Lecture Notes in Artificial Intelligence 3681

Edited by J. G. Carbonell and J. Siekmann

Subseries of Lecture Notes in Computer Science

Rajiv Khosla   Robert J. Howlett
Lakhmi C. Jain (Eds.)

Knowledge-Based Intelligent Information and Engineering Systems

9th International Conference, KES 2005
Melbourne, Australia, September 14-16, 2005
Proceedings, Part I

 Springer

Series Editors

Jaime G. Carbonell, Carnegie Mellon University, Pittsburgh, PA, USA
Jörg Siekmann, University of Saarland, Saarbrücken, Germany

Volume Editors

Rajiv Khosla
La Trobe University
Business Systems and Knowledge Modelling Laboratory
School of Business
Victoria 3086, Australia
E-mail: R.Khosla@latrobe.edu.au

Robert J. Howlett
University of Brighton
School of Engineering, Engineering Research Centre
Moulsecoomb, Brighton, BN2 4GJ, UK
E-mail: r.j.howlett@bton.ac.uk

Lakhmi C. Jain
University of South Australia
School of Electrical and Information Engineering, KES Centre
Adelaide, Mawson Lakes Campus
South Australia SA 5095, Australia
E-mail: Lakhmi.Jain@unisa.edu.au

Library of Congress Control Number: 2005932202

CR Subject Classification (1998): I.2, H.4, H.3, J.1, H.5, K.6, K.4

ISSN   0302-9743
ISBN-10   3-540-28894-5 Springer Berlin Heidelberg New York
ISBN-13   978-3-540-28894-7 Springer Berlin Heidelberg New York

Typesetting: Camera-ready by author, data conversion by Olgun Computergrafik
Printed on acid-free paper   SPIN: 11552413   06/3142   5 4 3 2 1 0

Preface

Dear delegates, friends and members of the growing KES professional community, welcome to the proceedings of the 9th International Conference on Knowledge-Based and Intelligent Information and Engineering Systems hosted by La Trobe University in Melbourne Australia.

The KES conference series has been established for almost a decade, and it continues each year to attract participants from all geographical areas of the world, including Europe, the Americas, Australasia and the Pacific Rim. The KES conferences cover a wide range of intelligent systems topics. The broad focus of the conference series is the theory and applications of intelligent systems. From a pure research field, intelligent systems have advanced to the point where their abilities have been incorporated into many business and engineering application areas. KES 2005 provided a valuable mechanism for delegates to obtain an extensive view of the latest research into a range of intelligent-systems algorithms, tools and techniques. The conference also gave delegates the chance to come into contact with those applying intelligent systems in diverse commercial areas. The combination of theory and practice represented a unique opportunity to gain an appreciation of the full spectrum of leading-edge intelligent-systems activity.

The papers for KES 2005 were either submitted to invited sessions, chaired and organized by respected experts in their fields, or to a general session, managed by an extensive International Program Committee, or to the Intelligent Information Hiding and Multimedia Signal Processing (IIHMSP) Workshop, managed by an International Workshop Technical Committee. Whichever route they came through, all papers for KES 2005 were thoroughly reviewed. The adoption by KES of the PROSE Publication Review and Organisation System software greatly helped to improve the transparency of the review process and aided quality control.

In total, 1382 papers were submitted for KES 2005, and a total of 688 papers were accepted, giving an acceptance rate of just under 50%. The proceedings, published this year by Springer, run to more than 5000 pages. The invited sessions are a valuable feature of KES conferences, enabling leading researchers to initiate sessions that focus on innovative new areas. A number of sessions in new emerging areas were introduced this year, including Experience Management, Emotional Intelligence, and Smart Systems. The diversity of the papers can be judged from the fact that there were about 100 technical sessions in the conference program. More than 400 universities worldwide participated in the conference making it one of the largest conferences in the area of intelligent systems. As would be expected, there was good local support with the participation of 20 Australian universities. There was a significant business presence, provided by the involvement of a number of industry bodies, for example, CSIRO Australia, DSTO Australia, Daewoo South Korea and NTT Japan.

KES International gratefully acknowledges the support provided by La Trobe University in hosting this conference. We acknowledge the active interest and support from La Trobe University's Vice Chancellor and President, Prof. Michael Osborne, Dean of the Faculty of Law and Management, Prof. Raymond Harbridge, Dean of the Faculty of Science and Technology, Prof. David Finlay, and Head of the School of Business, Prof. Malcolm Rimmer. KES International also gratefully acknowledges the support provided by Emeritus Prof. Greg O'Brien.

A tremendous amount of time and effort goes into the organization of a conference of the size of KES 2005. The KES community owes a considerable debt of gratitude to the General Chair Prof. Rajiv Khosla and the organizing team at La Trobe University for their huge efforts this year in bringing the conference to a successful conclusion. As the conference increases in size each year the organizational effort needed increases and we would like to thank Prof. Khosla and his colleagues for coping efficiently with the largest KES conference to date.

We would like to thank the Invited Session Chairs, under the leadership and guidance of Prof. Lakhmi Jain and Prof. Rajiv Khosla for producing high-quality sessions on leading-edge topics. We would like to thank the KES 2005 International Program Committee for undertaking the considerable task of reviewing all of the papers submitted for the conference. We express our gratitude to the high-profile keynote speakers for providing talks on leading-edge topics to inform and enthuse our delegates. A conference cannot run without authors to write papers. We thank the authors, presenters and delegates to KES 2005 without whom the conference could not have taken place. Finally we thank the administrators, caterers, hoteliers, and the people of Melbourne for welcoming us and providing for the conference.

We hope you found KES 2005 a worthwhile, informative and enjoyable experience.

July 2005

Bob Howlett
Rajiv Khosla
Lakhmi Jain

KES 2005 Conference Organization

General Chair

Rajiv Khosla
Business Systems and Knowledge Modelling Laboratory
School of Business
La Trobe University
Melbourne, Victoria 3086
Australia

Conference Founder and Honorary Program Committee Chair

Lakhmi C. Jain
Knowledge-Based Intelligent Information and Engineering Systems Centre
University of South Australia, Australia

KES Executive Chair

Bob Howlett
Intelligent Systems and Signal Processing Laboratories/KTP Centre
University of Brighton, UK

KES Journal General Editor

Bogdan Gabrys
University of Bournemouth, UK

Local Organizing Committee

Malcolm Rimmer – Chair, School of Business, La Trobe University
Rajiv Khosla, Selena Lim, Brigitte Carrucan, Monica Hodgkinson, Marie Fenton,
Maggie Van Tonder, and Stephen Muir
La Trobe University, Melbourne, Australia

KES 2005 Web Page Design Team

Joe Hayes, Anil Varkey Samuel, Mehul Bhatt, Rajiv Khosla
La Trobe University, Melbourne, Australia

KES 2005 Liaison and Registration Team

Rajiv Khosla, Selena Lim, Brigitte Carrucan, Jodie Kennedy, Maggie Van Tonder, Marie Fenton, Colleen Stoate, Diane Kraal, Cary Slater, Petrus Usmanij, Chris Lai, Rani Thanacoody, Elisabeth Tanusasmita, George Plocinski
La Trobe University, Melbourne, Australia

KES 2005 Proceedings Assembly Team

Rajiv Khosla
Selena Lim
Monica Hodgkinson
George Plocinski
Maggie Van Tonder
Colleen Stoate
Anil Varkey Samuel
Marie Fenton
Mehul Bhatt
Chris Lai
Petrus Usmanij

International Program Committee

Toyoaki Nishida, University of Tokyo, Japan
Vasile Palade, Oxford University, UK
John A. Rose, University of Tokyo, Japan
Rajkumar Roy, Cranfield University, UK
Udo Seiffert, Leibniz Institute of Plant Genetics and Crop Plant Research, Germany
Flavio Soares Correa da Silva, University of Sao Paulo, Brazil
Von-Wun Soo, National Tsing Hua University, Taiwan
Sarawut Sujitjorn, Suranaree University of Technology, Thailand
Takushi Tanaka, Fukuoka Institute of Technology, Japan
Eiichiro Tazaki, University of Yokohama, Japan
Jon Timmis, University of Kent, UK
Jim Torresen, University of Oslo, Norway
Andy M. Tyrrell, University of York, UK
Eiji Uchino, University of Yamaguchi, Japan
Jose Luis Verdegay, University of Granada, Spain
Dianhui Wang, La Trobe University, Australia
Junzo Watada, Waseda University, Japan
Xin Yao, University of Birmingham, UK
Boris Galitsky, University of London, UK
Bogdan Gabrys, Bournemouth University, UK
Norbert Jesse, Universität Dortmund, Germany
Keiichi Horio, Kyushu Institute of Technology, Japan
Bernd Reusch, University of Dortmund, Germany
Nadia Berthouze, University of Aizu, Japan
Hideyaki Sawada, Kagawa University, Japan
Yasue Mitsukura, University of Okayama, Japan
Dharmendra Sharma, University of Canberra, Australia
Adam Grzech, Wroclaw University of Technology, Poland
Mircea Negoita, Wellington Institute of Technology, New Zealand
Hongen Lu, La Trobe University, Australia
Kosuke Sekiyama, University of Fukui, Japan
Raquel Flórez-López, Campus de Vegazana, Spain
Eugénio Oliveira, University of Porto, Portugal
Roberto Frias, University of Porto, Portugal
Maria Virvou, University of Piraeus, Greece
Daniela Zambarbieri, Università di Pavia, Italy
Andrew Skabar, La Trobe University, Australia
Zhaohao Sun, University of Wollongong, Australia
Koren Ward, University of Wollongong, Australia

Invited Session Chairs Committee

Krzysztof Cios, University of Denver, USA
Toyoaki Nishida, University of Tokyo, Japan
Ngoc Thanh Nguyen, Wroclaw University of Technology, Poland
Tzung-Pei Hong, National University of Kaohsiung, Taiwan
Yoshinori Adachi, Chubu University, Japan
Nobuhiro Inuzuka, Nagoya Institute of Technology, Japan
Naohiro Ishii, Aichii Institute of Technology, Japan
Yuji Iwahori, Chubu University, Japan
Tai-hoon Kim, Security Engineering Research Group (SERG), Korea
Daniel Howard, QinetiQ, UK
Mircea Gh. Negoita, Wellington Institute of Technology, New Zealand
Akira Suyama, University of Tokyo, Japan
Dong Hwa Kim, Hanbat National University, Korea
Hussein A. Abbas, University of New South Wales, Australia
William Grosky, University of Michigan-Dearborn, USA
Elisa Bertino, Purdue University, USA
Maria L. Damiani, University of Milan, Italy
Kosuke Sekiyama, University of Fukui, Japan
Seong-Joon Yoo, Sejong Univwersity, Korea
Hirokazu Taki, Wakayama University, Japan
Satoshi Hori, Institute of Technologists, Japan
Hideyuki Sawada, Kagawa University, Japan
Vicenc Torra, Artificial Intelligence Research Institute, Spain
Maria Vamrell, Artificial Intelligence Research Institute, Spain
Manfred Schmitt, Technical University of Munich, Germany
Denis Helic, Graz University of Technology, Austria
Yoshiteru Ishida, Toyohashi University of Technology, Japan
Giuseppina Passiante, University of Lecce, Italy
Ernesto Damiani, University of Milan, Italy
Susumu Kunifuji, Japan Advanced Institute of Science and Technology, Japan
Motoki Miura, Japan Advanced Institute of Science and Technology, Japan
James Liu, Hong Kong Polytechnic University, Hong Kong, China
Honghua Dai, Deakin University, Australia
Pascal Bouvry, Luxembourg University, Luxembourg
Gregoire Danoy, Luxembourg University, Luxembourg
Ojoung Kwon, California State University, USA
Jun Munemori, Wakayama University, Japan
Takashi Yoshino, Wakayama University, Japan
Takaya Yuizono, Shimane University, Japan
Behrouz Homayoun Far, University of Calgary, Canada
Toyohide Watanabe, Nagoya University, Japan
Dharmendra Sharma, University of Canberra, Australia
Dat Tran, University of Canberra, Australia

Kim Le, University of Canberra, Australia
Takumi Ichimura, Hiroshima City University, Japan
K Yoshida, St. Marianna University, Japan
Phill Kyu Rhee, Inha University, Korea
Chong Ho Lee, Inha University, Korea
Mikhail Prokopenko, CSIRO ICT Centre, Australia
Daniel Polani, University of Hertfordshire, UK
Dong Chun Lee, Howon University, Korea
Dawn E. Holmes, University of California, USA
Kok-Leong Ong, Deakin University, Australia
Vincent Lee, Monash University, Australia
Wee-Keong Ng, Nanyang Technological University
Gwi-Tae Park, Korea University, Korea
Giles Oatley, University of Sunderland, UK
Sangkyun Kim, Korea Information Engineering Service, Korea
Hong Joo Lee, Daewoo Electronics Corporation, Korea
Ryohei Nakano, Nagoya Institute of Technology, Japan
Kazumi Saito, NTT Communication Science Laboratories, Japan
Kazuhiko Tsuda, University of Tsukuba, Japan
Torbjørn Nordgård, Norwegian University of Science and Technology, Norway
Øystein Nytrø, Norwegian University of Science and Technology, Norway
Amund Tveit, Norwegian University of Science and Technology, Norway
Thomas Brox Røst, Norwegian University of Science and Technology, Norway
Manuel Graña, Universidad Pais Vasco, Spain
Richard Duro, Universidad de A Coruña, Spain
Kazumi Nakamatsu, University of Hyogo, Japan
Jair Minoro Abe, University of Sao Paulo, Brazil
Hiroko Shoji, Chuo University, Japan
Yukio Ohsawa, University of Tsukuba, Japan
Da Deng, University of Otago, New Zealand
Irena Koprinska, University of Sydney, Australia
Eiichiro Tazaki, University of Yokohama, Japan
Kenneth J. Mackin, Tokyo University of Information Sciences, Japan
Lakhmi Jain, University of South Australia, Australia
Tetsuo Fuchino, Tokyo Institute of Technology, Japan
Yoshiyuki Yamashita, Tohoku University, Japan
Martin Purvis, University of Otago, New Zealand
Mariusz Nowostawski, University of Otago, New Zealand
Bastin Tony Roy Savarimuthu, University of Otago, New Zealand
Norio Baba, Osaka Kyoiku University, Japan
Junzo Watada, Waseda University, Japan
Petra Povalej, Laboratory for System Design, Slovenia
Peter Kokol, Laboratory for System Design, Slovenia
Woochun Jun, Seoul National University of Education, Korea
Andrew Kusiak, University of Iowa, USA
Hanh Pham, State University of New York, USA

IIHMSP Workshop Organization Committee

General Co-chairs

Jeng-Shyang Pan
National Kaohsiung University of Applied Sciences, Taiwan
Lakhmi C. Jain
University of South Australia, Australia

Program Committee Co-chairs

Wai-Chi Fang
California Institute of Technology, USA
Eiji Kawaguchi
Keio University, Japan

Finance Chair

Jui-Fang Chang
National Kaohsiung University of Applied Sciences, Taiwan

Publicity Chair

Kang K. Yen
Florida International University, USA

Registration Chair

Yan Shi
Tokai University, Japan

Electronic Media Chair

Bin-Yih Liao
National Kaohsiung University of Applied Sciences, Taiwan

Publication Chair

Hsiang-Cheh Huang
National Chiao Tung University, Taiwan

Local Organizing Chair

R. Khosla
La Trobe University, Australia

Asia Liaison

Yoiti Suzuki
Tohoku University, Japan

North America Liaison

Yun Q. Shi
New Jersey Institute of Technology, USA

Europe Liaison

R.J. Howlett
University of Brighton, UK

IIHMSP Workshop Technical Committee

Prof. Oscar Au, Hong Kong University of Science and Technology, Hong Kong, China
Prof. Chin-Chen Chang, National Chung Cheng University, Taiwan
Prof. Chang Wen Chen, Florida Institute of Technology, USA
Prof. Guanrong Chen, City University of Hong Kong, Hong Kong, China
Prof. Liang-Gee Chen, National Taiwan University, Taiwan
Prof. Tsuhan Chen, Carnegie Mellon University, USA
Prof. Yung-Chang Chen, National Tsing-Hua University, Taiwan
Prof. Yen-Wei Chen, Ritsumeikan University, Japan
Prof. Hsin-Chia Fu, National Chiao Tung University, Taiwan
Prof. Hsueh-Ming Hang, National Chiao Tung University, Taiwan
Prof. Dimitrios Hatzinakos, University of Toronto, Canada
Dr. Ichiro Kuroda, NEC Electronics Corporation, Japan

KES 2005 Reviewers

H. Abbass, University of New South Wales, Australia
J.M. Abe, University of Sao Paulo, Brazil
Y. Adachi, Chubu University, Japan
F. Alpaslan, Middle East Technical University, Turkey
P. Andreae, Victoria University, New Zealand
A. Asano, Hiroshima University, Japan
K.V. Asari, Old Dominion University, USA
N. Baba, Osaka-Kyoiku University, Japan
R. Babuska, Delft University of Technology, The Netherlands
P. Bajaj, G.H. Raisoni College of Engineering, India
A. Bargiela, Nottingham Trent University, UK
M. Bazu, Institute of Microtechnology, Romania
N. Berthouze, University of Aizu, Japan
E. Bertino, Purdue University, USA
Y. Bodyanskiy, Kharkiv National University of Radioelectronics, Ukraine
P. Bosc, IRISA/ENSSAT, France
P. Bouvry, Luxembourg University, Luxembourg
P. Burrell, South Bank University, UK
J. Cao, La Trobe University, Australia
B. Chakraborty, Iwate Prefectural University, Japan
Y.-W. Chen, Ryukyus University, Japan
Y.-H. Chen-Burger, University of Edinburgh, UK
V. Cherkassky, University of Minnesota, USA
K. Cios, University at Denver, USA
C.A. Coello, LANIA, Mexico
G. Coghill, University of Auckland, New Zealand
D. Corbett, SAIC, USA
D.W. Corne, University of Exeter, UK
D. Cornforth, Charles Sturt University, Australia
F.S.C. da Silva, University of Sao Paulo, Brazil
H. Dai, Deakin University, Australia
E. Damiani, University of Milan, Italy
M.L. Damiani, University of Milan, Italy
G. Danoy, Luxembourg University, Luxembourg
K. Deep, Indian Institute of Technology Roorkee, India
D. Deng, University of Otago, New Zealand
V. Devedzic, University of Belgrade, Serbia and Montenegro
D. Dubois, Université Paul Sabatier, France
R. Duro, Universidad de A Coruña, Spain
D. Earl, Oak Ridge National Laboratory, USA
B. Far, University of Calgary, Canada

D.H. Kim, Hanbat National University, Korea
S. Kim, Korea Information Engineering Service, Korea
T.-H. Kim, Security Engineering Research Group (SERG), Korea
L.T. Koczy, Budapest University of Technology and Economics, Hungary
P. Kokol, Laboratory for System Design, Slovenia
A. Konar, Jadavpur University, India
I. Koprinska, University of Sydney, Australia
H. Koshimizu, Chukyo University, Japan
S. Kunifuji, Japan Advanced Institute of Science and and Technology, Japan
A. Kusiak, University of Iowa, USA
O. Kwon, California State University, USA
W.K. Lai, MIMOS Berhad, Malaysia
P.L. Lanzi, Polytechnic Institute, Italy
K. Le, University of Canberra, Australia
C.H. Lee, Inha University, Korea
D.C. Lee, Howon University, Korea
H.J. Lee, Daewoo Electronics Corporation, Korea
R. Lee, Hong Kong Polytechnic University, Hong Kong, China
V. Lee, Monash University, Australia
Q. Li, La Trobe University, Australia
C.-P. Lim, University of Science, Malaysia
S. Lim, La Trobe University, Australia
J. Liu, Hong Kong Polytechnic University, Hong Kong, China
I. Lovrek, University of Zagreb, Croatia
H. Lu, La Trobe University, Australia
B. MacDonald, Auckland University, New Zealand
K.J. Mackin, Tokyo University of Information Sciences, Japan
L. Magdalena-Layos, EUSFLAT, Spain
D.C. Marinescu, University of Central Florida, USA
F. Masulli, University of Pisa, Italy
J. Mazumdar, University of South Australia, Australia
B. McKay, University of NSW, Australia
S. McKinlay, Wellington Institute of Technology, New Zealand
R. Mesiar, Slovak Technical University, Slovakia
J. Mira, UNED, Spain
Y. Mitsukura, University of Okayama, Japan
M. Miura, Japan Advanced Institute of Science and Technology, Japan
J. Munemori, Wakayama University, Japan
H. Nagashino, University of Tokushima, Japan
N. Nagata, Chukyo University, Japan
K. Nakajima, Tohoku University, Japan
K. Nakamatsu, University of Hyogo, Japan
R. Nakano, Nagoya Institute, Japan
T. Nakashima, Osaka University, Japan
L. Narasimhan, University of Newcastle, Australia

V.E. Neagoe, Technical University, Romania
C.D. Neagu, University of Bradford, UK
M.G. Negoita, WelTec, New Zealand
W.-K. Ng, Nanyang Technological University, Singapore
C. Nguyen, Catholic University of America, USA
N.T. Nguyen, Wroclaw University of Technology, Poland
T. Nishida, University of Tokyo, Japan
T. Nordgård, Norwegian University of Science and Technology, Norway
M. Nowostawski, University of Otago, New Zealand
Ø. Nytrø, Norwegian University of Science and Technology, Norway
G. Oatley, University of Sunderland, UK
Y. Ohsawa, University of Tsukuba, Japan
E. Oliveira, University of Porto, Portugal
K.-L. Ong, Deakin University, Australia
N.R. Pal, Indian Statistical Institute, India
V. Palade, Oxford University, UK
G.-T. Park, Korea University, Korea
I.C. Parmee, University of the West of England, UK
G. Passiante, University of Lecce, Italy
C.-A. Peña-Reyes, Swiss Federal Institute of Technology - EPFL, Switzerland
H. Pham, State University of New York, USA
D. Polani, University of Hertfordshire, UK
T. Popescu, National Institute for Research and Development Informatic, Italy
P. Povalej, Laboratory for System Design, Slovenia
M. Prokopenko, CSIRO ICT Centre, Australia
M. Purvis, University of Otago, New Zealand
G. Resconi, Catholic University, Italy
B. Reusch, University of Dortmund, Germany
P.K. Rhee, Inha University, Korea
J.A. Rose, University of Tokyo, Japan
T.B. Røst, Norwegian University of Science and Technology, Norway
E. Roventa, York University, Canada
R. Roy, Cranfield University, UK
D. Ruan, Belgian Nuclear Research Centre, Belgium
A. Saha, NCD, Papua New Guinea
K. Saito, NTT Communication Science Laboratories, Japan
T. Samatsu, Kyushu Tokai University, Japan
E. Sanchez, Université de la Méditeranée, France
B.T.R. Savarimuthu, University of Otago, New Zealand
H. Sawada, Kagawa University, Japan
M. Schmitt, Technical University of Munich, Germany
M. Schoenauer, INRIA, France
U. Seiffert, Leibniz Institute of Plant Genetics
    and Crop Plant Research Gatersleben, Germany
K. Sekiyama, University of Fukui, Japan

D. Sharma, University of Canberra, Australia
H. Shoji, Chuo University, Japan
A. Skabar, La Trobe University, Australia
B. Smyth, University College Dublin, Ireland
V.-W. Soo, National Tsing Hua University, Taiwan
A. Stoica, NASA Propulsion Jet Laboratory, USA
M.R. Stytz, Yamaguchi University, Japan
N. Suetake, Yamaguchi University, Japan
S. Sujitjorn, Suranaree University of Technology, Thailand
Z. Sun, University of Wollongong, Australia
A. Suyama, University of Tokyo, Japan
H. Taki, Wakayama University, Japan
T. Tanaka, Fukuoka Institute of Technology, Japan
M. Tanaka-Yamawaki, Tottori University, Japan
E. Tazaki, University of Yokohama, Japan
S. Thatcher, University of South Australia, Australia
P. Theodor, National Institute for Research and Development Informatics, Romania
J. Timmis, University of Kent at Canterbury, UK
V. Torra, Artificial Intelligence Research Institute, Spain
J. Torresen, University of Oslo, Norway
D. Tran, University of Canberra, Australia
K. Tsuda, University of Tsukuba, Japan
C. Turchetti, Università Politecnica delle Marche, Italy
A. Tveit, Norwegian University of Science and Technology, Norway
J. Tweedale, Defence Science and Technology Organization, Australia
A.M. Tyrrell, University of York, UK
E. Uchino, University of Yamaguchi, Japan
A. Uncini, University of Rome, Italy
P. Urlings, Defence Science and Technology Organization, Australia
M. Vamrell, Artificial Intelligence Research Institute, Spain
J.L. Verdegay, University of Granada, Spain
M. Virvou, University of Piraeus, Greece
S. Walters, University of Brighton, UK
D. Wang, La Trobe University, Australia
L. Wang, Nanyang Technical University, Singapore
P. Wang, Temple University, USA
K. Ward, University of Wollongong, Australia
J. Watada, Waseda University, Japan
K. Watanabe, Saga University, Japan
T. Watanabe, Nagoya University, Japan
T. Yamakawa, Kyushu Institute of Technology, Japan
Y. Yamashita, Tohoku University, Japan
A. Yang, University of New South Wales, Australia
X. Yao, University of Birmingham, UK
S.-J. Yoo, Sejong University, Korea

KES 2005 Keynote Speakers

1. Professor Jun Liu, Harvard University, MA, USA
   Topic: From Sequence Information to Gene Expression

2. Professor Ron Sun, Rensselaer Polytechnic Institute, New York, USA
   Topic: From Hybrid Systems to Hybrid Cognitive Architectures

3. Professor Jiming Liu, Hong Kong Baptist University, Hong Kong, China
   Topic: Towards Autonomy Oriented Computing (AOC):
      Formulating Computational Systems with Autonomous Components

4. Professor Toyoaki Nishida, Kyoto University and Tokyo University, Japan
   Topic: Acquiring, Accumulating, Transforming, Applying,
      and Understanding Conversational Quanta

5. Professor Marimuthu Palaniswami, University of Melbourne, Australia
   Topic: Convergence of Smart Sensors and Sensor Networks

Table of Contents, Part I

Soft Computing Techniques in Stock Markets

Intelligent Network Based Education

Maintenance and Customization of Business Knowledge

Intelligent Data Processing in Process Systems and Plants

Intelligent Agent Technology and Applications I

Intelligent Design Support Systems

Data Engineering, Knowledge Engineering and Ontologies

Knowledge Discovery and Data Mining

Advanced Network Application

Information Hiding and Multimedia Signal Processing

Soft Computing Techniques and Their Applications I

Intelligent Agent Technology and Applications II

Smart Systems

Knowledge – Based Interface Systems

Intelligent Information Processing for Remote Sensing

Intelligent Human Computer Interaction Systems

Experience Management and Knowledge Management

Network (Security) Real-Time and Fault Tolerant Systems

Advanced Network Application and Real-Time Systems

Approaches and Methods of Security Engineering II

Soft Computing Techniques and Their Applications II

Author Index

Table of Contents, Part II

Medical Diagnosis

Intelligent Hybrid Systems and Control

Emotional Intelligence and Smart Systems

Context-Aware Evolvable Systems

Intelligant Fuzzy Systems and Control

Knowledge Representation and Its Practical Application in Today's Society

Approaches and Methods into Security Engineering III

Communicative Intelligent I

e-Learning and ICT

Logic Based Intelligent Information Systems

Intelligent Agents and Their Applications I

Innovations in Intelligent Agents

Ontologies and the Semantic Web

Knowledge Discovery in Data Streams

Computational Intelligence Tools Techniques and Algorithms

Approaches and Methods to Security Engineering IV

Watermaking Applications I

Watermaking Applications II

Multimedia Retrieval I

Soft Computing Approach to Industrial Engineering

Experience Management and Information Systems

Table of Contents, Part III

Intelligent Agent Ontologies and Environments

Intelligent Multimedia Solutions and the Security in the Next Generation Mobile Networks

Intelligent E-Mail Analysis, News Extraction and Web Mining

Semantic Integration and Ontologies

Computer Vision, Image Processing and Retrieval

Communicative Intelligence II

Approaches and Methods to Security Engineering V

Multimedia Retrieval II

Multimedia Compression

Multimedia Signal Processing

Emergence and Self-organisation in Complex Systems

Soft Computing Techniques and Their Applications III

Information Engineering and Ubiquitous Computing

Location and Context-Based Systems

e-Based Systems in Education, Commerce and Health

Communicative Intelligent III

Speech Processing and Robotics

Stegnography

Soft Computing Approach to Industrial Engineering II

Medical Text Mining and Natural Language Processing

Knowledge Based Intelligent Systems for Health Care

Intelligent Learning Environment

Intelligent Data Analysis and Applications

Table of Contents, Part IV

Innovations in Intelligent Systems and Their Applications

Data Mining and Soft Computing Applications II

Skill Acquisition and Ubiquitous Human Computer Interaction

Soft Computing and Their Applications – IV

Agent-Based Workflows, Knowledge Sharing and Reuse

Multi-media Authentication and Watermarking Applications

Knowledge and Engineering Techniques
for Spatio-temporal Applications

Intelligent Data Analysis and Applications II

Creativitiy Support Environment and Its Social Applications

Collective Intelligence

Computational Methods for Intelligent Neuro-fuzzy Applications

Evolutionary and Self-organizing Sensors, Actuators and Processing Hardware

Knowledge Based Systems for e-Business and e-Learning I

Multi-agent Systems and Evolutionary Computing

Ubiquitous Pattern Recognition

Neural Networks for Data Mining

Intelligent Systems for e-Business and e-Learning II

Knowledge-Based Technology in Crime Matching, Modelling and Prediction

Soft Computing Applications

Multi-branch Neural Networks
and Its Application to Stock Price Prediction

Takashi Yamashita, Kotaro Hirasawa, and Jinglu Hu

Waseda University, 2-7 Hibikino, Wakamatsu-ku,
Kitakyushu, Fukuoka, Japan
`yamasita@ruri.waseda.jp`, {`hirasawa,jinglu`}`@waseda.jp`
`http://www.hflab.dyndns.org/index.html`

Abstract. Recently, artificial neural networks have been utilized for financial market applications. We have so far shown that multi-branch neural networks (MBNNs) could have higher representation and generalization ability than conventional NNs. In this paper, a prediction system of a stock price using MBNNs is proposed. Using the stock prices in time series and other information, MBNNs can learn to predict the price of the next day. The result of our simulations shows that the proposed system has better accuracy than a system using conventional NNs.

1   Introduction

Artificial neural networks (ANNs) can model complex nonlinear systems by learning from sampling data that describe inputs and outputs relations. With such capability, ANNs are being applied to various applications, and time series prediction is one of them.

Although predicting a stock market is a difficult task, a lot of research concerning the stock market prediction using ANNs [1] [2] [3] has been done. In these research, authors used a hybrid algorithm, genetic algorithm with ANNs, or Radial Basis Function (RBF) networks for the prediction because conventional Multi-Layer Perceptrons (MLPs) can not meet the requirements.

In this paper, Multi-Branch Neural Networks (MBNNs) [4] are applied to the prediction of stock prices. MBNNs are constructed using multi-branches between nodes, which adds additional nonlinear functions to the branches of the networks.

A benefit of MBNNs is gained by the smaller network size with smaller number of hidden nodes. We have shown that MBNNs could have higher representation and generalization ability than conventional NNs, even if the number of hidden nodes is reduced. Therefore, MBNNs are considered suitable to apply to real world problems such as time series predictions of the stock market. In this paper we examine the usefulness of application of MBNNs to real world problems.

Against this background, we predicted the stock price using MBNNs, and investigated the accuracy of the prediction comparing to conventional NNs.

The outline of this paper is as follows. In section 2, we briefly summarize the structure, effectiveness and learning method of MBNNs. Section 3 presents the

R. Khosla et al. (Eds.): KES 2005, LNAI 3681, pp. 1–7, 2005.

methodology for the time series prediction using MBNNs. In section 4, we show the simulation results. Finally, a brief summary of the research and conclusions are presented in section 5.

2 Multi-branch Neural Networks

In this section, we describe Multi-Branch Neural Networks (MBNNs). The basic concept of the multi-branch structure comes from Universal Learning Networks (ULNs) [5] [6]. ULNs provide a generalized framework to all kinds of structures of neural networks with supervised learning, and one of whose characteristics is a multi-branch structure. It connects nodes using some branches with arbitrary number.

2.1 Structure of MBNNs

Conventional NNs have a single branch structure for connections with each branch having a weight parameter. The extension of this is a multi-branch structure between nodes as shown in Fig.1, which could be realized easily by ULNs. Conventional NNs could make the number of nodes in the network increase in order to obtain higher representation ability. However, a large number of nodes lead to higher costs of space and time for training especially in backpropagation algorithm [7].

It is desirable for neural networks to reduce the costs of space and time when they are applied to many kinds of problems in fact. One of the points of MBNNs is that increasing the number of branches in order to obtain higher performances could keep the costs lower than increasing the number of nodes. So, it is expected that compact neural networks could be realized using MBNNs.

Eq.(1) shows the conventional processing of the branch which connects to node j; α_j in Eq.(1) is the input of node j.

$$\alpha_j = \sum_{i \in JF(j)} w_{ij} h_i + \theta_j, \tag{1}$$

where $JF(j)$ denotes the set of suffixes of nodes which are connected to node j, w_{ij} is the weight parameter of the branch from node i to node j, h_i is the output value of node i and θ_j is the threshold parameter of node j.

In the case of MBNNs, Eq.(2) is used instead.

$$\alpha_j = \sum_{i \in JF(j)} \sum_{p \in B(i,j)} w_{ij}(p) g_{ij}(p, h_i) + \theta_j, \tag{2}$$

where $B(i, j)$ denotes the set of suffixes of branches from node i to node j, $w_{ij}(p)$ is the weight parameter of pth branch from node i to node j and $g_{ij}(p, h_i)$ is the modified output value of node i used for pth branch from node i to node j.

The modified function g_{ij} is to be a certain nonlinear function in order to obtain nonlinear effects from the multi-branch structure. When $g_{ij}(p, h_i) = h_i$, Eq.(2) becomes Eq.(1).

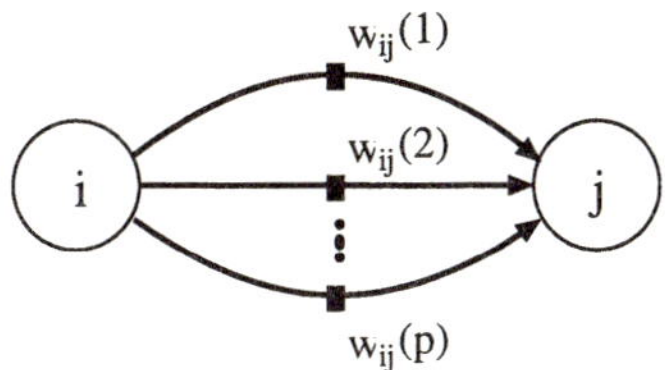

Fig. 1. Multi-branch structure between nodes

2.2 Localized Property of MBNNs

MBNNs using a gaussian function have been proposed, where $g_{ij}(p, h_i)$ in Eq.(2) is expressed by Eq.(3).

$$g_{ij}(p, h_i) = \exp\left(-\frac{(h_i - c_{ij}(p))^2}{\sigma_{ij}^2(p) + L}\right), \tag{3}$$

where $c_{ij}(p)$ is the center parameter of pth branch from node i to node j and $\sigma_{ij}(p)$ is the width parameter of pth branch from node i to node j. L is a positive constant for the adjustment of generalization ability.

By using the gaussian function, the MBNNs can utilize localized property. This causes each multi-branch between nodes to have a different function to the input. In other words, each parameter of the branch can make the network respond to a part of the inputs, not the whole range of them. In addition, it is easy to adjust their parameters, resulting in better representation and generalization ability.

The above multi-branch structure also means the modularized structure for signal processing between nodes. Generally a task can be dealt with easily when it is divided into a small and simple piece of tasks. This principle is accepted in various fields.

Radial Basis Function (RBF) networks [8] [9] have a similarity to the multi-branch structure, that is, RBF networks have some RBFs on the input space, while MBNNs have some gaussian functions on multi-branches between nodes.

2.3 Learning Parameters of MBNNs

MBNNs have unconventional learning parameters $c_{ij}(p)$ and $\sigma_{ij}(p)$. So, they need to adjust as well as weight $w_{ij}(p)$. Learning of MBNNs is realized by minimizing a criterion function E based on the following gradient method:

$$\lambda_m \leftarrow \lambda_m - \gamma\frac{\partial^+ E}{\partial \lambda_m}, \tag{4}$$

$$\lambda_m \in \{w_{ij}(p), c_{ij}(p), \sigma_{ij}(p), \theta_j\},$$

where γ is the learning coefficient assigned a small positive value and λ_m is the learning parameter. Here, the ordered derivative $\frac{\partial^+ E}{\partial \lambda_m}$ is calculated by backpropagation algorithm [5].

The point is that each learning parameter has different coefficient γ. We make the coefficients of $c_{ij}(p)$ and $\sigma_{ij}(p)$ smaller than that of $w_{ij}(p)$ because the change of $c_{ij}(p)$ or $\sigma_{ij}(p)$ could have larger effects than that of $w_{ij}(p)$. The coefficients of $c_{ij}(p)$ and $\sigma_{ij}(p)$ can be set so that the ratio of the absolute maximum update $(\Delta w_{max} : \Delta c_{max} : \Delta\sigma_{max})$ becomes $1 : r_c : r_\sigma$. We used 0.4 and 0.1 as r_c and r_σ, respectively.

3 Application of MBNNs to Prediction of Stock Price

In this section, we describe the methodology of the prediction of stock prices using MBNNs.

3.1 Model for Prediction

It is important for the application of ANNs to generate the appropriate training data, that is, inputs and outputs suitable for the learning. In this paper, the target value of the prediction is the rate of deviation of 5 days moving average, that is, NNs learn the behavior of the rate of deviation of the moving average in a time series. The rate of deviation of the moving average $v(t)$ is calculated by the following equation:

$$v(t) = \frac{C(t) - MA_m(t)}{MA_m(t)}, \tag{5}$$

where $C(t)$ denotes the closing price at time t and $MA_m(t)$ is the m days moving average at time t. $MA_m(t)$ is a function of $C(t)$; $MA_m(t) = \frac{\sum_{u=t-m+1}^{t} C(u)}{m}$. Therefore, the predicted next day's $\tilde{C}(t)$ can be calculated by Eq.(6) using the predicted $\tilde{v}(t)$.

$$\tilde{C}(t) = \frac{(\tilde{v}(t) + 1)\sum_{u=t-m+1}^{t-1} C(u)}{m - \tilde{v}(t) - 1}. \tag{6}$$

Using Eq.(6) we predict the next day's price value. We used the following function form for predicting the price at time $(t + 1)$ in time series.

$$v(t + 1) = f(v(t), v(t - 1), v(t - 2), v(t - 3), v(t - 4),$$
$$\bar{y}(t), \bar{y}(t - 1), \bar{y}(t - 2), \bar{y}(t - 3), \bar{y}(t - 4)), \tag{7}$$

where $v(t)$ is the rate of deviation of 5 days moving average at time t, and $\bar{y}(t)$ is the external information vector at time t. Five data points in time series are used for predicting the next data point.

We have to use not only $v(t)$ but also other external information vector because the stock market has the structure of great complexity. The external information vector of $\bar{y}(t)$ is as follows.

- Rate of the deviation of stock price moving average in terms of 25 and 75 days;
- Rate of the deviation of turnover moving average in terms of 5 days;
- Value of (closing price - opening price) / closing price;
- Value of (closing price - high price) / closing price;

- Value of (closing price - low price) / closing price;
- Change of closing price;
- Change of turnover;
- Psychological line in terms of 12 days;

Turnover means the amount of dealt stocks and psychological line means the rate of days increasing the stock price in terms of certain of the days.

These v and $\bar{y}$ are input variables and $v(t+1)$ is the output of the network. Therefore, we used these 50 input variables.

3.2   Learning of MBNNs

The data set is to be divided into three subsets: training, validation and testing data sets.

First, the network is trained with the training data. Then, the trained network is evaluated with the validation data in order to check its over-fitting. In this prediction problem, over-fitting would occur because NNs learn based on the limited information. This process is iterated for sufficient learning steps. Finally, the trained network acquiring the best evaluation with the validation data is tested using the testing data.

The number of branches between nodes in MBNNs can be chosen by the system designer. Architecture of MBNNs for the prediction of the stock price possesses both single-branch and multi-branch structure, that is, single-branch between input-hidden layer, multi-branch between hidden-output layer. The multi-branch structure was used only between hidden-output layer because the number of input variables are quite so many. In addition, only 3-branch is used as the multi-branch between hidden-output layer of the network, which is enough for our problems.

4   Simulations

4.1   Simulation Conditions

Simulations were carried out in order to compare the predictor using MBNNs and conventional NNs. Here, several conditions for simulations are described.

The criterion function E is usually defined as follows.

$$E = \sum (h_o - h_o')^2 + \eta \left(\sum_{w_{ij} \in W_s} w_{ij}^2 + \sum_{w_{ij}(p) \in W_m} \frac{w_{ij}^2(p)}{\sigma_{ij}^2(p)} \right), \tag{8}$$

where W_s denotes the set of weights of single-branch, while W_m denotes the set of those of multi-branch, h_o is the output value of the network, h_o' is the desired output value of the network and η is the regularization coefficient in weight decay [10]. Here, η was determined adaptively so that the penalty term in Eq.(8) becomes 0.2% of the error term. As a result, the generalization ability becomes effective.

In the parameter update, moment term [11] is introduced: The increment of the parameter update is written by $(1 - \alpha)\Delta\lambda_m(s) + \alpha\Delta\lambda_m(s - 1)$, where $\lambda_m(s)$ is a parameter at step s, and $\alpha = 0.95$. In the multi-branch NNs, training parameter λ_m includes $c_{ij}(p)$ and $\sigma_{ij}(p)$ as well as $w_{ij}(p)$ and θ_j. The parameters $w_{ij}(p)$ and θ_j are initialized randomly in the range of [-0.1:0.1], $c_{ij}(p)$ is initially located uniformly in the output range of the nodes [-1:1] and $\sigma_{ij}(p)$ is initialized in the range of [0.4:0.6].

The constant L was set at 0.16 in order to obtain enough generalization ability. The learning coefficient γ initialized at 0.001 changes adaptively: When the value of E decreases, $\gamma \leftarrow 1.05\gamma$, while $\gamma \leftarrow 0.95\gamma$ when it increases. In addition, if $E(s)/E(s - 1) > 1.05$, then the parameter update is not executed, where $E(s)$ is the criterion value at step s.

Sigmoidal function $\frac{1-e^{-\alpha_j}}{1+e^{-\alpha_j}}$ is used as the node function in the simulations. The learning is carried out by BP algorithm.

In these conditions, simulations were carried out 30 times with different initial parameters with each learning being done for 2000 iterations. Conventional NNs have 12 hidden nodes and 612 branches (which include 612 parameters with the exception of θ_j), while MBNNs have 10 hidden nodes and 530 branches (which include 590 parameters with the exception of θ_j).

We used 660 daily time series data of the stock price; 480 for training, 60 for validation and 240 for testing in sequential order.

4.2   Simulation Results

In the simulations, NNs were trained in order to predict the next day's price from the rate of deviation of 5 days moving average using Eq.(7). We adopted Nikkei-225 for this simulation, that is one of the representative stock price indexes of Japan. The results described in the following are the evaluations using testing data.

The predicted prices can be measured in various ways. Here, we measured the correct up-down sign of the predicted next day's price. Table 1 shows the rate of correct up-down sign of predicted Nikkei-225 at time $t + 1$. Table 1 includes the rate of correct up-down sign for three years, from 2002 to 2004. The average of MBNNs have better accuracy in any periods than conventional NNs.

Table 1. Rate of correct up-down sign of predicted Nikkei-225 at time $t + 1$

Year	Conventional NNs (%)			MBNNs (%)		
	Ave.	Max.	Min.	Ave.	Max.	Min.
2002	63.3	74.2	55.0	71.3	81.9	52.5
2003	62.0	69.6	57.1	66.9	73.1	56.6
2004	61.8	69.6	56.1	66.8	76.7	55.7

5   Conclusions

Multi-branch neural networks (MBNNs) are utilized for the prediction of stock prices in this paper. Simulations are carried out in order to investigate the accuracy of the prediction. The results show that MBNNs with fewer parameters could have better accuracy than conventional NNs when predicting Nikkei-225 at time $t + 1$. The prediction system with high profitability using MBNNs suggests that MBNNs are applicable for real world problems such as predicting stock market prices.

However, what kinds of information for the network inputs should be selected remains as one of the future research topics. It is very important for the predictor using NNs to select effective information. In fact, the accuracy of the prediction depends on these selections.

In addition, we have not investigated yet if the prediction itself could be useful for dealing the stock. We construct the dealing system using MBNNs in a near future.

References

1. N. Baba and M. Kozaki, "An intelligent forecasting system of stock price using neural network", Proceeding of *IJCNN*, Vol. 1, pp. 371–377, 1992.
2. A. Lendasse, E. De Bodt, V. Wertz and M. Verleysen, "Non-linear financial time series forecasting - Application to the Bel 20 stock market index", *European Journal of Economic and Social Systems*, Vol. 14, No. 1, pp. 81–91, 2000.
3. N. Baba, N. Inoue and Y. Yanjun, "Utilization of Soft Computing Techniques for Constructing Reliable Decision Support Systems for Dealing Stocks", Proceeding of *IJCNN*, pp. 2150–2155, 2002.
4. T. Yamashita, K. Hirasawa and J. Hu, "Multi-Branch Structure and its Localized Property in Layered Neural Networks", Proceeding of *IJCNN*, pp. 1039–1044, 2004.
5. K. Hirasawa, X. Wang, J. Murata, J. Hu and C. Jin, "Universal learning network and its application to chaos control", Neural Networks, Vol. 13, pp. 239–253, 2000.
6. K. Hirasawa, M. Ohbayashi, H. Fujita and M. Koga, "Universal Learning Network Theory", Trans. of Institute of Electrical Engineers of Japan, Vol. 116-C, No. 7, pp. 794-801, 1996.
7. D. E. Rumelhart, G. E. Hinton and R. J. Williams, "Learning internal representations by error propagation", Parallel distributed processing, Vol. 1, The MIT Press, pp. 318–362, 1986.
8. M. T. Musavi, W. Ahmed, K. H. Chan, K. B. Faris and D. M. Hummels, "On the Training of Radial Basis Function Classifiers", Neural Networks, Vol. 5, No. 4, pp. 595–603, 1992.
9. S. Chen, C. F. N. Cowan and P. M. Grant, "Orthogonal Least Squares Learning Algorithm for Radial Basis Function Networks", *IEEE Trans. Neural Networks*, Vol. 2, No. 2, pp. 302–309, 1991.
10. G. E. Hinton, "Connectionist Learning Procedure", Machine Learning, ed. J. Carbonell, The MIT Press, 1990.
11. Sejnowski, T. J. and C. R. Rosenberg, "Parallel networks that learn to pronounce English text", Complex Systems 1: pp. 145–168, 1987.

An Intelligent Utilization of Neural Networks for Improving the Traditional Technical Analysis in the Stock Markets

Norio Baba and Toshinori Nomura

Information Science, Osaka Kyoiku University,
Asahiga-Oka, 4-698-1, Kashiwara City, Osaka Prefecture, 582-8582, Japan

Abstract. The use of soft computing techniques such as neural networks and fuzzy engineering in the financial market has recently become one of the most exciting and promising application areas. In this paper, we propose a new decision support system (DSS) for dealing stocks which improves the traditional technical analysis by using neural networks. In the proposed system, neural networks are utilized in order to predict the "Golden Cross"("Dead Cross") several weeks before it occurs. Computer simulation results concerning the dealings of the Nikkei-225 confirm the effectiveness of the proposed DSS.

1 Introduction

During the last several decades, the soft computing techniques such as Neural Networks (NNs) & Genteic Algorithms (GAs) have been studied quite extensively by many researchers [1],[2] and have been applied to various real world problems. They have been successfully utilized for constructing a large number of intelligent systems. One of the most remarkable application areas may be the financial market. Nowadays, comparatively large numbers of researchers show keen interest in the application of soft computing techniques to the financial market [3]-[10].

In this paper, we shall propose a DSS for dealing stocks which improves the traditional technical analysis by utilizing NNs. In the proposed system, NNs are utilized in order to detect "Golden Cross" and "Dead Cross" several weeks before real crossing occurs. The outline of this paper is as follows. In the second section, we will briefly touch upon the traditional technical analysis using the measure "the long term moving average (LTMA) versus the short term moving average (STMA)". In the third section, we shall propose a modified DSS for dealing stocks which utilizes NNs in order to detect the current tendency of the stocks and predict "Golden Cross"("Dead Cross") several weeks before it actually occurs. In the fourth section, we shall show several computer simulation studies which confirm the effectiveness of the proposed DSS. Finally, the paper concludes with a brief summary.

2 Traditional Technical Analysis Utilizing the Measure "Long Term Moving Average Versus Short Term Moving Average"

In order to detect the current tendency of the stock market, many stock traders have so far used the traditional technical analysis which utilizes the measure "long term

R. Khosla et al. (Eds.): KES 2005, LNAI 3681, pp. 8–14, 2005.

moving average (LTMA) versus short term moving average (STMA)". Fig.1 and Fig.2 illustrate the golden cross and the dead cross of the two moving averages, respectively. As is well known among many stock traders, "Golden Cross" ("Dead Cross") is one of the most familiar indicators that predict that the stock price will become higher (lower). Therefore, many people have often utilized these measures for dealing stocks.

The prototype DSS utilizing the traditional technical analysis can be described as follows:

1) If the sign of (the LTMA (such as 13 weeks moving average) – the STMA (such as 5 weeks moving average)) changes from positive to negative, then buy stocks.
2) If the sign of (the LTMA (such as 13 weeks moving average) – the STMA (such as 5 weeks moving average)) changes from negative to positive, then sell stocks.

Remark 2.1: As the measure "LTMA versus STMA", "13 weeks moving average (13 MA) versus 5 weeks moving average (5 MA)" has often been utilized. In the simulations (concerning dealings in the Nikkei-225) done in this paper, 13 MA versus 5 MA has been used for the measure "LTMA versus STMA". As is well known among many traders, the measure "26 MA versus 13 MA" has also been utilized for detecting the tendency of the stock market.

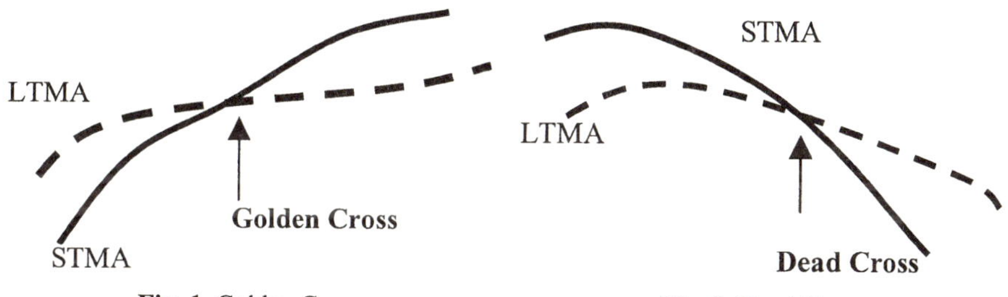

Fig. 1. Golden Cross   Fig. 2. Dead Cross

3 A New Decision Support System for Dealing Stocks Which Improves the Traditional Technical Analysis

The traditional technical analysis utilizing the measure LTMA versus STMA has been widely recognized as one of the most reliable techniques for dealing stocks. However, we consider that the traditional technical analysis could further be improved by the use of NNs. The prototype DSS utilizing the traditional technical analysis executes dealings after golden cross or dead cross has been found. However, such a way of dealings often loses an important chance to execute dealings in a reasonable price in the stock market. If one could deal stocks several weeks before real crossing occurs, one could save considerable amount of money. We shall explain this by using the real data which illustrates the changes of the price of "Nissan Motors" during the period January 2001 – May 2002. In the Fig.3, the solid line and the broken line indicate the STMA (13 MA) and the LTMA (26 MA), respectively. Further, the point A and the point B express the dead cross and the golden cross, respectively. If we utilize the traditional technical analysis, we have to sell the stock with the price around 750 yen and buy it with the price around 730 yen. On the other hand, we can sell the stock with the price higher than 800 yen and buy it with around 700 yen if we could predict the dead cross and the golden cross several weeks before those crossings occur.

In this section, we shall propose a decision support system (DSS) for dealing stocks which utilizes information (concerning the possibility of the "Golden Cross" ("Dead Cross")) having been obtained by the use of NNs effectively.

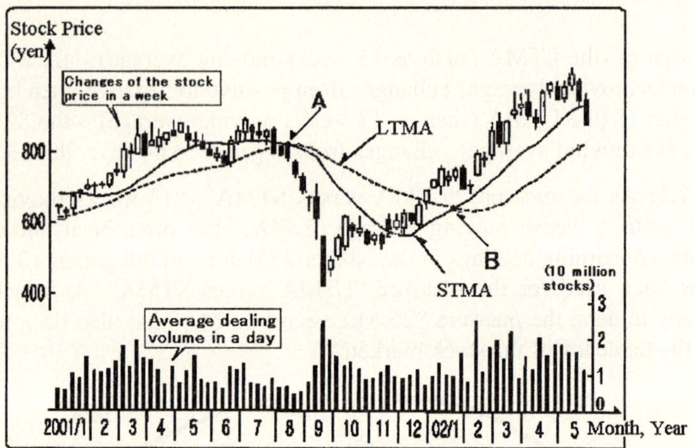

Fig. 3. Golden Cross and Dead Cross of the two moving averages of the stock price of Nissan Motors

3.1   Prediction of the Crossings Between the STMA and the LTMA by NNs

3.1.1   Choosing the Input Variables into the Neural Network Model

In order to construct a DSS using NNs which makes appropriate prediction of the crossings between the STMA and the LTMA, one has to choose input variables carefully which may give significant effect upon the changes of the two moving averages of the stock price. We have utilized the sensitivity analysis for this objective [11],[12].

3.1.2   Teacher Signals

The proposed decision support system (DSS) has been constructed in order to predict crossing (Golden Cross or Dead Cross) several weeks before it occurs. In order to let the DSS have such an intelligent property, we have prepared the teacher signals for neural network training as shown in the Fig.4. (Due to limitation of space, we don't touch upon teacher signals concerning Dead Cross. Interested readers are kindly asked to attend our presentation.)

3.1.3   Neural Network Training

Neural network training has been carried out by using MATLAB Toolbox in order to make a prediction of the "Golden Cross" and "Dead Cross" several weeks before they occur.

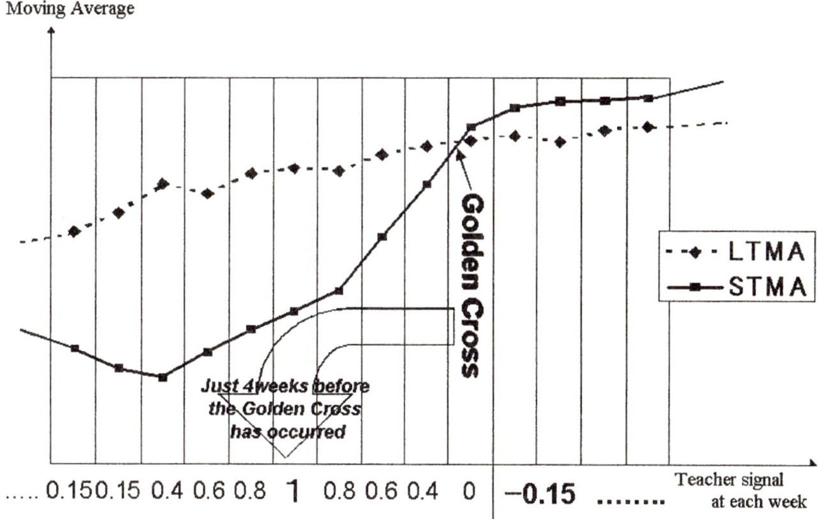

Fig. 4. Teacher signals near to the week when the Golden Cross has been observed

3.2 Decision Support System for Dealing Stocks

By utilizing outputs from the trained neural network, one can make a prediction of the crossings "Golden Cross" and "Dead Cross" several weeks before they occur.

This means that one can use the information concerning outputs obtained from the neural network (NN) effectively in order to carry out dealing stocks.

Below, let us propose a new decision support system (DSS; Fig.5) which utilizes this information effectively for dealing stocks:

Remark 3.1: Zmax (Ymax) denotes the maximum difference between LTMA and STMA having been observed after the recent Dead Cross (Golden Cross).

Remark 3.2: The condition Zmax $> \alpha$ (Ymax $> \beta$) in the Fig.5 is used in order to guarantee the maximum gap having been observed after the recent Dead Cross (after the recent Golden Cross) between LTMA and STMA to be large enough. Further, the condition LTMA – STMA $<$ Zmax $*\theta_1$ (STMA – LTMA $<$ Ymax $*\theta_2$) is introduced in order to guarantee the difference between LTMA and STMA to have been shortened considerably.

Remark 3.3: If the two conditions in the Remark 3.2 are satisfied and the output from the trained neural network surpasses the threshold γ, then the proposed DSS makes the decision "Buy". If the two conditions in the parentheses in the Remark 3.2 are satisfied and the output from the trained neural network falls below the threshold - δ, then the proposed DSS makes the decision "Sell".

4 Computer Simulations

We have carried out computer simulations for dealings in the Nikkei-225 from 1997 to 1999. Table 1 shows the changes of the initial money (10 billion yen) by the dealings in the Nikkei-225 utilizing the proposed DSS, the traditional technical analysis, and the Buy-and-Hold method. These simulation results demonstrate that the proposed DSS outperforms the traditional technical analysis and the Buy-and-Hold method considerably. (Although the traditional DSS and Buy-and-Hold method have not been able to produce return, the proposed DSS has succeeded in yielding around 15% average return each year; Fig.6 illustrates rather in detail how the dealings in 1998 have been carried out using the proposed DSS and the traditional technical analysis.)

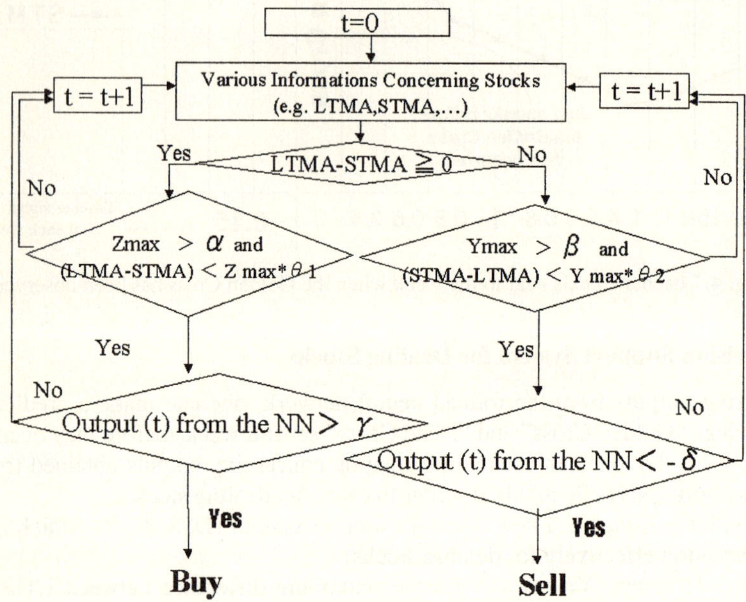

Fig. 5. Decision support system for dealing stocks which utilizes NNs

Table 1. Computer Simulation Results (Nikkei-225; From 1997 to 1999)

	Proposed DSS	Traditional DSS	Buy&Hold
Initial Amount of Money	10.000	10.000	10.000
Final Amount of Money	14.470	9.726	9.584
Returns	+4.470	-0.270	-0.415

(Billion yen)

Remark 4.1: In the above computer simulations, we have taken the charge for dealing into account by subtracting {0.001 * (total money used for dealing) + 250,000} yen from the total fund for dealing.

Remark 4.2: In the above computer simulations, the following parameter values have been used:

$$\alpha=300, \beta=300, \gamma=0.65, \delta=0.65, \theta_1=0.85 \ \theta_2=0.85$$

Remark 4.3: In the above computer simulations, we have not taken "selling on credit" into account. We have just followed the rule: "Buy first, and then sell"

We have also followed the following rules:

1) Buy stocks by using all of the money on hand when the DSS has made the decision "buy".
2) Sell all of the stocks on hand when the DSS has made the decision "sell".

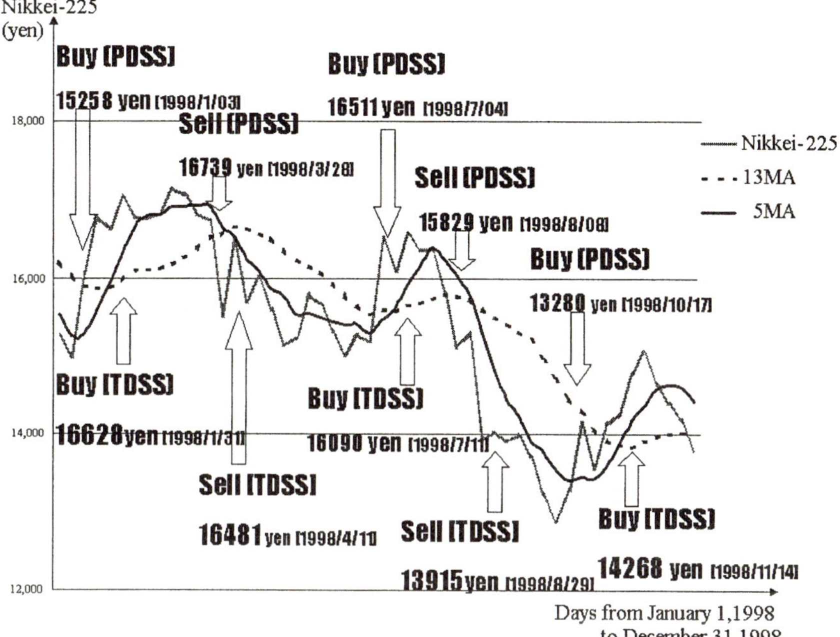

Fig. 6. Dealings by the proposed DSS(PDSS) and the traditonal DSS(TDSS). (Year:1998)

5 Concluding Remarks

A decision support system for dealing stocks which improves the traditional technical analysis by utilizing NNs has been proposed. Computer simulation results concerning Nikkei-225 suggest the effectiveness of the proposed DSS.

However, these simulations have been done during a rather short range of years. In order to execute full confirmation concerning the developed DSS, we need to carry out simulations through a rather long range of years. We also need to check whether the proposed DSS can be successfully applied for dealing in the other indexes such as TOPIX and/or for dealing each listed stock in various stock markets.

Acknowledgements

The authors would like to thank QUICK Corporation for their kind support in giving them various financial data. The authors also gratefully acknowledge the partial financial support of the Grant – In –Aid for Scientific Research (C) by Ministry of Education, Science, Sports and Culture, Japan and the Foundation for Fusion of Science & Technology (FOST).

References

1. Rumelhart, D.E. et al.: Parallel Distributed Processing. MIT Press(1986).
2. Goldberg, D.E.: Genetic Algorithms in Search, Optimization & Machine Learning. Addison-Wesley(1989).
3. Refenes, A.P., Editor: Neural Networks in the Capital Markets. Wiley(1995).
4. Weigend, A.S. et al.: Decision Technologies for Financial Engineering. World Scientific(1997).
5. Baba, N. and Kozaki, M.: An intelligent forecasting system of stock price using neural network. in Proceedings of IJCNN 1992, (1992)371-377.
6. Baba, N. et al.: A hybrid algorithm for finding the global minimum of error function of neural networks and its applications. Neural Networks 7(1994) 1253-1265.
7. Baba, N.: Construction of the decision support system for dealing stocks which utilizes neural networks. MTEC Journal 11(1998)3-41.
8. Baba, N. and Suto, H.: Utilization of artificial neural networks and TD-Learning method for constructing intelligent decision support systems. European Journal of Operational Research 122(2000)501-508.
9. Baba, N. et al.: Utilization of soft computing techniques for constructing reliable decision support system for dealing stocks. in Proceedings of IJCNN 2002, (2002)2150-2155.
10. Baba, N. and Kawachi, T.: A new trial for improving the traditional technical analysis in the stock markets. in Proceedings of KES2004, (2004)434-440.
11. Zurada, J.M. et al.: Sensitivity analysis for minimization of joint data dimension for feedforward neural network. in Proceedings of the IEEE International Symposium on Circuits and Systems.(1994)447-450
12. Viktor, H.L. et al.: Reduction of symbolic rules from artificial neural network using sensitivity analysis. in Proceedings of the IEEE International Symposium on Circuits and Systems. (1995)1788-1793.

Minority Game and the Wealth Distribution in the Artificial Market

Mieko Tanaka-Yamawaki

Department of Information and Knowledge Engineering
Tottori University, Tottori 680-8552, Japan
mieko@ike.tottori-u.ac.jp

Abstract. We focus on the wealth-distribution problem in an artificial market in the context of multi-agent modeling in which trading activity is represented by the minority game. This model is widely investigated in the community of computational intelligence and econo-physics, due to its rich contents including the emergence of self-organization among agents, natural incorporation of limited rationality of agents, and good affinity to evolutionary computation. However, the major defect of this kind of models is the difficulty in quantitatively explaining the constitution of the society in comparison to the real society. In particular, the wealth distribution among agents does not follow the scale-invariant distribution known as Pareto's law. We consider two new elements to be added to the standard minority game in order to make the resulting wealth distribution of the society to fit the real-world value.

1 Introduction

We consider an artificial society in the context of multi-agent system whose trading activity is represented by a simple model known as the minority game (MG) [1,2,3]. Minority game attracts much attention for its simplicity as well as its potential ability to explain self-organized property observed in the real society in spite of the fact that the agents are heterogeneous and act and learn independently each other.

Agents' intelligence is incorporated into the model in order to represent so-called limited rationality, namely the agents are moderately intelligent and they can learn to some extent, yet they have access to a very limited part of the entire information about the system. In order to clarify this point, we will spend some moment considering how traders think and behave in the market.

Traders always try to outwit others. They would think of buying while others are selling, and selling while others are buying. They are content if such a choice was a success. Seemingly a result of much deliberation, a trader's satisfaction simply comes from the action he took was indeed the right one. Traders also know the history is useful if it is not too long. Contrary to other games such as chess, a long process of thinking is meaningless for the game of trading. What is important is when the current trend ends and the market enter the reverse trend cycle. Those who can judge the moment of turnaround would win by outwitting the others.

Traders usually do not exchange ideas each other. In this sense they are independent agents. They all have a common knowledge of price history of the market in the

R. Khosla et al. (Eds.): KES 2005, LNAI 3681, pp. 15–20, 2005.

near past, e.g., whether many traders sold last weekend or not, and use it to make their own decisions. The market trend is formed when the bulk of traders make the same choice. This occurs as a stochastic process. In other words no one can predict based on any deterministic logic exactly when it occurs and how. The asset price is another stochastic process in the market. Those two processes are correlated each other. One drastic assumption is to regard those two processes as a single stochastic process of the system. MG is then naturally derived as a result of such assumption.

2 Minority Game (MG)

The game is played by odd numbers of (N) independent agents, who decide which of two actions (e.g., buy or sell) to make. The agents who have chosen the minority action get rewarded by one unit of wealth, while the majority agents loose one unit each. Agents do not communicate each other and they use only the public information of which choice won. In order to reflect reality, agents are given only limited intelligence in terms of time as well as space. All the agents use the same length of memory (M) and the same fixed number (S) of strategies. Individual agent uses the same strategy given at the beginning of the repeated games and different agents are given different set of strategies.

Fig. 1. shows an example of strategy set owned by an agent for the case of M=3 and S=8. Note that M and S are common to all the agents. Each agent chooses one action out of 0 or 1 at each match based on the best-scored strategy out of S strategies in its own bag. All the agents have 8 strategies although the content of the bag is different for a different agent. The history column shows the array of winning choices for M consecutive steps. There are $L=2^M$ possible arrays of history for the memory length M, for each of which there are two choices of action, 0 or 1. Thus the total number of possible strategies mounts to 2^L=256 for M=3.

History of winning side	Strategy 1	Strategy 2	...	Strategy 8
000	1	0	...	0
001	0	0		0
010	0	0		1
011	0	1		1
100	1	0		0
101	1	1		1
110	1	1		1
111	1	1		0
Scores →	5	0	...	-2

Fig. 1. Example of strategy table held by one agent for the case of M=3, S=8. The bottom row contains scores given to each strategy as a result of games. For example, when 0-1-0 is winning record of the past three matches, strategy1 having the best winning score tells you to vote for the action 0 in the next match

The reason to call MG as a game lies in the following elements:
(1) Each agent chooses one out of multiple choices of actions
(2) Each agent has a set of strategies and uses the best-scored one
(3) The winners of the game are rewarded and the losers pay penalties/commissions

3 Spontaneous Cooperation

The advantage of the Minority Game is not only its simplicity but also the potentiality to clarify the origin of self-organized behavior of agents [4]. This property is shown in many computer simulation results [5] that the number of winners tends to be maximized while the agents is never taught to be cooperative each other. Note that the number of winners cannot exceed N/2, namely one half of the total agents, since the winners belong to the minority side. The appearance of such spontaneous cooperation prompts us to explore its origin.

We first check how the agents cooperate when no incentive of cooperation is given to the agents in the model. A result of simulation is shown in Fig. 2., where the average number of winners vs. memory length (M) for different numbers of strategies (S), for M=1-12, S=2, 4, 8, 16, 32, that correspond to the 5 lines from top to bottom for M>1. Those data are the average over 100 runs, where each run has 10000 matches of games. The horizontal line around 95 corresponds to the random strategy, which indicates that 95 wins out of 201 total agent on the average if no strategy is employed. Note that the usage of some sort of strategy makes more agents win in many cases. Especially, the two lines corresponding to S=4 and S=8 stay over the random line for all the values of M=1-12. This fact tells us one can make more agents win if a little knowledge is thrown into the playing strategy. However, too much effort of having a long memory size of M>10 or large size of knowledge S>10 simply worsen the result. Considering the fact that the minority is a winner in this game, the maximum number of the winners is (N-1)/2 where N is the total number of agents. For the memory length M larger than 2, the smaller the number of strategies makes more agents to win. This is probably because many agents have chances to share the same strategies since the choice is done within very limited alternatives.

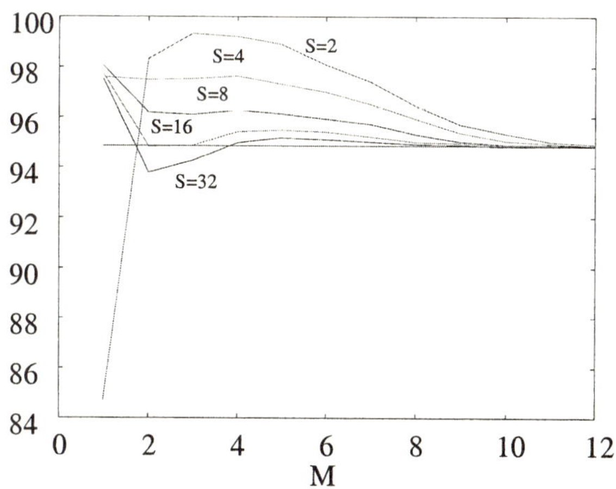

Fig. 2. Number of winning agents vs. memory length M for S = 2, 4, 8, 16, 32 for total number of agents N is 201 in the original MG

In short, MG employs a table of strategy set combined with the learning ability by rewarding outperformed strategies without changing the initially selected set of strategies throughout the game. The reason why the agents appear to be cooperative seems to lie in the severely limited intelligence in terms of space (small S) and of time (small M), although more work is called for in order to establish the reason to explain the result shown in Fig.2.

4   Wealth Distribution of the Society: Pareto's Law

In the real world, asset distribution among people is not random but is known to follow a statistical rule called Pareto's law, discovered by Italian economist Pareto about a century ago [6]. Roughly speaking this law means that a handful (20%) of the total population of a society holds major part (80%) of the total asset and the handful rich members of the society obeys a fractal structure. More precisely, the law is stated that the accumulated wealth distribution density P(x) follows the power law as a function of wealth x

$$P(x) = Ax^{-\alpha}$$

where A is a positive constant and the index α is called Pareto's index. The wealth is more evenly distributed among agents for smaller value of α and the other way around for larger value of α. Montroll and Shlesinger [7] showed that this rule indeed holds for very rich side of the society by using the U.S. statistics, while income distributeon of the working-class people follows the normal distribution. If we compute P(x) in the artificial society of agents playing MG defined in the past two chapters, the Pareto's index amounts to $\alpha = 48$, much larger than the real value of 1-3 [8,9].

This large deviation from the reality seemingly comes from the fact that the agents in MG society are treated even and do not receive any incentive to win and get rewarded.

In order to make the game more attractive, we attempt to change the way of rewarding by giving +2 for each winner and -1 for each looser. By encouraging the winners we expect more selfish competition among agents. As a result of simulations we observed more cooperation among agents, and less dependence on the size of intelligence S. Also for small memory length (M=1-3) the average winners are close to the maximum (101 for N=201), for which the society members enjoy the maximum profits as a whole. However, the number of winners monotonically decreases as the memory length M grows as the wealth distribution becomes biased [10].

Toda and Nakamura [11] considered a modified version of MG in which the agents invest a fixed percentage Y out of the current wealth and receive the profit in proportion to the invested amount only when it wins. In doing so they obtained much smaller value of α and argued that the real society can be simulated by this kind of modification. However, the accumulated distribution of wealth does not have wide enough range of power law behavior to see the fractal property as shown in Figure 3.

We have combined this idea with biased reward system of giving +2 for winners and -1 for losers. The result is shown in Figure 4 for the value of investment rate Y=0.001. We observe wide enough region of power-law dependence from 100 to 10000, where we estimate the power as the Pareto's index to be approximately $\alpha = 2.4$.

This value is comparable to the measured index values in the statistical report published by the Government of Japan [12], and other research using (unpublished) asset data of large companies.

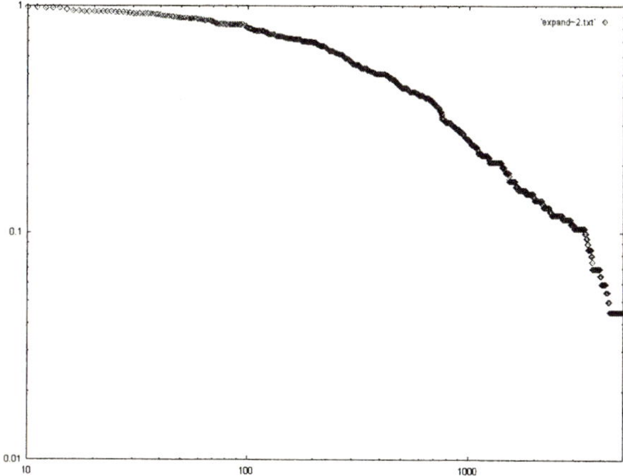

Fig. 3. Accumulated distribution of wealth obtained in the investors model. Pareto's index obtained from the straight line is about 2 which is close to the real value for Y=0.01

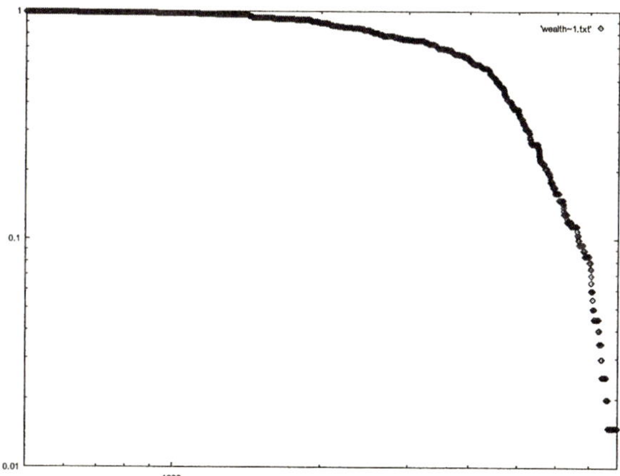

Fig. 4. Accumulated distribution of wealth obtained in our model for investment rate Y=0.001. The value of Pareto's index obtained from the straight line is about 2.4 which is close to the real value

5 Conclusion

We have discussed in this paper a possibility of constructing an artificial society by means of MG and its variances. We have also attempted to describe the essence of MG in an introductory manner. Then we dealt with the problem of wealth distribution in such a society beeing too even compared to the real world situation where Pareto's law holds with index α being close to unity. Following the idea of Toda and Naka-mura who proposed an inventing model we have combined this with a reward-driven model, we have sucessfully derived a fractal structure in the wealth distribution of agents in this MG society.

Acknowledgments

The author thanks Mr. Seiji Tokuoka for helping simulations used in this work. This work is supported in part by the Scientific Grant in Aid by the Ministry of Education, Science, Sports, and Culture of Japan (C2:14580385).

References

1. Damien Challet and Yi-Cheng Zhang, Physica A 246 (1997) 407-418
2. Ton C. Coolen, "Mathematical Theory of Minority Games" (Oxford, 2005)
3. Damien Challet, Yi-Cheng Zhang, Matteo Marsili, "Minority Games: Interacting Agents in Financial Markets" (Oxford, 2004)
4. Mieko Tanaka-Yamawaki, Physica A 324 (2003) 380-387
5. Minority Game web page: http://www.unifr.ch/econophysics/minority/
6. V. Pareto, "Cours d'Economique Politique" (McMillan, London 1897)
7. E.W. Montroll and M.F. Shlesinger, Journal of Statistical Physics (1984)
8. H. Aoyama, W. Souma, Y. Nagahara, H.P. Okazaki, H. Takayasu, and M. Takayasu, Fractals 8 (2000) 293. W. Souma, Fractals 9 (2001) 463.
9. A. Ishikawa, Physica A 349 (2005) 597.
10. Ministry of Health, Labour and Welfare of Japan (2003) Report on Income Distribution in Japan, 2002,15, 5 (Nov. 2003), 795-825.
11. Mieko Tanaka-Yamawaki and Seiji Tokuoka, submitted to AESCS2005
12. K. Toda and Y. Nakamura, "Wealth Dynamics in the Minority Game" (2005) to appear in IPSJ Journal.

A R/S Approach to Trends Breaks Detection

Marina Resta

University of Genova, DIEM sez. di Matematica Finanziaria,
via Vivaldi 5,16126 Genova, Italy
`resta@economia.unige.it`

Abstract. This note focuses on a pointwise estimation of the Hurst exponent H, and on its reliability as a method to detect breaking signals inside a given financial time–series. The idea is that, although classical H can give information about the average scaling behavior of data, its estimation over proper subsamples of the original time–series can reveal variability which is perfectly compliant with sudden changes in direction that are typical of financial markets. The behavior of the pointwise estimation of H is then analyzed on different proxies of market price levels (namely: log–returns, squared log–returns, and the absolute value of log–returns), focusing on the relationships existing with bursts in the markets and those observable in such indicators as well. In this context we find that breaks in the upward/downward tendency of financial time–series are generally anticipated by analogous movements in the estimated H values given on the squared log–returns.

1 Outline of the Work

As widespread known, the rescaled range analysis (or R/S analysis) recalls a method originally introduced by the hydrologist Harold Hurst in 1951. In its classical form, given a time–series of length n, the method provides the statistics:

$$R/S(n) = \frac{1}{s_n} \left\{ \max_{1 \leq k \leq n} \left[\sum_{j=1}^{k} (X_j - \bar{X}_n) \right] - \min_{1 \leq k \leq n} \left[\sum_{j=1}^{k} (X_j - \bar{X}_n) \right] \right\} \quad (1)$$

where s_n is the sample standard deviation, while the first and second terms on the right hand side of Eq.(1) refer, respectively, to the maximum and the minimum of the partial sums of the first k deviations of X_j from the sample mean.

Mandelbrot and Wallis (1969) employed the R/S analysis to detect long range dependence, highlighting the existence in random processes of a scaling relationship between the rescaled range and the number of observations n:

$$R/S(n) \propto nH$$

where H is the so called *Hurst exponent*. Additionally, the same authors provided evidence via Montecarlo simulations that, for sufficiently large n, a proxy of the

R. Khosla et al. (Eds.): KES 2005, LNAI 3681, pp. 21–26, 2005.
© Springer-Verlag Berlin Heidelberg 2005

Hurst coefficient may be given through the regression of the logarithm of $R/S(n)$ against $log(n)$.

For a white noise process we will have $H = 0.5$, whereas $H > 0.5$ generally accounts for a persistent, long memory process (and, conversely, $H < 0.5$ is generally intended as a signal of antipersistence inside the data).

More recently, however, various contributions (see among the others: Lo (1991), Moody et al. (1997)) pointed on the sensitivity of such estimated values to short range dependence, and suggested some modifications of the above procedure. Moving accordingly, we have chosen to refer to the Hurst estimator suggested by Fleming et al. (2001) who derive H through the variance of the set of wavelet detail coefficients.

The major question we are now trying to address is about the reliability of H to provide useful information about financial time–series dynamics. To the best of our knowledge, the Hurst exponent has been mainly considered as an indicator of the *global* behavior of assigned data (see for instance: Fang (1994), Fung (1994), Geweke et al. (1983) and Peters (1994), but the documented literature forms a very huge *apparatus*).

In a recent contribution Los and Yalamova (2004), observed that multifractal spectra of various financial time–series show non–stationary behavior, and they linked such exhibits to the appearance of extreme price movements.

We will try to move one step further, searching for correlations among a pointwise version of the Hurst exponent and the behavior of financial data. In particular, our aim is primarily that to verify if some evidence can be provided that upward/downward movements in prices are in somewhat fashion mirrored by analogous movements on H, or, better, to detect if exists a significant threshold of H fluctuations that can be used to monitorize inversions in consolidated trends.

What remains of the paper is organized as follows. Section 2 briefly introduces the methodology used and the sets of data. Section 3 discusses the results obtained, and Section 4 gives some conclusions and perspectives for future works.

2   Methodology

We derive H through the variance of the set of wavelet detail coefficients $\{d_{j,r}\}$ based on dyadic scaling, being:

$$Var\{\{d_{j,r}\}\} \propto 2^{-j\gamma} \tag{2}$$

with $\gamma = 2H$. This method has been applied to subsamples of the original time–series, obtained running over the data through a rolling window with amplitude τ and shift length of δ time units: in particular in this study we will provide results for the cases of $\tau = 500, 750$, and $\delta = 25$.

Table 1 summarizes the main features of financial time–series we have used.

Table 1. Basic features of data.

Name	Symbol	First date	Last Date	Overall Data Length
Cac40	35000	02/01/1985	25/02/2005	5214
Dax30	846900	14/08/1996	25/02/2005	2202
Nasdaq Combined	$CCO	06/05/1985	25/02/2005	5117
Nikkei 225	N225	02/01/1985	25/02/2005	5214
S&P 500	$INX	20/04/1982	25/02/2005	5673

For each dataset, the available price levels have been transformed according to the following general formulas:

$$r(t) = log\frac{X(t)}{X(t-1)} \quad \text{log–returns}$$

$$a(t) = |r(t)| \qquad \text{absolute value of log–returns}$$

$$s(t) = [r(t)]^2 \qquad \text{squared log–returns}$$

Such transformations are of widespread use in financial econometrics, since they are particular reliable to identify statistical features of underlying data.

While generally the Hurst exponent is evaluated on the $r(t)$'s of a given time–series, we are interested to capture the differential information (whether existent) which may be kept when H is estimated on other proxies of market dynamics.

3 Discussion of the Results

Once the data have been transformed according to the three formulas given in the previous section, the pointwise estimation of H has been performed, thus obtaining three different time–series H_r, H_a, and H_s.

In order to choose among them the most significant one, with respect to the dynamics of the underlying financial time–series, the classical *rho* statistics has been evaluated, opposing to the series of price levels those of H's, obtained for both $\tau = 500$, and $\tau = 750$. The results are reported in Table 2.

Table 2. Rho statistics for H estimated on different proxies of price levels.

rho	Cac40	Dax30	Nasdaq Combined	Nikkei 225	S&P 500
$H_r(\tau = 500)$	−0.2405	−0.3688	-0.2741	0.2438	0.1152
$H_a(\tau = 500)$	0.1688	−0.3509	0.3909	0.3976	0.5463
$H_s(\tau = 500)$	0.3828	−0.0879	0.4550	0.1854	0.6127
$H_r(\tau = 750)$	0.0185	−0.3892	−0.2714	0.3370	−0.1007
$H_a(\tau = 750)$	0.1729	−0.7597	0.3711	0.5806	0.4321
$H_s(\tau = 750)$	0.4892	−0.6289	0.4805	0.3906	0.6234

Two considerations are noteworthy at this point.

Firstly, the *rho* statistics has generally provided results closer to $|1|$ when $\tau = 750$. However, this result is in accordance with the sample size requirements needed for the significance of R/S test.

As second remark, we observe that the *rho*'s associated to H_s are on average the closest to $|1|$: this probably makes the H_s's the most significative to provide alert signals on the underlying price levels.

With those motivations we are now going to test the above conjecture, putting together the behavior of price levels and that of both log–returns and pointwise H_s.

Such comparison is made graphically in Figure 1 (shown at the end of the paper): just for readibility sake, instead of the behavior of H_s its corresponding first difference is shown.

The vertical lines running through the 3–plots blocks are at the occurence of crosses between the $H_s(\tau = 750)$ and the confidence (dotted) band lines at $\mu_{H_s(\tau=750)} \pm 0.75 \cdot \sigma_{H_s(\tau=750)}$.

We search for correspondence between such crosses in pointwise H and bullish/bearish trend breaks in the original time–series.

Note that in all the examined cases there is a reasonably good correspondence between downward/upward movements in $H_s(\tau = 750)$ and in price levels (and obviously, this corresponds to strong fluctuations in the corresponding log–returns).

4   Conclusions

Some preliminary results have been shown about the power of pointwise estimations of the Hurst exponent as indicators of trend breaks in financial time–series. In particular, it has been shown that such indicators are particularly succesfull when the Hurst exponent is evaluated on the squared log–returns of the original data.

This is not really surprising, since the concept of squared log–returns is intimately linked to that of volatility: studying H_s means to have an in–deep look at the volatility features of data.

However, a number of steps still need to be performed. The first question to be addressed is related to find the *ideal* size τ of blocks the original time–series has to be divided in. Here we have considered $\tau = 500$ and $\tau = 750$, since according our experience they provide the most affordable results, but we are perfectly aware that more investigations are necessary, since the choice of τ could be dependent on the kind of data under examination. A second point that should be object of deeper studies is related to the creation of alert signals based on the information provided by the H sequences, i.e. to what extent it is possible to use H's like *typical* technical analysis indicators.

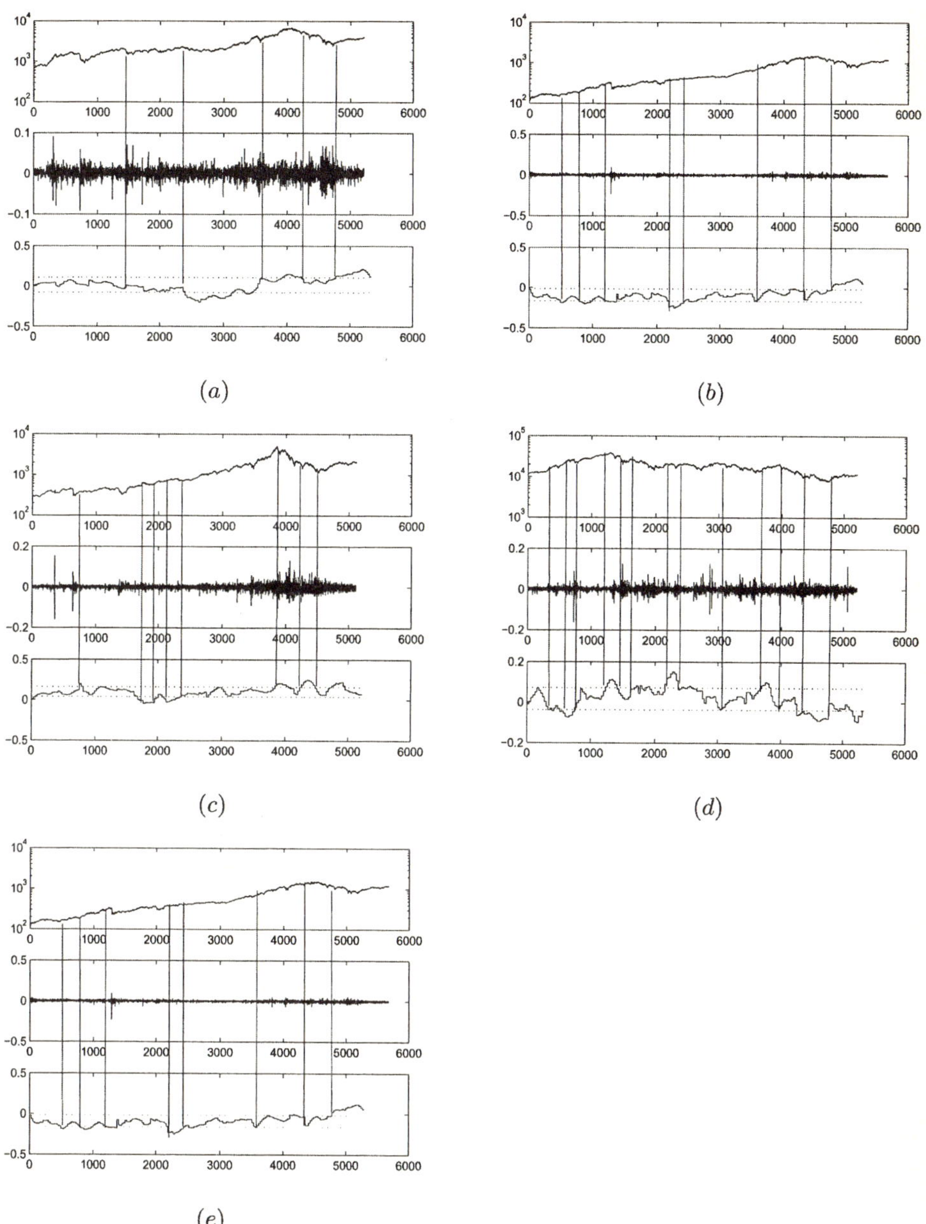

(a)

(b)

(c)

(d)

(e)

Fig. 1. From top to bottom, and from left to right: price levels (on semi–logarithmic scale), log–returns, and $H_s(\tau = 750)$ for the Cac40 (a), Dax30 (b), Nasdaq Combined (c), Nikkei 225 (d), and S&P 500 indexes.

References

1. Fang, H., Lai K. S., and M. Lai (1994): "Fractal Structures in currency futures price dynamics", *Journal of Futures Markets*, **14(2)**, 169-181.
2. Fleming, B. J. W., Yu, D. , Harrison, R. G., and D. Jubb (2001),"Wavelet-based detection of coherent structures and self- affinity in financial data", *European Physical Journal B*, **20(4)**: p. 543-546.
3. Fung, H., Lo, W., and J. E. Peterson (1994), "Examining the dependency in intraday stock index futures", *Journal of Futures Markets*, **14(4)**, 405-419.
4. Geweke, J., and S. Porter-Hudak (1983), "The estimation and application of long memory time–series models", *Journal of time–series Analysis*, **4(4)**, 221-238.
5. Hurst, H. E. (1951), "Long term storage capacity of reservoirs", *Transactions of the American Society of Civil Engineers*, **116**: 770-799
6. Lo, A, (1991), "Long-Term Memory in Stock Market Prices", *Econometrica*, **59**: 1279-1313.
7. Los, C. A. and R. Yalamova (2004), "Multifractal Spectral Analysis of the 1987 Stock Market Crash", Kent State University Workin paper.
8. Mallat, S. (1999), *A Wavelet Tour of Signal Processing*, 2nd ed., Academic Press, Boston, MA
9. Mandelbrot, B. and J. R. Wallis (1969), "Robustness of the resaled range r/s in the measurement of noncyclic long run statistical dependence ", *Water Resources Research*, **5(5)**, 967-988
10. Muzy, J. F., Bacry, E., and A. Arnéodo (1991), "Wavelets and Multifractal Formalism for Singular Signals: Application to Turbulence Data", *Physical Review Letters*, **67-25**, December, 3515-3518.
11. Peters, E. E. (1994), *Fractal market analysis: applying chaos theory to investment and economics*, John Wiley & Sons, New York.

Applying Extending Classifier System to Develop an Option-Operation Suggestion Model of Intraday Trading – An Example of Taiwan Index Option

An-Pin Chen, Yi-Chang Chen, and Wen-Chuan Tseng

Institute of Information Management, National Chiao Tung University,
Hsinchu, Taiwan, 1001 Ta Hsueh Road, Hsinchu, Taiwan, R.O.C.
{apc,ycchen}@iim.nctu.edu.tw, zoet.iim91g@nctu.edu.tw

Abstract. This novel study developed an option-operation suggestion model by applying integrated artificial intelligence technique, extending learning classifier system (XCS), which incorporates reinforcement machine learning method to the dynamical problems to the behavior finance. Due to the history of Behavior Finance, many researches have found that the shape of stock trend is not following random walk model, but the repeated trading patterns exist which are referred to as investors experiences. Furthermore, some classical researches have been merely adopted traditional artificial intelligence to analyze the result. Those methodologies are not sufficiently to resolve the dynamical problem, such as economical trading behaviors. Therefore, the model has been proposed concerning intraday trading but avoiding the system risk in the short-term position to benefit investors. By dynamic learning ability of XCS and general population features, the output operation suggestions could be obtained as a reference strategy for investors to predict the index option trend. As an example of Taiwan Index option, the results of the accuracy and accumulative profit have been exhibited remarkable outcome, and so as the simulations of short term prediction with 10-minute and 20-minute tick data.

1 Introduction

In 1980s, most researches have followed with Fama (1970)[4] who proposed Efficient Market Hypothesis and shown the economic trend pattern as random walk model, so that no approach could be utilized for the prediction about economics. Until NYSE market crashed in 1987, many economists have indicated that the investment behaviors in the market have been found repeated patterns even but against random walk model. Recently, Behavioral Finance theory incorporates the investors' psychological factors into trading behavior analysis, such as that Bondt et al. (1985)[1] have pointed out that investors incline to overreact from news. Shleifer (2000)[14] also has addressed that the repeated patterns exist in the investor trading behaviors based on declination. But, in the dynamical economic environment, those trading behavior analysis just focused on statistics-based result, of which prerequisite was not rational at all. By the way, those researches could not be workable on prediction for investors to accumulate profit.

Since financial prediction studies of stock market have been applied artificial intelligence techniques to predict the future or analyze the data. In the past experience most of them, by neural network for example, were the learning models. In reality, these traditional AI approaches could not produce the adequate solution to the dynamical economic data. Thus, the classifier systems, integrated AI approaches, are developed by the chance. The original ones, Learning Classifier System (LCS), was brought up by Holland (1996)[7]. For more accuracy, Wilson [16] proposed Extend CS (XCS) by modifying the fitness value function and only doing the GA function in the action-set. Even though, Liao and Chen (2001)[11] tried to apply XCS on stock forecasting and their research could be obtained the stable result and remarkable outcome. Besides, our previous research [13], applied LCS, has presented obvious conclusion in stock forecasting as well.

These studies have addressed that the stock market behavior analysis is not simply following as random walk model and the repeated patterns existed, so that the stock market trend is predictability. However, stock, just a simple financial commodity, is not sufficient to be formed as an optimal investment combination, but the derivative financial commodities, such as Index Future and Index Option, are provided specifics with lower cost and higher leverage. Recently, few forecasting researches have been focused on Index Option because it is indeed a complex financial commodity, except Yao (2000)[9] applying the simple back-propagation of neural network to option price forecasting. Due to the Index Option could also only avoid risks for investment but also keep investors away from economic system risk, we combined the integrated AI model, XCS, searched for the force relationship between financial commodities and concerned with the index option for hedge, to build an option operation suggestion model of intraday trading to provide short-term hedge investment strategy.

2   The Proposed Model

We proposed a scheme to develop a dynamical strategy model of put-call option to do financial product arbitrage, shown as Fig. 1. As the figure shown, this model has two strategies, put and call option. The put-option part was focused on put-index-option, and the other part, call-option part, was focused on call-index-option. Furthermore, the kernel of proposed model was applied Wilson's XCS [16], described as the following paragraph. As regards the relative inputs, which to be encoded, they would be detailed in section 2.2 and 2.3.

2.1 An XCS-Based Model

As for the XCS model, it would be divided into three steps, training, data-recall with testing, and testing. First, in the training step, learning from training set data is the main job which is employed the pay-off landscape function and dynamic GA parameters. It is by exploration-search approach to generate rule and randomize to match the action as output. Furthermore, the pay-off would be obtained and be utilized to modify the parameters, predicted-payoff, predicted-error and fitness etc, which are reinforced the rule. The other is applied GA to "evolve" the new rules, and GA operator, including duplication, crossover, and mutation, is employing tournament selection as selecting rule [2]. Continuously, new rules are survival or not, are decided by subsumption mechanism. Therefore, the training population is built up. The accuracy of training step was calculated by exploitation-search approach and to be utilized to the prepared population.

In the testing step, applying the training population as the initial one and determining the appropriate rule, according to the fitness and predicted pay-off of rules, are two main works. The accuracy of testing was estimated by analyzing the output actions. As for the data-recall step, as dynamic learning, it is also employed to recall the testing set data to reinforce the model. In addition to the three steps, the relational inputs would be detailed in the next section.

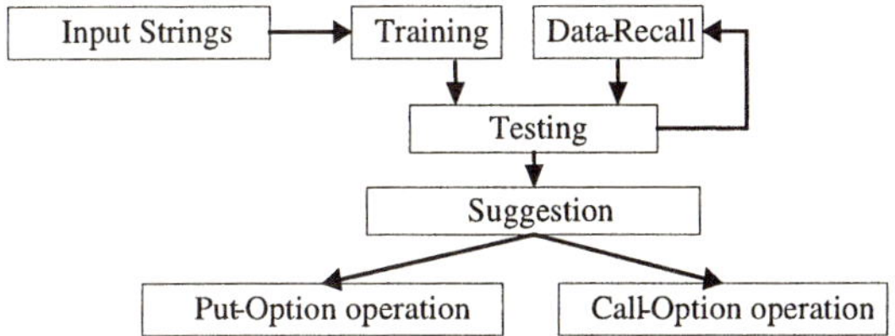

Fig. 1. A put-call option operation suggestion model based on XCS

2.2 Relative Factors

For the model developed, relative inputs that deeply affected the accuracy of outputs should be considered, and the past but classical literatures will be mentioned, such as put-call option, the price and volume fluctuation of future and the volume and implied volatility of put-call option should be taken into consideration. According to B-S model, these factors, which could determine option price, include the price of underlying, the exercise price, the mature term, the volatility of underlying, and the riskless interest ratio. In intraday trading, we first could ignore the riskless interest ratio, which is a constant, and the mature term after preprocess. As regards the other critical factors, some relative surveys are detailed as follow.

1. The relation between the option, future and spot. Lots of option pricing models were given to the assumption, that is, the commodities in related market would be in response to the news immediately. Actually, most of them have different abilities of response to the effects in real market. Kawaller et al. (1987)[13] and Hsieh (2002)[7] both have indicated that the future leads spot at least 5-mins. Whaley et al. (1996)[5] also have demonstrated that the future and option both lead spot. From these perspectives, the future could be evaluated the option price trend.

2. The relation between option volume and underlying price fluctuation. In Market-Microstructure, many researches investigated that the higher volume frequently effects the fluctuation of future. From the aspect, we could identify some market information from the commodity volume. French et al. (1986)[6] agreed with the future price fluctuation that could be predicted by the volume. Furthermore, Easley et al. (1998)[3] has showed that the option volume also obviously follows the same phenomenon. Thus, the volume, including option volume as well, should be taken into consideration to the future price fluctuation.

3. The correlation between option implied volatility and underlying volatility. The volatility was generally estimated by historical volatility and implied volatility. Lantane et al. (1976)[10] has demonstrated that to estimated the volatility from implied volatility is a proper approach. CBOE has been employed the market volatility index approach by Whaley (1993)[15] to calculate the VIX index for in-

vestors to predict the future price fluctuation. These researches all have exhibited that the implied volatility could be a measured factor for the future market. As the result, this study also has been concerned to the different exercise price, the different mature term, the put or call option, and even the implied volatility of different trading time.

As the three issues described, we will conclude some critical factors to the proposed model which will be detailed as below.

2.3 Input-Output Variables of the Structure

The inputs and outputs of the structure in the proposed model are divided into two parts, condition part and action part, which are both encoded by the binary code-the Gray code. About the coding method, the differential of interval states should be one bit to the reason, maximal generalization. In the previous section as mentioned, those mainly critical factors would be considered as condition part. Action part is rules generating, which consists of the operation suggesting by the model and the analysis of trend. The detailed structure is shown as Fig.2.

According to the structure, the raw data also needs to be translated into formatted data. The following steps are described as the translation processes:

1. For the performance, the minute-tick data is selected, which is Taiwan Index Option (trading date, exercise price, mature year-month, call-option or put-option, trading time, price, volume), and Taiwan Index Future (trading date, mature year-month, trading time, price, volume).
2. Recalculating the mature date of the trading option and the day-tick volume.
3. Applying the Newton-Raphson approach to estimate the option implied volatility.
4. Calculating 5-minute-tick data, including volume, highest price, lowest price, last price, and weighted-average of implied volatility.
5. Following the pre-4 steps, the relative input factors of condition part would be consequently obtained, for example, the price and volume moving-average of future and option.
6. Calculating the Taiwan Index Option price fluctuation, ΔV_m and θ_m .

    ΔV_m , the next m-mins valid fluctuation

$$\Delta V_m = \begin{cases} V_{m\_high} - V_{close} & \text{if } V_{m\_high} > V_{close} \\ \Delta V_m + (V_{close} - V_{m\_low}) & \text{if } V_{close} > V_{m\_low} \end{cases} \tag{1}$$

    θ_m , the ratio of next m-mins valid fluctuation to 5mins close price.

$$\theta m = \Delta V m \div V close \times 100\% \tag{2}$$

    Here, V_{close}, V_{m_high}, and V_{m_low} are respectively denoted to closing price of current day, the highest price in m minutes and the lowest price in m minutes.
7. From 1-6 steps and the encoding method, the relative variables of condition part are translated to the inputs of the designed XCS model. As regards to the action part, the rules would be still calculated by step 6.

All of the raw data should be translated by these seven steps. After that, those inputs would be ready for XCS model, and the outputs should be presented to the option-operation strategy as well.

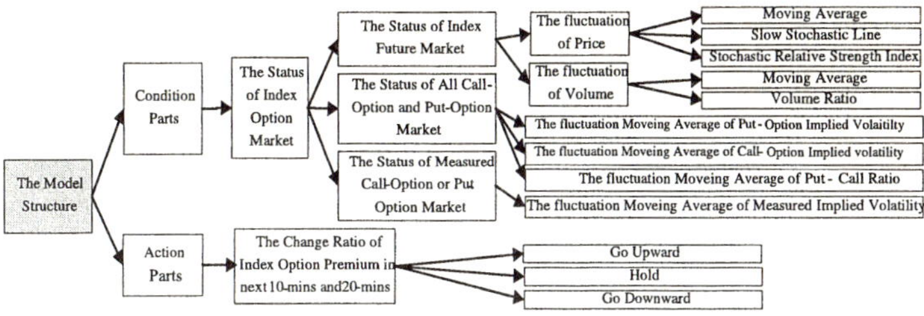

Fig. 2. The structure of inputs and outputs. The structure is consisted by the condition part and the action part. The condition part is designed to reveal the current market status. The action part is designed to the prediction of option price trend fluctuation, the strategy suggestion

3 Experiment and Result

The investment simulations are the experiments focused on Taiwan Index Option, including the call-option and put-option, which employed Taiwan Index Future to describe the trend of option price. All raw data are collected by Taiwan Future Exchange (TAIFEX), and the period is from 2003/12 to 2004/02. All data are divided into two parts. The training set is the daily data in Jan. 2004 and testing set is the daily data in Feb. 2004.

For the adjustment of accuracy, first of all, we respectively form four datasets, the highest-four volume of options within 5-minute-tick data in the two experiments. Then eight datasets by two kind options, which are put and call, are able to be predefined separately.

In the experiments, the best accuracy result has demonstrated to predict the next 10-mins could more than 70%. In call-option model, the accuracy of lower exercise price contract is better than the higher one. In put-option model, the accuracy of higher exercise price contract is better than the lower one. The eight accuracy results in testing step were shown as Fig.3.

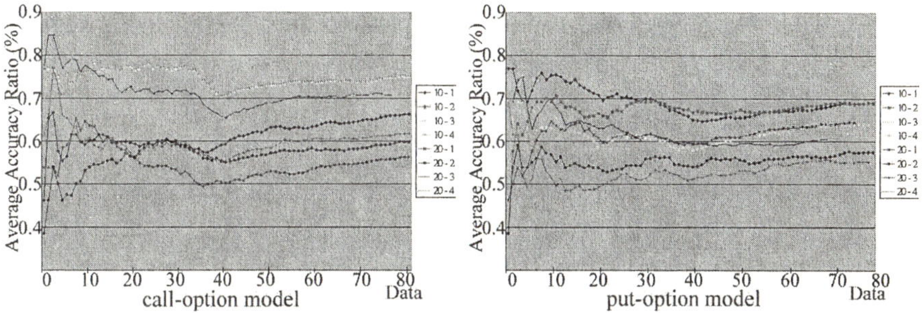

Fig. 3. The four accumulative accuracy ratios of call-option model and put-option model with two part dataset

In the second step, the simulation is the real investments according to the option operation by the proposed model which is suggested, go-upward signal or go-downward signal, and it is compared with Buy & Hold strategy. The results of 10-min and 20-min ticks both could be figured out that those accumulative profits are all better than Buy & Hold strategy and the eight kinds annual profit ratio of the proposed model are all at least 0.5, exhibited as Fig.4.

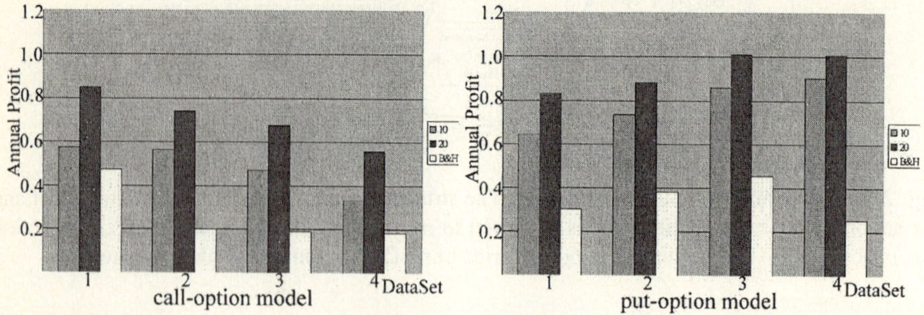

Fig. 4. The figure shows the comparison of annual profit ratio by the proposed model with m=10 and m=20, and Buy & Hold model. No matter which datasets, the results all show the better annual profit than Buy & Hold model. Besides, at least 0.5, the annual profit ratio has been exhibited that the proposed model could be given as a suggestion successfully to the short-term hedge investment

4   Conclusion

Few AI researches traditionally were focused on option forecasting model because of its complexity and unstable environment. However, we paid attention to this aspect and applied XCS developing an option operation suggestion model of intraday trading on Taiwan Index Option. Finally, the empirical results have been demonstrated that the average accuracy ratio of predicting next 10-min option trend could be more than 70% and the annual accumulative profit ratio could be achieved to at least 50%. The result has exhibited the remarkable outcome and is able to be as a referent suggestion for short-term hedge investments and concerning avoiding system risk as well. In the future, we suggest that data clustering process could be considered the level of exercise price related to spot price, and data coding process could be employed as continuous-valued representation to substitute the binary code. In that case, much better results and performance would be obtained.

References

1. Bondt, W.D., Thaler, R.: Does the Stock Market Overreact?. Journal of Finance. 40 (1985) 793-808.
2. Butz, M., Sasry, K., Goldberg, D.: Strong, Stable and Reliable Fitness Pressure in XCS due to Tournament Selection. IlliGAL Report No. 2003027 (2003).
3. Easley, D., O'Hara, M., Srinivas, P.: Option Volume and Stock Prices: Evidence on Where Informed Traders Trade. Journal of Finance. 53 (1998) 431-465.
4. Fama, E.: Efficient Capital Markets: a Review of Theory and Empirical Work. Journal of Finance. 25 (1970) 383-417.
5. Fleming, J., Ostdiek, B., Whaley, R.: Trading Costs and the Relative Rates of Price Discovery in Stock, Futures, and Option Markets. Journal of Futures Markets. 16 (1996) 353-387.
6. French, K., Roll, R.: Stock Return Variances: the Arrival of Information and the Reaction of Traders. Journal of Financial Economics. 17 (1986) 5-16.

7. Holland, J.: Escape Brittleness: the Possibilities of General Purpose Learning Algorithms Applied to Parallel Rule-based Systems. Machine Learning, an Artificial Intelligence Approach, Vol. 2, (1986) 593-623.
8. Hsieh, W.L.: Market Integration, Price Discovery, and Information Transmission in Taiwan Index Future Market. Journal of Financial Studies. 10 (2002) 1-31.
9. Yao, J., Li, Y., Tan, C.L.: Option Price Forecasting Using Neural Networks. The International Journal of Management Science. 28 (2000) 455-466.
10. Lantane, H., Rendleman, R.: Standard Deviations of Stock Price Ratios Implied in Options Price. Journal of Finance. 31 (1976) 361-381.
11. Liao, P.Y., Chen, J.S.: Dynamic Trading Strategy Learning Model Using Learning Classifier Systems. Proceedings of the 2001 Congress on Evolutionary Computation. 2 (2001) 783-789.
12. Lin, J.Y., Cheng, C.P., Tsai, W.C., Chen A.P.: Using Learning Classifier System for Making Investment Strategies Based on Institutional Analysis. Table of Contents AIA (2004) 765-769.
13. Kawaller, I., Koch, P., Koch, T.: The Temporal Price Relationship between S&P 500 Futures and the S&P 500 Index. Journal of Finance. 42 (1987) 1309-1329.
14. Shleifer, A.: Inefficient Market. Oxford University Press, New York. (2000).
15. Whaley, R.: Derivatives on Market Volatility: Hedging Tools Long Overdue. Journal of Derivatives. 1 (1993) 71-84.
16. Wilson, S.: Classifier Fitness Based on Accuracy. Evolutionary Computation. 3 (1995) 149-175.

Applying Two-Stage XCS Model
on Global Overnight Effect for Local Stock Prediction

An-Pin Chen, Yi-Chang Chen, and Yu-Hua Huang

Institute of Information Management National Chiao Tung University,
Hsinchu, Taiwan, 1001 Ta Hsueh Road, Hsinchu, Taiwan, R.O.C.
{apc,ycchen}@iim.nctu.edu.tw, alvin.iim91g@nctu.edu.tw

Abstract. This study applied an integrated artificial intelligence method, extend learning classifier system (XCS), to predict the stock trend fluctuation considering the global overnight effect. However, some researchers have already indicated that XCS model that is applied successfully to form a forecast model in local market. Based on those prediction models, we put more effort to focus on the financial phenomenon, overnight effect between each two global markets, and we developed a two-stage XCS model to forecast the local stock market. In the experiments, DJi and Twi are chosen as referent and predicted markets respectively, and the model is trained by their historical data. For its accuracy verified, the model is tested by recently data. Finally, we have concluded that the proposed model successfully simulates the phenomenon, and the high ratio of correctness is definitely figured out.

1 Introduction

Then overnight effect, existed in the global economic market, originates from the opening sequence of each market. Some relative researches focused on the relationship between markets. For instance, Chan et al. (1991) [2] firstly found the overnight effect existing by using return rates or volatilities on S&P500 index spots and futures. Moreover, Chan et al. (2000) [3] analyzed 30 NYSE stocks, 21 corporations in England and 7 corporations in Asia, by employing simple regression statistical model and shown that the overnight information strongly influences the open price of next trading day in first 30 minutes. Eun and Shim (1989) [6] obtained US stock market as the key impact in global by applying vector-auto-regression (VAR) model to analyze the correlation of 9 national stock markets, such as Europe, US, Asia, and so forth. Chou and Wu (1998) [5] founded that US stock market affects the 12% of daily volatility of return, especially the return of overnight, in Taiwan stock market. They used GARCH model to investigate the influences of returns and volatilities between Taiwan weighted index and US S&P500 index. Besides, Huang et al.(2000) [8] found that Dow Jones index obviously affecting Taiwan stock market. In those cases, the overnight effect should be indeed existed. However, many financial forecast models only focuse on single market so that they are almost mentioned simple factors. Thus, this is the first reason that we develop the model.

Besides, Liao and Chen (2001) [12] indicated that the traditional artificial intelligence(AI), such as neural network, is hard to be explained intutively and is unstable in some dynamic environment states until an novel integrated AI technique, Classifier System (CS), Holland (1986) [9]. CS is composed of the inductive

R. Khosla et al. (Eds.): KES 2005, LNAI 3681, pp. 34–40, 2005.

approach and evolutionary approach to be utilized for the dynamical problems, such as predication problem. Recently, CS is also applied in related financial issues, such as Mitlöhner et al. (1996) [13] and Beltrametti et al. (1997) [1]. Those results would conclude that CS has to classify the environment state, predict the appropriate output, produce, and adjust the current rule to respond new surrounding change under the dynamical state. Also, Lin et al. (2003) [11], our previous research, successfully indicated that stock trend is influenced by unknown and unpredictable surroundings and is applying a kind of CS, Learning Classifier System (LCS), to develop model for the capability to discover the patterns of future trends. Futhermore, Wilson (1995) [14] proposed an Extend Learning Classifier System (XCS) to improve the learning ability of LCS. Thus, considering our previous research, Liao and Chen [12] and the overnight effect, we tried to develop a two-stage XCS model to simulate the overnight effect to increase the investment profit.

2   A Prediction Model

For development of prediction model, those researches were almost successfully proven their models practicable for predicting the trend, but few of them were considered to the relative factors that was exactly affected the results. Therefore, in this paper, we not only concerned the relative factors from global economic market but also applied the integrated AI techniques to develop the model. In the following subsections, the relative factors would be firstly mentioned, and then the foundation of prediction model would be detailed.

2.1   The Relation of Input Factors and the Overnight Effect Theory

The overnight effect is simply verified by those researches [2], [3]. For the model development and simulation, we choose the Taiwan weighted index (Twi) as observed market and Dow-Jones index (DJi) as the overnight information referent market.

First, the overnight effect theory is figured out in part I, shown as Figure 1. Considering the phenomenon, the time series data of stock becomes non-continuous and some trading behaviors maybe exist [4], [7]. Owing to the restriction on trading time, the opening price of the observed market would not only be affected by the overnight effect and the closing price of the previous day, but also by the opening or the closing price of the referent market. As regards the time t, it is the proposed model starting to predict the trend of the next trading day by the current stock price and the following overnight information (roi_t) of non-trading period. As for part II, it is an inference from part I by DJi substituting for roi_t, and the following equations are the deduced steps.

1. As part I shown and only one observed market considered, the overnight effect could affect the next day opening price, which is exhibited by eq.1, while $P\_Return\ ^o_{t+1}$ and $Return^{co}_t$ are respectively denoted as the opening price of the next trading day and the trading return of the current day at the closing time.

$$P\_Return\ ^o_{t+1} = Return\ ^{co}_t + roi_t \tag{1}$$

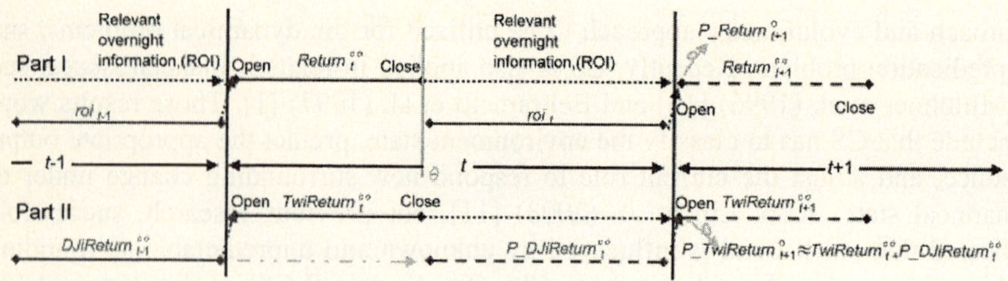

Fig. 1. The overnight effect theory

2. In this step, the referent market, DJi, is additionally involved. Owing to the trading time of DJi is just during the time between the closing time and the next day opening time of Twi, the trading return of DJi ($DJi\_Return_t^{co}$) is utilized to substitute the local overnight information (roi_t). In the meanwhile, eq. 1 would represent to eq. 2. $P\_TwiReturn_{t+1}^{o}$ is denoted the prediction return of the $t+1$ day opening of Twi. $TwiReturn_t^{co}$ is the Twi return of current day.

$$P\_TwiReturn_{t+1}^{o} = TwiReturn_t^{co} + DJiReturn_t^{co} \tag{2}$$

3. Actually, $DJi\_Return_t^{co}$ would not be obtained at the time θ, 5 minutes before the closing time, not the trading time in DJi market. For the reason, we designed the first-stage of proposed model to predict DJi market so that the predicted return of DJi ($P\_DJiReturn_t^{co}$) could be obtained while using two inputs, $DJi\_Return_{t-1}^{co}$, the return of day t-1 in DJi, and $TwiReturn_t^{co}$. After this, eq.2 could represent to eq.3.

$$P\_TwiReturn_{t+1}^{o} = TwiReturn_t^{co} + P\_DJiReturn_t^{co} \tag{3}$$

Finally, before the closing time, we still adopted the second-stage and two inputs, $TwiReturn_t^{co}$ and $P\_DJiReturn_t^{co}$, to obtain the output, $P\_TwiReturn_{t+1}^{o}$, meanwhile, $P\_TwiReturn_{t+1}^{o}$ would be immediately utilized as suggestion, the trend of next day, to investors making decision.

2.2 Two-Stage XCS Prediction Model

Due to the overnight theory and the different time zone between Taiwan and US, the trading time of the DJi between PM 22:30 and AM 05:00 (Taiwan time), which crosses two days, the model we proposed contains two stages. The first stage is to generate the predicted return of DJi, $P\_DjiReturn_t^{co}$. Second, $P\_TwiRetum_{t+1}^{o}$, the prediction trend of Twi, is the final output, shown as Fig. 2. Both stages are applying the methodology, XCS, which is detailed in [10], [13].

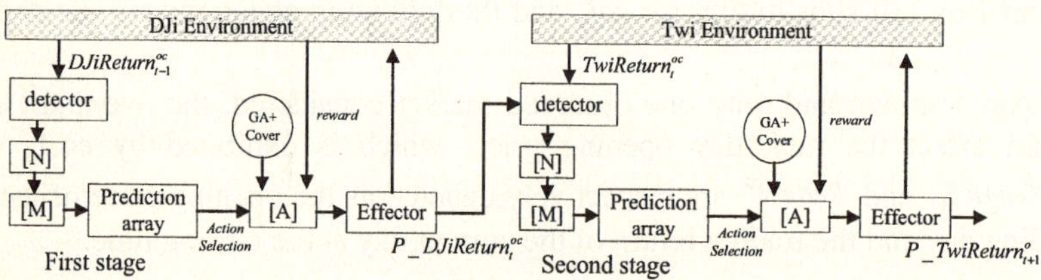

Fig. 2. Two-stage XCS prediction model

As Fig. 2 drawn, in each stage, the detector of XCS model transforms the outside environment information into the message for this classifier system. Comparing with the rule base [N], it is for sure whether the matching rules exist. Then, the match set [M] is formed from these classifiers which are matched the rule base. The prediction array is formed, based on each action of the match set from counting the weight average of fitness degree. While XCS starts action selection, setting action is by the best choice from the prediction array or randomly, if no fitting classifiers form the match set and covering rules. Finally, the effector takes the message from the action set [A]. In effector, the outputs or actions are corresponding to the content of the message. Besides, the outputs or actions based on the responses of the ordinary rule base producing are still necessary to be compared with the outside environment to do the feedback. According to the feedback of reinforcement program, the related parameters of XCS are renewed, such as the expected return, the expected return error, the accuracy and the fitness in the classifier. After executing the action, carrying out the GA in action set [A] includes duplicate, crossover and mutation on the most fitness outside environment population.

As regards the input variables, the historical moving average data of price and volume are required based on stock prediction. Therefore, the price-volume moving average data and the moving average convergence-divergence (MACD) extended by the price moving average are both utilized as the input variables which would be translated into bit-string type. In the first stage, predicting the DJi, about the price, the daily, weekly, monthly, and quarterly moving averages (1, 5, 20 and 60 days) are separately adopted as input variables (4 bits), down-trend (0) or up-trend (1), and their trend-permutation (4!=24 kinds) which would be encoded by 5 bits. As for the volume, its daily, weekly and monthly (1, 5 and 20 days) moving averages are utilized as the input variables (3 bits), down-trend (0) or up-trend (1), and their trend-permutation (3!=6 kinds) which would be encoded by 3 bits. We have already considered the quarterly moving averages (60 days) of volume as input, but it is insensitive. Also, 8 kind statuses that encoded by 3 bits are presented by 12 and 26 MACD patterns. Thus, the first-stage output $P\_DJiReturn_t^{co}$ (3 bits) would be obtained the corresponding input, 18 bits.

In the second stage, predicting the Twi, the input variables and process are the same as in the first stage. Additionally, this stage input bits (21 bits) need to be considered to $P\_DJiReturn_t^{co}$ (3 bits) from the first stage. After second stage, $P\_TwiReturn_{t+1}^{o}$ (3 bits) would be obtainded.

Figure 3 shows the distribution of the historical data of DJi and Twi. In order to analyze, this research separates the data into 8 groups equally, 4 up-trends and 4 down-trends. These 8 groups are encoded by c_1, c_2, c_3, detailed as table 1.

In this study, we assume that the critical time for making prediction is several minutes before the day t closing of Twi, which depends on the model performance. Predicting the day $t+1$ opening return of Twi is the purpose of proposed model. Once day $t+1$ opening, the accuracy of predication would be calculated that means investors will balance its investments at opening time to win the profit from the gap price because of overnight.

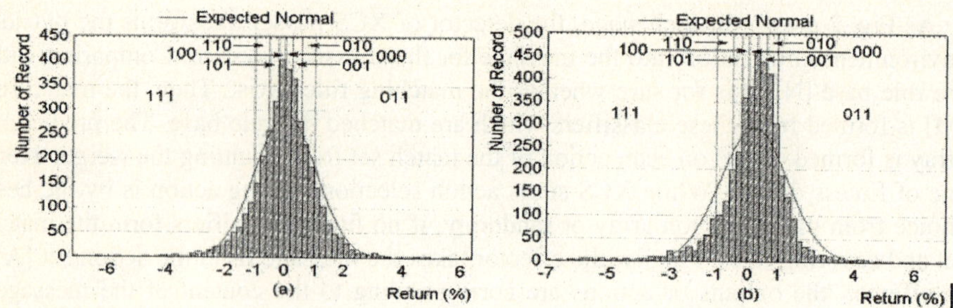

Fig. 3. (a) The distribution of the historical DJi return. (b) The distribution of the historical Twi return

Table 1. The predicted fluctuation of Twi Return and its accuracy indicator

| c_1~c_3 | ~predicted Up fluctuation | | c_1~c_3 | predicted Down fluctuation | |
	fluctuation (%)	accuracy indicator		fluctuation (%)	accuracy indicator
000	(0, 0.32]	(-0.5, ∞]	100	(-0.19, 0)	(-∞, 0.61)
001	(0.32, 0.61]	(-0.19, ∞]	101	(-0.5, 0.19)	(-∞, 0.32)
010	(0.61, 0.97]	(0, ∞]	110	(-1.04, -0.5)	(-∞, 0)
011	(0.97, ∞]	(0, ∞]	111	(-∞, -1.04)	(-∞, 0)

2.3 Simulation

All data is separated into the training period and the testing period. First of all, the data in the training period is applied to establish classifier rule populations. Second, the testing period data is applied to the opening price accuracy test. In the training stage, the data is established and trained the rules which is generated with the exploration, searching in the space. Afterward, the output is generated and presented the price trend-fluctuation. Moreover, the outputs that compared with the real fluctuation are still used to modify and replace the reinforcement learning rules which evolve naturally by GA. At last, applying the subsumption decides the evolved rules redundant or being set up. Following those sub-steps, the correct rule populations are generated.

In the testing stage, utilizing the rule population that verified in the training period treats as the initial one. The fitted rule populations depend on the conditions. According-ing to the fitness and return of fitted rule populations, the trend-fluctuation of Twi would be practicable. As to estimate the accuracy of the predicted fluctuation, the judgment is showed as table 1. The correctness (%) is calculated by observing states of Twi predicted fluctuation which is decoded by output codes.

Because of the trading day consistence of Twi and DJi, DJi daily data should be processed for synchronizing to predict Twi, which preprocesses are listed as follows.

1. In case of the day of Twi open but DJi non-open, we assume that DJi fluctuation is zero, which means $P\_DJiRetur\,n_t^{co} = DJiReturn\,_t^{co} = 0$.

2. In case of the day of Twi non-open but DJi open, we delete the DJi day (t to s-1) data and accumulate the day (t to s-1) Twi fluctuation to next trading day s, which denotes $TwiReturn\,_s^{co} = TwiReturn\,_t^{co} + \cdots + TwiReturn\,_{s-1}^{co}$.

3 Experimental Results

The daily data of Twi and DJi between Jan. 1990 and Sep. 2004 is utilized during the experiments. The data sizes of Twi and DJi are 4020 and 3717, respectively. The training set collects from Jan. 1990 to Dec. 2003, and the other period is the testing set.

Experimentally, 6 and 10 runs were separately executed in this study. Each one renews the training and reinforcing of predicting the DJi firstly. The predicting DJi is secondly used to predict the Twi. Figure 4 shows that each average correctness ratios of 6 and 10 runs are almost above 70% better than 34% [12]. In evidence, this proposed model works in stability. Consequently, the proposed two-stage XCS model is verified.

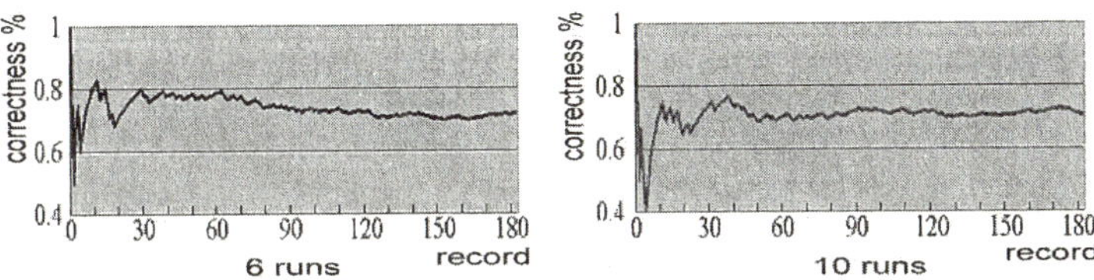

Fig. 4. The average correctness of 6, 20 runs of the predicted Twi

4 Conclusion

Traditional finance professionals hardly focus on AI methodology, and few advanced AI researchers concentrate on the brilliant financial theory. However, we pay attentions to this aspect and propose a mixture AI method, XCS, to predict the Twi stock trend considering the overnight effect while utilizing the DJi as referent market. Owing to the remarkable outcome, two conclusions are given that global overnight effect is indeed existed, and that we successfully develop this two-stage XCS prediction model. Thus, the international arbitrage by global foreign investors will be automatically made, and trend prediction will also be useful for normal investors.

In the future, some suggestions are able to be mentioned. More referent markets should be taken into consideration in the global. Furthermore, the strength of overnight effect should be done more tests. Even, the better detailed outputs should be desinged for the model, and the precise presentation of trend fluctuation needs to be more explicit as well.

References

1. Beltrametti, L., Fiorentini, R., Marengo, L., Tamborini, R.: A Learning-to-forecast Experiment on the Foreign Exchange Market with a Classifier System. J. of Economic Dynamics and Control. 21 (1997) 1543-1575.
2. Chan, K, Chan, K.C., Karolyi, G.A.: Intraday Volatility in the Stock Index and Stock Index Futures Markets. Review of Financial Studies. 5 (1991) 657-684.
3. Chan, K., Chockalingam, M., Lai, K.: Overnight Information and Intraday Trading Behavior: Evidence from NYSE Cross-listed Stocks and Their Local Market Information. J. of Multinational Financial Management. 10 (2000) 495-509.

4. Chen, S.H.: Lecture 7: Rescale Range Analysis and the Hurst Exponent, Financial Economics (I). Department of Economics. National Chengehi University, Taiwan. (2000).
5. Chou, L., Wu, C.S.: Modeling Taiwan Stock Market and International Linkages. NTU Int. Conf. on Finance. Department of Finance, National Taiwan University, Taiwan. (1998).
6. Eun, C.S., Shim, S.: International Transmission of Stock Market Movements. J. of Financial and Quantitative Analysis. 24 (1989) 241-256.
7. Hinich, M. J., Patterson, D.M.: Evidence of nonlinearity in Daily Stock Returns. J. of Business & Economic Statistics. 3-1 (1985) 69–77.
8. Huang, B.N., Yang, C.W., Hu, W.S.: Causality and Cointegration of Stock Market among the United States, Japan, and the South China Growth Triangle. Int. Review of Financial Analysis. 9-3 (2000) 281-297.
9. Holland, J.: Escape Brittleness: the Possibilities of General Purpose Learning Algorithms Applied to Parallel Rule-based Systems. Machine Learning, An Artificial Intelligence Approach, Vol. 2, (1986) 593-623.
10. Kovacs, T.: Evolving Optimal Populations with XCS Classifiers Systems. Technical Report CSR-96-17, School of Computer Science, University of Birmingham, UK. (1996).
11. Lin, J.Y., Cheng, C.P., Tsai, W.C., Chen A.P.: Using Learning Classifier System for Making Investment Strategies Based on Institutional Analysis. Proc. of AIA 2004. 765-769.
12. Liao, P.Y., Chen, J.S.: Dynamic Trading Strategy Learning Model Using Learning Classifier Systems. Proc. of the 2001 Congress on Evolutionary Computation. 2 (2001) 783-789.
13. Mitlöhner, J.: Classifier Systems and Economic Modeling. Proc. of the 1996 Conference on Designing the Future. (1996) 77 – 86.
14. Wilson, S.: Classifier Fitness Based on Accuracy. Evolutionary Computation. 3 (1995) 149—175.

Automatically Assembled Shape Generation Using Genetic Algorithm in Axiomatic Design

Jinpyoung Jung[1], Kang-Soo Lee[2,*], and Nam P. Suh[3]

[1] Samsung Corning Precision Glass Co., Ltd., Asan, Korea
`jinpyoung.jung@samsung.com`
[2] Hanbat National University, Daejeon, Korea
`kslee@hanbat.ac.kr`
[3] Massachusetts Institute of Technology, Cambridge, MA, USA
`npsuh@mit.edu`

Abstract. Axiomatic design is a powerful tool for in designing a product and describes functional requirements (FRs) and design parameters (DPs). In this paper, we extended axiomatic design by introducing shapes and databases. We used functional geometric features (FGFs) for the shape of DP. The databases store links of FR-DP-GE and their hierarchies. The retrieval of proper FGFs from the database is performed by matching a query FR with stored FRs linked to the corresponding FGFs by a lexical search based on the frequency of words and the sequence of the words in the FR statements using a synonym checking system. A computer algorithm automatically combines and assembles the retrieved FGFs. Genetic algorithm (GA) searches for the best matching of interface types between FGFs and generates the corresponding assembly sequences based on the codes of the chromosomes. Geometric interface-ability between FGFs is calculated as a value of INTERFACE_metric. The higher value of INTERFACE_metric means better the design solution for the given FRs that must be satisfied in the sense of language and geometric interface matching.

1 Introduction

The axiomatic design approach proposed by Suh [1, 2] is widely used in designing a product. According to axiomatic design approach, design is done by describing proper functional requirements (FRs) and design parameters (DPs). But it cannot handle the shape of the designed product and reuse previous designs. In this paper, we proposed the concept of functional geometric features (FGFs) to handle the shape of a product and used database to store a design to reuse previous designs. Genetic algorithm (GA) is used to generate an assembled shape from some parts that are stored in databases.

2 Attaching Shapes to DP in Axiomatic Design

2.1 Axiomatic Design

The Axiomatic design [1, 2] provides a scientific basis for design and improves design activities. Designers can be more creative, minimize trial-and-error processes, and determine the best design among the alternatives based on the axioms.

---

* Corresponding author

R. Khosla et al. (Eds.): KES 2005, LNAI 3681, pp. 41–47, 2005.

In axiomatic design theory, synthesized solutions that satisfy the highest-level FR can be created by a zigzagging decomposition process between the functional domain and the physical domain as shown in Figure 1. A designer can decompose a top-level FR into lower-level FRs and DPs until it reaches leaf level FRs and DPs, which are not decomposable any further. FRs are defined as a minimum set of independent requirements that completely characterizes the functional needs of the product in the functional domain. DPs are the key physical variables in the physical domain that characterize the design that satisfies the specified FRs. Both FRs and DPs have been expressed by language descriptions.

Once leaf level DPs are found or created, they have to be integrated into higher-level DPs, finally up to a top-level DP, which is mapped to a whole physical solution. Because the lower-level FRs have to be generated from their higher-level DP, the lower-level FRs depend on the designer's choice as the higher-level DP. For example, FR1, FR2, and FR3 are determined from DP0 as we can see in Fig. 1. Thus, different decompositions can be created to satisfy the top-level FR. The different decompositions result in different design solutions that satisfy the same set of the highest-level FRs.

Once all candidate DPs are found and integrated, the candidate design solutions have to be evaluated by two design axioms using axiomatic design. The first axiom is the independence axiom, which states that the independence of FRs must be maintained in design. The second axiom is the information axiom, which states that the information content of the design must be minimized. Those two axioms can be explained by starting from a mathematical representation of axiomatic design.

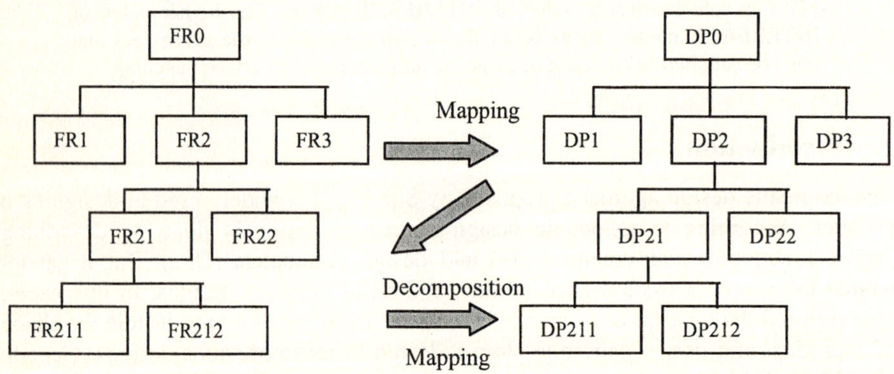

Fig. 1. Zigzagging decomposition/mapping/composition process of axiomatic design

2.2 Functional Geometric Feature (FGF)

The top-down decomposition process is the zigzagging decomposition process between FRs and DPs as shown in Fig. 1. It finally produces language descriptions of FRs and DPs. A DP is a language description of a proposed solution to satisfy the corresponding FR, and plays a role as a key design variable as a part of the whole design solution.

The mapping process is a process to create FGFs from the leaf level DPs. The FGFs mapped to the leaf level DPs are key geometric objects, which are created to satisfy the corresponding leaf level FRs. They are interfaced with each other and integrated into a complete geometric model in the physical domain to satisfy higher-level FRs through the bottom-up integration process.

There are mainly three steps in the bottom-up integration process. The first step is to establish interfaces of the identified FGFs created by the mapping process. The second step is to construct topologies by interfacing the FGFs into complete solids. It is performed by locating the FGFs and by connecting them in the physical space. The third step is to determine the final shapes. A constructed topology constrains the change of the FGFs and therefore affects the satisfaction of FRs.

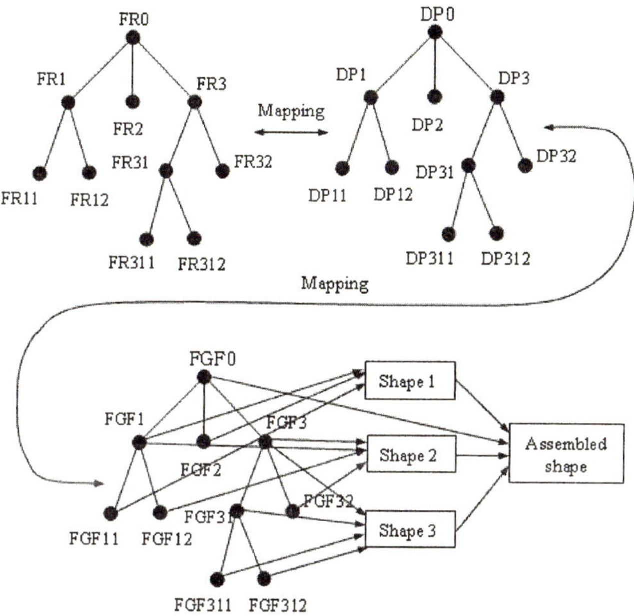

Fig. 2. Fundamental structure of decomposed FRs, DPs, and FGFs

In Fig. 2, FR1, FR2, and FR3 cannot be determined before a design concept on DP0 has been determined. For this reason, each DP in a certain level must be a sub-set or an element of its parent DP from one-level higher. For example, DP2 is an element of DP0, and DP1 and DP3 are sub-sets of DP0. DP11 and DP12 are elements of DP1, and DP31 is a sub-set of DP3. DP311 and DP312 are elements of DP31. All the FGFs have the same relationship between each other as the DPs have. In this sense, lower-level FGFs must be more fundamental and simple, and possibly integrated into higher-level FGFs based on the given FR hierarchies.

2.3 Storing a Design in the Database

The existing design knowledge should be stored and collected in the databases before the designers use the databases. Designers can use previous design like case based design (CBS) methods. [3, 4, 5] The design knowledge is stored as linked lists of FR-DP-FGF and the corresponding hierarchies.

3   Proposed Design Process

The proposed design process follows four steps, decomposing FRs-DPs, mapping FGFs into DPs at the leaf level, generating assembled shape, and determining shapes satisfying given FR. For explanation, a simple design problem to contain coffee keeping its temperature warm is illustrated.

3.1   Decomposing FRs-DPs

In this step, FR-DP hierarchies are produced by the zigzagging decomposition process. Language descriptions are used to express the meaning of FRs and DPs. In this example, DP0 for FR0 is a container; FR1 and FR2 are language descriptions, which are decomposed from DP0. DP1 and DP2 are found from the database, which means that they were created from other designs. Fig. 3 shows the decomposed FRs and DPs of the example design from top to bottom.

FR0 = contain coffee keeping its temperature warm   DP0 = container
FR1 = contain coffee   DP1 = water container
FR2 = cover DP1   DP2 = cover

Fig. 3. A simple example to show the proposed design process

3.2   Mapping FGFs into DPs at the Leaf Level

In this step, leaf level FGFs are created or found at the database, and linked to the corresponding language description of DPs. In this case, DP1 and DP2 are found at the database and corresponding FGFs are mapped to the DPs already. But, the position and scale of the FGFs may not adequate for assembly because they were created for different purposes at different designs. Fig. 4 shows the shapes created in I-DEAS^{TM} for the corresponding DPs

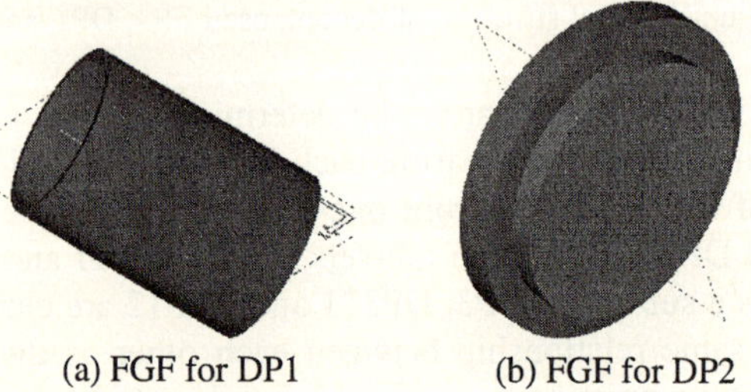

(a) FGF for DP1   (b) FGF for DP2

Fig. 4. Mapping FGFs into the corresponding DPs

3.3   Generating Assembled Shape

Integration of various FGFs to a complete shape is one of the key factors for geometry design with FRs. In the integration process, candidate FGFs are assembled and geometric interface-ability between FGFs is calculated as a value of INTERFACE_metric.

$$\text{INTERFACE\_metric} = \text{IT\_metric} * \text{Pos\_metric} \ . \tag{1}$$

$$\text{IT\_metric} = \sum^{\text{number of pairs}} \frac{\text{number of mated faces at each pair}}{\text{number of total faces at each pair}} \Big/ \text{number of pairs} \ . \tag{2}$$

The combination of FGFs is performed by genetic algorithm (GA) [6, 7] and geometric interface-ability is measured by an algorithm that automatically simulates assembling operations to integrate them in the physical space, which measures Pos_metric in Equation 1. Equation 1 is used as goal function in GA simulation.

$$\text{Pos\_metric} = \begin{cases} \text{AssyFactor} & \text{if assembly is successful} \\ 0.5 * \text{AssyFactor} & \text{if assembly is failed} \end{cases}$$

$$\text{AssyFactor} = \begin{cases} 1.2 & \text{if cylinder and hole are mated} \\ 1 & \text{else} \end{cases} \tag{3}$$

GA chromosomes represent assembly sequences of faces of FGF. In this research, we restricted the number of faces to be assembled as three, because the assembly more than three faces generally results in bad shape assembly. So, the two FGFs in Fig. 4 have five faces and have 86 chromosomes, one case for no mating face, five cases for one mating face, 20 cases for two mating faces, 40 cases for three mating faces. Fig.5 shows the whole process for generating assembled shapes.

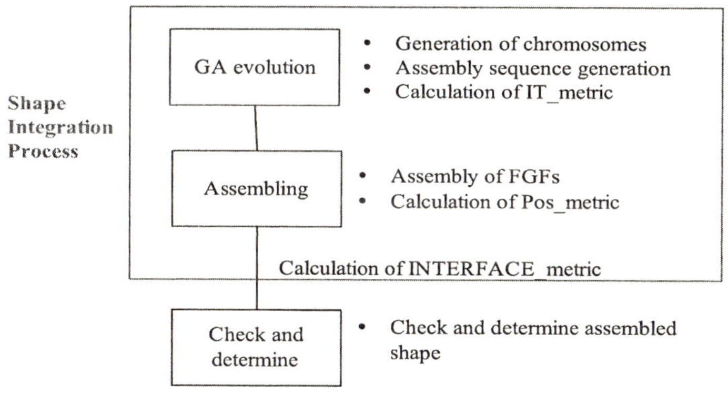

Fig. 5. Shape integration process of FGFs

Table 1. Summary of the design example

Pair	Pair1	
FGF	FGF1	FGF2
No. of faces	5	5
No. of codes	86	86
Chromosome of solution	38	51
Sequence of faces	2,1,3	3,1,5
IT_metric	0.6	
Pos_metric	1.2	
INTERFACE_metric	0.72	

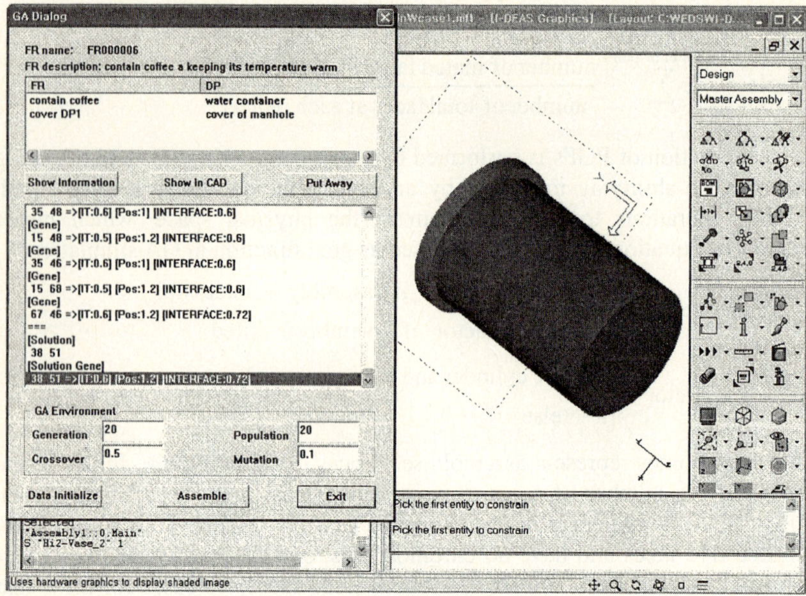

Fig. 6. Generation of assembled shape by GA simulation

Table 1 shows summary of GA simulation of the example design. There is only one mating pair because two parts are assembled. In each FGF, the value of chromosome will be between 0 and 85, which represents sequence of faces in generating assembled shape. In this simulation, 35 of FGF1 and 51 of FGF2 were selected as the chromosomes of solution, which gives the best INTERFACE_metric value. Fig. 6 shows the implemented program to simulate assembly-ability between FGFs.

3.4 Determining Shapes Satisfying Given FR

After simulation, Fig. 6 gives the best solution that can be selected as the assembled shape of the given FR and DP. Designer should determine the solved shape satisfies the given FR.

4 Conclusions

This research presented a method to generate assembled shapes from functional descriptions in axiomatic design approach. It can provide more convenient method to designers. It used GA simulation to determine the solution of assembled shape, which used the sequence of faces in FGFs as chromosomes.

Acknowledgement

This work was supported by the Post-doctoral Fellowship Program of Korea Science and Engineering Foundation. (KOSEF)

References

1. Suh, N.P.: The Principles of Design, Oxford University Press, New York (1990)
2. Suh, N.P.: Axiomatic Design: Advances and Applications, Oxford University Press, New York (2000)
3. Sycara K., NavinChandra D., Guttal R., Koning J., Narashimhan S.: CADET: A case-based synthesis tool for engineering design. International Journal of Expert Systems, Vol.4 No. 2 (1992) 157 – 188
4. Maher M.L., Chang D.M.: CADSYN: A case-based design process model. AI EDAM, Vol. 7 No. 2 (1993) 97 – 110
5. Goel A., Bhatta S., Strolia E.: KRIKIK: An early case-based design system. In Issues and Applications of Case-Based Reasoning to Design (Maher M.L., Pu P. Eds.) Hillsdale, NJ: Erlbaum (1996) ftp://ftp.cc.gatech.edu/pub/ai/goel/murdoc/kritik.ps
6. Michalewicz, Z.: Genetic Algorithms + Data Structures = Evolution Programs. 3rd edn. Springer-Verlag, Berlin Heidelberg New York (1996)
7. Wall, M.: GAlib: A C++ Library for Genetic Algorithm Components. (1996) http://lancet.mit.edu/ga/

Network Based Engineering Design Education

Kang-Soo Lee[1] and Sang Hun Lee[2]

[1] Hanbat National University, Daejeon, Korea
`kslee@hanbat.ac.kr`
[2] Kookmin University, Seoul, Korea
`shlee@kookmin.ac.kr`

Abstract. In graduate class, education on shortening the lead-time in designing and developing products by various computer-aided systems in a network and applying virtual engineering was carried out. To accomplish such a purpose, network based education about engineering design was applied. Some laboratories consisting of various computer-aided systems were constructed and connected by a network. A sample project to develop a radio controlled (R/C) was carried out in order to practice the design of a product based on network based technologies. While students were doing the sample project, they had experienced how to use various computer-aided systems efficiently in a network for concurrent engineering and how to apply the technology of virtual engineering.

1 Introduction

Manufacturing companies are trying to shorten the lead-time in developing products that satisfy customers' needs. [1, 2] They have many computer-aided systems such as a computer aided design (CAD) system, computer aided engineering (CAE) system, a computer aided manufacturing (CAM) system, a product data management (PDM) system and a digital mockup (DMU) system that will be more powerful if they are integrated by a network efficiently. In order to shorten the lead-time in developing a product, it is essential to integrate various computer-aided systems, which enables concurrent engineering. Therefore network based engineering design education is necessary to use various computer systems efficiently.

Some researches on shortening the lead-time in developing products by integrating computer-aided systems and applying virtual engineering are being performed. [3, 4] To do these kinds of researches, laboratories consisting of various computer-aided systems were constructed. The laboratories have 47 copies of Catia[TM] as a CAD system, which plays the most important role in design. We installed TeamPDM[TM] of Dassault Systems as a PDM system, VisMockup[TM] of EDS as a DMU system and Z-master[TM] of CubicTek as a CAM system. As CAE systems, ADAMS[TM], DFMA[TM], CFD-ACE+[TM], Star CD[TM], Ansys[TM], Nastran[TM], HyperMesh[TM], PamCrash[TM] and DADS[TM] were used. A sample project to develop a radio controlled (R/C) was carried out to practice the design of a product by network based technologies. While students were doing the sample project, they had experienced how to integrate computer-aided systems efficiently for concurrent engineering and how to apply the technology of virtual engineering.

The following researches were performed in this project.

First of all, we established a development process using a network for concurrent engineering. We defined the development process using CAD, CAE, CAM, PDM and DMU systems and did the project developing an R/C car based on that process.

R. Khosla et al. (Eds.): KES 2005, LNAI 3681, pp. 48–55, 2005.

Second, we tried to apply the technology of virtual engineering. Our research was particularly focused on the application of a DMU. During the development process, various DMUs were generated from a CAD model according to their application. The application of a DMU for visualization was well formulated compared with other applications. A DMU for visualization was tightly integrated with a PDM system, which enabled engineers to use the DMU data of any hierarchical level in the product structure at any place. It was used in the design review, collaborative design and the publication of various manuals. The application of a DMU for virtual testing was not well defined and is still under research. Generally, automotive companies spend a lot of time and cost to test strength, durability, riding/handling and crash/safety of a car. Although CAE activities were performed manually in this project, we were interested in formulating the process of various virtual tests using DMUs in the corresponding virtual test laboratories.

Finally, one of the purposes of this project was to train graduate students to utilize many computer systems efficiently in a network. They used various computer systems to do the sample project, or, the development of an R/C car. In the project, they experienced many kinds of computer-aided systems and the whole process to develop a product using computer-aided systems in a network.

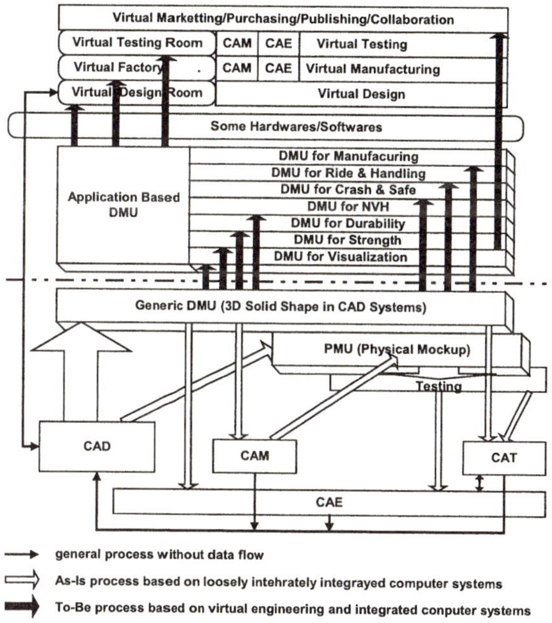

Fig. 1. System architecture for network based engineering design education

2 Computer Systems for Network Based Education

In order to apply network based technology in developing a product, we should integrate various kinds of computer-aided systems and customize many modules that have not been formulated. We constructed the computer systems based on Fig. 1.

The lower part of Fig. 1 shows the utilization of computer-aided systems in developing a product, which is easy to apply. Engineers design a product by using solid systems such as Catia™, analyze the product by using CAE systems with the solid data generated in solid systems, and generate NC data from the solid data. Although there are some interface problems, it is possible to use such systems seamlessly. Prototypes are produced after some activities such as manufacturing, purchasing and assembly. If some errors occur in manufacturing, purchasing, assembling and testing the product, the design will be modified. If necessary, the same process is applied to the second design, such as manufacturing, purchasing, assembling and testing.

The upper part of Fig. 1 represents the process of virtual engineering based on the current computer systems in a network explained previously. The prototypes generated on the computer systems are called digital mockups (DMUs) [5] in contrast to the physical prototypes. The data generated in the solid modeling system is a generic DMU, which will be used throughout the whole virtual process, that is, virtual design, virtual manufacturing and virtual testing. In contrast to the generic DMU, the DMUs used in the specific areas such as virtual design, virtual manufacturing and virtual testing are called application-based DMUs. The application-based DMUs retrieve data from the generic DMU and add more information for their engineering activities. Among application-based DMUs, a DMU for visualization is well defined to be used. It retrieves simple facet data from a solid model and adds some simple data on it. All members can check it easily with a simple viewer. The infrastructures of the computer systems shown in Fig. 1 are a PDM system and a network.

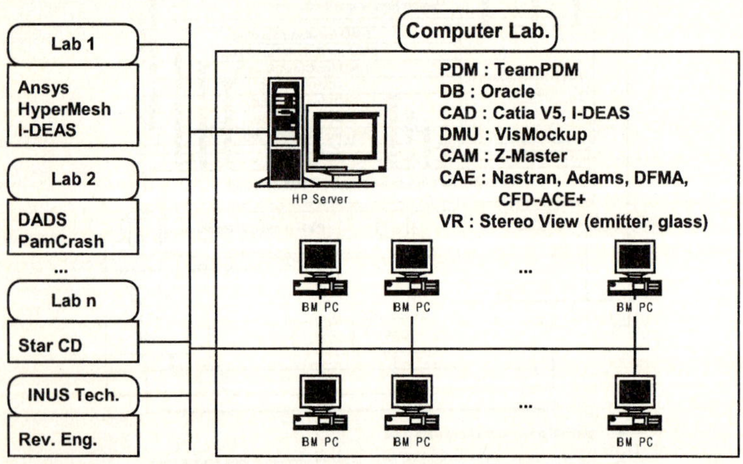

Fig. 2. System construction for network based engineering design education

3   Construction of Computer Systems in Laboratories

In order to verify the plan shown in Fig. 1, laboratories consisting of various computer-aided systems were constructed as shown in Fig. 2. Some basic systems such as CAD, CAM, PDM and DMU systems were installed. Some CAE systems such as Nastran™, ADAMS™, DFMA™ and CFD-ACE+™ were also installed in some laboratories.

4 Development of a Remote Controlled (R/C) Car Using Network Based Method

A sample project to develop an R/C car was accomplished at the constructed laboratory according to the process [6] given in Fig. 3. The purpose of this project was to simulate a development process based on the proposed method and give some intuitions to graduate students about network based engineering design. This project consisted on some engineering activities such as planning, design, analysis, manufacturing and service.

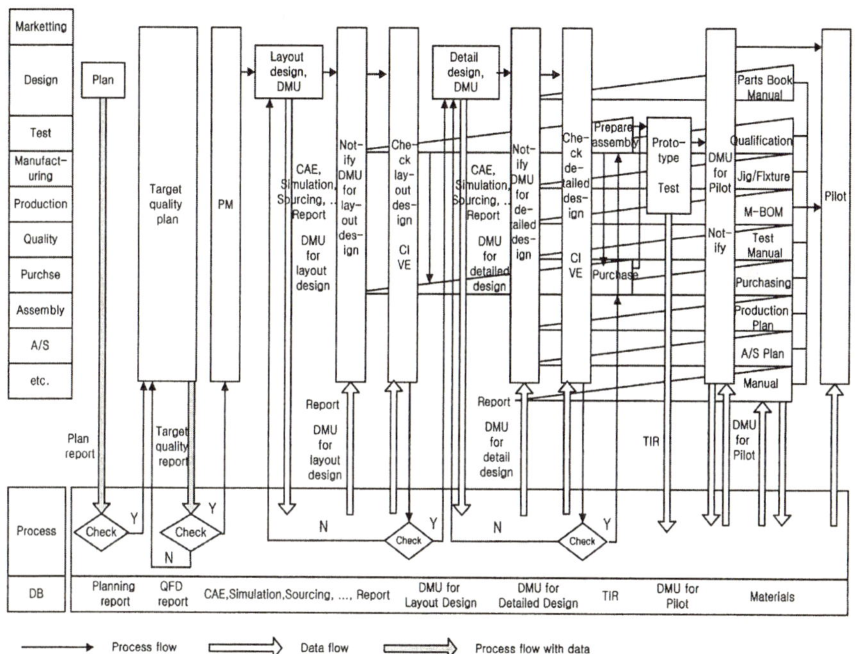

Fig. 3. Development process for concurrent engineering using networks

4.1 Planning

A project to develop an R/C car was planned by using the Quality Function Deployment (QFD) method. [1, 2, 7] Some customers' needs were converted as numeric target values that support the engineer's decision. The QFD tables prepared in planning were used and modified until the project was finished.

4.2 Design

In design stage, engineers used Catia^{TM} as a design system. The solid data were converted as visualization DMU data and stored on the PDM system. DMU data helped concurrent engineering by making it easy for engineers to check the solid shape of a part or an assembly.

Fig. 4 shows some design activities in developing an R/C car. Fig. 4 (a) shows the shape of the R/C car to be designed, Fig. 4 (b) is a solid shape designed in Catia^{TM}, Fig. 4 (c) shows the shape on DMU system, and Fig. 4 (d) is a PDM system that manages all data needed in development.

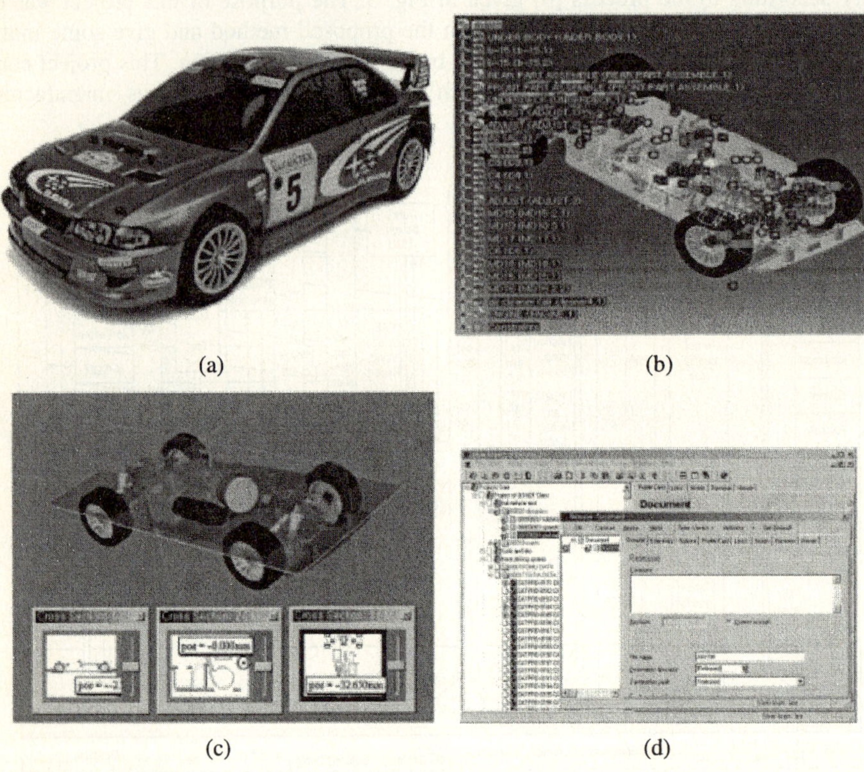

(a)   (b)

(c)   (d)

Fig. 4. Engineering design activities based on networks using various computer systems

4.3 Engineering Analysis

Various CAE systems shown in Fig. 2 were used to analyze the designed product. CAE analysis is very important for virtual testing. It should be performed from layout design to testing throughout all the development process. In this project, we applied CAE activities manually throughout the whole development process. If the interface between the CAD system and the CAE system is formulated, virtual testing will be possible. [1]

Fig. 5 (a) shows the result of strength analysis by using Ansys^{TM}. Strength analysis is the most basic work out of all the CAE activities. Fig. 5 (b) is the result of dynamic analysis by ADAMS^{TM}. The red circle means that there is interference between two parts during the steering motion. Fig. 5 (c) shows the simulation of lane change. It shows the behavior of a car when it changes its lane. This technique can be enhanced to a pothole test and a proving ground test with more research and work. Fig. 5 (d) shows the analysis of natural frequencies of under frame performed by HyperCam^{TM}.

A vibration test also carried out for correlation between CAE analysis and a real test. Fig. 5 (e) shows a front crash analysis by PamCRASH™. Offset and oblique crash analyses were also performed. Fig. 5 (f) shows the result of CFD analysis performed by StarCD™, pressure distribution when the R/C car drives at the velocity of 50 km/h. The surface of the car body was achieved by reverse engineering technique with laser-scanner. 3D meshes for CFD analysis were generated from the scanned surface.

From the results of CAE activities, some design changes were made. We replaced seven parts from the first prototype and made the second prototype, which showed better functions than those of the first one.

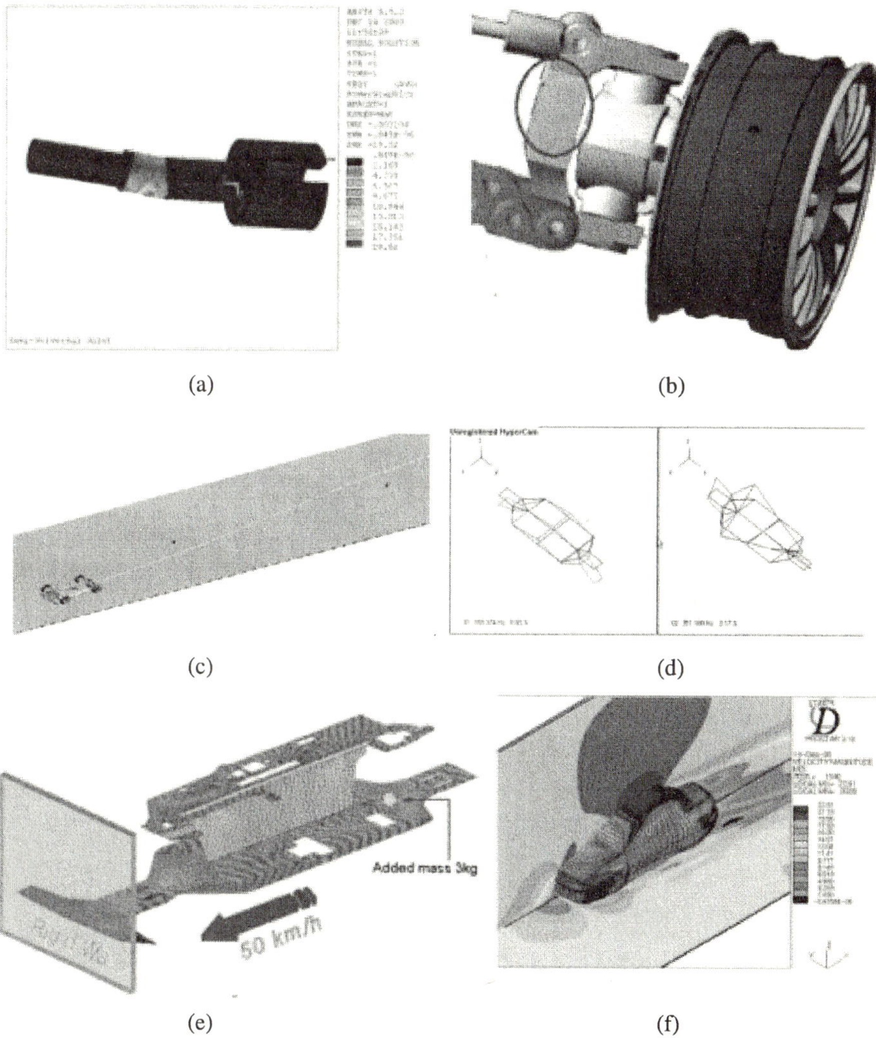

Fig. 5. Engineering analysis activities based on networks using various computer systems

4.4 Manufacturing and Service

Engineers at the departments of manufacturing and service reviewed the designed product carefully and suggested some design-changes reflecting the opinions of production and post sales service. In the manufacturing department, they generated NC data from the solid data created in the design department. They also used the DMU in making a parts list and a maintenance book, manufacturing bill of materials (M-BOM) was created based on engineering bill of materials (E-BOM) stored in the PDM system.

5 Conclusion

In this project, an R/C car was developed to simulate product development based on the proposed method using various computer systems integrated in a network. The following researches were carried out.

First, we established a development process various computer systems integrated in a network, particularly the PDM and DMU systems. The sample project progressed according to the defined process.

Second, although some activities were performed manually, the students became to know how to apply the technology of virtual engineering in the design of a product. We tried to formulate the interface between computer systems, which was thought to be a way toward virtual engineering.

Finally, the educational aspect of this project was emphasized. As the project progressed, graduate students became accustomed to the computer systems integrated in a network and the technology of virtual engineering.

Recently, manufacturers have been trying to introduce some computer-aided systems in developing products. In the future, we will do research into more the efficient methods of how to apply computer-aided systems using networks in developing products. The sample project presented in this paper gave a strong intuition to the students and researchers who previously had no experience in various computer systems integrated in a network.

Acknowledgement

This work was supported by the Post-doctoral Fellowship Program of Korea Science and Engineering Foundation. (KOSEF)

References

1. Prasad, B.: Concurrent Engineering Fundamentals, Prentice Hall (1996)
2. Hartley, J. R.: Concurrent Engineering: Shortening Lead Times, Easing Quality, and Lowering Costs, Productivity Press (1998)
3. Heo, S.J.: Survey on the application of virtual engineering to automotive development, Proceedings of a symposium on automotive technology (1999) (in Korean)

4. Han, S.H. and et al.: Report on the technology used for shortening the lead-time in developing new cars (1998) (in Korean)
5. Lee, K.: Principles of CAD/CAM/CAE Systems, Addison Wesley (1999)
6. Lee, K.-S. and et al.: A case of digital mockup system implementation for improving product development process, Proceedings of 2000 Korean CAD/CAM Conference (2000) 247-252 (in Korean)
7. Hauser J.R., Clausing D.: The House of Quality, Harvard Business Review May-June (1988) 63-73

Representative Term Based Feature Selection Method for SVM Based Document Classification

YunHee Kang

Department of Computer and Communication Engineering, Cheonan University
115 Anseo-dong, Cheonan 330-704, Korea
yhkang@cheonan.ac.kr

Abstract. This paper describes a document classifier for web documents in the fields of Information Technology and uses SVM to learn a model, which is constructed from the training sets and its *representative terms*. To reduce information overload, it needs to exploit automatic text classification for handling enormous documents. Support Vector Machine (SVM) is a model that is calculated as a weighted sum of kernel function outputs. The basic idea is to exploit the representative terms meaning distribution in coherent thematic texts of each category by simple statistics methods. Vector-space model is applied to represent documents in the categories by using feature selection scheme based on TFiDF. We apply a category factor which represents effects in category of any term to the feature selection. Experiments show the results of categorization and the correlation of vector length.

1 Introduction

As the growth of the Internet, to reduce information overload, it is necessary to exploit automatic text classification for handling enormous documents. Text classification or categorization is the process of organizing information logically. It can be used in many fields such as document retrieval, routing, and clustering. The document classifier is used for classifying documents based on category [1,5,8].

In this paper, a new document classification system is proposed for categorizing web documents in the fields of *Information Technology*. The basic idea is to exploit the representative terms meaning distribution in coherent thematic texts by simple statistics methods.

The SVM, which is based on minimizing the expected risk of making a classification error, finds an optimal separating hyperplane between binary classes[3,4,6]. Using this hyperplane, it is possible to discriminate text collection between a relevant documents set and an irrelevant documents set. To this end, the SVM has recently been applied to a number of applications, e.g., image classification[3,4] and text categorization[6].

In some document classification applications, it might be desired to pick a subset of the original features rather then select all of the original features. The benefits of finding this subset of features could be in cost of computations of unnecessary features. We applied these two methods into multi-class document categorization problems. We apply a category factor which represents effects in category of any term to the feature selection. In experiment, we show the result of classification and evaluate the results based on recall and precision rate.

R. Khosla et al. (Eds.): KES 2005, LNAI 3681, pp. 56–61, 2005.

The rest of the paper is organized as follows: Section 2 describes related works including feature selection and a machine learning method SVM. Section 3 describes the proposed document classification system. In section 4, we show that our proposed system outperforms in experiments. We conclude in section 5.

2   Related Works

Documents can be represented by sparse vectors, which consist of each unique words extracted from the documents. Typically, the information vector is very high dimensional, at least in the thousands for large collections. The dimensionality reduction is an example of "feature selection" which is a technique to extract the feature data from observed raw data and is an important issue, especially in the case of high dimensional data. Lewis has observed that "many algorithms scale poorly to domains with large numbers of irrelevant features"[8] and it is not uncommon to have thousands of terms in the vocabulary of a text filtering system[1].

The removed features are not used in the computation anymore. The aim of feature selection method usually determines a low dimensional d features from the set of features m, for which a criterion J will be maximized.

There are many methods to make a feature vector from keyword vector [5]. In this paper, the threshold methods are applied for term-number reduction in text area. Threshold methods are based on removal features, whose frequencies are greater than a defined threshold value. These methods are currently very popular because they are reasonably fast and efficient. On the other hand, they completely ignore the existence of other features and evaluate every feature with its own.

One of the well-established techniques for text in the field of IR(Information Retrieval) is to represent each document as a TF*IDF-vector in the space of keywords that appeared in training documents [2,8], sum all interesting document vectors and use the resulting vector as a model for classification. Each component, $d(i)$ of a document vector is calculated as follows:

$$d(i) = TF(w_i,d) * IDF(w_i) , \qquad (1)$$

where $TF(w_i,d)$ represents the number of times word w_i occurred in document d and $IDF(w_i)= \log D/ DF(w_i)$ where D is the number of documents and document frequency $DF(w_i)$ is the number of documents, where a word w_i occurred in at least once.

A number of statistical and machine learning techniques have been applied to text categorization [5,6,8]. Among those techniques for classification problem, SVM has been shown to be efficient and effective for text categorization [6]. The SVM is a learning method introduced by Vapnik [3] and has been applied efficiently to various problems. This method has a good property that the learning is independent of the dimensionality of the problem, while text categorization problem has very high dimensional data.

A SVM is a classifier based on structural risk minimization [3]. A linear SVM is a hyperplane that separates positive training data from negative training data with maximum margin which minimizes the structural risk. First, SVM is summarized as follows:

Consider a set of data points, $\{(x_1, y_1),..., (x_N, Y_N)\}$, where x_1, y_1 is an input and target data, respectively. An SVM is a model that is calculated as a weighted sum of kernel function outputs. The kernel function of an SVM is written as $K((x_1, y_1)$, which is considered as an inner product in feature space, that is, K is a Gram matrix in feature space. This means that we do not need to transfer all data into feature space. All what we need to calculate is the kernel function such as polynomial, radial basis function and other kernel functions which obey Mercer's condition. We do not cover kernel functions in more detail.

3   The Overall Document Classification System

For the construction of the classification system, this paper employs preprocessing on the documents selected from the web. In preprocessing, input feature vector is constructed for training (eliminating the unnecessary terms from the terms selected from the data) according to the frequency. The TF-IDF vector is inputted to a SVM learner, and the learner creates the models for classification [7].

After learning the data set of each categories using SVM, the learner generates a model for classification. It is concluded that the categories of the test documents are determined by the value which is generated by each classifier of a learning model. The number of SVM models is equal to that of categories to be created the classification models. Fig. 1 shows the procedure of the document classification.

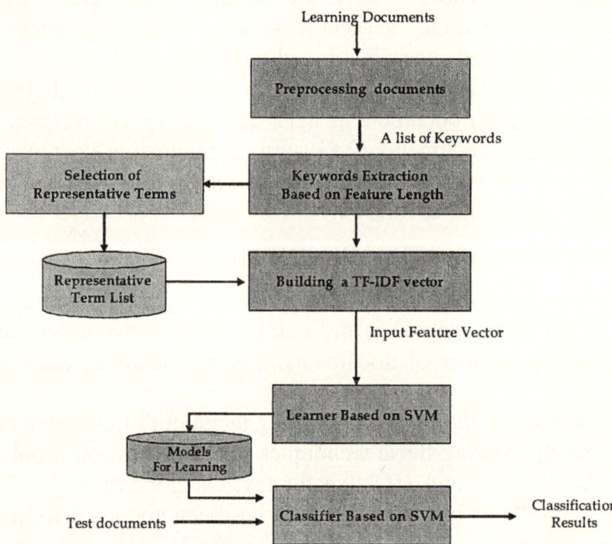

Fig. 1. The Document Classification procedure

4   Experiments and Evaluation

For this experiment, the SVMs model is constructed by using SVM^{Light} [7] Individual SVMs create models for document classification using training documents. Each

SVM is used to learn the training set for the models using both positive examples and negative examples. This experiment draws up the scenario according to two criteria as follows:

- Performance evaluation when a new category is to be added
- Performance evaluation according to the length of vector space

In this experiment, the feature selection module selects the terms which are based on frequency and category relationship. To select the appropriate terms, Eq. 2 is applied for calculating category factor which represents effects in category of any term T_j. It can be adjusted the weight value of terms based on frequency and category factor.

$$\text{if } (CDF_{ij}/TDF_j >= \tau) \text{ then } RT_i \leftarrow RT_i + T_j , \tag{2}$$

where CDF_{ij} represents frequency of a term i in a category j, TDF_j represents total frequency of term i and $\tau = 0.25$. RT_i denotes a set of representative terms of category j.

To construct input feature vector and to improve the feature selection, we apply a procedure based on heuristic TF*IDF expression as follows:

1. Sort n terms by frequency order, where n is the number of terms in learning set except for stopword
2. Compute TF * iDF value of input feature vector
3. Adjust TF * iDF value by using category factor

In the first experiment, it is evaluated the performance of precision and recall for classifiers formed when a new category is added. In the next experiment, it is evaluated that the performance of precision and recall for classifiers formed by restricted vector length.

Fig. 2 shows performance of each category, where the length of the feature vector is 500 and the number of test documents is 126. We observe that the two classes 1 and 4 have similar features. We evaluate the result of the experiment that two categories are more correlated each other. The other categories except for 1 and 4 have the precision rates, which are greater than 70%. In most cases precision levels reached more than 90% by refining sample or more training examples.

In order to evaluate performance of classification it is compared with it both the original category set and other category set which is added to a new category represents a noise category. As shown in Fig. 3, the overall performance is lower than the first case as shown in Fig.2. But we observe that most of categories have more than high precision reached more than 80 %. Also, we observed that the classifiers based on SVM are robust when the category set has noise documents for learning. In this case we got the performance that the average precision is 85.7%, the average recall is 78.8%.

Fig. 4 and Fig.5 show that the performance of categorization depends on the length of feature vector. As shown in Fig.5, as the length of feature vector is increased, the performance with respect to recall and precision is also increased.

5 Conclusions

In this paper, a document classification method is proposed, which is based on the extended TF*IDF feature selection for SVM. The basic idea is to exploit the repre-

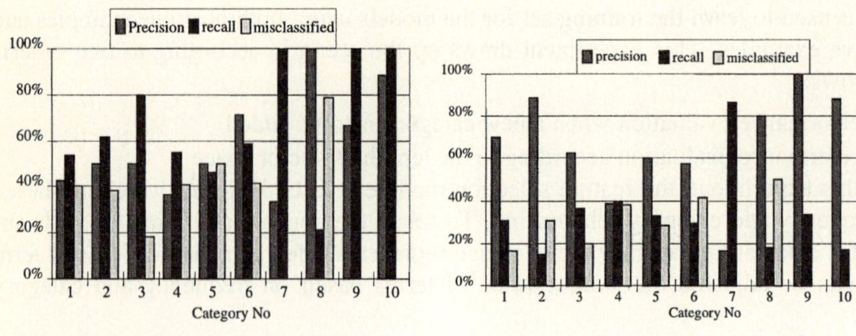

Fig. 2. Experiment-10 Categories   **Fig. 3.** Experiment -11 Categories

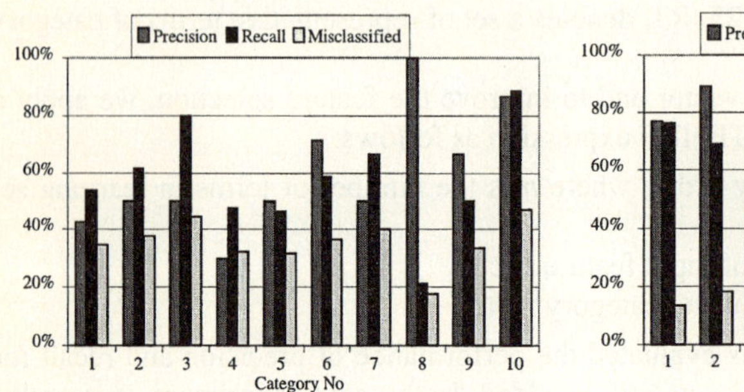
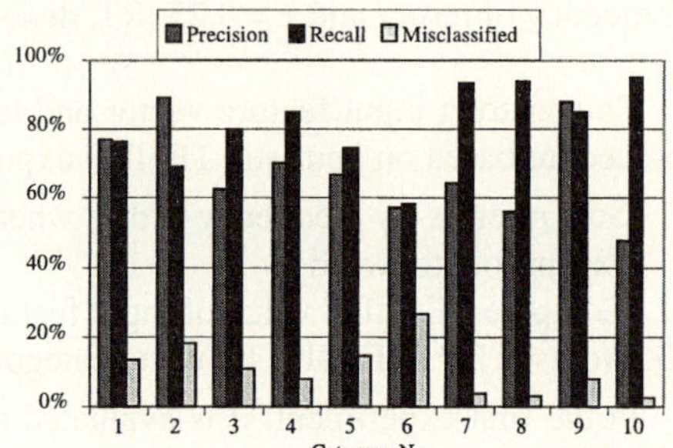

Fig. 4. Vector Length – 500   **Fig. 5.** Vector Length – 2000

sentative terms meaning distribution in coherent thematic texts by simple statistics methods. Also we described the feature selection for document classifier of web documents in the fields of Information Technology. In the first experiment, we evaluated the performance of precision and recall for classifiers formed, which is added to new category. In the next experiment, we evaluated the performance of precision and recall for classifiers formed by vector length. We evaluated the result of experiments that two categories are more correlated each other. The other categories except 1 and 4 have the precision, which is more than 70%. In most cases effectiveness, precision levels reached more than 90% by refining sample or more training. For these experiments we showed that the feature length affects highly performance of classification.

References

1. Tak W.Yan and Hector Garcia-Molina, Sift - a tool for wide-area information dissemination. In *Proceedings of the 1995 USENIX Technical Conference*, pages 177–186, 1995.
2. Salton, G. Automatic Text Processing: The Transformation, Analysis and Retrieval of Information by Computer, Addison-Wesley, 1989
3. V. Vapnik.Statistical Learning Tehory. John Wiley and Sons, Inc., New York, 1998.
4. Chapelle O., Haffner P., and V Vapnik. Svm for histogram-based image classification. *IEEE Trans. on Neural Networks*, 10(5):1055--1065, 1999.

5. Yang Y., J.O. Pdedersen, "A Comparative Study on Feature Selection in Text Categorization," Proc. Of the 14th Internatinal Conference on Machine Learning ICML-97, pp.412-429, 1997.

6. Joachims, T., "Text categorization with support vector machines: Learning with many relevant features," Proc. European Conference on Machine Learning (ECML), pp. 137-142,1998

7. Joachims, SVM^{Light}, http://ais.gmd.de/~thorsten/svm_light, 1998.

8. D. Lewis, W. A. Gale, A sequential algorithm for training text classifiers, Proc. SIGIR '94, pp. 3-12. Dublin, Ireland. 1994.

The Design and Implementation
of an Active Peer Agent
Providing Personalized User Interface

Kwangsu Cho[1], Sung-il Kim[2],*, and Sung-Hyun Yun[3]

[1] Learning Research and Development Center, University of Pittsburgh, USA
kwangsu@pitt.edu
[2] Dept. of Education, Korea University, Seoul, Korea
sungkim@korea.ac.kr
[3] Div. of Information and Communication Engineering, Cheonan University, Korea
shyoon@cheonan.ac.kr

Abstract. This study describes the design and implementation of a naive computer peer agent called KORI-2. Unlike prevalent intelligent tutoring systems that implement a machine tutor replacing a human expert, teachable agent systems try to implement a machine peer tutee. Human student tutors are motivated to learn domain knowledge effectively while tutoring a machine tutee. With KORI-2, a human student who plays a tutor role teaches KORI-2 by constructing concept maps. In this tutoring process, KORI-2 actively initiates and guides tutoring interactions by raising (re-)questions or refuting tutor's explanations. In addition, the KORI-2 system implements interface to contextualize the navigational situations of human tutors.

1 Introduction

Although the importance of motivation to learning in education contexts is evident, most learning systems do not focus on this issue [1]. Rather, they bypass the issue by including external motivators such as rewards and games, assuming that those components will motivate users who would otherwise grow bored. Although learners enjoy games and rewards, they often overlook learning content, which leads to poor learning. Therefore, it seems important to integrate motivation learning activities.

To serve the goal, we have developed KORI-2, an intelligent peer tutee agent that is taught by a human student tutor. Previous research supports the potentials of our approach, showing that similarities between peers such as status and knowledge motivate learners to be deeply engaged in peer tutoring activities [2]. It was found that novice students may learn better from peers than from adults or experts [3]. Although both tutor and tutee can benefit from the participation of peer tutoring [4], tutors learn better than tutees [5] by responding or elaborating their knowledge to tutees' questions [14].

---

* Corresponding Author.

R. Khosla et al. (Eds.): KES 2005, LNAI 3681, pp. 62–68, 2005.

There have been many attempts to develop tutee agents, peer agents, or teachable agents. One of the well-known systems is Betty [6]. The system is a multi-agent system consisting of a machine tutee agent called and an expert tutor agent as major components. Betty interacts with a human tutor by responding to questions Forbes' qualitative reasoning algorithm. However, it seems questionable what aspects of Betty define the system a teachable agent. For example, while a student who takes the tutor role tutors Betty by constructing a concept map, Betty is simply assume to learn the content on the map. However, Betty does not have a machine learning algorithm unlike a general expectation. Instead, it simply accepts content from student teaching. More importantly, Betty is a passive learner in that it reacts to questions raised by humans. It does not have actively engaging learning behaviors such as initiation, question, requestion, and rebuttal like human peer tutees.

We try to advance the research and development of teachable agents by designing KORI-2 to have more human peer tutee-like behaviors. The goal of this study is to describe the design and development of KORI-2 that interacts with fifth to seventh graders in order to learn geology concepts. As an updated version of KORI [13], KORI-2 is active in that KORI-2 may initiate and shape interactions and also that KORI-2 is personalized according to the level of student tutors. In this paper, we describe a major improvement of the system focusing on the activeness and personalized interface of the system. The other major components can be referred to the previous study [13].

2   The Internal Structure

The brain of KORI-2 consists of inference engines and a knowledge base. The classes of concept maps and inference engines are coded to represent and compute the knowledge of KORI-2. The system also has two independent user interface modules: a knowledge sharing module and a resource module. In the knowledge-sharing module, human tutors teach knowledge to KORI-2 through concept maps with the help of an action-language translator component, which translates between actions on the concept map and structured written language. In the test module, KORI-2's learning is evaluated with a built-in expert module. The resource module consists of web site links or documents related to the learning content. Figure 1 shows the user interface for human tutors. Human tutors teach KORI-2 which makes its own knowledge and updates it through the feedback mechanism. Concept maps are used to structure and organize knowledge into objects and relations.

Concept maps take a form of diagram consisting of nodes and arcs which play a role of external representations. Concept maps have been used in many disciplines such as education, policy studies and organizational studies to visualize knowledge structures and arguments. In educational psychology, Novak developed a system of concept maps [7, 8]. Toumin [9] developed scientific arguments based on concept maps. In AI, Quillian developed concept maps called semantic networks [10]. Along with the development of formal semantics, networks have been widely used for knowledge representations [11].

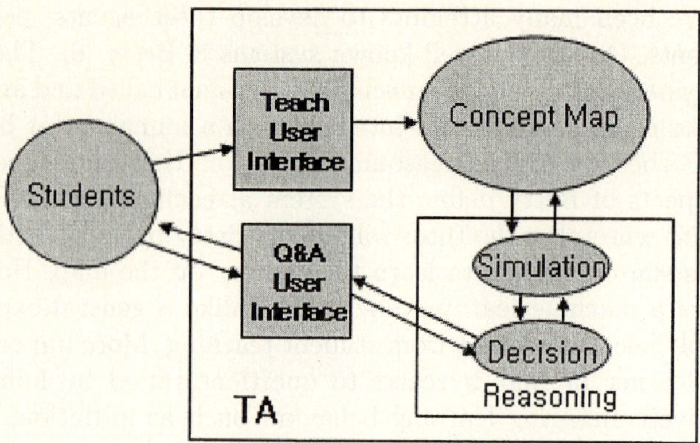

Fig. 1. Knowledge Acquisition and Representation Interface between Students and the TA.

In the concept map of KORI-2, a node takes the form of a rectangle which represents an instance or a concept. An arc stands for the causality between two nodes. Labels on arcs provide relational specifications between two nodes. KORI-2 represents inheritance between objects as in semantic networks [12]. Domain knowledge is represented in two layers. On a bottom layer, domain ontology is represented in an object-oriented scheme. The top layer is domain task-specific representations using a semantic network. As a form of the semantic network, we use a concept map consisting of nodes and arcs that play a role of external representations. As in Figure 2, for example, the object sedimentary rock has relation to the object deposit. To represent relation between these two objects, an arc is used with the label 'be weathered'. Users can represent their knowledge into concept maps using this method.

Individual nodes and arcs have a threshold value, which indicates the importance of the concept or learning difficulty. The threshold value ranges from 0 to 1, and its default value is 1. Concepts deemed important or known to be difficult to understand will be assigned a smaller threshold value to prompt further questioning from the tutee. The value interacts with a tutee activation algorithm, determining whether KORI-2 initiates questions.

3   Active Tutee Behaviors

Since we have developed a computer tutee agent to motivate human tutors to more deeply engage with tutoring, it is important to know what kinds of tutee behaviors prompt or discourage desirable tutor behaviors. To build guidelines to model effective peer tutee behaviors, we observed six pairs of 5th graders interact in peer tutoring situations. Qualitative protocol analyses revealed the following design guidelines of peer tutee behaviors:

1) A student tutor engages in tutoring when a student tutee takes a role of a less capable peer relative to the student tutor.
2) A student tutor engages in deeper levels of learning when a student tutee asks questions about the tutor's explanations.
3) A student tutor appears to lose motivation when a student tutee is unresponsive.
4) A student tutee accepts wrong explanations from a student tutor rather than correct wrong explanations.

Following these guidelines, KORI-2 can operate in a naive mode where the agent assumes the role of a less capable peer. The tutee can ask the human tutor questions such as "How sedimentary rock is formed?" or "Are you sure?". It asks these questions based on both threshold values assigned to the concepts in the domain map and tutee activation values. If the concept the student is currently focused on has a concept threshold value smaller than a tutee activation value, then the system is prompted to question and keeps questions until the activation value is smaller than the threshold value.

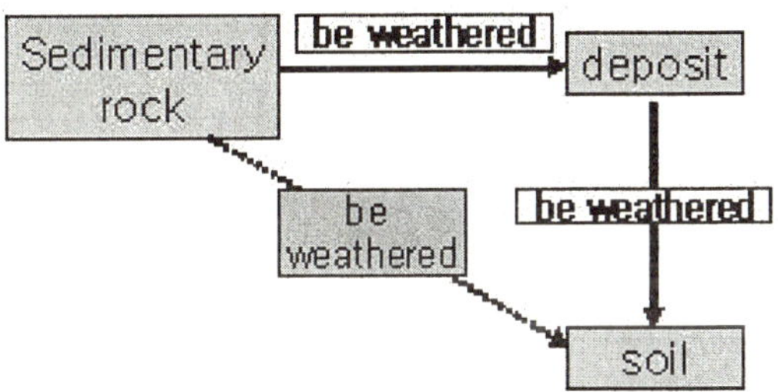

Fig. 2. Representation of Objects and their Relations.

As in Figure 2, we use rock cycles as an example of learning contents. The object sedimentary rock has an arc to the object deposit. To represent the relation between these two objects, the arc is labeled with 'be weathered'. Students can represent their knowledge into concept map by using this method. The inference engine uses forward and backward to infer about new relations between objects. Its reasoning result is reflected in the concept map window. After the tutor and tutee finish the concept map, the system activates the inference engine with predetermined rules and tries to find missing or incorrect relations between the sedimentary rock object and the soil object. The dotted arrow with 'be weathered' label shows this reasoning result.

4   Personalized User Interface

"Personalization is a toolbox of technologies and application features used in the design of an end-user experience. Features classified as 'personalization' are wide-ranging, from simple display of the end-user's name on a web page, to complex catalog navigation and product customization based on deep model of user's needs and behaviors." [15]. As an initial stage of implementing personalized user interface, we took the approach of personalizing navigational context [16]. The KORI-2 system is designed to allow human tutors to select their learning options before tutoring KORI-2. Mainly, it would be effective to vary the form of concept maps with which human tutors teach KORI-2: A null map, a partial map and an incorrect map. The partial map shows parts of a full solution. The null map interface presents a blank space without any node or arc. The incorrect map includes partially wrong solutions.

Table 1. Measurement Results of Students' Interest and Understanding to Learn

	grade	competence	null map	partial map	incorrect map
Interest	4th	low	2.10	3.30	2.85
Interest	4th	high	3.38	3.53	3.30
Interest	5th	low	3.11	2.92	2.94
Interest	5th	high	4.05	3.53	3.81
Understanding	4th	low	2.50	3.23	3.05
Understanding	4th	high	3.69	4.32	3.43
Understanding	5th	low	3.54	3.25	3.29
Understanding	5th	high	4.36	3.46	3.65

In table 1, to support the personalized user interface, we collected empirical data for the level of competence of 4th and 5th grade elementary students. The participants were instructed how to use the concept map. Then, the participants used concept map in forms of a null map, a partial map and an incorrect map. The participants were asked to indicate which type of concept maps they wanted to work on. Results reveal that the participants with high learning motivation and competence tended to prefer the null map or the incorrect map to the partial map. By contrast, the users with low learning motivation and competence prefer the partial map to the other types. A more interesting finding is that users' interest on the types of concept maps tend to have a high correlation with their understandings, $r = .87$. Therefore, the results support the personalized user interface approach.

The following packages are used to create the interface. KORI-2 is implemented on JAVA platform. For GUI user interface, we use JAVA Swing and Jgraph components.

conceptmap - knowledge representation and acquisition using a concept map
editor - to manage concept maps in the main window with objects and relations
ask_answer - to manage Q&A interface using inference modules
main - frame structure and main function
To simulate the knowledge of KORI-2, the following classes are used.
ConceptMap class - manage objects, relations, link paths between objects
InferenceEngine class - inference rules, reasoning process, inherit methods
from ConceptMap class

The reasoning is processed based on concept map knowledge structure. In InferenceEngine class, the following three methods are used to implement reasoning process.

get_reasoning() - acquisition of new knowledge through reasoning process
generate_reasoning() - make reasoning result to human tutors' question
ask_question() - the agent asks a question to a human tutor in case that there exist several paths to go to the next state during reasoning processes.

The ConceptMap class is used to structure and organize the tutee's knowledge. It gets knowledge from both human tutors and reasoning. During the reasoning process, the following search methods are used to get right results from concept map representations.

SearchPath() - the method to access and get knowledge from concept maps
buildLinkPath() - set link path between objects
addNextPath() - set next object and its relationship

5   Conclusion

This study describes an approach to an intelligent peer agent called KORI-2 that actively interacts with human student tutors and provides the personalized user interface to enhance learner's motivation and cognition. Since research and development of teachable agents are still in an embryonic stage, we face various technical and educational challenges. However, the system offers various assets that traditional intelligent tutoring systems and teachable agent systems do not provide. In addition, the design and development of intelligent peer tutee systems may inform learning theories as to the necessity of understanding under what conditions tutors learn better and what kinds of tutee behaviors are effective for human tutor learning.

Acknowledgments

This research was supported by Brain informatics Research Program sponsored by Korea Ministry of Science and Technology.

References

1. Barth, S., "KM horror stories.," Knowledge Management, 3, pp.36-40, 2000.
2. Cazden, C. B, "Classroom discourse. In M. C. Wittrock (ed.)," Handbook of reseach on teaching, 3rd Ed. New York: MacMillan, 1986.
3. Cho, K. & Schunn, C. D. (in press), "Scaffolded writing and rewriting in the discipline," Computers & Education, 2005.
4. Devin-Sheehan, L., Feldman, R., & Allen, V., "Research on children tutoring children: A critical review," Review of Educational Research, 46 (39), pp.355-385, 1976.
5. Webb, N., "Peer interaction and learning in small groups," International Journal of Educational Research 13, pp.21-40, 1989.

6. Biswas, G., D. Schwartz, K. Leelawong, N. Vye, and TAG-V, "Learning by Teaching: A New Agent Paradigm for Educational Software," Applied Artificial Intelligence, special issue on Educational Agents, vol. 19, no. 3, March 2005.
7. Novak, J. D, "Concept mapping as a tool for improving science teaching and learning," In:D. F. Treagust; R, Duit; and B.J. Fraser(Eds), Improving Teachable and Learning in Sceince and Mathematics, London: Tearchers College Press, pp. 32-43, 1996.
8. Novak, J.D., Gowin, D.B., "Learning How To Learn," New York, Cambridge University Press, 1984.
9. Toulmin, S., "The Uses of Argument," Cambridge, UK, Cambridge University Press, 1958.
10. Quillian, M.R., "Semantic Memory," In: Minsky, M., Ed.: Semantic Information Processing, Cambridge, Massachusetts, MIT Press, pp. 216-270, 1968.
11. Gaines, B.R., "An Interactive Visual Language for Term Subsumption Visual Languages," IJCAI'91: Proceedings of the Twelfth International Joint Conference on Artificial Intelligence, San Mateo, California, Morgan Kaufmann, pp.817-823. 1991.
12. Ames, M., Charles, N., "Snitch: Augmenting Hypertext Documents With A Semantic Net," 1993.
13. Kim, S., Yun, S.H., Yoon, M., So, Y., Kim, W., Lee, M., Choi, D. & Lee, H., "Design and Implementation of the KORI: Intelligent Teachable Agent and Its Application to Education," Lecture Notes in Computer Science 3483, Springer-Verlag Berlin Heidelberg, pp.62-71, 2005.
14. Chi, M. T.H., Siler, S. A., Jeong, H., Yamauchi, T., Hausmann, R. G., "Learning from Human Tutoring," Cognitive Science, 25, pp.471-533, 2001.
15. Kramer, Joseph et al. "A User-Centered Design Approach to Personalization," CACM, Vol. 43, No. 8, pp.45-48, 2000.
16. Gustavo, R., Daniel, S., Robson, G. "Designing Personalized Web Applications," ACM, pp.275-284, 2001.

Using Bayesian Networks
for Modeling Students' Learning Bugs and Sub-skills

Shu-Chuan Shih[1] and Bor-Chen Kuo[2]

[1] Department of Mathematics Education, National Taichung Teachers College,
Taichung, Taiwan
`ssc@mail.ntctc.edu.tw`
[2] Graduate School of Educational Measurement and Statistics,
National Taichung Teachers College, Taichung, Taiwan
`kbc@mail.ntctc.edu.tw`

Abstract. This study explores the efficiency of using Bayesian networks for modeling assessment data and identifying bugs and sub-skills in addition and subtraction with decimals after students have learned the related contents. Four steps are involved in this study: developing the student model based on Bayesian networks that can describe the relations between bugs and sub-skills, constructing and administering test items in order to measure the bugs and sub-skills, estimating the network parameters using the training sample and applying the generated networks to bugs and sub-skills diagnosis using the testing sample, and assessing the effectiveness of the generated Bayesian network models work in predicting the existence of bugs and sub-skills. The results show that using Bayesian networks to diagnose the existence of bugs and sub-skills of students can get good performance.

1 Introduction

In recent years, psychological and educational researchers, frustrated by the lack of diagnosticity of traditional tests, had become increasingly interested in building new models to diagnose students' cognitive skills at a more specific level, which could explain, for instance, individual student's problem-solving strategies, bugs, or knowledge-states in the sub-skills of the domain.

This paper is focused on the diagnosis of students' bugs and sub-skills. However, it has been known that manifestations of the bugs are often "unstable" within an individual student's performance, meaning that the bugs appear inconsistently during a-few-days-apart testing sessions or even during a single-day session [2]. This instability contributes significantly to the overall complexity and unpredictability in bug diagnosis [8]. So, we need to find a way to assess and diagnose these bugs with their instability and uncertainty. Several techniques have been used to implement cognitive models that incorporate uncertainty. One of the best-known uncertainty modeling techniques is Bayesian networks, which incorporate uncertainty by using a probability-based approach [5,6,14]. By means of Bayesian network technique, instability of bug occurrences can be efficiently coped with by use of probability. Hence, the goal of this paper is to build a diagnostic model of students' performances based on Bayesian networks which can be use to predict the occurrences of students' bugs and sub-

R. Khosla et al. (Eds.): KES 2005, LNAI 3681, pp. 69–75, 2005.
© Springer-Verlag Berlin Heidelberg 2005

skills in the domain of addition and subtraction with decimals, and to evaluate the effectiveness of the proposed diagnostic model. Besides, two factors are considered in this study. One is the input data type. The performances of applying Binary-scored (right/wrong) and the multiple-choice answer test data are evaluated. The other is the selection of cut-point. Teachers need to have dichotomous measures of bugs and sub-skills in order to proceeding appropriate remedial instruction. It's necessary to set a cut-point for the posterior probabilities that we can decide the bug manifestations of each student. Consequently, we want to assess the effects of various cut-points in diagnosing bugs.

2 Bayesian Networks

Bayesian networks have become one of the most widely used tools for managing and assessing uncertainty in a number of fields [5,6,14]. A Bayesian network is a pair (D,P) which combine graph theory with conditional probabilities, where D is a directed acyclic graph (DAG), consisting of a set of nodes and directed arcs. The nodes represent random variables, and each node can take on a set of possible values. The arcs signify direct dependence between the connected nodes.

In addition to the graphical structure, $\mathbf{P} = \left\{ p(x_1 | \pi_1), \cdots, p(x_n | \pi_n) \right\}$ is a set of conditional probability distributions (CPDs) associated with each node in the network, π_i is the set of parents of node X_i in D. By computing the CPDs of all nodes in the network, we can represent the joint distribution of all variables in the network compactly and economically. That is to say, the set P defines the joint probability distribution (JPD) as

$$p(x) = \prod_{i=1}^{n} p(x_i | \pi_i) \tag{1}$$

As formula (1) states, we can compute any desired probabilistic information with a given Bayesian network.

Evidential reasoning - using some evidence of causal events so as to update and draw inferences about the likelihood of other events - is a capability of Bayesian networks [8]. How to update the probability distribution of states of a node when it receives a piece of evidence from one or more than one of its neighboring nodes? This mechanism is called Bayesian Network Propagation. Various algorithms have been proposed to perform this probability updating [5,7,14]. In this study, the method of trees of cliques [5] is used to calculate all posterior probabilities.

The computation of network propagation in most cases can be very computationally demanding. So, many computer programs have been developed from which we can get help for this message propagation calculation. We used Bayes Net Toolbox for Matlab [13] to compose our program and to calculate all posterior probabilities of bugs and sub-skills.

In the last few years, researchers of educational assessment had also studied the applications of Bayesian networks in education [1,8,9,11,12,15]. One line of such efforts was involved it's application to cognitive diagnosis [8,11]. According to the literatures[11,12], Mislevy suggested a Bayesian network framework for proceeding probability-based inference in cognitive diagnosis. Three key points were described in this framework: building Bayesian networks for a student model (SM-BIN); constructing tasks that students can be provided an opportunity to reveal their performance in targeted knowledge and skill; and creating Bayesian networks for evidence models (EM-BINs) that describes how to extract the key items of evidence from what a student does in the context of a task, and pointers to parent student-model variables. This is where the inference network and local computation come into play.

3   The Domain Area

Before constructing Bayesian networks and preparing the items according to them, we have to do an analysis of the domain of decimal addition and subtraction. A number of studies had been carried out on this area [3,4,10]. According to the literatures, five of the sub-skills chosen as possible causes of the addition bugs and the subtraction bugs separately. The sub-skills of decimal addition are: "Line up the decimal points and the digit with the same place value"(aSkill1), "Carry"(aSkill2), "Use the decimal points exactly"(aSkill3), "Single-digit addition"(aSkill4), "Construct an accurate mathematics model for a addition word problem"(aSkill5). The sub-skills of decimal subtraction are: "Line up the decimal points and the digit with the same place value"(sSkill1), "Borrow"(sSkill2), "Use the decimal points exactly"(sSkill3), "Single-digit subtraction"(sSkill4), "Construct an accurate mathematic model for a subtraction word problem"(sSkill5). In addition, the six bugs for decimal addition and the seven bugs for decimal subtraction that were found to be most common are selected for this study. Descriptions of these selected bugs follow.

- The bugs in decimal addition
    - abug1. Don't line up the decimal points, and line up the numerals on the right or the left (e.g. 12.4+1.35=2.59).
    - abug2. Line up the decimal points, and line up the decimal digit on the right (e.g. 12.4+1.35=13.39).
    - abug3. Fail to carry (e.g. 12.4 + 10.84=22.24).
    - abug4. Fail to insert the decimal point into the answer (e.g. 12.4 + 10.84=2.324).
    - abug5. Don't understand the meaning of a word problem or use wrong operation.
    - abug6. The computational procedure in addition is wrong or careless.

- The bugs in decimal subtraction
    - sbug1. Don't line up the decimal points, and line up the numerals on the right or the left (e.g. 14.8-1.35=0.13).
    - sbug2. Borrowing for the ones column appears to be done correctly, except that there is no decrement in the tens column which should have been decremented (e.g. 12.9-10.84=2.16).
    - sbug3. Do correctly in the borrow-only-once problem, but fail in double-borrow problem (e.g. 12.4-10.84=2.16).
    - sbug4. Take liberties to change unequal-digit decimal subtraction for equal-digit decimal subtraction (e.g. 7.84-2.4 □ 7.84-0.24).
    - sbug5. Fail to insert the decimal point into the answer (e.g. 14.18-14.15=0.3).
    - sbug6. Don't understand the meaning of a word problem or use wrong operation.
    - sbug7. The computational procedure in subtraction is wrong or careless.

4 Method

The goal of the empirical work reported here is to evaluate how well Bayesian networks can be used to diagnose specific bugs and sub-skills in decimal addition and subtraction. We construct the test that the items are designed to elicit bug symptoms. The test administers in the sample of fourth grade students. The presence or absence of the bugs and sub-skills is identified in each student by experts' judgement using his/her actual answers to test items. These experts' diagnosis serves as the external criterion variables. Then, four Bayesian network architectures are proposed and tested. The performance of each network in diagnosing the presence of each bugs and sub-skills is evaluated.

The set of test items contains seven decimal addition problems and eleven decimal subtraction problems, which were carefully constructed so that the bugs and sub-skills can appear in various types of tasks. All items are 4 options (A, B, C, D) multiple-choice format and no response item is coded N.

The test are administered to 816 fourth grade students. The responses of 408 samples are used as a training data set for building Bayesian networks and others are treated as a testing data set for evaluating the holdout classification accuracies of the trained Bayesian networks.

4.1 Building the SM-BIN and EM-BINs

Four Bayesian network models are proposed and constructed in this study, varying with contents and input data types. Fig.1 and Fig.2 show the networks of decimal addition and subtraction respectively.

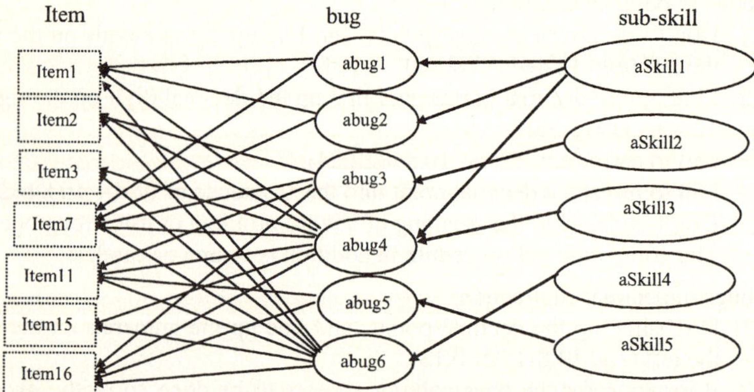

Fig. 1. Bayesian network in decimal addition

In the above figures, the input variables are the item responses based on the items. If "correct" or "incorrect" information are adopted, it's called the binary-answer network. Otherwise, if "specific multiple-choice answer" information is adopted, it's called the multiple-choice-answer network. Thus, the four proposed networks were: a binary-answer decimal addition network, a multiple-choice-answer decimal addition network, a binary-answer decimal subtraction network, and a multiple-choice-answer decimal subtraction network.

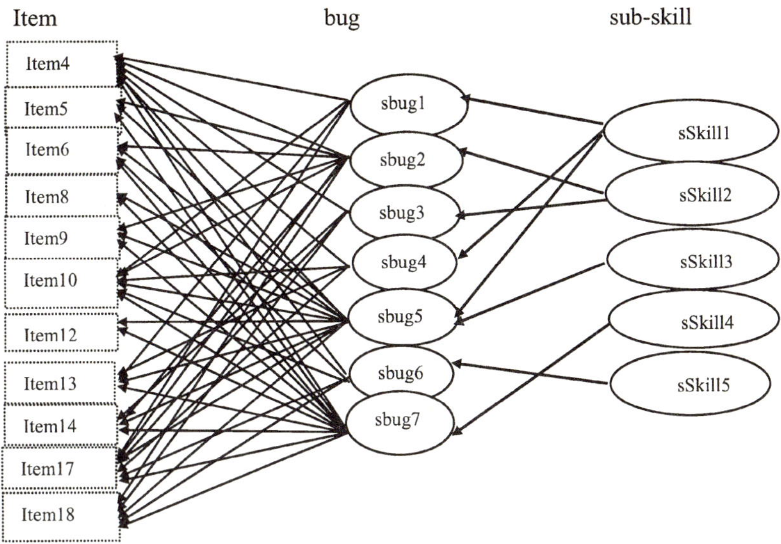

Fig. 2. Bayesian network in decimal subtraction

4.2 Computing Classification Accuracy

Computing the correct prediction rate (classification accuracy) is to assess how well the generated Bayesian network models would work in terms of predicting bugs and sub-skills in each individual student. The computing formula as follows:

BN classification Experts' criterion	Yes(1)	No(0)
Yes(1)	f_{11}	f_{10}
No(0)	f_{01}	f_{00}

$$\text{the correct prediction rate} = \frac{f_{11} + f_{00}}{N}$$

N : the number of testing samples   (2)

5 Evaluations and Discussions

As Table 1 and Table 2 shown below, overall classification results for each network are reported. We can find that the classification accuracies of all bugs are greater than 90% in all decimal addition networks. The classification performances carried out by the decimal subtraction network also are greater than 80% for all bugs. On the whole, the results of using Bayesian network for modeling assessment data and identification of bugs in addition and subtraction with decimals show good performance in predicting bug manifestations in individual students. By contrast, the prediction rates for all sub-skills are not as good as the bugs. However, except the aSkill1 and the cut-point 0.8, all of the prediction rates also exceed 70%.

Both Table 1 and 2 show that the performances of networks using two input data types are similar. This may be due to the distracters of our test are not designed by specific bug carefully. Classification accuracy is influenced by cut-point and different networks, and layers need different cut-points.

Table 1. Classification accuracy (%) of decimal addition network

binary-answer(0,1) decimal addition network								multiple-choice-answer(A, B, C, D, N) decimal addition network						
	Cut-points								Cut-points					
	0.2	0.3	0.4	0.5	0.6	0.7	0.8	0.2	0.3	0.4	0.5	0.6	0.7	0.8
abug1	93.7	93.5	92.7	94.2	95	95	95.1	94.7	94.6	94.1	94.1	94.9	95.1	95.1
abug2	97.3	98.6	98.6	98.7	98.7	98.7	98.7	97.4	97.9	98.1	98.5	98.7	98.7	98.7
abug3	97.4	97.7	98	98.5	98.7	98.7	98.8	98	98.7	98.7	98.7	98.7	98.7	98.7
abug4	96.1	97.1	97.6	97.6	97.6	97.7	97.7	96.4	96.6	96.9	97.3	97.4	97.8	97.8
abug5	97.6	97.8	98.3	98.4	98.4	98.4	98.4	97.2	98	98.1	98.1	98.3	98.3	98.3
abug6	93.9	94.6	95.3	95.5	95.9	96.1	96.1	95.9	95.9	95.9	96.1	96.1	96.2	96.2
average	96	96.6	96.8	97.2	97.4	97.4	97.5*	96.6	97.85	96.97	97.13	97.35	97.5*	97.5*
aSkill1	53.2	58.2	62.1	62.6	47	46.8	46.8	55.5	58.9	61.3	61.7	47.3	46.8	46.8
aSkill2	92.3	92.3	92.3	92.3	92.3	92.5	93.2	92.3	92.3	92.3	92.3	92.3	92.4	92.4
aSkill3	92.4	92.4	92.4	92.4	92.5	92.5	92.3	92.4	92.4	92.4	92.4	92.3	92.1	91.7
aSkill4	79.8	79.8	80.3	81.1	82.1	83.2	83.8	79.8	79.8	79.9	80.4	80.8	81.5	83.3
aSkill5	85.7	85.7	85.7	85.7	85.7	86.2	87.3	85.7	85.7	86.2	86.3	86.4	86.4	89.3
average	80.7	81.7	82.6	82.8*	79.9	80.2	80.7	81.1	81.8	82.4	82.6*	79.8	79.8	80.7

An asterisk(*) indicates the best classification accuracy in the row

Table 2. Classification accuracy (%) of decimal subtraction network

binary-answer(0,1) decimal subtraction network								multiple-choice-answer(A, B, C, D, N) decimal subtraction network						
	Cut-points								Cut-points					
	0.2	0.3	0.4	0.5	0.6	0.7	0.8	0.2	0.3	0.4	0.5	0.6	0.7	0.8
sbug1	94.7	95.4	96.2	96.3	96.3	96.8	96.9	96.8	97.1	96.8	96.9	97.1	96.9	96.8
sbug2	85.9	87.7	88.5	88.7	89	89.5	90	88.1	88.4	89.3	89.7	89.3	89.7	89.7
sbug3	98	98	98	98	98	98	98	96.9	97.5	97.5	97.5	97.5	97.5	97.5
sbug4	97.5	98.1	98.4	98.4	98.4	98.4	98.4	98.1	98.1	98.1	98.3	98.3	98.4	98.4
sbug5	90.5	92.6	93.9	94.3	94.7	95.2	95.4	93.9	94.2	94.7	94.9	94.7	94.9	95
sbug6	80	80.6	82.6	82.9	83.6	83.6	83.6	84.5	85	84.7	83.7	83.7	83.9	84.1
sbug7	88.6	89.8	91.1	91.6	92.2	92.8	92.8	91.7	92.1	92.5	92.6	92.9	92.6	92.9
average	90.6	91.7	92.7	92.9	93.2	93.5	93.6*	92.9	93.2	93.4	93.4	93.4	93.4	93.5*
sSkill1	79	79	79	80	81.5	85.4	81.8	79.1	79.3	79.4	80	80.9	81.5	73.2
sSkill2	79	79	79	79	79.8	83.7	83.2	79.2	79.2	79.2	79.2	80.6	81.5	72.5
sSkill3	83	83	83	83	83	83	86.5	83.1	83.1	83.1	83.1	83.1	83.3	83.3
sSkill4	77.8	77.8	77.8	78.1	80.5	84	32.2	77.9	77.9	77.9	78.5	79.6	80	26.7
sSkill5	72.3	72.3	72.4	76.1	82.1	86.6	27.7	72.4	72.4	73.3	79.5	81.1	83.1	27.7
average	78.1	78.1	78.2	79.2	81.4	84.5*	62.3	78.3	78.4	78.6	80.1	81.1	81.9*	56.6

An asterisk(*) indicates the best classification accuracy in the row.

6 Conclusions

Based on the above results, we can find that

- Using Bayesian networks for modeling assessment data and identifying of students' bugs and sub-skills is a powerful and practical way and can provide useful information for follow-up remedial instruction.
- Dynamic cut-point selection is a way to increase classification performance.
- Theoretically, multiple-choice input data type can provide more information, but real data experimental result shows that it does not help in this study.

References

1. Almond, R. G., & Mislevy, R. J.: Graphical models and computerized adaptive testing. Applied Psychological Measurement, 23, (1999) 223-237
2. Brown, J. S., & Burton, R. R.: Diagnostic Models for Procedural Bugs in Basic Mathematical Skill. Cognitive Science, Vol.2, 2 (1978) 155-192
3. Chiang, Ai-Hua.: The study of diagnostic instruction of the fifth graders in decimal number. Unpublished master's thesis, National Pingtung Teachers College, Taiwan (2002)
4. Hiebert, J. & Wearne, D.: A model of students' decimal computation procedures. Cognition and Instruction, 2(3&4), (1985) 175-205
5. Jensen, F. V.: An introduction to Bayesian networks. New York: Springer (1996)
6. Jensen, F. V.: Bayesian networks and decision graphs. New York: Springer (2001)
7. Lauritzen, S. L., & Spiegelhalter, D. J.: Local computations with probabilities on graphical structures and their application to expert systems. Journal of the Royal Statistical Society, Series B, 50(2), (1988) 157-224
8. Lee, Jihyun.: Diagnosis of bugs in multi-column subtraction using Bayesian networks. Unpublished Ph. D. thesis, Columbia University (2003)
9. Liu, C. L.: A Bayesian network-based simulation environment for investigating assessment issues in intelligent tutoring systems. Int. Computer Symposium. Taipei (2004)
10. Liu, M-L.: An exploration on the teaching of decimals. The academic journal of National Pingtung Teachers College, Taiwan, 16, (2002) 319-354.
11. Mislevy, R.J.: Probability-based Inference in Cognitive Diagnosis. In P.Nichols, S.Chipman & R. Brennan. (Eds.) Cognitively Diagnostic Assessment. Hillsdale, NJ: Lawrence Erlbaum Associates (1995)
12. Mislevy, R.J., Almond, R.G., Yan, D., & Steinberg, L.S.: Bayes nets in educational assessment: Where do the numbers come from? In K. B. Laskey & H. Prade (Eds.), Proceedings of the Fifteenth Conference on Uncertainty in Artificial Intelligence (437-446). San Francisco: Morgan Kaufmann Publishers, Inc (1999)
13. Murphy, K.: Bayes Net Toolbox for Matlab. [online]. Available: http://www.ai.mit.edu/ ~murphyk/Software/BNT/bnt.html (2004)
14. Pearl, J.: Probabilistic Reasoning in Intelligent Systems: Networks of Plausible Inference. San Mateo, CA: Morgan Kaufmann (1988)
15. Vomlel, J.: Bayesian networks in educational testing, International Journal of Uncertainty, Fuzziness and Knowledge Based Systems, Vol. 12, Supplementary Issue 1, (2004) 83-100.

A Case-Based Reasoning Approach to Formulating University Timetables Using Genetic Algorithms

Alicia Grech and Julie Main

La Trobe University, Bundoora 3086, Australia
{a.grech,j.main}@latrobe.edu.au

Abstract. This paper presents a technique to construct generic university timetables using case-based reasoning (CBR) with genetic algorithms (GAs). The case-based reasoning methodology allows a past memory of timetables to be stored and accessed via retrieval mechanisms, finding a past solution most fitting to the new timetable input problem. In the instance that a past solution is not well suited to the new timetable requirements, a genetic algorithm is employed to adapt the past timetables in the case memory. The hybrid technique used implements a learning mechanism to aid in the revision and adaptation of new timetable solutions. The focus of this technique is a feedback mechanism which allows the system to diagnose the violation of hard-constraints fitted to timetable creation.

1   Introduction

The problem of creating a valid educational timetable involves scheduling lessons, teachers and rooms into a fixed number of periods, in such a way that no teacher, class or room is used more than once per time period [1]. When creating a new timetable, previous parts of timetables are often re-used, based on the fact that constraints in a new timetable problem do not usually change significantly from a past timetable [2]. Case-based reasoning (CBR) is a methodology that finds solutions to new problems, using solutions from previous problems as well as the experienced gained from solving those problems. This is a motivating factor for applying CBR to timetable creation. In CBR, past solutions related to a new problem are retrieved and if required, past solutions can then be adapted to meet the demands of a new problem. The central processes to be executed in all case-based reasoning methods have been identified by Aamodt and Plaza [3] as: identify the current problem; find a past case similar to the new case (*retrieve*); use the past case to suggest a solution to the current problem (*reuse*, also known as *adaptation*); evaluate the proposed solution (*revise*); and update the system by learning from experience (*retain*).

Though the use of past experience in solving problems is an advantage of CBR, a few phases are difficult to engineer in the CBR cycle, especially in the retrieve and reuse phases. Genetic algorithms (GAs) are adaptive methods

R. Khosla et al. (Eds.): KES 2005, LNAI 3681, pp. 76–83, 2005.

applied to solve search and optimisation problems, based on genetic processes. Applying GAs in the CBR adaptation stage across varied domains has been shown to be a successful innovation over traditional CBR adaptation techniques (see section 2). This is likely to be so as CBR adaptation tends to be very domain specific [4]. The hybrid CBR-GA approach is well fitted to a timetabling domain, as using a GA to adapt previous cases can make the output case more novel than previous solutions which may not fit the new case requirements closely. We apply a GA at the adaptation stage of the CBR cycle, exploiting the mutation operation in order to find more novel solutions. After applying GA adaptation, we collect knowledge at the revise stage of the CBR cycle, with the intent to feed this knowledge back into adaptation to aid in finding healthier mutations to apply. We also perform case to chromosome mapping maintenance, including repairing solutions that are in an incorrect format due to unequal chromosome lengths.

2   Past Applications Using GAs in CBR

Louis and Xu approached open shop scheduling and re-scheduling problems with a hybrid CBR-GA system, injecting cases into a genetic algorithm's population to speed up and augment genetic search. The preliminary results indicated that the combination of GAs with CBR quickly finds better solutions than CBR on its own [5]. Domains covered using the GA component within retrieval or adaptation in case-based reasoning include: tablet formulation [6, 7]; modelling a Checkers game [8]; Open Shop Scheduling and Rescheduling [5]; developing lay out design of residences so they conform to the principles of feng shui [4]; and estimating the flow rates in Estuaries [9].

3   Application of the Timetabling Problem Using CBR-GA

The goal of this work was to implement a case-based reasoner to solve new problems involving University timetables, then adapt unsatisfactory retrieved solutions using a GA, feeding knowledge learnt from the GA adaptation back into future adaptations. Our implementation domain focuses on creating timetables for La Trobe University's Department of Computer Science and Computer Engineering. Due to the size and complexity of the timetabling problem, the scope and size of the problem domain has been reduced using some constraints. A prototype case-based reasoner named CBR-GAALTA (CBR Genetic Algorithm Adaptive Timetabling Learning Application) was created that handles timetable schedule cases.

The requirements for a new timetable arrives in the form of a 'new case' to the CBR system, specifying the number of labs, lectures, and problem classes required plus the expected class sizes. The purpose of the case-based reasoner is to retrieve a timetable from a case-base of previous timetables, that matches the new case as closely as possible. If this solution is insufficient it should be adapted, creating a timetable that is adjusted to suit the input requirements.

4   Creating Timetables with CBR and GAs

Many issues are related to implementing the central processes executed in the general four step CBR cycle. The major issues associated with executing the central processes include: appropriately representing a case; properly indexing and storing cases in a *case memory* (case base); and assigning relevance functions. Each process includes a range of tasks that must be engineered towards the specifications of a timetabling domain.

The reasoning process is heavily dependent on the structure and content of the collection of cases stored [3, 10]. For our application to the timetabling domain, a case represents an entire solution to a timetable, for all subjects on all days of the week. This implies that the content of a case is the *Subject* name, *Lecturer*, *Room* and *Lesson* for as many subject Lessons that exist within a timetable for a regular school week. Based on this, timetable objects are composed of these four other objects containing the information most relevant to a timetable. Each of the objects contain attributes that make the timetable robust, and are paramount to a well-defined case using relevant, traceable data. Figure 1 is a UML representation of a timetable case, displaying all objects and attributes stored within a timetable.

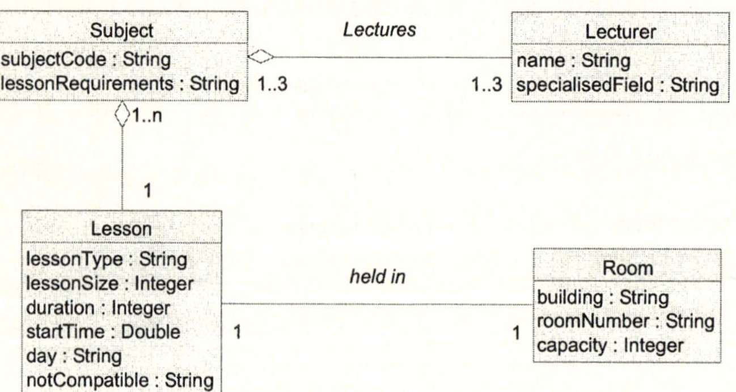

Fig. 1. Object Oriented Case Format.

To retrieve past cases, a partial problem description enters the system in the form of a new case. Retrieval ends when a past case has been found that best matches the new case. Each case in the case memory is tested for a relevance factor with respect to the new input problem, where the relevance factor indicates the closeness in similarity between cases in case memory and the new input case.

A point awarding scheme was developed to evaluate a relevance factor. To award relevance, ten relevance metrics were developed and are calculated across the new case, y, and for the comparison past case, x. This scheme rewards past cases with very similar content to the new case, and punishes past cases with less similar content.

The relevance metric can be mathematically explained as:

$$Relevance = \sum_{p=1}^{10} PM_p$$

Where PM is the amount of points awarded for the performance metric, p.

If the retrieved solution is not detailed enough to be a solution for the new problem, then modification (adaptation) may be needed. In this work, the use of a GA to adapt cases will minimise the burden on retrieval techniques, as well as create novel cases that are unique and not repetitive [8].

Revision of cases allows learning from failure. If an incorrect case solution is generated during the reuse stage, the solution can be revised and fixed [11]. At retain the user decides whether to add a solution to the case base, or whether to reject it. Retain also give the user an opportunity to add any extra domain rules to the system that went unnoticed before evaluation of the current solution. This allows new rules to be used for validating solutions next time the system is run.

4.1   Using a Genetic Algorithm for Case Adaptation

A genetic algorithm mimics genetic processes of biological organisms, using principles of natural selection and 'survival of the fittest' to evolve solutions to real world problems, when suitably encoded [12, 13]. General issues associated with integrating a CBR cycle with a GA for adaptation are encoding a case to a GA *chromosome*; determining a parent selection scheme; determining a fitness function correlating to relevance in the CBR process; and setting GA main parameters. The GA implemented for case adaptation follows the steps of a simple GA. Details of this cycle can be found in Mitchell [14]. For our application, the main parameters to run a GA have been set at the values listed in Table 1. The crossover probability is set to a standard value for a GA [14], though the mutation probability is set at a higher value than usual (0.001 is a standard value for P_m [14]). This probability has been set higher to allow the application to fully test the potential for mutations to thoroughly examine the search space. This gives our application the opportunity to greater test learning through mutations, allowing learning to be fed back into case reuse.

Table 1. GA Main Parameters.

GA Parameter	Assigned Value
Size of Population	75
Number of Generations	100
Crossover Probability (P_c)	0.7
Mutation Probability P_m)	0.01

In order to translate a CBR case to a GA chromosome, a mapping needed to be established to allow genetic operations (*crossover* and *mutation*)to be performed. When mapping the timetable case to a chromosome, the typical bit string or integer representation was not sufficient, as the timetable case uses rich symbolic notation. The chromosome mapping needed *genes* (encoding 'traits' within the problem) and *alleles* ('settings' for each of the traits) to be extracted from the case information, then encoded into consecutive blocks of vector spaces. Subject lessons were the genes for our chromosome mapping, with the alleles in the genes being the attributes within a subject lesson. With reference to Figure 1, each gene contains Subject, Lesson, Lecturer and Room information. The alleles were set to each attribute of this object, such as the subjectCode, lessonType, building etc. The sequence of genes continues until all subject lessons within a timetable case are encoded into the chromosome. A decision was made to allow chromosomes to be of unequal length instead of a fixed length, allowing for more open-ended evolution and also accommodating the timetabling domain. In the timetabling domain, different timetables are likely to have a different number of subject lessons, so having chromosomes of unequal length is generally unavoidable. The initial population of chromosomes for the GA is a set of candidate solutions fed from the CBR case memory into the initial GA population.

After encoding a CBR case into a GA chromosome, the GA fitness function was evaluated. The GA fitness is mapped back to CBR relevance, ensuring that relevance is maintained if a solution chromosome is accepted at the CBR retain phase. The selection scheme implemented for the adaptation GA in CBR-GAALTA is the Tournament Selection algorithm [14]. Tournament Selection is the least computationally expensive selection method, and was chosen to alleviate some of the computational costliness of the CBR adaptation task. The replacement policy employed in CBR-GAALTA is Steady State replacement [14].

5   Research Objectives

This work introduces a scheme in which information is gathered at the CBR revise phase, and is fed back into the CBR reuse phase. The aim of our research is to collect knowledge at the revise stage of the CBR cycle, and then feed this knowledge back into GA adaptation to aid in finding healthier mutations to apply. To successfully mesh the CBR case mapping to chromosome mappings within the GA, we needed to develop a 'repair' mechanism to maintain genes within chromosomes remained in the required case format. We also evaluate detrimental mutation values and avoid illegal mutation parameters by placing bounds on mutable values during adaptation. Any other illegal mutation problems are monitored through the feedback repair procedure.

5.1   Feedback from Revise into Reuse

The objective of this research is to implement a new learning method that filters information learnt in revise back into reuse. In order for initial learning to arise, rules are input to the system explaining illegal values and circumstances within a timetable. Examples of the types of rules are: the class size must be less than the room size; a class should not start before 8am or after 9pm; a lesson duration

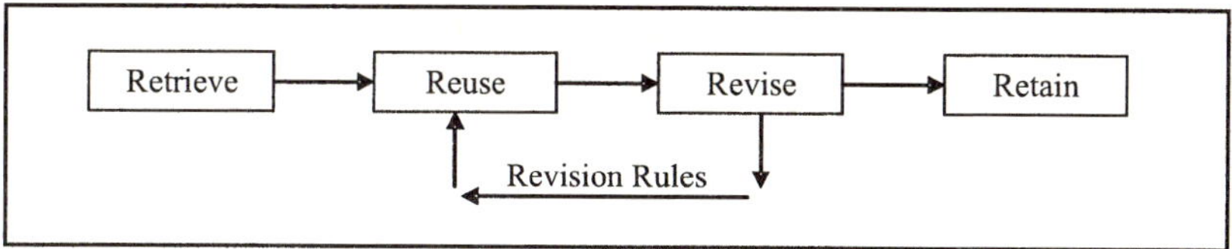

Fig. 2. Revision Feedback flow.

can not be equal to zero hours, etc. These revision rules are provided and input by system users.

Revision rules are maintained at the revision stage of the CBR cycle, but are fed back into the application at the GA adaptation stage, allowing the system to learn which mutations are not favourable to use. The rules are also used to check that core subjects are not running concurrently, and that any requirements within an incompatible subject list are considered, before allowing the experts to assess the timetable. When the chromosome repair mechanism is applied, system users are alerted to problems remaining within the chromosome structure. At this point, it is up to the user to evaluate and trial the solution in the real world. Dependent on the application of the solution, the user will choose to retain the solution or not.

6   Experiments with Varied Case Bases

The CBR system was tested using three different case bases, each containing 8 cases. The three case bases are called Case Base 1 (containing low relevance cases), Case Base 2 (containing average relevance cases) and Case Base 3 (containing high relevance cases).

Within these case base data sets, four different types of tests were conducted to evaluate the effect of mutations on fitness. All tests were conducted over 10 runs of the GA adaptation, with averages taken across the runs to yield results. The tests were run with the feedback mechanism (from section 5.1) activated (**WL**) and deactivated (**WOL**). The user can decide whether to run the system with or without learning. Table 2 shows all results for all experiments. The GA parameters remain the same as in Table 1 (except when mutation values are being changed in Test 3). Test 1 tests general adaptation with the GA, while in Test 2 we vary the probability of mutation, testing how well the GA reacts to varying the P_m. In Test 3, we test if the relevance is better when mutations are actually present, where Test 4 concentrates on how relevant cases are when we retain an adapted case.

6.1   Discussion of Results

The results displayed in Table 2 are the grounds for discussion of the learning feedback built into CBR-GAALTA. For general production of fitness while running the GA adaptation, the feedback mechanism has no direct effect on

Table 2. Results of GA adaptation using three separate case bases.

Test Number	Case-Base 1		Case-Base 2		Case-Base 3	
	WOL	WL	WOL	WL	WOL	WL
Test 1 - General CBR adaptation	27.35	26.00	30.90	31.15	33.70	33.10
Test 2 - Effect of changing GA Mutation Probability						
Mutation Probability = 0.1	25.55	28.05	26.45	29.60	30.75	36.50
Mutation Probability = 0.01	27.75	26.90	31.00	30.80	29.55	32.15
Mutation Probability = 0.001	27.15	26.50	30.55	32.15	32.75	32.15
Test 3 - Relevance based on presence of mutations						
Average Relevance With Mutations	24.85	29.35	29.62	29.00	33.00	35.13
Average Relevance Without Mutations	30.00	26.90	29.67	29.58	33.40	34.91
Test 4 - Relevance with retaining adapted solutions						
Average Relevance With Mutations	28.25	29.35	33.87	33.00	36.50	36.00
Average Relevance Without Mutations	23.00	27.83	34.00	35.00	37.20	37.20

producing chromosomes that exhibit improved fitness. However when testing the viability of chromosomes with varied mutations during test two, results became fitter as P_m increased. We found that when there is a higher chance of mutations occurring, a mechanism that learns which mutations adversely affect chromosome fitness is desirable. These results are consistent across all case base implementations.

Test three indicates how well the learning mechanism operates when mutations are present in the adaptation. Interesting results occur when comparing the three case bases. When the learning mechanism is activated, mutations generally produced higher fitness solutions, with the exception of Case Base 2. This demonstrated to us that a learning mechanism benefits average fitness, maintaining that mutation does not adversely affect the relevance of adapted solutions.

Test four records fitness based on feeding learnt information from an adapted solution back into the reuse. Feeding previously repaired solutions back into a learning mechanism allowed the solutions to continue repairing, improving over solutions where no mutations occurred.

7   Conclusions

This paper has shown that if using a GA for adaptation in a CBR cycle, feeding knowledge from the revision to adaptation stages can benefit the health of solutions. To create more novel solutions during adaptation, we focused on the importance of mutations and learning which mutations are best to use. A learning feedback mechanism was developed to feed knowledge learnt in the revise stage of the CBR cycle, into a CBR-GA adaptation stage. We found that when there is a higher chance of mutations occurring, using a mechanism that learns which mutations adversely affect chromosome fitness is desirable.

References

1. Abramson, D., Abela, J.: A parallel genetic algorithm for solving the school timetabling problem. In: Division of Information Technology, Melbourne (1992)
2. Burke, E.K., MacCarthy, B., S.Petrovic, R.Qu: Case-based reasoning in course timetabling: An attribute graph approach. In Aha, D., I.Watson, eds.: ICCBR. Volume 2080 of LNAI., Berlin Heidelberg, Springer-Verlag (2001) 90–104
3. Aamodt, A.: Case-based reasoning: Foundational issues, methodological variations, and system approaches. In: AICOM. (1994) 39–58
4. de Silva Garza, A.G., Maher, M.L.: An evolutionary approach to case adaptation. In: Proceedings of Third International Conference on Case-Based Reasoning. (1999) 162–172
5. Louis, S., Xu, Z.: Genetic algorithms for open shop scheduling and re-scheduling. In Cohen, M., Hudson, D., eds.: 11th International Conference on Computers and their Applications. (1996)
6. Jarmulak, J., Craw, S., Rowe, R.: Self-optimising CBR retrieval. In: Proceedings 12th IEEE International Conference on Tools with Artificial Intelligence. (2000) 376–383
7. Jarmulak, J., Craw, S., Rowe, R.: Genetic algorithms to optimise CBR retrieval. In Blanzieri, E., Portinale, L., eds.: EWCBR. Volume 1898 of LNAI., Springer-Verlag Berlin Heidelberg (2000) 136–147
8. Oppacher, F., Deugo, D.: Integrating case-based reasoning with genetic algorithms. In Cercone, N., Gardin, F., eds.: Computational Intelligence III, Elsevier Science Publishers (1991) 103–114
9. Pasone, S., Chung, P.W., Nassehi, V.: Case-based reasoning for estuarine model design. In Craw, S., Preece, A., eds.: ECCBR. Volume 2416 of LNAI. (2002) 590–603
10. Main, J., Dillon, T.S., Shiu, S.C.: A tutorial on case based reasoning. Soft Computing in Case Based Reasoning (1999) 1–28
11. Leake, D.B., ed.: CBR in Context: The Present and Future. In: Case-based Reasoning: Experiences, Lessons and Future Directions. Menlo Park: AAAI Press MIT Press (1996) 1–25
12. Beasley, D., Bull, D., Martin, R.: An overview of genetic algorithms: Part 1, fundamentals. University Computing (1993) 58–69
13. Wah, B.W., Ieumwananonthachai, A.: Teacher: A genetics based system for learning and generalizing heuristics. In Pal, S., Dillon, T., Yeung, D., eds.: Soft Computing in Case-Based Reasoning, Springer-Verlag London, UK (2001) 179–211
14. Michell, M.: An Introduction to Genetic Algorithms. MIT Press (1996)

A Computational Korean Mental Lexicon Model for a Cognitive-Neuro Scientific System*

Heui Seok Lim[1]

Dept. of Software, Hanshin University, Korea
limhs@hs.ac.kr
http://nlp.hs.ac.kr

Abstract. In this paper, we review major issues that have been addressed in Korean word recognition research, and we propose a computational model to explain Korean word recognition. The organization of the chapter is as follows: First, we briefly introduce some experimental results in Korean word recognition and morphological processing research. In this chapter, we will focus on studies regarding word frequency, word length, neighborhood effects, form priming, and morphological processing. Second, a computational model for Korean word recognition will be presented. The computational model was proposed to explain the characteristics of Korean morphological representation and processing. However, this computational model can explain other lexical effects such as word frequency, word length, and form priming effects. Third, the simulation results of the computational model will be shown and discussed. There will be comparison of the simulation results with human data: how well the model simulates Korean word processing and how much the simulated results coincide with that of human processing. We also discuss both the strength and weakness of the computational model.

1 Introduction

Computational modeling of cognitive process is an approach to simulate and to understand the central principles involved in human cognitive process mechanism. Some advantages of using computational models in researches on cognitive mechanism in human brain is as follows. First, it gives us detailed explanation of the cognitive process by mimicking results which are already known. Second, it enable us to perform lesion study of human brain function which is impossible with human subjects. Furthermore, it can predict some unknown phenomena which are worthy of investigating with human subjects.

Researches by using the computational models in cognitive language process are rapidly growing. Many computational models are proposed and implemented to explain many important principles involved in human language processing[1], [2], [3]. The processing mechanism of different languages have not only universal characteristics but only very peculiar ones of each language. However, many

---

* This Work was Supported by Hanshin University Research Grant in (2005).

R. Khosla et al. (Eds.): KES 2005, LNAI 3681, pp. 84–89, 2005.

of the models are related with lexical access and knowledge representations of English human mental lexicon. There rarely have been researches on developing computational models of Korean language processing. This paper proposes a computational model of Korean mental lexicon which can explain and simulate some aspects of Korean lexical access process.

The recent models of the lexical representation of the morphologically complex words can be divided into three types: the full-list representation model, the decomposition representation model, and the hybrid representation model. The supporters of the full-list model propose that full forms are stored separately in the mental lexicon and these full forms are linked together with the related forms. On the other hand, in the decomposition model, the stored representations are composed of roots or stems and affixes, and these morpheme representations are combined to make an inflected word and a derivation word. In the decomposition model, there should be decomposition of the morphologically complex words into the component morphemes before the lexical access and there should be morpheme-based lexical access. The hybrid model proposes that some word representations are stored in the full form and others in the decomposition form.

There have been a few experiments regarding the representation of the morphologically complex Korean words and Eojeols. Nam and his colleagues made several experiments to dig out representation unit of Korean Eojeol in the mental lexicon and they found some consistent evidences supporting full-list representation[4], [5], [6].

As main purpose of this paper is building a computational model which can simulate major characteristics of Korean lexical access, we propose a computational model which explain several basic finding that are consistently obtained using visual Korean lexical decision task(LDT)[4], [5], [6], [7]. LDT(lexical decision task) is most frequently adopted in the study of lexical access[7]. In a lexical decision experiment, subjects are required to discriminate words from nonwords by pressing one button if the presented stimuli is a word and another button if it is not. The LDT has proven to be a very fruitful method for exploring the nature of the mental lexicon.

1) **Frequency effect.** The time taken to make a lexical decision response to a word of high frequency in the language is less than that to a word of low frequency in the language.
2) **Length effect.** The time taken to make a lexical decision response to a long word is less than that to a short word.
3) **Word similarity effect.** When a legal nonword is sufficiently similar to a word, it is hard to classify as a nonword. While there are many other effects observed using LDT, the above are perhaps the most fundamental in Korean LDT and the target effects which our proposed model try to simulate.

2   Backgrounds

It is very helpful to find out how mental lexicon is organized and how a lexical item is accessed to understand human lexical processing. It is also very important to make a human-oriented efficient machine readable dictionary(MRD) for natural language processing. Efficient memory structure and fast accessibility are ultimate goals in building the MRD.

There have been many structures proposed to make efficient MRD such as B-tree structure, hashing, and Trie. We can guess some principles on the mental lexicon through investigation of the previous MRDs' structure and their operating algorithms. A B-tree is an m-way search tree to which we can make easy modification through insertion or deletion of an item whereas maximal search time is increased, proportional to the total number of items in the dictionary. In hashing, we store items in a fixed size table called a hash table. An arithmetic function, f which is called a hashing function is used to determine the address of an item when inserting and searching the item. If we use perfect hashing function, we can get information of an item by a direct access regardless of the total number of items in the dictionary. However, we are bothered to construct a new hashing table whenever an item is inserted or deleted. A trie is for storing strings, in which there is one node for every common prefix. The strings are stored in extra leaf nodes. The origin of the name is from the middle section of the word "reTRIEeval", and this origin hints on its usage. The trie data structure is based on two principles: a fixed set of indices and hierarchical indexing. The first requirement is usually met when you can index dictionary items by the alphabet. For example, at the top level we have a 26-element array or a linked list which we call a node from now on. Each of the nodes' elements may point to another 26-element node, and so on. Figure 1 represents structures of a 26-element array and a linked list. One of the advantages of the trie data structure is that its search time depends on only the length of an item, not on the number of stored items. Another advantage is that we can easily get a set of words with the same prefix, which is not easy in other dictionary structures presented in above.

Table 1 shows the MRDs' capacities of the characteristics of Korean mental lexicon and lexical recognition. As shown in 1, it is impossible to simulate frequency effect and length effect by using conventional hashing or B-tree index structure. In the trie structure, it can simulate length effect due to the principle of hierarchical indexing though frequency effect and word similarity effect are hard to be simulated.

Table 1. Data Structure of a Node of a FB-Trie.

	B-Tree	Hashing	Trie
Freq. Effect	X	X	X
Length effect	X	X	○
Sim. effect	X	X	X

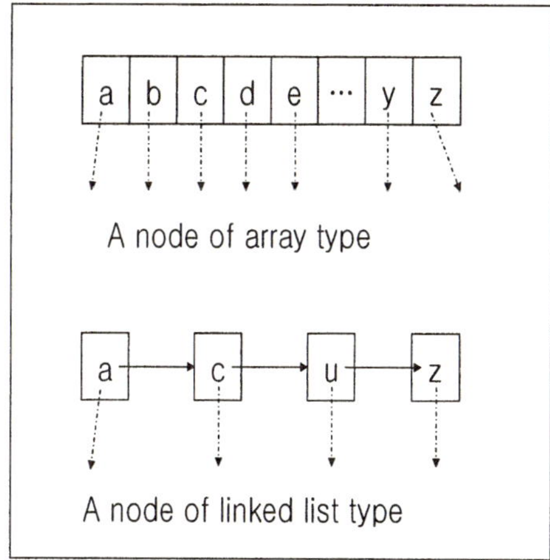

Fig. 1. Two Types of Trie nodes.

3   Frequency-Based Trie Model

Our goal of modeling Korean mental lexicon is to build a model which can simulate or reflect phenomena induced from Korean lexical recognition; full-form representation, length effect, and frequency effect. To do that, we decided that the previous trie structure is the most promising. If we make some modification to reflect frequency effect in trie structures, the trie structure can satisfy our requirements. We propose a frequency-based trie(FB-trie) to make the simple trie structure reflect frequency effect. The proposed FB-trie is a modified trie structure which can satisfy the following requirements. 1) A full-form of Korean words is stored in the trie. 2) The structure of a node is linked list structure. 3) The alphabet in a node is a set of Korean phonemes. 4) The alphabet elements in a node are sorted by descending order of frequencies of the alphabets in a Korean corpus.

Whereas in the general trie structure, the alphabets in a node are sorted alphabetically, the 4th requirement is needed to model frequency effect, which means more frequent words are accessed faster by visiting minimal elements and nodes. We tried to model language proficiency by adjusting the size of the corpus in indexing; larger for proficiency of an adult or an expert and smaller for that of an infant or novice. Data structure for a node and an indexing algorithm of FB-trie are presented in table 2 and table 3 respectively.

In table 2, 'ch' is a variable to store a phoneme and 'freq' is a variable to store frequency of the phoneme. Variable 'next' points to the next element in a node in which the current element exists and variable 'child' points to a hierarchically next node. Variable 'isfinal' is a flag to indicate whether a node is a final node of a word or not.

Table 2. Data Structure of a Node of a FB-Trie.

```
typedef struct node {
char ch;
int freq;
struct node *next;
struct node *child;
short int isfinal;
} nodetype;
```

Table 3. Indexing algorithm of FB-trie.

```
1) Get a set of unique words in a corpus and count
the frequency of each unique word
2) For each unique word
   2.1) Convert the word into sequences of phonemes
   2.2) Insert the sequences of phonemes into a trie.
   2.3) Increase 'freq' of each phoneme by frequency of the current word
3) For each unique word 4) Traverse all nodes of the trie
made in step 2) and sort phoneme elements of each node by
descending order of frequency of the phoneme
4) Output the trie made in step 3)
```

4   Experimental Results of the Proposed Model

We constructed two FB-tries with two training corpora of size about 12 million words and 7 million words. We also made two tries with both training corpora for the purpose of comparison with FB-tries. Table 2 shows the experiment results of the correlation between frequency and human reaction time and between the length and human reaction time in word recognition. In table 2, Trie1 and Trie2 represent tries trained with 7 million word size corpus and 12 million word size corpus. Simillary, FB-trie1 and FB-trie2 represent FB-tries trained with 7 million word size corpus and 12 million word size corpus. Human result of the second row is from [6].

Table 4. Comparison of Correlations with Human Data.

	Corr. of Frequency	Corr. of Length	Corr. of Similarity
Human	**-0.22200**	**0.22700**	**0.25500**
Trie1	-0.03258	0.58241	0.62241
Trie2	-0.06674	0.497034	0.51254
FB-Trie1	-0.15169	0.460327	0.44926
FB-Trie2	**-0.21718**	**0.337082**	**0.28731**

The experiment results were very promising in that correlations of frequency and length with bigger training corpus are more similar to those of human recognition than with the smaller. We aimed to model language ability or proficiency with the size of training corpus. Correlations of frequency of Trie1 and Trie2 are very small while those of FB-trie1 and FB-trie2 are relatively large and very similar to those of Human recognition. The reason of high correlations of length of Trie1 and Trie2 is due to inherent charactcristic of trie indexing structure requiring more time for retrieving a longer string. We can see that this characteristic is alleviated in FB-trie resulting in more similar correlation with human word recognition.

5   Conclusion

In this paper, we introduced some aspects in Korean lexical and morphological processing and a computational model for human mental lexicon. The proposed model is based on trie composed of nodes of linked list type and elements in each node is sorted by frequency of phonemes in corpus. We showed with some experimental results that the model reflects frequency effect and length effect which matter in Korean word recognition. In addition, we found some interesting phenomena when simulating FB-trie. First, words with high frequent neighborhood of syllables are likely to be accessed faster than with less frequent ones. Second, the difference between frequencies of phonemes which is the boundary between morphemes are very high and prominent. We guessed that this would be a clue to acquire a morpheme with many experiences with words with the morpheme. We are preparing a subsequent study on building a computational model which can encompass the neighborhood effect and the morpheme acquisition.

References

1. Bradley, A. D. Lexical representation of derivational relation. In M. Aronoff and M. L. Kean(Eds.), Juncture, 37-55. Cambridge, MA: MIT Press, 1980.
2. Caramazza. A., Laudanna, A., Romani, C. Lexical access and inflectional morphology. Cognition, 28, 207 - 332, 1988.
3. Foster, K. I. Accessing the mental lexicon. In R. J. Wales, E. Walker (Eds.), New approches to language mechanisms, 257-287. Amsterdam : North-Holland, 1976.
4. Jung, J., Lim, H., Nam, K., Morphological Representations of Korean compound Nouns. Journal of Speech and Hearing Disorders, 12, 77-95, 2003.
5. Kim, D., Nam. K., The Structure an Processing of the Korean functional category in Aphasics. Journal of Speech and Hearing Disorders, 12, 21-40, 2003.
6. Nam, K., Seo, K., Choi, K., The word length effect on Hangul word recognition. Korean Journal of Experimental and Cognitive Psychology, 9, 1-18, 1997.
7. Taft, M., Reading and the mental Lexicon. Hillsdale, NJ : Erlbaum, 1991.

An Application of Information Retrieval Technique to Automated Code Classification

Heui Seok Lim[1] and Seong Hoon Lee[2]

[1] Dept. of Software, Hanshin University, Korea
limhs@hs.ac.kr
http://nlp.hs.ac.kr
[2] Dept. of Information and Communications, Cheonan University, Korea
shlee@cheonan.ac.kr

Abstract. This paper describes an application of information retrieval techniques to automated industry and occupation code classification for Korean Census records. The purpose of the proposed system is to convert natural language responses on survey questionnaires into corresponding numeric codes according to standard code book from the Census Bureau. The system was experimented with 46,762 industry records and occupation 36,286 records using 10-fold cross-validation evaluation method. As experimental results, the system showed 87.08% and 66.08% production rates when classifying industry records into level 2 and level 5 codes respectively. In semi-automated mode, it showed 99.10% and 92.88% production rates for level 2 and level 5 codes respectively.

1   Introduction

The Korean National Statistical Office(KNSO) collects industry and occupation information of individuals in annual census. This task has been usually done by human coders who read the description written in informal natural language by individuals and decide corresponding codes with the help of coding guidelines and standard code book. This manual coding is laborious, time-consuming, and error prone.

There have been several researches on automated coding systems since early 1980s in U.S., France, Canada, Japan[1], [3], [5], [10]. AIOCS(Automated Industry and Occupation Coding System) was developed and used for the 1990 U.S. Census [1], [5]. Since then, the AIOCS has been improved with new approaches: Eli Hellerman algorithm, a self-organizing neural network, the holograph model, nearest neighbor and fuzzy search techniques, etc. [1], [5]. Memory Based Reasoning system was one of the successful system among the efforts to improve the AIOCS [4]. The ACTR(Automated Coding by Text Retrieval) system is the generalized automated coding system developed by Statistics Canada. It is based on the Eli Hellerman Algorithm[10], similar to AIOCS and is also designed for a wide range of coding application. ACTR is however unsuitable for Census Bureau industry and occupation coding since it is designed to assign a code to a single text string and weighting scheme cannot be altered. At this time, many

R. Khosla et al. (Eds.): KES 2005, LNAI 3681, pp. 90–96, 2005.

of the systems are at preliminary stages and the Census Bureau is conducting research on a wide range of different automated coding and computer-assisted coding systems.

While there have been many researches on automated coding system as described above, researches on automated coding system for Korean have been rarely studied. Since Korean is agglutinative language and very different from foreign languages like English, research on developing an automated coding system considering Korean characteristics is very important.

In this paper, we propose an Automated Korean Industry and Occupation Coding System(AKIOCS) using information retrieval and automatic document classification techniques. We abstract the coding problem as a code retrieval problem that the most highly ranked codes are classified into Korean Standard Industry and Occupation codes.

2   Overview

Figure 1 shows an overview of the system. The proposed system consists of three main modules: Indexing module, candidate codes generator module, and code generation module. The indexing module extracts index terms by using morphological analyzer and noun extractor[8], calculates weight of each term using conventional TF/IDF weighting scheme and makes an index database of which structure is inverted file. The Candidate codes generator retrieves a set of candidate codes by selecting several similar codes with input record. The code generator module selects more than one final codes from the candidate set by using a code generation function defined in this paper. The details of each module are explained in the following sections.

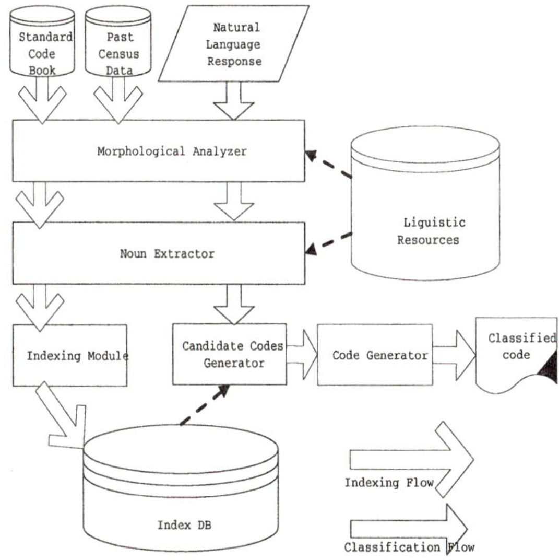

Fig. 1. The overall architecture of the proposed system.

3   Indexing

The standard code book provided by the KNSO describes each classification code
with four fields: code name, short description of the code, several examples and
exceptional cases. Each code is treated as a document in conventional information retrieval system. Unfortunately, the description of code book is not enough
to represent each code completely. The current code description for each code
provides very limited information of only 40 to 50 word size. Another problem
with the standard code book is that many individual responses usually contain
ambiguous, incorrect and ill-formed data. Variability of the terms and expressions between many respondents is also very serious: a equivalent occupation
can be described in so many different ways with so many different terms while
the standard code description contains very small number of fixed terms. To alleviate the problem of limited information available in the standard code book,
it is desirable to use Korean thesaurus which is not available for the present.
Instead, we propose to augment code description by adding nouns and phrases
in the records of which classification code were manually assigned in the past
census.

The indexing module extracts index terms by using morphological analyzer
and noun extractor[8] and calculating weight for each indexing term. Then, it
makes an inverted file including code ID, starting position of the posting file,
and the number of the posting files. Code ID is a classification code in a code
book. The posting file includes the code ID in which a term is occurred, weight
for the term. The starting position of the posting file is a starting address in the
posting file for a term.

Weight for a term indicates how the term is discriminative and it is calculated
by the following modified TF/IDF weighting scheme[2, 9].

$$w_{ij} = \frac{f'_{ij}}{max_l f_{lj}} \times log\frac{N}{n_i} \qquad (1)$$

where w_{ij} is the weight of the ith term in the jth code, N is the total number
of codes in a code book, and n_i is the number of code descriptions in which the
term appeared. Modified frequency, f'_{ij} is equal to $5^1 \times f_{ij}$ if term i occurred
in past census data, otherwise, equal to the conventional term frequency, f_{ij}.
Through indexing and weighting terms, each code is represented as vector in the
n-dimensional term space. The dimension of the code vector is the total number
of the valid terms. Value of each component of the vector is the weight of the
corresponding term. Later in code generation step, input record is also represented as a vector and distance between the vectors is calculated as a measure
of similarity.

---

[1] The constant 5 is acquired heuristically through many experiments

4 Candidate Code Generation

The individual response is converted into query which is set of nouns of the company name, business type and job description. The candidate code generator retrieves a set of relevant codes with input query by using *vector space model.* The vector space model uses vector representation for each code description and user query, and retrievcs codes by calculating cosine similarity between the code vectors and query vector. Cosine similarity between a jth code vector, $\overrightarrow{C_j} = (w_{1j}, w_{2j}, \ldots, w_{tj})$ and query vector, $\overrightarrow{Q} = (w_{1q}, w_{2q}, \ldots, w_{tq})$ is defined in equation 2.

$$sim(C_j, Q) = \frac{\overrightarrow{D_j} \cdot \overrightarrow{Q}}{|\overrightarrow{C_j}| \times |\overrightarrow{Q}|} = \frac{\sqrt{\sum_{i=1}^{t} w_{ij} \times w_{iq}}}{\sqrt{\sum_{i=1}^{t} w_{ij}^2} \times \sqrt{\sum_{i=1}^{t} w_{iq}^2}} \tag{2}$$

In the equation (2), $\sqrt{\sum_{i=1}^{t} w_{ij}^2}$ is calculated in indexing time for each code and $\sqrt{\sum_{i=1}^{t} w_{iq}^2}$ is not calculated because it is same for all the codes.

5 Classification Code Generation

The code generator module select one or more classification codes by corresponding classification modes: fully automated mode and semi-automated mode. The proposed system assigns a classification code in fully automated mode. In semi-automated mode, it assigns p classification codes and human export selects a final code among them. The human export can assign a correct code with less difficulty by only searching p classification codes than fully manual coding. We define two target functions, DVF(discrete-valued function) and SF(similarity-based function) as in equation 3 and 4.

$$\bar{f}(candidate set, l)_p = argmax_{c \in C_l} \sum_{i=1}^{k} \zeta(c, candidate_i) \tag{3}$$

where $\zeta(a, b) = 1$ if $a_l = b_l,$ 0 otherwise

$$\bar{f}(candidate set, l)_p = argmax_{c \in C_l} \sum_{i=1}^{k} \zeta(c, candidate_i) \tag{4}$$

where $\zeta(a, b) = sim(\overrightarrow{(q}, \overrightarrow{candidate_i})$ if $a_l = b_l,$ 0 otherwise.

In euations 3 and 4, *candidate set* is the result of the candidate code generator and l indicates classification level, 1-5. a_l or b_l represents a substring consisting of l digits from the first digit of string a or b. For example, 12345_3 means string 123. p indicates the number of codes made by code generator. If p is 1, the system operates as a fully automated system while it makes m final codes as a semi-automatic system when p is m. DVF of equation 3 assigns the most common c_l

code in the candidate set. Although the DVF is simple and easy to implement, it rarely uses similarity information between code vector and query vector and it is not robust to noisy data. The SF is an alternative way to consider the similarity information, it selects classification codes by sum of similarities between query vector and code vectors which have the sam c_l. Assuming that candidate code set is like in table 1, $p = 3$, and $l = 4$, table 2 shows the results ranked by DVF and SF.

Table 1. Example of candidate code set.

Code	Sim	Code	Sim	Code	Sim	Code	Sim
11110	0.5	11115	0.1	11119	0.6	33337	0.2
11118	0.2	1111	0.2	12345	0.5	3333	0.3
34567	0.3	22225	0.3	2222	0.1	22222	0.3
12347	0.1	4444	0.4	11117	0.5	1114	0.4
11141	0.5	44444	0.4	34560	0.6	34568	0.4

Table 2. Example of candidate code set.

Code	Num of codes	Rank by DVF	Sum of sim	Rank by SF
1111	6	1	2.1	1
2222	3	2	0.7	5
3456	3	3	1.3	2
1234	2	4	0.6	6
3333	2	5	0.6	7
4444	2	2	0.8	4
1114	2	7	0.9	3

6   Experimental Results

We used 46,762 records of which industry code were manually assigned and 36,286 records for occupation records to evaluate our proposed system. Performance of the system is evaluated by 10-fold cross-validation.

We defined a evaluation measure, *production rate* as in equation 5. We assume a code is correctly assigned in p mode if the system makes p best candidate codes which include correct code for the input case.

$$PR_p = \frac{\sharp \ of \ correctly \ assigned \ cases}{\sharp \ of \ input \ cases} \times 100 \qquad (5)$$

Table 3 shows results with some different p value. The results of *after augmenting* represent the efficacy of using past census data as to alleviate term inconsistency problem between code book and respondents.

As we can see in table 3, using past census data is very effective to improve production rate. In fully automatic mode, production rate for classifying level 2

Table 3. Production rate of Industry Code in p mode(%).

	p	level2	level3	level4	level5
before augmenting	1	70.03	66.16	60.23	57.08
	2	81.04	80.91	75.02	67.23
	3	83.52	81.10	77.02	72.98
	10	84.55	83.02	80.70	78.38
after augmenting	1	87.08	82.46	73.12	66.08
	2	95.16	90.91	85.14	77.01
	3	97.22	94.16	90.03	82.65
	10	99.10	98.22	95.80	92.88

Table 4. Production rate of Job Code in p mode(%).

	p	level2	level3	level4	level5
before augmenting	1	66.03	60.25	55.83	47.30
	2	71.04	65.72	58.45	57.23
	3	73.52	70.84	62.71	61.72
	10	74.55	73.99	78.80	73.27
after augmenting	1	75.08	71.36	68.12	64.08
	2	84.67	78.56	85.14	77.01
	3	89.14	83.65	90.03	82.65
	10	96.20	94.86	93.29	92.88

code was 87.08% and 66.08% for level 5 code. Though the performance of fully automatic mode is not very high, the system showed very promising performance in semi-automatic mode($p \geq 2$). It showed 99.10% and 92.88% production rate for level 2 and level 5 code classification in semi-automated mode.

Table 4 shows results with some different p value. The production rate of job code is rather lower than that of industry code for most cases. This is because the average number of words included in input records for job code is more smaller than that for industry code. This means the input for industry code has many distinguishable index terms and information between different codes in industry code classification. It is very promising that the production rate after augmenting code set are also much improved. This means that our proposed method to alleviate inconsistency between sets of index terms from code book and respondents are very effective.

Most failure of classification of the system resulted from spacing errors of input records and 1-syllable sized noun. There were much spacing errors in input records made in entering descriptions of respondents into computer. Most of errors resulted from spacing errors would be addressed, if there were Korean word-spacing system which could deal with spacing errors in short phases or short sentences. Most of current Korean word-spacing system works well on normal sized sentences which have enough context information. The another problem of 1-syllable sized noun resulted from noun extraction system of indexing module.

Most of 1-syllable sized Korean syllables are nouns. So, usual Korean noun extractor extracts nouns of which size are more than 2 syllables. It is one of future works to address the problem.

7 Conclusions

This paper describes development of the automated industry and occupation coding system for Korean using information retrieval and automatic document classification technique. We used 46,762 manually assigned records of industry code and 36,286 records for occupation records to evaluate our system. The 10-fold cross-validation is used to evaluated the system. The experimental results shows that using the past census data is very successful in increasing production rate and target function, SF is more effective than DVF to increase performance. Although production rate of the proposed system as a semi-automatic system is pretty good, the performance as a fully automated coding system is not satisfactory yet. This unsatisfactory result resulted from variability of the terms and expressions used to describe same code and spacing errors in input data. The system has much room to be improved as a fully automated system such as in dealing with spacing errors in input records and 1-syllable sized noun problem. Nevertheless, we expect that the system is very successful and enough to be used as a semi-automate coding system which can minimize manual coding task or as a verification tool for manual coding results.

References

1. Apeel, M. V. and Hellerman, E.: Census Bureau Experiments with Automated Industry and Occupation Coding. Proceedings of the American Statistical Association, 32-40, 1983.
2. Baeza-Yates and Ribeiro-Neto.: Modern Information Retrieval. Addison-Wesley, 1999.
3. Chen, B., Creecy, R. H., and Appel, M.: On Error Control of Automated Industry and Occupation Coding. Journal of Official Statistics, Vol. 9, No. 4, 729-745, 1993.
4. Creecy, R. H., Masand, B. M., Smith, S.J., and Walts, D. L.: Trading MIPS and Memory for Knowledge Engineering. Communications of the ACM, Vol. 35, No.8, 48-64, 1992.
5. Gilman, D. W. and Appel, M. V.: Automated Coding Research At the Census Bureau. U.S. Census Bureau, http://www.census.gov/srd/papers/pdf/rr94-4.pdf
6. Korean Standard Industrial Classification. National Statistical Office, January 2000
7. Korean Standard Classification of Occupations. National Statistical Office, January 2000
8. Lee, D.G.: A High Speed Index Term Extracting System Considering the Morphological Configuration of Noun. M.S. Thesis, Dept. of Computer Science and Engineering, Korea Univ., Korea, 2000.
9. Salton, G. and McGill, M.J.: Introduction to Modern Information Retrieval. McGraw-Hill, New York, 1983.
10. Rowe, E. and Wong, C.: An Introduction to the ACRT Coding System. Bureau of the Census Statistical Research Report Series No. RR94/02 (1994)

A Framework for Interoperability in an Enterprise

Antonio Caforio[1], Angelo Corallo[1], and Danila Marco[2]

[1] e-Business Management School – ISUFI, University of Lecce
Via per Monteroni – 73100 Lecce, Italy
{antonio.caforio,angelo.corallo}@ebms.unile.it
[2] Avio S.p.A., V.le I Maggio 99 – Torino, Italy
danila.marco@aviogroup.com

Abstract. This study highlights benefits to the interoperability between heterogeneous and geographically distributed workgroups, generated by the adoption of an Enterprise Architecture Framework. These groups need to collaborate in order to execute business transactions scheduled by common activities, for both complex activities that involve a lot of people like the concurrent design of a mechanical component and activities that involve communication just between software applications. The problem of interoperability is part of a more complex scenario provided by the Business Process Management which allows, through the use of a suitable framework, to describe and implement enterprise business processes in order to get a faster process execution and a more efficient and robust process management. But to obtain advantages from this kind of management it is necessary to identify and to be able to describe adequately which is the useful process knowledge and who are the actors of the process. So you need methodologies and tools to capture and to implement process knowledge to share and make it available for people who need it. These methodologies and tools should be provided inside an Enterprise Architecture Framework, based on Enterprise Application Integration techniques using a common schema for data, which offers business process modelling capabilities.

1 Introduction

Companies and organizations are constantly engaged in the effort of providing better products and services at lower costs, reducing the time to market, improving and customizing their relationships with customers and, ultimately, increasing customer satisfaction and the company's profits. To achieve these objectives companies should start to concentrate on their main business processes, on gearing all functions and resources to their processes and on improving the communication and interoperability by widely sharing information within processes. The integration of separated functions, the optimisation of the main business processes and the specification of a suitable information flow require a higher degree of transparency within the organization. In consideration of the complex relationships <u>models or modelling methods</u> have to be applied in order to support, to ease and to systematize the planning and integration of functions into business processes and to describe the related organizational structure. Moreover these models could enable a quick and easy implementation of the enterprise information system and help to identify processes to be automated.

Hence the toughest, but mostly rewarding task is analyzing business processes and creating a formal description. Such a formal description can be utilized to generate the

R. Khosla et al. (Eds.): KES 2005, LNAI 3681, pp. 97–103, 2005.
© Springer-Verlag Berlin Heidelberg 2005

"plumbing technology" that governs the flow of information between processes and people and the integration between software applications allowing to increase an organization's agility in responding to customer, market, and strategic requirements. This is the most important task required to automate this information flow as much as possible in order to streamline operations and give to people a mechanism to collaborate efficiently.

Business process automation [1], meaning the process of integrating enterprise applications, reducing human intervention wherever possible, and assembling software services into end-to-end process flows, can provide:

- **Improved efficiency** – automation of many business processes results in the elimination of many unnecessary steps
- **Better process control** – improved management of business processes achieved through standardizing working methods and the availability of audit trails
- **Improved customer service** – consistency in the processes leads to greater predictability in levels of response to customers
- **Flexibility** – software control over processes enables their re-design in line with changing business needs
- **Business process improvement** – a focus on business processes leads to their streamlining and simplification

In section 2 we'll describe the rationale for using Business Process Modelling, in section 3 we'll analyse new approaches in the application integration that benefits from business process orientation and in section 4 we'll summarily describe layers constituting a framework for emerging real-time enterprises.

2 Business Process Modelling

In general Business Process Modelling is used for two main objectives:

- to create an understanding of business processes;
- to assist the design and implementation of software systems.

The choice to use business modelling, and the choice of the type of modelling to use, require a clear understanding of the objectives of the modelling exercise. Business modelling is an area where excessive effort and cost can be expended for little benefit if those objectives are not properly understood.

2.1 What Is a Model?

A model is an abstract view of something that exists in reality, but it is quite different from what it models because some details are left out, it depends on the level of abstraction that the model contains. In the business domain a model could describe a business or a company itself (or a part of it), it should allow to model business goals, business processes, stakeholders, departments, dependencies between processes and so on.

A business model [2] does not necessarily say anything about the software system used within a company, whenever may (and should) serve as a basis for the information system model, ensuring consistency and accurate requirements being passed on to the software design. Therefore it is also called a Computational Independent Model

(CIM). A CIM is a software independent model used to describe a business system. Certain parts of a CIM may be supported by software systems, but the CIM itself remains software independent.

Finally, a model must have a purpose, for a business this may be understanding its structure, improving it or re-engineering it. The details of the model will depend on its objectives.

2.2 Why a Business Process Model

There are several reasons for using Business Process Models. Mainly they allow a reduction of complexity and a clear understanding of process features through the use of a graphical representation; they also provide a common and shared agreement between stakeholders; they give an unambiguous explanation of interactions between processes and description of interfaces.

Business Process Models offer also the opportunity to reach the following goals:

1. to state how value-creating activities are carried out;
2. to create a common approach for work to be carried out;
3. to improve incrementally processes;
4. to analyse properties of a process.
5. to support processes by workflow management systems;

More than a method to have a clear understanding of the process that allows to precisely assign task, to underline issues and further improvement, a business process model could be very useful to define requirements for the workflow management systems (as stated above). In the last years a big effort has been done by many companies and organizations to define a formal language (a meta-model) to describe business processes in order to achieve the automatic implementation of a business process management system. Results of such work could extremely speed up the building of the organizations' information systems, provide a framework to easily and rapidly manage changes in the organization that impact on IT systems and allow a more efficient collaboration between partners. These formal languages are near to become mature standards that could revolutionise the business process management.

3 Enterprise Application Integration

Application integration is a strategic approach to binding many information systems together, at both the service and information levels, supporting their ability to exchange information and leverage processes in real time [3]. While this sounds like a pure technology play, the resulting information and process flow between internal and external systems provides enterprises with a clear strategic business advantage: the ability to do business in real time, in an event-driven atmosphere, and with reduced latency. The business value of this is apparent.

3.1 Business Process Integration-oriented Approach

The goal of Business Process Integration-oriented Application Integration (BPIOAI) [4] is to bring together relevant processes found in an enterprise or trading community to obtain the maximum amount of value, while supporting the flow of information and control logic between these processes.

In reality, business process integration is another layer of value resting upon existing application integration solutions, solutions that include integration servers, application servers, distributed objects, and other middleware layers. Business process integration offers a mechanism to bind disparate processes together and to create process-to-process solutions that automate tasks once performed manually.

It is clear that the binding of middleware and process automation tools represents the future of application integration.

Business process integration is a strategy which strengthens your organization's ability to interact with disparate applications by integrating entire business processes, both within and between enterprises. Indeed, business process integration delivers application integration by dealing with several organizations using various metadata, platforms, and processes. Thus, business process integration technology must provide a translation layer between the source and target systems, and the business process integration engine.

There are many differences between traditional application integration and business process integration:

- a single instance of business process integration typically spans many instances of traditional application integration;
- application integration typically means the exchange of information between two or more systems without visibility into internal processes;
- Business Process Integration leads with a process model and moves information between applications in support of that model;
- application integration is motivated by the requirement for two or more applications to communicate.
- Business Process Integration is strategic, leveraging business rules to determine how systems should interact and better leverage the business value from each system through a common abstract business model.

There are three main services that business process integration provides:

- *the visualization of processes contained within all trading partner systems* (by visualizing enterprise and cross-enterprise processes contained within trading partners, business managers are able to become involved in enterprise integration. The use of graphics and diagrams provides a powerful tool for communication and consensus building. Moreover, this approach provides a business-oriented view of the integration scenarios, with real-time integration with the enabling middleware or points of integration.
- *interface abstraction* (refers to the mapping of the business process integration model to physical system interfaces and the abstraction of both connectivity and system integration solutions from the business analyst. Business process integration exists at the uppermost level in the application integration middleware stack. Those who use business process integration tools are able to view the world at a logical business level and are not limited by physical integration flows, interfaces, or adapters)
- *the real-time measurement of business process performance* (by leveraging tight integration with the process model and the middleware, business analysts are able to gather business statistics in real time from the trading community; for example, the performance of a supplier in shipping goods to the plant, and the plant's ability to turn those raw materials into product.)

The use of a common process model that spans entire company for application integration provides many advantages, including:

1. The ability to create a common, agreed-upon process between companies automating the integration of all information systems to react to business events such as increased consumer demand, material shortages, and quality problems in real time.
2. The ability to monitor all aspects of the business and trading community to determine the current state of the process in real time.
3. The ability to redefine the process at any given time in support of the business, and thus makes the process more efficient.
4. The ability to hide the complexities of the local applications from the business users and to have the business user work with a common set of business semantics.

4 Enterprise Architecture Framework

An enterprise which aims to reduce costs, to rapidly identify problems, to faster take decisions and act, should move to the Real-Time Enterprise concept. Real-Time Enterprise is able to do things faster, this means faster response to problems, customers and opportunities.

To achieve such objectives, enterprise needs a business and information technology integrated environment which allows

– to integrate several existing applications
– to model business processes that spans across the whole enterprise and his customers and suppliers (even assigning human and application resources to specific process tasks)
– to establish an information flow and a control flow between processes
– to automatically activate processes through business events
– to monitor and manage systems and activities.

Such environment could be provided by the following framework.

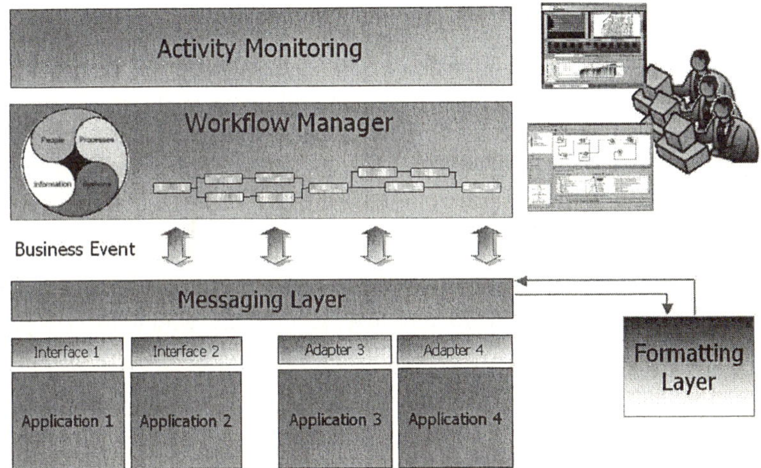

Fig. 1. Enterprise Architecture Framework

As showed in the Figure 1, the overall framework is based on a Messaging Layer (the Information Bus) that provides a transportation mechanism for messages between the source and target applications, the new and already existing ones, the open and proprietary ones; it also provides a fast and efficient event handling by passing messages to the Workflow Manager. Through the Information Bus, a flow of timely information across the business could be established and managed exploiting messaging capabilities for enabling real-time information flow between unlimited number of globally distributed endpoints, to perform application monitoring and management to keep enterprise systems up-and-running at all time and application connectivity.

The Formatting Layer allows data transformation for the reformatting of information as it's being moved from one system to another. It is particularly in this layer that the use of ontologies could be very useful to grant interoperability between software applications.

The Business Process and Workflow Layer (Workflow Manager) provides modeling tools and mechanisms that help people to define how information is moved throughout and between enterprises and how the movement of information relates to Business Events, to manage and monitor business process integration and execution. These service layer provide the ability to model and manage the flow of activities and interactions across systems, people and even organizations through workflow management and process automation capabilities allowing to get more value out of existing systems by tying them together in real-time. Moreover at this level it could be provided B2B integration improving collaboration and information exchange with trading partners of all sizes.

The Activity Monitoring layer provides tools that give a better visibility on the ongoing activities, allowing organization, by measuring performances and analyzing current and historical information, to drive immediate actions.

5   Conclusions

This study contributes to highlight the core components needed by an enterprise to succeed in providing better products and services at lower costs, reducing the time to market, exploiting at a maximum level information technology services.

The framework resulting from this research is based on a layer in which specific tools are provided to model business processes and on an application layer that exploits these models to assign task to people when it is required and to bind together software applications to improve collaboration and to achieve fast and efficient execution of activities.

Acknowledgements

This study was conducted in collaboration with Avio S.p.A. within the *Design for Six Sigma* Project, supported by Regional Administration of Puglia.

References

1. "Business Process Automation - ARIS in Practice", A.W. Scheer, F. Abolhassan, W. Jost, M. Kirchmer, Springer, 2004

2. "MDA Explained, The Model Driven Architecture, Practice and Promise", A. Kleppe, J. Warmer, W. Bast, Addison-Wesley, 2003
3. "Enterprise Modelling and Integration", F.B. Vernadat, Chapman & Hall, 2005
4. "Next Generation Application Integration", D. Linthicum, Addison-Wesley, 2003
5. "Planning and Building an Architecture that Lasts: The Dynamic Enterprise Reference Architecture", Doculabs, 2003

Detecting is-a and part-of Relations
in Heterogeneous Data Flow

Paolo Ceravolo[1] and Daniel Rocacher[2]

[1] Università di Milano - Dipartimento di Tecnologie dell'Informazione
Via Bramante, 65 - 26013 Crema, Italy
`ceravolo@dti.unimi.it`
[2] IRISA/ENSAT
BP 805018, 22305 Lannion cedex, France
`daniel.rocacher@enssat.fr`

Abstract. The `KIWI` project developed a platform called *OntoExtractor* enabling semi-automatic extraction of ontologies. This platform is composed of different modules. A first module extract typical instances from a data flow. A second module propose some topological relations among the detected classes. A third module assist the knowledge engineer in transforming classes and relations tentatively proposed by the system in a standard knowledge representation format. This paper address the specific problem of detecting `is-a` and `part-of` relations among typical instances of a document base.

Keywords: Knowledge Representation, Ontology, Bottom-up design, Fuzzy Logic

1   Introduction

This paper is inserted in the general framework of bottom-up knowledge representation design. The work introduced in these pages is a part of a more extended research worked out in the activities of the `KIWI` project [1]. The `KIWI` project is developing a technological platform supporting knowledge extraction, profilation, and navigation. The extraction technique proposed in the project is described in other works, such as [2] and [1]. In this paper we address the exposition only on a specific task of the general extraction technique. The first extraction phase is a clustering process selecting from a document base a set of typical instances. These representatives are view as semi-structured descriptions of domain documents. The second phase is focused on the organization of these representatives into a hierarchy. We do not claim to provide a completely automatized technique, our aim is to test some possible relations and to provide to the knowledge engineer a set of candidates relations, organized according to a satisfaction degree. The knowledge engineer will use this information in support to his design work.

---

[1] `KIWI` is a project supported by the Basic Research Fund (`FIRB`) of the Italian Research Ministry. This work was partially funded by this project.

R. Khosla et al. (Eds.): KES 2005, LNAI 3681, pp. 104–111, 2005.

The aim of the paper is to describe a technique for detecting `is-a` and `part-of` relations. This technique is based on fuzzy logic notions and use fuzzy predicates as parameters for adapting the detection algorithm to the domain features.

2   Ad-Hoc Ontology Construction

The aim of the `KIWI` project is to support knowledge engineer in building a representation of the domain, using some bottom-up instruments. For this purpose our system can provide a list of *candidate relations* interconnecting *representatives* extracted from the document base.

Basically, we deal with two types of relations: `is-a` and `part-of`. A class A `is-a` a class B if it is a specialization of B. Therefore a description of A must contain the same attributes than B, and at least one more. On the contrary A is a `part-of` B if it is (strictly) included in B, i.e. A has a subset of the attributes of B. Nesting is also important: as illustrated in Fig. 1, in case of `is-a` relation, we expect nodes in A to have the same nesting level that the corresponding nodes in B. In case of a `part-of` relation the nodes of A appear in B at a lower level.

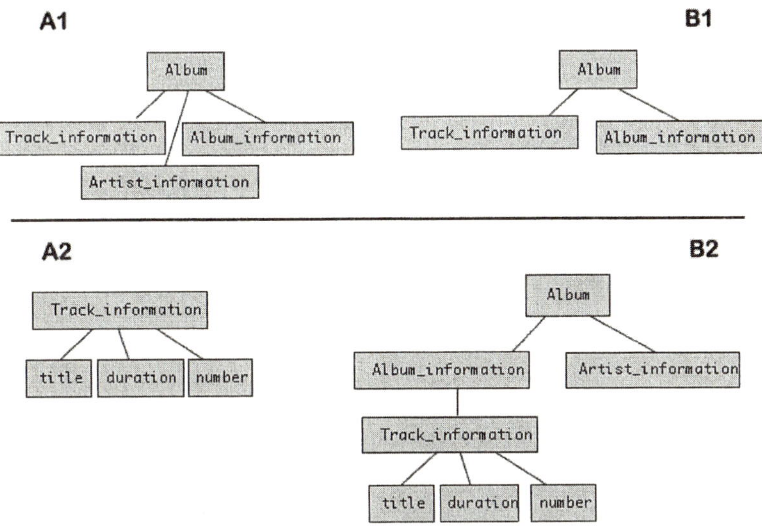

Fig. 1. $A1$ is a $B1$, $A2$ is a part-of $B2$.

2.1   Conditions for Detecting `is-a` and `part-of` Relations

We have now to formalize conditions for detecting `is-a` and `part-of` relations. From this list of conditions we can obtain a set of function allowing us to evaluate how much the comparison of two objects satisfies the conditions.

It is well known that tree matching problem have a computational complexities in the range from linear to NP-complete [3]. In this work we want to avoid hight computational costs and we prefer to accept a degree of incertitude. In fact the scenario we are drawing provides to organize a set of candidate classes and a set of candidates relations among classes, according to a relevance degree. For this reason we are selecting a set of conditions quite easy to compute and sufficient for distinguish is-a relations from part-of, but not enabling an exact matching. Actually, using our functions, trees with different topology can produce the same matching degree. This is acceptable, because the topological information we are losing is not so informative in order to distinguish is-a and part-of relations.

The fact that we express matching in a degree drives us in inserting our evaluation function into a fuzzy logic framework. Hence, if a condition is satisfied the function must to return a value equal to 1.

We isolated three conditions. Let A be a tree we are comparing to another tree called B. The first condition is about the rational between the intersection of A B, and the tree we expect to be the smaller. The second one is about the delta between the root of A and the root of B in B. The third one is about the structure cohesion between A and the subtree rising from the root of A in B.

A is part-of B *Conditions:*

– Intersection:

$$|A| = |A \cap B| \tag{1}$$

– Depth level:

$$(B_r \triangle A_r) \ in \ B > x \tag{2}$$

– Cohesion of intersection:

$$\forall \, m \, (\, A_r \triangle A_m) \ in \ B = (\, A_r \triangle A_m) \ in \ A_B \tag{3}$$

A is-a B *Conditions:*

– Intersection:

$$|A \cap B| < |A| \wedge |A \cap B| = |B| \tag{4}$$

– Depth level:

$$(B_r \triangle A_r) \ in \ B = 0 \tag{5}$$

– Cohesion of intersection:

$$\forall \, m \, (\, A_r \triangle A_m) \ in \ B = (\, A_r \triangle A_m) \ in \ A_B \tag{6}$$

Where $\triangle$ represents the number of edges dividing two elements of a tree. A_r is the root node of the tree A, m are all nodes of $|A \cap B|$, A_B the subtree rising from the root of A in B.

The conditions described above are not used in our system as boolean criteria, where the truth values are 0 or 1: satisfaction or non-satisfaction. Our aim is to return a gradual evaluation of conditions. For this reason the satisfaction is expressed in term of a degree into a range from 0 to 1. This degree is used to propose some candidate relations between representatives and to organize relations according to an estimated relevance.

2.2   Gradual Evaluation of the First Condition

In the `part-of` case the first condition can be measured dividing $|A \cap B|$ by $|A|$: $\frac{|A \cap B|}{|A|} = t$. If this two crisp cardinalities are equal their quotient is 1. How much the division result is close to 1, how much the condition 1 is satisfied. Filtering this result into a fuzzy predicate expressing the linguistic variable "close to 1", we obtain a degree of satisfaction expressing the extent to which the cardinality of A is close to the cardinality of $A \cap B$. Such a fuzzy predicate is a parameter of our process. For example it may correspond to the fuzzy predicate described by Fig. 2. Along the x axis we ordered the inputs of the function: in our case all the allowed values of the division. Along the y axis we have the output of the function: $[0, 1]$. In brief, the fuzzy predicate is a function associating some domain values to a degree of membership to the notion expressed by a linguistic variable.

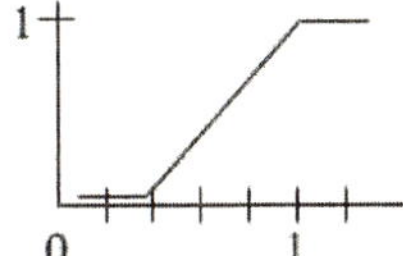

Fig. 2. The fuzzy predicate "close to 1".

For example evaluating this condition for the trees $A2$ and $B2$ in Fig. 1 we have $\frac{4}{4} = 1$ and $\mu(1) = 1$.

In the `is-a` case the first condition can be evaluated putting two sub-conditions. If the boolean condition $|A \cap B| < |A|$ is satisfied then the second can be evaluated otherwise no. The second sub-condition is: $\frac{|A \cap B|}{|B|} = t$. This division returns a result that can be used for assigning to the condition 4 a degree of satisfaction. In this case the fuzzy function we selected is equal to the one used in the `part-of` case (Fig. 2).

2.3   Gradual Evaluation of the Second Condition

The second condition can be measured computing in the tree B the number of edges dividing the root of A to the root of B: $B_r \in A \triangle A_r \in B = t$. For example comparing the tree $A2$ with the tree $B2$ in Fig. 1, the $\triangle$ between the root of $B2$ `Album`, and the root of $A2$ `Track information` is equal to 2, because two edges divide `Album` from `Track information`.

This distance can not to be too low, because a `par-of` relation presupposes the nesting of the included object in one node of the containing object. By means of a fuzzy predicate we can transform this distance in a value into the range from 0 to 1. Fig. 3a shows an example of a predicate setting a distance of 2 edges as threshold for having a degree of satisfaction greater than 0. It is clear how the fuzzy predicate can be used as a parameter for fixing degrees of satisfaction. In the `is-a` case, the second condition can be evaluated setting a

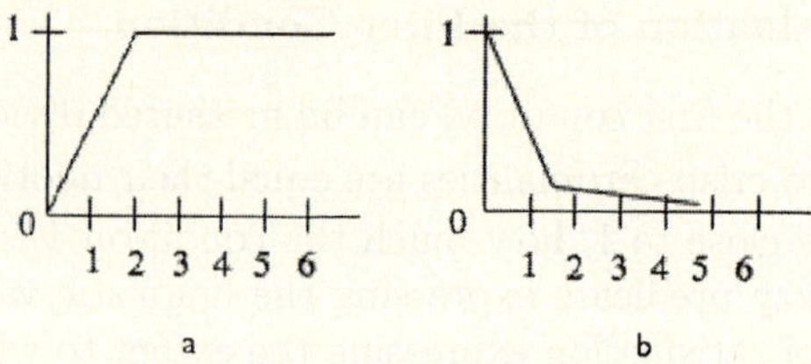

Fig. 3. The fuzzy predicates: a) "at least 2" b) "close to 0".

predicate returning hight values only if the roots distance is near to 0. Fig. 3b shows a predicate enabling these features.

2.4 Gradual Evaluation of the Third Condition

The third condition can be measured comparing the distance from the root of all the nodes of A to the distance from the root of all the nodes of the subtree of B rising from the first node of B equal of the root of A. In this case the fuzzification process in more complex. The result of the objects comparison in not a single value but a set of values. Our aim is to compare this set with the ideal set of values that should result if the comparison between objects was exact. In a crisp framework for comparing two set we have to compare their cardinality. The method for evaluating if two quantities are equal is to divide them and to estimate how much the result is close to 1. In a fuzzy framework this operation requires to compute the fuzzy cardinality of fuzzy sets and to execute a division on two fuzzy cardinalities. The mathematical notions for enabling exact divisions on fuzzy numbers are provided in [5]. Because of space limits, we are not detailing theoretical aspects of our computation. In the following we provide only a general description of the process of computation explaining it by means of a brief example. In the sequel we will use these notions for executing divisions on fuzzy integers and for approximating the result of this division to a real value.

To do that we take all the nodes distances (Eq. 7) and we filter them into a fuzzy predicate like the one in Fig. 3b. This operation returns a fuzzy set associating each node of the sub tree A_B to a degree of satisfaction of the predicate in Fig. 3b. Computing the cardinality of this fuzzy set we get a first fuzzy integer. A second fuzzy integer is taken assuming the nodes distance equal to 0; this fuzzy integer represents the cardinality of a set satisfying the requirement of our condition. Dividing the first fuzzy integer to the second one we have a fuzzy rational expressing how much the fuzzy set t is near to the satisfaction of condition 3.

$$(\forall\, m\, ((A_r \triangle A_m) \in B) - ((A_r \triangle A_m) \in A_B) = t \tag{7}$$

For the sake of clarity now we will transpose in an example the exposition of the previous paragraph. Fig. 4 show two trees, M and N, where N is near to be a part of M but with some topological differences. The node B is the root of N.

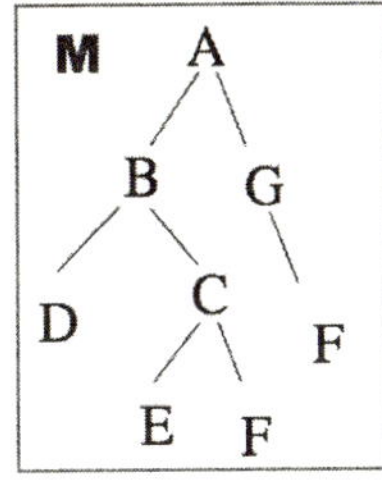
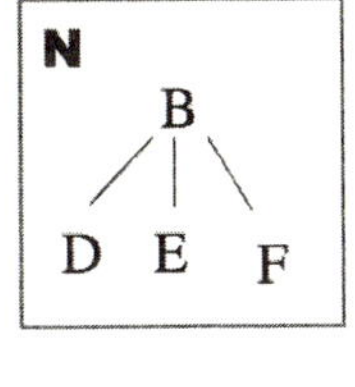

Fig. 4. N share all its nodes with the sub tree of M rising in the node B, but the topological organization of N differ form the sub tree in B of M.

In the tree M a sub tree rising in B exist. Computing the $\triangle$ between the nodes of N and the sub tree rising in B of M we obtain this set of values.

$$((1-1)/D, (2-1)/E, (2-1)/F) = (0/D, 1/E, 1/F)$$

Filtering this crisp set with the fuzzy predicate in Fig. 3b we obtain the following fuzzy set:

$$X = \{1/D, 0.7/E, 0.7/F\}$$

The fuzzy cardinality of this set can be viewed in terms of the following fuzzy integer: $\{0/1, 1/1, 0.7/2\}$, where, for example, the degree 0.7 is the extent to which the fuzzy set X contains at leat two elements.

Because the intersection between N and M is equal to 3 the complete satisfaction of the third condition should be realized with a fuzzy integer such as the following: $\{1/0, 1/1, 1/2, 1/3\}$.

For evaluating the distance from our fuzzy integer to the one allowing the complete satisfaction, we compute the ratio between them. This operation can be done using the mathematical notions provided in [5]. The division on these two fuzzy integer returns the following fuzzy rational: $\{1/0.33, 0.7/0.66\}$.

In the final step an approximation in the set of real values $\mathbb{R}$ of this resulting fuzzy rational number is performed. This approximation is computed thanks to a Lebesgue integral of the characteristic function of the fuzzy rational. Fo our example we obtain: $(1 * 0.33) + (0.3 * (0.66 - 0.33)) = 0.423$.

The procedure to be executed in case of `is-a` is the same we exposed for the `part-of` case.

2.5 Aggregating Conditions

When a value representing the degree of satisfaction is obtained for each condition the last step is to aggregate those values in a unique result. The choice of the aggregation function depends on the relevance we want to assign to each condition. The average function is the first option we can choose. In this case each condition have the same relevance. If we want to assign different relevances to the conditions we can use a weighted average or more complex function such

Is a	A1	B1	A2	B2
A1	1	1	0.28	0
B1	0	1	0	0
A2	0.37	0.37	1	0
B2	0.89	0.89	0.88	1

Part of	A1	B1	A2	B2
A1	0.68	0.52	0.24	0.6
B1	0.68	0.68	0.28	0.6
A2	0.24	0.28	0.68	1
B2	0.6	0.6	0.4	0.68

Fig. 5. Is-a and part-of values between representatives shown in Fig. 1.

as OWA and WOWA. A rich description of these operator can be found in [7] and [6].

In order to simplify the exposition, here we will develop our example using a weighted function.

$$\bar{x} = \frac{\sum_{i=1}^{n} p_i x_i}{\sum_{i=1}^{n} p_i}$$

In function 2.5, x_i are the values we are aggregating and p_i the weights associated to each value. In our example we take the weights values (p) into a range from 0 to 1. We assign to the condition 1 and 4 a weight equal to 1; to the condition 2 and 5 a weight equal to $0,8$; and to the condition 3 and 6 a weight equal to $0,7$.

3   An Example of Ad-Hoc Ontology Construction

We are now ready to build our bottom-up ontology; referring to representatives of the document base A1, B1, A2 and B2 in Fig. 1, we can connect one each other in a cyclic graph, labeling edges with the degrees of is-a (red arrows) and part-of (black arrows) relations obtained applying the process previously introduced (results are shown in Fig. 5). We represent classes as vertexes of a direct graph. Initially we obtain a cyclic graph. Cutting the less relevant edges we convert it into a directed acyclic graph, expressing a partial order. This information can be used during the construction of a hierarchy of classes, ordering instances of the domain. In Fig. 6 the result of this process is shown.

4   Conclusions

The paper issue is inserted in the general framework of bottom-up generation of ontologies. The paper discussed the role of fuzzy techniques in detecting topological relations among representatives of a document base. Each relation is described by means of a set of conditions. Fuzzy predicates are proposed as a tool for translating the conditions evaluation in a degree of satisfaction. This allow us to support the knowledge engineer with a set of candidate relations, that can be used in the design of a knowledge schema. Fuzzy predicates can also be used for parametrizing the detection process, tuning the condition evaluation functions according to the domain requirements.

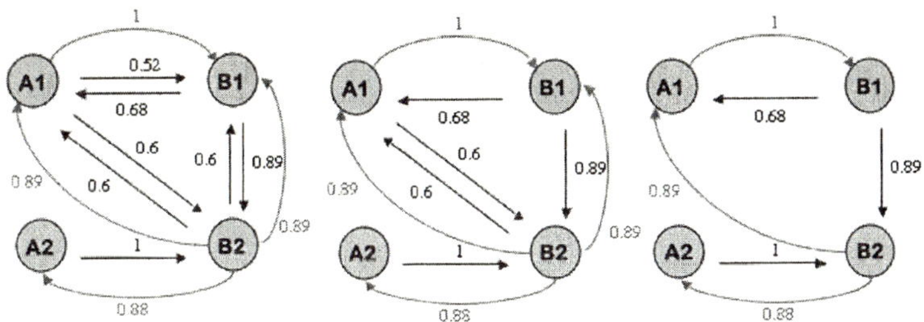

Fig. 6. The direct cyclic graph (representing classes and relations between them) is converted in a directed acyclic graph.

References

1. Ceravolo P., Nocerino M. C., Viviani M.: Knowledge extraction from semi-structured data based on fuzzy techniques. Eighth International Conference on Knowledge-Based Intelligent Information & Engineering Systems (KES 2004), Wellington, New Zealand, (2004) 328–334.
2. Damiani E., Nocerino M. C., Viviani M.: Knowledge Extraction from an XML Data Flow: Building a Taxonomy based on Clustering Technique. EUROFUSE Workshop on Data and Knowledge Engineering (EUROFUSE 2004), Warszawa, Poland, (2004) 22-25.
3. Kilpelinen P.: Tree Matching Problems with Applications to Structured Text Databases, PhD Thesis, Department of Computer Science, University of Helsinki, (1992).
4. Rocacher D.: On fuzzy bags and their application to flexible querying. Fuzzy Sets and Systems, (2003) 93-110.
5. Rocacher D., Bosc P.: The set of fuzzy rational numbers and flexible querying. Fuzzy Sets and Systems, to appear, (2005).
6. Torra, V.: The weighted owa operator. International Journal of Intelligent Systems, (1997) 153–166.
7. Yager, R.: On ordered weighted averaging aggregation operators in multicriteria decision making. IEEE Transactions on Systems, Man and Cybernetics , (1988) 183–190.

OntoExtractor:
A Fuzzy-Based Approach in Clustering Semi-structured Data Sources and Metadata Generation

Zhan Cui[2], Ernesto Damiani[1], Marcello Leida[1], and Marco Viviani[1]

[1] Università degli Studi di Milano,
Dipartimento di Tecnologie dell'Informazione
Via Bramante, 65 – 26013 Crema (CR), Italy
{leida,viviani}@dti.unimi.it
http://ra.crema.unimi.it/kiwi
[2] Intelligent Systems Research Centre, BT Group
Orion Building - Adastral Park - Martlesham Heath
IP5 3RE Ipswich - Suffolk, UK
zhan.cui@bt.com

Abstract. This paper describes a theoretical approach on data mining, information classifying and a global overview of our `OntoExtractor` application, concerning the analysis of incoming data flow and generate metadata structures.

In order to help the user to classify a big and varied group of data, our proposal is to use fuzzy-based techniques to compare and classify the data.

Before comparing the elements, the incoming flow of information has to be converted into a common structured format like XML.

With those structured documents now we can compare and cluster the various data and generate a metadata structure about this data repository.

1 Introduction

The need to have an easy access and a reliable classification of the amount of data available nowadays on the web, is one of the challenges of the future directions of data mining.

The community of users have to confront with a huge amount of digital information such as text or semi-structured documents in the form of web pages, reports, papers and e-mails. In order to improve the effectiveness of community-wide co-operation, the capability of extracting sharable knowledge from heterogeneous documents produced by multiple sources and in a timely fashion is therefore crucial. Our answer to this kind of demand is to supply a suite of tools for the end user and for the developer as well, to create, maintain and navigate an ontology generated from an varied amount of data.

R. Khosla et al. (Eds.): KES 2005, LNAI 3681, pp. 112–118, 2005.

The tool described in this paper is the `OntoExtractor`, an ontology generator. This tool wrap the set of data sources in a structured format, such as XML and, using fuzzy techniques in order to compare the incoming data flow, cluster them in a semi-automatic way.

The final purpose is to generate a browsable ontology of the processed resources.

This paper is a description of the tool `OntoExtractor` and the theory behind it. In the section describing our tool will be described the implementation of the fuzzy elements and algorithms used, for who wants to use them for his own fuzzy-based applications.

The following sections will be discussed without going in the deeper, for details refer to [4][3].

2 Data Wrapping

The first step in order to cluster a documents repository is to create a homogeneous set of semi-structured data objects. The idea is to use ad-hoc developed wrappers to process different kind of data types (pdf, html, emails, DB records,...) and convert them into a coherent structured XML documents.

We developed different types of wrappers for each kind of document to wrap, following a reverse-pyramid schema: using external licensed libraries to read and convert the document into a handling object. All the useful data will be extracted from the documents, using the respective libraries. After that the data needs to be cleaned out and converted into an unified data object which gives more importance to the document structure, simplify the XML generation process and have a common layer which simplify the next operations and adds more flexibility to the wrappers' structure: everyone can write its own wrapper for other document types, the only thing to take care about is to implement the specific interface which returns the unified data object. Once all the documents are converted, cleaned and translated into structure-oriented data object, it is recommended to save them in a common repository to avoid lost of data.

The repository initially will be a directory containing all the data objects saved using XML format[1]. This repository will be used by the `OntoExtractor` software as the document source, from this step all the documents contained into the repository are generated following a common XML schema and any information about the previous format of the document is needed, the only information that the `OntoExtractor` needs is about the structure of the document.

3 Fuzzy Techniques

After the wrapping process, the varied flow of incoming data is rendered in a coherent and homogeneous data source in a known repository. Now the goal is to

---

[1] This format is used only during the development step. Once the tool is stable and correct others solutions can be tested (i.e. DataBases).

analyze the documents in order to find some similarity between them. As XML documents are nested representation of data, we can use Fuzzy techniques to encode a nested object structure in a flat representation, keeping some memory of the structure in the membership values. The clustering process goes on comparing all the *FuzzyBags* using Fuzzy logic techniques described in [3] and putting the comparison value into a matrix representation. Once all the documents are compared, the similarity matrix is fully generated. The documents now are ready to be clustered. Using an α cutting value the similar documents will be clustered into the same cluster. This α value is used as a threshold for the clustering process: the documents that have resemblance value bigger than α are clustered together.

3.1   From Semi-structured Datasource to FuzzyBag

In order to compare two documents and obtain a resemblance value, first we need to convert the tree-structure of the XML document to a flat mathematical representation.

The XML documents are composed of a sequence of nested tags, each containing another tag, a value, or both of them. In our approach we only take into consideration tag names and not their value. It is possible to use a vocabulary where a domain expert associates a membership degree (between 0 and 1) to each tag, depending on its importance in the domain. Given an XML document like:

```
<Student>
    <Personal_data/>
    <University_data/>
</Student>
```

Its FuzzyBag representation will looks like the following set of elements:

$$S2 = \{\{M_1\}*\texttt{Student}, \{M_2\}*\texttt{Personal\_data}, \{M_3\}*\texttt{University\_data}\}$$

Where M_n is the membership value of the element. The membership value is an index of the position of the element in the document, it is used to keep memory of the document structure and the nesting level of the tag.

3.2   FuzzyBag Comparision

With the FuzzyBag representation of the document now is possible to perform mathematical operations (addiction, product, union, intersection, ...). We use easy-to-compute fuzzy norms to calculate the similarity between fuzzyBags representing documents. The norm we used [5][3] produces a value in $R_{[0,1]}$ which is an index of the resemblance of two documents.

Basically this norm compares the intersection of two fuzzyBags with their union, so if two documents are identical we obtain 1 as a result, on the other hand, if two documents doesn't have anything in common the intersection and the result of the norm is 0.

3.3 Clustering Using Alpha-Cut Values

Once applied the norm to all possible documents combination (actually, the procedure used into OntoExtractor is a bit different, as described in 5.4, but it comes from the same reasoning process) the results are used to build a matrix of resemblance values. This matrix is used during the clustering process: the clustering process uses an α cut-off value as a threshold in order to select which documents will be considered similar. The procedure is quite simple and is widely described in [3].

The result of clustering procedure is a set of containers with, inside of them, the documents which have a mutual resemblance value bigger than the cut-off value α. It is also important to generate or select a representative element for each cluster generated, this element is called *Cluster Head* [4].

4 Ontology Generation

The aim of the clustering process is to generate an ontology of the concepts isolated during the clustering process. The tool implements a set of functions for detecting typical instances from a document corpus and producing assertions in RDF format [6]. The detection process allows, as a final result, to associate a resource (a document of the corpus) to a class (a typical instance of the domain). The outcomed ontology can be browsed using RDF navigation tools such as our Semantic Navigator[2].

5 OntoExtractor

The practical translation of the previous theorical notions is the `OntoExtractor` tool. This tool is a step-by-step tool with the final purpose of generate the RDF ontology.

The current version of the tool works with a repository of XML documents, clusters those documents and select a Cluster Head for each cluster of documents.

We used different algorithms for each step of the clustering process, so it's possible to analyze the behavior of the tool using different techniques.

5.1 Project

The `OntoExtractor` is project oriented, this means that when the tool is started the user can choose to open an existing project file or create a new one. The operation of opening a project puts the tools at the situation when the project was closed (i.e. if the project was closed after FuzzyBag generation, the user now can go on with the others operations). When the user selects to create a

---

[2] Semantic Navigator is a ontology navigation tool developed by the Knowledge Management Group of the Department of Information Technologies of the University of Milan. `http://ra.crema.unimi.it/kiwi`

new project, a dialog box appears for collecting informations about the project, such as: the *name* of the project, the *folder* where the project description file will be saved, the *repository of the documents*, and the folders where the *data object* used by the tool will be saved (FuzzyBag, vectors and clusters). Once a project is ready to be used, is possible to associate a vocabulary, where a domain expert can assign a value between 0 and 1 to each tag extracted from the source documents, this vocabulary will be used in FuzzyBag generation process, to product the membership value of each tag.

5.2   Documents Selection

The first step to cluster the documents is to select a representative sub-set of the source documents. This procedure is introduced because, with a big amount of documents to process, the similarity matrix generation process has a big computational weight. The user can choose between two different selection algorithms:

- *Default Selection Algorithm:* this algorithm analyses the first tag of each document, considering all the documents with different root. If a document with that tag is already taken the algorithm considers also the second tag.
- *Node Number Algorithm:* this is a simple algorithm which selects the documents considering their number of nodes. The algorithms considers documents with 5, 10 and 15 nodes.

The first algorithm is slower, but is better to discriminate different types of documents (i.e. documents from different sources such as PDF, HTML, DB,...).
The second algorithm is faster and is suitable for homogeneous document source.

5.3   FuzzyBag

Once the selection algorithm produces a sub-set of the initial documents set, the next step is to generate the FuzzyBag of each selected document. Not only the selected documents are converted to FuzzyBags. The remaining documents will be converted later. The procedure is the same for both of the generation processes.
The user can choose between two different FuzzyBag generation algorithms to calculate the membership value (section 3.1) of each tag element in the document:

- *Nested Algorithm:* in this algorithm the value of M is:

$$M = V/L$$

- *MV Algorithm:* in the MV algorithm the value of M is:

$$M = (V + M_p)/L$$

Where V is the weight of the tag in the vocabulary, M_p the membership value of the parent tag in the bag and L the nesting level of the tag (with $L_{root} = 1$).

The difference between the two algorithms is that MV Algorithm considers also the membership value of the parent tag, bringing more meaning to the structure of the document, respect to the Nested Algorithm which considers more the nesting level of the tag.

5.4   Vectors

This section implements the similarity matrix generation process. For memory occupancy reasons in the tool we use a vector data structure for each row of the matrix, instead of a proper matrix representation. Even in this step the user can choose between two algorithms. Considering the following FuzzyBag:

$$N = \{(M_1/a_1), (M_2/a_2), ..., (M_n/a_n)\}$$

- *Default Approximation Algorithm:* This approximation algorithm computes the formula:

$$A = \sum_{i=0}^{n}(M_i - M_{i+1}) * a_i \ (with \ a_{n+1} \ = \ 0)$$

- *Average Approximation Algorithm:* This approximation algorithm computes the formula:

$$A = \frac{\frac{\sum_{i=0}^{n}(M_i)}{n}}{\frac{\sum_{i=0}^{n}(a_i)}{n}}$$

Those algorithms are used for the operation of approximation of the norm used to calculate the resemblance values between two documents as described in section 3.2. The tool is designed also to support different algorithms to each mathematic operation in the norm (union, intersection, division and cardinality).

5.5   ClusterSet

Once the similarity matrix is generated it is possible now to cluster the documents. The user has to select an α_{min} and an α_{max} and the number of samples k. Using those values the tool will generate k different clusters sets one for each value of α. Besides the previous values the user needs to select which algorithm will be used for the generation of the representative cluster head of each cluster.

- *Minimum Size Cluster Head Algorithm:* this algorithm, for each cluster, takes the document with the less number of tags as the representative cluster head.
- *Generic Cluster Head Algorithm:* this algorithm generates an ad-hoc cluster head, produced by the union of all the tags of the documents contained in that cluster.

5.6   Cluster

Browsing the generated cluster set, the user selects his preferred α cut-off value.
With this value the tool clusters all the documents that was not selected during the selection process.

Those documents are converted in FuzzyBags, and compared only with the cluster head of each cluster related to the α value selected by the user. Before this process the user can select one of the following cluster generation algorithms, the clustering procedure is the same described in [3], the algorithms decide only where to put the unclustered documents:

- *Default Clustering Algorithm:* this algorithm puts the documents that has not been clustered in the cluster where its cluster head has the biggest resemblance value in respect to the document to cluster.
- *Cluster not Clustered Documents Algorithm:* this algorithm creates a new cluster and puts all the unclustered documents into this cluster.

Once the clustering process is finished the domain expert can drag and drop documents from a cluster to another.

Acknowledgements

This work was partly funded by the Italian Ministry of Research Fund for Basic Research (FIRB) under project RBAU01CLNB_001 "Knowledge Management for the Web Infrastructure" (KIWI) and also partially supported by British Telecom (BT) research and venturing.

References

1. B. Bouchon-Meunier, M. Rifqi, S. Bothorel: Towards general measures of comparison of objects, Fuzzy Sets and Systems, Vol. 84, 143–153 (1996)
2. P. Ceravolo: Extracting Role Hierarchies from Authentication Data Flows, Computer Systems Science & Engineering Journal (IJCSSE), Vol. 19, No. 3, 121–127 (2004)
3. P. Ceravolo, M. C. Nocerino, M. Viviani: Knowledge Extraction from Semi-structured Data Based on Fuzzy Techniques, Knowledge-Based Intelligent Information and Engineering Systems, Proceedings of the 8th International Conference, KES 2004, Part III, 328–334 (2004)
4. E. Damiani, M. C. Nocerino, M. Viviani: Knowledge Extraction from an XML Data Flow: Building a Taxonomy based on Clustering Technique, Current Issues in Data and Knowledge Engineering, Proceedings of EUROFUSE 2004: 8th Meeting of the EURO Working Group on Fuzzy Sets, 133–142 (2004)
5. M. Leida: Structural information extraction techniques from semi-structured data flows, coming from differents data sources, Università degli Studi di Milano, DTI – Note del Polo – Research, No. 70 (2005)
6. RDF W3C Recommendation: http://www.w3.org/TR/rdf-primer/

Generate Context Metadata
Based on Biometric System

Antonia Azzini[1], Paolo Ceravolo[1], Ernesto Damiani[1],
Cristiano Fugazza[1], Salvatore Reale[2], and Massimiliano Torregiani[3]

[1] University of Milan - Department of Information Technologies
Via Bramante, 65 - 26013 Crema - Italy
`{azzini,ceravolo,damiani,fugazza}@dti.unimi.it`
[2] Siemens Mobile Communications S.p.A.
Via Monfalcone, 1 - 20092 Cinisello Balsamo, (MI) - Italy
`salvatore.reale@siemens.com`
[3] Siemens Informatics S.p.A.
Via Pirelli, 10 - 20126 Milano - Italy
`massimiliano.torregiani@siemens.com`

Abstract. This paper presents the design and the architecure implementation of a system, which provides an user access control framework
by means of multiple authentication methods. The user description is
provided according to the semantic web standard architecture, based on
two different kind of metadata: profile and context. Biometric technology
guarantees the maximal level of security, and it allows metadata definition with a personal authentication agent, based on knowledge, token
and biometric-based processes.

1 Introduction

Traditionally, access control relies on profile information associated to users and
resources in a given domain. In some scenarios it is necessary to downgrade or
even negate access privileges according to context information, like the authentication method used for accessing the system.

Our context-based architecture [3] makes resources accessible to users by
means of multiple authentication methods. Biometric technology can guarantee
the highest security level, but this technology is only available in a trusted environment. For this reason, remote access is provided by means of other authentication methods. In such a scenario, the `Access Control Framework` (ACF)
must distinguish between the user profile, by statically determining the set of
resources to be made accessible, and the access context, by dynamically limiting
the actions that can be performed on resources according to the authentication
method trustfulness.

In our system users are described by means of `profile metadata` and
`context metadata`, according to a Semantic Web standard format [9]. The
choice of such a standard is due to the capabilities provided by these languages
to validate metadata by means of an associated schema. Also, schemata can
be generated by means of a bottom up acquisition process, detecting system
evolution, monitoring the authentication data flow.

R. Khosla et al. (Eds.): KES 2005, LNAI 3681, pp. 119–126, 2005.

2   Authentication Techniques

Authentication is becoming an increasingly important issue in modern society. In this topic it is extremely important to allow a secure authentication and to prove that a person is who he claims to be. There are several methods that can be implemented, but the choice depends mainly on the considered environment, the system that it should be implemented, and different parameters like the percentage of accuracy and security that we want to obtain, together with costs and power consumption. The level of security to be associated to each method depends on these features and can change when the domain evolves. We also notice that some authentication methods are not available in all environments. For this reason, in some scenarious it is required to manage multiple authentication methods in the same application.

In a generic security access system a personal identification takes the form of **knowledge-based methods**, such as secret passwords and personal identification numbers (PINs), or **token-based methods**, such as keys, magnetic or electronic smart cards, ID and SIM cards, etc. In the first case, cryptographic keys are based on encryption standards, but they are difficult to memorize. Hence they are stored somewhere (i.e. on a computer or smart card) and released based on some alternative authentication method, like passwords. Most of them are easy and simple, but, for this reason, possible to crack, to share and then it is possible to compromise security. On the other hand, complex passwords are difficult to remember and, thus, expensive to maintain, so usually they are used toghether with a token-based hardware.

The lastest devices can store and process data electronically. The portability and size make them suitable for small personal data records. In addition, storing informations on a smart card provides more privacy than storing them in a central database.

Moreover, if we consider security in mobile systems, we have to underline that, while wireless communications provide great flexibility and mobility, they often come with a price in security. Indeed, important data are often placed on a mobile device that is vulnerable to theft and loss. For this scope, GSM systems are designed considering authentication and security processes. GSM services receive a Subscriber Identity Module (SIM) card, specifying an International Mobile Subscriber Identity (IMSI) that can be used worldwide for identification, billing, and security checking purposes. The SIM is a removable security module which is issued and managed by the user home service operator (even when the user is roaming) and is independent from the terminal. SIM-based authentication does not require any user action, except entering PIN into terminal.

In some of these technologies, the main problem is that user authentication processes determine whether a user is authorized to access the content, but they do not identify the user personally. Furthermore, they are no longer considered reliable enough to satisfy security and stability requirements, since the stored data can be stolen, forgotten, easily misplaced or shared.

Another valid choice considers **biometric-based methods**. These systems exploit automated methods of individual recognition based on the unique physiological or behavioral characteristics, called features. It is inherently more reliable than knowledge or token-based authentication. As biometric characteristics cannot be lost or forgotten or easily forged, they are extremely difficult to copy, share, distribute, and they require the person being authenticated to be present at the time and point of authentication, like in [6].

Physical biometrics include fingerprints, hand or palm geometry and retina, iris or facial characteristics, while behavioral features are based on signature, voice (which has also physical component), keystroke pattern and gait. The choice for each biometric depends on the application problem. The match between a specific biometric and an application is determined giving the requirements of the application itself and the properties of the biometric characteristics. Biometric representations are also depending on the acquisition method, environment and user interaction with the acquisition device.

3   Context-Based Access Control Framework

In comparison to traditional solutions our architecture introduces a new task in the system. The aim is to support all the different authentication technologies above introduced. During the authentication phase the user is associated with a role-context pair. The role defines the user's profile, determining invariant information such as the resources the user is granted access to, or the preferences set by the user in previous sessions. The context defines a temporary status of the user, determining the pool of actions that the user is allowed to. On the basis of this information, access control policies are filtered according to the level of security the authentication method can guarantee. It is straightforward to extend this model to encompass other aspects of user's interaction, for instance the connection protocol or the user agent issuing the request.

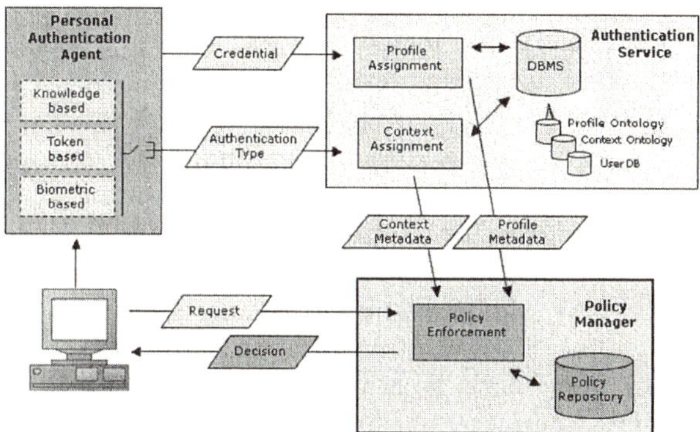

Fig. 1. Context-Based Authentication Architecture.

Fig. 1 describes the main modules that are composing our architecture. The *Personal Authentication Agent* allows the user to identify himself and keeps track of the authentication method used. The *Authentication Service* is a trusted server that processes user information to generate metadata describing both user profile and access context. Finally, the *Policy Manager* enforces the access control policies according to the metadata provided. In principle, profile metadata can be generated by different sources; for example, an authority can certify the user role by means of a standard certificate. For this reason we separate the Profile Assignment process, which aggregates profile metadata from different sources, from the Context Assignment, the centralized process that generates context metadata.

4   Biometric Design

In our work, the maximum security level is guaranteed by an authentication method relied on a biometric module. We can provide to the user maximum access rights only if the user is authenticated to the system by means of the biometric module.

Although biometric technologies may differ in several ways, their basic operations are very similar and their structures are shown in Fig.2.

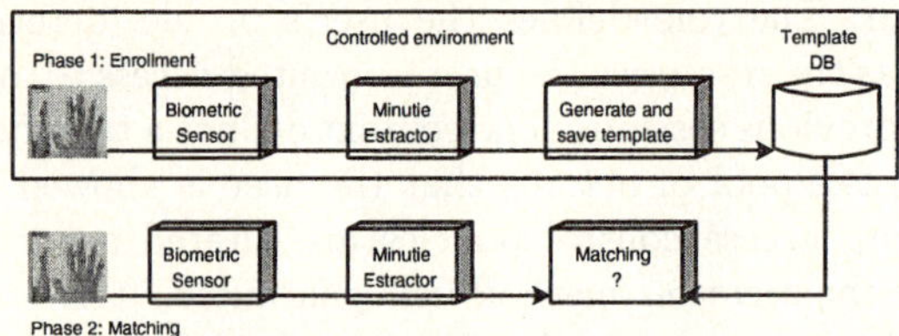

Fig. 2. General biometric structure.

First, a sensor acquires a sample of the user presented to the biometric system (fingerprint, hand geometry, face image). The sample can be trasmitted, eventually exploiting compression/decompression techniques. Some systems store the complete sample data in the storage unit. Storing samples is often deprecated in the literature due to privacy and security issues. More correctly, a biometric system uses and stores only a mathematical representation of the information extracted from biometric samples by the signal processing module: the biometric feature. Examples are minutie coordinates. If the extracted feature is stored (enrolled) into the biometric system, a template for future identification or verification (matching) is added. Each biometric system has a measure of the similarity between features derived from a presented input sample and a stored template. The measure produces the matching score. Hence, a match/non-match decision may be made according to whether this score exceeds a decision threshold or not. Then, the final step in the verification process is the decision to accept or reject the user, and it is based on the security threshold.

Biometric systems can be composed by one or more sensors included into an embedded system, or connected to a PC, or to a distributed system. Examples of embedded biometric systems are cellular phones with biometric authentication systems for credit card transactions, laptops with biometric user authentication biometrics, implemented with a fingerprint reader, or just simples biometric smart card readers. Examples of distributed biometric systems are authentication processes with many sensors units and a shared biometric traits database placed in another location. The design process looks for a trade-off among requirements, like accuracy, system cost, and computational complexity, that are competitive against each others, like in [4].

The architecture that we realize in this work considers a biometric approach based on the fingerprint verification, since for its relatively low cost, small size, and for the easily of integration of fingerprint authentication devices into an architecture design, like in [5].

5   Context Metadata Principles

Traditionally, the profile information associated to users can be exemplified by an Access Control List (ACL), which tuples specify: (i) the subject owning the access right, (ii) the action the subject is allowed for, and (iii) the target object for the given right. Example tuples in the profile ACL are the following:

```
user-xyz (can) update (on) chart-001
user-xyz (can) append (on) note-002
```

The ACF implemented in our system combines this information with context metadata. This sidekick context allows to model the different ways a user can authenticate with the system. Therefore, this second ACL has the following general form: (i) the target object for the given right, (ii) the action requested on the object, and (iii) the authentication method required. Example tuples in the context ACL are the following:

```
(on) chart-001 (users can) update (if authenticated with) knowledge-based
(on) note-002 (users can) append (if authenticated with) biometric-based
```

Advanced access control languages, such as XACML [10], provide means for extending the schema of the vocabulary used for referring subjects, objects, actions, as well as the set of evaluation functions that can be applied to them. As an example, the XACML policy in Fig. 3 conveys the semantics of the tuples related to chart-001 from both the example ACLs shown above.

6   Semantics-Aware Access Control

The state of the art access control we can express with plain XACML refers to entities (e.g., users, resources, access modes, and authentication schemes) as independent attribute values; instead, the ACF semantics-aware knowledge base represents them according to an ontology-based representation. To understand the role of semantic Web techniques it should be noted that, as soon as the

```
<Rule RuleId="urn:...:rule-id:1" Effect="Permit">
    <Condition>
        <Apply FunctionId="urn:oasis:...:function:and">
            <Apply FunctionId="urn:oasis:...:function:string-equal">
                <AttributeValue DataType="http://www.w3.org/2001/XMLSchema#string">
                    user-xyz
                </AttributeValue>
                <SubjectAttributeDesignator AttributeId="urn:...:user-id"
                    DataType="http://www.w3.org/2001/XMLSchema#string"/>
            </Apply>
            <Apply FunctionId="urn:oasis:...:function:string-equal">
                <AttributeValue DataType="http://www.w3.org/2001/XMLSchema#string">
                    knowledge-based
                </AttributeValue>
                <SubjectAttributeDesignator AttributeId="urn:...:authentication-scheme"
                    DataType="http://www.w3.org/2001/XMLSchema#string"/>
            </Apply>
            <Apply FunctionId="urn:oasis:...:function:string-equal">
                <AttributeValue DataType="http://www.w3.org/2001/XMLSchema#string">
                    chart-001
                </AttributeValue>
                <ResourceAttributeDesignator AttributeId="urn:...:resource-id"
                    DataType="http://www.w3.org/2001/XMLSchema#string"/>
            </Apply>
            <Apply FunctionId="urn:oasis:...:function:string-equal">
                <AttributeValue DataType="http://www.w3.org/2001/XMLSchema#string">
                    update
                </AttributeValue>
                <ActionAttributeDesignator AttributeId="urn:...:access-mode"
                    DataType="http://www.w3.org/2001/XMLSchema#string"/>
            </Apply>
        </Apply>
    </Condition>
</Rule>
```

Fig. 3. A XACML rule expressing policy contraints on resource `chart-001`.

entities involved are given fine-grained descriptions, the system administrator effort for producing consistent policies must increase consequently.

The different access modes, a subject could request on a resource, configure different degrees of criticality: for instance, to delete a document is a far more critical action than simply downloading it. Moreover, a document is supposed to be downloaded first in order to be modified; then `user-xyz` should also be allowed to read `chart-001`, since this is implied by the first tuple in the ACL profile. On the contrary, `user-xyz` cannot read on `note-002` since she is only allowed to append data, not implying read access to the resource. This criticality degree can then be used to induce a partial order on the set of possible access modes.

On the other hand, different degrees of trustfulness can also be associated with authentication methods provided by a system. Documents allowed for download with the most basic authentication scheme envisaged in our model, the `knowledge-based` authentication exemplified by a username-password pair, should then be available with any stronger authentication scheme. Therefore, `chart-001` should be also made accessible with `token-based` and `biometric-based` authentication schemes, while `note-002` cannot be accessed with any authentication schema other than biometric. This trustfulness degree induces a partial order on the set of possible authentication schemes.

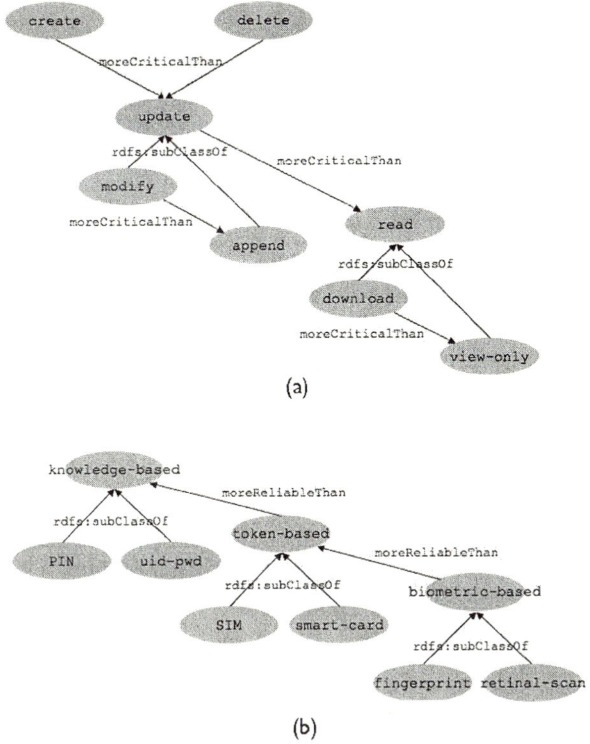

Fig. 4. The ontology fragment arranging entities according to partial order relationships.

The key point of our model is defining a partial order on both authentication schemes and access modes so that the ACLs shown above can be enriched with tuples implied by these relationships. Fig. 4 arranges the entites corresponding to access modes and authentication methods according to properties `moreCriticalThan` and `moreReliableThan` respectively; also taking advantage of is-a subtyping to define abstractions.

The expanded profile ACL is the following:

```
user-xyz (can) update (on) chart-001
user-xyz (can) read (on) chart-001
user-xyz (can) append (on) note-002
```

The expanded context ACL is the following:

```
(on) chart-001 (users can) update (if authenticated with) knowledge-based
(on) chart-001 (users can) update (if authenticated with) token-based
(on) chart-001 (users can) update (if authenticated with) biometric-based
(on) chart-001 (users can) read (if authenticated with) knowledge-based
(on) chart-001 (users can) read (if authenticated with) token-based
(on) chart-001 (users can) read (if authenticated with) biometric-based
(on) note-002 (users can) append (if authenticated with) biometric-based
```

The encoding of these structures rely on a RDFS schema [7], where classes model the entities described so far and properties represent the partial order relationships. A custom rule set is applied to the knowledge based for deriving the information enriching the ACLs.

7   Conclusions

This paper addresses the issue of integrating biometric-based authentication method in a multiple authentication-base system. The proposed solution introduces the `Context Based Access Control Framework`. This framework uses semantic-aware metadata specification for enabling a policy enforcement mechanism that is able, together with biometric process, to distinguish the user profile from the context of access.

References

1. P. Ceravolo, E. Damiani, S. De Capitani di Vimercati, C. Fugazza, P. Samarati: Advanced Metadata for Privacy-Aware Representation of Credentials. ICDE 05 (to appear)
2. P. Ceravolo M. Viviani M. C. Nocerino: Knowledge extraction from semi-structured data based on fuzzy techniques. Lecture Notes in AI, Springer, September 2004, pp 328-334.
3. E. Damiani, S. De Capitani di Vimercati, C. Fugazza, P. Samarati: Extending Policy Languages to the Semantic Web. Proceedings of the International Conference on Web Engineering, Munich, Germany, July 28-30, 2004
4. M. Gamassi, M. Lazzaroni, M. Misino, V. Piuri, D. Sana, F. Scotti: Accuracy and Performance of Biometric Systems. IMTC 2004: Instrumentation and Measurement Technology Conference, Como, Italy, 18-20 May 2004
5. A. K. Jain, L. Hong, S. Pankanti, R. Bolle: An Identity-Authentication System Using Fingerprints. Proceedings of the IEEE, Vol. 85, No. 9, September 1997
6. U. Uludag, S. Pankati, S. Prabhakar, A. K. Jain: Biometric Cryptosystems: Issues and Challenges. Proceedings of the IEEE, Vol. 92, No. 6, June 2004
7. World Wide Web Consourtium: RDF Vocabulary Description Language 1.0: RDF Schema, http://www.w3.org/TR/rdf-schema/, December, 2003.
8. World Wide Web Consortium: Biometric-based personal identification and verification technologies. http://www.biometrics.org
9. World Wide Web Consourtium: Semantic Web, http://www.w3.org/2001/sw/
10. eXtensible Access Control Markup Language, http://www.oasisopen.org/committees/tc_home.php?wg_abbrev=xacml

Complex Association Rules for XML Documents

Carlo Combi, Barbara Oliboni, and Rosalba Rossato

Dipartimento di Informatica
Università degli Studi di Verona
Ca' Vignal 2 — Strada le Grazie, 15
37134 Verona — Italy
{combi,oliboni,rossato}@sci.univr.it

Abstract. In the XML data context, documents may have a "similar" content but a different structure, thus a flexible approach to identify recurrent situations is needed. In this work we show how it is possible to compose structural and value association rules (simple rules) in order to describe complex association rules on XML documents. Moreover, we propose an approach to quantify *support* and *confidence* both for simple and complex rules in the XML data context.

1   Introduction

XML (eXtensible Markup Language) is spreading out as a standard for representing, exchanging, and publishing information [2]. XML is a markup language used for representing semistructured data, i.e. data that have no absolute schema fixed in advance, and whose structure may be irregular or incomplete [1]. Even without a document with a fixed schema, in this context it could be important to identify possible recurrent situations within data: for example, in a document related to hospital patients, it could be useful to discover that a "patient" element usually contains "therapy" elements with some specific drug.

To this aim, a recent research direction is related to the extraction of association rules from XML documents ([5], [6], [7], [8]).

In general, association rules describe the co-occurrence of data items in a large amount of collected data [4]. Agrawal et al. [3] introduced the formal definition of an *association rule* for the problem of mining association rules between sets of items $\mathcal{I} = \{I_1, I_2, \ldots, I_m\}$ in a large database of customer transactions T, where each transaction t is represented by means of a binary vector, with $t[k] = 1$ if t bought the item I_k, and $t[k] = 0$ otherwise. An association rule is an implication in the form $X \Longrightarrow I_j$, where X is a set of items in $\mathcal{I}$, and I_j is a single item in $\mathcal{I}$ that is not present in X.

In the classical context, the quality of an association rule is described by means of *support* and *confidence*. Given an association rule $X \Longrightarrow Y$, the *support* corresponds to the frequency of the set $X \cup Y$ in the dataset, while the *confidence* corresponds to the conditional probability $p(Y|X)$, i.e. the probability of finding Y having found X.

In [5], the authors describe association rules from XML documents by introducing an XML-specific operator, called XMINE RULE, which is based on

R. Khosla et al. (Eds.): KES 2005, LNAI 3681, pp. 127–133, 2005.

the use of XML query languages. In [6], the authors propose a flexible approach to evaluate simple association rules for XML documents. In [7] and [8] the authors show a technique which allows one to extract association rules from XML documents by using XQuery.

As for mining association rules from XML documents, in order to introduce flexibility in the definition of association rules and thus in the description of structures and values to verify, in this work we show how it is possible to compose the simple association rules, described in [6], into *complex association rules*. An example of simple association rule, is Symptom → Drug, which means that the node Symptom must be connected, by means of a path, to the node Drug. A complex association rule allows us to express more complex condition about the structure of the document to consider, i.e. more complex paths on the graphs.

In this work, we discuss how to quantify *support* and *confidence* in the XML data context, both for simple and complex association rules.

The structure of the paper is as follows. In Section 2 we briefly recall simple association rules (*structural* and *value* rules) for XML documents, and in Section 2.1 we introduce support and confidence for them. In Section 3 we show how it is possible to compose the previously mentioned simple rules to obtain *complex association rules* and we define support and confidence for complex rules. In Section 4 we report the conclusions and draft some future research topics.

2   Simple Association Rules

In this section we briefly describe the flexible approach, proposed in [6], to evaluate different kinds of simple association rules on XML documents.

In our approach, we represent XML documents as *XML graphs*, and thus we represent XML elements by means of nodes, and their containment relationships by means of non-labeled edges. In Figure 1 we show the graphical form of an XML document containing information about some patients, their symptoms and the assumed drugs.

We consider general XML documents, thus we allow the presence of mixed XML elements. For this reason, we have *mixed XML graphs* composed by three kinds of nodes: *simple*, *complex*, and *mixed* nodes. A *simple* node is a leaf and has a specific value (see node Drug in Figure 1), while a *complex* node has at least one outgoing edge and it may have a specific value. When a complex node has a value, then it is called *mixed* (*mixed complex* node). In the example of Figure 1, the nodes Patient, Visit and Therapy are mixed nodes, while the node Hospital is a *complex* node.

Simple association rules over a (set of) XML document(s) can be *structural* association rules and *value* association rules. A *structural association rule* allows one to evaluate the similarity of an XML document with respect to a given structure. For example, with respect to the XML document of Figure 1, a structural association could be Symptom → Drug. This association means that the node Symptom must be connected, by means of a path, to the node Drug. A *value association rule* allows one to check the similarity of the information con-

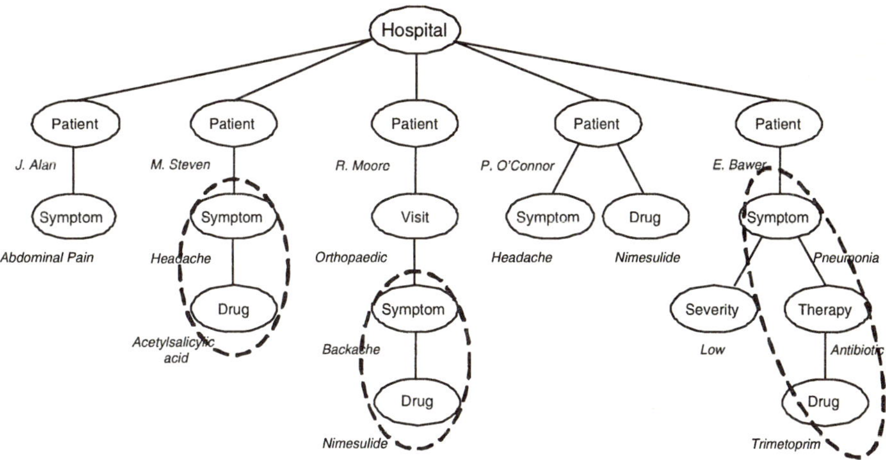

Fig. 1. An example of an XML graph.

tained in an XML document with respect to a given scenario. For example, a value association rule, on the XML document shown in Figure 1, could be the association Symptom $\rightarrow$ Drug(Acetylsalicylic acid) describing that, starting from a node with name Symptom, it is possible to reach a node Drug having value Acetylsalicylic acid.

The graphical representation of a rule is a directed graph showing the fact that the destination node (with or without a specific value) can be reached by the starting node. The graphical representations of the structural and value association rules mentioned above are reported in Figure 2.

Fig. 2. The graphical representations of (a) a structural rule and (b) a value rule.

2.1   Support and Confidence in the XML Context

The concepts of transaction and dataset, common for a database, cannot be applied directly to XML documents, thus we have to suitably model these definitions for the XML context. We first introduce the concepts of support and confidence for structural and value association rules and then, in Section 3.1, we will combine them to define the support and the confidence of a complex association rule.

Structural Rules. A structural association rule in the form $A \to B$, is used to describe the fact that the element A is "connected" to the element B by means of a path; different kinds of evaluation techniques, such as direct or undirect reachability, have been proposed in [6] to characterize the "connection" between two elements included in a structural rule. The direct reachability allows one to evaluate whether the starting node (in this case the node A) is directly connected to the destination node (B). The undirect reachability allows one to evaluate whether the destination node can be reached from the starting node with a path composed by more than one edge.

In this work, we consider as *transactions* related to an XML document, the set of paths extracted from the associated XML graph. In Figure 1 some transactions are shown in dashed regions.

The number of transactions related to an XML document can be computed in different ways by using the direct or undirect reachability. We define a function $Desc^n(A, B)$ that returns the number of transactions in which an element B is reached from an element A with a path having length n; if $n = 1$, we can compute the direct reachability, while in the case of undirect reachability we compute the function $Desc^*(A, B) = \sum_{n=1}^{l\_max} Desc^n(A, B)$, where l_max is the maximum length of a path in the XML graph. For example, the function $Desc^1(\mathsf{Symptom}, \mathsf{Drug})$ returns the value 2 because only in the first and in the second transactions of Figure 1, the node Drug is a child of the node $\mathsf{Symptom}$. In the third transaction, the path between $\mathsf{Symptom}$ and Drug has length 2.

In the XML context, the notions of support and confidence can take into account the strategy used to evaluate the rule.

The *supports* of a structural rule $A \to B$ w.r.t. the direct reachability ($supp_D$) and the undirect reachability ($supp_U$) are defined respectively as:

$$supp_D(A \to B) = \frac{Desc^1(A,B)}{Desc^1(X,Y)} \qquad supp_U(A \to B) = \frac{Desc^*(A,B)}{Desc^*(X,Y)}$$

where $Desc^1(X, Y)$ and $Desc^*(X, Y)$ are computed for each couple of nodes X and Y in the XML graph. The support of the structural rule $\mathsf{Symptom} \to \mathsf{Drug}$ w.r.t. the XML document of Figure 1 assumes value $\frac{2}{17}$ if it is computed with direct reachability, while in the case of undirect reachability the support of the rule is $\frac{3}{37}$.

The confidences of a structural rule $A \to B$ w.r.t. the direct reachability ($conf_D$) and the undirect reachability ($conf_U$) can be defined as:

$$conf_D(A \to B) = \frac{Desc^1(A,B)}{Desc^1(A,Y)} \qquad conf_U(A \to B) = \frac{Desc^*(A,B)}{Desc^*(A,Y)}$$

In the case of indirect reachability, the confidence is defined as the ratio between the number of transactions in which the node B is connected to the node A (with a path of any length) and the number of transactions in which A is connected to any other node Y ($Desc^*(A, Y) \geq Desc^*(A, B)$).

The confidence of the structural rule $\mathsf{Symptom} \to \mathsf{Drug}$ w.r.t. the XML document of Figure 1 is $\frac{2}{4}$ in the case of direct reachability, while in the case of undirect reachability the confidence of the rule is $\frac{3}{5}$.

Value Rules. A value association rule in the form $A \rightarrow B(Val)$, is used to describe the fact that the element A is "connected" by means of a path to the element B having value Val; also in this case, different kinds of evaluation techniques have been proposed to characterize the satisfiability of a value association rule with respect to an XML graph [6].

We define a function $Desc^n(A, B)|_{\rho(Val)}$ which returns the number of transactions in which an element B, having a value satisfying the relation $\rho(Val)$, can be reached from an element A by means of a path having length n; moreover we define $Desc^*(A, B)|_{\rho(Val)} = \sum_{n=1}^{l\_max} Desc^n(A, B)|_{\rho(Val)}$.

For example, if in a value rule we require that the value of the node B is equal to 100, then the function ρ will represent the equality relation, while if we require that the value of B is greater than 100, then the function ρ will be the relation $>$.

The supports of a value rule $A \rightarrow B(Val)$ w.r.t. the direct reachability and undirect reachability are defined as:

$$supp_D(A \rightarrow B(Val)) = \frac{Desc^1(A,B)|_{\rho(Val)}}{Desc^1(X,Y)} \quad supp_U(A \rightarrow B(Val)) = \frac{Desc^*(A,B)|_{\rho(Val)}}{Desc^*(X,Y)}$$

The confidences of the rule $A \rightarrow B(Val)$ w.r.t. the direct reachability and undirect reachability are defined as:

$$conf_D(A \rightarrow B(Val)) = \frac{Desc^1(A,B)|_{\rho(Val)}}{Desc^1(A,Y)} \quad conf_U(A \rightarrow B(Val)) = \frac{Desc^*(A,B)|_{\rho(Val)}}{Desc^*(A,Y)}$$

For example, the support of the value rule Symptom $\rightarrow$ Drug(Acetylsalicylic acid) evaluated with an exact value technique w.r.t. the XML document of Figure 1 and w.r.t. the direct reachability is $\frac{Desc^1(\text{Symptom},\text{Drug})|_{=\text{Acetylsalicylic acid}}}{Desc^1(X,Y)} = \frac{1}{17}$ while in the case of undirect reachability the support of the rule is $\frac{1}{37}$. The confidence of this rule w.r.t. the direct reachability is $\frac{1}{4}$, while by using the undirect reachability the confidence is $\frac{1}{5}$.

3   Combining Association Rules on XML Documents

In the XML data context, the same information can be represented in different ways. Thus, it could be useful and interesting to evaluate relationships, by means of particular association rules, between the structure of the XML document and its content.

For this reason, in this work, we define complex association rules in which both the antecedent and the consequent could be either a structural or a value association. The formal definition of a *complex association rule* is:

Definition 1. *Let* s, s$_1$ *and* v, v$_1$ *be structural associations and value associations, respectively. A complex association rule is an implication in one of the following forms:* s $\Longrightarrow$ v, s $\Longrightarrow$ s$_1$, v $\Longrightarrow$ v$_1$, *or* v $\Longrightarrow$ s.

For example, a complex rule in the form v $\Longrightarrow$ v$_1$ could be

$$(\text{Patient} \rightarrow \text{Symptom}(\text{Headache})) \Longrightarrow (\text{Patient} \rightarrow \text{Drug}(\text{Nimesulide}))$$

3.1   Support and Confidence for Complex Rules

In order to define the concept of support and confidence for complex rules, we need to extend our notation. In particular, with abuse of notation, we will denote with $Desc(R_1)$ either the function $Desc(A, B)$ if R_1 is the structural rule $A \to B$, or $Desc(A, B)|_{\rho(Val)}$ if R_1 is the value rule $A \to B(Val)$.

Moreover, we define $Desc(R_2)|_{R_1}$ the result of the function $Desc(R_2)$ applied to the XML graph satisfying the rule R_1. In particular, given an XML graph G and a complex rule $R_1 \Longrightarrow R_2$, with $R_1 = A \to B$ (or $A \to B(Val)$), in order to compute the function $Desc(R_2)|_{R_1}$, we need to obtain all the subgraphs $s_G(s)$ formed by the transactions satisfying R_1. Then, we need to construct the XML graph G' composed by the extensions of $s_G(s)$, i.e. the subgraphs starting from the root and ending into the leaves, and extended with the ancestors of A and the descendants of A and B. The function $Desc(R_2)|_{R_1}$ returns the result of $Desc(R_2)$ computed on G'.

We define the *support* of a complex association rule $R_1 \Longrightarrow R_2$, where R_1 and R_2 are structural or value association rules, as:

$$supp(R_1 \Longrightarrow R_2) = \frac{Desc(R_2)|_{R_1}}{Desc(X, Y)}$$

Note that, the proposed definition of support is consistent w.r.t. the traditional concept of this measure for classical association rule, i.e. it is equal to the ratio between the transactions where both the antecedent and the consequent of the rule appear simultaneously and the total of the transactions. For example, we now consider the complex rule described above which is composed by the rules $R_1 = $ Patient $\to$ Symptom(Headache) and $R_2 = $ Patient $\to$ Drug(Nimesulide), each of them evaluated by using an exact value technique. The support of $R_1 \Longrightarrow R_2$ is $\frac{1}{17}$.

The *confidence* of a complex rule $R_1 \Longrightarrow R_2$ is defined as:

$$conf(R_1 \Longrightarrow R_2) = \frac{Desc(R_2)|_{R_1}}{Desc(R_1)}$$

In Figure 3 we show the graph G' obtained from the XML graph G of Figure 1 by considering the rule $R_1 = $ Patient $\to$ Symptom(Headache). Note that the nodes Drug are added because descendants of the node Symptom and Patient. The result of $Desc(R_2)$ on G' with $R_2 = $ Patient $\to$ Drug(Nimesulide) is 1; we conclude that the confidence of the proposed rule is $\frac{1}{2}$.

4   Conclusions and Future Work

In this work we have proposed a possible approach to describe complex association rules on XML document. Complex association rules are composed by simple rules, i.e. *structural* and *value* rules. Moreover, we have defined *support* and *confidence* of simple and complex association rules.

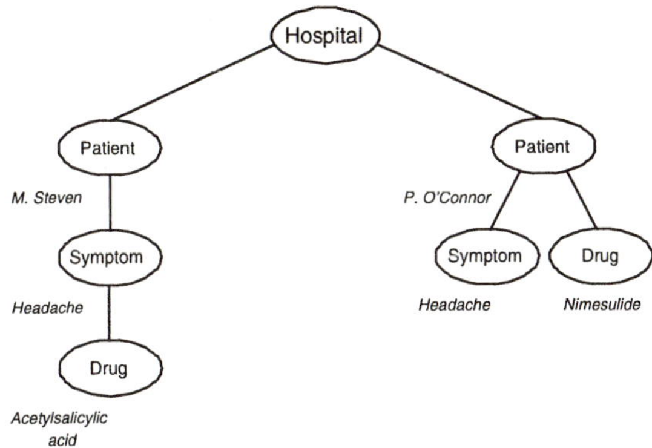

Fig. 3. The XML graph extracted.

We plan to extend the notion of value rules in order to allow the definition of the content also for the starting node, as well the extention of complex association rules to consider both in the antecedent and in the consequent a set of structural or value rules.

References

1. S. Abiteboul. Querying Semi-Structured Data. In *Proceedings of the International Conference on Database Theory*, volume 1186 of *LNCS*, pages 262–275, 1997.
2. S. Abiteboul, P. Buneman, and D. Suciu. *Data on the Web: from relations to semistructured data and XML*. Morgan Kaufman, 2000.
3. R. Agrawal, T. Imielinski, and A. Swami. Mining association rules between sets of items in large databases. In P. Buneman and S. Jajodia, editors, *Proceedings of the 1993 ACM SIGMOD International Conference on Management of Data*, pages 207–216, 1993.
4. R. Agrawal and R. Srikant. Fast algorithms for mining association rules in large databases. In J.B. Bocca, M. Jarke, and C. Zaniolo, editors, *Proceedings of the 20th International Conference on Very Large Data Bases*, pages 478–499. Morgan Kaufmann, 1994.
5. D. Braga, A. Campi, S. Ceri, M. Klemettinen, and P. Lanzi. Discovering Interesting Information in XML Data with Association Rules. In *SAC 2003*, pages 450–454. ACM Press, 2003.
6. C. Combi, B. Oliboni, and R. Rossato. Evaluating fuzzy association rules on xml documents. In *Proceedings of FUZZY DAYS 2004*, 2005. To appear in Springer series "Advances in Soft Computing".
7. J. W.W. Wan and G. Dobbie. Extracting association rules from XML documents using XQuery. In *Proceedings of DASFAA 2004*, pages 110–112, 2003.
8. J. W.W. Wan and G. Dobbie. Mining association rules from XML data using XQuery. In *Proceedings of DMWI 2004. To appear*, 2004.

A Rule-Based and Computation-Independent Business Modelling Language for Digital Business Ecosystems

Maurizio De Tommasi, Virginia Cisternino, and Angelo Corallo

e-Business Management School - ISUFI
University of Lecce
via per Monteroni sn, 73100 Lecce, Italy
{maurizio.detommasi,virginia.cisternino,angelo.corallo}@ebms.unile.it

Abstract. This paper aims at defining the Business Modelling Language main characteristics in the context of the Digital Business Ecosystem Project. One of the main obstacles in the adoption of e-business technologies from organizations and communities depends on the divide between information technologies and methodologies enabling business knowledge modeling. These methodologies are, on one side, not sufficiently connected with software development and management processes and, on the other, poorly usable and understandable by business users. BML main purpose is to allow simple communication among business people and IT system designers through a rule-based and natural language approach, granting at the same time strong expressiveness and formal logic mapping.

1 Introduction

The fast growth and the diffusion of the World Wide Web, its effect on trade, on business transactions and on communication systems, boosted the Information Technology Infrastructure creation and reinforcement. The technological convergence enlarged the connection option range, while the decreasing of the connection costs, allowed the exceptional diffusion of the access nodes to the network. These phenomena allowed companies and organizations to grow globally, increasing efficiency and reducing coordination costs. Organisations geographically distributed started to interact and trade globally, enhancing the growth of networks of distributed companies. Despite this growth, e-business still accounts for a relatively small share of total transactions. Using the widest range of B2B transactions, including established EDI (Electronic Data Interchange) as well as Internet transactions, it emerged that in 2000 total on-line transactions were generally 8% or less of total business transactions. The situation is still worst for Small and Medium-sized Enterprises (SMEs), since they lag behind larger firms in Internet transactions [1]. E-Business adoption represents a unique opportunity for SME to: reduce transaction costs and increase the speed and reliability of transactions; reduce inefficiencies resulting from lack of co-ordination between

R. Khosla et al. (Eds.): KES 2005, LNAI 3681, pp. 134–141, 2005.

firms in the value chain; reduce information asymmetries between buyers and suppliers [2]. SMEs that usually remain in local and regional market can expand their markets geographically, joining and competing in a wide variety of supply chains, including those previously inaccessible because of the use of costly closed EDI networks. Nevertheless, small organisations are not ready to use Internet extensively as a business tool. With the exception of few companies that lead the edge of the IT revolution, most of SMEs are not able to overcome the barriers that prevent them to use e-business technologies [3]. The main barriers to e-business for SMEs are represented by lack of connectivity, lack of attitude to innovation, shortage of ICT and management competencies, high cost for adopting and maintaining e-business solutions or lack of technological solutions and interoperability that could grant transactions with more different partners. We can add other difficulties as creating trust infrastructure, overcome the complexity of regulation or finding capital to invest in such kind of modernisation.

In this paper we present an approach to business modelling through a rule-based and computation-independent language, named Business Modelling Language (BML) enabling rich semantic description of business in all its aspects and explicitly devoted to sustain e-business adoption by SMEs. Such language aims at enabling business people to exchange business knowledge in a formal and machine readable way though with a natural language approach, providing, at the same time, formulation of meaning independently from expression in order to achieve reliable communication among and across communities in a Digital Business Ecosystem perspective.

2   The Digital Business Ecosystem Project

The high degree of changeability characterizing new markets context forces firms to transform themselves, becoming extremely reactive, flexible and focused on core competences. To obtain this result, during last years several organizations tended to adopt the *networked enterprise model*, that exploits new digital technologies in order to overcome organizational boundaries and focus on value creating activities. Nevertheless, even if practice is demonstrating the actual effectiveness of this model, its adoption seems to be very slow and limited only to those organizations able to strongly invest in IT systems. The main barriers limiting the diffusion of the networked enterprise model can be summarized as following:

1. Most of methodologies supporting business transformation toward this new paradigm are essentially thought for large organizations and efforts to adapt them to SMEs' needs seem to be very limited and ineffective.
2. In order to reach positive results the technological infrastructure and the organizational structure of a firm have to be integrated and strictly related each other; the actual problem is represented by a wide gap between IT and methodologies/technologies to represent business knowledge. The latter are usually separated from models for software development and management or they are barely understandable and usable by business users.

3. There is not enough care about development of network-based technological frameworks, essential to enable virtual collaborations among globally spread partners, who use very different technologies.
4. To enable transformation of traditional enterprises into digital enterprises, core organizational processes and routines must be transformed into a digital form. Since this transformation needs a technological framework flexible enough to not alter these processes, on-the-shelf products cannot be a suitable choice; on the other side, ad hoc solutions are often inaccessible to SMEs.

The DBE is an EU VI framework Integrated Project, whose main aim is to support the transaction of the Small and Medium Enterprise toward e-business, fostering local economic growth through new forms of dynamic business interaction and global co-operation among organizations and business communities enabled by digital ecosystem technologies.

Digital Business Ecosystems are entangled networks (of groups, departments, alliances, peripheral organisations, suppliers, collaborative joint ventures) where keystone species develop and sustain a technology platform on which a variety of organisations thrive, much in the same way as a variety of species depend on a coral reef substrate.

The DBE project aims at overcoming the aforementioned difficulties by creating a new way of conceiving co-evolution among organization and technology that shifts from:

– a mechanistic way of organizing business based on static view of the market to a new organicistic approach based on Mathematics, Physics and biological Science models;
– an approach for technology development unrelated to inter organizational issues to new paradigms in which technology and organization are related variables enabling innovative way of collaborating and competing;

In particular, focus of the DBE projects is on empowering the EU competitiveness in software ICT service industry in Europe and sustaining SMEs in ICT adoption to increase their productivity. To achieve such goals the DBE project aims to:

– Propose a new paradigm of software production based on open standards and Model Driven Architecture [4] through new forms of cooperation, developing reusable components Europe-wide, accessing multi-revenue models and sustaining software developer in creating and deploying low cost e-business application;
– Foster public-private cooperation among national and international research communities with a specific regard to SMEs software producer involvement;
– Support the growth of new multidisciplinary profiles at European level, capable to conjugate business, organisational, regulatory and technological competencies;
– Propose a technological infrastructure, which enables the evolution of the concept of business-ecosystems, enabling access to global markets by SMEs, through software and networking services.

By achieving such objectives the DBE will increase both the efficiency of business transactions and processes for SMEs, facilitating their ICT adoption and the market for software services and applications supporting SME software producers.

3   Defining BML

BML is conceived as a general framework enabling business people to represent business knowledge related to organization, its services, products, business processes, in order to allow trade mechanisms and search of potential e-business partners based on semantically rich information models. Moreover, BML allows to describe information exchanges between enterprises about volume, discount policy, rules and parameters by which potential suppliers are admitted, payment methods, contract duration, warranty, import/export rules and limitations.

Following the DBE approach, the BML architecture is based on MDA which motivates the paradigm switch (from object-oriented business modelling techniques to a rule-based and fact-oriented paradigm founded on natural language) introduced by BML.

3.1   The Model Driven Architectural Context

It has been estimated that more than 72% of software projects fail or fall short of their goals[1] and the worldwide cost of these failures is on the order of US$350 billion per year, of which about $200 billion per year is attributable to inadequate consideration of the business concerns in the system design[2]. The MDA mechanism for bridging the business-IT gap is the Computation Independent Business Model, or CIM.

As stated before, BML is focused on business modelling and, as such, it is a CIM language within the MDA framework. The CIM "represents the concerns of business owners, managers, agents and employees in the models. Anything business people have to say in the models goes in the CIM. [...] Using a CIM a business can capture, manage, and better use some of its most valuable assets: knowledge of its resources, policies, rules, terminology and processes. [...] Above all, CIM modeling must be easy for non-technical business people to understand and use. The salient feature of CIM languages is their reliance on and formal use of natural language."[5].

Since a Business Model is a "model of the business in its environment, by the business, in the language of business people, for business (not necessarily IT) purposes" [6], BML is focused on the description of business concepts without regarding to which parts of the business model are subject to automation in the underlying IT infrastructure. Obviously, the advantages coming from the adoption of a DBE-based platform are widely amplified if it is more accessible

---

[1] The Standish Group, CHAOS 2000.
[2] Tony Morgan, Brunel University, Presentation at Business Rules Forum, November 2002.

to business people and, in particular, to those involved in SMEs. A direct action
in business modelling by organizational people allows to:

- bridge the existing gap between technology and organization: business models could be designed in a business-oriented perspective, obtaining actual
  CIMs that preserve formality and rigorousness needed by technology;
- reduce costs related to ad hoc solutions development: if part of the development process is performed by business people IT professionals could focus
  on different activities, saving time and effort and overcoming communication
  problems.

In order to achieve direct involvement of business people in creating their own
business descriptions, BML introduces a new paradigm for business modelling
based on natural language. Moreover, inasmuch a CIM is also a formal model,
the BML language, mapped to formal logic, can drive automatic generation of
Platform Independent Models (PIM) and Platform Specific Models (PSM) for
the IT system supporting the business.

3.2   Natural Language Business Modelling

Methodologies used in software development (e.g. OOD) are typically applied
only when a problem is already formulated and well described. Starting from this
point, software developers transform requirements into code with a relatively
mechanical process. Nevertheless, the actual difficulty lies in the previous step,
that is describing problems and expected functionalities. Stakeholders involved
in software development can express their ideas using a language very close
to them, but they usually are not able to formalize these concepts in a clear
and unambiguous way. Indeed, the richness of structures and meanings of a
natural language can provide a great expressiveness, but it determines a lack
in terms of formalism. As a consequence, a very important role is played by
requirements analysts that act as a sort of translator between stakeholders and
software developers. Obviously, this implies a large effort in order to interpret
and understand real meanings and concepts hidden among stakeholders words.
Special constraints on syntax or predefined linguistic structures can be used
in order to overcome this problem, enabling natural language to well represent
and formally define problems and requirements. The main purpose of *natural
language modelling* is hence to make natural language suitable for conceptual
modelling. In other words, it aims at designing analytic processes able to produce
a simple syntax and to reduce ambiguity and vagueness, preserving language
completeness and essential meaning [7]. The focus is on semantic aspects and
shared meanings, while syntax is thought in a perspective based on formal logic
mapping.

Natural language is generally used by business organizations in order to describe themselves and their rules. Nevertheless, even if complex constructs and
ambiguous forms of expression provide a great communicative power, they usually make this description unclear and informal. This problem is largely am-

plified considering that involved people often do not share concepts and meanings. Conversely, as stated above, system requirements gathering and creation of machine-readable documents need a higher degree of precision and formality, with a consequent loss in richness of meaning and expressions. Bridging the existing gap between business people language and other formal languages, used for software development and document interchange capabilities, represents a fundamental issue for BML. In this perspective, a modelling approach based on natural language could be a very interesting choice in order to balance these opposite needs. Indeed, this approach can provide BML users with a powerful means allowing them to use their own language in order to create consistent, unambiguous and formal business models.

3.3   The Rule-Based Approach

Rules play a very important role in defining business semantics: they can influence or guide behaviours and support policies, responding to environmental situations and events. This means that rules represent the primary means by which an organization can direct its business, defining the operative way to reach its objectives and perform its actions. "A business rule is a statement that defines or constrains some aspect of the business" [8]

Therefore, BML is founded on a *rule-based approach*. It can be described as a way "for identifying and articulating the rules which define the structure and control the operation of an enterprise" [8].

This approach did not arise as a response to any emerging new class of software tools or as an academic conjecture. "Rather, the business rule approach is a real-world movement whose driving force is business success, not technology. It arose from the vision of dedicated professionals with many years of experience in the trials and challenges of business software", whose goal is "to offer companies the best possible approach to developing business solutions involving automated systems" [9].

Consequently, the rule-based approach aims to address two different kinds of users: from one side, it addresses business communities, in order to provide them with a structured approach, based on a clear set of concepts and used to access and manage business rules; from the other side, it addresses IT professionals, in order to provide them with a deep understanding about business rules and to help them in models creation. However, the rule-based approach aims to create enterprise models oriented toward business rules identification, formalization and management.

Over the years, a great number of analysing and modelling methodologies have been introduced in order to describe an enterprise in terms of data structure and performed activities. Even if these traditional approaches are able to describe the enterprise in terms of data flows and operative behaviours, they tend to neglect the constraints under which an organization operates. This implies that only those rules embedded in organizational structure and behaviours are documented, formalized and represented within the enterprise systems. Rules that prevent, cause, or suggest things to happen are not covered by these approaches. Moreover, business rules represented in this way can be understood only by technical experts and by automated systems; business experts are not able to review, correct and change them, in order to meet new needs and react to market changes. Due to these difficulties, a new way to think about enterprise and its rules is needed, in order to enable a complete business representation made by and for business people. The rule-based approach just meets this need.

The previous considerations allow to highlight some specific characteristics needed by rules describing business semantics: these constraints should be explicit, to be understood and directly managed by business people, and dynamic, to answer to the changeability of the external and internal environments. The rule-based approach aims at providing the means to formalize rules compliant with these purposes.

3.4   A Meta-language for BML

In order to be a CIM language, BML models (expressed in natural language through business rules) have to be mapped to formal logic (enabling automatic generation of Platform Independent system Models) and to interchangeable format (e.g. MOF and XMI, granting interoperability among different SMEs and communities).

To fulfil these requirements the BML abstract syntax and semantics is formally defined through an emerging standard, *Semantics of Business Vocabulary and Business Rules* (SBVR) which is a submission by the Business Rule Team to an OMG Request For Proposal called *Business Semantics of Business Rules* (BSBR). OMG issued the BSBR RFP in order to create a standard "to allow business people to define the policies and rules by which they run their business in their own language, in terms of the things they deal with in the business, and to capture those rules in a way that is clear, unambiguous and readily translatable into other representations" [10]. In particular, SBVR enables:

– business vocabularies construction, whose definitions represent shared understanding among a community of business people and whose contents are uniquely identified by this community;
– rules formalization, based on vocabularies; these rules should be expressed in a language close to natural (business) language, called Structured English but, at the same time, representable from an information technology point of view, in order to allow their sharing and transferring.

Building BML on top of SBVR provides linguistic support to business men in order to formally define and share business semantics in terms of business facts and rules represented in Structured English notation.

Furthermore, the SBVR approach provides BML with means (i.e. MOF/XMI mapping rules defined by SBVR in [11]) to translate natural language artifacts into MOF-compliant artifacts; this allows to exploit all the advantages related to MOF (repository facilities, interchangeability, tools, ...), satisfying BML requirements related to its use within a distributed environment.

Due to its compliance with the MDA approach, SBVR can offer support in automating software production. This implies that BML models based on SBVR can be used in order to generate models for IT systems (PIMs, PSMs, diagrams, classes an code) in an automated way, thus supporting SMEs producing software to exploit reuse and automation and allowing business people to play a central role in system models creation.

4 Conclusions

In this paper we presented an approach to the preliminary work for the creation of BML based on business rules and natural language modelling paradigm. BML will provide the DBE with an interoperable business layer needed to support interactions between SMEs in a digital business ecosystem context where business people can describe their business and where companies can search for e-business partners. In this scenario, business people can use simple and natural language based tools to represent their business knowledge and rules; information produced could be used as input, in an MDA scenario, for software development and at the same time to populate business registries with company description.

Nevertheless, many problems are currently unsolved and will be faced in the future of the project such as privacy and legal validity of information in the BML framework, organizational and technological mechanisms through which BML will be maintained and evolved in time, and technological tools that could allow BML use by SMEs. Other concerns relate building industrial or domain models on top of which BML specific business models can be easy created.

References

1. OECD Report: ICT e-business and SME (2005)
2. OECD Report: Science, Technology and Industry Outlook: Drivers of Growth: Information Technology, Innovation and Entrepreneurship (2001)
3. F. Nachira: Toward a Network of digital Business Ecosystem fostering the local development. (2002)
4. Joaquin Miller and Jishnu Mukerji: MDA Guide Version 1.0.1 (2003)
5. Stan Hendryx, Hendryx and Associates: Architecture of Business Modeling (2003)
6. Stan Hendryx, Hendryx and Associates: Integrating Computation Independent Business Modeling Languages into the MDA with UML 2 (2003)
7. Nik Boyd: Using Natural Language in Software Development, Journal Of Object Oriented Programming (1999)
8. The Business Rules Group: Defining Business Rules - What Are They Really?, Final Report, revision 1.3 (2000)
9. Ronald G. Ross: Principles of the Business Rules Approach, Addison-Wesley (2003)
10. OMG: Business Semantics of Business Rules - Request For Proposal (2004)
11. Business Rules Team: Semantics of Business Vocabulary and Business Rules (SBVR), ver6.0, bei/2005-01-01 (2005)

Building Bottom-Up Ontologies for Communities of Practice in High-Tech Firms

Marina Biscozzo, Angelo Corallo, and Gianluca Elia

e-Business Management School – ISUFI, University of Lecce
Via per Monteroni – 73100 Lecce, Italy
{marina.biscozzo,angelo.corallo,gianluca.elia}@ebms.unile.it

Abstract. In high-technology environments, where New Product Development (NPD) is an essential activity, providing effective decision support by making knowledge about development efforts readily available and accessible can contribute to the survival and growth of organizations. Often, the lack of a shared understanding leads to poor communication between people; moreover the co-existence of different software applications, languages and practices in people documentation efforts makes it hard to manage and exploit firm's knowledge base. This paper identifies in Community of Practice (CoP) and Ontology a possible successful solution to these problems. The bottom-up approach proves to be suitable to ontology building for a CoP. Taking into account the characteristics of the context, a methodology for building bottom-up ontologies for CoPs is proposed.

1 Introduction

Shorter product life cycles, globalization and rapid environmental changes are some of the factors inducing organizations to innovate faster, at lower costs and better than competitors. In many industries, competitive advantage comes from being *the first* and survival often depends on the speed at which new products can be developed.

In this context, New Product Development (NPD) becomes an essential activity for organizations. This can be especially found in high-technology environments, where the cost of development can get 85% of the total product cost[1] [1].

Providing effective decision support by making knowledge about past and current development efforts readily available and accessible can contribute to the reduction of costs and time required for development.

Product innovation and NPD usually require the integration and collaboration of people from different functional areas (e.g. marketing and R&D) or different geographic locations (like in the networked enterprise) and from many external entities (e.g. suppliers, customers and other organizations).

The lack of a shared understanding due to different needs and background contexts can be easily overcome when people work together, sit close to each other, and interact with each other on a daily basis. But when people/teams have very little contact with other people/teams or are isolated for long periods, they face serious problems in sharing and exchanging information.

---

[1] Source: N.R. Council, Improving Engineering Design: Designing for Competitive Advantage, National Academy Press, Washington, DC, 1991

R. Khosla et al. (Eds.): KES 2005, LNAI 3681, pp. 142–148, 2005.

As result, while on the one hand a considerable part of tacit knowledge is lost, on the other hand the firm's Knowledge Base (KB) is enriched with multiple versions of information, often without references to the context in which they were generated or in which they will be used. The coexistence of different software applications, languages and practices in people documentation efforts makes it hard representing the relationships between information in the KB, the integration of data and knowledge across the product development cycle and the transfer of existing knowledge into other parts of the organization. As consequence, this amount of explicit knowledge often cannot be used because of the ineffective instruments for its management. To make it valuable, firms need a new approach in order to transform their knowledge resource in an usable asset.

Two questions arise at this point: how to overcome people barriers to share and exploit knowledge? How to create a common language and shared understanding of context and data?

A possible successful solution is to combine cross-functional teams and communities of practice (CoPs) in order to make an organization simultaneously oriented to output and learning. Then, it is indispensable to provide CoPs with an appropriate language in order to create a shared understanding and capture tacit aspects of knowledge.

2 CoP as Vehicle for the Creation and the Diffusion of Knowledge in High-Tech Firms

The concept of CoP was introduced in 1991 by Lave and Wenger and developed by other researchers during last years. According to Etienne Wenger, "communities of practice are everywhere. They exist within businesses and across business units and company boundaries". Even though they are informally constituted and reside within a specific area of practice, these self-organising systems share the capacity to create and use organizational knowledge through informal learning and mutual engagement (Wenger, 2000). Communities of practice are particularly useful in high-tech industry, where people from different disciplines frequently meet each others. Communities of practice are a way to connect people with technical peers while maintaining the focus on cross-functional teams. The community provides information and insight on tools, analyses, and approaches current in the discipline. Community coordinators know the experts in each topic area and who is working on a specific technical problems. So, they can quickly link individuals from different cross-functional teams to peers in their discipline. In this multidisciplinary context the use of a common language is fundamental to promote collaboration, knowledge creation and knowledge sharing dynamics [2]. The high penetration of ICTs in today's landscape and the consequent process of automation require a language that is shared among community's members, and also readable and understandable by machines.

3 Ontology as Common Language for Community Members

What constitutes a basic element for a community – the negotiated meaning – is the goal of the process of creating an ontology. At the same time, ontology represents the common machine-readable and machine-understandable language that enables CoPs to interact with firm's knowledge base in an effective way.

Far from its original meaning[2], the term ontology is used here with reference to Gruber definition: *"An ontology is an explicit specification of a formal and shared conceptualization"* (Gruber, 1993). In other words, an ontology is an abstract model of the reality whose concepts, relations and rules are defined in an explicit and machine-readable way and accepted by a community. Ontology can be seen as a common vocabulary, a common structure of concepts that all community's members can use to communicate, interact and exchange knowledge. Moreover, the use of ontologies offers an opportunity to improve knowledge management capabilities in large organizations, since they represent a sort of *interface* between CoPs and the firm's KB. An ontology serves as a mean to structure a firm's KB and to implement its inference mechanisms; it enables the navigation and browsing of a KB and provides a global schema for the integration of the different knowledge objects coming from heterogeneous sources.

How CoP is successful in managing and exploiting firm's KB mainly depends on the acceptance degree of the ontology in the whole CoP. The choice of the right approach to ontology building can increase consensus of community members.

4   A Methodology for Building Ontologies for CoPs in High-Tech Firms

The introduction of ontology as language shared by all community's members implies replacing of personal conceptualization of the domain with a common one and changing of own language into a formal one. How to build the ontology in order to increase its acceptance? Following a *top-down approach*, an ontology is created by knowledge engineers and domain experts and implemented into the CoP's dynamics without any previous feedback, suggestions and contributions provided by its members. On the one hand, this approach allows community's members to reduce their effort in knowledge sharing process and to save time in having an ontology. On the other hand, this could cause the not-completely acceptance of the language by all members. In the context of high-tech firms, the advantage of following a top-down approach in ontology construction is to mitigate the competition between communities of practice and cross-functional teams for people's time. However, such an approach could make problems we are facing here unsolved.

A *bottom-up approach* can be adopted. CoP's members participate in the construction of taxonomies or ontologies, proposing their personal domain conceptualization or modifications to the community's language. The concept structures proposed by each member are compared with those of other members through the use of collaboration tools. Since the building process is collaborative and involves all CoP members, the acceptance of the ontology will be wide. The resulting ontology will reflect the "operative knowledge" coming from the concrete job experiences of participants.

In the context of high-tech firms, the adoption of a bottom-up approach can contribute to the construction of a robust backbone for the organization's KB. However, building an ontology following a bottom-up approach usually is time-consuming and

---

[2] In philosophy, ontology is the study of the kinds of things that exist

requires very high efforts by each member, because of the negotiation and agreement process between experts. This is particularly true in the case of high-tech firms, given the large amounts of unstructured and (semi-) structured information available on organizational KB. Due to the dimension and the complexity of the domain to model, building an ontology manually is not feasible in terms of time and resources. This implies the adoption of an automatic approach, using tools that take documents as input and generate an ontology from them. In most cases, it will not be a real ontology, but a very simple structure with a small variety of relation types as compared to a manual development process. Therefore, the outcome of an automatic approach will rather represent a useful basis for a successive manual development.

Here we propose a methodology for building bottom-up ontologies for CoPs in high-tech firms:

- automatic ontology/taxonomy extraction from documents present in the firm's KB;
- ontology/taxonomy evaluation made by community members that propose some modifications (elimination of duplicated concepts, addition of new concepts or relations) through the use of collaborative tools;
- the proposed changes are discussed by community members by using the collaborative tools;
- the ontology is populated with instances from documents

What we expect from the application of this methodology is a knowledge representation that almost quite reflects community members shared conceptualization, since:

- most concepts and conceptual structures of the domain as well the company terminology are described in documents produced by community members themselves, and put in the KB;
- the result of the automatic approach is further refined through the active involvement of community members in a collaborative environment.

In the following sections, we give a general description of technologies on which ontology extraction tools are based and of collaborative tools supporting ontology construction.

4.1 Ontology Learning

Various technologies and techniques can be applied to reduce the cost of identifying and manually defining several thousands of concepts. They aim at the discovery of terms, the description of their taxonomic relations and related terms and their enrichment with semantic information. All these technologies and techniques- such as *Natural Language Processing* (NLP) technologies - play an important role in supporting *Ontology learning*, an emerging field aimed at assisting a knowledge engineer in ontology construction and semantic page annotation with the help of machine learning techniques [3]. Based on the assumption that most of concepts and conceptual structures of the domain as well the company terminology are described in documents, acquiring knowledge from text can have good results in ontology design.

The semi-automatic ontology acquisition process starts with the selection of input data (Figure 1): a generic core ontology as a top level structure for the domain-specific ontology; a dictionary that contains important corporate terms described in natural language; a domain-specific and a general corpus of texts to remove concepts that are not domain-specific [4].

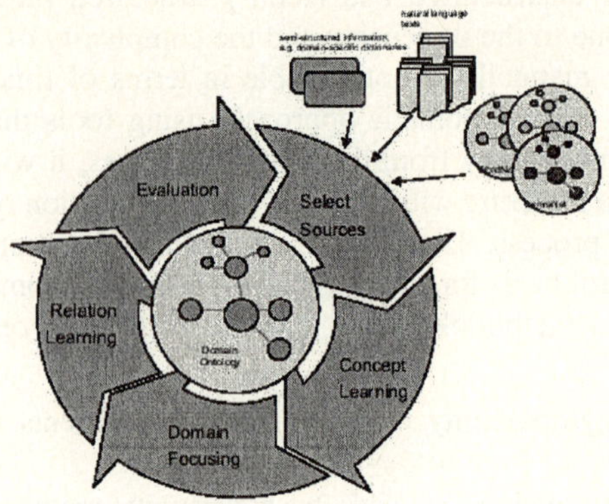

Fig. 1. Semi-Automatic Ontology Acquisition Process

Through a linguistic analysis, a glossary of terms is extracted from the dictionary of corporate terms or from domain-specific textual documents. These steps enrich the ontology with both domain-specific and domain-unspecific concepts. The latter group of concepts must be removed from the ontology. Therefore the user must identify a domain-specific and a general corpus of texts to be used for the removal of concepts, since domain-specific concepts usually are more frequent in a domain-specific corpus than in generic texts.

Terms are grouped together if they represent very similar concepts as in a classic thesaurus. These conceptual units, or sets of semantically similar terms, can be organized on the base of their specificity, giving rise to a taxonomy. New conceptual relations are introduced applying multiple learning methods to the selected texts. Two approaches can be used for this purpose:

- a statistical approach, based on the idea that frequent couplings of concepts in sentences can be regarded as relevant relations between concepts;
- a pattern-based approach, in which a regular expression captures re-occurring expressions; then it maps the results of the matching expression to a semantic structure, such as taxonomic relations between concepts.

The resulting domain-specific ontology can be further refined and improved by repeating the acquisition cycle. It is important to notice that this cyclic process occurs in cooperation with users that have to validate the results.

Nowadays several tools for automatic ontology extraction or taxonomy generation are available, both commercial products and prototypes developed by universities or other research institutes. Existing surveys or benchmarking can help CoPs evaluate the degree to which such tools are able to reduce manual efforts in ontology construction.

4.2 Collaborative Ontology Building

Community's members have to use tools that help them collaborate in order to build ontologies that reflect their domain conceptualization. These tools enable people to:

- propose their modifications, providing them with the ability to publish, browse, create, and edit ontologies;
- inform each user of the changes other users made;
- facilitate discussions about the proposed changes.

Support for distributed and collaborative development of ontologies is usually provided through the web interface by multi-user sessions.

When a user opens a session, any other members of that group can join the session and work simultaneously on the same set of ontologies. A notification mechanism can inform each user of the changes that other users have made. Notification could be an hyperlink to the changed definitions and describe changes in terms of basic operations such as add, delete, and modify. Moreover, users' annotations can be recorded and attached to any concept or instance.

The importance of informal communication has led to a variety of tools designed to stimulate casual conversation among CoP's members in different sites.

5 Building a Bottom-Up Ontology for a CoP in High-Tech Firms: The Avio S.P.A Case Study

The application and validation of the methodology for building a bottom-up ontology in high-tech firms is an ongoing work. The interested CoPs are the communities of CAD designers, CAE users, specialists in realizing prototypes, and experts of other disciplines in Avio SPA.

Avio SPA is an aerospace company, involved in the design, testing, manufacturing, maintenance, revision and overhaul of parts of aircraft and rocket engines. It has several centres of excellence geographically distributed. Furthermore, Avio works in partnership with a large number of aerospace companies (customers and suppliers).

Aerospace context is highly complex and requires a wide spectrum of technical and scientific knowledge to be managed during product development. To remain competitive Avio needs to reduce time and effort in product development, improving the sharing and reuse of technical knowledge and transforming people's practices in robust and standardized processes.

In this context, the use of an ontology enables CoPs of designers and specialists to manage Avio's KB in a efficient and effective way. The methodology described in the previous section seems to be appropriate to satisfy the needs of Avio's CoPs.

At the present firm's Knowledge Management (KM) team is involved in:

- testing a tool for automatic taxonomy generation from Technical Memory, Regulations, etc.
- promoting a complete KM initiative whose scope is to support and motivate people to participate in the construction of the ontology.

The results of this process will be an ontology (accepted by all communities' members) that allows the management of a part of Avio's KB.

Firm's KB includes not only the Technical Memory and Regulations, and thus HTML and XML documents, or other textual resources. An important part of the KB consists of artefact models, CAD control structures, 2D geometries, STEP files, etc. Therefore, an ultimate goal of KM team is to provide a common representation for such resources. Future development of this work can be oriented towards the realization or choice of a modelling language to represent archetypes and rules. This representational model will be integrated or merged with the results of the current ontology building process.

6 Conclusions

This work described the main issues concerning the building of an ontology for CoPs in high-tech firms. In particular, it showed how the bottom-up ontology is more appropriate than the top-down one, since the former reflects the "operative knowledge" coming from the concrete job experiences of participants and its acceptance is wide.

The paper proposed a methodology for building bottom-up ontology, taking into account problems to be faced in the management of large and heterogeneous KBs. Great attention is posed on tools for automatic ontology extraction from texts, since they can reduce CoP's efforts in ontology building. Such tools are not immune to mistakes; therefore, their results have to be refined by community members through the use of collaborative tools. The proposed methodology is currently applied into the CoPs of Avio, an aerospace firm. The experimentation is ongoing yet, and new challenges are emerging. Therefore, the results of this research will be extended to face such challenges.

Acknowledgments

This work is supported by *Knowledge-based Innovation for the Web Infrastructure (KIWI)* research project. We would like to acknowledge the valuable contributions of Avio staff, and in particular, Roberto Merotto for his challenging questions and assistance.

References

1. Ramesh, B., Tiwana, A., (1999). Supporting Collaborative Process Knowledge Management in New Product Development Teams.
2. Mc Dermott, R., (1999). Learning across teams: the Role of Community of Practice in Team organization. Published in Knowledge management Review
3. Navigli, R., Velardi, P. Gangemi, A. (2003). Ontology Learning and Its Application to Automated Terminology Translation. Published by the IEEE Computer Society.
4. Kietz, J., Maedche, A., Volz, R.,(2000). A Method for Semi-Automatic Ontology Acquisition from a Corporate Intranet.

Decision Support System for Atmospheric Corrosion of Carbon Steel Pipes

Kazuhiro Takeda[1], Yoshifumi Tsuge[1],
Hisayoshi Matsuyama[2], and Eiji O'shima[3]

[1] Department of Chemical Engineering,
Kyushu University, Fukuoka, 812-8581, Japan
[2] Graduate School of Information, Production and Systems,
Waseda University, Kitakyushu, 808-0135, Japan
[3] Tokyo Institute of Technology, Yokohama, 226-8503, Japan

Abstract. Pipes of a plant should be properly managed, since corrosion
of pipes cause leakage of dangerous material. But, the management is a
hard and costly job. Though the pipes are maintained costly, the ratio
of really corroded points to inspection points is 5-10 %. Then, if max
of corrosion rate is find out using condition of the inspection point, inspection points can be reduced and much cost can be cut. In this work,
corrosion rate model of atmospheric corrosion of a carbon steel pipe is
proposed. Furthermore, decision support system for atmospheric corrosion of carbon steel pipes is developed. Using database collected from
companies, investigation of database healthy is reported.

1   Introduction

Pipes of a plant should be properly managed, since corrosion of pipes cause
leakage of dangerous material. But, the management is a hard and costly job.
The reasons are as follows.

(1) Thousands of points are inspected, because the pipes are very long (ex. over
100 km per plant).
(2) For safety, all of the suspicious points are inspected.
(3) The pipes are placed at hard maintenance position (ex. under equipment).
(4) The pipes may not be directly observed, since the pipes may be insulated for
heating or cooling. To observe, the insulators should be removed in a regular
maintenance.

Though the pipes are maintained costly, the ratio of really corroded points
to inspection points is 5-10 %. Then, if max of corrosion rate is find out using
condition of the inspection point, inspection points can be reduced and much
cost can be cut.

Gerven *et al.* [1] has proposed the support system to prevent pitting corrosion.
Farina *et al.* [2] has proposed the diagnosis system using corrosion phenomena
to identify the cause of boiler failure. Cristaldi *et al.* [3] has proposed the expert system for the process engineer to select suitable materials during chemical

R. Khosla et al. (Eds.): KES 2005, LNAI 3681, pp. 149–154, 2005.

plant design. Pieri *et al.* [4] has proposed the data and knowledge-based system for supporting the maintenance of chemical plant considering economic factors. However, no support system for atmospheric corrosion of carbon steel pipes has been proposed.

In this work, corrosion rate model of atmospheric corrosion of a carbon steel pipe is proposed. Furthermore, decision support system for atmospheric corrosion of carbon steel pipes is developed. Using database collected from companies, investigation of database healthy is reported.

2 Corrosion Rate Model of Atmospheric Corrosion of Carbon Steel Pipes

Main atmospheric corrosion modes of carbon steel pipes are as follows.

- Uniform corrosion
- Oxide concentration cell corrosion
- Bimetallic corrosion

In the modes, atmospheric corrosion rate ACR [mm/y] of carbon steel pipe is formalized as follows. $ACR = ICR \times WCR$. Where, ideal corrosion rate ICR [mm/y] is corrosion rate under assumption that the carbon steel pipe is always covered by water, and water covered rate WCR [-] is the period per year that the pipe is covered by water. Therefore, estimation of atmospheric corrosion rate is divided to estimation of ICR by analyzing corrosion phenomena and estimation of it WCR.

ICR are affected by some factors of pipe surface, which are (a) temperature, (b) pH, (c) chlorine ion concentration and (d) diffusion rate of dissolved oxygen. ICR by oxide concentration cell corrosion are affected by (e) distribution of concentration of dissolved oxygen of pipe surface. ICR by bimetallic corrosion are affected by (f) contact with bimetal. The factors (a) - (f) are called electrochemical factors.

WCR is affected by (a), (L01) mechanism of water covered and (L02) mechanism of keep water. A default value of (a) is temperature in the pipe. (a) is affected by (L03) mechanism to raise temperature and (L04) mechanism to lower temperature. A default value of (b) is 7. (b) is affected by (L05) mechanism to make acid. (c) is affected by (L06) mechanism to feed chlorine ion. (d) is affected by two factors on pipe surface, which are (L07) mechanism to stream and (L08) mechanism to generate water membrane. (e) is affected by (L09) mechanism to generate distribution of concentration of dissolved oxygen. (f) is caused by (L10) adjacency of bimetal and (L11) lack of insulator. The factors (L01) - (L11) are called local circumstance factors.

From a global viewpoint, WCR is affected by (G01) site, (G02) location to wet, (G06) rip, (G07) mechanism gathering water to rip and (G08) mechanism to transport water from rip to corrosion point. (b) is affected by (G01) and (G03) location to lower pH. (c) is affected by (G01) and (G04) location to higher chlorine concentration. The factors (G01) - (G08) are called global circumstance factors.

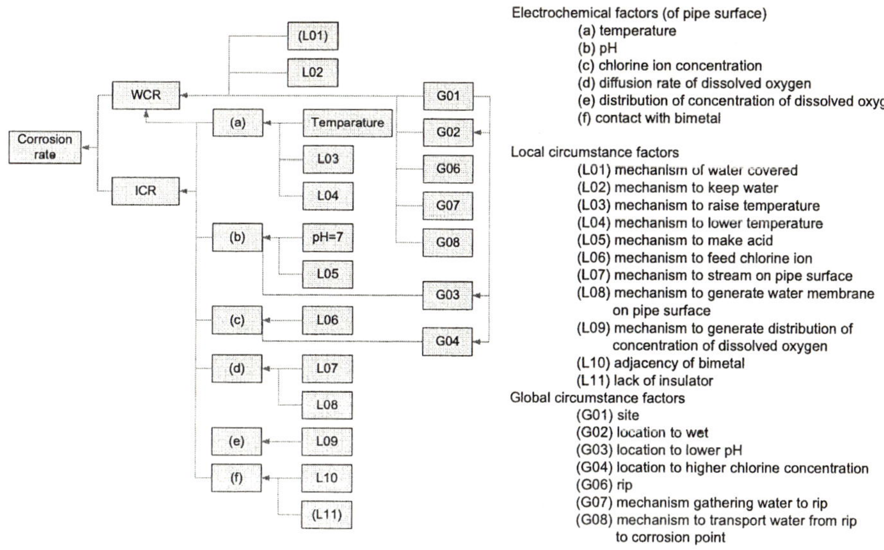

Fig. 1. The corrosion rate model.

The corrosion rate model using the factors is shown in Fig. 1.

3 Decision Support System for Atmospheric Corrosion of Carbon Steel Pipes

Based on the corrosion rate model, the decision support system for atmospheric corrosion of carbon steel pipes is developed. The system consists of search and estimation subsystem, database, and database maintenance subsystem.

The search and estimation subsystem outputs range of corrosion rate (the maximum and minimum corrosion rate) from investigation requirement.

In the database, each electrochemical, local circumstance and global circumstance factors is called 'term', an attribute value of each term is called 'element term', and a set of corrosion rate and element terms for the data is called 'case'.

The database maintenance subsystem registers the cases and estimated result in database.

For correct search and estimation, the element terms of the mechanisms (L01) - (L04), (L07) - (L09), (G06) - (G08) are selected from predefined examples. The element terms of the other terms are corrected automatically or manually.

The search and estimation subsystem processes along the following steps.

Step 1. Search the cases, whose element terms are the same as investigation requirement.

If the cases are found out, then output the corrosion rates and terminate, else output the message 'there is no required case' and go next step. The outputted corrosion rates are the minimum and maximum corrosion rate of searched cases.

Step 2. Collect a set of the closest cases to investigation requirement.
One term of the closest cases is dissimilar from investigation requirement and all the other terms of the closest cases are the same as investigation requirement.

Step 3. Determine the order among element term intensity of the dissimilar term from investigation requirement.
Collect a set Dc of cases, whose term C is dissimilar from each other and all the other terms are the same. For Element terms x and y of term C of the set Dc, search the minimum corrosion rate $Min(x)$ and $Min(y)$, and the maximum corrosion rate $Max(x)$ and $Max(y)$, respectively.

If $Min(x) = Min(y)$ and $Max(x) = Max(y)$, or $Min(x) < Min(y)$ and $Max(x) > Max(y)$, then define as $x = y$ and express as $x \leftrightarrow y$. Else if $Min(x) \leq Min(y)$ and $Max(x) \leq Max(y)$, then define as $x < y$ and express as $x \leftarrow y$.

The whole order among element terms intensity (effect to the corrosion rate) of term C can be presented using a graph. In the graph, intensity of all element terms in a strong connected component is of equal order. Intensity of all element terms in the connected component can be ordered. The orders among intensity of the element terms in unconnected components are not defined.

Step 4. Estimate the corrosion rate by interpolation or extrapolation using the order.
At first, the corrosion rate is interpolated for set Dc. It is assumed that the element term of term C of investigation requirement is y, the element terms of C of the cases of Dc are x and z.

If the order among x, y and z is $x < y < z$, then it is estimated that $Min(y) = Min(x)$ and $Max(y) = Max(z)$. Else if the order is $x = y = z$, then it is estimated that $Min(y) = min[Min(x), Min(z)]$ and $Max(y) = max[Max(x), Max(z)]$.

If x or z is not exist, the corrosion rate is extrapolated. If x is not exist, then $Min(y) = 0$. If z is not exist, then $Max(y)$ is equal to the maximum corrosion rate in the database.

Furthermore, in the same manner, $Min(y_1)$, $Min(y_2)$, ..., $Min(y_n)$ and $Max(y_1)$, $Max(y_2)$, ..., $Max(y_n)$ for all terms n are estimated.

Then, the minimum and maximum corrosion rate of investigation requirement is estimated to $Max[Min(y_1), Min(y_2), ..., Min(y_n)]$ and $Min[Max(y_1), Max(y_2), ..., Max(y_n)]$.

Step 5. Output the estimation result and register the result to database if the result is satisfied.

The decision support system is constructed on the Web. When a user inputs the investigation requirement, the system outputs the searched or estimated result along with estimated process. To avoid an unhealthy input, retry is made for a wrong input or no input to indispensable term.

4   Results and Discussion

Over 500 cases were stored in the database from corporate companies. The healthy of the database was examined using cross-validation method. The cross-validation method was processed as follows.

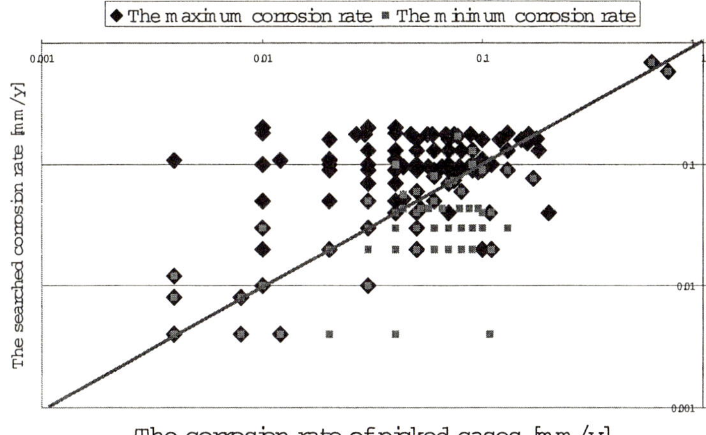

Fig. 2. The corrosion rate of picked cases VS. searched result.

Step 1. Pick one case Cp from the database.
Step 2. Search and estimate corrosion rate for the element terms of Cp.
Step 3. Compare the result with the corrosion rate of Cp.
Step 4. Repeat from Step 1 to 3 for all of the cases.

In ideal situation, the outputted corrosion rates will be the same as the corrosion rates of picked cases. When the outputted corrosion rate is slower than the corrosion rate of picked case, the investigated point is not corroded so much contrary to expectations. When the outputted corrosion rate is faster than the corrosion rate of picked case, the investigated point is much corroded contrary to expectations. Therefore, it is safety side when the outputted corrosion rate is slower than the corrosion rate of picked case.

The corrosion rates of picked cases and the searched results are displayed in Fig. 2. The axes of this figure have logarithmic scales. The proposed system outputted useful information, since most of the maximum outputted corrosion rates are in safety side. But, some of the maximum outputted corrosion rates are not in safety side. The reasons are that many cases of no element term or error element terms are stored in the database.

Next, the estimation results are displayed in Fig. 3. The cases are not found the same case in the database. All of the maximum outputted corrosion rates are in safety side. However, some of the estimation results are the maximum corrosion rate of the database. The reasons are that the orders of element terms cannot be determined, since the amount of cases in the database is insufficient.

5   Conclusion

The corrosion rate model of atmospheric corrosion of carbon steel pipes was proposed. The Factors used in the model were listed. Based on the model, the decision support system was developed. The system searches and estimates the

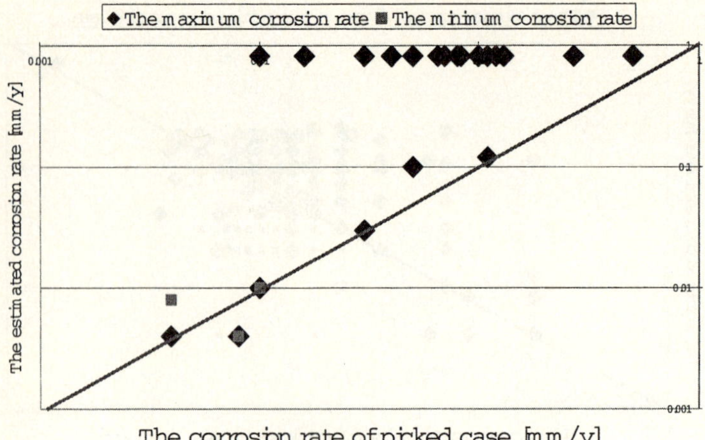

Fig. 3. The corrosion rate of picked cases VS. estimated result.

corrosion rate from investigation requirement. The real cases were stored in the database from corporate companies. The system was evaluated using cross-validation method. Most of the maximum searched corrosion rates were in safety side, but some of them were not in safety side. The reasons are that many cases of no element term or error element terms stored in the database. And all of the maximum estimated corrosion rates are in safety side. However, some of the estimation results are the maximum corrosion rate of the database. The reasons are that the orders of element terms cannot be determined.

In the future works, unhealthy cases will be eliminated from the database automatically, and how to treat the cases with no element term will be determined to be able to search and estimate in safety side.

Acknowledgment

It is great appreciation that the support for this work and the permission to present in this conference was done by members of 2nd committee, society of equipment maintenance technology, institute of research and innovation.

References

1. Gerven, T., Gurnia, H., Schlagner, W.: Wissenbasiertes System zur Lochkorrosion. Werkstoffe und Korrosion. **44** (1993) 426–430
2. Farina, C., Mininni, S.: DOCES: An expert system for material damage in boilers. Proceedings of the 11th International Corrosion Congress. (1990)
3. Cristaldi, L., Orsini, G.: Sistema esperto materiali. Metallurgica Italiana. **84** (1992) 249–254
4. Pieri, G., Klein, M. R. and Milanese, M.: MAIC: A data and knowledge-based system for supporting the maintenance of chemical plant. International Journal of Production Economics. **79** (2002) 143–159

Analogical Reasoning Based on Task Ontologies for On-Line Support

Takashi Hamaguchi[1], Taku Yamazaki[1], Meng Hu[1], Masaru Sakamoto[1],
Koji Kawano[2], Teiji Kitajima[3], Yukiyasu Shimada[4],
Yoshihiro Hashimoto[1], and Toshiaki Itoh[1]

[1] Nagoya Institute of Technology, Gokiso-cho, Showa-ku, Nagoya, Japan
`hamachan@nitech.ac.jp`
[2] Advanced Solutions Division, Mitsubishi Chemical Engineering Corporation, Japan
[3] Tokyo University of Agriculture and Technology, Japan
[4] National Institute of Industrial Safety, Japan

Abstract. In this paper, it is discussed how to search the previous reports, which serve as useful references. Usefulness depends on the scopes or aims. In order to express the scope or the knowledge to achieve aims, ontology is applied. Even if the objects are identical, the expressions of various task ontologies might be different from each other. The ontologies can be linked via the common parts of the object. By utilizing the links effectively, similar items can be searched. As an example scenario, search problem of useful previous reports about similar machines for on-line support was shown.

1 Introduction

The term in which a new product can maintain its competitive strength in the market is becoming shorter. Under this situation, the improvement of after-sales service is considered important as one of the strategies to differentiate from other companies. When good on-line support service is applied to such industries as machine tool manufacturer, it is possible to heighten the productivity not only by reducing the cost of maintenance between manufacturers and customers, but also by shortening the MTBF of production line.

However, it is difficult to maintain a high customer satisfaction because new products are frequently developed and the kinds of products are increased. Even if there were on-line support persons who had much knowledge and experience, the increase of the necessary information is too rapid. In addition, if the trouble reports for the maintenance of each product are accumulated, the amount of information for a product becomes small because of the short product life. It is desired to utilize the experience on similar products in order to obtain knowledge for new products. If on-line support persons can get pervious reports, which serve as useful references, and if they can understand their analogy, they might be able to find a good advice to customers.

Therefore, in this paper, it is aimed to develop a methodology for analogical reasoning to maintain the database for on-line customer support.

R. Khosla et al. (Eds.): KES 2005, LNAI 3681, pp. 155–161, 2005.

While the database for on-line support contains rich information about troubles, such as observed phenomena, faults origins, countermeasures and so on, the information about design or procurement is not contained. There are various similarities. Similarity depends on the scope or aims.

For evaluation of analogy, other knowledge than user support might be necessary. If similar component parts, which can realize the identical function, want to be found, the relationships between functions and component parts are necessary. They must be considered in design activity. Even if the objects are identical, the databases are different between the different activities. Their own database contains enough information for its activity. It is impossible to build a unified database or knowledge base which contains every kind of information. In order to obtain various information, it is necessary to utilize plural database or knowledgebase. At first, the expression of scope and knowledge for each activity is discussed.

2   Combining Various Task Ontologies

2.1   Database and Task Ontology

Database is a model which expresses the objective world to fulfill a certain purpose through the attribute data and item data. Thus naturally data themselves are different according to the purpose. But it is not obvious to locate the items of data because only the staff who structured the database knows how and why the model is created. So it is not easy to connect the different database. It is the objective of this study to remove this.

Ontology contributes to the understanding of the content of knowledge which should be expressed in the object world, and to heightening the sharing of knowledge and reusing of it [1],[2],[3],[4],[5],[6].

So it is made up of the needed concept which expresses the object world and the relation among the concepts. We can think of it as ontology when the model is shown which expresses the location of data items in database clearly through meta information and connection. Ontology differs in the objective of database.

There are various activities in pro, order processing, production scheduling, production control, material and energy control, procurement, quality assurance, product inventory control, product cost accounting, product shipping a maintenance management, research development and engineering, marketing and sales et al. Therefore the classification of useful information differs at each task in the same company and it is natural that each activity uses its own database. Each activity has its own "task ontology".

The information in one database is not enough to exatract the information for on-line support. Therefore, diffrent task ontologies must be connected to make search beyond plural task ontologies possible.

2.2   Ontologies in On-Line Support Center

In this paper, a case study examined the usefulness of the proposed method by assuming the scenario of customer support to a new product which premises to structure the database and Task Ontology including the following information besides design, procurement and customers support activity.

Customer Service Ontology. Reports about the trouble cases are registered in the database of customer service. Generally such information as reported date, the type, responding person, and the contents of the trouble is stored as a table. As shown in Fig. 1, such information as appliance by the customers, reports of service, arrangements of parts, responding history, logs of countermeasures to equipments is already arranged. It is our theme how to link the unlinked data in this database to other databases. In Fig.3, a part of task ontology for customer service is shown. The relationships are "is attribute".

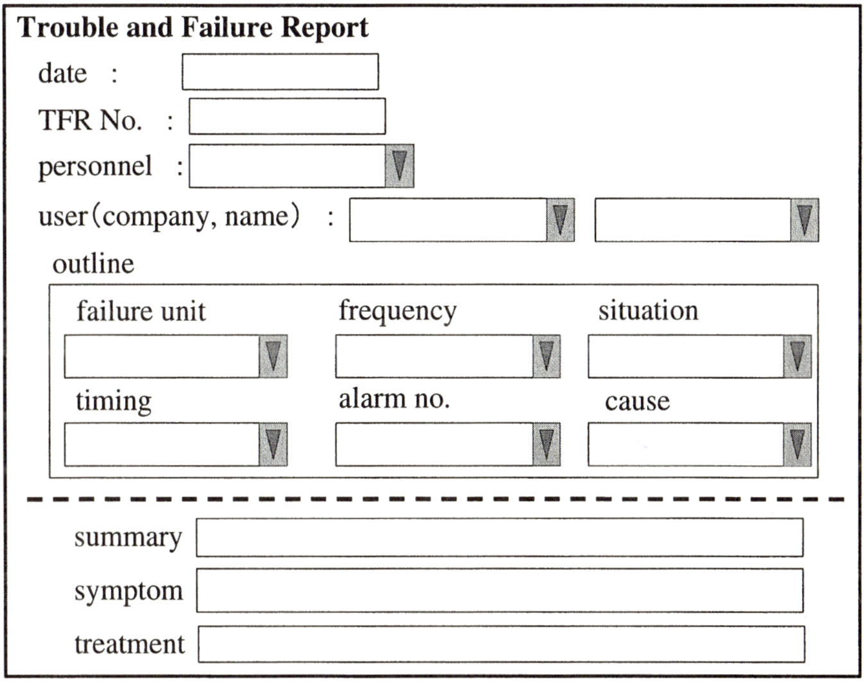

Trouble and Failure Report

date :

TFR No. :

personnel :

user (company, name) :

outline

| failure unit | frequency | situation |
| timing | alarm no. | cause |

summary

symptom

treatment

Fig. 1. Human Interface of customer service center database.

Design Ontology. The design activity can be regarded as the conversion of product specification into component specification. The design task ontology contains the relationships between functions realized the demanded specification. By analysing sub-functions, the necessary component parts for realizing the functions can be found. The specifications of the parts are the output of design task.

Product BOM Ontology. Fig.2 also shows the product BOM ontology. The structures of the part modules are shown in it. It is shown that turning machine has headstock and its motor has an inverter.

Procurement Ontology. Specifications of component parts are given by the design section to procurement section. Procurement section does not care which machine contains the parts. The same parts might be utilized in different machines. Procurement section persons search the catalogues of the parts, which satisfy the given specification. They purchase the parts considering their prices, shipping and so on besides their specification from design section. Although Fig.2 does not show the information, purchase ontology contains data of purchase, the distribution and so on which realize traceability.

3   Example Scenario

The user, who purchased the TL3 type turret lathe, made contact with the customer support center. The user reported that the shaft of headstock couldn't spin although he turned the start switch on. There is little information about TL3 in customer support database because the TL3 type is quite new product. The operator tried to obtain the useful information through the past similar reports in the database.

Firstly, the operator checked the expression of the task ontology for design about TL3. Fig.2 shows the relationships between functions and component parts

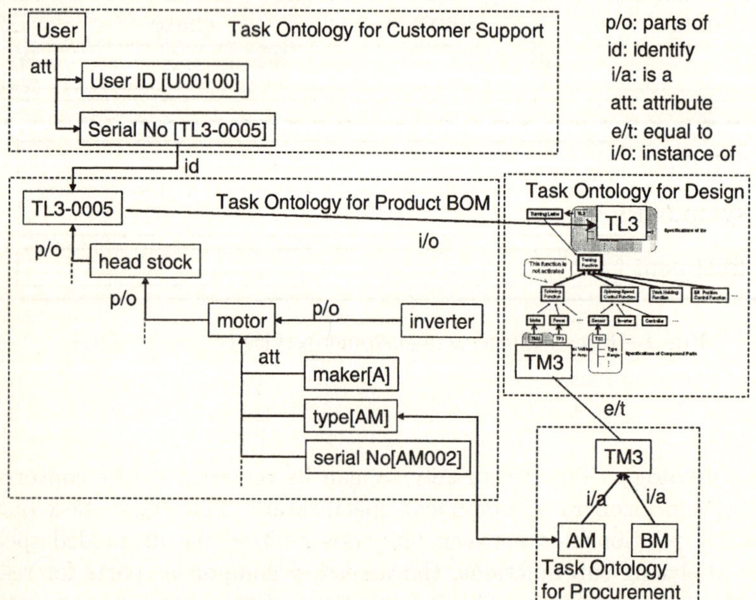

Fig. 2. Relationships among different task ontologies.

to satisfy the specification of the product. Design tasks can be regarded as the conversion of the product specification component into parts specification. Because the user reported that the spinning function of the headstock did not work, the component parts relating to the function can be found in the design task ontology shown in Fig.2

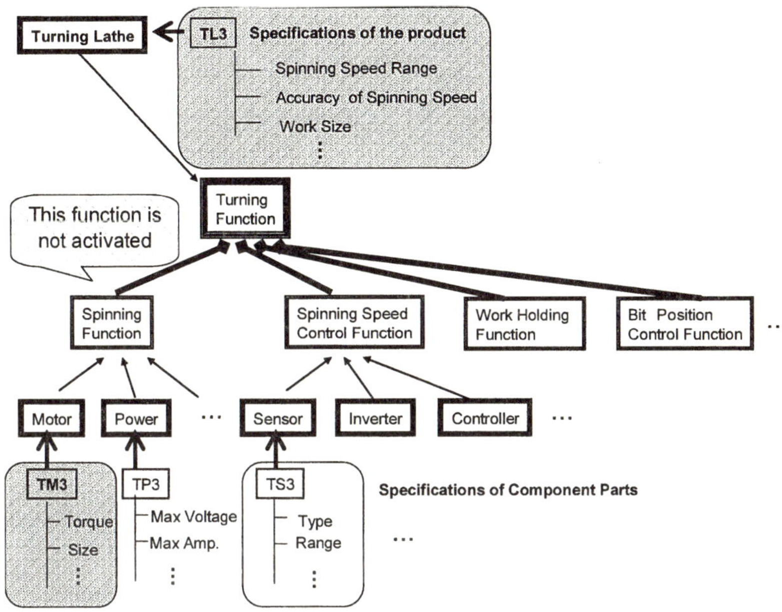

Fig. 3. The schematic of the expression of task ontology for design about TL3.

To realize the specifications of TL3, the specifications of each sub function are determined in the design stage. TM3 is the class of the specifications of the spinning motor.

Fig.2 shows the relationships among the different task ontologies.

The customer support person checked the machine ID of the customer's machine. He found that the class of the turning lathe was TL3 by using the product BOM. From the design ontology, he can detect that the component part to realize the specification TM3 might have a relationship to the reported trouble. By using procurement ontology, the both parts AM and BM satisfy the specification TM3. The product BOM shows the turning lathe TL3-0005 has an AM type motor.

The id link is used the connection between the expression of task ontology for customer support and the expression of task ontology for product BOM. To access the product BOM for TL3-0005, the operator used the id link. In the expression of task ontology for product BOM, the p/o link is prepared to express the unit composition. Hence he reached the motor, which is component of headstock easily. The motor instance has some attribute link att, which are connected with maker, type, serial number and so on. The motor has the p/o link, which is connected inverter circuit, too. From this search based on the expression of task ontology for product BOM, the operator identified the troubled TL3's motor information. It had implemented the type AM motor for headstock.

The operator explores AM in the customer support database. The searched result about AM is shown in Fig.4.

Type	General Area	Part of Unit	Name	Failure mode	Cause	Effect	Countermeasure
TL1	tool rest	motor	AM	Programming miss	Overload by tool	Motor can't run	To reduce load

Fig. 4. The searched result about AM.

The failure mode, cause, effect and countermeasure about the type AM motor were found from trouble report database as shown in Fig.4. The searched trouble was caused the bit control failure. Therefore, the operator asked the user whether the bit was touched. The user answered the bit didn't affect shaft in the situation. The operator judged that the found trouble report could not be applicable to the user's trouble. The trouble report is not always useful even if the trouble report concerned with the same parts.

The trouble was the ill condition of the spinning function. The spinning function is realized by the motor, which satisfies the specification TM3. The ontology for procurement shows that not only AM but also BM satisfy TM3. BM can be regarded as a similar motor to AM. Then, the search of trouble reports concerned with BM was executed. Another trouble report was found as shown in Fig.5. The operator could find that inverter circuit failure was another option of the trouble cause and its countermeasure is replacement of the power supply unit.

Type	General Area	Part of Unit	Name	Failure mode	Cause	Effect	Countermeasure
TL2	headstock	motor	BM	Power Supply Failure	Inverter Circuit Failure	Motor can't run	Power Supply Replacement

Fig. 5. Search Result using BM.

4   Conclusion

There are many kinds of databases for various kinds of activities. Even if they deal with the identical object, the data structures or technical terms might be different from each other. It is almost impossible to build unified data structure, which is available to any activities. In this paper, it was discussed how to extract useful information by combining different kinds of databases. It was assumed that the ontology of each activity was expressed by using meta-data and relations. It was not necessary that the expression ways of the ontologies are unified. Because every activity concerns with an existing machine, there are many relationships between diffrent ontologies via the existing object. By utilizing the relationships among different task ontologies, the search of similar object could be realized. Because the product life becomes shorter and shorter, the accumulation of trouble reports of an identical product becomes difficult. Therefore, it is important to utilize the experience for similar products effectively. In this paper, a method to combining diffrent ontologies was shown. Not only implementation of the proposed idea but also expression of other ontologies such as fault diagnosis and so on are our further works.

Acknowledgements

This research was partially supported by the Ministry of Education, Culture, Sports, Science and Technology, Grant-in-Aid for Scientific Research (B) 16310115, 2004 and 16360394, 2004.

References

1. Uschold, M., Gruninger, M.: Ontologies : Principles, methods and applications. Knowledge Engineering Review, 11(2)(1996)
2. Gruninger, M., Fox, M. S.: Methodology for the Design and Evaluation of Ontologies, Workshop on Basic Ontological Issue in Knowledge Sharing, IJCAI-95, Montreal(1995)
3. Fernandez-Lopez, M., Gomez-Perez, A., Pazos Sierra, J. : Building a Chemical Ontology Using Methontology and the Ontology Design Environment, IEEE Intelligent Systems, 14(1)(1999) 37-46
4. Staab, S. H., Schunurr, P., Studer, Sure, Y. : Knowledge Processes and Ontologies, IEEE Intelligent Systems, Special Issue on Knowledge Management, 16(1) (2001) 26-34
5. Mizoguchi, R., Ikeda, M., Seta, K., Vanwelkenhuysen, J.: Ontology for Modeling the World from Problem Solving Perspectives, Proc. of IJCAI-95 Workshop on Basic Ontological Is-sues in Knowledge Sharing (1995) 1-12
6. Mizoguchi, R., Kozaki, K., Sano, T., Kitamura, Y.: Construction and Development of a Plant Ontology, in Knowledge Engineering and Knowledge Management - Methods, Models and Tools -, The 12th International Conference, EKAW2000, Lecture Notes in Artficial Inteligence 1937, Juan-les-Pins, France(2000)113-128

Development of Engineering Ontology
on the Basis of IDEF0 Activity Model

Tetsuo Fuchino[1], Toshihiro Takamura[1], and Rafael Batres[2]

[1] Tokyo Institute of Technology, Department of Chemical Engineering, Tokyo 152-8552, Japan
fuchino@chemeng.titech.ac.jp
[2] Toyohashi University of Technology, Department of Production Systems Engineering,
Toyohashi 441-8580, Japan
rbp@pse.tut.ac.jp

Abstract. The chemical process industry which aimed at the high increase in efficiency so far has been facing the issue of the technology transfer newly and its technology should be defined explicitly. However, in general, technology is mutually related, and the relation depends greatly on its engineering process. Therefore, technology and its mutual relation are to be defined within engineering process where the technology is applied. In this study, engineering ontology for chemical process industry to define its technology is developed. Recently, the trial to describe the engineering process as an activity model has started. The PIEBASE (Process Industry Executive for achieving Business Advantage using Standard for data Exchange), which is an international group of process industry, has provided an activity model describing enterprise activities of process industry. The PIEBASE activity model was applied as the reference model to provide the engineering ontology here.

1 Introduction

Technology is an important element which decides competitiveness in the chemical industry. The technology belongs to people, and it is not an exaggeration that its transfer has so far been performed with the form of the initiation to people from a man. While competition intensifies, when movement of people became intense, the traditional technology transfer, which is time conscious, has been stopping fully functioning. To maintain competitive power, technology should be made to belong to a company organization not to a person, and a system environment to support it is necessary. For this purpose, technology itself should be defined.

Technology is strongly related with the engineering processes in which the technology is applied. However, the engineering, such as process design, is carried out implicitly, in general. In recent years, researches to describe the engineering process as IDEF0 (Integrated Definition Language) activity model have been found [1], [2]. The purpose of these studies is to clarify the engineering process explicitly. The PIEBASE, which is the global consortium of companies in the process industry, developed IDEF0 activity model (PIEBASE model [3]) to provide an overview of the business activities and information flows that constitute the process industries. This model was made to confirm the common understandings on the engineering process within the consortium in the STEP (Standards for the Exchange of Product Data, ISO standard 10303) project.

R. Khosla et al. (Eds.): KES 2005, LNAI 3681, pp. 162–168, 2005.

It is clear that the engineering process can be described explicitly by using IDEF0 activity model. However, it is not the purpose of IDEF0 to retrieve the information. Therefore, the relation between technology and the engineering process should be found out by hand, and it is difficult to define technology comprehensively in the engineering process on the IDEF0 activity model. In order to realize the retrieving function to the IDEF0 activity model, IDEF0 activity model is represented into ontology in this study.

In IDEF0 activity model, the information to perform the activities is categorized into four types; 'Input', 'Control', 'Output' and 'Mechanism', which is called the ICOM for short. On the basis of the feature of ICOM, the generic property classes between ICOM and activities are defined. The contents of ICOM and the activities are made into objects and their class definition was performed from the consideration of their meanings. This research proposes the development rule of engineering ontology by making an IDEF0 activity model based on PIEBASE model into a reference model.

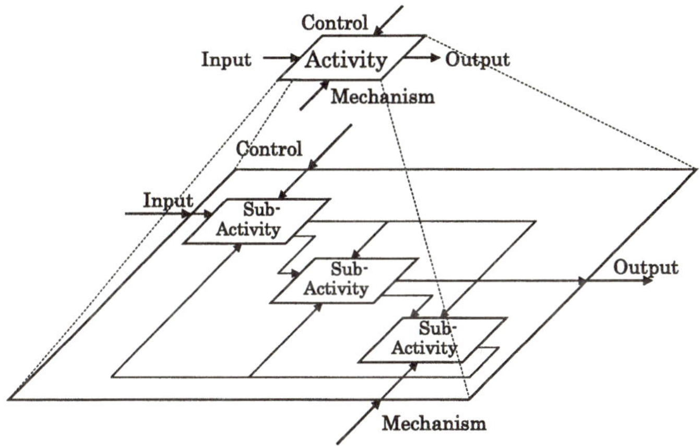

Fig. 1. Property of IDEF0 activity model

2 IDEF0 Activity Model

IDEF0 is a method to describe business and/or engineering process, where the rectangle represents activity, and the arrows describe the ICOM, where 'Input' is the information to be changed by the activity, 'Control' is the information to constraint the activity, 'Output' is results of the activity and 'Mechanism' is the resources for the activity. Each activity is expanded into sub-activities hierarchically, and the ICOM is also made in detail, as shown in Fig. 1. Although the description form of IDEF0 is specified in this way, the model creation method is not specified. Therefore, the consistency of the model had depended greatly upon the author(s) of the model.

PIEBASE developed an IDEF0 activity model, called PIEBASE model [3], to specify the business activities and information flows that constitute the process industries. In developing the activity model, PIEBASE proposed a template [4] used as the norm of activity model development. This template consists of three activity classes, i.e. 'Manage', 'Do' and 'Provide Resources', as shown on Fig. 2. The 'Manage' activity receives 'Directives' from the upper activity. These 'Directives' are interpreted, and the transferred sub-'Directives' are informed to 'Do' and/or 'Provide Resources' activities. The 'Do' activities perform anything on the constraint of the sub-'Directives', and their status and the 'Resource requirements' are informed to the 'Manage' activity. The requirements are specified by the 'Manage' activity, and the 'Directives' to prepare necessary resources are informed to the 'Provide Resources' activity. The 'Provide Resources' activity supplies the 'Do' activity with resources fit for the purpose. The most characteristic point of the PIEBASE template is that all control information is 'Directive'. This template was followed across all principal activities in the PIEBASE model, and the consistent activity model was developed by making 'Conduct Core Business' into the top activity. Fig. 3 shows the sub-model, which is expanded one step from the top activity. The PIEBASE model consists of 46 sub-models, and in this study, this activity model is converted into ontology.

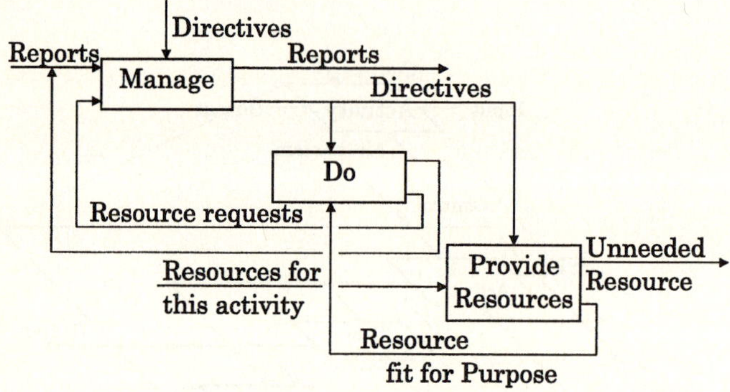

Fig. 2. Template of PIEBASE model

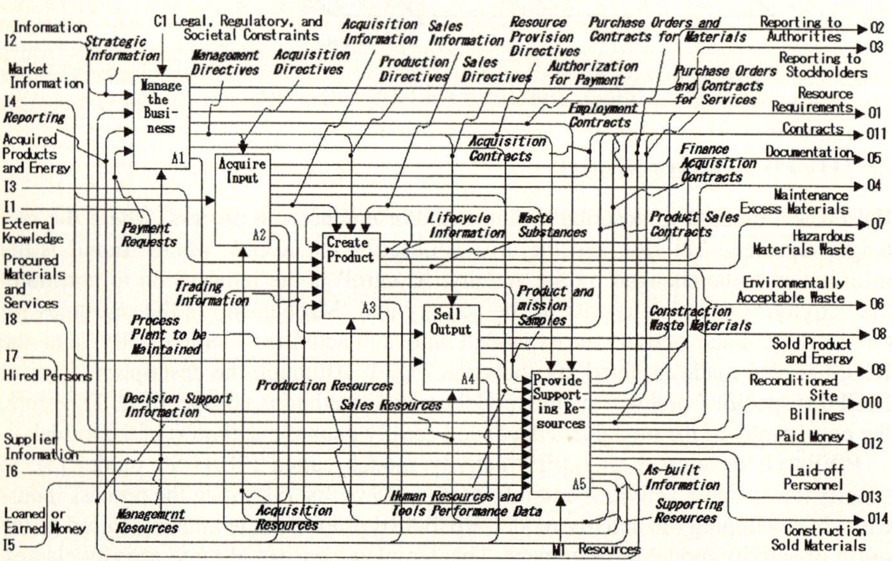

Fig. 3. One step expanded sub-activity from the top activity of PIEBASE model

3   Developing Ontology Based on IDEF0 Activity Model

In order to develop engineering ontology, in consideration of the feature of the domain, it is necessary to perform the definition of classes, instances, and properties. In this study, IDEF0 activity model, which is according to the PIEBASE template, is considered as the reference model for the engineering domain, and the feature of the PIEBASE model is considered.

3.1   Definition of Objects and Properties

Although ICOM can be regarded also as the property between activities, there are two aspects for respective ICOM information, i.e. outputted information and inputted information. Furthermore, the inputted information is divided into three cases; inputted as 'Input', inputted as 'Control' and inputted as 'Mechanism'. When ICOM is set to property, relations between activities equal to the number of ICOM(s) are to be necessarily defined. From the view point of information retrieval, it is not convenient. Therefore, we defined both Activity and ICOM as object, and decided to express the relation between an activity and ICOM as property. For example, a simple IDEF0 activity model consisting two activities as shown on Fig. 4 is considered, Activity-1, Activity-2, ICOM-1 to ICOM-7 are defined as objects here as shown on Fig. 5. Then, it is clarified that the relations between an activity and ICOM(s) can be categorized into five types from the feature of the PIBASE model, i.e. when (1) ICOM is inputted to an activity as 'Input', (2) ICOM is inputted to an activity as 'Control', (3) an activity outputs ICOM to be inputted to another activity as 'Input', (4) an activity outputs ICOM to be inputted to another activity as 'Control', and (5) ICOM is inputted to an activity as 'Mechanism'. In the PIEBASE model, all the 'Control' information is 'Directive', and the 'Directive' makes an activity start or stop under its condition. Therefore, two properties; 'Beginning' and 'Ending' are defined to describe the relation between 'Directive' and an activity. Moreover, when an activity outputs 'Control' information ('Directive'), the activity behaves as a resource for the 'Directive'. In order to describe such a relation between an activity and 'Directive', 'Causes' is defined as property. For the rest of the above three types of relations, still more three properties are introduced considering the meaning of 'Input', 'Output' and 'Mechanism' in the IDEF0 activity model respectively; 'has_consumable', 'has_product' and 'has_performable', as shown on Fig. 5.

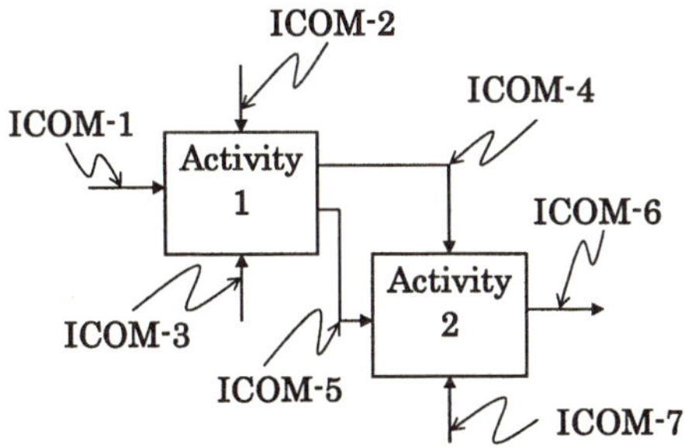

Fig. 4. Simple example IDEF0 model

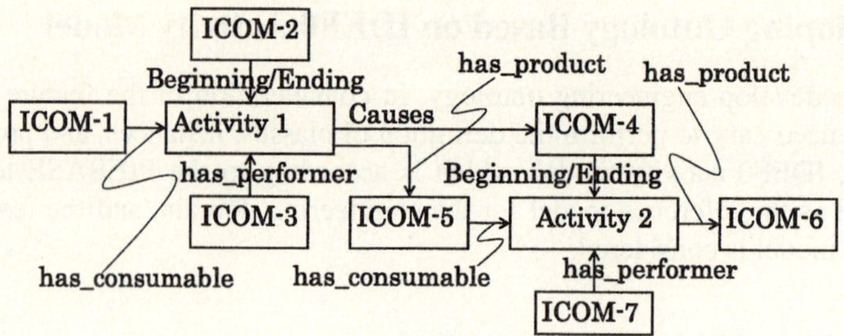

Fig. 5. Relation between activity and ICOM

3.2 Object Class Definition

In order to share the developed ontology through the lifecycle, the object class definition is performed on the basis of ISO 15926 data model [5]. In the ISO 15926, 'Thing' is defined as the top class, and has two sub-classes; 'Abstract_object' and 'Possible_Individual'. In the 'Possible_Individual', furthermore two sub-classes; 'Physical_object', 'Arranged_object', are originally defined, and these sub-classes are furthermore decomposed into sub-classes as shown on Fig. 6. On the other hand,

form the view point of the PIEBASE and/or IDEF0 activity model, four types of objects are to be defined; 'Input' (or 'Output'), 'Directive' (or 'Output'), 'Mechanism' and activity. The objects for 'Input' and/or 'Mechanism' can be classified into the sub-classes under the 'Physical_objectat' and/or 'Arranged_object'. However, sub-classes to categorize the objects for 'Directive' and/or activities are not introduced in ISO-15926, because it is to define the lifecycle data for process plant including oil and gas production facility. Therefore, 'Event' and 'Activity' sub-classes are newly introduced here. The 'Activity' sub-class has the activities as instance, and the 'Event'

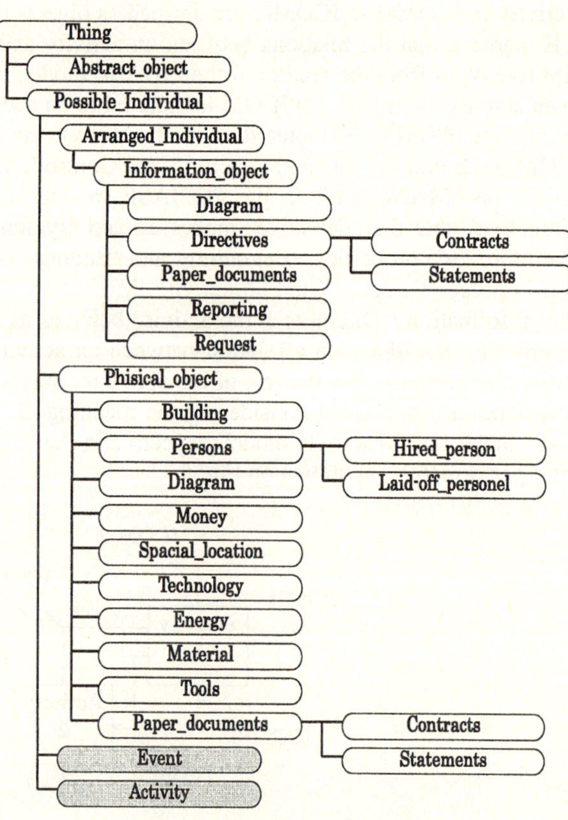

Fig. 6. Object Class hierarchy

sub-class has directives as instance. In order to divide the physical entities form the activities explicitly, 'Activity' and 'Event' sub-classes are defined under the 'Possible_Individual' sub-class as shown on Fig. 6.

4 Developed Ontology

According to forgoing object class, object and property definitions, the PIEBASE model is defined as engineering ontology on Protégé 2.1.2 [6]. Fig. 7 shows the interface image of Protégé 2.1.2. The class hierarchy is appeared on the left window, and the instances are appeared on the center window here. As explained in the previous section, activities in the PIEBASE model are defined in the 'Activity' sub-class.

Since PIEBASE model is expressed as engineering ontology, it becomes possible to retrieve the information on PIEBASE model freely through a search engine. For example, when activities outputting PFD (Process Flow Diagram) are retrieved, then four activities are enumerated; 'define_processes', 'perform_process_design', 'create_process_flow_diagram' and 'design_and_engineer_process_plant_in_consept'. As the same manner, it is possible to retrieve the related technology and/or activities, necessary resources, necessary conditions, and so on against any engineering action. Therefore, the technology can be defined by developing the engineering ontology on the basis of IDEF0 activity model.

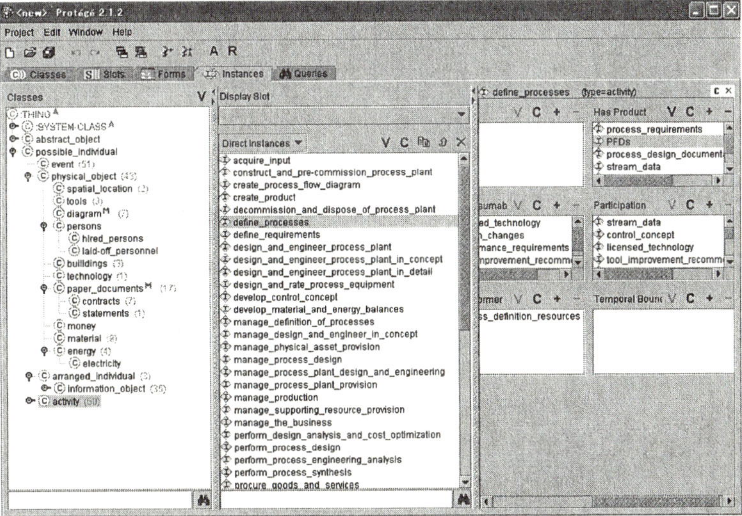

Fig. 7. Developed ontology on Protégé 2.1.2

5 Conclusion

In order to provide technology transfer supporting environment, technology should be defined within the engineering activity, where the technology is applied. In this study, on the basis of the PIEBASE activity model, engineering ontology is developed to define technology. The ontology developing rule used here would be applicable for any other IDEF0 activity models based on the PIEBASE template.

References

1. Fuchino, T., Shimada, Y.: IDEFO Activity Model Based Design Rationale Supporting Environment for Lifecycle Safety, Lecture Notes in Computer Science, 2773 (2003) 1281-1288
2. Fuchino, T., Wada, T., Hirao, M.: Acquisition of Engineering Knowledge on Design of Industrial Cleaning System through IDEFO Activity Model, Lecture Notes in Computer Science, 3214 (2004) 418-424
3. PIEBASE: Activity Model Diagrams and Definitions, http://www.posc.org/piebase/
4. PIEBASE: PIEBASE Activity Model Executive Summary, http://www.posc.org/piebase/
5. ISO 15926-2, Industrial automation system and integration –Integration of lifecycle data for process plants including oil and gas production facilities- Part2: Data model (2003)
6. Protégé, http://protege.stanford.edu/index.html

A Novel Approach to Retrosynthetic Analysis Utilizing Knowledge Bases Derived from Reaction Databases

Kimito Funatsu

Department of Chemical System Engineering, School of Engineering
The University of Tokyo, 7-3-1Hongo, Bunkyo-ku,
Tokyo 113-8656, Japan

Abstract. Novel approach to empirical retrosynthetic analysis of computer-assisted synthesis design is proposed using the KOSP (Knowledge base-Oriented system for Synthesis Planning) system. KOSP has four functions: (1) strategic site pattern perception, (2) precursor skeleton generation, (3) retrosynthetic scheme evaluation, and (4) retrosynthetic analysis termination. This system is based on knowledge bases in which reactions are abstracted by structural and/or stereo characteristics of reaction sites and their environments. This paper describes the framework of KOSP and results of its execution.

1 Introduction

We began to develop a novel empirical synthesis design system by employing knowledge bases which are free from the disadvantages of transform-knowledge base, called KOSP (Knowledge base-Oriented system for Synthesis Planning). We have concretized a similar concept in the reaction prediction system SOPHIA [1,2]. However, this approach is a first attempt, to our knowledge, for retrosynthetic analysis. Here, the knowledge bases of AIPHOS (Artificial Intelligence for Planning and Handling Organic Synthesis) [3] in which reactions are abstracted by structural and/or stereo characteristics around reaction sites are applied to this approach because these knowledge bases can be automatically derived from reaction databases. AIPHOS is designed to propose suitable retrosynthetic routes from the standpoints of both novelty and practicality based on knowledge bases. Since KOSP proposes empirically more practical retrosynthetic schemes than those of AIPHOS, the concept of KOSP is different from that of AIPHOS.

2 Retrosynthetic Analysis Procedures in KOSP

The KOSP system is roughly divided into four stages, as shown in Fig. 1. First, KOSP automatically acquires all possible strategic site patterns for an input target molecule by utilizing the RKB. In the next step, bonds for constructing each strategic site pattern are cut, and all possible reconnections of cut patterns and/or addition of appropriate atoms or atomic groups to them are performed to generate all possible precursors. Then, each generated retrosynthetic scheme is evaluated to determine whether it occurs by the RKB. Finally, the system judges whether retrosynthetic analyses can be terminated by comparing precursor structures with the starting material library of AIPHOS [4]. The details are described below.

R. Khosla et al. (Eds.): KES 2005, LNAI 3681, pp. 169–175, 2005.
© Springer-Verlag Berlin Heidelberg 2005

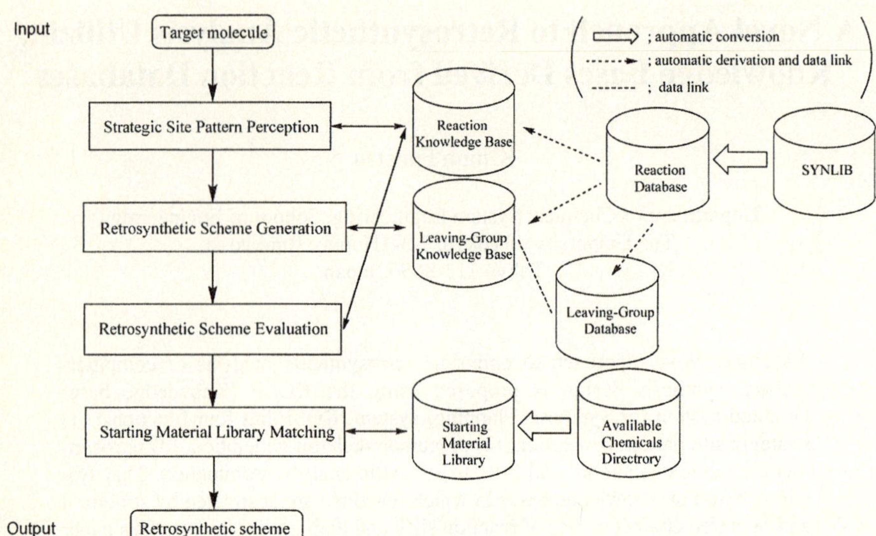

Fig. 1. The outline of KOSP

2.1 Automatic Perception of Strategic Site Pattern

Acquisition of strategic site patterns is important for retrosynthetic analysis because it determines whether or not proposed retrosynthetic schemes are prospective ones. In order to make detection of attractive and plausible strategic site patterns from the standpoint of chemists possible utilizing the RKB, in the KOSP system a strategic site pattern is defined as a set of bonds which is similar to characterized reaction schemes in the Reaction Knowledge Base (RKB). With this definition, KOSP can acquire prospective strategic site patterns automatically. The actual procedure for strategic site pattern perception is as follows: Input of a desired target structure with a graphical editor is performed, the user specifies whether or not stereoselective retrosynthetic analyses are considered, and basic structural information (e.g.; types of atoms and bonds, connectivity) is computed. The SSSR (smallest set of smallest ring) [5,6] and the aromaticity based on Hückel's rule are determined. With this information, the structural and stereo characteristics are recognized, and the bit sequences for every bond in the target molecule are obtained. Next, every combination of the bit sequences from one-bond to six-bonds of the target molecule is generated in turn, and the bit sequences of the generated combination are compared with the bit sequences in the reaction site file by one-to-one correspondence. In this method of comparison, five levels can be considered. The system judges a match according to one of five levels: (1) minimum matching level 1 is a match of type of bond and type of atoms belonging to this bond; (2) minimum matching level 2 is a match of Structural Characteristics of a reaction site and its Environments (SCE) of the atoms of level 1; (3) minimum matching level 3 is a match of SCE of the atoms of level 1 including the information around their α position; (4) minimum matching level 4 is a match of SCE of the atoms of level 1 including the information around their β position; (5) minimum matching level 5 is a match of SCE of the atoms of level 1 including the information around their γ position. Level 2 is default, though the user can select another level if necessary. If the user specifies consideration of stereoselective retrosynthetic analyses, the system combines these five levels and stereo characteristics into a criterion for this matching procedure.

This comparison procedure continues until all possible combinations up to six-bonds of the target molecule are generated. By this procedure, sets of all bonds judged to be disconnective are automatically perceived as strategic site patterns. Here, if necessary, acquired strategic site patterns can be ordered with the depth of matching level and/or number of original data in the Reaction Database (RDB) corresponding to the characterized reaction schemes of the RKB that were the basis for acquiring each strategic site pattern.

2.2   Retrosynthetic Scheme Generation

This procedure is required for KOSP to obtain precursor structures as concrete forms in utilizing the RKB in which reactions are described as abstract forms by Structural and Stereo Characteristics of a reaction site and its Environments (SSCE). From the perceived strategic site pattern, all possible structure of precursors are generated on the basis of the cyclic permutation also used in the reaction generator of AIPHOS. This generating procedure is continued for all acquired strategic site patterns.

2.2.1   Precursor Skeleton Generation

KOSP cuts all bonds involving the strategic site pattern and generates all possible precursor skeletons by reconnecting free bonds and by adding dummy atoms to the free bonds. If the user specifies consideration of stereoselective retrosynthetic analyses at the strategic site pattern perception, the system generates all possible precursor skeletons corresponding to reactions which proceed with retention and inversion. If an aromatic bond is cut, the system considers this bond both as a single bond and double bond. Furthermore, if a multiple bond is cut, recombination can also be considered.

2.2.2   Determination of SLG

KOSP determines Suitable Leaving Groups (SLG) for the proposed precursor skeletons by using the Leaving-group Knowledge Base (LGKB). SLG determination is composed of three steps: (1) matching original data in the RDB corresponding to characterized reactions of the RKB which were the basis for proposal of the strategic site pattern, (2) matching the precursor skeletons, and (3) determination of SLG. Details of these steps are described below.

(1) **Matching Original Data in the RDB Corresponding to Characterized Reactions of the RKB Which Were the Basis for Proposal of the Strategic Site Pattern.** The system judges whether original reaction data in the RDB corresponding to characterized reactions of the RKB which were the basis for acquiring the strategic site pattern have the leaving groups, by comparison with the Leaving-group Database (LGDB). If the data are involved in the LGDB, the precursor skeletons are sent to the next step. If not, the SLG determination procedure is discontinued, and only precursor structures which have no dummy atoms can be sent to the retrosynthetic scheme evaluation step (section 2.3).

(2) **Matching the Precursor Skeletons.** The system compares SCE of reaction site of reactant concerning the LG elimination of the free bonds which were involved in the generated skeletons with those in the LGKB. Here, only SCE of reaction site of reactant related to LG elimination in the LGKB, corresponding to the ID nos. of original data which were the basis for proposing the strategic site patterns, are considered. This matching procedure is performed for both cases of the pairing LG being considered and those not being considered. This procedure also considers the five levels of matching used for the acquiring strategic site patterns (section 2.1).

(3) **Determination of SLG.** SLG for the free bonds of each precursor skeleton are determined by three levels of priority. They are

Priority 1: LG of a matched characterized reaction in the LGKB are pairing, and they were extracted from a common individual reaction.

Priority 2: LG of a matched characterized reaction in the LGKB are pairing, and the depth of matching level is high and the number of matchings is maximum as computed in matching of the precursor skeletons.

Priority 3: The depth of matching level is high, and the number of matchings is maximum as computed in the matching of the precursor skeletons. Pairing LG is not considered.

2.2.3   Addition of SLG to Precursor Skeleton

The system replaces the corresponding dummy atoms of the precursor skeletons with the substructures of the determined SLG obtained from the LGDB. Thus, the system generates the precursor structures.

2.3   Retrosynthetic Scheme Evaluation

It is determined whether each of the retrosynthetic schemes corresponding to all of the generated precursor structures occurs. The evaluation procedure is as follows: First, basic structural information for a set of precursors in a retrosynthetic path is obtained, and SSSR and aromaticity are recognized in the same manner as for strategic site pattern perception (section 2.1). As mentioned above, generated precursor skeletons which have dummy atoms are ignored in this evaluation procedure; this is because computation of SSCE is incomplete for a bond to which a dummy atom is attached.

2.4   Matching Precursor with Starting Material Library

The system judges whether the proposed retrosynthetic analysis can be terminated or not by comparison with the starting material library [4] of AIPHOS which is retrieved from the chemical inventory database, Available Chemicals Directory (ACD) [7]. The library is comprised of a hierarchical structure, and four levels of abstract structures are used. (1) The S.M. Data level consists of completely specified starting materials. (2) The low level contains information about positions and kinds of functional groups. (3) The high level contains only information about positions of functional groups. (4) The top level comprises basic skeletons of starting materials. Data for about ten thousand are registered in the current library. The matching operation is performed as described below. In the first stage, the precursor structure involved in the evaluated retrosynthetic scheme is changed to abstract graphs corresponding to those of the library by recognizing functional groups with the indicated structural characteristics. Next, the system determines whether the abstract graphs of the evaluated precursor structure exist by set reduction [8,9]. If the evaluated precursor exists in the S.M. Data level of the library, the retrosynthetic analysis of this retrosynthetic scheme is terminated.

3   Execution Example

Two examples of KOSP executions are given below. These executions employ the RKB and LGKB prepared from 28000 reactions of the RDB which was constructed from SYNLIB containing 81000 reactions. The current KOSP system can be executed without consulting with the user after the input to the system is the desired target compound. This system has been written in FORTRAN77 and is currently running on a SGI-INDIGO2 system (R4400, 250MHz).

The input target molecule is (*S*)-7-amino-5-azaspiro[2.4]heptane, which is the C-7 substituent, not considering the protecting group, of a new-generation antibacterial quinolone carboxylic acid, DU-6859a [10]. One of the proposed routes with KOSP is shown in Fig. 2.

It was about thirty minutes to find this retrosynthetic route. This route is attractive, because it involves the synthesis of a single enantiomer by means of asymmetric microbial reduction. The retrosynthetic paths were also suggested by AIPHOS, and one of the authors has developed an efficient and stereoselective route for the C-7 substituent of DU-6859a based on this proposal. This reported procedure is illustrated in Fig. 3 [11].

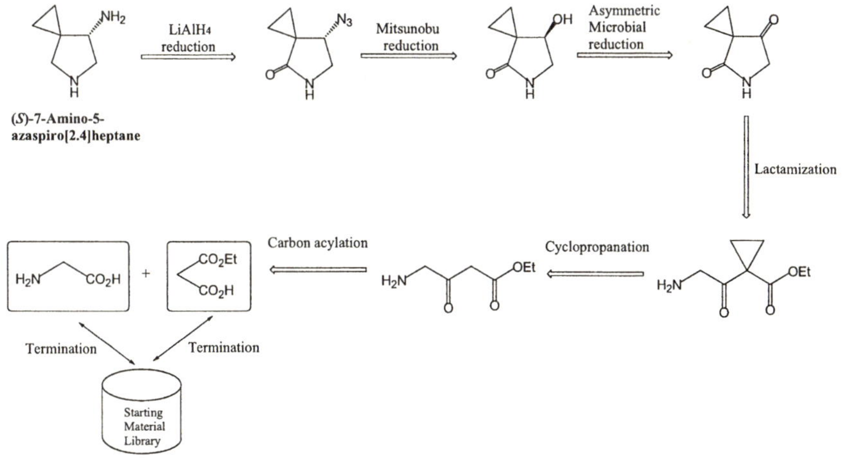

Fig. 2. One of the proposed retrosynthetic routes of (*S*)-7-amino-5-azaspiro[2.4]heptane from KOSP

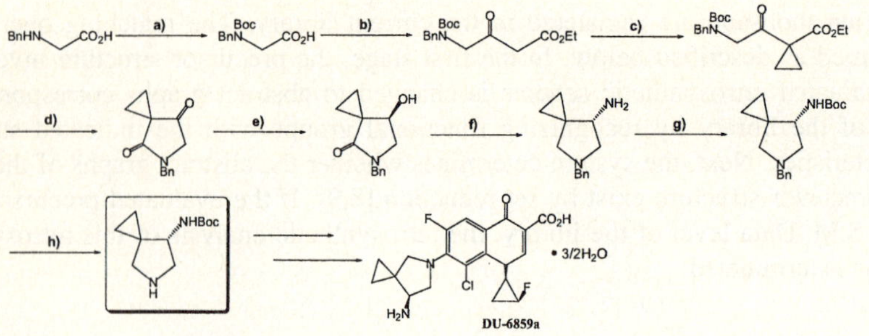

(a) Boc$_2$O; (b) CO(Im)$_2$; KO$_2$CCH$_2$CO$_2$Et, MgCl$_2$; (c) (CH$_2$Br)$_2$, K$_2$CO$_3$;(d) CF$_3$CO$_2$H; then toluene, reflux
(e) *Phaeocreopsis* sp.; (f) Ph$_3$P, EtO$_2$CN$_2$CO$_2$Et, DPPA; then LiAlH$_4$; (g) Boc$_2$O; (h) 5%Pd-C, H$_2$

Fig. 3. A reported synthetic method of the C-7 substituent of DU – 6859a based on the retro-synthetic route

4 Conclusions

The aim of the KOSP system is to adjust the reaction knowledge base, which is represented by information about structural and stereo characteristics of a reaction site and its environments (SSCE), and the leaving-group knowledge base, which is represented by information about structural characteristics of a reaction site and its environments (SCE), to an empirical retrosynthetic approach.

To achieve this purpose, four functions are required: strategic site pattern perception, retrosynthetic scheme generation, retrosynthetic scheme evaluation, and retrosynthetic analysis termination. By employing these functions, we have established a novel approach to empirical retrosynthetic analysis. One of the advantages of KOSP is that knowledge bases can be immediately derived from reaction databases in cases in which reaction data increases; KOSP can thereby consider novel and effective reactions which are being developed in the organic synthesis field. KOSP can thus utilize reaction data efficiently.

One plan to enhance KOSP is noninteractive generation of retrosynthetic sequences. For noninteractive generation of retrosynthetic sequences, a method of pruning of a synthesis tree is required to prevent increase in retrosynthetic proposals. As noted in section 3.1, KOSP orders strategic site patterns with the depth of matching level and/or the number of original data in the RDB suggesting each strategic site pattern. However, this function is not sufficient for ordering retrosynthetic paths from the chemist's point of view. Hence, we are now working on the development of a novel ordering technique of proposals by KOSP.

References

1. Satoh, H.; Funatsu, K. SOPHIA, a Knowledge Base-Guided Reaction Prediction System—Utilization of a Knowledge Base Derived from a Reaction Database. *J. Chem. Inf. Comput. Sci.* **1995**, *35*, 34—44.
2. Satoh, H; Funatsu, K. Further Development of a Reaction Generator in the SOPHIA system for Organic Reaction Prediction. Knowledge-Guided Addition of Suitable Atoms and/or Atomic Groups to Product Skeleton. *J. Chem. Inf. Comput. Sci.* **1996**, *36*, 173—184.

3. Funatsu, K.; Sasaki, S. Computer-Assisted Synthesis Design and Reaction Prediction System AIPHOS. *Tetrahedron Comput. Method.* **1988**, *1*, 27—38.

4. Satoh, K.; Azuma, S.; Satoh, H.; Funatsu, K. Development of the Program for Construction of a Starting Material Library for AIPHOS. *J. Chem. Software*, **1997**, *4*, 101—112.

5. Wipke, W. T.; Dyott, T. M. Use of Ring Assemblies in a Ring Perception Algorithm. *J. Chem. Inf. Comput. Sci.* **1975**, *15*, 140—147.

6. Schmidt, B.; Felischauser, J. Fortran IV Program for Finding the Smallest Set of Smallest Rings of a Graph. *J. Chem. Inf. Comput. Sci.* **1978**, *18*, 204—206.

7. *ACD: Available Chemicals Directory*; MDL Information Systems, Inc.: San Leandro, CA.

8. Susseguth, E. H. Graph-Theoretic Algorithm for Matching Chemical Structures. *J. Chem. Doc.* **1965**, *5*, 36—43.

9. Funatsu, K.; Miyabayashi, N.; Sasaki, S. Further Development of Structure Generation in the Automated Structure Elucidation System. *J. Chem. Inf. Comput. Sci.* **1988**, *28*, 18—28

10. Kimura, Y.; Atarashi, S.; Kawakami, K.; Sato, K.; Hayakawa, I. Fluoro-cyclo-propyl.quinolones. 2. Synthesis and Stereochemical Structure-Activity Relationships of Chiral 7-7-Amino-5-azaspiro[2.4]heptane-5-yl.-1-2-fluorocyclo-propyl.quinolone Antibacterial Agents. *J. Med. Chem.* **1994**, *37*, 3344—3352.

11. Satoh, K.; Imura, A.; Kanai, K.; Miyadera, A.; Yukimoto, Y. An Efficient Synthesis of a Key Intermediate of DU-6859a *via* Asymmetric Microbial Reduction. *Chem. Pharm. Bull.* **1998**, *46*, 587—591.

Reconfigurable Power-Aware Scalable Booth Multiplier

Hanho Lee

School of Information and Communication Engineering, Inha University,
253 Yonghyun-Dong, Nam-Gu, Incheon, Korea
`hhlee@inha.ac.kr`

Abstract. An energy-efficient power-aware design is highly desirable for digital signal processing functions that encounter a wide diversity of operating scenarios in battery-powered intelligent wireless sensor network systems. To address this issue, we present a reconfigurable power-aware scalable Booth multiplier designed to provide low power consumption for DSP applications in highly changing environments. The proposed Booth multiplier makes use of a dynamic-range detection unit, a shared radix-4 Booth encoder, a shared configurable partial product generation unit, 8-bit/16-bit optimized Wallace-trees, a 4-bit array-based adder-tree and a shared final carry-lookahead adder.

1 Introduction

Energy-efficient reconfigurable digital signal processing modules are becoming increasingly important in intelligent wireless sensor networks, where from tens to thousands of battery-operated microsensor nodes are deployed remotely and used to relay sensing data to the end-user [1]-[3]. Given the constantly changing environments of portable devices and the extreme constraints placed on their battery lifetimes, the considerations for power-aware design need to be taken into account [2]. Therefore, it is advantageous to design power-aware signal processing modules with power scalability hooks, such as variable bit-precision and variable digital filters, so that they can be used for a variety of scenarios and the changing operating conditions of each individual sensor node. Multipliers are a very important part of digital filters, and there is a present need for digital filters that can be automatically adapt. In practical wireless sensor network applications, the actual inputs often have a small range although the datapath hardware is designed to accommodate the maximum possible precision. The multiplication bit size is a source of operational range and, thus, large multipliers are power-inefficient under these conditions. For example, performing an 8-bit multiplication on a 16-bit multiplier can lead to serious power inefficiencies, due to the unnecessary signal switching that this causes. Thus, it is highly desirable to use common functional units that can be shared and reused over a range of scenarios.

In this paper, we present a reconfigurable power-aware scalable Booth multiplier, which detects the dynamic range of the input operands and implements 16-bit, 8-bit or 4-bit multiplication operations accordingly. The multiplication mode is determined by the dynamic-range detection unit, which generates and dispatches the control signals used in the pipeline stages. The proposed reconfigurable scalable Booth multiplier proves to be globally 20% more power efficient than a non-scalable Booth multiplier, as well as offering high speed operation due to pipelining.

R. Khosla et al. (Eds.): KES 2005, LNAI 3681, pp. 176–183, 2005.

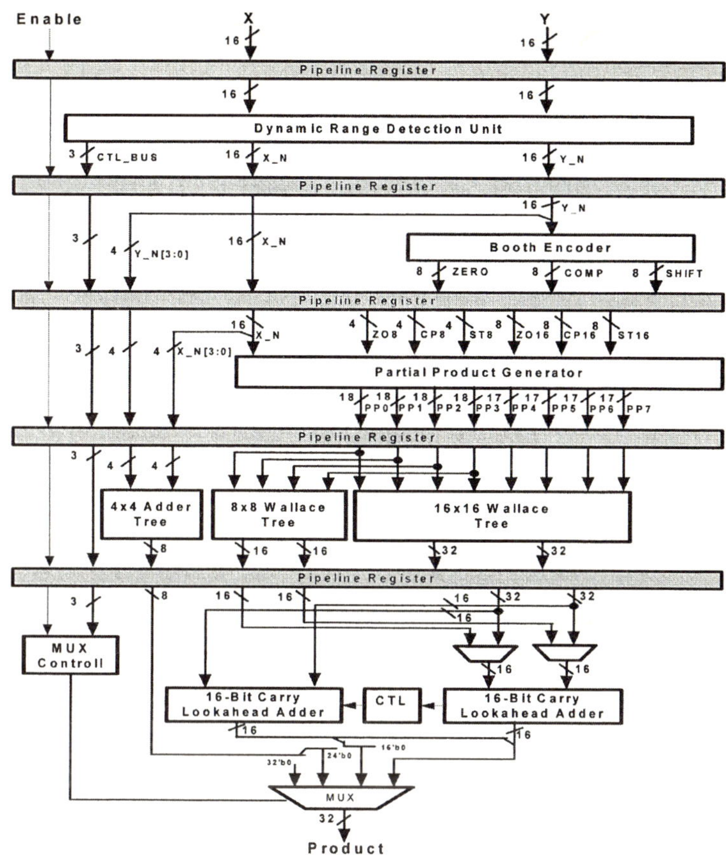

Fig. 1. Reconfigurable power-aware scalable Booth multiplier

2 Reconfigurable Power-Aware Scalable Booth Multiplier

The proposed reconfigurable Booth multiplier consists of a dynamic-range detection unit, a shared radix-4 Booth encoder, a shared configurable partial product generation unit, 16-bit/8-bit optimized Wallace-trees for partial product summation, a 4-bit array-based adder-tree and a shared final carry-lookahead adder, as shown in Fig. 1. The dynamic-range detection unit detects the effective dynamic ranges of the input data, appropriately selects the multiplier and multiplicand operands, and then generates control signals to deactivate the parts of the Wallace-trees and the array-based adder-tree to match the desired run-time data precision. Depending on the number of multiplier bits, the Booth encoder and partial product generator adjust the number of partial products generated, while maintaining the unused partial product generator sections in the static condition. The control and enable signals generated by the dynamic-range detection unit select either the Wallace-tree or the 4-bit array-based adder-tree for a given single multiplication operation. Thus, the components of the unused circuits are able to maintain their static state. In the static state, the previous values are held, so as to avoid any switching from occurring in the unused part of the structure. To take advantage of short precision, signal gating can selectively deactivate those parts of the Wallace-trees and 4-bit array-based adder-tree, which are not currently in use, and make the proposed Booth multiplier behave like a variable size multiplier. In the final stage, the product is generated from the outputs of the active Wallace-tree and carry-lookahead adder. Thus, the Booth encoder, partial product generator and carry-lookahead adder can be shared and reused to process 16 or 8-bit multiplications. For 4-bit multiplications, the Booth encoder and partial product generator units are unused and deactivated. The proposed Booth multiplier is implemented with five pipeline stages, with an input enable signal being used as a power-down switch for each stage.

2.1 Pipeline Gating Techniques

The latched-based clock gating technique used in each of the five pipeline stages enables the multiplier circuit to deactivate the unused part of the logic at runtime, thereby avoiding excessive power-consumption in each multiplication operation. The Synopsys Power Compiler feature was utilized to generate a latch-based clock gating circuit [9]. The top-level pipelined configuration for the multiplier is shown in Fig. 2. Each pipeline stage in the proposed multiplier is optimized in order to minimize the amount of switching that occurs in the logic between the pipeline stages, as well as within the pipeline registers. The enable signal, "ENABLE", is used to indicate the validity of the input operands for the multiplication operation. Thus, the first stage pipeline registers utilize a 16-bit clock gating D-FF with the enable signal used as the control signal for the clock gating circuit. The pipeline registers are divided into two types, namely Type-I and II. The only difference between the Type-I and II pipeline register units is the existence of an extra control unit, which is employed in the Type-II pipeline register units to eliminate any dependencies between the three multiplication modes of the proposed multiplier. The 3-bit control signal, "CTL_BUS", is generated from the dynamic-range detection unit using the dynamic range of the input operands, where the most significant bit is used to indicate a 16-bit operand, the second bit is used for an 8-bit operand and the least significant bit is used to indicate a 4-bit operand. The second Type-II pipeline register stage is partitioned into three configuration states, that is, the most significant 8-bits of the input operand are selected only if the input range is greater than 8-bits, the middle 4-bits are selected if the input range is 16-bits or 8-bits, and finally the least 4-bits are selected if the input range is 16-bits, 8-bits or 4-bits. All of these control signals for each gated pipeline register are generated if the enable signal is asserted high. The third pipeline stage after the Booth encoder utilizes a Type-II pipeline register. As a function of the "CTL_BUS" signal, 2-way partitioning is done to generate the control signals for the eight partial product generator units. The first four partial product generators are split into two modes (16-bit and 8-bit modes), whereas the last four partial product generators are only activated when 16-bit operands are detected. The fourth pipeline stage utilizes a Type-I pipeline register, in which the appropriate partial products and the complement values, "COMP", are distributed to the 16-bit and 8-bit Wallace-tree structures. In the case of the 4-bit unsigned multiplier, the least significant 4-bits of both the multiplier and multiplicand are used. Similarly, the fifth pipeline stage utilizes the Type-I pipeline register before the final addition.

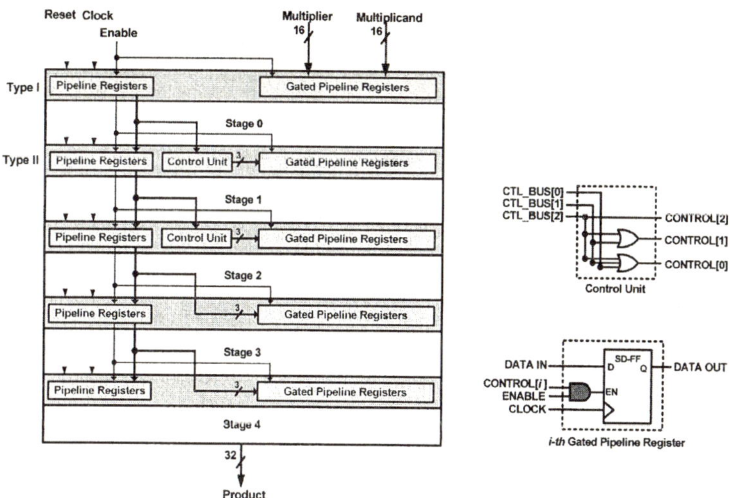

Fig. 2. Top-level pipelined configuration for the multiplier

2.2 Dynamic Range Detection Unit

The dynamic-range detection unit detects the effective dynamic ranges of the input data, and then generates the control signals that are used to deactivate the appropriate parts of the Wallace-trees and array-based adder-tree, so as to match the run-time data precision. In the proposed multiplier, the control signals not only select the data flows, but also manipulate the pipeline register units, in order to maintain the non-effective bits in their previous states and, thus, ensure that the functional units addressed by these data do not consume switching power. Additionally, these control signals are used to control the Wallace-trees and array-based adder-tree.

Fig. 3 shows the functional blocks of the dynamic-range detection unit that includes one-detect circuit (OR gate), comparator, multiplexers and logic gates. The OR gate is used, which asserts the output to be high if any of the inputs are high. Both the 16-bit wide multiplier and the multiplicand input operands are sectioned into three parts, where the detection is done for the 16-bit, 8-bit and 4-bit ranges. The proposed multiplier supports three multiplication modes that are 16-bit, 8-bit and 4-bit multiplication modes. Both the 16-bit and 8-bit multiplication modes are implemented using Booth multipliers with a shared Booth encoder, partial product generator and final adder. Since the 8-bit Booth multiplier is a signed multiplier, the maximum range of the 8-bit multiplication is $7F_{HEX}$. Thus, any input values greater than $7F_{HEX}$ utilize the 16-bit multiplication mode. This problem is avoided in the 4-bit multiplication mode by using a 4-bit unsigned array-based adder-tree. The output of the three one-detect circuits of both the multiplier and multiplicand are grouped into a 3-bit bus, where the most significant bit represents the 16-bit multiplication mode, the second bit represents the 8-bit multiplication mode and the least significant bit represents the 4-bit multiplication mode. The two 3-bit buses are compared using a 3-bit comparator. This comparison detects if the input multiplicand operand is greater than the input multiplier operand. The output of the comparator is used to switch the input multiplicand and multiplier operands.

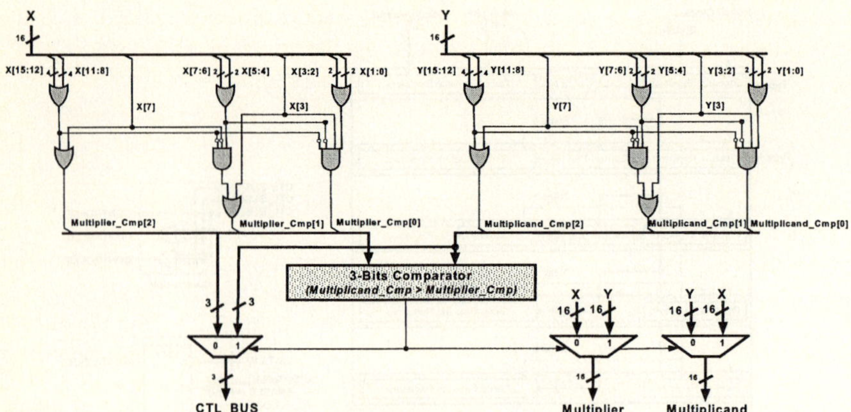

Fig. 3. Dynamic range detection unit

2.3 Booth Encoder and Partial Product Generator

The radix-4 Booth encoder can generate five possible values, viz. -2, -1, 0, 1 and 2 times the input datum [5][6][8]. The Booth encoder is used to generate the control signals, which are used in the partial product generator to determine the appropriate operation (OP_i, $i = 0 \dots 7$). The configurable partial product generator is designed to be configurable so that it can be shared between the 16-bit and 8-bit multiplication modes. The total number of partial products (PP) generated is $N/2$ (where N = max. number of multiplier bits), where $PP_i = OP_i \bullet M_i$ (where $M_i = i^{th}$-bit of the multiplicand, $i = 0 \dots 15$). The 8-bit multiplication mode only requires the first four of the partial products. Thus, in the pipeline stage between the Booth encoder and the partial product generator, all the necessary control signals are generated in order to activate (deactivate) the circuit in use (not in use). There are three types of configuration modes for the partial product, which are consistent with the operation modes of the power-aware multiplier. In the 16-bit multiplication mode, all eight of the partial product (PP_j, where $j = 0 \dots 15$) generator rows are active, while in the 8-bit multiplication mode only the first four of the partial product (PP_j, where $j = 0 \dots 7$) generator rows are active, and the rest of the circuit is in the static condition. Finally, in the 4-bit multiplication mode, all eight of the partial product generator rows are put in the static condition.

2.4 8-Bit/16-Bit Wallace-Trees and Final Carry-Lookahead Adder

The partial product summation is done in an optimized Wallace-tree structure. The Wallace-tree is the only structure in the 16-bit and 8-bit multipliers that is not shared. The 8-bit Wallace-tree has a maximum of five equally weighted inputs, which are the four partial products and the COMP input from the 4th partial product generator. The output of this tree structure is two 16-bit numbers, which are then added together using the final carry-lookahead adder (CLA).

The final CLA is shared between the 16-bit and 8-bit Wallace-trees. Two 16-bit CLAs are used, where the "CTL_BUS" control signal indicating the multiplication mode determines the order of selection of these structures for the final addition. If the 8-bit multiplication mode is utilized, then the right most 16-bit CLA is used and the left most CLA is deactivated. However, if the 16-bit multiplication mode is activated, then both CLA structures are utilized, acting as a single 32-bit CLA structure. The 32-bit output of the 16-bit Wallace-tree is distributed between the two adder structures. The input to the least significant 16-bit CLA structure comes from the output of the multiplexer that chooses the 16-bits from the 8-bit Wallace-tree for the 8-bit multiplication mode and chooses the least 16-bits of the 32-bits from the 16-bit Wallace-tree for the 16-bit multiplication mode. Both CLA circuits keep all of their internal values static during the 4-bit multiplication mode operation. However, in order to fully exploit the power-efficiency of this structure, it is important to join the two CLA structures together using the carry control unit. The carry control unit enables the activation and deactivation of the carryout from the least significant 16-bit CLA structure, as shown in Fig. 4. During the operation of the 16-bit multiplication mode, the carryout from the least significant CLA structure is passed to the carryin of the most significant 16-bit CLA. During the operation of the 8-bit multiplication mode, this carry is disabled, since a carryout generated by the least significant 16-bit CLA can cause the internal signals of the most significant 16-bit CLA to change unnecessarily. When either the input multiplicand or the multiplier operand consists of all zeros, the detection unit disables all of the multiplication modes. Therefore, all of the logic is by-passed during the operation of the multiply-by-zero mode, in order to avoid unnecessary toggling. The final multiplexer makes the final selection, choosing from the 16-bit, 8-bit, 4-bit or the "multiply-by-zero" mode to yield the final product.

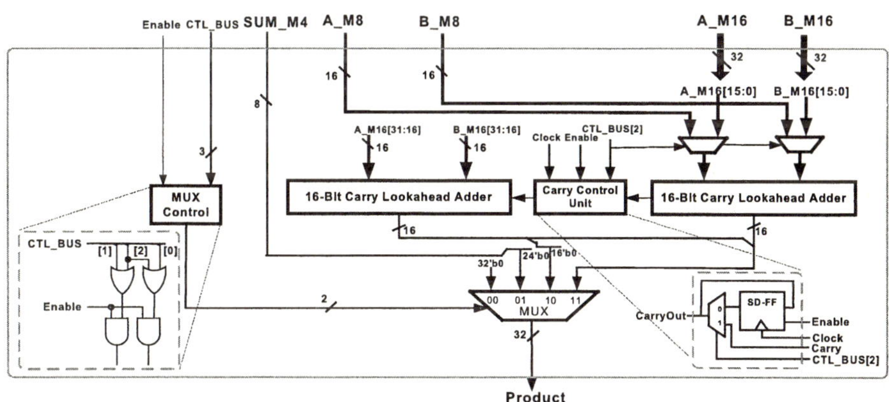

Fig. 4. Final carry-lookahead adder

3   Results

The Verilog HDL simulations for the proposed Booth multiplier were conducted using uniformly distributed random input test vectors with a supply voltage of 1.2V under nominal conditions. Following its functional validation, the architecture was synthesized for appropriate time/area constraints and 0.13-μm CMOS technology using the SYNOPSYS Design Compiler. Table 1 shows a comparison of the hardware complexity and power consumption for different Booth multipliers in the case of the different input precisions (16-bit, 8-bit, 4-bit). The power analysis of the gate level structure was conducted using the Synopsys PrimePower tool. For the 8-bit and 4-bit computations, the proposed reconfigurable scalable Booth multiplier affords reductions in power consumption of 29% and 58%, respectively, in comparison with a non-scalable multiplier. However, for the 16-bit computations, the power consumption of the proposed Booth multiplier is higher than that of the non-scalable Booth multiplier, due to the overhead logic used to render the reconfigurable power-aware. Globally, however, the proposed Booth multiplier is 20% more power-efficient than the non-scalable Booth multiplier, as well as offering high speed operation due to pipelining.

Table 1. Comparison of power consumption for different Booth multipliers (Clock frequency: 100 *MHz*, Latency: 5 clock cycles)

	16-bit Non-scalable multiplier	4, 8, 16-bit Scalable multiplier
Gate count	*5,800*	*7,850*
Power (16-bit)	*900 μW*	*1,035 μW*
Power (8-bit)	*740 μW*	*526 μW*
Power (4-bit)	*665 μW*	*282 μW*
Total	*2,305 μW*	*1,843 μW*

4 Conclusion

In this paper, we propose the reconfigurable power-aware scalable Booth multiplier designed to provide low power consumption in highly changing environments. The proposed Booth multiplier makes use of a dynamic-range detection unit, a shared radix-4 Booth encoder, a shared configurable partial product generation unit, 8-bit/16-bit optimized Wallace-trees, a 4-bit array-based adder-tree and a shared final carry-lookahead adder. In the preliminary experiments, power savings were observed in comparison with the non-scalable Booth multiplier. As a result, the proposed Booth multiplier has improved power-scalability characteristics and, thus, is globally more power efficient than non-scalable Booth multipliers in a variety of scenarios. The proposed multiplier architecture can be applied to other signal processing computations for different applications with a variety of scenarios.

Acknowledgment

This work was supported by Inha University Research Grant. (INHA-32739).

References

1. Wang, A. and Chandrakasan, A. P.: Energy-Efficient DSPs for Wireless Sensor Networks. IEEE Signal Processing Magazine (2002) 68-78.
2. Bhardwaj, M., Min, R., Chandrakasan, A. P.: Quantifying and Enhancing Power-Awareness of VLSI Systems, IEEE Transactions on Very Large Scale Integration (VLSI) Systems (2001), Vol. 9, No. 6, 757-772.

3. Min, R., Bhardwaj, M., Cho, S-H., Ickes, N., Shih, E., Sinha, A., Wang, A., Chandrakasan, A.: Energy-Centric Enabling Technologies for Wireless Sensor Networks, IEEE Wireless Communications (2002), Vol. 9, No. 4, 28-39.
4. Liao, Y. and Roberts, D. B.: A High-Performance and Low-Power 32-bit Multiply-Accumulate Unit With Single-Instruction-Multiple-Data (SIMD) Feature, IEEE Journal of Solid-State Circuits (2002), Vol. 37, No. 7, 926-931.
5. Elguibaly, F.: A Fast Parallel Multiplier-Accumulator using the Modified Booth Algorithm, IEEE Transactions on Circuits and Systems (2000), Vol. 47, No. 9, 902-908.
6. Chen, O., Wang, S., Wu, Y-W.: Minimization of Switching Activities of Partial Products for Designing Low-Power Multipliers, IEEE Transactions on Very Large Scale Integration (VLSI) Systems (2003), Vol. 11, No. 3, 418-433.
7. Jou, S-J., Tsai, M-H., and Tsao, Y-L.: Low-Error Reduced-Width Multipliers for DSP Applications, IEEE Transactions on Circuits and Systems-I (2003), Vol. 50, No. 11, 1470-1474.
8. Wang, A., and Chandrakasan, A. P.: Energy-Aware Architectures for a Real-Valued FFT Implemenattion, Proceedings of ISLPED (2003), 360-365.
9. Emnett, F., and Biegel, M., Power Reduction Through RTL Clock Gating, SNUG, San Jose (2000).

Base Reference Analytical Hierarchy Process for Engineering Process Selection

Elina Hotman

Knowledge-based Systems Group
Department of Computer Science
University of Paderborn, 33098 Paderborn, Germany
elina@uni-paderborn.de

Abstract. Recent developments in material and manufacturing technology have brought many exciting changes in the field of engineering. Engineers can benefit from wider selection of materials and processing techniques. The Analytical Hierarchy Process (AHP) is a multi attribute decision method (MADM) that utilizes structured pairwise comparisons. This method has been used widely in industry to aid selection process. The drawback of AHP is that the exhaustive pairwise comparison is tiresome and time consuming when there are many alternatives to be considered. In this paper, we propose a new approach to improve this limitation of AHP, the so-called Base Reference Analytical Hierarchy Process (BR-AHP). This study is accompanied by a case study to show the application of BR-AHP in engineering process selection.

1   Introduction

The Analytical Hierarchy Process (AHP) [4] is commonly used in engineering and industry to aid in selection problems. The AHP has been applied to a wide variety of engineering selection problems [1, 2, 7], including machine tool selection, plant selection, project selection, site selection and many other selection problems.

The method involves the comparison of several candidate alternatives, which are pairwise compared using several different criteria. Pairwise comparisons can be tiresome and time consuming, especially if there are many pairs. Methods for reducing the number of pairwise comparisons have been proposed. One of the ideas is to use hierarchical decomposition [5]. A large number of criteria may be collected into smaller group. Each group is compared, and then the groups themselves are compared to each other. However, this method only works for reducing pairwise comparison among the criteria and it does not work for reducing the number of pairwise comparisons of the alternatives. Another approach is to use the duality approach, which has been proposed by Triantaphyllou [6], to reduce the total number of pairwise comparisons when the number of alternatives is larger than the number of criteria plus one. However, with this approach the total number of pairwise comparisons is still quadratic with the number of alternatives.

In engineering, most of the engineers are mostly familiar only with one technique and they work by comparing their own technique with other techniques.

R. Khosla et al. (Eds.): KES 2005, LNAI 3681, pp. 184–190, 2005.
© Springer-Verlag Berlin Heidelberg 2005

If, for example, they want to compare two techniques they are not familiar with, first they will compare the first technique with their own technique, then the second technique to their own. And afterwards they can deduce the comparison of the two techniques. So they would usually have a base that they refer to, i.e. their own technique. A new method, the so-called Base Reference Analytical Hierarchy Process (BR-AHP) draws upon this idea.

In this paper, a modified AHP, called BR-AHP, is proposed to reduce the number of pairwise comparison of the alternatives. The rest of this paper is organized as follows. The next section gives a detailed description of the proposed BR-AHP. The third section presents the application of BR-AHP in engineering process selection. The final section is the conclusions.

2   Base Reference Analytical Hierarchy Process

The Base Reference Analytical Hierarchy Process (BR-AHP) is proposed to overcome the limitation of AHP when there are a large number of pairwise comparisons of the alternatives. The following subsections provide more detailed descriptions about BR-AHP.

2.1   BR-AHP Algorithm

The BR-AHP algorithm consists of four steps:

1. The construction of the hierarchy structure of the selection problem.
2. The evaluation of the relative importance of the criteria using pairwise comparison method.
3. The selection of the base alternative and the evaluation of the base alternative relative to each other alternatives on the basis of each selection criterion using base pairwise comparison method.
4. The integration of the ratings derived in steps 2 and 3 to obtain an overall relative ranking for each potential alternative.

In BR-AHP most procedures follow the original architecture of AHP. The difference is only in step 3 where it uses a base pairwise comparison method. By using the pairwise comparison method it needs to estimate $n(n-1)/2$ judgements for n alternatives, but with the base pairwise comparison method it only needs $n-1$ judgements.

We assume that there are m criteria and n alternatives. In the AHP, for each criteria it needs to determine $n(n-1)/2$ pairwise comparisons. So for m criteria it needs to determine $mn(n-1)/2$ pairwise comparisons. Total pairwise comparisons for the goal are $m(m-1)/2$ pairwise comparisons. Total pairwise comparisons for the conventional AHP are $mn(n-1)/2 + m(m-1)/2 = m/2\left(n(n-1) + (m-1)\right) = m/2\left(n^2 - n + m - 1\right).$

In the BR-AHP, for each criteria it needs only to determine $n-1$ pairwise comparisons. So for m criteria it needs to determine $m(n-1)$ pairwise comparisons. The total pairwise comparisons for the goal are $m(m-1)/2$ pairwise comparisons. Total pairwise comparisons for the BR-AHP are $m(n-1) + m(m-1)/2 = m/2\left(2(n-1) + (m-1)\right) = m/2\left(2n+m-3\right)$.

The different of total pairwise comparisons of AHP and BR-AHP is $m/2\left(n^2 - n + m - 1\right) - m/2\left(2n + m - 3\right) = m/2\left(n-1\right)\left(n-2\right)$.

2.2   Pairwise Comparison Method

We can construct the pairwise matrix between goal and criteria layer, which is shown in Eq. 1. In the same way, the matrices of criteria layer vs. sub-criteria layer can be considered.

$$A = \begin{bmatrix} \frac{w_1}{w_1} & \frac{w_1}{w_2} & \cdots & \frac{w_1}{w_n} \\ \frac{w_2}{w_1} & \frac{w_2}{w_2} & \cdots & \frac{w_2}{w_n} \\ \vdots & \vdots & \ddots & \vdots \\ \frac{w_n}{w_1} & \frac{w_n}{w_2} & \cdots & \frac{w_n}{w_n} \end{bmatrix} = \begin{bmatrix} 1 & a_{12} & \cdots & a_{1n} \\ 1/a_{12} & 1 & \cdots & a_{2n} \\ \vdots & \vdots & \ddots & \vdots \\ 1/a_{1n} & 1/a_{2n} & \cdots & 1 \end{bmatrix} \tag{1}$$

where A is the pairwise comparison matrix (size $n \times n$); $\frac{w_i}{w_j}$ represents the relative importance of the i-th criteria over the j-th criteria ($i, j \in 1, 2, \ldots, n$). In general the value of $\frac{w_i}{w_j}$ is given by a decision maker subjectively. There are $n(n-1)/2$ judgements which are required to develop the set of matrices. Reciprocals are automatically assigned in each pairwise comparison shown in Eq. 1.

2.3   Base Pairwise Comparison Method

In the construction of the pairwise matrix we collect much more information than we need. In fact, in order to fill the right upper corner of a pairwise matrix, in a problem with n alternatives, the decision maker has to carry out $n(n-1)/2$ basic comparisons, whereas $n-1$ properly chosen judgements should be sufficient. However this major amount of information could be used in the case where we need to manage base pairwise comparison matrix. The base pairwise matrix comparison between criteria layer and alternative layer is thus defined:

$$B = \begin{bmatrix} \frac{w_1}{w_1} & \frac{w_1}{w_2} & \cdots & \frac{w_1}{w_k} & \cdots & \frac{w_1}{w_n} \\ \frac{w_2}{w_1} & \frac{w_2}{w_2} & \cdots & \frac{w_2}{w_k} & \cdots & \frac{w_2}{w_n} \\ \vdots & \vdots & \ddots & \vdots & & \vdots \\ \frac{w_k}{w_1} & \frac{w_k}{w_2} & \cdots & \frac{w_k}{w_k} & \cdots & \frac{w_k}{w_n} \\ \vdots & \vdots & & \vdots & \ddots & \vdots \\ \frac{w_n}{w_1} & \frac{w_n}{w_2} & \cdots & \frac{w_n}{w_k} & \cdots & \frac{w_n}{w_n} \end{bmatrix} = \begin{bmatrix} 1 & b_{12} & \cdots & b_{1k} & \cdots & b_{1n} \\ b_{21} & 1 & \cdots & b_{2k} & \cdots & b_{2n} \\ \vdots & \vdots & \ddots & \vdots & & \vdots \\ b_{k1} & b_{k2} & \cdots & 1 & \cdots & b_{kn} \\ \vdots & \vdots & & \vdots & \ddots & \vdots \\ b_{n1} & b_{n2} & \cdots & b_{nk} & \cdots & 1 \end{bmatrix} \tag{2}$$

where B is the pairwise comparison matrix (size $n \times n$); k denotes the base; $\frac{w_k}{w_j}$ represents the relative importance of the k-th alternative over the j-th alternative ($k, j \in 1, 2, \ldots, n$).

In this case, a decision maker only needs to determine the base alternative and then compare the base alternative with other alternatives (b_{kj}). Next, we have to evaluate the missing values using the information derived from the available base elements of the base pairwise matrix. Here, the value of the base elements of the base pairwise matrix represents the ratio between the weights of two different alternatives in comparison with a defined criteria. For example b_{kj} is the ratio between the weight of the k-th alternative (w_k) and the weight of the j-th alternative ($w_j; j \in 1, \ldots, n; j \neq k$) in comparison with a common overall criteria. As a consequence we can write: $b_{kj} = w_k/w_j$.

We can automatically calculated the missed element b_{ij} by using information derived from the base elements without making any assumption on the value of the missed element. For example, if we do not have any evaluation for b_{ij}, but we do have an evaluation for b_{ki} and b_{kj}, then:

$$b_{ij} = w_i/w_j = \frac{w_k/w_j}{w_k/w_i} = \frac{b_{kj}}{b_{ki}} \tag{3}$$

In using Eq. 3 the reciprocals of base elements are also automatically assigned, here shown in Eq. 4.

$$b_{ik} = \frac{b_{kk}}{b_{ki}} = \frac{1}{b_{ki}} \tag{4}$$

The judgement scale used here is the nine-point scale, which was proposed by Saaty [4]. By using Saaty's nine-point scale the range of judgement is from $\frac{1}{9}$ (extremely less important) to 9 (extremely more important). To ensure the consistency with the nine-point scale, we have to modify the automatic value assignment in base pairwise matrix. To avoid the value higher than 9, we use the min operator to choose the lower one. And to avoid the value lower than $\frac{1}{9}$, we use the max operator to choose the higher one. Then, the modified base pairwise comparison matrix is expressed as follows:

$$B = [b_{ij}], \quad b_{ij} > 0, \quad i,j = 1, 2, \ldots, n$$
$$b_{ij} = \begin{cases} 1 & i = j \\ \min\left\{9, \frac{b_{kj}}{b_{ki}}\right\} & \frac{b_{kj}}{b_{ki}} > 1, i \neq j \\ \max\left\{\frac{1}{9}, \frac{b_{kj}}{b_{ki}}\right\} & \frac{b_{kj}}{b_{ki}} < 1, i \neq j \end{cases} \tag{5}$$

3 A Case Study of BR-AHP in Engineering Process Selection

To give an illustration of applying BR-AHP to an engineering process selection, we shall use it to select the best microencapsulation method.

Microencapsulation is the name given to a novel technique for the preparation of small substances, which began to develop about 50 years ago. Many different techniques have been proposed for the production of microcapsules by academics and industrial researchers. Nowadays more than 1000 methods can be identified in the patent literature [3]. In this case study we will only consider the case of selecting the best microencapsulation technique from six alternative methods.

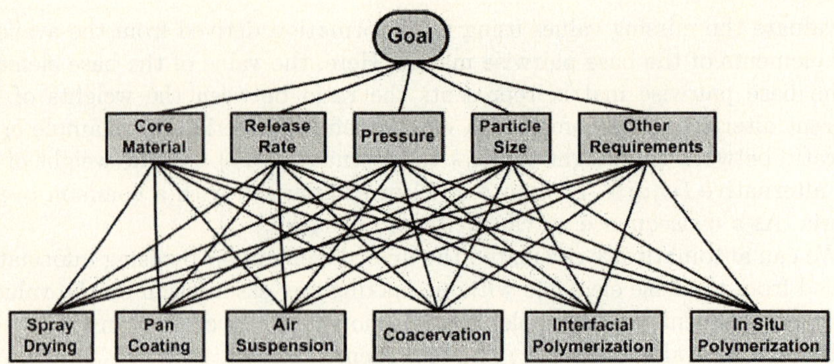

Fig. 1. Microencapsulation Process Selection Hierarchy.

The first step in BR-AHP is to construct the hierarchy structure of microencapsulation selection problem as shown in Fig. 1. Having defined the selection criteria, the next step in BR-AHP is the pairwise comparison of the importance of the microencapsulation criteria. This is done by assigning a weight to each of the criteria. The results of this operation are presented in Fig. 2. Here we have a comparison of 5 criteria, so the decision maker needs to determine 10 pairwise comparisons.

Goal	CoreMaterial	ReleaseRate	Pressure	ParticleSize	OtherRequire...	Weight
CoreMaterial	1.0	2.0	3.0	4.0	5.0	42%
ReleaseRate	0.5	1.0	2.0	3.0	4.0	26%
Pressure	0.33333	0.5	1.0	2.0	3.0	16%
ParticleSize	0.25	0.33333	0.5	1.0	2.0	10%
OtherRequirements	0.2	0.25	0.33333	0.5	1.0	6%

Fig. 2. Pairwise Comparison of Microencapsulation Criteria.

CoreMaterial	SprayDrying	PanCoating	AirSuspension	Coacervation	InterfacialPol...	InSituPolyme...	Weight
SprayDrying	1.0	1.0	0.5	0.25	1.0	1.0	10%
PanCoating	1.0	1.0	0.5	0.33333	1.0	2.0	12%
AirSuspension	2.0	2.0	1.0	0.5	2.0	3.0	21%
Coacervation	4.0	3.0	2.0	1.0	4.0	5.0	39%
InterfacialPolymeriz...	1.0	1.0	0.5	0.25	1.0	1.0	10%
InSituPolymerization	1.0	0.5	0.33333	0.2	1.0	1.0	8%

Fig. 3. Base Pairwise Comparison of Microencapsulation Alternatives.

The next step is the base pairwise comparison of the microencapsulation alternatives. Figure 3 shows this operation. First, the decision maker had to select the base alternative. In Fig. 3, the base alternative is highlighted horizontally. In this case study, the decision maker selected an alternative 'Coacervation' as the base alternative. In this step, the decision maker only needed to determine

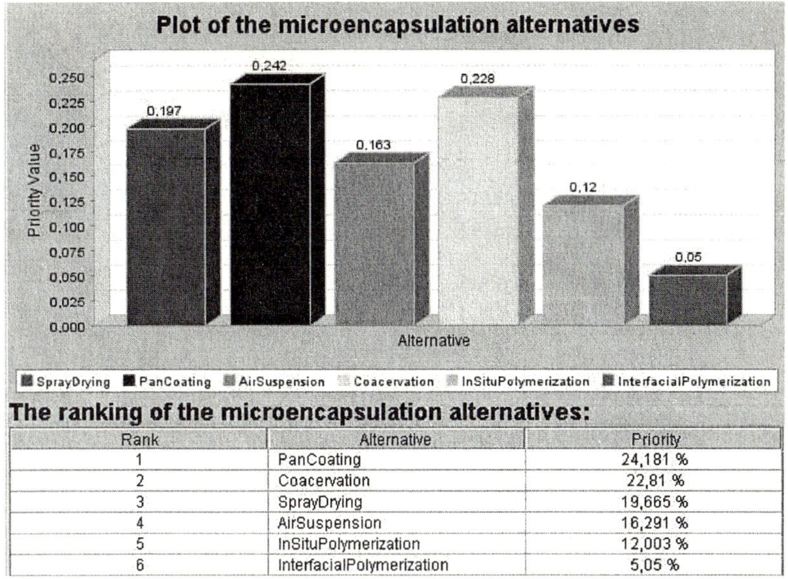

The ranking of the microencapsulation alternatives:

Rank	Alternative	Priority
1	PanCoating	24,181 %
2	Coacervation	22,81 %
3	SprayDrying	19,665 %
4	AirSuspension	16,291 %
5	InSituPolymerization	12,003 %
6	InterfacialPolymerization	5,05 %

Fig. 4. Microencapsulation Process Selection Result.

5 judgements by comparing the base alternative 'Coacervation' with the other five alternatives.

The final step in BR-AHP is to combine all the weights derived in the previous steps to obtain the overall ranking for the alternatives. The result of the ranking of microencapsulation alternatives is shown in Fig. 4.

For the case with 5 criteria and 6 alternatives, by using BR-AHP we only need 35 pairwise comparisons. If we use AHP, then we need 85 pairwise comparisons. In this case, we can reduce 50 pairwise comparisons by using BR-AHP. In many situations, in the engineering selection we have hundreds or even thousands alternatives. Here then we can reduce pairwise comparisons even more by using BR-AHP. Table 1 shows the total pairwise comparisons needed for both methods and the reduced number of pairwise comparisons by using BR-AHP in the case of m criteria and n alternatives. Δ is the total number of reduced pairwise comparisons by using BR-AHP. From Table 1, we see that when the number of alternatives grows the BR-AHP reduces more (Δ becomes larger).

4   Conclusions

The selection process is a very important issue in engineering field but is extremely complex since it usually involves many criteria and alternatives. The AHP method makes the selection process transparent, but it can be tiresome and time consuming if there are many alternatives to consider. This work presents a BR-AHP algorithm, which enhances the AHP by using base pairwise comparison method. The BR-AHP method draws upon ideas from engineering, where

Table 1. Number of Pairwise Comparisons by AHP and BR-AHP.

m	5			10		
n	AHP	BR-AHP	Δ	AHP	BR-AHP	Δ
2	15	15	0	55	55	0
3	25	20	5	75	65	10
4	40	25	15	105	75	30
5	60	30	30	145	85	60
6	85	35	50	195	95	100
7	115	40	75	255	105	150
8	150	45	105	325	115	210
9	190	50	140	405	125	280
10	235	55	180	495	135	360
100	24760	505	24255	49545	1035	48510
1000	2497510	5005	2492505	4995045	10035	4985010

most of the engineers are only familiar with one technique and they are used to comparing the other techniques with their own technique. A prototype has been developed to show the effectiveness of BR-AHP algorithm. The results demonstrate that BR-AHP method enhances AHP method by reducing the number of pairwise comparison to be performed.

Acknowledgement

The author would like to thank the two anonymous reviewers for their valuable comments. This work was supported by the International Graduate School of Dynamic Intelligent Systems, University of Paderborn.

References

1. Akash, B.A., Mamlook, R., Mohsen, M.S.: Multi-criteria selection of electric power plants using analytical hierarchy process. Electric Power Systems Research 52 (1999) 29 - 35
2. Golden, B.L., Wasil, E.A., Harker, P.T. (Eds.): The Analytic Hierarchy Process - Applications and Studies. Springer-Verlag (1989)
3. Gouin, S.: Microencapsulation - industrial appraisal of existing technologies and trends. Trends in Food Science & Technology, 15, 330 - 347 (2004)
4. Saaty, T. L.: The Analytic Hierarchy Process. McGraw-Hill (1980)
5. Scott, M. J.: Quantifying Certainty in Design Decisions - Examining AHP. Proceeding of DETC 2002 ASME Design Engineering Technical Conferences (2002)
6. Triantaphyllou, E.: Reduction of Pairwise Comparisons in Decision Making via Duality Approach. Journal of Multi-Criteria Decision Analysis 8 (1999) 299 - 310
7. Yurdakul, M.: AHP as a strategic decision-making tool to justify machine tool selection. Journal of Materials Processing Technology 146 (2004) 365 - 376

A Multiagent Model
for Intelligent Distributed Control Systems

José Aguilar, Mariela Cerrada, Gloria Mousalli,
Franklin Rivas, and Francisco Hidrobo

CEMISID, Dpto. de Computación, Facultad de Ingeniería,
Av. Tulio Febres. Universidad de los Andes, Mérida 5101, Venezuela
{aguilar,cerradam,mousalli,rivas}@ula.ve

Abstract. One approach for automating industrial processes is a Distributed Control System (DCS) based on a hierarchical architecture. In this hierarchy, each level is characterized by a group of different tasks to be carried out within the control system and by using and generating different information. In this article, a reference model for the development of Intelligent Distributed Control Systems based on Agents (IDCSA) inspired by this hierarchical architecture is presented. The agents of this model are classified in two categories: control agents and service management environment agents. To define these agents we have used an extension of the MAS-Common KADS methodology for Multi-Agent Systems (MAS).

1 Introduction

Presently, there is great interest in the development of integrated automation systems that permit monitoring different plant operation variables in a broad and dynamic way and to transform such variables in control commands that are later integrated into the plant through actuators. Moreover, they should take into account production and economic criteria that can be applied as control commands or as part of a plant programming function. The integrated structure should permit the flow of information at all levels (management, operation, etc.) concerning the plant, the products obtained, and all the relevant information. On the other and, a MAS can be defined as a network of "problem- solvers" that work together to solve problems that could not be worked out individually [1]. The main preoccupation of the MAS is the coordination between the groups of autonomous agents, perhaps intelligent, to coordinate their objectives, skills, knowledge, tasks, etc.

In this work, a reference model for Intelligent Distributed Control System (IDCS) based on Agents is proposed. Our model will be made up of entities called agents, that work together dynamically to satisfy the control systems local and global objectives and whose design can be made completely independent of the system could be developed. The description of the agents in this reference model is based on the MASINA methodology [5] which has an extension of MAS-Common KAD methodology to incorporate other characteristics of agents such as emerging behavior, the reasoning, and the possibility of using intelligent techniques (expert systems, artificial neuronal networks, genetic algorithms, fuzzy logic, etc.) for carrying out their jobs.

R. Khosla et al. (Eds.): KES 2005, LNAI 3681, pp. 191–197, 2005.

2 Reference Model for Intelligent Distributed Control System Based on Agents (IDCSA)

The control jobs and information management needed in automation processes can be distributed and expressed through a hierarchical logical structure. Figure 1 shows hierarchical reference architecture to develop a Distributed Control System (SCD), that permits the automation of an industrial plant [3].

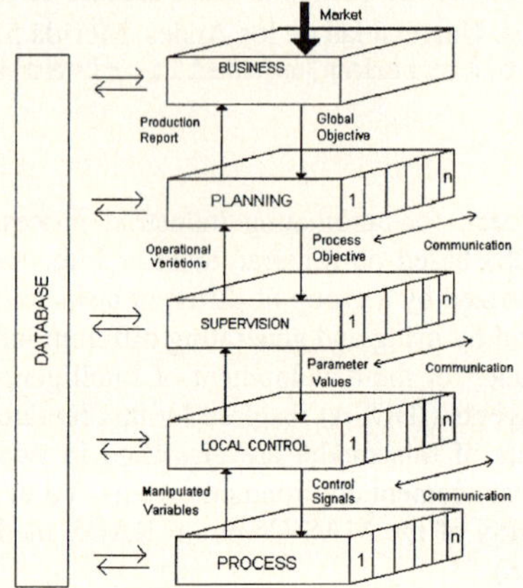

Fig. 1. Reference Model for Distributed Control Systems

At the business and planning levels decisions are made at the managerial level, and the control process jobs are carried out at the lowest levels. The Supervisory level adjusts the parameters of the controllers, the control signal is obtained at the Local Control level, to later be incorporated at the plant at the Process level. This SCD model is complemented with a group of agents in each one of the levels of the hierarchy, this is the IDCSA model shown in Figure 2. These agents carry out diverse jobs looking towards reaching the specific control objective.

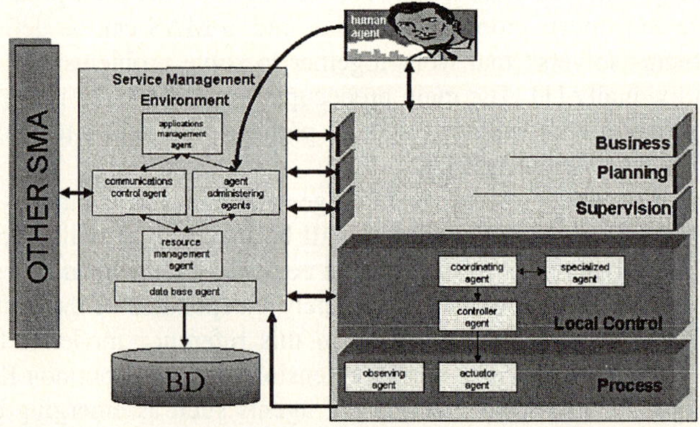

Fig. 2. IDCSA Reference Model

The incorporation of the MAS to the reference model permits the control to emerge from the interactions of those entities (agents). The IDCSA is seen as a network of autonomous agents, with different responsibilities according to the level to which it belongs. As such, the details of each agent (objectives, communication, jobs, intelligence, etc.) come according to the levels to which the given agents belong in the IDCSA model. This way, the agents are distributed through the control hierarchy and can be geographically dispersed. The dynamic interaction of the multiple agents happens between levels and in the interior of the levels. In the IDCSA model two categories of agents exist:

Control Agents: Carry out the jobs of control, measurement, control decision making, and putting the decisions into practice, among others. The agents are:

- Coordinating Agent: make decisions and plan control jobs
- Specialized Agent: carry out specific jobs that serve to support the coordinating agent
- Controller Agent: obtains control action
- Actuator Agent: executes the control action
- Observation Agent: measures and processes the variables of the plant
- Human Agent: supervises the SCIDA

Agents of the Service Management Environment (SME): It is the base of the distribution system given that it manages the communication of the IDCSA model and permit the distribution of the control system and heterogeneity among geographically dispersed agents. The agents of the SME are [4]:

- Agent Administering Agents: coordinates the multi-agent society
- Resource Management Agent: administers the use of the resources.
- Applications Management Agent: administer the use of applications (software) of the system
- Data Base Agent: manages the information storage means.
- Communication Control Agent: establishes communication with other MAS.

3 Specifications of the IDCSA Reference Model

The models proposed in the MASINA methodology have been used according to the needs of IDCSA [5].

3.1 Organization Model

The main objective of the organization model is to specify the environment in which IDCSA will be developed. In this stage of modeling, a human organization is analyzed to determine the areas susceptible to the introduction of the IDCSA model. The model is conceived in three stages:

- Modeling of the organization environment: Describes the environment where we are going to introduce the IDCSA model.
- Evaluating the organization environment: The viability of introducing the IDCSA model to the organization is evaluated.

- Model of the Multi-Agent Society: In this stage, the following activities are developed:
  - Identify the IDCSA levels in the organization: some applications can consider the levels proposed in the IDCSA; others only require a group of them [2].
  - Agent's proposal at each IDCSA level: after identifying the levels needed, the group of IDCSA agents adapted to the functions that are carried out at each level is defined.
  - Identify the environment objects: in each one of the levels, the surrounding objects with which agents will interact to achieve their objective should be identified.

3.2 Agent Model

The agent model serves to specify the generic characteristics of the IDCSA agents. The agent model proposed is oriented towards services. Agents are capable of carrying out jobs which are offering to other agents (called services). An agent will be selected, the Observation Agent, to discuss the templates of the Agent Model (to see the remaining templates, refer to [6]).

1. General Information: i) Name: Observation Agent, ii) Type: Software Agent and Physical Agent, iii) Paper: Environmental, iv) Position: Lowest level agent of the Multi-agent System, v) Description: The Observation Agent makes up the data acquiring system: the sensor, the conditioning system and the transmitter. Through them changes in the process can be detected.
2. Agent Objectives: i) Name: Process Variable Measurement, ii) Type: Persistent, direct objective, iii) Entry Parameters: Variables that need to be measured, repository where data is stored, iv) Exit Parameters: Measured variable, v) Activation Condition: Order of measurement (non-periodic case) or period of time available for measurement, vi) End Condition: Measured variable, vii) Representation Language: Natural language, vii) Ontology: Control ontology, viii) Description: is associated with the capture of process variable values through carrying out direct or indirect measurements on given variables. The obtained signals should be filtered to eliminate measurement errors, and transported and stored in systems.
3. Agent Services: Process measurement variables, Conversion of variables, and Detection of Quick Changes
4 Agent Capabilities: contact with the process, make available to the rest of the MAS changes occurring in the environment

3.3 Job Model

The elements of this model are ingredients, capacity, environment and method. A job has associated entry ingredients that permit the job to be executed and at the same time, can produce exit ingredients as a product of the execution job. Jobs can be made using a specific technique (classic, intelligent) or hybrid techniques. Below a list of IDCSA jobs is presented. In [6] the use of each one of the agent's cases is presented, in which the jobs that are carried out is inferred.

1. Measurement Jobs: i) Measure: Sense and condition, ii) Process the variable
2. Control Jobs: i) Decision-making, ii) Obtain control action, iii) Process control action, iv) Execute control action
3. Specialized Jobs: i) Process specialized agent
4. Information Management Job: i) Information storage, ii) Information up-date, iii) Look for information, iv) Management of the BD
5. Location Jobs: i) Locate resources, agents or applications, ii) Assign agent/resource/application

The Observation agent will continue to be utilized to describe the templates that specify the jobs (to see the remaining refer to [6]).

Sensing: i) Objective: Process variables are registered in a device sensor, ii) Precondition: The sensor is an element of the lowest level; the only thing that is required is to have an associated process variable, iii) Control Structure: Sensing does not have sub-jobs but rather belongs to a group of jobs that correspond to measuring agents, iv) Execution Time: Continue, v) Frequency: Absolute frequency (constant), vi) Content: The variable is a signal coming from the process, which can be measured and is associated with the state of the process, vii) Surrounding: Process level, viii) Regulations: the sensor can be calibrated, ix) Method: Classic techniques for capturing data are used.

3.4 Intelligence Model

This model is implemented in those intelligent agents which have reasoning capabilities to decide about the jobs they are going to carry out to solve a situation. For that, the agent can use previous experience or accumulated knowledge through the learning process. According to agent jobs and services defined by IDCSA, their intelligent agents are: Coordinating Agents, Specialized Agents and Controller Agents. The intelligent model for these agents is generically defined around three elements:

- Experience: i) Representation: Rules, ii) Type: Based on cases
- Learning Mechanism: i) Type: Adaptive, ii) Representation Technique: Rules, iii) Learning Source: Process, exit and or failure conditions occurring in the proposal of a control action, proposal of new control and decision-making parameter values in coordination jobs, iv) Up-Dating: Feedback (using previous experiences for up-dating knowledge)
- Reasoning Mechanisms: i) Information source: Previous results obtained for system control agents, ii) Activation Source: control and coordination jobs, iii) Type of Inference: Based on rules, iv) Strategic Reasoning: Coordinate strategies for the proposed selection of appropriate control algorithms and for decision-making in the supervisory and planning level. Confront unknown situations to enrich the experience.

3.5 Coordination Model

In general, this model describes the coordinating mechanisms between agents. Basically, this model describes the way activities are organized in the IDCSA to reach the objectives. It deals with the planning process and conflict resolution mechanisms that occur between agents. The model is complemented with a description of protocols, ontologies, and communication mechanisms (direct and indirect). In the IDCSA, only predefined planning mechanisms are used. The proposed coordinating model is oriented to services. A service can have associated specific properties (cost, duration, etc). For the IDCSA case, conversations generated from the predefined planning process are: i) Obtaining control action, ii) Manage level objectives, iii) Sound alarm, iv) Coordinate humans. Below, the first conservation alone is described:

1 When communication exists with the coordinator, the control agent is only in charge of generating the control action (the associated interactions are shown in Table 1)

2 When no communication exists with the coordinator, the control agent has the possibility of auto-regulating itself. In this case, the control agent, making use of the intelligent model, can identify parameters of the current situation and attempt to define the parameters that remedy said situation.

Table 1. Interactions of the Conversation "Obtaining Control Action"

Interactions	Services
Request: Control-Data Base	Search for information
Transmit: Control-Data Base	Store decisions
Transmit: Control-Coordinator-Inferior Levels	Coordinate goals
Transmit: Current Control (level of local control)	Conversion of Control Signals to physical values

3.6 Communication Model

This model describes each interaction between agents (speech acts). In the case of IDCSA, a predefined planning process was used, which determines direct communication carried on between agents. Each interaction between two agents is carried out through sending a message, and has a speech act associated with it (see table 1).

4 Conclusions

This work proposes a reference model for distributed control systems based on MAS whose architecture is inspired in a hierarchical scheme. The use of MAS incorporates collaborative, organizational and intelligent aspects that permit the control system to have emerging behavior. The use of MASINA methodology permitted the specification of the IDCSA model. The IDCSA model allows model distributed control system in a way that we can include intelligent components in our system. Two main aspects can be considered in our model: we can distribute the different control tasks among the control agents of IDCSA, and the intelligence is part of the control agents.

References

1. Wooldridge M., Jennings N.: Intelligent Agents: Theory and Practice. Knowledge Engineering Review (1994).

2. Aguilar J., Cerrada M., Hidrobo F., Mousalli G., Rivas F., Díaz, A: Application of the Agents Reference Model for Intelligent Distributed Control Systems. Advances in Systems Sciences Measurement, (Ed. N. Matorakis, L. Pecorelli), WSES Electrical and Computer Engineering Series (2001) 204-210.
3. Franek B., Gaspar C.: SMI++ object Oriented Framework for Designing and Implementing Distributed Control Systems. IEEE Trans. on Nuclear Science, Vol. 45 (1998), 1946-1950.
4. Haring G., Kotsis G., Puliafito A., Tomarchio O.: A Transparent Architecture for Agent Based Resource Management. Proc. Of IEEE Intl. Conf. On Intelligent Engineering (1998) 338-342.
5. Hidrobo F., Aguilar J.: Resources Management System For Distributed Platforms Based on Multi-Agent Systems. Proc. of 5th International Conference on Artificial Intelligence, Simulation and Planning-AIS' 2000 (2000) 60-69.
6. Aguilar J., Cerrada M., Hidrobo F., Mousalli G., Rivas F., Díaz, A: Especificaciones de los Agentes del SCDIA usando MASINA. Technical Report # 23-2004, FONACIT, Ministerio de Ciencia y Tecnologia, (2004).

Study on the Development of Design Rationale Management System for Chemical Process Safety

Yukiyasu Shimada[1], Takashi Hamaguchi[2], and Tetsuo Fuchino[3]

[1] Chemical Safety Research Group, National Institute of Industrial Safety,
Tokyo, 204-0024, Japan
shimada@anken.go.jp
[2] Department of Engineering Physics, Nagoya Institute of Technology,
Nagoya 466-8555, Japan
hamachan@nitech.ac.jp
[3] Department of Chemical Engineering, Tokyo Institute of Technology,
Tokyo 152-8552, Japan
fuchino@chemeng.titech.ac.jp

Abstract. Chemical process design is one of the most important activities in the plant life cycle (PLC). To manage the safety of process plant through the PLC, the system environment to enable recording and accessing design rationale (DR) of the current process and/or plant is indispensable. Theoretical DR information which is discussed on process design activity is very important for safety management activities. However, conventional process risk assessment and safety design mainly depends on engineer's experience, know-how, sense of value etc, and it has not been recorded explicitly and could not be shared for other PLC activities. This paper proposes a DR management model (DR model) for the chemical process design and a simplified DR search system for managing the process design information.

1 Introduction

Process structure is decided so as to maintain the consistency with process behavior including material property and operation based on the result of risk assessment. Information on technical and theoretical thinking process in the process design affects the safety of subsequent plant life cycle (PLC) activities such as operation and maintenance. Previously, such information has not been recorded explicitly and could not be shared for other PLC activities. This will entail the problem in which other engineer, plant operator or maintenance person cannot understand the technical background and the thinking process behind the design, even if they can interpret the design specifications. For safer plant management, it is important to understand not only design results but also designer's thinking process as well as design intents and constraints. The reasoning embedded in the design can be explicitly expressed as *design rationale* (DR). In order to accomplish integrated process safety environment through the PLC, it is necessary to manage the design information, including the DR effectively.

This paper proposes a DR management model to enable link with the DR and the design result information. For supporting the safer PLC activities, it is important to reuse the information on safety management technique, including the process DR information. A simplified search system to get the DR information on the background knowledge of process design is proposed.

R. Khosla et al. (Eds.): KES 2005, LNAI 3681, pp. 198–204, 2005.

2 Conventional Approach of Design Rationale Management

Researches related to the DR have been very active during past decade in many engineering disciplines [1]. The Issue-based Information System (IBIS) uses an *issue-based* approach that has been in architectural design, city planning and public-policy discussion. The Question, Option and Criteria (QOC) is one semi-formal notation of *design space analysis*, which focuses on the three basic concepts indicated in its name. The QOC representation emphasizes the systematic development of a space of design options structured by questions, whose purpose is to capture the history of design deliberation.

Some representation schemes of the DR have been attempted in the chemical process/plant design. Bañares-Alcántara et al. (1997) proposed the Knowledge Based Design System-IBIS (KBDS-IBIS), which is the result of integrating a representation of DR to the design alternative history maintained by a design support system [2]. Chung et al. (1999) proposed an Integrated Design Information System (IDIS) that supports the design of chemical plants [3]. IDIS provides an integrated framework for recording three different aspects of DR: exploration of design alternative, reasons for design decision and design constraints. Batres et al. (2002) proposed a two-layer methodology approach to DR by combining activity modeling and IBIS-like methods [4]. Fuchino et al. (2003) proposed an environment to support the DR approach based on a feature-oriented approach [5]. Shimada et al. (2004) proposed a method of retrieving DR and reusing of it for supporting the plant safety operation [6, 7].

3 Safety Management Environment
by Using Plant Design Information

Previously, information on the design result such as PFD (Process Flow Diagram) and P&ID (Piping and Instrument Diagram) and the result of risk assessment has been tried to be used for the process safety management. On the other hands, it is also important to capture the DR such as design thinking processes and design intents and reuse them for supporting the theoretical and reliable process safety management. By offering the DR information, engineer, plant operator and maintenance personnel can understand the safety design process and the background knowledge of design, and carry out each activity correctly. Fig.1 shows a concept of proposed safety management environment based on the plant design information.

(1) PLC Activity Model: To-be model which manages the information on process design and other PLC activities and the flow of them.
(2) Process Information Model: Information on plant structure, process behavior and operation procedure.
(3) DR Management Model (DR mode): Information on theoretical thinking process and background knowledge of process design.
(4) Design Information Database: Database of information on design result and the DR.
(5) Design Information Management System: To get the information on process design and send it to each activity support system.

(6) Safety Management System for Operation and Maintenance: To support safer operation and more effective maintenance based on the theoretical design information.

(7) MOC Support System: To ensure consistency against the change of process design, operation, or social requirement, etc.

(8) Design Support System: To support HAZOP, safety design for other plant design.

Within this environment, information on the process design including DR and their flows can be managed according to the PLC activity model and reused for supporting the PLC activities such as the safer operation, the management of change (MOC) of objective plant and the safer design for other plant. This paper focuses on the DR management model and the design information management system.

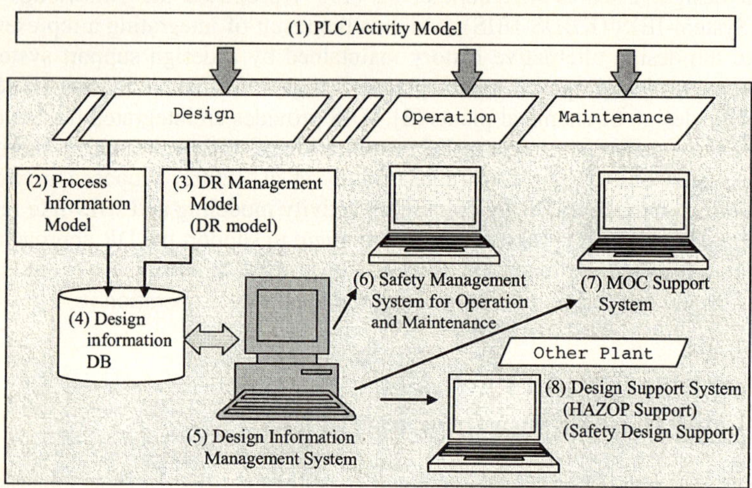

Fig. 1. Safety management environment based on plant design information

4 Design Rationale Management Model (DR model)

Previous studies on the chemical process design have focused on how to record the design history and apply IBIS as representation scheme in process-oriented approach. The IBIS can represent the design alternative and its designer's thinking process against the problem. However, it cannot record the relationships between alternative, evaluation term and final decision making which is the result of design such as PFD, P&ID and SOP (Standard Operating Procedure) clearly.

This paper proposes a DR management model (DR model) based on the IBIS method to record the DR information and enable the link with final decision making. Fig.2 shows the basic DR model. Original IBIS representation scheme is enclosed by dot line. This basic DR model consists of five types of objects: *issue, position, argument, basis* and *decision*, and their interrelated arrows. Table 1 shows the meanings of each object. Issue *is suggested by* positions and position *responds to* issue. Argument *is based on* one or more basis and either *supports* or *objects to* position. Some positions *are selected as* final decisions. Issue and decision *are generalized* or *specialized* into new issues.

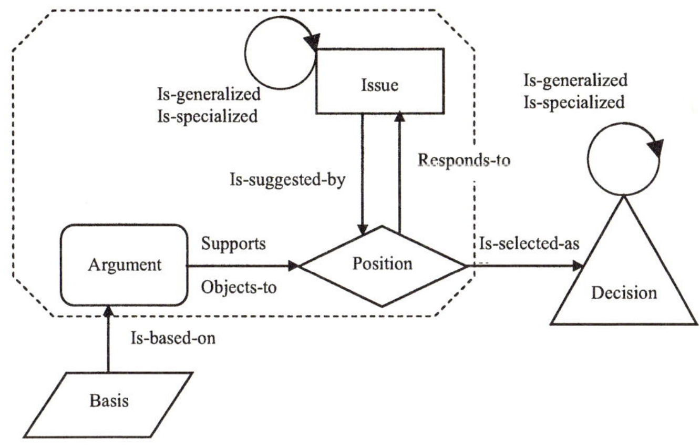

Fig. 2. DR management model (Basic DR model)

Table 1. Meanings of objects in the basic DR model

Object	Meanings
Issue	Question or subject on process design
Position	Solutions against Issue (Alternatives)
Argument	Discussion based on Basis (How considered?)
Basis	Evaluation item, standard, criteria for evaluation, design intent, policy, etc.
Decision	Positions selected finally (Design result)

Fig.3 shows an example of DR model for designing the countermeasure against an issue; 'Loss of EC (loss of effective cooling)' [8]. To prevent high temperature reaction, two kinds of positions: 'BSC (Back-up Source of Cooling)' and 'ERD (Emergency Relief Device)', respond to this issue. Each position is evaluated based on the general or specific bases and selected as final countermeasures or not. In the process design, plant structure is decided to satisfy with following design bases.

(1) 'SQCD (Safety, Quality, Cost and Delivery)': Prioritization of countermeasures considered from SQCD viewpoints and balances between them.

(2) 'Reg': Compliance with regulations and standards required minimally.

(3) 'Other': Requirements or constraints from process operations and behaviors, including material properties, owner's and social requirements, and so on.

The bases from SQCD viewpoints include the designer's intent for plant operation such as safety-conscious or quality-conscious operation. Other engineers can understand which design intent effects mainly designed countermeasure (position). In this example, both of positions are selected to satisfy with the safety and the delivery requirements and to comply with some regulations and standards such as IEC61508. Furthermore, each decision is specialized into new issues for detail design (ex. Selection of type of 'BSC' in Fig.3). The DR model is represented graphically for each design process and makes the relationships between the issue on process design and the decision making based on the argument and the basis clear. It becomes possible to unify the management of DR information which leads the final design result from design issue by selecting some alternatives based on evaluation items or criteria.

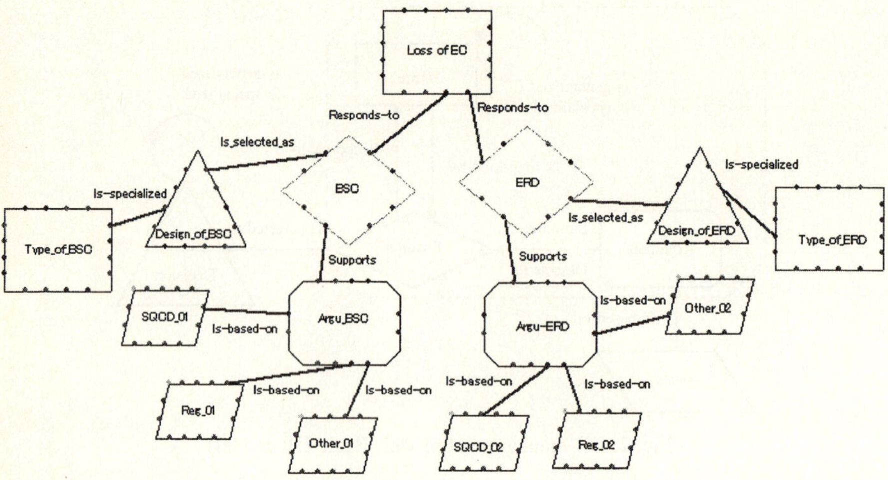

Fig. 3. DR model for issue 'Loss of EC'

5 Design Information Management System

In conventional support system, only information on the result of risk assessment and the process structure such as PFD and P&ID has been used for appropriate decision making for the following PLC activities. However, for example, plant operator cannot understand the logic on why such operation should be done or what the intent of operation is even if they know what to do. As shown in Fig. 1, the design information including the DR can be useful for carrying out the plant safety management on ensuring the consistency between the design intent and the process structure since the DR information also consists of process design intent, operation policy, and so on. The safer plant operation, the more effective maintenance activity and MOC of the objective plant, or the more efficient risk assessment and the safer process design for other plant can be realized by reusing the DR information.

In this paper, a simplified DR search system is proposed as part of design information management system. The graphical DR model can be transferred to Prolog fact data and is stored into the design information database. Process specific information, which is the design result such as plant structure, process behavior and operation procedure, is also represented by Prolog fact data. Such approach makes possible the link between the design results and the DR information. Part of the DR model in Fig.3 is described as following Prolog fact data.

```
topology('MC1','BSC','MP6','Loss of EC','MP3').
topology('MC3','Argu_BSC','MP9','BSC','MP2').

obj_unit('Loss of EC',issue).
obj_unit('BSC',position).
```

These relations are represented by objects and associations, and the bidirectional search between the design intention and the process structure via the DR information is enabled. The search of DR information is done by using recursive rules of Prolog, and the design intent and the background mechanism are outputted based on the process specific information. Final decision-making by the safety management system can be justified based on the DR information since the design intent as well as the reasoning behind the design is explicitly offered.

```
Q- Why is the emergency relief device 'ERD' selected?
    ?-why_selected ['ERD'].

A- To prevent high temperature reaction due to 'Loss of EC'.
    to_solve_the_issue ['Loss of EC'].
    to_be_supported_by_argument ['Arg-ERD'].
```

Fig. 4. Example of searching the DR information

Fig.4 shows an example of search. When user inputs a question why the 'ERD' is selected, the system outputs contents of issue 'Loss of EC' on design problem and argument 'Arg-ERD' as answers according to DR model shown in Fig.3. Furthermore, user can understand the background knowledge, that is the DR on deciding the process structure so as to satisfy SQCD requirements, regulation or standards and the requirements from the process behaviors, including material properties and the operations by searching basis of this argument 'Arg-ERD'. In this case, it could be clarified that 'ERD' is designed to intend the safety-conscious operation and to comply with some standards, and operators can carry out their activities correctly based on this information. In this way, these outputs can be reused as useful information for supporting the plant safety management.

6   Conclusion

To clarify the process design rationale (DR) information, a DR management model (DR model) is proposed based on the original IBIS method. This DR model can represent the relationship between the design intent, the thinking process and the design result (process information). Furthermore, a simplified search system is proposed as a part of the plant safety management system. This system can give useful information for the plant safety by the easy question through the Prolog inference engine.

As next steps, it is important to develop the system which can support the record of process design process based on proposed DR model and store them in the database. Then, the operation support system, the maintenance support system and the management of change (MOC) support system based on the DR information and the safety management technical information, etc will be developed.

Acknowledgement

The authors would like to gratefully acknowledge the support of the Secom Science and Technology Foundation and the Grant-in-Aid Scientific Research (16310115) from Japan Society for the Promotion of Science. And the authors also thank to Dr. C.Zhao and Prof. V.Venkatasubramanian in Purdue University (USA) for discussion.

References

1. Regli,W.C., Hu,X., Atwood,M. and Sun,W.: A Survey of Design Rationale Systems: Approach, Representation, Capture and Retrieval. Eng Comput. 16 (2000) 209-235
2. Bañares-Alcántara,R. and King,J.M.P.: Design Support Systems for Process Engineering iii – Design Rationale as a Requirement for Effective Support. Comput. Chem. Eng. 21 (1997) 263-276
3. Chung,P.W.H. and Goodwin,R.: An Integrated Approach to Representing and Accessing Design Rationale. Eng. Appl. Artif. Intell. 11 (1998) 149-159
4. Batres,R., Aoyama,A. and Naka,Y.: Design Rationale as an Enabling Factor for Concurrent Process Engineering. Proc. of WWDU2002, Berchtesgaden, Germany (2002) 612-614
5. Fuchino,T. and Shimada,Y.: IDEF0 Activity Model based Design Rationale Supporting Environment for Lifecycle Safety. Proc. of KES2003 1 (2003) 1281-1288
6. Shimada,Y. and Fuchino,T.: Safer Plant Operation Support by Reusing Process Design Rationale. Proc. of 11th Int. Symp. on Loss Prevention and Safety Promotion in the Process Industries D (2004) 4352-4357
7. Shimada,Y. and Fuchino,T.: Retrieval and Reuse of Design Rationale for Supporting Safer Plant Activities. Proc. of PSAM7 & ESREL'04 6 (2004) 3447-3452
8. AIChE/CCPS: Guidelines for Process Safety in Batch Reaction Systems. AIChE/CCPS (1999)

A Multiple Agents Based Intrusion Detection System

Wanli Ma and Dharmendra Sharma

School of Information Sciences and Engineering
University of Canberra, Canberra, Australia
`{Wanli.Ma,Dharmendra.Sharma}@canberra.edu.au`

Abstract. Three reasons, in our opinion, are responsible for the high false alarm rate of current intrusion detection practice. They are: (i) only single information source is analysed by an intrusion detection system, (ii) only a single method is used for the analysis, and (iii) there is no distinction of vulnerability, threat, attack, and intrusion. This paper first studies the dynamics of attackers and defenders and then lists all possible information sources. A multiple agents based integrated intrusion detection system (IIDS) is then proposed. The status of our current work is also discussed.

1 Introduction

A plethora of intrusion detection systems (IDS) has been proposed. In general, there are two types of intrusion detection systems: signature matching based intrusion detection systems and anomaly behaviour analysis based intrusion detection systems. Many different types of technology have been proposed as the detection engines, which analyze the collected information to find the signs of intrusions. Using probability distribution to detect intrusions was proposed by Eskin [1]. Hidden Markov Model was studied by Ourston et al [2]. Anderson and Khattak proposed to use information retrieval technology to comb through computer system logs to find clues of intrusions [3]. Autonomous agents and distributed intrusion detection were proposed by Balasubramaniyan et al [4] and Bradshaw [5]. Jai Sunder et al also proposed an intrusion detection architecture using autonomous agents [4]. Crosbie and Spafford proposed an intrusion detection system with many "small, independent agents monitors the system" [6]. A good survey paper was written by Sherif et al, Part I [7] and Part II [8]. In practice, there are several open source and commercially available intrusion detection systems, for example Snort [9] and cisco IDS [10].

The goal of an intrusion detection system is to efficiently and effectively detect intrusions. The requirements are so basic, yet they are far from achievable. If an intrusion detection system raises alarm while there is no successful intrusion, it is called *false positive*. On the other hand, if there is a successful intrusion, the intrusion detection system fails to raise the alarm, it is called *false negative*. The rate of failure, both false positive and false negative, is very high. Some even estimate of over 99% [11].

In our opinion, there are three reasons responsible for the high failure rate. First, almost all existing intrusion detection systems rely on limited single information source, for example, network traffic content for pattern matching, program execution paths, or system logs. There is no attempt to establish a coherent big picture of the activities of all computer and network systems; neither the attempt to study the intrusion dynamics.

R. Khosla et al. (Eds.): KES 2005, LNAI 3681, pp. 205–211, 2005.
© Springer-Verlag Berlin Heidelberg 2005

Second, only a single method is used in an intrusion detection system to analyse a piece of information. It is in contrary of human intelligence analysis, where the same piece of information is always analysed by experts of different skills. We call the analysis module of an intrusion detection system *analysis engine*. There is no reason why a piece of information can only be analysed by a single analysis engine.

Finally, there is no distinction of vulnerability, threat, attack, and intrusion [12]. A threat does not necessarily lead to attack, and an attack does not necessarily succeed and hence lead to intrusion. In comparison with the number of attack attempts, the attack success rate is extremely low. Attack attempts should only raise the attention of an intrusion detection system instead of alarm!

In this article, we would like to go back to the basics, starting from the dynamics of both attackers and defenders on the fields and then listing all available information sources. Based on the observation, we then propose a multiple agents based Integrated Intrusion Detection System (IIDS).

The paper is organized as following. We first study the dynamics of both attackers and defenders in Section 2, and we further look at all possible information sources in Section 3. In Section 4, we provide details of IIDS. The current status of IIDS implementation work is discussed in Section 5. Section 6 concludes the paper.

2 The Dynamics of Attackers and Defenders

To research, design, and implement an intrusion detection system which can efficiently and effectively detect intrusions, it is essential to understand the dynamics and the technology used by both attackers and defenders.

In general, there are three steps an attacker may take:

1. **Reconnaissance:** attacking on computer systems is the same as battle field attacking. The first step is to gather intelligence. In the context of computer system reconnaissance, the intelligence includes (i) the basic computer system information, such as manufacturers, models, hardware and software version etc.; (ii) the local configuration, for example, the topological structure of device, services provided by the computer systems, and device population etc.; (iii) based on the intelligence collected, work out the possible vulnerabilities or security holes of this particular site.

2. **Attacking:** After reconnaissance, the attacker now fully understands the vulnerabilities of the site. He or She can start attack. Depending on the skill level of the attacker, the nature of the vulnerabilities, and the availability of the attacking tools, the attacker may take attacking actions ranging from as simple as just running an existed tool, orchestrating a number of different tools, to crafting a new tool and then attacking. As the result of the attack, some computer systems are penetrated. In reality, reconnaissance and attack are often combined together, especially for these not-so-skilled attackers. That is why we see many attack attempts, but little success. An intrusion detection system should take the phenomenon into consideration when analysing potential intrusions.

3. **Covering up:** After a successful attack, the attacker will install some kind of back door so that he or she can come back without going through the process of attacking the vulnerabilities again. After setting up the proper back door access, the experienced attacker will cover up the intrusion trace by wiping out possible log en-

tries, hoodwinking intrusion detection systems, and even changing computer system monitoring programs so that they won't display or report any activities done by this attacker. Intrusion detection becomes much more difficult after this covering up operation.

On defending site, a defender operates constantly in two steps and has to be more diligent than an offender (attacker).

1. **Gathering information:** The defender normally subscripts to a number of security related email services and visits several respectable security related web sites. Upon receiving any security alert or even just a tip off, he/she has to do thorough investigation to see if the security threat is real. The other way to gather information about the security status is to do the probing himself/herself. He/She will use the same tools as an attacker does.

2. **Managing security threats:** if a threat is real, the system administrator will have to get it fixed or mitigated. If the vulnerability is fixed completely, any attack based on it will not cause any damage, and the attack attempts should not set off alarms. Instead, alert would be more appropriate.

3   Available Information

Intrusion detection is like a detective work: assembling the whole picture of activities based on small pieces of available information. There are many different information sources. No single information source alone can construct the big picture. An intrusion detection system should take full advantage of all available information sources. At least, it should not preclude useful information, as in the worst case, relying only on single source of information.

Network traffic is the primary information source for intrusion detection, especially for network intrusion detection systems. System status information is also very important. Any computer device, being a computer, a printer on the network, or a network switch etc., has its special characters. It is a certain type of hardware with its version of firmware; it is running an operating system of its own version; it is patched to certain patch level; it provides certain types of services via some active network ports. This information should be taken into consideration when analysing a possible intrusion.

Any type of modern systems has extensive logs. The logs provide vital information about the activities of the systems. Similarly, applications have extensive logs as well.

Finally, other information, such as fingerprints of computer system files and open Internet ports of a computer etc., may not be readily available. It may take quite effort to generate and collect this type of information; however, it provides vital evidence, sometimes, positive proof, for system tampering.

4   Proposed Agents Approach

A good intrusion detection system should not ignore the operation dynamics of both attackers and defenders, the current status of the underlying computer and network infrastructure, and all available information. Based on the observation in the previous sections, we propose an integrated intrusion detection system (IIDS) using multiple distributed agents technology.

A software agent is *"a software entity which functions continuously and autonomously in a particular environment"* [5]. In IIDS, an agent has a special skill (analysis engine) to monitor certain types of computer or network activities. It can be regarded as a traditional intrusion detection system by itself, be it host based or network based, signature matching based or anomaly analysis based. However, an agent will not raise any alarm by itself, as it can only monitor part of the computer or network activities. It does not have the whole picture. When there are suspicious activities, the agent will pass the information back to the assessment centre, unlike other standalone intrusion detection systems. An agent also has the knowledge of the local system, which helps it to make better decision. Instead of alarming the assessment centre, it will just pass on the failed attempt as alerting information. The assessment centre, upon learning the failed attack attempt, may further instruct the other agents to pay attention to this potential attack, Figure 1.

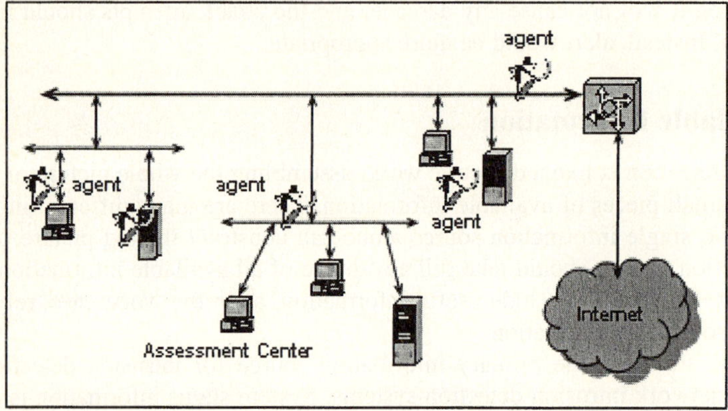

Fig. 1. The Architecture of Integrated Intrusion Detection System

An agent may also have the skills to take counter attack or mitigation measures. If an attacker successfully penetrates a system, there is virtually no time for a human operator to take action manually. The penetration could happen at anytime, while the human operator cannot be on standby 24 hours a day and 7 days a week. If an intrusion is positively identified, the agent should be able to take some kinds of counter measures, for example, changing firewall rules to stop traffic going to a remote host or stopping a particular type of traffic completely.

5 Current Work

IIDS has an assessment centre and many agents, Figure 1. We have already installed Snort on two hosts. They are Snort agents and specialized in network traffic content monitoring. Our current priority is to implement the network monitoring agent, Windows Autopilot agent, and Wigo agent. We leave the assessment centre for manual operation at this stage. After we have some experience of running multiple agents simultaneously, we will design and implement this assessment centre.

5.1 Network Traffic Monitoring Agent

Network traffic monitoring agent is specialized in analysing network traffic. It analyses network traffic by doing:

- simple statistics,
- temporal analysis: analysing the traffic patterns along time line. The analysis will take the dynamics of attackers and defenders into consideration.
- network packet content monitoring: searching the content of network packets against a set of signatures.

In most case, the agent just builds up a watching list. The watching list is then sent to the assessment centre for further analysis. The watching list may be passed to the other agents, and then they will pay more attention to any activity involving the IP addresses on the list.

An alert may raised by the temporal analysis. If a sequence of network communication involves a port of a local host, and the port is not supposed to provide Internet service, an alert is sent to the assessment centre. The assessment centre, based on the information from the other agents, it will:

- ignore the alert if there is no enough evidence to warrant further actions,
- instruct the relevant agents to scrutinize the activities more carefully at all stages, or
- raise intrusion alarm and instruct the relevant agents to take counter attack measures.

The bulk part of the network traffic monitoring agent is its temporal analysis engine. We are currently building it up.

5.2 Windows Autopilot: The Agent for System Status Information Collection

Windows Autopilot is an agent which can collect the system status information of a Windows based computer. The agent is part of the security frame work being developed at the University of Canberra [13].

5.3 Wigo (What is going on): A Windows Forensics Analysis Agent

Wigo (What is going on) is another agent. It is responsible for collecting fingerprints of the underlying computer system files and the information of open Internet ports of the computer. The fingerprints of a file are different types of message digests of that file, for example, MD5 and SHA/SHS [12]. If the content of a file changes, its fingerprints change. It is almost impossible to alter the content of a file while keep two or more types of fingerprints of that file unchanged. By keeping a baseline fingerprints database for the system files of a computer, it is easy to verify if a file is changed. On the other hand, the open Internet ports on a computer reflect the services provided by the computer, for example, TCP/80 for HTTP connection, TCP/25 for email connection, and TCP/161 for SNMP connection. If a computer suddenly opens a new port, which is not arranged by the local system administrator, it is almost certain the computer is compromised by invaders.

5.4 Assessment Centre and Agent Communication Framework

With all these agents running on different computers and different locations of the underlying network, an assessment centre is needed to coordinate the activities of the agents, analyse the collected information, raise alarms, and perhaps, and take counter attack or mitigation actions. The assessment is also responsible for the communications among these agents. An agent will not communicate directly to another agent. It will just pass the information back to and take instructions from the assessment centre. It does not have to be aware of the existence of other agents.

6 Conclusion

In this paper, we suggest three reasons responsible for high false alarm rate in the current intrusion detection practice. The three reasons are: (i) only single information source is analysed by an intrusion detection system, (ii) only a single method is used as detection engine, and (iii) there is no distinction of vulnerability, threat, attack, and intrusion.

We base our approach on the dynamics of attackers and defenders and then listing all possible information sources. Armed with the study, we propose a multiple agents based integrated intrusion detection system (IIDS). These agents are deployed in the strategically important locations of the underlying computer and network systems. Each agent has specialized skills for analysis. Agents with different specialties may work on the same piece of information. The reasoning of these agents are passed to an assessment centre for further analysis.

To implement IIDS, we are currently building several different skilled agents. We have already installed Snort on two hosts. They will be Snort agents and specialized in network traffic content monitoring. Our current priority is to build the network monitoring agent, Windows Autopilot agent, and Wigo agent. The assessment centre is currently manually operated, but we expect to fully automate these tasks in the near future.

References

1. Eskin, E. Anomaly Detection over Noisy Data Using Learned Probability Distributions. in The 17th International Conference on Machine Learning. 2000. Morgan Kaufmann, San Francisco, CA, USA.
2. Ourston, D., et al., Coordinated Internet attacks: responding to attack complexity. Journal of Computer Security, 2004. **12**: p. 165-190.
3. Anderson, R. and A. Khattak. The use of Information Retrieval Techniques for Intrusion Detection. in First International Workshop on Recent Advances in Intrusion Detection (RAID'98). 1998. Louvain-la-Neuve, Belgium.
4. Balasubramaniyan, J.S., et al. An Architecture for Intrusion Detection using Autonomous Agents. in 14th IEEE Computer Security Applications Conference (ACSAC '98). 1998. Scottsdale, AZ, USA: IEEE Computer Society.
5. Bradshaw, J.M., An Introduction to Software Agents, in Software Agents, J.M. Bradshaw, Editor. 1997, AAAI Press/The MIT Press. p. 3-46.
6. Crosbie, M. and G. Spafford. Defending a Computer System using Autonomous Agents. in 18th National Information Systems Security Conference. 1995. Baltimore, Maryland, USA.

7. Sherif, J.S., R. Ayers, and T.G. Dearmond, Intrusion Detedction: the art and the practice, Part 1. Information Management & Computer Security, 2003. **11**(4): p. 175-186.
8. Sherif, J.S. and R. Ayers, Intrusion detection: methods and systems, Part II. Information Management & Computer Security, 2003. **11**(5): p. 222-229.
9. Snort, Snort web site, http://www.snort.org.
10. cisco, Cisco IOS Firewall Intrusion Detection System, http://www.cisco.com/en/US/products/sw/secursw/ps2113/products_white_paper09186a00 8010e5c8.shtml.
11. Julisch, K. Mining Alarm Clusters to Improve Alarm Handling Efficiency. in 17th Annual Computer Security Applications Conference (ACSAC'01). 2001. New Orleans, Lousiana, USA: IEEE.
12. Pfleeger, C.P. and S.L. Pfleeger, Security in Computing. Third ed. 2003: Prentice Hall.
13. Sharma, D., W. Ma, and D. Tran. On a Computer Security Framework. in Ninth International Conference on Knowledge-Based Intelligent Information & Engineering Systems (KES2005). 2005. Melbourne, Australia: Springer-Verlag.

Agent-Based Software Architecture
for Simulating Distributed Negotiation

V. Kris Murthy

School of Business Information Technology
RMIT University, Melbourne, Victoria 3000, Australia
kris.murthy@rmit.edu.au

Abstract. A cooperative multi-agent model is proposed for simulating distributed negotiation for military and government applications. This model uses the condition-event driven rule based system as the basis for representing knowledge. In this model, the updates and revision of beliefs of agents corresponds to modifying the knowledge base. These agents are reactive and respond to stimulus from the environment in which they are embedded. The distributed agent-based software architecture will enable us to realise human behaviour model environment and computer-generated forces (or computer-generated actor) architectures.

1 Introduction

A multi-agent model is proposed using object-based systems that can support a distributed virtual environment for simulation of negotiation in military and government. The distributed agent-based software architecture will enable us to realise intelligent military operations planning systems and computer-generated forces (also called computer-generated actor (CGA) systems. These CGA systems will be useful in modeling cyberwarfare, or information warfare.This model also provides an insight into the self-organized criticality in a network of agents [1-9]. The cooperative software agents are useful in realizing distributed military operations planning systems [10-12].

2 Agent Model

A multi-agent system consists of the following subsystems [6]:

(1) Environment U: Those states which completely describe the universe containing all the agents.
(2) Input (Percept): This is an input from the environment. Depending upon the sensory capabilities (input interface to the universe or environment) an agent can partition U into a standard set of messages T, using a sensory function Perception (PERCEPT). PERCEPT is interpreted by an agent and involves various senses.
(3) Mind M: The agent has a mind M (namely, a problem domain knowledge consisting of an internal database for the problem domain data and a set of problem domain rules) that can be clearly understood by the agent without involving any sensory function. The database D sentences are in first order predicate calculus (also known as extensional database) and agents mental actions are viewed as inferences arising from the associated rules that result in an intentional database, that changes (revises or updates) D.

R. Khosla et al. (Eds.): KES 2005, LNAI 3681, pp. 212–218, 2005.

The beliefs are first order logic sentences resulting from information about the environment at a certain time. These beliefs can be of three types:

 (i)   Elementary belief: This is assumed or self supported,
 (ii) Derived belief: This is got from perception and communication.
(iii) Inferential belief: This is got through analysis.

An agent therefore knows what the belief is, how it was arrived at and why it is true. A distributed belief is composed of the union of the beliefs of all agents. Thus M can be represented by an ordered pair of elements (D, P). D is a set of beliefs about objects, their attributes and relationships stored as an internal database and P is a set of rules expressed as preconditions and consequences (conditions and actions). When T is input, if the conditions given in the left-hand side of P match T the elements from D that correspond to the right-hand side are taken from D and suitable actions are carried out locally (in M) as well as on the environment.

(4) Organizational Knowledge (O): Since each agent needs to communicate with the external world or other agents, we assume that O contains all the information about the relationships among the different agents. For example, the connectivity relationship for communication, the data dependencies between agents, interference among agents with respect to rules, information about the location of different domain rules are in O.

(5) INTRAN: M is suitably revised or updated by the function called Internal transaction (INTRAN). Revision means acquisition of new information about the world state, while update means change of the agent's view of the world. Revision of M corresponds to a transformation of U due to occurrence of events and transforming an agent's view due to acquisition of new information that modifies rules in P or their mode of application (deterministic, nondeterministic or probabilistic) and corresponding changes in database D (e.g modifying the tax-rules).

(6) EXTRAN: External action is defined through a function called global or external transaction (EXTRAN) that maps an epistemic state and a partition into an agent action.

(7) EFFECT: The agent also has an effectory capability on U by performing suitable actions.

Thus, an agent is defined by: (U, T, M, (P, D), O, A, PERCEPT, INTRAN, EXTRAN, EFFECT).

The nature of internal production rules P, their mode of application and the action set A determines whether an agent is deterministic, nondeterministic, probabilistic or fuzzy. Rule application policy in a production system P can be modified by:

(1) Assigning probability fuzziness for applying a rule.
(2) Assigning strength to a rule by using a measure of its past success.
(3) Introducing a support for a rule by using a measure of its likely relevance to the current situation.

The above factors provide for competition and cooperation among the different rules. Such a model is useful for collaboration and cooperation involving many agents.

3   Agent-Based Negotiation

"Negotiation" is an interactive process among a number of agents that results in varying degrees of cooperation, competition and ultimately leads to a total agreement, consensus or a disagreement. It has many applications, including Economics, Psychology, Sociology and Computer science.

A negotiation protocol is a set of public rules that dictate the conduct of an agent with other agents to achieve a desired final outcome. A negotiation protocol among agents involves the following actions or conversational states:

1. Propose: one puts forward for consideration a set of intentions called a proposal.
2. Accept: The proposal is accepted for execution into actions.
3. Refuse: The proposal is rejected for execution into actions.
4. Modify: This alters some of the intentions of the proposer and suggests a modified proposal- that is at the worst it can be a Refuse and a New proposal; or a partial acceptance and new additions.
5. No proposal: No negotiation.
6. Abort: Quit negotiation.
7. Report agreement: This is the termination point for negotiation in order to begin executing actions.
8. Report failure (agree to disagree): Negotiation breaks down.

Note that the above actions are not simple exchange of messages but may involve some intelligent or smart computation.

Multiagents can cooperate to achieve a common goal to complete a transaction to aid the customer. The negotiation follows rule-based strategies that are computed locally by its host server. Here competing offers are to be considered; occasionally cooperation may be required. Special rules may be needed to take care of risk factors, domain knowledge dependencies between attributes, positive and negative end conditions. When making a transaction several agents have to negotiate and converge to some final set of values that satisfies their common goal. Such a goal should also be cost effective so that it is in an agreed state at the minimum cost or a utility function. To choose an optimal strategy each agent must build a plan of action and communicate with other agents.

The negotiation process is usually preceded by two cooperating interactive processes: Planning and reasoning. To solve a problem through negotiation, we start with a set of desired properties and try to devise a plan that results in a final state with the desired properties. For this purpose, we define an initial state where we begin an operation and also define a desirable goal state or a set of goal states. Simultaneously, we use a reasoning scheme and define a set of intended actions that can convert a given initial state to a desired goal state or states. Such a set of intended actions called the plan exists if and only if it can achieve a goal state starting from an initial state and moving through a succession of states. Therefore to begin the negotiation process, we need to look for a precondition that is a negation of the goal state and look for actions that can achieve the goal. This strategy is used widely in AI and forms the basis to plan a negotiation. The same approach is used for devising a negotiation protocol.

3.1 Negotiation Termination

For the negotiation protocol to be successful, the process must terminate. We now describe an algorithm that can detect the global termination of a negotiation protocol. Let us assume that the N agents are connected through a communication network represented by a directed graph G with N nodes and M directed arcs. Let us also denote the outdegree of each node i by Oud (i) and indegree by Ind(i). Also we assume that an initiator or a seeding agent exists to initiate the transactions. The seeding agent (SA) holds an initial amount of money C. When the SA sends a data message to other agents, it pays a commission:

C/(Oud (SA) + 1) to each of its agents and retains the same amount for itself. When an agent receives a credit it does the following:

a. Let agent j receive a credit C(M(i)) due to some data message M(i) sent from agent i. If j passes on data messages to other agents j retains C((M(i)) / (Oud(j)+1) for its credit and distributes the remaining amount to other Oud(j) agents. If there is no data message from agent j to others, then j credits C(M(i)) for that message in its own savings account; but this savings will not be passed on to any other agent, even if some other message is received eventually from another agent.

b. When no messages are received and no messages are sent out by every agent, it waits for a time-out and sends or broadcasts or writes on a transactional blackboard its savings account balance to the initiator.

c. The initiator on receiving the message broadcast adds up all the agents' savings account and its own and verifies whether the total tallies to C.

d. In order to store savings and transmit commission we use an ordered pair of integers to denote a rational number and assume that each agent has a provision to handle exact rational arithmetic. If we assume C=1, we only need to carry out multiplication and store the denominator of the rational number.

We prove the following theorems to describe the validity of the above algorithm:

Theorem 1: If there are negotiation cycles that correspond to indefinite arguments among the agents (including the initiator itself) then the initiator cannot tally its sum to C.

Proof: Assume that there are two agents i and j are engaged in a rule dependent argument cycle. This means i and j are revising their beliefs forever without coming to an agreement, and wasting the common resource C. Let the initial credit of i be x. If i passes a message to j, then i holds x/2 and j gets x/2. If eventually j passes a message to i ,then its credit is x/4 and i has a credit x.3/4 ; if there is continuous exchange of messages for ever then their total credit remains (x - x/2^k) with x/2^k being carried away by the message at k th exchange. Hence the total sum will never tally in a finite time.

Theorem 2: The above algorithm terminates if and only if the initiator tallies the sum of all the agents savings to C.

Proof: If part: If the initiator tallies the sum to C this implies that all the agents have sent their savings and no message is in transit carrying some credit and there is no chattering among agents.

Only if part: The credit assigned can be only distributed in the following manner:

a. An agent has received a message and credit in a buffer; if it has sent a message then a part of the credit is lost; else it holds the credit in savings.
b. Each message carries a credit; so, if a message is lost in transit or communication fails then total credit cannot be recovered.

Thus termination can happen only if the total sum tallies to C, i.e., the common resource is not wasted and all the agents have reached an agreement on their beliefs.

We illustrate the above algorithm using the E-auction in the English auction scenario in which there is a single auctioneer and a set of clients participating.

3.2 Example

Auction is a controlled competitive peocess among a set of agents (clients and auctioneer) coordinated by the auctioneer. In this example, the belief is first obtained from the auctioneer and other clients through communication and these are successively updated. Finally, the distributed belief among all participants is composed of all the existing beliefs of every agent involved in the process.

The rules that govern the English auction protocol are as follows:

1. At the initial step the auctioneer-agent begins the process and opens the auction.
2. At every step, decided by a time stamp, only one of the client-agent is permitted to bid and the auctioneer relays this information. The bidding client agent is called active and it does not bid more than once and this client becomes inactive until a new round begins.
3. After the auctioneer relays the information a new client becomes active and bids a value strictly greater than a finite fixed amount of the earlier bid.
4. When at a given time-out period no client-agent responds, the last bid is chosen for the sale of the goods and the auction is closed.

3.3 Auction Protocol

Let us assume that there are three clients (A, B, C) and an auctioneer G. The auctioneer G initiates the auction. Then each of the clients A,B and C broadcasts their bid and negotiates, and the auctioneer relays the information. The bidding value is known to all the clients and the auctioneer. When the bid reaches a price above a certain reserve price, and no bid comes forth until a time-out, G terminates the auction and the object goes under the hammer for that price. In E-auction the above scenario can be realised and the negotiation termination algorithm can be used.

At initiation, the node G is the seeding agent (auctioneer). It transmits the information to each client the beginning of the E-auction. Also it starts with a credit 1 and retains a credit of 1 /(Oud (SA)+ 1 to itself, and transmits the same amount to its neighbours (A, B, C) which in this case is 1/4. The retained credit for each transmission is indicated near the node. To start with the agent-client A bids a value. Then all clients and G get this information and the credits. Then agent-client node B updates its earlier belief from the new message received from G; but the other nodes A, C do not update their initial beliefs and remain silent .The agent-client node C then bids. Finally as indicated in the rules described above, we sum over all the retained credits after each transmission.

4 Battlefield Modelling and Simulation

The agent negotiation system can model a distributed battlefield simulation system. In this system, the military operations are modelled as a distributed process among many soldiers (agents) coordinated by the group commander (controlling agent).

In this simulation system, the domain data D, rules P and organizational knowledge O are based on three factors:

1. The experience and knowledge of a soldier is based totally on his criteria (elementary belief)
2. The soldier acquires knowledge through communication with other other soldiers and commanders; such a soldier is called a fundamentalist (derived belief).
3. The soldier acquires knowledge by observing the behavior of other soldiers and commanders; such a soldier is called a trend chaser (inferential belief). In practice a soldier is influenced by the above factors and the modified knowledge is incorporated in D, P and O.

In a battlefield (military) simulation system, a soldier or a commander (an agent) can perform various actions including Attack and Retreat. Each agent can communicate with other agents and this creates a bond among them. This results in the modification of the organizational knowledge O. This bond is created with a certain probability determined by a parameter, which characterises the willingness of a soldier or commander to comply with others.

We can assume that any two soldiers or commanders (agents) are randomly connected with a certain probability. This divides the agents into clusters of different sizes whose members are linked either directly or indirectly via a chain of intermediate agents. These groups are coalitions of military participants who share the same opinion about their activity. The decision of each group is independent of its size and the decision taken by other clusters.

Using percolation theory [11] it can be shown that when every agent is on average connected to another, more and more agents join the spanning cluster, and the cluster begins to dominate the overall behaviour of the system. This gives rise to 'Offensive Action' (if all the soldiers and commanders decide to attack) and a 'Defensive Action' (if all the soldiers and commanders decide to retreat). Accordingly, an analogy exists between Offensive Action or Defensive Action and critical phenomena or phase transitions in physics. Thus a distributed agent system can eventually enter into a phase-transition like situation [1, 2, 8, 9, 11].

When soldiers and commanders (agents) collaborate in a battlefield simulation system, the collaboration graph consists of many nodes and edges. As more military participants join and their collaboration increase, the number of links increase and the collaboration graph grows. The links among the agents can be established in a certain preferential manner rather than a uniform distribution. Recently, Barabasi and Albert [2] observe that the growth and preferential attachment leads to a power-law distribution, namely, the probability $P(k)$ that each agent has k links is k^{-x}, where $x = 2.3$. Thus the development of distributed complex systems is governed by robust self-organizing phenomena that go beyond the particulars of the individual systems. Therefore complex systems involving a large number of agents will self organize into a scale-free state. This phenomenon will be useful in complex distributed military operations planning systems involving many smart agents.

5 Conclusion

A cooperative multi-agent model was described for supporting distributed simulation technologies for military applications. The distributed collaborative agent-based software architecture based on Java tools will enable us to realise human behaviour model environment and computer-generated forces architectures. Also this model will provide an insight on the self-organized criticality among a network of agents.

References

1. Bak, P.: How Nature works: The Science of Self-organized criticality, Springer, New York (1996).
2. Barabasi, A and Albert, R.: Emergence of scaling in random networks, Science 286 (1999) 509-512.
3. Chen, Q. and Dayal, U.: Multi agent cooperative transactions for E-commerce, Lecture Notes in Computer Science 1901 (2000) 311-322.
4. 4. DeLoach, S.A.: Multiagent Systems Engineering, International Journal of Software Engineering and Knowledge Engineering 11 (2001) 231-258.
5. Dignum, F. and Sierra, C.: Agent Mediated E-Commerce, Lecture Notes in Artificial Intelligence 2003 (2000).
6. Fisher, M.: Representing and executing agent-based systems, in Intelligent Agents, Woolridge, M. and Jennings, N.R (Eds.), Lecture Notes in Computer Science,Vol.890, Springer-Verlag, New York (1995) 307-323.
7. Jennings, N.R.: On agent-based software engineering, Artificial Intelligence, Vol.117 (2000) 277-296.
8. Kirkpatrick, S. and Selman, B.: Critical behaviour in the satisfiability of random boolean expressions, Science, Vol.264 (1994) 297-1301.
9. Lloyd, A. and May, R.M.: How viruses spread among computer and people, Science, Vol.292 (2001) 1316-1317.
10. Nagi, K.: Transactional agents, Lecture Notes in Computer Science. New York, Springer Verlag (2001).
11. Paul, W. and Baschnagel, J.: Stochastic Processes, Springer Verlag, New York (2000).
12. Winikoff, M. et al: Simplifying the development of intelligent agents, Lecture Notes in Artificial Intelligence, Vol. 2256 (2001) 557-568.

A Multi-agent Framework for .NET

Naveen Sharma and Dharmendra Sharma

School of Information Sciences and Engineering
University of Canberra, Canberra ACT, Australia
N.Sharma@student.canberra.edu.au
Dharmendra.Sharma@canberra.edu.au

Abstract. In this paper an approach is presented to modeling Multi Agent Systems (MAS) as a framework and its implementation as an extension to .NET. The MAS characteristics are identified in the context of the proposed MAS framework. The framework is developed and implemented in VB.NET language based applications. The relevant .NET-based technologies (i.e. Remoting, Services, and Web Services) are assessed for the implementation. A prototype application for dynamic truck scheduling is presented to test theproposed MAS framework and some future research and development work is outlined.

1 Introduction

Design and development of intelligent Multi Agent System (MAS) for distributed computing is one of the main potential area of development, as it is fairly new area of research and much work is still required to achieve any major milestones [1, 6]. Agent is an acting entity and system is a bunch of supportive functions in MAS. However there is not any universally accepted definition of agent, but following definition describes all the aspects of agent in the context of MAS well.

An agent is a computer system that is *situated* in some *environment*, and that is capable of *autonomous action* in this environment in order to meet its design *objectives* [9]. The proposed MAS is an environment where agents act to achieve their design objectives. MAS implements many of agent properties, which distinguish them from an object. MAS captures a working behaviors of anonymous agents, which might be residing in same application domain or in different environments on a network. Agents working under MAS cooperate with each other and work like a team rather an stand alone unit in order to achieve their common goals [1]. They facilitate their community members (other agents) to share their resources to gain their own individual goals. MAS agents can sense their environment, their goals, and services they offer and are available for public consumption. They can sense other agents around and negotiate their services for their purpose of life (objectives). They can gradually learn to do things (task) in better ways with the time. The main motivation for MAS is to equip them with collaborative and cooperative features to distribute and perform their tasks intelligently. The collective approach is known to be better because of limitations on individual agents.

A complete MAS environment has been proposed which can facilitate almost all the pre-established characteristics of an Agent [3, 11]. It also allows the rapid development of MAS application by giving an application developer a choice to develop the application in an interactive and user friendly environment within .NET. There are many advantages of using .NET as the development environment for the MAS framework and is also discussed in this paper.

R. Khosla et al. (Eds.): KES 2005, LNAI 3681, pp. 219–225, 2005.

2 Characteristics of MAS

The purpose of MAS environment is to make all the applications in the environment to jointly come up with dynamically coordinated computational power to perform some special sort of task which cannot be done by using normal software (individual software). To perform this team work there are some special characteristics, which must be implemented in the system [11]. Following are the minimum basic characteristics.

- **Autonomous** – Agent should be able to act independently without any external support to achieve its individual goals.
- **Communication** – An agent should be able to communicate **and** coordinate with other agents, services etc residing with in the environment.
- **Learning** – Learn from previous experience and adapt with time so the agent would have its own experience of work as the time pass.
- **Goal Orientation** – Agent behaviors should be directed towards achieving the purpose of its life and helping other agents to achieve their purpose of life.
- **Mobility** – Should be able to travel from one environment to another according to their need and to achieve their life goals.
- **Persistence** – Should be able to execute uninterruptedly even when their supporting services are not available.

All the agents must be designed to achieve their purpose of life (objectives) at their own; [11] it means without any external support and any special runtime service a proposed MAS agent should work like an individual piece of software. When it senses the existence of other agent applications with in the environment than it should start working as part of the team to achieve their common goals but still agent should coordinate its efforts to achieve its own purposes of life. The main advantage of application based on this design (proposed MAS Framework) is to ensure the survival of the system despite the failure of one or more agents. This is major cause of concern in centralized computing. In other words MAS is a network of loosely coupled software application (modeled as software agents) which can work individually as well as jointly with other software applications (agents, services etc) and can deliver the results beyond their individual software and hardware limits. In order to coordinate their work agents need to have some sort of communication to send and receive information and data. In order to make this communication effective some protocols are required so when one agent sends some data to other agent in different process or network which has different definition and understanding of data storage, the data must be read and used as it should be, So that application doesn't produce any unexpected or unreliable result. It is very crucial and critical characteristic of the agent as if it is not effectively implemented; the whole purpose of MAS environment will be lost. Agent's capability of learning from its own work experience takes it miles ahead than any other distributed application system. With this unique capability agents will become more valuable as the time go pass like experienced human practitioner. This functionality requires artificial intelligence. As there are some purposes of every software application, MAS applications also have purposes, the only difference is the approach to the problem and dynamic strategy to achieve the purposes of their life.

3 MAS

MAS framework facilitates the rapid development of MAS applications. It is incorporated in Integrated Development Environment (IDE) of .NET and it uses the default Microsoft wizard engine of Visual studio. That gives it very familiar look and helps the developer to predict the behavior of the IDE. As .NET application developers are familiar with the working patterns of IDE so they do not need any special training other than reading the Developer manual before they start developing the MAS applications. MAS Framework consists of many services, utilities and tools other than Development environment. Each one of them has their unique functionality, so the significance in the framework. Following is a diagrammatic explanation of the MAS framework.

3.1 Components of MAS Framework

Figure 1 summarises the various components of the MAS framework. The components are described below.

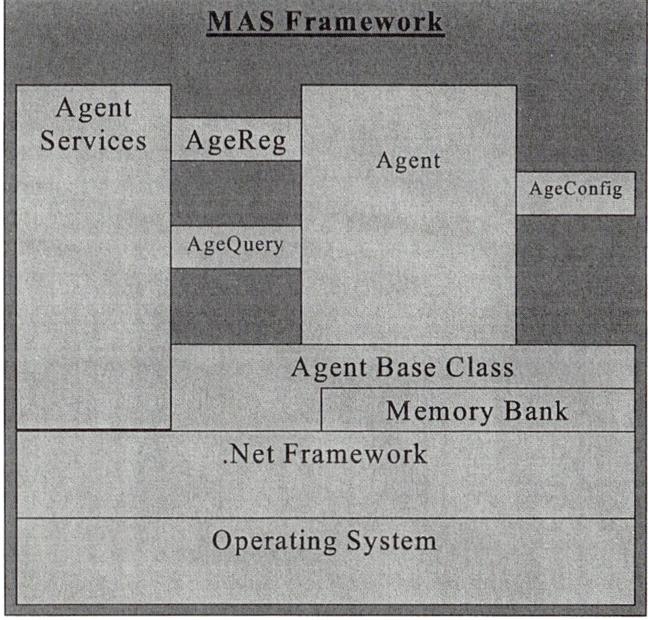

Fig. 1. Summary of MAS Framework components

1. **Agent Base Class** – This is Base class of all Agents application, all agent's applications must inherit from this class. It facilitate the understanding of communication protocols, social awareness (sensing other agents), exposing memory bank for fail safe data and more importantly it implements the functionality required to facilitate the cross domain and cross computer communications.
2. **Agent Services** – It is a key component of the Framework it is a sort of information meeting place of agents in the system. It provides the information to agents

about other agents, their public library and facilitates the use of public function of the agents. It also plays the role of life supporter by rescuing the agents from any unexpected crash and bring them back nearly to their last state of mind

3. **Agent Config** – It is used to change the configuration information of the agent application i.e. communication port number, Agent Service name, URI of the service etc.

4. **Agent Reg** – It is used to register the agents as a public agent and expose their public library and generate the configuration information file based on XML standards.

In the proposed MAS framework every agent is an extension of Agent Base Class (ABC). It registers itself in to local system's Agent Services where all local agents with in the system are registers and agents over the network can be contacted using standard functions of the services. Data is passed between agents using XML standards, Agents Services make this communication possible but it stays invisible from the process. Port number and URI sort of configuration can be changed using agent configuration files. Role of ABC is important as it registers the agent in the environment and become wrapper of the agent for dynamic connection to existing agents. Communication between agents is facilitated using Agent Service (AS). It also makes the agent a robust piece of code.

3.2 Major Advantages

1. **Rapid Development** – In MAS framework, roles and significance of each component has been defined clearly and all these components jointly implement all the pre-understood and pre-established characteristics of Multi Agent System. This clarity of roles facilitates the rapid development MAS applications.

2. **Standardized Protocols** – uses of framework facilitate the implementation of reliable working model for communication based on standardized protocols regardless of application domain or network.

3. **Agent's System Definition** – roles of every component of MAS framework are clearly defined. So we can differentiate that which function is part of the system and which is to be part of the agent. All the functions which are suppose to be available with in the system need not to be developed separately, moreover that system functions are also exposed so the user know how to use the function with in the agents.

4. **User-Friendly Practice of INTERFACE Exposure** – MAS framework consists of numbers of different tools and services, all of them are exposed. XML is usual standard of exposure. This entire system is developed in VB.NET so the source code can be retrieved back and examined as per need.

5. **Centralized Development** – MAS framework is the key for the agents to implement most of its features. So agents are heavily depends on framework. Therefore any possible future enhancement can centrally be implemented.

6. **Single Approach to All Agents** – all agent applications are required to extend "AgentBase" class, which retains all the functions to communicate with the "AgentService" and tools. So all agents have equal approach to the framework and it is up to the "AgentService" to grant the amount of resources and access to the resources.

4 Why .NET

1. **User Friendly** – Visual Studio .NET is regarded as one of the best user-friendly environment for software development. It allows using least keystrokes to develop utmost customized application. IDE behavior is highly predictable across all project types. MAS framework is also developed using .NET language therefor there are no issues regarding incompatibility with .NET environment.

2. **Platform Independence** – .NET offers most extensible range of platform independence which is extremely useful for MAS, as some point of time in near future it has to go inter platform to make it widely usable. Applications written in .NET environment identically behaves in other platforms so that it reduce the development time and increase the viability of the application.

3. **Language Bar** – any development in .NET does not require learning a special language, in fact developer can chose his own preferred language from hundreds of available languages, this characteristic gain the recognition of .NET as widely acceptable Application Development Environment.

4. **Open Source** – .NET environment let the user to see the code written in the application that facilitate the user to understand what is going on in the application more accurately.

5. **Existing Library Options** – .NET contains library of more than 2000 classes, which are very useful for developers. All what user has to do is extend these classes and incorporate the functionality they additionally required and it is all done. It also has in built support of Remoting, windows and web services which allows integrating the inter-platform communication with in the application with out much hassle.

6. **Extensible** – every agent application is a extension of agent base class which facilitate the use of all tool and services available in the environment. MAS are new piece of technology and it is very likely to extend to cope up with new future requirements. It can easily be done in .NET environment as this entire system is developed in .NET environment.

5 Application: Dynamic Truck Scheduling

The motivation for current work is to provide a solution for the dynamic truck scheduling problem from [12]. The solution of DTS system is designed and implemented in the MAS framework. Sample parameters are summarized below.

Players	Capabilities	Descriptions
Central Depot	Schedule Making	It provides the final schedule of truck to Regional Report.
	Schedule posting	It optimize the newly added schedules in context of all existing schedules
Regional Depot	Proposed transportation form	It accepts data input from the user
	Proposed transportation poster	It posts the proposed transportation data to central Agent.
Trucks	Availability checker	It checks the availability of the truck for a span of time in order to add new schedule

Players	Capabilities	Descriptions
Destination	Destination constraints container	It contains all the constraint information in context of destination
Drivers	Availability checker	It checks the availability of the drivers for the span of time to assign to schedule trucks.

Agent Central Depot

Name: Central Depot
Description: this agent is like heart of the system. It takes the schedule requests from Regional depots and exercises the schedules Regional Agents and keeps them posted.
Cardinality Minimum: 1
Cardinality Maximum: 1
Life Time: ongoing
Goal: Make schedule and keep regional depots posted
Functionality: Schedule making, schedule posting

6   Conclusion

We have designed a leading multi agent framework for .NET the results determined so far is promising. MAS Framework is at its primary stage of development but shows a good potential. Most of the basic functionality has been achieved or path has been laid down to achieve them. Framework has been developed in .NET based language and it is completely extendable so the future version can and most like will be based on current version. Existing framework encapsulate the most of the known characteristics of the agent. It facilitates the rapid development of the Agents. Agents can work as stand alone pieces of code for which Agent do not required any supported system but to work as part of team agents depends on framework services. Dynamic truck scheduling problem has answered using this framework which established a role model of using Agent framework to implement agents.

Future work includes development of artificial inteeligence components for MAS including capability to learn from the environment, platform independence, ensuring robustness of the framework, memory bank for fail safe.

References

1. Adriani, D. http://www.dsl.uow.edu.au/dsl/techreports/deby/
2. Brooks, Rodney A. (1990), "Elephants Don't Play Chess," In Pattie Maes, ed., Designing Autonomous Agents, Cambridge, MA: MIT Press.
3. Brustoloni, Jose C. (1991), "Autonomous Agents: Characterization and Requirements," Carnegie Mellon Technical Report CMU-CS-91-204, Pittsburgh: Carnegie Mellon University.
4. Sharma, D., *An intelligent agent planning architecture – IAPA*, submitted for publication, 2003 – http://resources.dmt.canberra.edu.au/sharma/iapa.htm.
5. Dethie, Vincent G. (1986), "The Magic of Metamorphosis: Nature's Own Sleight of Hand," *Smithsonian*, v. 17, p. 122ff
6. Dethie, Vincent G. (1986), "The Magic of Metamorphosis: Nature's Own Sleight of Hand," *Smithsonian*, v. 17, p. 122ff

7. Etzioni, Oren, and Daniel Weld (1994), A Softbot-Based Interface to the Internet. *Communications of the ACM,* 37, 7, 72p;79.
8. Franklin, Stan (1995), Artificial Minds, Cambridge, MA: MIT Press
9. Hayes-Roth, B. (1995). "An Architecture for Adaptive Intelligent Systems," *Artificial Intelligence:* Special Issue on Agents and Interactivity, 72, 329-365, .
10. Kautz, H., B. Selman, and M. Coen (1994), "Bottom-up Design of Software Agents." *Communications of the ACM*, 37, 7, 143-146
11. Langton, Christophcr, cd. (1989), Artificial Life, Redwood City, CA: Addison-Wesley
12. Sharma, D., Dynamic Scheduling using Multiagent Architecture, *KES'04 International Conference*, Wellington, 2004.

On an IT Security Framework

Dharmendra Sharma, Wanli Ma, and Dat Tran

School of Information Sciences and Engineering
University of Canberra, ACT 2601, Australia
{Dharmendra.Sharma,Wanli.Ma,Dat.Tran}@canberra.edu.au

Abstract. IT security is becoming an area of increasing importance especially with the increasing dependency of humans and businesses on computers and computer networks. This paper outlines a generic framework to deal with security issues. The framework proposed is based on a multi agent architecture. Each specialist task for security requirement is modeled as a specialist agent task and to address the global security tasks an environment is invoked in which the multiple agents execute their specialist skills and communicate to produce the desired behavior. This paper presents the framework and its constituent parts and discusses the characteristics of the security problems, the agent roles and how they link up. The security related areas investigated in the current study are discussed and modeled in the proposed framework.

1 Introduction

Nowadays, computers pervade every aspect of human life, ranging from personal entertainment, office operation, corporate business, finance hubs, such as banking and stock exchange systems, and critical infrastructure control to defense and weaponry systems. These computers are connected through wired or wireless connection to generate networks for resource sharing through which processing power or typically large amounts of data are exchanged.

The management of the computers, network devices, and other IT equipment is increasing in complexity with time. The development and implementation of computer systems far out pace computer security needs. Many of the systems currently in place are running critical business, yet little research and development work have been done to secure these systems. Recent trends show a sharp increase in computer security breaches [1] and incidences of virus and worm attacks [2]. The costs incurred in dealing with these security problems are phenomenal. In 2004, in Australia, on average, IT crime and IT security related incidents cost $116,212 per organization [1], while the potential of damage would be even bigger if we take the loss of productivity, loss of confidence in IT, and the resulting negative impact on business.

Despite these difficulties, we are not in a position to readily answer questions such as *"What defenses do I have that will be effective against this attack?"* or *"How do I prevent such attacks in the future?"* [3]. Much work needs to be done to develop security technology and market them to gain user confidence for their uptake of information technology for their secure needs.

R. Khosla et al. (Eds.): KES 2005, LNAI 3681, pp. 226–232, 2005.

IT security covers many issues such as security policy development and implementation, user education, encryption, system administration, network firewall, intrusion detection, and programming practice etc [4]. To secure an IT infrastructure, which consists of the involved computer systems and network devices, many efforts from different areas should be taken. The IT infrastructure is an integrated entity and so it should have security management.

In this paper, we propose an IT security framework. We attempt to systemically integrate research work in several areas of security: system administration assistance, biometrics authentication, intrusion detection, and computer forensics. The proposed security framework integrates the various tasks of gathering security information, analysis of this information using experiential knowledge, generating alert and actions to respond to any security breaches or attempts on breaches. The functionality is integrated in a multi agent distributed environment.

2 Software Agents

A software agent is *"a software entity which functions continuously and autonomously in a particular environment"* [5]. The continuous and autonomous nature of agents and the communication among these agents make them a primary candidate as the underlying framework for IT security. There are many advantages of using multiple agents:

- **An Agent Has Local Knowledge**
  Any security related operation requires the knowledge of IT devices, such as firmware version; operating system type and version, patched levels, and the role of the devices etc. It is virtually impossible to collect this information in a central repository. The information changes constantly. On the other hand, some of the information is only useful to a local device, for example the virus signature data on a particular computer, open ports and provided services on another computer, and remote access to a network router etc. It makes perfect sense to have this information collected, kept, and used by a local software agent. A software agent thus has the local knowledge. It uses or provides the knowledge when needed.

- **An Agent Can Specialize in One Specific Area**
  An agent can also be specialized in a specific area, for example, an intrusion detection agent, an authentication agent, and a traffic analysis agent etc. We can also have different skilled agents to work on the same task. For a piece of information, we can have agents with different skills to examine if there is any trace of intrusion. An agent may be programmed with pattern-matching skills, another agent may be programmed with neural network skills, yet the third agent may be programmed with Hidden Markov Model. This simulates how humans analyse and solve problems, where the same piece of information maybe analyzed by experts with different skills for different purposes.

- **Survivability**
  Due to the distributed and autonomous nature of agents, an agent system can easily survive the loss of some agents. From a system point of view, a large system consists of my components. These components have certain degree of reliability. Such systems have to detect and deal with the possibility of component failure.

- **Scalability**
  Agents are autonomous. Therefore, cooperation among the agents may have varying levels of coupling. An agent may join in or drop out of cooperation at any time. This arrangement gives scalability to an agent system. Within the system, an agent only contacts the other agents when it is necessary. An agent may not be aware of the existence of every other agent. Adding more agents into the agent system does not have to notify every single agent in the system. Only the involved agents are aware of the entrance of the new agents.

- **Extendibility: Multiple Skills, Easy to Introduce New Technology**
  Agents are heterogeneous. Different agents may be programmed with different skills. It is easy to program an agent with new technology and then introduce the agent into the agent system. To program a new agent, one does not have to make changes to other agents.

- **Communication**
  For agents to interact efficiently, their mental states (society layer) should be as convergent as possible ie it would be ideal if agents share the same beliefs and intentions. In order to synchronize their tasks and mental states, the agents must share knowledge. Knowledge sharing is not a trivial task and it leads to a number of potential problems. The knowledge sharing problem in a heterogeneous multi agent system can be stated as: *"Expressions in given agent's native language should be understood by some other agent that uses a different implementation and domain assumptions"* [6]. Any communication language that aims to solve this problem should guarantee the following:

  Syntactic Translation – This caters for a uniform mechanism of sharing the propositions – or to share the knowledge and the knowledge about knowledge (meta-knowledge)

  Semantic Translation – Each agent maintains its view of its environment. This collection of terms, definitions and the relationships between them is referred to as ontology. Semantic translation ensures that ontology context is preserved.

  Communication Models – Agents must be able to communicate using various means of communication (one-to-one, one-to-many, subscription to new knowledge etc.) and to request services from each other. Communication models must define these tasks.

3   Current Work

Our current research work concentrates on a designing and developing a multi agent framework to address the above IT security areas. The proposed framework is summarized in Figure 1, where many agents with different skills are clustered into groups to perform the tasks such as System Administrator Assistant, Integrated Intrusion Detection, Bio-authentication, and Computer Forensics. These tasks together form this proposed IT Security Framework.

The main security problems considered are summarized below.

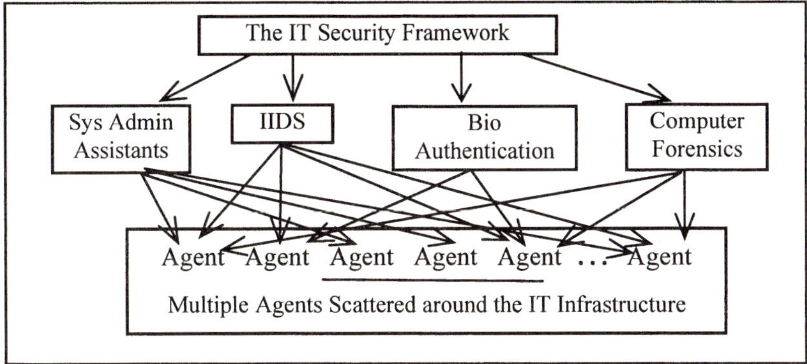

Fig. 1. The Architecture of the IT Security Framework

3.1 System Administration Assistant – Windows Autopilot

Personal computers seem to be a household necessity. The management of these computers especially OS and application configuration, security updating, software upgrading, system fine tuning, routine maintenance, and critical data backup pose major technical challenges. These are beyond the scope of ordinary users. Even IT professionals, who are not experienced in the field, feel incapable of managing these issues. Our *Windows Autopilot* project is aiming at using agent technology to provide computer users with auto management facility. An agent, which has the knowledge of an experienced system administrator, is running on a computer and performing all the required system management and administration tasks. It will also remind the users with good practice and answer the queries. Window Autopilot agent can serve 3 different purposes:

- To help end users to manage the Windows computers.
- To be part of Integrated Intrusion Detection System for intrusion detection purpose.
- To be part computer forensics investigation, if needed.

3.2 Authentication and Authorization

User authentication (human-by-machine authentication) is the process of verifying the identity of a user: is this person really who he/she claims to be? User authentication has become more complicated and difficult with the onset of the computer age. An entity cannot be seen on the remote end of a computer network. The other entity could be the genuine recipient, a machine or an attacker. It is apparent that attackers can use the Internet to access other user's records without the need for physical presence.

Authentication systems can be categorized into three main types: knowledge-based authentication (what you know, e.g., passwords and PIN), Object-based authentication (what you have, e.g., physical keys), and ID-based authentication (who you are, e.g., biometric ID such as voiceprint and signature, and physical ID such as passport and credit card) [7].

Passwords are excellent authenticators, but they can be stolen if recorded or guessed. Physical keys or other physical devices have similar advantages and disadvantages. Biometrics are useful to establish authenticity and for non-repudiation of a transaction, wherein a user cannot reject or disclaim having participated in a transaction. There has been a significant surge in the use of biometrics for user authentication in recent years because of the threat of terrorism and the Web-enabled world [8]. However, biometrics can be copied or counterfeited, so they cannot ensure authenticity or offer a guaranteed defense against repudiation. Different types of authenticators should be combined to enhance security and performance [9].

We propose a multiple agents based user authentication and authorization system based on knowledge and biometric ID features, which will operate on standard computers providing a high level of security. Our proposed paradigm is a process of verifying the claimed identity of a user based on his/her handwriting and voice characteristics as well as the information content of the spoken and written phrases. Therefore the proposed multiple agents based user authentication and authorization system will consist of the following agents: voice verification (VV), verbal information verification (VIV), signature verification (SV), and written information verification (WIV). The VV (or SV) agent is to verify the user based on his/her voice (or signature). The VIV (or WIV) agent is to verify spoken (or written) words against personal information, such as the date of birth or the mother's maiden name, which is stored in a given personal data profile.

The proposed agents are loosely coupled. They may join in or drop from the cooperation at any time. For example, the VV agent may individually join in and it may not be aware of the existence of the other agents. For voice-based authentication, the involved agent is the VIV agent.

The proposed system can be used either in Tablet PCs and future-generation mobile phones equipped with both microphones and pen, or in standard PCs where a pen is unavailable. The proposed system can achieve a considerably higher level of security.

3.3 Intrusion Detection

Three reasons, in our opinion, are responsible for the high false alarm rate of current intrusion detection practice. They are:

- only a single information source is analyzed by an intrusion detection system,
- only a single method is used for the analysis, and
- there is no distinction of vulnerability, threat, attack, and intrusion.

We propose a multiple agents based integrated intrusion detection system (IIDS) [10]. There are many agents in an IIDS, and they are deployed on the strategic points of the underlying computer and network systems. An agent of IIDS has a special skill to monitor certain types of computer or network activities. It can be regarded as a traditional intrusion detection system by itself. A single agent will not raise alarm by itself. It just passes information back to an assessment center. The assessment center builds up the whole picture of the current activities based on the information from these agents. It will further instruct some agents to pay attention to certain type of activities or raise alarm. On the other hand, an agent also has the knowledge of the local system, which helps it to make better decision. For example, if an agent is running on a particular host, it would have the knowledge of the host: its operating system, patch level, and services etc. If it observes an attack attempting on a vulnerability that has already been patched on the system, it knows that there will be no damage to the system at all. With multiple agents scattered all over the underlying systems gathering information and the assessment center evaluating the situation, IIDS is anticipated to detect intrusion efficiently and effectively.

3.4 Computer Forensics

Computer Forensics (CF) techniques are used to discover evidence for a variety of crimes relating to illegal use of a computer, or computer network. They are usually developed by the 'try and fix' method which means that there is little academic or industrial research literature available. However, these ad-hoc methods are problematic if the gathered information is to be used as evidence in a prosecution. Also the knowledge that is acquired during an investigation is often lost over time due to long times, and loss of experienced staff, between each case. What is needed is a clear methodology for finding, analyzing and recording forensic evidence that will increase the validity of the evidence in a court. To increase the accuracy of the evidence collected a wide variety of background knowledge is needed: computer protocols, information about the target network, the location of the logs for the system, and knowledge of CF tools and experience in their use. The ability to find and use this knowledge will also determine the speed with which the evidence can be gathered. Both of these factors are critical in providing a timely and useful response to a compromise or to a prosecution. CF is a complex collaborative problem where many different skill sets (including investigator, system administrator, and researcher) are needed. Automation of the collaborative tasks is amenable to multi-agent technology.

It is the aim of this project to use the multi-agent framework to facilitate the gathering of computer forensic evidence. It is expected that the developed agent system will increase the validity, accuracy and speed of gathering computer forensic evidence.

4 Conclusion

The paper summarizes the need for a better security technology for IT and presents a design for a multi agent based framework. Several security problems that have been discussed are being considered as applications for the framework. The framework is being implemented and tested for the identified problem and the framework will be enhanced into a model to cater for security needs for data and applications shared on networks.

References

1. AusCert, Australian Computer Crime & Security Survey, http://www.auscert.org.au/download.html?f=114. 2004, AusCERT.
2. Kienzle, D.M. and M.C. Elder. Recent Worms: A Survey and Trends. in ACM Workshop on Rapid Malcode, WORM'03. 2003. Washington, DC, USA: ACM.

3. Saydjari, O.S., Cyber Defense: Art to Science. Communications of ACM, 2004. **47**(3): p. 53-57.
4. Pfleeger, C.P. and S.L. Pfleeger, Security in Computing. Third ed. 2003: Prentice Hall.
5. Bradshaw, J.M., An Introduction to Software Agents, in Software Agents, J.M. Bradshaw, Editor. 1997, AAAI Press/The MIT Press. p. 3-46.
6. Finin, T., Y. Labrou, and J. Mayfield, KQML as an Agent Communication Language, in Software Agents, J. Bradshaw, Editor. 1995, MIT Press.
7. O'Gorman, L., Comparing Passwords, Tokens, and Biometrics for User Authentication. Proceedings of the IEEE, 2003. **91**(12): p. 2021-2040.
8. Bolle, R.M., et al., Biometrics 101, IBM Research Report. 2002, IBM: IBM T. J. Hawthorne, New York, USA.
9. Namboodiri, A.M. and A.K. Jain. On-line Script Recognition. in Proceedings of the Sixteenth International Conference on Pattern Recognition. 2002. Quebec City, Canada.
10. Ma, W. and D. Sharma. A Multiple Agents Based Intrusion Detection System. in Ninth International Conference on Knowledge-Based Intelligent Information & Engineering Systems (KES2005). 2005. Melbourne, Australia: Springer-Verlag.

Investigating the Transport Communicating Rates of Wireless Networks over Fading Channels

Xu Huang[1] and A.C. Madoc[2]

[1] School of Information Sciences and Engineering, University of Canberra,
ACT 2601, Australia
Xu.Huang@canberra.edu.au
[2] Center for Actuarial Studies, Department of Economics, University of Melbourne,
VIC 3051, Australia
a_c_m81@hotmail.com

Abstract The large wireless networks under fading, mobility, and delay constraints are further considered under more generic conditions comparison with previous investigating are discussed. The previous implicit restrictions are relaxed, together with more realistic assumptions, make the conclusions, the transport capacity of wireless networks over fading channels more useful for the relevant protocol designs for the wireless networks.

1 Introduction

In recent years, one of the popular topics in wireless networks consisting of nodes with radios has been involved [1-2]. Wireless ad hoc networks consist of mobile nodes communicating over a wireless channel. Contrary to cellular networks, where the nodes are restricted to communicate with a few strategically placed base stations, in wireless ad hoc networks nodes are allowed to communicate directly. Such networks have no wired backbone network and all communication must take place only over the shared wireless medium. However, because of the nature of the wireless channel, each node can effectively communicate with only some of the others, typically those that lie in its vicinity. On the other hand, the traffic requirements are taken to be arbitrary and all notes hear a superposition of the attenuated signals transmitted by all other nodes, thus, one would like to have an information theoretic basis for organizing such information transfer.

Current protocol development effects [3] are aimed at realizing the following strategy. Packets are relayed from node to node until they reach their intended destination. At each hop a packet is fully decoded, thus digitally generated, and then retransmitted to the next node on its path. The decoding of a packet at a node is done by treating all interference from concurrent transmissions as useless noise. We may call this as the "multi-hop" strategy. To make sure this strategy works well, it requires a suite of protocols. For example, a medium access control protocol is needed to avoid excessive interference at receivers, since transmitters in their vicinity need to be silenced. This is the goal of the IEEE 802.11 protocol in Distributed Coordination Function (DCF) mode [4], also the proposals such as Dual busy tone multiple access (DBMTA) [5], and SEEDEX [6]. A power control protocol is needed not just look after the saving battery life, but also to regulate the power of transmissions which

R. Khosla et al. (Eds.): KES 2005, LNAI 3681, pp. 233–239, 2005.

need only traverse a short hop while avoiding creating unnecessary interference for other concurrent transmissions [7]. A routing protocol is vital for a network to determine the path to be followed by packets from a source to its destination [8].

It is obvious and important to characterize how much has been traded for different models from the so-called idea case, for example, how much we lose in capacity if we use point-to-point coding rather than network coding, how much will affect the capacity maximum estimation if we use Gaussian noise model rather than other statistic or stochastic models.

In the rest of the paper, the organization is as follows. In section 2, we focus on one of our key issues of this paper, namely, what the model we are going to discuss and what else we need to consider for the wireless networks under fading, mobility, and delay constraints. In the section 3, we are going to extend the discussion of paper [9] to more generic situation. We also conclude with some remarks about this extension.

2 The Model

The usual theoretic model was enriched by taking into account the distances between nodes located on a plane with a minimum separation distance between nodes. Each node is power limited, and a performance measure C, called the transport capacity, was investigated, which is the supremum of the distance weighted sum of rates taken over all feasible rate vectors. The wireless medium itself was very simplistically modeled. An attenuation of the form $e^{-xr/rd}$ was presumed, as a signal traveled a distance r, here $x \geq 0$ is the absorption constant, and d is the path loss exponent.

But there is a very important issue in a real application needs to considered, which is the presence of multi-path fading. In a wireless network, due to the real environment, the electromagnetic waves travel to receivers along a multitude of paths, encountering delays and suffering gains that vary with time. This was not particularly taken into account for in [9].

It is well known that depending on the frequency bandwidth used, and how fast the environment changes, the fading can be classified into four types. In other words, we can have a close look the fading natures both in frequency and time aspects. If the bandwidth W of the signal is much smaller than the channel coherence bandwidth B_{ch}, namely $W << B_{ch}$, then the channel is called frequency non-selective or flat fading. In this case, the channel only has a multiplicative effect on the signal. If the case is with $W \geq B_{ch}$, the receiver will have several resolvable signal components; therefore such a channel is titled frequency selective. Besides the frequency natures, we can also characterize a fading channel by comparing the time duration T_{sig} of a signal symbol with the channel coherence time T_{coh}. A channel is titled as slow fading if $T_{sig} << T_{coh}$ or as fast fading if $T_{sig} \geq T_{coh}$. Summarizing the above descriptions, we have four combinations, which formed four types of fading channels, i.e. flat slow, flat fast, frequency selective slow, and frequency selective fast fading channels [10].

Following [9-12], we consider a collection of n nodes, representing the immobile devices, X_1, X_2, ..., X_n, placed randomly, uniformly and independently, in the two dimensional area, $\{(x, y) : -\frac{1}{2} \leq |x|, \ |y| \leq \frac{1}{2}\}$. We also assume that each node is the source of a single stream, and the destination of a single stream. A node cannot be both source and destination of the same stream, in other words we are not going to discuss the case for self transition. Each node can transmit with a power $P_i \leq P_{max}$, where P_{max} is the maximum power at this node. If we assume the received power at node X_j is $G_{ij}P_i$, where $G_{ij} = Kd_{ij}^{-\alpha}$, K is a constant, the same for all nodes, d_{ij} is the distance between nodes X_i and X_j, and $\alpha > 2$ is defined as decay exponent factor. Let $\{X_t : t \in \mathfrak{J}\}$ be the set of transmitting nodes at a given time, each node X_t transmitting with power P_t. Then, let X_j, $j \notin \mathfrak{J}$ is receiving information from X_i, $i \in \mathfrak{J}$. We use the signal to interference and noise ratio (SINR) at node X_j, as

$$C_j = \frac{G_{ij}P_i}{\psi + \sum_{k \in \mathfrak{J}, k \neq i} G_{kj}P_k} \tag{1}$$

where ψ is the thermal noise power at the receiver, which is assumed the same for all nodes. Therefore, we have the fact that the transmission of the packet will be successful if and only if the transmission rate used, R_j, meets the following inequality:

$$R_j \leq f_R(C_j) = W \log_2(1 + \frac{1}{\Gamma}C_j) \tag{2}$$

where $\log_2(x)$ is defined as the logarithm of x, of base 2. If $\Gamma = 1$, the receiver achieves Shannon's capacity. In fact it is important to note that this is linking with the channel bandwidth, for example if we take a channel with infinite bandwidth, with the white noise, we shall have limit capacity, $C_j = 1.44$ (signal/noise) [16]. When $G > 1$, equation (2) approximates the maximum data rate under a given bit of error (BER).

Let us focus on the case that node mobility with a fixed delay constraint in [13]. In order to discuss the nodes with fading, it was introduced a factor f_{ij}, called fading coefficient, a non-negative random variable that models fading and does not change with time. Also in [9], it was assumed that the expectation of a factor, $E[f_{ij}] = 1$ and is permutable, i.e. $f_{ij} = f_{ji}$. The paper makes a very mild assumption that there is a median value $f_m > 0$ such that the probability $P[f_{ij} \geq f_m] \geq (1/2)$. Under those assumptions, the results [9] shows for (mobile nodes with a polynomial upper bound on delay) under node mobility, fading, and an upper bound on the acceptable packet delay $d_{max} \leq (4Ns)n^d$, each node can communicate with its destination with a rate equal to equation (3). Here $C^F{}_{min}(n)$ is the capacity under this case, where "F" means "fading", k_2 is a constant. From equation (3) we can see that the rate is proportional to f_m, which was proven that it existed for the most three popular distributions, i.e. the Nakagami, Ricean, and Rayleigh [12].

It is well known that for investigating the natures of multiple-channel fading one of the important tools is using the Nakagami distribution [13-15]. Hence, let us focus on the Nakagami distribution and try to extend to the results obtained by [9] to more generic situation.

$$\lambda_4(n) = R_{min}(n)\frac{1}{18}[Q(n)k_3 \log n]^{-1}$$

with

$$R_{\min}(n) = W \log_2 [1 + \Gamma^{-1} C_{\min}^{F}(n)]$$

$$C_{\min}^{F}(n) = \frac{f_m}{k_2 \log n} C_{\min}(n) \tag{3}$$

$$C_{\min}(n) = \frac{3\alpha - 6}{3\alpha - 5} 10^{-\frac{\alpha}{2}-1}$$

3 Communicating Rates Under the Nakagami Distributions

We are going to discus two issues in this section, firstly we shall very briefly introduce Nakagami distribution related to our focus in this paper, and then we relax the condition in [9].

The Nakagami model is a model purely based on empirical. But it is well known that it provides a closer match to some experimental data than the Rayleigh, Ricean or log-normal distributions. The Nakagami distribution describes the received envelope by the distribution [11]:

$$p_{A_k}(A_k) = \frac{2}{\Gamma(m_k)} \cdot [\frac{m_k}{\Omega_k}]^{m_k} \cdot A_k^{2m_k-1} \cdot e^{\frac{m_k}{\Omega_k} A_k^2}, \quad k = 1,2,...,M \tag{4}$$

where $\Gamma(\cdot)$ is the Gamma function, $\Omega_k = <A^2_k>$ is the average power on k^{th} branch, m_k is the fading parameter. As m_k becomes smaller, the degree of fading becomes more severe. The Rayleigh distribution and the one-side Gaussian distribution are special cases with $m_k = 1$ and $m_k = 0.5$, respectively. The fading parameter m_k can be any real value greater than or equal to 0.5. Sometimes it would be simple for the case that it only takes integer values of m_k, considering that the measurement accuracy of the channel is typically only of an integer order. The performance for non-integer m_k can be obtained by interpolating from two integer m performance curves. Therefore, we restrict our analysis to the case of identical branch fading parameters, i.e. $m_k = m$ for $k = 1, ..., M$, and denoted the Nakagami distribution as NK(m,Ω). Hence, we may use the probability density function (PDF) of NK(m,Ω) to obtain the moment as below:

$$E[z^\alpha] = \frac{2}{\Gamma(m)} (\frac{m}{\Omega})^m \int_0^\infty z^{2m+\alpha-1} \exp(-\frac{m}{\Omega} z^2) dz \tag{5}$$

which can be used to obtain the following equation as shown in equation (6). It is well know that the Gamma function $\Gamma(m)$ is defined by equation (7), which , for a positive integer m, can be simplified to $\Gamma(m) = (m-1)!$. Therefore, the average power Ω is an intermediate parameter, which we do not need to specify. Within this expression, it is easy to obtain the equations (8) and (9).

$$E[z^\alpha] = \frac{\Gamma(m+\frac{\alpha}{2})}{\Gamma(m)} (\frac{\Omega}{m})^{\alpha/2} \tag{6}$$

$$\Gamma(m) = \int_0^\infty x^{m-1} e^{-x} dx \tag{7}$$

Now if we let $y = z^2$, we have the following results from equations (8) and (9), which shows the relation between var$[y]$ and var$^2[z]$ in equation (10) and can be used to determine the variance of squared Nakagami signal.

$$\text{var}[z] = E[z^2] - (E[z])^2 = [1 - \frac{1}{m} \frac{\Gamma^2(m+\frac{1}{2})}{\Gamma^2(m)}]\Omega \tag{8}$$

$$\text{var}[z^2] = \frac{\Omega^2}{m} \tag{9}$$

$$\text{var}[y] = \frac{\text{var}^2[z]}{m}[1 - \frac{1}{m} \frac{\Gamma^2(m+\frac{1}{2})}{\Gamma^2(m)}]^{-2} \tag{10}$$

Now let us have a look at the assumptions made by the [9] for the Nakagami distribution. As we mentioned that in [9] there is assumption $E[f_{ij}] = E[R^2] = \Omega = 1$, now submitting this into equation (9) with $z = R$, and $m \geq 0.5$, we have var$[R^2] \leq 2$, which implies that the Gaussian components have lower standard deviations for the results obtained from [9]. In fact that we can see that the different m values affect the shapes of their pdf, which is shown in the left hand side of Figure 1, where we assumed the $\Omega = 1.0$ and allowed m changes from 0.6 to 5.0. It is obvious from the diagram (left hand side of Figure 1) we have an observation of large m, say $m = 5.0$ with the same Ω value, is almost symmetric shape but narrower than lower m, say $m = 0.6$, which effectively caused the Gaussian components have lower variance for a large m values. As mentioned before, since the paper [9] also assumed that $\Omega = 1$, which, upon closer inspection, needs to be more carefully investigated, as we first show that the Ω values do affect the distribution bands as shown in the right hand side of Figure 1, where we let the $m = 0.7$, and Ω takes values from 0.7 to 5.0. When Ω values are smaller, say $\Omega = 1$, the symmetric shape is more than that in larger values, for example when the Ω takes larger value, say $W = 5.0$, at the same m-value, its shape would be more non-symmetric shape than smaller one, say $\Omega = 1.0$, as shown in the right hand side of Figure 1. Since the core factor defined by [13] is f_m, which

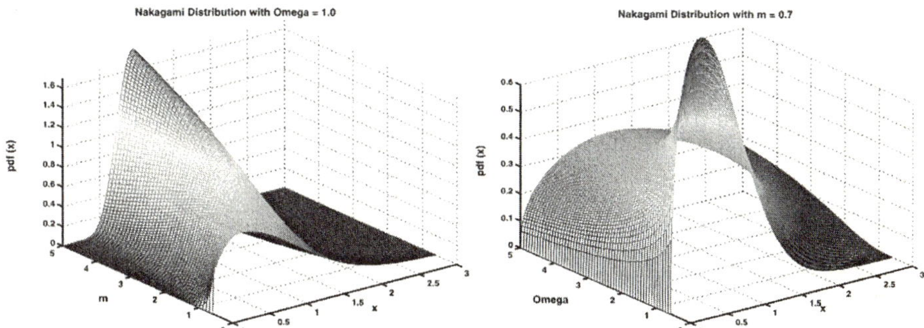

Fig. 1. Left hand side is Nakagami-m distribution with $\Omega = 1.0$ and different m values. The right hand side is Nakagami-m distribution with $m = 0.7$ and different Omegas values. We can see that the Gaussian components have lower variance for a larger m values and the different band with different deviation-like values.

directly affects the capacity of wireless networks as well as the communication rates as presented in equation (3). Therefore, we can immediately obtain, from the Figures 1, that the f_m will be affect by the key parameters, namely m and Ω.

It is interesting to note that the numerical results show that even the parameters m and Ω are affecting the Nakagami distributions, in terms of symmetric shapes and bandwidths, in different directions, we cannot expect them simply "compensate", which is shown in Figure 2, which will be discussed in another paper.

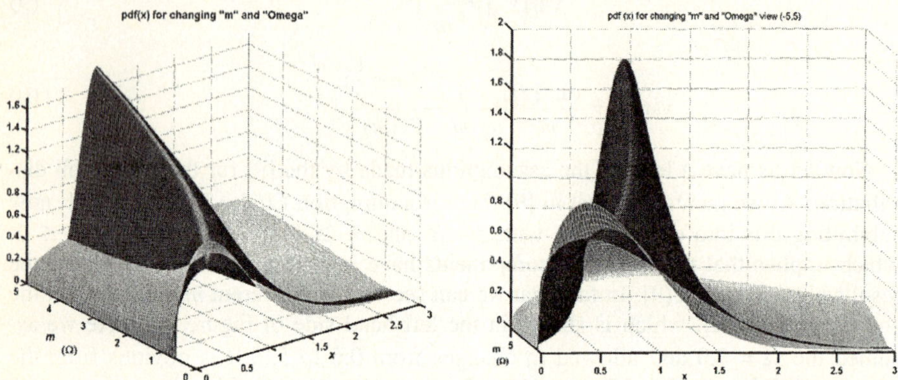

Fig. 2. Left hand side of the diagram is simply put both diagrams in Figure 1 together and the right hand side of the diagram is a view turning a degree for the left hand side diagram. It shows that the parameter m and Ω can not be simply compensated due the non-linear natures.

4   Conclusions

We have investigated the transport communicating rates of wireless networks over fading channel. For the fixed constant Γ^{-1}, we knew the rate is proportional to $R_{\min}(n)$ in equation (3), hence we have the conclusions that if we use larger m and smaller Ω values or either one case, we may have smaller communication rates for the same situations with smaller m and larger Ω values. The previous implicit restrictions described by Nakagami distributions, which is powerful tools for investigating multi-channel fading, are examined via two key parameters, m and Ω. The effects made by those two parameters to the communication rates are also discussed, which offers a correct direction for designs of protocols for wireless networks.

References

1. "The Fourth ACM International Symposium on Mobile Ad Hoc Networking and Computing Mobihoc 2003," June 1-3 2003. Annapolis, MD, USA, http://www.sigmoblile.org/mobihoc/2003/.
2. Jie Wu and Ivan Stjmenovic.: Recent developments offer potential solutions to problems encountered in ad hoc network." IEEE Computer Society, Computer, February (2004) 29.
3. "Mobile Ad-hoc networks (manet). IETF Secretariat," October 10 2003. http://www.ietf.org/html.charters/manet-charter.html.

4. Mustafa Ergen, Duke Lee, Raja Sengupta, and Pravin Varaiya: Wireless Token Ring Protocol-Performance Comparison with IEEE 802.11, Proceeding of the Eighth IEEE International Symposium on Computers and Communication (ISCC'03), (2003), 1-6.
5. Z. J. Haas and J. Deng: Dual busy tone multiple access (DBMTA) – performance evaluation, in Proceedings of the IEEE VTC'99, (Houston, Texas), May, (1999) 17-21
6. S. Narayanaswamy, V. Kawadia, R. S. Sreenivas, and P. R. Kumar: Power control in ad-hoc networks: Theory, architecture, algorithm and implementation of the COMPOW protocol, in European Wireless Conference-Next Generation Wireless Networks: Technologies, Protocols, Services and Applications, (Florence, Italy), (2002) 156-162.
7. V. Kawadia, S. Narayanaswamy, R. Rozovsky, R. S. Sreenivas, and P. R. Kumar: Protocols for Media Access Control and Power Control in Wireless Networks, Proceeding of the 40th IEEE Conference on Decision and Control, Orlando, FL, Dec (2001), 1935-1940.
8. X. Hong, K. Xu, and M. Gerla: Scalable routing protocols for mobile ad hoc networks, IEEE Network Magazine, (2002) 11-21.
9. S. Toumpis and A. J. Goldsmith: Large wireless networks under fading, mobility, and delay constraints," *IEEE INFOCOM,* Hong Kong, China, Mar. 2004.
10. Feng Xue, Liang-Liang Xie, and P.R. Kumar: The Transport Capacity of Wireless Networks over Fading Channels, *IEEE Transactions on Information Theory* on 7, (2004), 1-14.
11. B. P. Lathi: Modern Digital and Analog Communication Systems, third edition, Oxford University Press (1998).
12. S. Toumpis: Capacity of Wireless Ad Hoc Networks, Ph.D. thesis, Stanford University, 2003.
13. Sang Wu Kim: Adaptive Rate and Power DS/CDMA Communications in Fading Channels, IEEE Communications Letters. Vol. 3 No.4 (1999), 85-87.
14. Marc Torrent-Moreno, Daniel Jiang, and Hannes Hartenstein: Broadcast reception rates and effects of priority access in 802.11-Based vehicular ad-hoc networks, International Conference on Mobile Computing and Networking Proceeding of the First ACM workshop on Vehicular ad hoc Network, Philadephia, PA, U.S.A. (2004) 10-18.
15. George K. Karagiannidis, Dimitris A Zogas, and Stavros A. Kotsopoulos: An Efficient Approach to Multivariate Nakagami-m Distribution Using Green's Matrix Approximation, IEEE Transactions on wireless communications vol. 2, no.5 (2003), 883-889.
16. M. Nakagami: The m-distribution, a General Formula of Intensity Distribution of Rapid Fading, in Statistical Methods in Radio Wave Propagation. W. G. Hoffman, Ed. Oxford, England: Pergamon, 1960.

Agent Team Coordination in the Mobile Agent Network

Mario Kusek, Ignac Lovrek, and Vjekoslav Sinkovic

University of Zagreb, Faculty of Electrical Engineering and Computing
Department of Telecommunications, Unska 3, HR-10000 Zagreb, Croatia
{mario.kusek,ignac.lovrek,vjekoslav.sinkovic}@fer.hr

Abstract. The paper deals with coordination of an agent team with mobile agents. A formal model based on the Mobile Agent Network is proposed. It elaborates a shared plan through the process that allows team formation according to tasks complexity and characteristics of a distributed environment where they should be performed. A case study describing the multi-agent system with mobile and intelligent agents for software management is included.

1 Introduction

A multi-agent system containing mobile and intelligent agents is a promising paradigm for network and distributed system management, especially for software operation and configuration in a large environment, such as a mobile telecommunication network or Grid. Benefits expected from the agent-based approach are the following ones: human operations reduced to specification and supervision of management tasks to be executed by agents, and remote operations treated as the local ones because of the agent mobility [1, 2].

The paper discusses a multi-agent system organized as an agent team and coordinated according to the shared plan that defines each agent's role (operations and where to do them). A shared plan is elaborated through the process that allows team formation according to the complexity of a management task and characteristics of the environment in which it should be performed. A formal model is based on the Mobile Agent Network where the multi-agent system resides in a network with agents migrating from node to node and performing the operations [3].

The rest of the paper is organised as follows: Section 2 deals with an agent team in the Mobile Agent Network and elaborates the shared plan. A case study describing agent-based software operation is given in Section 3. Performance evaluation of the agent coordination schemes is described in Section 4, while Section 5 concludes the paper.

2 An Agent Team in the Mobile Agent Network

The Mobile Agent Network (MAN) is used for modelling agent coordination in the agent team. MAN is represented by triple $\{A, S, N\}$, where A represents a multi-agent system consisting of cooperating and communicating mobile agents that can migrate autonomously from node to node; S is a set of n_s nodes in which the agents perform operations; and N is a network that connects nodes and allows agent mobility.

R. Khosla et al. (Eds.): KES 2005, LNAI 3681, pp. 240–246, 2005.
© Springer-Verlag Berlin Heidelberg 2005

An agent $agent_k$ is defined by the triple {$name_k$, $address_k$, $task_k$}, where $name_k$ defines a unique agent identification, $address_k$ the list of nodes to be visited by the agent, and $task_k$ functionality it provides, i.e. the set of assigned elementary operations, s_i. When hosted by the node $S_i \in address_k$, $agent_k$ performs operations, $s_i \in task_k$. If an operation requires specific data, the agent carries them during migration ("loaded agent").

In organization of the multi-agent system suitable for network and distributed system management, the agent team that uses shared plan is considered [4-7]. An intelligent stationary agent is used for decomposing a complex management task into n_t elementary operations and ordering of operations. The same agent also collects and interprets data which describe characteristics of the nodes and network in order to define a suitable agent team.

The following assignments of elementary operations are considered as the basis for identification of the agents-shared plan:

- R1: a single agent executes all operations on all nodes;
- R2: an agent executes a single operation on one node only;
- R3: an agent executes all operations on one node only;
- R4: an agent executes a specific operation on all nodes;
- R5: an agent executes a specific operation only once on all nodes;
- R6: operations are assigned to the agents in order to exploit maximal parallelism of operation. Mutually independent operations are assigned to different agents, in order to execute them simultaneously on nodes with parallel execution supported;
- R7: hybrid solution combining R4 and R3. An agent is responsible for a specific operation on all nodes; all other agents execute all other operations, each on a different node;
- R8: hybrid solution combining R5 and R3 (specialization of R7 in a way R5 is specialization of R4).

Table 1. Characteristics of assignment schemes

Scheme	Operation/Agent	Agent/Node	Agent/Team	Coordination
R1	n_t	1	1	I
R2	1	n_t	$n_t \times n_s$	I, L and G
R3	n_t	1	n_s	I
R4	1	n_t	n_t	I and L
R5	1	n_t	n_t	L
R6	1 to n_t	1 to n_t		I and L
R7	1 or n_t -1	2	$1 + n_s$	I and L
R8	1 or n_t -1	2	$1 + n_s$	I and L

The assignment scheme defines complexity of agents (elementary operations per agent), their number in the agent team and the coordination mechanism required. Table 1 describes characteristics of the proposed assignment schemes. Column Operation/Agent shows the number of operations executed by an agent; column Agent/Node shows the number of agents needed to execute all operations on each node, and column Agent/Team shows the number of agents in a team. Three types of coordination are considered:

a.) internal (I), when the operations are performed by the same agent;

b.) local (L), when the operations are performed by different agents on the same node;

c.) global (G), when the operations are performed by the agents on different nodes.

Figure 1 shows lifecycle of an agent. Agent creation is characterized by its birth. After birth the agent migrates to the first node where it has to execute its operation. If the agent carries data its migration time is longer. Having arrived to the first node it executes operation and via dialogue informs other agents of the result. Dialog depends upon coordination (local or global). Then the agent takes over next operation. If this operation has to be executed at the same node the migration is skipped. If, however, next operation is to be executed on another node the agent migrates to it, executes the operation and performs a dialogue. This is repeated for all operations from the task list ($task_k$). The last operation is the agent's death by which the agent is disposed.

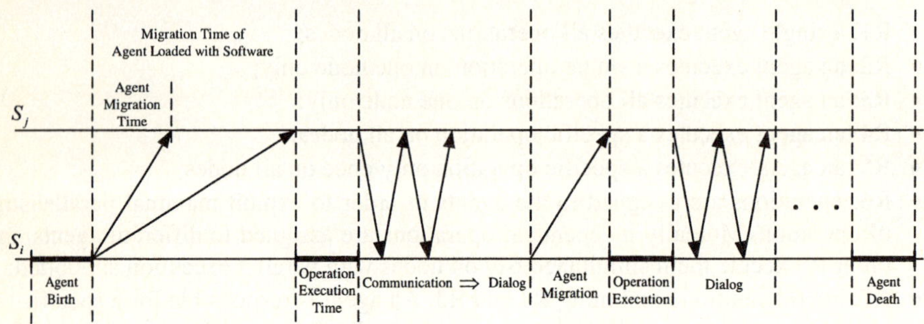

Fig. 1. Agent's lifecycle

Assignment scheme efficiency depends on the task submitted (number, ordering and complexity of elementary operations) and environmental characteristics. Basic parameters describing environment are the following ones: operation execution time, agent migration time, loaded agent migration time, time required for a dialogue between the agents residing at different nodes, and the type of agent communication, which can be either direct or indirect. When the agent sends a message, the dialogue fails if the receiving agent is not at the expected destination. In direct communication the sender waits for some time and retries a dialogue until it succeeds. In indirect communication the sender creates a transport agent which migrates to destination, and delivers a message to the receiving agent when it arrives [6].

3   Case Study: An Agent Team for Network Software Operations

Software management in large environments includes introduction and modification of software products on many nodes in order to correct the faults, to improve performance or other attributes, to adapt the products to a changed environment or to improve their maintainability.

A multi-agent system called MA-RMS (Remote Maintenance Shell) serves for the case study [8, 9]. It offers the following software operations: migration, installation/uninstallation, starting/stopping, running according to the specified execution parameters and tracing. Testbed which allows software verification at the target node is provided as well. Graph shown in Figure 2 describes the tasks defined as "introduction of a new software" and "replacement of the existing one by the new one".

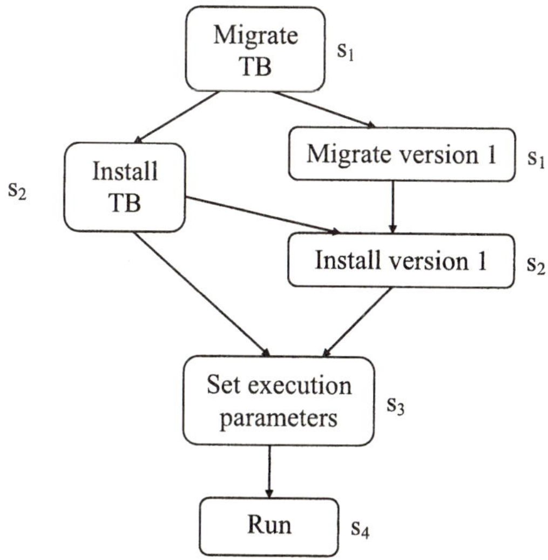

Fig. 2. Software operation graph

Six operations are used to introduce/replace and verify software at a single node. Because the required final state is "run software at the specified node", the agents can reduce the number of operations by checking the initial state and skipping already completed operations. For instance, if a testbed (TB) is preinstalled, only operations s_1 – s_4 related to software (version) will be executed.

4 Performance Assessment of an Agent Team

Performance assessment of the agent team [10, 11] for different plans is based on the following assumptions:

a.) agent migration and global dialogue time are the same;
b.) migration time for a loaded agent is multiplied by 1-10 for small and by 10-1000 for large amount of data;
c.) agent migration and dialogue time depend on the available network transmission rate;
d.) operation execution time depends on the node capacity and is the same for all operations;
e.) agent creation and agent death consumes time equals to operation execution time;
f.) time is measured in a discrete time unit Δt.

Figure 3 shows total execution time for the assignments R1-R8 with direct agent communication in a small network having less than 10 nodes. The origin of graph represents a fast network with high capacity nodes.

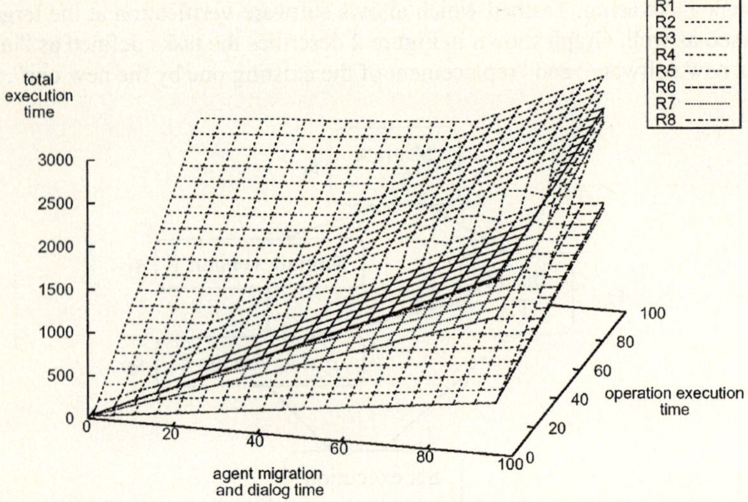

Fig. 3. Execution graph for direct communication between the agents

Assignments R1 and R3 will be discussed with some details shown in Figure 4. R1 gives better results in Area A, while R2 is more effective in Area B. There is only one agent in R1 and only $2\Delta t$ are needed for its birth and death. In R3, there are as many agents as network nodes and, therefore, $2n_s\Delta t$ is spent for their birth and death. That is why assignment R1 is better than R3 in the fast networks where agent migration time is practically negligible (Area A). On the other hand, agent migration has more influence in slower networks, when R3 gives better results (Area B). Agent teams always perform better in the environments with high number of widely distributed nodes.

Another important question is related to the agent load i.e. software carried by an agent during migration. Two examples are marked with big black dots in the Area A and Area B (Figure 4).

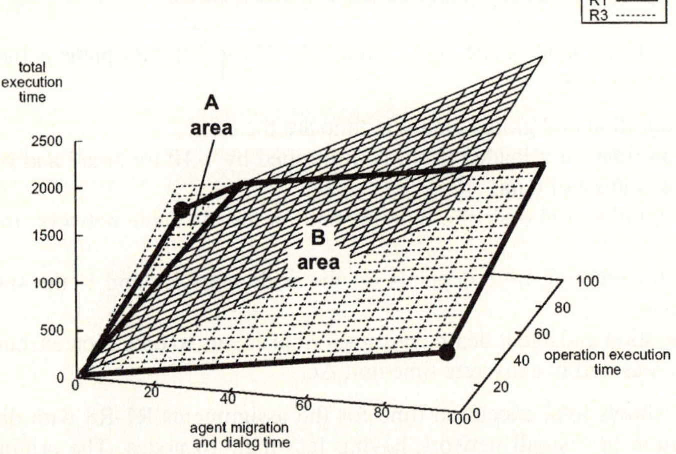

Fig. 4. Execution graph for assignments R1 and R3

Total execution time for different sizes of software (ratio 1:1 to 1:1000 between the agent and its load) is shown in Figure 5. In Area A (Figure 5a) the assignment R1 performs better when the ratio is below 1:300, while R3 is preferable for all other loads. In Area B (Figure 5b) the assignment R3 performs better for all loads.

The comparison between direct and indirect agent communication has shown that direct communication is less time consuming in most cases. It can be concluded that agent teams with a single agent responsible for all operations on a node lead to better performance.

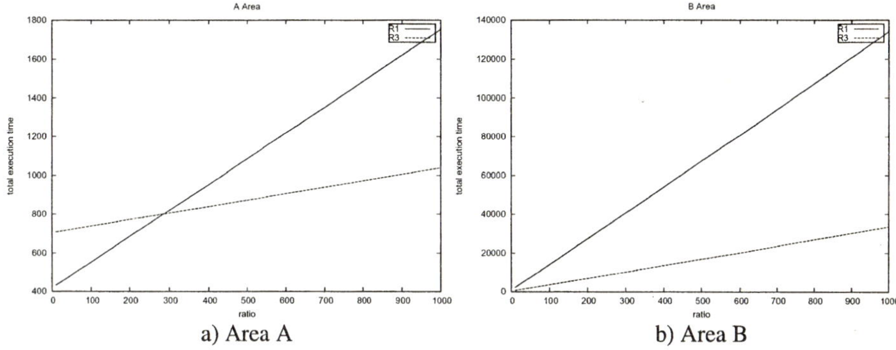

a) Area A   b) Area B

Fig. 5. Total execution time for different software sizes

5   Conclusion

Agent team coordination in the Mobile Agent Network based on a shared plan is elaborated. Selection of the plan shared by the agents is based on characteristics of the given task and environment, notably of the nodes and network. Different schemes for assigning operations to the agents are discussed from performance point. A case study describing multi-agent system for software management in the communication network and Grid is used to analyze the proposed approach. Future work will include investigation of the agent competition.

References

1. Lovrek, I., Carić, A., Huljenić, D.: Remote Maintenance Shell: Software Operations Using Mobile Agents, Proceedings of the International Conference on Telecommunications ICT 2002, pp. 175-179, Beijing, 2002.
2. Jezic, G., Kusek, M., Marenic, T., Lovrek, I., Desic, S., Trzec, K., Dellas, B.: Grid Service Management by Using Remote Maintenance Shell, Lecture Notes in Computer Science, LNCS 3270 (2003), 1143-1149.
3. Lovrek I., Sinković V.: Performance Evaluation of Mobile Agent Network, Frontiers in Artificial Intelligence and Applications 69 Part 2 (2001) 924-928.
4. Weiss, G.: Multiagent Systems, MIT Press, Cambridge, 1999.
5. Grosz, B.J., Kraus, S.: Collaborative plans for complex group actions, Artificial Intelligence 86(1996) 269-357.
6. Arisha K. A., Ozcan F., Ross R., Subrahmanian V. S., Eiter T., Kraus S. and Ilan B.: Impact: A Platform for Collaborating Agents, IEEE Intelligent Agents, 14 (1999) 64-72.

7. Silva L. M., Batista V., Martins P., and Soares G.: Using Mobile Agents for Parallel Processing, Proceedings of the International Symposium on Distributed Objects and Applications, pp. 34-43, Edinburgh, UK, 1999.
8. Lovrek I., Sinković V., Jezic G.: Communicating Agents in Mobile Agent Network, Frontiers in Artificial Intelligence and Applications, 82 (2002) 126-130.
9. Lovrek I., Jezic G., Kusek M., Ljubi I., Caric A., Huljenic D., Desic S., Labor O.: Improving Software Maintenance by using Agent-based Remote Maintenance Shell, Proceedings of International Conference on Software Maintenance, pp. 440-449, Amsterdam, 2003.
10. Jezic G., Kusek M., Desic S., Caric A., Huljenic D.: Multi-Agent Remote Maintenance Shell for Remote Software Operations", Lecture Notes in Computer Science - Lecture Notes in Artificial Intelligence, LNAI 2774, (2003) 675-682.
11. Kusek M.: Coordination of Mobile Agents for Remote Software System Operations, Ph.D. Thesis, University of Zagreb, 2005 (in Croatian).
12. Ismail, L., Hagimont, D.: A performance evaluation of the mobile agent paradigm, ACM SigPlan Notices, 34 (1998) 306-313.

Forming Proactive Team Cooperation by Observations

Yu Zhang[1] and Richard A. Volz[2]

[1] Computer Science Department, Trinity University, San Antonio, TX 78212
Tel:(210)999-7399, Fax(210)999-7477
yzhang@cs.trinity.edu
[2] Computer Science Department, Texas A&M University, College Station, TX 77843
Tel:(979)845-8873, Fax(979)862-4813
volz@cs.tamu.edu

Abstract. Proactivity is the ability to take initiative by exhibiting goal-directed behavior. Agents with proactivity can respond to dynamic environment in a timely way. Hence the ability to anticipate the information needs of teammates and assist them proactively is highly desirable. This research is directed toward helping agents produce effective communication through observations. We show how agents can use observation to estimate the teammates' beliefs and proactively provide the information that the teammates need. We present two experiments that explore: 1) effectiveness of different aspects of observability; 2) scalability of using observability with respect to the number of agents in a team.

Keywords: Multi-Agent Teamwork, Agent Communication, Observation

1 Introduction

In this paper, we focus on how to represent observability in the description of a teamwork, and how to include it into the basic reasoning for proactive communication. We define several different aspects of observability (e.g., seeing a property, seeing another agent perform an action, and believing another can see a property or action are all different), and propose an approach to the explicit treatment of an agent's observability that aims to achieve more effective information exchange among agents. We employ the agent's observability as the major means for individual agents to reason about the environment and other team members. We deal with communication with the 'right' agent about the 'right' thing at the 'proper' time in the following ways:

- Reasoning about what information each agent on a team will produce. This is achieved through: 1) analysis of the effects of individual actions in the specified team plans; 2) analysis of observability specification, indicating what and under which conditions each agent can perceive about the environment as well as the other agents.

- Reasoning about what information each agent will need in the process of plan execution. This is done through the analysis of the preconditions of the individual actions involved in the team plans.

- Reasoning about whether an agent needs to act proactively when producing some information. The decision is made in terms of: 1) whether or not the information is mutable according to information classification; 2) which agent(s) needs this in-

R. Khosla et al. (Eds.): KES 2005, LNAI 3681, pp. 247–254, 2005.

formation; and 3) whether or not an agent who needs this information is able to obtain the information independently according to the observation of environment and other agents' behaviors.

2 Agent Observability

The syntax we use for observability is given in Fig. 1.

```
<observability>  ::= (CanSense <viewing>)*
                     (BEL <believer> (CanSense <viewing>))*
<viewing>        ::= <observer><observable> <cond>
<believer>       ::= <agent>
<observer>       ::= <agent>
<observable>     ::= <property>|<action>
<cond>           ::= (<property>|<action>)*
<property>       ::= (<property-name> <object> <args>)
<action>         ::= (DO <doer> (<operator-name> <args>))
<object>         ::= <agent>|<non-agent>
<doer>           ::= <agent>
```

Fig. 1. The synatx of observability

The following shows the observability specification for a carrier in Multi-Agent Wumpus World (refer to section 4 for details). Here ca, rca, fi, rfi represent carrier, carrier's detection radius, fighter and fighter's detection radius, respectively.

```
;The carrier can see the location property of an object.
(CanSense ca (location ?o ?x ?y)
   (location ca ?xc ?yc)(location ?o ?x ?y)
   (inradius ?x ?y ?xc ?yc rca))

;The carrier can see the shootwumpus action of a fighter.
(CanSense ca (DO ?fi (shootwumpus ?w))
   (play-role fighter ?fi)(location ca ?xc ?yc)
   (location ?fi ?x ?y)  (adjacent ?xc ?yc ?x ?y))

;The carrier believes the fighter is able to see the location of an object.
(BEL ca (CanSense fi (location ?o ?x ?y))
   (location fi ?xi ?yi)(location ?o ?x ?y)
   (inradius ?x ?y ?xi ?yi rfi))

;The carrier believes the fighter is able to see the shootwumpus action of
 another fighter.
(BEL ca (CanSense fi (DO ?f (shootwumpus ?w)))
   (play-role fighter ?f)(≠ ?f fi)(location ca ?xc ?yc)
   (location fi ?xi ?yi)(location ?f ?x ?y)
   (inradius ?xi ?yi ?xc ?yc rca)(inradius ?x ?y ?xc ?yc rca)(adjacent ?x ?y
   ?yi))
```

Next we discuss the semantics. Let $\text{Sense}_t(a, \psi)$ express that agent a observes ψ at time t. There are two cases to consider, first where ψ is a property, and secondly, where ψ is an action. When ψ is a property, sensing ψ means determining the truth value of ψ, with unification of any free variables in ψ. If ψ is an action, sensing ψ means that the agent believes the doer believed the precondition of ψ immediately before the action occurred and the doer believes the effect of ψ immediately after performing the action. We use the predicate $\text{Hold}_t(c)$ to mean c holds in the world at time t. We make the assumption below:

$$\forall a, \psi, c, t, \text{CanSense}(a, \psi, c) \wedge \text{Hold}_t(c) \rightarrow \text{Sense}_t(a, \psi),$$

which means that if the condition c holds at time t and agent *a* has the capability to observe ψ under c, then agent *a* actually does determine the truth value of ψ at time t.

Next, we consider the relation between sensing something and believing it. Belief is denoted by the modal operator BEL and for its semantics we adopt the axioms K, D, 4, 5 in modal logic [2]. We made an analogous assumption to the one that "seeing is believing" [1]. It is stated separately for properties and actions. In the case of properties, it is formalized in this axiom:

$$\forall a, \varphi, t, \text{Sense}_t(a, \varphi) \rightarrow [\text{Hold}_t(\varphi) \rightarrow \text{BEL}_t(a, \varphi)] \wedge [\neg \text{Hold}_t(\varphi) \rightarrow \text{BEL}_t(a, \neg\varphi)],$$

which says that for any property φ sensed by agent *a*, if φ holds, *a* believes φ; if φ does not hold, *a* believes not φ ($\neg\varphi$).

Agent *a*'s belief is more complex when an action, ϕ, is observed. Let Doer(ϕ), Prec(ϕ), Efft(ϕ) denote the doer, the precondition, and the effect of action ϕ. We assume that the precondition of the action must be believed by <doer> before the action can be performed and the effect must be believed after the action is performed. When agent *a* senses action ϕ performed by some agent, agent *a* believes that the agent believed the precondition and effect. This process is expressed by the following axiom:

$$\forall a, \phi, t, \text{Sense}_t(a, \phi) \rightarrow$$
$$\text{BEL}_t(a, \text{BEL}_{t-1}(\text{Doer}(\phi), \text{Prec}(\phi))) \wedge \text{BEL}_t(a, \text{BEL}_t(\text{Doer}(\phi), \text{Efft}(\phi))).$$

From the belief update perspective in our current implementation where beliefs are assumed persistent, for any $p \in \text{Prec}(\phi)$, agent *a* believes that Doer(ϕ) still believes p at time t unless $\neg p$ is contained in Efft(ϕ). This is similar for "BEL CanSense".

An agent's belief about what another agent's sensing is based on the following axiom:

$$\forall a, b, \psi, c, t, t', \text{Believe}(a, \text{CanSense}(b, \psi, c)) \wedge$$
$$\text{BEL}_t(a, \text{BEL}_{t'}(b, c)) \rightarrow \text{BEL}_t(a, \text{Sense}_{t'}(b, \psi)),$$

which means that if agent *a* believes that agent *b* is able to observe ψ under condition c, and agent *a* believes c at time t', then agent *a* believes at time t that agent *b* sensed (t'<t), senses (t'=t), or will sense (t'>t, which requires some prediction capability for agent *a*) ψ at time t'. In our approach, each agent focuses on the reasoning about current observability, not in the past or in the future. Therefore, the axiom above can be simplified as follows:

$$\forall a, b, \psi, c, t, \text{Believe}(a, \text{CanSense}(b, \psi, c)) \wedge \text{BEL}_t(a, c) \rightarrow$$
$$\text{BEL}_t(a, \text{Sense}_t(b, \psi)).$$

Combining this with the previous assumption that "seeing is believing," we extend this to belief. Again we have two separate cases for properties and actions. When agent *a* believes agent *b* senses a property φ, *a* believes that *b* believes φ:

$$\forall a, b, \varphi, t, \text{BEL}_t(a, \text{Sense}_t(b, \varphi)) \rightarrow$$
$$\text{BEL}_t(a, \text{BEL}_t(b, \varphi)).$$

When agent *a* believes agent *b* senses an action ϕ, *a* believes that *b* believes the doer believed the precondition at the previous time step and believes the effect at the current time step. This consequence is expressed by the following:

$$\forall a, b, \phi, t, \mathrm{BEL}_t(a, \mathrm{Sense}_t(b, \phi)) \rightarrow$$
$$\mathrm{BEL}_t(a, \mathrm{BEL}_t(b, \mathrm{BEL}_{t-1}(\mathrm{Doer}(\phi), \mathrm{Prec}(\phi)))) \wedge \mathrm{BEL}_t(a, \mathrm{BEL}_t(b, \mathrm{BEL}_t(\mathrm{Doer}(\phi),$$
$$\mathrm{Efft}(\phi)))).$$

3 Observation-Based Proactive Communication

To generate inter-agent communications when information exchange is desirable, we developed two protocols named *ProactiveTell* and *ActiveAsk*. They are designed based on following three types of knowledge:

- Information needers and providers. In order to find a list of agents who might know or need some information I, we analyze the preconditions and effects of operators and plans and generate a list of needers and a list of providers for every piece of I [4]. The providers are agents who might know I, and the needers are agents who might need to know I.

- Relative frequency of information need vs. production. For any piece of information I, we define two functions, f_C and f_N. $f_C(I)$ returns the frequency with which I changes. $f_N(I)$ returns the frequency with which I is used by agents. We classify information into two types – static and dynamic. If $f_C(I) \leq f_N(I)$, I is considered static information; otherwise, I is considered dynamic information. For static information we use *ProactiveTell* by providers, and for dynamic information we use *ActiveAsked* by needers.

- Beliefs generated after observation. Agents use these beliefs to track other team members' mental states and use beliefs of what can be observed and inferred to reduce the volume of communication. For example, if a provider believes that a needer senses or infers I, the provider will not tell the needer.

```
/* Let T be the time step when the algorithm is executed.*/
/* Independently executed by each agent (self) when it needs the value of I.*/
ActiveAsk(self, I, KB_self, T){
        candidateList=null;
    if (I is dynamic and (I t) v (notI t) is not true in KB_self for any t≤T)
            if there exists a x≥0 such that
            ((BEL Ag I T-x) v (BEL Ag notI T-x)) is true in KB_self
                    let xs be the smallest such value of x;
                for each agent Ag≠self
                        if ((BEL Ag I T-xs) v (BEL Ag notI T-xs)) is true in KB_self
                                add Ag to candidateList;
                    randomly select Ag from candidateList;
                    ask Ag for I;
        else randomly select a provider
            ask the provider for I;}


/* Independently executed by each agent (self), after it executes updateKB.*/
ProactiveTell(KB_self, T){
    for each conjunct I for which (I, T) is true in KB_self and I is static
    ∀ Agn∈ needers
            if (BEL Agn I T) is not true in KB_self
                tell Agn I;}
```

Fig. 2. Proactive communication protocols

An algorithm for deciding when and to whom to communicate for *ActiveAsk* and *ProactiveTell* is shown in Fig. 2. For *ActiveAsk*, an agent requests *I* from other agents who may know it, having determined it from the information flow. The agent selects a provider among agents who know *I* and ask for *I*. For *ProactiveTell*, the agent tells other agents who need *I*. An agent always assumes others know nothing until it can observe or reason that they do know a relevant item. Information sensed and beliefs about others' sensing capabilities become the basis for this reasoning. First, the agent determines what another agent needs from the information flows. Second, the observation rules are used to determine whether or not one agent knows that another agent can sense the needed information.

4   Experiments

Multi-Agent Wumpus World [3] is an extension to the classic Wumpus World problem [4]. The world is 20×20 cells and has 20 wumpuses, 8 pits, and 20 piles of gold. The goals of the team, four agents, one carrier and three fighters, are to kill wumpuses and get the gold. The carrier is capable of finding wumpuses and picking up gold. The fighters are capable of shooting wumpuses. When a wumpus is killed, agents can determine whether the wumpus is dead only by getting the message from others, who kill wumpus or see shooting wumpus action. We used two teams defined below. Each team was allowed to operate a fixed number of 150 steps. Except for the observability rules, conditions of both teams were exactly.

- Team A: the carrier can observe objects within a radius of 5 grid cells, and each fighter can see objects within a radius of 3 grid cells.
- Team B: none of the agents have any seeing capabilities beyond the basic capabilities described at the beginning of the section.

There are two categories of information needed by the team: 1) an unknown conjunct that is part of the precondition of a plan or an operator (e.g., "wumpus location" and "wumpus is dead"); 2) an unknown conjunct that is part of a constraint (e.g., "fighter location", for selecting a fighter closest to wumpus). The "wumpus location" and "wumpus is dead" are static information and the "fighter location" is dynamic information. Agents use *ProactiveTell* to impart static information they just learned if they believe other agents will need it. For example, the carrier *ProactiveTell*s the fighters the wumpus' location. Agents use *ActiveAsk* to request dynamic information if they need it and believe other agents have it. For example, fighters *ActiveAsk* each other about their locations and whether a wumpus is dead.

4.1   Evaluating Different Beliefs Generated from Observability

The first experiment tested the contribution of different aspects of observability to the successful reduction of the communication. These aspects are belief about observed property, belief about the doer's belief about preconditions of observed action, belief about the doer's belief about effects of observed action and belief about another's belief about observed property. For simplify, we call them belief1, belief2, belief3 and belief4 correspondently. We used Team B, as reference to evaluate the effectiveness of different combinations of observability with Team A. We named this test combination 0, since there is none of such four beliefs involved in. For Team A, we tested another 4 combinations of these beliefs to show the effectiveness of each, in terms of the average number of communications per wumpus killed (ACPWK). These combinations are:

0. Team B, which involves none of beliefs.
1. In Team A, for each agent, leave off BELCanSense rules and do not process belief2 and belief3 when maintaining beliefs after observation. Therefore every agent only has belief1 about the world.
2. Keep every condition in comb. 1, except for enabling the belief2 process.
3. Enabling the belief3 process in comb. 2.
4. Add BELCanSense rules into comb. 3. This combination tests the effect of belief4 as well as show effectiveness of the beliefs as a whole.

Each combination is run in 5 randomly generated worlds. The average results of these runs are presented in Fig. 3, in which one bar shows ACPWK for one combination.

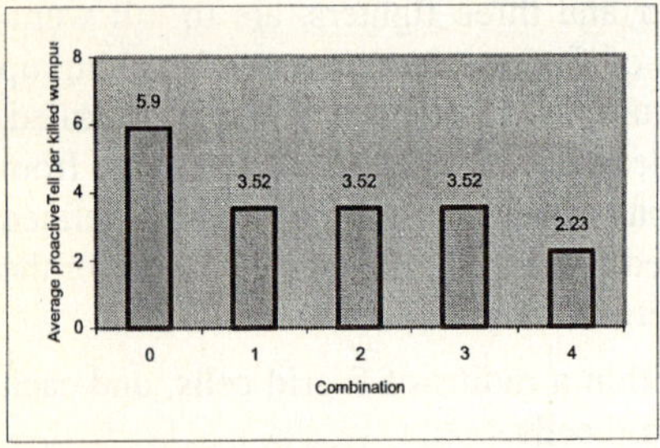
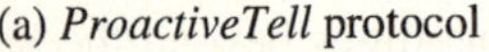

(a) *ProactiveTell* protocol

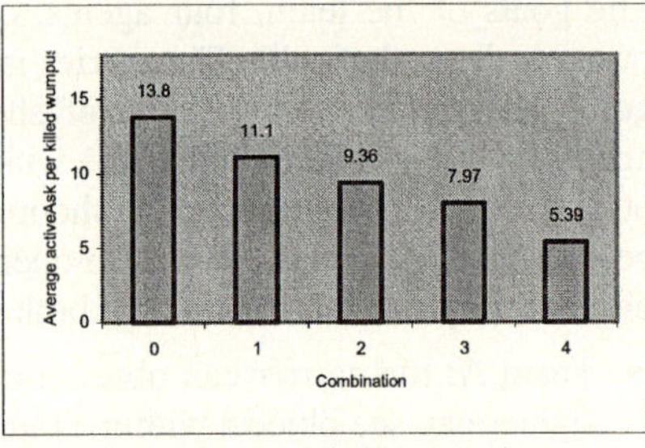

(b) *ActiveAsk* protocol

Fig. 3. Average communication per killed wumpus in different combinations

The result indicates three things. First, belief1 and belief4 have a strong effect on the efficiency of both *ProactiveTell* and *ActiveAsk*. Therefore, CanSense / BELCanSense a property can be generally applied to dual parts communication involving both Tell and Ask. Second, belief2 and belief3 have weak influence on the efficiency of *ProactiveTell*, this suggests that CanSense an action may be applied to communication which incurs more Ask than Tell, such as goal-directed communication. Third, these beliefs work best together, because each of them provides a distinct way for agents to get information from the environment and other agents. Furthermore, they complement each other's relative weakness, so using them together better serves the effectiveness of communication as a whole.

4.2 Evaluating Scalability of Observation-Based Proactive Communication

We designed the second experiment to show how communication load scales with increased team size. Based on the assumption that *ProactiveTell* brings more communication into play than *ActiveAsk*, we choose to test *ProactiveTell* protocol. ActiveAsk is directed to only one provider at certain time, while *ProactiveTell* goes to all needers who do not have the information. Therefore if the test results are good for *ProactiveTell*, we can expect that they are valid for *ActiveAsk* as well.

We used the same sensing capabilities for Teams A and Team B as in the first experiment. However, we increased the number of team members by 1, 2 and 3, in two tests that we ran. In the first test, we increased the number of needers, (i.e. fighters) and kept the same number of providers, (i.e. carriers). In the second test, we did it the other way around. In each test, for each increment and each team, we ran 5 randomly generated worlds and used the average value of ACPKW produced in each world.

Fig. 4 shows the trend of ACPKW as a function of increasing team size. In (a), Team B has an obvious increase in ACPKW with increasing the team size. However, Team A keeps the same ACPKW. The cause can be attributed to two factors: first, the amount of the increasing *ProactiveTell*s is held down because if the carrier believes the fighters can see wumpus, the carrier does not perform *ProactiveTell*; second, the more fighters there are, the more wumpuses will be killed, which enlarges the numerator of ACPKW. In (b), increasing the number of providers breaks the constant trend in Team A and shows an increased ACPWK. However, comparing this increase to that of Team B, it is a moderate number. In Team B, every provider increment means almost double the number of *ProactiveTell*s. The carriers lack an effective way to predict when a piece of information is produced and by whom, which is our main concern of future work. This experiment shows that the team empowered with observability has a slower growth of ACPWK with increase of team size, which may indicate that observability will improve team scalability in some sense.

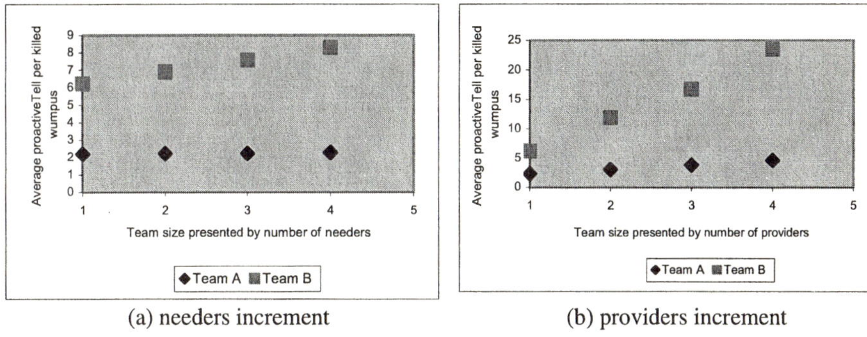

(a) needers increment   (b) providers increment

Fig. 4. The comparison of *ProactiveTell* with different team size

5 Conclusion

Our present proactive information algorithm analyzes the preconditions and effects of operators for which each agent is responsible in the team plan. The purpose of doing so is to determine potentially useful information flow among agents. In the future, we would like to make the recognition of needed information more dynamic. One way to solve this problem is to recognize the plans of other agents by observing actions of the other agents, and tracking the sequence of sub-goals on which they are working dynamically. Using this information together with the action an agent has most recently performed, the most likely information needs of other agents can be dynamically estimated over a finite time horizon. Then we can send other agents only unknown information that will be needed in the near future.

References

1. Bell, J. and Huang, Z., 1998. Seeing is Believing. Prod. of Common Sense 98', pp. 391-327.
2. Fagin, R., Halpern, J., Moses, Y., and Vardi, M, 1994. Reasoning About Knowledge, MIT.
3. Russell, S. and P. Norvig, 1995. Artificial Intelligence: A Modern Approach, Prentice Hall.
4. Zhang, Y., Volz, R.A., Ioerger, T.R., Cao, S. and Yen, J., 2002. Proactive Information Exchange During Team Cooperation. pp. 341-346, ICAI'02.

Microscopic Simulation in Decision Support System for the Incident Induced Traffic Management

Hojung Kim[1], S. Akhtar Ali Shah[1], Heyonho Jang[2], and Byung Ha Ahn[1]

[1] System Integration Lab., Department of Mechatronics,
Gwangju Institute of Science and Technology,
Oryong-dong, Buk-gu, Gwangju, 500-712, Republic of Korea
{snoopy,shahg,bayhay}@gist.ac.kr
[2] Highway and Transportation Technical Institute, Korea Highway Corporation,
Sancheok-li, Dongtan-myen, Hwasung-si, Gyenggi-do, 445-812, Republic of Korea
nettrek@freeway.co.kr

Abstract. This paper is a part of a comprehensive study in which a decision support system is developed to facilitate traffic mangers in tackling the post incident traffic flows on the freeway network in Korea. The system employs Cellular Automata, which is a microscopic simulation platform. This paper suggests two additional rules to in the modified version of the conventional model. This has enabled us to model more realistically the dynamic flow parameters in post incident scenario. A comparative study of the proposed model with the conventional and modified models suggests that our model displays more realistic results.

1 Introduction

Incident induced congestion on freeways is of particular importance due to its greater impacts on the regional economies and involvement of large numbers of vehicles because of control exit points. Nonetheless, as argued by Catherine and Demetsky [1], the incident related congestion is one of the most disruptive but least understood problems of the traffic management. Considering the gross severity of the non-recurrent congestion problem on freeways and in the perspective of higher traffic growth in near future, Korea Highway Corporation launched a project named as "Development of Dynamic Freeway Simulation Model for Incident Management" in year 2002 with the broader objective of making mobility on freeways safe, efficient and predictable particularly in post-incident scenario.

In this study we have developed a Freeway Incident Analysis System (FIAS), which is a microscopic simulation based tool to assist traffic managers in the Traffic Management Centers (TMCs)[2]. However, the theme of this paper is just to focus the microscopic simulation platform and discuss the adaptation in the conventional model of cellular automata (CA) developed by Nagel and Schreckenberg [3]. We have further extended the model by adding two additional rules to accommodate the typical freeway incident dynamics.

2 Traffic Modeling

Road traffic is influenced by a number of factors such as the driver's behavior, road geometry, and land use pattern of the abutting property. To depict these flows

R. Khosla et al. (Eds.): KES 2005, LNAI 3681, pp. 255–260, 2005.

through general mathematical expression or models is both difficult as well as time consuming because of the complexity, randomness and interdependency of these parameters.

Traffic flow simulators rather than analytical models are often employed to simulate traffic on various spatio-temporal scales. Traffic flow models can conventionally be classified as microscopic and macroscopic in nature. In microscopic modeling the movement of individual vehicle is focused that yields a more detailed description of the vehicle. Whereas macroscopic modeling describes aggregate variables of the traffic dynamical process behind traffic flow flows like traffic density, mean speed and volume. The later approach considers traffic flow as a compressible fluid and employs fluid-dynamical partial differential equations to propagate flow and density between large discrete time and space section. The use of these parameters in the model reduces the computation time considerably. However researchers like Xaichen and Daniel [4] argue that although the macroscopic traffic model reduces the computation time but it cannot estimate travel time, turning movement and other control parameters on a short time scale. Thus the macroscopic model in incident scenario cannot be regarded as the best option wherein short time scale information becomes significantly important to take immediate relieve measure to limit the spreading of congest effects.

2.1 Cellular Automata

Cellular Automata (CA) is an artificial approach in which physical quantities are transformed into a finite set of discrete units. Von Neumann and Ulam originally pioneered this simulation modeling approach in 1960s for the representation of biological self-reproduction phenomenon [4]. Since then, CA has been regarded as a promising tool for high-speed micro simulation and is used to model a variety of physical scenarios [4].

Nagel and Schreckenberg [3] employed CA to depict traffic flow on a single lane of a highway. In this model, which is also referred as "NaSch model" in [5], they proposed a simple rule-based approach to simulate the driver's behavior [4] in free flow conditions. Car following model is a key element in modeling driver's behavior and is also employed in "Nasch Model". For an elaborate review, the model is briefly described for single lane traffic. The full stretch of road is subdivided into 'n' boxes ('cells') of 7.5 meters length which can be either occupied or be empty by one vehicle at any discrete instant of time 't'. The velocity v of the vehicles is an internal parameter that characterizes its state. It can take on only integer values, i.e., $v = 0, 1, 2,\ldots,$ v_{max}. The dynamics of the model is described by speed update and movement rules for the velocities, which are given below:

1. Speed update step

    Acceleration: IF $(v_i < g_i)$ THEN $v_{i+1} = \min[v_{max}, v_i +1]$

    Deceleration: IF $(v_i \geq g_i)$ THEN $v_{i+1} = \min[v_i, g_i]$

    Randomization: IF $(p_{noise} >$ random value) THEN $v_{i+1} = \max[v_{i+1}-1, 0]$

2. Movement step $x_{i+1} = x_i + v_{i+1}$

Where, $v_i (0 \leq v_i \leq v_{max}$, cell/timestep) is a velocity of car at time step t, and g_i is the number of unoccupied cells to the leading the car at time step t. p_{noise} is random noise parameter (0~1) and illustrates the deviation from the optimal way approach encompassed in acceleration and deceleration rules. Thus the stochastic nature of traffic flows is incorporated in the model that explains the phantom jam dynamics. Without this parameter the dynamics would be purely deterministic as the system shows strong dependence on the initial condition. The randomization or the movement step takes into account natural fluctuations in the driver's behavior.

2.2 Velocity Dependent Randomization

Real traffic delivers strong hysteresis effects near maximum flow due to the existence of metastable states in certain density regimes [3, 5]. Traffic stays laminar when approaching from the low densities till a threshold point 'ρ2' and then breaks down into start-stop traffic. [6] describes that it does not become laminar in reverse direction until 'ρ1', which is about 30% smaller than ρ2. [5] incorporates this behavior into the Nash model and term it as Slow-to-Start rule. It more realistically captures the breakdown flow behavior. To describe the acceleration rate of stopped vehicles, STS rule uses random noise parameter p_s that has higher value than p_{noise}. If $p_s \gg p_{noise}$ then drivers escaping out of a jam will hesitate so that a jam grows and moves backward. STS rule is directly added to NaSch model, and an integrated NaSch model with STS rule is called as the modified NaSch model. And if $p_s = p_{noise}$ then the modified NaSch model has the same function as the NaSch model.

However the NaSch model and modified model has following common shortcomings in view of microscopic behaviors:
a. Unrealistic braking capability to avoid back collision in one time step, generally 1 second, from v_{max} to 0.
b. Not distinguished drivers' behavior between staying in a jam and escaping from it.

2.3 Additional Movement Rules for More Realistic Behaviors

In this paper, two additional movement rules, **Stopping Maneuver Rule (SMR)** and **Low Acceleration Rule (LAR)**, will be introduced to tackle explained-above two defects and to capture velocity quantities more realistically under heterogeneous traffic flow conditions. SMR is to describe a decelerating vehicle to arrive and stop in the tail of a jam, and LAR is to capture an accelerating vehicle, which follows a stopped leading vehicle in a jam. And the two rules are easily and directly added into the modified NaSch model.

An approaching vehicle, with velocity (v_i >0, cell/sec), to the back of a stopped leading vehicle can't help slowing down to protect itself from a rear end collision or a crash, and needs enough **Stopping Distance (SD)** to accomplish a smooth safe stop. Let us assume followings (**NUC** stands for the **Number of Unoccupied Cells** to the first stopped vehicle.):

$$SD = \sum_{i=1}^{v_i} i + v_i \text{ for all vehicles with } v_i > 0$$

Using SD and NUC, SMR is applied with random noise (p_{sm}, $p_{noise} \leq p_{sm} \leq 1$) as following ($v_{lead}$ is the velocity of the leading vehicle.):

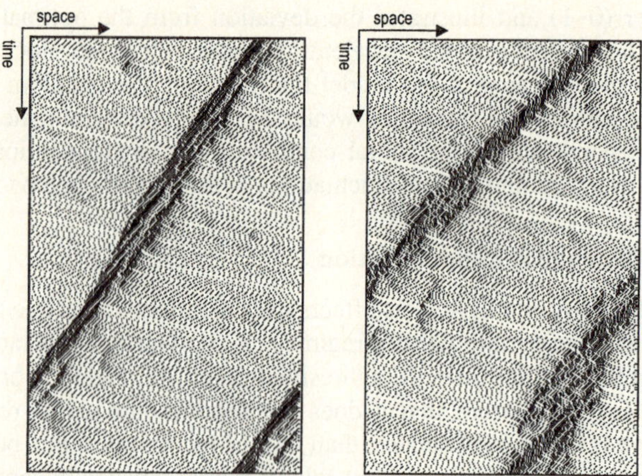

Fig. 1. Vehicle trajectories on a 1-lane periodic system with parallel update: dots represent vehicles that move to the right. The horizontal direction is discrete space to the right and the vertical direction to the down is discrete time. Left: NaSch model (p_{noise}=0.135, v_{max}=5). Right: the integrated model with SMR and LAR (p_{noise}=0.135, v_{max}=5, p_{sm}=0.8, p_{la}=0.8)

> **IF** ((SD $\geq$ NUC or $(v_i -1 \geq v_{lead}$ and $v_i \geq g_i)$) and $(p_{sm} >$ random value))
> **THEN** $v_{i+1} = \min[v_i -1, g_i]$
> **ELSE** the speed update step of the modified NaSch model

In the modified NaSch model, noise parameter values, namely p_{noise} and p_s are used for both the acceleration of stopped vehicles in a wide moving jam (1-1/(v_{max} +1) $<<$ $\rho \leq 1$) and the acceleration of escaping vehicle from a jam. But an aggressive accelerating maneuver with high acceleration rate to escape a jam is different from accelerating maneuver in a jam. Because, in a jam, an accelerating maneuver is physically accomplished with low acceleration rate and drivers psychologically hesitate about accelerating. To describe a low acceleration rate in a jam, LAR is introduced. Let us make assumption about followings:

1. A stopped vehicle with g_i =1, following a stopped vehicle, stays in a jam.
2. A stopped vehicle with g_i =1, following a running vehicle, escapes from a jam.

And then LAR are modeled with random noise (p_{la}, $p_{noise} \leq p_{la} \leq 1$ or $p_s \leq p_{la} \leq 1$) as following:

> **IF** (v_i =0, g_i =1, v_{lead} =0, $p_{la} >$ random value) **THEN** v_{i+1}= 0
> **ELSE** v_{i+1}= 1

SMR and LAR are directly integrated into the modified NaSch model as following.

1. Velocity update step

> **IF** ((v_i >0, SD $\geq$ NUC) or $(v_i -1 \geq v_{lead}, v_i \geq g_i)$) **THEN** SMR
> **ELSE** the speed update step of the modified NaSch model

> **IF** (v_i =0, g_i =1, v_{lead} =0) **THEN** LAR
> **ELSE** the speed update step of the modified NaSch model

2. Movement step $x_{i+1} = x_i + v_{i+1}$

Above the integrated model, the modified NaSch model with SMR and LAR, has the convertibility. If $p_{noise} = p_s = \rho_{sm} = p_{la}$ then SMR and LAR have the same function as the NaSch model, and if $p_{noise} \ll p_s$, $p_{noise} = \rho_{sm}$, $p_s = p_{la}$ then as the modified NaSch model.

3 Results and Discussion

The random noise parameter p_{sm} is used with very higher values, because the deceleration of a vehicle, arriving to the end of a jam, is mandatory and more or less deterministic. And to describe low deceleration rate in a jam, the random noise parameter ρ_{la} is considered with high values, for example if $p_{la} = 0.8$ and cell length 7.5 then acceleration rate is $1.5 \approx 7.5 * (1-0.8)$. For heterogeneous flow ($\rho > \rho_c$), in Fig. 2, NaSch model shows a more or less convex flow-density relationship, but the integrated model does a concave flow-density relationship according to increasing the values of ρ_{sm} and ρ_{la}.

The modified NaSch model explains the reversed λ-shaped flow-density curve with two maximum flow, q_{max} and q_{min}, and the integrated model generates the same relationship. The modified NaSch model describes heterogeneous flow as straight-line relation, whereas the integrated model does as concave-curve relation according to increasing the values of p_{sm} and p_{la}.

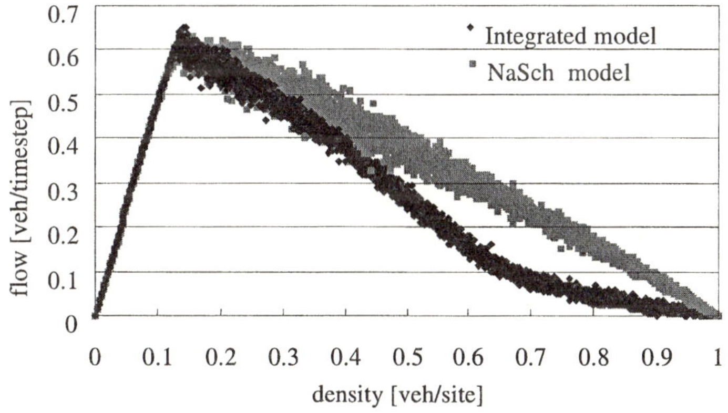

Fig. 2. Fundamental diagram of the NaSch model *vs.* the integrated model: 5-min polling average. NaSch model with $p_{noise}=0.135$, $v_{max}=5$, and the integrated model with $v_{max}=5$, $p_{sm}=0.8$, $p_{la}=0.8$

4 Conclusions

FIAS is a decision support system that employs microscopic simulation model to forecast post incident traffic volume and speed. This paper discussed the limitations of the conventional approach to tackle the post incident traffic flows and suggest a model which more versatile then the existing versions of CA. The distinguishing

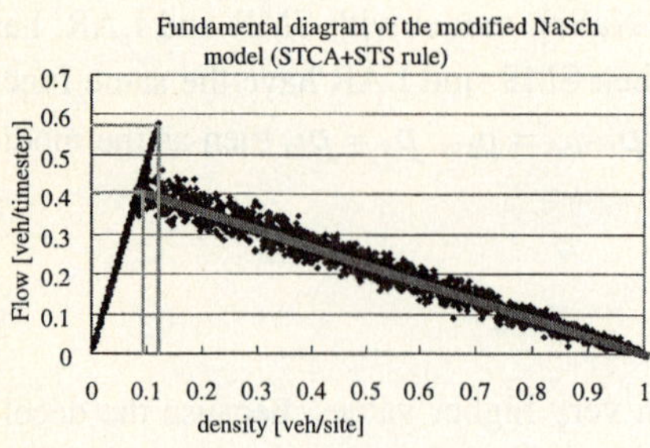

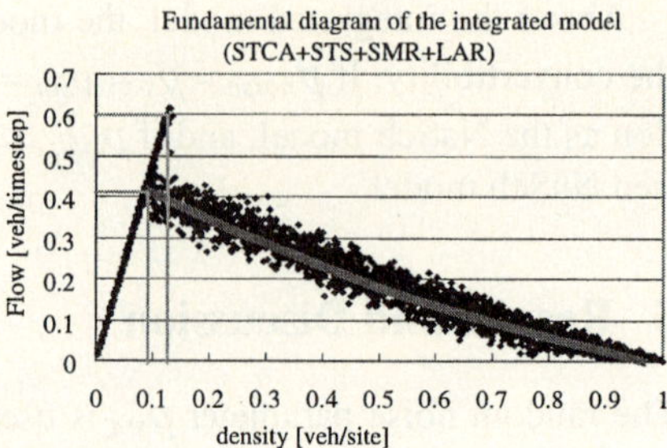

Fig. 3. Fundamental diagram of the modified NaSch model *vs.* the integrated model: 5-min polling average. The modified NaSch model with p_{noise} =0.135, p_s =0.5, v_{max} =5, and the integrated model with p_{noise} =0.135, p_s =0.5, p_{sm} =0.8, p_{la} =0.8, v_{ma}=5

feature of our proposed model is the incorporation of two additional rules to the conventional model so that it can simulate congested flows more elastically. It can also simulate breakdown on the bottleneck sections along with simulation of Stop/Go phenomena more realistically.

References

1. Catherine A. Cragg, Michael J. Demetsky, "Simulation analysis of route diversion strategies for freeway incident management. Final Report" VTRC Report No. 95-R11. Prepared for Virginia Department of Transportation, Virginia Transportation Research Council, Charlottesville (1995)
2. Hojung Kim, S.Akhtar Ali Shah, Seungkirl Baek, Byungha Ahn, "Architecture of decision support system for incident management – A case study of Republic of Korea", Proc. Of the 11th World Congress on ITS (2004)
3. Nagel K., and Schreckenberg M., "A cellular automaton model for freeway traffic", J. Phys. I France 2:2221 (1992)
4. Xaichen Xu, Daniel J. Dailey, "Real Time Traffic Simulation and Prediction Using Inductive Loop Data", Vehicle Information and Navigation System Conference Proceedings IEEE Service center, Piscataway, NJ, USA (1995) pp 194-199.
5. Barlovic R., Santen L., Schadschneider A., Schreckenberg M. (1998), "Metastable States in Cellular Automata for Traffic Flow", European Physical Journal B, 5:793.
6. Kai Nagel Cellular Automata models for transportation applications Cellular automata 5th International Conference on Cellular Automata for ACRI (2002), Geneva, Switzerland

Development and Implementation
on a Fuzzy Multiple Objective Decision Support System

Fengjie Wu, Jie Lu, and Guangquan Zhang

Faculty of Information Technology, University of Technology Sydney
PO Box 123, Broadway, NSW 2007, Australia
`{fengjiew,jielu,zhangg}@it.uts.edu.au`

Abstract. A fuzzy-goal optimization-based method has been developed for solving fuzzy multiple objective linear programming (FMOLP) problems where fuzzy parameters in both objective functions and constraints and fuzzy goals of objectives can be in any form of membership function. Based on the method, a fuzzy multiple objective decision support system (FMODSS) is developed. This paper focuses on the development and use of FMODSS in detail. An example is presented for demonstrating how to solve a FMOLP problem by using the FMODSS interactively.

1 Introduction

Many realistic decision-making problems are featured by the existence of multiple and conflicting objectives that should be considered simultaneously and are subject to constraints. Multiple objective linear programming (MOLP) is popularly used to deal with such complex and ill-structured decision-making [1].

Due to the unavoidable influence of uncertainty, MOLP models must take into account vague information, imprecise requirements, and modifications of the original input data from the modeling phase. Therefore, MOLP with fuzzy parameter called fuzzy multiple objective linear programming (FMOLP) would be more suitable to reflect the reality [2]. For the same reason, uncertainty is also reflected in the problem solution process since decision makers often have imprecise goals for their objectives.

Many optimization methods and techniques for modeling and solving FMOLP problems have been proposed [2-5]. Wu et al. [6, 7] also developed some algorithms and methods solving FMOLP problems.

Especially, Lu et al. [8] developed an interactive fuzzy goal optimization-based method for solving FMOLP problems with fuzzy parameters and fuzzy goals under the different satisfactory degree α. In this paper, this method is implemented in the form of decision support system (DSS) called fuzzy multiple objective DSS (FMODSS) under Windows environment. The main aim of the FMODSS development is to help the decision makers gather the knowledge about the fuzzy problem in order to make a better-informed decision.

2 A Fuzzy Goal Optimization-Based Method
for FMOLP Problems

In this paper, the FMOLP problems in which all coefficients of the objective functions and constraints are real fuzzy numbers are considered. Therefore, the FMOLP problems can be formulated as follows:

R. Khosla et al. (Eds.): KES 2005, LNAI 3681, pp. 261–267, 2005.

$$(\text{FMOLP}) \begin{cases} \text{Max} \quad \tilde{f}(x) = \tilde{C}x \\ \text{s.t.} \quad x \in X = \left\{ x \in R^n \mid \tilde{A}x \preceq \tilde{b}, x \geq 0 \right\} \end{cases} \tag{1}$$

where $\tilde{C}$ is an $k \times n$ matrix, $\tilde{A}$ is an $m \times n$ matrix, $\tilde{b}$ is an m-vector, each element of these matrices is a fuzzy number, and x is an n-vector of decision variables, $x \in R^n$.

An interactive fuzzy goal optimization-based method has been developed for solving FMOLP problems with fuzzy parameters and fuzzy goals under the different satisfactory degree α. The main steps of the method are listed as follows:

Step 1: Ask the decision maker to select satisfactory degree α $(0 \leq \alpha \leq 1)$, and input the membership functions of $\tilde{c}$ for $\tilde{f}(x) = \tilde{c}x$, $\tilde{a}$ and $\tilde{b}$ for $\tilde{a}x \preceq \tilde{b}$.

Step 2: Under the current degree α, solve FMOLP problem. If the Pareto optimal solution with the optimal decision variable x^* and $\tilde{f}(x^*)$ exist, go forward to the next step. Otherwise, go back to Step 1 to reassign degree α

Step 3: Ask the decision maker whether to be satisfied with the initial fuzzy Pareto optimal solution calculated in Step 2. If satisfied, the whole interactive process stops, and the initial Pareto optimal solution is the final satisfying solution. Otherwise, go to Step 4.

Step 4: Specify the new fuzzy goals $\tilde{g} = (\tilde{g}_1, \tilde{g}_2, ..., \tilde{g}_k)^T$ for the objective functions based on the current fuzzy Pareto optimal solution and new degree α if needed.

Step 5: Calculate the fuzzy Pareto optimal solution based on the current fuzzy goals of objective functions and degree α specified in Step 4.

Step 6: If the DM is satisfied with the solution calculated in Step 5, the whole interactive process stops, and the current fuzzy Pareto optimal solution is the final satisfying solution of FMOLP problem. Otherwise, go back to Step 4.

3 Implementation

According to the FMOLP model (1) and the method in Section 2, the following common data need to be input from the user for setting up model (1).

- The number of decision variables, the number of objective functions, and the number of constraints,
- The coefficients of objective functions, the max/min for individual objective function, as showed in Figure 1,
- The coefficients of constraints, the coefficients of rhs, the relation sign of individual constraint, as showed in Figure 1,
- Especially, as the coefficients of objective functions and constraints and fuzzy goals are represented by fuzzy numbers. A special Dialog Box as showed in Figure 2 is designed for entering these fuzzy numbers. The forms of left and right function of membership function can be selected as linear, quadratic, cubic, exponential, logarithmic, etc., and four end-points of left and right function of fuzzy numbers defined in Definition 1 are entered simultaneously.

In the window of Figure 3, there are 4 FlexGrids on the left side of window used to entering or displaying different sort of data. FlexGrid 1 is for entering the different weights for fuzzy objective functions. The degree of membership functions of all fuzzy numbers can be set by the slider on the right side of window.

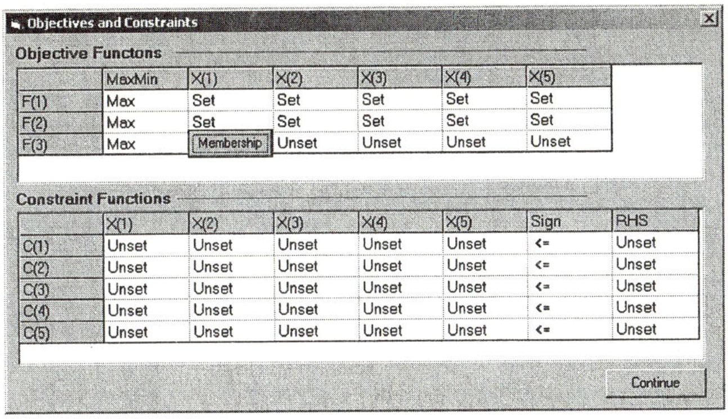

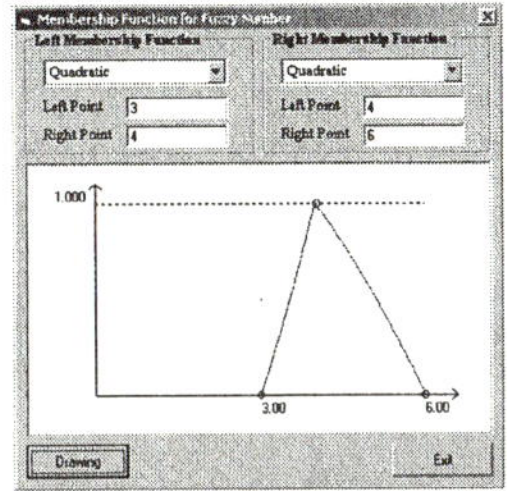

Fig. 1. Input window for fuzzy objective functions and constraints

Fig. 2. Input window for the membership function of a fuzzy number

Following the method described in Section 2, click Button "Initiate", the initial solution of the problem will be shown as decision variables in FlexGrid 2 and fuzzy objective functions in FlexGrid 3. To display the membership functions of the fuzzy objective function output, click the corresponding grid in the first row "Objectives" of FlexGrid 3, then another window will be pop up as shown in Figure 4.

When not satisfied with the solution calculated in the previous trial, the decision maker can specify the fuzzy goals to be achieved in the next trial by two ways. One way is by increasing or decreasing the previous individual fuzzy objective function solution by percentage through clicking the corresponding grid in the second row "By %" of FlexGrid 3. The other is by inputting the new fuzzy goals. In order to input the fuzzy goals which are represented by fuzzy numbers, click the corresponding grid in the third row "By value" of FlexGrid 3, then another Window "Fuzzy Goal Input" similar as shown in Figure 2 is pop up. After the new fuzzy goals being created, click on Button "Continue" for getting new solution of the current trial.

The optimal solutions for each trial during the interactive decision-making process are recorded and listed in FlexGrid 4.

4   An Illustrative Example

In order to show an outline of the basic analysis performed with FMODSS, a small sample decision problem [9] in a fictitious international toy manufacturing company is used here. The FMOLP problem is modeled as below.

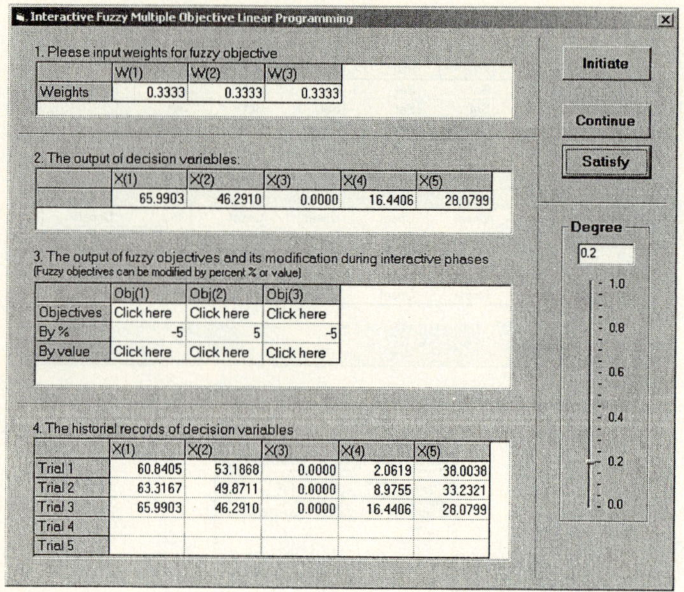

Fig. 3. Main window with method for solving FMOLP problem interactively

$$\text{Max} \begin{pmatrix} \widetilde{f}_1(x) \\ \widetilde{f}_2(x) \\ \widetilde{f}_3(x) \end{pmatrix} = \text{Max} \begin{pmatrix} \widetilde{c}_{11}x_1 + \widetilde{c}_{12}x_2 + \widetilde{c}_{13}x_3 + \widetilde{c}_{14}x_4 + \widetilde{c}_{15}x_5 \\ \widetilde{c}_{21}x_1 + \widetilde{c}_{22}x_2 + \widetilde{c}_{23}x_3 + \widetilde{c}_{24}x_4 + \widetilde{c}_{25}x_5 \\ \widetilde{c}_{31}x_1 + \widetilde{c}_{32}x_2 + \widetilde{c}_{33}x_3 + \widetilde{c}_{44}x_4 + \widetilde{c}_{55}x_5 \end{pmatrix} = \text{Max} \begin{pmatrix} \widetilde{1}x_1 + \widetilde{9}x_2 + \widetilde{10}x_3 + \widetilde{1}x_4 + \widetilde{3}x_5 \\ \widetilde{9}x_1 + \widetilde{2}x_2 + \widetilde{2}x_3 + \widetilde{7}x_4 + \widetilde{4}x_5 \\ \widetilde{4}x_1 + \widetilde{6}x_2 + \widetilde{7}x_3 + \widetilde{4}x_4 + \widetilde{8}x_5 \end{pmatrix} \quad (2)$$

$$\text{s.t.} \begin{cases} \widetilde{a}_{11}x_1 + \widetilde{a}_{12}x_2 + \widetilde{a}_{13}x_3 + \widetilde{a}_{14}x_4 + \widetilde{a}_{15}x_5 = \widetilde{3}x_1 + \widetilde{9}x_2 + \widetilde{9}x_3 + \widetilde{5}x_4 + \widetilde{3}x_5 \preceq \widetilde{b}_1 = 1039 \\[6pt] \widetilde{a}_{21}x_1 + \widetilde{a}_{22}x_2 + \widetilde{a}_{23}x_3 + \widetilde{a}_{24}x_4 + \widetilde{a}_{25}x_5 = -\widetilde{4}x_1 - \widetilde{1}x_2 + \widetilde{3}x_3 - \widetilde{3}x_4 - \widetilde{2}x_5 \preceq \widetilde{b}_2 = 94 \\[6pt] \widetilde{a}_{31}x_1 + \widetilde{a}_{32}x_2 + \widetilde{a}_{33}x_3 + \widetilde{a}_{34}x_4 + \widetilde{a}_{35}x_5 = \widetilde{3}x_1 - \widetilde{9}x_2 - \widetilde{9}x_3 - \widetilde{4}x_4 - \widetilde{0}x_5 \preceq \widetilde{b}_3 = 61 \\[6pt] \widetilde{a}_{41}x_1 + \widetilde{a}_{42}x_2 + \widetilde{a}_{43}x_3 + \widetilde{a}_{44}x_4 + \widetilde{a}_{45}x_5 = \widetilde{5}x_1 + \widetilde{9}x_2 + \widetilde{10}x_3 - \widetilde{1}x_4 - \widetilde{2}x_5 \preceq \widetilde{b}_4 = 924 \\[6pt] \widetilde{a}_{51}x_1 + \widetilde{a}_{52}x_2 + \widetilde{a}_{53}x_3 + \widetilde{a}_{54}x_4 + \widetilde{a}_{55}x_5 = \widetilde{3}x_1 - \widetilde{3}x_2 + \widetilde{0}x_3 + \widetilde{1}x_4 + \widetilde{5}x_5 \preceq \widetilde{b}_5 = 420 \\[6pt] x_1 \geq 0; \quad x_2 \geq 0 \end{cases} \quad (3)$$

In this example, the unified form of all membership functions of the parameters of the objective functions and constraints is assumed to be as following:

$$\mu_{\widetilde{c}_{11}}(x) = \begin{cases} 0 & x < a \text{ or } d < x \\ (x^2 - a^2)/(b^2 - a^2) & a \leq x < b \\ 1 & b \leq x \leq c \\ (d^2 - x^2)/(d^2 - c^2) & c < x \leq d \end{cases} \quad (4)$$

where left and right function of membership functions of a fuzzy number are quadratic. For simplicity, we will represent the above form of membership function (4) as quadruple pair (a, b, c, d) in latter on. All membership functions of the fuzzy parameters of the objective functions and constraints are listed in Table 1, 2 and 3.

Table 1. The membership functions of objective functions' parameters

$\tilde{c}_{ij}$	1	2	3	4	5
1	(0.5,1,1,2)	(8,9,9,11)	(9,10,10,12)	(0.5,1,1,2)	(2,3,3,5)
2	(8,9,9,11)	(1,2,2,4)	(1,2,2,4)	(6,7,7,9)	(3,4,4,6)
3	(3,4,4,6)	(5,6,6,8)	(6,7,7,9)	(3,4,4,6)	(7,8,8,10)

Table 2. The membership functions of constraints' parameters

$\tilde{a}_{ij}$	1	2	3	4	5
1	(2,3,3,5)	(8,9,9,11)	(8,9,9,11)	(4,5,5,7)	(2,3,3,5)
2	(-6,-4,-4,-3)	(-2,-1,-1,-0.5)	(2,3,3,5)	(-5,-3,-3,-2)	(-4,-2,-2,-1)
3	(2,3,3,5)	(-11,-9,-9,-8)	(-11,-9,-9,-8)	(-6,-4,-4,-3)	(0,0,0,0)
4	(4,5,5,7)	(8,9,9,11)	(9,10,10,12)	(0.5,1,1,2)	(-4,-2,-2,-1)
	(2,3,3,5)	(-5,-3,-3,-2)	(0,0,0,0)	(0.5,1,1,2)	(4,5,5,7)

Table 3. The membership functions of rhs's parameters

$\tilde{b}_i$	1
1	(1038, 1039, 1039, 1041)
2	(93, 94, 94, 96)
3	(60, 61, 61, 63)
4	(923, 924, 924, 926)
5	(419, 420, 420, 422)

Following the method described in Section 2, first of all, the FMOLP model (2) (3) needs to be entered. Based the description of the problem above, the number of decision variables, objective functions, and constraints are set to 5, 3, and 5 respectively. The window for setting up the model including the information of the objective functions and constraints is showed as in Figure 1.

Also in Figure 1, in order to enter fuzzy coefficients of the objective functions and constraints, click on the corresponding grid in the table, and then another Dialog Box will pop up as showed in Figure 2. For example, for the membership function of coefficient $\tilde{c}_{31}$, the forms of left and right function of membership function are selected as quadratic, and four end-points are set to 3, 4, 4, and 6 respectively.

After finishing entering FMOLP model, the decision maker will switch to the window as shown in Figure 3 to solve the problem. Suppose the satisfactory degree α is set to 0.2. Click Button "Initiate", the initial solution of the problem can be generated. The decision variables are: $x_1^* = 60.84$, $x_2^* = 53.19$, $x_3^* = 0$, $x_4^* = 2.06$, $x_5^* = 38.00$, and the fuzzy objective functions are

$$\begin{cases} \tilde{f}_1^*\left(x_1^*,x_2^*,x_3^*,x_4^*,x_5^*\right) = 60.84\tilde{c}_{11} + 53.19\tilde{c}_{12} + 2.06\tilde{c}_{14} + 38.00\tilde{c}_{15} \\ \tilde{f}_2^*\left(x_1^*,x_2^*,x_3^*,x_4^*,x_5^*\right) = 60.84\tilde{c}_{21} + 53.19\tilde{c}_{22} + 2.06\tilde{c}_{24} + 38.00\tilde{c}_{25} \\ \tilde{f}_2^*\left(x_1^*,x_2^*,x_3^*,x_4^*,x_5^*\right) = 60.84\tilde{c}_{31} + 53.19\tilde{c}_{32} + 2.06\tilde{c}_{34} + 38.00\tilde{c}_{35} \end{cases} \tag{5}$$

The membership functions of (5) are shown as in Figure 4.

Suppose the decision maker is not satisfied with the initial solution, and will specifie new fuzzy goals for the fuzzy objective functions to be achieved. Let these new

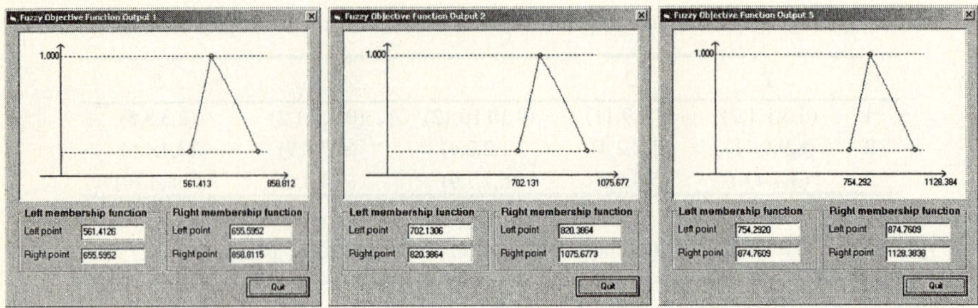

Fig. 4. The membership functions of $\widetilde{f}_1^{*}$, $\widetilde{f}_2^{*}$ and $\widetilde{f}_3^{*}$ at Trial 1

fuzzy goals be assigned by decreasing both of the first and third fuzzy objective functions by 5% as the first and third fuzzy goals respectively, and increasing the second fuzzy objective function by 5% as the second fuzzy goals based on the initial solution. That is:

$$(\widetilde{g}_1,\widetilde{g}_2,\widetilde{g}_3)=\left(0.95*\widetilde{f}_1^{*}(x_1^{*},x_2^{*},x_3^{*}),\ 1.05*\widetilde{f}_2^{*}(x_1^{*},x_2^{*},x_3^{*}),\ 0.95*\widetilde{f}_3^{*}(x_1^{*},x_2^{*},x_3^{*})\right) \tag{6}$$

Pressing Button "Continue", the new solution based on achieving the fuzzy goals (6) is generated. Consequently, the decision variables are $x_1^{*}=63.32$, $x_2^{*}=49.87$, $x_3^{*}=0.0$, $x_4^{*}=8.98$, $x_5^{*}=33.23$. And the membership functions of the fuzzy objective functions are shown as in Figure 5. Comparing the fuzzy optimal objective functions $\widetilde{f}_1^{*}$, $\widetilde{f}_2^{*}$ and $\widetilde{f}_3^{*}$ in Figure 4 with the ones in Figure 5, $\widetilde{f}_1^{*}$ and $\widetilde{f}_3^{*}$ obtained some decrement, and $\widetilde{f}_2^{*}$ got some increment. That is the purpose of the fuzzy goals (6).

If the decision maker is also not satisfied with the solution calculated in previous step, the interactive process will carry on. Suppose the decision maker creates some new fuzzy goals as followings:

$$(\widetilde{g}_1,\widetilde{g}_2,\widetilde{g}_3)=\left(6\widetilde{0}0,9\widetilde{0}0,8\widetilde{5}0\right), \tag{7}$$

the membership functions of which in quadruple pair format are listed as:

$$\begin{cases} u_{g_1}=(590,600,600,615) \\ u_{g_2}=(890,900,900,915) \\ u_{g_3}=(840,850,850,865) \end{cases}, \tag{8}$$

then the solution based on the fuzzy goals (7) (8) is generated. The decision variables are $x_1^{*}=65.99$, $x_2^{*}=46.29$, $x_3^{*}=0.0$, $x_4^{*}=16.44$, $x_5^{*}=28.08$. The membership functions.

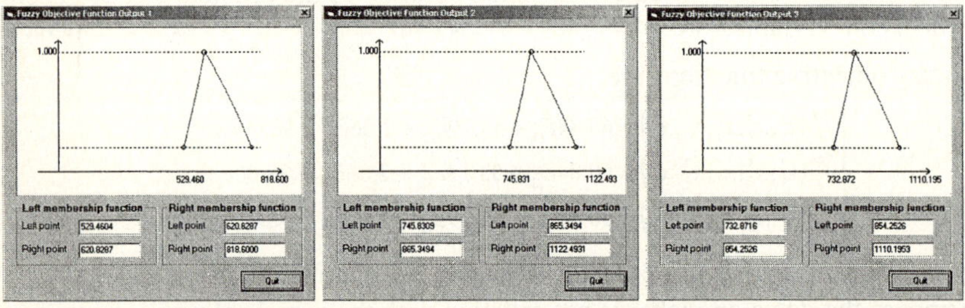

Fig. 5. The membership functions of $\widetilde{f}_1^{*}$, $\widetilde{f}_2^{*}$ and $\widetilde{f}_3^{*}$ at Trial 2

5 Conclusions

In this study, we develop FMODSS a decision support system for solving fuzzy multiple objective linear programming in which the fuzzy parameters in both the objective functions and the constraints can be represented by any form of membership functions, and the decision makers are allowed to provide their fuzzy goals by any form of membership functions as well. This system has been initially tested by a number of examples and results are pretty positive. This system will be developed further and put online in order to attract more applications to use the system.

Acknowledgements

This research is partially supported by Australian Research Council (ARC) under discovery grant DP0211701.

References

1. C. L. Hwang and A. S. Masud, *Multiple Objective Decision Making: Methods and Applications*. Berlin: Springer-Verlag, 1979.
2. M. Sakawa, *Fuzzy sets and interactive multiobjective optimization*. New York: Plenum Press, 1993.
3. Y. J. Lai and C. L. Hwang, *Fuzzy Multiple Objective Decision Making: Methods and Applications*. Berlin: Springer-Verlag, 1994.
4. J. Ramik and H. Rommelfanger, "Fuzzy mathematical programming based in some new inequality relations", *Fuzzy Sets and Systems*, vol. 81, pp. 77-87, 1996.
5. H. Rommelfanger, "Interactive decision making in fuzzy linear optimization problems", *European Journal of Operational Research*, vol. 41, pp. 210-217, 1989.
6. F. Wu, J. Lu, and G. Q. Zhang, "A new approximate algorithm for solving multiple objective linear programming with fuzzy parameters", presented at The Third International Conference on Electronic Business (ICEB 2003), Singapore, Dec 9-13, 2003.
7. F. Wu, J. Lu, and G. Q. Zhang, "A fuzzy goal approximate algorithm for solving multiple objective linear programming problems with fuzzy parameters", presented at FLINS 2004: 6th International Conference on Applied Computational Intelligence, Blankenberghe, Belgium, Sep 1-3, 2004.
8. J. Lu, F. Wu, and G. Q. Zhang, "On Generalized Fuzzy Goal Optimization for Solving Fuzzy Multi-Objective Linear Programming Problems", *International Journal of Uncertainty, Fuzziness and Knowledge-Based Systems*, pp. Submitted, 2004.
9. C. E. Downing and J. L. Ringuest, "Implementing and testing a complex interactive MOLP algorithm", *Decision Support Systems*, vol. 33, pp. 363-374, 2002.

IM3: A System for Matchmaking
in Mobile Environments

Andrea Calì

Faculty of Computer Science
Free University of Bozen-Bolzano
piazza Domenicani 3 – I-39100 Bolzano, Italy
ac@andreacali.com

Abstract. In this paper we present IM3 (Intelligent Mobile Match-Maker), a system that recommends tourist events to users. The system, based on a centralised server and accessible through mobile devices, matches user preferences (demand) with descriptions of tourist events (supply), in order to provide the user with information about the events that he/she is likely to be interested in. User and event descriptions in IM3 are represented with a formalism borrowed from AI and based on Description Logics; such descriptions are allowed to be incomplete in the parts that are considered irrelevant by the demander/supplier. The algorithms implemented in IM3 are able to deal with both conflicting and missing information between user and event profiles. Motivated by the nature of its application domain, IM3 can operate in a location-based fashion: it can provide the user with information related to events close to the user terminal, allowing the clients to connect via Bluetooth to registered local servers. Finally, also mobile terminals, running suitable software included in IM3, can match demand and supply profiles among them via Bluetooth, without relying on any centralised server.

1   Introduction

Matching demands and supplies of goods or services arises in several fields, including real estate agencies, recruitment agencies, dating agencies and advertisement in general. The problem of matchmaking consists in general in matching a set of demand profiles with a set of supply profiles; the match should be the *best possible* one, since perfect matches are unlikely to be possible. Different approaches to matchmaking can be found in the literature [6, 7, 10–12], based on bipartite graph matching, vector-based techniques, and record matching in databases, among others.

The formalism we adopt for representing demand and supply profiles, first presented in [5], differently from known approaches, allows both for the specification of incomplete profiles, and to reason on profiles in the presence of conflicting and missing information; such formalism is borrowed from AI, and in particular from Description Logics. In our approach, a demand profile and a supply profile are considered, and they are somehow "adjusted", so that the supply profile

R. Khosla et al. (Eds.): KES 2005, LNAI 3681, pp. 268–275, 2005.

fully satisfies the demand profile. At each adjustment a *penalty* is associated; the overall penalty is the sum of all penalties generated during the matchmaking process: the larger the penalty, the less the demand and supply match. If the supply fully satisfies the demand, the penalty is zero.

The matchmaking technique described above is incorporated in IM3 (Intelligent Mobile MatchMaker), a system for matchmaking that recommends tourist events to users. There are several known recommendation and matchmaking systems [3, 8, 9]; IM3 adopts the novel approach cited above [5] for modelling and reasoning about profiles, and it is accessible through mobile devices, in particular smartphones running J2ME. Users can store in a centralised server their demand profile, which describes the kind of tourist events they are interested in; supply profiles, each describing a tourist event, are published also by users.

Besides the obvious advantage of allowing users to access information about tourist events while carrying just a cellphone, there is one more reason to offer access to the system through mobile terminals: IM3 offers the possibility to connect to local servers via Bluetooth. The local servers offer information only about events happening nearby, thus providing a location-based service.

The matchmaking comes into play when the user accesses the system to retrieve data about tourist events he/she is likely to be interested in. IM3 presents the events sorted according to the penalty returned in the match with the demand profile, so that the events that should be of interested for the user are presented first. This is important when the user is using a small-screen and small-keyboard device: in fact in this case a minimisation of the steps required to navigate to the relevant information is desirable [4].

The paper is organised as follows. In Section 2 we briefly illustrate the formalism for representing profiles and the matchmaking algorithms adopted in IM3. In Section 3 we present the architecture of IM3; in Section 4 we describe how the system works. Section 5 concludes the paper.

2   Matchmaking Techniques in IM3

In this section we briefly describe how demand and supply profiles are represented in IM3, and how IM3 matches profiles in order to recommend events to the users according to their requests. In our case the problem is, given a demand profile and a set of supply profiles, to sort the supply profiles according to their capability of satisfying the demand profile.

When we come across the description of an event, we may face two different problems: *(i) Inconsistencies.* Some of the characteristics of the events are in contrast with our requirements. *(ii) Missing information.* The description (profile) of some event does not have information about some of the characteristics specified by our requirements.

In IM3 we adopt techniques that are capable of dealing with the above problems in a clear and formal approach, in particular by using an approach based on *Description logics* [1]. We adopt a special Description Logic, tailored for our needs: such formalism is able to *(i)* represent also quantitative information;

(ii) deal with conflicting and incomplete information with suitable reasoning services. The system IM3 manipulates profiles that are descriptions of tourist events; each event supplier gives a description of the supplied event, while each user specifies the kind of events he/she is interested in.

Now we briefly describe the formalism we use for modelling event profiles. Due to lack of space, we refer the reader to [5] for the details about syntax and semantics of this particular language. Description Logics are based on concepts, representing sets of objects, and roles, representing binary relations on sets of objects. The Description Logic adopted in IM3 is enriched with *concrete domains* [2]; such concrete domains, e.g. real numbers or intervals in the set of integers, are used to capture quantitative information such as price, date and/or time of beginning and end of events, and so on. We use *features*, assuming values in concrete domains, to represent such quantitative information; moreover, a special feature level is used to represent the level of relationship that an event has with a certain topic. The feature level assumes values in the domain $[0,1] \subset \mathbb{R}$.

A profile is a conjunction of constructs of four possible types:

1. *Atomic concepts.* Such concepts are used to represent atomic properties associated to an event; for example, concert denotes that the event is a concert.
2. *Features* of the form $p(f)$, where f is a feature, the (unary) predicate p can be one of the predicates $\geqslant_\ell(\cdot)$, $\leqslant_\ell(\cdot)$, $=_\ell(\cdot)$, and ℓ is a value of the concrete domain associated to f, or any logical conjunction of them. For example, $\leqslant_{20,00}(\text{price})$ denotes that the price of the event is less or equal to $20,00$ Euros.
3. *Relationship concepts* of the form $\exists\text{isRelated.}(C \sqcap \geqslant_x(\text{level}))$, where C is a conjunction of concept names, *isrel* is a special role denoting relationship with topics, and $0 \leqslant x \leqslant 1$. This kind of concepts denote that the event is related to a topic denoted by C with a level x. For example, $\exists\text{isRelated.}$ $(\text{japaneseCuisine} \sqcap \geqslant_{0,8}(\text{level}))$ denotes that the event has to do with Japanese cuisine with a high degree of relevance $(0,8)$.
4. *Non-relationship concepts* of the form $\forall\text{isRelated.}(\neg C \sqcup \leqslant_x(\text{level}))$, where C is a conjunction of concept names, and $0 \leqslant x \leqslant 1$. Concepts of this form denote that the level of relationship of the event with topic C is at most x.

Note that in a demand profile features like price or beginTime are expressed with *range predicates*, for instance $(\geqslant_{20:00} \wedge \leqslant_{22:00})(\text{beginTime})$, while in supply profiles there will usually be *equality predicates* like $=_{12,00}(\text{price})$.

When matching a demand profile and a supply profile, IM3 returns as a result a *penalty* value in $\mathbb{R}$; the greater the penalty, the less the two profiles fit each other. In the matchmaking process the system performs operations that can be thought as steps of a negotiation; we can think of the system changing the two profiles (note that there the profiles are not actually changed) in several steps, at each of which a partial penalty is added to the final penalty; at the end of the process the supply profile completely satisfies the demand, at the price of the penalties added during the matchmaking steps. There are two ways of executing a step in the negotiation, called *contraction* and *abduction* according to the corresponding logical operations, for which we refer to [5]. *(i) Contraction.* When the demand profile requests something that is in contrast with the properties of the supply, the demander partially *gives up* his request in order to avoid the conflict. *(ii) Abduction.* When the demand profile requests something that is left unspecified by the supply profile, we specialise the supply profile until the demander's request is satisfied.

The penalties are described by means of *penalty functions* defined by domain experts; penalty functions depend on the importance of the corresponding information and also on quantitative information: for example, if the demander willing to spend at most 10 Euros for an event, and the participation to the supplied event costs $10, 50$ Euros, the penalty will be much less than if the price of the supplied event was $100, 00$ Euros.

All the matchmaking techniques have to take into account the relationships that exist between concepts. For instance, if a user is not interested at all in sports events, and we have a supply profile describing a football match, the system needs to know that football is a sport. In order to allow the system to properly reason about concepts, in the server (and, when necessary, in the clients) a *domain ontology* represents inclusion and disjunction relations among concepts. For example, if the demand profile contains the atomic concept culturalEvent and the supply profile containts the atomic concept artExhibition, the system will recognise that the type of the supplied event satisfies the type required by the demand, since the assertion artExhibition $\sqsubseteq$ culturalEvent in the ontology specifies that an art exhibition is a cultural event.

3   Architecture of IM3

In this section we describw the architecture of IM3, shown in Figure 1.

The core of the whole system is the server, that stores supply and demand profiles, and runs the matchmaking algorithm on them. The server uses Java Server Pages technology and runs a MySQ relational DBMS for storing all the profiles. Also a Java server runs on the server, accepting connections via sockets from client software running on the mobile clients. Periodically, the server com-

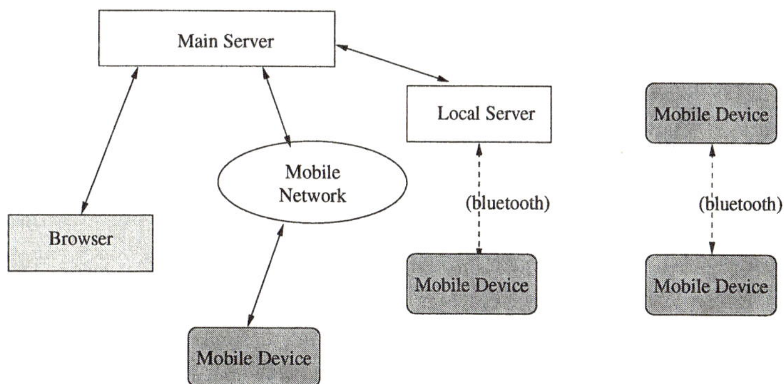

Fig. 1. Overall architecture of IM3.

putes the penalty between all demand/supply profile pairs; in this way, when a query is issued to the system, the answer can be computed quickly and the results presented to the user ordered according to the penalty (obviously, events returning the smallest penalty are returned first). The server runs on an Intel machine with Windows XP operating system.

Local servers do the same job as the main server, but with two significant differences: *(i)* each of them stores only events located in the area where the local server is based; *(ii)* due to the local nature of the information stored in them, the local servers are accessible by mobile clients only via Bluetooth. The local servers do not store any user profile, since the client software sends its demand profile on-the-fly to the local server when needed.

The central server is accessible on the Web by means of an ordinary browser. The mobile clients have the possibility of accessing the central server via Web through a browser capable of processing XHTML Mobile Profile markup. Alternatively, they can run a Java Midlet on their J2ME Virtual Machine that is able to connect to the central server, the local servers or other mobile clients running the same software. A mobile client can store a demand profile and a limited number of supply profiles.

4   Mobile Matchmaking at Work

In this section we describe how IM3 works in its interaction with the user. We will refer to the four ways in which the users interact with IM3.

1. The first way of connecting to the server is via its Web interface, implemented with Java Server Pages (JSP). Through the Web interface, the user is able, by using a browser, to configure his/her own profiles and issue specific queries with particular search criteria, such as location where events have to take place. When the result of a query is presented to the user, the events are sorted and displayed starting from the one that returns the smallest penalty when matched with the demand profile, which is stored in the database. Users can register also as event suppliers, and enter profiles of events they wish to post on IM3.

2. Another Web interface is offered through the wireless network; the JSPs of IM3 are capable of presenting to the mobile browser of the user XHTML pages that offer the same functionalities as the HTML ones. The only requirement for the mobile terminal in this case is a mobile browser capable of interpreting XHTML Mobile Profile. In this case, presenting the top-ranked events according to the matchmaking becomes more important than in the previous case; in fact, the small size of the screens with which mobile devices are equipped allows only a small amount of information to be displayed at once; it is crucial here to minimise the steps performed by the user to navigate the result [4]. In Figure 2 some screenshots are shown, generated using the Openwave mobile browser emulator.

3. Now we come to the location-based interaction with IM3. Users can install a Java Midlet and run it on their mobile terminals. Such a Midlet is a text-

based Java application that offers all the functionalities available through the Web interface; the connection to the main server is through sockets (notice that this requires MIDP 2.0 to ba available on the mobile device). What is relevant about the Midlet is that it is able to connect to local servers via Bluetooth, thus avoiding the access to the Internet through the mobile network. Since the local servers store only profiles of events located nearby, the users has immediate access to local information. The local servers do not store demand profiles; when a user queries the system, he/she has two possibilities: *(i)* he/she sends his/her ID, and the local server connects to the main server, thus being able to answer the query and sorting the results without running the matchmaking algorithm; *(ii)* if the user wants to provide a fresh profile, not yet stored in the main server, he/she can send such profile together with the query; in this case the local server runs the matchmaking algorithm on-the-fly; obviously, the query processing time here is longer than in the previous case.

4. The last possibility of interaction is among mobile terminals without any local or centralised server. The Midlet running on mobile terminals is capable of storing a limited number of supply profiles (at the moment the user has a limit of five supply profiles that can be published), and act as a local server, so that other terminals nearby can access and retrieve such profile, in order to run the matchmaking algorithm. In this terminal-to-terminal interaction, the matchmaking algorithm is always executed on-the-fly on the terminal acting as server. Notice that this requires such terminal to store the domain ontology.

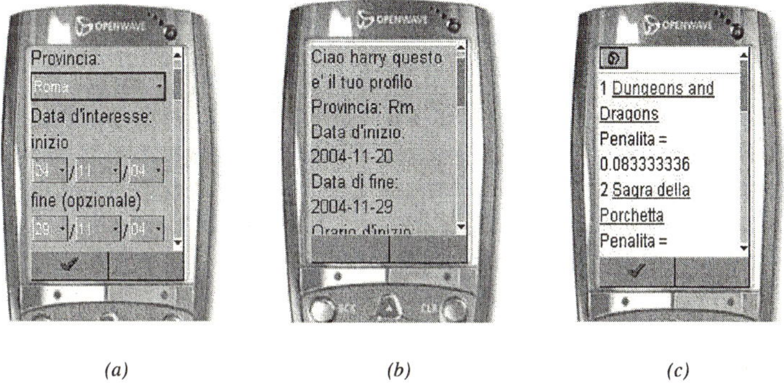

(a)   (b)   (c)

Fig. 2. Interaction with IM3 when accessed through a mobile browser. *(a)* profile input; *(b)* profile edit *(c)* browsing of results.

5 Conclusions and Future Work

In this paper we have presented IM3, a system for recommending tourist events that adopts matchmaking techniques to reason about demand and supply profiles; the techniques used in IM3 to represent and reason about profiles are borrowed from Description Logics. The reasoning tasks in IM3 are able to manage quantitative information, conflicting information between demand and supply, and missing information in profiles. The typical user of the system is likely to be a traveller equipped with a mobile phone; therefore, IM3 IM3 is fully accessible via mobile devices, both via a mobile browser or a Java Midlet. In the latter case, the mobile terminal can connect via Bluetooth to local servers, that store only profiles of events that are located nearby. Finally, each mobile terminal can access other mobile terminals via Bluetooth, each acting as a local server and storing profile of events.

Recommending the "best" events for the user is important when the system is accessed via a mobile phone or a PDA; in such a case, it is desirable to minimise the steps of the navigation of the information. To this aim, IM3 is being integrated with the system presented in [4], that adapts the information presented to the user according to usage data.

There are several directions in which IM3 can be extended. First, we are making IM3 server capable of providing local information, based on the location of the terminal and of the events; a simple ad-hoc GIS has been developed to this aim, but at the moment the geographic location has to be entered manually; integration with GPS terminals (e.g., accessed via Bluetooth) is under development. Another way to determine the location of terminals is to make them able to determine the GSM cell number they are connected to; unfortunately, there are no standard Java APIs for accessing cell information, so this feature will be implemented later (when APIs will be available or after rewriting the application in C++ on Symbian OS).

Aknowledgements

This work has been supported by Project MAIS (*Multichannel Adaptive Information Systems*), funded by the Italian Ministry of Instruction, University and Research (MIUR). The author wishes to thank Diego Calvanese, Simona Colucci, Tommaso Di Noia and Francesco M. Donini, with whom the formalisms and techniques for matchmaking were deveoped. The author is also grateful to Andrea Capata, Marco Cerrocchi, Matteo Berghella and Marco Oppedisano, who put a huge effort in the implementation of the system.

References

1. Franz Baader, Diego Calvanese, Deborah McGuinness, Daniele Nardi, and Peter F. Patel-Schneider, editors. *The Description Logic Handbook: Theory, Implementation and Applications*. Cambridge University Press, 2003.
2. Franz Baader and Philipp Hanschke. A schema for integrating concrete domains into concept languages. In *Proc. of IJCAI'91*, pages 452–457, 1991.
3. Chumki Basu, Haym Hirsh, and William Cohen. Recommendation as classification: Using social and content-based information in recommendation. In *Proc. of AAAI'98*, 1998.
4. Enrico Bertini, Andrea Calì, Tiziana Catarci, Silvia Gabrielli, and Stephen Kimani. Interaction-based adaptation for small screen devices. In *Proc. of UM 2005*, 2005.

5. Andrea Calì, Diego Calvanese, Simona Colucci, Tommaso Di Noia, and Francesco M. Donini. A logic-based approach for matching user profiles. In *Proc. of KES 2004*, 2004.

6. F. S. Hillier and G. J. Lieberman. *Introduction to Operations Research*. McGraw-Hill, 1995.

7. D. Kuokka and L. Harada. Integrating Information Via Matchmaking. *J. of Intelligent Information Systems*, 6:261–279, 1996.

8. David W. McDonald and Mark S. Ackermann. Expertise recommender: a flexible recommendation system and architecture. In *Proc. of CSCW 2000*, 2000.

9. Stuart E. Middleton, Nigel Shadbolt, and David De Roure. Ontological user profiling in recommender systems. *ACM Trans. Inf. Syst.*, 22(1):54–88, 2004.

10. Clinton Smythe and David Poole. Qualitative probabilistic matching with hierarchical descriptions,. In *Proc. of KR 2004*, 2004.

11. K. Sycara, S. Widoff, M. Klusch, and J. Lu. LARKS: Dynamic Matchmaking Among Heterogeneus Ssoftware Agents in Cyberspace. *Autonomous agents and multi-agent systems*, 5:173–203, 2002.

12. D. Veit, J.P. Muller, M. Schneider, and B. Fiehn. Matchmaking for Autonomous Agents in Electronic Marketplaces. In *Proc. AGENTS 2001*, pages 65–66. ACM, 2001.

Towards Using First-Person Shooter Computer Games as an Artificial Intelligence Testbed

Mark Dawes and Richard Hall

Dept. Computer Science & Engineering,
La Trobe University, Melbourne,
3086, Victoria, Australia
dawesm@cs.latrobe.edu.au, richard.hall@latrobe.edu.au

Abstract. One trend in first-person shooter computer games is to increase programmatic access. This allows artificial intelligence researchers to embed their cognitive models into stable artificial characters, whose competitiveness and realism can be evaluated with respect to human players. However, plugging cognitive models in is non-trivial, since games are currently not designed with AI researchers in mind. In this paper we introduce an intermediate architecture that fits between these game and the AI, and assess its feasibility by implementing it within the game Unreal Tournament 2004. Making such an architecture publicly available may potentially lead to improved quality of game AI.

1 Introduction

A game consists of the following: purpose, procedure for action, rules governing action, number of required participants, roles of participants, results or payoff, abilities and skills required for action, interaction patterns, physical setting and environmental requirements, and required equipment [1]. Many games exist; some (eg. dice rolling games), have been around for a thousand years, while others are relatively recent [2]. In computer games, not only is the equipment simulated, but its participants can also be simulated. To assist such simulation the computer games industry is increasingly turning to artificial intelligence techniques [3]. For example, in the 1990s the so-called first person shooters (FPS) emerged, where a player's view of the game world equates to the view of their character. Players controls a character (often humanoid), interact with the environment, and win by shooting other characters while keeping theirs alive [4].

The computer game industry has allowed programmatic access to games because it has been seen to greatly extend game shelf-life [5]. Many players are capable enough to use or even make game editing tools, and millions of copies of resulting game spin-offs have been sold [6, 7]. Increasing programmatic access to games for the gaming public incidentally creates an environment where various academic disciplines (such as artificial intelligence) can play a part.

While various definitions for artificial intelligence exist [8], the one pursued in the context of computer games seems to parallel the original McCarthian vision

R. Khosla et al. (Eds.): KES 2005, LNAI 3681, pp. 276–282, 2005.

of so-called *weak* AI; 'making a machine behave in ways that would be called intelligent if a human were so behaving' [9]. In this paper we require no definition, since we develop a testbed for implementing any AI which meet three criteria. First, the AI perception should receive no more information than does a player. Specifically, as client machines are told which objects should be drawn, the AI is informed as to which objects are visible (eliminating image processing requirements). Second, the AI implementation can only interact with the world via the same mechanisms that human players can interact with the world. Finally, its objective is to win the game through normal game play (without cheating).

Networked computer games approximate Turing's ideal arrangement for his intelligence test, where a teleprinter separates the interrogator I from the other two players (whose identity I attempts to determine), thus hiding irrelevant features [10]. Most FPS provide text support for conversation, which was originally proposed as the most demanding test of human mentality by Descarte in his *Discourse on Method* (1637). FPS thus provide an environment in which the Turing test, regarded as a sufficient test of intelligence [11], can be implemented.

The idea of implementing AI in computer games has been around since the dawn of computing and has met with varying success (see [12]). A number of artificial intelligence researchers have embedded cognitive models in games such as Pengi, Simcity, and Robocup [13–15]. With FPS moving increasingly towards photorealistic models of the world, AI researchers supposedly can focus on cognitive capabilities without needing to create these worlds [16]. Having the capability to easily evaluate an AI technique against potentially millions of enthusiastic networked game players (cast in the role of survey subjects) is certainly appealing. However, while networked computer games approximate Turing's test environment to a degree, differences must also be considered. In its original formulation, participants are limited to conversation. However in the FPS context, the goal is to win. Players may have no time to talk when action is intense, but failure to answer a question tells us nothing about intelligence. On the other hand, if all the AI did was communicate, it could not be evaluated within the FPS context since it must, at the very minimum, participate in the game.

Consequently, the Turing test is diluted twofold. First, the AI must prioritise on game play as opposed to conversation. But even if it can consistently win, it doesn't necessarily mean that it has achieved intelligence [11]. Second, the interrogation capacity is outside the FPS context. The interrogator must thus be positioned as an observer; their task becomes studying bot behaviour. They must try to infer from this behaviour who they are controlled by (assuming control cannot be shared by both human players and programs), which assumes that game play skill is correlated with intelligence. As such, an essential aspect of the Turing test; evaluating whether the interrogated knows things that one would expect people to know about the *real* world, is completely lost.

In addition, the assumption that millions of enthusiastic game players will become test subjects must be considered, since a primary reason for game play is supposedly peer acknowledgement [6]. Human players, by practicing against other human players, often develop their skills to a level that is much higher than the current computer-generated characters, and it is doubtful that these olympic-level competitors will be interested in playing AI that is no competition. However, if the AI really was competitive (without cheating), and no other strong human players were available, these super-players might use the AI for practice. Novices might also want to practise against this strong AI, since at the moment even if novices can beat the computer-generated characters they may not be guaranteed a level of skill that allows them to be competitive with the average super-player. On the other hand, novices might be so disappointed by being continuously destroyed by the AI that they give up entirely, thus a dynamic adjustment of the strength of computer-controlled characters could be necessary.

Nonetheless, some argue that game AI will improve, 'with or without input from the academic artificial intelligence community' [3, 16]. Three main reasons are given: the games industry is competitive (eg. better AI can be marketed); more AI cycles are available as graphics gets its own CPU; and the games industry is ahead of academia in some areas (eg. path planning). Clearly, AI research outside academic circles is ongoing and opportunities do exist to get involved.

The organisation of this paper is as follows. Section 2 describes the architecture of our intermediate layer which positions bot cognition on the client side and game recording in a server-side database. Section 3 demonstrates that our architecture functions by comparing recorded with play-back games. Section 4 concludes the paper.

2   Method

We initially built a very simple neural network into the FPS Unreal Tournament 2003 using UnrealScript. Unreal Tournament 2003 was originally chosen due to its player base, and its tight integration with its internal scripting engine and language, UnrealScript. Another key factor was the ease of modifying bot behaviours. Its AI hooks gave access to the bot's cognitive model (a form of finite-state machine), allowing us to map player controls to state transition triggers. We were dissatisfied with the result for three reasons: we were confined within the limits of the state representation and became responsible for a vast amount of mapping between player controls and triggers; the implementation occurred on the server-side thus increasing server load and limiting scalability; and because we required rather intimate knowledge of UnrealScript before we could even think about the AI.

Consequently, we stepped back from developing the AI and pseudo-Turing tests and focused on the precondition for these to proceed; developing an intermediate layer between a FPS and cognitive models, that provides client-side bot plug in support without requiring knowledge of UnrealScript. Our architecture does three things: records what individual game players do in such a way that machine learning techniques can be applied to these records; allows the external cognitive models to control characters in the game; and evaluates the competitiveness of players by comparing match statistics.

Taking what we had learnt from our earlier experiments, we came up with a list of the four primary requirements for our architecture: to minimise game play disturbance; to be able to record any data which we believed was important for training a cognitive model in a non-game-specific language, including adding data items after deployment; to be able to monitor games as they were being played to check that what we were recording actually matched the game play; and transparency, neither the server or a client should be aware that any given client is a bot or a human player. The architecture shown below in Figure 1 shows the interaction between the components that were necessary to represent this intermediate layer. UT2004 server game events are sent to and recorded in the MySQL database server, which holds not only all the game data but also the AI cognitive models (note that we have proposed to use a neural network, continuing on from our initial experiments). The database server is managed by an administrator client which provides monitoring capabilities for games. This server is also connected to an AI player UT2004 Client, which, from the point of view of the UT2004 game play engine, is a human player. This architecture overcomes the problems associated with the original approach; it is highly extensible and shifts the focus to creating cognitive models as opposed to having to understand the environment language.

In order to show that our intermediate layer is independent of which cognitive model is plugged in, we evaluated it without plugging in a cognitive model. For this evaluation, we created a synchronicity based testing method. This testing method involves comparing recorded games with play-back games. We recorded a game between multiple players (an individual player's records might become the inputs to a cognitive model). We then play-back these records simultaneously (matching the output of a cognitive model), and observe whether a similar game is reproduced, by matching positions, deaths, end of game statistics, etc with the original game.

UT modifications involved creating classes (called mutators) to replace UT classes. Server side, one mutator sends information to the database server. Client side, another mutator acts as a player. This client-side mutator reads data from configuration files detailing the cognitive models (eg. a set of neural networks linked to specific data streams) and selected game play data, all of which is converted to cognitive model inputs as defined by the configuration files.

The data in the MySQL database tables contains three main event types: server events, game events, and player events. *Server events* are simply recorded for administration purposes, including IP, player name, and session details. *Game events* describe game state, including item pickup, flag capture, and player death. *Player events* precisely record what a player is doing, such as moving, shooting, changing weapons, etc. These events will be used to create cognitive models of tactics and strategy.

Our intermediate layer largely satisfied our requirements. Server scalability is unaffected as bot clients appear to be player clients. Since the cycles for the AI are all client side, from a server point of view the load is identical to having the equivalent number of human players. The database server runs on a different machine to the game engine such that the game engine is basically unaffected in terms of extra load.

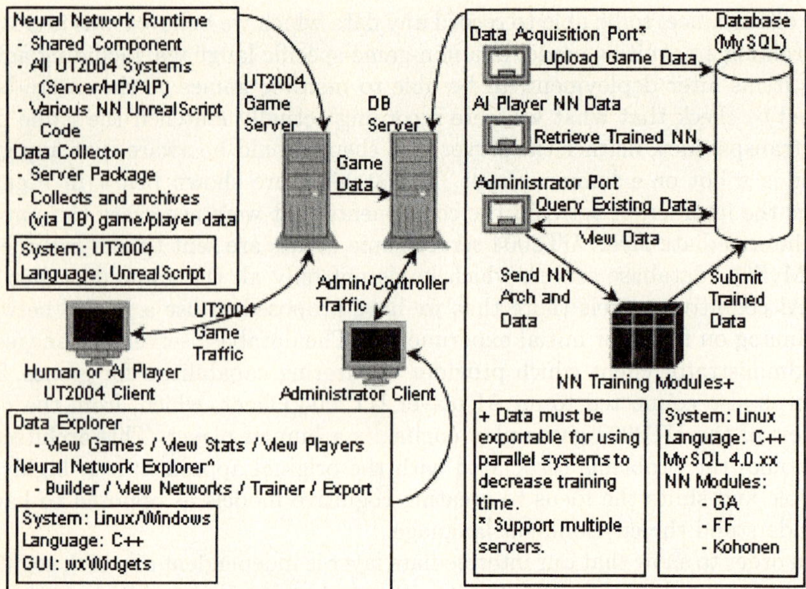

Fig. 1. FPS AI-Intermediate Architecture.

Data capture is generalised and flexible. Events are stored such that adding new event types is simply a matter of adding the code in UnrealScript to expose the data. No modification of the Administrator or Database clients are required.

The administrator application allows real time viewing of a game that is currently in progress. Seeing kills/deaths and other statistics, as well as seeing a 3D representation of where players are in the world was very helpful during development, as well as helping an administrator pick games for cognitive modelling by getting a simple representation of the game at any given time.

3   Results

To evaluate our intermediate architecture without a cognitive model, we recorded a game between multiple players, then played-back the player events out of the database on the same world map while observing whether a similar game was reproduced. An example game was recorded with the following features and a screen grab was taken halfway through. Game Type was One on One Deathmatch, map was Deck17, other miscellaneous settings where 15 minutes maximum time limit and 40 maximum kill limit.

After the game had finished, two separate computers were started with the Data Replay mutator and told to play the specified games back, each computer taking the role of one of the participants of the match. A screen grab was taken roughly halfway through playback (this grab was done manually thus the slight timing and positional differences) as shown in Figure 2.

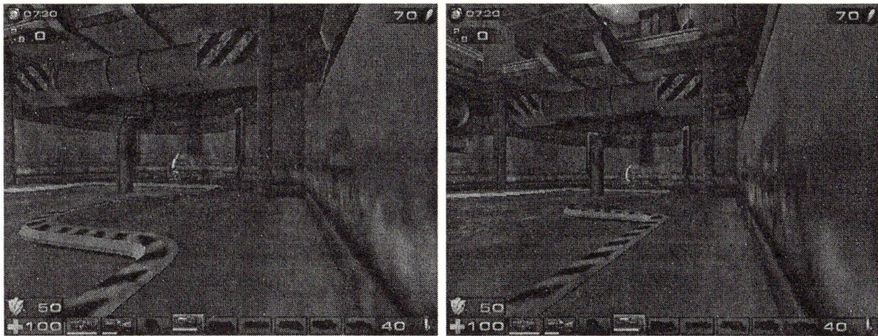

Fig. 2. Example Test Result.

This simple test demonstrates that our architecture records properly via the database server and plays back games via the AI Plugin UTServer Client. We ran many of these synchronicity tests, using many varied source games and containing players of different skill levels. These tests revealed tiny deviations between the games due to the random effects of some weapons, which might do less or more damage when used in the same circumstances but with a different random seed in playback. We believe that such randomness should not adversely affect the results of any training in any significant way, since any cognitive model used would be taking large amounts of data into account, and would be generalising behaviour based on many factors, not just damage.

4   Conclusion

In this paper we developed an intermediate layer between a first-person shooter and an AI cognitive model which uses UT class overrides to mimic client-side players and to store and retrieve game events using a database. We believe that such an architecture has five main advantages for AI researchers: it is independent of the specific programmatic language required by a FPS (in this case UnrealScript); many game servers could potentially be leveraged to quickly and cheaply generate large datasets for cognitive model development; different types of bots can be plugged in, allowing specialised bots for individual, team, or goal based play, or context-specific combinations; bots are told which objects they can see by the game server, eliminating the task of image-processing from the AI development process; and it allows an AI researcher to focus on the cognitive modelling rather than on modelling worlds and interactions within these worlds.

There are several other advantages that this intermediate layer provides for cognitive model development. A FPS servers can run stably for months, detailed data sets can be collected on individual players, potentially allowing the development of bots with distinctly different levels of skill. Thus, it allows the possibility of bots that are based exclusively on a particular player to participate in pseudo-Turing tests, where observers try to determine whether characters are controlled by computer or human. Regardless of the success of modelling the Turing test, we hope that the trend of allowing programmatic access to games continues and that the AI research community is given a chance to help make better bots. This architecture allows cognitive modelling to occur with only an understanding of the principles of the game being played (Capture the flag, DeathMatch, etc), no programming knowledge, no gaming skills, and no game specific implementation details are required. This should allow cognitive modellers to quickly develop and test models of various complexities, quickly and with minimal overhead.

Acknowledgement

Our thanks must go to the School of Engineering and Mathematical Sciences at LTU for the FBIS grant, and to all the UT players (FrEaKs and SPRIGers) who helped create our datasets.

References

1. Avedon, E.: The Study of Games. John Wiley and Sons Inc., New York (1971)
2. Jay, R., Wolff-Purcell, R.: Dice: Deception, Fate, and Rotten Luck. Quantuck Lane Press (2002)
3. Woodcock, S.: Game ai: The state of the industry. Game Developer 6 (1999)
4. Yerrick, D., et al.: First-person shooter: From wikipedia, the free encyclopedia. http://en.wikipedia.org/wiki/First-person shooter (2005)
5. Nieborg, D.: Who put the mod in commodification? - a descriptive analysis of the first person shooter mod culture. http://www.gamespace.nl/content/CommodificationNieborg2004.pdf (2004)
6. Herz, J.: Harnessing the hive: How online games drive networked innovation. Esther Dyson's Monthly Report 20 (2002)
7. Au, J.: Triumph of the mod. http://www.salon.com/tech/feature/2002/04/16/modding/ (2004)
8. Boden, M., ed.: The philosophy of artificial intelligence. Oxford University Press (1990)
9. Ross, D., et al.: Artificial intelligence: From wikipedia, the free encyclopedia. http://en.wikipedia.org/wiki/Artificial intelligence (2005)
10. Turing, A.: Computing machinery and intelligence. Mind 59 (1950) 433–460
11. Dennett, D.: Can machines think. In: How We Know. San Francisco: Harper & Row (1985)
12. Woodcock, S.: Classic Games & AI. http://www.gameai.com/clagames.html (2005)
13. Agre, P., Chapman, D.: Pengi: An implementation of a theory of activity. In Proceedings of the Sixth National Conference on Artificial Intelligence (1987) 268–272
14. Fasciano, M.: Real-Time Case-Based Reasoning in a Complex World. Technical Report, TR-96-05, Computer Science Department, University of Chicago. (1996)
15. Asada, M., Veloso, M., Tambe, M., Noda, I., Kitano, H., Kraetzschmar, G.: Overview of robocup-98. AI Magazine 21 (2000) 9–19
16. Laird, J., Lent, M.: Human-level ai's killer application: Interactive computer games. AAAI/IAAI (2000) 1171–1178

Structured Reasoning
to Support Deliberative Dialogue

Alyx Macfadyen, Andrew Stranieri, and John Yearwood

Centre for Informatics and Applied Optimization
University of Ballarat, Victoria, Australia

Abstract. Deliberative dialogue is a form of dialogue that involves participants advancing claims and, without power plays or posturing, deliberating on the claims of others until a consensus decision is reached. This paper describes a deliberative support system to facilitate and encourage participants to engage in a discussion deliberatively. A knowledge representation framework is deployed to generate a strong domain model of reasoning structure. The structure, coupled with a deliberative dialogue protocol results in a web based system that regulates a discussion to avoid combative, non-deliberative exchanges. The system has been designed for online dispute resolution between husband and wife in divorce proceedings involving property.

1 Introduction

Walton and Krabbe [17] classified six basic types of human dialogue: a) Information seeking (b) Inquiry (c) Persuasion (d) Negotiation (e) Deliberation and (f) Eristic. Deliberative dialogue involves two or more participants seeking to agree upon a course of action or decision. Participants to a deliberative dialogue advance claims and, without power plays, dishonesty or posturing deliberate on the claims of others until a consensus solution or decision is reached. The advantages of this type of human dialogue have been well documented and include improved outcomes and a sense of engagement in the process.[2, 6, 11]

McBurney and Parsons[10] cite the absence of hierarchy in deliberative dialogue as conducive to a sense of equality between participants. The non-combative and inclusive nature of a deliberative discussion with many participants with diverse points of view enables access to information otherwise not available. Traditional power based dialectic outcomes are frequently unsatisfactory leaving participants disgruntled and cynical. Consultative and participatory skills are developed by deliberative dialogues and can be carried across into other decision-making processes.

Information communication technologies have recently emerged as a convenient environment for deliberative dialogues[20]. Online deliberative discussions involve a community of participants from diverse locations using web technology. Participants to on an online discussion can deliberate on their own position and that of others posting a response.

R. Khosla et al. (Eds.): KES 2005, LNAI 3681, pp. 283–289, 2005.

The aim of this paper is to illustrate that knowledge based systems can facilitate online deliberative dialogue. For this to occur, the knowledge underpinning a dialogue must be structured so that vital issues are not overlooked and professional opinion, precedent and policies support all arguments. Argumentation theories have often been advanced as approaches to structuring reasoning.[1, 3, 9, 10] For example, Toulmin[16] concluded that most arguments, regardless of the domain, have a structure that consists of six basic invariants: claim, data, modality, rebuttal, warrant and backing. Yearwood and Stranieri [19] have varied this structure to yield a model that can more readily be applied to structuring reasoning for deliberative dialogues. Their model, called the Generic Actual Argument Model (GAAM) has been deployed in knowledge based systems in family law[14], nursing[15] and sentencing[4].

A deliberative dialogue support system not only requires a method to structure knowledge but also requires a way to regulate discourse. In their desiderata for a dialectical system McBurney and Parsons[11], advance specifications for a dialogue game protocol: (a) Set of topics for discussion (b) The syntax for a set of defined locutions for the topics (c) A set of rules that govern the utterance of these locutions (d) A set of rules which establish what commitments are created by the utterance of each locution (e) A set of rules governing the circumstances of the dialogue termination.

In this paper a dialogue game protocol is advanced that uses the GAAM to define topics for discussion and define a set of locutions for the topics. A protocol described in[13] is used to define a set of rules that govern utterance of these locutions and a set of rules to govern dialogue termination. The application domain is family law property proceedings. The participants to a deliberative discourse are the husband and wife of a failed marriage who are motivated to deliberate rather than combat in order to achieve a fair outcome for themselves and their children and to avoid legal costs.

The GAAM is briefly described in the next section with examples drawn from the family law domain. Following that is a discussion of two implementations for deliberative discussion between husband and wife using the family law structure and dialectical protocol in development.

2   The GAAM

The Generic Actual Argument Model is advanced by Yearwood and Stranieri[21] is a two tiered framework for structured reasoning. Firstly, a Generic Argument Structure, GAS, is used to represent the domain of the dialectic. An actual argument is a dialectical instantiation of a selected generic argument where values have been assigned by participatory deliberation. Subtrees are formed by child nodes that support decision making for ancestor nodes. Actual rule-based locutions occur within this domain.

Argument nodes are referred to as *claims*. Attributes of claims are the definition of the particular matter to be resolved, a set of possible values that each party may choose to represent their point of view, justification values and inference values.

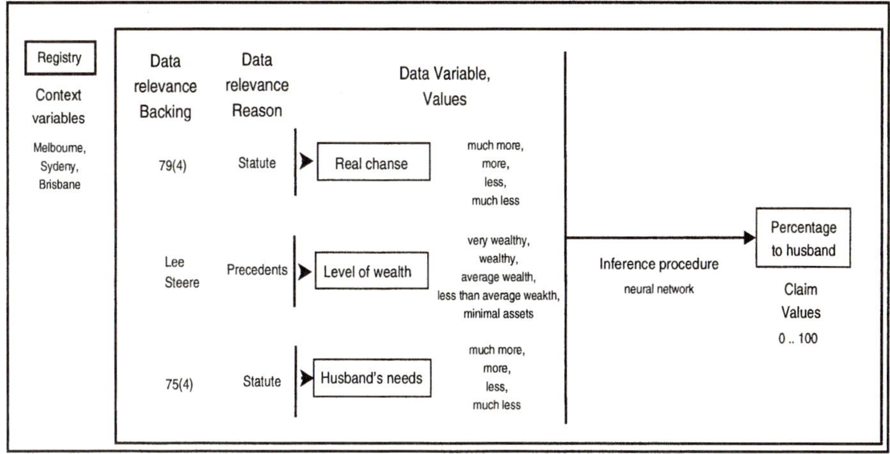

Fig. 1. A generic argument in Split-Up.

Split-Up, developed using the GAAM, models the reasoning used by Family Court judges in Australia in predicting the percentage of property that a Court is likely to award both parties to a marriage following divorce. It is based on a generic argument structure and its inference values determine a likely legal ruling. Figure 1 illustrates a generic argument from Split Up. The top level claim is the percentage split of assets. This is the main issue in dispute. The three key features relevant in an inference of percentage split are the husband's contribution to the marriage relative to the wife's, the level of wealth of the marriage and the husband's future needs relative to the wife's. In the knowledge based system, values on these three are inputs into neural network that has been trained with past cases. In the deliberation game here, the neural network is not used.

Figure 1 illustrates one generic argument. Each leaf node in that argument is a claim of another generic argument. For example, there is another generic argument that has, the *Husband's Contribution* as a claim and factors, *Husband's Direct Contribution, Husband's indirect Contribution, Husband's Negative Contribution* and *Marriage Length,* as factors. These four factors are used to infer a value on the *Husband's Contribution.* In this way a tree representing a hierarchy of factors is identified. The tree for Split Up comprises 94 factors in 35 generic arguments and was elicited from family law experts who verified the relevance of and position of each factor.

In the next section the way in which the tree underpins the facilitation of deliberative dialogue between husband and wife is illustrated.

3   Dialogue

A deliberative support system is needed to manage dialogue between participants. In its present state, Split-Up is operational for one participant. Two solutions are in development that manage the dialectic process for two players. These are (a) a two-dimensional web-based solution and (b) a three-dimensional solution using a three-dimensional environment. Both solutions use the same structured reasoning tree. The next sub-sections discuss these solutions.

3.1   Two Dimensional Dialogue

In the web-based solution, a dialogue agent manages interaction between participants who are using the system simultaneously via web browsers from separate remote locations. Dialogue interaction occurs by a series of prompts to each user in turn. The dialogue agent is registered for user interface events and outputs *string* user prompts. Values selected by participants are stored and compared by the *GAAM* engine. Agreement is defined as an assignment of the same value to the same factor. For example if both parties beleive the *Wife's Contribution* was *much more* than that of the husband, then there is agreement. Agreement between the two participants is represented as a point in common:

$$PIC = \{C \in C : C_v^{P1} = C_v^{P2}\} \tag{1}$$

Where an issue is not agreed, dialogue continues with arguments (child nodes) relating to the matter in question (represented as a subtree in the generic argument structure). The dialogue game presents both parties with prompts for the child nodes. Where there is no subtree and a disagreement, the argument is considered irreconcilable on data and listed as a point of disagreement.

$$POD = \{C \in C : C_v^{P1} \neq C_v^{P2}\} \tag{2}$$

Transformation of a wife or husband's assertion for a matter in question is possible through a retraction dialogue. Having accessed the assertions and reasoning process of the spouse, the wife or husband may decide to change a prior assertion. The store of claims is then modified.

3.2   Dialogue Using SAM

Using the UnrealScript game engine[1] a Simulated Argument Model represents the nodes of a structured argument tree by a set of zones in a three dimensional world. Participants may virtually interact by moving through the three sequence levels of the game making decisions for each argument zone. In level one each player must make a decision for each zone. At level two the players must negotiate points of differences to find agreement where possible. Level three predicts a possible legal ruling based on the information and context of the information provided by the two participants. A dialogue D is represented by a vector containing two sub-sets *zone* and *player*.

$$\{Z, P\} \in D \tag{3}$$

---

[1] Epic Games.

Sub-sets of Zone are directly mapped as instantiated interactive objects.

$$Z = \{states_i, argument, values_i, justification, inference\} \qquad (4)$$

The simulated model presents ludic processes for decision making. Participants can choose which matter in question to address by entering the *zone* for the matter in question. Each zone's status is indicated by lighting. Players can interact within each zone or choose not to interact. Assertions for matters in question are made by selecting the appropriate object in the zone.

The ultimate goal is agreement about the top level matter in question, the division of common assets. Rewards are earned for agreement about matters in question and for resolution of points of differences. When agreement cannot be reached, the mentor game character (an intelligent agent) makes a ruling. Retractions are made by simply revisiting the zone and selecting another object representing the revised assertion. Both solutions discussed here provide rule-based environments where dialogue cannot occur beyond the structured argument template and both solutions store and compare the player's chosen values for each argument node. In the next sub-section differences in implementation and user interaction between the two methods of dialectic are discussed.

3.3   Discussion

According to the dialogue game described by [21], a dialogue commences with either party selecting any matter from the *Issue list* for discussion. In the SAM model any party may select any matter in question at any time by selecting an object in a zone. In the web-based scenario, both parties are prompted in turn to assert a claim about the same matter in question.

In solution *a* both parties may choose from a list of assertions and in solution *b* they can agree or disagree with the statement represented in each zone. The objective in both solutions is to identify the key issues they agree on and those they disagree on.

Virtual face-to-face interaction of player avatars may occur in solution *b* while in solution *a*, interaction occurs using text messaging within a webform. The next section concludes the discussion.

4   Conclusion

The deliberative dialogue program is currently under development as an on-line web based application. Current research is in progress to extend the application to deploy a 3-D game engine environment. In this way a realistic simulated environment can be created and the effects of scenario playing in an emotionally charged domain such as divorce proceedings can be studied. Research is also in progress toward extending the deliberative discourse protocol from a 2 person game to an n-person game. This paves the way forward for electronic democracy applications where many individuals can deliberate on current affairs issues.

References

1. Amgoud, L., Maudet, N., Parsons, S.: Modelling Dialogues Using Argumentation. IEEE (2000) 1–8
2. Baas, N. (ed.): Summary. In: Mediation in civile en bestuurs-rechtelijke zakon. WODC Onderzoeksnototies. (2004) 109–111)
3. Belluci, E., Lodder, A., Zeleznikow, J.: Integrating Artificial Intelligence, Argumentation and Game Theory to Develop an Online Dispute Resolution Environment. In: Proceedings of the 16th IEEE International Conference on Tools with Artificial (ICTAI 2004). IEEE
4. Hall, M. J.J , Calabro, D. , Sourdin, T. , Stranieri, A . and Zeleznikow, J. Supporting discretionary decision making with information technology: a case study in the criminal sentencing jurisdiction In: (to appear) in University of Ottawa Law and Technology Journal. 2005-03-07
5. Fatima, S., Wooldridge, M., Jennings, M.: Optimal Agendas for Multi-Issue Negotiation. In: Proceedings of AAMAS'03. July 14-18, Melbourne, Australia. ACM Press
6. Hitchcock, D., McBurney, P., Parsons, S.: A Framework for Deliberation Dialogues. In: Hanson, H., Tindale, C., Blair, J., Johnson, R. (eds.) Proceedings of the 4th Biennial Conference. Ontario Soc. Study of Argumentation (OSSA-2001). Windsor, Ontario, Canada.
7. King, D.: Internet Mediation - A Summary. Australian Dispute Resolution Journal. (2000). 11:3, 180–186
8. McBurney, P., Parsons, S.: A Denotational Semantics for Deliberation Dialogues. In: Proceedings of the Third International Conference on Autonomous Agents and Multi-Agent Systems. New York, New York, IEEE Computer Society. (2004) 1, 86–93
9. McBurney, P., Parsons, S.: Locutions for Argumentation in Agent Interaction Protocols. In: Proceedings of AAMAS'04. New York, New York, ACM Press. (2004) July 19-23.
10. McBurney, P., Parsons. S. Democracy in Open Agent Systems. In: Proceedings of AAMAS'03. (2003) July 14–18. Melbourne, Australia.
11. McBurney, P., Parsons. S., Wooldridge, M, Desiderata for Agent Argumentation Protocols. In: Proceedings of the First International Joint Conference on Autonomous Agents and Multiagent Systems: part 1. Bologna, Italy, ACM Press. (2002) 402–409
12. Stranieri, A., Zeleznikow, J.: Copyright regulation with argumentation agents In: Information and Communication Technology Law, 2001. Vol 10, No. 1. p123-137.
13. Stranieri, A., Zelenikow, J., Yearwood, J.: Argumentation Structures that Integrate Dialectical and Non-Dialectical Reasoning. In: The Knowledge Engineering Review. (2001) 16:4 331–348
14. Stranieri, A., Zeleznikow, J., Gawler, M., and Lewis, B.: A Hybrid rule- neural approach for the automation of legal reasoning in the discretionary domain of family law in Australia. In: Artificial Intelligence and Law Vol (1999) 7(2-3). Pp153-183
15. Stranieri,A., Yearwood, J., Gervasoni, S., Garner, S., Deans, C and Johnstone, A. Web-based decision support for structured reasoning in health. In: Twelth National Health Informatics Conference. Health Informatics Society of Australia. pp61.
16. Toulmin, S. The Uses of Argument. Cambridge, Cambridge University Press. (1958)

17. Walton, D., Krabbe, E.: Commitment in Dialogue: Basic Concepts of Interpersonal Reasoning. State University of New York Press. Albany, NY., USA. (1995)
18. Yearwood, J., and Stranieri, A. The integration of retrieval, reasoning and drafting for refugee law: a third generation legal knowledge based system In: Seventh International Conference on Artificial Intelligence and Law. 1999. (199) ICAIL'99 ACM Press. Pp 117-137.
19. Yearwood, J., Stranieri, A. The Generic Actual Argument Model of Practical Reasoning. In: To appear in Decision Support Systems
20. Yearwood, J., Stranieri, A. Generic Arguments: A Framework for Supporting Online Deliberative Discourse. In: Enabling Organizations and Society Through Information Systems. Proceedings of the Thirteenth Australasian Conference on Information Systems (ACIS'2002). (2002) December 4–6 337–346
21. Yearwood, J., Stranieri, A. Deliberative Discourse and Reasoning from Generic Argument Structures. In: Technical Report (2004) School of Information Technology and Mathematical Sciences, University of Ballarat, Australia.

A New Approach
for Conflict Resolution of Authorization

Yun Bai

School of Computing and Information Technology
University of Western Sydney
Penrith South DC, NSW 1797, Australia
ybai@cit.uws.edu.au

Abstract. Authorization rules provide access control to information system in order to protect the system from unauthorized, malicious attempt. However, authorization rules need to be updated to reflect system and client's changing requirements. Update causes conflict. In this paper, we propose a new approach to solve conflict by using prioritized logic program. Addressing conflict resolution from logic programming point of view provides a novel, efficient approach for authorization conflict resolution.

Keywords: authorization rule, formal specification, conflict resolution, prioritized logic programming.

1   Introduction

Authorization rules provide the ability to control access to information system, and to limit what entities can do what kind of operations on the information and the resources of the system. To ensure the security of the system, authorization rules need to be specified in such a way that they not only have a powerful expressiveness to accommodate user and system requirements, but also they need to be flexible enough to capture the changing needs of the environment. Logic based specification approaches provide an appropriate level of such requirements. Therefore, more and more researchers are focusing on the issue of logic based authorization specification. jodia *et al* [4] proposed a logic language for expressing authorizations. They used predicates and rules to specify the authorizations; their work mainly emphasizes the representation and evaluation of authorizations. The work of Bertino *et al* [1] describes an authorization mechanism based on a logic formalism. It mainly investigates the access control rules and their derivations. In their recent work [2], a formal approach based on C-Datalog language is presented for reasoning about access control models. Li *et al* [5] developed a logical language called *delegation logic* to represent authorization policies, credentials in large-scale, distributed systems. The work emphasizes the delegation depth and a variety of complex delegation principals. Chomicki *et al* [3] discussed security policy management using logic program approach. Woo and Lam proposed a formal approach using default logic to represent and evaluate authorizations [6].

R. Khosla et al. (Eds.): KES 2005, LNAI 3681, pp. 290–296, 2005.
© Springer-Verlag Berlin Heidelberg 2005

This paper is to address high level authorization specification and conflict resolution by using prioritized logic programs. We first propose a logic language by using logic program to specify authorization rules, and then solve its conflict by using the concept and techniques of prioritized logic programs.

The paper is organized as follows. Section 2 describes authorization rules, its specification and evaluation. Section 3 investigates authorization conflict issue and proposes a new approach to solve it. We introduce prioritized logic programs for efficient conflict resolution. Section 4 concludes the paper with some future work.

2 Authorization Specification and Evaluation

2.1 Authorization Specification

We define that all the authorizations rules forms an *authorization domain*. The individual rule is specified by a language $\mathcal{L}$. Language $\mathcal{L}$ includes the following six disjoint sorts for *subject, group-subject, access-right, group-access-right, object, group-object* together with predicate symbols *holds*, $\in$, $\subseteq$ and logic connectives.

The six disjoint sorts and the predicate symbols are defined as follows:

1. Sort *subject*: with subject constants $S, S_1, S_2, \cdots$, and subject variables $s, s_1, s_2, \cdots$.
2. Sort *group-subject*: with group subject constants $G, G_1, G_2, \cdots$, and group subject variables $g, g_1, g_2, \cdots$.
3. Sort *access-right*: with access right constants $A, A_1, A_2, \cdots$, and access right variables $a, a_1, a_2, \cdots$.
4. Sort *group-access-right*: with group access right constants $GA, GA_1, GA_2, \cdots$, and group access right variables $ga, ga_1, ga_2, \cdots$.
5. Sort *object*: with object constants $O, O_1, O_2, \cdots$, and object variables $o, o_1, o_2, \cdots$.
6. Sort *group-object*: with group object constants $GO, GO_1, GO_2, \cdots$, and group object variables $go, go_1, go_2, \cdots$.
7. A ternary predicate symbol *holds* which takes arguments as *subject* or *group-subject, access-right* or *group-access-right* and *object or group-object* respectively.
8. A binary predicate symbol $\in$ which takes arguments as *subject* and *group-subject* or *access-right* and *group-access-right* or *object* and *group-object* respectively.
9. A binary predicate symbol $\subseteq$ whose both arguments are *group-subjects, group-access-rights* or *group-objects*.

In language $\mathcal{L}$, the fact that a subject S has access right R for object O is represented using a ground atom $holds(S, A, O)$. The fact that a subject S is a member of G is represented by $S \in G$. Similarly, we represent inclusion relationships between subject groups such as $G_1 \subseteq G_2$ or between object groups such as $GO_1 \subseteq GO_2$. In general, we define a *literal* which represents a *fact F* to be an atomic formula of $\mathcal{L}$ or its negation, while a *ground fact* is a fact without variable occurrence. We view $\neg\neg F$ as F. A *rule* is an expression of the form:

$$F_0 \leftarrow F_1, \cdots, F_m, notF_{m+1}, \cdots, notF_n, \tag{1}$$

where each F_i $(0 \leq i \leq n)$ is a literal. F_0 is called the *head* of the rule, while $F_1, \cdots, F_m, not\, F_{m+1}, \cdots, not\, F_n$ are called the *body* of the rule. Obviously, the body of a rule could be empty. In this case, it represents an authorization fact. A rule is *ground* if no variable occurs in it.

All the rules required to specify the access control of a system or an organization form an *authorization domain*. It is formally defined as:

Definition 1. *An authorization domain is a finite set $D = \{R_i\}$, $(i=1,2, ...k)$ where R_i is a rule of the form $F_0 \leftarrow$ or $F_0 \leftarrow F_1, \cdots, F_m, notF_{m+1}, \cdots, notF_n$ where $m>0$, $n>m$.*

The following is an example of an authorization domain.

Example 1. $D = \{R_1, R_2, R_3\}$, where
R_1: $holds(S, R, O) \leftarrow$
R_2: $holds(S_1, W, O) \leftarrow \neg holds(S_2, W, O)$
R_3: $holds(S_3, R, O) \leftarrow holds(S_3, R, O_1), O \in O_1, not\neg holds(S_3, R, O)$

This domain represents the current authorization information about the system: subject S has read right on object O; if subject S_2 does not have write right on object O, then S_1 can write on O; if S_3 can read O_1, O is a member of O_1 and there is no information stating that S_3 cannot read O, then S_3 has read right on O.

2.2 Authorization Evaluation

Evaluation is the other aspect of authorization apart from its specification. Once an authorization domain is properly specified and applied to a system, its task is to make the decision as which access request is to be granted and which is to be denied according to the specification of the authorization domain. That is, given an authorization domain and an access request, how to decide either to grant or deny such an access request?

Authorization evaluation is to answer this question. It first calculates all the authorization facts from the authorization domain. Then when an access query is proposed, it can make proper decision by checking the calculated set of authorization facts.

Some facts are explicitly represented, such as $holds(S, R, O)$ in example 1. This is the only authorization fact we can get from this domain. Let's have a look of another example:

Example 2. $D = \{R_1, R_2, R_3, R_4, R_5\}$, where
R_1: $holds(S, R, O) \leftarrow$
R_2: $holds(S_3, R, O_1) \leftarrow$

R_3: $O \in O_1 \leftarrow$
R_4: $holds(S_1, W, O) \leftarrow \neg holds(S_2, W, O)$
R_5: $holds(S_3, R, O) \leftarrow holds(S_3, R, O_1), O \in O_1, not\ \neg holds(S_3, R, O)$

This authorization domain says that currently subject S has read right on object O; subject S_3 has read right on object O_1; object O is a member of O_1; if subject S_2 does not hold write right on object O, then subject S_1 can have the write right on object O; if subject S_3 has read right on object O_1, and object O is a member of O_1, and it is not specified that subject S_3 does not hold read right on object O, then subject S_3 has read right on object O. In this domain, authorization facts $holds(S, R, O)$, $holds(S_3, R, O_1)$ and $O \in O_1$ are explicit facts; $holds(S_3, R, O)$ is implicit fact since it is implied in the domain and deduced from it.

In an authorization domain, the explicit facts are obvious. How can we get these implicit facts? Here we use answer set of extended logic program to find the implied access facts.

The set of rules of the authorization domain form an extended logic program Π. Here we use the answer set semantics proposed by Gelfond and Lifschitz to evaluate the extended logic program.

To simplify the procedure, we treat a rule r in Π with variables as the set of all ground instances of r formed from the set of ground literals of the language of Π. Let Π be an extended logic program not containing not and Gl the set of all ground literals in the language of Π. The *answer set* of Π, denoted as $Ans(\Pi)$, is the smallest subset S of Gl such that (i) for any rule $F_0 \leftarrow F_1, \cdots, F_m$ from Π, if $F_1, \cdots, F_m \in S$, then $F_0 \in S$; and (ii) if S contains a pair of complementary literals, then $S = Gl$. Now let Π be an arbitrary extended logic program. For any subset S of Gl, let Π^S be the logic program obtained from Π by deleting (i) each rule that has a formula $not\ F$ in its body with $F \in S$ (since $not\ F$ is not true, this rule will no longer take effect), and (ii) all formulas of the form $not\ F$ in the bodies of the remaining rules (for similar reason, such formulas will no longer take effect). We define that S is an *answer set* of Π, denoted by $Ans(\Pi)$, iff S is an answer set of Π^S, i.e. $S = Ans(\Pi^S)$. An extended logic program may have one, more than one, or no answer set at all.

3   Conflict Resolution

3.1   Prioritized Logic Program

Now, let's consider the following authorization domain.

Example 3. $D = \{R_1, R_2, R_3, R_4\}$, where
R_1: $holds(S_1, R, O_1) \leftarrow$
R_2: $\neg holds(S_1, R, O) \leftarrow$
R_3: $O \in O_1 \leftarrow$
R_4: $holds(S_1, R, O) \leftarrow holds(S_1, R, O_1), O \in O_1, not\ \neg holds(S_1, R, O)$

Obviously, rules R_2 and R_4 conflict with each other as their heads are complementary literals, and applying R_2 will defeat R_4 and *vice versa*. However, we can assign preference ordering among the conflict rules. If we define $R_2 < R_4$, we expect that rule R_4 is preferred to apply first and then defeat rule R_2 after applying R_4 so that the solution $holds(S_1, R, O)$ can be derived. On the other hand, if we define $R_4 < R_2$, we expect that rule R_2 is preferred to apply first and then defeat rule R_4 after applying R_2 so that the solution $\neg holds(S_1, R, O)$ can be derived.

We call the logic program with partial ordering $<$ on the rules *prioritized logic program* $\mathcal{P}$ [7]. $\mathcal{P}$ is defined to be a triplet $(\Pi, \mathcal{R}, <)$, where Π is an extended logic program, $\mathcal{R}$ is a naming function mapping each rule in Π to a name, and $<$ is a strict partial ordering on names. The partial ordering $<$ in $\mathcal{P}$ plays an essential role in the evaluation of $\mathcal{P}$. We also use $\mathcal{P}(<)$ to denote the set of $<$-relations of $\mathcal{P}$. Intuitively $<$ represents a preference of applying rules during the evaluation of the program. In particular, if $\mathcal{R}(r) < \mathcal{R}(r')$ holds in $\mathcal{P}$, rule r' would be preferred to apply over rule r during the evaluation of $\mathcal{P}$.

3.2 Evaluation of Prioritized Logic Program

The evaluation of a PLP will be based on its ground form. It is to find the answer set of the authorization domain. Given a PLP $\mathcal{P} = (\Pi, \mathcal{R}, <)$. We say $\mathcal{P}$ is *well formed* if there does not exist a rule r' that is an instance of two different rules r_1 and r_2 in Π and $\mathcal{R}(r_1) < \mathcal{R}(r_2) \in \mathcal{P}(<)$. In the rest of this paper, we will only consider well formed PLPs in our discussions, and consequently, the evaluation for an arbitrary program $\mathcal{P} = (\Pi, \mathcal{R}, <)$ will be based on its ground instantiation $\mathcal{P}' = (\Pi', \mathcal{R}', <')$. Therefore, in our context a ground prioritized (or extended) logic program may contain infinite number of rules. In this case, we will assume that this ground program is the ground instantiation of some program that only contains finite number of rules.

Definition 2. *Let Π be a ground extended logic program and r a rule with the form $R_0 \leftarrow R_1, \cdots, R_m, \text{not } R_{m+1}, \cdots, \text{not } R_n$ (r does not necessarily belong to Π). Rule r is* defeated *by Π iff Π has an answer set and for any answer set $Ans(\Pi)$ of Π, there exists some $R_i \in Ans(\Pi)$, where $m + 1 \leq i \leq n$.*

Let us consider program example 3 once again. If we choose $R_2 < R_4$ and R_2 is defeated by $\mathcal{D} - \{R_2\}$, rule R_2 should be ignored during the evaluation of $\mathcal{D}$. We will get the unique answer set $\{holds(S, R, O_1), O \in O_1, holds(S_1, R, O)\}$.

To calculate the set of access facts of an authorization domain, we need to evaluate its corresponding extended logic program. That is, to find the answer set of prioritized logic program $\mathcal{P}$. Now, we present the procedure for finding the answer set. We start from a reduced set or the reduct of $\mathcal{P}$.

Definition 3. *Let $\mathcal{P} = (\Pi, \mathcal{N}, <)$ be a prioritized extended logic program. $\mathcal{P}^<$ is a reduct of $\mathcal{P}$ with respect to $<$ if and only if there exists a sequence of sets Π_i ($i = 0, 1, \cdots$) such that:*

1. $\Pi_0 = \Pi$;
2. $\Pi_i = \Pi_{i-1} - \{r_1, r_2, \cdots \mid$ (a) there exists $r \in \Pi_{i-1}$ such that
   for every j $(j = 1, 2, \cdots)$, $\mathcal{N}(r) < \mathcal{N}(r_j) \in \mathcal{P}(<)$ and
   $r_1, \cdots$, are defeated by $\Pi_{i-1} - \{r_1, r_2, \cdots\}$, and (b) there
   does not exist a rule $r' \in \Pi_{i-1}$ such that $N(r_j) < N(r')$
   for some j $(j = 1, 2, \cdots)$ and r' is defeated by $\Pi_{i-1} - \{r'\}\}$;
3. $\mathcal{P}^< = \bigcap_{i=0}^{\infty} \Pi_i$.

In Definition 3, $\mathcal{P}^<$ is a ground extended logic program obtained from Π by eliminating some *less preferred rules* from Π. In particular, if $\mathcal{R}(r) < \mathcal{R}(r_1)$, $\mathcal{R}(r) < \mathcal{R}(r_2)$, $\cdots$, and $\Pi_{i-1} - \{r_1, r_2, \cdots\}$ defeats $\{r_1, r_2, \cdots\}$, then rules $r_1, r_2, \cdots$ will be eliminated from Π_{i-1} if no less preferred rule can be eliminated (i.e. conditions (a) and (b)). This procedure is continued until a fixed point is reached. It is worth to note that the generation of a reduct of a PLP is based on the ground form of its extended logic program part. Furthermore, if $\mathcal{R}(r_1) < \mathcal{R}(r_2)$ holds in a PLP where r_1 or r_2 includes variables, then $\mathcal{R}(r_1) < \mathcal{R}(r_2)$ is actually viewed as the set of $<$-relations $\mathcal{R}(r_1') < \mathcal{R}(r_2')$, where r_1' and r_2' are ground instances of r_1 and r_2 respectively.

Definition 4. *Let $\mathcal{P} = (\Pi, \nabla, <)$ be a PLP and Gl the set of all ground literals in the language of $\mathcal{P}$. For any subset S of Gl, S is an answer set of $\mathcal{P}$, denoted as $Ans^P(\mathcal{P})$, iff $S = Ans(\mathcal{P}^<)$ for some reduct $\mathcal{P}^<$ of $\mathcal{P}$. Given a PLP $\mathcal{P}$, a ground literal L is entailed from $\mathcal{P}$, denoted as $\mathcal{P} \models L$, if L belongs to every answer set of $\mathcal{P}$.*

Using Definitions 3 and 4, it is easy to conclude that in example 3, if we assign $R_2 < R_4$, $\mathcal{P}$ has a unique reduct as follows:

$$\mathcal{P}^< = \{holds(S_1 R, O_1) \leftarrow, O \in O_1 \leftarrow,$$
$$holds(S_1, R, O) \leftarrow holds(S_1, R, O_1), O \in O_1, not\neg holds(S_1, R, O)\}$$

from which we obtain the following answer set of $\mathcal{P}$:

$$Ans^P(\mathcal{P}_1) = \{holds(S_1, R, O_1), O \in O_1, holds(S_1, R, O)\}$$

Example 4. Now we consider another authorization doamin D, it's corresponding program $\mathcal{P}$ is:

$$R_1 : holds(S, W, O_3) \leftarrow,$$
$$R_2 : holds(S, W, O) \leftarrow not\ holds(S, W, O_1),$$
$$R_3 : holds(S, W, O_2) \leftarrow,$$
$$R_4 : holds(S, W, O_1) \leftarrow not\ holds(S, W, O),$$
$$R_1 > R_2, R_3 > R_4.$$

According to Definition 3, it is easy to see that $\mathcal{P}$ has two reducts:

$$\{holds(S, W, O_3) \leftarrow, \quad holds(S, W, O_2) \leftarrow,$$
$$holds(S, W, O_1) \leftarrow not\ holds(S, W, O)\},$$
and
$$\{holds(S, W, O_3) \leftarrow, \quad holds(S, W, O) \leftarrow not\ holds(S, W, O_1),$$
$$holds(S, W, O_2) \leftarrow\}.$$

From Definition 4, it follows that $\mathcal{P}$ has two answer sets:
$\{holds(S, W, O_3), holds(S, W, O_1), holds(S, W, O_2)\}$ and
$\{holds(S, W, O_3), holds(S, W, O), holds(S, W, O_2)\}$.

4   Conclusion

In this paper, we proposed a new approach to solve conflicts in authorizations. Either by update or in the initial specification, conflicts in authorization is an issue needs to be solved. In our work, we employed a prioritized logic program to resolve authorization conflicts in an authorization domain specified by a logic language. By assigning each rule a name representing its preference ordering, using a fixed point semantics to delete those less preferred rules (the rules will not take effect under current state), then using answer set theory to evaluate the authorization domain to get the preferred authorizations.

From example 4, some authorization domain returns more than one answer sets. In reality, we expect only one answer set. This is our future work to investigate the uniqueness of the answer set. We also will consider its implementation issue with authorization evaluation and dynamic policy update. A related work using logic program for conflict resolution in reasoning has been implemented in [8]. It is our future work to use logic program (stable model semantics) to implement the approach for authorization conflict resolution presented in this paper.

References

1. E. Bertino, F. Buccafurri, E. Ferrari and P. Rullo, "A Logic-based Approach for Enforcing Access Control". *Computer Security*, vol.8, No.2-2, pp109–140, 2000.
2. E. Bertino, B. Catania, E. Ferrari and P. Perlasca, "A Logical Framework for Reasoning about Access Control Models". *ACM Transactions on Information and System Security*, Vol.6, No.1, pp71–127, 2003.
3. J. Chomicki, J. Lobo and S. Naqvi, "A Logical Programming Approach to Conflict Resolution in Policy Management". *Proceedings of International Conference on Principles of Knowledge Representation and Reasoning*, pp121–132, 2000.
4. S. Jajodia, P. Samarati, M.L. Sapino and V.S. Subrahmanian, "Flexible Support for Multiple Access Control Policies". *ACM Transactions on Database Systems*, Vol.29, No.2, pp214–260, 2001.
5. N. Li, B. Grosof and J. Feigenbaum, "Delegation Logic: A Logic-based Approach to Distributed Authorization". *ACM Transactions on Information and System Security*, Vol.6, No.1, pp128–171, 2003.
6. T.Y.C. Woo and S.S. Lam, "Authorization in Distributed systems: A Formal Approach". *Proceedings of IEEE Symposium on Research in Security and Privacy*, pp33-50, 1992.
7. Y. Zhang and Y. Bai, "The Characterization on the Uniqueness of Answer Set for Prioritized Logic Programs". *Proceedings of the International Symposium on methodologies on Intelligent Systems*, pp349–356, 2003.
8. Y. Zhang, C.M. Wu and Y. Bai Implementing Prioritized Logic Programming, *AI Communications*, Vol.14, No. 4, pp183–196, 2001.

Knowledge-Based System for Color Maps Recognition

Serguei Levachkine, Efrén Gonzalez, Miguel Torres,
Marco Moreno, and Rolando Quintero

Geoprocessing Laboratory, Centre for Computing Research (CIC) –
National Polytechnic Institute (IPN)
{sergei,mtorres,marcomoreno,quintero}@cic.ipn.mx
efren@esfm.ipn.mx
http://geo.cic.ipn.mx, http://geopro.cic.ipn.mx

Abstract. In this paper, we describe the Fine-to-Coarse Scale Method in which the knowledge of cartographic patterns into small-scale map aids to recognize the corresponding patterns into large-scale map of the same territory. This approach exploits the user's experience providing the knowledge domain in the form of the prescribed feature-attribute set. The cartographic patterns are presented in raster maps. A map is composed by thematic layers that contain cartographic patterns. The knowledge to recognize cartographic objects in raster fine scale maps is based on the information about the objects of the coarse scale map. These recognized objects are removed from the map, reaching a simplified representation of the map. Then, the rest of objects are recognized as new cartographic material in this simplified map. The goal of these representations is to obtain a GIS database.

1 Introduction

Color cartographic pattern recognition by a computer system remains a hard task [1]. This is closely related to the spatial data vectorization for GIS-ready information [2]. In this context, the *coarse-to-fine scale* method, developed herein, represents a promising alternative to carry it out. Indeed, a typical situation in GIS development is that spatial data in a coarse scale are available, and GIS-developer only needs to transform (recognize, vector) them into finer scale. For example, Mexico has full-territory coverage by recognized (vector) topographic maps in scale 1:50,000, but not in 1:25,000; this is a commonality for many countries [1]. The method incorporates external (interpretative) knowledge to the segmentation and recognition processes; an early approach with similar core was described in [3]. Moreover, the coarse to fine scale method is, in essence, a simultaneous segmentation-interpretation-recognition system [4-6].

We approach the problem of cartographic pattern recognition in fine scale[1] maps, using information that comes from coarse scale[2] maps. The maps are raster-scanned color maps of different thematic, representing the same territory in coarse and fine scale respectively. Coarse to fine scale method is defined in terms of *means*: coarse scale maps and their *information*; *concepts*: image associated function, *cartographic*

---

[1] Inferior to 1:200,000
[2] Superior to 1:200,000

R. Khosla et al. (Eds.): KES 2005, LNAI 3681, pp. 297–303, 2005.

knowledge domain and cartographic pattern; and *tools*: a set of clustering criteria of the *Logic Combinatorial Pattern Recognition* [7]. In the present paper, we describe step-by-step the main ideas of the coarse-to-fine scale method of color cartographic pattern recognition. This method has been originated from unsolved problem of the vector description of raster objects [1][2].

Formal Statement of the Problem and System Description. Let us suppose a vector[3] image I_1 (or already recognized raster image) in scale s_1 of territory T and, a raster image I_2 of T to be converted to be vectored (recognized) in scale s_2, and $s_1 > s_2$ (e.g. $s_1 = 1:100,000$ and $s_2 = 1:50,000$). Our goal is to use the information from I_1 in vectorization of I_2. Note that I_1 can be considered as a "generalization"[4] of I_2: $I_1 = G (I_2)$, i.e. if an object $O_2 \in I_2$, then there can exist $O_1 \in I_1$, such that $O_1 = G (O_2)$. We denote Ω - the set of all such objects O_2 from I_2 and Θ - the compliment of Ω in I_2: $I_2 = \Omega \cup \Theta$. We also put $\omega = G (\Omega)$ and note that $\omega \subseteq I_1$. Obviously, to vector objects from Ω and Θ we need two different strategies. The objects from Ω can be vectored, using the features (position, color or colors, shape, etc.) of the vector objects from ω. After, the objects of Ω have being vectored we can vector the objects from Θ by one of the recognition modules [2][5] as a "new" cartographic material. Figure 1 shows a chart-flow with general system description.

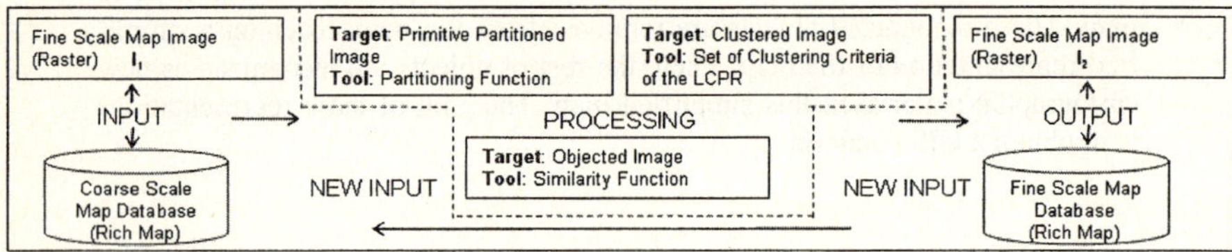

Fig. 1. System Description

In this work, we present a method to recognize cartographic patterns on geo-images. In Section 2 we present the theoretical background. The Fine-to-Coarse Scale Method is described in Section 3. Section 4 outlines our conclusions.

2 Theoretical Background

Definition 1. A digital image is a matrix $\|M (i, j)\|_{n \times m}$ such that $0 \leq i \leq n, 0 \leq j \leq m$; and $M(i, j) = (r, g, b)$, where r, g, b are elements of the set $\{0,1,..., 255\}$.

Definition 2. Let *I* be a digital image according to the definition 1, then *Image Associated Function* f_I: $Z \times Z' \to M$ to I is defined as follows: (a) image associated function domain is $Z \times Z'$ where $Z=\{0,1,..., n\}$ and $Z'=\{0,1,..., m\}$; (b) image associated function co-domain is a set M *without repeated elements*, which is composed of the elements of $M(i, j) \in I$; (c) image associated function is assigned to each pair (i, j) in $Z \times Z'$ the corresponding value given by the matrix $\|M (i, j)\|$. In other words, $f_I (i, j) = M(i, j), \forall (i, j) \in Z \times Z'$.

---

[3] See [2] for definition of raster and vector images.
[4] We do not discuss here what this generalization is.

Definition 3. *Cartographic Knowledge Domain (CKD)* is a finite space of attributive, topological, logical and spatial data, which are associated with raster cartographic objects presented in maps and represented by a set of concepts [8]. For example, attributive data like type of border, name of state, surface, population, administrative unit, among others; geometric data like contours, coordinates, etc. [2].

Definition 4. Let I be a digital image according to Definition 1 and let f_I: $Z{\times}Z' \to M$ be image associated function. *Cartographic Pattern P* in *I* is a function such that (1) $P = f_I|_{Z_p}$: $Z_p \to M_p$, where $Z_p \subseteq Z{\times}Z'$ and $M_p \subseteq M$. In other words, P is the restriction of image associated function f_I to some set Z_p; (2) a concept from *CKD* can be assigned to the set Z_p or M_p. Whereas *P* only fulfills condition 1, it is a candidate to be a Cartographic Pattern.

The definitions above are served as a compliment to clustering criteria coming from Logic Combinatorial Pattern Recognition: *β_0-Connected*, *β_0-Compacted*, *β_0-Complete* Maximal [7] to segment and recognize the image.

2.1 Clustering Criteria

Next, we analyze mentioned above clustering criteria and the similarity measures in application to color cartographic image processing. To our knowledge, this is one of the first works in this direction; see [2][9] for belief.

We use the following notations. Let I be a digital cartographic image; f_I be the Image Associated Function, f_I: $Z{\times}Z' \to M$; Γ: $U{\times}U \to [0,1]$ be a *Similarity Function*, where U is a finite space of objects to classify; β_0 be nonnegative real number.

A subset G_i in U is called *β_0-Connected*, if $\forall O_r$, $O_s \in G_i$, then there exist $O_{i_1},...,O_{i_q} \in G_i$ such that $O_r = O_{i_1}$, $O_s = O_{i_q}$ and $\forall p \in \{1,..., q-1\}$ $\Gamma(O_{i_p}, O_{i_{p+1}}) \geq \beta_0$. Moreover, if $O_k \in U$ and $\exists O_j \in G_i$ such that $\Gamma(O_k, O_j) \geq \beta_0$ then $O_k \in G_i$.

In cartographic digital images, this criterion can be used to find objects in co-domain M, defined by the elements that allow a gradual transition in terms of the similarity function. While for cartographic patterns, this can be used when the patterns are defined by color ramps, for instance, from "clear" blue to "dark" blue. A problem comes when there are gradual transitions between colors in the image and each transition is used to represent a different cartographic pattern. Notice that this problem depends on the image context and can be solved by using other two criteria that are defined in the following.

A subset G_i in U is called *β_0-Compacted*, if $\forall O_j \in G_i \exists O_i \in G_i$ such that $O_i \neq O_j$ and $\Gamma(O_i, O_j)$ takes the maximum value. In other words, the objects in a β_0-Compacted set are most similar objects with respect to the similarity function. To make it context–independent, we can successively apply the β_0-Compacted criterion and/or the following criterion.

A subset G_i in U is called *β_0-Complete Maximal* iff $\forall O_i$, $O_j \in G_i$, the similarity function Γ computed for both elements is greater or equal than β_0, and if $\Gamma(O_k, O_p) \geq \beta_0$, where $O_k \in G_i$, then $O_p \in G_i$. According to our experiments, this criterion has a disadvantage; if it is applied to the digital image before to the other two criteria it can generate unnecessary objects in the co-domain.

Summing-up the clustering criteria analysis, we note the following: (1) Elements of the image associated function co-domain can be grouped/structured in different objects depending on the clustering criteria and (2) objects/structures obtained by the application of a criterion Π (Π and Π' are one of the three criteria) can result in the union of objects/structures formed by another criterion Π'. In other words, successive application of clustering criteria leads to gradually finer structuring of the image associated function co-domain. Thus, a hierarchy under *inclusion* relation between the objects is settled down, in which "general" ("coarse") objects are at superior level (β_0-Connected criterion) and "specific" ("fine") objects are at inferior level (β_0-Compacted and/or β_0- Complete Maximal criteria) [7].

2.2 Similarity Criterion and Similarity Function

Let **I** be a cartographic digital image. Let f_I: $Z \times Z' \rightarrow M$ be image associated function. The *Similarity Criterion C*: $M \times M \rightarrow [0,1]$ is defined as follows: let p, q $\in$ M; we denote by I(k), S(k) and H(k) the intensity, saturation and hue respectively of some point k $\in$ M, then C(p, q):

$$C(p,q) = \begin{cases} C_1(p,q) = \begin{cases} C_3(q) \ if \ I(p) \leq N \\ C_4(p,q) \ otherwise \end{cases} & if \ |I(p) - I(q)| \leq C \\ 0 & otherwise \end{cases} \tag{1}$$

where

$$C_3(q) = \begin{cases} 1 \ if \ I(q) \leq N \\ 0 \ otherwise \end{cases} \tag{2}$$

$$C_4(p,q) = \begin{cases} C_5(q) \ if \ I(p) \geq B \\ C_6(p,q) \ otherwise \end{cases} \tag{3}$$

$$C_5(q) = \begin{cases} 1 \ if \ I(q) \geq B \\ 0 \ otherwise \end{cases} \tag{4}$$

$$C_6(p,q) = \begin{cases} C_7(q) \ if \ S(p) \leq R \ y \ I(q) \neq 0 \\ 1 - |I(p) - I(q)| \ If \ S(p) \leq 0.03 \ and \ I(q) = 0 \quad or \quad S(q) \leq 0.03 \\ 1 - \sqrt{(S(p)\cos(H(p)) - S(q)\cos(H(q)))^2 + (S(p)\sin(H(p)) - S(q)\sin(H(q)))^2} \ otherwise \end{cases} \tag{5}$$

Using this *Similarity Criterion*, it is possible to define the *Similarity Function* only in terms of this criterion. Unique feature used in definition of *C* is color. However, we can define another criterion C_a and consider other features besides of color, constructing C_a and corresponding similarity function. According to the similarity criterion above, the similarity function coincides with the criterion C(p, q). C(p, q) for elements p, q from the image associated function co-domain are computed as follows; refer the formulas above.

a) If the absolute value of difference between their intensities is more than some threshold *C* the value of similarity function is equal to zero.

b) Otherwise, if this value is less than *C*, then if both intensities are less or equal than another threshold *N* (average of all colors close to "black") or the difference is more or equal than another threshold B (average of all colors close to "white"), then the value of similarity function is equal to one (This can be interpreted as averaging of clear and dark tonalities).

c) If $\neg a \wedge \neg b$, then the saturation of *p* and *q* are considered as follows: (1) if both saturation are less or equal than some threshold R, then the value of similarity function is equal to one; (2) if one of saturation is less or equal than R, then the value of similarity function is equal to one minus the absolute value of the differ-ence of intensities; (3) if $\neg c1 \wedge \neg c2$, then the saturation as well as hue are determinate, forming two vectors for p and q in polar coordinates: saturation (radii), hue (angle).

3 Application of the Coarse to Fine Scale Method to Cartographic Pattern Recognition

Let **I** be a digital cartographic image $\|M\,(i,\,j)\|_{n \times m}$. We should compute the Image Associated Function as follows. Let $Z = \{0,...,\,n\}$, $Z' = \{0,...,\,m\}$, consider the Cartesian product $Z \times Z'$ and the difference of the sets A and B: $A\text{-}B = \{x \in A \mid x \notin B\}$. Next the following steps are employed:

1. Let $p_0 = 0$, $q_0 = 0$, then the sets are defined:
   $G_0 = \{\,M(\,x,\,y) \mid M(x,\,y) = M(p_0,\,q_0)\}$ and $G'_0 = \{(x,\,y) \mid M(\,x,\,y) \in G_0\}$.
2. Let $(p_1,\,q_1) \in Z \times Z' - G'_0$, then the sets are defined:
   $G_1 = \{\,M(\,x,\,y) \mid M(x,\,y) = M(p_1,\,q_1)\}$ and $G'_1 = \{(x,\,y) \mid M(\,x,\,y) \in G_1\}$.
3. Let $(p_2,\,q_2) \in Z \times Z' - G'_0 \cup G'_1$, then the sets are defined:
   $G_2 = \{\,M(\,x,\,y) \mid M(x,\,y) = M(p_2,\,q_2)\}$ and $G'_2 = \{(x,\,y) \mid M(\,x,\,y) \in G_2\}$.
4. The process continues until the first natural number k such that $Z \times Z' - G'_0 \cup G'_1 \cup G'_2 \cup ... \cup G'_{k+1} = \varnothing$ is found. It can be demonstrated that sets G'_i, $i \in \{0,...,k\}$ generate a partition of set $Z \times Z'$.

To relax the equality condition that is required in the definition of sets G_i, $i \in \{1,...,\,k\}$, we set up the generalized sets G_i^{ε} as follows:

$$G_i^{\varepsilon} = \{\,M(\,x,\,y) \mid \|M(x,\,y) - M(p_i,q_i)\| \le \varepsilon\}\,,^{5} \qquad (7)$$

where $\|\bullet\|$ could be, generally speaking, any metric. In this work we use the "Manhattan" metric, i.e. $\|(x,\,y,\,z,) - (p,\,q,\,r)\| = |x - p| + |y - q| + |z - r|$, where $|\bullet|$ denotes the absolute value.

Suppose that a cartographic pattern P_{TE} already recognized by a computer system in a cartographic map in a coarse scale exists. The recognition of P_{TE} implies that corresponding attributive, geometric, topologic data are known and, therefore, form the *Cartographic Knowledge Domain (CKD)*. Information of special interest is its location with respect to some coordinate system, color or colors, shape, etc. We use this information to recognize or assign a concept from *CKD*, e.g. "Lerma River" to P_{FE} presented in the fine scale map. Note that the characteristics of P_{TE} and P_{FE} are conceptually similar, e.g. they have the same name, although they can be distinct as functions according to definition 4. Information about the location of P_{TE} is used to find a candidate to be "fine" cartographic pattern P_{FE}. Notice that P_{FE} is generated by the application of the clustering criteria. This way, once P_{FE} has been located the system is assigned to it the concept defined by "coarse" cartographic pattern P_{TE}, finally reaching P_{FE} recognition under that concept.

---

[5] Notice that when $\varepsilon = 0$, the sets G_i^{ε} are converted in the sets G_i; that is why we call them "generalized".

302   Serguei Levachkine et al.

3.1 Examples: Recognition of Punctual, Linear and Area Patterns

Recognition of Punctual Patterns. Suppose that a cartographic map in coarse scale provides information to construct the *cartographic knowledge domain* as: {name: "Palm"; location: (84,101)}; see Fig. 2.

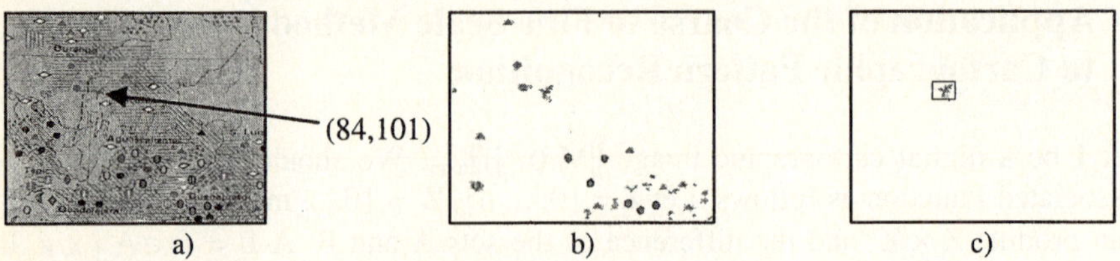

Fig. 2. Recognition of a cartographic pattern classified as punctual. (a) Original image. (b) Image obtained after applying the criterion β_0-Connected (c)Recognition of the pattern "Palm"

Recognition of Linear Patterns. Suppose that a cartographic map in coarse scale provides information to construct the *cartographic knowledge domain* as: {name: "River_ 2"; location: (76,113)}; see Fig. 3.

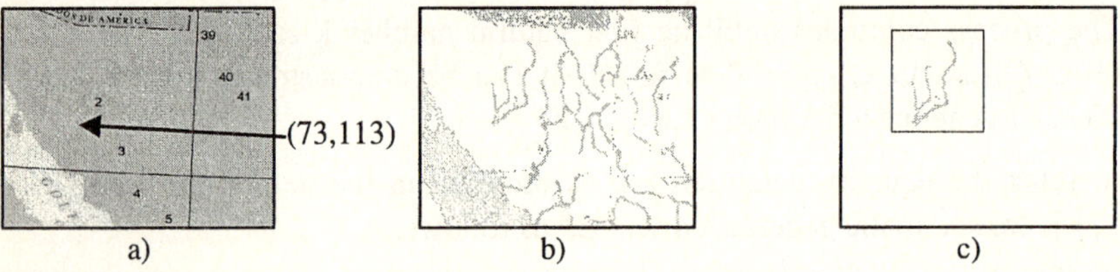

Fig. 3. Recognition of a cartographic pattern classified as linear. (a) Original image. (b) Image obtained by applying the criterion β_0-Complete Maximal, $\beta_0=0.9$. (c) Recognition of the pattern (function) "River_2"

Recognition of Area Patterns. Suppose that a cartographic map in coarse scale provides information to construct the CKD as {name: "Lake G"; location: (315,239)}. Note that the location, extraction and recognition are immediate because the classification has been employed over all image pixels by means of the Image Associated Function co-domain; see Fig. 4.

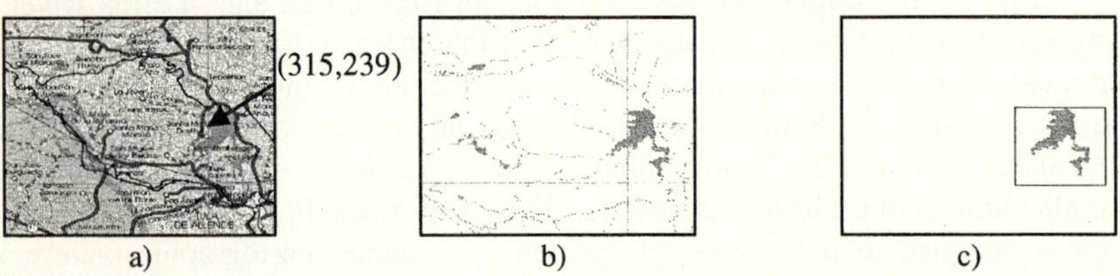

Fig. 4. Recognition of a cartographic pattern classified as area. (a) Original image. (b) Image obtained by applying the criterion β_0-Complete Maximal, $\beta_0=0.9$. (c) Recognition of the pattern (function) "Lake G"

4 Conclusion

The method herein presented possess some "universality"; it was designed to recognize cartographic patterns into fine scale maps, however, further development shown that it is possible to recognize patterns independently to the map scale by modifying the *cartographic knowledge domain.*

In this approach, the user must define the clustering criteria parameters. However, it is not a big obstacle. To our knowledge, this is one of the first attempts to design a segmentation-recognition computer system for complex color images of arbitrary type. We believe that this is a kind of advanced simulation of the human's visual perception. However, it is necessary to emphasize that for optimization of labor-intensive program training a strong formalization of composite image technique is now required

This approach is based on pattern clustering of one feature (color). In the future other features such as *shape* and *size,* can improve the performance of this method.

References

1. Levachkine, S. P., Polchkov, E.A.: Integrated Technique for Automated Digitization of Raster Maps, *Revista Digital Universitaria*, Vol. 1, No. 1. (2000). Available on-line at http://www.revista.unam.mx/vol.1/art4/index.html
2. Levachkine, S.: Raster to Vector Conversion of Color Cartographic Maps. Graphics Recognition: Recent Advances and Perspectives, *5th International Workshop, GREC 2003*, Barcelona, Spain, July 30-31, 2003, Revised Selected Papers, Editors: Josep Lladós, Young-Bin Kwon. Lecture Notes in Computer Science. Vol. 3088, Springer-Verlag (2003) 50-62
3. Meyers G.K., Chen, C.-H.: Verification–based Approach for Automated Text and Feature Extraction from Raster-scanned Maps. Lecture Notes in Computer Science, Vol. 1072. Springer-Verlag (1996) 190-203
4. E. Gonzalez-Gómez, S. Levachkine. Color cartographic pattern recognition using the coarse to fine scale method. Lecture Notes in Computer Science, Vol. 3287, Springer-Verlag (2004) 540-547
5. Levachkine, S., Velázquez, A., Alexandrov, V., Kharinov, M.: Semantic Analysis and Recognition of Raster-scanned Color Cartographic Images. Graphics Recognition. Algortihms and Applications: *4th International Workshop, GREC 2001*, Kingston, Ontario, Canada, September 7-8, 2001, Selected Papers, Editors: D. Blostein, Y.-B. Kwon. Lecture Notes in Computer Science, Vol. 2390. Springer-Verlag (2002) 178-189
6. Levachkine, S., Torres, M., Moreno, M., Quintero, R.: Knowledge-Based Method to Recognize Objects in Geo-Images. Knowledge-Based Intelligent Information and Engineering Systems: *8th International Conference, KES 2004*, Wellington, New Zealand, September 20-25, 2004, Proceedings, Part III, Editors: Mircea Gh. Negoita, Robert J. Howlett, Lakhmi C. Jain. Lecture Notes in Artificial Intelligence, Vol. 315. Springer-Verlag (2002) 718-725
7. Martínez-Trinidad, J.F., Guzmán-Arenas, A.: The Logical Combinatorial Approach to Pattern Recognition, an Overview through Selected Works. *Pattern Recognition*, Vol. 34, No. 1. (2001) 741-7519.
8. Torres-Ruiz, M., Levachkine, S.: Semantics Definition to Represent Spatial Data. In: Levachkine, S., Ruas, A., Bodansky, E. (eds.), *Proc. International Workshop on Semantic Processing of Spatial Data (GEOPRO 2002)*, 3-4 December 2002, Mexico City, Mexico (2002)
9. Lazo-Cortés, M., Ruiz-Shulcloper, J., Alba-Cabrera, E.: An Overview of the Evolution of the Concept of Testor. *Pattern Recognition*, Vol. 34, No. 4. (2001) 753-762

Constructing an Ontology
Based on Terminology Processing

Soo-Yeon Lim, Seong-Bae Park, and Sang-Jo Lee

Department of Computer Engineering, Kyungpook National University,
Daegu, 702-701, Korea
nadalsy@hotmail.com

Abstract. An ontology consists of a set and definition of concepts that presents the characteristics of a given domain and relationship between the elements. This paper proposes a semiautomatic method to construct a domain ontology using the results of text analysis and applies it to a document retrieval system. An experiment domain used to construct an ontology was selected by the pharmacy field in which the types of ontology appeared in a document were analyzed. By using these results, a processing method of the terminologies, which are combined with some specific nouns of suffices, and uses the semantic relation to construct an ontology. In order to present usefulness for retrieving a document using the hierarchical relations in an ontology, this study compares a typical keyword based retrieval method with an ontology based retrieval method, which uses related information in an ontology for a related feedback. As a result, the latter shows the improvement of precision and recall.

1   Introduction

This paper proposes the method of constructing domain ontology with the semantic relation that exists in an ontology by extracting a semantic group and hierarchical structure after classifying and analyzing the patterns of terminology that appeared in a Korean document as a type of compound nouns [8]. A built ontology can be used in various fields. This paper proposes an ontology that classifies words related to a specific subject by using a hierarchical structure for a method to improve the performance of retrieving a document. A retrieval engine can be used as a base of inference to use the retrieval function using the concept and rule defined in the ontology. In order to experiment this retrieval engine, the texts that exists in a set of documents related to the pharmacy field are used as an object of the experiment.

The existing methods for extracting terminology can be largely classified by a rule based method and statistics based method. A rule-based method [2,3,5,6] constructs a configuration pattern of terminology by hand or learning corpus and recognizes terminology by constructing a recognition pattern automatically by using it. Here, it presents a relatively exact result because people directly describe the rules using a noun or suffice dictionary. A statistics based method [4,9] uses some kind of knowledge, such as the hidden Markov model, maximum entropy model, word type, and vocabulary information in order to learn the knowledge of the recognition from a learning corpus. This paper extracts terminology by using a rule based method and configures a rule to extract terminology by analyzing the appearance patterns of the terminology.

R. Khosla et al. (Eds.): KES 2005, LNAI 3681, pp. 304–310, 2005.
© Springer-Verlag Berlin Heidelberg 2005

2   Constructing an Ontology

In order to perform this process, it is a definition of the concepts and structure through a conference with the specialists of domain is required and constructs by using these results. In an actual application system, it is necessary to an ontology that includes a specific knowledge for each domain. This paper selected the pharmacy field as an experiment domain and limited a document that existed in the pharmacy domain for learning. We call this constructed ontology Pharm-Ontology..

2.1   Process of Constructing Pharm-Ontology

The constructing process of the proposed ontology consists of four steps as follows. First, the web documents that exist in the related web will be collected to create a corpus and structurized through a document transformation process. Second, the verbs that will express the relation between the nouns that become concept and the extracted concepts will be extracted after passing a simple natural language process. Third, the terminology will be extracted from the extracted concepts and produce a hierarchical structure from the results of the analysis of the structure. Finally, the extracted relations will be added to the existing ontology with the concepts. The relation that will append the concepts of the Pharm-Ontology, which will be built using the results of the documents that exist in the pharmacy database (http://www.druginfo.co.kr) and the results will be configured. The collected documents will form concepts according to the tag that is attached after the transformation process to fit the configured structure using a semi-structurized (tagging) document. The concepts that existed in the built ontology and its relation is represented as OWL.

2.1.1   Constructing of Base Ontology

For constructing an ontology, a designer decides a minor node, which is located in the upper level, and constructs an ontology based on it and extends it. It is called by a base ontology. This paper configures a base ontology using 48 words. In order to configure a base ontology, the concepts of the name of a disease, symptoms, drugs are defined as the highest node and its 45 hyponym nodes. The hyponym nodes consist of each node group; 20 nodes, which are defined by following the classification of a specific noun or suffices that form the name of a disease or symptom that exist in the pharmacy domain, and 15 nodes for the configured structures, and 10 nodes, which express a general noun that has a high frequency of appearance.

2.1.2   Extracting Concepts and Adding Relations

The stop words included in a document will be removed after the processes of morpheme analysis and tagging. In order to perform this process, the stop words list was made by using 181 words by considering the morphological characteristics of Korean language in which a part of suffices was excluded in the process of stemming because these suffices will be nicely used to extract terminology. The extracted nouns from an ontology that is a kind of network with a lot of words mean the concept of ontology. The tags and verbs present the relation between the concepts and act as a network, which plays a role in the connection of the concepts to each other. There are some proper nouns and compound nouns in the documents what we have an interest in, such as the name of a disease, symptoms, ingredients, and other things. The proper nouns that present a major concept are processed the same way as a general noun. In addition, the terminology that appeared as a compound noun in the domain is extracted and hierarchicalized, and then is added into the ontology.

We use two methods in order to give a relation for the extracted concepts. The one is using the value of tag attached at the front of text, and the other is using the verb by extracting it from the text. The left part of Table 1 shows 15 types of semantic relations according to the value of the attached tag.

Table 1. Types of the semantic relations

Semantic Relations based on Tag Value		Semantic Relations based on Extracting Verb	
productedBy	hasContra	appear	use
hasInsCode	hasSideEffect	beWorse	cause
hasComCode	byMean	inject	accompany
hasClsCode	byAmount	noTake	prevent
hasColor	byUnit	reduce	control
hasKind	byAge	return	infect
hasForm		rise	cure
hasEffect		take	improve
hasMethod		relax	maintain

Moreover, the verbs, which connect the nouns, those are appeared in the surrounding area are also extracted to define the relation between the concepts. As a result, the frequency of 35 verbs that have a frequency of over 200 was 11,250. It covers 47.97% of the frequency of the entire verbs. This paper classifies these results as a semantic pattern and defines it 18 semantic relations as presented in the right part of Table 1. The relationship between the extracted verbs and nouns can be verified by the co-occurrence information. If there is a relationship that exists between the nouns and verbs, otherwise it will compare a relationship between other nouns and verbs.

2.1.3 Extracting Terminology

Terminology is a set of words that has a specific meaning in a given domain and means a lexical unit that characterizes a subject by expressing the concept used in a domain. This paper analyzes an appearance type of terminology in order to extract the information automatically. The shape combining of terminology shows very varied ways. Almost all of the terminology appeared in an appropriate domain presented as a type of compound noun and can be classified as two types as follows. The one is a singleton term, that is, it has a simple shape of combining with one word that has no spacing words. The other is a multi-word term that has spacing words and is a kind of compound noun with two more words in which it has a semantic relation with the front element of a word. The nouns and suffices that are configured by a singleton terminology are classified by 20 kinds. A multi-word terminology has almost a relation of modifier and keyword like "acute bronchitis" in which there are many cases that a keyword consists of terminology, which has a singleton term. This paper configures 5 semantic patterns and defines the semantic relation of an ontology according to these patterns [8].

2.2 Extending an Ontology

A constructed ontology can be extended by other resources, such as other ontologies, thesaurus, and dictionaries. It is possible to reduce the time and cost to extend an ontology by applying the predefined concepts and rules by using the existing resources. The process consists of 3 steps, such as the import, extract, and append. This paper uses two external resources. (http://www.nurscape.net/nurscape/dic/frames.html, http://www.encyber.com/)

3 Application to Document Retrieval

This paper proposes a method that will improve the effect of retrieving a document using this constructed ontology. A process of ontology selects a major set of documents in a specific field and extracts the concepts by analyzing these documents, and then appends these concepts by using a link. The objective of concept extraction is extracting the nouns that will represent the documents as well as possible.

A relevance feedback is well known as an effective way to process a reformation of a query that is an important part to access information. In the case of applying the relevance feedback to improve the traditional method of $tf \cdot idf$, it is well recognized that it can improve a precision rate if it is applied to a set of sample document of small scale.

This paper uses a hierarchical relation for the process of user relevance feedback in the ontology. The query will be extended by using the terminologies, which appeared as a hyponym information in the ontology that related to the inputted queries, and calculates the weights for the rewritten query. In this process, the hyponym retrieval level to retrieve the nodes in the ontology was set by 2. Figure 1 presents the configuration of a document retrieval system mentioned above in which it largely consists of a preprocessing module and retrieval module.

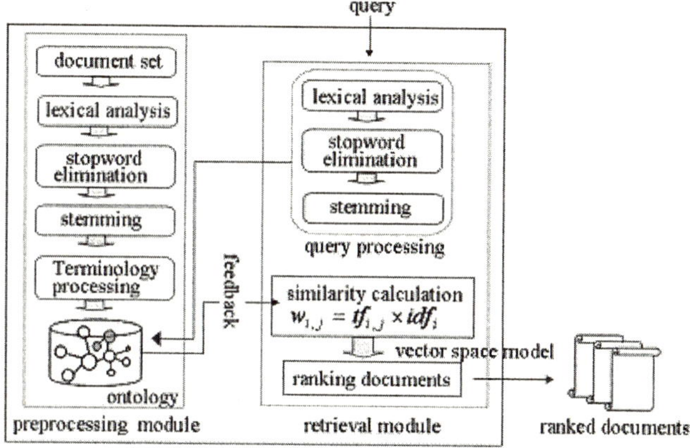

Fig. 1. Configuration of the document retrieval system

In this system, the ontology will be used as a set of index terms. In order to compare the retrieval performance, a set of the upper 30 correct answered documents for the 10 queries by introducing the 5 specialists' advices about 430 documents was configured. Then, the rates of recall and precision for each question were produced based on this configuration.

4   Results of Experiments

From the results of the analysis of the morpheme of text in the formed corpus, the terminologies that were applied by specific expressions or patterns will be extracted from the extracted nouns.

4.1   Recognizing Terminology

The experiment was applied by the extraction method proposed in this paper for the text in the pharmacy domain. The experiment documents used in this experiment are 21,113. From the results of the expression analysis, the total numbers of the extracted nouns are 78,902. The numbers of terminology for the entire extracted nouns of 78,902 are 55,870. It covers about 70.8% for the total nouns. This means that the rate of terminology is very high in a specific domain. There is good recognition for the terminologies that appeared and addition of 2,864 hyponym concepts after applying the extraction algorithm proposed in this paper in which the average level of the nodes, which existed in the ontology, is 1.8. The extracted terminologies can be investigated by three specialists by hand and evaluated by the precision. The precision of the extraction shows the ratio of the terminologies that are related by the correct relation for the extracted terminologies.

As the results of experiments, the precision of the singleton term is 92.57% that shows a relatively good performance compared with the proposed algorithm. And the average precision of the multi-word term terminologies was 79.96%. In addition, 574 concepts were added.

4.2   Performance of Retrieval

To show the effectiveness of the constructed ontology, a keyword based document retrieval system that gives weights by using the traditional method of $tf \cdot idf$ will be compared and analyzed with an ontology based document retrieval system that uses a hyponym information that exists in the ontology to a relevant feedback and recalculates the weights. An experiment reference collection and evaluation scale is used to evaluate an information retrieval system. An experiment reference collection consists of a set of literatures, information query examples, and set of relevant literatures for each information query. This paper collects 430 health/disease information documents from the home page of Korean Medical Association (http://www.kma.org) in order to configure a reference collection and configures an information query using 10 queries as follows. The experiment was carried for the 430 documents extraction. The objective of the experiment was to produce the recall and precision for the 10 queries in which a set of correct answers for each inputted question was defined in order of the documents set by the specialists.

Query 1:   {otitis media, symptom, kinds}   Query 2:   {otitis media, cure}
Query 3:   {otitis media, cure, drug}   Query 4:   {otitis media, property}
Query 5:   {otitis media}   Query 6:   {chronic otitis media diagnosis, cure}
Query 7:   {acute otitis media}   Query 8:   {otitis media, infection route}
Query 9:   {fever, disease   Query 10:   {ear, disease}

An index for the texts was produced to increase the retrieval speed. It consists of two elements, such as the vocabulary and frequency. The vocabulary is a set of all words that exist in the text in which it has an appeared document vector for each word. An appeared document vector stores the location of the appeared document including the weights considered by the frequency. We compared of the distributions of the precision and recall for the inputted queries using the two methods mentioned above. The average recall and precision for the two retrieval methods were respectively produced. The results are shown in Table 2. From the results, the ontology based document retrieval system that uses a hyponym information existed in the ontology to extend a question and grants the weights presented a high recall and precision by 0.78% and 4.79% respectively compared with the traditional method of $tf \cdot idf$. This means that a hierarchical relation in the ontology that is used as a relevant feedback in a retrieval system will not largely affect to the recall but will affect the increase of the precision.

Table 2. Average recall and precision

	Keyword based Retrieval	Ontology based Retrieval
Recall	46.34%	47.12%
Precision	47.26%	52.25%

5 Conclusions

This paper proposes a semiautomatic method to construct a domain ontology using the results of text analysis and applies it to a document retrieval system. The experiment domain defined by the field of pharmacy. In addition, a corpus was formed by collecting a relevant document for drugs from the web. The structure of ontology can be configured by analyzing the structure of the text in the corpus and sets the types of relation to extract the concepts and relations. This paper especially proposes a method to process the ontologies that were combined with a specific nouns or suffices in order to extract the concepts and relations for constructing the pharmacy ontology after analyzing the types of ontologies appeared in the relevant documents. The proposed method is a kind of semiautomatic method using a text mining and can reduce the time and cost for constructing an ontology.

In addition, a keyword based document retrieval system that gives weights by using the frequency information compared with an ontology based document retrieval system that uses relevant information existed in the ontology to a relevant feedback in order to verify the effectiveness of the hierarchical relation in the ontology. From the evaluation of the retrieval performance, it can be seen that the precision increased by 4.97% while the recall was maintained as similar to other one. This means that if the hierarchical relation in the ontology is used as relevant feedback information, the precision will be improved.

It is necessary to modify and extend the ontology in order to process a more precise processing of an query. On the other hand, 33 semantic relations defined in the present pharmacy ontology may be short. And it is necessary for constructing an ontology for various domains and to study for a method that applies the proposed method of constructing ontology to the average domain.

References

1. Baeza-Yates, R. and Robeiro-Neto, B.: Modern Information Retrieval. ACM Press, New York, NY, USA, 1999.
2. Gyeong-Hee Lee, Ju-HO Lee, Myeong- Choi, Gil-Chang Lim, "Study on Named Entity Recognition in Korean Text," Proceedings of the 12th Conference on Hangul and Korean Information Processing, pp. 292-299, 2000.
3. Hyo-Shik Shin, Young-Soo Kang, Key-Sun Choi, Man-Suk Song, Computational Approach to Zero Pronoun Resolution in Korean Encyclopedia, Proceedings of the 13th Conference on Hangul and Korean Information Processing, pp. 239-243, 2001.
4. JongHoon Oh, KyungSoon Lee, KeySun Choi, "Automatic Term Recognition using Domain Similarity and Statistical Methods," Journal of the Korea Information Science Society, Vol. 29, No. 4, pp. 258-269, 2002.
5. Jung-Oh Park, Do-Sam Hwang, "A Terminology extraction system," Proceedings of Korea Information Science Society Spring Conference(2001),Vol 27, No 1, pp. 381-383, 2000.
6. Klavans, J. and Muresan, S., "DEFINDER:Rule-Based Methods for the Extraction of Medical Terminology and their Associated Definitions from On-line Text," Proceedings of AMIA Symposium, pp. 201-202, 2000.
7. Missikoff, M., Velardi, P. and Fabriani, P., "Text Mining Techniques to Automatically Enrich a Domain Ontology," Applied Intelligence, Vol. 18, pp. 322-340, 2003.
8. Soo-Yeon Lim, Mu-Hee Song, Sang-Jo Lee, "Domain-specific Ontology Construction by Terminology Processing," Journal of the Korea Information Science Society(B), Journal of the Korea Information Science Society Vol. 31, No. 3, pp. 353-360, 2004.
9. Yi-Gyu Hwang, Bo-Hyun Yun, "HMM-based Korean Named Entity Recognition," Journal of the Korea Information Procissing Society(B), vol.10, No. 2, pp. 229-236, 2003.

A Matrix Representation and Mapping Approach to Knowledge Acquisition for Product Design

Zhiming Rao and Chun-Hsien Chen

School of Mechanical & Aerospace Engineering, Nanyang Technological University
50 Nanyang Avenue, Singapore 639798
{raoz0001,mchchen}@ntu.edu.sg

Abstract. In constructing a knowledge-based system, knowledge acquisition plays a critical role. It is the most time-consuming phase and has been recognized as the bottleneck in the development of a knowledge-based system. This paper describes a matrix representation and mapping (MRM) approach to facilitate the knowledge acquisition process in building the knowledge bases. The MRM approach involves six consecutive steps: knowledge sources identification, taxonomic tree generation, identification of linkages between conditions and conclusions sets, generation of sub-clauses to form IF and THEN clauses, identification of logical interrelationship among sub-clauses and rule generation. The proposed approach has established an integral and complete process for knowledge acquisition.

1 Introduction

In constructing a knowledge-based system, knowledge acquisition plays a critical role. It covers the activities of collecting knowledge and transforming it into a form that can be processed by a computer. It is the most time-consuming phase and has been recognized as the bottleneck in the development of a knowledge-based system [1-2].

For the knowledge acquisition bottleneck, much research has contributed to resolve the problem. As summarized by Chen & Occena [3], there are two common approaches to knowledge acquisition: (1) acquire knowledge directly from the domain expert(s); (2) acquire knowledge through the historical records, i.e. by rule induction. One of the machine learning techniques, ID3 algorithm is a popular tool used for rule induction. Several studies on ID3 algorithm for rule induction can be found in the literature [2, 4]. However, as pointed out in Gonzalez & Dankel [1], although these inductive tools for knowledge acquisition can often assist in the development of a knowledge-based system, they are not useful in all cases. Inductive techniques are appropriate for classification tasks only and often most helpful in the development of small systems. In addition, Wong [5] summarized that the typical existing learning systems such as CART, C4.5, ASSISTANT, AQ15, and CN2 used attribute-value language for representing the training examples. Those representation limits them to learn only propositional descriptions in which concepts are described in terms of a fixed number of attributes. Chen & Occena [3] proposed a knowledge sorting process to facilitate the efficiency of knowledge acquisition. The process consists of three stages: identification of knowledge sources, generation of taxonomic trees, and organization of acquired knowledge. Another knowledge acquisition tool using a goal-

R. Khosla et al. (Eds.): KES 2005, LNAI 3681, pp. 311–317, 2005.

driven strategy was developed by Wu & Kao [6]. However, these approaches did not consider the quantitative characteristics of the design information. In this study, a knowledge acquisition approach using matrix representation and mapping for the construction of knowledge-based systems is proposed. The knowledge acquisition approach proposed in this study is mainly for dealing with product design problems.

2 Overview of the MRM Approach

In order to construct a knowledge-based system, knowledge is elicited from domain expert(s) or related literature. The MRM approach to knowledge acquisition begins by identifying the appropriate knowledge sources, and then preliminarily organizes the acquired knowledge in a taxonomic manner. Based on the taxonomic trees generated, a series of matrix representation and mapping operations will be performed to extract the rules automatically. The MRM approach involves six consecutive steps, namely, knowledge sources identification, taxonomic tree generation, identification of linkages between conditions and conclusions sets, generation of sub-clauses to form IF and THEN clauses, identification of logical interrelationship among sub-clauses and rule generation. Figure 1 illustrates the overall procedure of the MRM approach. As shown in the figure, the solid lines with arrows indicate the major procedures of the approach. The dashed lines with arrows indicate the sub-procedures.

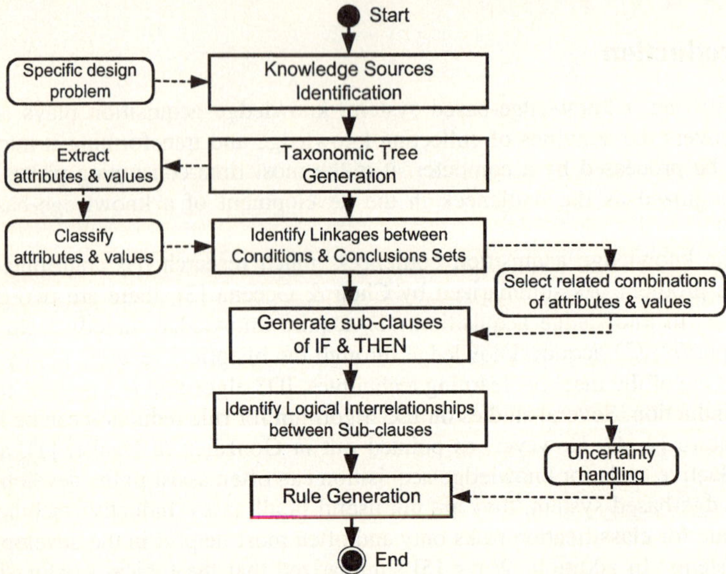

Fig. 1. Overall procedure of the MRM approach

3 Fundamental Concepts of the MRM Approach

In building a knowledge-based system, rule is an important and widely used knowledge representation scheme. People find it is natural to express knowledge using the

IF-THEN format. The IF portion of a rule is a condition, which tests the truth value of a set of facts. If these are found true, the THEN portion of a rule is inferred as a new set of facts. Figure 2 shows a conceptual process for the rule formation.

$$\begin{array}{ccc} \text{Rules} & \text{IF} & \text{THEN} \end{array}$$

$$\begin{bmatrix} R_1 \\ R_2 \\ \vdots \\ R_n \end{bmatrix} = \begin{bmatrix} X_1 \\ X_2 \\ \vdots \\ X_n \end{bmatrix} + \begin{bmatrix} Y_1 \\ Y_2 \\ \vdots \\ Y_n \end{bmatrix}$$

Fig. 2. Conceptual process for rule formation

In the MRM approach, the rule generation is defined as the mapping process between the conditions domain and conclusions domain. This process can be characterized mathematically. The characteristics of the conditions domain are represented by a set of IF clauses, these may be treated as a vector with n elements. Similarly, the THEN clauses in the conclusions domain also constitute a vector with n elements. As a result, the rule generation process involves selecting a set of the right conditional IF clauses to deduce a set of corresponding conclusive THEN clauses, which can be expressed as:

$$Y_{THEN} = MX_{IF}. \tag{1}$$

where Y_{THEN} is the vector comprising the THEN clauses in the conclusions domain; X_{IF} is the vector comprising the IF clauses in the conditions domain; M is the mapping matrix. Equation (1) may be written as:

$$[Y_1\ Y_2 \dots Y_n]^T = M[\ X_1\ X_2 \dots X_n]^T \tag{2}$$

where $Y_{THEN} = [\ Y_1\ Y_2 \dots Y_n]^T$, and $X_{IF} = [\ X_1\ X_2 \dots X_n]^T$. The mapping matrix comprises a relationship array and each element of the matrix relates an element of the IF vector to an element of THEN vector.

4   Description of the MRM Approach

4.1   Knowledge Sources Identification

For design problems, knowledge sources may include domain experts, historical records, and related literature, etc. During the construction of a specific knowledge-based system, the knowledge engineers should identify the appropriate knowledge sources required and figure out how to acquire knowledge from these sources.

4.2   Taxonomic Tree Generation

After the knowledge is elicited from a number of different knowledge sources, the acquired knowledge is organized in a taxonomic manner to facilitate the subsequent activities for rule generation. A variety of terms used to define a specific design problem are elicited from different knowledge sources. Then the elicited terms are divided and assorted into different groups according to certain domain knowledge. Thus, a hierarchical taxonomic tree illustrating the knowledge structure for a specific design problem can be formed.

4.3 Identify Linkages Between Conditions and Conclusions Sets

Based on the taxonomic tree generated in the above steps, the design attributes and their corresponding values can be extracted and then classified. These extracted attributes and values are classified into two categories: conditions set and conclusions set. For a specific design problem, the attributes and corresponding values representing the design information such as the specifications and design parameters can be classified into the conditions set. Meanwhile, the attributes and corresponding values representing the design task or objectives can be classified into the conclusions set. The attributes and corresponding values in the conditions set are used to represent the IF clauses for rule generation. Similarly, the attributes and their corresponding values in the conclusions set are used to represent the THEN clauses for rule generation. Let S_{CD} and S_{CC} denote the set of arrays representing design attributes and their corresponding values in the conditions set and conclusions set respectively.

$$S_{CD} = \{[a_{Xp}, u_{Xp}] \mid p = 1,2,\cdots,m\}; S_{CC} = \{[b_{Yq}, v_{Yq}] \mid q = 1,2,\cdots,n\} \qquad (3)$$

where a_{Xp} and u_{Xp} denote the attributes and their corresponding values in the conditions set respectively; b_{Yq} and v_{Yq} stand for the attributes and their corresponding values in the conclusions set respectively. According to Equations (1) and (2), in order to generate the rules, the relationship and linkage between the conditions set and conclusions set need to be identified. Let D represent the array set expressing linkages of attributes and values between the conditions set and conclusions set:

$$\begin{aligned}
D &= \{ S_{CD} : S_{CC} \} \\
&= \{ [a_{Xp}, u_{Xp} : b_{Yq}, v_{Yq}] \mid p = 1,2,\cdots,m ; q = 1,2,\cdots,n \} \qquad (4) \\
&= \{ [a_{Xp}, b_{Yq} : u_{Xp}, v_{Yq}] \mid p = 1,2,\cdots,m ; q = 1,2,\cdots,n \}
\end{aligned}$$

$$
\begin{array}{c|ccccc}
 & b_{Y1} & b_{Y2} & \cdots & & b_{Yn} \\
\hline
a_{X1} & 0 & 1 & \cdots & 0 & 1 \\
a_{X2} & 1 & 0 & \cdots & 1 & 0 \\
\vdots & \vdots & \vdots & \cdots & \vdots & \vdots \\
 & 0 & 0 & \cdots & 0 & 1 \\
a_{Xm} & 1 & 1 & \cdots & 0 & 0 \\
\end{array}
\qquad
\begin{array}{c|ccccc}
 & v_{Y1} & v_{Y2} & \cdots & & v_{Yn} \\
\hline
u_{X1} & 0 & 1 & \cdots & 0 & 1 \\
u_{X2} & 1 & 0 & \cdots & 1 & 0 \\
\vdots & \vdots & \vdots & \cdots & \vdots & \vdots \\
 & 0 & 1 & \cdots & 1 & 0 \\
u_{Xm} & 0 & 0 & \cdots & 1 & 0 \\
\end{array}
$$

(a) P_{mxn}: attributes relationship   (b) Q_{mxn}: values relationship

Fig. 3. Relationship matrices for attributes and values

To obtain the relationship between S_{CD} and S_{CC}, the corresponding attributes and values in the conditions set and conclusions set are compared respectively. Those attributes and lnalues which have significant relationship between the conditions set and conclusions set will be selected to generate the rules. The relationships are determined by taking into account the related design knowledge or expert opinions, which are elicited from the selected knowledge sources for a specific design problem. The two matrices, P_{mxn} and Q_{mxn}, illustrated in Figure 3, are defined to represent the relationships of the attributes and values between the conditions set and conclusions set. The choice of the value "1" or "0" for the elements in both of the matrices represents the strength of the relationship.

4.4 Generate Sub-clauses of IF and THEN

After those related design attributes and values that have significant relationships between the conditions set and conclusions set are selected, the conditions set and conclusions set can be linked via the attributes and values selected. Those combinations of the selected attributes and values are expressed as:

$$D_{AB} = \{[a_{Xp_a}, u_{Xp_a} : b_{Yq_b}, v_{Yq_b}] | p_a \in p, q_b \in q\} \tag{5}$$

where D_{AB} represents the set of combinations of the selected attributes and values; p and q denote the set of array numbers in the conditions set S_{CD} and conclusions set S_{CC} respectively, $p=1,2,...,m$, $q=1,2,...,n$. Since the elements of D_{AB} are selected from D, D_{AB} is a subset of D and thus can be expressed as: $D_{AB} \subseteq D$.

Rules of knowledge-based systems are represented using the IF and THEN clauses. These clauses can be divided into a group of sub-clauses that are composed of the selected attributes and their corresponding values. Let A_X and B_Y denote the set of the combinations of the attributes and values used to represent the IF and THEN clauses, respectively.

$$A_X = \{A_{Xp_a} | p_a \in p\} = \{[a_{Xp_a}, u_{Xp_a}] | p_a \in p\}$$
$$B_Y = \{B_{Yq_b} | q_b \in q\} = \{[b_{Yq_b}, v_{Yq_b}] | q_b \in q\} \tag{6}$$

The elements of A_X and B_Y are those related design attributes and values selected from the conditions set and conclusions set, hence, $A_X \subseteq S_{CD}$ and $B_Y \subseteq S_{CC}$ can be derived. According to Equation (6), the sub-clauses used to represent the IF clauses can be generated through the combinations of $[a_{Xp_a}, u_{Xp_a}]$. Similarly, the sub-clauses used to represent the THEN clauses can be generated through the combinations of $[b_{Yq_b}, v_{Yq_b}]$. Consequently, the representation of the sub-clauses of the IF and THEN clauses are defined as follows:

$$X = \{X_i = \{A_{Xij_1} | j_1 = 1,2,\cdots,m_1\} | i = 1,2,\cdots,n\}$$
$$Y = \{Y_i = \{B_{Yij_2} | j_2 = 1,2,\cdots,m_2\} | i = 1,2,\cdots,n\} \tag{7}$$

where X and Y denote the vector sets of IF and THEN clauses for rule generation, respectively; j_1 and j_2 are the number of sub-clauses in the IF and THEN clauses, $j_1=1,2,...m_1$ and $j_2=1,2,...m_2$; A_{Xij_1} and B_{Yij_2} denote the individual sub-clauses in the IF clauses and THEN clauses.

4.5 Identify Logical Interrelationships Between Sub-clauses

As aforementioned, the rules are usually generated using the IF and THEN clauses, which in turn can be produced by combining the sub-clauses according to certain logical interrelationships. Moreover, due to the uncertain and imprecise information is often involved in knowledge-based systems, a uncertainty handling mechanism should be provided to deal with the confidence level of the generated rules.

(i) *Mechanism for uncertainty handling.* In product design, the knowledge that has vague, imprecise and probabilistic descriptions is inevitably involved. In order to acquire such knowledge, the empirical associations and heuristic relationships are oftentimes modeled in building the knowledge-based system. This results in that the knowledge-based system should be able to reason with uncertainty. A number of methods such as MYCIN-based certainty factors (CF) [7], Bayesian probability [8], Dempster Shafer method [9], and Zadeh's fuzzy logical [10] have been developed to handle uncertainty in knowledge and data. Compared with other methods, the CF formalism is a popular tool for uncertainty handling in knowledge-based system with specific advantages [1]. Those advantages are especially suitable for design process in which many factors and complex design knowledge are involved. It appears that the CF formalism is a good choice for the uncertainty handling in the MRM approach. As such, the CF formalism is integrated in the proposed MRM approach to handle uncertainty.

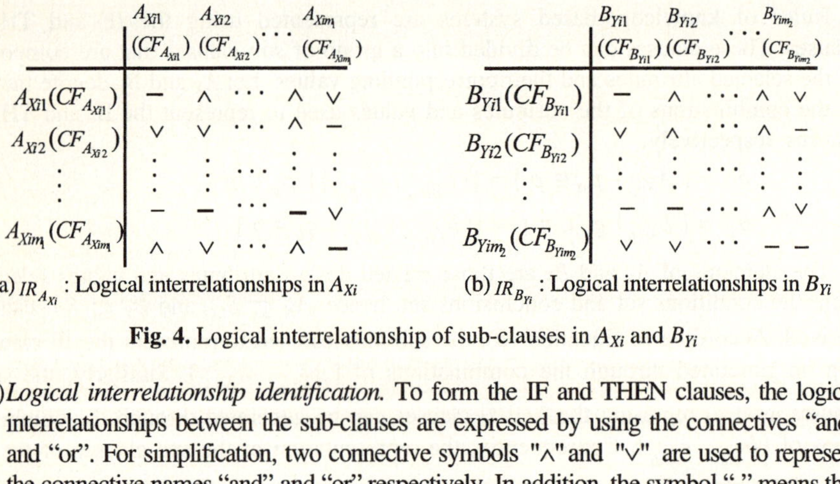

$$
IR_{A_{Xi}} = \begin{array}{c|ccccc}
 & A_{Xi1} & A_{Xi2} & & A_{Xim_1} & \\
 & (CF_{A_{Xi1}}) & (CF_{A_{Xi2}}) & \cdots & (CF_{A_{Xim_1}}) & \\
\hline
A_{Xi1}(CF_{A_{Xi1}}) & - & \wedge & \cdots & \wedge & \vee \\
A_{Xi2}(CF_{A_{Xi2}}) & \vee & \vee & \cdots & \wedge & - \\
\vdots & \vdots & \vdots & \cdots & \vdots & \vdots \\
 & - & - & \cdots & - & \vee \\
A_{Xim_1}(CF_{A_{Xim_1}}) & \wedge & \vee & \cdots & \wedge & -
\end{array}
$$

(a) $IR_{A_{Xi}}$: Logical interrelationships in A_{Xi}

$$
IR_{B_{Yi}} = \begin{array}{c|ccccc}
 & B_{Yi1} & B_{Yi2} & & B_{Yim_2} & \\
 & (CF_{B_{Yi1}}) & (CF_{B_{Yi2}}) & \cdots & (CF_{B_{Yim_2}}) & \\
\hline
B_{Yi1}(CF_{B_{Yi1}}) & - & \wedge & \cdots & \wedge & \vee \\
B_{Yi2}(CF_{B_{Yi2}}) & \vee & \wedge & \cdots & \wedge & - \\
\vdots & \vdots & \vdots & \cdots & \vdots & \vdots \\
 & - & - & \cdots & \wedge & \vee \\
B_{Yim_2}(CF_{B_{Yim_2}}) & \vee & \vee & \cdots & - & -
\end{array}
$$

(b) $IR_{B_{Yi}}$: Logical interrelationships in B_{Yi}

Fig. 4. Logical interrelationship of sub-clauses in A_{Xi} and B_{Yi}

(ii) *Logical interrelationship identification.* To form the IF and THEN clauses, the logical interrelationships between the sub-clauses are expressed by using the connectives "and" and "or". For simplification, two connective symbols "$\wedge$" and "$\vee$" are used to represent the connective names "and" and "or" respectively. In addition, the symbol "-" means that there are no direct logical relationships between the corresponding elements. As a result, the logical interrelationships between the different sub-clauses can be illustrated in matrices $IR_{A_{Xi}}$ and $IR_{B_{Yi}}$ shown in Figure 4, according to the sub-clause representations in Equation (7). The elements of $IR_{A_{Xi}}$ and $IR_{B_{Yi}}$ denote the logical interrelationships among the sub-clauses listed along the rows and columns in the matrices. The certainty factors associated with each sub-clause are illustrated in the parenthesis. Subsequently, all the rules can be generated through combining the different sub-clauses according to the logical interrelationships illustrated in the matrices.

4.6 Rule Generation

According to the two matrices $IR_{A_{Xi}}$ and $IR_{B_{Yi}}$, the IF and THEN clauses can be determined and consequently the rules for building the knowledge-based system can be generated.

For instance, rule i can be expressed as: IF $A_{Xi1}(CF_{A_{Xi1}})$ and $A_{Xi2}(CF_{A_{Xi2}})$ or $A_{Xim_1}(CF_{A_{Xim_1}})$ THEN $B_{Yi1}(CF_{B_{Yi1}})$ and $B_{Yi2}(CF_{B_{Yi2}})$ ...or $B_{Yim_2}(CF_{B_{Yim_2}})$. Meanwhile, the CF_{ri} of rule i can be obtained using the reasoning algorithm based on the CF formalism.

5 Conclusions

In this paper, a preliminary knowledge acquisition approach, so called MRM, is proposed to facilitate rule generation in building a knowledge-based system. The activities involved in this approach are elaborated. The MRM approach begins by identifying the appropriate knowledge sources. After relevant knowledge is acquired by the knowledge engineer(s) from a variety of knowledge sources, the acquired knowledge is organized in a taxonomic manner. Based on the taxonomic trees generated, a series of matrix representation and mapping activities are performed to extract the rules. The procedure of the approach has been demonstrated using a case study on the process of diagnosing automotive problems. The result of the case study is promising and reveals the potential of the proposed MRM approach. In future work, this approach will be improved and possibly computerized to automate the process of knowledge acquisition.

References

1. Gonzalez, A. J., Dankel, D.D.: The Engineering of Knowledge-based Systems. Theory and Practice. Prentice-Hall, Inc (1993)
2. Shao, X., Zhang, G., Li, P., Chen, Y.: Application of ID3 algorithm in knowledge acquisition for tolerance design. Journal of Materials Processing Technology. 117 (2001) 66-74
3. Chen, C.-H., Occena, L. G.: Knowledge sorting process for a product design expert system. Expert Systems. 16 (1999) 170-182
4. Wu, C.H., Kao, S.C.: Induction-based approach to rule generation using membership function. International Journal of Computer Integrated Manufacturing. 15(2002) 86-96
5. Wong, M.L.: An adaptive knowledge-acquisition system using generic genetic programming. Expert Systems with Applications. 15(1998) 47-58
6. Wu, C.H., Kao, S. C.: GDKAT: A goal-driven knowledge acquisition tool for knowledge base development. Expert Systems. 17 (2000) 90-105
7. Qiu, D.: Certainty factor theory and its implementation in an information system-oriented expert system shell. Applied Artificial Intelligence. 6 (1992) 147-166
8. Guan, J.W., Guan, Z., Bell, D.A.: Bayesian probability on Boolean algebras and applications to decision theory. Information Sciences. 97(1997) 267-292
9. Shafer G.: A Mathematical Theory of Evidence. Princeton University Press (1976)
10. Zadeh, L.A.: Fuzzy Sets. Information and Control. 8(1965) 338-353

A Unifying Ontology Modeling
for Knowledge Management

An-Pin Chen and Mu-Yen Chen

Institute of Information Management, National Chiao Tung University
Hsinchu, 300, Taiwan, ROC
{apc,mychen}@iim.nctu.edu.tw

Abstract. The knowledge-based economy is approaching rapidly, and knowledge management (KM) has disseminated in leaps and bounds in academic circles as well as in the business world. This paper develops a unifying framework to evaluate KM activities for supporting intelligent knowledge-based system (KBS) using web interface, and expert system technology to help inexperienced administrators in insuring the smooth operation of KM performance. This paper makes three important contributions: (1) it proposes an efficient KM Ontology Construction Algorithm to fast conceptualize KM domain concept; (2) it provides a hybrid model used for knowledge acquisition through skeletal concept model and IDEF (Integrated DEFinition function modeling) analysis; and (3) it presents a methodology for using KM ontology in building a unifying framework and evaluation guideline for KM that works well and effective.

1 Introduction

Recent surveys indicate that issues such as 'measuring the value of knowledge management (KM)' and 'evaluating KM performance' are of great importance to managers in Asia [1], the United States [9] and the United Kingdom [10]. Given the increasing role of KM in upgrading business competitiveness, the wide interest of managers in measuring and evaluating both KM performance and benefit is not surprising. This creates an important research issue: how do most firms that initiated KM develop appropriate metrics to measure the effectiveness of their initiative? We need a more rigorous metric to assess KM performance with the ability to explain it and to suggest future strategic actions that the firms should take to improve KM performance.

Our research objective was therefore to propose a new ontology construction algorithm to evaluate knowledge management performance. This paper is organized as follows: we start by giving an overview of prior research on KM evaluation, structured analysis, and ontology in Section 2. We then integrate the framework of structured analysis and KM ontology in Section 3. Section 4 explains how the knowledge acquisition for ontology construction algorithm in KM. Finally, the conclusion and future work are discussed in Section 5.

2 Preliminary

In this section, we provide an overview of prior research on KM evaluation approaches, software engineering with structured analysis, and ontology function.

R. Khosla et al. (Eds.): KES 2005, LNAI 3681, pp. 318–324, 2005.

2.1 The Methodology of KM Evaluation

While authors differ in the terminology used in describing the KM process, the aggregate of their works can be described as a simple KM process. We generalized a conclusion from a collection of related KM research and defined the "4C" process of KM activities: creation, conversion, circulation and carry out [4].

Knowledge creation relates to knowledge addition and the correction of existing knowledge. Nonaka and Takeuchi [8] suggest four modes of knowledge creation: socialization, externalization, internalization and combination.

Knowledge conversion relates to individual and organizational memory. While organizational memory reflects the shared interpretation of social interactions, individual memory depends on the individual's experiences and observations.

Knowledge circulation is the didactic exchange of knowledge between a source and a receiver. An import aspect of the knowledge carry out is that the source of competitive advantage resides in the knowledge itself. Here a major challenge is how to integrate internal knowledge and the knowledge gained from outside. The creation, conversion, and circulation process of new knowledge without its exploitation (i.e., carry out) leads to its underutilization as a driver of performance [2].

2.2 Overview of Structured Analysis

The Unified Modeling Language (UML) is the industry-standard language for specifying, visualizing, constructing and documenting the artifacts of software systems [7]. It simplifies the complex process of software design, making a "blueprint" for construction. An important part of the UML is the facilities for drawing structured analysis and design. IDEF (Integrated DEFinition function modeling) modeling is a powerful structured analysis tool for analysts to identify and evaluate activities functionality. The goal of IDEF is to apply a structured method to better contain a high level of support for process understanding, communication, and improvement. In this paper, we applied IDEF modeling to help extract KM entity while building the KM ontology.

2.3 The Needs of Ontologies

Ontology analysis clarifies the structure of knowledge. Given a domain, its ontology forms the heart of any system of knowledge representation for that domain. Without ontologies, or the conceptualizations that underline knowledge, there cannot be a vocabulary for representing knowledge [3]. Accordingly, a reusable task/domain model can be represented, and a computer program code is generated in that ontology for knowledge acquisition, reuse and heuristic learning. Gomez-Perez et al. believes that ontological engineering is a craft rather than a science [6]. Each development team usually follows its own set of principles, design criteria, and phases in the ontology development process.

3 System Architecture of the Unifying Framework

The following top-level phases and sub-level phases are modified and defined for construction knowledge management according to the above-mentioned references.

Top-level phases are knowledge creation; knowledge conversion, knowledge circulation, and knowledge carry out.

3.1 Structured Analysis with KM

The concept of IDEF modeling is proposed and used for designing construction knowledge management systems. The IDEF suite of structured analysis approaches, consisting IDEF0 (Function Modeling), IDEF1 (Information Modeling), IDEF4 (Object-oriented Design), and IDEF5 (Ontology Description Capture) has been applied extensively in support of large industrial engineering projects. The information/knowledge contents and processes managed in the KM system are identified through the IDEF0 function modeling approach.

The development of construction knowledge management encompasses all activities from ''knowledge creation'' to ''knowledge carry out''. Figure 1 presents the top-level IDEF0 context diagram to represent the scope of the process. In this paper, the context diagram is decomposed into four sub-functions—''knowledge creation'', ''knowledge conversion'', ''knowledge circulation'', and ''knowledge carry out''. Figure 2 shows the "knowledge creation" sub-functions respectively.

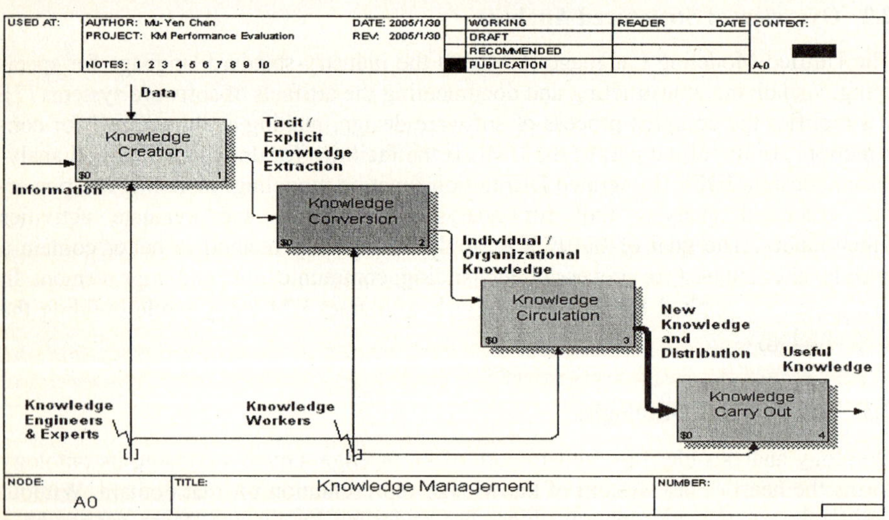

Fig. 1. Top level of construction knowledge management

3.2 Building of a KM Ontology

Ontology provides the framework for the domain knowledge base. With the help of ontology, the knowledge is not only human-readable but also machine-readable. In the following, we will explain some of the main points of the KM ontology.

In general we can say that there are eight main types of relationships: (1) "isa" is a generalization relationship, which can be used to describe the relationship in the class hierarchy. For example, collect data is an activity in knowledge creation; (2) "part of" is generalization relationship, which can be used to denote the "belong to" relation-

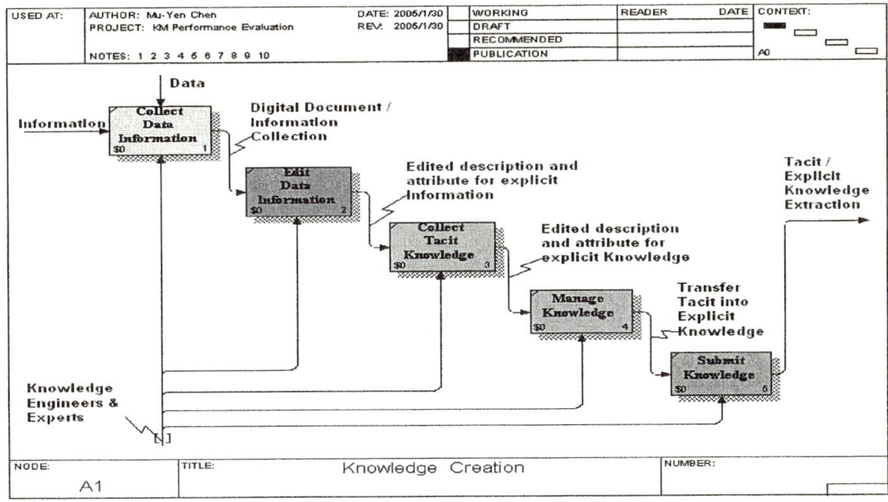

Fig. 2. Sub-level of knowledge creation

ship. For example, a digital document belongs to the "collect data" activity; (3) "conceptual part of" is logic or a non-material relationship. For example, knowledge creation belongs to KM; (4) "connected to" is a "related to" relationship, which denotes that there exists some substance relationship between these terms. For example, "edit data" activity is connected to "collect data" activity, because these activities are belonging to the knowledge creation process; (5) "form of" is describing the evolvement relationship, which means that there exists an inheritance relationship between these terms. For example, knowledge creation is form of whole KM process; (6) "location of" indicates a derivative knowledge relationship. For example, a "manage knowledge" activity will transfer tacit knowledge into explicit knowledge. So explicit knowledge is location of "manage knowledge" activity; (7) "issue in" shows a special knowledge relationship; (8) "evaluation of" represents a verification knowledge relationship. For example, the most common method to verify a "collect data" activity is by using the quantitative analysis method. So the quantitative analysis method is an evaluation of gastric "collect data" activity.

4 Knowledge Acquisitions and Evaluation

It is important to bear in mind that knowledge acquisition (KA) is an independent activity in the ontology development process [5]. Most of the acquisition is done simultaneously with the requirement specification phase, and decreases as the ontology development process moves forward.

4.1 KA Process

In this paper, to help extract knowledge from experts and construct the KM ontology hierarchy, we propose an efficient KA scheme (e.g., KM Ontology Construction

Scheme) to quickly conceptualize KM domain knowledge by using a hybrid method consisting of employees' interviews, expert interviews, and structured analysis by IDEF modeling. A simplified scheme is illustrated as shown in Figure 3.

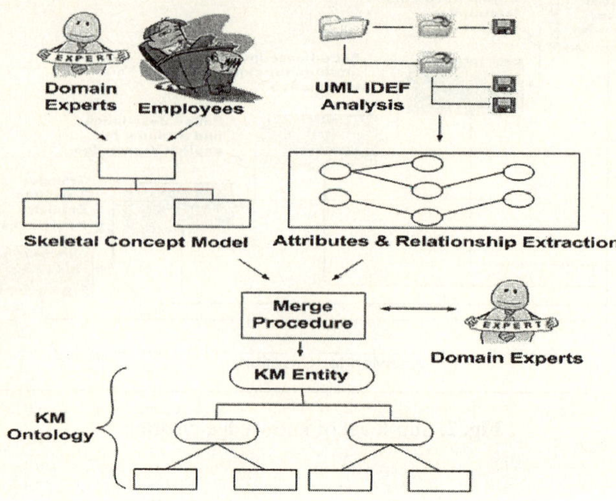

Fig. 3. KM knowledge Acquisition Process

4.2 KM Ontology Construction Algorithms

The whole KM KA process can be roughly divided into six steps. Each step can be described as follows:

Step 1: Build the Skeleton KM ontology by top-down approach.
Step 2: Initiate IDEF modeling
Step 3: Conduct the attributes and relationships extraction by the bottom-up approach
Step 4: Merge the findings from step 1 and step 3
Step 5: Experts verify the ontology under construction
Step 6: The ontology generated is confirmed OK

In the following, we give an example for helping to demonstrate the operation details of the KA process during Steps 2–4.

Example: We have an IDEF modeling consisting of "Knowledge Creation" sub-level that contains many issues including: Collect Data / Information and Submit Knowledge, etc.

Step 2: Figure 4 shows the original skeletal concept model of the "Knowledge Creation" concept. We use a hybrid method to construct the original skeletal concept through employees' interviews and expert interviews. Therefore, there are existing gaps between the skeletal concept model and the IDEF modeling.

Step 3: As shown in Figure 2, we can easily extract other knowledge creation concepts, including "Edit Data / Information", "Collect Tacit Knowledge", and "Management Knowledge". In addition to the concepts of extraction, we can find that the "is-a" relationship exists between the "Knowledge Creation" concept and five other concepts. Therefore, by following our former heuristics, we can construct a part of the ontology hierarchy concerning the "Knowledge Creation" sub-level, as shown in Figure 5.

Step 4: Finally, we find that the "Knowledge Creation" concept is just one part of the "Knowledge Management" concept through our algorithm. Hence, we can connect these two concepts directly. After the operation, the whole KM ontology can be illustrated as shown in Figure 6.

Fig. 4. An initial skeletal concept model of Knowledge Creation

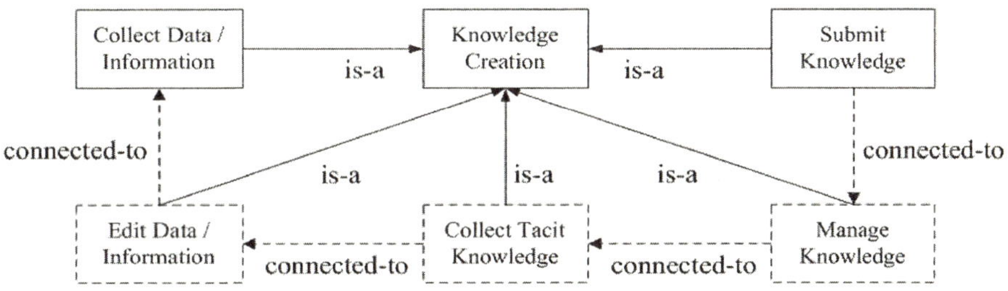

Fig. 5. A concept model of the Knowledge Creation

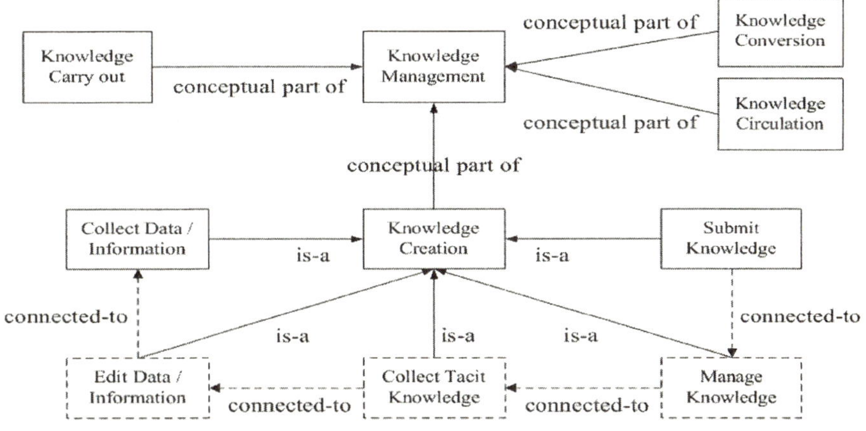

Fig. 6. The KM ontology after the adding of the knowledge creation concept

4.3 KM Evaluation

In our study, we evaluated KM activities by questionnaire from the following perspectives: knowledge creation, knowledge conversion, knowledge circulation and knowledge carry out. The respondents completed a self-administered, 17-item questionnaire. The respondents indicated their agreement or disagreement with the survey instruments via a five-point Likert-type scale, with anchors ranging from "strongly disagree" to "strongly agree". Pre-testing and pilot testing of measures were conducted by selected users from the KM field, as well as by experts in this area.

5 Conclusion

In this paper, we proposed a KM ontology construction algorithm to help knowledge engineers construct the KM ontology. During computational design application, the ontology can be used to determine the suitability of problem solvers for the KM including: (1) to re-organize the knowledge hierarchy of the KM, (2) to upgrade the scale of the KM when the environment is changed, (3) and most important, to provide expertise to evaluate KM activities and its performance.

References

1. Ahn, J.H., and Chang, S.G.: Assessing the Contribution of Knowledge to Business Performance: the KP3 Methodology, Decision Support Systems, Vol. 36 (2004) 403 – 416
2. Alavi, M.: Knowledge Management and Knowledge Management Systems: Conceptual Foundations and Research Issues, MIS Quarterly, Vol. 25, No. 1 (2001) 107-136
3. Chandrasekaran, B., Josepheson, J. and Benjamins, V.R.: What are Ontologies, and Why Do We Need Them? IEEE Intelligent Systems, Vol. 14, No 1 (1999) pp.20-26
4. Chen, A.P., Chen, M.Y.: Measurement Practices for Knowledge Management: An Option Perspective, accepted to appear in Proceedings of the 17th Conference on Advanced Information Systems Engineering (CAiSE'05), Porto, Portugal (2005)
5. Fernandez, M.L., Gomez-Perez, A., Juristo, N.: METHONTOLOGY: From Ontological Art Towards Ontological Engineering, Workshop on Ontological Engineering, AAAI97 Stanford, U.S.A (1997)
6. Gomez-Perez, A., Fernandez, M.L., De Vicente, A.J.: Towards a Method to Conceptualize Domain Ontologies, In Workshop on Ontological Engineering (1996)
7. Kobryn, C.: A Standardization Odyssey, Communications of the ACM, Vol. 42, No. 10 (1999) 29–37
8. Nonaka, I and Takeuchi, H.: The Knowledge Creating Company, New York, NY: Oxford University Press (1995)
9. Schultze, U. and Leidner, D.E.: Studying Knowledge Management in Information Systems Research: Discourses and Theoretical Assumptions, MIS Quarterly, Vol. 26, No. 3 (2002) 213-242
10. Shin, M., Holden, T., and Schmidt, R.A.: From Knowledge Theory to Management Practice: Towards an Integrated Approach, Information Processing and Management, Vol. 37 (2001) 335-355

A Model Transformation Based Conceptual Framework for Ontology Evolution

Longfei Jin, Lei Liu, and Dong Yang

College of Computer Science and Technology, Key Laboratory of Symbolic Computation
and Knowledge Engineering of Ministry of Education of P.R. China
Jilin University, Changchun, 130012, P.R. China
jinlongfei@vip.163.com

Abstract. Ontology evolution in the Model Driven Semantic Web can be looked as a process of model transformations, so a model transformation based conceptual framework for ontology evolution is presented in this paper. The applications of model transformations in every phase of ontology evolution process are descript respectively. The framework combines technologies of ontology evolution, ontology definition metamodel and model transformations, and it can be looked as a method for ontology evolution in the Model Driven Semantic Web.

1 Introduction

The OMG's Model Driven Architecture (MDA) initiative and the W3C's Semantic Web project initially proceeded in parallel with no coordination. The marriage of these two technologies can dramatically improve current approaches to ontology and knowledge base development and, at the same time, enable next generation enterprise computing by automating business semantics interchange and execution. This new domain is named Model Driven Semantic Web (MDSW) [1]. In MDSW, ontologies are represented as models derived from Ontology Definition MetaModel (ODM) [2] – a metamodel of MDA, and many problems of ontology engineering can be solved by model technologies.

Ontology evolution is a very complex problem of ontology engineering, and it is very important for semantic web. As ontologies are represented as models in MDSW, ontology evolution can be looked as a process of model transformations. In this paper, ontology evolution process was analyzed in a model transformation viewpoint, and a conceptual framework for ontology evolution is presented. The framework uses an OMG's QVT proposal [3], it can act as the foundation of ontology management, and it can also benefit other aspects of ontology engineering.

The paper is organized as follows. In section 2, we introduce the OMG's QVT proposal. Section 3 overviews the conceptual framework for ontology evolution and the applications of model transformations in every phase of the ontology evolution process are descript respectively. Section 4 introduces related works. Section 5 concludes the paper and discusses the further works.

2 Introduction of QVT

MOF 2.0 Query/View/Transformation (QVT) is currently undergoing standardization through the OMG.

R. Khosla et al. (Eds.): KES 2005, LNAI 3681, pp. 325–331, 2005.

We now present high level definitions of the QVT's subjects, and more details can be seen in the OMG's QVT proposal.

- **Queries** take as input a model, and select specific elements from that model.
- **Views** are models that are derived from other models.
- **Transformations** take as input a model and update it or create a new model.

The kernel of the QVT proposal is a model transformation language named QVT language. Main ideas of definition of QVT language are followed.

1. Transformation is a super-type of Relation and Mapping.
2. Relations are multi-directional transformation specifications. Relations are not executable in the sense that they are unable to create or alter a model: they can however check two or more models for consistency against one another.
3. Mappings are transformation implementations. Unlike relations, mappings are potentially unidirectional and can return values. Mappings can refine any number relations, in which case the mapping must be consistent with the relations it refines.
4. QVT language allows model fragments to be matched against meta-model patterns and used in transformations.
5. Domains in QVT language are comprised of patterns and conditions. By utilizing conditions, arbitrarily complicated expressions can be specified to augment patterns.

3 The Conceptual Framework and the Ontology Evolution Process

Our conceptual framework for ontology evolution is shown in Fig. 1.

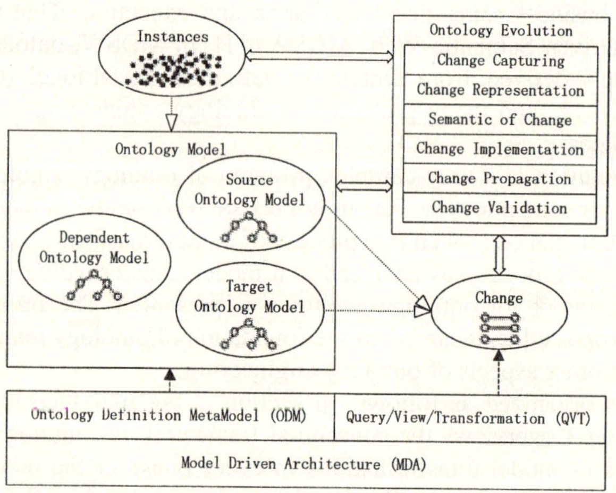

Fig. 1. A conceptual framework for ontology evolution

Figure 1 shows the framework with an ontology evolution process comprised of six phases [4]: (1) change capturing, (2) change representation, (3) semantic of change, (4) change implementation, (5) change propagation, (6) change validation. QVT language based model transformation technologies are used in every phase of the process.

3.1 Change Capturing

The process of ontology evolution starts with capturing changes either from explicit requirements or from the result of change discovery methods, which induce changes from existing data. In the conceptual framework, explicit requirements are represented in model transformation programs. Ontology engineers can transform informal requirement specifications derived from domain analysis to this kind of formalization. For an example, a informal requirement specification: adding two conceptions "Man" and "Woman" as sub conceptions of "Person" into an ontology can be represented in a model transformation program followed.

```
relation R{
domain{(Old.RDFSClass)[localName="Person", ...]}
domain{(New.RDFSClass, c1)[localName="Person", ...]}

domain{(New.RDFSClass)[localName="Man",
RDFSsubClassOf=c1, ...]}

domain{(New.RDFSClass)[localName="Woman",
RDFSsubClassOf=c1, ...]} }
```

There are three kinds of implicit requirements resulted from change discovery methods: structure-driven changes, data driven changes and usage-driven changes. Structure-driven changes can be deduced from the ontology structure itself. In the conceptual framework, we can get them through a model analysis process, in which queries and views of QVT language are used. Data-driven changes are generated by modifications to the underlying dataset. We can get such changes represented in model transformation programs through a dataset mining process. Usage-driven changes result from the usage patterns created over a period time. They are counterparts of design patterns of software engineering. In the conceptual framework, Usage-driven changes are represented as model transformation programs, too. An example of such change named "pull concept up" can be represented in a model transformation program followed (We extend the QVT language in this example).

```
relation PullConceptUp(Old.RDFSClass c /*an extended
template parameter*/){

domain{(Old.RDFSClass, c)[localName=n3,
RDFSsubClassOf=(Old.RDFSClass)[localName=n2,
RDFSsubClassOf=(Old.RDFSClass)[localName=n1, ...], ...],
...]}

domain{(New.RDFSClass, c1)[localName=n1, ...]}

domain{(New.RDFSClass)[localName=n2,
RDFSsubClassOf=c1, ...]}

domain{(New.RDFSClass)[localName=n3,

RDFSsubClassOf=c1, ...]} }
```

Besides these requirement driven ontology changes, most popular ontology changes are those caused by edit operations in the IDE of ontology. In these cases, model transformation programs are generated by the IDE automatically.

3.2   Change Representation

Change representations mainly refer to the granularity of changes. There are two granularities representing ontology changes in the conceptual framework: elementary changes and composite changes. Elementary changes cannot be decomposed into simpler changes. A composite change is an ontology change that modifies (creates, removes or changes) one and only one level of neighborhood of entities in the ontology. The QVT language permits recursions of patterns, operations for transformations – NOT, AND, and OR, and calls among transformations. This makes it easier to represent rich ontology changes.

3.3   Semantic of Change

The semantics of change phase is the phase in the ontology evolution process that enables the resolution of ontology changes in a systematic manner by ensuring the consistency of the ontology. In the conceptual framework, conditions of model transformation programs are used to keep consistency of ontology models. Checks are invoked before every transformation. For example, only a RDFS class with URI that does not exist in current ontology model can be added. This can be represented in a model transformation program followed.

```
relation AddConcept{
domain{(Old.RDFSClass, parent)[...] }

domain{(New.OWLClass)[URI=uri,
RDFSsubClassOf=parent, ...]}

when{allElement->forAll(element, element.URI<>uri)} }
```

There are many evolution strategies encapsulating policies for evolution with respect to user's requirements in the conceptual framework. When ontology models are inconsistent, users can select and execute those strategies. For an example, when a user deletes a concept, he can select whether to delete its entire offspring, or only to pull its sub concept up, or to make its sub concept as a child of the top concept. In the conceptual framework, evolution strategies are represented as different domains with same patterns and conditions. When different domains are satisfied at the same time during a model transformation process, execution is interrupt and will resume after user select a special strategy. For convenient, user can set default evolution strategies for a special transformation process.

3.4   Change Implementation

Three phases are required in change implementation: (1) change notification, (2) change application, (3) change logging. In order to avoid performing undesired changes, a list of all implications to the ontology and dependent artifacts should be generated and presented to the ontology engineer, who should then be able to accept or abort these changes. The conceptual framework uses views of QVT to provide change impact analysis views to users. The application of a change should have transactional properties. The conceptual framework realizes this requirement by the strict separation between the request specification and the change implementation. This allows to easily treating the set of change operations as one atomic transaction, as all the changes are applied at once. The conceptual framework uses model transformation programs as logs. Through computing inverse transformations, models can be traced back. As relations are multi-directional transformation specifications, models can be traced back easily through execution of inverse mappings.

3.5   Change Propagation

Ontologies often reuse and extend other ontologies. Therefore, an ontology update might also corrupt ontologies depending on the modified ontology (through the inclusion, mapping integration, etc.) and consequently, all the artifacts based on these ontologies. Ontology changes can impact: (1) local ontology in which elements changed exist, (2) dependent ontologies which depend on local ontology, (3) instances of ontologies, (4) applications using ontologies. Change propagations can be computed recursively. As change impact analysis of applications relate to the implements of those applications, we will not discuss them in this paper.

Ontology changes can impact local ontology. There are many relations among elements, such as inherit, equal, different, etc. We can construct a dependency graph of local ontology by computing those relations. When a change occurs in local ontology, we can get elements impacted by the change through the dependency graph. In the conceptual framework, implement of the change impact analysis comprised of five steps: (1) to transform local ontology to a dependency graph, (2) to get impacted elements from the dependency graph through queries of the QVT language, (3) using those elements as inputs to execute step 2 recursively, and to get change impact views, (4) to check inconsistency and generate model transformation programs, (5) to execute those programs to keep consistency of local ontology.

Ontology changes can impact dependent ontologies. We can get change impact views by querying the dependency graph comprised of local ontology and dependent ontologies. As elements not changed in local ontology will not impact any elements of dependent ontologies, we can optimize the construction of dependency graph through replacing the whole local ontology by a change impact view of it.

Ontology changes can also impact instances. There are two kinds of implementations in the conceptual framework. One is to generate date transformation programs from model transformation programs and to keep consistency of ontology model and instances by data transformations. The other is to keep instances unchanged, and to send model transformation programs as additional inputs to an extended reasoner. Because the reasoner knows changes of ontologies, it can give a transparent reasoning - ability of a reasoner that can give same reasoning results as with old ontologies when instances changed.

3.6   Change Validation

Change validation enables justification of performed changes and undoing them at user's request. As we have provided formal requirement specifications and change impact analysis views, many validations have been done in other phases of ontology evolution process.

4 Related Works

In [4] the authors identify a possible six-phase evolution process, on which our framework is built. [5] defines three types of change discovery. An implementation of data-driven change discovery is included in the KAON tool suite [6]. [7] describes a set of changes for the OWL ontology language, based on an OWL meta-model. [5] proposes an evolution log based on an evolution ontology for the KAON ontology model. [8] presents an approach for evolution in the context of dependent and distributed ontologies. [9] concentrates on the benefits of modular ontologies with respect to local containment of terminological reasoning. [10] discusses OntoView, a web-based change management system for ontologies.

The OMG's QVT proposal introduces a general model transformation language. EMF [11] is a model transformation tool built on the Eclipse [12] platform, and it can transform models to java codes. GMT [13] is general model transformer built on the Eclipse platform, and its basic model transformation language is named ATL [14]. YATL [15] is another QVT language, and some examples are discussed.

Ontology Definition MetaModel (ODM) is currently undergoing standardization through the OMG. It is a standard of ontology representation. UML, Entity Relationship, Description Logic, RDFS/OWL, Topic Map and Simple Common Logic metamodel are including in ODM. Our conceptual framework is built on ODM, QVT and ontology evolution.

5 Conclusion and Further Works

Our work combines research results of ontology evolution, model transformation and ontology definition metamodel. A model transformation based conceptual framework for ontology evolution is presented in this paper. The framework resolves the problem of ontology evolution in the domain of Model Driven Semantic Web, and it can benefit other researches of ontology engineering. As the QVT proposal and the ODM proposal are not standards, and ontology evolution is far from manual technology, our framework is still conceptual. With researches of related technologies going on, the framework will develop to a model transformation based ontology management system.

There are a lot of works need to do, to name just a few. To refine the framework, to cut and extend the model transformation language, to design and implement an interpreter of the model transformation language, to add ontology versioning functions to the framework, to investigate how ontology evolution impact ontology dependencies, to apply other model technologies to ontology evolution, etc.

References

1. David F., Pat H., Elisa K., Deborah M.: The Model Driven Semantic Web. in Proceedings, the 8th International IEEE EDOC Conference, 1st International Workshop on the Model Driven Semantic Web (MDSW 2004), Monterey, CA, September 21, (2004)
2. Daniel T. C., Elisa K.: Major Influences on the Design of ODM. in Proceedings, the 8th International IEEE EDOC Conference, 1st International Workshop on the Model Driven Semantic Web (MDSW 2004), Monterey, CA, September 21, (2004)
3. MOF 2.0 Query/View/Transformation Revised Submission, ad/04-04-01, April 5, (2004)

4. Ljiljana S., Alexander M., Boris M., Nenad S.: User-driven ontology evolution management. In European Conf. Knowledge Eng. and Management (EKAW 2002). Springer-Verlag, (2002) 285–300
5. Ljlijana S.: Methods and Tools for Ontology Evolution. PhD thesis, University of Karlsruhe, (2004)
6. FZI Karlsruhe and AIFB Karlsruhe. KAON the Karlsruhe ontology and semantic web framework—developer's guide for KAON 1.2.7, (2004)
7. Michel K.: Change Management for Distributed Ontologies. PhD thesis, Vrije Universiteit Amsterdam, (2004)
8. Alexander M., Boris M., Ljiljana S.: Managing multiple and distributed ontologies in the semantic web. VLDB Journal, 12(4):286–302, (2003)
9. Heiner S., Michel K.: Integrity and change in modular ontologies. In Proceedingso of the 18th International Joint Conference on Artificial Intelligence, Acapulco, Mexico, August (2003)
10. Michel K., Atanas K., Damyan O., Dieter F.: Finding and characterizing changes in ontologies. In Proceedings of the 21st International Conference on Conceptual Modeling (ER2002), Lecture Notes in Computer Science, Vol. 2503. Springer-Verlag, Berlin Heidelberg (2002) 79-89
11. EMF Developer FAQ. www.eclipse.org/emf
12. Eclipse.org: Eclipse platform technical overview. Technical report, http://www.eclipse.org/whitepapers/eclipse-overview.pdf (2003)
13. GMT, General Model Transformer Eclipse Project, http://www.eclipse.org/gmt/
14. ATL, ATLAS Transformation Language Reference site http://www.sciences.univnantes.fr/lina/atl/
15. Patrascoiu O.: YATL:Yet Another Transformation Language. In Proc. of First European Workshop MDA-IA, University of Twente, the Nederlands, (2004)

On Knowledge-Based Editorial Design System

Hyojeong Jin and Ikuro Choh

Global Information and Telecommunication Studies, Waseda University.
A209, 1011 Okuboyama Nishi-Tomida Honjo-shi Saitama-ken, Japan
`jin@moegi.waseda.jp`, `choh@giti.waseda.ac.jp`
`http://www.media.giti.waseda.ac.jp/`

Abstract. The ultimate goal of this research is to create an intelligent editorial design system integrating knowledge from expert designers. In this research, a grid system of layout theories and data mining are used to clarify the relation between design elements and their mechanisms from real design products by expert designers. Subsequently, design logics are theorized and knowledge-base of the editorial design system is created. When using our system, even computer or design novices can effectively and effortlessly perform design works. Furthermore, the result of scientific systemization of layout design process can contribute to the efficient design knowledge management.

1 Introduction

One way to enhance the efficiency of design work is to utilize knowledge of expert. This research concerns supporting creation of new designs in documents by formalizing expert designers' tacit knowledge.

Till now, almost design works has depended on the sense of an individual designer. And the ambiguity of design process makes difficult to manage design works and to understand. In order to solve such problems, design processes should be theorized. Hence, in this paper, we theorize expert designers' tacit knowledge from practical layout designs, and those outcomes are utilized to create automatic layout design system to support new editorial design environments.

2 Background

2.1 The Necessity of Design Knowledge Management

The goal of this research is to support the management of design process by integrating design knowledge to computer system.

There are 2 types of knowledge, tacit and explicit knowledge. Tacit knowledge is the knowledge gaining from experience or intuition and it can not be represented by words, formulas or pictures. Explicit knowledge is the knowledge that can be represented by words, formulas or pictures as objects or reasons. The creation of knowledge can be explained by 4 modes - Externalization, Combination, Internalization and Socialization - which are circulation of tacit and explicit knowledge [2]. In the present, almost design knowledge is tacit knowledge while only a few are explicit knowledge. Therefore, it has to spend too much time to practically learn or acquire the knowledge in design works. Since tacit knowledge is shapeless and no-pattern,

R. Khosla et al. (Eds.): KES 2005, LNAI 3681, pp. 332–338, 2005.

development is limited or can not be done effectively. Furthermore it is difficult not only to lucidly explain to other people but also to make effective collaborative works in multimedia contents production such as making agreement among designers. Thus, in this paper, in order to perform efficient knowledge management in the major of design, we propose the Editorial Design System by using the concepts of transforming tacit knowledge from designers to explicit knowledge.

2.2 The Analysis Based on Gestalt Theory

In this research, the Analysis for theorizing designers' tacit knowledge is based on Gestalt Theory, which is the theory of visual perception.

According to this theory, when human perceive a page of a book, they perceive a whole including the relation of their elements or characteristics, not recognition of each separate part. And, when human perceive an object, they analyze "figure" and "ground" in every areas. Subsequently, they bind or separate them mutually. At that time, the principle of perceived formation, the influence elements of this kind of human perception are "distance", "size", "brightness" and "position".

In this research, in order to analyze layout design, characteristics of all elements are utilized to prioritize their relation and create the fixed form. Besides, the influence elements for the grouping, distance, size (width, height, and area), grayscale (brightness) and position are established to measure the parameters.

In order to operate all objects in a page as a part of grid system, unified valuation basis is necessary. Hence, we assume that the shape of objects in a page is square. In addition, distance, size (width, height, and area), grayscale (brightness) and position are established and measured as valuation basis. Consequently, the measured data are grouped and the results are analyzed thoroughly.

2.3 Layout Methods Based on Grid System

Grid system is a theory of layout method. The grid is useful in planning and designing layout of pages and utilizing principles and elements of visual design. The elements on the page are placed on the cell border lines and overall aligned on vertical and horizontal lines. The grid creates systematic and consistent rule for placing objects. It creates visual rhythms and makes it easier and more pleasant for the eye to scan the objects on the page [3]. Thus, layout design based on grid system are utilizing in our proposed system.

3 The Proposed Editorial Design System (EDS)

Editorial Design System (EDS) provides users environment supporting a design process from drawing up a document to making and modifying layout design based on Knowledge-base. EDS is operated on the environment of "Legible Design System (LDS)". The LDS projects the image by using a video projector to be editable on the table acting as a tablet computer. LDS is show in Figure 1.

In this section, firstly, we describe the method of creating knowledge-base, and how this system supports users through editorial design process, respectively.

Fig. 1. The Legible Design System

3.1 Creating Knowledge-Base of Expert Designers

We gather practical design samples designed by using grid system. Subsequently, we utilize kohonen-network and decision tree which are techniques for data mining to discover and extract the knowledge from expert designers by using real world design products. In the other words, the tacit knowledge is transformed to explicit knowledge of design. Consequently, extracted knowledge is used to build a knowledge-base system. This knowledge-base is main logic of EDS. The detail of the process is clarified as below (Fig.2).

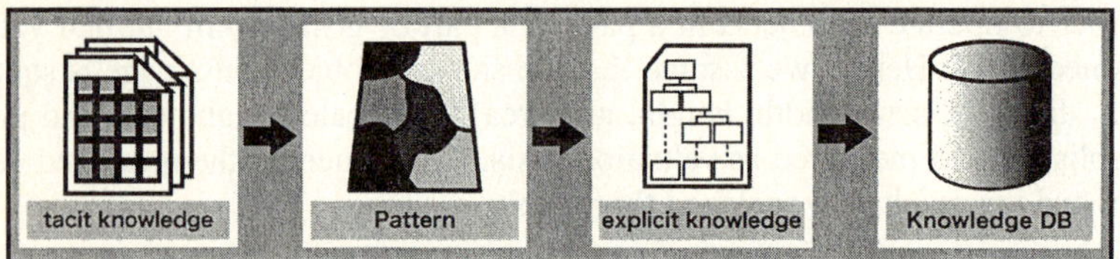

Fig. 2. The process of building knowledge-base

3.1.1 The Measurement of Parameters

First, we scan books which are designed by using grid system and objects in each page are transformed to square shape of gray-scale. Subsequently, all pages are recorded and variables shown in table 1 are measured.

3.1.2 The Classification of Objects to Contents Structure

After finished the process of measuring parameters, all objects are classified to each separated content structure. Moreover, in order to prioritize objects, the levels of independency are weighted. In this research, units of content structure are classified to "Title Head" and "Detail Body". One of content structure units are continued part to lower layer of the structure. The structure and example is shown in Fig.3.and 4.

3.1.3 Discovering Rules for Classification of Clustering Layout Data

The data of initial classified object positions are clustered by using Kohonen network. Consequently, grid structures of all clusters are analyzed based on the results from clustering. An example is shown in Table 2. Moreover, the rules of layout design are created by using analytical results and stored in knowledge base respectively. An example of layout method is shown in Figure 5.

Table 1. The measurement of parameters

	Measure Item		
Entire page	Size	Width(Page),Height(Page)	
	Grid	A (column) * B (line)	
	Margin	Left, Right, Top, Bottom	
object	Size	Width, Height	
	Degree of Grayscale	Equality	Level 1, 2, 3, …
		Inequality	Average : 5Levels
			Principle position : 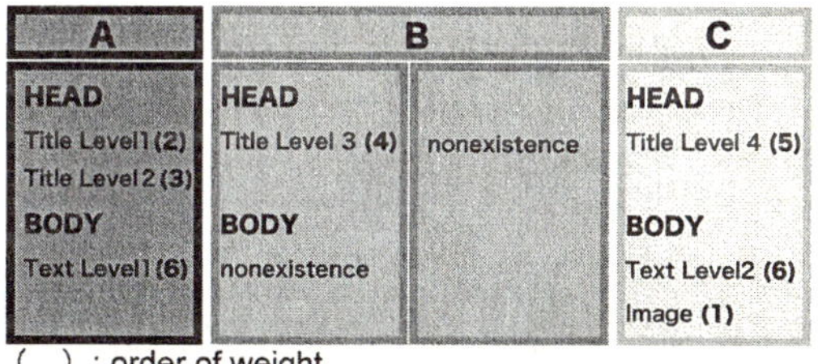
	Position	coordinate of upper left point	

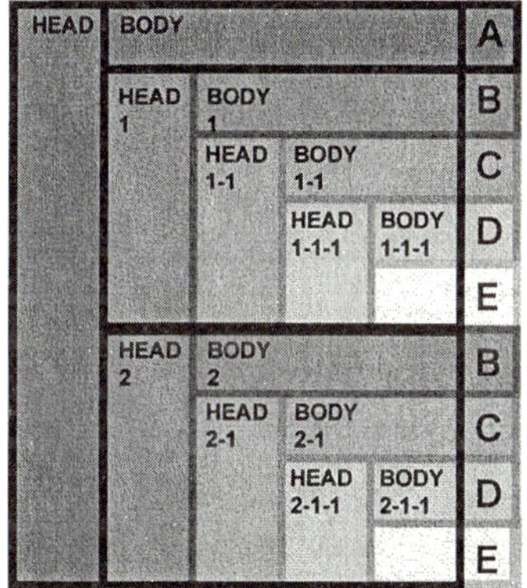

Fig. 3. Content structure

Fig. 4. An example of classification of objects to contents structure (A subject of analysis: Grid Systems in Graphic Design [3])

3.1.4 Building System's Knowledge-Base

In this process, analyzed layout methods are applied to the system's knowledge-base. The structure of this knowledge-base is classified to genre (book, magazine, newspaper, etc.) and content structure based on Fig.3.

3.2 The Structure of Editorial Design System

First, users only reconstruct information from various sources to create their own document. Subsequently, they choose prefer layouts provided by the proposed system and the system will layout the document automatically. After that, users can modify the outcomes produced by the system. The detail of work flow is clarified as below.

Table 2. The example of classification of clustering layout data (A subject of analysis: Grid Systems in Graphic Design [3])

	Cluster1	Cluster2	Cluster3	Cluster4
composition	Title(Level1) Text(Level1)	Title(Level2) Title(Level4) Image Text(Level2)	Title(Level2) Title(Level3) Title(Level4) Image Text(Level2)	Image
Grid				

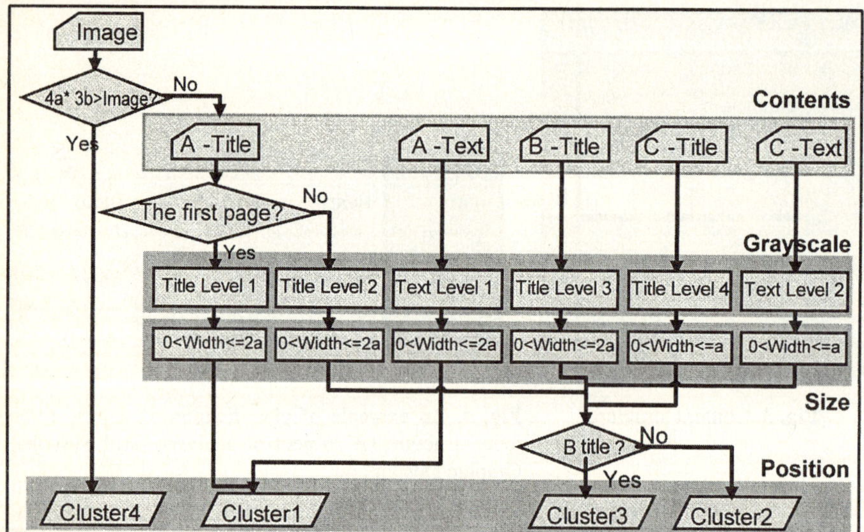

Fig. 5. The example of layout method (A subject of analysis: Grid Systems in Graphic Design [3])

3.2.1 Reconstruction of Contents and Documentization

In this process, users select and reconstruct contents from various sources of digital or analogue format and reconstruct contents of new documents. At this process, system displays data source and saves the selected contents, if the data source is analogue format, the selected contents are scanned (Fig.6). After that, users create the document from reconstruction of contents. In the detail of document, objects which have to put in to same page are grouped (Fig.7).

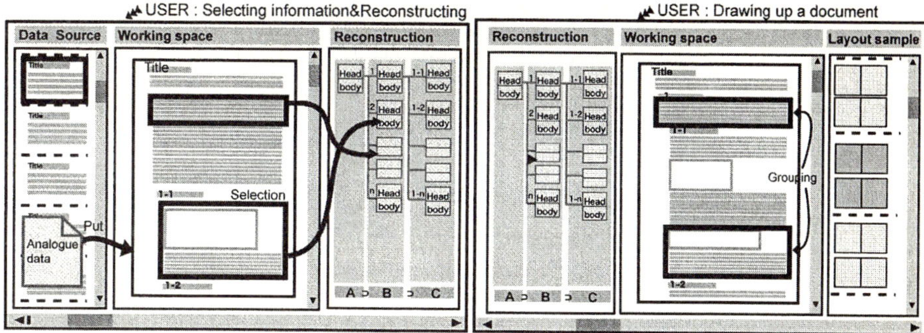

Fig. 6. Reconstruction of contents   **Fig. 7.** Documentization

3.2.2 Layout Sampling and Its Procedure

In this process, user will choose prefer layout provided by knowledge-base in the system, and the document will be separated to pages by considering the quantity of objects. Subsequently, layout of all documents will be automatically created based on selected layout samples. To be noticed that, our layout samples contained design knowledge. An example of creating layout process is shown as Fig.8.

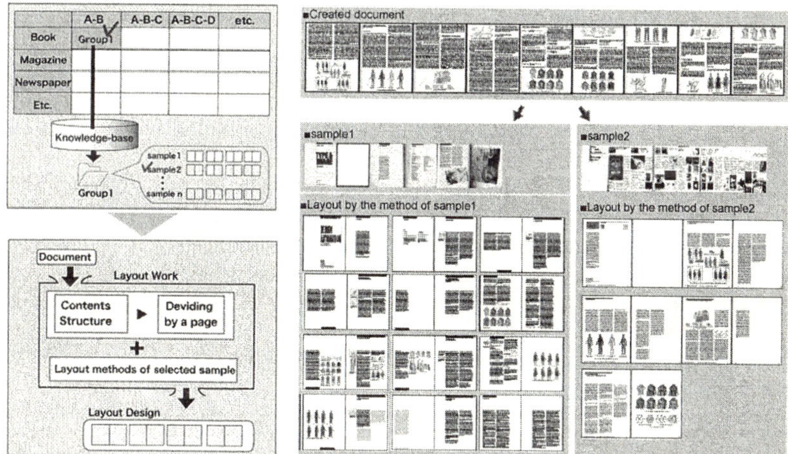

Fig. 8. Layout work flow and the example of Re-Layout

3.2.3 Modification of Layout

In this process, users can modify created layouts by editing, deleting or inserting the objects and their position in the layouts (Fig.9). When a layout is modified and the result influences to the layout of next page, influenced layout will be also automatically updated based on selected knowledge-base.

4 Conclusion

In this research, the tacit knowledge from expert designers and their experience are transformed to explicit knowledge. Subsequently, the results are applied to our system

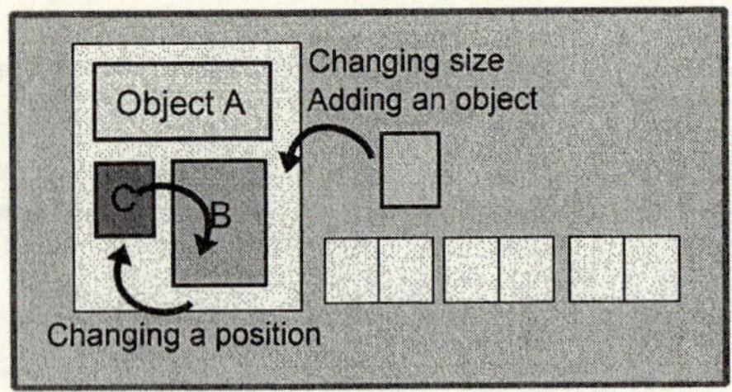

Fig. 9. Modification of layout

to build automatic layout design system which we call Editorial Design System (EDS). Moreover, users can also effectively modify the results created from EDS.

Our results show that not only the proposed system can reduce expense, processes and labors, but also increase design productions with constrain environments for design producers. Furthermore, even design and computer novices can perform expert design works, effortlessly. In addition, new expert design knowledge can be created in EDS, because new tacit knowledge from users can be added to transformed explicit knowledge by EDS.

References

1. D.C.L.Ngo, L.S.Teo, J.G.Byrne, "Modeling interface aesthetics", Information Sciences 152, Elsevier Science Publishers, pp 25-46, 2003.
2. Ikujiro Nonaka, Hirotaka Takeuchi, The Knowledge-Creating Company, Oxford University Press, 1995.
3. Josef Muller-Brockmann, Grid Systems in Graphic Design, Arthur Niggli, 1996.
4. Koffca. K., Principles of Gestalt Psychology, Routledge and Kegan, London, 1955.
5. Michael Polanyi, "Tacit Dimension", Peter Smith Publisher Inc, 1983.
6. Rudolf Arnheim, Art and Visual Perception: A Psychology of the Creative Eye, University of California Press, 1974.
7. Tom, M. Mitchell, Machine Learning, McGraw-Hill, Inc, 1997.
8. Usama M. Fayyad, Gregory Piatetsky-Shapiro, Padhraic Smyth, Ramasamy Uthurusamy (ed.), Advances in Knowledge Discovery and Data Mining, AAAI Press, 1996.

Knowledge Representation
for the Intelligent Legal Case Retrieval

Yiming Zeng[1], Ruili Wang[1], John Zeleznikow[2], and Elizabeth Kemp[1]

[1] Institute of Information Sciences and Technology, Massey University,
Palmerston North, New Zealand
{y.zeng,r.wang,e.kemp}@massey.ac.nz
[2] School of Information Systems, Victoria University, Melbourne MC,
Victoria, 8001, Australia
john.zeleznikow@vu.edu.au

Abstract. In this paper, we develop a knowledge representation model for the intelligent retrieval of legal cases, which provides effective legal case management. Examples are taken from the domain of accident compensation. A new set of sub-elements for legal case representation has been developed to extend the traditional representation elements of issues and factors. In our model, an issue may need to be further decomposed into sub-issues, and factors are categorized into pro-claimant, pro-responder and neutral factors. These extensions can effectively reveal the factual relevance between legal cases. Based on the knowledge representation model, we propose the IPN algorithm for intelligent legal case retrieval. Experiments and statistical analysis have been conducted to demonstrate the effectiveness of the proposed representation model and the IPN algorithm.

1 Introduction

In legal proceedings under the common law system, past legal cases (precedents) are frequently followed, analogized, distinguished or overruled for arguments and judicial opinions [1]. The reason for using precedents is because of the *open textured* nature of Law. Some commentators view open texture as the uncertainty about whether given legal terms match specific facts [2]. Lawyers and judges often use precedents in open textured domains, to see how legal terms will apply to the current case facts [3].

Past legal cases can be seen as sources of legal knowledge [4], because they contain the tacit knowledge about how open-textured legal terms were once applied. The process of using precedents for making arguments and giving judicial opinions employs the tacit knowledge contained in the precedents. Developing a knowledge representation model for the intelligent retrieval of legal cases can greatly facilitate the users' ability to correctly access tacit legal knowledge, which is an important goal of legal knowledge management [5].

The need to manage legal knowledge effectively for lawyers and judges to locate knowledge and information becomes urgent, due to the rapidly growing volume of the reported legal precedents [4]. The traditional retrieval and management methods for legal precedents, such as alphabetical indexing and keyword-based search, return many irrelevant cases or miss the relevant cases.

R. Khosla et al. (Eds.): KES 2005, LNAI 3681, pp. 339–345, 2005.

Case-based retrieval is an appropriate solution because of the nature of use of legal precedents in arguments and judicial opinions [6]. For legal tasks, which generally have a weak or no domain model [6], decision-makers are known to rely heavily on their knowledge of past cases, since verdicts must be consistent and transparent [2].

In Ashley's version of legal case-based reasoning [10], *issues* and *factors* are the basic elements to represent legal cases. Lawyers frequently make arguments by analyzing and interpreting the similarities and differences between cases [7]. Successful arguments depend on whether the cited precedents are factually relevant to the current case, i.e., share issues and factors with the current case.

We shall extend Ashley's work by developing a set of sub-elements (sub-issues, pro-claimant factors, pro-responder factors and neutral factors) that better reveal factual relevance between cases. In this paper, we focus upon the domain of accident compensation. We believe this representation model has the potential to be used in other areas.

The remainder of the paper is organized as follows. In Section 2, we will introduce the new representation sub-elements. In Section 3, we will illustrate a new knowledge representation model and IPN algorithm for the retrieval of legal precedents. In Section 4, we will examine our intelligent case-based retrieval system, and demonstrate the effectiveness of the algorithm. Finally, in Section 5, we will provide a summary of our research and indicate our future directions.

2 Sub-elements for Case Representation

2.1 Issues and Sub-issues

Legal issues rather than factual issues are often the focus of the disputes between parties. Even simple legal rules generally contain some open texture [5]. For example, the use of the apparently simple term *"personal injury"* can generate conflicts as to whether a mental injury belongs to the class *"personal injury"*. Lawyers identify the issues in dispute, identify the relevant precedents for these issues, and organize their arguments according to these issues [8]. Courts explain their decisions according to the disputed issues [8].

Many issues focus on the open textured nature of law. For example, an issue that frequently appears in the context of accident compensation is "whether the victims suffered from the personal injury caused by the accident". Logically, this compound issue also has several subordinate issues, such as "whether the accident was unexpected", and "whether the accident was identifiable". To precisely retrieve the factually relevant precedents, the compound issues can be necessarily decomposed into more refined sub-issues. Fig. 1 shows the sub-issues of the compound issue "Personal_Injury_By_Accident".

2.2 Factor Classification

A factor is abstract knowledge related to a stereotypical fact pattern of a case [8]. In the HYPO and CATO models [7, 8], the factors are classified into *pro-plaintiff* (pro-p) and *pro-defendant* (pro-d), favorable to plaintiff and defendant respectively. The pro-p and pro-d factors support one side and potentially influence the outcome of the case. A case is represented simply as a set of applicable factors [11].

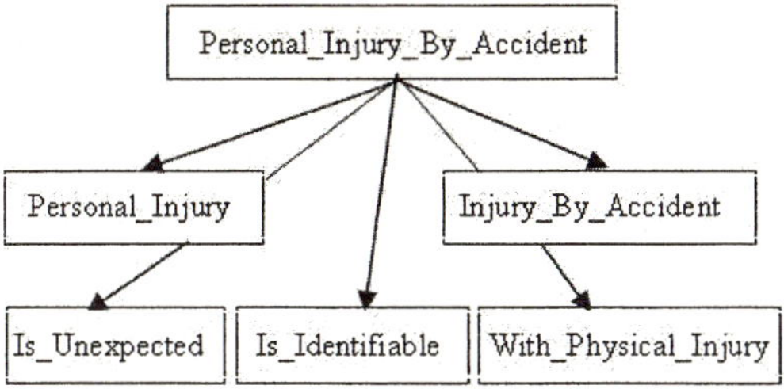

Fig. 1. Sub-issues of "Personal_Injury_By_Accident". These sub-issues focus on more detailed open texture and dispute, and facilitate the retrieval of the relevant precedents

The classification of factors in a case into pro-p and pro-d is argument-oriented. It facilitates the organization of arguments for the new case. However, it is less knowledge representation-oriented [11]. Also, in practice, the roles of one side as plaintiff or defendant are not fixed. An applicable factor with the same underlying fact can be pro-p in one case and pro-d in another one. The lack of a uniform predefined standard to classify factors in precedents makes case retrieval more difficult.

We propose a scheme to classify all factors into three categories to represent and retrieve cases: *pro-claimant* (pro-c), *pro-responder* (pro-r), and *neutral* factors, in order to give a uniform predefined standard to label the factors and at the same time to keep the argument-oriented characters of the pro-p and pro-d factors.

In an accident compensation case, the *claimant* is the *VICTIM* who makes a claim for damages, and the *responder* is possibly liable for the accident or damage. The *pro-c* and *pro-r* factors are those important facts favorable to the claimant and responder respectively. Similar with pro-p and pro-d factors, whether an applicable factor is pro-c or pro-r depends on the underlying facts (data), i.e., evaluating the evidence. For example, the factor *speeding* in traffic infringement law does not support either the driver or the police. It depends on the underlying data. However, the advantage to categorize the factors as pro-c/r rather than pro-p/d is that: once both the past case and current case share a factor with the same label as pro-claimant (or pro-responder), we can easily know that these two cases share this factor, no matter the victims in both cases are both plaintiffs, both defendants, or plaintiff and defendant respectively.

Neutral factors are those important facts neutral to both parties. However, these factors strongly characterize certain features of a case, and potentially imply the appearance of certain legal issues. The neutral factors are different from the pro-c or pro-r factors that are favorable to one side or the other. From the viewpoint of argument, neutral factors can be ignored. Thus, traditionally all factors are classified positive or negative to one side and there is no such concept of neutral factors. However, from the viewpoint of knowledge management, neutral factors should not be ignored as they can provide some additional information that is useful for case retrieval. Neutral factors are not surface features, but factual and contextual knowledge related to the evidence. Neutral factors strongly characterize certain kind of cases and potentially imply the presence of certain legal issues. Therefore, neutral factors are useful for case retrieval. For example, Fig. 2 shows the pro-c factors, pro-r factors, neutral factors and issues in a case *Re Chase* (Court of Appeal Wellington, [1989] 1 NZLR 325).

Admittedly neutral decisions are rarely given in litigated cases, which is one of the reasons disputants often prefer Alternative Dispute Resolution Techniques. In some commercial and family law cases, the plaintiff may win on some issues but not on others. In any criminal case, the onus is on the crown to prove the case beyond reason-able doubt. In taxation cases, the onus is on the taxpayer to prove her claim is legitimate. Nevertheless, we believe that some factors in a legal case support neither the claimant nor the responder. Hence we have introduced the concept of a neutral factor.

> *Re Chase*
> The <u>police</u> [**neutral: role of responder**] armed offenders squad <u>executed a search warrant</u> [**pro-r**] at a flat occupied by a gang member… In the course of executing the search warrant an armed offenders squad <u>shot and killed Chase apparently but mistakenly</u> [**pro-c**] <u>in reasonable self-defence</u> [**pro-c**]… The proceeding alleged <u>assault, battery, negligence and trespass to property</u> [**neutral: injury reason**]… The pleading expressly disavowed <u>a claim for damages</u> [**neutral: claim type**] arising <u>directly or indirectly out of personal injury by accident</u> [**Issue**]…

Fig. 2. The pro-c, pro-r, and neutral factors in *Re Chase*. The neutral factor *Claim Type* (CT) refers to the type of compensation the claimants claim for. For example, those cases with the "backdated payments" as the value of CT will probably have a common issue, e.g., whether the delay makes the claim still time effective. Therefore, these cases are considered relevant

3   Representation Model and Algorithm

The purpose of our model for intelligent legal case retrieval is to effectively represent and precisely position knowledge and information contained in the precedents; based on the concept of introduced sub-elements. Fig. 3 shows the four levels of the model. At the top is the source level containing the plain-text legal cases. The second level is the representation level, where the representation elements, such as issues, sub-issues, pro-c, pro-r and neutral factors, are drawn from each precedent and the new case. The third level is the repository level, where the legal precedents are stored in a case base according to the abstract representation. On the bottom is the output level, where the most factually relevant precedents are captured with a retrieval algorithm.

It is impossible to find a 100% similar precedent with the new case. However, those precedents sharing one or more representation elements with the new case are desired as they are potentially useful for making legal arguments. We propose the following algorithm to retrieve those relevant precedents.

- First, those precedents sharing more issues are considered as more factually relevant. The precedents sharing the issue(s) with the new case will be checked further to determine how many sub-issues are shared with the new case. We use I_Score to count the number of shared issues and sub-issues between a precedent and the new case. When I_Score is calculated, issues are treated differently depending on whether the issues have sub-issues or not. We calculate I_Score by adding 1 point for each shared issue that has no sub-issue; when a shared issue has sub-issues, adding 1 point for each shared sub-issue; otherwise if no shared sub-issue between the precedent and the new case, adding only 0.5 point for the shared issue.

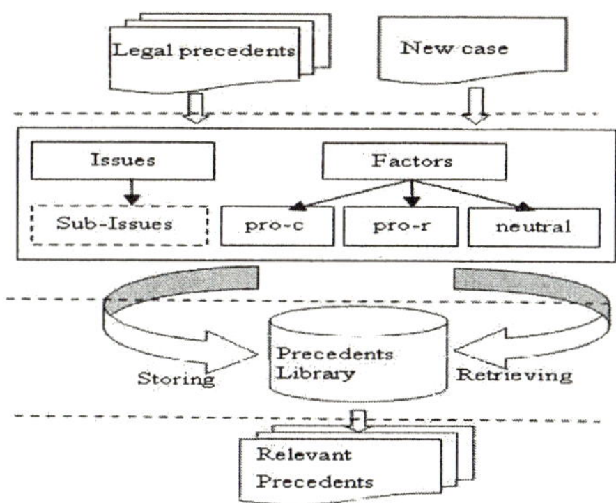

Fig. 3. The knowledge representation model

- Second, within the precedents retrieved in the first step, those that share more pro-c and pro-r factors with the new case are more convincing due to their factual relevance. We use P_Score to count the number of shared pro-c and pro-r factors. We calculate P_Score by adding 1 point when a precedent shares a same pro-c or pro-r factor with the new case; otherwise adding 0 points.
- Finally, within the precedents retrieved in the first two steps, those that share more neutral factors with the new case will demonstrate additional conviction that the precedent has the similar context with the new case. We use N_Score to count the number of shared neutral factors. We calculate N_Score by adding 1 point when a precedent shares a same neutral factor with the new case; otherwise adding 0 points.

We rank the precedents in the order of three scores: I_Score, P_Score, and N_Score. For simplicity, we refer to this algorithm as the *IPN* algorithm. Note that whether a precedent will be selected also depends on other considerations like time and citation frequency. The algorithm intends to locate the most useful cases, rather asking experts to spend their time performing this tedious task. Also the abstracted representation elements can provide additional knowledge for users to explain why a precedent is considered factually relevant. Of course the N_Score is a much weaker indicator of significance than either the I_Score or the P_Score.

4 Experiments and Discussion

4.1 An Example in the Domain of Accident Compensation

We have built a legal case base of 152 accident compensation cases reported by Court of Appeal or High Court of New Zealand since 1958. Here we give an example of how the IPN algorithm works. When a new case arises, users will study the fact of the case and extract the factors and the possible issues and sub-issues. For experiments, we use the already reported legal cases as "new" cases to test whether the retrieved precedents with the IPN algorithm have actually been used in the reported case. Take the case *Harrild v Director of Proceedings* (Court of Appeal Wellington, [2003] 3 NZLR 289) for example. The scores returned by IPN algorithm are shown in Table 1 (Those retrieved cases with I_Score less than 3 are ignored).

From the scores, we can see the most relevant precedents are Cases 105, 114 and 111. Actually, Case 114 was used in *Harrild*. The reason why Case 111 and 105 were not used is that Case 114 is the most recent and both Case 111 and 105 were already used in Case 114.

Table 1. The relevant precedents for *Harrild*

Case	I_Score	P_Score	N_Score
105	*3.5*	*0*	*1*
114	*3*	*1*	*0*
111	*3*	*1*	*0*

4.2 Statistical Analysis

We first examine the cover rate, i.e., the ratio of retrieved citations to total occurred citations. For example, when a case cites another two cases, *two* citations appear. Thus, altogether *109* citations appear in the 152 cases. IPN retrieves *105* of them. The cover rate is (*105/109* =) 96.3%. This means almost all useful precedents can be retrieved by IPN. Reasons for the omission of four cases are: two cases are just mentioned as exceptions which should not be followed; and the other two are mentioned in *obiter dicta* (things said by judges optionally [9]) and too flexible to predicate.

A high cover rate only is not enough. We also examine how precise IPN is by testing the rank of the retrieved 105 citations. For example, Case 114 is cited in *Harrild* and is second in the ranking list (see Table 1). Thus, the percentage of citations which are in the first, second, third and fourth place in the ranking list are *43.8%, 30.5%, 11.4%, 10.5%*, respectively. The total percentage for citations within top four is (*43.8% + 30.5% + 11.4%+ 10.5%* =) *96.2%*. This means generally users can find useful cases within the top four retrieved cases. We can see that the retrieval system can greatly free users from the burdensome searching on precedents.

5 Summary

A knowledge representation model for effective and precise legal case retrieval is needed to aid the everyday work of the lawyers and judges. Based on existing representation elements, we have proposed a set of sub-elements to better reveal factual relevance between cases. First, we have introduced the concept of sub-issues. Second, we have adapted the HYPO-style factors to our pro-claimant and pro-responder factors. Also, we have introduced a new category of factors, neutral factors. Based on this set of sub-elements, we have proposed a new model for legal precedent retrieval and the IPN retrieval algorithm to rank the precedents. Experiments and statistical analysis have shown that the model and IPN algorithm can achieve satisfying results.

Future research might focus upon giving weightings to each of the issues used in I_Score and P_Score so that these two measures can give a better measure of the similarity between the retrieved precedent and the new case. This will lead to an improved IPN algorithm. We will also conduct a rigorous evaluation of the system constructed, including user evaluation surveys.

References

1. Prakken, H., Sartor, G.: Modelling Reasoning with Precedents in a Formal Dialogue Game. Artificial Intelligence and Law. 6 (1998) 231-287
2. Stein, G.C.: Common-Sense Reasoning about Beliefs. Ph.D. Thesis. New Mexico State University (1996)
3. Porter, B.W., Bareiss, E.R., Holte, R.C.: Concept Learning and Heuristic Classification in Week-theory Domains. Artificial Intelligence. 45 (1990) 1-2
4. Oskamp, A., Tragter, M.W., Lodder, A.R.: Mutual Benefits for AI & Law and Knowledge Management. In Proceedings of the 7th International Conference on Artificial Intelligence and Law (ICAIL-99). (1999) 126-127
5. Zeleznikow, J.: Using an Argumentation Based Approach to Manage Legal Knowledge. in D. G. Schwartz (ed) Encyclopedia of Knowledge Management, Idea Group Inc. Hershey PA. (2005) in press
6. Zeleznikow, J., Stranieri, A., Hunter, D.: Beyond Rule Based Reasoning – the Meaning and Use of Cases. In Proceedings of the 11th Conference on Artificial Intelligence for Applications. (1995)
7. Ashley, K.D., Rissland, E.L.: Waiting on Weighting: A Symbolic Least Commitment Approach. In Proceedings of AAAI-88, St. Paul, MN. AAAI Press/MIT Press, Cambridge, MA. (1988) 239–244
8. Aleven, V.: Using Background Knowledge in Case-based Legal Reasoning: A Computational Model and an Intelligent Learning Environment. Artificial Intelligence. 150 (2003) 183–237
9. Waddams, S.M.: Introduction to the Study of Law. Carswell/ Thomson Professional Publishing, Ontario (1992) 77-84
10. Ashley, K. D.: Case-based Reasoning and Its Implications for Legal Expert Systems. Artificial Intelligence and Law. 1(2) (1992) 113-208
11. Rissland, E.L., Ashley, K.: A Note on Dimensions and Factors. Artificial Intelligence and Law. 10 (2002) 65-77

An OCR Post-processing Approach Based on Multi-knowledge

Li Zhuang and Xiaoyan Zhu

Department of Computer Science and Technology, Tsinghua University
State Key Laboratory of Intelligent Technology and Systems
Beijing, 100084, P.R. China
zhuangli98@mails.tsinghua.edu.cn, zxy-dcs@tsinghua.edu.cn

Abstract. This paper proposes an OCR post-processing approach based on multi-knowledge, which integrates language knowledge and candidate distance information given by the OCR engine. In this approach, statistical language model and semantic lexicon are combined, and candidate distance information is used to reduce the size of the search space. The experimental results show that this approach is very effective. After post-processing, the recognition accuracy rate on the test set increases from 58.45% to 83.73%, which means 60.84% error reduction.

1   Introduction

In most OCR systems, independent character recognition engine is often used to recognize each segmented part of an image, where only shape and structure of the character are considered. In order to improve the recognition accuracy rate, it is necessary in post-processing to use language knowledge, which introduces context information, to correct the image recognition results. Post-processing approaches based on language knowledge include using a lexicon [1][2] or some syntax and semantic rules [3] to correct the spelling of words, and using some statistical language models (SLM) [4][5] to select out the best sequence from the candidate characters given by the OCR engine. Because of the complexity of language, all kinds of language knowledge sometimes are used together to obtain better performance [6].

An OCR engine outputs not only candidate characters, but also candidate distance information of each candidate character, which is also important in OCR post-processing. Currently, candidate distance is usually transformed to reliability of the corresponding candidate character to be utilized. Generally speaking, the bigger the reliability of a candidate character, the smaller the corresponding candidate distance. In early period, the reliability was calculated by using some empirical formulas [7][8]. Afterwards, a statistical approach was proposed [9], which calculates the reliability according to the distribution of candidate characters and correct characters with different candidate distances. It reflects some statistical characteristics, and its complexity is low, therefore it achieves good results in some applications. However, the use of candidate distance is still limited in OCR post-processing.

R. Khosla et al. (Eds.): KES 2005, LNAI 3681, pp. 346–352, 2005.

This paper proposes an OCR post-processing approach based on multiknowledge, which integrates language knowledge and candidate distance information. In this approach, statistical language model and semantic lexicon are combined, and candidate distance information is used to reduce the size of search space. The experimental results show that this approach is very effective. After post-processing, the recognition accuracy rate on the test set increases from 58.45% to 83.73%, which means 60.84% error reduction.

This paper is organized as follows. In Section 2, the multi-knowledge based approach that integrates language knowledge and candidate distance information is introduced. The experimental results are given in Section 3. In Section 4, some conclusions and future works are given.

2　The Multi-knowledge Based Approach

In the proposed approach, first the candidate distance information is used to construct the search space, where only the candidates whose distances satisfy a specific condition are added into the space; and then, statistical language model and semantic lexicon are used on the search space, with Viterbi algorithm to find the best sequence as the result as usual. In the following, statistical language model, semantic lexicon and use of candidate distance information will be introduced respectively.

2.1　Statistical Language Model

The most widely used statistical language model in OCR post-processing is n-gram model, which supposes that the occurrence probability of a word w_i is only related with the previous $n - 1$ words, i.e.

$$P(w_i|w_1^{i-1}) = P(w_i|w_{i-n+1}^{i-1}) \tag{1}$$

There are some other models based on n-gram [10], one of which is distance-m n-gram model [11]. This model is the same as n-gram except using the history information with an interval of $m - 1$ to the word w_i, i.e.

$$P(w_i|w_1^{i-1}) = P(w_i|w_{i-m-n+2}^{i-m}) \tag{2}$$

Distance-m n-gram model can utilize long-distance history information. However, it is reported that its precision decreases quickly along with the increase of the value of m [11]. Therefore, it is often combined with conventional n-gram model to maintain the performance as well as to use more contextual information.

In our experiment, six different models were used, including bigram, trigram, bigram+trigram, bigram+distance-2 bigram, trigram+distance-2 trigram and bigram+trigram+distance-2 bigram+distance-2 trigram. The last four models are combined via linear interpolation. For example, trigram+distance-2 trigram model can be described as

$$P(w_i|w_{i-3}w_{i-2}w_{i-1}) = \lambda_1 P_1(w_i|w_{i-2}w_{i-1}) + \lambda_2 P_2(w_i|w_{i-3}w_{i-2}) \tag{3}$$

where $P_1(w_i|w_{i-2}w_{i-1})$ and $P_2(w_i|w_{i-3}w_{i-2})$ correspond to the trigram model and the distance-2 trigram model respectively, and λ_1, λ_2 are the weight parameters which satisfy $\lambda_1 + \lambda_2 = 1$.

2.2   Semantic Lexicon

Many applications in natural language processing require extensive common language knowledge. Recently, there are many attempts to overcome the lack of available large knowledge bases by using semantic lexicon (SL). The most famous semantic lexicon is WordNet [12], which was developed by Princeton University. In WordNet, entries are sets of synonyms, called synsets, each of which represents a concept. Between the concepts there are semantic relations. For example, the most common relation between nouns is the ISA (is-a) relation that organizes the noun synsets into hierarchies.

Similar semantic lexicons for other languages have been developing by many countries, such as EuroWordNet for Italian, Spanish etc [13], KoreaNet for Korean [14], and HowNet [15] for Chinese. In our experiment, a sub-lexicon of HowNet was used, which describes relations between nouns.

2.3   Use of Candidate Distance Information

The analysis of distribution of candidate distance shows that it is almost impossible for a candidate to be the correct character when its candidate distance is too much bigger than the distance of the corresponding first candidate. On the other hand, the first candidate distance reflects the reliability of the first candidate character. The bigger the first candidate distance, the more possible that the correct character is not the first candidate, and the distance difference between the correct character and the first candidate may be more.

Based on the above analysis, an approach to reduce the size of search space by using the candidate distance information can be proposed. For the candidates of the same character, the first candidate distance d_1 is used to obtain the threshold $th = f(d_1)$, where $f(d)$ is called threshold function. If and only if another candidate distance d_i satisfies the condition $d_i - d_1 < th$, the corresponding candidate is added into the search space. Here the threshold function can be selected differently, but it should be an increment function so that when the first candidate distance increases, the number of candidate characters added into the search space may increase, which accords with the analysis proposed previously.

Generally speaking, the size of the reduced space is much smaller than that of the original space that contains all the candidates given by the OCR engine. Therefore it can speed up the search process significantly.

Figure 1 is an example of Chinese text, where each line contains six candidates for the same character and the number following every candidate is the corresponding candidate distance. When the threshold function is selected as $f(d) = 0.1d$, only underlined candidates are in the reduced space, while all the candidates are in the original space. The difference between them is obvious.

河 385.86 向 421.72 何 428.82 白 433.18 角 434.86 份 441.87

北 312.53 业 325.92 兆 375.34 扎 383.92 址 385.00 壮 386.47

省 294.28 有 331.72 荷 342.91 肩 353.12 雀 354.71 备 364.01

$$\vdots$$

金 250.52 全 271.03 釜 305.59 盆 324.36 重 329.74 皇 331.51

西 258.11 酉 286.36 两 291.86 酋 338.89 面 363.00 丙 365.83

街 332.51 衔 341.73 衙 356.18 做 358.91 御 376.96 衍 381.40

Fig. 1. Comparison of the reduced space and the original space.

3   Experiment

3.1   Data

We put the multi-knowledge based approach into Chinese address OCR post-processing. A training corpus containing 196,009 Chinese addresses is used to train the character-based n-gram and distance-2 n-gram models. Another corpus containing 20,000 Chinese addresses is used as a held-out data to optimize the weight parameters of linear interpolation language models. The test set is a corpus containing 15,000 handwritten Chinese addresses. After OCR, ten candidates for each character and the corresponding candidate distances are given. The evaluation criterion is the recognition accuracy rate of whole address, and the baseline on the test set is the recognition accuracy rate when results consist of the first candidates, which is 58.45% (8768 correct addresses).

3.2   Results

The results are shown in Table 1, where B, T, DB and DT denote bigram, trigram, distance-2 bigram and distance-2 trigram respectively, and the threshold function was selected as $f(d) = 0.1d$.

Table 1. The results with different n-gram types.

N-gram type	SLM with original space	SLM with reduce space	SLM-SL with original space	SLM-SL with reduce space
B	54.87%	70.49%	61.33%	79.64%
T	68.17%	**73.92%**	75.91%	83.13%
B+T	67.73%	73.63%	75.80%	82.97%
B+DB	55.07%	70.37%	62.41%	80.51%
T+DT	68.23%	73.23%	77.29%	83.71%
B+T+DB+DT	**68.23%**	73.25%	**77.29%**	**83.73%**

The advantages of the multi-knowledge based approach are shown from two sides. Firstly, the comparison between column 2 and column 4, or column 3 and column 5 in Table 1 shows that for every SLM, the recognition accuracy rate with semantic lexicon used is higher than that without semantic lexicon used. The reason is that the semantic lexicon introduces more effective language knowledge. Especially, APO (a-part-of) relation between nouns is mainly used here. In Chinese, there is such a relation between place-names in an address (a latter place-name is a part of a former place-name in geographical meaning). The relation can help to correct the search result very effectively, so that the recognition accuracy rate increases. Secondly, the comparison between column 2 and column 3, or column 4 and column 5 in Table 1 shows that in every instance, the recognition accuracy rate on the reduced space is higher than that on the original space. The fact indicates that the search space can influence the performance of a language model significantly. Because there is only one or no correct candidate for each character, if all the candidates are used, there must be lots of wrong candidates in the search space. The effect of them is like much noise in speech recognition, and these wrong candidates will probably lead the search along a wrong direction. Among the language models we used, bigram model and bigram+distance-2 bigram model are the most imprecise, so they are influenced badly. When using SLM only, the results of them are even worse than the baseline. However, when using candidate distance to reduce the size of search space, many wrong candidates are deleted, thus the "noise" in recognition decreases, and the recognition accuracy rate is improved clearly.

In addition, use of the reduced space can speed up the post-processing significantly. For example, with trigram model, when using the original space, 6 hours is needed for the test set on a computer with 3.2GHz CPU, i.e., 1.44 seconds for per address in average. While using the reduced space, only 3 minutes is needed on the same computer, i.e., 0.012 seconds for per address in average. The speed is improved by 120 times or so.

From the above results, it can be seen that the multi-knowledge based approach (column 5 in Table 1) integrates the advantages of using both semantic lexicon and reduced search space, so that it is very effective. On the test set, the best recognition accuracy rate increases from 58.45% to 83.73%, which means 60.84% error reduction.

3.3　Discussion

When using candidate distance to reduce the size of search space, it can be noticed obviously that different threshold function can make different search space and then influence the result of post-processing. Here we used positive proportion function $f(d) = ad$ as the threshold function, and observed the results on the same test set with changing the proportion coefficient a. The results when using trigram and semantic lexicon are shown in Figure 2.

From Figure 2, it can be seen clearly that along with the increase of the proportion coefficient a, the recognition accuracy rate first increases, and then decreases. When $a = 0.1$, the best recognition accuracy rate 83.13% is achieved. The reason is that along with the increase of the proportion coefficient, more and more candidates are added into the search space. At the beginning, the candidates added can bring correct characters, so that the chance of selecting

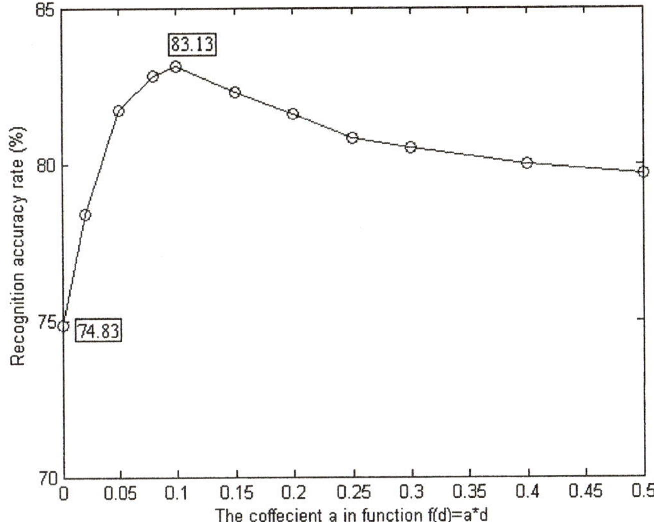

Fig. 2. Recognition accuracy rate to different proportion coefficient a.

out the correct result increases, so does the recognition accuracy rate. However, after adding an amount of candidates, the "noise" in the search space increases quickly, which results in the decrease of the recognition accuracy rate.

Besides the influence of different threshold function, it also can be noticed that the precision of the reduced space depends on the original image recognition result. If the precision of the original result is poor, most correct characters are in the back of the candidate list, and the distance difference between correct character and the first candidate may be much. Thus many correct candidates will not be in the reduced space, and the post-processing result may be very poor. Therefore, this approach cannot be used for a bad OCR engine. Fortunately, most applied OCR engines are good enough to satisfy the requirement of this approach, which provides good condition for using it widely.

4   Conclusion and Future Work

This paper proposes an OCR post-processing approach based on multi-knowledge, in which statistical language model and semantic lexicon are combined, and candidate distance information is used to reduce the size of search space. In this approach, more semantic information and a smaller and more precise search space are used, which makes it very effective. In an application of Chinese address OCR post-processing, the recognition accuracy rate on the test set increases from 58.45% to 83.73%, which means 60.84% error reduction.

There are two key points in this approach: the language model and the threshold function. In this paper, only the simplest forms of them are discussed. In the future, more research about them will be carried out, including using some more precise language models, and looking for a proper function according to some analysis of original recognition results. In addition, we will test the approach in processing some general contents.

Acknowledgements

The authors are thankful to Fujitsu Laboratories Ltd. for providing the experimental data and partly supporting the work. Moreover, this work is supported by the Natural Science Foundation of China (Grants No. 60272019 and 60321002).

References

1. Wimmer Z., Dorizzi B., "Dictionary preselection in a neuro-Markovian word recognition system", Proceedings of the 5th International Conference on Document Analysis and Recognition, Bangalore, India, 1999: pp.539-542.
2. Procter S., Illingworth J., "Mokhtarian F. Cursive handwriting recognition using hidden Markov models and a lexicon-driven level building algorithm", Proceedings of the IEE Vision, Image and Signal Processing 2000, 147(4): pp.332-339.
3. Marti U., Bunke H., "A full English sentence database for off-line handwriting recognition", Proceedings of the 5th International Conference on Document Analysis and Recognition, Bangalore, India, 1999: pp.705-708.
4. Brakensiek A., Willett D., Rigoll G., "Unlimited vocabulary script recognition using character n-grams", Proceedings of the 22nd DAGM Symposium, Tagungsband Springer-Verlag, Kiel, Germany, 2000: pp.436-443.
5. Zhuang L., Bao T., Zhu X.Y., "A Chinese OCR spelling check approach based on statistical language models", Proceedings of the IEEE International Conference on System, Man and Cybernetics, Hague, Netherlands, 2004: pp.4727-4732.
6. Golding A.R., Schabes Y., "Combining trigram-based and feature-based methods for context-sensitive spelling correction", Proceedings of the 34th Annual Meeting of the Association for Computational Linguistics, Santa Cruz, CA, 1996: pp.71-78.
7. Lee H.J., Tung C.H., Chang Chien C.H., "A Markov language model in Chinese text recognition", Proceedings of the 2nd International Conference on Document Analysis and Recognition, Tsukuba, Japan, 1993: pp.72-75.
8. Tung C.H., Lee H.J., "Increasing character recognition accuracy by detection and correction of erroneously identified characters", Pattern Recognition, 1994, 27(9): pp.1259-1266.
9. Li Y.X., Zhu X.Y., "Post-processing for handwritten Chinese address recognition", Proceedings of the International Conference on Intelligent Information Technology, Beijing, China, 2002.
10. Goodman J.T., "A bit of progress in language modeling", Computer Speech and Language, 2001, 15(4): pp.403-434.
11. Rosenfeld R., "A maximum entropy approach to adaptive statistical language modeling", Computer Speech and Language, 1996, 10(3): pp.187-228.
12. http://wordnet.princeton.edu/
13. http://www.dcs.shef.ac.uk/research/groups/nlp/funded/eurowordnet.html
14. Moon Y., "Design and implementation of WordNet for Korean nouns", Journal of the Korea Information Science Society, 1996, 2(4): pp.437-445.
15. http://www.keenage.com/

Data Mining in Parametric Product Catalogs

Lingrui Liao and Tianyuan Xiao

Department of Automation, Tsinghua University,
100084 Beijing, P.R. China
llr98@mails.tsinghua.edu.cn, xty-dau@tsinghua.edu.cn

Abstract. Parametric product catalogs provide unique context for data mining techniques. The specific user access mode is analyzed. The availability of structured content data is taken into consideration. Three mining scenarios (category association, feature evaluation and product recommendation) are discussed. Related mining techniques are introduced and adjusted to the context. A system framework is put forward to show how these techniques can support each other efficiently to build a more integrated and intelligent system.

1 Introduction

WWW has become a major vehicle for accessing online information in the Internet era. Through web browsers, parametric product catalogs are bringing the power of enterprise information systems to users around the world. This type of system is usually built on a structured data repository. How numerous contents in these catalogs can be made more accessible using data mining techniques is the focus of this article.

1.1 Defining P.P.C.

Namely, a parametric product catalog supports parametric search, which allows users to specify the product features and corresponding values they are interested in. The catalog system is responsible for retrieving available products that meet these criteria.[1]. Since products in different categories have different features, users need to choose a category at first to find out applicable features for specifying. Due to the great number of categories that may exist in a catalog, product taxonomy is provided to users in the form of category hierarchy, as such as at www.globalspec.com.

Although parametric search can narrow down the search space efficiently, there are still improvements to be considered. First, how can a user quickly find out the right category to search in? Secondly, which features really make differences among the products in the chosen category? Further, how to find related products without explicitly search for them? For the first question, a peer-category relationship is desirable. This knowledge of how categories are pragmatically related may help users jump to "also-wanted" categories directly.

R. Khosla et al. (Eds.): KES 2005, LNAI 3681, pp. 353–358, 2005.

This relationship can be mined from previous user behaviors. We call it **category association**. For the second one, **feature evaluation** is a quantitative description of how discriminative a feature is in a chosen category. It provides helpful hints when users are organizing search criteria. For the last one, **product recommendation** is a solution. Automated recommendation is no so novel, but there are interesting differences in the context of parametric catalogs.

As we can see, data mining techniques can bring intelligent improvements to a parametric product catalog. Following paragraphs will further explain the ideas and system framework. Before that, we shall see what kinds of data are available.

1.2   Relevant Data

In a P.P.C., product is described in terms of category-dependent features. A product in category C is described as a vector of feature values:

$$P^i = \langle f_1^i, f_2^i, \cdots, f_{n_C}^i \rangle \tag{1}$$

n_C is the number of applicable features for C, f_j^i is value description for the jth feature. Feature descriptions can be either numerical or literal. These structured data make products in the same category easily comparable, which is only true for P.P.C.

Besides content data, user access history is also a useful data source. Extracted from server logs, access behaviors can be divided into three phases: **category navigation**, **criteria organization** and **results browsing**. Category navigation is the process of choosing categories to search in. Possible actions include choosing a subcategory, returning to the parent category and selecting a category to query etc. In criteria organization, users specify features and corresponding values to filter products in a chosen category. A query in category C can be formalized as:

$$Q = \langle C, (o_1, f_{j_1}), (o_2, f_{j_2}), \cdots, (o_m, f_{j_m}) \rangle \tag{2}$$

$o_{1 \cdots m}$ are boolean operators on feature values. For numerical values the operators could be "=", ">", etc. For literal ones they could be "contains", "exactly match". Criteria organization is an iterating interaction as users may refine criteria based on retrieved results. Since most catalogs display search results in a list&details mode, result browsing means which products in a result list are "seen for details" by the user.

2   Mining Scenarios

2.1   Category Association

As mentioned earlier, category association is meant to find sets of categories that are not directly connected in the category hierarchy but have certain pragmatic relations. Clearly, previous category navigation behavior is a source for this knowledge.

In [2], 3 kinds of navigational actions are defined; "Predefined Interaction Sequences" are given to determine whether restructuring may improve the catalog's organization. If some PIS's frequency and relevance have reach the thresholds given by the administrator, corresponding restructuring operations will take place. [2]'s work gives clues on how to extract category relationship from the navigation behaviors. However, detailed navigational actions are not always observable. For example, the category hierarchy may be shown in one page as a DHTML multi-level menu. In this case, "referer" extracted from HTTP head is of little use to determine a user's category navigation path. So the only reliable observation is categories the user has selected to query. Seeing this, we turn to consider association rule mining.

Association rule mining was used to generate page-wised recommendation in the work of [3]. Borrowing this idea to the context of category association, queried categories comprise the itemset of a user session (transaction). Using the last w queried categories in current session as the condition part, association rules with better confidence are chosen to generate category relationships. With associate rules pre-computed, recommendation engine can infer related categories on-line. Since items (categories) have an inherent taxonomy in a P.P.C., generalized associate rule mining[4] is promising to improve flexibility for category recommendation.

2.2   Feature Evaluation

When a user searches in a chosen category, he will be prompted with applicable features. Unfortunately, not all features are good as criteria. Some features have long spans of possible value; give users a hard time adjusting criteria before a result list of desirable length can be retrieved. To address this problem, system in [1] gives marketing people the control on whether each feature is displayed on the search page. Cardinality of feature value is the major basis for their decision. In our opinion, more informative hints should be given to users who actually do the search.

There are two methods to reach this goal. First, content mining techniques can be used to give an overview of the feature value's distribution. For example, with a histogram of numeric feature values, users can estimate the length of result list before submitting the query. Secondly, "criteria organization" behaviors can be recorded and benefit users in the future. When a user stops submitting queries and begins to check product details, it stands a good chance that the last query generated a desirable result. If these successful queries are summarized, evaluations could be given on features' usability as search criteria. For instance, the relative frequency of a feature's occurrence in successful queries:

$$Ev(f_j) = \frac{|\{Q_s : \exists o, s.t.(o, f_j) \in Q_s\}|}{|\{Q_s : C \in Q_s\}|} \tag{3}$$

These evaluations not only suggest features' discrimination, but also imply their importance to the users' practical needs. Therefore, besides to be helpful hints in criteria organization, results of feature evaluation are also useful in other mining scenarios, as will be shown in the following section.

2.3   Product Recommendation

Clearly, product recommendation can save users' time in finding related products or provide chances for cross-selling. Collaborative Filtering is one of the most popular techniques in recommendation systems. CF method has the ability of adapting to group preferences and generating recommendation without domain knowledge; it also has some well-known problems, such as the sparsity of user-item rating matrix and first-rater problem. In [5], content-based methods are incorporated into a CF framework, trying to solve these problems. First, a classifier is trained with the content descriptions of items rated by a user. This user profile is then used to predict the same person's ratings on other items. In this way, a much denser pseudo-rating matrix can be got. Finally collaborative filtering is carried out on this matrix using neighborhood-based algorithm.

Due to the application domain of [5], rating prediction is a text-categorization problem. In a parametric catalog, more complex methods of rating prediction need be contrived. Here is the reason. In parametric catalogs, products in different categories have different features, that is, different forms of descriptions. Therefore, content-based prediction is limited to the scope of each single category. To learn a user's preferences, we need a profile for each category. What's more, if a user has not rated any product in a category, no training set is available to learn from. Considering the large number of categories in a real world catalog, this situation will not be rare.

Facing this difficulty, we are inspired by the work from [6]. Categories form a kind of class hierarchy; subcategories inherit features from their parents. If a user profile is represented in a tree structure, preference in a parent category can be deduced from its subcategories using inherited features. This generalized preference can be used to predict ratings in its unrated subcategories. In more details, user preference is described using a list of taste-weight pairs.

$$Pref_C = \langle (T_1, w_1), (T_2, w_2), \cdots, (T_{n_C}, w_{n_C}) \rangle \qquad (4)$$

Each pair corresponds to a feature of the category. While weights (w_i) are scalar values specified by experts, tastes (T_i) have different forms. For a numerical feature, the taste is a user's preferred range of value. For a literal feature, vector space model is used to represent the taste. A parent category's taste-weight list is determined by aggregating the projected lists of its subcategories. Similarity between products and user's preference profile can be evaluated as:

$$Sim\left(Pref_C, P^i\right) = \sum_{j=1}^{n_C} \left(TasteSim(T_j, f_j^i) \times w_j\right) \qquad (5)$$

Using this similarity measurement to predict product ratings, collaborative filtering can be boosted with a denser rating matrix. Moreover, since feature evaluation results indicate pragmatic importance. They can be used as the weights for tastes, instead of experts specified ones.

3   System Framework

The parametric characteristic of catalogs lends us a good opportunity to bring content mining and usage mining together. The mining methods for different scenarios should be integrated and support each other. Fig. 1 illustrated our system framework. Preference Learning Module learns from previous ratings and product descriptions to generate a hierarchical profile for each user. This profile is then used by Rating Prediction Module to rate unrated products in the catalog. Feature Evaluation results act as weights for similarity measurement in the rating process. Supplied with a denser rating matrix and current user's browsing behaviors, Collaborative Filtering is carried out with the KNN method. To obtain higher efficiency, relationships inferred by Category Association Module are also used in Collaborative Filtering. A limited number of categories are selected from the taxonomy, based on whether they are related to the categories within user's current interest. Then rating vectors are projected to the items from selected categories. Since vector length is dramatically reduced, distances between rating vectors can be computed much faster.

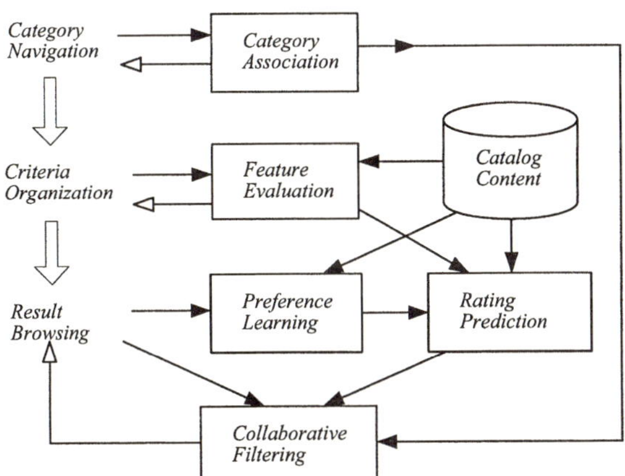

Fig. 1. System framework of a data mining system for parametric product catalog.

4   Conclusions and Perspectives

Although parametric product catalogs are implemented as web applications, they have specific user access mode, which distinguishes them from other web sites. This access mode and the availability of structured content data created a unique context for data mining. In this paper, we discussed promising ways to improve intelligence in different phases of user access. They are category association, feature evaluation and product recommendation. For each of them, related works are introduced and techniques are adjusted to be used in parametric catalogs. We also put forward a harmonized system framework, in which modules provide valuable knowledge for each other. However, this framework has not been tested within a real world system, more detailed consideration and evaluations should be carried out.

References

1. J. J. Rofrano. IBM WebSphere Commerce Suite Product Advisor. IBM Systems Journal, Vol. 40 Issue 1. (2001)
2. Hye-young Paik, Boualem Benatallah. Building Adaptive E-Catalog Communities Based on User Interaction Patterns. IEEE Intelligent Systems, Vol. 17 Issue 6. (2002)
3. Bamshad Mobasher, Honghua Dai, Tao Luo, Miki Nakagawa. Effective Personalization Based on Association Rule Discovery from Web Usage Data. Proceedings of the Third International Workshop on Web Information and Data Management(WIDM), (2001)
4. R. Srikant, R. Agrawal. Mining generalized association rules. Future Generation Computer Systems 13. (1997) 161-180
5. Prem Melville, Raymond J. Mooney, Ramadass Nagarajan. Content-Boosted Collaborative Filtering for Improved Recommendations. Proceedings of the 18th National Conference on Artificial Intelligence (AAAI). (2002)
6. Li Niu, Xiao-Wei Yan, Cheng-Qi Zhang, Shi-Chao Zhang. Product hierarchy-based customer profiles for electronic commerce recommendation. Proceedings of 2002 International Conference on Machine Learning and Cybernetics, Vol. 2. (2002)
7. Anindya Datta, Kaushik Dutta, Debra VanderMeer, Krithi Ramamritham, Shamkant B. Navathe. An architecture to support scalable online personalization on the Web. The VLDB Journal, Vol. 10. (2001) 104-117
8. Dimitrios Pierrakos, Georgios Paliouras, Christos Papatheodorou, Constantine D Spyropoulos. Web usage mining as a tool for personalization: A survey. User Modelling and User-Adapted Interaction, Vol. 13 Issue 4. (2003)

Mining Quantitative Data Based on Tolerance Rough Set Model

Hsuan-Shih Lee[1], Pei-Di Shen[2], Wen-Li Chyr[3], and Wei-Kuo Tseng[1]

[1] Department of Shipping and Transportation Management
National Taiwan Ocean University
Keelung 202, Taiwan
[2] Department of Information Management
Ming Chung University
Taipei 111, Taiwan
[3] Graduate School of Management
Ming Chung University
Taipei 111, Taiwan

Abstract. Rough set theory has been widely used in knowledge acquisition. However, In conventional application of rough set to numeric data, data must be pre-classified. In this paper, a different approach is introduced to deal with numeric data. We develop a mining algorithm based on fuzzy sets and tolerance rough set model, which offers a way of relating data in their semantics.

1   Introduction

The concept of the rough set is a new mathematical approach to imprecision, vagueness and uncertainty in data analysis beside fuzzy set theory. Rough sets were first introduced at the beginning of the eighties by Z. Pawlak [3, 4, 6] and belong to the family of concepts concerning the modeling and representing of incomplete knowledge [5, 7]. Since then researches related to rough sets were booming [2].

The origin of the rough set philosophy is the assumption that with every object we associate some information. Objects can be something like patients, while the symptoms of the disease are the information employed to characterized patients. Objects are similar or indiscernible, if they are characterized by the same information. In contrast to the theory of fuzzy set which models the uncertainty with membership function, rough sets approach vague concept by charactering the vagueness with a pair of crisp sets, the lower and the upper approximation of the vague concept.

Rough set is usually applied to nonnumeric data for data mining. In this paper we are going to develop a mining algorithm for numeric data based on rough set. To deal with numeric data, all numeric data are transformed into fuzzy sets. For the time being, we assume only condition attributes are numeric.

R. Khosla et al. (Eds.): KES 2005, LNAI 3681, pp. 359–364, 2005.

2 Tolerance Rough Set Model

Let U be a nonempty set and R be an indiscernibility relation or equivalence relation on U. Then (U, R) is called a Pawlak approximation space. Let the concept X be a subset of U. Then the lower approximation of X in (U, R), denoted as $LA_R(X)$, is defined to be

$$LA_R(X) = \{x | [x]_R \subseteq X\}$$

and the upper approximation of X in (U, R), denoted as $UA_R(X)$, is defined to be

$$UA_R(X) = \{x | [x]_R \cap X \neq \phi\},$$

where $[x]_R$ is an equivalence class of R containing x. The equivalence classes of R and the empty set ϕ are called elementary or atomic sets in the approximation space (U, R). The union of one or more elementary sets is called a composed set. The family of all composed sets, including the empty set, is denoted by $Comp((U, R))$, which is a Boolean algebra and a subalgebra of Boolean algebra 2^U. Pawlak regards the group of subsets of U with the same upper and lower approximations in (U, R) as a rough set in (U, R). Using lower and upper approximations, an equivalence relation $\approx_R$ can be defined on the powerset of U:

$$X \approx_R Y \Leftrightarrow LA_R(X) = LA_R(Y) \text{ and } UA_R(X) = UA_R(Y),$$

where $X, Y \in 2^U$ and R is an equivalence relation on U.

This equivalence relation induces a partition on the power set 2^U. An equivalence class of such partition is called a P-rough set. The set of all P-rough set is denoted by $2^U / \approx_R$. More specifically, a P-rough set can be defined as follows:

Definition 1. *Given the approximation space (U, R) and two sets $A_1, A_2 \in Comp((U, R))$ with $A_1 \subseteq A_2$, a P-rough set is the family of subset of U described as follows:*

$$< A_1, A_2 >= \{X \in 2^U | LA_R(X) = A_1, UA_R(X) = A_2\}.$$

Equivalently, a P-rought set containing $X \in 2^U$ can be defined as:

$$[X]_{\approx_R} =$$
$$\{Y \in 2^U | LA_R(Y) = LA_R(X), UA_R(Y) = UA_R(X)\}.$$

In other words,

$$[X]_{\approx_R} =< LA_R(X), UA_R(X) > .$$

A member of $[X]_R$ is also referred to as a generator of the P-rough set [1].

Pawlak [8] defined the rough membership function as

$$\frac{|[x]_R \cap X|}{|[x]_R|}.$$

Based the definition of rough membership function, we introduced the concept of coverage and certainty which is defined in the subsequent section.

3 Mining Algorithm

3.1 Fuzzy Indiscernibility Relation

Given an information table such as table 2 of which the condition attributes are fuzzy sets, we extend the concept of indiscernibility relation to fuzzy indiscernibility relation. Take table 2 as an example. Consider one condition attribute, say $Hemoglobin$. Then the attribute induces the following fuzzy indiscernibity relation:

$$B_R =$$
$$\{((\{O_1, O_3, O_4, O_5, O_8\}, Normal * 0.2), (\{O_1, O_2, O_3, O_8\}, High * 0.5),$$
$$(\{O_4, O_5, O_6, O_7\}, Low * 0.7)\}.$$

There are three classes in the fuzzy indiscernibility relation. The first is induced by $Normal$. Since O_1, O_3, O_4, O_5, and O_8 all contain $Normal$, they are in the first class. The degree of membership of this class or $Normal$ is 0.2, which is obtained by taking the minimum of each object's degree in $Normal$ that is the minimum of $\{0.4, 0.4, 0.2, 0.2, 0.2\}$.

Consider two condition attributes, say $Hemoglobin$ and $Hematocrit$. The fuzzy indiscernibility relation induced is:

$$\{((\{O_1, O_5, O_8\}, (Normal, Normal) * 0.2), (\{O_3\}, (Normal, High) * 0.4),$$
$$(\{O_4\}, (Normal, Low) * 0.2), (\{O_1, O_8\}, (High, Low) * 0.5),$$
$$(\{O_2, O_3\}, (High, High) * 0.5), (\{O_5, O_7\}, (Low, Normal) * 0.7),$$
$$(\{O_4, O_6\}, (Low, Low) * 0.7)\}.$$

There are 7 classes in the relation. The first class is induced by ($Hemoglobin = Normal, Hematocrit = Normal$). The degree of membership, 0.2, is obtained by taking the minimum of $0.4 * N, 0.7 * N, 0.2 * N, 0.8 * N, 0.2 * N, 0.8 * N$.

3.2 Notation Definition

The notations used in our algorithm are defined as follows:

X: The set of objects sharing the same value of decision attributes.
X_t: A subset of X that have not yet been classified.
A: The set of condition attributes.
B: A subset of A.
D: Value of decision attributes.
B_R: Fuzzy indiscernibility relation induced by B.
$L(B)$: The set of linguistic values for condition attributes B.
l: An element of $L(B)$.
$[l]_{B_R}$: The set of objects that share the same linguist value in condition attributes B.

$Cov(l)$: The coverage of l in X_t, which is defined to be

$$\frac{[l]_{B_R} \cap X_t}{[l]_{B_R}}$$

$Cer(l)$: The certainty of l that is defined to be

$$\frac{[l]_{B_R} \cap X}{[l]_{B_R}}$$

$M(l)$: The degree of membership of l in B_R.

3.3 Algorithm

Step 1 The quantitative values of condition attributes of objects in the information table are transformed into fuzzy set by the membership function of linguistic values.

Step 2 Let T be set to the information table given.

Step 3 If T is empty, then stop. Otherwise proceed to next step.

Step 4 Let X denote the set of objects in T share the same value of decision attribute.

Step 5 Set $i = 1$ and $X_t = X$.

Step 6 If $i > |A|$ then go to step 13.

Step 7 Choose i condition attributes from A. Let it be denoted as B. If all options have been chosen, then go to step 12

Step 8 If all elements for X_t have been chosen, then goto step 11. Otherwise, choose one element from X_t. Let it be x.

Step 9 If $[x]_{B_R^{-1}} \subseteq X_t$, then generate certain rule:

$$B(x) \xrightarrow{1} D(X_t)$$

and set

$$X_t = X_t - [x]_{B_R^{-1}}.$$

Step 10 Go to step 8.

Step 11 Go to step 7.

Step 12 Set $i = i + 1$. If $X_t \neq \phi$, go to step 6.

Step 13 If $X_t = \phi$ then go to step 17.

Step 14 Let l^* and B^* be the optimal solution of (1).

$$\max_{l \in L(B), B \subseteq A} Cov(l) \tag{1}$$

Generate possible rule:

$$(B = l^*)_{M(l^*)} \xrightarrow{Cer(l^*)} D(X_t).$$

Step 15 Set $X_t = X_t - [l^*]_{B_R^*}$.

Step 16 Go to step 13.

Step 17 Set $T = T - X$.

Step 18 Go to step 3.

Table 1. Information table with numeric data in condition attributes.

| | Condition attributes | | Decision attribute |
Objects	Hemoglobin	Hematocrit	Anemia
O_1	15	40	No
O_2	16	48	No
O_3	15	60	No
O_4	12	35	Yes
O_5	12	44	No
O_6	10.5	30	Yes
O_7	11	45	Yes
O_8	15.5	41	Yes

Table 2. Information table with fuzzy data in condition attributes.

| | Condition attributes | | Decision attribute |
Objects	Hemoglobin	Hematocrit	Anemia
O_1	$0.4 * N + 0.5 * H$	$0.7 * N$	No
O_2	$1 * H$	$0.3 * N + 0.6 * H$	No
O_3	$0.4 * N + 0.5 * H$	$1 * H$	No
O_4	$0.2 * N + 0.7 * L$	$1 * L$	Yes
O_5	$0.2 * N + 0.7 * L$	$0.8 * N$	No
O_6	$1 * L$	$1 * L$	Yes
O_7	$1 * L$	$0.7 * N$	Yes
O_8	$0.2 * N + 0.7 * H$	$0.8 * N$	Yes

3.4   Example

Assume there is an information table as shown in table 1. The linguistic values for *Hemoglobin* are defined as fuzzy sets by following membership functions:

$$\mu_{Normal}(x) = \begin{cases} \frac{x-11.5}{2.25}, & 11.5 \leq x \leq 13.75, \\ \frac{16-x}{2.25}, & 13.75 \leq x \leq 16, \\ 0, & otherwise, \end{cases}$$

$$\mu_{High}(x) = \begin{cases} \frac{x-14.5}{1.5}, & 14.5 \leq x \leq 16, \\ 1 & 16 \leq x, \end{cases}$$

and

$$\mu_{Low}(x) = \begin{cases} 1, & 0 \leq x \leq 11.5, \\ \frac{13-x}{1.5}, & 11.5 \leq x \leq 13. \end{cases}$$

The linguistic values for *Hematocrit* are defined as fuzzy sets by following membership functions:

$$\mu_{Normal}(x) = \begin{cases} \frac{x-35}{2.5}, & 35 \leq x \leq 40, \\ \frac{45-x}{2.5}, & 40 \leq x \leq 45, \\ 0, & otherwise, \end{cases}$$

$$\mu_{High}(x) = \begin{cases} \frac{x-40}{5}, & 40 \leq x \leq 45, \\ 1 & 45 \leq x, \end{cases}$$

and

$$\mu_{Low}(x) = \begin{cases} 1, & 0 \le x \le 35, \\ \frac{40-x}{5}, & 35 \le x \le 40. \end{cases}$$

Then we go through the algorithm with information table (1) and the association rules mined from table 1 are summarized as follows:

certain rule 1

$$(Hematocrit = Low)_1 \xrightarrow{1} (Anemia = Yes).$$

certain rule 2

$$(Hematocrit = High)_{0.6} \xrightarrow{1} (Anemia = No).$$

possible rule 1

$$(Hematocrit = Normal)_{0.3} \xrightarrow{\frac{2}{5}} (Anemia = Yes).$$

possible rule 2

$$(Hemoglobin = Normal, Hematocrit = Normal)_{0.2} \xrightarrow{\frac{2}{3}} (Anemia = No).$$

4 Conclusions

In this paper, we proposed an algorithm for numeric data where rough set acts awkward in the past. we have incorporated fuzzy set theory to deal with numeric data. A mining algorithm based on rough set has been proposed to deal with numeric data such that certain rules and possible rules may be generated for the information table contains numeric data.

Acknowledgement

This research work was partially supported by the National Science Council of the Republic of China under grant No. NSC93-2416-H-019-004-.

References

1. S. Chanas, D. Kuchta, Further remarks on the relation between rough and fuzzy sets, Fuzzy Sets and Systems 47 (1992) 391-394.
2. H.-S. Lee "A fuzziness measure of rough sets", Lecture Notes in Artificial Intelligence 2715 (2004) 64-70.
3. Z. Pawlak, Rough sets, Report No. 431, Polish Academy of Sciences, Institute of Computer Science (1981).
4. Z. Pawlak, Rough sets, Internat. J. Inform. Comput. Sci. 11(5) (1982) 341-356.
5. Z. Pawlak, Rough classification, Internat. J. Man-Machine. Stud. 20 (1984) 469-483.
6. Z. Pawlak, Rough set and fuzzy sets, Fuzzy Sets and Systems 17 (1985) 99-102.
7. Z. Pawlak, Rough sets: A new approach to vagueness, in Fuzzy Logic for the Management of Uncertainty (L. A. Zadeh and J. Kacprzyk, Eds.), Wiley, New York (1992) 105-118.
8. Z. Pawlak and A. Skowron, Rough membership functions, in Fuzzy Logic for the Management of Uncertainty (L.A. Zadeh and J. Kacprzyk, Eds.), Wiley, New York (1994) 251-271.

Incremental Association Mining
Based on Maximal Itemsets

Hsuan-Shih Lee

Department of Shipping and Transportation Management
Department of Computer Science
National Taiwan Ocean University
Keelung 202, Taiwan
Republic of China

Abstract. Incremental association mining refers to the maintenance and utilization of the knowledge discovered in the previous mining operation for later association mining. In paper, we propose a notion called maximal itemset based on which large itemsets with dynamic minimum support can be identified easily. When new transactions are inserted into a database, the maximal itemsets of the new database can be generated from previous maximal itemsets and new transactions.

1 Introduction

Mining association rules from transaction databases has been one of the most interesting research topics in data mining [1–5]. Various methods and new theory theories such rough sets [10, 11] have been proposed for KDD. When databases are dynamic, how to utilize previously mined patterns for latter mining process has become important research topics. Incremental data mining methods are introduced to deal such problems [6–9, 12].

The problem of mining association rules of a database can be decomposed into two subproblems. Identifying large itemsets and generating associating rules from large itemsets. The later can be accomplished easily once large itemset are identified. Hence the focus of mining rules becomes the identification of large itemsets. Incremental mining methods based on Apriori [3] need the minimum support be known in advance. In this paper, we propose a notion called maximal itemset to keep track of the occurrences of itemsets. An incremental method to construct maximal itemsets is proposed. Large itemsets can be identified from the maximal itemsets with dynamic minimum support. That is, mining information abstracted by maximal itemsets can be used for any minimum support in generating large itemsets. Moreover, the maximal itemsets can be easily updated when new transactions are inserted into a database.

2 Maximal Itemset

Definition 1. *Given a set of items $I = \{i_1, i_2, \ldots, i_3\}$ and a database $B = \{t_1, t_2, \ldots, t_n\}$ where B is a multiset and $t_i \subseteq I$, for itemset $X \subseteq I$, the num-*

R. Khosla et al. (Eds.): KES 2005, LNAI 3681, pp. 365–371, 2005.

ber of the occurrences of X, denoted as $O_B(X)$, is defined to be the number of transactions in database B that contain X.

Definition 2. Given $X \subseteq I$, X is a maximal itemset in B if and only if $X = I$ or $O_B(X) > O_B(Y)$ for all $Y \subseteq I$ and $Y \supset X$. That is, an itemset is a maximal itemset if its number of occurrences is greater than occurrence of all its proper supersets or it has no proper superset.

Lemma 1. For $X, Y \subseteq I$, if $X \subset Y$, then $O_B(X) \geq O_B(Y)$.

Proof: The lemma follows directly from that if Y appears in a transaction so does X. $\qquad\square$

Lemma 2. Let $X_1, X_2 \in M(B)$. If $X_1 \subset X_2$, then $O_B(X_1) > O_B(X_2)$.

Proof: By the definition of maximal itemset, since X_1 is a maximal itemset and $X_1 \subset X_2$, we have $O_B(X_1) > O_B(X_2)$. $\qquad\square$

Lemma 3. Let $X_1, X_2 \in M(B)$. If $X_1 \cap X_2 \neq \emptyset$, $X_1 - X_2 \neq \emptyset$ and $X_2 - X_1 \neq \emptyset$, then there exists $Y \in M(B)$ such that $X_1 \cap X_2 \subseteq Y$ and $O_B(Y) \geq O_B(X_1) + O_B(X_2)$.

Proof: Let $Z = X_1 \cap X_2$. The number of the occurrence of Z in the context X_1 is $O_B(X_1)$ and the occurrence number of Z in the context X_2 is $O_B(X_2)$. Since $X_1 - X_2 \neq \emptyset$ and $X_2 - X_1 \neq \emptyset$, the contexts X_1 and X_2 are different. Therefore, the occurrence number of Z in B is at least $O_B(X_1) + O_B(X_2)$. Hence, there exists $Y \in M(B)$ such that $Z \subseteq Y$ and $O_B(Y) \geq O_B(X_1) + O_B(X_2)$. $\qquad\square$

Lemma 4. Let B_1 and B_2 be two databases. Assume $B_1 \subseteq B_2$. If $X \in M(B_1)$, then $X \in M(B_2)$. That is if X is a maximal itemset in current database, it is also a maximal itemset after new transactions are inserted.

Proof:

Case 1. X has no proper superset.
Then X is a maximal itemset in B_2 by definition.
Case 2. X has a proper superset.
Let it be Y. That is, $X \subset Y \subseteq I$. Since $X \in M(B_1)$, by definition, we have

$$O_{B_1}(X) > O_{B_1}(Y). \tag{1}$$

Since $X \subset Y$ and $B_1 \subseteq B_2$, by lemma 1, we have

$$O_{B_2 - B_1}(X) \geq O_{B_2 - B_1}(Y). \tag{2}$$

That is,
$$O_{B_2}(X) - O_{B_1}(X) \geq O_{B_2}(Y) - O_{B_1}(Y). \tag{3}$$

Combining (1) and (3), we have

$$O_{B_2}(X) > O_{B_2}(Y). \tag{4}$$

Hence, following the definition of maximal itemsets, we can conclude that $X \in M(B_2)$.

$\qquad\square$

Definition 3. *A maximal itemset Y is said to be a supreme cover of itemset X in database B if $O_B(X) = O_B(Y)$ and $X \subseteq Y$.*

Lemma 5. *If $t \in B$, then $t \in M(B)$. That is, every transaction in a database is a maximal itemset.*

Proof: Let $B' = B - \{t\}$.

Case 1. $O_{B'}(t) = 0$:

For $Y \supset t$, $O_{B'}(Y) = O_B(Y) = 0$. Since $O_B(t) = 1$, t is a maximal itemset in B.

Case 2. $O_{B'}(t) > 0$:

Case 2.1. $t \in M(B')$:

Following lemma 4, we have $t \in M(B)$.

Case 2.2. $t \notin M(B')$: Since $t \notin M(B')$, the proper supersets of t can be categorized into two classes. Let Y be the proper superset of t. Then either

$$O_{B'}(t) > O_{B'}(Y) \tag{5}$$

or

$$O_{B'}(t) = O_{B'}(Y). \tag{6}$$

If (5) or (6) holds, then

$$O_B(t) = O_{B'}(t) + 1 > O_{B'}(Y) = O_B(Y). \tag{7}$$

Hence, $t \in M(B)$.

$\square$

Lemma 6. *For each itemset X in database B, there is one and only one supreme cover for X.*

Proof: First we show that every itemset X in B has a supreme cover. According to the definition of supreme cover, if X is a maximal itemset, then X is a supreme cover of itself. If X is not a maximal itemset, then let X' be the largest proper superset of X such that $O_B(X) = O_B(X')$. If $Y \supset X'$, $O_B(Y) < O_B(X')$. That is, X' is a maximal itemset and hence a supreme cover of X in B.

Assume the supreme cover of X is not unique. Let Y and Y' be two different supreme covers of X. Hence,

$$O_B(X) = O_B(Y) = O_B(Y'). \tag{8}$$

If $Y \subset Y'$, following lemma 2, we have $O_B(Y) > O_B(Y')$, which is contradictory to (8). Hence Y is not a proper subset of Y'. With the same argument, we can also conclude that Y' is not a proper subset Y. However, both Y and Y' have X in common. That is, $Y \cap Y' \neq \emptyset$, $Y - Y' \neq \emptyset$ and $Y' - Y \neq \emptyset$. Following lemma 3, there exists $Z \in M(B)$ such that $Y \cap Y' \subseteq Z$ and $O_B(Z) \geq O_B(Y) + O_B(Y')$. Hence,

$$O_B(Z) > O_B(X). \tag{9}$$

Z is a superset of X. By the definition of maximal itemset, we have $O_B(X) > O_B(Z)$, which is contradictory to (9). Therefore, the assumption about the supreme cover of X is incorrect.

$\square$

Theorem 1. *The number of the occurrences of itemset X can be determined as follows:*

$$O_B(X) = \begin{cases} \max_{Y \in M(B) \text{ and } X \subseteq Y} O_B(Y) & \text{if } X \text{ occurs in } B \\ 0 & \text{otherwise.} \end{cases}$$

Proof: Following lemma 6, if X occurs in the database B, then there exist one supreme cover for X, which is a maximal itemset in B. Let it be Y. If $Y = X$, then $O_B(Z) < O_B(Y)$ for all $Z \in M(B)$ and $X \subset Z$. If $X \subset Y$, then Y will be the only maximal itemset in $M(B)$ that contains X. $\square$

3 Updating Maximal Itemsets and Identifying Large Itemsets

Lemma 7. *If $X \in M(B)$ and $t \subseteq I$, then $X \cap t \in M(B \cup \{t\})$. That is, $X \cap t$ is a maximal itemset in $B \cup \{t\}$.*

Proof: Let $Z = X \cap t$. If $Z = X$, according to lemma 4, Z is a maximal itemset in $B \cup \{t\}$. If $Z \subset X$, we want to show that for every proper superset of Z has fewer occurrences than Z in $B \cup \{t\}$ and hence Z is a maximal itemset in $B \cup \{t\}$.

Let $B' = B \cup \{t\}$ and Y be a proper superset of Z. That is, $Z \subset Y$.

Case 1: If Y does not occur in B, then

$$O_{B'}(Y) = 1 < 2 \le O_{B'}(Z). \tag{10}$$

Case 2: Assume Y occurs in B. Then either $Y \subseteq t$ or $Y \nsubseteq t$.

  Case 2.1: If $Y \nsubseteq t$, then $O_B(Y) = O_{B'}(Y)$. Since $Z \subset Y$, by lemma 1, we have $O_B(Y) \le O_B(Z)$. Therefore,

$$O_{B'}(Y) = O_B(Y) \le O_B(Z) < O_{B'}(Z). \tag{11}$$

  Case 2.2: Assume $Y \subseteq t$. By lemma 1, we obtain $O_B(Z) \le O_B(Y)$ because of $Z \subset Y$. We want to show that $O_B(Z) = O_B(Y)$ is impossible. We show this by contradiction. Assume

$$O_B(Z) = O_B(Y). \tag{12}$$

Since Z is a proper subset of Y and they have the identical number of occurrences in database B, both Z and Y must appear simultaneously in the same transaction of database B. Following lemma 6, the supreme cover is unique. Therefore, Z and Y have the same supreme cover in B. Let L be the supreme cover of Z and Y in B. Since $Z \subset Y$, we have

$$Y - Z \ne \emptyset. \tag{13}$$

Since $Y \subseteq t$, $Z \subset Y$ and $t \cap X = Z$, we have

$$Y \cap X = Z.$$

That is,

$$Y - Z \nsubseteq X. \tag{14}$$

Since $Z \subset Y \subseteq L$, we have

$$Y - Z \subseteq L. \tag{15}$$

Following (13), (14) and (15), for those items in $Y - Z$ will appear in L but not in X. Hence, we have

$$L - X \neq \emptyset. \tag{16}$$

If $X \subset L$, then $O_B(X) > O_B(L)$ because of X is a maximal itemset in B. This would yield that X, instead of L, is the supreme cover of Z. Therefore, $X \not\subset L$. Hence

$$X - L \neq \emptyset. \tag{17}$$

Since $Z \subset X$ and $Z \subset L$, we have

$$X \cap L \neq \emptyset. \tag{18}$$

Following lemma 3, (16), (17) and (18) would yield the following:

$$\exists L' \in M(B), \text{ such that } X \cap L \subseteq L' \text{ and } O_B(L') > O_B(L). \tag{19}$$

Since $Z \subset X$ and $Z \subset L$, we have

$$Z \subseteq X \cap L. \tag{20}$$

By (19) and (20), L' contains Z and $O_B(L') > O_B(L)$. Hence L is not the supreme cover of Z. Assumption (12) is incorrect. Therefore, $O_B(Z) > O_B(Y)$. Since $O_{B'}(Y) = O_B(Y) + 1$ and $O_{B'}(Z) = O_B(Z) + 1$, we have

$$O_{B'}(Z) > O_{B'}(Y) \tag{21}$$

(10),(11) and (21) yield that Z is a maximal itemset in database B.

$$\square$$

Assume the maximal itemsets of database B, denoted as $M(B)$ and the number of the occurrences of the maximal itemsets in B are known. If a new transaction t is inserted into B, the maximal itemsets of database $B \cup \{t\}$, $M(B \cup \{t\})$, can be derived by following procedure:

Maximal_insert;
    $B' = B \cup \{t\}$;
    $M(B') = M(B)$;
    Let $S = \{t \cap X | X \in M(B)\}$.
    For each $Z \in S$ do
        $O_{B'}(Z) = 1 + O_B(Z)$;
        $M(B') = M(B) \cup \{Z\}$

If $t \notin M(B')$ do
$$M(B') = M(B') \cup \{t\} \text{ and } O_{B'}(t) = 1;$$
End of Maximal_insert;

Theorem 2. *The Maximal_insert procedure maintains the maximal itemset of database $B \cup \{t\}$ and the number of the occurrences of the maximal itemsets in $B \cup \{t\}$ correctly. The time of updating is linear to the number of maximal itemsets.*

Proof: Following lemma 4, lemma 5 and lemma 7, the Maximal_insert procedure adds maximal itemsets in $M(B)$, itemsets in S and t into $M(B')$. Moreover, no maximal itemset is missed in $M(B')$. For simplicity, we omit the proof of this part. $\square$

Having maximal itemsets determined by the procedure in the preceding section, we can employ the following algorithm to generate large itemsets for a give minimum support s.

Large_of_maximal;
　　//input: $M(B)$ and minimum support s
　　//output: large itemsets L
　　$L = \emptyset$;
　　For each $X \in M(B)$ do
　　　　If $O_B(X) \geq s * |B|$ and $X \notin L$ do
　　　　　　For each $Y \subseteq X$ and $Y \neq \emptyset$ do
　　　　　　　　$L = L \cup Y$;
End of Large_of_maximal;

4   Conclusions

In this paper, we have proposed a notion called maximal itemset based on which we proposed an incremental method to construct the maximal itemsets of a database. Once maximal itemsets are identified, the large itemsets of different minimum support can be easily derived.

Acknowledgement

This research work was partially supported by the National Science Council of the Republic of China under grant No. NSC93-2416-H-019-004-.

References

1. R. Agrawal, T. Imielinksi and A. Swami, "Mining association rules between sets of items in large database," ACM SIGMOD conference, pp. 207-216, Washington DC, USA, 1993.
2. R. Agrawal, T. Imielinksi and A. Swami, "Database mining: a performance perspective," IEEE Transactions on Knowledge and Data Engineering, Vol. 5, No. 6, pp. 914-925, 1993.

3. R. Arawal and R. Srikant, "Fast algorithm for mining association rules," ACM International conference on Very Large Data Bases, pp. 487-499, 1994.
4. R. Agrawal and R. Srikant, "Mining sequential patterns," IEEE International Conferences on Data Engineering, pp. 3-14, 1995.
5. S. Brin, R. Motwani, J.D. Ullman and S. Tsur, "Dynamic itemset counting and implication rules for market basket data," ACM SIGMOD Conference, pp. 255-264, Tucson, Arizona, USA, 1997.
6. D.W. Cheung, J. Han, V.T. Ng and C.Y. Wong, "Maintenance of discovered association rules in large databases: an incremental updating approach," IEEE International Conference on Data Engineering, pp. 106-114, 1996.
7. D.W. Cheung, S.D. Lee, and B. Kao, "A general incremental technique for maintaining discovered association rules," In Proceedings of Database Systems for Advanced Applications, pp. 185-194, Melbourne, Australia, 1997.
8. R. Feldman, Y. Aumann, A. Amir, and H. Mannila, "Efficient algorithms for discovering frequent sets in incremental databases," ACM SIGMOD Workshop on DMKD, pp. 59-66, USA, 1997.
9. T.P. Hong, C.Y. Wang and Y.H. Tao, "A new incremental data mining algorithm using pre-large itemsets," International Journal on Intelligent Data Analysis, 2001.
10. H.-S. Lee "A fuzziness measure of rough sets", Lecture Notes in Artificial Intelligence 2715 pp. 64-70, 2004.
11. Z. Pawlak, Rough sets, Report No. 431, Polish Academy of Sciences, Institute of Computer Science, 1981.
12. N.L. Sarda and N.V. Srinvas, "An adaptive algorithm for incremental mining of association rules," IEEE International Workshop on Database and Expert Systems, pp. 240-245, 1998.

Information-Based Pruning
for Interesting Association Rule Mining
in the Item Response Dataset

Hyeoncheol Kim and Eun-Young Kwak

Department of Computer Science Education,
College of Education, Korea University,
Anam-Dong 5-Ga, Seoul, 136-701, Korea
{hkim,key}@comedu.korea.ac.kr

Abstract. Frequency-based mining of association rules sometimes suffers rule quality problems. In this paper, we introduce a new measure called *surprisal* that estimates the informativeness of transactional instances and attributes. We eliminate noisy and uninformative data using the *surprisal* first, and then generate association rules of good quality. Experimental results show that the surprisal-based pruning improves quality of association rules in question item response datasets significantly.

1 Introduction

Association rule mining is one of the most successful data mining methods. One of the issues is how to control quantity and quality of the generated rules with respect to specific application domains. Apriori-like algorithms generate association rules based on co-occurrence of items in the transactional datasets. The quantity and quality of the generated rules are controlled by support and confidence measure [5]. However the measures are not good enough to solve the rule quality problems [1, 2, 6, 8]. The *quality* of rules means how surprising and valuable those rules are. We also call it the *interestingness* of the rules. Not all of the association rules provide interestingness to us. Various research effort has been done on the interestingness measures to extract implicit and significant association rules [3, 7, 9, 10]. Some approaches first generate association rules and then apply interestingness measures to the rules. Other approaches incorporate their measures into rule generation algorithms to generate interesting rules [3, 9]. There are still many debates on effectiveness of the interesting measures and what the definition of interestingness is.

In this paper, we apply the association mining to analyze associations between question items in an item response dataset, which is quite different from market-basket-type dataset. Item response dataset is illustrated in Table 1. Attributes represent question items and each transaction represents a student instance. Value of each cell in the data table tells whether the student made correct answer for the question item; 1 for correct answer and 0 for wrong answer. By mining the transaction data, we can extract very useful and valuable patterns

R. Khosla et al. (Eds.): KES 2005, LNAI 3681, pp. 372–378, 2005.

such that one question item is associated to another. When we create a question item set to test or evaluate students' achievement, it is very difficult to find out one question item logically related to another.

In the item response dataset, each transaction (i.e., student) has its own score and is considered differently. That is, very high scored student making correct answer on item I_j should be considered differently from very low scored student making correct on the same item. Even for market-basket-type domains, a shopaholic buying MP3 and DVD player should be considered differently from thrifty customer buying them when we analyze item associations based on the transactional dataset. If not, uninteresting or wrong association rules might be generated.

We can not apply frequency-based association rule generation to the item response analysis directly because of the following aspects:

- Frequency-based association does not mean dependency or logical relationship among items. For example, frequency-based mining will generate association rules of very easy question items regardless of their domain relationship. Frequency of easy questions will be very high and frequent-based association represents their co-occurrence chance only, but not their logical associations.
- Item response dataset provides us with the information about how many items each student have made correctly. Even for a same question item I_j, meaning of the answer correctness of low scored students will be different from the one of a very high scored student.

By analyzing the item response dataset, we want to mine the patterns among question items regardless of question item easiness and student scores. In this paper, we propose the approach which removes uninteresting instances and attributes from the dataset first, and then generate association rules. We introduce an information-based measure, *surprisal*, which estimates the interestingness of data instances and attributes.

2   Interestingness Measures and Information-Based Pruning

Apriori-based association mining uses support and confidence thresholds to control quantity and quality of rules generated. They are based on frequency of

Table 1. Item Response Data. The "1" means correct answer and the "0" means wrong.

		I_1	I_2	...	I_j	...	I_{m-1}	I_m
	S_1	1	1	...	1	...	1	1
	S_2	1	1	...	1	...	0	0
Students	...			...				
	S_i	1	1	...	0	...	0	0
	...			...				
	S_n	1	0	...	0	...	0	0

(Question Items)

Table 2. Definition of support, confidence, correlation and conviction.

Measure	Definition
$Support(A, B)$	$P(A \wedge B)$
$Confidence(A \Rightarrow B)$	$\frac{P(A \wedge B)}{P(A)}$
$Corr(A, B)$	$\frac{P(A \wedge B)}{P(A) \cdot P(B)}$
$Conviction(A \Rightarrow B)$	$\frac{P(A) \cdot P(\neg B)}{P(A \wedge \neg B)}$

itemsets, and simple and useful. Even though frequency-based measures are good enough for market basket analysis, they are not sufficient to explain meaningful relationship among items. They generate rules in which items are associated by their frequency, not by their semantic relationship or dependency. Thus, the frequency-based association mining sometimes generates uninteresting rules that are already known and unrelated. A bigger problem is that the support-confidence framework does not work well when correlation is the appropriate measure. For example, consider items A and B with $P(A) = 80\%$, $P(B) = 90\%$ and $P(A \wedge B) = 72\%$. Then the *support* and *confidence* of rule $A \Rightarrow B$ are as high as 72% and 90% respectively. However, their correlation is 1.0 which means no dependency each other. Many other interestingness measure have been proposed to solve the rule quality problem, including correlation, lift, conviction, Gini index, IS, PS, etc [3, 4, 7, 9]. For our experiments, we investigated support, confidence, correlation and conviction, which are defined in Table 2. Support and confidence are chosen as they are regarded as basic tools in mining association rules. Correlation is employed to find out the relationships between question items and conviction is selected as a measure to assess cause-effect relationships.

Frequency-based measures can generate uninteresting or wrong association rules if a dataset includes uninformative instances. For example in the question item response dataset, it might generate a rule such as $A \Rightarrow B$ where frequencies of A and B are the highest. The rule means "Students tend to make correct on the two easiest questions together", but not "the two questions are related". They are frequency-associated, but not semantically-associated. To mine out dependency between question items, we introduce an information-based pruning which eliminates uninformative instances first and then does association rule mining.

The surprise that we get when we see the symbol was called the "surprisal" by Tribus [11] and is defined by

$$u = -log_2(P)$$

For example, if P approaches 0, we will be *very surprised* to see the symbol since it should almost never appear, and the u approaches ∞. On the other hand, if $P = 1$, then we won't be surprised at all to see the symbol because we expected it there and $u = 0$.

For an item response dataset in Table 1, we suppose columns (i.e., question items) are sorted in descending order by their correctness probabilities (i.e., easiness). For each student instance (i.e., transaction) S_i, we can calculate the surprisal for his answers (i.e., responses) to question items. If we know that he made k question items correct, we can expect that he made the first k items among total m items. If he really did, it is no surprise since we expected it. If he did not, it is surprise to us. We can measure the surprise by Tribus [11]'s surprisal.

The $P(B_{i,j})$ is defined by the probability that student S_i makes correct on the question item I_j. The $P(S_i)$ is defined by average of the $P(B_{i,j})$s for all j, and the $P(I_j)$ by the average of $P(B_{i,j})$s for $1 \leq i \leq n$. The $P(I_j)$ represents easiness of a question item I_j and the $P(S_i)$ represents scores of a student S_i in probability scale.

$$P(S_i) = \frac{\sum_j P(B_{i,j})}{m}, \quad P(I_j) = \frac{\sum_i P(B_{i,j})}{n}$$

The $P(M)$ is defined by the average of $P(B_{i,j})$s for all i and j.

$$P(M) = \frac{\sum_i \sum_j P(B_{i,j})}{n \cdot m} = \frac{\sum_j P(I_j)}{m}$$

We want to get $P(B_{i,j})$ to calculate the surprisal u. To obtain the $P(B_{i,j})$, we divide the $P(S_i)$ by $P(M)$ as follows.

$$\frac{P(S_i)}{P(M)} = \frac{\sum_j P(B_{i,j})/m}{\sum_j P(I_j)/m}$$

Then

$$\sum_j P(B_{i,j}) = \frac{P(S_i) \cdot \sum_j P(I_j)}{P(M)}$$

Since $P(B_{i,j}) \propto P(I_j)$ for i fixed, we can define $P(B_{i,j}) \propto \frac{P(S_i)}{P(M)} \cdot P(I_j)$ for i fixed. If $\frac{P(S_i) \cdot P(I_j)}{P(M)} \leq 1$, then we define $P(B_{i,j}) = \frac{P(S_i) \cdot P(I_j)}{P(M)}$ for i fixed. However, note that the $\frac{P(S_i) \cdot P(I_j)}{P(M)}$ could be greater than 1. If it is so, then we define $P(B_{i,j}) = 1$ for i fixed, and we donate the amount of $(1 - \frac{P(S_i) \cdot P(I_j)}{P(M)})$ to other $P(B_{i,j})$s that are less than 1 to make their sum is equal to $P(S_i)$. The amount is donated to each $P(B_{i,j})$ that is less than 1 proportionally to their probability values. The bigger one gets more. In this way, we can define $P(B_{i,j})$ for every i and j.

Then we define the surprisal u that student S_i makes question item I_j as follows:

$$u(B_{i,j}) = -log_2 P(B_{i,j})$$

The surprisal for each student S_i or each question item I_j is defined by averaging $u(B_{i,j})$ for i or j.

$$u(S_i) = \frac{\sum_j u(B_{i,j})}{m}, \quad u(I_j) = \frac{\sum_i u(B_{i,j})}{n}$$

3   Experimental Results

For our experiments, we used an artificial dataset to demonstrate the effectiveness of the surprisal u. The surprisal u measures interestingness or informativeness of each instance or attribute. We prune uninteresting or uninformative instances and attributes from the original dataset and then do association rule mining. We demonstrate the surprisal-based pruning improves the quality of association rules significantly, compared to no-pruning.

The artificial dataset is generated by random number generator and includes $10,000$ student instances and 10 question items. Correct answers for each question item are randomly generated by 90% through 10%, and student instances are also randomly generated with $P(S_i)$ according to normal distribution. Table 3 shows the distribution of the original dataset. We intentionally set up 2-pair associations between question items so that we can validate our experimental results. We set 90%-association between item I_6 and I_2. That is, 90% of students who made correct answers for I_6 also made correct answers for I_2. We also set another 90%-association between item I_{10} and I_4. The rest of the data is randomly generated. Thus, our experiment will be validated by finding the two rules, $I_6 \rightarrow I_2$ and $I_{10} \rightarrow I_4$.

Table 3. Summary of artificial dataset. The first table shows the distribution of easiness of 10 question items and the second table shows the distribution of scores of $10,000$ students.

Question Items	I_1	I_2	I_3	I_4	I_5	I_6	I_7	I_8	I_9	I_{10}
Number of students who made correct	9018 (90%)	6913 (70%)	7010 (70%)	6004 (60%)	5992 (60%)	4920 (50%)	5064 (50%)	2888 (30%)	1063 (10%)	996 (10%)

Scores	0	1	2	3	4	5	6	7	8	9	10
Number of students	200	200	200	1,400	1,600	2,300	2,700	900	100	200	200

Our experiment consists of the following steps: (1) We first prune the original dataset by minimum surprisal; (2) We generate association rules from the pruned datasets by interesting measures such as confidence, correlation and conviction; (3) We compare the association results from pruned datasets and unpruned original dataset.

We generated 2-pair association rules with Apriori algorithm and ranked them with the measures such as support, confidence, correlation and conviction. Table 4 shows the generated 2-pair rules by the different measures. The intentionally inserted rules (I_6, I_2) and (I_{10}, I_4) are ranked relatively low by the frequency-oriented measures such as support and confidence, while they are ranked high by the measures such as correlation and conviction. However many other uninteresting rules are also generated from the original un-pruned dataset.

Surprisal-based pruning is applied to the original dataset. We experimented two types of pruning: instance pruning and attribute(i.e., question item) pruning. Surprisal values $u(S_i)$ are calculated for every student instance and if the

Table 4. Generated association rules ranked by different measures. The rules are generated from the original un-pruned dataset. Many uninteresting rules are also generated as well, in addition to the intentionally inserted rules, (6,2) and (10, 4).

Rank	Support		Confidence($\geq$ 80%)		Correlation($\geq$ 1.0)		Conviction($\geq$ 1.0)	
	item	value	Item	value	Item	value	Item	value
1	2,1	67	7,1	94	**10,4**	1.43	9,3	3.40
2	3,1	60	8,1	93	9,3	1.37	**10,4**	2.80
3	7,1	47	6,1	91	10,3	1.30	10,3	2.38
4	6,1	40	3,1	91	9,5	1.30	**6,2**	2.20
5	**6,2**	39	9,3	90	8,3	1.25	8,3	1.97
6	8,2	27	2,1	89	**6,2**	1.18	9,5	1.95
7	8,3	24	**6,2**	89	7,1	1.06	7,1	1.83
8	10,3	12	10,3	86	8,1	1.05	8,1	1.60
9	**10,4**	12	**10,4**	86	3,1	1.02	6,1	1.21
10	9,5	9	8,3	83	6,1	1.02	3,1	1.21
11	9,3	9	9,5	80	2,1	1.02	2,1	1.03

value is less than 0.5, then the instance S_i is eliminated from the dataset. We eliminated 2155 instances from the set leaving 7845 instances. Attribute pruning is done in the same way. That is, if surprisal value $u(I_j)$ for each attribute is less than minimum value, we eliminate the attributes I_j from the dataset. Question items I_1 and I_3 were eliminated from the set by the minimum value 0.3 in our experiments. Table 5 shows the generated 2-pair rules from the surprisal-based pruning. It demonstrates that the surprisal-based pruning eliminates noisy data instances or attributes, and thus only interesting rules are generated no matter what measures are used.

Table 5. 2-pair association rules generated from the surprisal-based pruning. The number of uninteresting rules are dramatically reduced, compared to the rules from original dataset. The first table lists the rules after instance pruning, and the second table lists the rules after attribute pruning. The surprisal-based pruning improves the quality of a rule set significantly.

Rank	Support		Confidence($\geq$ 80%)		Correlation($\geq$ 1.0)		Conviction($\geq$ 1.0)	
	Item	value	Item	value	Item	value	Item	value
1	**6,2**	30	**6,2**	86	**10,4**	1.45	9,3	2.53
2	10,3	9	9,3	86	9,3	1.34	**10,4**	2.39
3	**10,4**	9	10,3	82	10,3	1.28	10,3	1.99
4	9,3	6	**10,4**	82	**6,2**	1.16	**6,2**	1.86

Rank	Support		Confidence($\geq$ 80%)		Correlation($\geq$ 1.0)		Conviction($\geq$ 1.0)	
	Item	value	Item	value	Item	value	Item	value
1	**6,2**	39	**6,2**	89	**10,4**	1.43	**10,4**	2.80
2	**10,4**	12	**10,4**	86	**6,2**	1.19	**6,2**	1.20

4   Conclusions

We introduced an information-based measure, *surprisal*, which estimates informativeness of instances or attributes. Surprisal-based pruning eliminates noisy or uninformative data and thus improves the quality of dataset to be mined. Experimental results show that the surprisal-based pruning improves the quality of association rules significantly. In the problem domain in which probabilities of every instances and attributes are known, the surprisal measure will improve the quality of association rule mining.

References

1. T. Brijs, K. Vanhoof, and G. Wets. Defining interestingness for association rules. *Intelligent Data Analysis*, pages 229–240, Apr 2000.
2. S. Brin, R. Motwani, and Silverstein. C. Beyond market baskets: Generalizing asso-ciation rule to correlation. *Data Mining and Knowledge Discovery*, 2, 1998.
3. A.A. Freitas. On rule interestingness measures. *Knowledge-Based Systems*, 12, 1999.
4. J. Han and M. Kamber. *Data Mining Concepts and Techniques*. Morgan Kaufmann, 2000.
5. F. Hussain, H. Liu, E. Suzuki, and H. Lu. Exception rule mining with a relative interestingness measure. *Proceedings of the 4th Pacific-Asia Conference on Knowledge Discovery and Data Mining, Current Issues and New Applications*, 2000.
6. T. Imielinski, L. Khachyan, A. Abdulghani, and Cubergrades. Generalizing association rules. Technical report, Dept. Computer Science, Rutgers Univ., Aug 2000.
7. A. Silberchatz and A. Tuzhilin. On subjective measures of interestingness in knowledge discovery. *Poc. of the 1st Int. Conf. on Knowledge Discovery and Data Mining*, 1995.
8. C. Silverstein, S. Brin, and B. Motwani. Beyond market baskets-generalizing association rules to dependence rules. *Data Mining and Knowledge Discovery*, 1997.
9. P. Tan and V. Krumar. Interestingness measures for association pattern. *Technical Report TR00-036, Department of Computer Science, University of Minnestota*, 2000.
10. P. Tan, V. Krumar, and J. Srivastava. Selecting the right interestingness measure for association patterns. 2002.
11. M. Tribus. Thermostatics and thermodynamics. *D. van Nostrand Company, Inc.*, 1961.

On Mining XML Structures Based on Statistics

Hiroshi Ishikawa, Shohei Yokoyama, Manabu Ohta, and Kaoru Katayama

Graduate School of Engineering, Tokyo Metropolitan University

Abstract. We propose an approach to dynamically generate database schemas for well-formed XML data. Our approach controls the number of tables to be divided based on statistics of XML so that the total cost of processing queries is reduced. We devise schemas appropriate for complex data such as text formatting and child elements with the small maximum number of occurrences in order to reduce the number of tables. To this end, we define three functions NULL expectation, Large Leaf Fields, and Large Child Fields for controlling the tables to be divided. We evaluated typical XML queries over the generated schemas and normalized schemas and measured and compared both of the costs. Through this, we successfully validated our approach.

1 Introduction

Now we are surrounded by a lot of XML data. Therefore we must design database schemas to accommodate such voluminous XML data. Approaches to storing XML data into databases are generally divided into the following categories:

(1) To store XML data into predefined schemas of databases
(2) To store XML data into dynamically generated schemas from XML schemas such as DTD

The first category which we call a static schema approach includes some variations. Zhang et al.[Zhang et al. 01] take an approach to store elements and strings (PCDATA) into separate tables together with locations within XML data. Florescu et al.[Florescu et al.99] propose three approaches; an approach to edges and values of XML trees into separate tables, an approach to cluster data according to the kinds of edges into the same tables, and an approach to store all elements into a universal relations by viewing them as having the same set of tags virtually. Both Jiang et al.[Jiang et al.02] and [Yoshikawa et al.01] decompose XML data into values, elements, and paths and store them into separate tables.

All these approaches have the following merits:

(1) It is possible to accommodate any XML data into the uniform schemas.
(2) It is easy to preserve the ordering of various structures of XML data.
(3) It is easy to automatically translate queries including hierarchical relationships (i.e., ancestor and descendants).

On the other hand, all these have the following demerits:

(1) It is rather difficult to distinctively define basic data structures such as numerals and characters of XML values.
(2) It is rather complex to formulate queries involving aggregate functions such as summation and average.

R. Khosla et al. (Eds.): KES 2005, LNAI 3681, pp. 379–390, 2005.

Note that Tian et al. [Tian et al.02] did a comparative study about the above methods.

The second category which we call a dynamic schema approach allows basic data structures to be represented according to each element value in a straightforward manner. In general, however, as the number of tables increases according to the number of element kinds, it is often necessary to join tables in queries. Therefore, the cost of query processing may be very high.

The solution to the above problem is to reduce the number of tables constituting database schemas. The works along this line include methods of Deutsch et al. [Deutsch et al.99] and of Shanmugasundaram et al. [Shanmugasundaram et al.99]. However, both of them assume the existence of DTD (Document Type Definitions), which not every end-user is expected to be able to describe. The approach of Klettke et al. [Klettke et al.01] supports generation of ORDB (Object-Relational Databases) schemas by analyzing the number of occurrences of elements in DTD and XML data and analyzing frequent queries, but it takes no account of the efficient way of decomposing tables.

Further, since DTD are not intended to be used to succinctly describe relational schemas, they are insufficient to generate efficient schemas since it is not possible to determine appropriate data types and maximum number of occurrences of child elements in parent elements. In other words, XML data generated from DTD only are likely to be diverse and it is difficult to determine optimal database schemas from such XML data.

On the other hand, XML Schemas [XML Schema 04] are more expressive than DTD and are richer in information for generating database schemas. Therefore, it seems more promising to generate efficient database schemas by using XML Schemas, but not every end-user is expected to describe XML Schema like DTD.

In this paper, we propose an approach to generate database schemas based on statistics of XML data only. In other words, we don't assume the existence of XML schemas. We assume the following criteria:

(1) Databases (i.e., tables) containing unnecessary null values are redundant and undesirable.

(2) Database schemas containing many tables are undesirable because many joins consuming large processing costs are needed.

Therefore, our objective is to dynamically generate as little number of tables and as little redundant tables as possible. Salient features of our approach are that it generates database schemas based on statistics such as number of occurrences of tags and maximum number of occurrences of child elements in parent elements, but not DTD nor XML Schema. Our approach generates database schemas by combining the following methods:

(1) Data resulting in complex models such as text formatting and leaf elements more often than a predetermined threshold occurring in parent elements are stored as texts in order to reduce the number of referenced tables.

(2) An evaluation function to determine the ratio of null values based on statistics of XML data is defined. Database schemas producing small value of the function are generated in order to reduce redundancy resulted from reduction of the number of tables.

We introduce general concepts and definitions needed to describe our approach in Section 2. In Section 3, we discuss our approach in detail. In Section 4, we explain our experiments to validate the effectiveness of our approach.

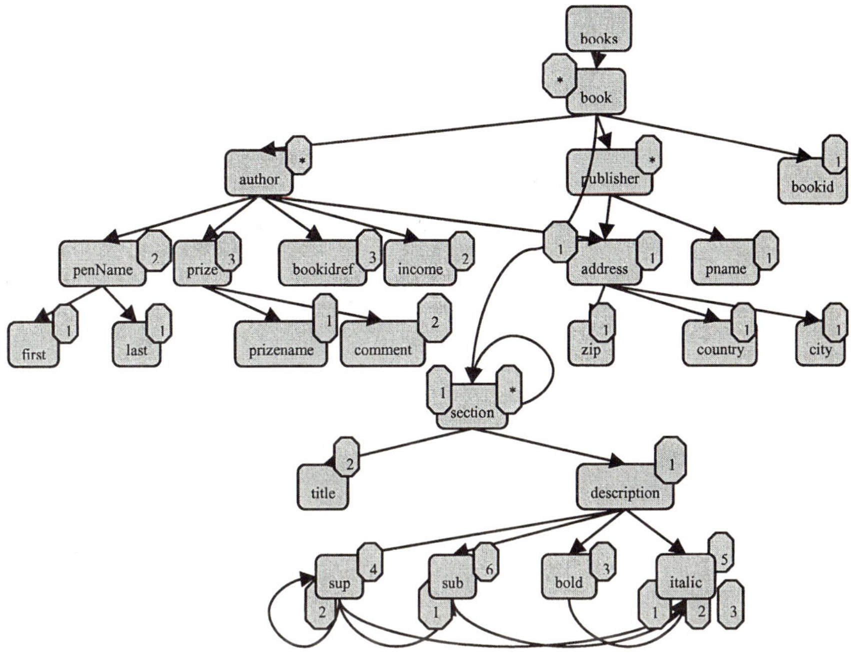

Fig. 1. Model for XML data

2 Concepts and Definitions

2.1 Concepts

Ideally, good database schemas contain little redundancy and little number of tables to be joined. However, reducing redundancy of databases will usually increase the number of tables and the times of joining such tables. On the other hand, reducing the number of tables for reducing the times of joining them will usually increase redundancy. There is a trade-off between these two objectives. One possible way is to apply normalization used for logical design of relational databases [Elmasri et al.99]. Recently, ORDB systems allow advanced data types such as arrays and objects to represent partially-normalized data. In other words, our proposed approach is to discover latent structures corresponding to such data types from XML data. However, availability of such data types depends on specific database systems (i.e., object relational databases or object-oriented databases). So we use data types only supported by pure relational database systems. We assume that it is more efficient to stores enumerated leaf elements and non-leaf elements occurring for times less than a prescribed threshold into one table. So we try to reduce the total number of tables by using the maximum number of occurrences of parent and child elements that are not expressed by DTD. We also try to reduce the number of tables by storing rather complex data such as text formatting as character strings. To reduce redundancy caused by storing different type of elements into one table, we define a function for evaluating redundancy related to the maximum number of occurrences.

2.2 Definitions

We call rules expressing XML elements and parent-child relationships a model of XML data. We represent a model as a graph by corresponding elements and parent-child relationships to nodes and links in the graph, respectively. Schema definitions by DTD or XML Schema are examples of models. For example, a graph induced by DTD has links labeled "?", "*", "+" expressing the number of occurrences of parent and child elements. Instead, we represent the number of occurrences explicitly (See Figure 1). Please not that if the number of occurrences is larger than or equal to ten, the link is labeled "*" similarly for the moment. A model which we mention hereafter in this paper means a model in our proposal. We call XML elements simply nodes. They are divided into non-leaf nodes for elements having any descendants and into leaf nodes for elements having no descendants.

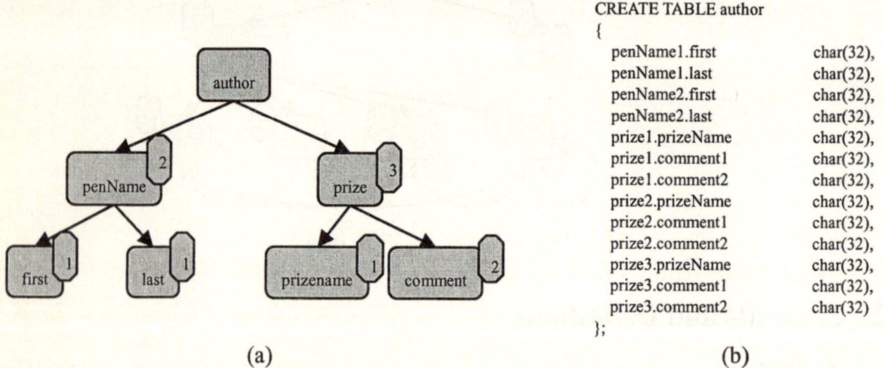

Fig. 2. (a) Sub-tree of a model. (b) Table schema

We use the following notations for a model M and a node A:

|M|: the number of nodes contained by M
R (M): the root node of M
P (A): parent nodes of A
p (A): a parent node of A having a shortest path to the root of the model. If there are more than one, a node occurring most often is chosen among them.
C (A): child nodes of A
D (A): descendant nodes of A
L (A): leaf and descendant nodes of A

For example, if we take the model illustrated in Figure 1 as M, the above parameters have the following values:

|M|=24
R (M)=book
P (address)={author, publisher}

p (address)={author}; We assume that publisher elements occurs more often than address elements.

C (author)={penName, prize, bookdlref, income}

D (author)={penName, first, last, prize, prizename. comment, bookdlref, income, address, zip, country, city}

L(author)={first, last, prizename. comment, zip, country, city}

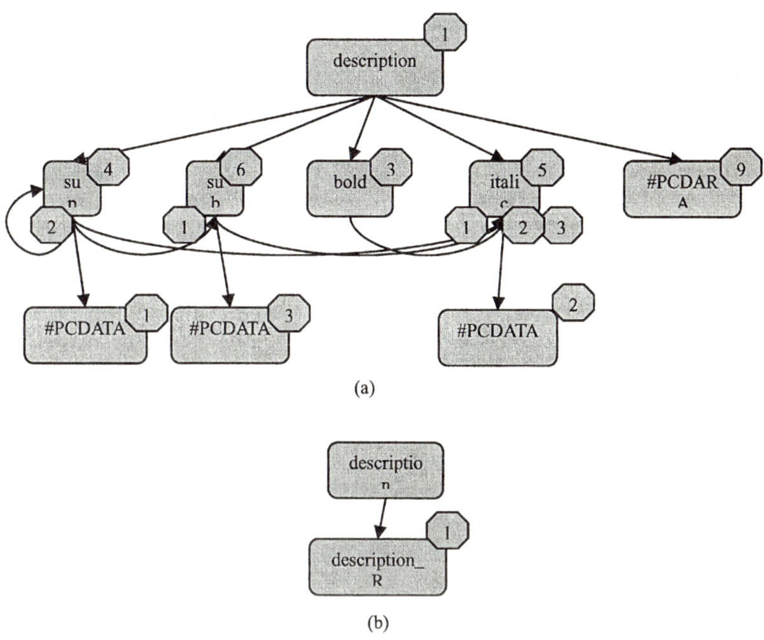

(a)

(b)

Fig. 3. (a) Complex data. (b) Transformed tree

Now we describe XML statistics. We calculate statistics for XML data by following rules:

(1) We view attribute names and values as element names and values, respectively.

(2) We views texts (i.e., character strings) not directly contained by elements as a child element <#PCDATA>.

(3) We view child elements having the same attribute name and value as the same element only if the parent elements have the same name.

By assuming that e denotes an element, we define sub functions used by our evaluation functions as follows:

cnt (e): the number of occurrences of the starting tag of e.

mo (p(e), e): the maximum number of occurrences of a child e in a parent p(e)

np (p(e),e): 1- cnt (e) / (cnt (p(e)) * mo (p(e), e)); Null probability

For example, assuming that cnt (author) is 100 and cnt (penName) for M illustrated in Figure 1, the Null probability is calculated as follows:

np (author, penname) = 1 − cnt (penname) / (cnt (author) * mo (author, penname)) = 0.25

Here we summarize relations between models and database schemas. We call a sub-tree of a model a tree model. The number of tree models is equal to the number of tables in the database schemas. The name of the root node of a tree model is equal to the name of a table. The size of a tree model is determined by the number of the fields of a table. The name of a field is a concatenation of a name of an element and its order. Names of a parent and a child are delimited by ".".

For example, Figure 2(a) depicts a tree model for a model M in Figure 1 with a node author as its root. From this tree model, table schemas expressed by SQL are generated as in Figure 2(b).

3   Our Approach

3.1   Discovery of Complex Data

We describe how we discover complex data such as text formatting. Complex data are elements where child elements such as <#PCDATA>, <sup>, <sub>, <italic>, <bold> occur recursively. Apparently, it is inefficient to mechanically divide such complex structures into different tables. Instead, we store them as character strings. We call such complex data raw data in this paper. We discover raw data as complex data and simplify a final model.

We illustrate how we discover raw data by taking a character string as an example. The model depicted in Figure 3(a) is a sub-tree of a model depicted in Figure 1. We describe the algorithm by using this example. Figure 3(a) illustrates the maximum number of occurrences, too because the number is considered as important.

(step1) Candidate roots of raw data are {description, sub, italic} as character strings occur for more than or equal to two. Among the three, description has a shortest path to the root of the model, so the description node is chosen as the root of raw data.

(step2) Among candidate raw nodes are sub because the maximum number of occurrences of character strings is more than or equal to one. {bold, #PCDATA} are also among candidates because the number of kinds of child elements is less than two.

(step3) From descendants of the raw root node description, candidate raw root nodes and candidate raw nodes are deleted.

(step4) A node description_R occurring only once at most is added as a child element of the raw root node description.

The resultant structure is illustrated in Figure 3(b). We have successfully discovered <text> from standard.xml in the XML Benchmark Project [Schmidt et. al 01] and <title> dblp.xml in DBLP [DBLP 04] as complex data.

3.2   Division of Tables

Here we describe how to determine whether we should store elements occurring less often into one table or into separate tables. If we store elements occurring less often into one table, we can reduce the number of tables and increase unnecessary NULL values, too. So it is important to determine the target database schemas so that the number of tables is as small as possible and the function for evaluating unnecessary NULL values satisfies the prescribed threshold.

author

authorID
1
2
3
4
5

penName

penNameID	authorID
1	1
2	3
3	3
4	4
5	4
6	5
7	5

first

penNameID	first
1	
2	
3	
4	
6	
7	

last

penNameID	last
1	
2	
3	
4	
5	

(a)

	penName1.first	penName1.last	penName2.first	penName2.last
1			NULL	NULL
2	NULL	NULL	NULL	NULL
3				
4			NULL	
5		NULL		NULL

(b)

Fig. 4. (a) Tables to be merged. (b) Merged tables

We describe XML structures to be divided when design database schemas. Non-leaf child and parent elements whose maximum number of occurrences is more than or equal to ten, annotated as "*" in our approach, are to be divided. Child elements with more than one parent element are also to be divided. Such structures are to be divided by other approaches, too. The resultant model is a forest (i.e., a set of trees).

For example, a parent element Book and a child element author in Figure 1 are designed as separate tables because their maximum numbers of occurrences are more than or equal to ten (i.e., "*"). An element section is also designed as a separate table because it has more than one parent (i.e., book and section itself).

We describe how to divide tree models taking redundancy into consideration. The number of possible cuts of tree models is equal to the number of links. We select a candidate division of sub-trees so that the following three statistical functions (NULL expectation, Large Leaf Fields, and Large Child Fields) satisfy the prescribed thresholds and the number of tables is the smallest. A candidate with the smallest summation of NULL expectations is chosen among all candidates. We determined the above thresholds by experiments.

(1) NULL expectation (NE)

NULL expectation is a function for counting NULL values incurred by packing XML data into one table. For example, if four tables such as author, penName, first, and last are stored into one table, unnecessary NULL values are generated (See Figure 4(b)). NULL expectation is calculated by dividing the number of such NULL values by the number of records. In this case, NULL expectation is 9/5 (1.8). In general, NULL expectation is calculated as follows:

$$NE(T) = \sum_{l \in L(T)} \prod_{n=l}^{p(n)=r(T)} mo(p(n),n) - \sum_{l \in L(T)} \prod_{n=l}^{p(n)=r(T)} mo(p(n),n)(1 - np(p(n),n))$$

where

$$\prod_{n=l}^{p(n)=r(T)} mo(p(n),n) = mo(p(l),l) \cdot mo(p(p(l)),p(l)) \cdot \ldots \cdot mo(p(n) = r(T),n)$$

A database schema which has a lot of NULL values is considered as inefficient, so candidates to be divided, which contain tree models with NULL expectation higher than the threshold, are deleted.

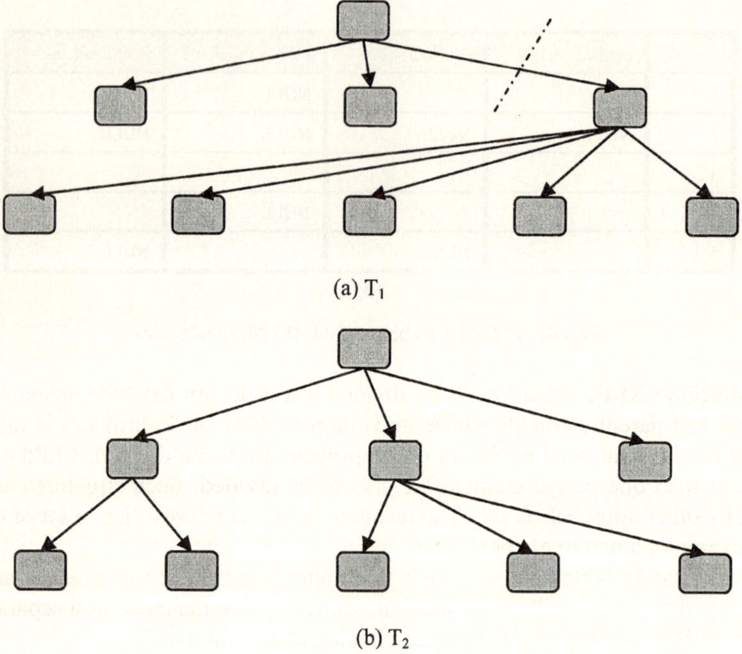

(a) T_1

(b) T_2

Fig. 5. Tree models

(2) Large Leaf Fields (LLF)

Large Leaf Fields is a function for counting most frequent leaf nodes, related to the cost of query processing. For example, if we retrieve comment nodes in the table structure described by Figure 2(b), we have to retrieve six fields such as prize1.comment1, prize1.comment2, prize2.comment1, prize2.comment2, prize3.comment1, and prize3.comment2.

The largest number of fields needed to retrieve a particular node is the function Large Leaf Fields. In general, the function is calculated as follows:

$$LLF(T) = \max_{l \in L(T)} \left(\prod_{n=l}^{p(n)=r(T)} mo(p(n),n) \right)$$

For example, we assume that T is a tree model in Figure 2(a). LLF(T)=max{(mo(penName,first)*mo(author,penName), mo(penName,last)*mo(author,penName)), (mo(prize,prizeName)*mo(author,prize), mo(prize,comment)*mo(author,prize))=(1*2, 1*2, 1*3, 2*3)}=6. In this case, 6 is the count for the comment node. A database schema which has a lot of fields necessary to retrieve one node is considered to produce a lot of "or" conditions in SQL commands. This can make the cost for query processing high. So it is considered as better that such a database schema is to be divided. Therefore, candidates to be divided, which contain tree models with LLF higher than the threshold, are deleted.

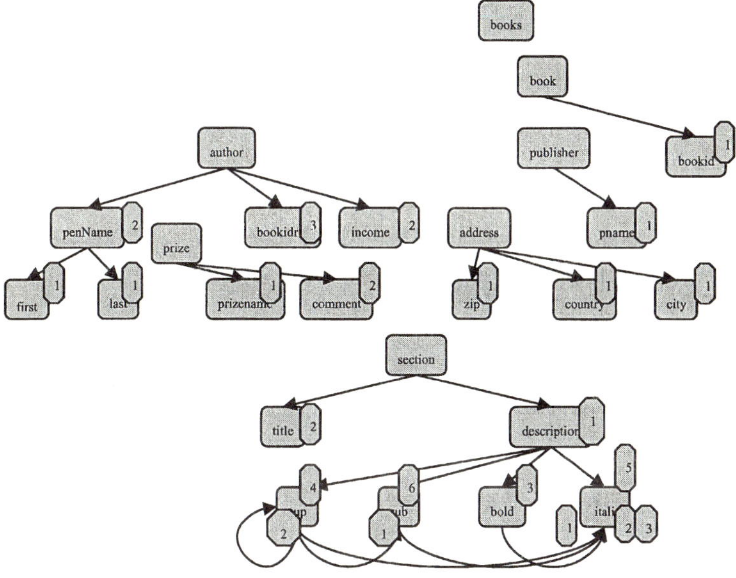

Fig. 6. Division of a model

(3) Large Child Fields

Large Child Fields is a function for counting the most frequent child nodes in a parent node, related to good division of tables. For example, in two tree models T1 and T2 (See Figure 5), we assume that |T1|=|T2| and NE(T1)=NE(T2) and LLF(T1)=LLF(T2). Then we will decide either of T1 or T2 to be divided. In this case, T1 is less balanced than T2. T1 is to be divided at a broken line.

To express such an imbalance of tree models, we count the most frequent child nodes in tree models. A tree model which has a lot of child nodes is considered to be unbalanced, so such child nodes are to be divided from the tree model. Therefore, candidates to be divided, which contain tree models with LCF higher than the threshold, are deleted. In general, LCF is calculated as follows:

$$LCF(T) = \max_{c \in C(T)} \left(\sum_{l \in L(c)} \prod_{n=1}^{p(n)=r(T)} mo(p(n), n) \right)$$

For example, LCF of a tree model T in Figure XXX is calculated by using the above formula.

LCF(T)=max{(mo(penName,first)*mo(author,penName)+ mo(penName,last)*mo (author,penName)), (mo(prize,prizeName)*mo(author,prize)+ mo(prize,comment)* mo(author,prize))=(1*2+1*2, 1*3+2*3)}=9.

In this case, the node prize is to be divided.

3.3 Resultant Database Schema

Database schemas generated by our approach consist of a master table for containing information about a XML document and elements and of a file table for containing XML instances. For example, we consider a model for book.xml (See Figure 1). We assume that book.xml is divided as in Figure 6. The resultant database schemas are depicted in Figure 7. The final design of schemas contains flags in each table for the match with the shortest path to the root. Such information helps to reduce the number of referenced tables and the number of joins.

4 Evaluation

We have done experiments in order to validate our approach. We have processed queries over database schemas generated both by our approach and by the usual normalization theory [Elmasri et al.99] and measured and compared the needed costs.

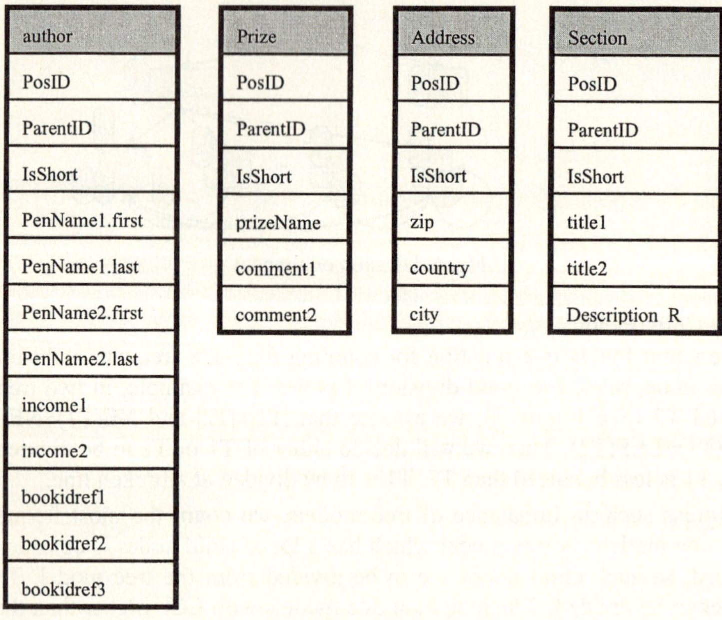

Fig. 7. Table schemas

In the normalized schemas, we have divided parent and child elements with the largest number of occurrences more than one. For experiments, we have generated XML data with the low largest number of occurrences. The generated XML documents have a size of 360MB and 10.2 millions of elements.

Figure 8(a) depicts statistics about the database schemas generated by our approach and normalization approach. The number of records corresponds to the total number of records contained by all the tables. Our approach has a smaller number of tables than the normalization approach. The number of records in our approach is also smaller. This can reduce both the database size and index size.

Our approach can reduce the number of tables and can increase the number of fields to retrieve at the same time as it normalizes database schemas partially. We have formulated three queries as follows:

(1) A basic query involving more than one field.
(2) A query returning all descendants of an element
(3) A query calculating an aggregate function

For all these queries, the costs of our approach are smaller than those of the normalization approach (See Figure 8(b)). This is because the number of referenced tables is smaller than that of the normalization.

5 Conclusion

We have proposed an approach to dynamically generate database schemas for well-formed XML data. Our approach controls the number of tables to be divided based on XML statistics so that the total cost of processing queries is reduced. We have devised schemas appropriate for complex data such as text formatting and child elements with the small maximum number of occurrences in order to reduce the number of tables. To this end, we have defined three functions NULL expectation, Large Leaf Fields, and Large Child Fields for controlling the tables to be divided. We have evaluated typical queries over the generated schemas and normalized schemas and measured and compared both of the costs. Through this, we have successfully validated our approach.

	Our approach	Normalization
# of tables	7	16
# of data	2,890,292	7,780,837
Data size (MB)	760	1801
Index size (MB)	71	296

(a)

	Our approach	Normalization
Basic query	0.99	26.28
Descendant	6.70	28.14
Aggregate	2.25	3.44

(b)

Fig. 8. (a) Statstics of database. (b) Query Processing Costs (second)

Acknowledgements

This work is partially supported by the Ministry of Education, Culture, Sports, Science and Technology, Japan under a Grant–in-Aid for Scientific Research on Priority Areas 16016273. We appreciate Takeyoshi Maku for his great efforts in the implementation and evaluation of the general ideas described in this paper.

References

[DBLP 04] DBLP (Digital Bibliography & Library Project): http://www.informatik.uni-trier.de/~ley/db/index.html Accessed 2004.

[Deutsch et al. 99] Alin Deutsch, Mary Fernandez, and Dan Suciu: Storing Semistructured Data with STORED. In *Proceeding of ACM SIGMOD Conference*, pp.431-442, June 1999.

[Elmasri et al. 99] Ramez Elmasri and Shamkant B. Navathe: *Fundamentals of Database Systems, Third Edition*, Pearson Addison Wesley, 1999.

[Florescu et al.99] Daniela Florescu and Donald Kossmann: Storing and Querying XML Data Using an RDBMS. *IEEE Data Engineering Bulletin*, vol.22, No.3, pp.27-34, 1999.

[Jiang et al. 02] Haifeng Jiang, Hongjun Lu, Wei Wang, and Jeffrey Xu Yu: Path Materialization Revisited: An Efficient Storage Model for XML Data. In *Proceedings of Thirteenth Australasian Database Conference*, pp.85-94, 2002.

[Klettke et al. 01] Meike Klettke and Holger Meyer: XML and Object-Relational Database Systems Enhancing Structural Mappings Based On Statistics. *Lecture Notes in Computer Science*, vol.1997, pp.151-170, 2001.

[Schmidt et. al. 01] A.R. Schmidt, F. Waas, M.L. Kersten, D. Florescu, I. Manolescu, M.J. Carey, and R. Busse: The XML Benchmark Project. *Technical report*, INS-R0103, CWI, 2001. http://monetdb.cwi.nl/xml/index.html. Accessed 2004.

[Shanmugasundaram et al. 99] Jayavel Shanmugasundaram, Kristin Tufte, Gang He, Chun Zhang, David DeWitt, and Jeffrey Naughton: Relational Databases for Querying XML Documents: Limitations and Opportunities. In *Proceedings of the VLDB Conference*, pp. 302-314, 1999.

[Tian et al. 02] Feng Tian, David J. De Witt, Jianjun Chen, and Chun Zhang: The Design and Performance Evaluation of Alternative XML Storage Strategies. *ACM Sigmod Record*, vol.31, no.1, pp.5-10, 2002.

[XML Schema 04] XML Schema: http://www.w3.org/TR/xmlschema-0/ Accessed 2004.

[Yoshikawa et al. 01] Yoshikawa, M. and Amagasa, T.: XRel: A path-based approach to storage and retrieval of XML documents using relational databases. *ACM Transactions on Internet Technology*, vol.1, no.1, pp.110-141, 2001.

[Zhang et al. 01] Chun Zhang, Jeffrey F. Naughton, David J. DeWitt, Qiong Luo, and Guy M. Lohman: On Supporting Containment Queries in Relational Database Management Systems. In *Proceedings of ACM SIGMOD Conference*, pp.425-436, 2001.

A New Cell-Based Clustering Method
for High-Dimensional Data Mining Applications

Jae-Woo Chang

Dept. of Computer Engineering and Research Center for Advanced LBS Technology
Chonbuk National University, Chonju, Chonbuk 561-756, South Korea
jwchang@chonbuk.ac.kr

Abstract. Many clustering methods are not suitable for high-dimensional data mining applications because of the so-called 'curse of dimensionality' and the limitation of available memory. In this paper, we propose a new cell-based clustering method for the high-dimensional data mining applications. The proposed clustering method provides efficient cell creation and cell insertion algorithms using a space-partitioning technique, as well as makes use of a filtering-based index structure using an approximation technique. In addition, we compare the performance of our cell-based clustering method with the CLIQUE method which is well known as an efficient grid-based clustering method for high-dimensional data. The experimental results show that our clustering method achieves better performance on cluster construction time and retrieval time.

Keywords: Cell-based clustering methods, high-dimensional data mining

1 Introduction

Data mining is concerned with the extraction of interesting knowledge from a large amount data, i.e. rules, regularities, patterns, constraints. One of the most important research topics in data mining is clustering. Clustering is the process of grouping data into classes or clusters, in such a way that objects within a cluster have high similarity to one another, but are very dissimilar to objects in other clusters [1]. The existing clustering methods have a critical drawback that they do not work well for clustering high-dimensional data because their retrieval performance is generally degraded as the number of dimension increases. In this paper, we propose an efficient cell-based clustering method for high-dimensional data mining applications. Our clustering method provides a cell creation algorithm to make cells by splitting each dimension into a set of partitions, and provides a cell insertion algorithm to construct clusters as cells with more density than a given threshold and insert them into an index structure. By using an approximation technique, we also propose a new filtering-based index structure to have fast accesses to the clusters.

The rest of this paper is organized as follows. The next section discusses related work on high-dimensional clustering methods. In Section 3, we propose a new cell-based clustering method. In Section 4, we analyze the performances of our cell-based clustering method. Finally, we draw our conclusion in Section 5.

2 Related Work

The existing clustering methods for high-dimensional data mining applications can be roughly classified into two groups, such as grid-based and partitioning [2].

R. Khosla et al. (Eds.): KES 2005, LNAI 3681, pp. 391–397, 2005.
© Springer-Verlag Berlin Heidelberg 2005

CLIQUE[3] and MAFIA[4] belong to the grid-based approach while PROCLUS[5], FINDIT[6], and DOC[7] belong to the partitioning approach. In this section, we discuss the typical clustering methods for each approach. First, CLIQUE(CLustering In QUEst) was proposed as a density-based clustering method. CLIQUE automatically finds subspaces(grids) with high-density clusters. CLIQUE produces identical results irrespective of the order in which input records are presented, and it does not presume any canonical distribution of input data. Input parameters are the size of the grid and a global density threshold for clusters. Next, CLIQUE scales linearly with the number of input records, and has good scalability as the number of dimensions in the data. PROCLUS was the first top-down clustering method. It samples the data, then selects a set of k medoids, and iteratively improves the clustering. PROCLUS is biased toward clusters that are hyper-spherical in space. While cluster may be found in differ subspaces, the subspaces must be of similar sizes since the user must input the average number of dimensions for the clusters.

3   An Efficient Cell-Based Clustering Method

Since the existing clustering methods assume that a data set is resident in main memory, they are not efficient for handling large amounts of data. As the dimensionality of data is increased, the number of cells increases exponentially, thus causing the dramatic performance degradation. To remedy that effect, we propose an efficient cell-based clustering method for handling large amounts of high-dimensional data. The overall architecture of our clustering method is shown in Figure 1.

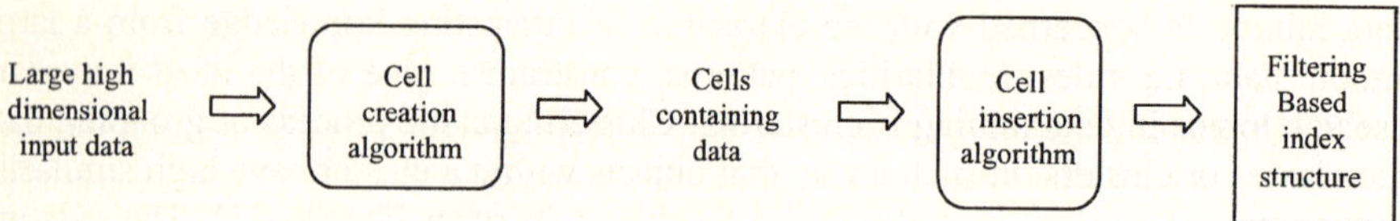

Fig. 1. Overall architecture of our clustering method

3.1   Cell Creation Algorithm

Our cell creation algorithm makes cells by splitting each dimension into a group of sections using a split index. Density based split index is used for creating split sections and is efficient for splitting multi-group data. Our cell creation algorithm first finds the optimal split section by repeatedly examining a value between the maximum and the minimum in each dimension. The split index value is calculated by Eq. (1) before splitting and Eq. (2) after splitting. Using Eq. (1), we can determine the split index value for a data set S in three steps: i) divide S into C classes, ii) calculate the square value of the relative density of each class, and iii) subtract from one all the square values of the densities of C classes. Using Eq. (2), we compute a split index value for S after S is divided into S_1 and S_2. If the split index value is larger than the previous value before splitting, we actually divide S into S_1 and S_2. Otherwise, we stop splitting. Secondly, our cell creation algorithm creates cells being made by the optimal split sections for n-dimensional data. As a result, our cell creation algorithm creates fewer cells than the existing clustering methods using equivalent intervals.

$$Split(S) = 1 - \sum_{j=1}^{C} P_j^{\;2} \tag{1}$$

$$Split(S) = \frac{n_1}{n} Split(S_1) + \frac{n_2}{n} Split(S_2) \tag{2}$$

If a data set has n dimensions and the number of the initial split sections in each dimension is m, the conventional cell creation algorithms make m^n cells, but our cell creation algorithm makes only $K_1*K_2*...*K_n$ cells ($1 \leq K_1, K_2, .. ,K_n \leq m$).

3.2 Cell Insertion Algorithm

For cell insertion, we first obtain the cells created using the cell creation algorithm. Secondly, we construct clusters as the cells with more density than a given cell threshold and store them into a cluster information file. Thirdly, we calculate the frequency of a section in all dimensions whose frequency is greater than a given section threshold. Finally, we set to '1' the bits corresponding to the sections with a high frequency in an approximation information file. We set to '0' the other bits for the remainder sections. we calculate the frequency of data in a cell. Finally, The cell threshold and the section threshold are shown in Eq. (3).

$$Section\ threshold = \begin{cases} \lambda = \dfrac{NR \times F}{NI} \\ NI : \text{the number of input data} \\ NR : \text{the number of sections per dimension} \\ F : \text{minimum section frequency being regarded as '1'} \end{cases} \tag{3}$$

$$Cell\ threshold(\tau) : \text{positive integer}$$

3.3 Filtering-Based Index Structure

In order to reduce the number of I/O accesses to the cluster information file, it is possible to construct a new filtering-based index scheme using the approximation information file. Figure 2 shows a filtering-based index scheme containing both the approximation information file and cluster information file. Let assume that K clusters are created by our cell-based clustering method and the numbers of split sections in X axis and Y axis are m and n, respectively. The following equation, Eq.(4), shows the retrieval times (C) when the approximation information file is used and when it is not used. We assume that $\square$ is an average filtering ratio in the approximation information file. D is the number of dimensions of input data. P is the number of records per page. R is the average number of records in each dimension. When the approximation information file is used, the retrieval time decreases as $\square$ decreases. For high-dimension data, our two-level index scheme using the approximation information file is an efficient scheme because the K value increases exponentially in proportion to dimension D.

  i) Retrieval time without the use of an approximation information file

$$C = \lceil K/P \rceil /2 \quad \text{(Disk I/O accesses)}$$

ii) Retrieval time with the use of an approximation information file

$$C = \lceil (D*R)/P \rceil * \alpha + (1-\alpha) \lceil K/P \rceil / 2 \text{ (Disk I/O accesses)} \qquad (4)$$

Figure 2 shows our filtering-based index scheme used to answer a query when a cell threshold and a section threshold are 1, respectively. For a query Q1, we determine 0.6 in X axis as the third section and 0.8 in Y axis as the fourth section. In the approximation information file, the value for the third section in X axis is '1' and the value for the 4-th section in Y axis is '0'. Because one of sections' values is '0', Q1 can be discarded without searching the corresponding cluster information file. For a query Q2, the value of 0.55 in X axis and the value of 0.7 in Y axis belong to the third section, respectively. Because the third bit for X axis and the third bit for Y axis have '1' in the approximation information file, we calculate a cell number and obtain its cell frequency by accessing the cluster information file. As a result, we obtain a cell number 11 and its frequency 3 for Q2.

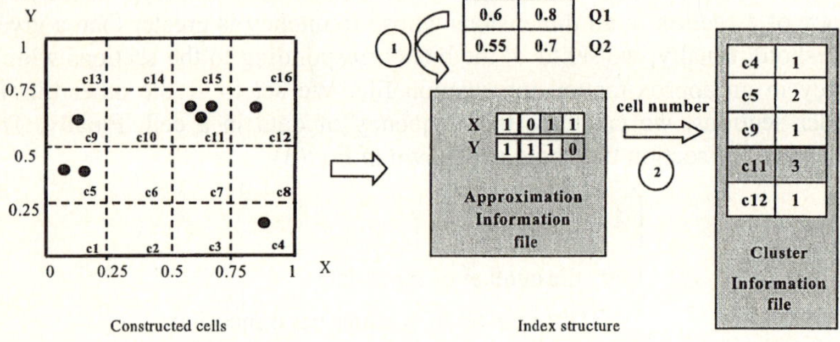

Fig. 2. Filtering-based index scheme

4   Performance Analysis

For our performance analysis, we implement our cell-based clustering method under Linux server with 650 MHz dual processor and 512 MB main memory. We also make use of one million 16-dimensional data which are made by Synthetic Data Generation Code for Classification in IBM Quest Data Mining Project [8]. A record is composed of attributes, such as salary, commission, age, elevel, zipcode hvalue, hyears, loan, area, children, ctype, job, tax, interest, cyear, and balance. The factors of our performance analysis are cluster construction time, precision, and retrieval time. Because our cell-based clustering methods belongs to a grid-based approach, we compare it with the CLIQUE method which is well known an efficient grid-based clustering method. Though MAFIA is a grid-based clustering method, we exclude it for our comparison because it is impossible to fairly compare MAFIA with our method due to MAFIA's parallel execution. For fair comparison, we also our method with the CLIQUE+A, which is made by applying an approximation information file to the CLIQUE method.

4.1   Performance Analysis Without Considering Section Thresholds

For cluster construction time, we set threshold values to 0 so as not to consider section thresholds. Figure 3 shows the cluster construction time of three methods. The experimental result shows that the CLIQUE and the CLIQUE+A need about 730 and 1,200 seconds, respectively. However, our cell-based clustering method needs only 100 seconds. Because our clustering method makes the much smaller number of cells than the CLIQUE, our method leads to 85% decrease in cluster construction time. Because the CLIQUE+A makes an approximation information file, it causes about 160% increase in cluster construction time, compared to the CLIQUE.

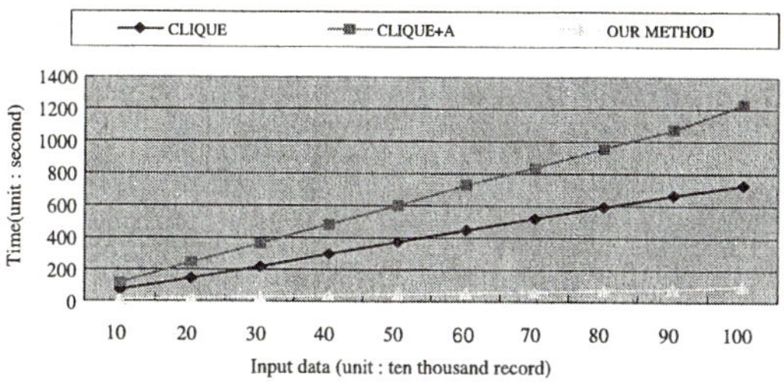

Fig. 3. Cluster construction time

Figure 4(a) shows the average retrieval time of three methods. The CLIQUE, the CLIQUE+A, and our clustering method need 31, 8, and 1 second, respectively. It is shown that our clustering method is much better on retrieval performance than the CLIQUE and the CLIQUE+A. This is because our clustering method creates the small number of cells by using our cell creation algorithm and achieves good filtering effect by using the approximation information file.

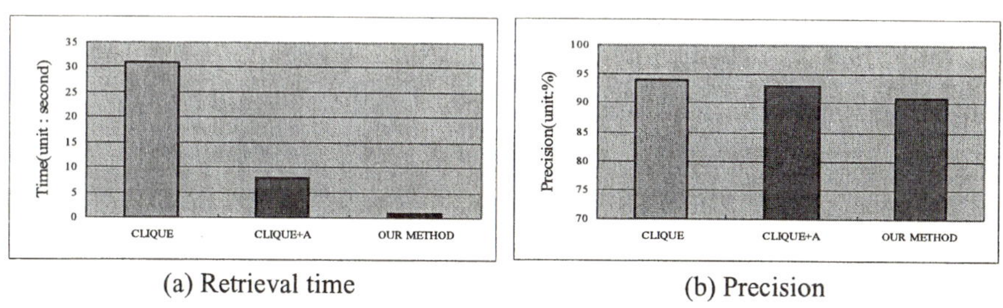

(a) Retrieval time   (b) Precision

Fig. 4. Retrieval time and precision

Figure 5 shows the performance of precision. The result shows that our clustering method and the CLIQUE+A achieve over 90% precision while the CLIQUE achieves about 94% precision. In the viewpoint of the precision, it is shown that the our clustering method is slightly worse than the CLIQUE. This is because our clustering method makes much smaller number of cells, compared with the CLIQUE.

Because the retrieval time and the precision have a trade-off relationship, we estimate a measure to combine them. For this, we define a system efficiency measure as Eq. (5). E_{MD} means the system efficiency of MD(method) and W_p and W_t mean the weight of precision and that of retrieval time, respectively. P_{MD} and T_{MD} mean the precision and the retrieval time of the three methods(MD). P_{MAX} and T_{MIN} mean the maximum precision and the minimum retrieval time, respectively, for all methods. When the weight of precision and that of retrieval time are assumed to be 0.5, respectively, our clustering method achieves about 1.0 on the system efficiency while the CLIQUE and the CLIQUE+A achieve about 0.5 and 0.6, respectively. Conclusively, our clustering method achieves the best performance on the system efficiency.

$$E_{MD} = W_p \bullet \frac{P_{MD}}{P_{MAX}} + W_t \bullet \frac{1}{\dfrac{T_{MD}}{T_{MIN}}} \tag{5}$$

4.2 Performance Analysis with Considering Section Thresholds

Because the CLIQUE method does not use an approximation information file, we only compare our clustering method with the CLIQUE+A. Figure 5(a) shows the retrieval times of the two methods. When the section threshold is 0.1, our clustering method requires about 8 seconds while the CLIQUE+A requires 1 second. It is shown that our clustering method is about one order of magnitude better than the CLIQUE+A. The result shows that retrieval time is decreased as the section threshold increases because we can reduce the number of records accessed by filtering. Figure 5(b) shows the precisions of the two methods. When the section threshold is 0.1, our clustering method achieves 92% precision while the CLIQUE+A achieves about 90% precision. This is because our clustering method makes the smaller number of cells than the CLIQUE+A. In addition, the precision is decreased as the section threshold increases. This is because as the section threshold increases, we can discard the more number of cells in the approximation information file, thus leading to low precision.

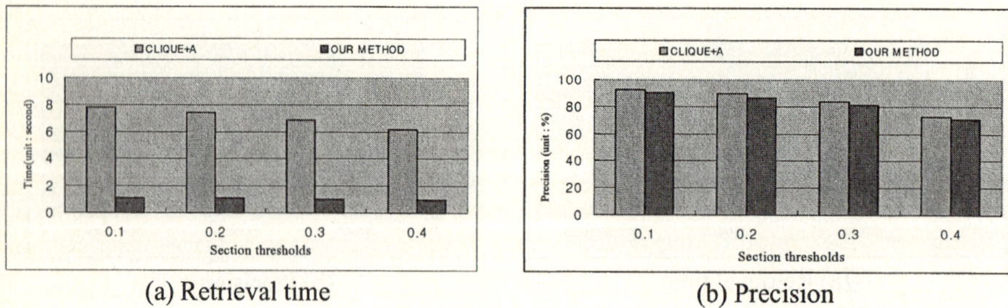

(a) Retrieval time   (b) Precision

Fig. 5. Precision and retrieval time with respect to section thresholds

We measure a system efficiency by using Eq. (6) in order to show an overall performance to combine retrieval time and precision. In Eq. (6), PMD(λ) and TMD(λ) denote the precision and the retrieval time of MD(method) when a section-threshold value is λ. Figure 6 shows the performance of system efficiency, EMD. Our cell-based clustering method is the most efficient when λ is 0.2 as well as Wp and Wt are 0.5.

$$E_{MD}(\lambda) = W_p \bullet \frac{P_{MD}(\lambda)}{P_{MAX}(\lambda)} + W_t \bullet \frac{1}{\dfrac{T_{MD}(\lambda)}{T_{MIN}(\lambda)}} \tag{6}$$

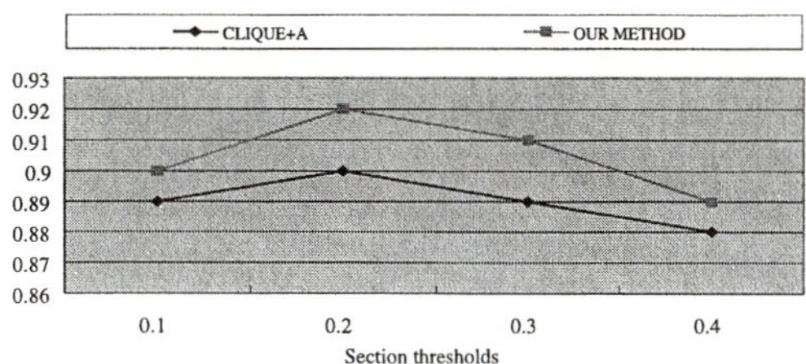

Fig. 6. System efficiency with respect to section thresholds

5   Conclusions

Because the existing clustering methods are not efficient for large amounts of high-dimensional data, we proposed an efficient cell-based clustering method so as to overcome the difficulty. It created the small number of cells, and it adopted an approximation technique for fast retrieval. For performance analysis, we compared our clustering method with the CLIQUE method. Our clustering method showed slightly lower precision, but it achieved good performance on retrieval time as well as cluster construction time. Finally, our method showed good performance on a system efficiency measure to combine both precision and retrieval time.

Acknowledgement

This research was supported by University IT Research Center Project.

References

1. Han, J., Kamber, M.: Data Mining: Concepts and Techniques. Morgan Kaufmann (2000).
2. Gan, G., Wu, J.: Subspace Clustering for High Dimensional Categorical Data. ACM SIGKDD Explorations networks, Vol. 6, Issue 2 (2004) 87-94.
3. Agrawal, R., Gehrke, J., Gunopulos, D., Raghavan, P.: Automatic Subspace Clustering of High Dimensional Data Mining Applications. Proc. of ACM SIGMOD (1998) 94-105.
4. Nagesh, H., Goil, S., Choudhary, A.: A Scalable Parallel Subspace Clustering Algorithm for Massive data Sets. Proc. of Int. Conf. on parallel Processing (2000) 477-486.
5. Aggrawal, C., Wolf, J., Yu, P., Procopiuc, C., Park, J.: Fast Algorithms for Projected Clustering. Proc. of ACM SIGMOD (1999) 61-72.
6. Woo, K, Lee, J.: FINDIT: A Fast and Intelligent Subspace Clustering Algorithm using Dimension Voting. PhD thesis, Korea Advanced Institute of Sci.&Tech, Dept. of CS (2002).
7. Procopiuc, C, Jones, M., Agrawal, P., Murali, T.: A More Carlo Algorithm for Fast Projective Clustering. Proc. of ACM SIGMOD (2002) 418-427.
8. http://www.almaden.ibm.com/cs/quest

An Application of Apriori Algorithm
on a Diabetic Database

Nevcihan Duru

Department of Computer Eng., University of Kocaeli, 41440, Izmit, Kocaeli, Turkey
nduru@kou.edu.tr

Abstract. In recent days, mining information from large databases has been recognized by many researchers and many data mining techniques and systems have been developed. In this study, a software (DMAP), which uses Apriori algorithm, was developed. Apriori is an influential algorithm that used in data mining. The name of the algorithm is based on the fact that the algorithm uses prior knowledge of frequent item set properties. The software is used for discovering the social status of the diabetics. A diabetic database that belongs to faculty of medicine of Kocaeli University has been used. The software was executed on a database which has records of 66 patients for test purpose. In the literature, diabetic databases have been often analyzed by rough sets. In this paper, Apriori algorithm, which has been usually used for the market basket analysis, was used for analyzing a diabetic database.

1 Introduction

The explosive growth in databases has generated an urgent need for new techniques and tools that can intelligently and automatically transform the processed data into useful information and knowledge [1]. In fact, as data volumes grow dramatically, this type of manual data analysis is becoming completely impractical in many domains. Databases are increasing in size in two ways: (1) the number N of records or objects in the database and (2) the number d of fields or attributes to an object [2]. Therefore, data mining has become a research area with increasing importance [3,4]. Although data mining and knowledge discovery in databases are often treated as synonym, data mining is actually part of the knowledge discovery process. There have been many advances on researches and developments of data mining, and many data mining techniques and systems have recently been developed. Different classification schemes can be used to categorize data mining methods and systems based on the kinds of databases to be studied, the kinds of knowledge to be discovered, and the kinds of techniques to be utilized [5]. In this approach, Apriori algorithm, on a diabetic database, has generated association rules.

There have been a lot of works on diabetic databases for different purposes. Micheal and Beguin have used a database to query for diabetes mellitus [6]. Kopelman and Sanderson used a database to provide continuous quality improvement in diabetes care [7]. Breault used rough sets on Pima Indian diabetic database which has become a standard for testing data mining algorithms to see their accuracy in predicting diabetic status [8]. According to Knowler and Bennett et al., the Pima Indians may be genetically predisposed to diabetes [9]. For this reason, there have been many studies on data mining techniques to the Pima Indian database.

R. Khosla et al. (Eds.): KES 2005, LNAI 3681, pp. 398–404, 2005.

Diabetes is suitable for applying data mining technology, for a number of reasons. First, there is tremendous amount of data. Second, diabetes is a disease that can cause many complications of blindness, kidney failure, amputation, premature cardiovascular death and so on [8]. Third, it can be decided that a person's predisposition to diabetes by examining these type of database.

1.1 A Brief Look at Data Mining and the Models

Data mining is a step in the knowledge discovery process. However, in industry, in media and in the database research area, the term data mining has become more popular than the term of knowledge discovery in databases. It means a process of nontrivial extraction of implicit, previously unknown and potentially useful information (such as knowledge rules, constraints, regularities) from data in databases [10]. Data mining should be applicable to relational databases, data warehouses, World Wide Web and advanced database systems like object oriented and object relational databases.

There have been many methods for mining different kinds of knowledge, including association rules, characterization, classification, clustering, etc. In general, the models that are used in data mining can be classified into two categories: predictive and descriptive [1]. Descriptive mining tasks characterize the general properties of the data in the database. Predictive mining tasks perform inference on the current data in order to make predictions. Classification and regression are predictive; association rules and clustering are descriptive models [11].

Association analysis is the general process of determining which things go together. That means it is the process of discovering association rules showing attribute-value conditions that occur frequently together in a given set of data. Association rules are unlike traditional classification rules in that an attribute appearing as a precondition in one rule may appear in the consequent of second rule. In addition, traditional classification rules usually limit the consequent of a rule to contain one or several attribute values [12]. It is widely used for market basket analysis. In this approach it is applied to a diabetic database by means of developed software.

A mathematical model was proposed in [13] to address the problem of mining association rules. Let $I=\{ i_1, i_2, \ldots, i_m\}$ be a set of literals, called items. Let D be a set of transactions, where each transaction T is a set of items such that $T \subseteq I$. Note that the quantities of items bought in a transaction are not considered, meaning that each item is a binary variable representing if an item was bought. Each transaction is associated with an identifier, called TID.

Let X be a set of items. A transaction T is said to contain X if and only if $X \subseteq T$. An association rule is an implication of the form $X \Rightarrow Y$, where $X \subset I$, $Y \subset I$ and $X \cap Y = \emptyset$. The rule $X \Rightarrow Y$ holds in the transaction set D with confidence c if c% of transactions in D that contain X also contain Y. The rule $X \Rightarrow Y$ has support s in the transaction set D if s % of transactions in D contains $X \cup Y$.

Confidence denotes the strength of implication and support indicates the frequencies of the occurring patterns in the rule. It is often desirable to pay attention to only those rules, which may have reasonably large support. Such rules with high confidence and strong support are referred to as strong rules. The task of mining association rules is essentially to discover strong association rules in large databases [5].

When several attributes are present, the association rules generating process becomes difficult to deal with because of the large number of possible conditions for the consequent of each rule. To generate the rules efficiently, special algorithms have been developed. One such algorithm is the Apriori algorithm [14]. Apriori algorithm is one of the most known algorithm used to generate association rules.

This algorithm generates item sets, which are attribute-value combinations. Those attribute-value combinations, which do not meet the coverage requirement, are discarded. Because of this, the rule generation process can be completed in a reasonable amount of time.

Apriori association rule-generation is a two-step process. The first step is item set generation and the second step is to generate a set of association rules by using the generated items.

2   Implementation of the Algorithm

In the approach, a software, which uses Apriori algorithm was developed. Apriori is an influential algorithm, which is used in data mining. The software is named as DMAP (**D**ata **M**ining with **Ap**riori). Here DMAP has been used for defining the diabetic persons' social status. A diabetic database, which has been formerly created by the Faculty of Medicine, Kocaeli University were used. The database originally was created in SPSS. It has n=66 patients each with 8 variables. These 8 variables are: 1) Family Type, 2) Age, 3) Career, 4) Date of diagnose, 5) Marital status, 6) Education, 7) Method of care type, 8) Sex. The available values for these variables (attributes) are given in Table 1.

Table 1. The available values for the attributes

DV	FmlyType	Age	Career	Beg.Date (years)	MaritalStatus	Education	MethodofCare	Sex
1	BasicFamily	30-39	Housewife	Less than 1	Married	Literate	Insulin	Male
2	LargeFamily	40-49	Retired	1-5	Unmarried	Primary	Oral	Female
3	Alone	50-59	Independent	6-10	Widow	Secondary	Diet	
4		60-69	Official	11+		University		
5		70+	Worker					
6			Unemployed					

The database file which has been built in SPSS, was converted to MS Access file. During this conversion process, only integer values between 1 and 6 were used. These six values are shown in the DV field of Table 1. For example, number 2 was used, if the patient was "female", and number 6 was used, if the patient was "unemployed". After this value conversion process, the database was shaped as shown in Figure 1.

DMAP was developed by using Borland Delphi 7.0. The first created item set table contains single-item set. This set is shown in Table 2. These values were determined by the DMAP interface which is shown in Fig. 2.

In the first iteration, Apriori simply scans all the transactions to count the number of occurrences for each item. The candidate 1-itemsets obtained is shown in Table 2.

FmlyType	Age	Career	MaritalStatus	Education	MethodofCare	BegDate	Sex
1	5	1	3	1	1	1	1
1	5	2	1	2	1	3	2
1	5	1	3	6	2	3	1
1	4	2	1	2	2	2	2
1	5	2	1	2	2	1	2
1	4	3	1	4	1	3	2
3	5	3	3	2	2	3	2
1	5	2	1	2	1	3	2
1	4	1	3	2	1	1	1
1	5	2	1	2	2	1	2

Fig. 1. Diabetic database for 10 samples

Table 2. Single-item set

Single-Item sets	Number of Items	Single-Item sets	Number of Items
Family Type=1	51	Education=1	10
Family Type=2	13	Education=2	36
Family Type=3	2	Education=3	6
Age=1	3	Education=4	4
Age=2	21	Education=5	3
Age=3	14	Education=6	7
Age=4	17	MethodofCare=1	22
Age=5	11	MethodofCare=2	42
Career=1	36	MethodofCare=3	2
Career=2	21	Sex=1	40
Career=3	9	Sex=2	26
MaritalStatus=1	54		
MaritalStatus=3	12		
BeGdate=1	28		
BeGdate=2	16		
BeGdate=3	22		

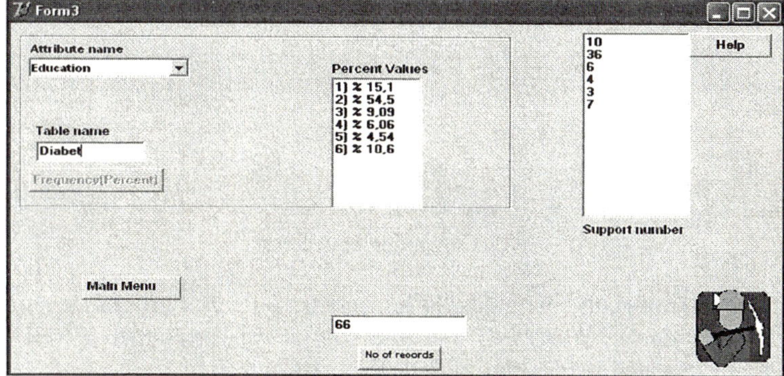

Fig. 2. DMAP interface (frequency values of the attributes)

Assuming that the minimum support required is 6, some items shown in Table 2 are discarded and new set is produced. This new set is shown in Table 3. As noticed,

Table 3. The set of frequent 1-itemset

Single-Item sets	Support	Single-Item sets	Support
Family Type=1	51	BeGdate=1	28
Family Type=2	13	BeGdate=2	16
Age=2	21	BeGdate=3	22
Age=3	14	Education=1	10
Age=4	17	Education=2	36
Age=5	11	Education=6	7
Career=1	36	MethodofCare=1	22
Career=2	21	MethodofCare=2	42
Career=3	9	Sex=1	40
MaritalStatus=1	54	Sex=2	26
MaritalStatus=3	12		

reproducing new n-itemsets is available with DMAP. By means of executing the interface shown in Fig.3. iteratively, the sets are generated. As seen on the top of the 2-item set in Fig. 3, our first possibility is:

***Family Type=1*(Basic Family) & *Career=1* (Housewife).** The rule confidence is obtained as 39.3%. The next step is to use the attribute-value combinations from the 2-item set table to generate 3-item sets.

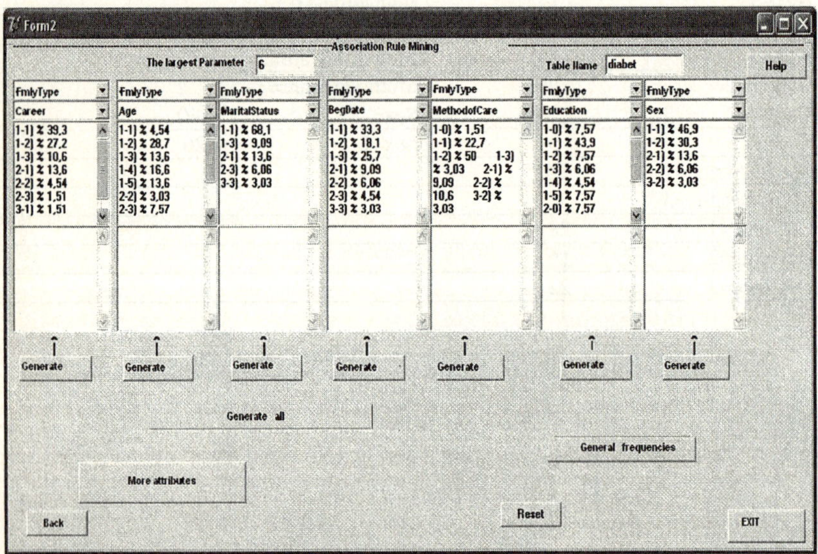

Fig. 3. Generating of 2-itemset

In Fig.4. it is shown another example for another 2-itemset groups. In *Age=2* **(40-49) & *MaritalStatus=1* (Married)**, the rule confidence is obtained as 30.3%. This process can be repeated for generating 3-itemset sets and so on.

Association rules are particularly popular because of their ability to find relationships in large databases without having the restriction of choosing a single dependent variable. However caution must be exercised in the interpretation of association rules since many discovered relationships turn out to be trivial [12].

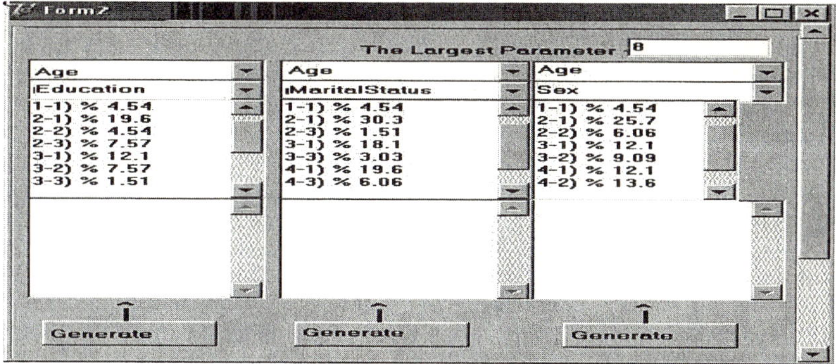

Fig. 4. Generating of 2-itemset for another items

3 Conclusions

In this approach, a software, which uses Apriori algorithm, was developed. Even there are many commercial tools that one can make data mining, they are costly and not available in our university. In this work, our purpose was to develop a software which could be used for data mining.The software is used for discovering the social status of the diabetics. A diabetic database that belongs to faculty of medicine of Kocaeli University has been used. It has 66 records of patients, each with 8 numeric variables. The purpose of this software is to serve to analyze the diabetics. Before the development of this software, SPSS database could only be surveyed by eye and only single-itemset relations could have been generated. The use of this software made possible to generation of two, three and even four itemsets. It concluded that, developed software and the methodology have served the purpose and worked well. A comparative analysis of different data mining techniques seems as an interesting work for the near future.

References

1. Han, J. and M. Kamber (2001). Data mining: concepts and techniques. San Francisco, Morgan Kaufmann Publishers.
2. Fayyad U.:From Data Mining to Knowledge Discovery in Databases, American Association for Artificial Intelligence, 1996.
3. U. M. Fayyad, G. Piatetsky-Shapiro, P. Smyth, and R. Uthurusamy. Advances in Knowledge Discovery and Data Mining. AAAI/MIT Press, 1996.
4. G. Piatetsky-Shapiro and W. J. Frawley. Knowledge Discovery in Databases. AAAI/MIT Press, 1991.
5. Chen M., Han J.: Data Mining: An Overview from Database Perspective, *IEEE* Transactions on Knowledge and Data Engineering, 8(6):866-883, 1996.
6. Michel, C. and C. Beguin: Using a database to query for diabetes mellitus, Stud Health Technol Inform 14: 1994, 178-182.
7. Kopelman, P. G. and A. J. Sanderson: Application of database systems in diabetes care, Med Inform (Lond) 21(4): 1996, 259-271.
8. Breault, J. L.: Data Mining Diabetic Databases: Are Rough Sets a Useful Addition?, Computing Science and Statistics, Vol:34, 2001.

9. Knowler, W. C., P. H. Bennett, et al. (1978). "Diabetes incidence and prevalence in Pima Indians: a 19-fold greater incidence than in Rochester, Minnesota." Am J Epidemiol 108(6): 497-505.
10. G. Piatetsky-Shapiro and W. J. Frawley. Knowledge Discovery in Databases. AAAI/MIT Press, 1991.
11. Berry, M.J.A., Linoff, G.S.:Mastering Data Mining:The Art and Science of Customer relationhip Management, John Wiley & Sons, 1 st Ed., 1999.
12. Roiger, R.J., Geatz M. W.: Data Mining, A Tutorial-based Primer, Addison Wesley,2003.
13. Agrawal, R., Imielinski, T., Swami, A.:Mining Association Rules between Sets of Items in Large Databases. Proceedings of ACM SIGMOD, pages 207-216, May 1993.
14. R. Agrawal and R. Srikant. Fast Algorithms for Mining Association Rules in Large Databases. Proceedings of the 20th International Conference on Very Large Data Bases, pages 478-499, September 1994.

Network Software Platform Design
for Wireless Real-World Integration Applications

Toshihiko Yamakami

ACCESS, 2-8-16 Sarugaku-cho, Chiyoda-ku, Tokyo, Japan
`yam@access.co.jp`

Abstract. The mobile Internet quickly penetrates the every-day life. After reaching the mobile multimedia stage, the real world integration is one of the next challenges in the mobile software frameworks. The author proposes an Internet-converged embedded web solution for the real world integration named *DirectConnect* in the analogy of the Live Connect in the Netscape Navigator. The author describes evaluation on five approaches for the enhanced real world integration and presents advantages of the minimum modification approach. The proposed approach is a browser-centric Internet-converged solution that enables the micro-browser on the mobile handsets to be an integration point of the real world integration. It uses the common Internet framework, and is adaptable to a wide range of web and real world integration. The scheme encapsulates the security issue in each *object* implementation.

1   Introduction

The embedded non-PC Internet and its real world integration is a challenge for the next decade [1]. The mobile Internet quickly penetrates the every-day life. The emerging mobile Internet is witnessed in Asia [2]. There were 73.5 million mobile Internet users using micro browsers in their mobile handsets in Japan at the end of 2004. Other digital appliances like game consoles, automotive systems and digital TV start to become ready for the Internet connection. The digital appliances have their own use scenes, which need more context-awareness. Context-aware computing has a fit with mobile computing, because *context-aware computing is characterized by the ability of a software system to continuously adapt its behavior to a changing environment* [3]. The context aware mobile application research started in the middle of 1990's [4]. In addition to the fundamental mobility in the mobile Internet, the demands for the integration with the physical environment are quickly growing. The 3G mobile phone in Japan reached 18 millions in May 2004. The 3G mobile Internet services have the two different directions:

- mobile multimedia [5], and
- real world integration.

The real-world integration quickly draws attention from the mobile carriers. The converged web and virtual-real linkage research includes HP Lab's Cooltown [6].

R. Khosla et al. (Eds.): KES 2005, LNAI 3681, pp. 405–411, 2005.
© Springer-Verlag Berlin Heidelberg 2005

Dating back the Internet history, the URI scheme "tel:" was the first example of the real-world integration in the mobile Internet. "tel:" is an example to integrate the mobile web and the wireless telephony environment. It is a real turning point for wireless telephony. The data communication reaches a certain stage where the fundamental nature of the wireless communication may change. Originally, the mobile data communication continues to seek the higher bandwidth in the restricted wave resource. However, with the introduction of flat rate data services, the new business models in non-human communication applications need inevitably to be explored to replace the per-packet charge business models. One example of the carrier's exploration of real world integration is the *wallet handset*, with the smart-card (*FeliCa* card) for e-money, ticketing and other handy mobile functions. This is another indication of the real world integration. The e-money function needs the peer-to-peer communication function. It accelerates the real world equipment that can communicate with e-money on the handset.

2   Challenges

In order to cover the variety of the real-world integration and enable easy integration, a common middleware platform available on the digital appliances are needed. There are four major requirements for real-world integration:

- Internet convergence,
- Rich Functionality,
- Availability, and
- Security.

The first requirement is Internet convergence. When the increasing digital appliances are connected to the Internet, the Internet convergence enables the efficient resource reuse and fit to the download mechanism. The mobile Internet is rapidly shifting toward Internet convergence in standardization [7]. The second requirement is rich functionality. The monitoring of the real world needs different operations with different granularity depending upon the use context. The flexibility of the capability is needed. The third requirement is availability. The function should be usable in a wide range of digital appliances and networks. The fourth requirement is security. When the Internet converged real world integration is attacked from the hackers in the Internet, it will cause a significant damage due to the integration capability that impacts the real world. The secure method with the sufficient granularity is needed. These requirements sometimes contradict. Therefore, the trade-off design is needed.

3   DirectConnect Design

The authors propose the DirectConnect service that enables the secure device control/monitoring with digital appliances. In order to enable the real-world integration, the mechanism to invoke the real world integration needs in the mark up language or scripting languages. The approaches include:

- Create a new language,
- Create a new URI scheme,
- Create a new attribute in a markup language, and
- Use programming mechanism e.g. Live Connect in Java.

Our approach is based on the 4th approach to enhance the available programming components, avoiding creating new things in the embedded web. It is an analogy of Live Connect in Netscape Navigator [8]. When the Live Connect was coined, the real world Internet integration was still in a very early stage, therefore, the Live Connect was targeted to the multiple browser integration. Live Connect facilitated the plug-in program control from Java and JavaScript (equivalent to ECMAScript). ECMAscript [9] provides the scripting ability embedded in the XHTML web pages. XHTML is an XML-cast HTML [10]. It will be a standard content language for the mobile Internet [11] in the mobile Internet industrial standards. In DirectConnect, the principle components are XHTML+ECMAscript in align to the Internet-convergence in the mobile Internet. The security level can be set per method and per attribute. In order to facilitate the real-world integration, it can accept the external triggered event. An example of DirectConnect is depicted in Fig. 1. The XHTML+ECMAscript program is downloaded from a HTTP server with optional SSL encryption. The NetFront 3 Browser executes the ECMAscript program to realize the NetFront DirectConnect interaction with devices. When the security is required, the NetFront browser implementation checks each execution status.

A device definition example based on XHTML is described in Fig. 2. Each device has the corresponding object definition with an object element with type attribute describing the MIME type for each device.

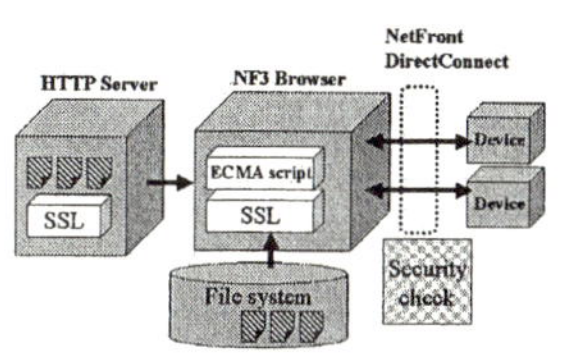


```
<html>
<head>
<object name=xxx
  type="application/x-nf3-zzzzz">
</object>
</head>
...
```

Fig. 1. Example of DirectConnect.

Fig. 2. A device definition example based on XHTML.

A script access example based on ECMAscript is shown in Fig. 3. When the button is clicked, switch-on operation described in the *onclick* attribute is performed.

A device-triggered operation example is depicted in Fig. 4. The usual XHTML script is user-operation-oriented. The *onValueChange* is used to overwrite the device-triggered function. The f function is called when the object changes its value using *onValueChange* function registration. This enables the device to browser event handling in an Internet-compliant way.

```
<html>
<head>
<script>
document.write(xxx.value);
</script>

<button onclick="xxx.switch(true)">
switch on
</button>
</head>
...
```

Fig. 3. A script access example based on ECMAscript.

```
<html>
<head>
<object name=xxx
 type="application/x-nf3-zzzzz">
</object>

<script>
function f() {
    window.alert("sensor detects value-change");
}
xxx.onValueChange = f;
</script>
</head>
...
```

Fig. 4. A device-triggered operation example.

Usability including interactions with helpdesk can be achieved using common XHTML.

The security level can be controlled on attribute-based or property-based. Each implementation of method on each property is implemented in a separate code. In each code implementation, the security policy is embedded. The security policy includes:

- restriction of URI scheme, like *file:*
- restriction of Internet domains
- restriction of server certificate

The security policy is determined by device identifier or user identifier. The appropriate identifier is verified by:

- network identifier like those in the mobile telephony
- device identifier like in *utn* attribute in i-mode (utn denotes Universal Telephone Number)
- user identifier like those in the mobile Internet proprietary deployments

4   Discussion

The digital appliance real world integration covers a wide range of applications. Examples are depicted in Table 1.

The evaluation of Internet-converged integration is depicted in Table 2. The new language is easy to coin for each application domain, however, it is difficult and time-consuming to increase user agents to support the language in a standardized manner. The new attribute is a more practical approach, however, the expressing ability of the programming features are quite restricted especially in the programmable features like event-handling. The new URI scheme is a new way to provide activation in the Internet framework, however, it is difficult to maintain the security level because the interactions between two different user agents are not protected. It also difficult to provide a universal method to provide programmable features. The programming language like Java is a more feasible approach than creating a new language from the scratch. It is a more Internet

Table 1. Digital appliance integration examples.

telephone	call control
	telephone book control
voice guidance	voice synthesis control
printer	print control
indicator	display control
soft-key	display control
GPS	current position acquisition
meter	current position acquisition
device configuration	setup control
	reservation confirmation

Table 2. Evaluation of Internet-converged integration.

	Internet Convergence	Functionality	Availability	Security
DirectConnect	○	○	○	○
New Language	×	○	×	△
New Attribute	△	△	×	×
New URI scheme	△	△	×	△
Programming Language	△	○	×	×

converged approach, however, it needs more memory than DirectConnect approach which implements the minimum code in the real-world integration part. It is coded by the native language like C, however, the operation interface is encapsulated by the object interface, which provides more security-protected than the arbitrary programming language.

This object approach can encapsulate the real world event handling in an Internet-compliant manner. The object based encapsulation protects the scripting part compared to the arbitrary programming language implementation. The scripting part reuses the Internet-converged ECMAScript which minimizes the footprint increase. When the scripting language is re-invented separately, the footprint is increased by 50% or even doubled because scripting language implementation needs detailed event-handlings and their resource handling. The footprint increase from this approach is significantly small compared to the other non-Internet-compliant approaches. Our implementation shows that the increase of the DirectConnect footprint is less than 10 Kbytes, depending upon the complexity of the real world integration exchange data size. The proposed approach combined with the NetFront browser architecture can be applied to a wide range of the execution platforms, Palm OS, Symbian, BREW, Wireless T-Engine, Linux and iTRON. Java is still on the way for the real world integration language. It has the restricted upload and download target that needs relaxation when the real world integration is aimed. The new programming language will duplicate the device handling in the existing ECMAScript. In order to enhance the security level, the standardized SSL can be used to secure the connection using enciphering. The event handling is a key for the bilateral real world integration. It can be implemented in

- operating system layer,
- browser application environment layer, and
- application layer.

The operating system layer approach is promising when the execution environment leads to the uniform one. Otherwise, the porting and testing cost is prohibiting. The application layer approach is flexible and easy to port, however, it does not provide the footprint-optimized solution. As the XHTML and ECMAscript become common in the embedded browser environment, the reuse of the ECMAscript event handling execution environment is easy to implement and provides the footprint-optimized solution.

In the past research, the context-aware software platform varies from network-specific middleware approaches to centralized message passing approaches. When the context-aware network is built from the scratch, hard-wired middleware can be build, but it lacks the network-wide applicability. An example of multiple sensor network architecture is [12]. On the other hands, in the pervasive software environment, Java is preferred for its execution neutrality. Application framework is build upon it like [13]. There are hundred millions of Java-enabled mobile handsets available, however, the execution environment for browsers and Java VM are not well integrated. In the footprint-optimized solutions, the XHTML-ECMAScript solution provides the simple Internet-converged approach with environment encapsulation in each *object* element implementation. In the past, Java was the only choice to install a standardized execution environment on mobile handsets or other non-PC devices. As the micro-browser becomes a standard component for the mobile Internet, it provides another candidate for the integration point for the mobile and real-world integration. It provides a standard approach to embed real world integration functions in non-PC devices.

5   Conclusion

The advanced wireless data communication quickly turns the corner from the high bandwidth applications to the real world integration applications. There is a need for the Internet-compliant real world integration solution. Especially, a compact and secure middleware platform that can be run on a wide variety of digital appliance is needed for easy real-world integration. The author proposes an object and event based approach to include the real world integration in a secure manner in the W3C-compliant Internet converged solution. When used with the Internet-compliant browser, this approach avoids the duplicate implementation of the scripting engine as well as the secure encapsulation of real-world operation interfaces. It facilitates the XHTML-ECMAscript-based footprint optimized solutions. The real world empowered embedded browsers will play an important role in the emerging ubiquitous environment with its simple and secure real world context aware capability. As the micro-browser becomes a standard component for the mobile Internet, it provides another candidate for the integration point for the mobile and real-world integration. This enables a wide variety of *embedded intelligence* and *embedded guidance* in the collaborative real-world integration applications to embed the monitoring/triggering functions in the real world environment.

References

1. Cerf, V.: Beyond the post-PC internet. CACM **44** (2001) 34–37
2. Barnes, S.J., Huff, S.L.: Rising sun: imode and the wireless internet. CACM **46** (2003) 78–84
3. Roman, G.C., Julien, C., Huang, Q.: Network abstractions for context-aware mobile computing. In: Proceedings of the 24th international conference on Software engineering, ACM Press (2002) 363–373
4. Abowd, G., Atkeson, C., Hong, J., Long, S., Kooper, R., Pinkerton, M.: Cyberguide: a mobile context-aware tour guide. Wireless Networks **3** (1997) 421–433
5. Yamakami, T.: Leveraging information appliances: A browser architecture perspective in the mobile multimedia age. In Chen, Y.C., Chang, L.W., Hsu, C.T., eds.: Advances in Multimedia Information Processing - PCM 2002, Third IEEE Pacific Rim Conference on Multimedia. Volume LNCS 2532., Hsinchu, Taiwan (2002) 1–8
6. Kindberg, T. et al: People, places, things: web presence for the real world. Mobile Networks and Applications **7** (2002) 365–376
7. Baker, M., Ishikawa, M., Matsui, S., Stark, P., Wugofsky, T., Yamakami, T.: XHTML™ basic. W3C Recommendation 19 December 2000, available at http://www.w3.org/TR/xhtml-basic (2000)
8. Bouvin, N.: Designing open hypermedia applets: experiences and prospects. In: Proceedings of the Ninth ACM conference on Hypertext and hypermedia, ACM Press (1998) 281–282
9. ECMA: Standard ecma262 ecmascript language specification 3rd edition. available at http://www.ecma-international.org/publications/files/ECMA-ST/Ecma-262.pdf (1999)
10. Altheim, M. et al: Modularization of XHTML™. available at http://www.w3.org/TR/2002/REC-xhtml-modularization-20010410/ (2001)
11. Open Mobile Alliance™: XHTML™ mobile profile v1.1. http://www.openmobilealliance.org/ (2002)
12. Gellersen, H., Schmidt, A., Beigl, M.: Multi-sensor context-awareness in mobile devices and smart artifacts. Mobile Networks and Applications **7** (2002) 341–351
13. Samulowitz, M., Michahelles, F., Linnhoff-Popien, C.: Adaptive interaction for enabling pervasive services. In: Proceedings of the 2nd ACM international workshop on Data engineering for wireless and mobile access, ACM Press (2001) 20–26

Development of Ubiquitous Historical Tour Support System

Satoru Fujii[1], Yusuke Takahashi[1], Hisao Fukuoka[1],
Teruhisa Ichikawa[2], Sanshiro Sakai[2], and Tadanori Mizuno[2]

[1] Matsue National College of Technology, 14-4 Nishi-Ikuma Matsue 690-8518, Japan
{fujii,fukuoka}@matsue-ct.ac.jp,
s0513@stu.cc.matsue-ct.ac.jp
[2] Faculty of Information, Shizuoka University, 3-5-1, Johoku Hamamatsu 432-8011, Japan
{ichikawa,sakai,mizuno}@inf.shizuoka.ac.jp

Abstract. We have developed the historical tour support system that provides its users with appropriate information according to each phase of the tour ubiquitously. There are many systems to serve sightseeing information used WWW, but most contents of those systems are not so interesting. In addition, there are few systems to support the lifecycle of historic tour. Our system has interesting display such as 3D graphics and multilingualism for users before sightseeing. And this system serves information about the historic spot where they are now with a cellular phone and GPS. After the tour user can enjoy his picture albums that have been produced automatic by the server during tour. We have tried this system and have gotten a good evaluation by 10 user's questionnaire.

1 Introduction

We can see some virtual space systems used 3D graphics used WWW [1][2]. And recently we can see some kind of ubiquitous computing [3][4]. We are interested in historical tour support systems used 3D graphics and ubiquitous computing. There are many historical tour support systems to serve sightseeing information to their users on the Web in Japan, but most of their contents are not so attractive [5]. One of the reasons may be the lack of the lifecycle support for the tour. The system should provide some kind of support in each phase of the tour such as pre-tour, on-tour and post-tour. Other reason may be the lack of support to foreign visitors, because most of the information is provided only in Japanese. We have developed a historical tour support system to address these problems. Our system offers the necessary information to its users according to the phase of the tour. It provides the information on HORIKAWA sightseeing tour - one of the famous historical tours in Japan that gives travelers a sightseeing tour around the moat of the Matsue castle onboard a pleasure boat. In the preparation phase, it gives users the 3D virtual HORIKAWA sightseeing tour in multi languages. During the tour, each visitor has a cellular phone with GPS function and is given the location-based information on each historic spot through his/her cellular phone [6]. After the tour, it supports maintaining the memories of the tour by providing the electronic album function.

2 System Structure

Fig. 1 shows the structure of our historical tour support system[7]. The system is on the Internet and can be accessed using PCs and cellular phones. The system consists

R. Khosla et al. (Eds.): KES 2005, LNAI 3681, pp. 412–417, 2005.

of three parts: virtual world manager, data manager and cellular phone manager. The virtual world manager gives users 3D virtual space and interacts with them. The data manager handles all tour information that will be provided to the users. The cellular phone manager administers users' cellular phones. The 3D virtual space model is written in Java 3D [8] and downloaded to PCs on the Internet. The tour information is written in XML and XSLT and processed by the data manager [9]. The cellular phone manager is implemented as a CGI program. The main part of the system is written in Java and HTML which controls the interaction among those three parts mentioned above.

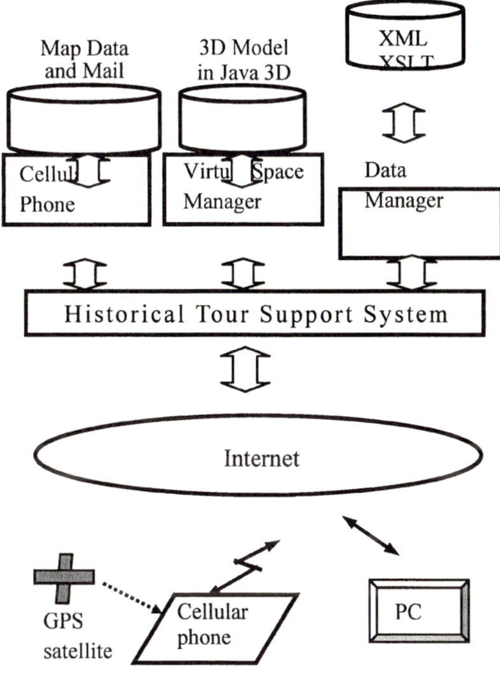

Fig. 1. System Structure

3 System Function

Some current screen shots of our system for pre-tour support are in Fig. 2. An avatar of a user is a virtual pleasure boat immersed into the 3D virtual space and can move around the virtual castle (Fig.2(a)). The user can see the motions of other pleasure boats in the moat and the motions of a sightseeing bus around the moat, which are robots in the virtual world. If the user clicks on the castle gate, the information of the castle retrieved from the XML database are displayed in another window (Fig.2(b)). The information is presented textually or audibly in his/he preferable language selected among Japanese, English and Chinese(in this case written in Chinese).

Some screen shots of cellular phone for on-tour support are in Fig.3. During the actual tour, the location of the user can be determined using GPS function of the user's cellular phone (Fig.3(a)). The location information is sent to a server in real time and collated with the map. On the server, the information on the historic site, which can be identified by the user's location information, is retrieved from the database and sent back to the user via e-mail (Fig.3(b)). By the function of reading out e-mail of the cellular phone, a user can go sightseeing while hearing the explanation about the historical site in Japanese or English (Fig.3(c)).

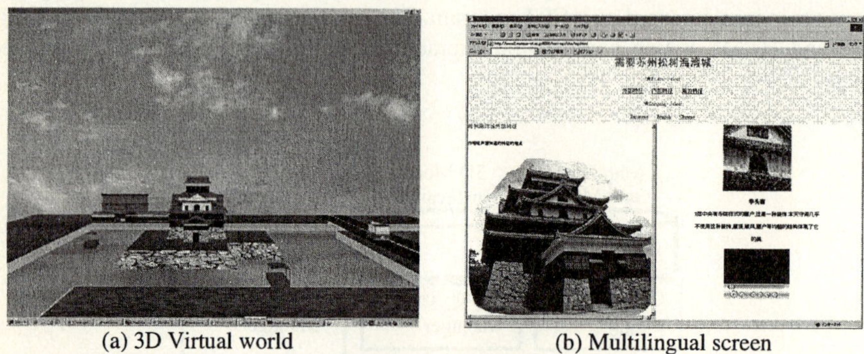

(a) 3D Virtual world   (b) Multilingual screen

Fig. 2. Screen shots of pre-tour support

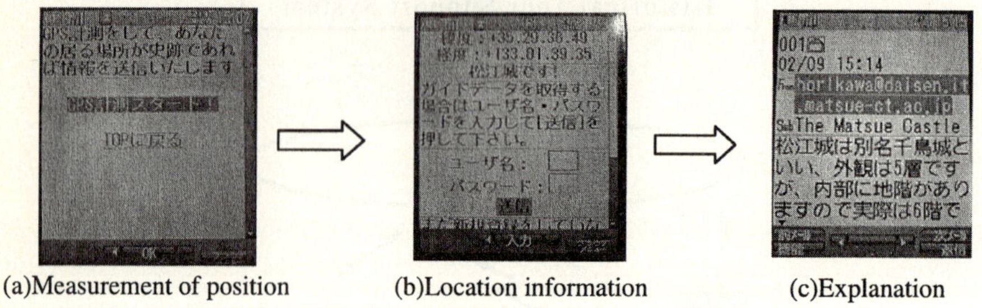

(a)Measurement of position   (b)Location information   (c)Explanation

Fig. 3. Screen shots of on-tour support

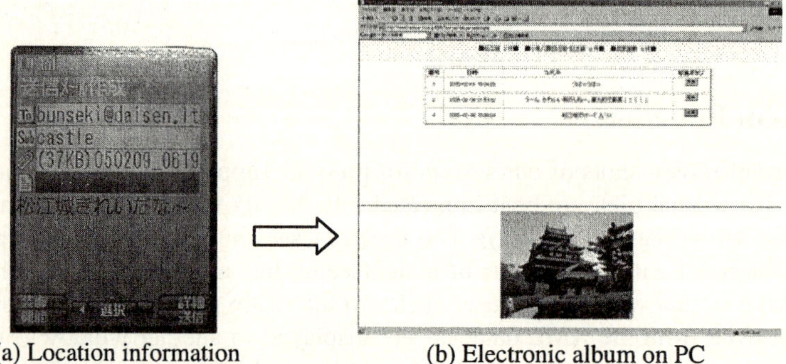

(a) Location information   (b) Electronic album on PC

Fig. 4. Screen shots of post-tour support

Screen shots of cellular phone and electric album for post-tour support are in Fig.4. The user sends the photographs taken with the cellular phone and the notes associated with them to the server (Fig.4(a)). The server creates an electronic photo-album automatically, and the user can enjoy it with PC or cellular phone later (Fig.4(b)).

Thus this system can support historical tour ubiquitously according to its user's situation: pre-tour, on-tour, and post-tour.

4   Implementation

We have developed 3D graphics of HORIKAWA sightseeing in Java3D. The 3D virtual space is constructed with Scene Graph, and has two views; avatar's view and external view. List.1 is a program to set avatar's view. The statement "SetFrontClip-Distance(0.05);" sets minimum distance and the statement "setBackClipDistance(150.0);" sets maximum distance to be seen by the avatar. The unit of both arguments is 'meter'. The statement "BranchGroup scene = createSceneGraph (canvas);" defines branch group 'scene', "getUniverse();" calls Universe, and "add-BranchGraph(scene);" adds branch group 'scene' as a child of Universe.

```
List. 1  Setting code of avatar's view

GraphicsConfiguration config =
SimpleUniverse.getPreferredConfiguration();
Canvas 3 D canvas = new Canvas 3 D (config);
this.add(canvas, BorderLayout.CENTER);
    .setUniverse(new SimpleUniverse(canvas));
getUniverse().getViewer().getView()
    .setFrontClipDistance(0.05);
getUniverse().getViewer().getView()
    .setBackClipDistance(150.0);
BranchGroup scene = createSceneGraph(canvas);
getUniverse().addBranchGraph(scene);
```

We have implemented the administration functions for cellular phone in PHP and MySQL. The system registers a user by analyzing the empty e-mail from the user's cellular phone. The system analysis used 'mimeDecode.php' of PEAR that has the function to make association array from e-mail automatically[10]. After the analysis, the system creates user's name and password, stores them with the user's mail address in DB, and sends back the e-mail with result to the user.

The server has a HTML file including the following statement.

<a href="device:gpsone?url=fileurl">

When a user clicks this reference in his/her screen of cellular phone, the function to get position data is invoked. Then the cellular phone communicates with the location server (gpsOne server), which is operated by the cellular phone carrier and is out of our system, to obtain its own position data and put it into 'fileurl'. It leads that our server can get latitude and longitude by 'GET' function of PHP. Then the server compares this data with the position data of historic point stored in DB, and sends the information of the historic point to the user via e-mail.

Our server maintains the file which contains the picture data taken by the user at the historic spots. In order to allow the user to access this file, the server maintains the table in DB which consists of the file name, the name of historic spot the user visited, the time the user visited there, the notes the user took then. The user can access this table through the Internet and can retrieve the memorial record after the tour.

5 Evaluation of the System

Ten students at our school used this system by PC and cellular phone and evaluated by the following nine questions.

(1) Were you interested in 3D graphics display?
(2) Did the 3D graphics image change smoothly?
(3) Did you think it useful to support multilingual historic information in this system?
(4) Did you feel convenient for the location-based information service using GPS?
(5) Do you want to see your photo-album not only by PC but also by cellular phone?
(6) Do you think it convenient for you to produce your own photo-album using cellular phone?
(7) Were you able to operate this system smoothly?
(8) Was the virtual space displayed clearly?
(9) Is this system useful for your own historical tour?

Fig. 5 shows result of the evaluation. Results are almost good;(1)~(4) and (7)~(9). Result of (9) hows that all users evaluated this system was useful for historic learning. But we must improve function of GPS and album, because result of (5) and (6) were not so good. In addition, some users had complaint about few contents of 3D and e-mail. After that, we must evaluate this system again, because we did not use and evaluate at the real historic spots, and we had no foreigners to evaluate.

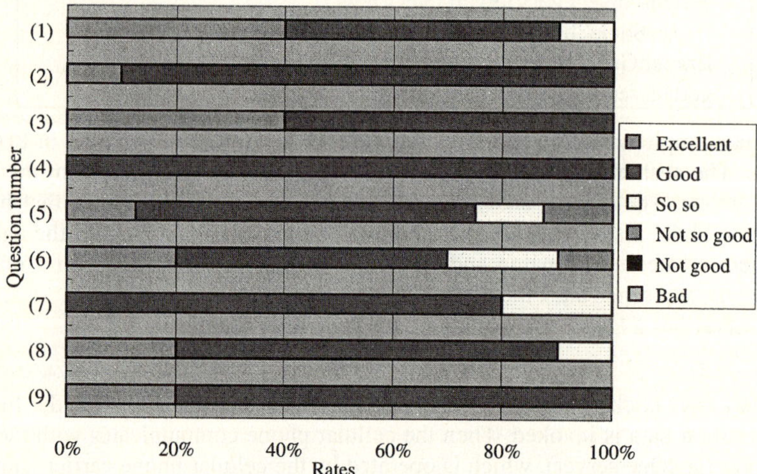

Fig. 5. Result of the evaluation

6 Conclusion

We rarely see the system which supports the lifecycle of historical tour. We have developed the historical tour support system that provides its users with appropriate information according to each phase of the tour ubiquitously. In the pre-tour phase, it gives the users 3D virtual tour graphic in multi languages. During the tour, the user's cellular phone with GPS offers the information about the historic spot based on the user's location. In the post-tour phase, it offers the electronic photo-album which records the experience of the tour visually.

We had ten users to use this system and got a good response from them according to the results of their answers to the questionnaire. From their response to the function for the pre-tour phase, we found that the users enjoyed this system because of the effect of 3D graphical presentation of the virtual tour.

We must evaluate this system again, because we did not evaluate it at the actual historic spots or had no foreigners in the evaluation. After we complete implementing the function to offer the information at actual historic spots with a cellular phone and the function to create the photo-album, we will evaluate the entire system.

References

1. Murray,T. and Hilz,S.,R.,: "Software Design and the Future of the Virtual Classroom",Journal of Information Technology for Teacher Education, Vol.4,No.2,pp.197-215(1995).
2. Matsuura,K.,Ogata,H.and Yano,Y.: "Agent-based Asynchronous Virtual Classroom",Proc. Of ICCE99, Vol.1.1, pp.133-140(1999).
3. Abowd,G.D. and Mynatt,E.D.: "Charting Past, Present and Future Research in Ubiqutous Computing",ACM Trans. Computer Human Interaction, Vol.7,No.1,pp.29-58(2000).
4. Lyytien, K. and Yoo,Y.: "Issues and Chalanges in Ubiqutous Computing", Comm.ACM, Vol.45,No.12, pp.63-65(2002).
5. Matsue Jozan Koen Administration Office, Special Features Of Matsue Castle Pamphlet (2004).
6. K.Tarumi, S.Tokuda, T.Yasui and K.Matsubara: "A Virtual City for Mobile Terminals based on Locations", IPSJ SIG Technical Report, 2004-GN-52, pp.37-42(2004).
7. Y.Takahashi, M.Koike and S.Fujii: "Development of Ubiqutous Historical Learning System", The Third International Conference on Active Media Technology (AMT2005), Takamatsu (2005).
8. Java3D: http://developer.java.sun.com/developer/onlineTraining/java3d/, Sun Microsystems (2004).
9. S.Fujii, J.Iwata, Y.Miura, K.Yoshida, S.Sakai and T.Mizuno: "Development of a System for Learning Ecology Using 3D Graphics and XML", 8th International Conference on Knowledge-Based Intelligent Information and Engineering Systems(KES2004), LNAI 3214, pp.912-919 (2004).
10. Manual PEAR: http://www.xml.jp/php/manual_pear/ (2004).

Capturing Window Attributes for Extending Web Browsing History Records

Motoki Miura[1], Susumu Kunifuji[1], Shogo Sato[2], and Jiro Tanaka[3]

[1] School of Knowledge Science, Japan Advanced Institute of Science and Technology
{miuramo,kuni}@jaist.ac.jp
[2] Master's Program in Science and Engineering, University of Tsukuba
satosho@iplab.cs.tsukuba.ac.jp
[3] Graduate School of Systems and Information Engineering, University of Tsukuba
jiro@iplab.cs.tsukuba.ac.jp

Abstract. We propose a method to enhance web browsing history records by considering browsing window attributes. The browsing window attributes means properties of web browser window such as size, location, z-order, activated status, etc. We often open multiple web documents for each window and browse the contents through handling these windows at the same time. Since a series of the window attributes describe the user's browsing activities, we exploit the attributes to enrich the browsing records to help the user find previously visited web pages.

1 Introduction

Most web browsers have a built-in history mechanism to help users find previously visited web pages. Cockburn et al. pointed out that approximately 81% of pages are previously visited by the user[3]. Therefore, there is a need to develop techniques and tools to efficiently revisit those web pages.

Basically, bookmarks are managed by the user, whereas the built-in history mechanism retrieves data without any explicit operations. Consequently, many built-in history mechanisms only use limited attributes such as the title, visited time, and visit counts. Furthermore, the history contains a large number of entries, even those made several weeks before, making the user struggle a lot to narrow down his/her search to a certain target page.

We consider the lack of attributes of entries to be one of the main reasons for struggling with many history entries. A history mechanism should provide richer attributes by attaching more contexts to the history entry. A richer attribute will help the user to efficiently filter the entries and find the target URLs.

2 Display-Oriented Attributes
for Extending Web History

In this paper, we propose a method to enhance web browsing history records by considering browsing window attributes. By browsing window attributes, we

R. Khosla et al. (Eds.): KES 2005, LNAI 3681, pp. 418–424, 2005.

mean the display-oriented properties of web browser window such as size, location, z-order and status of the window. We often open multiple web documents and handle several windows while browsing the page contents. The browsing window attributes describe how the web documents were displayed and managed on a desktop screen. Therefore those attributes can be used to precisely infer the user's browsing activities.

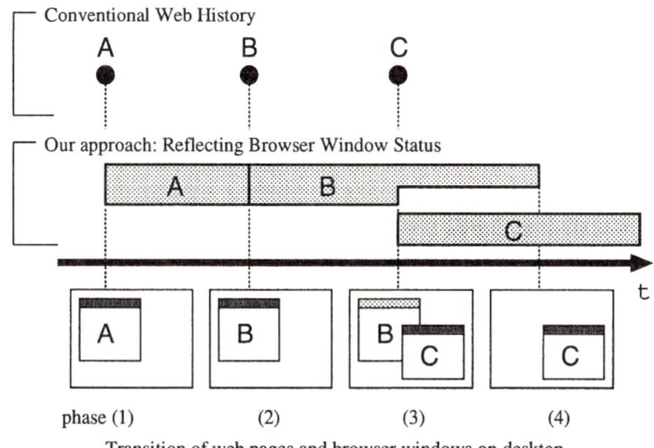

Fig. 1. Basic concept of our approach.

2.1   Basic Concepts

Our approach is originally motivated by a hypothesis that "the appearance of web documents shown on a desktop screen describes the significance of the document." Figure 1 illustrates a basic concept of our approach comparing it with the conventional web history model. The four boxes below the time-line represent transition of desktop screen, and the inner box represents a browser window. In this scenario, each phase of the transition corresponds to the following actions: (1) the user opened page [A], (2) the user opened page [B] in the same window, (3) the user opened page [C] as a new window (the first window is deactivated), and (4) the user dismissed the first window.

The conventional web history only retrieves opened time (shown on the top of Figure 1). Thus the opened time can be illustrated as a "point." Our model retrieves both the "opened time" and the "closed time." The duration of appearing of a web document on the computer screen can be represented as a "line." Our model represents browsing activity with high accuracy by introducing the notion of "duration."

In addition to the concept of duration, our model also considers how a web document is displayed on the screen. The size and location of the browser window can be captured to add the importance of URL. When a window overlaps other windows (see phase (3) of Figure 1), the importance of overlapped web document ([B]) is reduced. By employing our model, the importance of web document can be calculated by a definite integral of document appearing area.

2.2   Extracting User's Browsing Activity from Window Properties

In addition to the above basic model, we try extracting user's browsing activities from window events and status. We have currently focused on the following window attributes; (1) window ID, (2) window class, (3) location and size of the window, (4) z-order, (5) title, (6) URL, and (7) combination of events, and those events are: opened, closed, iconified, de-iconified, maximized, de-maximized, activated, de-activated, moved, resized, and reordered. We utilize "Window ID" to track a window attributes. "Window class" represents the type of application. The "Window class" property can be used to distinguish web browser window from other applications.

Rating Records. As we described in 2.1, the precise records in both temporal and spatial factors based on window appearance can be retrieved by considering the location, size, and z-order. These factors can be used to evaluate importance of web pages. Other attributes such as active/inactive status and iconified/deiconified events can supplement these factors for ratings. For example, the page which the user activates the window for a long time will be noted. Also the numbers of events such as activated, maximized and resized are informative factors because they indicate positive stance for the page.

We can infer some browsing situations and intentions by tracking the attributes. Suppose that the user dismissed a page. By considering the attributes, we can distinguish the following cases, (1) when, the window remains, the page was altered by following link but the browsing session continued. (2) When the window is destroyed, the user wanted not only to leave the page but also to quit browsing session. (3) When the window is iconified (stored in a task bar), the user had some reasons to interrupt browsing but he/she was still keeping attention to the page, and he/she will continue browsing.

Relating Records. Tools such as Footprints[8] and Browsing Icons[6] visualize page trails as graphs. These tools consider "explicit" relationship describing link structure among pages. The explicit relationship based on the "link" is promising. However, we focus on "implicit" relationship to vary the patterns of web revisit. We consider the window attributes as one of the implicit relationships.

The implicit relationship can be extracted by browsing actions, which include interaction between pages. Suppose that the two pages are displayed in mutually-adjoined windows. If the user moves a window near the other one, and if the edges of windows snap, the user intends to browse these pages concurrently. If the user activates these windows alternately, these pages can be related.

This relationship can be used to provide an associative search function. Suppose that a user, who cannot recall a keyword of target page directly but does remember some keywords of the page, opened either temporally or spatially adjoined windows. The user can reach the target page by following the implicit relationship.

The window attributes can be used to extract semantic aspects of web browsing activities. We consider the effect of this approach to become higher as the display size becomes wider. A large display enables the user to freely layout the web windows; thus SXGA (1280×1024) or a larger display is appropriate for this approach.

3 System Implementation

In order to investigate the effects of extended web browsing records, we have developed two applications; a web-browsing window attributes collector for logging and a web browsing history viewer for searching.

3.1 Collecting the Attributes

We have designed the collector to run on Microsoft Windows environment, which is one of the most commonly distributed platform for web browsing tasks. The collector watches browser window attributes every second, and records the attributes if the value changes. The record is stored as a csv file. The current implementation of the collector supports Internet Explorer, Opera, and Mozilla.

The collector application initially appears as a tray icon. When the user clicks the icon, status window (Figure 2 top) appears, and lists current logging attributes. Figure 2 shows the situation when the user activates the left browser window. As a result, the left window (ID:1115800) becomes active and is displayed in the front. Thus the z-order of the right window (ID:6162018) changes to two. The status window is helpful in understanding how the collector works, but users are free from the window in daily use.

The additional merit of the collector is that the system can integrate web history records performed by several web browser applications. Typical web history records are closed by each browser. Of course a situation of handling several

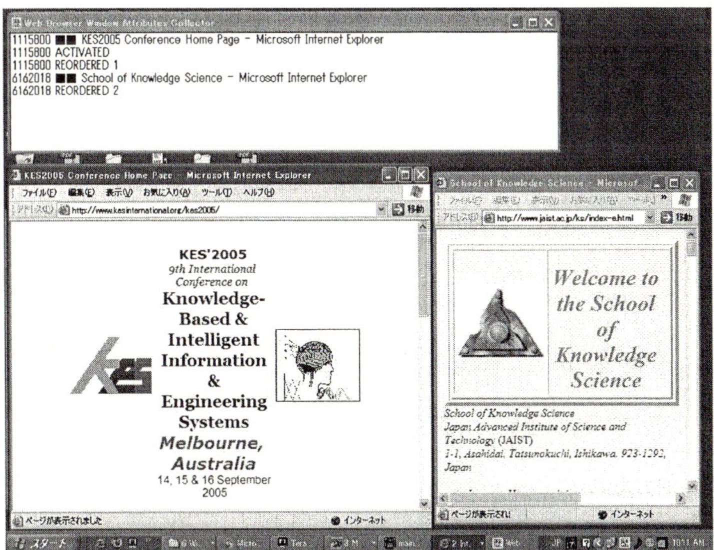

Fig. 2. The collector status window.

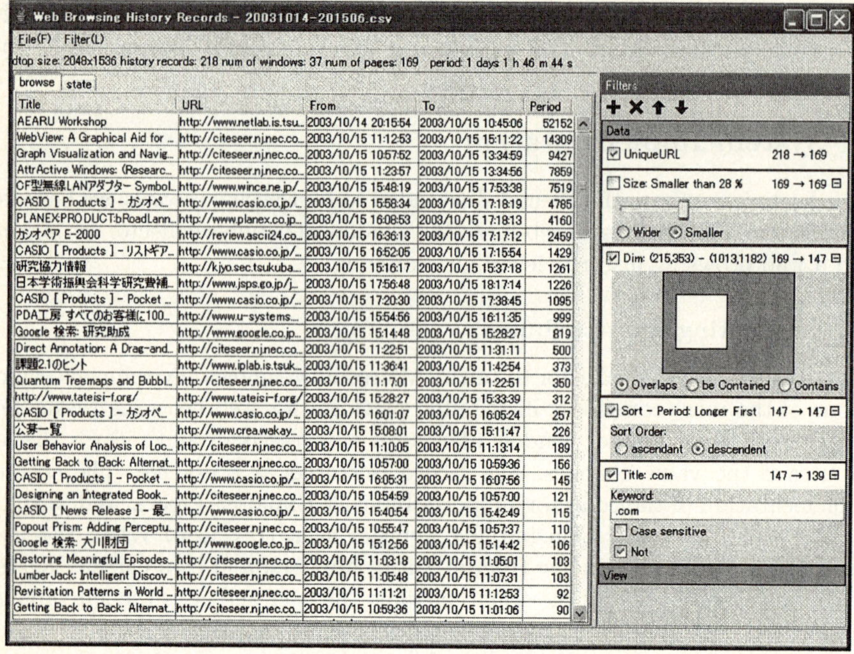

Fig. 3. Web History Records Viewer.

browsers at the same time is not popular. However, there are cases where people have to change their browsers to check a page, which recommends a particular browser environment. The collector unites multiple browser attributes held on the user's machine and associates them for a wider use.

3.2   Searching the Records to Re-visit Sites

We have implemented a prototype of the web browsing history viewer as a Java application (Figure 3). The viewer loads a log file generated by the collector, and lists the records in a table view. The initial table view shows the following five attributes; page title, URL, open time, close time, and duration. The user can revisit a page by double-clicking on the URL.

The user can sieve the records by specifying filters. The primary filters are; left unique by title/url, and can be searched by title/url, and sorts by title/url/date/duration/etc. When the user chooses a filter from a filter list, the filter appears in a filter dock (Figure 3 right-pane) and updates the list. The filters in the dock are applied in a top-down order. The user can easily mark a checkbox to validate/invalidate the filters.

4   Informal Observation

We conducted an informal observation to reveal the characteristics of our method. We collected several history records for two month, which included 11,476 items.

Out of that the number of unique title items was 2,505. We calculated two different kinds of measures such as visited times and exposures for every items. Incidentally, we adopted only foremost and activated windows for accumulation of the exposures.

The top two results were common in measures; a start (portal) page and a search engine site like Google. According to our data, Wiki editable pages held the top of the exposures indexed list. The visited times tends to rise up a major "hub" page and scheduler page.

We assessed the distribution of these measures in Figure 4. The visited times was biased. More than half (59.6%, 1,492) pages were visited once, and 96.7% of pages were visited less than 10 times. The distribution of exposures were flatten than the visited times. The main purpose of re-visitation is to find a page, which was infrequently visited. Therefore the measures of exposures have advantages in finding lost pages.

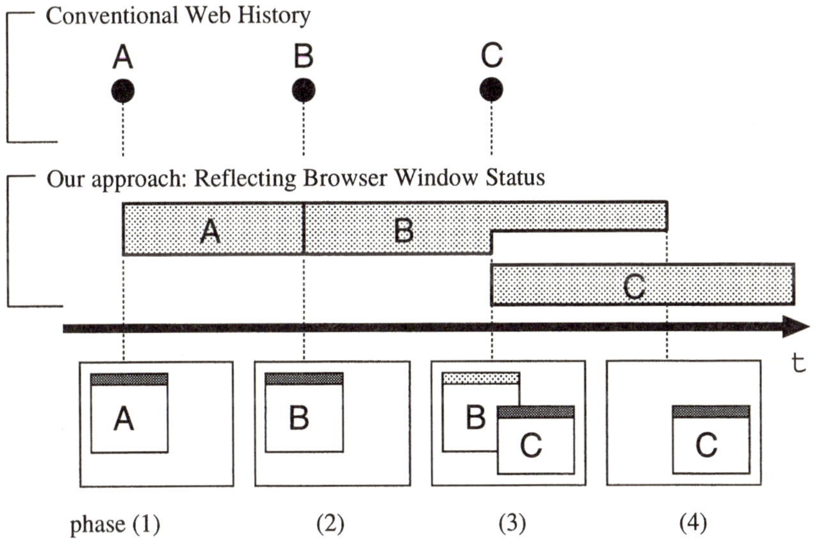

Fig. 4. Distributions of Visit Counts and Proposed Measure.

5   Related Works

Many researches have been carried out to visualize the Web in a zoomable two-dimensional pane [4] or a three-dimensional space [1, 2]. Systems which visualize web history such as WWW3D, The HyperVIS system, and WebPath are also summarized in different research papers. These systems can help a user to understand the path and its structure by mapping history records in a flexible virtual space. Our system aims to add extra factors which are extracted from window attributes to history records.

Regarding usability of web history, JasonSmith et al. [5] compared a linear temporally ordered list typified Netscape Navigator's history with a hierarchical structured list typified by Internet Explorer's history. Nadeem et al. [7] also compared a simple listed history with a visual graph history which is similar to PadPrints. These experiments mainly focus on the effectiveness of the interface with traditional history framework. In this paper we endeavor to promote a framework to enhance typical history records by capturing window attributes.

6   Concluding Remarks

We propose a method to enrich web history records by capturing attributes of web browser windows. The method focuses on "appearance" of web documents on the desktop screen. We believe that the attributes naturally reflect the user's intention of web browsing activities.

To realize the proposed method, we have developed tools for collecting window attributes, and for revisiting via history records. No operation is necessary for collecting and enhancing the history records; the window attributes are automatically captured, and are associated with URLs. The history viewer enables the user to search the records with a flexible filtering function. These tools can be helpful in finding valuable pages from huge history records. As a future work, we assess the effectiveness of this method using our system.

References

1. S. Benford, I. Taylor, D. Brailsford, B. Koleva, M. Craven, M. Fraser, G. Reynard, and C. Greenhalgh. Three dimensional visualization of the world wide web. *ACM Computing Surveys*, 31(4es):25, Dec. 1999.
2. E. H. Chi. Improving Web Usability Through Visualization. *IEEE Internet Computing*, 6(2):64–71, 2002.
3. A. Cockburn and B. McKenzie. What Do Web Users Do? An Empirical Analysis of Web Use. *Int. J. Human-Computer Studies*, 54(6):903–922, June 2001.
4. R. R. Hightower, L. T. Ring, J. I. Helfman, B. B. Bederson, and J. D. Hollan. Graphical Multiscale Web Histories: A Study of Padprints. In *Proceedings of the ninth ACM conference on Hypertext and hypermedia*, pages 58–65, 1998.
5. M. JasonSmith and A. Cockburn. Get a Way Back: Evaluating Retrieval from History Lists. In *Proceedings of the Fourth Australian User Interface Conference (AUIC2003)*, pages 33–38. Australian Computer Society, Inc., 2003.
6. M. Mayer and B. B. Bederson. Browsing Icons: A Task-Based Approach for a Visual Web History. Technical Report CS-TR-4308, Computer Science Department, University of Maryland, 2001.
7. T. Nadeem and B. Killam. A Study of Three Browser History Mechanisms for Web Navigation. In *Proceedings of the Fifth International Conference on Information Visualisation (IV'01)*, pages 13–21, 2001.
8. A. Wexelblat and P. Maes. Footprints: History-Rich Tools for Information Foraging. In *Proceedings of the SIGCHI conference on Human Factors in computing systems*, pages 270–277, May 1999.

Development of an Intercultural Collaboration System with Semantic Information Share Function

Kunikazu Fujii, Takashi Yoshino, Tomohiro Shigenobu, and Jun Munemori

Wakayama University
930 Sakaedani, Wakayama 640–8510, Japan
{s075065,yoshino,s020055,munemori}@sys.wakayama-u.ac.jp

Abstract. The Internet provides the opportunity to implement the real seamless communications over different geographic domains. However, when they cope with the network level challenges, they encounter the context and cultural level challenges. Therefore, we have developed Intercultural Collaboration System with Semantic Information Share Function. This system is a multilingual conferencing system. This system consists of three functions Spark2 for shared workspace, AnnoChat for chat communication tool and TalkGear2 for audio and video communication tool. The system has a semantic information share function and a translation function.

1   Introduction

Asian Internet population surpassed the US or Europe in 2003. The wide range of people can use the network for various purposes. However, the differences of language exist as a barrier of information exchange. Generally, the information on nonnative languages makes sufficient understanding difficult.

Global collaboration is ongoing in English speaking countries. European Union encourages Europe wide projects. However, a serious language barrier exists in Asian countries. This is because neighboring languages are not taught in Asia. Asian people cannot think in English and want to make interim documents in our first languages. Some collaboration studies have been conducted in other countries [1]-[3].

We have developed Intercultural Collaboration System with Semantic Information Share Function. This system is a multilingual conferencing system. This system consists of three functions Spark2 for shared workspace, AnnoChat for chat communication tool and TalkGear2 for audio and video communication tool. The system has a semantic information share function and a translation function. The use field of the system can be a business area where the exchange of intercultural is active. Moreover, recently, the exchange of intercultural is rapidly active in the field of the show business. It is possible to spread from such a field.

R. Khosla et al. (Eds.): KES 2005, LNAI 3681, pp. 425–430, 2005.

2   Design Policy

(1) Machine translation function
     When a machine translation function is not provided, it is necessary to communicate in a common language (e.g. English). It may disturb smooth communications because a psychological barrier is high in communications with nonnative language. Then, this system supports to communicate with each other only by the mother language. The machine translation cannot completely prevent the mistranslation. This system has a turn back translation function and an annotation added function for improving the machine translation.
(2) Annotation added function
     One word often has two or more meanings. This may occur misunderstandings between different languages, or the persons among other languages might be incomprehensible. For the users among the same language, there are a lot of requirements to add various explanations.
(3) Shared workspace tool
     PC has a lot of content, such as text files, image files, movie files, voice or sound files, and document files (PowerPoint files, MS Word files and MS Excel files). We think sharing those files is useful in an actual collaboration.
(4) Replay function
     We think operation processes show a part of thinking processes. Then, we think that operation processes can help to realize a certain collaboration result. In addition, a record of audio and video is effective for participants who are absent from an electronic meeting.

3   Intercultural Collaboration System with Semantic Information Share Function

Intercultural Collaboration System consists of three tools, such as shared workspace tool, a chat communication tool and an audio and video communication tool. In this system, the size of a group is about 5~6 persons.

3.1   Spark2: A Tool for Providing Shared Workspace

Spark2 is a tool for providing shared workspace. The tool can outline personal or group knowledge through illustrations. Figure 1 shows a screenshot of Spark2. Texts, Images, files on PC, are arranged freely on a workspace. Those objects can be grouped and be arranged hierarchical. PowerPoint file and Movie file can be seen on a screen without other software. Text data can be translated among Chinese, Japanese, Korean, and English by Machine Translation. Figure 2 shows an example of translated texts. Text is translated automatically by the background processing. For instance, the sentence that the Japanese input in Japanese is displayed in Chinese on Chinese's screen. Annotations can be added to those objects. Annotation is shown in the balloon shape in Figure 1. One

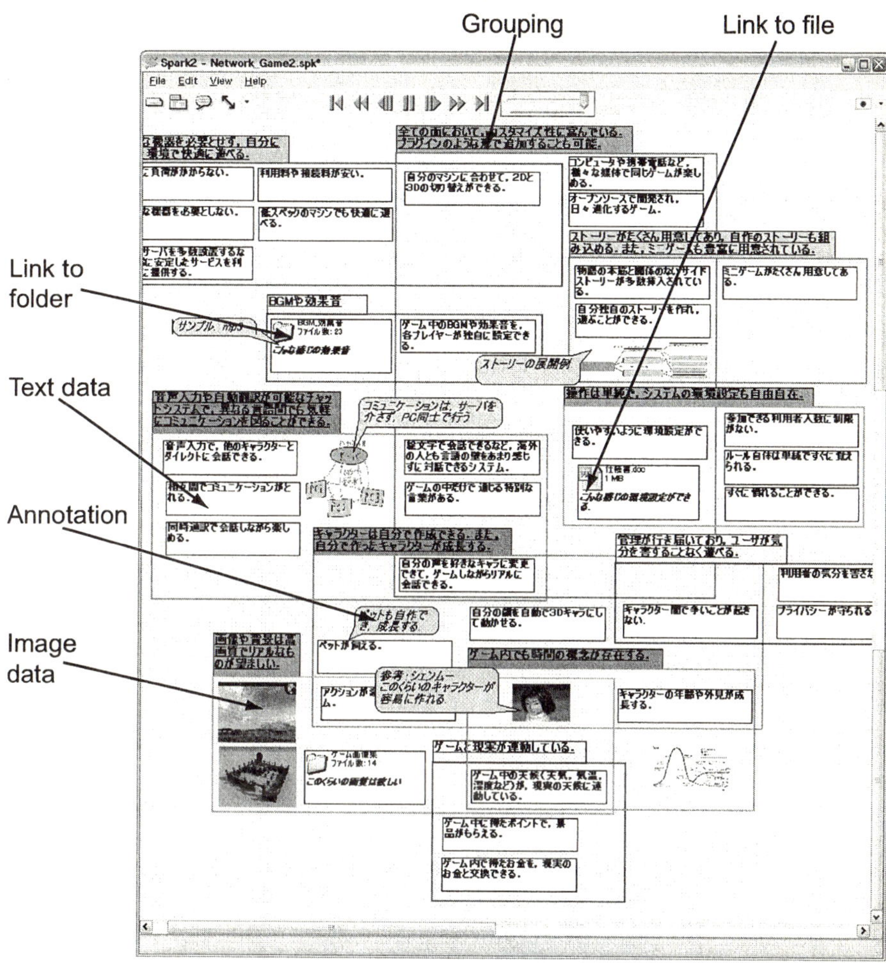

Fig. 1. Screenshot of Spark2.

object can have some explanations. The annotation function can be added to the content of the other annotation further. Figure 3 shows the hierarchized annotations. Spark2 has a replay function to show operations on the workspace. A user can see the history of the operations on the workspace at any time.

3.2   AnnoChat: Chat Communication Tool

AnnoChat is a chat communication tool. Figure 4 shows a screenshot of AnnoChat. The tool has a function to input together with a chat text and pictographic characters like emoticon. An inputted text sentence is translated and displayed on a user's native language. The annotation function can be added to the content of the other annotation further.

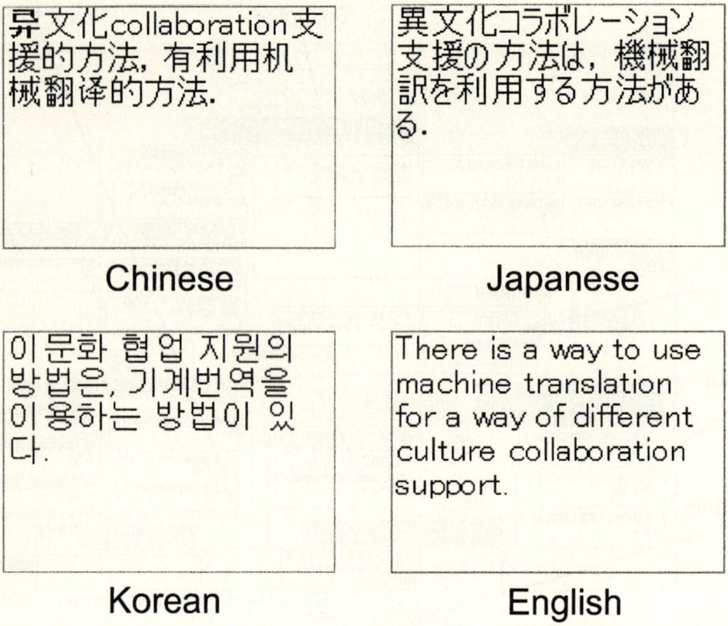

异文化collaboration支援的方法，有利用机械翻译的方法.
Chinese
異文化コラボレーション支援の方法は，機械翻訳を利用する方法がある.
Japanese
이문화 협업 지원의 방법은, 기계번역을 이용하는 방법이 있다.
Korean
There is a way to use machine translation for a way of different culture collaboration support.
English

Fig. 2. Translated texts.

無線送受信の機器は購入済み
振動で壊れないようする必要がある.
無線方式で,データのやりとりを行う.
クッションでくるむ
スリッパに内蔵する方式
無線を使うがRS232Cと同じで可能

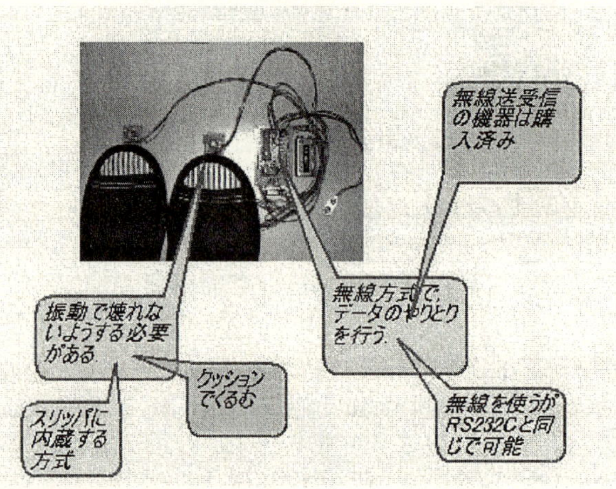

Fig. 3. Hierarchized annotations.

3.3 TalkGear2: Audio and Video Communication Tool

TalkGear2 is an audio and video communication tool. Figure 5 shows a screenshot of TalkGear2. The tool has a function to communicate voice and video among multi-points on the Internet. The tool can use simultaneously two sets of the video capture devices connected to one set of PC. For example, the first set

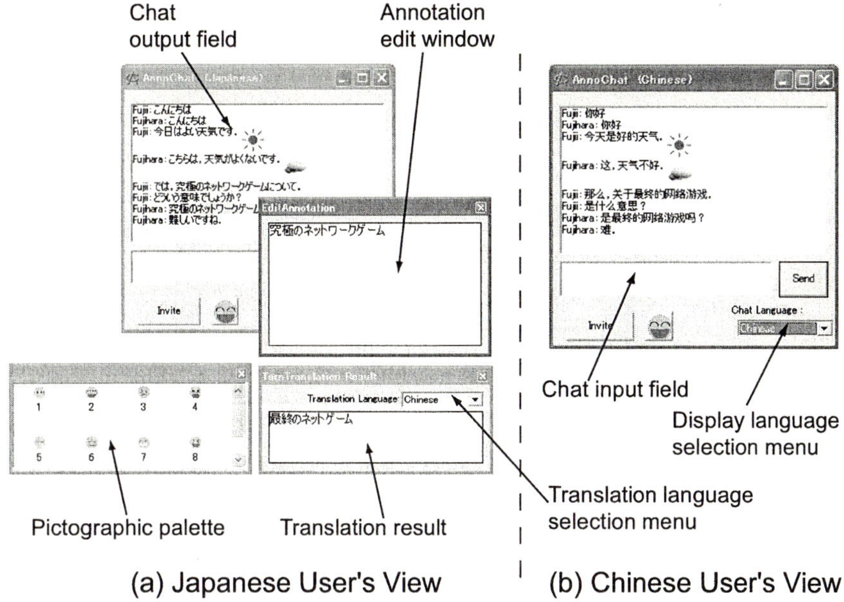

Fig. 4. Screenshot of AnnoChat.

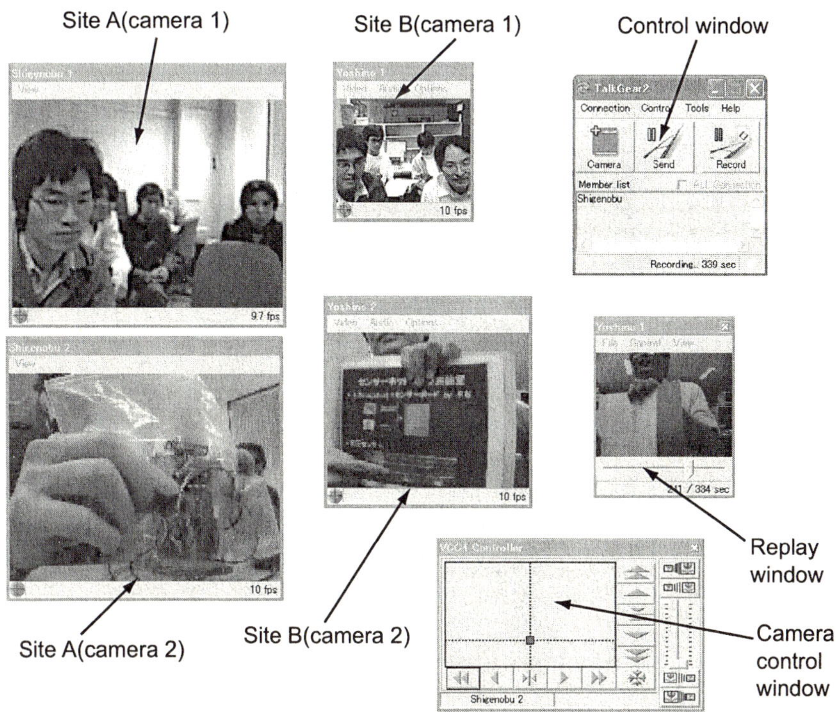

Fig. 5. Screenshot of TalkGear2.

is used as background/audience video and the other set is used as a presenter's video. Participants can operate the local cameras and the partner's cameras in remote place mutually. TalkGear2 can store the sent and received audio and video data. The tool can also replay the stored data.

4   Conclusion

We have developed Intercultural Collaboration System with Semantic Information Share Function. We will plan to use our system among some countries.

References

1. Takashi Yoshino, Tomohiro Shigenobu, Shinji Maruno, Hiroshi Ozaki, Sumika Ohno, Jun Munemori: Development and Application of an Intercultural Synchronous Collaboration System, Proceedings of Eighth International Conference on Knowledge-Based Intelligent Information Engineering Systems & Allied Technologies (KES 2004) (LNAI 3214, Springer Verlag), pp.869-875, Sept. 2004.
2. Milam Aiken: Multilingual Communication in Electronic Meetings, SIGGROUP Bulletin ACM, 2002.
3. Kaname Funakoshi, Akishige Yamamoto, Saeko Nomura, and Toru Ishida: Lessons Learned from Multilingual Collaboration in Global Virtual Teams. 10th International Conference on Human - Computer Interaction (HCII2003), Crete, Greece, June 2003.

Position Estimation for Goods Tracking System Using Mobile Detectors

Hiroshi Mineno[1], Kazuo Hida[1], Miho Mizutani[1], Naoto Miyauchi[2],
Kazuhiro Kusunoki[3], Akira Fukuda[4], and Tadanori Mizuno[1]

[1] Shizuoka University, 3-5-1 Johoku, Hamamatsu-shi, Shizuoka 432-8011, Japan
mineno@cs.inf.shizuoka.ac.jp
[2] Mitsubishi Electric Corporation, 5-1-1 Ofuna, Kamakura-shi, Kanagawa 247-8501, Japan
[3] Nagoya Works, Mitsubishi Electric Corporation,
5-1-14 Yadaminami, Higashiku, Nagoya-shi, Aichi 461-8670, Japan
[4] Kyushu University, 6-1 Kasuga-Koen, Kasuga-shi, Fukuoka 816-8580, Japan

Abstract. Determining physical location of indoor objects is one of the key issues in ubiquitous computing. Although there are many proposals to provide physical location tracking, those have some restrictions such as a dependence on the type and size of objects and a trade-off between position accuracy and the number of sensing devices. In this paper we present a hierarchical approach to conquer these restrictions using mobile detectors. We describe a system called MobiTra that estimates the position of indoor any and all objects using mobile detectors' detection histories. The effect of position estimation is evaluated through prototype testbed and simulation.

1 Introduction

Determining the location of indoor objects is one of the fundamental issues in ubiquitous computing. The importance and promise of location-aware applications has led to the design and implementation of systems for providing location information, particularly in indoor environments where the Global Positioning System (GPS) does not work well [1] [2] [3] [4]. Although many systems are proposed, there is no system that is widely used for the indoor any and all goods tracking system. One reason is that the system requires a special device to track the object, such as ultrasonic sensor and wireless LAN interface. The special device limits the type and size of target objects. The other reason is that there is a trade-off between position accuracy and the number of sensing devices. In general, to achieve more accurate location information, we have to deploy sensing devices in the high density. It increases the total cost for installing the system.

This paper addresses the problems of tracking any and all goods using Radio Frequency Identification (RFID) tags and mobile detectors, or RFID tag readers in this case. The architecture of a location system influences its scalability, its ease of deployment, and its object-tracking accuracy. We propose a hierarchical tracking system called MobiTra that conquers these restrictions using mobile detectors' detection histories. The object is indirectly detected by mobile detectors so that the object needs not to install a special device to be able to detect directly from the long distance. To track the

R. Khosla et al. (Eds.): KES 2005, LNAI 3681, pp. 431–437, 2005.
© Springer-Verlag Berlin Heidelberg 2005

location of mobile detectors, existing proposed systems can be applied to the MobiTra according to the required location accuracy. This hierarchical approach provides scalability and ease of deployment for the any and all goods tracking system. The location accuracy depends on the detection range of mobile detector because of the objects are indirectly detected by mobile detectors. Therefore, position estimation is applied to the detection histories from the latest one so that the location accuracy is improved.

The rest of the paper is organized as follows. Section II outlines related work. In section III, we describe the features of MobiTra in more detail and the implementation architecture. Section IV provides the evaluation results of prototype testbed and simulation. Section V ends the paper with a brief summary and some concluding remarks.

2   Related Work

The past few years have rapidly seen growing interest in location-aware applications and in systems such as Active Badge, Active Bat, Cricket, and RADAR that provide location information in indoor environments.

Active Badge uses infrared signal, which has dead spots in some locations. Room level granularity can be obtained by the Active Badge. Active Bat and Cricket both use time-difference-of-arrival (TDoA) between RF and ultrasound. A Bat node carries an ultrasound transmitter whose signals are picked up by an array of receivers mounted on the ceiling. The location of a Bat can be calculated via multilateration with a few centimeters of accuracy. An RF base station coordinates the ultrasound transmissions such that interference from nearby transmitters is avoided.

RADAR uses received-signal-strength-indicator (RSSI) of 802.11 RF from three fixed base stations in two phases. First, a comprehensive set of received signal strength measurements is obtained in an offline phase to build a set of signal strength maps. The second phase is an online phase during which the location of users can be obtained by observing the received signal strength from the user stations and matching that with the readings from the offline phase. This process eliminates multipath and shadowing effects at the cost of considerable preplanning effort. RADAR is not as accurate as the systems that use TDoA, but does not require any infrastructure other than 802.11 access points.

An RF based proximity method is presented in [6], in which the location of a node is given as a centroid. This centroid is generated by counting the beacon signals transmitted by a set of beacons pre-positioned in a mesh pattern. A different approach is taken in the Pico Radio project at UC Berkeley. It provides a geolocation scheme for an indoor environment [7], based on RF received signal strength measurements and pre-calculated signal strength maps.

A survey of alternative localization techniques for a range of applications is provided in [5]. Researchers have also investigated Bayesian and Kalman filter methods for tracking users [8] [9]. There is an extensive and growing body of work on improving the performance of GPS using various filtering and statistical techniques, some of which might also apply to indoor location systems.

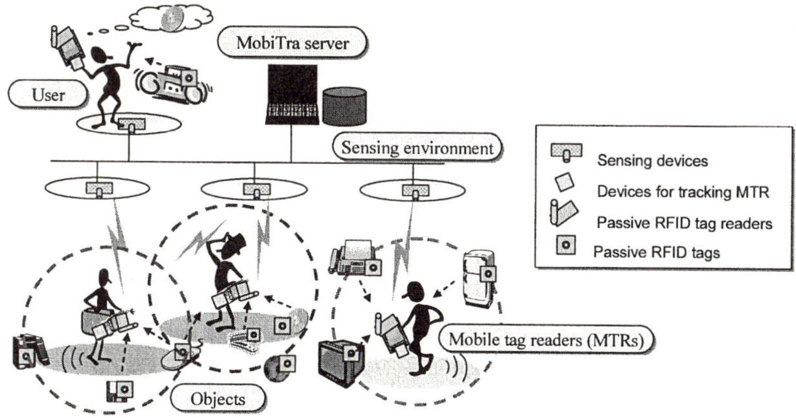

Fig. 1. System overview.

3 MobiTra Overview

3.1 Hierarchical Tracking

Although there has been an increasing interest for indoor localization systems, there is no system that is widely used for the indoor any and all goods tracking. The special devices like ultrasonic sensor and wireless LAN limit the type and size of target objects. Therefore, the device for tracking should not be too large, expensive, and hard to install.

We think the solution is RFID tag. RFID tags are categorized as either passive or active. Passive RFID tags operate without a battery. They reflect the RF signal transmitted to them from a reader and add information by modulating the reflected signal. Passive RFID tags are mainly used to replace the traditional barcode technology and are much lighter and less expensive than active RFID tags, offering a virtually unlimited operational lifetime. However, their read ranges are very short.

Active RFID tags contain both a radio transceiver and a button-cell battery to power the transceiver. Since there is an onboard radio on the RFID tag, active RFID tags have more range than passive tags. Active tags are ideally suited for the identification of high-value products moving through a tough assembly process.

As described above, passive RFID tags are suited for installing a large variety of goods. Passive RFID tags reduce the limitation of the type and size of target objects. In order to resolve the limited read range of passive RFID tag, tag readers are moved as mobile detectors. Since small and lightweight tag readers have been developed, tag readers will be commonly mobile and can be called mobile tag readers (MTRs). The goods tracking system discover the MTRs' location using described techniques in section II. We call this hierarchical tracking system MobiTra that estimates the location of indoor any and all objects using MTRs' detection histories.

Figure 1 shows the MobiTra system overview. It consists of users, a MobiTra server, sensing environment, mobile tag readers, and target objects. The MobiTra server provides the tracking service to users, and the location of target object is estimated by using MTRs' detection histories. Sensing environment can be deployed by the existing

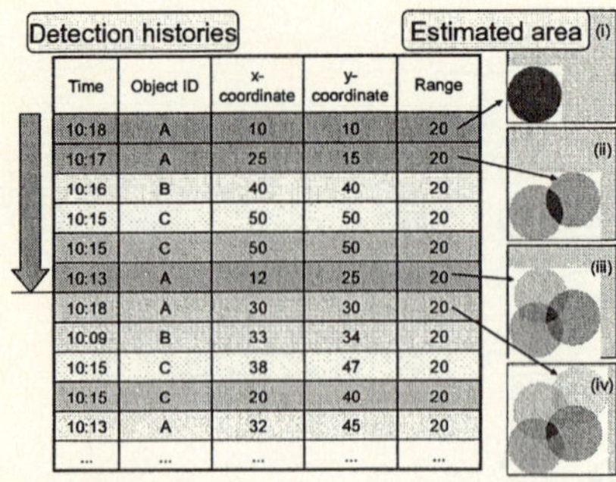

Time	Object ID	x-coordinate	y-coordinate	Range
10:18	A	10	10	20
10:17	A	25	15	20
10:16	B	40	40	20
10:15	C	50	50	20
10:15	C	50	50	20
10:13	A	12	25	20
10:18	A	30	30	20
10:09	B	33	34	20
10:15	C	38	47	20
10:15	C	20	40	20
10:13	A	32	45	20
...	...	...	...	...

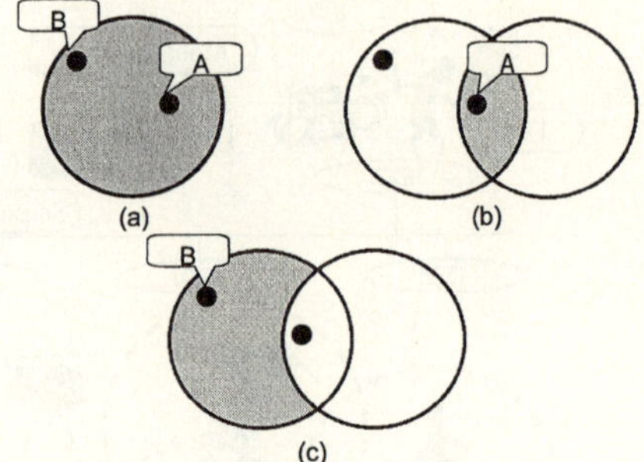

Fig. 2. Analyzing detection histories.   **Fig. 3.** Position estimation methods.

several techniques as described above. The infrastructure for collecting the detected location and tag information of MTR from the sensing environment is provided by such as power line communications (PLC), which is suited for exchange of small control data with no-new-wire.

3.2 Position Estimation

The location accuracy depends on the read range of MTR, since the objects are indirectly detected by MTRs. Therefore, position estimation is applied to the detection histories from the latest one so that the location accuracy is improved. As illustrated in Fig. 2, this simple iteration scheme leads more accurate localization of target object. If the area of under analyzing detection history does not overlap previous estimated area (see Fig. 2(iv)), the move of target object is detected, and the position estimation is finished.

In fact, the read range of MTR is affected by absorption, reflection, and interference. We approximate the range of MTR by a circle (assumed to be the same for all MTRs). Although this approximation overestimates the uncertainty of the detected object's position, it guarantees that the actual position of the object is always within the circle.

Figure 3 shows some applicable position estimation methods, for illustrative purposes, we assume that the target's exact position is known. (a) and (b) in Fig. 3 are already depicted in position estimation shown in Fig. 2. In some cases if there are histories that a MTR does not detect a target whereas its neighbor-MTRs' histories do, it may be able to use this information to reduce its estimated area uncertainty. This situation is depicted in Fig. 3(c). If the target object's actual position was in the shaded region within the circle, it would have detected target. Note, that we should use this *negative* information constraint only when the probability of not existence is high, since the approximation of read range overestimates.

Position estimation is done by applying these methods as shown in Fig. 4. Firstly, a target object's history is searched from detection histories. The estimated area, Q_i, is initialized to the circle with the barycentric coordinate (x,y) and radius r, where i gives the target object identification. There are three distinct procedures a position estimation uses to update its estimated area Q_i: (i) using sensing constraints imposed by

```
search a target object's history from the latest one
initialize Qi to the circle
iterate if there are detection histories
    if the target object's history is detected
        if the area is overlapped to the Qi
            update Qi using connectivity constraints
        else
            exit from a loop
        end if
    else
        if the area is overlapped to the Qi
            update Qi using negative information
        end if
    end if
end for loop
```

Fig. 4. Pseudocode of position estimation algorithm.

the MTR (see Fig. 3(a)), (ii) using connectivity constraints when the target object's history is detected and the area is overlapped to the Q_i (see Fig. 3(b)), and (iii) using negative information constraints when the target object's history is not detected and the nonexistent area is overlapped to the Q_i (see Fig. 3(c)).

4   Tracking Performance

4.1   Experimental Result

The effect of position estimation was evaluated through prototype testbed and simulation. First of all, we developed MobiTra prototype testbed to evaluate the effect of proposed position estimation as shown in Fig. 5. In this prototype testbed, we used active RFID tags to detect target objects. To get a certain level of read range using passive RFID tags, UHF RFID tags using the 950 -956 MHz band are suited. However RFID tags using the 950MHz - 956MHz band is still not available (in Japan, this frequency band will likely be available for RFID systems next spring), we chose active RFID system, Spider System manufactured by RF Code [10], alternatively for detecting the target objects. To deploy the tracking environment for MTRs, we chose the proximity technique using passive RFID tags, V700-D13P21 manufactured by OMRON. User with MTR wears sandals that installed downward a small active RFID tag reader, V700-HMD11 manufactured by OMRON. Since the read range is 50mm, the move of MTR can be detected by the passive RFID tags deployed in the high-density. Although it took so much work to deploy the tracking environment, the purpose was to evaluate the effect of position estimation. We can choose other effective techniques like TDoA or signal pattern matching.

We approximate the detected area by a rectangular bounding box. Although this approximation overestimates the uncertainty of the object's position, it guarantees that the actual position of the object is always within the bounding box and considerably simplifies the computation of the position estimation.

This prototype testbed is only implemented two position estimation methods, using sensing constraints (see Fig. 3(a)) and using connectivity constraints (see Fig. 3(b)), so that the proposed hierarchical approach is useful for a goods tracking system. The extinction ratio of estimated area is shown in Fig. 6. As the number of applied detection

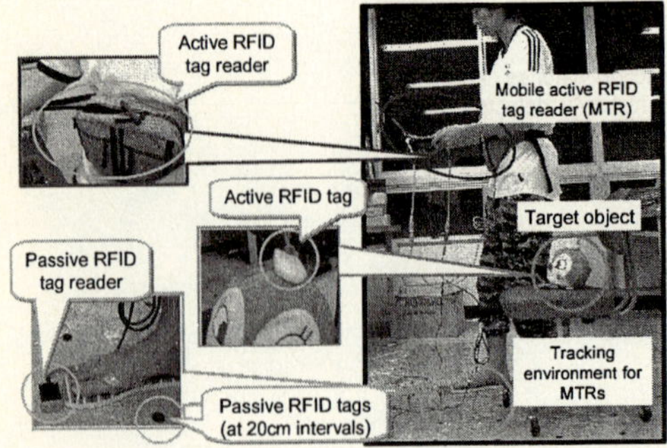

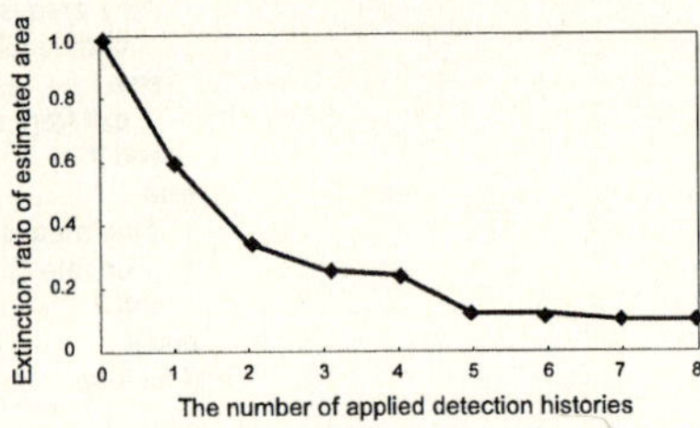

Fig. 5. Prototype testbed environment.

Fig. 6. Experimental result.

histories increased, the estimated area was reduced exponentially. This result shows the availability of proposed hierarchical approach.

4.2 Simulation Result

More detailed effect of position estimation is evaluated through simulation. In this simulation, we compared the effect of using negative constraints (see Fig. 3(c)). 25 tags were distributed at $1.25m$ static intervals in a $5m \times 5m$ square area. We consider one MTR moving in the area according to random waypoint mobility model [11]. The random way point model breaks the entire movement of a mobile node into alternating motion and rest periods. MTR moves to its randomly chosen destination point with a speed 1m/s. When MTR reaches its destination, it does not pause , then chooses a new destination, and continues moving. The read range of MTR is 1m with a radial pattern. Although the radial range model may not be a realistic description of wireless communication in physical environments, it is a valid starting point for evaluating purposes and it guarantees that the actual position of the object is always within the area.

We approximate the estimated area by a rectangular bounding box as well as previous experimental prototype. We compared average estimated area of tags at four corners and other tags to investigate the effect of using negative constraints. The result is shown in Fig. 7. The estimated area was dramatically reduced by using negative constraints than not using in the case of tags at four corners. Since the tags at four corners are low probability of detection by MTR, additional using of negative constraints information is more effective than using only connectivity constraints. We confirmed the effect of using negative constraints through simulation.

5 Conclusion

We presented a discussion on determining the location of indoor any and all goods. We focused position estimation for MobiTra, which is a hierarchical goods tracking system using RFID tags and mobile detectors. The effect of position estimation is evaluated through prototype testbed and simulation. From the experimental result, the availability of proposed hierarchical approach was investigated. Simulation result showed the

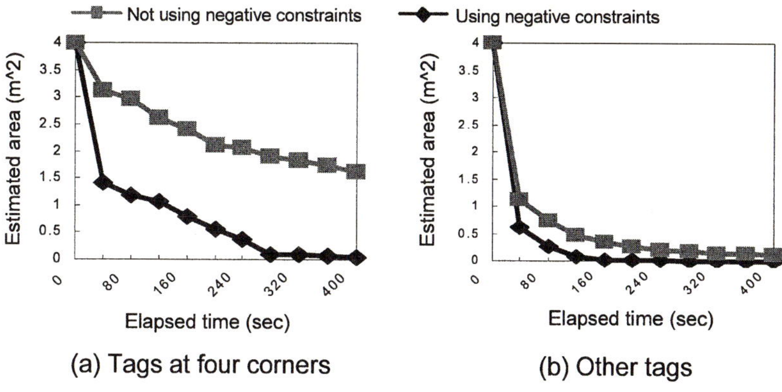

Fig. 7. Estimated area through simulation.

additional using of negative constraints information is more effective to the position estimation at four corners than using only connectivity constraints.

We have to consider the effect of position estimation when more costless way like TDoA or signal pattern matching is applied to the deployment of tracking environment for MTRs. Then, we have a plan for deploying MobiTra system in our laboratory, and evaluate the accuracy of estimated location. Furthermore, if MTRs can control their read range freely, torus-shape shaded region is estimated by controlling read range of MTRs. The effect of this additional method will be applied to the positoin estimation algorithm.

References

1. R. Want, A. Hopper, V. Falcao and J. Gibbons, "The Active Badge Location System," *ACM Transactions on Information Systems*, Vol. 40, No. 1, pp. 91–102, 1992.
2. A. Harter, A. Hopper, P. Steggles, A. Ward and P. Webster, "The Anatomy of a Context-Aware Application," *IEEE Computer*, Vol. 34, No. 8, pp. 50–56, 2001.
3. N. B. Priyantha, A. Chakraborty and H. Balakrishnan, "The Cricket location-support system," *ACM/IEEE MobiCom 2000*, pp. 32–43, 2000.
4. P. Bahl and V. N. Padmanabhan, "RADAR: An In-Building RF-Based User Location and Tracking System," *INFOCOM 2000*, pp. 775–784, 2000.
5. J. Hightower and G. Borriella, "Location Systems for Ubiquitous Computing," *IEEE Computer*, Vol. 34, pp. 57–66, 2001.
6. N. Bulusu, J. Heidemann and D. Estrin, "GPS-less low cost outdoor localization for very small devices," *IEEE Personal Communications*, Vol. 7, No. 5, pp. 28–34, 2000.
7. J. Beutel, "Geolocation in a Pico Radio Environment," *Master Thesis*, UC Berkeley, 2000.
8. J. Krumm, "Probabilistic Inferencing for Location", *Workshop on Location-Aware Computing*, pp. 25–27, 2003.
9. D. Fox, J. Hightower, L. Liao, D. Schultz and G. Boriello, "Bayesian Filtering for Location Estimation," *IEEE Pervasive Computing*, Vol. 2, No. 3, pp. 24–33, 2003.
10. RF Code, Inc., http://rfcode.com/ProductsFrame.asp.
11. J. Broch, D. Maltz, D. Johnson, Y. C. Hu, and J. Jetcheva, "A performance comparison of multi-hop wireless ad hoc network routing protocols," *ACM/IEEE Mobicom*, pp. 85–97, 1998.

Dual Communication System Using Wired and Wireless Correspondence in Home Network

Kunihiro Yamada[1], Kenichi Kitazawa[2], Hiroki Takase[2], Toshihiko Tamura[2],
Yukihisa Naoe[2], Takashi Furumura[2], Toru Shimizu[3], Koji Yoshida[4],
Masanori Kojima[5], and Tadanori Mizuno[6]

[1] Tokai University
[2] Renesas Solutions Corporation
[3] Renesas Technology Corporation
[4] Shonan Institute of Technology
[5] Osaka Institute of Technology
[6] Shizuoka University

Abstract. A system of simultaneously communicating of the same data in a home or relatively small home-like relatively small space using different communication methods, wireless and wired, was tried. As a result, 100% successful data communication was achieved without taking specific measures in making any changes to the communication environment. By a single means of wireless or wired communication, the message transfer success rates of message transfer were about 80% and about 70%, respectively, without changes to no particular measures taken in the communication environment, and 96% success was achieved in a mathematical simulation by the simultaneous communication method. However, the result of the above trial encouraged us to explore the find possibility of installing a network in a home or relatively small home-like relatively small space without giving making any specific improvements to the communication environment.

1 Introduction

The Internet, mailing systems, bath and hot water supply systems, audio-visual systems, intercom systems and security systems are operated independently inside homes and information exchange between them information exchange is limited [1].

A Desirable "home information community" should realize security (crime and disaster prevention) and convenience, and ensure that the condition of those home appliances the householder considers necessary is detectable and controllable. Controlling and detecting the condition of home appliances should be easy.

Networks should be easily installable in homes and other small spaces. Easy installation means installation without improving the any environmental communication conditions and without the assistance of engineers to operate the networks being required. A single communication medium is generally used for communication between nodes in these networks. However, we intend to improve communication quality by using plural types of communication media which supplement each other. This paper describes our study of using two types of communication media, wired and wireless, and how the improvement was realized.

We planned to use Power Line Carrier Communication (PLC) using electric power lines as the wired medium and IEEE 802.15.4 as the wireless medium, which are already, or about to be, in practical use as independent communication media [2-4].

R. Khosla et al. (Eds.): KES 2005, LNAI 3681, pp. 438–444, 2005.

2 Communication Quality in Wired and Wireless Single Communication

2.1 Quality of Wired Communication

The results of measuring the PLC communication quality in a common family house are shown in Figure 1. The X axis indicates packet error rate (PER), and a value of 0% means the best condition. The Y axis indicates the frequency distribution as a percentage (%). A PER of 0% existed at a frequency of 70%, and a PER of 100%, or totally impossible to communicate, existed at 20%. The cause of impossible communication was an improper phase combination in the power cable placement [5].

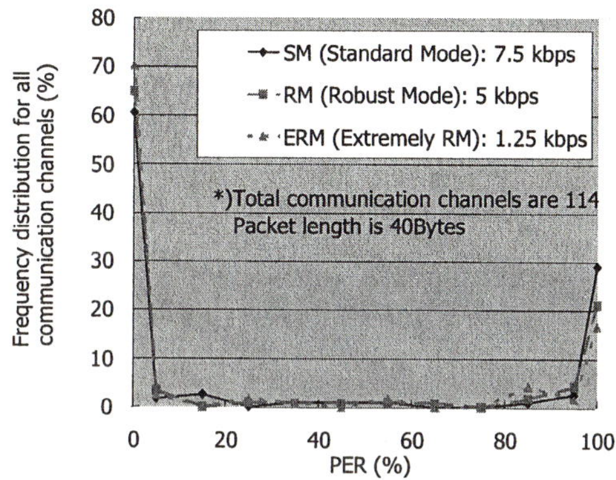

Fig. 1. Testing PLC (100 kHz–400 kHz) in a typical Japanese home

2.2 Quality of Wireless Communication

Figure 2 shows the results of a simplified experiment to measure how distance changes the level of received signals and PER in a certain number of packets, using IEEE 802.15.4. When the distance is 30 m, PER was found to be 1.5% or so. However, IEEE 802.15.4 is easily affected by obstacles such as metal items or reinforced concrete (rebar) as are other wireless communication methods, and sometimes it is not possible to achieve stable communication.

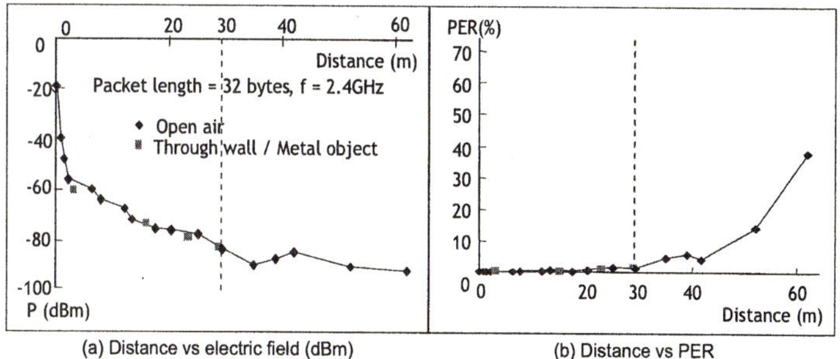

Fig. 2. IEEE 802.15.4 Test result data

3   Mutually Supplementary Wired and Wireless Network System

Our intention is to develop networks for relatively small spaces such as a detached three-story house within a 300 m^2 floor area, which do not require any changes in the communication conditions or assistance by engineers to operate the networks. This is based on the idea that low cost networks which are easy to operate are essential for home networks. We planned to use two or more different communication media and have the media supplement each other. In this study we used two communication media, wired PLC and wireless IEEE 802.15.4 to evaluate the network communication quality.

3.1   Wired and Wireless Network Unit Cells

We called our communication system with wired and wireless media a "network unit cell". Figure 3a shows how the home appliances are interfaced with the network unit cell and Figure 3b explains the symbols used. The network unit cell has three sections, the wireless section, the wired section, and the data processing section.

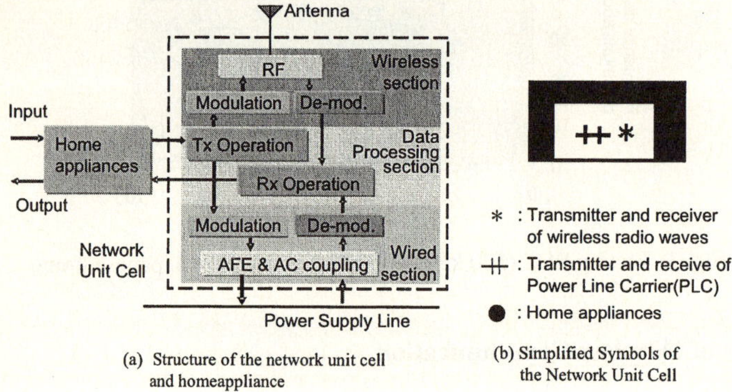

(a) Structure of the network unit cell
and homeappliance

(b) Simplified Symbols of
the Network Unit Cell

Fig. 3. Network Unit Cell

3.2   Simulation of Communication Quality

We defined the evaluation criteria for communication quality in modeling the networks as follows. When packet data of about 40 bytes is sent 10 times successively, we can classify the received data into correct data (Truth: the same data is received in the correct condition twice successively), possibly correct data (Truth?: the correct data is received twice, but not successively), and false data (False: conditions other than above). The communication quality in the data transmitted and received is classified into three ranks in the following examination.

The following section discusses the case of a three-story detached house, using the data in Section 2. In the case of PLC, we assume the frequency of Truth is 70%, Truth? is 10%, and False is 20% from the data in Figure 1. In the case of IEEE 802.15.4, we assume PER on the same floor, between adjacent floors, between the 1st and 3rd floors to be 2%, 50% and 70%, respectively, from the data in Figure 2, and the frequency of the weighted average for the conditions above is 82% for Truth, 14% for Truth?, and 4% for False, as shown in Table 1.

Table 1. The values indicated for the Truth, Truth?, and False categories for IEEE 802.15.4 are in %

Condition		Events	PER	Truth	Truth	False
On the same floor	1F to 1F, etc	45 routes	2	100	0	0
Between adjoining floors	1F to/from 2F, etc.	72 routes	50	86	13	1
Between floors separated by more than 2 floors	1F to/from 3F	36 routes	70	50	35	15
		Weighted average		82	14	4

Shown in Table 2 are the communication reliabilities when the wired and wireless media are used simultaneously. The communication quality is improved from 70% to 94.6% for the independent wired communication and from 82% to 94.6% for the independent wireless communication, which can be interpreted as an improvement to 96% if the Truth and Truth? categories are considered as the Truth category.

Table 2. The communication quality in % when the wired and wireless communication media are used simultaneously

		Wireless		
		Truth 82	Truth? 14	False 4
Wired	Truth 70	Truth 57.4	Truth 9.8	Truth 2.8
	Truth? 10	Truth 8.2	Truth?.Truth? 1.4	Truth?.False 0.4
	False 20	Truth 16.4	Truth?.False 2.8	False.False 0.8

4 Verification Experiment

Next, a verification experiment of the mutually complementing wired and wireless network system was conducted in a three-story detached house (Fig. 4). In this system, measures were taken so that, if communication could not take place between the transmission unit cell and the reception unit cell, the message was relayed via another unit cell and then sent to the target unit cell [5]. It is deemed successful if two consecutive correct data packets can be received 20 times when 38 bytes of data was sent 100 times.

4.1 Results of Demonstration Test

In the demonstration test, a success rate of 100% was obtained in data communication for all combinations from the 1st floor to the 3rd floor, as shown in Table 3. The table also shows the success rate between single unit cells for wired and wireless communication.

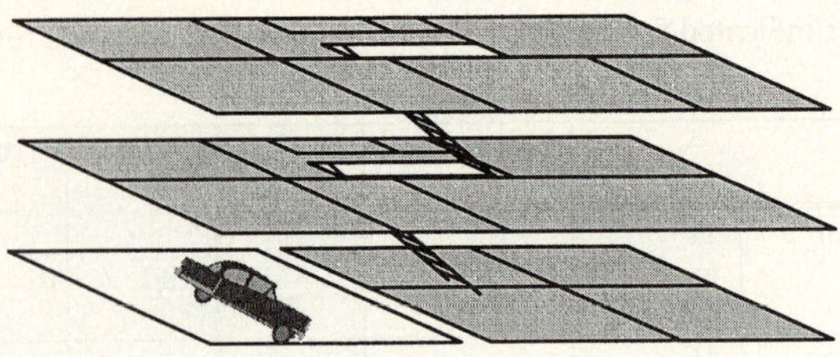

Fig. 4.

Table 3. Result of verification experiment

	Floor of reception	Overall success rate			Wired single success rate			Wireless single success rate		
		1F	2F	3F	1F	2F	3F	1F	2F	3F
Floor of transmission	1F	100%	100%	100%	98%	80%	74%	100%	83%	55%
	2F	100%	100%	100%	82%	81%	75%	81%	87%	84%
	3F	100%	100%	100%	75%	76%	76%	60%	83%	93%

The average success rate was 80% for wired single, and 81% for wireless single. In wired single, a combination showed a success rate of about 70% or lower even on the same floor, which was supposed to be coming from the wrong phase connection in the power cable, and this agreed with the result of an actual check of the power distribution on each floor (Fig. 5).

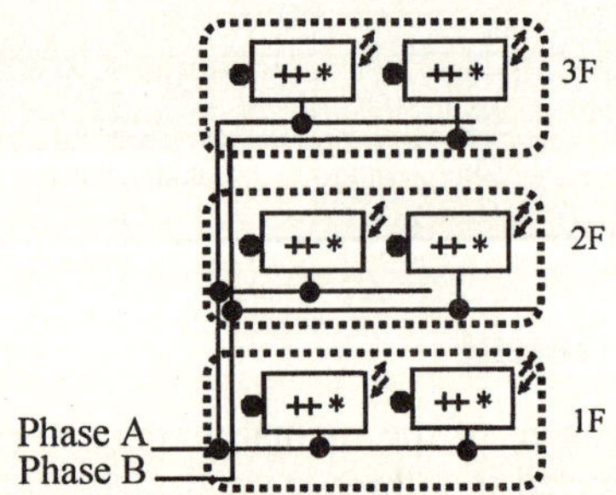

Fig. 5. Power distribution condition

For wireless communication, the success rate was 93% between two unit cells on the same floor, about 83% between adjacent floors, and about 58% between two floors separated by two or more floors in between, and thus the range of transmission decreased as the difference in floors increased. If backup for a communication error arising during wired single communication or wireless single communication is done by selecting wired and wireless in the unit cells, the error rate is:

Wired single communication error rate × Wireless single communication error rate

According to the rate of a single medium (Table 3), the success rate is as shown in Table 4.

Table 4. Assumed success rate between unit cells

		Floor of reception		
		1F	2F	3F
Floor of transmission	1F	100%	97%	88%
	2F	97%	97%	96%
	3F	90%	96%	98%

The average success rate in Table 4 is 95%, which almost agrees with the simulation value, 96%. Since we used the another unit cell to relay the message and then send to the target unit cell, the rate is actually 100% as shown in Table 3.

4.2 Consideration of Power Distribution Pattern

In residential houses, power distribution combinations are available such as in (a), (b) and (c) in Figure 6 in addition to the cable pattern in the house used in our test.

Here (a) is a common power distribution, which has two phases on all floors. According to our test result, a success rate of 100% can be expected by a combination of wired and wireless. Unusual distributions are (b) and (c). Here (b) is a power distribution with a single phase, a success rate of 100% can be expected with the wired method, while (c) shows a case in which different phases are used for power distribution in adjacent floors, and wired communication cannot be used between 1F and 2F, and between 2F and 3F, for which we have to rely on wireless communication only. In such a situation, the success rate between two adjacent floors is about 83% from the result of our test, and the average rate is 92%. This is considered to be the severest power distribution pattern.

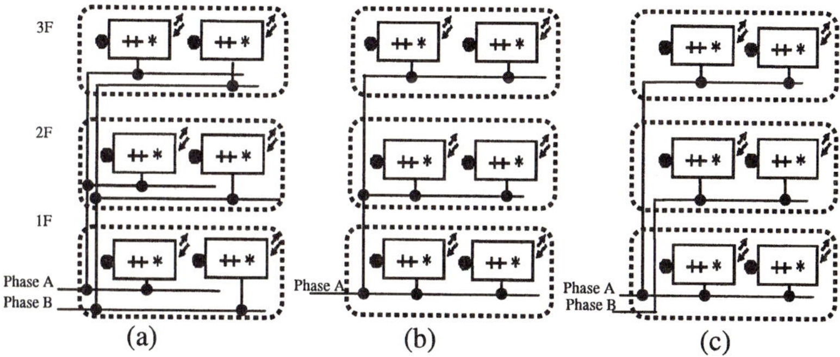

Fig. 6. Power distribution pattern

5 Conclusion

A system of simultaneous communicating the same data in a three-story detached house using a mutually complementing wired and wireless network system was tried.

As a result, 100% successful data communication was achieved without making any changes to the communication environment. The peculiar distribution is inconvenient to operate the network because the advantage of wired and wireless complimentary system is not available, yet 92% of success rate is assured.

By means of wireless or wired communication, the message transfer success rates were about 80% and 70%, respectively, with no changes made to the communication environment. However, the above trial encouraged us to find possibility of installing a network in a home or home-like relatively small space without giving specific improvement to the communication environment.

References

1. H. Fujimura, Y. Ogawa, H. Kawana, A. Kuwayama, "Home Safety Station "Beruhi"" Matsushita Technical Journal Vol. 47, No. 47, Feb. 2001
2. G. Evans. "The CEBus Standard User's Guide" Section 5. The Training Dept. Publications, 1996
3. URL http:/www.itrancom.com
4. E. Takagi, Y. Takabetake, "Current Status of and Future Trends in Wireless Technologies" Toshiba Review Vol. 58, No. 4, Apr. 2003
5. K. Yamada, et al. "Dual Communication System Using Wired and Wireless Correspondence in a Small Space" KES2004, Proc-II, pp. 898-904.
6. T. Furumura, Y. Hirata, S. Kuroda, K. Yamada, M. Kojima, T. Mizuno, "Power Line communication System for Home Network" pp. 93-96 Multimedia, Distributed, Cooperative and Mobile Symposium, IPSJ Symposium Series Vol. 2003, No. 9, ISSN 1344-0640, June 2003
7. Y. Okada et al. "Evaluation Method for High-Reliability Spread Spectrum Communication on Power Line" Matsushita Electric Works Technical Report, March 2003
8. URL http://www.zigbee.org/

The Architecture to Implement
the Home Automation Systems with LabVIEW™

Kyung-Bae Chang, Jae-Woo Kim, Il-Joo Shim, and Gwi-Tae Park

ISRL, Korea Univ, 1, 5-ka Anam-dong Sungbuk-ku 136-701 Seoul, South Korea
{lslove,kjwoosh,ijshim,gtpark}@korea.ac.kr
http://control.korea.ac.kr

Abstract. It becomes a topic that constructional environment could be controlled at any place and anytime in the ubiquitous era. It demands much expenses and technology to control everything. Therefore, we presented construct environment that could control home from outside using inexpensive USB web camera, we suggested that we could control it quickly by PDA. Proposed program could be used to many applications including home automation, monitor system and was verified through the experiment. [1] For implementation LabVIEW™ was chosen as software along with USB web camera and a wireless client (PDA).

1 Introduction

Ubiquitous computing aims to "enhance computer use by making many computers available throughout the physical environment, but making them effectively invisible to the user". Thus, it is a desirable system for home networks where every home owner acts as an administrator. [2]

In a home environment, some devices might be controlled by a PDA and Web Server for Smart home. In efficient aspect, a user can use his/her PDA as a universal communication. A user can use his/her PDA as a universal communication and control agent, not only for communication, but also as a controller of network-connected appliances. Furthermore a representative application field for security and monitoring is the home network service, which is an emerging topic in the IT field.

We use inexpensive web camera instead of expensive devices, and are going to propose and verify a scenario that can embody home monitoring systems through control signal of wireless devices. The digital home service becomes popular and hence more affordable. We can use such inexpensive equipment to monitor our home for intruders or nursery, operating room in the hospital, building entrance and so on. [3] Therefore, in this paper, we embodied program that can send control signal as real time after accept the information of someone in far away place using by USB web camera and also can watch with direct control to Internet.

The Internet leads to the revolution of the Internet-enabled instrumentation using programs like LabVIEW™. As LabVIEW™ is based on graphical programming, users can build instrumentation called "Virtual instruments" (*VIs*) using software objects. VIs built using LabVIEW™ are the easiest to be controlled and shared over Internet as LabVIEW™ has a number of VIs such as VI server, data Socket server

R. Khosla et al. (Eds.): KES 2005, LNAI 3681, pp. 445–451, 2005.

and visual basic and active X, Java and Java Script that can be used to design and develop Inter-net-enabled virtual instruments. [4] This is the reason that we selected LabVIEW™.

We implemented home automation using LabVIEW™ in home network technology (*using centered home server that control, utilize, and monitoring some electronic devices*) and are going to propose digital home system that uses web server as well.

2.1 Architecture discusses advantages and drawbacks of using technologies for communication with LabVIEW™ and LabVIEW™ PDA module and discusses to analysis front panels and block diagrams in a server and a client program. [5] We confirmed proposed home model through 3 experiments in 2.2 home controls. Lastly, 2.3 applications discuss about useful functions in a building and about home controls and conclude the paper.

2 Implementation and Proposal of Home Control

2.1 Architecture

Because LabVIEW™ is a graphical program language, the user can make programs easy and fast. It is easy to learn and understand. LabVIEW™ displays strong power when is used to control. In the communication field, LabVIEW™ is developing to be worth comparing with functions which are realizable in the C++. LabVIEW™ PDA Module is one of options among other application modules and it was chosen for the development. Figure 1 shows overview of technologies for communication with Lab-VIEW™ over a network. [6]

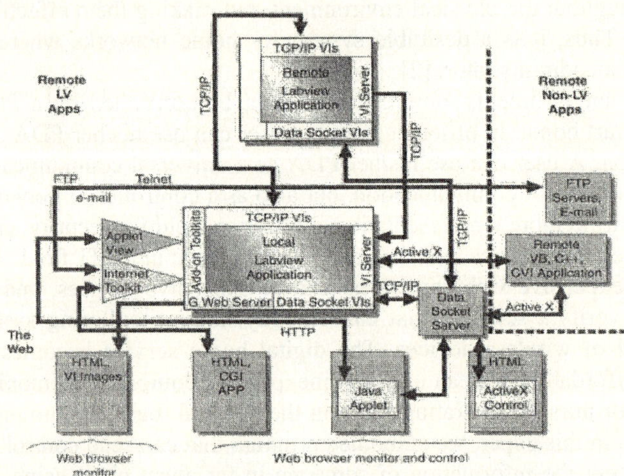

Fig. 1. Technologies for communication with LabVIEW™ over a network

We can confirm application parts of LabVIEW™ in the figure. First of all, we must compose server and client's communication program using TCP/IP function to use communication applications. Figure 2 is block diagrams that indicate the way of mak-ing a data for communication using TCP/IP. The main characteristic of Lab-

VIEW™ is that it first sends the size of the letter and second it sends the content. The Receiver has a structure that receives the size first and then it receives the content.

When "send key" is pressed, the program would change the sender's name and size and context of data to the proper format for transfer. If you "send key" is not pressed, the program will send empty message continuously. If the receiver does not receive any context during a certain time period, a time out error will occur, which is the reason why empty messages are being sent continuously.

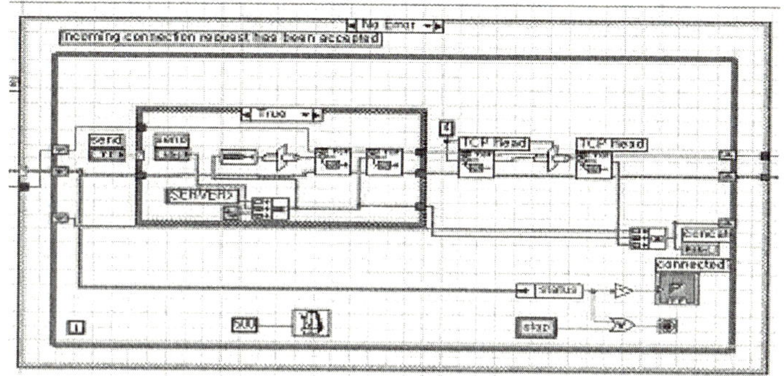

Fig. 2. TCP/IP Communication block diagram (*If send Boolean values is true*)

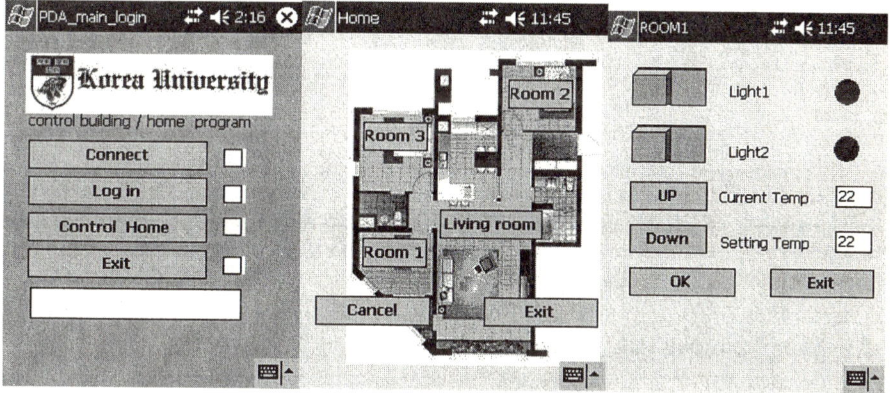

Fig. 3. PDA Main and Home and Room

With TCP/IP programs, we can divide to server and client program. Then After executing the program in the PDA(*client program*), pressing on the "connect" button makes a connection with the server. After successful login, some control is enabled, like in the previous concept described. Figure 3 shows screens that show from first page to next. Room 1 is connected with lighting temperature. Following (Room 2, Room 3.) is also prepared to have additional functions such as controlling the light or the door. The third picture of figure 3 is screens that show what happens when click is made in Room 1. Using the program, you can turn-on or off the lighting, and control the temperature. PDA transmits key value and according to signals ON/OFF, the light will function and send the response back to the PDA program.

2.2 Home Control

There are three different experiments of the home control. First control is made by using a PDA, second control by using the PDA or a cell phone after monitoring program operated by USB web camera, lastly a controlling through the Internet by using program that is consisted of below three parts (*monitoring part, sound part, and control part*). Whole concept is same as in Figure 4 and we need to experiment equipment such as Table 1.

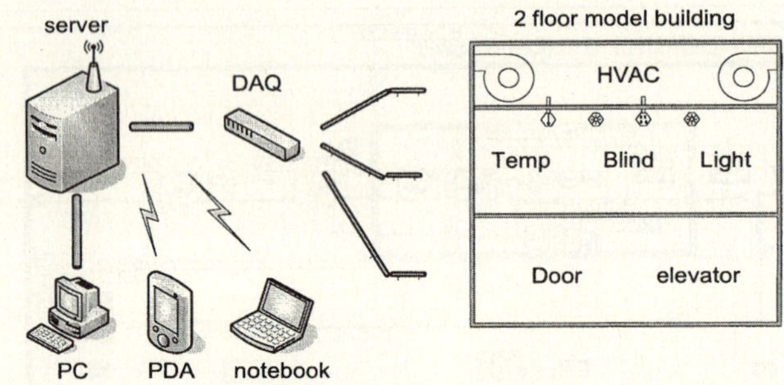

Fig. 4. Example model

Table 1. Experiment appliance specification

Computer - P4 2.4G, RAM 512M
PDA - HP 5450
Samsung USB web camera(*OS Windows Pocket PC 2003*)
AP(Access Point) - LINKSYS WRT54G
DAQ – NIDAQ (*PCI 6040E*)(*Data Acquisition System*)
Valve-valcon AT12-3T,   Lamp-300w,   Fan

We can see that there are some network appliances that can be controlled from mobile environment in the world and software to develop the remote control application.

2.2.1 Home Control Using PDA

We composed a model of the home network consisting of the lighting, a fan, a temperature indicator and a valve like in Figure 2. So we created a PDA client program that has five on-off is switches available in order to enable the control as shown in the Figure 5. PDA program enables the control of the each device pictured in the Figure 6

2.2.2 Home Control Using Monitoring System

Second experiment was done by using an USB web camera. Web camera is connected to server and when motion is detected, program developed in LabVIEW™. Since, when the E-mail arrives, using a service that sends such message to portable phone in case when we are not at the house, we can be notified when someone enters our house. Since we can confirm someone's identity by looking at the picture, a user can control device in user's house (door, window, temperature) using client program through wireless LAN. Therefore, we are proposing a scenario as in Figure 7. Program that is stored on a server sends the email, and when the email arrives, we can confirm someone visit in real time using service that sends the messages. Figure 7 shows downloading JPEG files from Internet in a PDA, and log in pages. After log-in process, we can control other device like in the first experiment (*send control signal by PDA*).

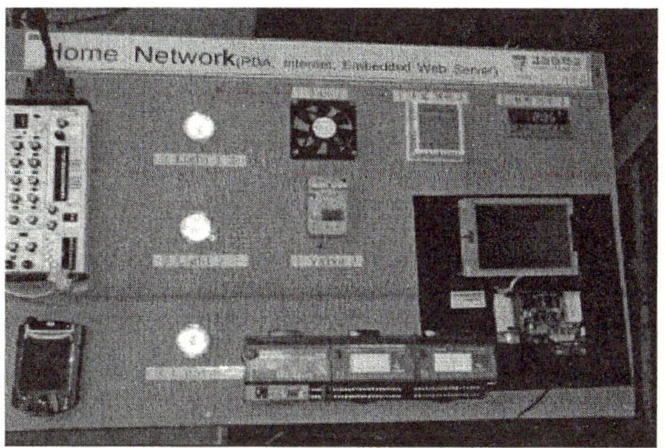

Fig. 5. Home control system

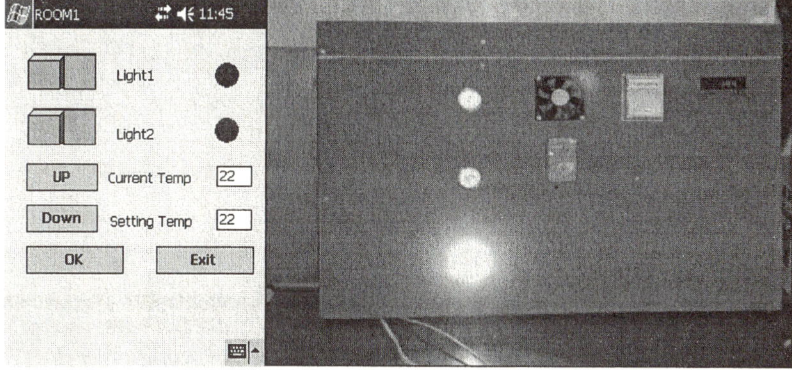

Fig. 6. PDA client program for experiment (*left*) and Operating Home Control System(*right*)

2.2.3 Home Control Through the Web-Based Interfaces

It is important for us to enable the control by a web interface, to allow the control to be possible anytime and anywhere. So we developed the application for the control in run-time using the LabVIEW™ Run-time Module. Figure 8 shows the control using the Web .[7] It can be controlled from Web environment in the world.

2.3 Applications

For example, to inform the home owner about a person entering or trying to enter the home and even allow entrance via a certification process through a mobile device such as a cell-phone or PDA. After analyzed information, we can do the second control. Further, it can allow to remotely monitoring young children or a patient in a house or in the hospital while being away enhancing personal safety and comfort.

Furthermore, it may be possible to take a photograph through sounds (*crash spot*) and to monitor financial institution or jail with web server, and to analyze degree of congestion of complicated place. And with log files, it is possible to predict the lighting, fan, motor's total time of use, etc. So we can know time of exchange and equipment and this program is also capable of informing administrator about the time of managing.

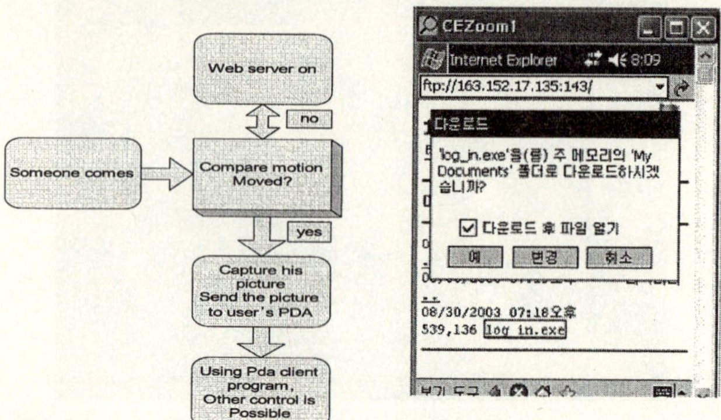

Fig. 7. Scenario and Connecting pages after downloading the picture

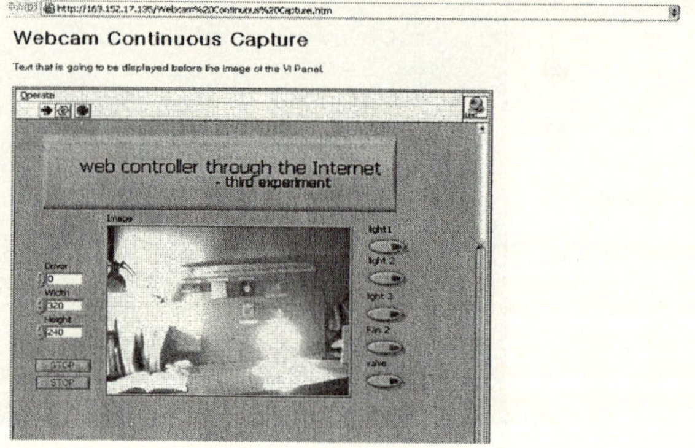

Fig. 8. Control panel on web pages

3 Future Work

For future work, we will enforce program's security and check its stability in a long time period. Further, when multiple users connect to this program, there is a priory problem that has to be settled. And need research about link and implementation of embedded web server in order to apply in additional application.

4 Conclusion

For home networks, many network appliances that can be controlled from mobile environment.We made a control program in a server and a wireless client (PDA) using WLAN and LabVIEW™. And we also made a smart home model to propose useful functions in building and home controls. Since control program is always processing signals of sensors, to make a program using LabVIEW™ with the support of measurement for safety and accuracy is profitable as communication program, since when we make programs in a PDA, to use LabVIEW™ helps us to shorten a developing time and gains in accuracy measurement values. Lastly, we implemented useful functions in building and home controls using PDA and web. Proposed monitoring system can be used conveniently without expensive image acquisition equipment and can be applied to other applications as well.

References

1. Alheraish, A. "Design and implementation of home automation system" Volume: 50, Issue: 4 1087 - 1092 IEEE Transactions Nov. 2004
2. Schulzrinne, H. Xiaotao Wu Sidiroglou, S. Berger, "Ubiquitous computing in home networks" Volome: 41, Issue: 11 128 - 135 Communications Magazine, IEEE 2003
3. Yuan-Hsiang Lin I-Chien Jan "A wireless PDA-based physiological monitoring system for patient transport" Volume: 8, Issue: 4 439 - 447 IEEE Transactions 2004
4. Nikunja K. Swain, "Remote data acquisition, control and analysis using LabVIEW™ front panel and real time engine", Proceedings IEEE southeast Con, 2003.
5. Kuo, S. Salcic, Z. Madawala, U. "A real-time hybrid Web-client access architecture for home automation" 1752 - 1756 vol.3 Processing 2003.
6. Jeffrey travis "INTERNET APPLICATION IN LabVIEW™", PHPTR
7. M. Joler and C. G. Cristodoulou, "Virtual laboratory instruments and simulations remotely controlled via the Internet," in Proc. IEEE Antennas Propagat. Soc. Int. Symp., vol. 1, pp. 388–391 Boston, MA, July 2001

Separated Ethernet Algorithm for Intelligent Building Network Integration Using TCP/IP

Kyung-Bae Chang, Il-Joo Shim, and Gwi-Tae Park

School of Electrical Engineering Korea University,
1, 5-ka, Anam-dong, Seongbuk-gu, Seoul, 136-701, South Korea
{lslove,ijshim,gtpark}@korea.ac.kr

Abstract. Problems like poor security, transfer delay or packet loss occur while building network systems that are applied with TCP/IP integrate with data network systems. To solve this problem, this paper proposes the Separated Ethernet, which can give logically separation and priority to the system, and by using the OPNET Modeler simulator, we will verify its performances.

1 Introduction

In a building, networks are divided by its facility's functions into control networks and data networks. And the control networks and data networks are completely separated, as shown in Fig. 1, for security reasons.

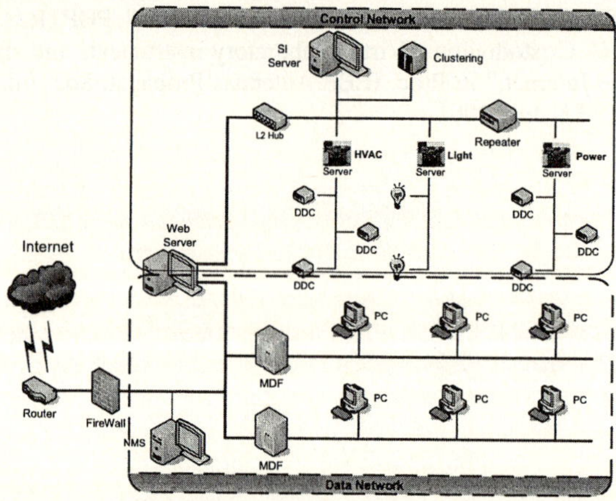

Fig. 1. The control networks and data networks are completely separated in building system

This paper proposes the Separated Ethernet, which can integrate the entire network in the building into TCP/IP protocol, and verifies its performance with a network simulator.

2 Separated Ethernet

2.1 Dual Queue Structure and Queue Priority

In building network systems, security is improved by integrating the two networks logically. We propose a method, which has independent upper layers of the control network and data network for improved security.

R. Khosla et al. (Eds.): KES 2005, LNAI 3681, pp. 452–458, 2005.
© Springer-Verlag Berlin Heidelberg 2005

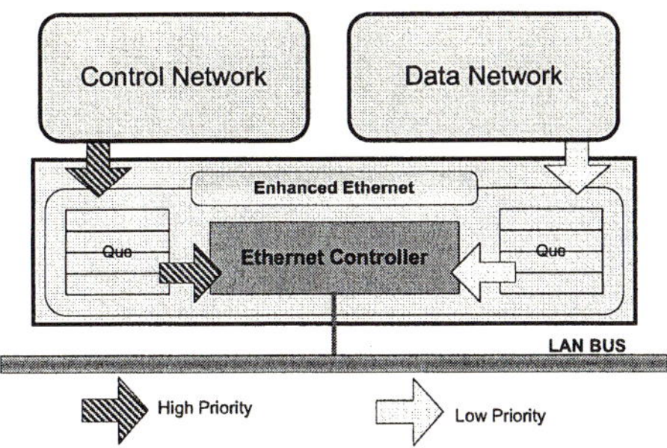

Fig. 2. The dual structure Queue and Priority

Fig. 2 shows a dual structure Queue. If the data network's Queue is disabled, and the data network frame is received from the Separated Ethernet, as the data is not transmitted to the upper layers, malignancy codes and virus's attacks have no effect on the station. If there is the same data in both the data network's and control network's Queue, priority must be given to the control network by extracting, fragmentation and transmitting the data from the control network.

2.2 Backoff Priority

Not like the MAC protocols that act in order, like Token Bus or Token Ring, Ethernet is a competitive MAC protocol, which gives priorities according to its mechanism. Ethernet gives priority to the station that attempts transmission first, and if a collision occurs by transmission delay or by transmitting at the same time, priority is decided by a BEB (Binary Exponential Backoff) algorithm. Sometimes, like with a wireless environment IEEE standard 802.11, priorities must be given. Research on giving priority by improvement of the BEB algorithm is progress, as BEB is used in deciding priorities on the collision of wireless LAN too. [1] These, Backoff-based priority schemes are methods in which the stations with a high priority of deciding the Backoff slot time, compete in a shorter slot time.

When a collision occurs on the improved BEB algorithm, the re-transmission time is decided by the number of collisions. If the fixed window size in a competitive situation is W, and r is the random slot time, the station's re-transmission times i and j can be shown as Ti and T_j. If i station's priority is higher than station j's, the equation can be shown as equation 1.

$$T_i = W * r_i < T_j = W * r_j. \tag{1}$$

Window size W is fixed, it can be shown as in equation 2.

$$r_i < r_j. \tag{2}$$

In the proposed S Ethernet, the control network transmitting station's random slot time value is r_c, and the data network transmitting system's is r_d. The random slot time is decided by the number of re-transmissions n, and is shown as equation 3 and equation 4.

$$r_c = random(\ ) \tag{3}$$
$$r_d = random(\ ). \tag{4}$$

Namely, the control network with a higher priority gets the lower competitive section. Fig. 3 shows the giving of priority, to the Backoff algorithm applied to the building network system.

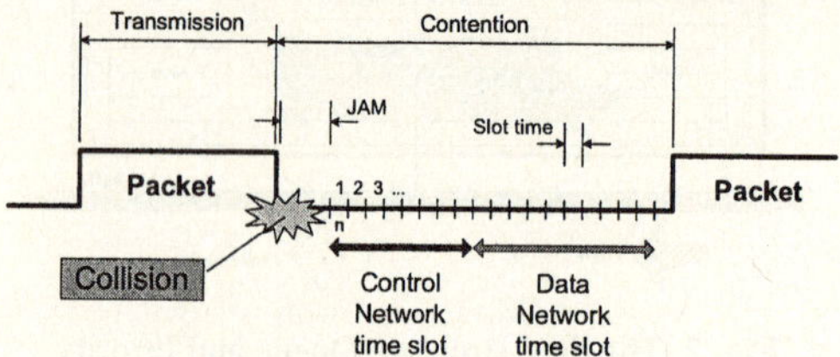

Fig. 3. Competive state resulting from collision of Separated Ethernet

2.3 Separated Ethernet Frame

All the frames transmitted on the channel for the integration of building networks must be classified into data network frame and control network frame.

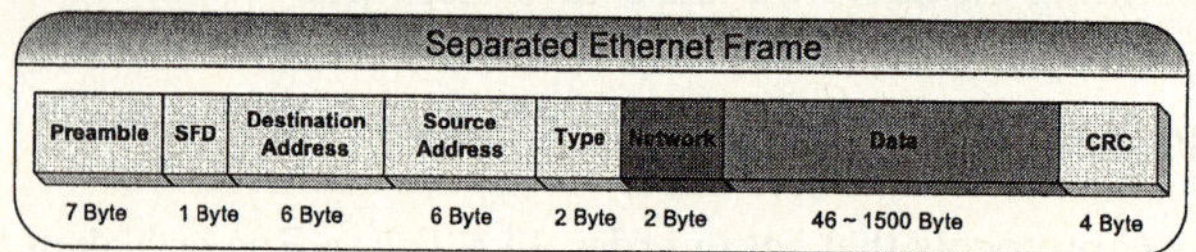

Fig. 4. The structure of Separated Ethernet

Fig. 4 shows the Separated Ethernet frame's structure. It is a form of which, a 2 byte field called Network is added to the existing Ethernet. The contents of this field are shown in Table 1.

Table 1. Network field values of the Separated Ethernet

Name	Code	Description
Network 1	0x00	Control Network
Network 2	0x01	Data Network
Control	0x10	Ext - Ethernet

The values of Network 1 and 2 are used to discriminate control network and data network. The control value is a designated value, which is used to add the MAC algorithm, an algorithm that can form the most suitable network by establishing the Separated Ethernet's constant value.

3 Simulation

The proposed Separated Ethernet is verified using OPNET's OPNET Modeler simulator. A LAN environment, which is consisted by 10 stations of 10MBps Link Models, is simulated under different scenarios of node model and created packet amount settings. To understand and valuate the performances of the Separated Ethernet, 4 scenarios have been made. Fig. 5 shows the simulating process of OPTNET simulator.

One model from the existing Ethernet and three models of Separate Ethernet applied ones were used in the node model simulation. Fig.6 shows the node models in simulation.

Separate Ethernet applied DDC(Direct Digital Controller) indicates the control facilities used in buildings, and the Separated Ethernet applied Client PCs indicates the data processing facilities. Web Server is the device that connects these two networks.

In the results estimated in the simulation, the values are numbers of transmitted packets, and of collisions in each station, and the mean value of the End-to End Delay, which is the time during which packets of two networks are transmitted.

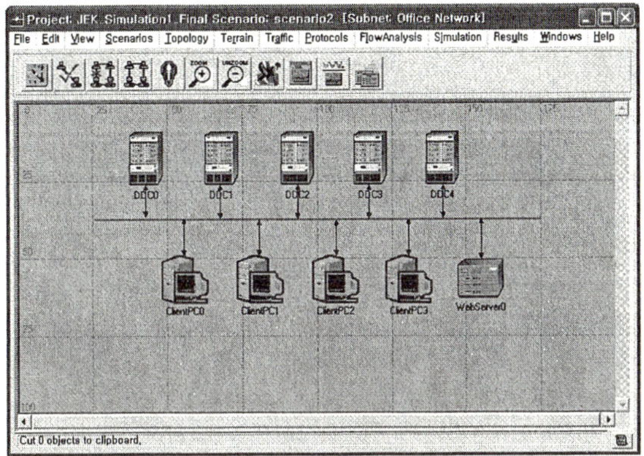

Fig. 5. Simulation environment

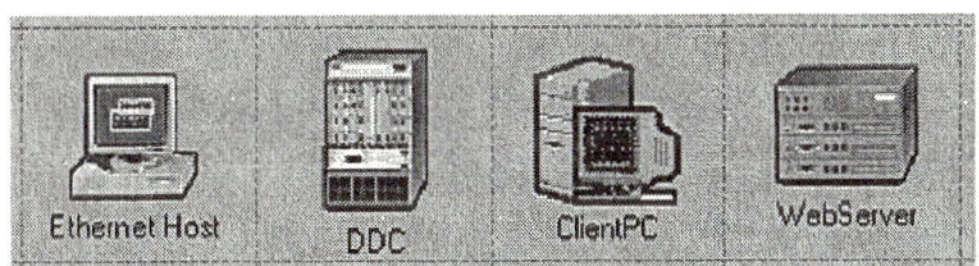

Fig. 6. Node models used in simulation

3.1 Confirming the Queue Priority

In the scenario of confirming the Queue's priority, packets are generated from the upper two layers of the Web Server and is input to Queue, from a network environment consisted of five DDCs, four Client PCs and one Web Server. The packets are generated every 5 seconds for 100Bytes, from both the control network packet and data network packet. This scenario's result is shown in Fig. 7.

The control network's End-to-End Delay average is 0.103[ms], the data network's End-to-End Delay average is 0.215[ms]. The control network's packet has been transmitted twice faster than the data network packet. The two packets have a regular Delay value, which is because there were no collision on the LAN, due to the packet's being generated from a single station.

3.2 Confirming the Backoff Priority

In the scenario of confirming the Backoff's priority, each packet is generated every second for 1024 Bytes, from the stations of control facilities and data facilities of a

LAN environment of DDC and five Client PCs. The End-to-End Delay average is shown on Fig. 8.

These results confirm that the control network's transmission time is always shorter than that of the data network, as the control network's Delay value is 1.602[ms] and the data network's Delay value is 3.595[ms].

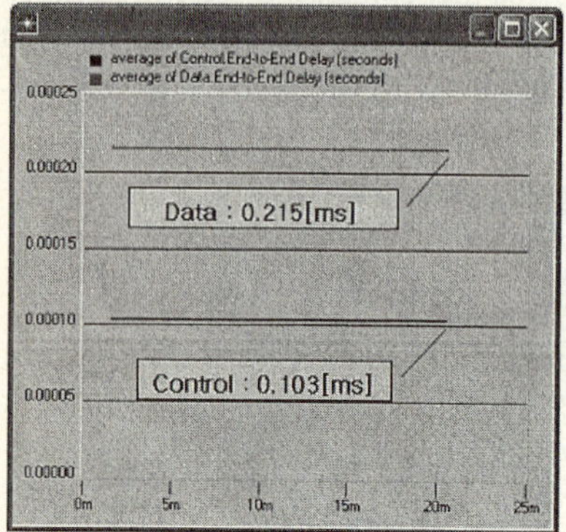

Fig. 7. Results of scenario 1

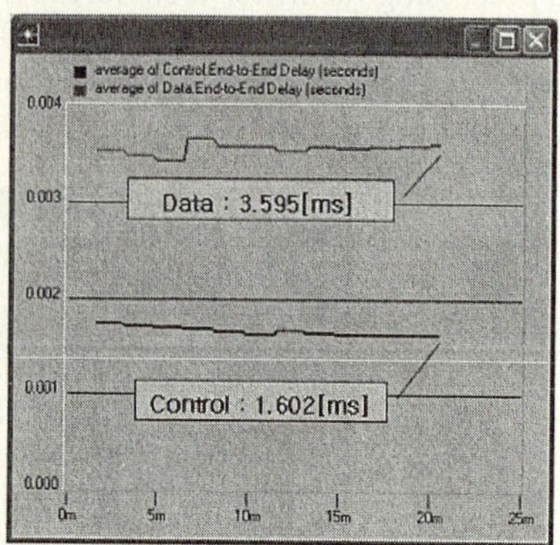

Fig. 8. Results of scenario 2

3.3 Separated Ethernet's Performances Under Overload

To study the performances of the Separated Ethernet under a sudden, big amount of traffic on the data network, five control facilities and 5 data facilities are connected to one bus. The control facility station generates 100 byte packets every 0.1[sec], and beginning at 600 seconds all the stations of the data networks generate 1024 byte packets every 0.1[ms]. The End-to-End Delay average is shown in Fig. 9, and the amount of packet loss is shown on Fig. 10.

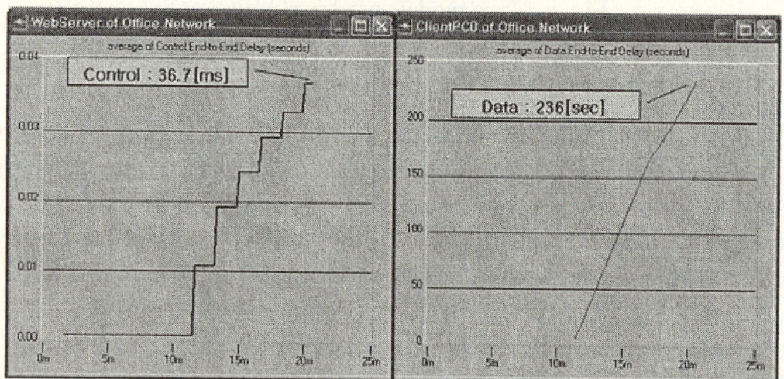

Fig. 9. Results from scenario 3's transfer delay2

As shown in Fig. 9, the control network's delay increases as the data network's packets become generated. But the results show a big distinction of 36.7[ms] and 236[sec]. Also, the data network's packet loss remains but the control network does not have any packet losses.

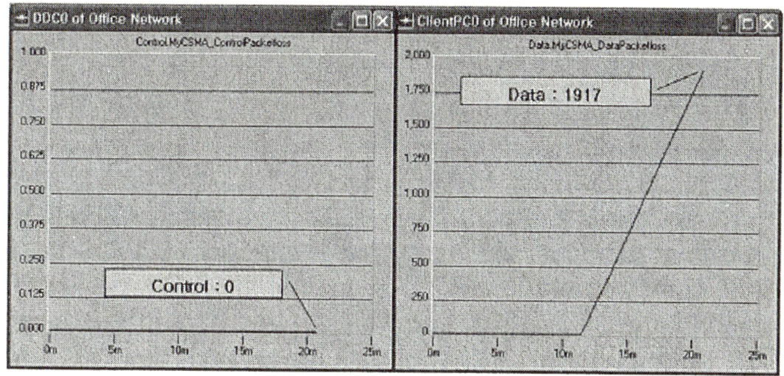

Fig. 10. Packet loss from scenario 3

3.4 Comparison Between Ethernet and Separated Ethernet

To compare the performances of the existing Ethernet and Separated Ethernet, the same amount of data has been generated for the same amount of intervals from the two LANs, each composed of Ethernet and Separated Ethernet, and the End-to-End Delay of the two LANs have been measured.

Two kinds of simulations were conducted, one where all the stations generate packets of a low, 1024Byte data every second, and one where packets of 1024Byte data are generated every 6 seconds.

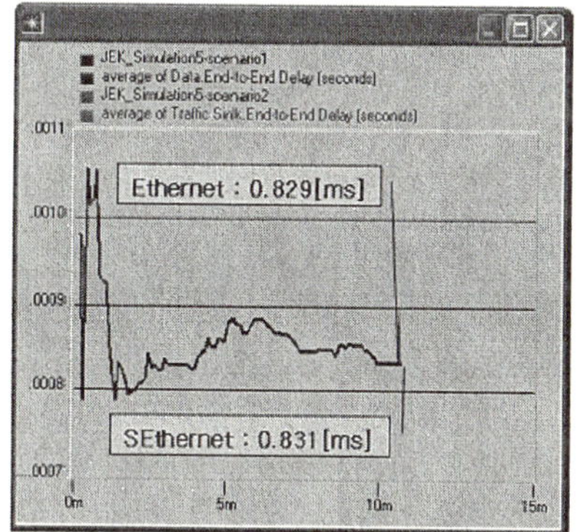

Fig. 11. Results of scenario 4 under low traffic

Fig. 12. Results of scenario 4 under high traffic

Fig. 11 shows the results of scenario 4 under low traffic. The existing Ethernet's transmission delay is smaller than the proposed S-Ethernet, but its differences are much smaller than the total transmission delay.

Fig. 12 shows the results of scenario 4 under high traffic. As the traffic increases, the transmitting delay of the two Ethernet methods become bigger, but this difference is too small comparing to the total transmission delay. In other words, there is little difference in the performances of the existing Ethernet and the Separated Ethernet.

By these results, we could confirm that the control network's packets have a fast and stable transmitting speed, even when a lot of packets are generated from the data network by the priority given to the Queue and Backoff algorithm. Also, there were little differences with the performances of the existing Ethernet.

4   Conclusion and Future Works

This paper proposed the integration of the building's control network and data network. Problems that occur with the integration of the two networks, like transmitting delay, packet loss and security matter. We proposed the Separated Ethernet to solve these problems. Ethernet MAC Protocol, which is in charge of layer1 and layer2 of the OSI 7 layer has been logically divided with the network, and the control network system has been prioritized on the Separated Ethernet.

The Separated Ethernet has been designed by adding Backoff Priority, double Queue and Queue priority functions to the existing Ethernet. The designed Separated Ethernet's logical dividing and priority giving functions have been verified by OPNET Modeler, by separate scenarios.

Reference

1. Yang Xiao, "Backoff-based priority schemes for IEEE 802.11", ICC, 2003. pages: 1568-1572 vol. 3

Operational Characteristics of Intelligent Dual-Reactor with Current Controlled Inverter

Su-Won Lee[1] and Sung-Hun Lim[2]

[1] BK21 Kunsan National University, San 68, Miryong-dong, Kunsan, Korea, 573-701
Fax:82-63-469-4699
lsw91@kunsan.ac.kr
[2] Research Center of Industrial Tech., Chonbuk National Uni., Chonju, Korea

Abstract. The intelligent dual-reactor with the current controlled inverter, which was operated based on the current controlled algorithm, was suggested in this paper. It could limit the fault current from the fault accident as well as compensate the reactive power of the grid, irrespective of the non-linear load type or its variation. The suggested equipment consisted of the dual-reactor with a power switch and the full bridge inverter. The operation of this equipment could be divided into the normal state and the short-circuit state. The operational characteristics of the proposed intelligent dual-reactor connected with the current controlled inverter for both the normal state and the short-circuit state were analyzed from the simulation results.

Keyword: dual-reactor, current controlled inverter, fault current, reactive power

1   Introduction

With the growth of industry and the appearance of various industrial electric appliances, the recognition for the importance of power quality, especially the need for its improvement, has been increased [1-2]. Due to these necessities, the various topologies and algorithms to improve the power quality of grid have been suggested [2-5]. However, the topologies to protect the system from the short-circuit effectively with the improvement of the power quality have been rarely developed. To protect the power system from the short-circuit, most of researches have been focused on the circuit breaker or series reactor, not the multi-functional equipment using the intelligent technology [6-7].

The suggested intelligent dual-reactor with current controlled inverter in this paper can protect the system from the over-current in case of a short-circuit, and compensate the reactive power of the grid due to non-linear load during a normal time as well by controlling the current from the inverter to the non-linear load. The operational principles for the suggested equipment were described and its usefulness was confirmed through the simulation.

2   Intelligent Dual-Reactor with Current Controlled Inverter

Fig. 1 shows a schematic diagram of the intelligent dual-reactor with the current controlled inverter, which is constructed between the grid and the load through the link

R. Khosla et al. (Eds.): KES 2005, LNAI 3681, pp. 459–464, 2005.

inductor(L_M). This equipment is composed of the dual-reactor with a power switch and the shunt power conditioner. The former consists of the primary winding (L_P) and the secondary winding (L_S) wound in parallel through the same iron core with a power switch (SW). The latter is composed of series inductor (L_C), bi-directional inverter and dc capacitor (C_{DC}). The fault current from a short circuit can be limited by keeping the power switch open in the dual-reactor, and the reactive power due to non-linear load can be compensated by the shunt power conditioner controlled based on the current controlled algorithm.

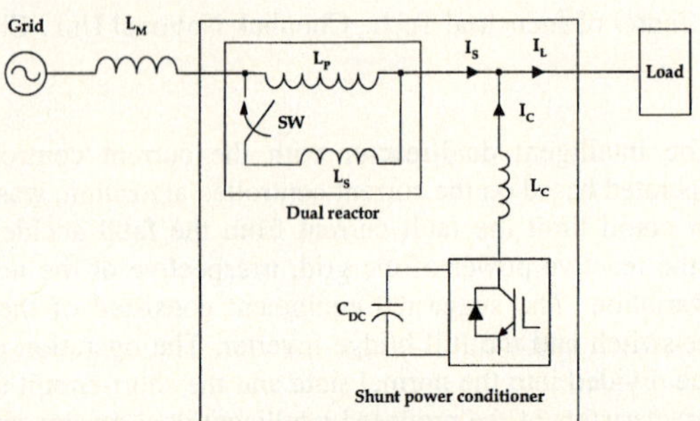

Fig. 1. Schematic diagram of intelligent dual-reactor with the current controlled inverter

2.1 Operational Principle of the Dual-Reactor with a Power Switch

Fig. 2 shows conceptual configuration of the dual-reactor with a power switch. i_1, i_2 and i_{In} are the primary, the secondary and the line currents, respectively. The primary and the secondary windings are wound to counteract each other's flux, which the flux generated from the primary winding is cancelled out by one from the secondary winding if the power switch keeps close. Therefore, the voltages across the primary winding and the secondary one are zeros. In fault condition like line-ground fault, if the power switch is open, the voltage induced in the primary winding is not equal to the voltage in the secondary one any more, which the fluxes generated by each winding are not cancelled out each other, and the fault current can be limited by the impedance of the dual-reactor.

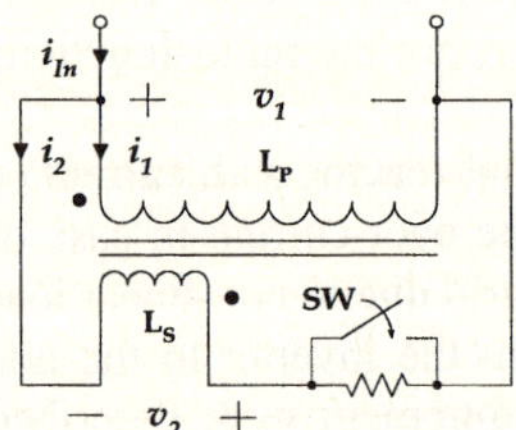

Fig. 2. Conceptual configuration of the dual-reactor with a power switch

The operational waveforms of the dual-reactor in case that the line-ground fault happened were shown in Fig. 3. As described above, the voltages in each winding kept zeros by closing the switch during a normal time. However, the voltages in each winding were generated by opening the switch when the fault happened.

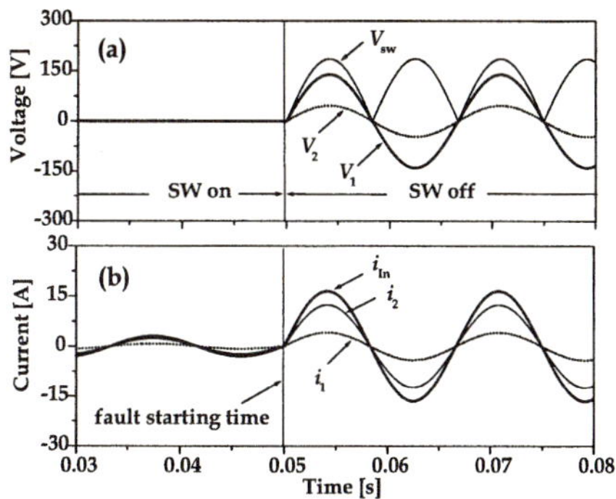

Fig. 3. Operational waveforms of the dual-reactor in case that the line-ground fault happens. (a) Voltages across each winding and the switch (b) Transporting currents of line and each winding

2.2 Operational Principle of the Shunt Power Conditioner

The basic configuration of the shunt power conditioner is shown in Fig. 4. Assuming that the fundamental component of the system current supplied by the power source can be expressed as equation (1), the amplitude of the fundamental component (I_{Sm}) can be derived as equation (2).

$$i_s(t) = I_{Sm} \sin wt \tag{1}$$

$$I_{Sm} = I_1 \cos\theta_1 + \frac{C_{DC}V_{Cr}^2}{V_{Sm}T} \tag{2}$$

Where I_1 , V_{Sm} and θ_1 represent the amplitude of the fundamental current, the amplitude of the source voltage and the phase difference between the source voltage and the fundamental current of the load, respectively. C_{DC}, V_{Cr} and T are the capacity of dc capacitor, the reference voltage across the capacitor and the period of source voltage, respectively. As seen in equation (2), the fundamental component of system current supplied by the voltage source is dependent on the capacity of dc capacitor, the reference voltage across the capacitor and so on. If the shunt power conditioner can supply the current as shown in equation (3), the source voltage supplies the current described in the equation (1).

$$i_c(t) = i_L(t) - i_s(t) \tag{3}$$

If the inverter can supply the same reference current to the load, the power quality of the system can be improved. For the power conditioning operation, the signal for the reference current as shown in equation (3) which the current of the series inductor follows should be transmitted into the power switches of the inverter. As the control algorithm for the generation of the reference current signal, the polarized ramp time (PRT) current control algorithm, which uses the previous same-side excursion of the current error signal and can keep the constant switching frequency, was applied with. The detailed implementation of PRT current control algorithm was described in reference [8].

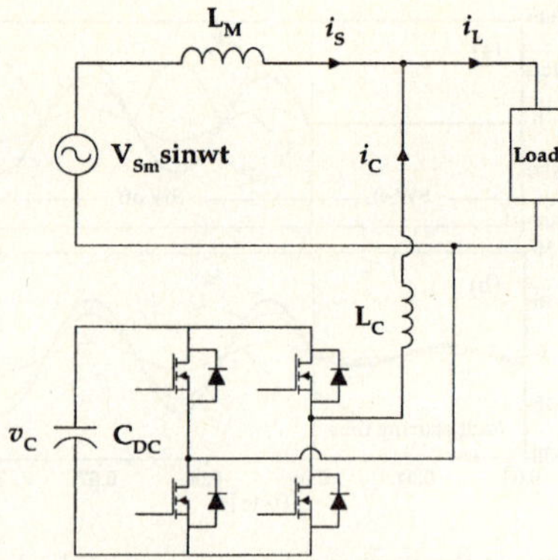

Fig. 4. Basic configuration of the shunt power conditioner

3   Simulation Results and Discussion

The simulation for the operational characteristics of the intelligent dual-reactor with the current controlled inverter was executed using PSIM simulator. The operational characteristics were analyzed into two states: the normal state and the line-ground fault state. The design parameters for the simulation were shown in Table 1.

Table 1. Design parameters for simulation

Parameter	Value	Unit	Parameter	Value	Unit
V_{Sm}	311	V_{peak}	V_{Cr}	600	V
L_M	1	mH	f_S	10	kHz
L_C	8	mH	L_P	42	mH
C_{DC}	330	μF	L_S	14	mH

Fig. 5 shows the simulation results in the normal state where the load current changes from peak value of 20 A to 40 A in the fifth period and then, recovers to 20 A after the fifth period. Though two step changes of load current from peak value of 20 A to 40 A and from peak value of 40 A to 20 A happen, the transport current from the source voltage to load responds immediately at the next period, and approaches to the new value. The voltage of dc capacitor, which decreases after the load current increases from 20 A to 40 A, converges on the reference voltage of 600 V. In case that the load current decreases from 40 A to 20 A, the voltage across dc capacitor stabilizes at the reference voltage after transient response with two periods.

Fig. 6 and Fig. 7 show the simulated voltage and current waveforms respectively, in case that the line-ground fault happens. With the turning-off operation of power switch in the dual-reactor in case of the line-ground fault, the voltages across the primary and the secondary winding can be induced, which leads to the fault current limiting operation as shown in Fig. 7(a).

During a fault period, the compensating operation was useless because the reactive component of the source current did not exist. Therefore, the voltage of dc capacitor kept constant as seen in Fig. 6(b).

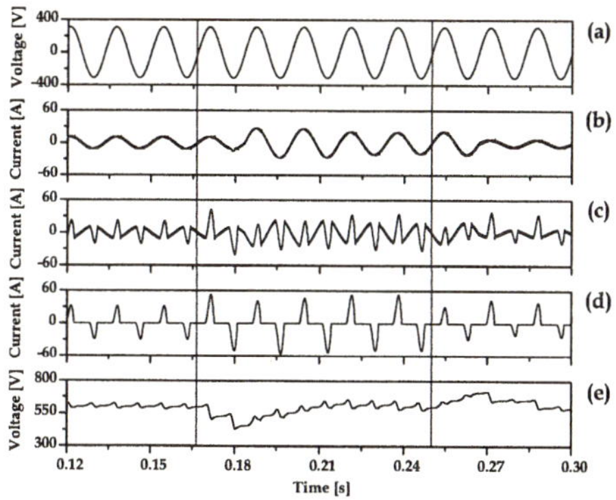

Fig. 5. Simulated current and voltage waveforms for steady state and transient response. (a) Source voltage (v_S) (b) Source current (i_S) (c) Compensation current (i_C) (d) Load current (i_L) (e) Voltage across dc capacitor (v_C)

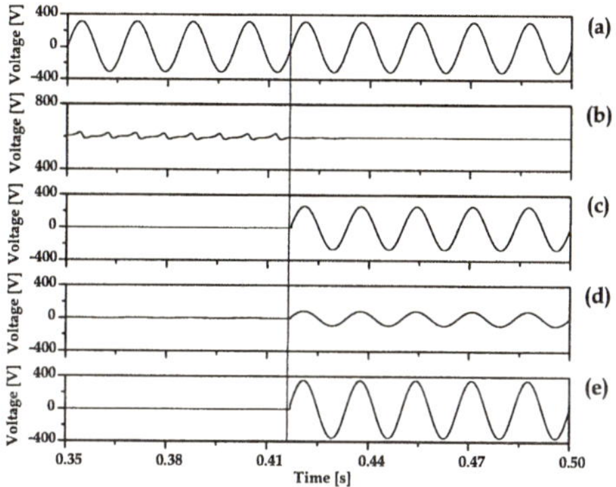

Fig. 6. Simulated voltage waveforms in case that the line-ground fault happens. (a) Source voltage (v_S) (b) Voltage across dc capacitor (v_C) (c) Voltage induced in the primary winding (v_1) (d) Voltage induced in the secondary winding (v_2) (e) Voltage across the power switch

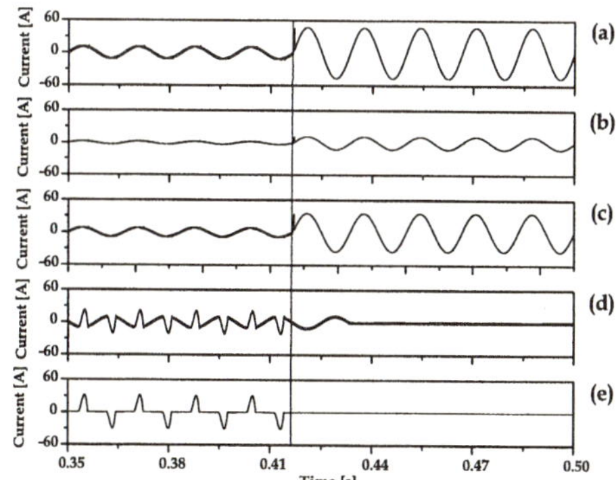

Fig. 7. Simulated current waveforms in case that the line-ground fault happens. (a) Source current (i_S) (b) Current of the primary winding(i_1) (c) Current of the secondary winding (i_2) (d) Compensation current (i_C) (e) Load current (i_L)

4 Conclusion

We suggested the intelligent dual-reactor with the current controlled inverter. The operation of the suggested equipment, which consisted of the dual-reactor with a power switch and the current controlled inverter, could be divided into the normal state and the short-circuit state. In the normal state, the reactive power of grid due to non-linear load or its variation could be compensated by the inverter controlled based on the algorithm of the polarized ramp time (PRT) current. In the short-circuit state such as line-ground fault, the fault current could be limited by controlling the power switch open. The operations such as UPS (uninterruptible power supply) and DSM (demand side management) will be realized in this intelligent equipment in the future.

Acknowledgment

"This work was partially supported by the Brain Korea 21 Project of Kunsan National University".

References

1. Kalyan K. Sen, "SSSC-Static Synchronous Series Compensator: Theory, Modeling and Applications", IEEE Trans. on Power Delivery, vol. 13, pp. 241-246, Jan. 1998.
2. M. EI-Habrouk, M. K., Darwish, P. Mehta, "A Survey of Active Filters and Reactive Power Compensation Techniques", Eighth International Conference on Power Electronics and Variable Speed Drives, 18-19 Sept. 2000, pp. 7-12.
3. B. M. Han and S. I. Moon, "Static Reactive-Power Compensator Using Soft-Switching Current-Source Inverter", IEEE Trans. on Industrial Electronics, vol. 48, pp. 1158-1165, Dec. 2001.
4. Duangkamol K., Mitani Y., Tsuji K. and Hojo M., "Fault Current Limiting and Power System Stabilization by Static Synchronous Series Compensator", Power System Technology, Proceedings Power Con 2000 International Conference, vol. 3, Dec. 2000, pp.1581-1586.
5. Lawrence B., "Four Quadrant Power Flow in a Ramp time Current Controlled Converter", Applied Power Electronics Conference and Exposition, Eleventh Annual Conference Proceedings, vol. 2, March 1996, pp. 898-904
6. Camilo Machado Jr., Nita Fukuoka, Eber A Rose, Airton Violin, Manuel Luis Barreira Martinez and Carlos Alberto Moura Saraiva, "Switching a series reactor - a concept developed to limit the level of short circuit currents", 2001 IEEE Porto Power Tech Conference, vol. 3, Sept. 2001.
7. Charles W. Brice, Roger A. Dougal and Jerry L. Hudgins, "Review of technologies for current-limiting low-voltage circuit breakers", IEEE Trans. on Industrial Applications, vol. 32, pp. 1005-1010, Sept./Oct. 1996.
8. Lawrence J. Borle and Chemmangot V. Nayar, "Zero Average Current Error Controlled Power Flow for AC-DC Power Converters", IEEE Trans. on Power Electronics, vol. 10, pp. 725-732, Nov. 1995.

Design and Evaluation of an SARHR Scheme
for Mobility Management in Ad Hoc Networks

Ihn-Han Bae

School of Computer and Information Communication Eng.,
Catholic University of Daegu, Gyeongbuk 712-702, Korea
`ihbae@cu.ac.kr`

Abstract. We present a scalable adaptive randomized home region scheme for mobility management in ad hoc mobile networks. In the proposed scheme, home regions are used to store the location of the network nodes and to manage the mobility of nodes. When a mobile's location changes, a number of randomly selected home regions considering the gravity of locality of the mobile node are updated. When a mobile's location is needed, such as a call arrival, a number of randomly selected databases are queried. The performance of the proposed scheme is evaluated through an analytical model, and it is compared with that of the conventional randomized database group (RDG) scheme. In the future, we will study the secure SARHR, which is a security enhancement to the original SARHR scheme.

1   Introduction

Advances in wireless communications and small, lightweight, portable computing devices have made mobile computing possible. One research issue that has attracted much attention recently concerns the design of a mobile ad hoc network (MANET) that consists of a set of mobile hosts that roam at will and communicate with one another. Communication takes place through wireless links among mobile hosts, using their antennas, but such as an environment supports no base stations. Further, the transmission distance limitation means that mobile hosts may not be able to communicate with one another directly. Hence, a multiple scenario occurs, and several hosts may need to relay a packet before it reaches its final destination. This situation requires each mobile host in a MANET to serve as a router [1].

Recent research in this field addresses ways of solving existing problems in MANETS by the use of node location information. However, maintaining node location information in the network is in itself a challenge, due to the frequently changing location of the nodes. Node location information may be used to provide various services such as location dependent query processing, navigation, geographic messaging and neighbor and service discovery [2].

In routing protocol of an Ad-Hoc network, the location service uses the location information of a node for packet routing. So, many researches for location management in an Ad-Hoc network were performed recently. In [3], based on the virtual backbone structure, randomized database group (RDG) method is proposed for Ad-Hoc mobility management. This method is doubly distributed at the point of view that both database

R. Khosla et al. (Eds.): KES 2005, LNAI 3681, pp. 465–471, 2005.
© Springer-Verlag Berlin Heidelberg 2005

allocation and database access are dynamic and non-deterministic. During the location update of a mobile host or when the call is delivered to a mobile host, the location of a mobile host is recorded to or read from some group of K databases selected randomly.

In [4], many different query methods for randomized database group scheme are studied and their performances are compared. Especially, the optimal update-query size and query-group size are determined. Furthermore, it shows the probability of the first query being successful, and the average query latency searching for the location of a mobile host, then evaluates the cost for implementing randomized database group scheme with different number of database functions.

In this paper, we design a scalable adaptive randomized home region (SARHR) scheme for mobility management in ad hoc mobile networks. In a proposed scheme, during the location update of a mobile host, the number of home regions that write the location of the mobile is adaptively determined by considering the gravity of locality of the mobile node. When a mobile's location is needed, a fixed number of randomly selected home regions are queried.

The rest of the paper is organized as follows. Section 2 contains details of the proposed SARHR scheme. Section 3 presents the performance of the SARHR that is evaluated through an analytical model and finally Section 4 concludes the paper.

2   SARHR Scheme

Given a square region of area A, GQS-based adaptive mobility management scheme divides the topography into G logical unit regions (referred to as Order-1 regions), where each node is aware of the size of the topography as well as the size of an Order-1 region. It then combines K^2 Order-1 regions to form Order-2 regions. Each node selects a home region in each Order-2 region via a function F that maps roughly the same number of nodes to each Order-1 region in an Order-2 region. Hence, every node has $O(\frac{A}{K^2})$ home regions in A [5]. The rest excepting home regions are considered far regions. Any node within every home region is selected randomly as a location server. At any given time, a relatively limited number of key locales in the network, where important events or activities are taking place, are referred to as focus or hot locales. The mobile nodes reside in these locales are called as focus or hot nodes. Otherwise, it is referred to as cold nodes respectively [6]. Figure 1 shows a sample square topography divided into Order-1 and Order-2 grids, where an Order-2 grid consists of 16 Order-1 grids. The shaded grid represents the home region in each Order-2 grid, and R_1 and R_2 represent the focus locale respectively.

We assume that each node can identify its location through a service like GPS. Furthermore, mobile node selected in each home region plays a role as a location server maintaining the location information of other nodes in a system. Each location data item is related to the timestamp representing the time obtained that data item.

In the MANET that consists of N home regions, when the information becomes available, K home regions are randomly chosen and the location information is updated. When the location information of the node is requested, Q home regions are randomly selected and queried. If the updated K home regions and the queried Q home regions have at least one common home region, the query will produce an answer and

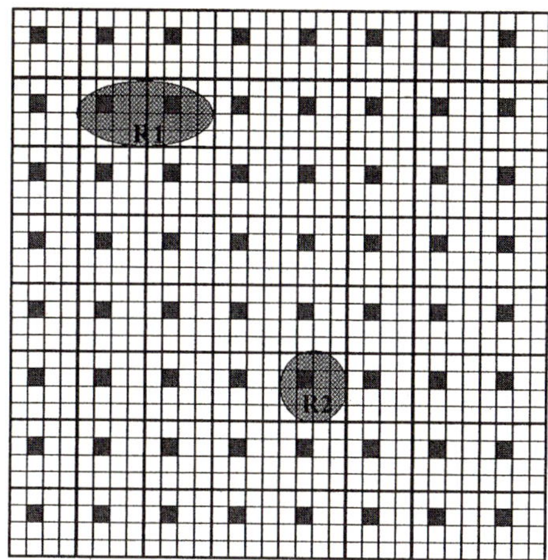

Fig. 1. The topolography of Ad-Hoc Network is divided into Order-2 regions.

is considered successful. If this is not the case, a new set of Q home regions is selected and queried. The process continues until in one of the rounds, a home region, which contains the sought location information, is found.

The operation of the SARHR scheme can be modeled as "Sampling from dichotomous population" [7], which states as follows: Consider a population, which consists of N balls, numbered 1 through K, are white and that the remaining balls, numbers $K+1$ through N, are black. Samples of Q balls are successfully drawn randomly from this population, until one of the samples contains at least one white ball. In our model, the ball population represents the nodes in the MANET; the white balls are the balls that were updated with the location information, while the drawn balls are those home regions that are queried. Let X denotes the number of drawings required to obtain a white ball

The probability that the first query will be successful is equal to one minus the probability that no such home regions are found in the first trial:

$$P(X=1) = \frac{\binom{N-K}{Q}}{\binom{N}{Q}} . \tag{1}$$

From the equation (1), the hot mobile's location information within focus locales is stored in K_1 home regions in order that the hot mobile's $P(X=1)$ can be got 0.9 from the queried Q home regions, and the cold mobile's location information within non-focus locales is stored in K_2 home regions in order that the cold mobile's $P(X=1)$ can be got 0.7 from the queried Q home regions, where $K_1 > K_2$.

The number of fixed queried home regions (Q), K_1 and K_2 for N are shown in Figure 2. In the case that N=36 and Q=5, if $P(X=1) \geq 0.9$, K_1 is 13, and if $P(X=1) \geq 0.7$, K_2 is 8. Also, in the case that N=144 and Q=11, if $P(X=1) \geq 0.9$, K_1 is 27, and if $P(X)=1 \geq 0.7$, K_2 is 15.

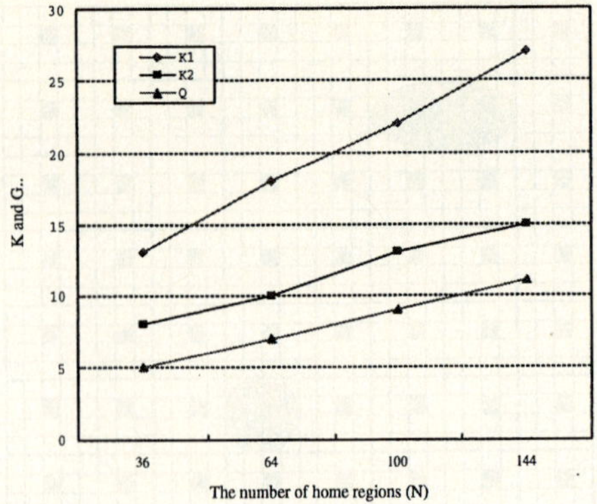

Fig. 2. K_1 and K_2 values according to $P(X = 1)$ for N.

There are three ways in which locations are updated in the home regions [4]:

1. Call-origination update: When a mobile host initiates a call, it queries an SARHR for the location of the destination and, at the same time, writes its current location into the queried home regions.
2. Location-change update: When a mobile host changes its location, it updates its new location in randomized home regions.
3. Periodic update: In order to avoid call loss during long time of lack of activity and immobility, a mobile host periodically sends its location information to randomized home regions.

To minimize the round of queries, we can use the following search strategies:

1. Strategy A: The first query uses a group size of Q; if the query is not successful, the second query uses a group size of $N - K_2 - Q + 1$. In this case, the maximum delay is two queries.
2. Strategy B: The first n queries use the same query group size of Q; if none is successful, the second query uses a group size of Q; if the second query is unsuccessful, the final query uses a group size of $N - K_2 - 2Q + 1$. In this case, the maximum delay is three queries.

3   Performance Evaluation

Let us label Y as the number of updated home regions within a sample of Q home regions $(Q > 1)$. Then the probability function of Y is given as:

$$P(X = 1) = \binom{Q}{k} \frac{K^{(k)}(N - K)^{(Q-k)}}{N^{(Q)}} = \frac{\binom{K}{k}\binom{N-K}{Q-k}}{\binom{N}{Q}}$$

$$\equiv P(k, Q, K, N), k = 0, 1, ..., Q \tag{2}$$

We are interested in the number of query rounds required; i.e., the probability function of X, which we obtain from equation (2) as follows:

$$P(X = 1) = 1 - P(0, Q, K, N)$$
$$P(X = 2) = P(0, Q, K, N)(1 - P(0, Q, N - Q))$$
$$P(X = 3) = P(0, Q, K, N)P(0, Q, K, N - Q)(1 - P(0, Q, M, N - 2Q))) \tag{3}$$

$$\cdots$$

The costs of the querying process for the different search strategies in Section 3 are as follows:

- Search Strategy A:

$$C_A = P(X = 1) \times Q + (1 - P(X = 1)) \times (N - K_2 - Q + 1) \tag{4}$$

- Search Strategy B:

$$C_B = P(X = 1) \times Q + P(X = 2) \times Q + (1 - (P(X = 1) + P(X = 2)) \\ \times (N - K_2 - 2Q + 1) \tag{5}$$

To evaluate the cost of such a dynamic process, the total cost of mobility management is given by the sum of the total update cost and the total query cost.

$$C_{total} = P_{h_1} K_1 C_u + (1 - P_{h_1}) K_2 C_u \\ + \frac{\lambda}{\mu} \{ P_{h_2}(C_{shm} C_q) + (1 - P_{h_2})(C_{scm} C_q) \} \tag{6}$$

In equation (6), C_u and C_q represent the cost per update and the cost per query respectively. P_{h_1} is the probability that the mobile updating location is hot, and P_{h_2} is the probability that the destination mobile is hot. Also, C_{shm} and C_{scm} are the query costs that find a hot mobile and a cold mobile through a searching strategy, respectively, and $\frac{\lambda}{\mu}$ represents call-mobility ratio (CMR). Figure 3 shows the results of the analytical performance evaluation using parameters in Table 1.

As shown in the Figure 3, the performance of the proposed SARHR scheme is better than that of RDG scheme regardless of the search strategy and the CMR. Especially, the performance of the SARHR scheme is getting better in high CMR. This result is obtained from the spatial locality of hot mobiles in location query procedure because the location information of hot mobiles is stored in more home regions. The total cost of the SARHR(B) scheme which uses the search strategy B is better than that of the SARHR(A) which uses the search strategy A.

Figure 4 demonstrates the average number of query rounds for mobility management schemes with different search strategy. From the Figure 4, we know that the average number of rounds of the SARHR scheme is superior to that of the RDG scheme regardless of the search strategy. The total cost of the SARHR(B) is better than that of the SARHR(A), but the average number of query rounds of the SARHR(B) is little worse than that of the SARHR(A). Generally, the performance of the SARHR(B) is better than that of the SARHR(A), but the SARHR(A) scheme is more applicable to real time applications.

Table 1. Parameters for analytical evaluation.

Parameters	Value
G	1600
R	4
N	100
K	16
K_1	22
K_2	13
Q	9
P_{h_1}	0.25
P_{h_2}	0.75
C_u, C_q	1

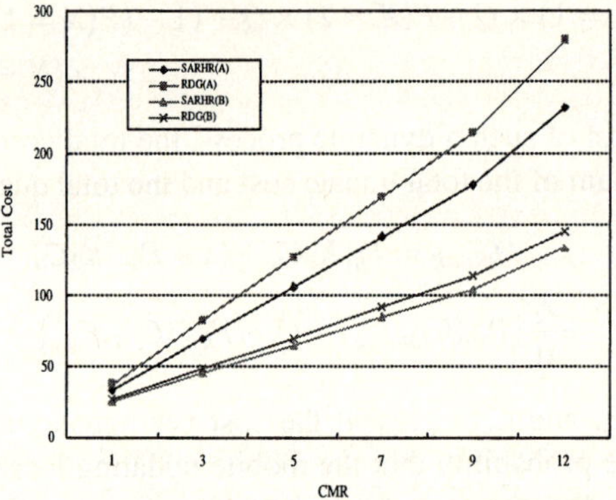

Fig. 3. The results of the total cost.

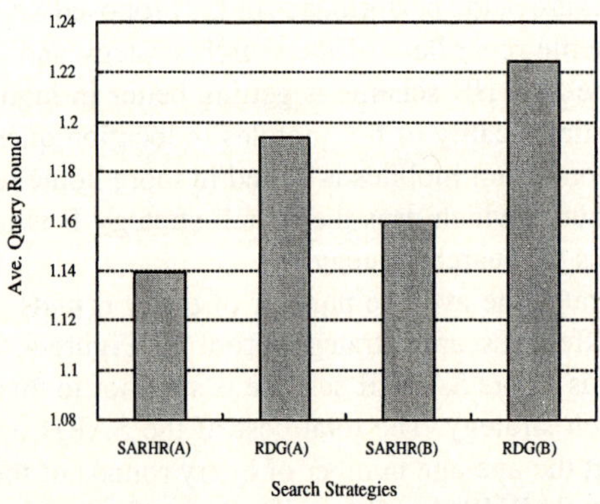

Fig. 4. The results of the average number of query rounds.

4   Conclusion

RDG scheme is a simple but robust and efficient method for implementation of mobility management in mobile ad hoc networks. In this paper, we have proposed an SARHR scheme considering the gravity of locality of a mobile node in a MANET. The performance of the proposed SARHR scheme is to be evaluated by an analytical model. Based on the results of the performance evaluation, we know that the performance of the proposed SARHR scheme is better than that of RDG scheme regardless of the search strategy and the CMR. Therefore, the proposed SARHR is the mobility management scheme that provides location information to mobiles in MANETs, efficiently and fast. In the future, we will study the secure SARHR, which is a security enhancement to the original SARHR scheme.

References

1. Yu-Chee, Shih-Lin Wu, Wen-Hwa Liao, Chih-Min Chao: Location Awareness in Ad Hoc Wireless Mobile Networks. IEEE Computer 34(6), (2001) 46–52
2. S. Bhattacharya: Randomized Location Service in Mobile Ad Hoc Networks. Proceedings of the 6th international workshop on MSWIM'03, (2003) 66–73
3. Z. J. Haas and B. Liang: Ad-Hoc Mobility Management with Randomized Database Group. International Conference on Communication (ICC'99), (1999) 6-10
4. J. Li, Z. J. Haas, and B. Liang: Performance Analysis of Random Database Group Scheme for Mobility Management in Ad hoc Network. IEEE International Conference on Communications (ICC2003), (2003) 11-15
5. S. J. Philip, C. Qiao: ELF : Efficient Forwarding on Ad hoc Networks. Proceedings of the IEEE Globecom, (2003)
6. S. Bhattacharya, H. Kim, S. Prabh, T. Abdelzaher: Energy-Conserving Data Placement and Asynchronous Multicast in Wireless Sensor Networks. The First International Conference on Mobile Systems, Applications, and Services (MobiSys), (2003)
7. E. Parzen: Modern Probability Theory and Its Applications. John Wiley and Sons Inc., (1992)

Reliability and Capacity Enhancement Algorithms
for Wireless Personal Area Network
Using an Intelligent Coordination Scheme

Chang-Heon Oh[1], Chul-Gyu Kang[1], and Jae-Young Kim[2]

[1] School of Information Technology, Korea University of Technology and Education
Byeoncheon-myeon, Cheonan-si, Chungcheongnam-do, 330-708 Korea
{choh,swing98}@kut.ac.kr
[2] Wireless Home Network Research Team Digital Home Research Division
ETRI, 161 Gajeong-dong, Yuseong-gu, Daejeon, 305-700 Korea
jyk@etri.re.kr

Abstract. In this paper, we propose reliability and capacity enhancement algorithms for high data rate wireless personal area network (HDR-WPAN) system using an intelligent coordination scheme. Conventional system (i.e., IEEE802.15.3, HDR-WPAN) supports maximum 55Mbps transmission rate at 15MHz bandwidth. Proposed intelligent coordination scheme is a modification of HDR-WPAN specification to enhance the reliability and transmission rate using an adaptive modulation and bandwidth expansion method. The proposed system is analyzed through a computer simulation in respect of reliability, transmission rate, and power spectrum density (PSD). From the results, the proposed system has maximum 110Mbps data rate and reliability of 10^{-5} at E_b/N_o=21dB in indoor radio channel when we employ a TCM-128QAM at 25MHz bandwidth.

1 Introduction

HDR-WPAN system is a wireless data communications system intended for operation in a small office/home office (SOHO) environment. The HDR-WPAN is intended for consumer application, therefore reliability and transmission rate are important. The HDR-WPAN system supports an ad-hoc networking, meaning little or no infrastructure (i.e., fixed access points) is required [1]. Conventional system (HDR-WPAN) supports maximum 55Mbps transmission rate. But it is not a suitable data rate for the applications of high speed data transmission, for example wireless home network applications. Therefore, in this paper, we propose reliability and capacity enhancement algorithms for HDR-WPAN system using an intelligent coordination scheme. And its performance is analyzed.

2 HDR-WPAN System

IEEE 802.15.3 HDR-WPAN system is single carrier system that use frequency band 2.4GHz. This clause specifies the PHY for a single carrier system that supports up to five modulation formats with coding at 11Mbaud to achieve scalable data rate [2]. Maximum transmission possibility frame size of HDR-WPAN system is maximum size of data frame that pass from higher level. Transmission possibility frame is 2044

R. Khosla et al. (Eds.): KES 2005, LNAI 3681, pp. 472–478, 2005.

octets. If security is enabled for the data connection, the upper layers should limit data frames to 2044 octets minus the security overhead. The QPSK and 16, 32, and 64-QAM formats shall use an 8-state 2-D trellis code. Non-coding shall be applied to the DQPSK modulation.

PHY frame contains four segments; preamble, a header, data, and a tail. The preamble is used to perform gain adjustment, symbol and carrier timing compensation, equalization, etc. The header is used to cover physical layer data necessary to process the data segment, such as modulation type and frame length. The tail is used to force the trellis code to a known state at the end of the frame to achieve better distance properties at the end of the frame. IEEE 802.15.3 PHY frame format is divided by 11Mbps mode and 22, 33, 44, 55 Mbps mode. PHY header, MAC header, and HCS are repeated two times in 11 Mbps.

3 Reliability and Capacity Enhancement Algorithms

3.1 TCM-128QAM Scheme

In 1980's, G. Ungerboeck published papers to demonstrate that a convolutional code could be integrated with a modulation scheme to achieve significant coding gain without reducing the data rate or requiring more bandwidth[3],[4].

When TCM is employed, the coding gain, the power saving given by the coding, is defined as equation (1).

$$\gamma = 10\log_{10}\left(\frac{d^2_{free}/E}{d^{2(u)}_{min}/E^{(u)}}\right)dB \tag{1}$$

where d^2_{free} is the MSED between two different sequences of coded signal points, $d^{2(u)}_{min}$ is the MSED between signal points in the uncoded system, and E and $E^{(u)}$ are the average signal energy of the coded and uncoded system, respectively.

The MSED between any two uncoded signals in the 2^6QAM constellation is $d^{2(u)}_{min} = d^2$, and the average signal energy of 2^6QAM system is given as equation (2) [5].

$$E^{(u)} = \frac{2^6 - 1}{6}d^2 = 10.5d^2 \tag{2}$$

For the TCM system, it is clear that

$$d^2_{free} = \min\left\{\sum_j d^2_{min}\left(S_j, S'_j\right), d^2_{min,\Lambda_i}\right\} = 8d^2 \tag{3}$$

Where $d^2_{min}\left(S_j, S'_j\right)$ is the MSED between the jth signal points which come from the different coded sequence and locate in subset S_j and S'_j, respectively. The average signal energy of 64QAM system is given as equation (4).

$$E^{(u)} = \frac{2^7 - 1}{6}d^2 = 21.17d^2 \tag{4}$$

Therefore, the coding gain of our example TCM-128QAM sequence is 3.97dB.

$$\gamma = 10\log_{10}\left(\frac{d_{free}^2/E}{d_{min}^{2(u)}/E^{(u)}}\right)dB = 10\log_{10}\left(\frac{8d^2/21.17d^2}{d^2/10.5d^2}\right) = 3.97\,dB \tag{5}$$

3.2 Pulse Shaping Method

The most popular pulse shaping filter used in mobile communications is the raised cosine filter. A raised cosine filter belongs to the class of filters which satisfy the Nyquist criterion (zero ISI at the sampling times). The transfer function of a raised cosine filter is given in equation (6).

$$H_{RC} = \begin{cases} 1 & 0 \le |f| \le \dfrac{(1-\alpha)}{2T_s} \\[2ex] \dfrac{1}{2}\left[1+\cos\left[\dfrac{\pi\left(|f|\cdot 2T_s - 1 + \alpha\right)}{2\alpha}\right]\right] & \dfrac{(1-\alpha)}{2T_s} \le |f| \le \dfrac{(1+\alpha)}{2T_s} \\[2ex] 0 & |f| \le \dfrac{(1+\alpha)}{2T_s} \end{cases} \tag{6}$$

Where α is the roll-off factor which ranges between 0 and 1. From equation (6), as the roll-off factor α increases, the bandwidth of the filter also increases, and the time side-lobe levels decrease in adjacent symbols slots. This implies that increasing α decreases the sensitivity to timing jitter, but increases the occupied bandwidth.

Starting with the Nyquist bandwidth constraint that the minimum required system bandwidth B for a symbol rate of R_s symbols/sec without ISI is $R_s/2$ hertz, a more general relationship between required bandwidth and symbol transmission rate involves the filter roll-off factor α [6],[7].

$$R_s = \frac{1}{T_s} = \frac{2B}{1+\alpha} \tag{7}$$

Band-pass modulation signals require twice the transmission bandwidth of the equivalent base-band signals. Such frequency-translated signals, occupying twice their base-band bandwidth, are often called double-sideband (DSB) signals. Therefore, for QAM and PSK modulated signals, the relationship between the required DSB bandwidth B and the symbol transmission rate R_s is expressed with equation (8).

$$R_s = \frac{1}{T_s} = \frac{B}{1+\alpha} \tag{8}$$

According to the equation (8), eHDR-WPAN system requires the bandwidth of 24.7495MHz to obtain maximum 110Mbps transmission rate using roll-off coefficient value α=0.35 with TCM-128QAM modulation.

3.3 Multipath Indoor Channel Models

In this paper, we consider the statistical channel model for indoor multipath, which is proposed by Saleh and Valenzuela in [8]. This statistical model is based on clustering phenomenon of rays observed in their experimental data. The clustering phenomenon is that rays arrive in several groups within an observation window, and the clusters are attenuated in amplitude. In addition, the magnitudes of rays within a cluster decay with time. The model proposes that both of these decaying patterns are exponential with time, and are controlled by the cluster arrival decay time constant Γ and the ray arrival decay time constant γ.

The cluster arrival times are modeled as a Poisson arrival process with the cluster arrival rate Λ. Within each cluster, subsequent rays also arrive according to a Poisson process with the ray arrival rate λ. Let the arrival time of the l-th cluster be denoted by T_l ($l = 0, 1, 2, \ldots$). Moreover, let tie arrival time of the k-th ray measured from the beginning of the l-th cluster be denoted by τ_{kl} ($k = 0, 1, 2, \ldots$). Thus, according to this model, T_l and τ_{kl} are described by the independent inter-arrival exponential probability density functions.

$$p(T_l \mid T_{l-1}) = \Lambda \exp[-\Lambda(T_l - T_{l-1})], \quad l > 0 \tag{9}$$

$$p(\tau_{kl} \mid \tau_{(l-1)l}) = \lambda \exp[-\lambda(\tau_{kl} - \tau_{(k-1)l})], \quad k > 0 \tag{10}$$

The impulse response of the channel is given by

$$h(t) = \sum_{l=0}^{\infty} \sum_{k=0}^{\infty} \beta_{kl} \delta(t - T_l - \tau_{kl}) \tag{11}$$

Where β_{kl} represents the complex gain of each arrival. The amplitude of β_{kl} is a rayleigh distributed random variable, while the phase of β_{kl} is a uniformly distributed random variable. The mean square value of β_{kl} is described as follows:

$$\overline{\beta^2}_{kl} = \overline{\beta^2(T_l, \tau_{kl})} = \overline{\beta^2(0,0)} e^{-T_l/\Gamma} e^{-\tau_{kl}/\gamma} \tag{12}$$

Where $\overline{\beta^2(0,0)}$ is the average power gain of the first ray of the first cluster. The characteristics of channels are changed according to the values of the channel parameters mentioned previously. Therefore, it is important to select appropriate values of the channel parameters.

4 Simulation and Results

In this section, we have analyzed the proposed eHDR-WLAN system through the computer simulation in AWGN and fading channel environments. The simulation parameters are listed in table 1.

Table 1. Simulation parameters

Parameters	Value
symbol rate	11Mbaud
Data rate	11/22/33/44/55/110Mbps
Trellis code	Variable, 8-state convolutional code
Transmit filter	Square-root raised cosine, =0.35
Channel bandwidth	15MHz, 25MHz
Frame length	1250bytes

The channel impulse response is plotted in Fig. 1 for the statistical indoor channel model of Saleh and Valenzuela, where $1/\Lambda = 15$ nsec, $1/\lambda = 0.5$ nsec, $\Gamma = 14$ nsec and $\gamma = 7.9$ nsec.

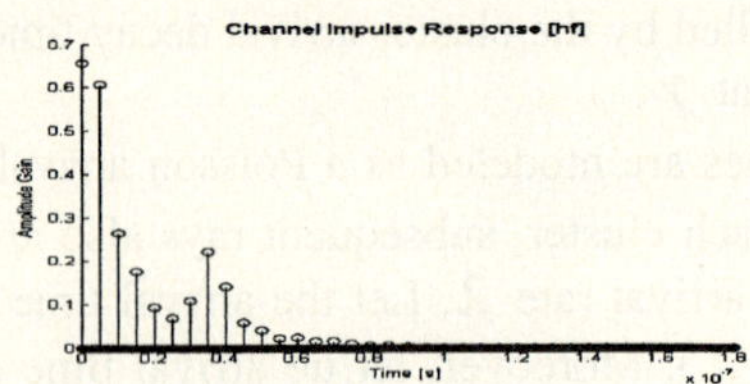

Fig. 1. Channel impulse response for NLOS

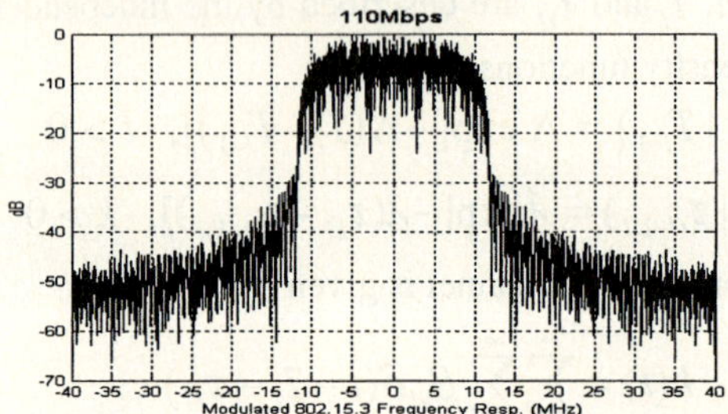

Fig. 2. Power spectrum density in 110 Mbps

Fig. 2 shows the power spectrum density characteristics of transmitted signals with TCM-128QAM scheme. Approximately 25MHz bandwidth is required at -30dBr point, which meets the IEEE802.15.3 specification. It is a comparable bandwidth with other systems, for example WLAN.

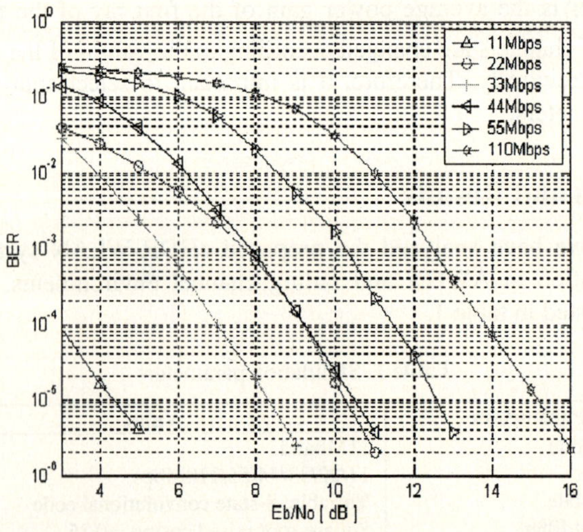

Fig. 3. BER performance in AWGN channel environment

Fig. 3 shows BER performance in AWGN channel environment. Required E_b/N_o is about 13dB at 55Mbps and about 15dB at 110Mbps to achieve the reliability of 10^{-5}. It is confirmed that to achieve the higher data rate, we need more transmission power at the same BER performance because of adopting the 128QAM scheme.

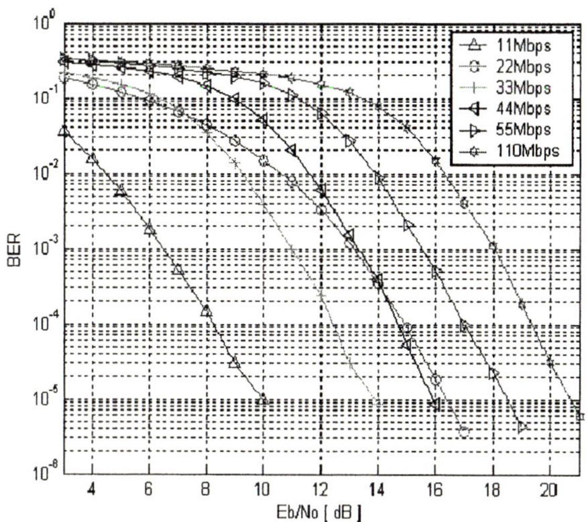

Fig. 4. BER performance in Fading channel environment

Fig. 4 represents BER performance in fading channel environment. Similarly the required power difference between the 55Mbps and 110Mbps system is 2dB. But   the required E_b/N_o is about 19dB at 55Mbps and about 21dB at 110Mbps to achieve the reliability of 10^{-5}. From the results, it is confirmed that we need more power to overcome the effect of multipath fading at the same BER performance.

5   Conclusions

In this paper, we proposed reliability and capacity enhancement algorithms for high data rate wireless personal area network using an adaptive modulation and bandwidth expansion scheme as an intelligent coordination technique. The proposed system is analyzed through a computer simulation in respect of reliability, transmission rate, and power spectrum density (PSD). From the results, the proposed system has maximum 110Mbps transmission rate and reliability of 10^{-5} at $E_b/N_o=21$dB in multipath indoor channel when we adopt a TCM-128QAM at 25MHz bandwidth. It is a reasonable performance for high data rate wireless personal area network applications.

References

1. Jeyhan Karaoguz: High-Rate Wireless Personal Area Networks. IEEE Communications Magazine, vol. 39, no. 12, pp. 96-102, Dec. 2001.
2. IEEE Std 802.15.3: Wireless Medium Access Control (MAC) and Physical Layer (PHY) Specifications for High Rate Wireless Personal Area Networks (WPANs), 2003.

3. G. Ungerboeck: Trellis Coded Modulation with Redundant Signal Sets, part I: Introduction. IEEE Communication Magazine, vol. 25, no. 2, pp.5-11, Feb. 1987.
4. G. Ungerboeck: Trellis Coded Modulation with Redundant Signal Sets, partⅡ: State of the Art. IEEE Communications Magazine, vol. 25, no. 2, pp.12-21, Feb. 1987.
5. John M. Cioffi: A Multicarrier Primer. Amati Communications Corporation and Stanford University, 1991.
6. John G. Proakis: Digital Communications. McGraw-Hill, Inc., 2001.
7. Theodore S. Rappaport: Wireless Communications Principles and Practice. Prentice-Hall, Inc., 2002.
8. A. M. Saleh and R. A. Valenzuela: A Statistical Model for Indoor Multipath Propagation. IEEE Journal on Selected Areas in Communications, vol. SAC-5, no. 2, pp. 128-137, Feb. 1987.

A Study of Power Network Stabilization
Using an Artificial Neural Network

Phil-Hun Cho[1], Myong-Chul Shin[1,*], Hak-Man Kim[2], and Jae-Sang Cha[3]

[1] Dept. of Electronic and Electrical Eng., SungKyunKwan Univ., Suwon, 440-746, Korea
{Cho,Shin} youinaru@kepco.co.kr
[2] KERI, 28-1 Sengju-dong, Changwon, Kyung-Nam, 641-120, Korea
kerihifi@naver.com
[3] Dept. of Inform. And Commun. Eng. Seokyeong Univ., Seoul, 136-704, Korea

Abstract. In power network, low frequency oscillation become a major concern for many years. In order to depress low frequency oscillation, the power system stabilizer parameters must be adjusted when there are changes in power network conditions. This paper presents the application of neural network to tune the power network stabilizer parameters. For training neural network, generator real power and reactive power are chosen as the input signals and the output are the desired power network stabilizer parameters. A popular type of neural network, the multi-layer perceptron with error-back-propagation training method, is employed. The neural network, once trained, can yield proper power network stabilizer parameters under any generator loading conditions. Simulation results show that the neural network based power system stabilizer yield better dynamic performance than conventional power system stabilizers in the sense of having large damping in responds to a step disturbance.

1 Introduction

Low frequency oscillation is observed in a special generator for a while when system conditions, load or a transmission line state, are changed in a steady state. This situation is happened because of lack of a damping torque element in the mechanical mode of the generator so that it harms stability of the power system [1][2]. When large scale power systems are mutually connected, an oscillation of low-frequency that is a number of cycle per a minute might be happened and disappeared in a few minute but in other case, it gradually increases so that causes collapse of the power system [3]. So these low frequency oscillation problems are important in stability, reliability, and high-quality of the power system. According to this, phase lead-lag compensator or Proportional-Integral Power System Stabilizer(PSS) is widely used because of utility of price, maintenance and repair to improve stabilization problem of power system. But operating conditions of synchronous generator are changed, it is deficient in adaptability and has inaccuracy problem of system modeling. Self-tuning regulator that changes PSS parameter into real-time by on-line measurement has difficulty to realize accurately mathematical modeling in practical application.[5~9]. Therefore Artificial Intelligent control techniques are used.[10~12]

In this paper, a neural network which is one of the Artificial Intelligent control technique is used so as to save calculation time and be convenient the mathematical modeling problem of PSS, which is a fault of a conventional power system stabilizer.

---

* Correspondence author

R. Khosla et al. (Eds.): KES 2005, LNAI 3681, pp. 479–484, 2005.

2 Power System Stabilizer

A cause of low frequency oscillation is lack of a mechanical damping torque of a synchronous generator that connected to power system. To control this low frequency oscillation, power system stabilizer is installed. Fig.1. is a block diagram that is widely used to study about low frequency oscillation of power system.

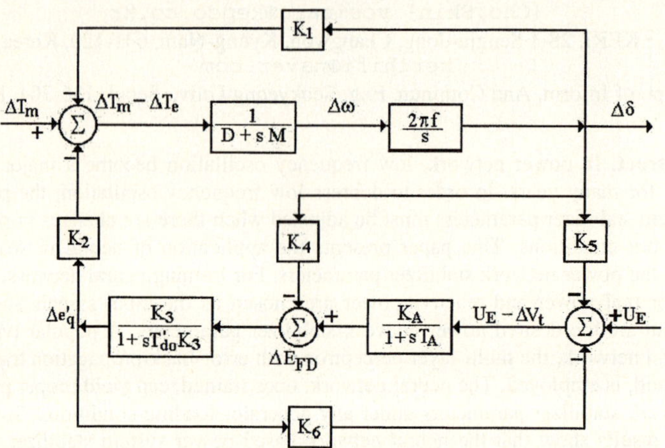

Fig. 1. Block diagram of synchronous generator

Power System Stabilizer is an equipment that supplies damping torque element through an excitation control. Fig. 2. is concept diagram of the power system stabilizer. When synchronous angle velocity are changed, PSS offers proper stable signal, Vpss.

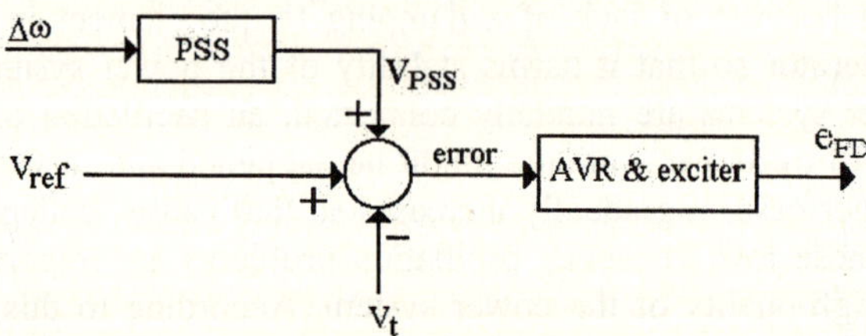

Fig. 2. Diagram of PSS

In this paper, parameters of Proportional-Integral PSS are properly changed using neural network when load conditions and transmission line states are changed. A diagram of the power system stabilizer using the neural network is figure 3.

Input parameters of neural network are active power(p) and reactive power(q) and output parameters (K_P, K_I) are calculated by eq.(1).

$$H(\lambda) = \frac{1}{C(\lambda I - A)^{-1} B} = \frac{\lambda T}{1 + \lambda T}(K_P + \frac{K_I}{\lambda}) \tag{1}$$

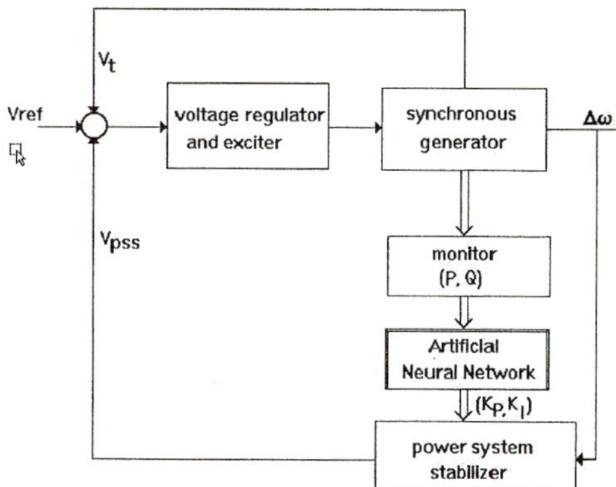

Fig. 3. A diagram of neural network PSS

3   Simulation and Results

3.1   Simulation

The system we considered in this paper is the synchronous generator that is connected to one-machine infinite-bus.

First of all, to make the learning data of neural network, we calculated the output (K_P, K_I) of the Proportional-Integral stabilizer for arbitrary input (P, Q) by using equation (1).

We used 99 learning data which are obtained by changing the reactive power(Q) and active power(P). The structure of the neural network requires the pre-process of the data. The data is normalized between 0.1 and 0.9 to adapt it to learning and action of neural network. We used the sigmoid function for activation function and 0.5 of bias for learning. In this paper, we performed the learning by the neural network with hidden layer of 1, 16 neurons and 0.3 of coefficient of momentum. Then the learned neural network underwent a deduction and we examined its error. To verify that the learned neural network reacts on the data that is not learned, we applied active power(P) and reactive power(Q). Here, we created 525 data by varying the active power and the reactive power.

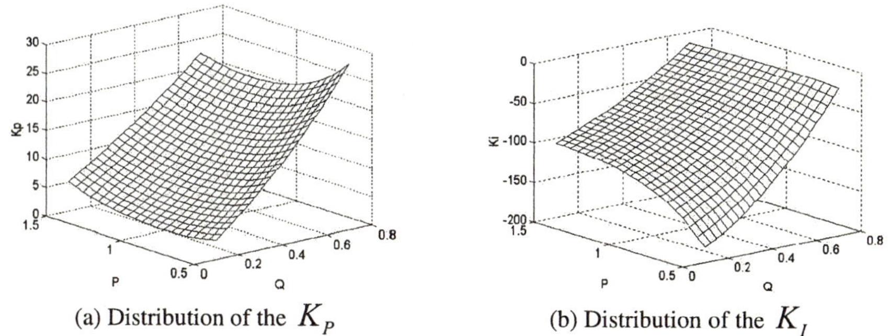

(a) Distribution of the K_P (b) Distribution of the K_I

Fig. 4. Output of the learned neural network

Fig 4. represents that the output of the neural network is nonlinear. The integration constant(K_I) is distributed over a wider range than the proportional constant(K_P).

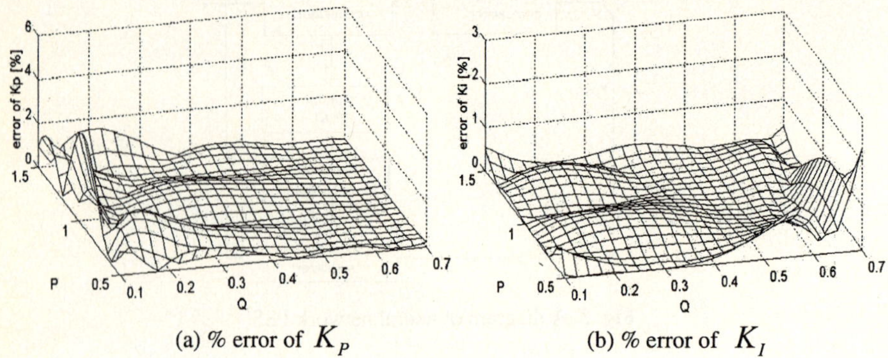

(a) % error of K_P (b) % error of K_I

Fig. 5. % error of the output of learned neural network

3.2 Case Study

These are the case studies of power system stabilizer using phase lead-lag compensator and the stabilizer that adopts neural network. Next figure show the responses of the system that is equipped with phase lead-lag stabilizer and neural network proportion integral stabilizer when a square wave torque is applied.

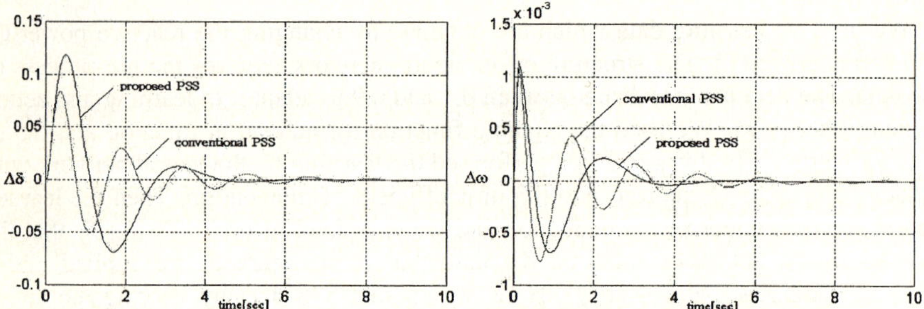

Fig. 6. Dynamic response of the generator with PSS at P=1.0 and Q=0.25

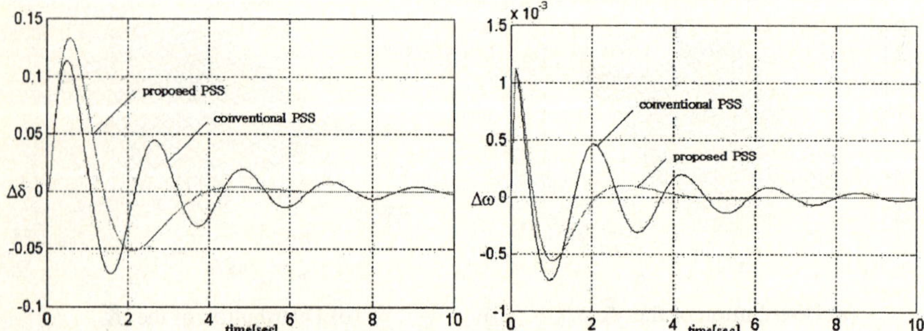

Fig. 7. Dynamic response of the generator with PSS at P=1.0, Q=0.55

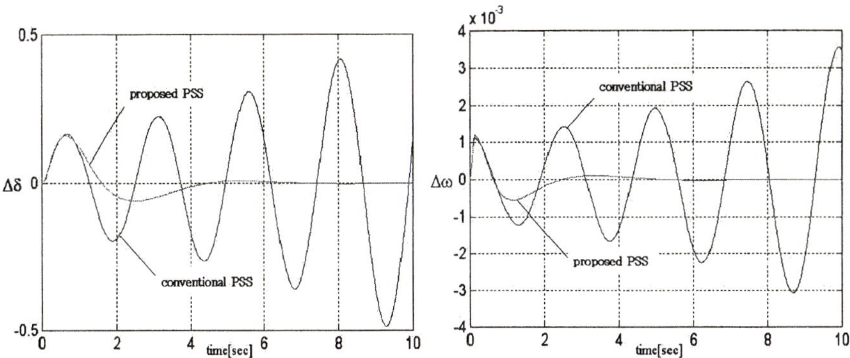

Fig. 8. Dynamic response of the generator with PSS at P=1.0, Q=0.62

When 0.1 [p.u] is applied to input torque for 0.1 sec, the neural network stabilizer is stabilized in 4.5 ~ 6.2 seconds. However, the phase lead-lag stabilizer required at least 9.5 seconds to be stabilized. Especially, when P=1.0 and Q=0.62, the system went into an unstable state on account of the emission. In fact, there occurred some cases where neural network stabilizer has a larger over suite than the phase lead-lag stabilizer, but it showed a lower oscillation and fast conversion.

Examining the case studies, the neural network stabilizer that was suggested in this paper showed a better damping characteristics under a wider range of operational condition than the conventional phase lead-lag stabilizer.

4 Conclusion

In this paper, we adopted an artificial neural network to control the parameter of power system stabilizer. The input values are the active power (P) and reactive power (Q) that represent the characteristics of load state of generator and the outputs are the appropriate parameter of P-I stabilizer. We used the back-propagation algorithm as the learning method of the neural network, and the adaptive method changed the learning rate during the learning process to enhance the learning speed.

The result showed that the neural network power system stabilizer didn't need the identification of the power system and more efficient than the conventional phase lead-lag power system stabilizer because it had fast responsibility and robustness which resulted in a good performance under a wide range of operational condition.

References

1. P.M.Anderson and A.A.Fouad, "Power System Control and Stability", Iowa State University Press, Ames, Iowa, 1977.
2. Yao-Nan Yu, "Electric Power System Dynamics", Academic Press, 1983.
3. Franscisco P. Demello and Charles Concordia, "Concepts of Synchronous Machine Stability as Affected by Excitation Control", IEEE Trans. on PAS, Vol. 88, No. 4, pp.316-329, April 1969.
4. Yuan-Yih and Chung-Yu Hsu, "Design of a Proportional-Integral Power System Stabilizer", IEEE Trans. on Power Systems, Vol. PWRS-1. No.2, pp.46-53, May 1986.

5. Chi-Jui Wu and Yuan-Yih Hsu, "Design of Self-Tuning PID Power System Stabilizer for Multimachine Power Systems." IEEE Trans. on Power Apparatus and Systems, Vol. PAS-101, No.9, pp.82-94, Sept. 1982.

6. Y.Y.HSU and K.L.Liou, "Design of Self-tuning PID Power System Stabilizers for Synchronous Generators", IEEE Trans. EC. Vol. 2, pp.343-348, 1987.

7. Yuan-Yih Hsu and Chi-Jui Wu, "Adaptive Control of a Synchronous Machine Using The Auto-Searching Method", IEEE Trans. on Power Systems, Vol. 3, No. 4, pp.1434-1440, November 1988.

8. Wenxin Liu, Dnald C.Wunsch II,"A Heuristic Dynamic Programming based Power System Stabilizer for Turbogenerator in a Single Machine Power System",

9. Yuan-Yih Hsu and Kan-Lee Lion, "Design of Self-Tuning PID Power System Stabilizers for Synchrous Generators", IEEE Trans. on Energy Conversion, Vol. EC-2, No.3, pp.343-348. September 1987.

10. Yuan-Yih Hsu and Chao-Rong Chen,"Tuning of Power System Stabilizers using an Artificial Neural Networks", IEEE Trans. on Energy Conversion, Vol. 6, No. 4, pp.612-619, December 1991.

11. D.K. Chaturvedi, O.P.Malik, "Experimental Studies With a Generalized Neuron-Based Power System Stabilizer" IEEE Trans. On Power Systems. Vol.19, N0.3, August 2004.

12. Y.L.Abdel-Magid, M.A.Abido,"Optimal Multiobjective Design of Robus Power System Stabilizers Using Genetic Algorithms", IEEE Trans. On Power System, vol.18, no.3 August 2003.

An Adaptive Repeater System for OFDM with Frequency Hopping Control to Reduce the Interference

Hui-shin Chae[1], Kye-san Lee[1], and Jae-Sang Cha[2]

[1] Kyunghee Univ., Seochunri, Kihung-eup, Yongin-si, Gyunggi-do, 449-701, Korea
{heesin,kyesan}@khu.ac.kr
[2] SeoKyeong Univ., Chongnung-dong, Songbuk-ku, Seoul,136-704, Korea
chajs@sku.ac.kr

Abstract. An adaptive repeater system for OFDM with frequency hopping control to reduce interferences is proposed. In this paper, we discuss a novel repeater to reduce the Inter Symbol Interference (ISI) caused by the delayed waves with Frequency Hopping controller in the frequency selective fading channel when the shadowing is occurred due to obstacles. The proposed system provides the solution for the shadowing, removing the interferences, results in improving the performance. An adaptive repeater system for OFDM with frequency hopping control improves the BER performance.

Keyword: Repeater System, ISI (Inter Symbol Interference), Frequency Hopping Controller

1 Introduction

OFDM is considered as a new technology for the mobile communication providing the high data transmission. It is effective in a frequency selective fading channel environment avoiding ISI (Inter Symbol Interference). OFDM is a transmission method of the multi-carrier transmission consisting of the number of sub-carriers. In this paper, a novel OFDM repeater with frequency hopping will be discussed. When the data are transmitted from a base station to a mobile station, there is a shadowing because of obstacles; the repeater is being used to overcome shadowing.

But, it is shown in Fig.1 that the conventional repeater uses the same frequency f0 when the base station transmits the signal using the frequency f0 initially. In this case, ISI (Inter Symbol Interference) is caused by the conventional repeater used by the same frequency f0, it deteriorates the signals. In order to solve this problem, Fig.2 show that the proposed repeater system uses a frequency hopping pattern controller, the proposed repeater change the frequency from f0 to f1.

2 The Proposed Adaptive Repeater System for OFDM with Frequency Hopping Control

2.1 Conventional Repeater System

The conventional repeater system have two types, one of them is re-transmitting all the signals which it received the transmitted signal, it amplify the received signal, and then a repeater transmit their signals. The other is the Regenerative Repeater devel-

R. Khosla et al. (Eds.): KES 2005, LNAI 3681, pp. 485–491, 2005.

oped recently. The regenerative repeater is used in digital transmission, it regenerates the signals which deteriorated and faded. When the conventional repeater transmits the signals to the mobile station, the delayed signal caused the interferences to an original signal at the repeater which it is delayed.

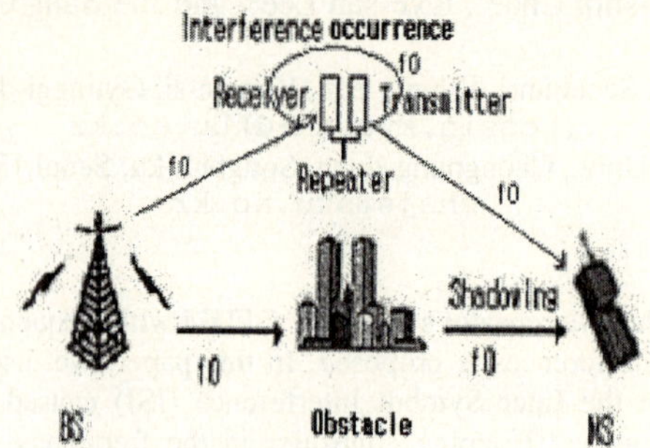

Fig. 1. Conventional system model

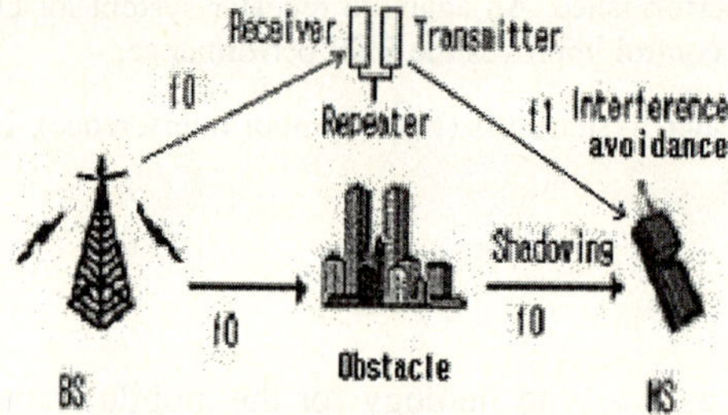

Fig. 2. Proposed repeater system model

But, the conventional Repeater contains the interference problem caused by the delayed signal which the repeaters transmit.

The signal from the base station to the repeater is transmitting through the frequency F0 first as shown in fig.3, The conventional repeater transmits the received signal faded through F0. For these reasons, the received signal at repeater contains the interference occurred by the transmitting signal at repeater because all the both of transmitting part and receiving part of repeater use the same frequency F0. It causes the more Interference result in worse performance.

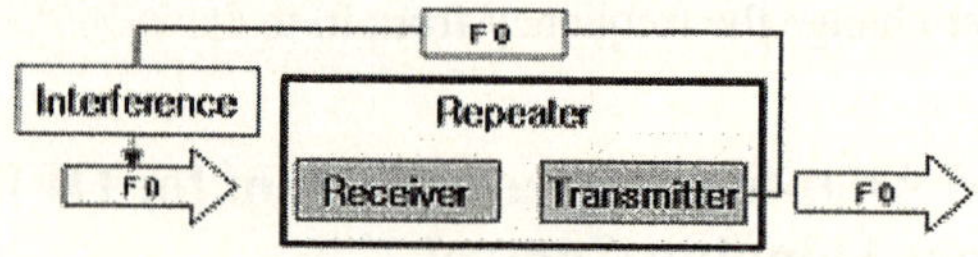

Fig. 3. Conventional Repeater system

2.2 Proposed Repeater System

It proposes from the dissertation and the system which does to sleep with Fig.4 uses a Frequency Hopping together in the Repeater and it changes an conventional frequency F0 with the F1 and it transmits it removes the interference which in receiver of the repeater appears it is a method.

Fig.5 shows a block diagram of the proposed repeater system.

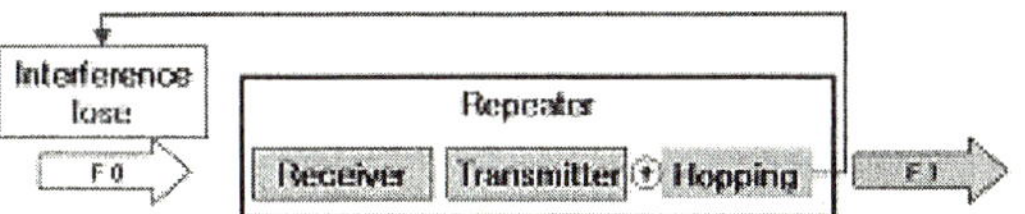

Fig. 4. Proposal Repeater system

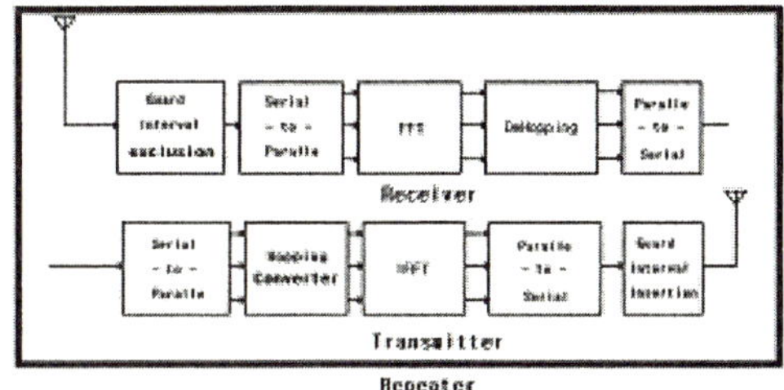

Fig. 5. Block diagram of Proposal Repeater system

Let's take a look repeater model. Guard interval is removed in the receiver. It converts the serial to parallel, FFT (Fast Fourier Transform) is utilized for modulation, De-hopping processing is done and converted parallel to serial. In the repeater transmitter, data converts in parallel to serial, IFFT (Inverse Fast Fourier Transform) is used, Guard interval is inserted.

Fig. 6 show Hopping converter of transmitter. In Hopping converter(as showed in Fig.6), it transmit the signal after changing the frequency from F0 to F6, from F1 to F7, it converts the frequency pattern. "+ 6" of Hopping converter is not fixed. It just used as an exam.

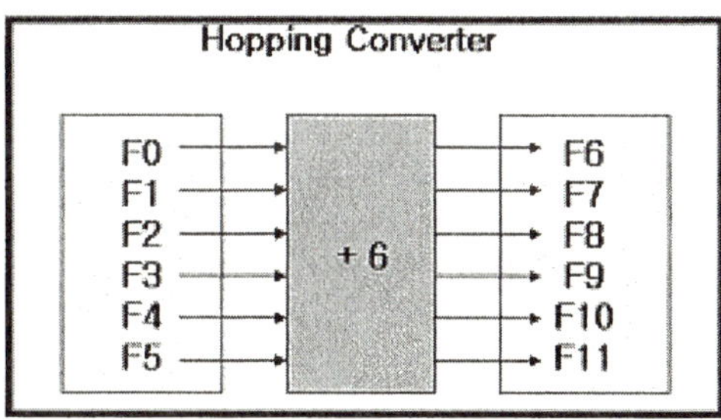

Fig. 6. Hopping Converter

3   Results

We performed simulation according to MATLAB SIMULINK, also compared to a conventional repeater with proposed repeater in OFDM. In OFDM transmitter, two channels would be used for repeater. On the other hand, for this simulation, we use one channel for repeater evaluation.

3.1 Simulation Environment Composition

The OFDM transmission method Parameter for the Simulation with table1 is same.

In the table 1, this parameter, especially 80ns can cope with delay spread in proportion to coding rate and modulation. To limit the relative amount of power and time spent on the guard interval to 1dB, the symbol duration chosen is 4 μs. This also determines the sub-carrier spacing at 312.5 kHz, which is the inverse of the symbol duration minus the guard interval. By using 48 data sub-carrier and In addition to the 48 data sub-carrier, each OFDM symbol contains an additional four pilot sub-carrier, which can be used to track the residual carrier frequency offset that remains after an initial frequency correction during the training phase of the packet.

Convolutional coding is used with the industry standard rate 1/2, constraint length 7 code with generator polynomials (133,171). Higher coding rates of 3/4 are obtained by puncturing the rate 1/2 code. The 3/4 rate is used together with 16-QAM only to obtain a data rate of 36 Mbps.

Multi-path degraded exponentially in Multi-path channel fig.7 shows exponential Multi-path channel.

Table 1. OFDM Parameter

Item	Value
Data rate	36 Mbps
Modulation	16-QAM
Coding rate	3/4
Number of sub-carriers	52
Number of pilots	4
OFDM symbol duration	4 μ s
Guard interval	800 ns
Sub-carrier spacing	312. kHz
-3-dB Band-with	16.56 MHz
Channel spacing	20 MHz
Doppler	40 Hz

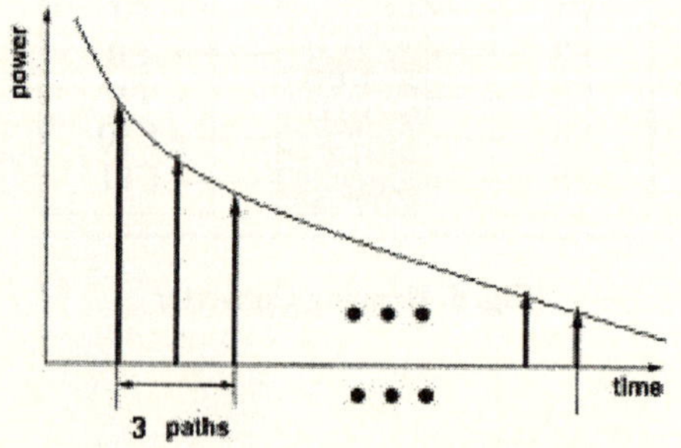

Fig. 7. Multi-path Channel model

In this simulation, we use 3path 1, 3, 5 delay and 0.2, 0.06, 0.01 gain.

3.2 Result of the Proposed Repeater Systems

This result comes from Repeater receiver which has passed AWGN channel, also added to conventional frequency. And also we get this result after Delaying old frequency in addition to Repeater receiver. To evaluate Repeater result, the channel which moves Repeater to OFDM Receiver isn't used.

Fig.8 shows (a) Conventional system Graph represents the result which comes from Repeater receiver which has passed AWGN channel, and added to initial frequency. (b) Proposed system Graph represents the result which transmitted OFDM Transmitter with Frequency Hopping, passed AWGN channel De-hopping at repeater, once again replaced frequency pattern by Frequency hopping.

According to Fig.8 E_b/N_0 loss is roughly 4.5dB in proposed Repeater system below BER 10^{-1} .1dB E_b/N_0 needs to get more the same BER in conventional system. To get a lower BER, E_b/N_0 needs 1.5dB below 10^{-2}, and 2dB between 10^{-3} and 10^{-6}.

Consequently, the proposed system has better performance than conventional system at this simulation.

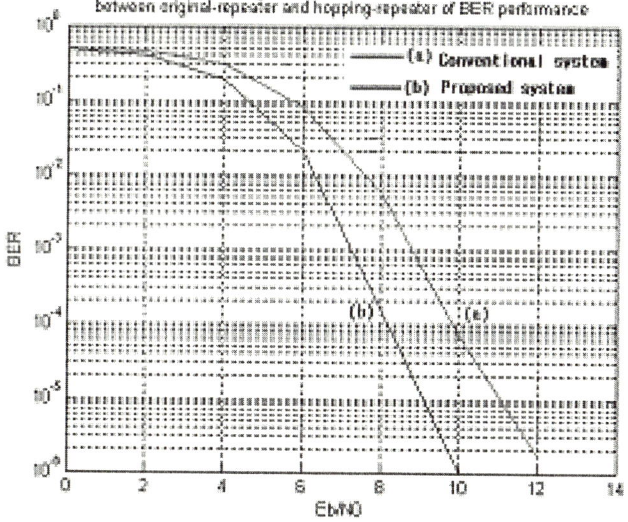

Fig. 8. The graph of the result as AWGN Channel Environment

Fig.9 shows The graph of the result as Multi-path Fading Channel result. It assume the 3 path fading model.

In fig.9, (a) indicates the conventional repeater system and (b) is meaning the proposed repeater system. The proposed system can achieves better performance to reduce the interference caused the delayed signal by converting the frequency hopping.

According to Fig.9, the BER E_b/N_0 difference of conventional Repeater system and proposal Repeater system is 2.5dB from 10^{-2}. The BER E_b/N_0 difference of conventional Repeater system and proposal Repeater system is 6dB with from 10^{-3}.

The novel repeater to reduce the Inter Symbol interference (ISI) with Frequency Hopping converter achieves the better BER according to increase of the E_b/N_0.

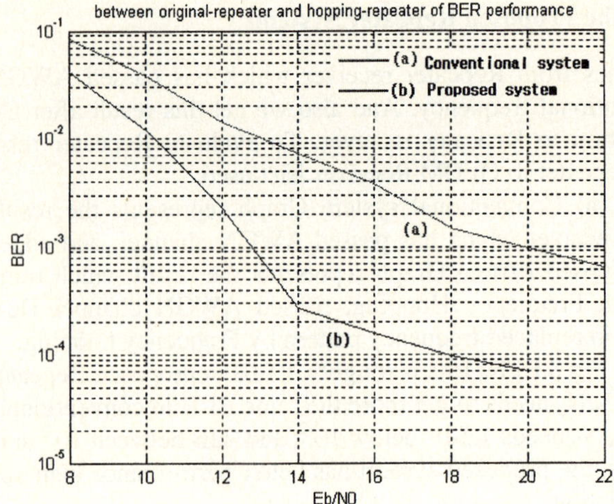

Fig. 9. The graph of the result as Multi-path Fading Channel Environment

4   Conclusions

An adaptive repeater system for OFDM with frequency hopping control to reduce interferences is proposed. The proposed adaptive repeater system is proposed to reduce the Inter Symbol Interference (ISI) caused by the delayed waves with Frequency Hopping controller in the frequency selective fading channel. In order to solve the shadowing problem, when the distortion of the signal is occurred by the shadowing due to obstacles, the proposed system provides the solution for the shadowing, removing the interferences, it results in improving the performance. The adaptive repeater to reduce the Inter Symbol interference (ISI) with Frequency Hopping converter achieves the better BER performance according to increase of the E_b/N_0. It was verified by the simulation results that the proposed schemes are effective, and practical in the frequency selective fading channel environment.

Acknowledgment

This research was supported by WIRI (Wireless Internet Research Institute).

References

1. E. Lawrey, "Multiuser OFDM," Fifth International symposium on signal processing and it's applications ISSPA '99, Aug.1999.
2. Rechard van Nee & Ramjee Prasad, "OFDM for Wireless Multimedia Communications,"
3. D.T. Magill et al. "Spread Technology for Commercial Applications," Proceedings of the IEEE Vol. 82, No.4, Apri 1994, pp.572-584
4. Barak, A.; Plotnik, E.; "An efficient baseband approach for simulation of frequency hopping spread spectrum systems," Electrical and Electronics Engineers in Israel, 1995., Eighteenth Convention of, 7-8 March 1995

5. Chanda, D.; Sesay, A.; Davies, B.; "Performance of clipped OFDM signal in fiber," Electrical and Computer Engineering, 2004. Canadian Conference on, Volume: 4, 2-5 May 2004 Pages:2401 - 2404 Vol.4
6. J. van de Beek, M. Sandell, P. O. Borjesson, "On Synchronization in OFDM Systems Using the Cyclic Prefix", Vehicular Technology Conference, 1995.
7. Cvetkovic, Z.; Tarokh, V.; Seokho Yoon; "Frequency synchronization in OFDM," Acoustics, Speech, and Signal Processing, 2004. Proceedings. (ICASSP '04). IEEE International Conference on, Volume: 4, 17-21 May 2004 Pages:iv-373 - iv-376 vol.4

Intelligent and Effective Digital Watermarking Scheme for Mobile Content Service

Hang-Rae Kim[1], Young Park[2], Mi-Hee Yoon[2], and Yoon-Ho Kim[3]

[1] Telecommunication R&D Center, SK C&C, 5F, Yonsei Severance B/D, 84, Namdaemunno, 5-ga, Chung-Gu, Seoul 100-753, Republic of Korea
hangrae@skcc.com
[2] Chungbuk Provincial University of Science & Technology, Okcheon-Gun, Chungbuk, 373-807, Republic of Korea
{py6363,mihee}@ctech.ac.kr
[3] Mokwon University, Doan-Dong 800, DaeJeon, 302-729, Republic of Korea
yhkim@mokwon.ac.kr

Abstract. In this paper, digital watermarking scheme that is intelligent and effective in protecting a copyright of image data under image transformation such as scaling, rotation, and JPEG lossy compression is proposed. Especially, the proposed scheme is a robust algorithm for mobile content service. It means that the proposed watermarking is robustness to protect impulse noise attacks. The watermark and inserted watermark are very different from each other. The original watermark is recomposed using the code granted to a copyrighter or customer. The proposed algorithm is an extension of spread-spectrum watermarking algorithm that obtains the convolution value of PN sequence. Correlator and decision circuit in our algorithm are required. Because just a little watermark is embedded in the original image, this decision circuit plays an important role for detecting this watermark and then the intelligent method for the decision must be required. Degradation of image quality between original image and watermarked image in order to confirm required invisibility in the digital watermarking system and watermark robustness to protect various attacks from the outside are analyzed using the value of PSNR and recovered rate of watermark image. The experimental result shows that image quality is less degraded as the PSNR of 93.75 dB. It is also observed that excellent watermark detection is achieved under an image transformation and JPEG lossy compression, and watermark can be effectively detected in case of the image inserted impulse noise, which PSNR is 7.97 dB.

1 Introduction

Recently, digital contents can be easily accessed by using computer networks and the problem of protecting multimedia information become more and more important. Furthermore, the mobile users using the mobile communication network have increased rapidly. Thus the problems of protecting multimedia information through wired and wireless channels become more and more important. As a solution to these problems, digital watermarking scheme is now drawing the attention as a new method of protecting copyrights for digital multimedia data [1, 2]. It is realized by embedding information data with an insensible form for human audio/visual systems. We call the embedded information data "watermark". Watermarking is defined to be a process that embeds data, called a watermark, into a multimedia object to help protect the

R. Khosla et al. (Eds.): KES 2005, LNAI 3681, pp. 492–497, 2005.

owner's rights to that object. The watermark may be either visible or invisible. A visible watermark typically consists of a conspicuously visible message or a company logo indicating the ownership of the image. On the other hand, an invisibly watermarked image appears visually very similar to the original. Therefore, visible watermarking is defined to be a process that embeds data that is intentionally perceptible to a human observer, and invisible watermarking is defined to be a process that embeds data that is not perceptible. Although "invisible" and "watermark" are visual terms, invisible watermarking is not limited to marking images. Several watermarking schemes for images have been presented which allow embedding of invisible and unremovable information into the image [3, 4].

Image watermarking schemes proposed so far can be generally divided into two groups, according to the type of feature set the watermark is embedded in the intensity value of the luminance in the spatial domain, or the image coefficients in a frequency domain (e.g. FFT, DCT, wavelet). To provide extra robustness against attacks, some algorithms recover the watermark by relying on the comparison between the marked and non-marked images. There have been many different watermarking schemes proposed in the literature and consequently there is a great deal of variation in how a watermark signal is embedded into an image. In spread spectrum method, the basic idea of watermarking for image is the addition of a pseudo-random signal to the image that is below the threshold of perception and that cannot be identified and thus removed without knowledge of the parameters of the watermarking algorithm.

In this paper, a new scheme to add a watermark to digital images is presented. The proposed algorithm is operated in Fast Fourier Transform (FFT) domain, which is widely used. Our approach is a direct extension of ideas from direct-sequence code division multiple access (DS-CDMA) communications [5]. Therefore, a pseudo-random sequence is not required and each pixel of watermark image is spread by the code of the copyrighter or customer. The proposed scheme represents a major improvement to methods relying on the comparison between the watermarked and original images [3, 4, 6].

2 Watermarking Algorithm

2.1 Watermark and Inserted Watermark

The original watermark $\mathbf{O}$ in spatial domain is defined as the following equation.

$$\mathbf{O} = \{o(x,y)|0<x\leq O_1, 0<y\leq O_2\} \tag{1}$$

where the original watermark $\mathbf{O}$ is a binary image of $O_1 \times O_2$.

The code $\mathbf{C}$ of the copyrighter or customer is defined as

$$\mathbf{C} = \{c(x,y)|0<x\leq C_1, 0<y\leq C_2\} \tag{2}$$

where the code $\mathbf{C}$ may be a binary image of $C_1 \times C_2$ and can be either a vector or matrix. In the watermarking algorithm we propose, it is impossible to detect watermark if there is no code $\mathbf{C}$ of the copyrighter or customer.

An inserted watermark will be an encoded watermark using the code $\mathbf{C}$ and is expressed as following expression.

$$\mathbf{W} = w(x,y) \times \mathbf{C} \tag{3}$$

where the original watermark **O** is a binary image of $O_1 \times O_2$. Eq. (3) means that a pixel of the original watermark is encoded by the code matrix **C**, which is used as the spread code.

2.2 Watermark Insertion

Original image **I** in spatial domain is defined as the following equation.

$$\mathbf{I} = \{i(x,y) | 0 < x \leq I_1, \ 0 < y \leq I_2\} \tag{4}$$

where the original image **I** is a gray level image of $I_1 \times I_2$. The original image and original watermark have to satisfy $I_1 \geq 2 \times O_1$ and $I_2 \geq 2 \times O_2$ for encoding. Also, the inserted watermark **W** and the original image **I** have to satisfy $I_1 \geq W_1$ and $I_2 \geq W_2$.

Frequency components of the inserted watermark are inserted to frequency components of the original image using two-dimensional Fast Fourier Transform (FFT) of watermark image **W** added in the code. FFT of the original image **I** and watermark image **W** can be written as eq. (5) and (6), respectively.

$$\mathbf{F_I} = FFT\{\mathbf{I}\} \tag{5}$$

and

$$\mathbf{F_W} = FFT\{\mathbf{W}\} \tag{6}$$

where $\mathbf{F_I}$ and $\mathbf{F_W}$ are a matrix form of $I_1 \times I_2$ and $W_1 \times W_2$, respectively. FFT{} denotes Fast Fourier Transform.

The watermark **W** can be generally embedded using the following equations:

$$\mathbf{F_E} = \mathbf{F_I} + \beta \ \mathbf{F_W} \tag{7}$$

and

$$\mathbf{F_E} = \mathbf{F_I} (1 + \beta \ \mathbf{F_W}) \tag{8}$$

where β is constant value.

Watermarked image is easily obtained from IFFT of $\mathbf{F_E}$ as the following equation.

$$\mathbf{I_W} = IFFT\{\mathbf{F_E}\} \tag{9}$$

where IFFT{} denotes Inverse Fast Fourier Transform.

Fig.1 shows insertion process as the above description that the watermark is inserted to the original image

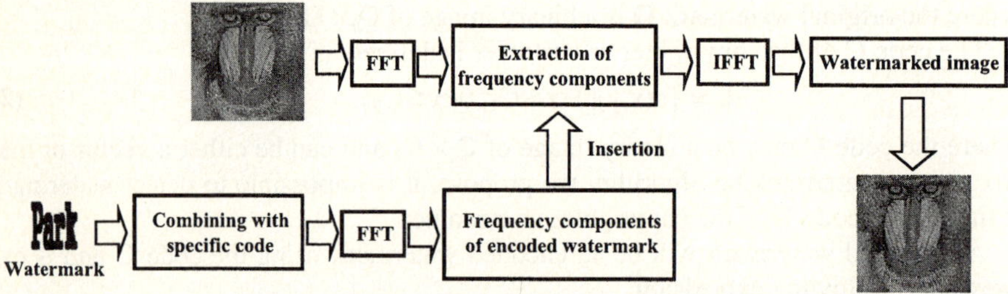

Fig. 1. Watermark insertion process

2.3 Watermark Detection

Watermarked image $\mathbf{I_W}$ is influenced by image transformation, JPEG lossy compression, or impulse noise and so on. If $\mathbf{I_A}$ is the transformed image as the signal processing, watermark can be detected according to the following equation.

$$\mathbf{F_D} = FFT\{\ \mathbf{I_A}\ \} - \mathbf{F_I} \tag{10}$$

where $\mathbf{F_D}$ denotes the detected watermark.

The detection process performs a spatial correlation operation to detect only the specific desired code. All codes appear as noise due to decorrelation. For detection of the watermark, detection process needs to know the code used by insertion process. In order to recover watermark, a pixel $w_D(x,y)$ of detected watermark $\mathbf{W_D}$ is decoded by the specific code.

$$w_D(x,y) = \sum IFFT\{\ \mathbf{F_D} \times \mathbf{C}\} \tag{11}$$

where the number of summation is $C_1 \times C_2$. The above equation means that the detection process is in reverse order of the insertion process.

Conformity degree between the detected watermark and original watermark was shown as recovered rate. Fig.2 shows detection process that the watermark is detected from the watermarked image.

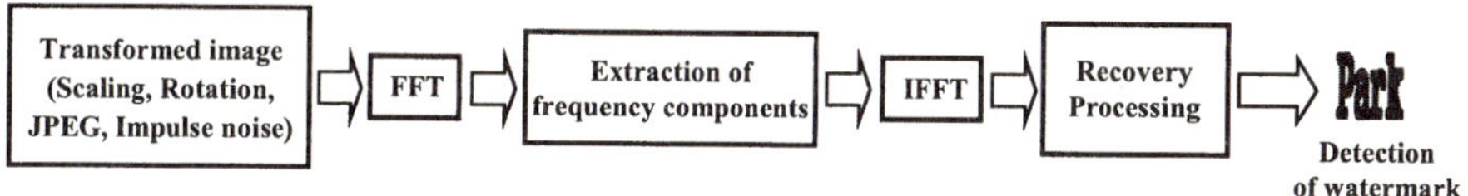

Fig. 2. Watermark detection process

3 Experimental Results

Baboon image of 256×256 gray level is used as a standard image for test of the proposed watermark scheme and the watermark consists of 64×64 binary image. Fig.3 shows original image, watermark and watermarked image, and a pattern with "Park" is used as the watermark. After inserted watermark image, PSNR for testing is 93.75dB. This result shows that it is difficult to find out inserted watermark with human visual system and that original image and the watermarked image are very similar.

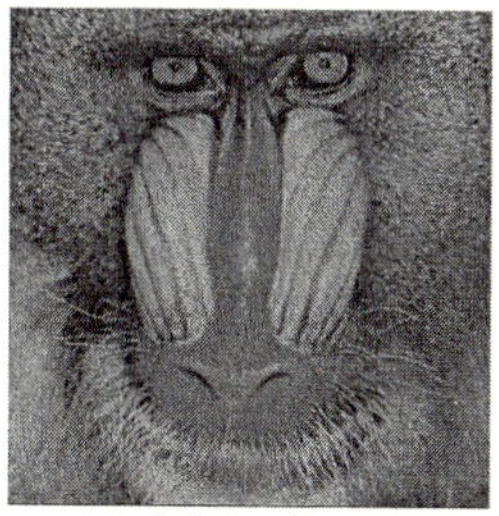

(a) Original image   (b) Watermark   (c) Watermarked image

Fig. 3. Experimental image

For testing robustness of the proposed watermark scheme, we firstly tested about scaling images. In case of scaling down the watermarked image, the recovered watermarks are shown in Fin. 4. As experiment results, it is observed that the watermarks are perfectly recovered about enlarged images and is recovered to 75.71% about 20% scaling down.

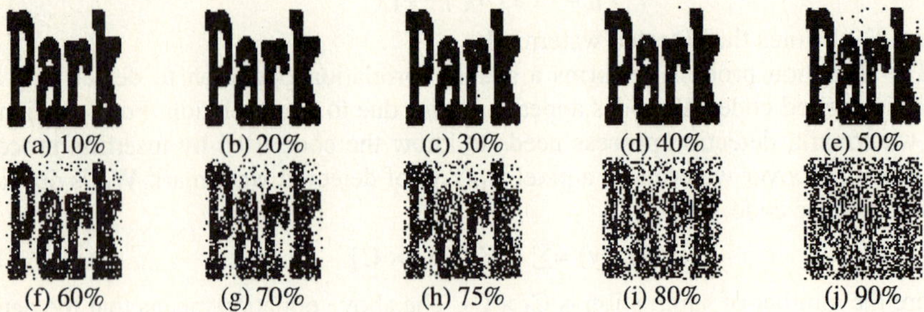

Fig. 4. The detected watermarks from scaling down images

In case of rotation, the recovered watermarks are illustrated in Fin. 5. It observed that the recovered rate in the proposed method is also very excellent about 45 degree and 90-degree rotation.

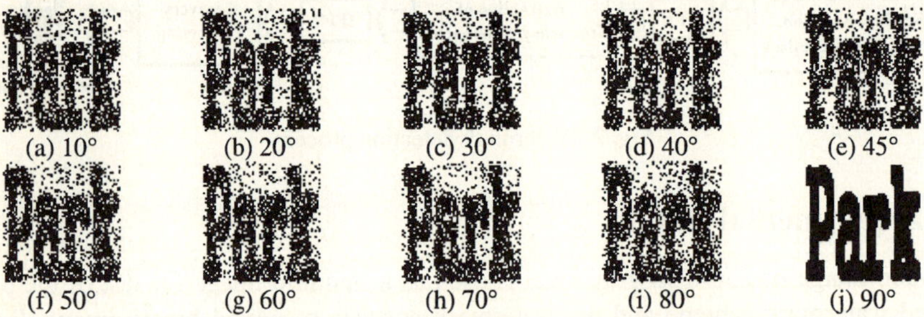

Fig. 5. The detected watermarks from the rotation images

Experiment images are tested with compression rate from 10% to 90% about JPEG lossy compression. Fig. 6 shows the detected results from JPEG lossy compressed version of the watermarked images according to compression rates.

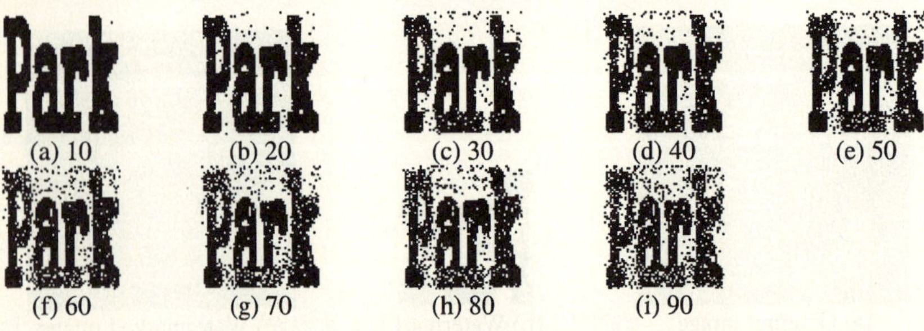

Fig. 6. The detected watermarks from JPEG lossy compression

In the case of image added impulse noise, the proposed scheme is tested according to noise density. Even though PSNR is 7.97 dB as in Fig. 7(e), the recovered rate is 87.57%. It is robust against impulse noise in this respect. Impulse noise can be interpreted as random noise occurring mobile environments.

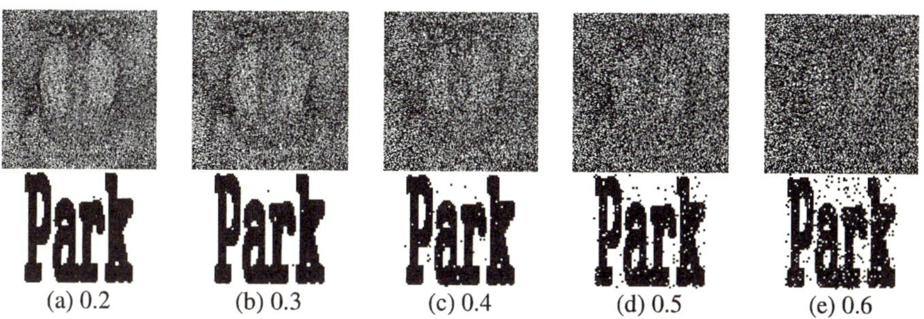

Fig. 7. The detected watermarks according to impulse noise density

4 Conclusions

In this paper a new watermarking algorithm for digital images is presented: watermark is encoded and decoded by the specific code of copyrighter or customer. A specific code is provided as a secret key that can be used to serve various embedding processes by using the same embedding scheme. Also, the proposed scheme is easily achieved by FFT, and then is hybrid method combining key with logo method. It was observed that invisibility of watermark is very effective since PSNR of the watermarked image is over 93.75 dB. It is confirmed that the proposed scheme is robust against several signal processing, such as scaling, rotation, and JPEG lossy compression. Also it observed that the proposed digital watermarking is very adequate for mobile image contents that are attacked by the random error due to mobile environment. Furthermore, we will investigate the tolerance of the other attacks and focus on the decision method in order to improve the performance of the proposed watermarking scheme. This decision can be achieved using algorithm such as neural or fuzzy.

References

1. W. Bender, D. Gruhl, N. Morimoto and A. Lu.: Techniques for data hiding," IBM System Journal, vol. 35, no. 3&4, pp. 313-336, 1996.
2. M. D. Swanson, M. Kobayashi and A. H. Tewfik, "Multimedia data embedding and watermarking technologies," Proceedings of the IEEE, vol. 86, no. 6, pp. 1064-1087, June 1998.
3. J. J. K. O'Ruanaidh, W. J. Dowling and F. M. Boland, "Watermarking digital images for copyright protection," IEEE Proceedings Vision, Image and Signal Processing, 143(4):250-256, August 1996.
4. I. J. Cox, J. Kilian, T. Leighton and T. Shamoon, "Secure Spread Spectrum Watermarking for Multimedia," IEEE Trans. on Image Processing, vol. 6, no. 12, pp. 1673-1678, 1997.
5. D. L. Nicholson, Spread Spectrum Signal Design-Low Probability of Exploitation and Anti-Jam Systems, Computer Science Press, 1988.
6. C. Podilchuk and W. Zeng, "Perceptual watermarking of still images," Proceeding The First IEEE Signal Processing Society Workshop on Multimedia Signal Processing, June 1997.

Performance Improvement of OFDM System Using an Adaptive Coding Technique in Wireless Home Network

JiWoong Kim[1] and Heau Jo Kang[2]

[1] FUMATE Inc., Korea
`kjwcomm@korea.com`
[2] Division of Computer & Multimedia Content Eng. 800, Doan-Dong,
Taejon Mokwon University, Taejon, 305-729, Korea
`hjkang@mokwon.ac.kr`

Abstract. Since OFDM method uses Fast Fourier Transform(FFT) algorithm,, it can make the size of circuits smaller and also can vary transmission capacity by adaptive changing the number of subcarrier. Therefore, OFDM method may be applied as digital broadcasting and physical layer transmission mode of intelligent system. In this paper, we analyzes how a synchronization error affects receiving system when using a adaptive coded OFDM(Orthogonal Frequency Division Multiplexing) transmission method in wireless LAN channel environment in which we can efficiently transmit wide-band information data. As a result, in OFDM system, we could see that the higher a frequency offset is, the worse performance will be, and we could see that there was performance improvement by applying adaptive coding technique in OFDM system in order to reach (BER=10^{-3}). However, when we use 64QAM(64Quadrature Amplitude Modulation), there was a huge influence between carriers by frequency offset at 0.05, 0.1.

1 Introduction

Recently wireless LAN which is based on the 802.11 standard has spread its domain in the U.S, Europe, England. In Korea, KT and Hanaro have commenced their wireless LAN service and SK telecom has just taken part in wireless LAN market. IEEE 802.11a, which is currently a standard of high-speed wireless LAN and HiperLAN/2 use wireless frequency 5GHz which is relatively wider than 2.4GHz and OFDM modulation and demodulation method in common[1].

Since OFDM method uses FFT(Fast Fourier Transform) algorithm, it can make the size of circuits smaller and also can vary transmission capacity by changing the number of subcarrier[2]. Besides it has an advantage that by installing a protective section we can completely remove the effect of ISI (Inter-Symbol Interference), which is occurred by delay elements[3].

However, OFDM system has a weakness that because of subcarrier's narrow band when it comes to frequency offset, it reduces SER (Symbol Error Rate) [4]-[6].

This paper shows how subcarrier frequency offset affects OFDM system and its performance reduction degree as well as performance improvement degree using convolution coding. As a result, we can decide maximum frequency offset in order to satisfy SER performance.

R. Khosla et al. (Eds.): KES 2005, LNAI 3681, pp. 498–503, 2005.

2 OFDM System Considering Subcarrier Frequency Offset

The model of OFDM which considers subcarrier frequency offset in AWGN(Additive White Gaussain Noise) channel is like the figure 1.

In this paper, we uses QPSK(Quadrature Phase Shift Keying), 16QAM and 64QAM as modulation methods. OFDM signal reduces system performance when orthogonality is corrupted between subcarriers while demodulating. Frequency offset is occurred by the influence of doppler shift.

If receiver generates sine wave $e^{-j2\pi(f_c-\Delta f)t}$ which has a frequency gap from f_c to $-\Delta f$, then baseband OFDM signal is like below. In this case in order to simplify we assumed transmission distortion is inexistent $(H(f)=1)$, in the equation we erased additional noises. We call Δf frequency offset.

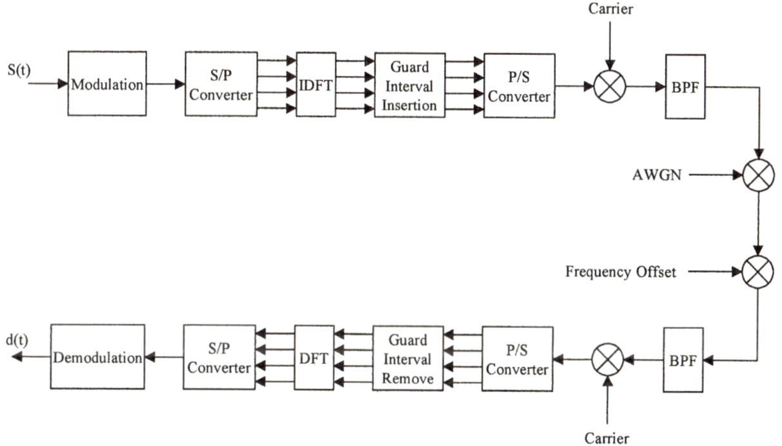

Fig. 1. The model of OFDM system which consider carrier frequency offset

$$s_B(t) = \sum_{m=-\infty}^{\infty} g(t - mT_s)$$
$$\sum_{n=0}^{N-1} d(m,n)e^{j2\pi nf_0(t-mT_s)}e^{j2\pi\Delta ft} \tag{1}$$

In equation (1), when there's no frequency offset, $e^{j2\pi\Delta ft}$ is multiplied. This signal can be sampled as equation (2).

$$s_B\left(kT_s + \frac{i}{Nf_0}\right) = \sum_{m=-\infty}^{\infty} g\left\{(k-m)T_s + \frac{i}{Nf_0}\right\}e^{j2\pi nf_0\left\{(k-m)T_s\frac{+i}{Nf_0}\right\}} \cdot e^{j2\pi\Delta(kT_s\frac{+i}{Nf_0})}$$
$$= \sum_{n=0}^{N-1} d(k,n)e^{j\frac{2\pi ni}{N}}e^{j2\pi\Delta f(kT_s\frac{+i}{Nf_0})} \tag{2}$$
$$= e^{j2\pi k\Delta fT_s}\sum_{n=0}^{N-1} d(k,n)e^{j\frac{2\pi(n+\alpha)i}{N}}$$

In equation (2), α is a parameter which is expressed by the formula below and the size of frequency offset. We call α regular frequency offset.

$$\alpha = \frac{\Delta f}{f_0} \tag{3}$$

As we can see from equation (4), demodulation $\hat{d}(k,l)$ includes expecting symbol $d(k,l)$ as well as all the subcarrier symbol in the same OFDM symbol, and there can be an interference between carriers.

$$\hat{d}(k,l) = \sum_{i=0}^{N-1} \hat{c}(k,l,n,)d(k,n) \tag{4}$$

$$\hat{c}(k,l,n) = \frac{1}{N} \cdot \frac{\sin\{\pi(n-l+\alpha)\}}{\sin\dfrac{\pi(n-l+\alpha)}{N}} \\ e^{j2\pi k\Delta fT_s} \cdot e^{j\frac{\pi(N-1)(n-l+\alpha)}{N}} \tag{5}$$

A coefficient $\hat{c}(k,l,n)(n \neq l)$ is n'th subcarrier and shows transferred symbol distortion. We can analyze symbol error by using this.

At first, search for a demodulation symbol $\hat{d}(k,l)$'s average power. However, make sure that k, l's symbol demodulation method is same and should be.

$$E[|\hat{d}(k,l)|^2] = \sigma_d^2$$

As we can see from above, demodulation symbol's average power is always same regardless of k and l. This shows that if a regular frequency offset exists, at random symbol different subcarrier's interference frequency and interference is same.

If multipass exists, this doesn't work out. At this point, l'th subcarrier's expecting symbol elements's average power is denoted as equation (6).

$$E\left[\left|\hat{c}(k,l,l)d(k,l)\right|^2\right] = \frac{\sigma_d^2}{N^2} \cdot \frac{\sin^2(\pi\alpha)}{\sin^2\dfrac{\pi\alpha}{N}} \tag{6}$$

From this, average power $P_I(l)$ of interference elements is denoted by

$$P_I(l) = \sigma_d^2 - \frac{\sigma_d^2}{N^2} \cdot \frac{\sin^2(\pi\alpha)}{\sin^2\dfrac{\pi\alpha}{N}} \tag{7}$$

At this point, we assume that AWGN is added on transmission path and each symbol's average power is $2\sigma_n^2$, l'th subcarrier's expecting SN ratio $\gamma(l)$ is defined by

$$\gamma(l) = \frac{\dfrac{1}{N^2} \cdot \dfrac{\sin^2(\pi\alpha)}{\sin^2\dfrac{\pi\alpha}{N}}}{1 - \dfrac{1}{N^2} \cdot \dfrac{\sin^2(\pi\alpha)}{\sin^2\dfrac{\pi\alpha}{N}} + (CNR)^{-1}} \tag{8}$$

It is possible to get symbol errors of each symbol when a frequency offset exists.

Table 1. OFDM System Parameter

Subcarrier number	1024
Subcarrier interval	4KHz
Symbol length	250
Guard interval length	31.25
Minimum Subcarrier frequency in transmit band	100MHz

3 Analyzing Figures

In this paper, We analyzes performance of OFDM System in which a subcarrier frequency offset exists and shows performance improvement by applying convolution coding.

Figure 2 analyzes performance of OFDM/QPSK system when 0, 0.05, 0.1 are give as a frequency offset figure. The higher a frequency offset is, the worse the system performance is, and we can see performance improvement by more than 4dB.

In addition, the higher frequency offset is, the better performance improvement is by applying convolution coding.

Figure 3, 4 analyze OFDM/MQAM system performance according to frequency offset in AWGN environment. From these, we can know OFDM/MQAM system is more sensitive to frequency offset than OFDM/QPSK and OFDM/64QAM can not be satisfactory when we have to reach (BER=10^{-3}). When a frequency offset figure is higher, SER rapidly decreases and more elaborate high frequency synchronization is needed. And we can find out satisfactory and acceptable frequency offset figures.

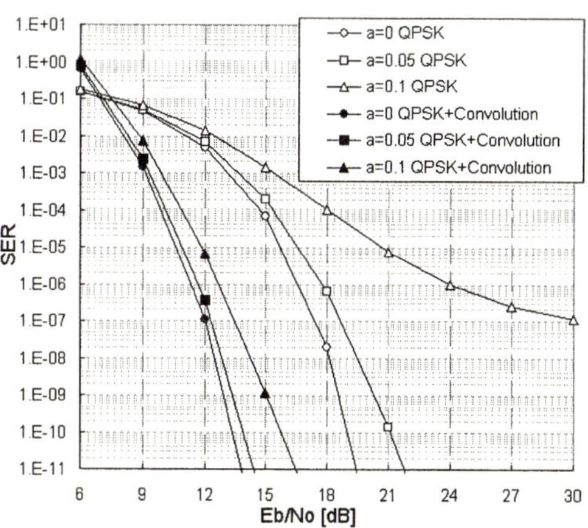

Fig. 2. SER performance of OFDM/QPSK system according to frequency offset

Figure 5 analyzes OFDM/MQAM system performance according to frequency offset. When convolution coding is used, we could see performance improvement. However, when frequency offset is set as 0.1 in OFDM/64QAM system, we can hardly see improvement, and when we tried to reach (BER=10^{-3}), we can not meet the conditions of SER. From this we can see interference by frequency offset between carriers is very important.

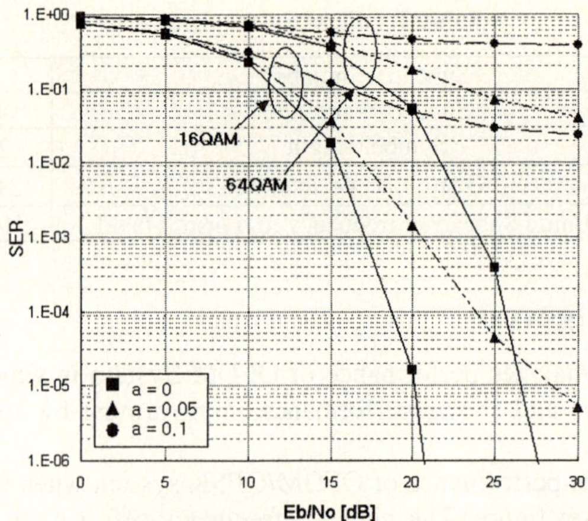

Fig. 3. SER performance of OFDM/QPSK system according to frequency offset in AWGN environment

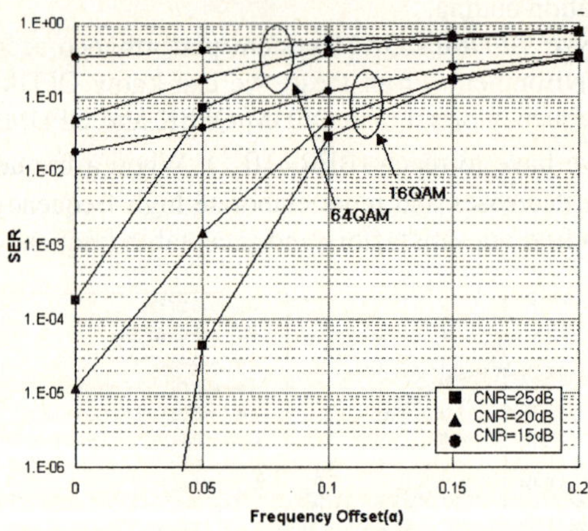

Fig. 4. SER performance of OFDM/QPSK system according to frequency offset (subcarrier=1024)

4 Conclusion

This paper showed effect of a receiver system when using OFDM transmission method so that we efficiently transmit wide band signal. In a way, by applying adaptive coding technique, we could improve performance reduction.

As a result, in OFDM/QPSK system the higher a frequency offset is, the better the system performance is, and by applying adaptive coding we could see performance improvement more than 4dB. Besides, in OFDM/QPSK system the higher a frequency offset is, the worse the system performance is, and performance reduction is bad even at high SNR. Lastly, wireless LAN system which aims to reach (BER=10^{-3}) needs additional research because interference effects between carriers is very important by 0.05, 0.1 frequency offset.

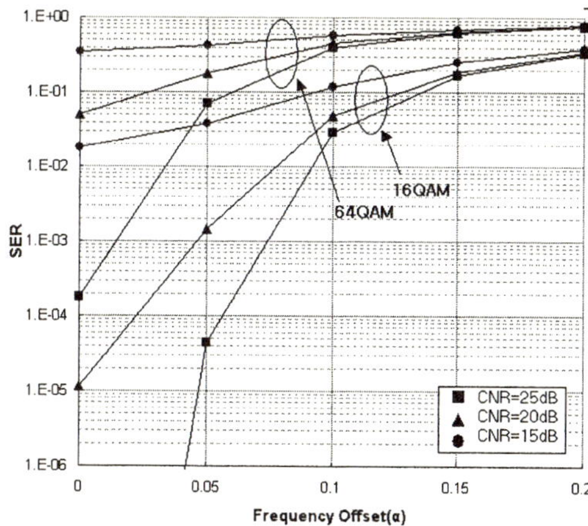

Fig. 5. SER performance of OFDM/QPSK system considering convolution coding

References

1. Y.W.Park, "The major issue and current event of WLAN markets," *Information and Communications Policy*, vol. 14, no. 8, May 2002.
2. J. A. C. Bingham, "Multicarrier modulation for data transmission: An idea whose time has come," *IEEE Commun. Mag.*, vol. 28, no. 5, pp. 5-14, May 1990.
3. S. Hara, M. Mouri, M. Okada, and N. Morinaga, "Transmission performance analysis of multicarrier modulation in frequency selective fast Rayleigh fading channel," *Wireless Personal Commun.*, vol. 2, pp. 335-356, Jan.-Feb. 1996.
4. H. Sari, G. Karam, and I. Jeanclaude, "Transmission techniques for digital terrestrial TV broadcasting," *IEEE Commun. Magazine*, pp. 100-109, Feb. 1995.
5. William D. Warner and Cyril Leung, "OFDM/FM frame synchronization for mobile radio date communication," *IEEE Transaction on Vehicular Technology*, pp. 302-313, vol. 42, no. 3, August 1993.
6. T. M. Schmidl and D.C. Cox, "Blind synchronization for OFDM," *Electronics Letters*, vol. 33, no. 2, pp. 113-114, Jan. 1997.

Arbitrated Verifier Signature
with Message Recovery for Proof of Ownership*

Hyung-Woo Lee

Dept. of Software, Hanshin University, Osan, Kyunggi, 447-791 Korea
`hwlee@hs.ac.kr`

Abstract. There are some problems in existing digital copyright mechanism as the rightful ownership cannot be resolved by using current knowledge based intelligent technology alone. Specially, some one can provide counterfeited watermark which can be performed on a watermarked image to allow multiple claims of rightful ownerships. Therefore, we propose *advanced arbitrated verification scheme* on the digitally watermarked contents by designated proving with message recovery function on trusted third party. In detail, the copyright owner of a watermark can sure that only the intended verifier(buyer) can be convinced about the validity of the watermark as anybody except for verifier can't produce identically distributed transcripts.

1   Introduction

A *digital watermark* is a signal added to digital data such that it could be used (1) to identify source of the data or uniquely establish ownership, (2) to identify its intended recipient, and (3) to check if the data has been tampered with. Within each class of applications, there could be variations on the requirement of the watermarking scheme[1]. Therefore, digital watermarks have been proposed as the means for *copyright protection of multimedia data* on network based digital image distribution structure. However, there are some problems in existing digital watermarking mechanism as the *rightful ownerships cannot be resolved by current watermarking schemes alone*. Specially, some one can provide counterfeited watermark which can be performed on a watermarked image to allow multiple claims of rightful ownerships[2].

We can define the *designated verifier property* in the following way. If Bob, after having received a proof (watermark or signature) from Alice, has a way to prove to Carol the truth of a given copyright statement, then he can produce indistinguishable verification process by his own. As a consequence, whatever Bob can do with the "real" transcript, he will be able to do with the "simulated" transcripts as well. Thus, Carol being aware of this fact, will never be convinced by Bob's proof, whatever the protocol that Bob initiates. So, proposed designated verifier signature provide authentication of a digital watermark message.

It has the property that it convince one specified recipient(Bob) that the watermark is valid, but unlike other digital watermark scheme, nobody else can

---

* This work was supported by Hanshin University Research Foundation.

R. Khosla et al. (Eds.): KES 2005, LNAI 3681, pp. 504–510, 2005.

be convinced about its validity or invalidity. The reason is that the designated verifier in these schemes is able to create a watermark intended to himself in an indistinguishable way. Therefore, when Bob receives a signature from Alice, he will certainly trust that it is originated from Alice upon verifying it.

As a solution, we propose *new designated verifier signature scheme with arbitrator* for providing the property that nobody else except buyer can be convinced about their validity or invalidity. Designated verifier signatures are very useful in situations where the signer(owner) of a message(digital watermark) should be able to specify who may be convinced by his/her signature. So, the signer can designate its verifier on signature verification steps. Verifiable watermark differ from ordinary digital watermark in that the verifier should be unable to distinguish between valid and invalid watermark. If the message in is copyright owner's information, we can also introduce zero-knowledge based watermarking mechanism for proving watermark message without revealing copyright owner's secret information.

2 Digital Watermark for Proof of Ownership

2.1 Watermark: Function and Mechanism

These watermarking schemes are designed mainly for two purposes : copyright protection and data authentication. In this paper we shall focus on the applicability of watermarking techniques for one instance of ownership verification, identification of an content's rightful owner(s)[2].

The watermarks embedded in an image have to be recoverable, despite intentional or unintentional modification of the image. They must also be invulnerable to deliberate attempts to forge, remove, or invalidate watermarks. Without any verification techniques or specification of certain requirements for digital watermark, anyone can claim ownership of any image by using his/her own verification methods. Therefore, it is crucial that watermark for copyright protection be able to be *verified by designated receiver(contents buyer) for proving his own usability* by interaction with original watermark embedder(copyright owner).

2.2 Digital Signature for Proof of Ownership

A watermark is inserted into an image, and the decoding process by which a watermark is recovered and then compared to the inserted watermark. We use to denote I an image, W a watermark consisting of a sequence of ownership labels $W = \{ w_1, w_2, ..., w_n \}$ and the watermarked image I^W . E is an encoder function if it takes an image I and a watermark W and generates a new image which is called the watermarked image I^W i.e., $E(I, W) = I^W$ [1]. We can construct both encoding and corresponding decoding processes.

We can consider three watermark mechanisms for digital data D', which are the (1) watermark(W) casting mechanism, (2) signature(D^*) signing mechanism, and (3) signature hiding and verification mechanism. We can embed some mark W on digital data D'. So, it is the common watermark embedding and verification(recovery) mechanism(*mark casting mechanism*). The other one is the cryptographic mechanism as we can attach digital signature D^* on digital data D' for authentication(*signing mechanism*).

From these two mechanism, we can consider new hybrid mechanism(*signature hiding and verification mechanism*). As we can think signature as its watermark information, we can hide this signature on its digital data for verifying ownership and authentication. So, we can construct hybrid mechanism on digital watermark verification and authentication system.

2.3   Attack and Its Possible Solutions

If there are multiple watermark in one image, how can we decide the original watermark ? And if we cannot find out which so-called *original* image is the *true* original, how can we decide which watermarked version of an image is the *truely watermarked* version in circulation?

These multiple watermarking problem can be solved easily if we consider it in *verification* process. In case of several attack, we only review its robustness on verification module. Therefore, we can solve proposed attack problems from different directions. In proposed system, the signer can receive the watermark (*pseudo-watermark*) from the trusted third party and embed it on contents. And the the owner(*pseudo-owner:prover*) designate buyer(*pseudo-buyer:verifier*) for proving the it's usability so that the signer(owner) can cooperate its verification process with the buyer(verifier). And in case that the watermark can also be matched with the registered message on TTP server, we can trust the pseudo-buyer as *really true-buyer*.

3   Overview on Existing Scheme

3.1   Designated Verifier Signature

A *digital signature* of a message is a number dependent on some secret known only to the signer, and, additionally, on the content of the message being signed. *Signatures must be verifiable*; if a dispute arises as to whether a party signed a document (caused by either a lying signer trying to repudiate a signature it did create, or a fraudulent claimant), an unbiased third party should be able to resolve the matter equitably, without requiring access to the signer's secret information (private key).

In 1996, Jakobsson, Sako and Impagliazzo[4] and Chaum[5] introduced designated verifier signatures and private signatures, respectively, that are based on the same idea. Designated verifier signatures provide authentication of a message, without however providing non-repudiation.

Although designated verifier signatures are *signer ambiguous*, in the sense that one cannot verify whether the real signer or the designated verifier issued the signature, they remain universally verifiable, i.e. everyone can convince himself that there are only two possible signers. We say that a proof is strong designated verifier if transcripts of a *real* proof may be simulated by anybody in such a way that they are indistinguishable for everybody other than *Bob*. So, accordingly to our definition of designated verifier proofs, we define the strongness as follows:

Definition 1 (Strong Designated Verifier Signature) *Let $P(A, B)$ be a protocol for Alice to prove the truth of the statement Ω to Bob. We say that $P(A, B)$ is a strong designated verifier proof if anybody can produce identically distributed transcripts that are indistinguishable from those of $P(A, B)$ for everybody, except for Bob.*

3.2 Existing Schnorr Based Designated Verifier Signature

A large prime p, a prime factor q of $p - 1$, a generator $g \in Z^*_p$ of order q and a one-way hash function h. Each user i chooses his secret key $x_i \in Z_q$ and publishes the corresponding public key $y_i = g_{x_i}(mod\ p)$. In order to sign a message m for Bob, Alice selects two random values k and t in Z_q and computes

$$c = y_b{}^k(mod\ p), \quad r = h(m, c), \quad s = kt^{-1} - rx_a(mod\ q). \tag{1}$$

The triple (r, s, t) is then the signature of the message m. Then knowing that a signature is originated from *Alice*, *Bob* may verify its validity by checking whether

$$h(m, (g^s y_a{}^r)^{tx_b}(mod\ p)) = h(m, (g^{kt^{-1} - rx_a} y_a{}^r)^{tx_b}(mod\ p)) = r. \tag{2}$$

Nobody else other than *Bob* can perform this verification, since his secret key is involved in the verification equation. This precisely means that our scheme verifies the strong designated verifier property[7].

4 Arbitrated Verifier Signature with Message Recovery

4.1 New Designated Verifier Signature with Message Recovery

Each entity creates a private key to be used for signing messages, and a corresponding public key to be used by other entities for verifying signatures. In this scheme, each entity creates a private key to be used for signing messages, and a corresponding public key to be used by other entities for verifying signatures. We can define the function of proposed signature as follows:

Definition 2 (Digital Signature with Message Recovery) *Entity A produces a signature $s \in S$ for a message $m \in M$, which can later be verified by any entity B and the message m is recovered from the entity A's signature s. Therefore, A priori knowledge of the message is not required for the verification algorithm.*

1. Each entity A should select a set $S_A = \{S_{A;k} : k \in R\}$ of transformations. Each $S_{A;k}$ is a 1-1 mapping from M_S to S and is called a *signing transformation*.
2. S_A defines a corresponding mapping V_A with the property that $V_A \cdot S_{A;k}$ is the identity map on M_S for all $k \in R$. V_A is called a *verification transformation* and is constructed such that it may be computed without knowledge of the signer's private key.
3. A's public key is V_A; A's private key is the set S_A.

We propose new ElGamal[10] like designated signature scheme with message recovery.

Definition 3 (Designated Verifier Signature with Message Recovery)
Let $P_{MR}(A, B)$ be a protocol for Alice to prove the truth of the statement Ω to Bob with message recovery. $P_{MR}(A, B)$ is a strong designated verifier proof if only Bob can recover identically distinguished transcripts.

Detailed procedure is as follows. A large prime p, a prime factor q of $p - 1$, a generator $g \in Z^*_p$ of order q and a one-way hash function h. Each user i chooses his secret key $x_i \in Z_q$ and publishes the corresponding public key $y_i = g^{x_i}(mod\ p)$. In order to sign a message m for Bob, Alice selects two random values k in Z_q and computes

$$m^* = R(m), r = y_b^{-k}(mod\ p), \ e = m^*r(mod\ p), \ s = x_a e + k(mod\ q). \quad (3)$$

The tuple (s, e) is then the signature of the message m. Then knowing that a signature is originated from *Alice*, *Bob* may verify its validity by checking whether

$$(g^s y_a^{-e})^{x_b} e(mod\ p) = g^{x_b k} m^* r(mod\ p) = m^* \quad (4)$$

Nobody else other than *Bob* can perform this verification, since his secret key is involved in the verification equation. This precisely means that our scheme verifies the strong designated verifier property.

4.2   New Arbitrated Verifier Signature Scheme

An arbitrated digital signature scheme is a digital signature mechanism requiring an unconditionally *trusted third party* (TTP) as part of the signature generation and verification[8]. Therefore, we can define new arbitrated verifier signature scheme as follows:

Definition 4 (Arbitrated Verifier Signature) *Let $P_{MR}^T(A, B)$ be a protocol for Alice to prove the truth of the statement Ω to Bob through the trusted third party (arbitrator) T. $P_{MR}^T(A, B)$ is a arbitrated verifier proof if only Bob can produce identically distinguished transcripts from intermediate process with unconditional arbitrator(TTP).*

Signature generation and verification for arbitrated signatures requires a symmetric-key encryption algorithm $E = \{E_k : k \in K\}$ where K is the key space. Assume that the inputs and outputs of each E_k are l-bit strings, and let $h : \{0, 1\}^* \to \{0, 1\}^l$ be a one-way hash function. The TTP selects a key $k_T \in K$ which it keeps secret. In order to verify a signature, an entity must share a symmetric key with the TTP.

For key generation, each entity selects a key and transports it secretly with authenticity to the TTP. Each entity A should do the following: (1) Select a random secret key $k_A \in K$. (2) Secretly and by some authentic means, make k_A available to the TTP.

For signature generation and verification, entity A generates signatures using E_{k_A}. Any entity B can verify $A's$ signature with the cooperation of the TTP. To sign a message m, entity A should do the following: (a) A computes $H = h(m)$. (b) A encrypts H with E to get $u = E_{k_A}(H)$. (c) A sends u along with some identification string I_A to the TTP. (d) The TTP computes $E_{k_A}^{-1}(u)$ to get H. (e) The TTP computes $m^* = E_{k_T}(H||I_A||T)$ on TTP's timestamp T and sends m^* to A. (f) $A's$ signature for m is m^*.

$$A \to TTP : u = E_{k_A}(H = h(m)), I_A \tag{5}$$

$$TTP \to A : m^* = E_{k_T}(E_{k_A}^{-1}(u)||I_A||T) = E_{k_T}(H||I_A||T) \tag{6}$$

$$A \to B : r = y_b^{-k}(mod\ p),\ e = (E_{k_T}(H||I_A||T)) \cdot r(mod\ p),$$
$$s = x_a \cdot e + k(mod\ q) \tag{7}$$

For verification, any entity B can verify A's signature m^* on m by doing the following: (a) B computes $v = E_{k_B}(m^*)$. (b) B sends v and some identification string I_B to the TTP. (c) The TTP computes $E_{k_B}^{-1}(v)$ to get m^*. (d) The TTP computes $E_{k_T}^{-1}(m^*)$ to get $H||I_A||T$. (e) The TTP computes $w = E_{k_B}(H||I_A||T)$ and sends w to B. (f) B computes $E_{k_B}^{-1}(w)$ to get $H||I_A||T$. (g) B computes $H' = h(m)$ from m. (h) B accepts the signature if and only if $H' = H[8]$.

$$B \to TTP : v = E_{k_B}(m^*), I_B \tag{8}$$

$$TTP \to B : w = E_{k_B}(E_{k_T}^{-1}(E_{k_B}^{-1}(v))) = E_{k_B}(H||I_A||T) \tag{9}$$

$$Verify\ B : E_{k_B}^{-1}(w) = H||I_A||T, (g^s y_a^{-e})^{x_b} e(mod\ p) = g^{x_b k} m^* r(mod\ p) = m^*,$$
$$H' = h(m),\ if\ ((m^* \equiv w)\&(H' \equiv H))\ accept\ m^*\ else\ reject\ m^* \tag{10}$$

5   Performance Comparison

5.1   Performance Comparison

In this section we give a performance comparison of our new scheme and the two existing schemes. Rivest, Shamir and Tauman introduced ring signatures[6]. Their scheme allows to generate a signature, with a group of potential signers. It merely requires 1 modular exponentiation at the generation, and only 4 modular multiplication to verify the signature. However, the exponentiation is computed with respect to a large RSA modulus, which results in a scheme that is less efficient than the discrete logarithm based scheme.

For this comparison we choose an implementation, setting p = 512 bits and q = 160 bits for the JSI[4], SKM[7] and our scheme. In order to have comparable security, we set the RSA modulus to 512 bits in the RST[6] scheme. For the comparison to be effective, we only counted the number of modular exponentiations.

In Table 1, we indicate the complexity of the these existing schemes as well as our new scheme(new dsg.ver.: designated verifier signature, new abt.ver.: arbitrated verifier signature scheme). We can see that our scheme is much more efficient than the existing scheme for both generation as well as verification. We also compared the size of the respective signatures, assuming that the hash function's output is of size 160 bits.

Table 1. Performance and Size Comparison.

	Gen.	Ver.	Total	Size(bits)	DW App.	Msg Rcv.	Abt.
JSI[4]	1200	1200	2400	2368	×	×	×
RST[6]	768	0	768	1536	×	×	×
SKM[7]	240	480	720	480	○	×	×
New Dsg.Ver.	240	480	720	480	○	○	×
New Abt.Ver.	360	720	1080	720	○	○	○

6 Conclusions

In current knowledge based intelligent network environment, digital watermarks have been proposed as the means for copyright protection of multimedia data. However, the rightful ownerships cannot be resolved from diverse attacks. In this study, we solve watermarking problem as we consider it by the *designated verification with arbitrator* process. Using this scheme, we can propose TTP(arbitrator) based designated verification mechanism when the confirmer of a watermark verify the validity of the watermark owner without revealing the secret of originator's true watermark information in further intelligent network.

References

1. Andre Adelsbach, Ahmad-Reza Sadeghi, "Zero-Knowledge Watermark Detection and Proof of Ownership", 4th International Information Hiding Workshop (IHW '01), LNCS 2137, Springer-Verlag, pp.273-288, 2001.
2. Andre Adelsbach, Birgit Pfitzmann, Ahmad-Reza Sadeghi, "Proving Ownership of Digital Content", 3rd International Information Hiding Workshop (IHW '99), LNCS 1768, Springer-Verlag, pp.117-133, 1999.
3. Scott Craver, Nasir Memon, Boon-Lock Yeo, Minerva M. Yeung, "Resolving Rightful Ownerships with Invisible Watermarking Techniques: Limitations, Attacks, and Implications", IEEE Journal on Selected Areas in Communications vol. 16, No. 4 pp.573-586. 1998.
4. Markus Jakobsson, Kazue Sako, and Russell Impagliazzo, "Designated Verifier Proofs and Their Applications", Advances in Cryptology - EUROCRYPT 96, LNCS 1070, Springer-Verlag, pp.143-154, 1996.
5. David Chaum, "Private signature and proof systems", United States Patent 5,493,614, 1996.
6. Ronald L. Rivest, Adi Shamir, and Yael Tauman, "How to leak a secret", Advances in Cryptology, ASIACRYPT 2001, LNCS 2248, Springer-Verlag, December 2001.
7. Shahrokh Saeednia, Steve Kremer, Olivier Markowitch, "Efficient Designated Verifier Signature Scheme", Technical Report, University Libre de Bruxelles, Belgium, 2001.
8. Alfred J. Menezed, Paul C. van Oorschot, Scott A. Vanstone (ed.), "Handbook of Applied Cryptography", CRC Press, (http://www.cacr.math.uwaterloo.ca/hac), 1996.
9. B. Schneier, "Applied Cryptography", Second Edition, Wiley, 1996.
10. T. ElGamal, "A Public Key Cryptosystem and a Signature Scheme based on Discrete Logarithm", IEEE Transactions on Information Theory, Vol. IT-30, No. 4, pp.469-472, 1985.

Human-Based Annotation of Data-Based Scenario Flow on Scenario Map for Understanding Hepatitis Scenarios

Yukio Ohsawa

Graduate School of Engineering, The University of Tokyo

Abstract. A human's process is proposed for data based scenario understanding, by integrating two existing tools in a new way of annotation. Scenario of a patient's hepatitis progress or recovery was obtained, by the presented process integrating a scenario map and a scenario flow diagram which are obtained from data on the patient's blood tests. The data was first visualized by KeyGaph as a scenario map, showing the rough view of event transitions. The user then discussed looking at the visualization, and wrote scenarios his/her thought of from the map. This text was visualized, again by KeyGraph, which externalizes the relations of events in his thought. On this visual output, the user came to be enabled to pay attention to potential chances existing at the cross points of scenarios. Based on this attention, the user annotated on the scenario flow diagram presented by the Discourse Structure Visualizer (DSV), showing the details of event transitions from the same data. As a result, the obtained annotations enabled hepatic experts to understand useful and novel scenarios underlying the patient's chronological history.

1 Introduction

Recently, data mining studies dedicated to acquiring "useful" knowledge from medical data stored in hospitals. The fruits of these efforts include patterns such as exceptional association rules [Suzuki et al 2003], inducted graphs of causality based on Graph based Induction (GBI) [Yoshida et al 2004], etc. Also, time-series of the changes in the values of blood attributes has been studied [Ho et al 2003, Ohsaki et al 2002]. Mixed presentation of results from various data visualization tools also aided discoveries in medical science [Ho et al 2002, Ohsawa et al 2005].

These methods have been evaluated on objective criteria such as confidence/support, or on comments from medical experts. However, it has not been clarified what a doctor understood when he/she says "this is interesting!" Without validating the understanding we can not trust the treatment based on such a discovery, because the reason for treatment should come from a sound understanding of underlying scenarios of recovery/progress.

Our problem in this paper is to aid a doctor's understanding of scenarios about recovery and progress of hepatitis. We define a scenario as an event sequence sharing a context. If the sequence has a critical change from/to scenario, the change should be included in the understood scenario, e.g., "the patient was treated by interferon, but it did not work. Iron reduction affected after that."

Due to the complexity of human body, the doctor must think about multiple possible scenarios of progress or recovery, from which to choose a desirable one at an appropriate moment. This choice of scenario is the essence of decision making. If the

R. Khosla et al. (Eds.): KES 2005, LNAI 3681, pp. 511–517, 2005.

doctor can understand useful scenarios, he/she can prepare for a forthcoming significant event and will not fail to catch a good timing to choose the way to a good scenario.

For keeping multiple scenarios in front of user's eyes, in the process of chance discovery, a "scenario map" has been shown where the closeness among events are visualized [Ohsawa et al 2005]. By tracing paths between events on this map, user understands there exist scenarios from which he/she can select one for optimizing the situation. In business domains, a scenario map made from data about customer behaviors, has been aiding marketers in finding a scenario leading to sales success [Usui and Ohsawa 2003]. In this paper, we present a method for chance discovery using the combination of tools KeyGraph and IDM for scenario map visualization, in the process of communicative human group to the discovery of chances and useful scenarios in the case of virtual decisions on medical treatment

2 The Data and Tools for Understanding Hepatitis Scenarios

Here let us introduce the target data, and the tools for visualizing scenario maps applied to these data in this paper. The data is from blood tests for 771 hepatitis patients, stored in a hospital from 1982 to 2001. In each blood test, the values of the tested part of the 456 given variables were recorded. We first cleaned these data (deleting meaningless symbols, unifying different attributes of the same meaning, etc), and obtained 71 variables of which the upper and the lower bounds of normal range were predefined, and translated the data into the form as D in Eq.(1) with including names of treatments e.g. "interferon." Here, each item represents an extraordinary value or the change in the value of a variable, and one line in Eq. (1) corresponds to one time of blood test. That is, item "X_Y" means an event where the value of variable X took the value or the change as Y shows in 'H' 'L' '+' or '-'. For example, D-BIL_H (L) means a state where the value of D-BIL, i.e., direct bilirubin, was higher (lower) than its predetermined upper (lower) bound of normal value range. And, D-BIL_+ (-) means the value of D-BIL increased (decreased) for two sequent tests. As a result, 284 unique items came to appear in the data as in Eq.(1). Then, items denoting treatments such as "interferon" were added in the lines of D corresponding to the periods of the treatment.

$$D = \text{CHE\_ - GPT\_H GOT\_H ZTT\_H ...,}$$
$$\text{Interferon ZTT\_H GPT\_H GOT\_H G-GPT\_-...,}$$
$$\text{G-GPT\_- GPT\_- CHE\_+,.} \tag{1}$$
$$\text{... ,}$$

From thus formed data D, a scenario map is created using KeyGraph (for details see [Ohsawa 2003b, Okazaki and Ohsawa 2003]) as follows.

[The Procedure of Making a Scenario Map on KeyGraph]

Step 1. (Show Basic Episodes as Island Clusters): The most frequent $M1$ events in D (e.g., "GPT_H" in Eq.(1)) are depicted with black nodes, and $M2$ pairs of such frequent events co-occurring the most frequently in the same set (i.e., the same line in Eq.(1)) get linked with solid lines. For example, GOT_H and ZTT_H in Eq.(1) are connected with solid lines in Fig.1. Each connected graph obtained here forms a cluster called an *island*, implying a frequent episode which can be a part of possible scenarios.

Step 2. (Show Bridges Between Episodes): Events co-occurring with items in multiple clusters, e.g., "interferon" in Eq.(1), are obtained as *hubs*. A path of links connecting islands via hubs is called a *bridge*. If a hub is rarer than black nodes, it is colored in red (converted to white in the figures of paper). Abridge may suggest a scenario with a decision to jump between islands.

A scenario map as in Fig. 1 is then obtained. A doctor understands that the large cluster at the bottom with black nodes and solid lines means an event-set in frequent episodes of progress, and the small part in the right corresponds to recovery. The dotted path bridging between these two parts may be regarded as a process of treatment, so the user can infer the underlying scenario if he/she had sufficient knowledge of hepatitis..

On the other hand, a doctor must understand the scenario of progress/recovery of one patient, in order to treat the patient. However, the data of one patient is too small to learn about the petinet's history. In order to approach this problem in this paper, we combine KeyGraph ,and DSV (Discourse Structure Visualizer) which takes advantage of the influence flows from events to events, on the influence diffusion model ([Matsumura et al 2002]) mentioned below.

[The Procedure of DSV (See Figure 3, Ignoring Annotations in Bold Letters and Lines)]

Step 1. Given a sequence of lines (as in Eq.(1)) $S1$, $S2$,..., Sm, obtain segments $X1$, $X2$,...Xn as follows: Set k to 1, and

For all j ($1<j<m+1$), for $i = j-1$

if sim(Si, Sj)> θ, $Xk = Si + Sj$ (combine the sentences),

Otherwise, increment k;

Here, sim(Si, Sj) stands for the similarity of Si and Sj, defined by the cosine of the word vectors of Si and Sj.

Step 2. Compute the response strength from each segment Xb to a preceding one Xa ($a < b$), as |common(Xa, Xb) | / |Xb|, where |X| denotes the number of key words (selected number of words with highest *tfidf* values [Salton and Buckley 1988]) in segment X and common(X, Y) means the set of words common to X and Y.

Step 3. Depict each segment as a node, and a response stronger than a predefined threshold as an arrow.

DSV has been applied so far for analyzing influence between messages in communication, regarding each message as a segment in the procedure above. There, the influence of each message to the communication was visually understood even if the communication had a few messages, i.e., for small data. And, DSV shows a word's various meanings by visualizing in different messages. Thus, DSV is expected to solve the two problems above of KeyGraph. In this paper, we apply DSV to blood-test data of individual patients, regarding the state of human body at a time as a response to previous states. Although DSV is potentially capable of visualizing the flow of scenarios, but a method for suitably focusing on meaningful items in the output with many items per segment is desired. Hence, we introduce a process for combining the scenario map on KeyGraph, the scenario flow on DSV, and, as we mentioned about chance discovery in Section 1, human's feeling about important events in the real world.

3 Scenario Understanding with Annotating the DSV Output on the Result of KeyGraph

Ohsawa proposed the process of chance discovery, starting from obtaining the target data (*object data* in [Ohsawa 2003a], but we call *external data* for readability in this paper), based on user's concern with a chance. Then, the data is visualized in a map of items. User(s) writes down his/her thoughts about scenarios readable from the map, as *internal data* (*subject data* in [Ohsawa 2003a], but here we call this day). Again, this data is visualized by a tool. By looking at the output, user can understand the most meaningful scenario to him/herself, and find chances. He/she makes an action on the scenario, acquires a new concern with a chance from the real world, which enables to run the process again to deepen the understanding.

In this paper, the users were the team formed by doctors and the person using the computer, with the concern about scenarios of each patient. This concern has been already reflected to our dealing with blood test data of each patient. The team is now at the stage of understanding scenarios with referring to the visual output of tools applied to this data. As we already mentioned in previous section, KeyGraph and DSV which have been applied to chance discovery has some defects for the current data, and they should compensate for each other's capability. However, this is not a fatal problem if we apply the robustness of chance discovery process. That is, the user is allowed to explain scenarios ambiguously, i.e., write/tell multiple possibilities of scenarios, and find new ideas occurring to his/her own mind by looking at the visualized thoughts of him/herself. The ambiguity of understanding is rather helpful for obtaining rich internal data coming from rich experience in the real world. The process below is a method for reflecting of awareness of one's own thought to the completion scenario understanding.

[The Process to Annotate DSV on the Result of KeyGraph]

Step 1. Apply KeyGraph to the blood test data of the target patient (i.e., in stead of a collection of patients). From the output, the user writes down scenarios found. Because the data is small, arrows on links not attached.

Step 2. Apply KeyGraph to the written document. The user looks at the output, and becomes aware of his/her own concern with some states of patient's blood.

Step 3. The user annotates on the output of DSV visualizing the flow of the changes in values of blood attributes. First, he/she chooses interesting items in the right half of DSV as in Figure 3, and then reflects the annotations in each segment in the right to the corresponding node in the left.

4 A Case Study of the Proposed Process

The user in this section had obtained generous scenarios of patients treated with interferon, but the requirement of the doctor urged to shift the concern to understanding the scenario of an individual patient for the purpose of treatment. The external data dealt with was on a patient of hepatitis C. The result of KeyGraph was as shown in Fig.1. The figure was hard to understand although the structure looked simple, because one cluster has opposite events, e.g. {GOT_- and GOT_+}, {GTP_- and GTP_+}, {FE_+ and FE_-} etc.

The thought in looking at the graph in Fig.1 was recorded in a text, as appearing in the left-bottom window of Fig.1. This text was then visualized as in Fig.2 with Key-Graph. The word "interferon" appeared as an important concern of the user, even though the word did not appear in the graph in Fig.1. From Fig.2, we find user's doubt how interferon really affected to the patient. He guessed this doubt came from the appearing side-effects in Fig.1, such as LDH_+. In addition, his concerns with "globurin" "LDH" and "CHE" known as enzymes relevant to destruction of liver cells, were externalized. Reflecting the concerns clarified so far to the annotations on DSV, we find a scenario flow in Fig.3. This means, according to the doctor, "This patient was first treated with interferon and iron-reduction. This worked, as the decrease in globurin shows, but strong side effects lasted, and the recovery effect stopped once. Iron reduction was applied again, and the recovery came back. I suggest (1) to check if the virus disappeared, (2) not to use interferon more, and to (3) keep observing the iron quantity and consider to do iron reduction if iron is increasing."

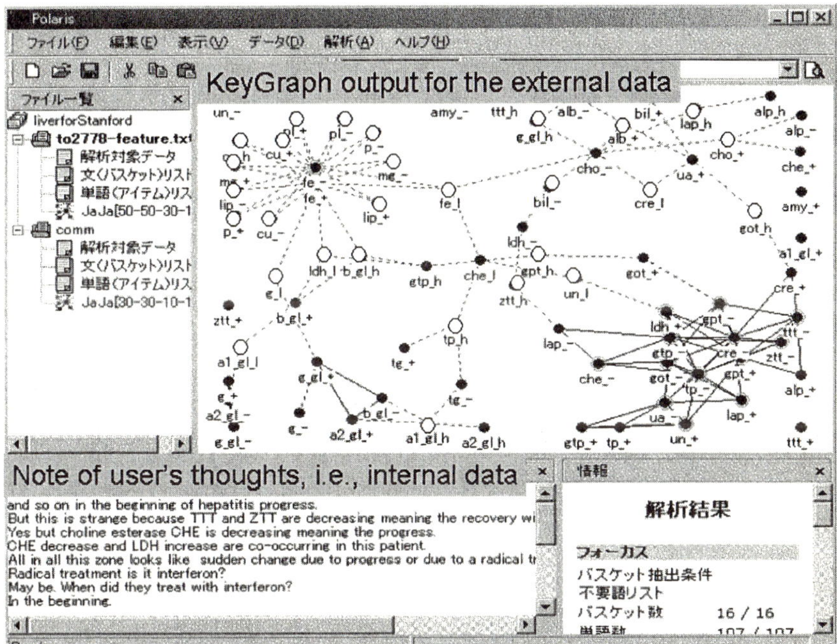

Fig. 1. The KeyGraph applied to a patient's blood test data. Polaris is this system's name on which user writes scenarios he/she thought, and the notes here are dealt as internal data

Note that we do not claim KeyGraph and DSV are the best combination for showing a scenario map and a scenario flow diagram. Although we assert a visualization with islands and bridges of events is required for a scenario map, we do not show the comparison of KeyGraph with other 2D visualization tools such as the corresponding analysis [Greenacre 1993], other co-occurrence graphs, or 3D visualization tools. And, although we assert DSV obtains scenarios even from small data, we do not prove this from the comparison of DSV and Bayesian networks. These are because our claim point is that understanding of scenarios can be enabled by the integration of a visualized scenario map and a scenario flow diagram, each of which may have some defects, following the process presented in Section 3.

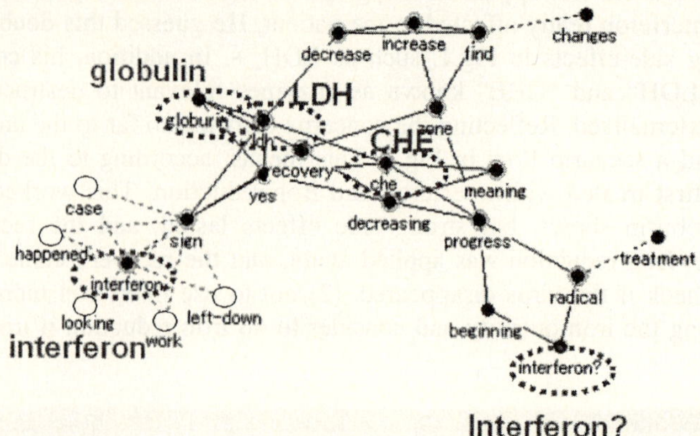

Fig. 2. The KeyGraph for the internal data from the thought in looking at Fig.1

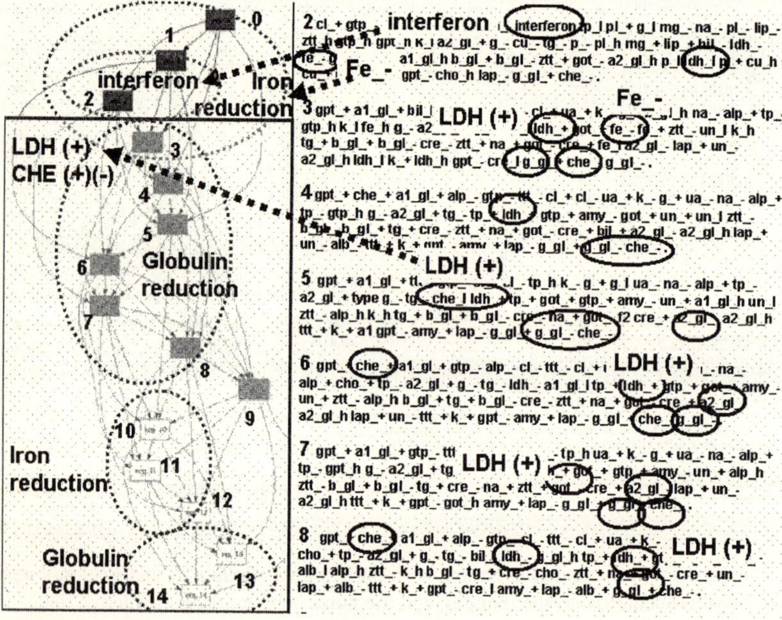

Fig. 3. The annotated scenario flow. The right hand list was annotated first, and this was reflected to the annotations in the left, considering the correspondence of segments between the right and the left. All annotated items were taken from figure 1 and figure 2

5 Conclusions

The chronological scenarios of individual patients of hepatitis are understood, from the presented process of chance discovery. Here, the internal data is obtained from the thoughts of non-experts who learned premature knowledge about hepatitis related blood attributes and overviewed the output KeyGraph applied to blood test data of individual hepatitis patients. This internal data was fed back to the awareness of the user him/herself about the significance of certain set of blood attributes. The patient's data was in parallel visualized using DSV, a visualizing tool of scenario flows, and the awareness above of the user was used to annotate on the output of DSV. This led to useful understanding of the patient's scenarios underlying the data.

References

[Greenacre 1993] Greenacre, M.J., *Correspondence Analysis in Practice*, London, Academic Press, 1993.

[Ho et al 2002] Ho T.B. et al, Visualization support for a user-centered KDD process, Proceedings of the Eighth ACM SIGKDD international conference on Knowledge discovery and data mining (KDD02), 519-524, 2002.

[Ho et al 2003] Ho T.B. et al, Mining Hepatitis Data with Temporal Abstraction, *Proc. The Ninth ACM Int'l Conf. on Knowledge Discovery and Data Mining* (KDD03), 519-524, 2003.

[Igata et al 2003] N.Igata, H.Tsuda, Y.Katayama, F.Kozakura: "Semantic groupware and its application to KnowWho using RDF", ISWC 2003 (2nd Intl. Semantic Web Conference) Poster, October 20th-23rd, 2003, Florida, USA

[Matsumura et al 2002] Matsumura, N., Ohsawa, Y., and Ishizuka, M., Influence Diffusion Model in Text-Based Communication, *Poster, The Eleventh Conf. World Wide Web*, 2002.

[Ohsawa and Usui 2005] Ohsawa, Y. and Usui, M. "Chance Discovery in Textile Market on Scenario Communications with Touchable KeyGraph, *Readings in Chance Discovery*, 2005.

[Ohsaki et al 2002] Ohsaki, M., et al, A Rule Discovery Support System for Sequential Medical Data – In the Case Study of a Chronic Hepatitis Dataset -, *International Workshop on Active Mining in IEES Int'l Conf. Data Mining*, 97-102, 2002.

[Okazaki and Ohsawa 2003] Okazaki, N. and Ohsawa, Y., Polaris: An Integrated Data Miner for Chance Discovery, In *Proceedings of The Third International Workshop on Chance Discovery and Its Management*, Crete, Greece, 2003.

[Salton and Buckley 1988]Salton, G. and Buckley, C. Term weighting approaches in automatic text retrieval., *Information Processing and Management*, 24(5):513-523, 1988.

[Suzuki et al 2003b] Suzuki,E., Watanabe, T., Yokoi, H., and Takabayashi, K.: Detecting Interesting Exceptions from Medical Test Data with Visual Summarization, *Proc. Third IEEE International Conference on Data Mining* (ICDM), 315-322, 2003.

[Ohsawa 2003a] Ohsawa, Y., Modeling the Process of Chance Discovery, in Ohsawa Y and McBurney P. eds, 2003, *Chance Discovery*, Springer Verlag: 2-15, 2003.

[Ohsawa 2003b] Ohsawa, Y., KeyGraph: Visualized Structure Among Event Clusters, in Ohsawa Y and McBurney P. eds., *Chance Discovery*, Springer Verlag, 262-275, 2003.

[Ohsawa et al 2005] Ohsawa, Y., et al, Mining Scenarios for Hepatitis B and C, in Paton, R. (ed), *Multidisciplinary Approaches to Theory in Medicine*, 2005.

[Yoshida et al 2004] Yoshida, T., et al, Preliminary Analysis of Interferon Therapy by Graph-Based Induction, in *Proc. International Workshop on Active Mining, Japanese Soc. AI*, 2004.

A Scenario Elicitation Method in Cooperation with Requirements Engineering and Chance Discovery

Noriyuki Kushiro and Yukio Ohsawa

Graduate School of Business Science, University of Tsukuba,
3-29-1 Otsuka Bunkyo Tokyo, Japan
{kushiro,ohsawa}@gssm.otsuka.tsukuba.ac.jp
http://www.gssm.otsuka.tsukuba.ac.jp

Abstract. The effective scenario elicitation method has been established by combining the chance discovery with the requirements engineering. The method consists of the multi-dimensional hearing and the hierarchical integration process newly combined on the double-helical model of chance discovery. The capability of the method to elicit a useful scenario has been confirmed through a theoretical analysis with a visualization tool named KeyGraph and an experiment of eliciting scenario on the method by using a prototype of an operation panel for electric appliances controller used in a home.

1   Introduction

Defining what stakeholders really ask for, that is called a scenario elicitation in this paper, is a very important activity for concept making of a new product and analyzing system requirements. To get useful scenarios is a way to lead product designer and system analysts to discovering a chance. We will use the term "stakeholders" to refer to all the people concerned with a product and a system, such as end users, customers, suppliers, and manufacturers. The term "scenario" is defined as a story explaining how a system or products satisfies several requirements among stakeholders. The scenario elicitation can be replaced by the requirements definition in the domain of requirements engineering. The requirements definition process is composed of the following three steps [1].

1. Requirements elicitation step: to elicit requirements from stakeholders.
2. Requirements analysis step: to analyze the requirements and develop solutions.
3. Requirements specified step: to integrate solutions into system specifications.

The requirements engineering has been focusing on the requirements analysis and specified steps and numerous methods for these phases have been established [2][3]. However, only a few studies have been made so far on the method for the elicitation step. Our aim is to establish a systematic method for the elicitation and analysis of requirements, with a multi-dimensional hearing and a hierarchical integration process, newly combined on the double-helical model of chance discovery [4].

2   Target Issues and Solutions

We begin from considering issues on scenario elicitation for a system development.
(1) Stakeholders don't always know requirements: Stakeholders often doses not understand what they really want. If they understood, they can only explain these requirements in abstract words.

R. Khosla et al. (Eds.): KES 2005, LNAI 3681, pp. 518–525, 2005.

(2) Analysts often misunderstand the requirements: Analysts cannot understand requirements because the requirements explained by the stakeholders are too vague.

We aim to establish a new scenario elicitation method to solve the above issues by combining the chance discovery and the requirements engineering methods. Table 1 summarizes the basic approaches of this study.

Table 1. Basic approaches

Issues	Solutions
How to get requirements	Multi-dimensional hearing using a prototype based on the requirement engineering
How to understand requirements	Hierarchical integration process based on the double-helical model of chance discovery
How to integrate requirements	

2.1 The Multi-dimensional Hearing Method

Analysts cannot use these raw demands elicited from stakeholders for making scenario [5]. Because raw demands reflect stakeholders' tacit knowledge such as premises and restrictions in the real environment usage of the system and the knowledge is not explained explicitly in raw demands. Approaches to derive such tacit knowledge are important to generate a scenario. There are some techniques for deriving tacit knowledge as follows:

- Laddering technique in the marketing research [6]
  Laddering is a technique to grasp demands by repeating "why" questions.
- Claim analysis in the requirements engineering [7]
  Claim Analysis is a technique for clarifying design premises by asking both positive and negative reason for the proposed design.

From the view of the object driven approach [8], a requirement consists of 3 layers, purpose, object and method [1]. Each requirement layer has premises and restriction. In this study, the claim analysis is reinforces to elicit a useful scenario. An extra dimension of knowledge corresponding to the purpose, the object and the method of requirement, is added to the positive/negative dimension of the original claims. The hearing method is called hereof the multi-dimensional hearing method and requirement elicited by the multi-dimensional hearing is called a requirement primitive.

A Scheme of multi-dimensional hearing method is shown in Fig.1.

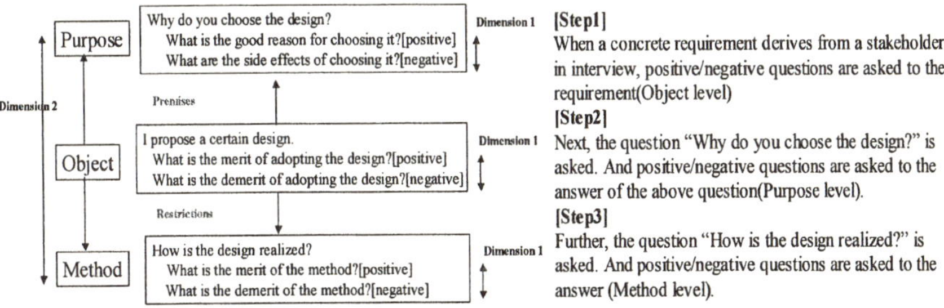

Fig. 1. Scheme of multi-dimensional hearing method

2.2   The Hierarchical Integration Process on the Double-Helical Model

We propose a requirement integration process for analyzing and integrating the various requirements primitives, on the double-helical model of chance discovery. The double helical model is a systematic approach to put stress on interaction among stakeholders on a cyclic process. On the other hand, the inspection test [9] on a prototype is an efficient way to elicit a useful scenario in the early stage of the system development[10].

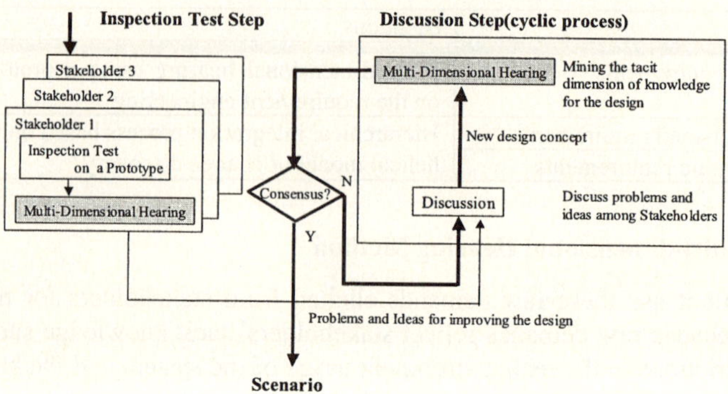

Fig. 2. The hierarchical integration process

We design an integration process composed of the inspection test at the front end of the process and the cyclic discussion step on the double-helical model (Fig.2). We call the process "hierarchical integration process". The following is the summary list of the hierarchical integration process.

[Inspection Test Step]
Step 1: Prototype of the design was shown to stakeholders. Moderator explains the concept and the features of the design. Each stakeholder writes requirements for the prototype on the paper.

Step 2: The Stakeholders express her/his requirements for the design as concrete as possible. The moderator interviews the stakeholders by the multi-dimensional hearing.

[Discussion Step]
Step 3: The stakeholders discuss about the problems and propose improvements of the prototype.

Step 4: The moderator interviews the stakeholders about the proposed improvements by the multi-dimensional hearing and go to Step3 until the consensus of improvements obtained.

3   How Primitives and Scenario Are Derived on the Process?

We explain how a requirement primitive and scenario are elicited through the hierarchical integration process by using a visualization tool named Keygraph[11]. KeyGraph is a tool for visualizing the map of word relations in requirements.

First, we explain how a requirement primitive is derived from the multi-dimensional hearing. The elicitation of primitive is monitored with KeyGraph by using an example that basic requirement "We want to utilize video information for a presentation".

[The Contents Summary of Multi-dimensional Hearing]

Object level: {We utilize video information for a presentation.}
  Positive: Video information appeals to user's impression directly.
  Negative: It is difficult to search and allows us only a sequential search.
Purpose level: {Why do you use video information for a presentation?}
  Positive: Video information includes visual information abundantly. Visual information has direct influence on drawing user's empathy.
  Negative: Video information is very difficult to search directly, because of the difficulty of encoding the information for indexing
Method level: {We use a fast forward function for retrieving video information.}
  Positive: We can estimate video clips quickly by the fast forward function.
  Negative: Intelligibility of sound and reality of time are lost.

The requirement primitive elicited by the multi-dimensional hearing are visualized by KeyGraph(Fig.3). The primitive consists of 3 layers of requirement (basic object, purpose and implementation method) and their premises and restrictions of the implementation. The shape of the primitive visualized by KeyGraph, shows a twin poles structure. The basic object of the primitive is placed at the center of the twin poles, one pole explains the purpose of the design, and the other pole shows the implementation methods. The bridge between the purpose and the basic concept means premises or design rationale. The other bridge is restrictions for the proposed implementation method.

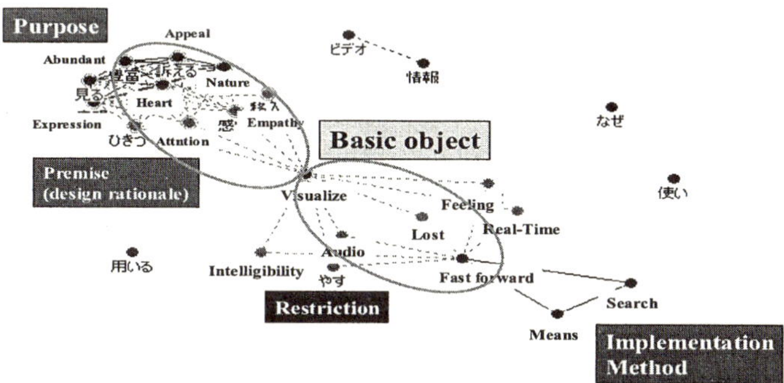

Fig. 3. The primitive derived the from multi-dimensional hearing

Let us consider the growth of primitives into scenario through the hierarchical integration process. Many Primitives are elicited in the inspection test step. Each primitive grows with respect to the correlation of primitives in the discussion step. When some primitives have common purpose, these primitives are connected by the common purpose. As result, more sophisticated premises or design rationales are obtained, where a combination of objects (object 1 and 2 in Fig4 (a)) is expected to fulfill a common purpose as in Fig.4 (a). When some primitives are in a hierarchical relationship, i.e. if the implementation in one primitive becomes the purpose of another primitive, the primitives are chained as in Fig.4 (b).

As mentioned above, by the repetitions of the multi-dimensional hearing in the hierarchical integration process, the primitives grow through these intersections and chains. Finally, the primitives are formed into a scenario, which explains how a system or products satisfies several requirements among stakeholders (Fig4(C)).

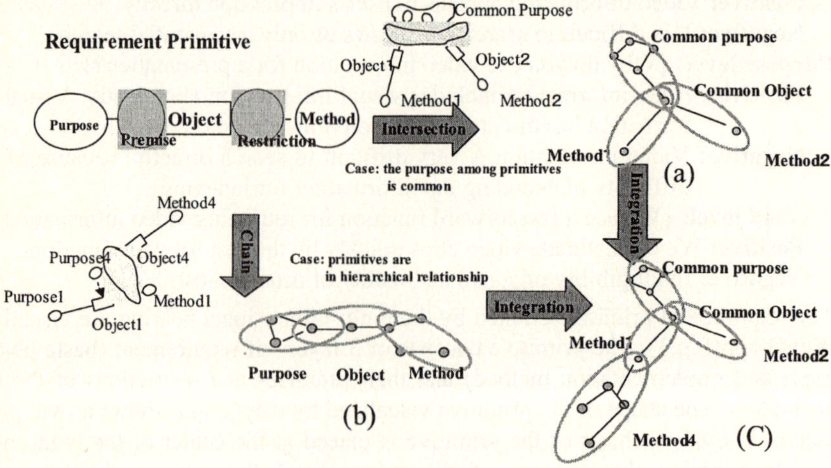

Fig. 4. The growth of scenario by intersections and crossing of primitives

4   The Experiment of the Elicitation of Scenario

An experiment of the elicitation of scenario was conducted on the proposed methods by using a prototype of home controller for household appliances.

4.1   Outline of the Experiment

Basic settings of the experiment are shown in Table 2. The experiment was conducted in the procedure shown in Chapter 2.2.

Table 2. Basic settings of the experiment

Items	Details
Stakeholders	Representatives of user(3), A representative designer(1), Human Interface design specialists(2), Moderator(1)
Task	4 tasks for operating of home appliances(ex. water heater) Task1: Set the temporary temperature of water Task2: Set schedule of machine operation Task3: Set the default temperature of water Task4: Operation when an error occurred
Examination Process of each group	Case1: Conventional inspection test (CIF)process Case2: CIF + the multi-dimensional hearing Case3: Proposed process in the paper

4.2 The Results of the Experiment

The results of the experiment are shown in Table 3. Scenario was elicited from every task trough the hierarchical integration process.

Table 3. The results of experiment

Items	Case1	Case2	Case3
Task Execution time(minutes)	9.0	16.6	31.0
Number of acquired primitives (for each task)	3.4	10.4	32.8
Number of elicited scenarios (for 4 tasks)	0	0	4

5 Evaluation

The proposed method was evaluated by two criteria: the efficiency of eliciting primitive and the contribution to constructing scenario.

(1) The efficiency of eliciting primitives

As the requirement elicitation method, it is a important to elicit request primitives efficiently in a limited time. The results are shown in Table 4. The proposed method exceeded the conventional inspection test in efficiency of eliciting primitives.

(2) Contribution to constructing scenario

In order to evaluate contribution of the multi-dimensional hearing and the hierarchical integration process toward constructing scenarios, the procedure of the hierarchical process for a scenarios is illustrated in Fig.5 and the growth of scenario also visualized in Fig.6.

Table 4. The efficiency of eliciting a primitive

Items	Group1	Group2	Group3
Time for eliciting a primitive (minutes)	2.6	1.53	0.96

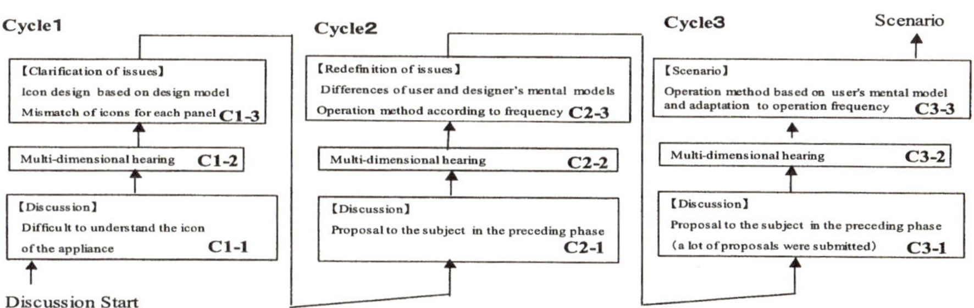

Fig. 5. The effect of the multi-dimensional hearing for rotating cycle of discussion

3 cycles of the discussion step were repeated. The multi-dimensional hearing was a driving force to rotate the cyclic process in Fig5. The following is the summary list step by step of each cycle in Fig5.

- **Basic Proposal:** A basic proposal that some icons on the panel should be changed to make it easier to operate (Fig.6 (a)).
- **From the Basic Proposal to Cycle 1:** The cause and the solution of the difficulty of operation were derived from the multi-dimensional hearing. Two primitives in hierarchical relationship were chained. The result of KeyGraph shows a twin poles structure (Fig.6 (b)).
- **From the Basic Proposal to Cycle 2:** Another cause was added to the first discussion. Two primitives newly derived from the hearing were chained by interdependency of the implementation method (Fig.6 (c)).
- **From the Basic Proposal to Cycle 3:** Two primitives derived from the multi-dimensional hearing were crossed and chained with the common purpose and the common implementation method (Fig.6 (d)).

Consequently, the scenario was formed by the intersections and chains of the primitives derived through the hierarchical integration process. When scenario was integrated, the answer to the multi-dimensional hearing was expressed recursively by the terms appeared in the former arguments.

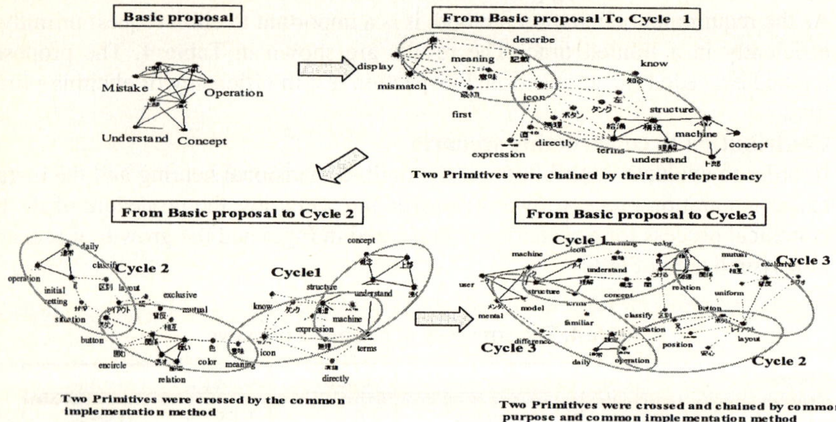

Fig. 6. Growth of scenario 2 visualized by KeyGraph

6 Conclusion

The efficient scenario elicitation method has been established by combining the chance discovery and the requirements engineering methods. The method consists of the multi-dimensional hearing and the hierarchical integration process. We evaluated its capability through the experiment using a proto type of home controller. The following have been confirmed: requirement primitives which consist of its purpose, object, method, premises and restrictions are elicited by the multi-dimensional hearing; the primitives grow by intersections and chains and are formed into a scenario through the hierarchical integration process; the proposed method exceeded the conventional inspection test in the efficiency of eliciting primitives.

We hope that the method will be widely used and open the way to lead us to discovering a chance in requirements analysis.

References

1. P.Loucopoulos and V.Karakostas, System Requirements Engineering, McGraw-Hill Book Company, 1995.
2. A.I.Anton, Goal-Based Requirements Analysis, ICRE, pp.136-144, 1996
3. E.S.K.Yu, Towards modeling and Reasoning Support for Early-Phase Requirement Engineering, ICRE, pp.194-203, 1997
4. Y.Ohsawa, Modeling the Process of Chance Discovery, in Y. Ohsawa and P.McBurney. Eds., Chance Discovery, Springer Verlag, pp.2-15 2003.
5. S. Robertson and J. Robertson, Mastering the Requirements Process, ACM press, 1999
6. C. Corbridge, G. Rugg, N. P. Major, N. R.Shadbolt and A. M.Burton, Laddering: technique and tool use in knowledge acquisition, Knowledge. Acquisition 6-3, pp.315-341, 1994.
7. J. M.Carrol, Making Use Scenario-based design of Human-computer interactions, The MIT press, 2000.
8. J.A. Bubenko and B.Wangler, Objectives driven capture of business rules and information systems requirements, IEEE Conference on SMC, 1993.
9. J.Nielsen and R.L.Mack, Usability Inspection Methods, John Wiley & Sons, 1994.
10. K.E. Wiegers, Software Requirements: Practical techniques for gathering and managing requirement through the product development cycle, Microsoft press, 2003.
11. Y.Ohsawa, KeyGraph: Visualized Structure Among Event Clusters, in Y. Ohsawa and P. McBurney. Eds., Chance Discovery, Springer Verlag, pp.262-275 2003.

The Interactive Evolutionary Computation Based on the Social Distance to Extend the KeyGraph

Mu-Hua Lin[1,2], Hsiao-Fang Yang[1,2], and Chao-Fu Hong[1,2,*]

[1] Department of Information Management, Altheia University
32 Chen-Li Street, Tamsui, Taipei, Tanwan 251
{fa925710,cfhong}@email.au.edu.tw
[2] Graduate School of Management Sciences, Altheia University
32 Chen-Li Street, Tamsui, Taipei, Tanwan 251
jimmy52@ms35.hinet.net

Abstract. In this study we combined the Kotler and Trias De Bes (2003) at Lateral Marketing had defined "what is the creativity" with the watt's Social Affiliation graph to extend the KeyGraph and discovered the chance (creative probability) or decreased the length of searching path. In our model, firstly we based on the KeyGraph chose the important keyterms as the selection in IEC. Secondly, the recombining mechanism as the crossover in IEC, it according to the merging probability extends the KeyGraph. And the mutation was as long distance changing mechanism in our Affiliation graph IEC (AGIEC) model. Finally we applied this model for cell phone design and from the interactive data found that the choosing and recombining mechanism as we expected the Key-Graph was based on the social distance extent to the preferable components and brought the effectively creative product. And the long distance mutating mechanism worked as lateral transmitting phenomenon and it could bring the spurring for the designer.

Keywords: KeyGraph, chance discovery, small world, interactive evolutionary computation (IEC), social affiliation network

1 Introduction

In recent years what the creativity was the one of hot researching topics, and the Kotler and Trias De Bes at 2003 in their lateral marketing tried to defined it. When the new purpose was born or the environment was changed, there will let new need was generated in customer's mind, then the product would be modified to suit it. It meant the lateral marketing was not only useful in new marketing but also suited to make a good product for consumer. Unfortunately, the Kotler and Trias De Bes only propose the conceptual model, did not supply the operating model. Here we believed only the dynamic and interactive evolutionary computational (IEC) model had a chance to merge the lateral transmitting into vertical transmitting and to supply the creative operation. Of course, the IEC also had many successful cases; there only relied on the designers' subjective value decided the evolving direction to solve their problems, but still had a fatiguing problem.

The other method was Ohsawa at 1998 combined the text mining with the graph theory to visualize the data and this method: at first the noise words were filtered out,

---

* Associate Proffessor

R. Khosla et al. (Eds.): KES 2005, LNAI 3681, pp. 526–532, 2005.

second analyzed the word's appearing frequency and the pair relation between words (Ohsawa, Benson, and Yachida, 1998; Ohsawa, 1999). Then the line was used to link these pair words. They found many clusters on the graph and discovered some words as the bridge connected the clusters. But the most important thing was some words appearing times were not very high and these words became the important keyterms (Llorà, Goldberg, Ohsawa, Ohnishi, Tamura, Washida, and Yoshikawa, 2004a; Llorà, Matsumura, Goldberg, Ohsawa, Ohnishi, and Gonzales, 2004b). Because these words was as the rare event and bridged two clusters. It meant these keyterms were defined as the chance to help the people cross over one cluster to the other cluster. And now they recombined this method into double helical process to discover the chance from business data (Ohsawa, and McBurney, 2003). The double helix: one helix was the human's decision process and the other helix was the computational data mining. When the human's subjective accessed the objective data to discover something new, in this work the KeyGraph was used to visualize the data, defined the clusters, bridge and important keyterm. Then the specialist could accord his experiments find the possibility for the new concerns or to evaluate there chances. This method also had many successful cases. But the problem is before analyzing they have to collect enough objective data, this is a big problem for personal design.

At IEC we can collect 6 data, and if we only use these data describe a meaningful KeyGraph is impossible. Therefore we believe that the people search on limited environment like in IEC to find his preferable product is a very difficult thing. But from Milgram's experiment we also can observe a few participants can successfully send the message to the targeting person and these participants can contact the number of persons as same as the designer in IEC environment (Milgram, 1992). It means the IEC may have a chance, only passes through about 5-7 times interactive processes and arrives the goal. But the problem is how to increase the successful probability, here we try from the result of Kleinberg's experiment; the man who has many social networks, and he relies on these networks to decide which person can go ahead and send the message to the goal. These social networks are as the geographic network and working network, these networks distance is useful to analyze the opportunity of chance (Kleinberg, 2000a; Kleinberg, 2000b; Wasserman, and Faust, 1994; Watts, 2003; Watts, Dodds, and Newman, 2002). Here we try to integrate the concept of distance with KeyGraph to find the chance (short-cut) for our AGIEC and help designer quickly find his preferable product.

2   Methodology

2.1   The Chromosome Design

Before we designed the chromosome, we had collected all cell phone from the market, and according to these samples designed the components of cell phone. The cell phone's chromosome was shown on table 1.

Table 1. The cell phone's chromosome

faceplate				handset				screen				function key				number key			
1	2	...	64	1	2	...	16	1	2	...	16	1	2	...	16	1	2	...	16

And we integrated the same characteristic genes in same allele's cluster under the attribute and let the designer had a chance depends on his subjective value (social distance) to choose the preferable cluster.

2.2   Based on the Social Distance to Extend the KeyGraph

Because the KeyGraph is according to the word's appearing frequency to decide its weight, and the line links the pair words to construct whole knowledge as a graph. From the KeyGraph we can find some important keyterms which words does not have high connecting times with cluster, but there as the bridge connect two clusters. This phenomenon means that the customer does not usually change his behavior, but these important keyterms have a chance to induce the customer changes his behavior's pattern. Therefore in environment of IEC, the population size is 6, and only 6 data can be collected. Of course these data is not enough to create a meaningful Key-Graph let the customer understand what is he want or find the chance. In addition, the Kleinberg's γ models really want to evidence how to find the short-cut path in the small world. Then he finds that if we want to connect any two points in the network, the connecting probability will decrease when the distance between these two points is increasing. But if we can arrange the data distribution at any distance has the same appearing probability, then on this environment we can easily find the short cut path and have a high effective searching. The other is that the man he is not usually measures the distance between any two individual persons, but he always relies on his social hierarchy to measure the distance between two individual clusters as A, B in Fig. 1. And this kind relationship map is as the apartment relationship or working relationship and the distance between cluster A and cluster B is 3 in Fig.1. In Watts's model (Watts, 2003), the multi-social distance was relied on the multi-social hierarchy and the minimum social distance would domain the searching direction. In his experiment only about 5-7 searching time the participant could arrive the goal. But for IEC it hardly create the whole hierarchy structure, here we implement the concept of partial KeyGraph to calculate the social distance and according to these distances extend the KeyGraph.

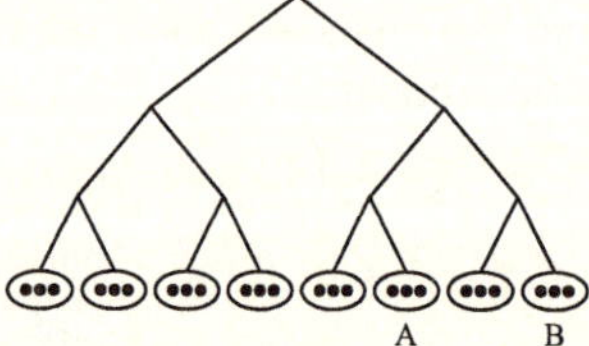

Fig. 1. One dimensional Social Affiliation Networks (Watts, 2003)

2.3   The KeyGraph Based Choosing Mechanism (Selection)

Because the AGIEC model's choosing mechanism relies on the designer's subjective value to estimates the product. The Fig.2 is a part (faceplate) of cell phone, firstly the system collect every individual product's score and calculates every cluster's score by Eq.1 (Fig.2a), this process as the KeyGraph pick up the keyterms and links pair words to create the graph as shown on Fig.2b. According to the cluster's score chooses the elitism cluster, here the a_{1j} is the C cluster and it also is an important keyterm. Finally we can build a social affiliation graph on every generation, but in this generation the KeyGraph is shown on Fig.2c. Of course when all attribute is calculated, we can find all best clusters for every attribute and built 5 affiliation graphs.

$$a_{1j} = (\frac{\sum_{k=1}^{g} SC_{jk(j=1..c)}}{\sum_{i=1}^{n} t_{ij(j=1..c)}}) \tag{1}$$

The n is the population size, the c is the number of cluster under the attribute, and the t is the cluster's presenting times on AGIEC and SC is the product's score.

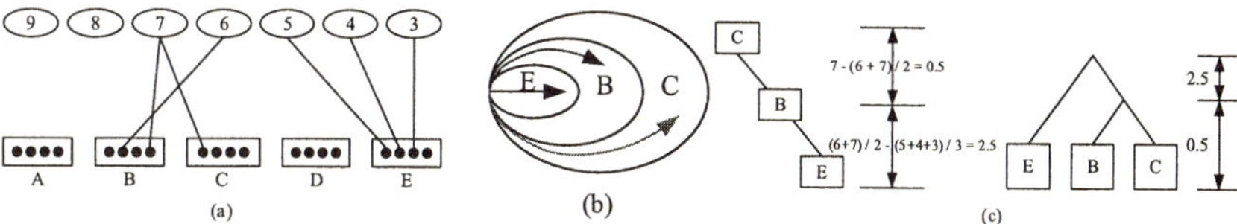

Fig. 2. (a) the relationship between the designer's subjective score (b) according to Eq.1 calculate the a_{1j} and find the cluster C (c) the a_{1j}'s social affiliation network

Even in this study we divide the cell phone into 5 parts, but for the man if the number of social dimension is larger than three, it will be too complexity for him to search his want product. That is why in this study we only consider the first and secondary dimension in our model.

$$a_j = f(a_{1j}, a_{2j}) \tag{2}$$

2.4 Using Recombination Mechanism Emerge the Chance to Extend the KeyGraph (Crossover)

The social distance in this study we use the social affiliation graph to define it as shown on Fig.3. On the more important two attributes we choose the best pair cluster like 1-1 and its distance is 1. If the first attribute is secondary cluster and the second attribute is the best cluster is the pair cluster 2-1 and its distance is 2, the other is shown on Fig.3.

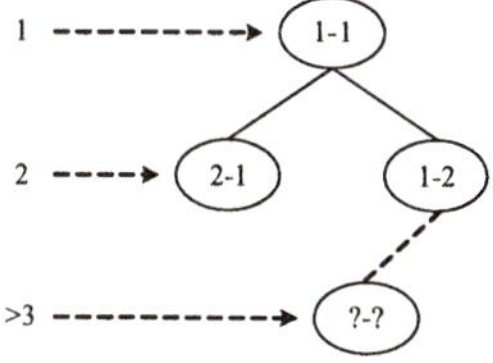

Fig. 3. The affiliation graph

Passed through the choosing mechanism we can find the strong tie clusters, and after the possibility analysis or analyzing the social distance, we find the best cluster.

Therefore in the recombination mechanism: first chance is we let non-present gene in best cluster has a high emerging possibility to recombine each other. The second chance belongs the social distance is 2 like the pair cluster 2-1 and pair cluster 1-2 emerges each other. Finally, the crossover is as the above choosing mechanism, selects the prefer a_j and connect with the $set(b_{ijk})$ as Eq. 3. The result of crossover is shown on Fig.4.

$$AGIEC_{crossover} = a_j + set(b_{ijk}) \tag{3}$$

And on next generation we want the system can supply the same numbers of the product in every social distance, let our model suits the Kleinberg's condition. As in Fig.3 the ?-? is a long distance pair cluster, it also has the same emerging possibility in our system. We induct this process is the mutation as in IEC.

2.5 Using Mutation Mechanism Emerge the Chance to Extend the KeyGraph (Mutation)

Because the long distance relationship is as the short cut path in small world, and from the theory of chance discovery, they define the rare event has a high possibility to be a chance. The specialist will focus on there and understand why there is a chance. Here we extend this concept to non-presenting cluster. As in Fig.2a we can find the clusters A and D do not present on first generation, we define these clusters (a_{new}) as the chance for the next generation. This also means the long social distance clusters is a chance to stimulate the designer and expands his designing space to these cluster.

$$AGIEC_{mutation} = a_{new} + set(b_{ijk}) \tag{4}$$

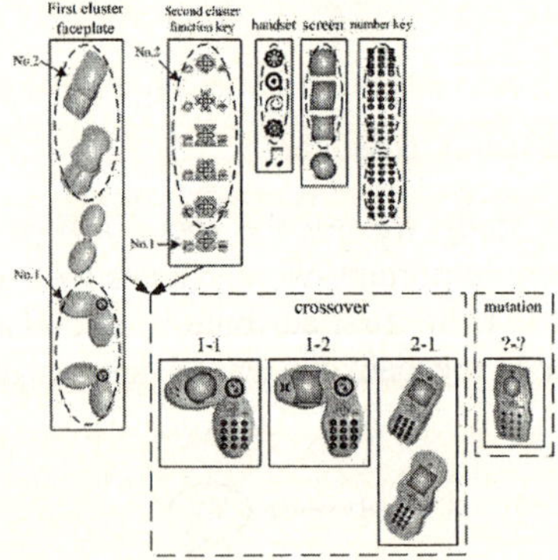

Fig. 4. According to the social distance to extend the KeyGraph

3   Experimental Results and Discussion

Here, we wanted to evidence the social distance can successfully extend the Key-Graph. After analyzed the interactive data, we could get the result of first generation as shown on Fig.5a. Firstly, the five dimensions were narrowed down to two dimensions by conjoin analysis and one of there was shown on Fig. 5b. Followed Eq.1, we could find the 4's cluster was the best cluster and secondary cluster was 3's cluster. Then after the crossover mechanism the searching space was extended to other gene in same cluster as shown on Fig.5c. And the long distance was depended on the mutation, from Fig.5c we could find the system had chosen a gene from 1's cluster. At second generation the 1's cluster was the best cluster and secondary cluster was 3's cluster. We could found that the product was supplied by the extending KeyGraph which graph was based on the social distance and it was chosen. This meant the designer also could satisfy this kind environment, and the subjective score increased step by step. In this case the designer at the fifth generation had already found his preferable product.

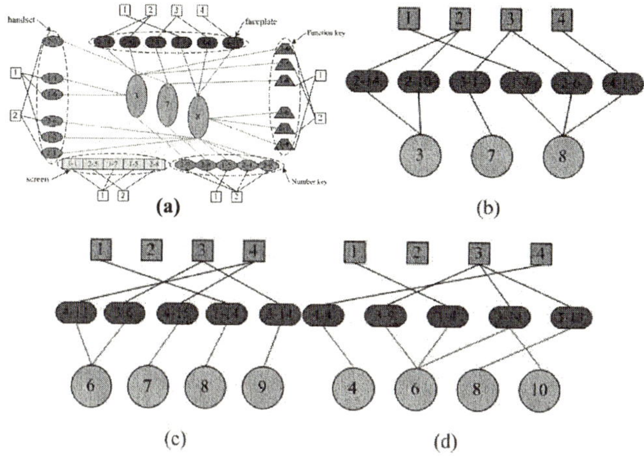

Fig. 5. (a) the first generation's KeyGraph (b) discover the most important attribute at first generation (c) the second generation (d) the fifth generation

4   Conclusion

How to increase the communicational ability between the designer and IEC is the important concept for creative design. Here our AGIEC model can depend on the distance of designer's social affiliation KeyGraph to choose the preferable cluster and under the extending KeyGraph to discover the chance. And the results of experiment also evidenced the operation of AGIEC could extend the KeyGraph and discovered the effective new gene or new cluster for designer to do the lateral transmitting as the Kotler and Trias De Bes (2003) said.

References

1. Kleinberg, J. (2000a), Navigation in a small world. Nature, 406, 845.
2. Kleinberg, J. (2000b), The small-world phenome-non: An algorithmic perspective, In Proceedings of the 32nd Annual ACM Symposium on Theory of Computing, Association of Computing Machinery, New York,, pp.163-170.
3. Kotler, P., and Trias De Bes, F. (2003), Lateral Marketing: New Techniques for Finding Breakthrough Ideas, John Wiley & Sons Inc.

4. Llorà, X., Goldberg, D. E., Ohsawa, Y., Ohnishi, K., Tamura, H., Washida, Y., and Yoshi-kawa, M. (2004a), Chances and Marketing: On-line Conversation Analysis for Creative Scenario Discussion, First European Workshop on Chance Discovery (EWCD'2004), pp.152-161, Valencia, Spain.
5. Llorà, X., Matsumura, N., Goldberg, D. E., Ohsawa, Y., Ohnishi, K., and Gonzales, A. (2004b), Discovering Chance Scenarios using Small-World KeyGraphs and Evolutionary Computation, First European Workshop on Chance Discovery (EWCD'2004), pp.51-61, Valencia, Spain.
6. Milgram, S. (1992), The Individual in a Social World: Essays and Experiments, 2d ed., McGraw-Hill, New York.
7. Ohsawa Y, Benson NE, and Yachida M (1998), Keygraph: Automatic Indexing by Co-occurrence Graph Based on Building Construction Metaphor, Proceedings of Advance in Digital Libraries Conference, pp. 12-18.
8. Ohsawa, Y. (1999), Get timely files from visualized structure of your working history, Knowledge-Based Intelligent Information Engineering Systems, pp.546-549.
9. Ohsawa, Y. and McBurney, P. (Eds.) (2003), Chance Discovery, Advanced Information Processing, Springer-Verlag.
10. Wasserman, S., and Faust, K. (1994), Social Network Analysis: Methods and Applications, Cambridge University Press, Cambridge.
11. Watts, D. J. (2003), Six Degrees: The Science of a Connected Age, New York: W.W. Norton & Company.
12. Watts, D. J., Dodds, P. S., and Newman, M. E. J. (2002), Identity and search in social networks. Science, 296, 1302-1305.

Chance Path Discovery: A Context of Creative Design by Using Interactive Genetic Algorithms

Leuo-hong Wang, Chao-Fu Hong, and Meng-yuan Song

Evolutionary Computation Laboratory, Department of Information Management,
Aletheia University, Taiwan
{wanglh,cfhong}@email.au.edu.tw

Abstract. Chance discovery is a research topic that emerges in recent years. To discover chances, a double helix model emphasizing human-computer interactions and a visual analysis tool named as KeyGraph have been proposed. In this paper, the mechanism of the double helix model is re-interpreted from the viewpoint of decision-making. The manipulation results of the double helix model are thus mapped to the objective space of the decision-making problem. In such a case, identifying the *chance path* to a "heaven" which is full of satisficing solutions becomes the key to success for the context of decision-making. A context of designing creative cell phones by using interactive genetic algorithms (IGA) has been discussed. To recognize the chance paths for IGA, an on-chance operator which evolves the individuals based on the historical data is massively used. The pilot experimental results, including satisfaction investigation, are all promising.

1 Introduction

Chance discovery, which emphasizes to discover *rare but significant* events from the *dynamic* data set, is a research topic emerged in recent years. A double helix model [1] has been proposed by Ohsawa to deal with the process of chance discovery. In this model, the human processes representing one helix will repeatedly interact with the other helix which consists of computer systems. More precisely, the data collected by each human process will be fed into the corresponding computer systems for further chance recognition. KeyGraph [2] is therefore one of the most important techniques used by these computer systems, since KeyGraph has been proven to be capable of discovering chances in many applications [3].

The intentions of the double helix model, in our opinion, can be also explained as searching creative targets in a dynamic context if we apply the model to solve decision-making problems. The result of each human process hence encloses a part of area of the objective space called the *state of context*, and the responsibility of the corresponding computer systems is then to figure out where the chances, or the *creative solutions* are in the state of context. In other words, the results of human processes at different time can be projected on a coordinate system whose vertical axis is time axis and without loss of generality, the

R. Khosla et al. (Eds.): KES 2005, LNAI 3681, pp. 533–539, 2005.

horizontal plane is the two dimensional objective space, as shown in Fig. 1. To understand the relationships between states of context in Fig. 1, each state is further projected onto the horizontal plane. The overlap occurs between consecutive states all the time, since the area focused currently are emerged from the chances found previously. Such emergences cause the initial state to gradually transfer to the final state. Meanwhile, a *chance path* that is linked by chances discovered at each state also appears to be the shortcut to creative solutions. In our opinion, generating a subtle chance path in the objective space is the reason that the double helix model can discover chances for many applications.

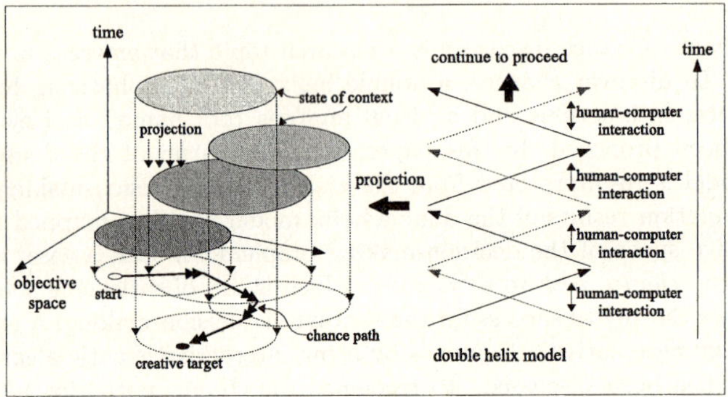

Fig. 1. The results of the double helix model processing are projected onto objective space to generate states of context. The states transfer in light of the results of chance discovery at each stage. The creative targets gradually emerge.

An idea to incorporate a chance path discovery mechanism with the interactive genetic algorithm (IGA) is inspired by the observation mentioned above. IGA [4], a variant of the genetic algorithm (GA), is an evolutionary computation framework that is suitable for solving optimization problems concerning user preferences. The major difference between IGA and GA is how to determine the fitness value of each evolvable individual. Generally speaking, GA assigns a fitness value to an individual via evaluating a pre-defined fitness function. However, formulating the fitness functions of the optimization problems concerning user preferences in advance is very difficult, humans are involved in IGA to interactively specify the fitness value of each individual. Humans, in other words, assist IGA in selection process which can't be done by computer automatically. Takagi [4], in his excellent survey, indicated that IGA performed well in the cases of art design, industrial design and product design.

Human-computer cooperation is the basis of IGA as well as the double helix model. More interestingly, if the time interval between stages of the model is very short, for example, just a few seconds, then IGA will be similar to the model. In other words, if chances can be identified in current generation of IGA and the population of next generation is then emerged from these chances, a chance path to creative solutions is possible gradually to appear. The context of IGA is, after all, slightly different from the canonical applications of chance discovery. The most important difference is that the execution time of IGA is restricted. Identifying a chance path correctly and *efficiently* is hence the underlying requirement for IGA.

Since efficiency is the most important issue, we abandon the canonical genetic operators which explore the solution space *by-chance* and define an *on-chance* operator. Unlike by-chance operators which generate new offsprings largely dependent on probability, the on-chance operator determinstically decide new offsprings based on the fitness value of each individual and *each gene*. A novel manipulation scheme of IGA, which extensively utilizes the *on-chance* operator to build chance paths, will be consequently presented in this paper.

The rest of paper is organized as follows. Section 2 introduces the on-chance operator and its underlying bipartite network model. Section 3 shows the experimental results and gives a brief discussion regarding on the effect of the on-chance operator. The last section summarizes our work and describes the future work.

2   The On-chance Operator and the Bipartite Network

To define an on-chance operator which can correctly and efficiently guide the population shifting to a state more closer to the creative solutions, the underlying manipulation model of IGA must be reformulated. Moreover, since the efficiency is the principal requirement, a processing model imitating the small world network, which is with a short characteristic path length, is probably suitable for IGA. A bipartite network model similar to the small world networks [5] is actually used by our IGA system to find the chance path to the creative solutions. The fitness values given by the users, which are scores, ranging from 1 to 10 and all possible genes (not individuals) are the two sets of nodes in the bipartite network respectively. The fitness values that a gene got make these two sets of nodes are connected. Since IGA is usually applied to combinatorial problems, manipulating the genes rather than individuals in our model benefits by easily to emerge new individuals. Figure 2 illustrates the bipartite network model.

A product design problem is used to demonstrate the concept of bipartite network model in this figure. The simplest formulation of the product design

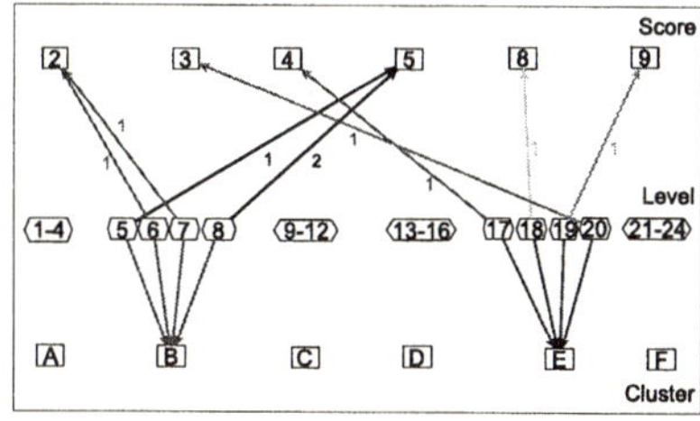

Fig. 2. The bipartite network model of IGA.

problem [6] considers a product consisting of k relevant attributes. Each attribute a_i $(i = 1, 2, \cdots, k)$ has further s_i different levels, $l_1^i, \cdots, l_{s_i}^i$, for instance. Only one attribute is shown in this figure for simplicity. There are 24 levels forming 6 clusters in this attribute. Levels in the same cluster are with similar part-worth utilities to users from the viewpoint of experts. These attribute levels are therefore in the neighborhood of *micro level*. In contrast, attribute levels belonging to different clusters are in the neighborhood of *macro level*.

The numbers on the upper side of the figure, which are enclosed by rectangles, are the fitness values (scores) given by the user of IGA systems. The link between a level and a score represents that the individual with this level is given such a score by the user. The weight of the link denotes the number of times that a level got the same score.

A bipartite graph will be created after each generation of IGA. The on-chance operator is consequently defined by using this graph. The on-chance value v_i of each level i is firstly calculated by applying the following equation:

$$v_i = \frac{w_{ij}}{\sum_{i \to j} w_{ij} \cdot s_j} \tag{1}$$

where w_{ij} is the weight of the link connecting level i and score j, $i \to j$ denotes level i and score j are linked, and s_j represents the value of score j which range from 1 to 10 in this study. The on-chance value calculated by (1) is the mean value of each *gene* that is a part of an *individual*. If the genes have appeared equally, this equation is actually a part-worth utility function which is commonly used in linear regression and is capable of predicting the importance of each gene. However, this equation is just served as an approximation function in IGA since the times of each gene that appears can not be equal.

The genes with *lowest* on-chance values will be the chances since individuals who have these genes got higher fitness values on average. These genes, in other words, are the key features attracting the user with higher probability and thus are selected by the on-chance operator to construct new individuals. The individuals emerging from the on-chance operator are subsequently placed into the population of next generation.

The rules concerning the construction of new individuals by the on-chance operator are twofold:

1. *The macro level transition.* The individuals with best on-chance value genes will be mutated and placed into next generation. However, the mutating parts inside these individuals are deterministically chosen by the on-chance value. The gene parts with worst on-chance values, in fact, will be replaced. Neighbors at the macro level, that is, levels of other clusters which are with better on-chance values or not yet appearing are the candidates for substitution. The selection probability of these candidates are equivalent.

2. *The micro level transition.* The individuals without best but moderate on-chance value genes will be mutated and placed into next generation as well. But unlike the case of macro level transition, levels in the same cluster are preferred to substitution.

Utilizing the on-chance operator described above, the new population is generated. When the population size is 9, for example, the new population consists of the elitist which is the best individual appeared so far, 5 new offsprings produced through means of the macro level transition and 3 new offsprings produced via the micro level transition. The macro level transition provides ways to explore the solution space as not appearing genes have probability to emerge. The micro level transition, on the other hand, possesses the capability of gene exploitation as only the local neighborhood is examined.

3   Experimental Results and Discussion

Two systems, an IGA with on-chance operator (IGA_{oc}) and a canonical IGA (IGA), are implemented for comparison. The simplified version of the product design problem mentioned previously, which is employed to model the creative cell phone design, are handled by these two systems respectively. More precisely, a cell phone is exactly composed of five components (attributes) in the experiments. The components are the appearance, display, numerical keypad, receiver and control button.

Each component, except the appearance attribute which has 24 alternatives, is divided into 16 different alternatives (levels) which are designed by experts. The alternatives are further grouped into clusters according to their similarities from the designer's viewpoint. The appearance attribute and its 24 levels, for example, are illustrated in Fig.3.

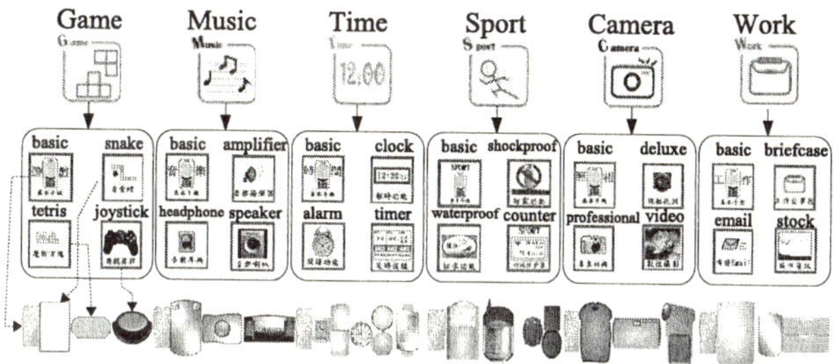

Fig. 3. The appearance attribute has 24 alternatives (levels) that are shown in the lowest part of the figure. Each shape has its own corresponding functionality. The alternatives are thus grouped into 6 clusters based on the functionality. The user will be shown the shape as well its corresponding functional icons when interacting with IGA.

The laboratory experiment was used to evaluate the performance of both IGA systems. Eighteen subjects were invited to manipulate IGA systems to identify their most favorite cell phones. The subjects were randomly divided into two groups and were requested to operate either one of the systems. Each subject was further asked to fill in a questionnaire which included topics related to the appearance, functionalities and desire to purchase of the derived cell phones.

The other questionnaire containing questions on innovation and correctness was also necessary to fill. Five-point scale, with *1= "very good"* and *5= "very bad"* was used for both questionnaires. The investigation results are listed in Table 1 and 2.

Table 1. Satisfaction investigation concerning appearance, functionalities and desire to purchase for the canonical IGA (IGA) and IGA with on-chance operator (IGA_{oc}) is listed. The three topics have 5, 5 and 4 questions respectively. Each score is an average over 9 subjects.

Topics	appearance					functionalities					desire to purchase			
IGA	2.44	2.11	2.00	2.33	2.33	2.33	2.44	2.44	1.78	2.00	1.67	2.22	2.11	2.33
IGA_{oc}	1.44	2.33	1.67	1.78	2.11	2.11	1.78	1.78	2.11	1.89	1.56	1.89	2.44	1.33

Table 2. Satisfaction investigation concerning correctness and innovation.

Topics	correctness			innovation		
IGA	2.78	2.11	2.33	1.67	3.11	2.44
IGA_{oc}	1.44	2.11	2.00	1.22	2.22	1.78

The Mann-Whitney procedure is used to test whether the observed means difference are significant. The directional hypothesis is that IGA_{oc} will prove the more effective. The probabilities to reject the hypothesis are 0.0084 for the data in Table 1 and 0.0274 for the data in Table 2. Consequently, when the significance level α equals to 0.05, it can be asserted with 95% confidence that IGA_{oc} is more effective than the canonical IGA for both investigations.

In addition, chance paths are examined using the IGA_{oc} manipulation data which are obtained from one of the subjects, as illustrated in Fig. 4.

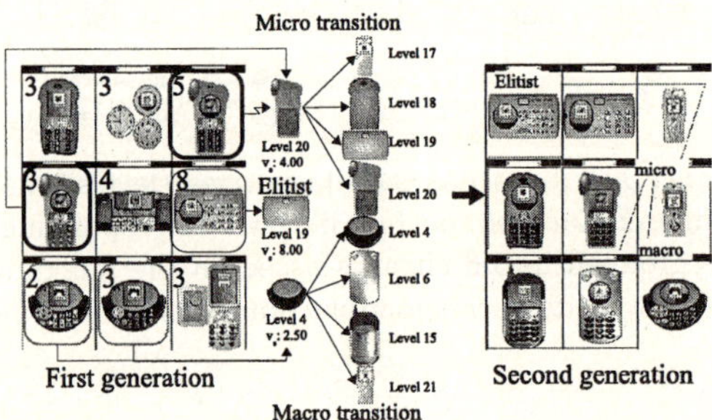

Fig. 4. The macro and micro level transitions in IGA with on-chance operator.

Only one attribute ("appearance") is shown in Fig. 4 for simplification. Level 19 is the best gene in the first generation and therefore will be reserved for the next generation. Level 20 is, moreover, one of the moderate genes and results in a micro level transition which introduces other levels belonging to the same cluster. Level 17 and 18 thus appear in the next generation. Level 4 and other worse genes, on the other side, consequently cause a macro level transition which makes these genes to be replaced. That is the reason why levels not yet appearing, such as level 6, 15 and 21, suddenly appear in the next generation, since not yet appearing levels have higher selection probabilities. The experiment finished after four generations. The most favorite level of the appearance attribute was the level 19.

The chance paths starting from level 19 and 20 through levels examined in the second and third generations and finally going back to level 19 are identified in this case. Exploring objective space in such a way is more efficient than the evolutionary approach of the canonical IGA since the exploration concentrates on the parts that the user is interested in. Moreover, the unattractive parts that usually appear repeatedly in the canonical IGA will be quickly replaced by the parts never appearing before. However, examining more experimental data are needed to confirm the existence of chance paths and the positive assistance for searching creative solutions.

4   Summary

An on-chance operator for IGA has been proposed in this paper. Identifying chances by using data cumulated so far and then exploring the objective space emerging from these chances provide an effective approach to search out creative solutions. Satisfaction investigation indicates preliminarily that this novel approach improves IGA in several aspects. However, more evidence is required to comprehensively prove the effectiveness of our approach.

References

1. Ohsawa, Y.: Modeling the Process of Chance Discovery. In: Chance Discovery. Springer (2003) 1-15
2. Ohsawa, Y., Benson, N.E., Yachida, M.: Keygraph: Automatic indexing by co- occurrence graph based on building construction metaphor. In: Proceedings of the Advances in Digital Libraries Conference. (1998)
3. Ohsawa, Y.: Keygraph as risk explorer from earthquake sequence. Journal of Contingencies and Crisis Management (2002) 119-128
4. Takagi, H.: Interactive evolutionary computation: Fusion of the capacities of ec optimization and human evaluation. Proceedings of the IEEE **89** (2001) 1275-1296
5. Newman, M.E.J., Strogatz, S.H.,Watts, D.J.: Random graphs with arbitrary degree distributions and their applications. Physical Review E **64** (2001)
6. Kohli, R., Krishnamurti, R.: A heuristic approach to product design. Management Science **33** (1987) 1523-1533

Supporting Exploratory Data Analysis
by Preserving Contexts

Mitsunori Matsushita

NTT Communication Science Labs., NTT Corp.
2-4 Hikaridai, Seikacho, Sorakugun, Kyoto, 619-0237, Japan
mat@cslab.kecl.ntt.co.jp

Abstract. The goal of the research presented in this paper is to support users in exploring a huge amount of data for the purpose of decision-making and problem-solving. Our approach is to design human-computer interaction as a natural discourse between the user who explores the data and the system that supports the user's exploration process. For that purpose, we developed a prototype system named InTREND that interprets the user's natural language query and presents statistical charts as a result of the query. InTREND encourages iterative exploration by maintaining the context of past interactions and uses this context to improve discourse with the user. This paper explains our research motivation and presents a framework for supporting exploratory data analysis. Our user studies evaluate the context preservation mechanisms of InTREND.

1 Introduction

Technological improvements, including larger databases, faster data retrieval, and various sensors, make more and more data on a variety of subjects (e.g., climate, economy, or population) available for our perusal. We use such data to make decisions and solve problems by finding and extracting useful information from the data. For instance, when a conpany manager wants to decide whether to open a new store, he/she may analyze the sales data of existing stores in the targeted area and data on population changes.

Prior to such analysis tasks, one does not usually have a clearly stated goal of how to use the data, only a high-level purpose or a problem context (such as, *"Shall we open a new store in this town?"*). This type of analysis task is *directed* toward decision-making and problem-solving, but it is also *arbitrary* in a sense that one does not initially know how to analyze the data in detail. What the person extracts from the data can be accounted for only a posteriori. This type of data analysis task is called *exploratory data analysis* [4]. Exploratory data analysis is not an assemblage of simple operation, but a very cognitive-intensive task, iterating through a series of "what-if" games through viewing data from multiple perspectives and examining data with a variety of emerging hypotheses by interacting with the data [5].

Exploratory data analysis would also be useful methodology for *Chance Discovery*. Chance Discovery means a process from discovery to utilization of *chance*,

R. Khosla et al. (Eds.): KES 2005, LNAI 3681, pp. 540–546, 2005.

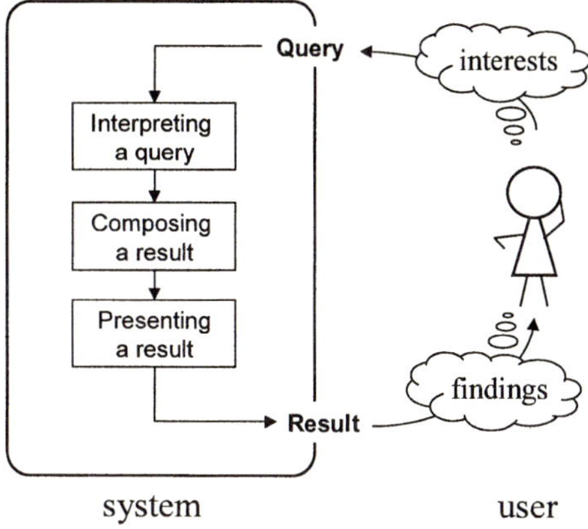

Fig. 1. A cycle of discourse as a cognitive process in exploratory data analysis.

an event or a situation significant for making a decision in a complex environment [7]. In Chance Discovery, it is quite important to comprehend data under
multiple perspectives and to depict a novel and compelling scenario based on
these understandings.

This paper discusses how a computer system supports such a data exploration task. In particular, we focus on the efficacy of context preservation of
past interaction conducted between a user who explores data and a system that
interprets the user's query, composes a result based on the query, and presents
the result.

2   Computational Support for Exploratory Data Analysis

An exploratory data analysis task does not begin with a clear goal of how to
use the data: In the process, a user initially has only a vague idea of what to
look for in the data and forms the first query based on this initial idea. The
system returns a result based on the query. After viewing the result, the user is
able to reformulate the query based on a better understanding of the information
need [8]. By repeating this process, the user gradually accumulates findings, from
which judgments can be made for problem-solving and decision-making.

Such exploratory data analysis has a different emphasis from that of data
mining [1] or knowledge discovery [3]. Data mining and knowledge discovery primarily aim at finding interesting phenomena and extracting "meaningful facts"
from a large amount of data on the initiative of calculation with a computer. In
contrast, exploratory data analysis is more guided by the user's decision-making
and problem-solving purposes, through cycles of discourse consisting of submitting a query and examining the result, as illustrated in Fig. 1. The primary goal
of the computational support for exploratory data analysis, therefore, is to make
the iteration take place as smoothly and as naturally as possible.

Our approach to achieve this goal is to use a natural language as an interface
for query articulation and to use statistical charts for presenting results. For that

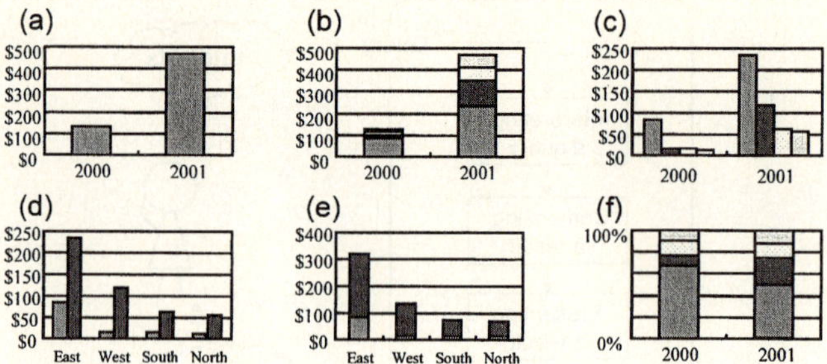

Fig. 2. Chart Examples. Chart (a) represents the sales amount of Kobe city in the last two years. Chart (b) represents the sales amount of each of the four areas of Kobe city in the last two years. Charts (c)-(e) all represent the same data content as that of (b). Chart (f) represents the proportion of the sales amount of each area.

purpose, we developed a prototype system named InTREND (an Interactive Tool for Reflective Exploration through Natural Discourse) [6]. InTREND supports this iterative exploration by (1) interpreting the user's query represented in a natural language (NL), (2) composing a chart, based on the interpreted query, for retrieved data, and (3) presenting an animated chart for the retrieval results. InTREND encourages such iterative exploration by maintaining the context of past interactions and using this context to improve discourse with the user. The system takes three types of contexts from previous interactions to support iterative cycles of discourse: in interpreting the NL query, in determining which chart to use to represent the retrieved data, and in presenting a result chart. The detail of each context is as follows.

1. *Natural Language Context (NL-Context)*

   In natural discourse, each NL query is interpreted against the prior ones. For instance, suppose a user asked the system, *"Show me the sales amount of Kobe city in the last two years"* and the system showed the chart depicted in Fig. 2-(a). A user might then want to see detailed data about Kobe city. In natural discourse with a person, the user would ask, *"How about each area?"* to obtain the chart shown in Fig. 2-(b), rather than having to say, *"Show me the sales amount of each area of Kobe city in the last two years."* This example shows that the system has to use the context of past queries (NL-Context) to interpret such partial queries.

2. *Graph Content Context (GC-Context)*

   Determining what chart (e.g., x- and y-axes, resolution, type of chart) is appropriate for a given query must consider the context as indicated by the previous interactions. When we asked people in a preliminary user study which chart was most appropriate among Fig. 2-(b), (c), (d), and (e) in response to the question, *"Show me the sales for each area of Kobe city over the last two years,"* the subjects' preference was for Fig. 2-(c). In contrast, when the initial question, *"Show me the sales for Kobe city over the last two years,"* lead to chart Fig. 2-(a), and that was followed by the follow-up

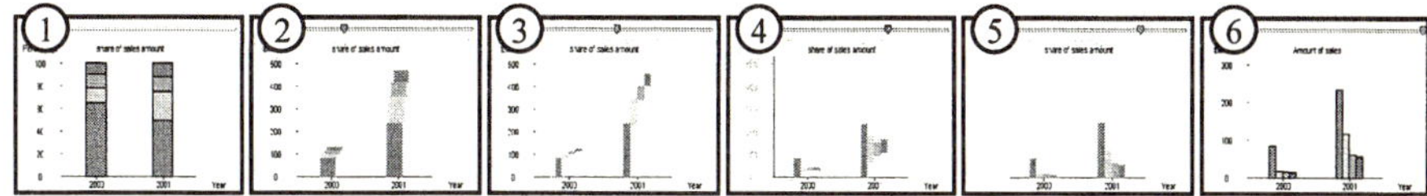

Fig. 3. Example of animation used when presenting a graph.

question of *"How about each area,"* they preferred Fig. 2-(b). This result indicates that the system needs to use the context of the previous chart (GC-Context) including the x- and y- axes attributes, resolution, and chart type to determine the most appropriate chart type.

3. *Presentation Continuity Context (PC-Context)*

Suppose a user is looking at the chart shown in Fig. 2-(f), which represents the proportion of the sales amount presented in the chart in Fig. 2-(b). If the user got interested in comparing the amount of each area, the chart shown in Fig. 2-(c) would be more appropriate. However, without maintaining correspondence between the two charts (PC-Context), this transformation would be very confusing. To avoid this confusion, the system must preserve visual features between the two charts (e.g., color, order, or shape of objects that are common in the two charts). Our system also animates the first chart to make the correspondence to the new chart easier to understand (e.g., Fig. 3). InTREND composes the animated transformation dynamically by taking the PC-Context into account.

Conventional dialogue systems that adopt NL input only take NL-Context into consideration [2]. The novel characteristic of our approach is to utilize GC-Context and PC-Context along with NL-Context. We assume that taking GC-Context and PC-Context into account contributes to the easiness of the user's exploration. See article [6] for the detailed mechanism of how the system manipulates these three contexts.

In the next session, we evaluate whether it is possible to respond users adequately by preserving the GC-Context and PC-Context.

3   User Studies

This section presents two kinds of user studies to evaluate the context preservation mechanisms of InTREND, that is, to evaluate the mechanism of taking a GC-Context and to evaluate that of a PC-Context.

3.1   User Study 1: Taking a GC-Context

We conducted a user study to evaluate a chart produced by InTREND taking the GC-Context into account.

We asked twenty-five subjects, both men and women in their twenties and thirties from our research institutes, to follow the following steps in six different tasks:

Table 1. Results of User Study 1

	Task 1			Task 2			Task 3			Task 4			Task 5			Task 6		
		GC	no-GC		GC	no-GC		GC	no-GC		GC	no-GC		GC	no-GC		GC	no-GC
Option1	6		x	2		x	18	x	x	7		x	1			21	x	
Option2	1			0			1			0			18	x		0		
Option3	8	x		0			0			7	x		3			0		
Option4	10			23	x		6			11			3		x	4		x
score		8	6		23	2		18	18		7	7		18	3		8	6

1. We showed each subject an NL query and presented a corresponding chart composed by InTREND (e.g., *"What was the sales amount in Kobe city in 2000 and 2001?"* and the chart in Fig. 2-(a)).
2. We showed each subject a follow-up query (e.g., *"How about in each area?"*).
3. We asked each subject to choose the best chart among four optional charts we presented to them (e.g., Fig. 2-(b), (c), (d), and (e)). These four charts are semantically equivalent but have different emphases.
4. We then submitted the same query to the InTREND system under two different conditions. In one condition, InTREND is used as a whole, and in the other condition, a part of the mechanism of a graph-planner component [6] has been turned off so that InTREND will not take the GC-Context into account.

Table 1 shows the distribution of the subjects' preference over four optional charts in each task, together with how InTREND produces charts. As can be seen, InTREND produces different charts in five out of six tasks with or without using a GC-Context. We counted the number of subjects who preferred the chart that InTREND produced as a score. We counted the scores for each task and calculated averages.

In summary, InTREND with the GC-Context gets 15.83 and without the GC-Context gets 6.67. By conducting a t-test, the with-GC condition is statistically significant compared to the no-GC condition ($t(10) = 2.551; p < 0.05$ (2-tailed)).

Table 1 shows that InTREND with the GC-Context could not produce a chart that was most preferred by the subject only in two tasks out of six. A perfect score would have been 16.83. Comparing this score with 15.83 demonstrates that the current context preservation mechanism of InTREND using a GC-Context can provide an adequate chart that meets user expectation.

3.2 User Study 2: Taking a PC-Context

This study evaluated the effect of animation for conveying the PC-Context appropriately to users. We used fifteen subjects, both men and women, also from our research institutes.

The study was conducted in two stages. The first stage was to examine if animation in general is helpful for understanding a chart and the second stage was to examine if animation taking a PC-Context into account aids understanding.

For the first stage, we prepared five sets of two charts: one for an initial query and the other for the follow-up query. We then prepared two conditions. In the first condition, the second chart was instantly replaced with the initial chart. In the second condition, the initial chart was first deleted and then the second chart was drawn using animation. This animation is generic in the sense that it does not take the context from the first chart, but simply drew elements of a chart step-by-step. For example, it might first draw the x-axis, then the y-axis, and then plot the values one by one. We asked the fifteen subjects which condition was most helpful for them to understand the second chart. We changed the order of the two conditions for each task.

The result is that 3.8 subjects on average preferred the instant-redrawn condition, whereas 11.2 subjects on average preferred the animation condition. This tells us that the use of animation in general is helpful for people to understand charts ($t(8) = 6.541$; $p < 0.001$ (2-tailed)).

For the second stage, the fifteen subjects followed the following steps in fifteen different tasks.

1. We showed each subject the first chart.
2. We verbally described the information that was extractable from the chart and told the subject to focus on a certain aspect of the information.
3. We replaced the initial chart with the second chart with three different conditions: instant-replace (for estimating the baseline of response time), generic-animation (the same one used in the first stage), and InTREND-animation.
4. We asked a question about a value of an object displayed on the second chart using characterization illustrated with the first chart.

In each of these tasks, in order to correctly answer the last question, the subject needed to both remember the first chart and understand the second chart. For instance, in one task, the chart shown in Fig. 2-(f) was given as the first chart. Then, we provided the following verbal description, "This chart shows the proportional sales amount of companies A, B, C and D. The vertical axis represents percentage. Please focus on how the share of each company changes." After a subject confirmed, the first chart was replaced with the second chart shown in Fig. 2-(c). After this, the experimenter asked the subject verbally: "I am now asking a question regarding two companies, one that has decreased its share in 2000 to 2001 and one that has the smallest share change. What is the difference in sales amount between the two companies?"

Each subject was asked to answer three questions in each of the three re-drawing conditions. The order of the conditions was altered for each subject.

For each subject, we tallied how many questions the subject answered correctly and how long it took the subject to answer.

The result shows that the average response time for (α) instant replacement, (β) general animation, and (γ) InTREND animation is 5.43, 4.86, and 4.20 seconds, respectively. We used statistical analysis to evaluate which animation is better for understanding the contents of the chart. The statistical analysis shows that the average response time for (γ) is significantly shorter than (α)

$(p < 0.05)$, but the average response time for (β) is not significantly shorter than (α) $(p > 0.1)$. From the post-experimental interviews with the subjects, seven subjects told us that it was easiest to answer the questions when presented with the (γ) option. On the other hand, one subject mentioned that the existence of animation itself is confusing and hinders him/her from focusing on answering the question.

4   Concluding Remarks

This paper presented our approach for supporting a user in performing exploratory data analysis. Since exploratory data analysis is a very cognitive-intensive task, the user must be able to focus on the task without being bothered by unsmart interface behaviour, for instance, being required to explicitly specify the previous context, which is in most cases obvious to the user. Our InTREND system prevents such an encumbrance and encourages iterative exploration by maintaining the context of past interactions, and uses this context to improve discourse with the user.

Our user studies so far have demonstrated that taking contexts into account helps the user at each step of expressing a query and understanding a graph. We continue further refinement of context preservation mechanisms and user studies to evaluate if the approach improves the quality of overall exploratory data analysis tasks.

References

1. Agrawal, R., Imielinski, T. and Swami, A. N.: Mining Association Rules between Sets of Items in Large Databases, *Proc. SIGMOD'93*, pp. 207–216 (1993).
2. Carbonell, J. G. and Hayes, P. J.: Recovery Strategies for Parsing Extragrammatical Language, *American Journal of Computational Linguistics*, Vol. 9, No. 3-4, pp. 123–146 (1983).
3. Fayyad, U. M., Piatetsky-Shapiro, G. and Smyth, P.: Knowledge Discovery and Data Mining: Towards a Unifying Framework, *Proc. KDD'96*, pp. 82–88 (1996).
4. Hartwig, F. and Dearing, B. E.: *Exploratory Data Analysis*, SAGE Publications (1979).
5. Mackay, W. E. and Beaudouin-Lafon, M.: DIVA: Exploratory Data Analysis with Multimedia Streams, *Proc. CHI'98*, pp. 416–423 (1998).
6. Matsushita, M., Nakakoji, K., Yamamoto, Y. and Kato, T.: InTREND: An Interactive Tool for Reflective Data Exploration Through Natural Discourse, *Proc. KES2004*, Vol. 2, pp. 148–155 (2004).
7. Osawa, Y. and McBurney, P.(eds.): *Chance Discovery*, Springer Verlag (2003).
8. Williams, M. D., Tou, F. N., Fikes, R., Henderson, A. and Malone, T.: RABBIT: Cognitive Science in Interface Design, *Proc. CogSci'82*, pp. 82–85 (1982).

Chance Discovery and the Disembodiment of Mind

Lorenzo Magnani

Department of Philosophy and Computational Philosophy Laboratory,
University of Pavia, 27100 Pavia, Italy
`lmagnani@unipv.it`

Abstract. The first part of the paper illustrates that at the roots of the chance discovery there is a process of disembodiment of mind that exhibits a new perspective on the cognitive mechanisms underling the emergence of scenarios productive for opportunities and risks. I will further illustrate the centrality to chance discoverability of the disembodiment of mind from the point of view of the cognitive interplay between internal and external representations. I consider this interplay critical in analyzing the relation between chance discovery and dynamical interactions with the environment. To the aim of explaining these relationships I conclude stressing the importance I attribute to the role of model-based and manipulative abduction, and of external representations and epistemic mediators.

1   From the Prehistoric Brains to Scenario Creation

In an interesting article of 1950, "Computing machinery and intelligence" [1], which adopts an evolutionary perspective, Turing maintains that a big cortex can provide an evolutionary advantage only in presence of a massive storage information and knowledge on external supports that only an already developed small community of human beings can possess. Evidence from paleoanthropology seems to support this perspective. Some research in cognitive paleoanthropology teaches us that high level and reflective consciousness in terms of thoughts about our own thoughts and about our feelings (that is consciousness not merely considered as raw sensation) is intertwined with the development of modern language (speech) and material culture. After 250.000 years ago several hominid species had brains as large as ours today, but their behavior lacked any sign of art or symbolic behavior. If we consider high-level consciousness as related to a high-level organization - in Turing's sense - of human cortex, its origins can be related to the active role of environmental, social, linguistic, and cultural aspects.

Handaxes were made by Early Humans and firstly appeared 1,4 million years ago, still made by some of the Neanderthals in Europe just 50.000 years ago. The making of handaxes is strictly intertwined with the development of consciousness. Many needed capabilities constitute a part of an evolved psychology that appeared long before the first handaxed were manufactured. It seems humans were pre-adapted for some components required to make handaxes [2]:

1. imposition of symmetry (already evolved through predators escape and social interaction). It has been an unintentional by-product of the bifacial knapping technique but also deliberately imposed in other cases. It is also well-known that the attention

R. Khosla et al. (Eds.): KES 2005, LNAI 3681, pp. 547–553, 2005.

to symmetry may have developed through social interaction and predator escape, as it may allow one to recognize that one is being directly stared at [3].

2. understanding fracture dynamics (for example evident from Oldowan tools and from nut cracking by chimpanzees today);
3. ability to plan ahead (modifying plans and reacting to contingencies, such unexpected flaws in the material and miss-hits), still evident in the minds of Oldowan tool makers and in chimpanzees;
4. high degree of sensory-motor control: "Nodules, preforms, and near finished artefacts must be struck at precisely the right angle with precisely the right degree of force if the desired flake is to be detached" [2, p. 285]. The origin of this capability is usually tracked back to encephalization - the increased number of nerve tracts and of the integration between them allows for the firing of smaller muscle groups - and bipedalism - that requires a more complex integrated highly fractionated nervous system, which in turn presupposes a larger brain.

The combination of these four resources produced the birth of what Mithen calls technical intelligence of early human mind, that is consequently related to the construction of handaxes. Indeed they indicate high intelligence and good health. They cannot be compared to the artefacts made by animals, like honeycomb or spider web, deriving from the iteration of fixed actions which do not require consciousness and intelligence.

1.1   Private Speech and Fleeting Consciousness

Two central factors play a fundamental role in the combination of the four resources above:

- the exploitation of private speech (speaking to oneself) to trail between planning, fracture dynamic, motor control and symmetry (also in children there is a kind of private muttering which makes explicit what is implicit in the various abilities);
- a good degree of fleeting consciousness (thoughts about thoughts).

In the meantime these two aspects played a fundamental role in the development of consciousness and thought:

> So my argument is that when our ancestors made handaxes there were private mutterings accompanying the crack of stone against stone. Those private mutterings were instrumental in pulling the knowledge required for handaxes manufacture into an emergent consciousness. But what type of consciousness? I think probably one that was fleeting one: one that existed during the act of manufacture and that did not the endure. One quite unlike the consciousness about one's emotions, feelings, and desires that were associated with the social world and that probably were part of a completely separated cognitive domain, that of social intelligence, in the early human mind (p. 288).

This use of private speech can be certainly considered a "tool" for organizing brains and so for manipulating, expanding, and exploring minds, a tool that probably evolved with another: talking to each other. Both private and public language act as tools for thought and play a fundamental role in the evolution "opening up our minds to ourselves" and so in the emergence of new meaning cognitive chances.

1.2   Material Culture as a Chance Scenario

Another tool appeared in the latter stages of human evolution, that played a great role in the evolutions of primitive minds, that is in the organization of human brains. Handaxes also are at the birth of material culture, so as new cognitive chances can co-evolve:

- the mind of some early humans, like the Neanderthals, were constituted by relatively isolated cognitive domains, Mithen calls different intelligences, probably endowed with different degrees of consciousness about the thoughts and knowledge within each domain (natural history intelligence, technical intelligence, social intelligence). These isolated cognitive domains became integrated also taking advantage of the role of public language;
- degrees of high level consciousness appear, human beings need thoughts about thoughts; - social intelligence and public language arise.

It is extremely important to stress that material culture is not just the product of this massive cognitive chance but also cause of it. "The clever trick that humans learnt was to disembody their minds into the material world around them: a linguistic utterance might be considered as a disembodied thought. But such utterances last just for a few seconds. Material culture endures" (p. 291).

In this perspective we acknowledge that material artefacts are tools for thoughts as is language: tools for exploring, expanding, and manipulating our own minds. They constitute the cognitive ancestors of those *information artefacts* [4] or *cognitive artifacts* [5] which constitute the external multiple tools - communication, context shifting, computational devices, like KeyGraph, etc. - recently analyzed by the researchers in the field of chance discovery [6].

Moreover, in this perspective the evolution of culture is inextricably linked with the evolution of consciousness and thought. Early human brain becomes a kind of universal "intelligent" machine, extremely flexible so that we did no longer need different "separated" intelligent machines doing different jobs. A single one will suffice. As the engineering problem of producing various machines for various jobs is replaced by the office work of "programming" the universal machine to do these jobs, so the different intelligences become integrated in a new universal device endowed with a high-level type of consciousness.

From this perspective the expansion of the minds is in the meantime a continuous process of disembodiment of the minds themselves into the material world around them. In this regard the evolution of the mind is inextricably linked with the evolution of large, integrated, material cognitive systems. In the following section I will illustrate this extraordinary interplay between human brains and the new cognitive systems they make.

2   Disembodiment of Mind: Chances That Cut Chains

A wonderful example of chance creation productive for the discovery of new concepts - through disembodiment of mind - is the carving of what most likely is the mythical being from the last ice age, 30.000 years ago, a half human/half lion figure carved from mammoth ivory found at Hohlenstein Stadel, Germany.

An evolved mind is unlikely to have a natural home for this being, as such entities do not exist in the natural world, the mind needs new chances: so whereas
evolved minds could think about humans by exploiting modules shaped by
natural selection, and about lions by deploying content rich mental modules
moulded by natural selection and about other lions by using other content rich
modules from the natural history cognitive domain, how could one think about
entities that were part human and part animal? Such entities had no home in
the mind (p. 291).

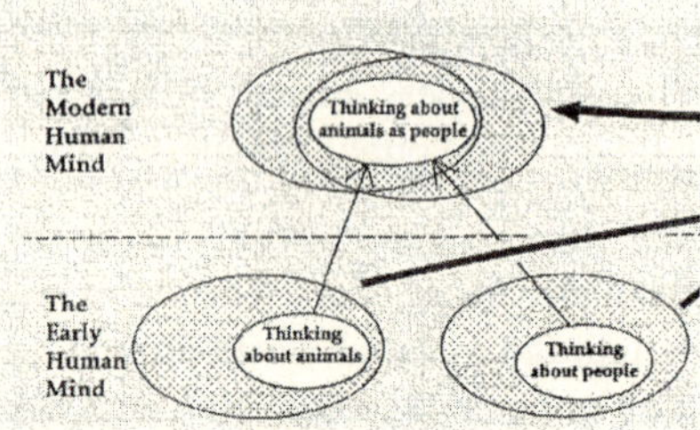

Fig. 1. In [2].

A mind consisting of different separated intelligences cannot come up with such
entity (Figure 1). The only way is to extend the mind into the material word, building
primitive *scenarios*[1] made by rocks, blackboards, paper, ivory, and writing, painting,
and carving: "artefacts such as this figure play the role of anchors for ideas and have no
natural home within the mind; for ideas that take us beyond those that natural selection
could enable us to possess" (p. 291).

In the case of our figure we face with an anthropomorphic thinking created by the
material representation serving to anchor the cognitive representation of supernatural
being. In this case the material culture disembodies thoughts (and provides chances potentially significant to adopt new inferences and new actions), that otherwise will soon
disappear, without being transmitted to other human beings. The early human mind possessed two separated intelligences for thinking about animals and people. Through the
mediation of the material culture the modern human mind can arrive to internally think
about the new concept of animal and people at the same time. But the new mean-ing
occurred over there, in the external material world where the mind picked up it.

Artefacts as external objects allowed humans to loosen and cut those *chains* on our
unorganized brains imposed by our evolutionary past. Chains that always limited the
brains of other human beings, such as the Neandertals. Loosing chains and securing
ideas to external objects was also a way to creatively re-organize brains as universal
machines for thinking.

---

[1] Ohsawa proposed this term to explain some aspects of change discovery. Cf. for example [7]
that stresses the attention on the activity of extracting chances through communication of scenarios: given that a scenario is a time series of events under a given context, during a communication scenario drawings are generated, externalized (that is disembodied) and subsequently
analyzed in the process of chance discovery.

I think the disembodiment of mind can nicely account for low-level cognitive processes of chance production, bringing up the question of how could higher-level processes be comprised and how would they interact with lower-level ones. To the aim of explaining these higher-level mechanisms I have provided a cognitive framework where model-based and manipulative abduction together with external representations and epistemic mediators play a central role (cf. the following sections).

3   Theoretical and Manipulative Reasoning

Abduction is the process of *inferring* certain facts and/or laws and hypotheses that render some sentences plausible, that *explain* or *discover* some (eventually new) phenomenon or observation; it is the process of reasoning in which explanatory hypotheses are formed and evaluated. There are two main epistemological meanings of the word abduction [8]: 1) abduction that only generates "plausible" hypotheses ("selective" or "creative") and 2) abduction considered as inference "to the best explanation", which also evaluates hypotheses. An illustration from the field of medical knowledge is represented by the discovery of a new disease and the manifestations it causes which can be considered as the result of a creative abductive inference. Therefore, "creative" abduction deals with the whole field of the growth of scientific knowledge. This is irrelevant in medical diagnosis where instead the task is to "select" from an encyclopedia of pre-stored diagnostic entities.

Theoretical abduction[2] certainly illustrates much of what is important in creative abductive reasoning, in humans and in computational programs, but fails to account for many cases of explanations occurring in science when the exploitation of environment is crucial. It fails to account for those cases in which there is a kind of "discovering through doing", cases in which new and still unexpressed information is codified by means of manipulations of some external objects (*epistemic mediators*). The concept of *manipulative abduction*[3] captures a large part of scientific and everyday thinking where the role of action is central, and where the features of this action are implicit and hard to be elicited:It at the level of manipulative abduction that various aspect of chance production and discovery can be illustrated.

4   Chance and Environment

Indeed in previous research in the field of chance discovery I have stressed the central role of manipulative abduction both in epistemic [10] and ethical setting [11], where prefiguring ethical chances becomes important. I maintain that a considerable part of the inference activity is model-based and many model-based ways of reasoning are performed in a manipulative way by using external tools and mediators.

---

[2] Magnani [8, 9] introduces the concept of theoretical abduction. He maintains that there are two kinds of theoretical abduction, "sentential", related to logic and to verbal/symbolic inferences, and "model-based", related to the exploitation of internalized models of diagrams, pictures, etc.

[3] Manipulative abduction and epistemic mediators are introduced and illustrated in [8].

For example, in the simple case of the construction and examination of diagrams in elementary geometrical reasoning [9], specific experiments serve as states and the implied operators are the manipulations and observations that transform one state into another. This process involves an external representation consisting of written symbols and figures that for example are manipulated "by hand". The cognitive system is not merely the mind-brain of the person performing the geometrical task, but the system consisting of the whole body (cognition is *embodied*) of the person plus the external physical representation where specific mind/brain activities are disembodied. In geometrical discovery the whole activity of cognition is located in the system consisting of a human together with diagrams, which consequently depicts a kind of *extended mind*. In this perspective chance discovery is seen in a perspective of a new materialistic concept of mind: what we usually call mind simply consists in the union of both the changing neural configurations of brains together with those large, integrated, and material cognitive systems the brains themselves are continuously building on external *scenarios*.

4.1 The Extra-Theoretical Dimension of Chance Discovery: Templates of Epistemic Acting and Epistemic Mediators

Various templates of manipulative behavior exhibit some regularities that are central to chance discovery. The activity of manipulating external things and representations is highly conjectural and not immediately explanatory: these templates are hypotheses of behavior (creative or already cognitively present in the scientist's mind-body system, and sometimes already applied) that abductively enable a kind of epistemic "doing". In the following paragraph I give some examples of manipulative templates active at the epistemological level of scientific reasoning.

Some common features of the tacit templates of manipulative abduction, that enable us to manipulate things and experiments in science are related to: 1. sensibility towards the aspects of the phenomenon which can be regarded as *curious* or *anomalous*; manipulations have to be able to introduce potential inconsistencies in the received knowledge and so to open new possible reasoning opportunities; 2. preliminary sensibility towards the *dynamical* character of the phenomenon, and not to entities and their properties, common aim of manipulations is to practically reorder the dynamic sequence of events into a static spatial one that should promote a subsequent bird's-eye view (narrative or visual-diagrammatic), fruitful for further outcomes; 3. referral to experimental manipulations that exploit *artificial apparatus* to free new possible stable and repeatable sources of information about hidden knowledge and constraints; 4. various contingent ways of epistemic acting: *looking* from different perspectives, *checking* the different information available, *comparing* subsequent events, *choosing*, *discarding*, *imaging* further manipulations, *re-ordering* and *changing relationships* in the world by implicitly *evaluating* the usefulness of a new order (for instance, to help memory). As I have illustrated more in detail in [10] The whole activity of manipulation is devoted to building various external *epistemic mediators* that function as an enormous new source of information and knowledge.

5 Conclusion

It is clear that the manipulation of external objects helps human beings in chance discovery and production and so in their creative tasks. I have illustrated the strategic role played by low-level cases of disembodiment of mind also relating it to the high-level cognitive processes of manipulative abduction, considered as a particular kind of abduction that exploits external models and epistemic mediators endowed with delegated explanatory cognitive roles and attributes, surely active in chance discovery and chance production.

References

1. Turing, A.: Computing machinery and intelligence. *Mind* **49** (1950) 433–460
2. Mithen, S.: Handaxes and ice age carvings: hard evidence for the evo-lution of consciousness. In Hameroff, A., Kaszniak, A., Chalmers, D., eds.: *Toward a Science of Consciousness III. The Third Tucson Discussions and Debates*, Cambridge, MIT Press (1999) 281–296
3. Dennett, D.: *Consciousness Explained. New York: Little, Brown, and Company.* Little, Brown, and Company, New York (1991)
4. Amitani, S., Hori, K.: A method and a system for supporting the process of chance discovery. In Abe, A., Oehlmann, R., eds.: *The First European Workshop on Chance Discovery*, Valencia (2004) 62–83
5. Shibata, H., Hori, K.: Toward and integrated environment for writing. In Abe, A., Oehlmann, R., eds.: *The First European Workshop on Chance Discovery*, Valencia (2004) 232–241
6. Oshawa, Y., McBurney, P., eds.: *Chance Discovery*, Berlin, Springer (2003)
7. Nara, Y., Ohsawa, Y.: Knowing melting pot. emerging topics based on scenario communication on technical foresights. In Abe, A., Oehlmann, R., eds.: *The First European Workshop on Chance Discovery*, Valencia (2004) 62–83
8. Magnani, L.: *Abduction, Reason, and Science. Processes of Discovery and Explanation.* Kluwer Academic/Plenum Publishers, New York (2001)
9. Magnani, L.: Epistemic mediators and model-based discovery in science. In Magnani, L., Nersessian, N.J., eds.: *Model-Based Reasoning: Science, Technology, Values*, New York, Kluwer Academic/Plenum Publishers (2002) 305–329
10. Magnani, L., Piazza, M., Dossena, R.: Epistemic mediators and chance morphodynamics. In Abe, A., ed.: *Proceedings of PRICAI-02 Conference, Working Notes of the 2nd International Workshop on Chance Discovery*, Tokyo (2002) 38–46
11. Magnani, L.: Moral mediators. prefiguring ethical chances in a human world. In Shoji, H., Matsuo, Y., eds.: *Proceedings of the 3rd International Workshop on Chance Discovery, HCI International Conference*, Greece (2003) 1–20

Modeling the Discovery of Critical Utterances

Calkin A.S. Montero[1], Yukio Ohsawa[2,3], and Kenji Araki[1]

[1] Graduate School of Information Science and Technology, Hokkaidō University,
Kita 14-jo Nishi 9-chome, Kita-ku, Sapporo City, 060-0814 Japan
{calkin,araki}@media.eng.hokudai.ac.jp,
http://sig.media.eng.hokudai.ac.jp
[2] Graduate School of Business Science, University of Tsukuba,
3-29-1 Otsuka Bunkyo, Tokyo, Japan
[3] Graduate School of Information Science an Technology, The University of Tokyo,
Tokyo, Japan
osawa@gssm.otsuka.tsukuba.ac.jp
http://www.gssm.de/comp/lncs/index.html

Abstract. The ubiquity of chance discovery can be seen in many areas of application, fundamentally when there are events that are difficult to find or predict. In this paper, we present our idea of applying a chance discovery tool, the *KeyGraph*, for identifying relationship of co-occurrence between the utterances of the interlocutors during a chat section.

1 Introduction

Despite of being a new field of research and still in a early stage of development, chance discovery is being applied to a wide range of branches of science. Chance discovery, as a field of investigation itself, is devoted to analyze the occurrence of events that might have important impact in their happening area. In this regard, the word "chance" has been defined as the information about an event or situation significant for making decisions, being *chance discovery* the discovery of a chance, *not* by chance [1]. This chance has been considered a rare opportunity - or risk - whose significance has been unnoticed, which may lead to an unexpected benefit.

Such chances have been studied in areas of application such as:
- earthquake prediction
- topic diffusion in communities
- genomes triggering diseases
- new products and consumer's behavior
- complex adaptive systems
- leading opinions in on-line communities
- data mining with visual interface
- discourse analysis

and so forth. In this regard, we applied techniques of chance discovery to process human-human chat[1].

---

[1] Friendly, informal conversation. Cambridge Advanced Learner's Dictionary.

R. Khosla et al. (Eds.): KES 2005, LNAI 3681, pp. 554–560, 2005.

During the analysis of human-human chat, a peculiar behavior of critical self-organization was observed [2]. It could be considered that the positive feedback effect that exists, from the interlocutors during the chat, triggers the self organization. We argue that for modeling a human-like computer chat this critical behavior should be taken in consideration.

Previous research has approached the problem of modeling computer dialogue [3]. In fact, three main approaches have been proposed to modeling dialogue: dialogue grammars approach, plan-based approach, and joint action or collaborative approach. The dialogue grammar approach is based on the observation that sequencing regularities appear in the dialogue [4] - so as questions are generally follow by answers, proposal by acceptances, etc -. Hence, it has been proposed that dialogues are a collection of such act sequences with embedded sequences of digressions and repairs. This kind of grammars can be constructed following the Chomsky hierarchy or finite state machines. The plan-based approach is based on the observation that the user's utterances are not simple string of words but communicative actions or speech acts [5]. This approach assumes that the user's speech acts are part of a plan, and the goal of the system would be to uncover and respond appropriately to this plan [6]. Finally, the joint action or collaborative approach views the dialogue as a collaboration between the interlocutors in order to achieve a mutual understanding [7]. All of those approaches focused on creating a dialogue in order to achieve certain specific goal, it is to say, domain specific dialogue approaches.

When it comes to modeling non-specific-goal computer dialogue[2], in spite of the increasingly development of such dialogue agents, there is a remarkable lack of research. Non-specific-goal dialogue agents are called chat robots or chatbots. A chatbot has been defined as a computer program that simulates human conversation through Artificial Intelligence (AI). The dialogue modeling of a chatbot is based on certain words combinations that it finds in a phrase given by the user, and in the case ALICE (Artificial LInguistic Computer Entity) [8] chatbot, AIML[3] is used to transcribe pattern-response relations. However, the chatbot often gets lost during chatting with a user since it is highly improbable to create all the possible pattern-response. In this research we applied a chance discovery tool, the *KeyGraph*, in order to observe the critical behavior and dynamics of the human chat, information that, we argue, is of great value for meliorating the modeling human-like computer chat.

2   Human-Human Chat Dynamics: A Chance

While researching on the structure of human-human chat, we observe a peculiar general behavior in the flow of the conversation. Let us explain our observations as follow: considering a trivial conversation - chat - between two interlocutors, it is noticeable that in the beginning the chat could be considered flat (greetings from each interlocutors). As the chat evolves toward one direction - some

---

[2] Open domain dialogue.

[3] Artificial Intelligence Markup Language, base on extended Markup Language, XML.

specific topic - certain utterances may cause the course of the chat to slightly chance the direction of the topic, although the initial topic still remains the same. Eventually, during the chat there is not anything else to utter about the initial topic, so this subject can not be discussed any further, either because the interlocutors have already agreed in their ideas regarding the subject or because the topic is not longer of interest for the interlocutors. At this point a single utterance given by any of the interlocutors can cause the whole direction of the chat to change toward a new topic and the whole process starts again. Making an analogy with self-organized criticality exemplified by the Bak's sandpile model [9], this behavior has been called *critically self-organized chat (CSOC)* [2]. The above described behavior can be seen in the following fragment of a real chat between two friends[4] [10]:

1A: Hello
1B: Hello A, ah
2A: Yes, B ah
2B: Ah
3A: Have you had your lunch
3B: No
4A: Uhm I haven't had my lunch also
4B: Uh busy uh
5A: Uhm busy playing with my nieces
5B: Both of them are in their own place eh at your home now
6A: Hah The younger one is always at my home mah
6B: Both of them – Orh
....
39A: You hear my little niece crying
39B: Heh uh Uhm so are you spring cleaning already
40A: Ah intend to do it today lor
40B: Orh

41A: So I will be home whole day lah but
41B: Uhm
42A: Uhm you know like can't get started like that do a little bit then no don't want to do uhm then
42B: Uh huh uh huh
....
52A: Ah – Then weekdays definitely can not – Weekdays I'm so busy in the office you know
52B: Ah hah
53A: This whole this week itself ah
53B: Ah hah
54A: Every night you know
54B: Ah
55A: Past nine o'clock then I leave the office
55B: Whoa everyday
....

In this example, the utterances 1A to 4B form the flat initial state of the system: greetings and topic-introductory utterances. At this point there are not big changes in the chat, it could be said the system is in equilibrium. The next utterance - 5A - initiates the chat toward a determined topic: A's nieces. This topic remains the main one, although it might get slightly changed, until the utterance 39B that "crashes" this main topic initiating a completely new one: "laundry on weekends". And then again, utterance 52A "crashes" the previous topic and started a new one: "busy at the office". This process repeats itself until the chat ends. The utterance that changes the direction of the chat - crashing one topic and starting another - is called "critical utterance" [2].

Considering the flow of a chat, it can be said that the chat is a dynamic system whose environment is formed by the interlocutor's utterances. In this

---

[4] The speakers names have been omitted.

dynamic system a *chance* could be seen as the moment when a critical utterance may come as to define the new direction of the chat.

3   Chance Discovery and Critically Self-organized Chat

It is precisely the dynamic structural behavior of human-human chat that allows the smooth transition from one topic to the other within a chat section. If we consider the time of transition between topics as potential chances, the analysis of critical utterances - when a topic changes - and their relationships becomes relevant. In this regard the application of chance discovery techniques, as data mining processing tool, are considered useful. However, it has been stated in [11] the difficulty of identifying a chance in complex systems - critically self-organized systems - due to the interrelation between the system and its environment. Therefore, a chance must be considered from the global collective nature of the dynamics of the system.

In order to analyze the dynamics of our system - a chat section - as to model how to identify potential chances - critical utterances or topic changing - a chance discovery graphical tool is needed. In this regard, the application of the *Key-Graph* becomes suitable. The *KeyGraph*, originally used for indexing documents [12], has developed as a data mining tool for extracting patterns of the appearance of chance events. As a data mining tool, it identifies relationship between terms in a document, particularly focusing on relationship of co-occurrence of both high probability and low probability events. A detailed description of the application of the *KeyGraph* for analyzing the chat section shown in Sect. 2 is given hereunder.

3.1   Design Experiment and Visual Results

We aim at visualizing potential chances during a human-human chat section in order to apply this knowledge to model computer chat. The used tool, the *Key-Graph*, has been applied to a variety of topics [13]:
- discovering deep building blocks for genetic algorithms
- discovering emerging topics from the World Wide Web
- discovering areas with high risk of near-future earthquakes
and so forth. In our research, it is used to visualize critical utterances.

The analysis was carried out as follow:
1) Each turn of the speakers - utterance - was considered one sentence.
2) Each sentence was segmented by words.
3) High frequency words were eliminated, i.e., I, you, are, my, and the like.
4) A vectorial representation of each sentence was created in order to find co-occurrence relation between them.
5) The co-occurrence document was analyzed using the *KeyGraph*

Let us suppose our chat section, C_S, as a matrix where columns represent words, w_n for $n = 1, 2, 3, ..., k$, and rows represent sentences, S_m for $m = 1, 2, 3, ..., l$, in the form:

$$C_S =$$

$$
\begin{array}{ccccc|c}
w_1 & w_2 & w_3 & \cdots & w_n & \\
0 & 0 & 1 & \cdots & 1 & S_1 \\
1 & 0 & 1 & \cdots & 1 & S_2 \\
0 & 1 & 0 & \cdots & 0 & S_3 \\
& & \vdots & & \vdots & \\
0 & 1 & 1 & \cdots & 0 & S_m
\end{array}
$$

where the zeros and ones represent the absence or presence of a word in each sentence. Transforming this document by obtaining the transposed matrix, it becomes:

$$(C_S)^T =$$

$$
\begin{array}{ccccc|c}
S_1 & S_2 & S_3 & \cdots & S_m & \\
0 & 1 & 0 & \cdots & 0 & w_1 \\
0 & 0 & 1 & \cdots & 1 & w_2 \\
1 & 1 & 0 & \cdots & 1 & w_3 \\
& & \vdots & & \vdots & \\
1 & 1 & 0 & \cdots & 0 & w_n
\end{array}
$$

In $(C_S)^T$ each row represents what words each sentence shares with each other, what we regard as the occurrence relationship between sentences. As a result, from the above matrix we obtain:

$$D =$$

$w_1 : S_2$

$w_2 : S_3, S_m$

$w_3 : S_1, S_2, S_m$

....

$w_n : S_1, S_2$

This document D is then analyzed using the *KeyGraph*.

Applying the above described algorithm to the chat section the graphical result is given in Fig. 1.

In this figure the results of the *KeyGraph* can be interpreted as follow:
- Each one of the dotes represent one sentence.
- Each cluster, formed by groups of dotes, represent the relationship between the sentences that form one specific topic.
- Each link, represent interconnection between topics.

In the analyzed chat there were eleven critical utterances as shown in Table 1. From those, ten (except 78) were identified in the graph - marked with arrows. Observing the clusters in the figure, it can be said that one critical utterance is leading to the other. This might be happening because of the interlocutor's desire (unconsciously) to organize his/her own mind (or schedule for doing things). This desire comes from a natural tendency of human self-discovery and coherence in their own chats. The time when a critical utterance arrives (when the topic under scope loses its novelty, so there is not more interest on it from the

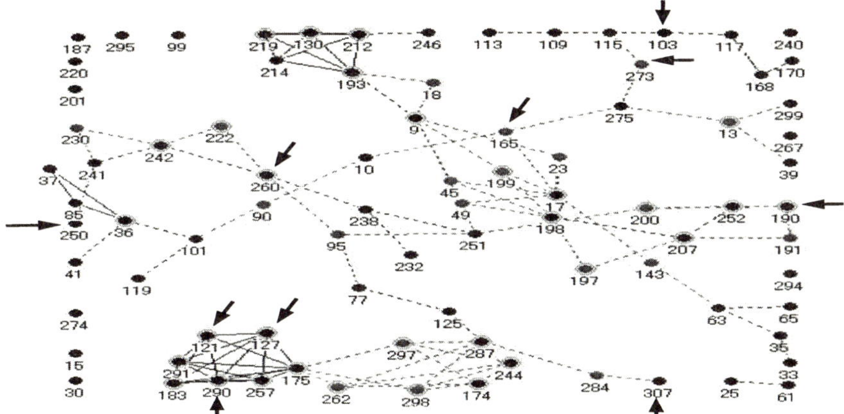

Fig. 1. Graphical View: Human-Human Chat.

participants) is triggering the different links and nodes given by the tool for the critical utterances. However, a topic may have lost its novelty temporarily, but later on the same topic may arise again from another point of view - what makes very difficult to distinguish what is a critical utterance and what is not- , giving birth to the clusters that contain more than one critical utterance all together with other utterances.

The links the graph are representing different ways of transition between one topic and the other. For example, the cluster containing the critical utterances 127 and 290 (concerning about New Year) is linked to utterance 287 (about bringing one of the speaker's niece out) through 297 (about "bring something"). Therefore, the bridges might be indicating transition between one topic and the other, information that can be used for smartly changing topics in a very smooth way when modeling computer chat, i.e., a topic regarding New Year can be shifted, although related, by asking or giving a question like "what are you bringing for New Year's party?".

Although the *KeyGraph* did not identified all of the critical utterances, the results shown by the graph reflect the dynamic behavior that characterize the tendency to critically self-organize of the human-human chat.

4   Conclusion

In this paper an algorithm for modeling the discovery of critical utterances and their relationship within a chat section, was described. Making used of the visual computer aid the *KeyGraph* the dynamics of the chat were appreciated graphically. In the graph, most of the utterances that caused a change in the topic during the chat section were observed. Those utterances were forming clusters of topics and the links between clusters were representing different ways of transition from one topic to the other. Future works are oriented toward the application of the described algorithm for analyzing and improving computer chat.

Table 1. Critical Utterances.

Old Topic	Critical Utterance	New Topic
Speaker's nieces	78:So are you spring cleaning already	laundry weekends
laundry weekends	103:weekdays.. I'm so busy in the office ..	busy in office
busy in office	121:the end of next week ..enjoy the New Year	relaxing at home
relaxing at home	127:Chinese New Year, who are you going visit	Chinese New Year
Chinese New Year	165:So you mean we have lunch at your place	meeting New Year
meeting New Year	190:Did I tell you I bought a game set	game
game	250:but the girls will dominate as usual	mix meeting
mix meeting	260:When you see C.. you know what to do right	bringing presents
bringing presents	273:lunch time is it	having lunch
having lunch	290: Okay, have you done New Year shopping	shopping
shopping	307: Okay, so talk again tonight	end of chat

References

1. Ohsawa Y: Modeling the Process of Chance Discovery. In: Ohsawa Y. and McBurney P (eds) Chance Discovery. Springer, Berlin Heidelberg New York. (2003)
2. Montero C.A.S., Araki K.: Discovering Critically Self-Organized Chat. The Fourth IEEE International Workshop on Soft Computing as Transdisciplinary Science and Technology. To appear. (2005)
3. Cohen Phil: Dialogue Modeling. In: Cole R, Mariani J, Uszkoreit H, Varile G, Zaenen A, Zampolli A, and Zue V (eds) A Survey of the State of the Art in Human Language Technology. Cambridge University Press, Cambridge. (1998)
4. Sacks H, Schegloff E, Jefferson G: A Simplest Systematics for the Organization of Turn-taking in Conversation. In: Schenkein J, (editor) Studies in the Organization of Conversational Interaction. Academic Press, New York. (1978)
5. Searle RJ: Speech Acts: An essay in the Philosophy of Language. Cambridge University Press, Cambridge. (1969)
6. Allen FJ, Perrault RC: Analyzing Intention in Dialogues. Artificial Intelligence, 15(3):143–178. (1980)
7. Clark HH, Wilkes-Gibbs D: Referring as a Collaborative Process. Cognition, 22:1–39. (1986)
8. Wallace RS: A.L.I.C.E. Artificial Intelligence Foundation. http://www.alicebot.org
9. Bak P: How Nature Works, Oxford University Press. (1997)
10. University of South California, Dialogue Diversity Corpus, Dialogue 91, http://www-rcf.usc.edu/~billmann/diversity/DDivers-site.htm
11. Jensen HJ: Self-organizing Complex Systems. In: Ohsawa Y, McBurney P (eds.) Chance Discovery. Springer, Berlin Heidelberg New York. (2003)
12. Ohsawa Y, Benson NE, Yachida M: KeyGraph: Automatic Indexing by Co-occurrence Graph Based on Building Construction Metaphor. Proc. of Advanced Digital Library Conference, pp. 12-18. (1998)
13. Ohsawa Y, McBurney P: Chance Discovery. Springer, Berlin Heidelberg New York. (2003)

Optimizing Interference Cancellation of Adaptive Linear Array by Phase-Only Perturbations Using Genetic Algorithms

Chao-Hsing Hsu[1], Wen-Jye Shyr[2], and Kun-Huang Kuo[1]

[1] Dept. of Electronic Engineering, Chienkuo Technology University
Changhua 500, Taiwan, R.O.C.
chaohsinghsu@yahoo.com

[2] Dept. of Industrial Education and Technology, National Changhua University of Education
Changhua 500, Taiwan, R.O.C.
shyrwj@cc.ncue.edu.tw

Abstract. An antenna array is often used as an adaptive antenna. In this paper, the pattern nulling of a linear array for interference cancellation is derived by phase-only perturbations using genetic algorithms. This design for radiation pattern nulling of an adaptive antenna can suppress multiple interferences by placing nulls at the directions of the interfering sources, i.e., to increase the Signal Interference Ratio (SIR). One example is provided to justify the proposed phase-only perturbations approach based on genetic algorithms. The simulation results show that optimizing interference cancellation of adaptive linear array has been achieved by phase-only perturbations using genetic algorithms.

1 Introduction

The modern wireless communication asks for higher quality and better efficiency. Pattern nulling techniques are very important to cancel undesired interference in wireless communication [1].

Due to the amazing development of computers, applying modern numerical optimization techniques to radiation pattern design is becoming possible. A perturbation method consists of small perturbations in the element phases or amplitudes to obtain the desired nulling pattern, which has got much attention. A search procedure based on the genetic algorithm is used to obtain the required perturbations for the designed patterns with null steering. Genetic algorithm possess the ability of global search can avoid being trapped in local optima by using mutation technique. So, it is suitable for optimizing interference cancellation of adaptive linear array by phase-only perturbations. The proposed genetic algorithm for optimizing interference cancellation of adaptive linear array by phase-only perturbations is presented. The genetic algorithm doesn't need to know these values of the input signal and input interference and it just need to know these directions of user and interferences.

The optimization design of an adaptive phase-only nulling with phase arrays was focus on the binary genetic algorithm and one interfering source nulling technique was investigated in 1999 [2]. In this article, we focus on suppressing the multiple different

R. Khosla et al. (Eds.): KES 2005, LNAI 3681, pp. 561–567, 2005.

interfering sources and the decimal genetic algorithm is investigated. The array antenna design optimization by the decimal genetic algorithm performs much better than that by the binary genetic algorithm, which has been proved in 1999 [3]. A trinary-phased array, in which a phase quantization unit of phase shifters is 120 degrees, was examined in 1999 [4]. In the article, the value of βn is set in any value between $-\pi$ and π. The genetic algorithm was used to synthesize the directional circular arc array as a sectored antenna. Then the performance of this sectored antenna in door wireless millimeter wave channels was investigated in 2000 [5]. In the article, the resulting weights adjustably place nulls in the far field pattern in the directions of interferences.

In this paper, nulling techniques for a linear antenna array are investigated. The technique features are phase-only perturbations. Compared to phase-amplitude or phase-position perturbations, the structure of phase-only perturbations can be simplified [6]. This procedure for the proposed nulling techniques provides an iterative solution. The interferences can be cancelled. The excellent nulling results are derived in this paper. For our paper, the proposed numerical method can be used in a real time signal processing and the mobile communication system.

2   Deduction of Radiation Pattern Formula Available for Searching Optimal Solutions

For a linear array of $2N$ equispaced sensor elements as Fig. 1, an interfering signal with wavelength λ impinges on any two adjacent sensor elements by a distance d and from a direction θ with respect to array normal as shown in Fig. 2. The ν is the propagation speed of radio wave. Then, there is a time delay τ as follows:

$$\tau = \frac{d\sin\theta}{\nu} \tag{1}$$

The τ corresponds to a phase shift of $\frac{2\pi}{\lambda}d\sin\theta$.

$$\psi = \frac{2\pi}{\lambda}d\sin\theta = kd\sin\theta \tag{2}$$

The adaptive array factor for far field [7] is given by

$$AF(\theta) = \sum_{n=1}^{2N} w_n e^{j(n-1)\psi} \tag{3}$$

If the reference point is at the physical center of the array, the array factor becomes

$$AF(\theta) = \sum_{n=1}^{2N} w_n e^{j(n-N-0.5)\psi} = \sum_{n=1}^{2N} \alpha_e^{j[(n-N-0.5)\psi+\beta_n]} \tag{4}$$

where $2N$=number of elements, $w_n = \alpha_e^{j\beta_n}$ complex array weights at element n α =constant amplitude weight for all elements, β_n =phase shifter weight at element n, $\psi = kd\sin\theta$, and θ =an incidence angle of interfering signal or desired signal.

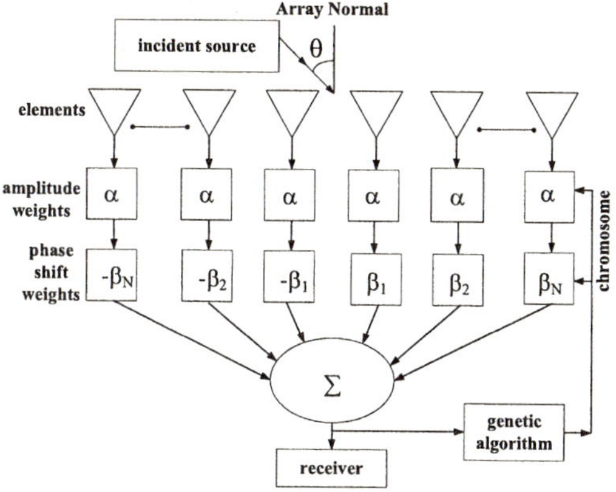

Fig. 1. Diagram of an adaptive linear array designed by phase-only perturbations using genetic algorithms.

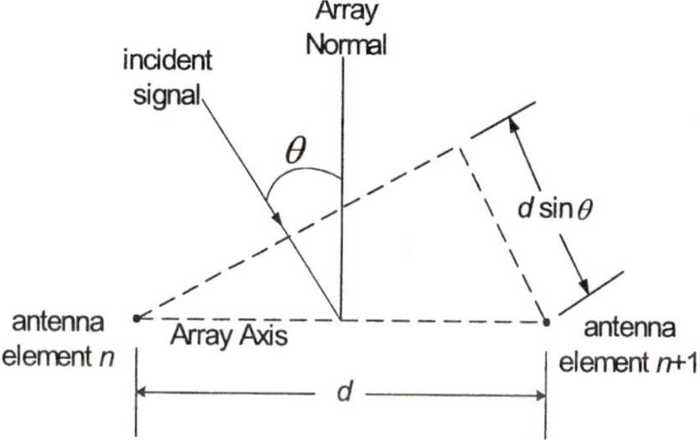

Fig. 2. Incident signal reaching any two adjacent elements.

If amplitude weights are constant and phase shift weights are odd symmetry, the equation (4) can be simplified to

$$AF(\theta) = 2\sum_{n=1}^{N} \alpha \cos[(n-0.5)\psi + \beta_n] \qquad (5)$$

The equation (5) is written in normalized form as follows:

$$AF_n(\theta) = \frac{1}{N}\sum_{n=1}^{N} \alpha \cos[(n-0.5)\psi + \beta_n] \qquad (6)$$

Using the optimization technique, the fitness function cannot exist the imaginary part. As the equation (6) only the real part is left, it is available for searching the optimal solutions using optimization technique. Genetic algorithm is a kind of optimization technique.

3   Genetic Algorithm for Pattern Nulling

In genetic algorithm, a set of solution to a problem is called chromosomes. A chromosome (a string of solution) is composed of genes (weights). The individual of the whole population is represented by a chromosome. The performance of the solution is called fitness. The fitness of a set of chromosome is evaluated. Then, new chromosomes are produced by using the selected candidates as parents and applying mutation and crossover operations. The new set of chromosomes is then evaluated again. This cycle continues till a suitable solution found. This suitable solution is called the optimal. A genetic algorithm adjusts the phase-only weights based on the goal of the power of the array in the interfering direction. The goal is to minimize total output power to the receiver. We need look for the relative power, whereas $P \propto V^2 \propto E^2 \propto AF(\theta)^2$. So, the fitness function is the square of $AF_n(\theta)$ in equation (6).

In this section we described the implementation of the genetic algorithm as follows [8-9]:

Step 1. Initialization: In this problem, a set of chromosomes is randomly generated. The problem is to minimize a function of β_n, The initial step may be to generate a collection of random matrix vectors (β_{nm}), $n=1,2,3,\ldots,N$, $m=1,2,3,\ldots,P$. The number of genes is $2\,N$. P is the number of chromosomes or population size.

Step 2. Evaluation: To evaluate each chromosome (a set of weights) fitness.

Step 3. Selection: Within the algorithm, population selection is based on the principle of survival of the fittest, which is based on the Darwinian's concept of "Natural Selection" [10]. The survivors of the current population are decided from the survival rate p_s. A random number generator is used to generate random numbers whose values are between 0 and 1. If the value of random number is smaller than p_s, this chromosome survives; otherwise, it dose not survive. The best fitness of the population always survives. If the best fitness of the population can not always survive, the fitness value can not be sure in a monotonous convergence (i.e., convergence is not for sure).

Step 4. Crossover: Pairs of parents are selected from these survivors. Single point crossover is selected to produce the next generation. The weight string from the beginning of chromosome to the crossover point is copied from one parent, the rest is copied from the other parent.

Step 5. Mutation: Some useful genes are not generated in the initial step. This difficulty can be overcome by using the mutation technique. The basic mutation operator randomly generates a number as the crossover position and then changes the value of this gene randomly.

Step 6. Stopping criteria: Steps 2-5 are repeated till the predefined number of generations has been reached. The best set of weights (phase shift weights) can be generated after termination.

The flow chart of genetic algorithms is given in Fig. 3.

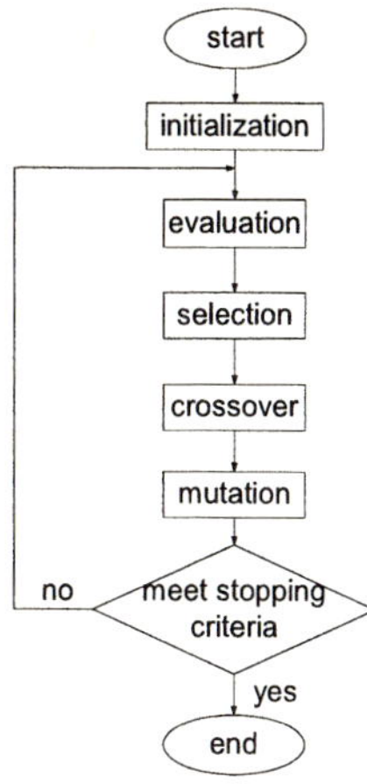

Fig. 3. Flow chart of genetic algorithms.

4   Results of Computer Simulations

The values assigned to the necessary variables used by genetic algorithms are as follows: population size P equals to 20; the selected survival rate p_s equals 0.5; the probability of crossover is 0.5; the proportion of mutation is 0.05.

In addition, in this problem, we assume a linear antenna array composing of 20 isotropic elements. So, $N = 10$. The distance d of any two adjacent elements is half of λ. The value of β_n is set between $-\pi$ and π. The unit of β_n is rad. The fitness function is the square of $AF_n(\theta)$ in equation (6). The value of α is constant and set between 0.1 and 1. In this case, let α equal 1. Pattern nulling skill can be used to suppress multiple interferences (including noises). This skill is able to do the cancellation of multiple interferences for different incident directions. One example is given as follows.

Example. The interfering signal's direction is from 40^0 with respect to the array normal. The pattern nulling is derived in the 40^0 interfering direction. The genetic algorithm stops after 250 generations. The result is listed in Table 1. The convergent map is shown in Fig. 4, and the radiation pattern is shown in Fig. 5. In our paper, we focus on the interference cancellation in order to increase the Signal Interference Ratio (SIR). As we know, when we optimized the signal to interference ratio, the noise can be ignored always. Actually, the main purpose of our paper is to propose a method of adjustable cancelling interfering signal in the mobile communication. By phase-only perturbation method, the nulling design of an adaptive antenna has been studied by the approach of the genetic algorithm. The excellent nulling results have been derived. In this example, whatever direction the desired signal is from, the SIR has 40 dB at least shown in Fig. 5. For the nulling design, the interferences can be suppressed effectively.

5   Conclusions

Pattern nulling design of adaptive antenna by phase-only perturbations using genetic algorithms is proposed and achieved. In the paper, first, the uplink output power pattern

Table 1. The weight vector $[\beta_n]$ for pattern nulling in the interfering directions.

Nulling direction at 40^0			
$\beta_1 =$	-0.364	$\beta_2 =$	-2.229
$\beta_3 =$	-1.394	$\beta_4 =$	0.195
$\beta_5 =$	-2.977	$\beta_6 =$	0.402
$\beta_7 =$	1.149	$\beta_8 =$	0.622
$\beta_9 =$	-2.016	$\beta_{10} =$	0.088

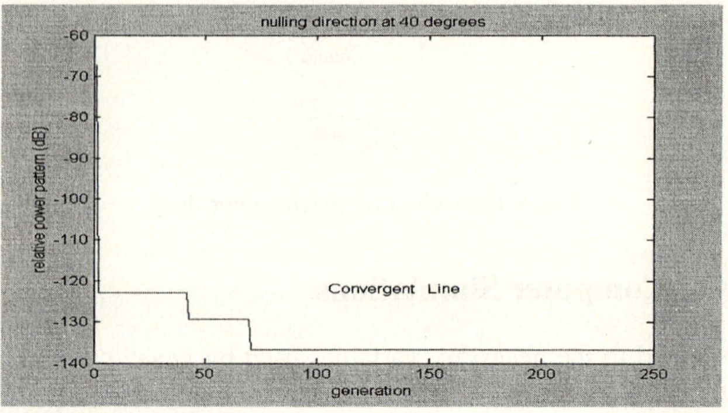

Fig. 4. Relative interference power getting convergent pattern for nulling direction at 40^0.

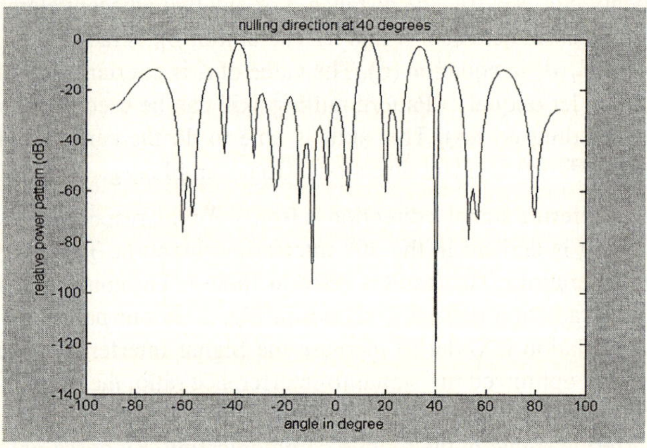

Fig. 5. Adaptive array pattern for single interfering source at 40^0.

formula based on phase composed by adaptive antenna array is deduced. In order to be able to adopt optimized techniques to search the optimal solution, the formula is reformed through assuming amplitude weights are constant and phase shift weights are in odd symmetry.

The genetic algorithm is applied to find the optimal weight vector of the proposed adaptive antenna. It is a useful and accurate search technique which is simple and easily implemented without gradient-based search procedure. Genetic algorithm possess the ability of global search can avoid being trapped in local optima by using mutation technique. The approach of genetic algorithms in linear array places null in the direction of interference with phase perturbations to the far-field pattern.

References

1. Roy, R. H.: Adaptive Antennas for Wireless Information Networks. IEEE Technical Application Conference, (1998) 60-65
2. Haupt, R. L.: Phase-Only Adaptive Nulling with a Genetic Algorithm. IEEE Transactions on Antennas and Propagation, Vol. 45(6) (1997) 1009-1015
3. Lee, Y. H., Marvin, A. C., Porter, S. J.: Genetic Algorithm using Real Parameters for Array Antenna Design Optimization. 1999 High Frequency Postgraduate Student Colloquium, IEE MTT/ED/AP/LEO Societies Joint Chapter, United Kingdom and Republic of Ireland Section, (1999) 8-13
4. Masaharu Fujita: A Trinary-Phased Array. IEICE Trans. Communications, Vol. E82-B(3) (1999) 564-566
5. Chen, C. H., Chiu, C. C.: Synthesizing Sectored Antennas by the Genetic Algorithm to Mitigate the Multipath of Indoor Millimeter Wave Channel. IEICE Trans. Fundamentals, Vol.E83-A(2) (2000) 350-356
6. Hsu, C. H., Babij, T. M.: Pattern Nulling of Adaptive Antenna by Phase and Amplitude Perturbations using Genetic Algorithm. KES-2001 Fifth International Conference on Knowledge-Based Intelligent Information Engineering Systems & Allied Technologies, Part 2, (2001) 1047-1051
7. Miller, M.: Introduction to Adaptive Arrays, Published by Wiley, New York, (1986)
8. Goldberg, D. E.: Genetic Algorithms in Search Optimization and Machine Learning, Addison- Wesley Publishing Company, (1989)
9. Pan, J. S., McInnes, F. R., Jack, M. A.: Codebook Design Genetic Algorithms. IEE Electronics Letters, Vol. 31(17) (1995)1418-1419
10. Dawkins, R.: The Selfish Gene, Oxford University Press (1976)

Optimizing Linear Adaptive Broadside Array Antenna by Amplitude-Position Perturbations Using Memetic Algorithms

Chao-Hsing Hsu[1] and Wen-Jye Shyr[2]

[1] Dept. of Electronic Engineering, Chienkuo Technology University
Changhua 500, Taiwan, R.O.C.
chaohsinghsu@yahoo.com
[2] Dept. of Industrial Education and Technology,
National Changhua University of Education
Changhua 500, Taiwan, R.O.C.
shyrwj@cc.ncue.edu.tw

Abstract. This paper presents an innovative implementation method of the broadside antenna by using adaptive array antenna. Compared with the conventional ones, its performance can be improved because it is possible to null the interfering signals adjustably and at the same time maximize the main lobe towards the array normal. Thus the optimal radiation pattern can be obtained. The conventional broadside array only can make a main lobe toward the array normal, but it can not cancel the interferences. So, it is easy to be seen for the significance of this proposed method. In many applications, it is required to have the maximum radiation of an array directed toward the array normal. An optimal radiation pattern design for an adaptive broadside array antenna is not only to derive the maximum power radiation at the array normal direction but also to suppress interference by placing nulls at the directions of the interfering sources. The signal to interference ratio (SIR) can be maximized. Memetic algorithm is used for the search of the optimal weighting of array factor of the optimal radiation pattern by amplitude-position perturbations.

1   Introduction

Conventional broadside antennas can form a main lobe always towards the array normal. Their structures are simple, but they are not able to suppress the interferences. In order to improve the performance of the broadside antenna, implementing the function of the broadside antennas with adaptive antenna array is a possible approach.

A broadside array antenna must keep maximum radiation toward array normal. In the previous papers [1-3], adaptive pattern nulling technique minimizes the power of the interfering signal coming from any direction by putting a null in its direction. In this paper, the optimal radiation pattern of adaptive broadside array will not only be able to suppress interferences by placing nulls, but also be able to derive the maximum power radiation in the array normal direction. A deduced formula of the radiation pattern of adaptive broadside array antennas is suitable for the optimal solution search. The formula can always keep maximum radiation of a broadside array antenna toward array

R. Khosla et al. (Eds.): KES 2005, LNAI 3681, pp. 568–574, 2005.

normal. At the same time, it can cancel all the interferences from the interfering sources. A perturbation method consists of small perturbations with the element parameters. The technique features are amplitude and position perturbations.

Memetic algorithm uses local search as improvement procedure. For many problems, there exists a well-developed, efficient search strategy for local improvement, e.g., hill-climber for optimization. These local search strategies complement the global search strategy of the genetic algorithm. The idea of combining genetic algorithm and local search heuristics for solving combinatorial optimization problems were proposed to improve the search ability of the genetic algorithm.

2 Deduction of Radiation Pattern Formula Available for Searching Optimal Solutions

For an adaptive linear array of $2N$ equispaced sensor elements as Figure 1, an interfering signal with wavelength λ impinges on any two adjacent element n and element $n + 1$ by a distance d from the direction θ with respect to array normal as shown in Figure 2.

To reach any two adjacent elements, there is a time delay τ as follows [4]:

$$\tau = \frac{d \sin \theta}{\nu} \tag{1}$$

The ν is the propagation speed of radio wave. The τ corresponds to a phase shift of ψ.

$$\psi = \frac{2\pi}{\lambda} d \sin \theta = kd \sin \theta \tag{2}$$

The array factor of adaptive linear array for far field using phase perturbations is given by

$$AF(\theta) = \sum_{n=1}^{2N} \alpha_n e^{j(n-1)\psi} \tag{3}$$

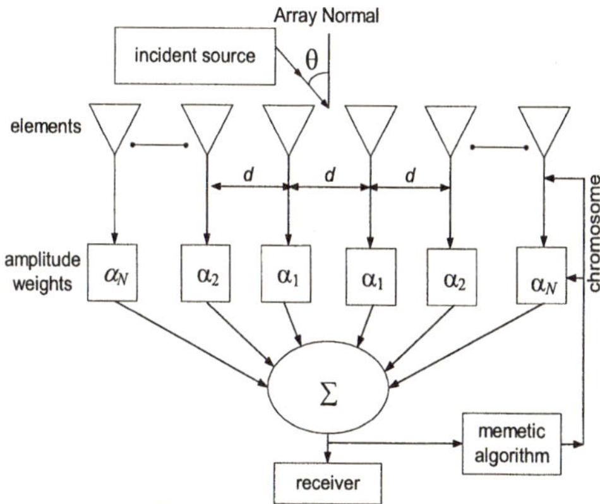

Fig. 1. Diagram of an adaptive linear array designed by amplitude-position perturbations using memetic algorithm.

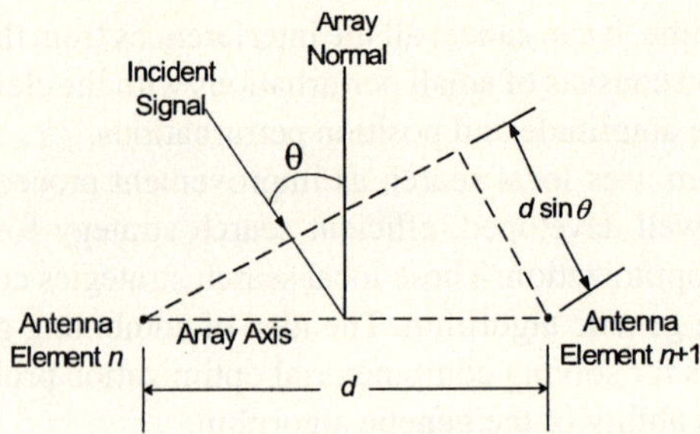

Fig. 2. The incident signal reaching any two adjacent elements.

The array factor of adaptive linear array for far field using phase perturbations is given by

$$AF(\theta) = \sum_{n=1}^{2N} \alpha_n e^{j(n-1)\psi} \tag{4}$$

where α_n = amplitude weight at element n, $2N$ = number of elements, θ = incident angle of interfering signal or desired signal.

If the reference point is at the physical center of the array, the array factor becomes

$$AF(\theta) = \sum_{n=1}^{2N} \alpha_n e^{j(n-N-0.5)\psi} \tag{5}$$

If amplitude weights are in even symmetry, the equation (5) can be simplified to

$$AF(\theta) = 2 \sum_{n=1}^{N} \alpha_n \cos[(n-0.5)\psi] \tag{6}$$

In order to increase the performance of adaptive array, the antenna position perturbations in even symmetry are added. The array factor becomes as

$$AF(\theta) = 2 \sum_{n=1}^{N} \alpha_n \cos[(n-0.5)k(d+(\Delta d_n/(n-0.5)))\sin\theta] \tag{7}$$

where Δd_n is the position weight at element n .

The equation (7) can always keep maximum radiation toward array normal no matter how the weights change. In addition, as only the real part is left, it is available for searching the optimal solutions using optimized technique.

3   Memetic Algorithm for Pattern Nulling

Memetic algorithm is a kind of optimization technique [5-6]. With fitness function given by the square of $AF(\theta)$ in equation (7), a memetic algorithm is used to adjust the position weights based on the power of the array in the interfering direction. The goals are to minimize the total output power of the interfering signal. During the process of memetic iteration, the weight vector kept for the next step iteration should make the output power of the interfering signal to be decreased monotonically. Owing to a broadside array, the main lobe is always toward 0^0 with respect to array normal using equation (7), the radiation pattern could not only make the main lobe is always toward 0^0 but also minimize the total output power of the interfering signal at same time. Obviously, this technique can maximize the signal to interference ratio (SIR) for broadside array antenna.

Steps of memetic algorithm are as follows:

Step 1. Initialization: A set of chromosomes is randomly generated. A chromosome (weight vector) is composed of genes (weights). For this problem, the genes are α_n and Δd_n. So, The initial step is generating a collection of random matrix vectors $[\alpha_n; \Delta d_n]$, $n=1,2,3\ldots N$. The number of genes is $2N$. Define a vector and variable to which the gradually optimized chromosome and its fitness are saved separately. Their initial values are the first chromosome of the generated chromosome set and its fitness.

Step 2. Evaluation: For every chromosome, two objective functions are calculated for evaluating its fitness. One is the output power of the desired signal; the other is the output power of the interfering signal. Check every chromosome's fitness step by step. Compared with the present best fitness, if one chromosome can not only give the higher output power of the desired signal but also the lower output power of the interfering signal, renew the value of the defined vector and variable with this chromosome and its fitness. Otherwise, keep their values unchanged.

Step 3. Selection: A random number generator is used to generate random numbers whose values are between 0 and 1. If the value of random number is smaller than p_s, this chromosome survives; otherwise, it dose not survive. The best fitness of the population always survives.

Step 4. Crossover: Pairs of parents are selected from these survivors. Single point crossover is selected to produce the next generation. The weight string from the beginning of chromosome to the crossover point is copied from one parent and the rest is copied from the second parent.

Step 5. Mutation: Mutation is operated at genes of chromosomes randomly. Selection and crossover effectively search and recombine the chromosomes, but occasionally they may lose some potentially useful genes and it is also possible that some useful genes are not generated in the initial step. The required result cannot be obtained or it is difficult to achieve due to the lack of these useful genes. This problem can be overcome by using the mutation technique.

Step 6. Local Search: A hill-climber memetic algorithm, which used as local search. Apply the local search procedure in the current population [7].

Step a: Specify an initial solution x.

Step b: Examine a neighborhood solution y of the current solution x.

Step c: If y is a better solution than x, replace the current solution x with y and return to step a.

Step d: If randomly chosen k neighborhood solutions of the current solution x have been already examined (i.e., if there is no better solution among the examined k neighborhood solutions of x), then end this procedure. Otherwise return to Step 2.

Step 7. Stopping Criteria: If the number of the current generation is equal to or larger than the predefined number of generations, end the algorithm. Otherwise return to the Step 2. The best set of weights (amplitude and position weights) can be generated after termination.

4   Simulations Results

For this problem, the necessary variables of memetic algorithm are defined as follows: population size p equals to 20; the selected survival rate p_s equals to 0.5; the probability of crossover is 0.5; the proportion of mutated genes is 0.5.

In this problem, assumed a linear antenna array is composed of 20 isotropic elements. So, N equals 10. The distance d of two adjacent elements is half of λ. The optimal radiation pattern technique features are by position perturbations. The fitness function is given by the square of $AF(\theta)$ in equation (7). Both amplitude weights and position weights are in even symmetry. The value of α_n is set between 0.1 and 1. The value of Δd_n is set between -0.24λ and 0.24λ.

The stopping criteria of memetic algorithm are no better solution among the examined 50 neighbor generations, and the number of the current generation is equal to or larger than 100 generations.

Case and Result: With respect to array normal, the interfering signal directions are from 60^0 and -60^0 respectively, and main lobe is toward 0^0. The memetic algorithm is going to stop if it meets the stopping criteria. Then, the weights for the optimal radiation of broadside array are derived from the memetic algorithm calculation as shown in Table 1. The relative power pattern is getting convergent as Figure 3. The optimal radiation pattern is derived and showed in Figure 4. The SIR is equal to 94 dB so that the interfering noise can be completely ignored. So, the efficiency of the memetic algorithm approach is very high, and the interference can be completely ignored in real time signal processing.

The interfering signal directions are from 60^0 and -60^0 respectively, and the maximum radiation toward 0^0. The radiation patterns are mapped from 90^0 to 270^0 as showed in Figure 5 in polar coordinate in dB scale.

Table 1. The weights for the optimal radiation pattern.

Main lobe toward 0^0 from 60^0 and -60^0	Nulling direction in
$\alpha_1 = 0.862$	$\Delta d_1 = -0.049$
$\alpha_2 = 0.680$	$\Delta d_2 = -0.173$
$\alpha_3 = 0.695$	$\Delta d_3 = 0.184$
$\alpha_4 = 0.822$	$\Delta d_4 = -0.238$
$\alpha_5 = 0.940$	$\Delta d_5 = -0.011$
$\alpha_6 = 0.768$	$\Delta d_6 = 0.173$
$\alpha_7 = 0.787$	$\Delta d_7 = 0.238$
$\alpha_8 = 0.770$	$\Delta d_8 = 0.060$
$\alpha_9 = 0.616$	$\Delta d_9 = 0.135$
$\alpha_{10} = 0.933$	$\Delta d_{10} = 0.209$

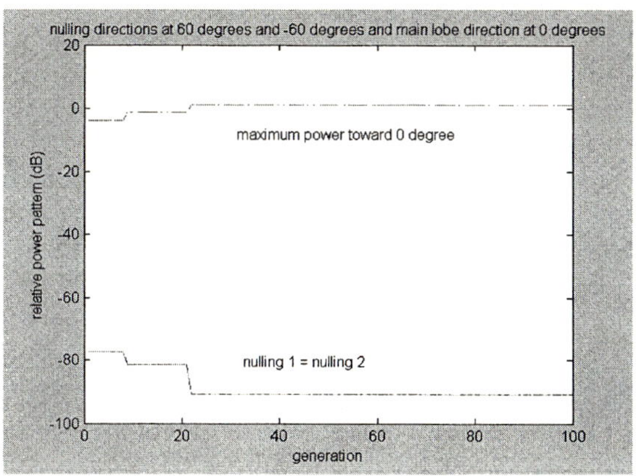

Fig. 3. Relative power pattern getting convergent using iteration.

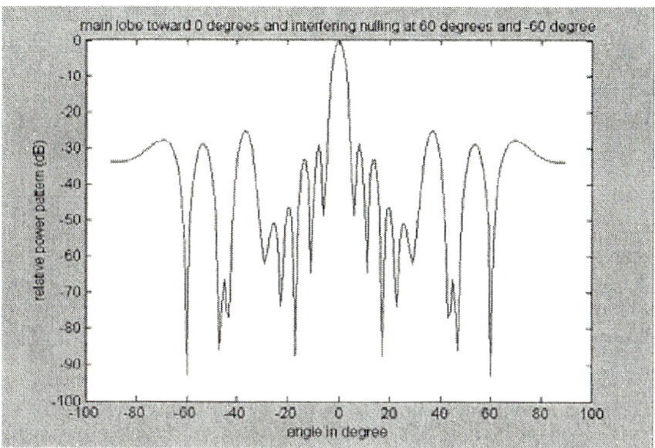

Fig. 4. Optimal radiation pattern of broadside array antenna for main lobe toward 0^0 and interfering source at 60^0 and -60^0 in Cartesian coordinate.

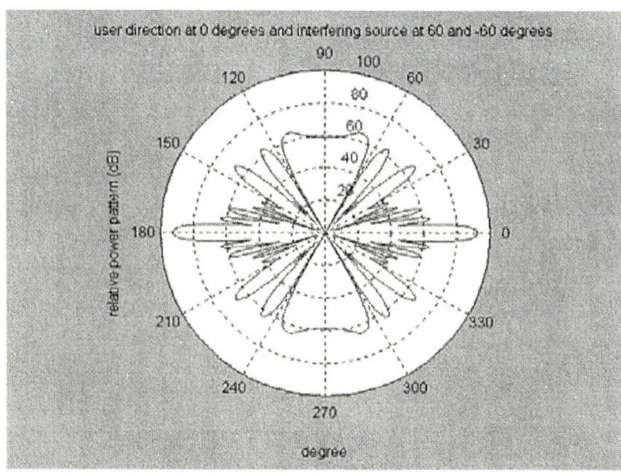

Fig. 5. Radiation pattern of adaptive broadside array antenna in polar coordinate.

5 Conclusions

The optimal radiation pattern design of adaptive broadside array antenna based on amplitude-position perturbations using memetic algorithm has been proposed and achieved. For a broadside array, the desired signal direction is always at 0^0 with respect to array normal. In the paper, first, the output power formula based on amplitude and position for broadside array antenna composed by adaptive antenna array was deduced. In order to be able to adopt memetic algorithm to search the optimal radiation pattern, the formula is reformed through assuming amplitude and position weights are in even symmetry. Memetic algorithms are applied to find the optimal radiation pattern of the proposed adaptive antenna. As the optimal radiation pattern means that the power of desired signal should be highest and the power of interfering signals should be lowest as much as possible at the same time.

In this paper, the optimal radiation pattern of adaptive broadside array was obtained. The convergence curves of memetic algorithm iteration show that it is effective for this problem. The approach of memetic algorithm in an adaptive linear array places nulls in the directions of interferences and forms maximum main lobe in the direction of array normal to the far-field radiation pattern. The optimal adaptive radiation pattern of broadside array antenna has been derived in this paper.

References

1. Hsu, C.H. and Babij, T.H. :Pattern Nulling of Adaptive Antenna by Phase and Amplitude Perturbations Using Genetic Algorithm. Knowledge-Based Intelligent information Engineering Systems & Allied Technologies, KES'2001, Part 2, (2001) 1047-1051
2. Liao, W.P. and Chu, F.L. :Array Pattern Nulling by Phase and Position Perturbations with the Use of the Genetic Algorithm. Microwave and Optical Technology Letters, Vol. 15(5) (1997) 251-256
3. Tennant, A., Dawoud, M.M. and Anderson, A.P. :Array Pattern Nulling by Element Position-Perturbations Using a Genetic Algorithm. Electronics Letters, Vol. 30(3) (1994) 174-176
4. Monzingo, M., :Introduction to Adaptive Arrays. Wiley, New York, (1986)
5. Moscato, P. and Norman, M.G. :A Memetic Approach for the Travelling Salesman Problem-Implementation of a Computational Ecology for Combinatorial Optimization on Message-Passing Systems. Proceedings of the International Conference on Parallel Computing and Computer Applications, IOS Press, 1992.
6. Knowles, J. D. and Corne, D. W. :M-PAES:A Memetic Algorithm for Multiobjective Optimization. Proceedings of the IEEE International Conference on Evolutionary Computation, Vol. 1 (2000) 325-332
7. Ishibuchi, H. and Murata, T. :Multi-Objective Genetic Local Search Algorithm. Proceedings of the IEEE International Conference on Evolutionary Computation, IEEE Press, (1996) 119-124

Stability in Web Server Performance with Heavy-Tailed Distribution

Takuo Nakashima and Mamoru Tsuichihara

Department of Information Science
Kyushu Tokai University, 9-1-1 Toroku Kumamoto-shi, Japan
{taku@ktmail,40mie102@stmail}.ktokai-u.ac.jp

Abstract. In this research, we focused on the heavy-tailed property on Web servers, and studied evaluation experiments by an active measurement method for Web servers. Our experiments have resulted in three conclusions; first, it has been shown that the total transmission time is dominated not only by the file size but also by first data transmission time. It implies Web data transmission has the robustness of heavy-tailed property. Second, we demonstrated that if servers have a heavy load on the initial data transmission phase, to find that the performance of server rarely have stable small variance. Third, the convergence property of performance evaluation measures of Web servers was examined.

1 Introduction

Since the seminal study of Leland et al.[1], self-similarity of network traffic has been widely adopted in the modeling and analysis of network performance. The properties of self-similarity are found in TCP/IP protocol stacks. Crovella et al.[2] showed that file sizes of Web servers had heavy-tailed distribution and Web file transmission times had appeared to show heavy-tailed characteristics. The relation between file sizes and self-similar traffic was explored in Park et al. [3]. Crovella et al.[2] described that self-similarity in Web traffic which might arise due to the heavy-tailed distribution of file sizes present on the Web. This manifestation of the application layer causes the self-similarity at multiplexing points in the network layer in the form of self-similar traffic[4]. In conjunction with this evidence, file size distribution in UNIX has heavy-tailed distribution, and ftp traffic has the heavy-tailed property of Pareto distribution with $0.9 \leq \alpha \leq 1.1$[5]. To analyze the cause of self-similarity in detail, one refined method of work is to characterize the model of user behavior, and the other method is to observe the effectiveness of mediation of TCP and flow controlled UDP.

Our purpose in this research is to find out the characters of Web servers to evaluate their systems working with the heavy-tailed property described above. Performance evaluation methods for Web servers are classified into two methods based on the measuring node. One method focuses on the local performance activities, e.g. CPU available ratio and memory access ratio, measured on the Web servers. The other method measures a remote server by sending measuring packets. Using this kind of methods, we can collect the data, e.g. response time including RTT (Round Trip Time).

R. Khosla et al. (Eds.): KES 2005, LNAI 3681, pp. 575–581, 2005.
© Springer-Verlag Berlin Heidelberg 2005

In this research, we stand on the client, i.e. a remote site and analyze packets by the active measurement method. Normal users are sensitive to the total response time of Web servers and Web applications are currently made using a lot of different technology, e.g. PHP, ASP, JSP and Servlet. As we want to compare different type servers, we will evaluate them with the same metrics, i.e. the total response time, on different type remote servers. That is because when users build a Web server, users cannot necessarily afford to construct high cost, performance machines. In many case, users tend to accept a trade-off between hardware cost and software development cost. That means that users can expect the server selection guideline to compare with other servers.

The construction of this paper is as follows. First we will define the heavy-tailed property, and explain a method of the evaluation experiments. Second we will show the results of the experiments, and illustrate stability of Web servers with heavy-tailed property.

2   Background

The heavy-tailed distribution is defined as follows[6]:

$$P[X > x] \sim x^{-\alpha}, \text{ as } x \to \infty, 0 < \alpha < 2. \tag{1}$$

Regardless of the behavior of the distribution for small values of the random variable, if the asymptotic shape of the distribution is hyperbolic, it is heavy-tailed.

The simplest heavy-tailed distribution is the Pareto distribution. The Pareto distribution is hyperbolic over its entire range; its probability density function is

$$p(x) = \alpha k^\alpha x^{-\alpha-1}, \alpha, k > 0, x \geq k$$

and its cumulative distribution function is given by

$$F(x) = P[X \leq x] = 1 - (k/x)^\alpha.$$

The parameter k represents the smallest possible value of the random variable, and is called the location parameter.

To assess the property of heavy-tailedness, we employ $log - log$ plots of the complementary cumulative distribution $\overline{F}(x) = 1 - F(x) = P[X > x] = (k/x)^\alpha$ for the random variable X. Plotted in this way, heavy-tailed distributions have the property that $d\log \overline{F}(x)/d\log x = -\alpha, x > \theta$ for some θ. To check the property of heavy-tailedness, we form $log - log$ plots, and look for approximately linear behavior in the tail. This heavy-tailed property causes the long-range dependence that is one of the properties of self-similarity. And Hurst parameter is related to the tail index by $H = (3 - \alpha)/2$.

On the other hand, if the distribution has the provability of Poisson Process, its cumulative distribution function is given by $\overline{F}(x) = P[X > x] = e^{-\lambda x}$, and has the property that $d\log_e \overline{F}(x)/dx = -\lambda$ When we employ log plots only for the distribution function, the slope becomes a linear line.

3   An Method of Evaluation Experiments

In these experiments, we implemented an original measuring program to collect the processing time at remote Web servers.

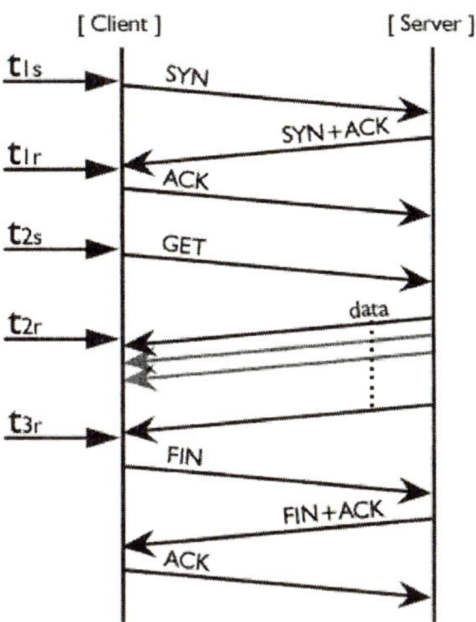

Fig. 1. This Figure illustrates the packet sequence of the measuring program. First the measuring host, client, stamps the time (t_{1s}) before the SYN packet of TCP at the connection establishment phase, and then the time (t_{1r}) is stamped after receiving the SYN+ACK packet. Just before sending data, i.e. GET request on HTTP, the measuring host stamps the time (t_{2s}), and then the measuring host stamps the time (t_{2r}) after receiving the first of the data from the Web server. Finally the time (t_{3r}) is stamped after receiving the final data. As a result, we can obtain five different time stamps.

We will regard $t1 = t_{1r} - t_{1s}$ as RTT, and will consider $t_2 = t_{2r} - t_{2s}$ as the initial processing time including RTT on the Web server, and $t_3 = t_{3r} - t_{2r}$ as the data processing time on the Web server, and $t'_2 = t_2 - t_1$ as the initial processing time without RTT, and $t = t_1 + t_2 + t_3 = t_{3r} - t_{1s}$ as the total transmission time. If we consider that the data processing time is affected by the file size, we propose the packet unit processing time (t'_3) with the assumption that one packet constitutes 1460 bytes, and we will compare the results relating to t_3 and t'_3. Both t'_2 and t'_3 are the packet unit processing time.

Using this program we sampled the top page data 3 times in experiment (I) and 20 times in experiment (II) for 1999 Web sites.

4   Property Evaluation Experiment

In this section, we argue the heavy-tailed property and its dominant factors based on data sets in experiment (I).

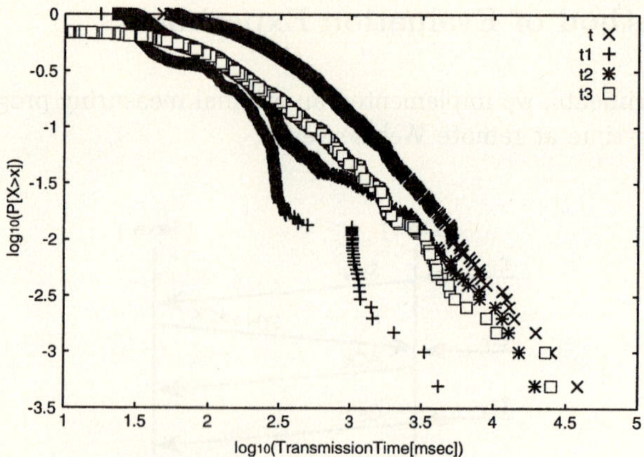

Fig. 2. This Figure shows the complementary cumulative distribution of the total transmission time (t) including RTT (t_1), t_2 and t_3.

4.1   Analysis of Server Transmission Time

This Fig. 2 implies that this distribution has the heavy-tailed property. The study of Crovella et al. [2] an estimate of $\alpha = 1.29$ and the current distribution has a slope of $\alpha = 1.34$. Both slopes have a similar curvature, but current distribution declines one digit less than the old text files. This means that current Web servers respond more quickly to requests from Web clients.

We show the relation between t and t_2 and t_3 illustrating that not only t_3 but also t_2 dominates the tail part of t. As a result of this evidence, first of all the total transmission time induces heavy-tailed distribution, and this causal mechanism is robust with respect to change in Web server overloads, i.e. the initial setting time or the file transmission time. Each Web server has the variation that the total transmission time is dominated by the initial setting time before data transmission or is dominated by file transmission time. In addition the heavy-tailed property is maintained irrespective of different dominated elements of Web performance.

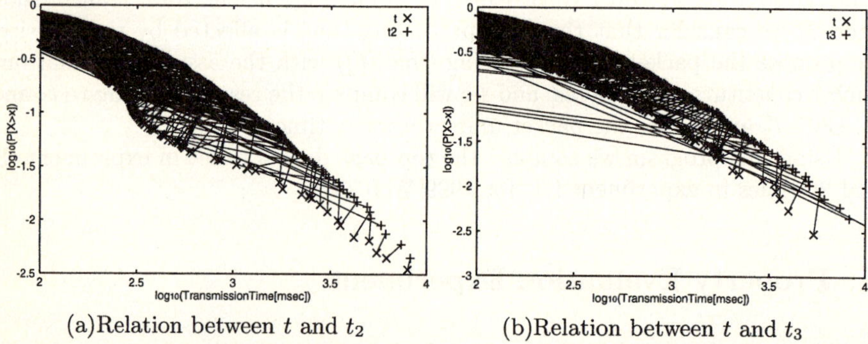

(a)Relation between t and t_2 (b)Relation between t and t_3

Fig. 3. We connected a plot of (a)t and t_2 and (b)t and t_3 with a straight line if data sets were observed on the same Web server.

4.2   Classification by the Dominant Factor for t

We discuss performance on classified Web servers as to whether the total transmission time is dominated by t_2 or t_3.

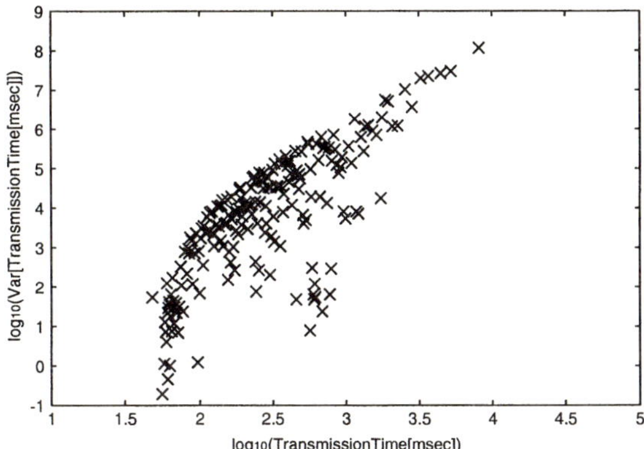

Fig. 4. The vertical axis shows the mean of $log(t_3)$ and the horizontal axis shows the variance of $log(t_3)$, and this figure shows performance distribution of Web server based on t_3.

This figure shows that the upper boundary of the variance gradually increases linearly and the lower boundary spreads widely to a small variance. As a whole, performance tends to grow to the bigger mean and variance, servers which have only a small variance might have the stable processing ability of the Web system.

Fig.5(a) shows the distribution of t_2 and t_3 for Web server performance dominated by t_2. All the plotted data covers the upper part of their performance, i.e. these servers wasted initial setting time and transmission time, and these servers could not operate in a stable state. These servers dominated by t_2 seemed to operate always at their upper limit.

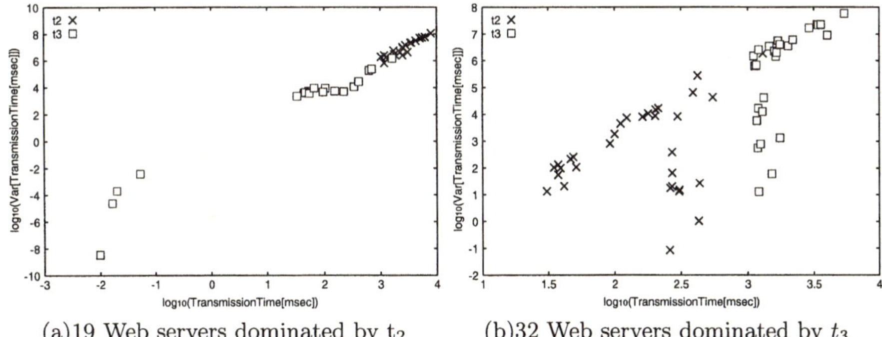

(a)19 Web servers dominated by t_2 (b)32 Web servers dominated by t_3

Fig. 5. Each Web servers have the total transmission time (t_3) more than 1 millisecond. The axes were constructed to depict the mean and variance with $log - log$ order.

On the other hand, Fig.5(b) illustrates the performance distribution of Web servers dominated by t_3. This data was plotted on a linearly increasing upper boundary and one part of lower variance compared with Fig.5(a). This lower degree of variance seemed to show that these servers operate under stable conditions.

5   Stability of Performance Metrics

In this section we discuss the convergence property of each element of Web performance metrics using the data sets in experiment (II). In this paper, convergence properties show the average of two sets of samples of 2 and 20 parameters on the same Web server respectively.

We adopted the measuring metrics of the two packet unit transmission times, t_2' and t_3' to eliminate the affection of file size, We have evaluated the Web server performance on the map of mean and variance plot.

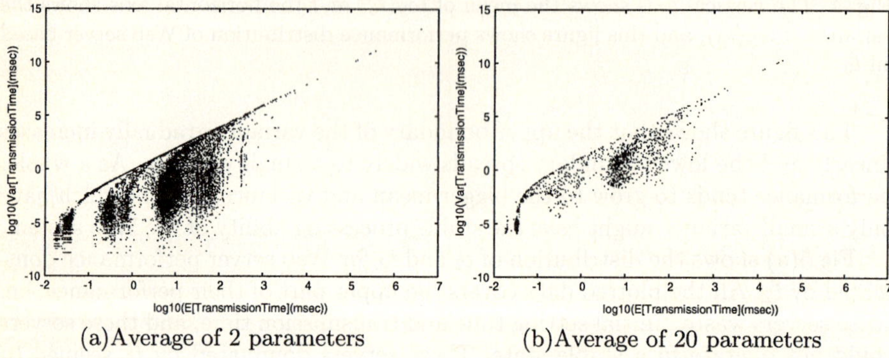

(a)Average of 2 parameters   (b)Average of 20 parameters

Fig. 6. Mean and variance plot in both axes are $\log - \log$ order for the packet unit transmission time t_3'.

Fig.6(a) to show t_3' performance for a set of 2 parameters indicates this performance has the upper boundary for the variance. In comparatively short mean time, under 100 millisecond, these values of variance spread to the small value. Compared to a set of 2 samples, this Fig.6(b) shows that second-order moment, namely variance, converged more rapidly than first-order moment, e.g., mean, and the mean value spreads widely over timescales. The plotted data can be considered to present the essential performance character of Web servers.

On the other hand, the plot of t_2' performance of a set of 2 samples scattered widely at small mean value in Fig. 7(a). The upper boundary for variance was also found over large mean value. Fig.7(b) illustrates t_2' performance of 20 samples, and it shows that the value of variance quickly converged to the upper boundary. The spread of mean is restricted to the plot between 1 and 10 millisecond for both mean and variance.

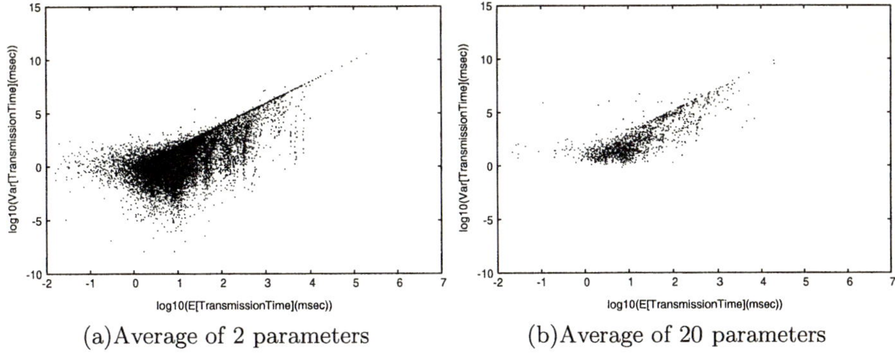

(a)Average of 2 parameters   (b)Average of 20 parameters

Fig. 7. Mean and variance in both axes are $\log - \log$ order for the packet unit transmission time t_2'.

6   Conclusion

In this research, we focused on the heavy-tailed property of Web servers on the Internet, and observed their performance using active measurement methods. The results of our experiments showed that the total transmission time was dominated not only by the file size but also by the first data transmission time. In addition these heavy-tailed distributions are robust with respect to the total transmission time.

We classified Web servers based on whether the tail part of heavy-tailed distribution is constructed by t_2 or t_3, and evaluated their performance. Servers which required on initial setting time had large variances in the transmission time. Whereas servers which stayed in a stable state for the initial setting time tended to have a smaller variance for one packet transmission time. Such stability as convergence property on the performance evaluation plots can be characterized.

References

1. Leland, W., Taqqu, M., Willinger, W. and Wilson, D.: On the Self-Similar Nature of Ethernet Traffic (Extended Version), IEEE/ACM Trans.Networking, **(2)**,1 (1994) 1–15
2. Crovella, M. and Bestavros, A.: Self-similarity in World Wide Web traffic: evidence and possible causes, Proc. of the 1996 ACM SIGMETRICS International Conference on Measurement and Modeling of Computer Systerms, (1996) 160–169
3. Park, K. Kim, G. and Crovella, M.: On the relationship between file sizes, transport protocols, and self-similar network traffic, Proc. IEEE International Conference on Network Protocols, (1996) 171–180
4. Park, K. ,Kim, G. and Crovella, M. E.: The Protocol Stack and Its Modulating Effect on Self-similar Traffic, Self-Similar Network Traffic and Performance Evaluation, K.Park and W.Willinger,Eds.,Wiley-Interscience, New York, (2000) 349–366
5. Paxson, V. and Floyd, S.: Wide-Area traffic: the failure of Poisson modeling, IEEE/ACM Trans. Networking, **(3)**, 3 (1995) 226–244
6. Crovella, M and Bestavros, A: Self-Similarity in World Wide Web Traffic: Evidence and Possible Causes, IEEE/ACM Trans. on Networking, **(5)**, 6 (1997) 835–845

Using Normal Vectors
for Stereo Correspondence Construction

Jung-Shiong Chang[1], Arthur Chun-Chieh Shih[2],
Hong-Yuan Mark Liao[2], and Wen-Hsien Fang[1]

[1] Department of Electronic Engineering,
National Taiwan University of Science and Technology, Taiwan
norris@iis.sinica.edu.tw, whf@et.ntust.edu.tw
[2] Institute of Information Science, Academia Sinica, Taiwan
{Arthur,Liao}@iis.sinica.edu.tw

Abstract. In scenes of an outdoor environment, depth recovery by matching a pair of stereo images is not very successful due to the strong effect of noise and changing illumination. In contrast, the surface normal is relatively robust against the influence of noise or changing illumination compared to other frequently used features. In this paper, we propose a two-step approach to solve the 3-D depth recovery problem. In the first step, we use the intensity feature to execute a rough comparison. We then use the surface normal vector, which is a much more discriminating feature, as the search basis for the second step. In addition, the 3-D invariant nature of a surface normal improves the accuracy of the stereo image matching results. The maps reconstructed in our experiments on images of outdoor scenes show that our approach is indeed more efficient and accurate than conventional methods.

1 Introduction

In two-frame stereopsis, solving the correspondence between two images is the most difficult issue [3]. There are several reasons for this difficulty. First, images taken from different viewing angles usually have different illumination, which makes the matching of corresponding points in two different images very hard. Second, stereo images are usually taken with different cameras without calibration; therefore, the average intensity levels of stereo images of the same object or scene may be different. Third, in some outdoor environments, such as overgrown fields, the content of images can be very complex and chaotic, so it is difficult to calculate the correct matching pairs in a local area. Finally, the correspondence problem may also occur in the homogeneous regions of an image, because the change of intensities within a smooth region is not sufficiently obvious. This makes the matching process even more difficult.

In this paper, we propose a two-step comparison process. In the first step, we use the intensity feature to select a set of candidates from the database. Since this feature exists naturally in an image, the computation does not require much time. In the second step, we calculate the surface normal vector (a more complicated feature) of every pixel in an image and use it to determine the best candidate from the candidate list obtained in the first step. Making comparisons based on the surface normal vector is more time-consuming; however, since the search space has been significantly reduced by the screening operation conducted in the first step, the more complicated comparison executed in the second step is actually tolerable. Our experiment results show that the proposed method is superior to the maximum-flow and the pixel-to-pixel methods for real outdoor scenes.

2 Calculating Surface Normal Vectors from Discrete Image Data

In differential geometry [4], the normal vector of a surface patch plays an important role because it represents the orientation of the surface patch in a three-dimensional space. A good property of the normal vector of a surface is its 3-D invariant nature when observed from different angles. Once the angle formed by the normal vectors of two adjacent surface patches is determined, it will be invariant from any viewpoint. Since the normal vector of a surface patch is so robust in computation, it is an important feature of our system.

We assume an image is composed of many digital surfaces and calculate the normal vectors of all pixels in the image. We call the calculated new image a *normal vector map*. Since the formed map is dense, we can use it to execute either point-wise matching or block-wise comparison, depending on the degree of efficiency and accuracy required. However, the normal vectors computed by a simple numerical method usually yield poor results due to the effects of noise and digitalization. Thus, we have to choose a continuous differentiable function to represent the discrete data of an image. Such continuous functions are used to calculate surface normals – a process generally known as *surface reconstruction*. Surface reconstruction based on local quadratic surface fitting is a popular method. We apply this method to reconstruct smooth image surfaces and calculate the normal vectors of these surfaces to form a feature map that can be used for further comparison.

In [4], a local surface patch is defined as $\bar{x}:(U \subset R^2) \to R^3$, where $\bar{x}(u,v)=[u\ v\ f(u,v)]^T$. U is an open set in R^2. This parametric form represents a general surface S with respect to a known coordinate system. Given a point $\bar{x}(u,v)$ on S, $\bar{x}_u(u,v)$ and $\bar{x}_v(u,v)$ represent its tangent vectors along the u direction and the v direction respectively. They are defined as

$$\bar{x}_u(u,v) = \frac{\partial \bar{x}(u,v)}{\partial u} = [1\ 0\ f_u(u,v)],\ \text{and}\ \bar{x}_v(u,v) = \frac{\partial \bar{x}(u,v)}{\partial v} = [0\ 1\ f_v(u,v)]. \tag{1}$$

$\bar{x}_u(u,v)$ and $\bar{x}_v(u,v)$ are considered as a pair of bases of the tangent plane at a point $\bar{x}(u,v)$. The unit normal vector of $\bar{x}(u,v)$ is defined as

$$\bar{n}(u,v) = \frac{\bar{x}_u \times \bar{x}_v}{\|\bar{x}_u \times \bar{x}_v\|}, \tag{2}$$

where $\times$ represents a cross product operator and $\|\cdot\|$ is a norm operator. Substituting Eq. (1) into Eq. (2), we have

$$\bar{n}(u,v) = \frac{[-f_u\ -f_v\ 1]^T}{\sqrt{1+f_u^2+f_v^2}} = [n_x\ n_y\ n_z]^T, \tag{3}$$

where $n_x = \dfrac{-f_u}{\sqrt{1+f_u^2+f_v^2}}$, $n_y = \dfrac{-f_v}{\sqrt{1+f_u^2+f_v^2}}$, and $n_z = \dfrac{1}{\sqrt{1+f_u^2+f_v^2}}$.

In this work, we use a least square surface fitting algorithm to compute the partial derivative of a surface point from its immediate neighbors within a local window. Using the window as an operator, we calculate a set of parameters for an approximated surface patch by minimizing the errors between the input data and the surface patch. For example, a 7×7 binomial smoothing window can be written as $[W] = \vec{w}\vec{w}^T$. We use the column vector, w, given in [4], i.e.,

$$\vec{w} = \tfrac{1}{64}[1\ 6\ 15\ 20\ 15\ 6\ 1]^T \tag{4}$$

Thus, a window operator that approximates the operation of an equally-weighted least squares derivative can be derived by

$$[D_u] = \vec{a}_1\vec{a}_2^T, \text{ and } [D_v] = a_2\vec{a}_1^T, \tag{5}$$

where the column vectors $\vec{a}_1$ and $\vec{a}_2$ are respectively calculated by

$$\vec{a}_1 = \tfrac{1}{7}[1\ 1\ 1\ 1\ 1\ 1\ 1]^T, \text{ and } \vec{a}_2 = \tfrac{1}{28}[-3\ -2\ -1\ 0\ 1\ 2\ 3]^T. \tag{6}$$

The partial derivative estimates of an image can then be computed via a set of 2-D convolutions as follows:

$$f_u(u,v) = (D_u * W * f)(u,v), \text{ and } f_v(u,v) = (D_v * W * f)(u,v). \tag{7}$$

3 Our Approach

We now describe the general design of the two-step search process. Let (I^L, I^R) be a pair of stereo images, where I^R is a reference image. In the first step, for each pixel in I^L, we compute the cross correlation based on the intensity of the pixel and the intensities of its neighbors along the epipolar line in I^R. Since these cross-correlations may have several very close values due to the effect of noise or illumination, the correct point may not receive the highest cross correlation value. Therefore, we make a further check of the candidate pixels that received a correlation value higher than a threshold T_C to find candidates for the second round check. T_C is determined by computing the maximum correlation value among all candidates, minus a default threshold difference, ΔT_C. In this step, we also calculate the normal vector maps for I^L and I^R, respectively, as described in Section 2.

In the second step, for each pixel in I^L, we calculate the similarity between the normal vector of the pixel and the normal vectors of its neighbors along the epipolar line in I^R. We then choose the matched pair with the maximum score as the final pair. If this score is less than a threshold T_m, we increase the threshold difference, ΔT_C, so that other possible disparity values in the candidate list have a chance to be further examined. This process continues until one of the following conditions holds: (a) the maximum score of the matched pairs is larger than or equal to T_m; or (b) all disparity values have been examined.

3.1 Deriving a Candidate List

Let $I_{i,j}^L$ and $I_{i+d,j}^R$ denote the intensity values of a point (i,j) in I^L and a candidate point $(i+d,j)$ in I^R respectively, where d is a potential disparity. Given a computing window of size $(2m+1) \times (2n+1)$, the cross correlation, $R(i,j,d)$, can be defined as follows:

$$R(i,j,d) = \frac{Cov(I_{i,j}^L, I_{i+d,j}^R)}{\sqrt{Var(I_{i,j}^L)} \cdot \sqrt{Var(I_{i+d,j}^R)}}, \tag{8}$$

where

$$Cov(I_{i,j}^L, I_{i+d,j}^R) = \sum_{k=-m}^{m} \sum_{l=-n}^{n} [I_{i+k,j+l}^L - \mu(I_{i,j}^L)] \cdot [I_{i+k+d,j+l}^R - \mu(I_{i+d,j}^R)], \tag{9}$$

$$Var(I_{i,j}^L) = \sum_{k=-m}^{m} \sum_{l=-n}^{n} [I_{i+k,j+l}^L - \mu(I_{i,j}^L)]^2, \tag{10}$$

$$Var(I_{i+d,j}^R) = \sum_{k=-m}^{m} \sum_{l=-n}^{n} [I_{i+k+d,j+l}^R - \mu(I_{i+d,j}^R)]^2 \tag{11}$$

$\mu(I_{i,j}^L)$ and $\mu(I_{i+d,j}^R)$ represent the average intensities obtained by applying convolution with the above mentioned window to a region centered at (i,j) in I^L and $(i+d,j)$ in I^R respectively. For each point (i,j) in I^L, we calculate the cross correlations with all potential candidate points in I^R, and sort their disparities by their correlation values from large to small. Unlike traditional block matching approaches, we record the sorted disparities and their associated correlations on the candidate list. These candidates are further examined with surface normal vectors in the next step.

3.2 Calculating the Normal Vector Matching Cost

In existing methods [1-3], the cross correlation function is designed to compute the degree of similarity, instead of making decisions in the vector space. We now introduce a vector map-based cost determination process that can be used to measure the degree of similarity between two normal vector maps. Let $(nx_{i,j}^L\ ny_{i,j}^L\ nz_{i,j}^L)$ and $(nx_{i+d,j}^R\ ny_{i+d,j}^R\ nz_{i+d,j}^R)$ be the normal vectors of a point (i,j) in I^L and its corresponding point $(i+d,j)$ in I^R respectively, where d has the highest priority on the disparity candidate list. The normal vectors are calculated by the quadratic surface fitting method mentioned in Section 2. For each point (i,j) in I^L, the normal vectors derived from the point and its eight nearest neighbors are combined to form a cubic matrix. Given a potential depth d on the disparity candidate list, a normal vector-map based matching cost of (i,j) can be defined as:

$$Cost(i,j,d) = \frac{1}{3} \left\{ Corr2(NX_{i,j}^L, NX_{i+d,j}^R) + Corr2(NY_{i,j}^L, NY_{i+d,j}^R) + Corr2(NZ_{i,j}^L, NZ_{i+d,j}^R) \right\}, \tag{12}$$

where

$$NX_{i,j}^L = \begin{bmatrix} nx_{i-1,j-1}^L & nx_{i,j-1}^L & nx_{i+1,j-1}^L \\ nx_{i-1,j}^L & nx_{ij}^L & nx_{i+1,j}^L \\ nx_{i-1,j+1}^L & nx_{i,j+1}^L & nx_{i+1,j+1}^L \end{bmatrix}, \; NX_{i+d,j}^R = \begin{bmatrix} nx_{i+d-1,j-1}^R & nx_{i+d,j-1}^R & nx_{i+d+1,j-1}^R \\ nx_{i+d-1,j}^R & nx_{i+d,j}^R & nx_{i+d+1,j}^R \\ nx_{i+d-1,j+1}^R & nx_{i+d,j+1}^R & nx_{i+d+1,j+1}^R \end{bmatrix},$$

$$NY_{i,j}^L = \begin{bmatrix} ny_{i-1,j-1}^L & ny_{i,j-1}^L & ny_{i+1,j-1}^L \\ ny_{i-1,j}^L & ny_{ij}^L & ny_{i+1,j}^L \\ ny_{i-1,j+1}^L & ny_{i,j+1}^L & ny_{i+1,j+1}^L \end{bmatrix}, \; NY_{i+d,j}^R = \begin{bmatrix} ny_{i+d-1,j-1}^R & ny_{i+d,j-1}^R & ny_{i+d+1,j-1}^R \\ ny_{i+d-1,j}^R & ny_{i+d,j}^R & ny_{i+d+1,j}^R \\ ny_{i+d-1,j+1}^R & ny_{i+d,j+1}^R & ny_{i+d+1,j+1}^R \end{bmatrix},$$

$$NZ_{i,j}^L = \begin{bmatrix} nz_{i-1,j-1}^L & nz_{i,j-1}^L & nz_{i+1,j-1}^L \\ nz_{i-1,j}^L & nz_{ij}^L & nz_{i+1,j}^L \\ nz_{i-1,j+1}^L & nz_{i,j+1}^L & nz_{i+1,j+1}^L \end{bmatrix}, \; and \; NZ_{i+d,j}^R = \begin{bmatrix} nz_{i+d-1,j-1}^R & nz_{i+d,j-1}^R & nz_{i+d+1,j-1}^R \\ nz_{i+d-1,j}^R & nz_{i+d,j}^R & nz_{i+d+1,j}^R \\ nz_{i+d-1,j+1}^R & nz_{i+d,j+1}^R & nz_{i+d+1,j+1}^R \end{bmatrix}.$$

The first and second elements in $(NX_{i,j}^L, NX_{i+d,j}^R)$ represent the x-component of the normal vectors of (i,j) in I^L and those of $(i+d,j)$ in I^R respectively. Similarly, $(NY_{i,j}^L, NY_{i+d,j}^R)$ represent the y-component of the normal vectors of (i,j) in I^L and those of $(i+d,j)$ in I^R respectively, and $(NZ_{i,j}^L, NZ_{i+d,j}^R)$ represent the z-component of the normal vectors of (i,j) in I^L and those of $(i+d,j)$ in I^R respectively. In addition, each pair of normal vectors can be substituted into a similarity function measurement, defined as:

$$Corr2(A,B) = \frac{\sum_{k=1}^{3} \sum_{l=1}^{3} (A_{kl} - \hat{A})(B_{kl} - \hat{B})}{\sqrt{\sum_{k=1}^{3} \sum_{l=1}^{3} (A_{kl} - \hat{A})^2} \sqrt{\sum_{k=1}^{3} \sum_{l=1}^{3} (B_{kl} - \hat{B})^2}}, \tag{13}$$

where $\hat{A}$ and $\hat{B}$ represent the means of all elements in matrix A and B, respectively. From Eq. (12), we calculate the normal vector matching costs of all candidates on the list whose correlations are larger than T_C. The candidate with the maximum cost is considered the correct depth for (i,j). If, however, the maximum matching cost is less than a default threshold, T_m, we decrease the threshold T_C by ΔT_C to recruit more candidates from the list. This process is repeated until the maximum matching score of a matching pair is larger than or equal to T_m, or all disparity values on the list have been examined. Then, the disparity with the maximum score is considered the correct depth for the pixel (i,j).

4 Experiment Results

In the experiments, the following test set of images was taken outdoors: Phone, left and right (Figures 1a and 1b). The test set was downloaded from the website of Point Grey Research Inc. [5]. For comparison, we used the disparity map shown in Figures 1c, which was also provided by Point Grey Research Inc. Because the map was reconstructed with triple images from the left, right, and top views, we assume they are close to the real ground truths. Based on the procedure described in Section 3, the matching costs can be calculated and used to determine the optimal true depths. Figure 1d shows the obtained disparity map of pair of Figures 1a and 1b. For comparison, we also applied the Maximum-Flow method [1] and the Pixel-to-Pixel (p2p) method [2] to estimate the disparity map shown in Figures 1e-f. In [1], the parameters of the program, called stereomf, are set as follows: drange=-20 to 20, dstep=81, smooth=10, reduce=1, and ref =2. In [2], the parameters of the program are set as follows: occlusion penalty=15 and reward=30. To evaluate the performance of the above algorithms, we adopted the Scharstein and Szeliski measurement, namely, the RMS (Root-Mean-Squared) error between the computed disparity map $d_C(x,y)$ and the ground truth map $d_T(x,y)$ [6]. The RMS is calculated as follows:

$$RMS = \left(\frac{1}{N} \sum_{(x,y)} \left| d_C(x,y) - d_T(x,y) \right|^2 \right)^{1/2}, \tag{14}$$

where N is the total number of pixels. The RMS error of our approach is equal to 3.4518, and that of the Maximum-Flow and the Pixel-to-Pixel methods are equal to 4.8948 and 4.6227 respectively.

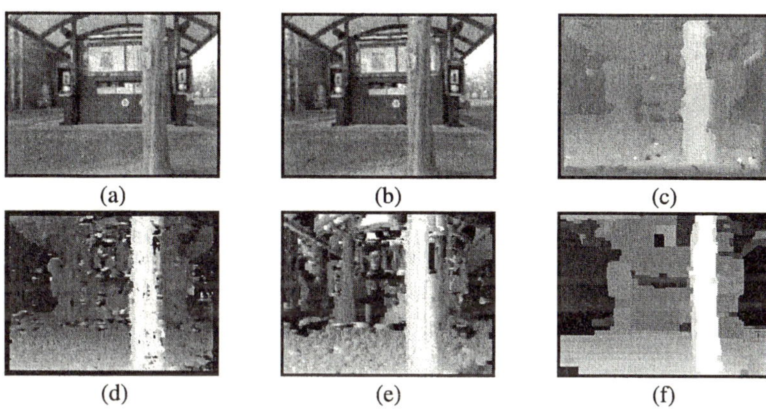

Fig. 1. Real outdoor stereo images: Phone: (a) Left image, (b) Right image, (c) Ground truth provided by Point Grey Research Inc [5], (d) Depth map obtained by our method, (e) Depth map obtained by the Maximum Flow method [1], and (f) Depth map obtained by the p2p method [2]

5   Conclusion

In this paper, we have proposed a novel approach to improve the accuracy of matching stereo images in an outdoor environment. Using an image's normal vectors as features, our approach overcomes the ambiguous correspondence problem in outdoor stereo images. The proposed method was applied to a set of outdoor stereo images. The experiment results show that our proposed approach is superior to the Maximum-Flow and the Pixel-to-Pixel methods for reconstructing 3-D outdoor scenes.

References

1. Sebastien Roy, "Stereo without epipolar lines: A maximum-flow formulation", IJCV, Vol. 34, No. 2/3, pp.147-161, 1999.

2. Stan Birchfield and Carlo Tomasi, "Depth Discontinuities by Pixel-to-Pixel Stereo", IJCV, Vol. 35, No. 3, pp. 269-293, 1999.
3. Carlo Tomasi and Roberto Manduchi, "Stereo Matching as a Nearest-Neighbor Problem", IEEE Transactions on Pattern Analysis and Machine Intelligence, Vol. 20, No.3, pp. 333-340, 1998.
4. John McCleary, Geometry from a differentiable viewpoint, Cambridge University Press, New York, 1994.
5. http://www.ptgrey.com/
6. D. Scharstein and R. Szeliski, "A Taxonomy and Evaluation of Dense Two-Frame Stereo Correspondence Algorithms", IJCV, Vol. 47, No. 1, pp. 7-42, 2002.

Color Image Restoration
Using Explicit Local Segmentation

Mieng Quoc Phu[1], Peter Tischer[1], and Hon Ren Wu[2]

[1] Faculty of Information Technology, Monash University, Clayton Campus,
Victoria 3800, Australia
Mieng.Quoc.Phu@infotech.monash.edu.au
Peter.Tischer@csse.monash.edu.au
[2] Software and Network Engineering, RMIT, City Campus, Victoria 3001, Australia
henry.wu@rmit.edu.au

Abstract. A local segmentation structure is proposed for color image denoising and restoration. Most state-of-the-art filters suppression impulse noise well but tends to destroy thin lines, edges and high image details. The proposed filters facilitate local segmentation technique to preserve image structures and noise reduction. Firstly, the K-VMF is used for local segmentation and then the selection of vector filters for reconstruction of the current pixel. The proposed filters demonstrated acceptable results for both objective and subjective assessments.

1 Introduction

Traditionally, in the area of image restoration noise removal is achieved by linear operations because of their simplicity and ease of implementation. Nonlinear filters generally perform better than linear filters due to the fact that most imaging systems produced non-stationary and nonlinear signals. The Vector Median Filter (VMF [2]) is one of the most popular nonlinear vector filters for impulse noise removal because of its simplicity. Fuzzy VMF (FVMF [3]) is an extended VMF. Other order statistic vector filters include the Vector Directional filter (VDF/GVDF [4]), Directional Vector filter (DDF [5]), Hybrid Directional filter (HDF [6]) and Adaptive Nearest Neighbor filter (ANNF [7]). These classic filters perform well in suppressing impulse noise, but they tend to degrade the high image details and introduce new artifacts such as blurring, smearing and shifting of the image structures.

The Multiple Window Configuration (MWC [8]) and Neighborhood adaptive vector (NAVF [9]) filters are detection and switch based filters which utilize a detector and switch the output between component filters according to the detection results. MWC is one of the most efficient impulse noise removal filter but it depends on the estimated reference image (corrupted image filtered by standard median filtering). Thus any distortion remaining in reference image will be introduced into the final reconstruction, resulting in image blurring, edge shifting and chromatic distortions [9].

Seemann and Tischer [1] proposed an explicit local segmentation filter for removal of Gaussian noise in grey-scale images. They used the k-means technique to segment pixels in the local window and utilized the estimated global noise variance in order to reconstruct the corrupted pixels. The Peer Group filter (PGF [10]) is another explicit local segmentation filter for impulse noise suppression in color images. It used the

R. Khosla et al. (Eds.): KES 2005, LNAI 3681, pp. 589–595, 2005.

Fisher's discriminant to classify the pixels into groups. It employs a standard filter to reconstruct the noisy pixel from the pixels in the peer group.

In this paper, the concept of local segmentation is incorporated into existing vector filters in order to suppress impulse noise in both natural and textured color images. Section 1.1 describes some limitations of the VMF. Section 2 presents the structure of the proposed filter. Section 3, describes the simulation results and in Section 4, some remarks and conclusion.

1.1 VMF Drawbacks

The VMF is regarded as one of the most popular vector filters and there are many extensions and variations of it in the literature. For example, detection based filters utilize it with a noise detector. Hybrid or switch based filters use it as a component filter for reconstruction. VMF can also be used as a noise detector and for reconstruction, as in [9]. Its definition is as follows:

Let $\underset{\sim}{x}$ be a multichannel sample vector and W be the window $(x_1, x_2, ..., x_n)$. For pixel x_i, its total distance to the other pixels in the window is given by (1).

$$d_w(x_i) = \sum_{j=1}^{n} d(x_i, x_j)$$

(1)

Often, the distance $d(x_i, x_j)$ is the Euclidean distance. Let the $d_w(x_i)$ be sorted in increasing order to produce $\{d_1, d_2, ..., d_n\}$. The pixel associated with d_1 has the smallest total distance to all the pixels in the window. In one sense, it is the most 'in lying' value and thus the best representative value. This value is chosen as the vector median (VM) and used in the VMF.

Although it has been used extensively, the VMF has its drawbacks as demonstrated in Fig.1. Even when there is no noise within an image the VMF will destroy thin lines and edges. The VMF and its variants may destroy underlying image structure when the mask contains pixels from more than 1 segment. For example, suppose the mask contains 9 pixels which belong to two segments. The VM will be chosen from the pixels which belong to the segment which has more members in the mask. Thus structure like corners and thin lines, which only have a small number of pixels in the mask, get erased. For a 3x3 mask, if the centre pixel is part of a segment whose intersection with the mask contains at most 4 pixels, the centre pixel will be replaced by a pixel from the other segment. Any feature whose intersection with the 3x3 mask never has more than 4 pixels will be completely erased. These drawbacks not only occur with the VMF but also with VDF, DDF, and ANNF, to name a few.

2 Formulation of the Local Segmentation Vector Filters

The motivation for the proposed filter is to overcome the drawbacks of the vector filters. Hence, it aims to preserve as much of the image structure as possible and to minimize chromatic distortions. To preserve local image structure, our filters must identify this structure and segments within the mask. In this paper we will only consider situations where the mask covers at most two segments.

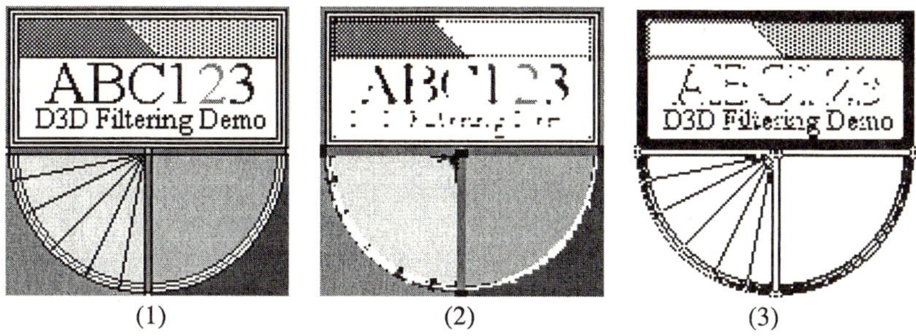

Fig. 1. (1) Noiseless *Letter* image, (2) Result of VMF and (3) Difference image

2.1 Homogeneity Detector

There are numerous ways to check whether a local window is homogeneous or not. If equation (2) is true, the region is classified as being homogeneous and unaffected by impulse noise and no filter is required. Otherwise, it will be treated as containing two segments and local segmentation. The threshold $\delta_R = 55$ is chosen from [9].

$$d_n - d_1 \leq \delta_R \tag{2}$$

2.2 Local Segmentation (LS)

In grayscale images there are many ways to cluster data. The K-means and K-median are among the popular choices because of their simplicity and low computation cost. K-means minimizes the mean square error, whereas K-median minimizes the mean absolute error. In color image processing, the K-median method can extended to K-VMF for local segmentation in the RGB color space.

Since we are only trying to find two classes, k=2. Our algorithm is as follows:

1. Choose two initial vectors to represent each class.
2. Assign each pixel in the mask to the representative vector which is closer.
3. Re-compute the representative vector for a class as the VM of the pixels in that class.
4. Repeat as long as the class representatives continue to change.

For initial vectors, we choose the two pixels in the mask which are the furthest apart.

2.3 Reconstruction Filters

If we have split the pixels into two segments our program allows any vector filter (E.g. VMF, VDF, DDF, ANNF, etc) to be applied, but it can only use pixels from the same segment as the pixel to the filtered. However, the vector filter will only be applied if the segment is considered to be 'valid'. If the pixel to be filtered is in the segment by itself, it could either be part of a one pixel feature or it could be adversely affected by impulsive noise. There is no way to distinguish between those two situations. Therefore, we need to set a limit on how small a valid segment can be. If the segment has 3 or more pixels, it is considered valid. If it has 2 pixels or fewer, it is considered invalid and the vector median of the other segment is used as the filtered value.

3 Simulation Results

In this section, the proposed local segmentation filter structure is evaluated and compared with existing filters for color image denoising. The performance analysis carried out in two ways:

1) A comparison of the classic vector filters (VMF and ANNF) with their proposed variants, LS-VMF and LS-ANNF.
2) A comparison of the proposed LS vectors with the state-of-the-art filters.

Several objective criteria are used to measure the distortion in image reconstruction, which includes the NMSE (normalized mean square error) and the NCD (normalized color difference). The NMSE and NCD are defined in [3].

3.1 Impulse Noise Corruption

In this paper, the corrupted noise model is assumed to be random impulse noise where the noise term is uniformly distributed over the range of all possible pixel values. All images in the simulation are corrupted by channel correlation method proposed by [4]. For example, a preset percentage is first corrupted by random impulse noise in an independent manner, then a correlation factor C=0.5 is used to introduce more noise into the other colour channels for each corrupt pixel. In another word, there is a 50% chance of further corruption if one channel has been already corrupted.

3.2 Experimental Performance

In the first experiment, the Mandrill image was corrupted by 5% and 10% of random impulse noise. In order to evaluate the fidelity of the LS vector filters, they are first compared with their corresponding classic vector filters. Table 1 shows the objective measures for VMF, ANNF and the proposed LS-VMF, LS-ANNF, respectively. It shows that the proposed filters have consistent improvement for both noise levels than their counterparts. Figure 2 demonstrates that the proposed filters perform better at preserving edges and at retaining image structures.

Table 1. Color Mandrill image corrupted by 5% and 10% random impulse noise. The measurements are scaled by 10e-2

Noise	Random Impulse (5%)		Random Impulse (10%)	
Filter	NMSE	NCD	NMSE	NCD
None	1.492	2.702	2.864	5.178
VMF	1.602	5.247	1.671	5.414
LS-VMF	**1.181**	**4.611**	**1.359**	**4.970**
ANNF	1.489	5.640	1.548	5.940
LS-ANNF	**1.114**	**4.954**	**1.279**	**5.440**

In the second experiment, the proposed filters are compared with the state-of-the-art filters for the Letter image corrupted by 10% and 20% of random impulse noise. In Figure 3, the proposed filters performed exceptionally well for noise suppression and detail preservation. Most state-of-the-art filters showed unacceptable results as they destroyed lines, edges, and high image detail. Moreover, GVDF, DDF and

MWC even have chromatic distortions along with structure degradations. Table 2 shows evidence of these degradations. The main reasons for such poor performance could be that the images are different to those in the training set used to tune the filters or the image model used by these filters is not appropriate. However, it is conclusive that these filters do not preserve image structure well and even further degrade the image quality even with or without noise. The PGF also used explicit local segmentation is one of the top performer for both natural and texture images.

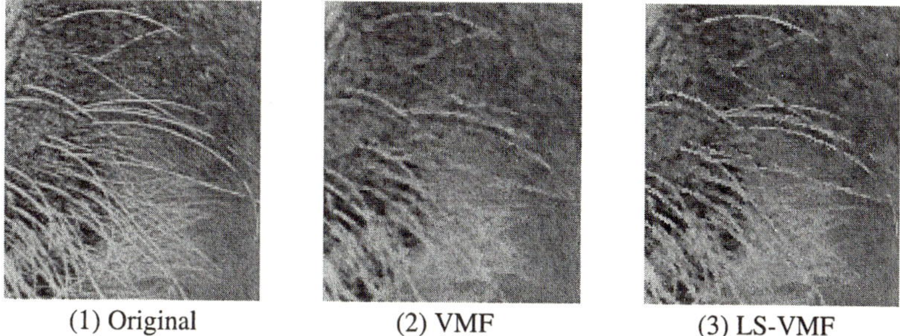

(1) Original   (2) VMF   (3) LS-VMF

Fig. 2. Filtered Mandrill image (portion) for the VMF and LS-VMF filters

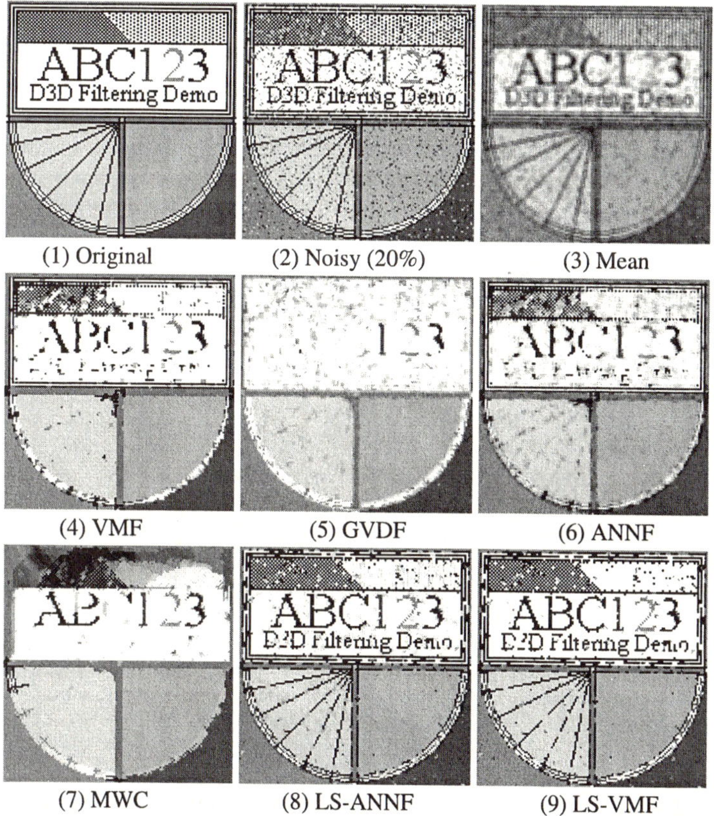

(1) Original   (2) Noisy (20%)   (3) Mean

(4) VMF   (5) GVDF   (6) ANNF

(7) MWC   (8) LS-ANNF   (9) LS-VMF

Fig. 3. Some resultant images of color *Letter* image corrupted by 20% random impulse

Table 2. Color *Letter* image corrupted by 10% and 20% random impulse noise. The measurements are scaled to 10e-2

Noise Filter	Random Impulse (10%)		Random Impulse (20%)	
	NMSE	NCD	NMSE	NCD
None	3.341	7.149	6.413	13.758
Mean	20.169	32.311	20.548	35.033
VMF	31.046	27.325	31.185	28.593
GVDF	29.165	25.437	28.415	26.325
DDF	28.971	25.706	29.357	26.495
HDF	30.477	26.400	30.552	27.278
AHDF	29.973	26.434	29.878	27.435
FVDF	28.415	25.920	27.773	26.902
ANNF	29.028	28.182	28.062	29.757
MWC	21.813	28.081	21.840	27.523
PGF	9.223	8.075	14.707	13.384
LS-ANNF	**5.383**	**6.914**	**7.412**	**10.935**
LS-VMF	**5.673**	**5.664**	**7.830**	**8.552**

4   Conclusions

In this paper, a local segmentation structure is proposed for use with existing vector filters for color image denoising and restoration. The proposed K-VMF technique is used for local segmentation and vector filters such as VMF and ANNF are used for reconstruction. However, the proposed structure is so diverse that it is not limited to these two filters. The proposed LS-VMF and LS-ANNF filters demonstrate greatly improved performance as compared to the VMF and ANNF filters. The proposed filters performed exceptional well for random impulse noise suppression and detail preservation, where most state-of-the-art filters showed unacceptable results as they destroy lines, edges, and high image detail.

References

1. T. Seemann and P. E. Tischer, Structure preserving noise filtering using Explicit Local Segmentation. In *Proc. IAPR Conf. on Pattern Recognition (ICPR'98)*, volume II, pages 1610.1612, August 1998.
2. J. Astola, P. Haavisto, Y. Neuovo, Vector median filters. *Proc. IEEE,* vol.78, 1990, pp.678-689.
3. K.N. Plataniotis and A. N. Venetsanopoulos, *Color Image Processing and Applications.* Berlin: Springer-Verlag Telos, ISBN: 3540669531, 31 July 2000.
4. P.E. Trahanias, A.N. Venetsanopoulos, Vector directional filters: A New class of multichannel image processing filters. *IEEE Trans. Processing,* vol.2, No.4, 1993, pp.528-534.
5. D.G. Karakos, P.E. Trahanias, Combining vector median and vector directional filters: The directional-distance filters. Proc. *ICIP-95,* vol.1, 1995, 171-174.
6. M. Gabbouj, F.A. Cheickh, Vector median-vector directional hybrid filter for color image restoration. *Proc. EUSIPCO-96,* vol. 2, 879-881, Trieste, Italy, Sept. 1996.
7. K.N. Plataniotis, et.al., An adaptive nearest neighbor multichannel filter. *IEEE Trans.,on Circuit, Syst., Video Tech.,* vol.6, no.6, pp.699-703, Dec. 1996.

8. E. S. Hore, B. Qiu and H. R. Wu, Prediction based image restoration using a multiple window configuration. *Optical Engineering*, vol.41, pp.1-11, Aug. 2002.
9. Z. Ma and H.R. Wu, A Neighborhood Adaptive Vector Filter for Color Image Restoration. *Proceedings of 2003 International Symposium on Intelligent Signal Processing and Communication Systems (ISPACS 2003),* Awaji Island, Japan, December 7-10, 2003.
10. C. Kenney, Y. Deng, B. S. Manjunath, and G. Hewer, Peer Group Image Enhancement. *IEEE Trans. Image Processing*, Vol. 6. No. 7, pp. 326-334, Feb. 2001.

Weight Training for Performance Optimization in Fuzzy Neural Network

Hui-Chen Chang[1] and Yau-Tarng Juang[2]

[1] Department of Electrical Engineering, National Central University, Chung-Li, Taiwan,
and the Department of Electronics Engineering,
Ta Hwa Institute of Technology, Hsin-Chu, Taiwan
[2] Department of Electrical Engineering, National Central University, Chung-Li, Taiwan

Abstract. An algorithm for improving performance by training a fuzzy neural network (FNN) based on the back-propagation (BP) algorithm and grey relations is proposed. This technique is developed by directly incorporating the grey relational coefficient (GRC) into the learning rule of the BP, and a BP with GRC technique is proposed in order to improve the performance of training the FNN. From the simulation results, we demonstrate this technique applied for controlling a nonlinear fuzzy model car system and find the result is better than the classical BP algorithm for training the FNN.

1 Introduction

In a fuzzy neural network approach, fuzzy rules are represented in the form of a network and its parameters are tuned by a similar method to that of the back-propagation algorithm [1] in neural networks. The advent of BP - an adaptation of the steepest descent method [6] - opened avenues for the application of multiplayer neural networks for many problems of practical interest [7-8]. In the standard learning scheme for BP algorithms, the weights of the network are iteratively adapted according to the recursion

$$\mathbf{w}(t+1) = \mathbf{w}(t) + \eta\mathbf{d}(t) \tag{1}$$

to find an optimal weight vector. The positive constant of η, which is selected by the user, is called the learning rate. The direction vector $\mathbf{d}(t)$ is the negative of the gradient of the output error function E

$$\mathbf{d} = -\nabla E(\mathbf{w}) \tag{2}$$

By inspecting Eq.(1), we see that the movable amount is determined by η and $\mathbf{d}(t)$. However, it seems that the movable amount can incorporate some important information, i.e., the relationships that actually exist between the inputs and the connection weights propagating to the neuron in a FNN. Indeed, there exist distinct relationships between any two subsystems in the real world [3,11], although we do not know exactly what these relationships are. Grey theory, as proposed by Deng [3], can perform grey relational analysis for these subsystems by dealing with finite and incomplete data series obtained from these subsystems [5]. Therefore, we consider that the BP algorithm can take into account the grey relation that actually exists between the inputs and the connection weights propagating to the neuron in each training iteration of the network. Such a significant relation is called the grey relational coefficient. We see that the movable amount of the weighting vector $\mathbf{w}$ is not determined only by η and $\mathbf{d}(t)$: it acquires more movement if there exists a GRC between the inputs and the connection weights.

2 Background

We use the following fuzzy rules whose consequent parts are assumed to be a linear combination of two fuzzy sets associated with an output linguistic variable [11].

$$R_i : \text{IF } x_1 \text{ is } A_{i1} \text{ AND...AND } x_n \text{ is } A_{in} \text{ THEN } y_i \text{ is } \omega_{i1} \cdot B_1 + \omega_{i2} \cdot B_2 \tag{3}$$

where n is the number of inputs $x_1 \ldots x_n$, y_i denotes the ith rule output, B_1 and B_2 are the output fuzzy sets, $A_{i1} \ldots A_{in}$ are labels of input fuzzy sets pertaining to the ith rule, ω_{i1} and ω_{i2} are the coefficients that form the consequent parameter set. Each linguistic term of input and output fuzzy sets is characterized by a triangular fuzzy set. These fuzzy rules are represented in a network, which is a 6-layer connective structure, as shown in Fig. 1, where the nodes in layer 1 are input nodes which represent input linguistic variables. The nodes in layer 2 are the input term nodes that stand for membership functions for the respective terms of the linguistic variables. The nodes in layer 3 are the rule nodes, which form the entire fuzzy rule base; layer 3 and layer 4 serve as the inference mechanism; the links of layer 3 define the preconditions of the fuzzy rule nodes, and the links of layer 4 define the consequent parts of the fuzzy rule nodes. Each fuzzy rule whose consequent parts appear in this article is a linear combination of two fuzzy sets associated with an output linguistic variable. Layer 5 has two nodes for the prearrangement of the defuzzification, and layer 6 is the last layer to perform the defuzzification. The activation function of node i in layer k can be represented as

$$f_i^k = f(f_1^{k-1}, f_2^{k-1}, \cdots, f_j^{k-1}, \cdots, f_n^{k-1}; \omega_{i1}^k, \omega_{i2}^k, \cdots, \omega_{ij}^k, \cdots, \omega_{in}^k) \tag{4}$$

where ω_{ij}^k is the connection weight between the ith unit in the kth layer and the jth unit in the $(k-1)^{th}$ layer, f_j^{k-1} is the jth input to a unit in the kth layer, f_i^k is the output of the ith unit in the kth layer. Next we describe the detailed functions of the nodes in the fuzzy logic controller structure.

Layer 1: The nodes in this layer just transmit input values to the next layer directly so the link weights at this layer are a unity and therefore,

$$f_i^1 = x_i \tag{5}$$

Layer 2: The nodes in this layer serve as membership functions, so the activation function in each neuron in this layer should be a membership function. A triangle-shaped membership function is used

$$f_i^2 = \begin{cases} 0 & ,\textit{if } f_i^1 < (c-L); \\ 1 - \dfrac{c - f_i^1}{L} & ,\textit{if } (c-L) \le f_i^1 < c; \\ 1 - \dfrac{f_i^1 - c}{L} & ,\textit{if } c \le f_i^1 < (c+R); \\ 0 & ,\textit{if } (c+R) \le f_i^1; \end{cases} \tag{6}$$

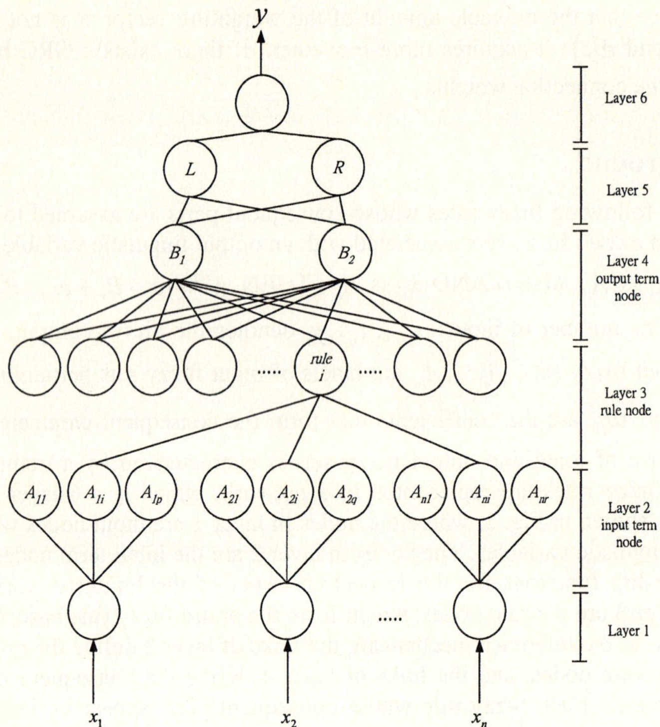

Fig. 1. Structure of the fuzzy neural network

where c is the center of a triangle-shaped membership function; L and R are the span of left side and right side of the function, respectively; and the weights in this layer are a unity also.

Layer 3: This layer is used to perform the precondition matching of fuzzy logic rules. Thus the rule should perform the fuzzy AND operation, and the link weights in this layer are also unity.

$$f_i^3 = \min(f_1^2, f_2^2, \cdots, f_n^2)$$
$$= \hat{\alpha}_i$$
(7)

Layer 4: This layer is called the sum of output term nodes, and each node in the 4th layer is a linear combination of two output fuzzy sets. Because the sup-product inference is adopted in this article, we have

$$f_j^4 = \sum_{i=1}^n (\omega_{ij}^4 \cdot \mu_{B_j}(y)) \cdot f_i^3 = \sum_{i=1}^n (\omega_{ij}^4 \cdot \mu_{B_j}(y)) \cdot \hat{\alpha}_i, \quad j = 1, 2$$
(8)

where ω_{ij}^4 is a weight of link between the node i in the layer 3 and the node j in the layer 4.

Layer 5: This layer functions as the prearrangement work of defuzzification. There are two nodes in this layer. Since we use the COS-HM defuzzifier, the node on the left is

$$f_L^5 = \sum_{i=1}^{m} c_j \cdot f_i^4 = \sum_{j=1}^{m} c_j \cdot [\sum_{i=1}^{n} \omega_{ij}^4 \cdot \hat{\alpha}_i \cdot \mu_{B_j}(c_j)] = \sum_{j=1}^{m} c_j \cdot \sum_{i=1}^{n} \omega_{ij}^4 \cdot \hat{\alpha}_i \tag{9}$$

where c_j is the center of output fuzzy set, and the node on the right is

$$f_R^5 = \sum_{j=1}^{m} f_j^4 = \sum_{j=1}^{m} \sum_{i=1}^{n} \omega_{ij}^4 \cdot \hat{\alpha}_i, \quad m = 1,2 \tag{10}$$

Layer 6: This layer performs the action of defuzzification. The function is used to represent the output of the COS-HM defuzzification method.

$$f^6 = \frac{f_L^5}{f_R^5} = \frac{\sum_{j=1}^{m} c_j \cdot \sum_{i=1}^{n} \omega_{ij}^4 \cdot \hat{\alpha}_i}{\sum_{j=1}^{m} \sum_{i=1}^{n} \omega_{ij}^4 \cdot \hat{\alpha}_i} = \frac{\sum_{i=1}^{n} \hat{\alpha}_i \cdot (\sum_{j=1}^{m} c_j \cdot \omega_{ij}^4)}{\sum_{i=1}^{n} \hat{\alpha}_i \cdot \sum_{j=1}^{m} \omega_{ij}^4} \tag{11}$$

3 Proposed Algorithm

The BP algorithm with GRC is proposed. The learning rule is as follows:

$$\mathbf{w}(t+1) = \mathbf{w}(t) - \eta(\xi)^k \nabla E(\mathbf{w}) \tag{12}$$

$$E = \frac{1}{2}[\hat{y}(t) - y(t)]^2 \tag{13}$$

where $\hat{y}(t)$ is the desired output, $y(t)$ is the current output, k is a pre-specified positive real number, and ξ is the GRC between the inputs and the connection weights propagating to the neuron. The algorithm proceeds as follows:

1. Set the pre-specified k and the learning rate η
2. Initialize connection weights corresponding to each node in layer 4 randomly.
3. Present a training input/output pair through the network
4. Compute ξ_j and ξ .
5. Update the new weights using Eq. (12).
6. Go to step 2 until the training iterations are finished.

We can see that the learning rule of the BP with GRC is determined not only by the learning rate and the gradient of the output error function but also by the ξ between the inputs and the connection weights propagating to the neuron. The computation of GRC is described in the next paragraph.

3.1 Grey Relational Coefficient (GRC)

Grey relational analysis is a method that can find the relationships between one major sequence and the other sequences in a given system [4]. Given the reference sequence $\mathbf{x} = (x_1, x_2, \ldots, x_n)$ and the comparative sequences $\mathbf{w}_i = (w_{i1}, w_{i2}, \ldots, w_{in})$ $(1 \leq i \leq m)$ with the normalized form, the GRC ξ_{ij} between x_j and w_{ij} can be computed as [2,4,5,11]

$$\xi_{ij} = \frac{\Delta_{\min} + \rho \Delta_{\max}}{\Delta_{ij} + \rho \Delta_{\max}} \tag{14}$$

where ρ is the discriminative coefficient $(0 \le \rho \le 1)$, and usually $\rho = 0.5$ is used [2,4]; and

$$\Delta_{\min} = \min_{i} \min_{j} \left| x_j - w_{ij} \right|, \, 1 \le i \le m, \quad 1 \le j \le n, \tag{15}$$

$$\Delta_{\max} = \max_{i} \max_{j} \left| x_j - w_{ij} \right|, \, 1 \le i \le m, \quad 1 \le j \le n, \tag{16}$$

$$\Delta_{ij} = \left| x_j - w_{ij} \right|, \tag{17}$$

4 Simulations

To illustrate and to examine the performance of the proposed technique, we apply it to control a nonlinear system, and then we compare the result of it with that of the classical BP algorithm. This is the well-known nonlinear fuzzy model car first analyzed by Sugeno [10,13]. As a performance index of a fuzzy rule-based system, we use the summation of square errors between the desired output y_p and the inferred output $y(\mathbf{x}_p)$ for each input-output pair $(\mathbf{x}_p ; y_p)$. The performance index (PI) is defined as

$$PI = \sum_{p=1}^{m} \left\{ y(\mathbf{x}_p) - y_p \right\}^2 / 2 \tag{18}$$

4.1 A Fuzzy Model Car

Figure 2 shows a fuzzy model car moving along a track with a rectangular turn. The input linguistic variables x_1 and x_2 represent the horizontal distance and the vertical distance of the car, respectively from the X-Y coordinate. x_3 is the steering angle. The output linguistic variable y is the next steering angle. Each linguistic variable of input is partitioned into five linguistic terms and the output space is set to 2 partitions whose membership functions are given in Fig. 3.

To gather the input/output data of the fuzzy model car for the training in the fuzzy rules, we have driven the model car [10,13] three times using a set of fuzzy control rules consisting of 125 rules and saved the data as $\hat{x}_1, \hat{x}_2, \hat{x}_3$ and $\hat{y}$ which will be used to give training in fuzzy rules. After the training is completed, three randomly selected initial starting points are chosen to test the resulting fuzzy rules. The specified parameter specifications are the same as the above example, and the ith control rule is of the following form:

$$R_i: \; IF \; x_1 \; is \; A_{i1}, \; x_2 \; is \; A_{i2}, \; x_3 \; is \; A_{i3}$$
$$THEN \; y_i \; is \; \omega_{i1} \cdot B_1 + \omega_{i2} \cdot B_2, \, i=1,\dots,125 \tag{19}$$

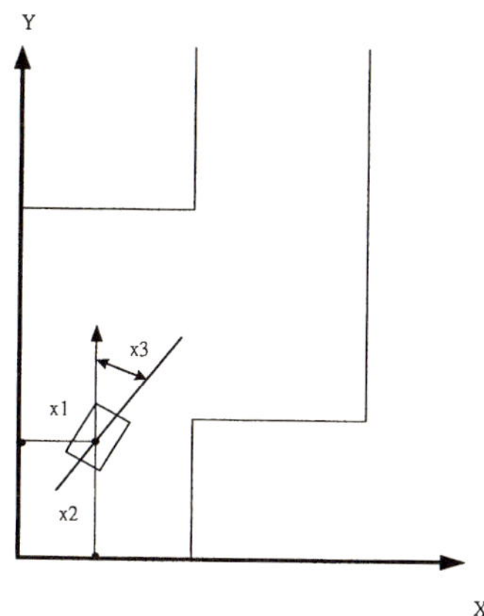

Fig. 2. A fuzzy model car

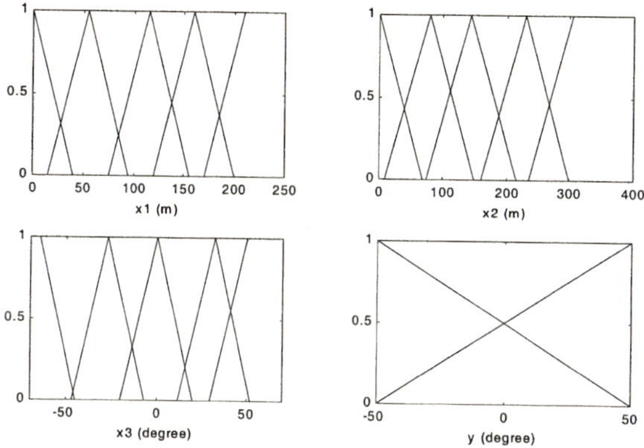

Fig. 3. Fuzzy partitions of input and output space in a fuzzy model car

As seen in Fig. 4, the dotted curves are the paths driven by the training data and the solid curves are the paths driven according to the resulting fuzzy rules trained following the classical BP, and it shows that the resulting fuzzy rules drive the car satisfactorily. The performance index is 0.1504.

Fig. 5 shows the results for BP with GRC. The best performance index is 0.0835, and it is smaller than the result obtained using classical BP.

5 Conclusions

We have presented in this paper a BP with GRC learning technique in training fuzzy neural networks. This technique is developed by directly incorporating the grey relational coefficient into the BP algorithm. The best result of the performance by the BP with GRC results in a smaller performance index than that of the classical BP from the simulation of this example. It thus demonstrates the feasibility and effectiveness of the BP with GRC for training fuzzy neural networks. Work is currently under way to apply the BP with GRC for finding the optimal partitions of the consequent part.

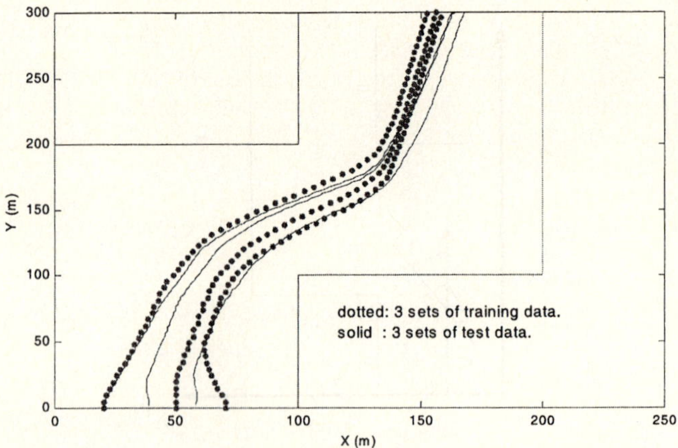

Fig. 4. Control result trained by the classical BP in a fuzzy model car

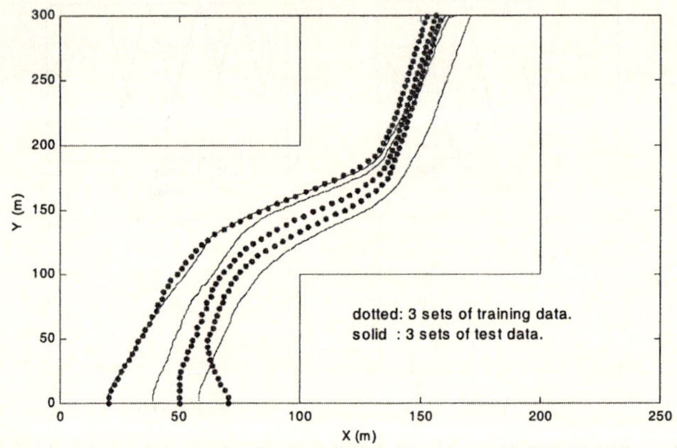

Fig. 5. Control result trained by the BP with GRC in a fuzzy model car

References

1. Rumelhart, D. E. and McClelland, J. L., Parallel Distributed Processing Vol. 1. MIT Press, Cambridge, MA, (1986).
2. Cheng, C. S., Hsu, Y. T., Wu, C. C., Grey neural networks, IEICE Transactions on Fundamentals Electronics, Commun. Comput. Sci. E81-A, 11, 1998, 2433-2442.
3. Deng, J. L., Control problems of grey systems, Systems Control Lett. 1,5, 1982, 288-294.
4. Hsu, Y. T. and Chen, C. M., A novel fuzzy logic system based on N-version programming, IEEE Trans. Fuzzy Systems 8, 2, 2000, 155-170.
5. Huang, Y. P. and Huang, C. H., Real-valued genetic algorithms for fuzzy grey prediction system, Fuzzy Sets Systems 87, 3, 1997, 265-276.
6. Bazaraa, M. S., Sherali, H. D. and Shetty, C. M., Nonlinear programming, theory and algorithms, New York: Wiley, 1993.
7. Salehi, F., Lacroix, R. and Wade, K. M., Effects of learning parameters and data presentation on the performance of back-propagation networks for milk yield prediction, Transactions of the ASAE 41, 1, 1998, 253-259.

8. Sejnowski, T. J. and Rosenberg, C. R., Parallel networks that learn to pronounce English text, Complex systems 1 (1) (1987) 145-168.
9. Lin, C. T., and Lee, C. S. George, Neural-Network-Based Fuzzy control and Decision System, IEEE transactions on computers 40 (12) (1991) 1320–1336.
10. Maeda, M., Maeda, Y. and Murakami, S., Fuzzy drive control of an autonomous mobile robot, Fuzzy sets and systems 39 (1991) 195-204.
11. Nozaki, K., Ishibuchi, H. and Tanaka, H., A simple but powerful heuristic method for generating fuzzy rules from numerical data, Fuzzy sets and Systems 86, 1997, 251-270.
12. Shi, K. Q., Wu, G. W. Hwang, Y. P., Theory of Grey Information Relation, Chuan Hwa, Taiwan, 1994.
13. Sugeno, M. and Murakami, K., An experimental study on fuzzy parking control using a model car, Industrial Application of Fuzzy Control (1985) 125-138.

Optimal Design Using Clonal Selection Algorithm

Yi-Hui Su[1], Wen-Jye Shyr[2], and Te-Jen Su[3]

[1] Dept. of Information Management, Tzu-Hui Institute of Technology
Pintung 926, Taiwan, R.O.C.
[2] Dept. of Industrial Education and Technology, National Changhua University of Education
Changhua 500, Taiwan, R.O.C.
shyrwj@cc.ncue.edu.tw
[3] Dept. of Electronic Engineering, National Kaohsiung University of Applied Sciences
Kaohsiung 807, Taiwan, R.O.C.
sutj@cc.kuas.edu.tw

Abstract. In this paper, the Clonal Selection Algorithm (CSA) is employed by
the natural immune system to define the basic features of an immune response to
an antigenic stimulus. This paper synthesizes the advantages of clonal selection
algorithm and proposed optimal design problem using clonal selection algorithm
which is a basis of the immune system. CSA, the essence of immune algorithm, is
effective to solve optimal problem. The clonal selection algorithm is highly parallel and presents a fine tractability in terms of computational cost. Like the genetic
algorithm, clonal selection algorithm is a tool for optimum solution. Clonal selection algorithm and genetic algorithm are used to reach the optimization performances for two numerical function. Then those results are compared each other.
These proposed algorithms are shown to be an evolutionary strategy capable of
solving optimal design problem.

1   Introduction

Many heuristic optimization algorithms have been developed and adapted to several
problems [1-3]. The problems of finding a maximum or a minimum of a function under
some constraint conditions are called optimization problem. Almost every engineering
evaluation problem can be formulated as an optimization problem. Optimization primarily aims to find the best possible combination of factors which can be defined as
evaluation parameters to maximize or minimize an objective function subjected to a set
of constraints.

The genetic algorithm paradigm is a stochastic optimization method for combinatorial optimization problems. Because genetic algorithm is based on multipoint search and
uses crossover operator, genetic algorithm can search large region of the solutions. But
genetic algorithm has two main disadvantages. The one is lack of the local search ability
and the other is the premature convergence. In order to overcome these drawbacks, several researchers have studied new optimization methods based on the immune system
[4, 5].

The immune system has two types of response: primary and secondary. The primary
response is reaction when the immune system encounters the antigen for the first time.
At this point the immune system learns about the antigen, thus preparing the body for

R. Khosla et al. (Eds.): KES 2005, LNAI 3681, pp. 604–610, 2005.

any further invasion from that antigen. This learning mechanism creates the immune system's memory. The secondary response occurs when the same antigen encountered again. This has response characterized by a more rapid and more abundant production of antibody resulting from the priming of the B-cells in the primary response [6]. By using this mechanism, immune algorithm shows a good performance as an optimization algorithm. Over the last few years, there has been an ever increasing interest in the area of artificial immune systems and their applications [7-10].

In this work, we use the clonal selection concept, together with the affinity maturation process, and demonstrate that these biological principles can lead to the development of powerful computational tools. The algorithm to be presented focuses on a systemic view of the immune system. It shows that some basic immune principles can help us not only to better understand the immune system itself, but also to solve engineering evaluation problems. This paper addresses comparison of optimal design using clonal selection algorithm and genetic algorithm.

2   The Human Immune System

The human immune system is a complex system of cells, molecules and organs that represent an identification mechanism capable of perceiving and combating dysfunction from our own cells and the action of exogenous infectious microorganisms. The human immune system protects our bodies from infectious agents such as viruses, bacteria, fungi and other parasites. Any molecule that can be recognized by the adaptive immune system is known as an antigen. The basic component of the immune system is the lymphocytes or the white blood cells.

Lymphocytes exist in two forms, B cells and T cells. These two types of cells are rather similar, but differ with relation to how they recognize antigens and by their functional roles, B-cells are capable of recognizing antigens free in solution, while T cells require antigens to be presented by other accessory cells. Each of this has distinct chemical structures and produces many Y shaped antibodies form its surfaces to kill the antigens. Antibodys are molecules attached primarily to the surface of B cells whose aim is to recognize and bind to antigens [11].

The immune system possesses several properties such as self/nonself discrimination immunological memory, positive /negative selection, immunological network, clonal selection and learning which performs complex tasks.

3   Artificial Immune System

3.1   Clonal Selection Theory

In order to explain how an immune response is mounted when a nonself antigenic pattern is recognized by a B cell, clonal selection theory [12] is been developed.

The main goal of the immune system is to protect the human body from the attack of foreign (harmful) organisms. The immune system is capable of distinguishing between the normal components of our organism and the foreign material that can cause us harm. These foreign organisms are called antigens. The molecules called antibodies play the main role on the immune system response. The immune response is specific to a certain foreign organism (antigen). When an antigen is detected, those antibodies that best recognize an antigen will proliferate by cloning. This process is called clonal selection theory.

The main features of the clonal selection theory are explored as follows [13]:

(1) generation of new random genetic changes subsequently expressed as diverse antibody patterns by a form of accelerated somatic mutation;
(2) phenotypic restriction and retention of one pattern to one differentiated cell (clone);
(3) proliferation and differentiation on contact of cells with antigens.

3.2   Clonal Selection Algorithm

The clonal selection algorithm developed on the basis of clonal selection theory of the immune system. It was proved that can perform and adapt to solve design of optimization tasks. In all runs of the algorithm, the stopping criterion is a predefined maximum number of generations.

The clonal selection algorithm can be described as follows [14]:

(1) Generate a set (P) of candidate solutions, composed of the subset of memory cells (M) added to the remaining (Pr) population (P = Pr + M);
(2) Determine (Select) the n best individuals of the population (Pn), based on an affinity measure;
(3) Reproduce (Clone) these n best individuals of the population, giving rise to a temporary population of clones (C). The clone size is an increasing function of the affinity with the antigen;
(4) Submit the population of clones to a hypermutation scheme, where the hypermutation is proportional to the affinity of the antibody with the antigen. A maturated antibody population is generated ($C*$);
(5) Re-select the improved individuals from $C*$ to compose the memory set M. Some members of P can be replaced by other improved members of $C*$;
(6) Replace d antibodies by novel ones (diversity introduction). The lower affinity cells have higher probabilities of being replaced.

4   Simulation Results

To verify the excellence of clonal selection algorithm, we considered three numerical objective functions and use for comparing the performance of genetic algorithms. The experiment 1 was run for 50 iterations and the experiment 2 was run for 100 iterations. In addition, the parameters are created and tested to ensure that results are listed in Table 1.

4.1   Simple Evaluation Function

Equation (1) is a simple evaluation function for clonal selection algorithm and genetic algorithm as follows:

$$f(x,y) = x^3 + y^3 \qquad (1)$$

where $[x_{min}, x_{max}] = [-2, 2]$, and $[y_{min}, y_{max}] = [-2, 2]$.

Table 1. The parameters created for clonal selection algorithm and genetic algorithm.

Clonal Selection Algorithm (CSA)	
Number of clones generated	100
Hypermutation probability	0.01
Scales of affinity proportion selection	100
Percentage of random new cells each generation	10%
Genetic Algorithm (GA)	
Population size	22
Crossover probability	0.5
Mutation probability	0.01

Figure 1 shows the fitness function obtained by clonal selection algorithm and Figure 2 represents the fitness function in genetic algorithm.

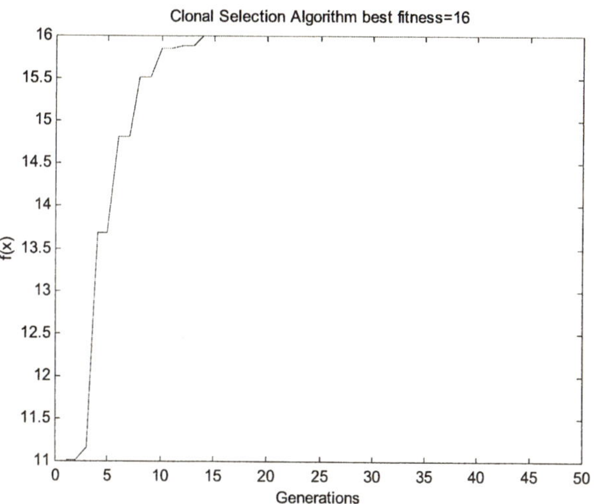

Fig. 1. Fitness function obtained by clonal selection algorithm (Simple evaluation function).

4.2   Rastrigrin Function

Equation (2) is the generalized Rastrigrin function for clonal selection algorithm and genetic algorithm as follows:

$$f(x,y) = x^2 - 10\cos(2\pi x) + 10 + y^2 - 10\cos(2\pi y) + 10 \qquad (2)$$

where $[x_{\min}, x_{\max}] = [-1, 2]$, and $[y_{\min}, y_{\max}] = [-1, 2]$.

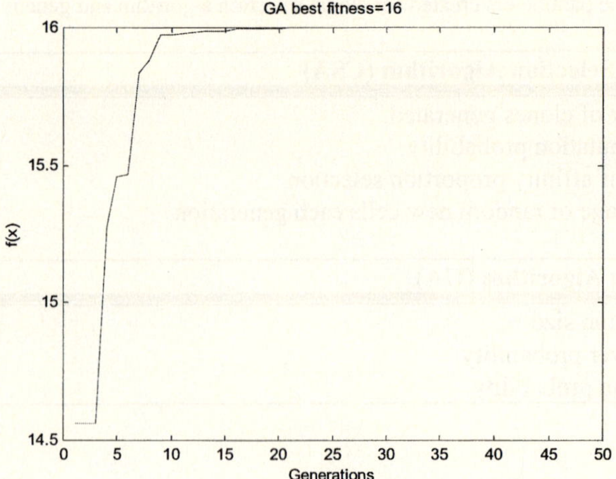

Fig. 2. Fitness variation in genetic algorithm (Simple evaluation function).

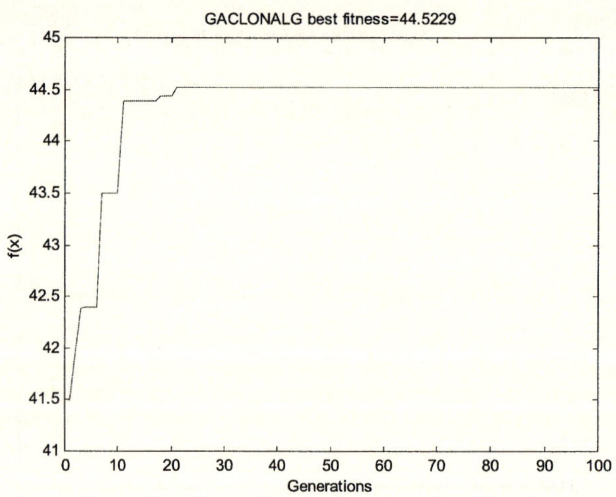

Fig. 3. Fitness function obtained by clonal selection algorithm (Rastrigrin function).

Figure 3 shows fitness function obtained by clonal selection algorithm and Figure 4 represents the fitness function in genetic algorithm.

Table 2 provides a set of results for the functions being optimized. Each algorithm is executed until the maximum value is found. As can be seen from Table 2, both of two algorithms perform well at finding optimal solution. Therefore, in terms of the metric for quality of solution there seems little to distinguish the two algorithms. However, when the number of generations takes into account, there are significant differences in the number taken to obtain the solution.

Like the genetic algorithm, the clonal selection algorithm is highly parallel and presents a fine tractability in terms of computational cost. But, we do not advocate that

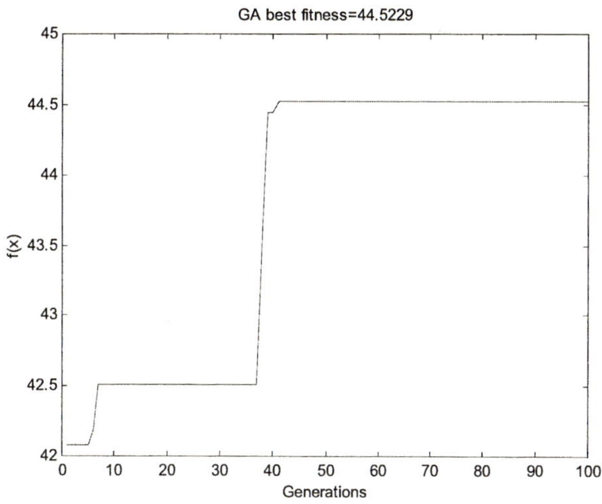

Fig. 4. Fitness variation in genetic algorithm (Rastrigrin function).

the clonal selection algorithm performs better than the genetic algorithm, on average, in all applications. Instead, we demonstrate that the proposed algorithm is also derived from a biologically inspired approach, which performs learning and optimal solutions search.

Table 2. Comparison of clonal selection Algorithm (CSA) and Genetic Algorithm (GA).

Optimal Value / Function	Fitness f(x,y)		x		y	
	CSA	GA	CSA	GA	CSA	GA
Simple Function	16	16	2	2	2	2
Rastrigrin Function	44.5229	44.5229	2	2	2	2

5 Conclusion

In this paper, genetic algorithm and clonal selection algorithm system are applied to two numerical functions for optimization. By comparing the proposed clonal selection algorithm with the standard genetic algorithm, we can notice that the clonal selection algorithm can reach the optima solution, while the GA tends to polarize the whole population of individuals towards the best candidate solution. It seems that little to distinguish the two algorithms. However, when the number of generations takes into account, there are significant differences in the number taken to obtain the solution.

Further work is clearly required. We have also done some work on this and it will be written in the future. Eventually, it is hopeful that this optimization approach of clonal selection algorithm can be helpful for the numerical optimization as a useful and effective alternative.

References

1. Goldberg, D. E. :Genetic Algorithms in Search, Optimization and Machine Learning. Addison-Wesley Publishing Company, Inc, 1989
2. Chun, J. S., Kim, M. K., Jung, H. K. and Hong, S. K. :Shape Optimization of Electromagnetic Devices Using Immune Algorithm. IEEE Transactions on Magnetics, Vol. 33 (2) (1997) 1876-1879
3. Forrest, S. and Perelson, A. S. :Genetic Algorithms and the Immune System. Proc. 1st workshop of PPSN, (1990) 320-325
4. Bersini, H. and Varela, F. J. :The Immune Recruitment Mechanism: A Selective Evolutionary Strategy. Proc. 4th Inter. Conf. on genetic algorithms, (1991) 520-526
5. Mori, K., Tsukiyama, M. and Fukuda, T. :Immune Algorithm with Searching Diversity and its Application to Resource Allocation Problem. T.IEE Japan, Vol. 113-C(10) (1993) 872-878
6. Kim, D. H. :Comparison of PID Controller Tuning of Power Plant Using Immune and Genetic Algorithms. IEEE Inter. Symposium on Computational Intelligence for Measurement Systems and Applications, (2003) 169-174
7. Ishida, Y. :The Immune System as a Self-Identification Process: A Survey and a Proposal. In Proc. of the IMBS'96, 1996
8. Dasgupta, D. :Artificial Immune Systems and Their Applications, Ed., Springer-Verlag, 1999
10. Hofmeyr S. A. and Forrest, S. :Immunity by Design: An Artificial Immune System. Proc. of GECCO'99, (1999) 1289-1296
10. de Castro, L. N. and Von Zuben, F. J. :Learning and Optimization Using the Clonal Selection Principle. IEEE Trans. on Evolutionary Computation, Special Issue on Artificial Immune Systems, Vol. 6(3) (2002) 239-251
11. Jerne, N. K. :The Immune System. Scientific American, Vol.229(2) (1973) 52-60
12. Burnet, F. M. :The Clonal Selection Theory of Acquired Immunity. Cambridge University Press, 1959
13. Burnet, F. M. :Clonal Selection and After. in Theoretical Immunology, (Eds.) G. I. Bell, A. S. Perelson & G. H. Pimbley Jr., Marcel Dekker Inc., (1978) 63-85
14. de Castro, L. N. and Von Zuben, F. J. :The Clonal Selection Algorithm with Engineering Applications. Proc. of GECCO'00, Workshop on Artificial Immune Systems and Their Applications, (2000) 36-37

Wrist Motion Pattern Recognition System by EMG Signals

Yuji Matsumura, Minoru Fukumi, and Norio Akamatsu

Faculty of Engineering, The University of Tokushima, Tokushima, 770-8506, Japan
Tel / Fax: +81-88-656 -7510
yuu@is.tokushima-u.ac.jp

Abstract. In this paper, we aim for construction of high-speed and high-accurate system using Fast Fourier Transform (FFT) for feature extraction, Simple-PCA (SPCA) for feature compression, and a neural network (NN) for recognition. In particular, we present a novel method based on Canonical Discriminant Analysis (CDA) to improve recognition accuracy for EMG. From results of computer simulation, it is shown that our approach is effective for improvement in recognition accuracy.

1 Introduction

Recently, information terminals such as a cellular phone have been widely used. According to this, industrial standard of radio communication such as Bluetooth has been established. As a result, it would be possible to combine and to perform various interfaces. However, now a day, the device that has various operations of portable machines and tools and can control networks (call "total operation device" for short) has not been provided yet. Moreover, the wristwatch type is preferable in the viewpoint of operationality. Therefore, we investigate ElectroMyoGram (EMG) [1]-[3] which is a signal generated from a living body with movement of a subject.

In this paper, we aim for construction of high-speed and high-accurate system using Fast Fourier Transform (FFT) [4] for feature extraction, Simple-PCA (SPCA) for feature compression, and a neural network (NN) [5] for recognition. In particular, we present a novel method based on Canonical Discriminant Analysis (CDA) to improve recognition accuracy for EMG. From results of computer simulation, it is shown that our approach is effective for improvement in recognition accuracy.

2 EMG Recognition System

A flowchart of construction of an EMG pattern recognition system that we propose in this paper is shown in Fig. 1. The proposed system is composed of an input, a signal processing, a data transform and learning-evaluation parts.

First, electrodes in the input part measure time series data for EMG. Second, this data are amplified and A/D transform is performed in the signal processing part. Next, this amplified data are converted to Fourier power spectra. Finally, we evaluate such a data in the learning-evaluation part. Fig. 2 shows the system developed in this research.

R. Khosla et al. (Eds.): KES 2005, LNAI 3681, pp. 611–617, 2005.
© Springer-Verlag Berlin Heidelberg 2005

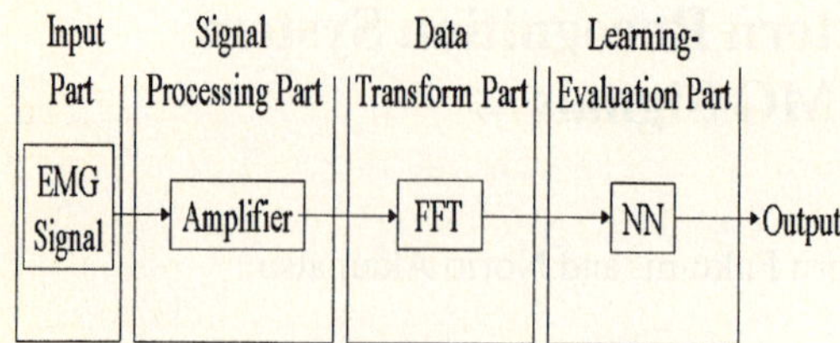

Fig. 1. A flowchart of the proposed system

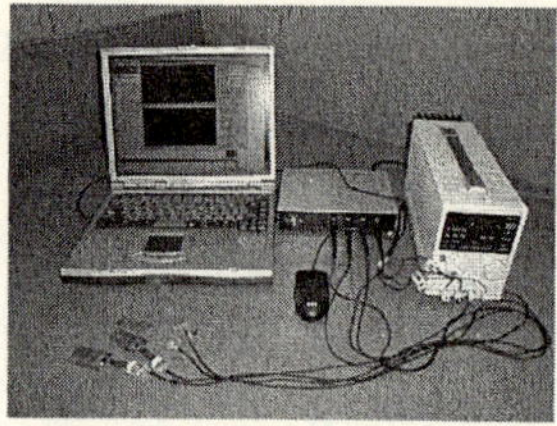

Fig. 2. The outside of the proposed system

2.1 Input Part

In this paper, we use the surface-type electrode as an input device because we try to recognize a lot of behavior from the various muscles. Furthermore, there are two types of surface electrodes. One is a wet type and the other is a dry type. From the viewpoint of easiness of operation, we adopt the dry type one.

Figure 3 illustrates a device for measuring EMG of the wrist. We measure 7 patterns of the wrist behavior using this device. The wrist behavior is neutral, up, down, bend to right, bend to left, twist of inside, twist of outside.

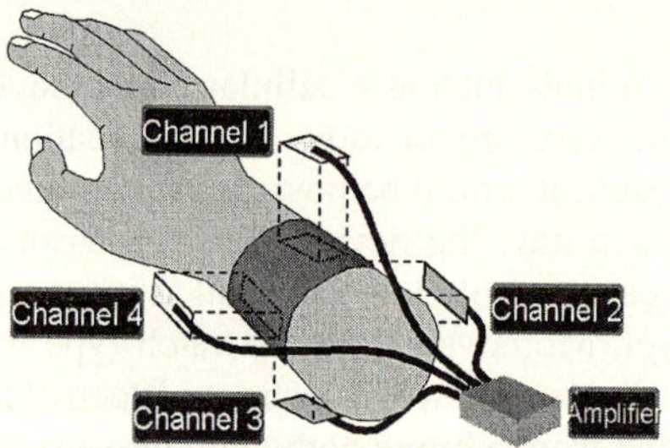

Fig. 3. The device of measuring

2.2 Signal Processing Part

It has been reported that the main frequency of EMG is generated between a few Hz and 2 kHz [6]-[8]. We extract frequency components between 70 Hz and 2 kHz to avoid the influence of the noise by the electromagnetic wave, that is, the commercial frequency of 60 Hz. As a result, the amplifier reduces the influence of the electromagnetic wave noise. The specification of the amplifier is shown in Table 1.

Table 1. The specification of the amplifier

Low Cut-off Frequency	70Hz (6th filter)
High Cut-off Frequency	2kHz (6th filter)
Gain	2000 times (65dB)
Output Voltage	±5V
Input Voltage	±2.5mV

2.3 Data Transform Part

The EMG output program yields 20,480 points time-series data for 2,048 milliseconds. And, FFT is performed against the data. The spectra are obtained by performing 1,024-points FFT from 10,000th data point for every channel, moving a data window every 512 data points, as shown in Fig.4. Using this operation, the average of the data generated by FFT can be obtained. Fig. 4 shows a schematic to yield learning data.

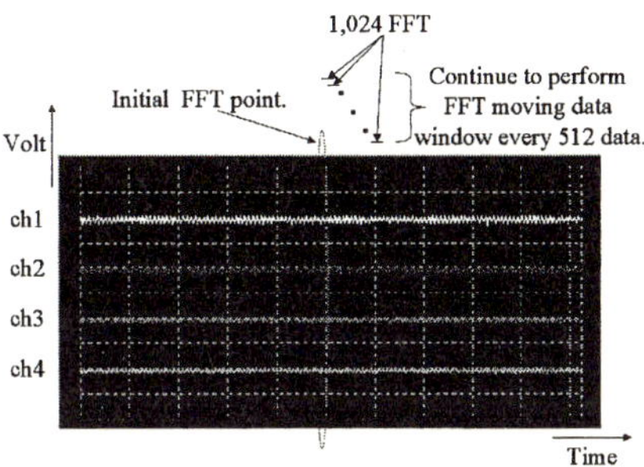

Fig. 4. A method to yield FFT spectra

Moreover, we used learning-evaluation data less than 1kHz according to the previous research [9]. But, the amount of feature data is about 400 and the construction of network becomes huge. For this reason, there may be problem in learning time. Therefore, we try to compress learning-evaluation data using Simple-PCA (SPCA).

2.3.1 Simple-PCA

SPCA has been proposed in order for PCA to achieve high-speed processing. It is effective for information compression at handwritten digits [10].

SPCA produces approximate solutions using a simple repetition operation without calculating a variance-covariance matrix. The SPCA algorithm is given as:

(i) Collect $\mathbf{V}$ set of an n dimensional data $\mathbf{v}_j$, $i = 1,2,\ldots,$m.

(ii) $\mathbf{X} = \{\mathbf{x}_1, \mathbf{x}_2, \cdots \mathbf{x}_m\}$, which are obtained by subtracting $\overline{\mathbf{v}}$ which is the average value of $\mathbf{V}$ from the center of gravity of set of the vectors, is used as input vectors.

(iii) The column vector a_1 is defined as connection weights between the inputs $\mathbf{x}_j$ and an output y. The first weight, a_1 is used to approximate the first eigenvector, α_1. The output function has a value given by:

$$y_1 = \alpha_1^T \mathbf{x}_j \tag{1}$$

(iv) Using the following equation (2), repetitive calculation of an arbitrary vector a_1 suitably given as an initial value is carried out. As a result, the vector a_1 can approach the same direction as α_1.

$$a_1^{k+1} = \frac{\sum_j \Phi_1(y_1, \mathbf{x}_j)}{\left\| \sum_j \Phi(y_1, \mathbf{x}_j) \right\|}; \quad y_1 = (a_1^k)^T \mathbf{x}_j \tag{2}$$

where Φ_1 is a threshold function given as

$$\Phi_1(y_1, \mathbf{x}_j) = \begin{cases} \mathbf{x}_j & \text{if } y_j \geq 0 \\ 0 & \text{otherwise} \end{cases} \tag{3}$$

(v) Using the following equation (4), we remove the first principal component vector from the dataset in order to find the next principal component.

$$\mathbf{x}_j' = \mathbf{x}_j - (a_1^T \mathbf{x}_j)a_1 \tag{4}$$

We obtain a_2 because we substitute a_1 with a_2 and $\mathbf{x}_j$ with $\mathbf{x}_j'$ in equation (2) and perform repetitive calculation once again.

If the same operation is being done, we obtain the principal component which is greater in contribution rate by turns.

In this paper, we calculate principal components until 20th vector that accumulated contribution ratio is more than 95% and obtain cosine distance between eigen vector and input data as new learning-evaluation data.

2.4 Learning-Evaluation Part

We perform learning and evaluation of EMG data using NN in the learning evaluation part. As the reason for using NN, EMG has a high nonlinearity and it is thought that we couldn't separate easily using a linear processing of threshold processing etc. Fig 5 shows a distribution of learning-evaluation data after SPCA.

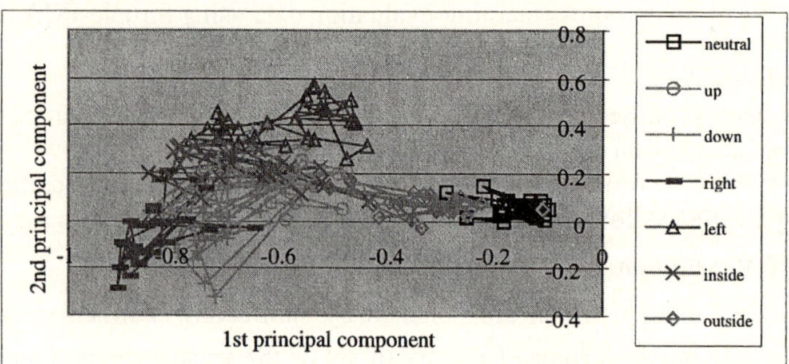

Fig. 5. Distribution of learning-evaluation data after SPCA

Moreover, a learning method we used in this paper is the BP method.

(1) The BP network

   The BP network is a three-layered feed-forward network that is composed of an input layer, a middle layer, and an output layer. The error between an output and a teacher signal is back-propagated in order of an output layer, the middle layer, and the input layer to update each weight w. It is also called the Back Propagation method.

(2) Double Judgment Network (DJN)

   Double Judgment Network is the method that NN recognizes using only selected data by Canonical Discriminant Analysis (CDA)[11]. CDA is a dimension-reduction technique related to principal component analysis and canonical correlation. Given a classification variable and several quantitative variables, CDA derives canonical variables (linear combinations of the quantitative variables) that maximize between-class variation in the same way that principal components maximize a total variation. In this occasion, Judgment is done in case that both outputs of CDA and NN are the same class. This will enable us to recognize without indefinite data, that is, hardly recognizable data. Therefore, it is expected to raise recognition accuracy. DJN's algorithm is shown below.

[STEP1]: First, PCA produces feature data for dimensional reduction of CDA's calculation amount.

[STEP2]: Canonical Discriment Analysis performs the first judgment.

[STEP3]: Neural Network performs the second judgment.

[STEP4]: If both outputs of NN and CDA are the same class, NN recognition is then carried out in easily recognizable data.

[STEP5]: If not, this data is regarded as hardly recognizable data. (Judgment: reject)

Moreover, Figs.6 and 7 show the outline of BP and DJN.

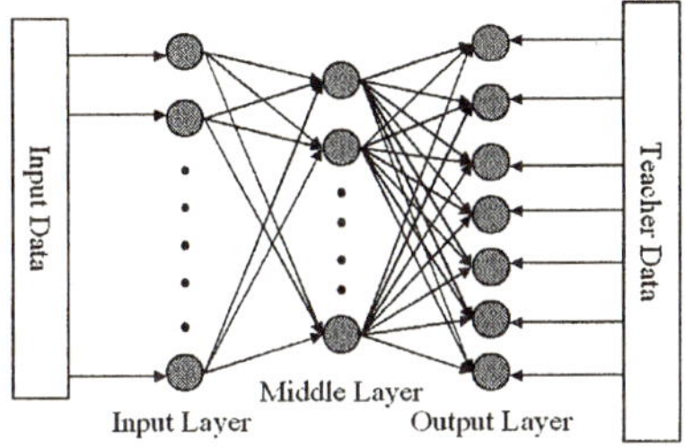

Fig. 6. The outline of BP **Fig. 7.** The outline of DJN

3 System Verification Experiments

In this paper, we verified the effectiveness of the wrist motion recognition system using 3 persons data of subject A (30 years old, male), subject B (22 years old, male), and subject C (21 years old, male).

The amount of data we used is 30(trials)×7(motions)×3(persons)=630. But, we use only the same subject data in learning-evaluation because we aim to construct an online-learning type recognition system.

3.1 Comparative Experiments

In this paper, we carried out comparative experiments between DJN and NN. Tables 2 and 3 show experimental conditions of NN and DJN, respectively. Table 4 shows experimental results.

According to results listed above, it is found that recognition accuracy of DJN is better than BP in every subject. Discriminant Analysis using primary judgment caused the results. Originally, there is a possibility where classification of nonlinear data isn't good because Discriminant Analysis is one of linear discriminant methods. By the use of its weaknesses, it is found that the same results data for both Discriminant Analysis and NN are easy to be classified. After all, it is thought that recognition accuracy rises because DJN recognizes using only this data.

Table 2. Experimental Conditions (BP)

No. of Unit's in the Input Layer	20
No. of Unit's in the Middle Layer	20
No. of Unit's in the Output Layer	7
No. of Learning Data	70
No. of Evaluation Data	140
Maximum Learning Cycles	20,000

Table 3. Experimental Conditions (DJN)

No. of feature data	20
No. of classification	7
No. of Learning Data	70
No. of Evaluation Data	140
No. of canonical variable	3

Table 4. Experimental results (BP & DJN) Recognition Accuracy (%)

	Subject A		Subject B		Subject C	
	BP	DJN	BP	DJN	BP	DJN
Up	79.0	99.1	63.0	96.0	84.0	94.7
Right	96.0	97.5	79.5	91.3	88.0	97.5
Left	90.5	100.0	94.0	98.8	97.5	100.0
Down	93.5	99.0	91.0	98.9	84.0	87.5
Neutral	86.0	96.5	75.0	93.6	82.5	93.8
Inside	88.0	91.1	73.0	91.8	86.0	96.6
Outside	70.0	69.3	75.5	84.9	90.5	94.8
Total	86.1	93.2	78.7	93.6	87.5	95.0

4 Conclusions

In this paper, we constructed the wrist motion pattern recognition system using wrist EMG, instead of brachial muscle using as the conventional methods. Furthermore, we constructed the new recognition system by a double judgment network that recognition is carried out using only easily recognizable data. As a result, some positive results were achieved through this verification. However, this led to create unused data (Judgment: reject) for recognition. In addition, the number of unused data is 30% of the number of evaluation data. Discontinuous motion systems such as the total operation device we described have to work without trouble. However, continuous motion systems such as pointing device are deadly in the number of rejects. Therefore, it is needed that we study how to deal with unused data in the next step. This research was partially supported by Suzuki Foundation and Grant-in-Aid for Scientific Research (B) 15300073 in MEXT.

References

1. K. Hashimoto, T. Endo: "How to Inspect to Vital Function", Japan publishing service, (1983), in Japanese

2. N. Hoshimiya: "Living Body Information Measure", Morikita publishing company, (1995), in Japanese
3. N. Hoshimiya, K. Akazawa: "[MBE topics series 3rd] EMG Control System", Sho-mei-do, (1993), in Japanese
4. J.W Cooley, J. W. Turkey: "An Algorithm for Machine Calculation of Complex Fourier Series", 19, pp. 297-301, (1965)
5. J. Hearts, A. Krogh, Richard. G. Palmer, "Introduction to The Theory of Neural Computation", Addison-Wesley Publishing Company, The Advanced Book Program, 350 Bridge Park0way, Redwood City, (1991)
6. Y. Ishioka, "Standard and Application of Stomatognathic Function Analysis", dental Diamond Company, pp.260-273, (1991), in Japanese
7. H.G Vaughan, Jr., L. D. Costa, L. Gilden and H. Schimmel: "Identification of Sensory And Motor Components of Cerebral Activity in Simple Reaction Time Tasks", Proc. 73rd, Conf. Amer. Psychol. Ass., 1, pp. 179-180, (1965)
8. C. J. De Luca: "Physiology and Mathematics of Myoelectric signals", IEEE Trans, Biomed. Eng., 26, pp. 313-325, (1979)
9. Y. Matsumura, M. Fukumi, Y. Mitsukura, "Recognition of EMG Signal Patterns by Neural Networks", ICONIP'02Workshop Proceeding, Singapore, (2002)
10. M.partridge, R.Calvo, "Fast dimensionality reduction and Simple PCA", IDA, Vol.2(3), pp. 292-298, (1997)
11. Anderson, T.W.: "An introduction to Multivariate Statistical Analysis", New York, John Wiley, (1958)

Analysis of RED with Multiple Class-Based Queues for Supporting Proportional Differentiated Services

Jahwan Koo[1] and Seongjin Ahn[2,*]

[1] School of Information and Communications Engineering, Sungkyunkwan Univ.
Chunchun-dong 300, Jangan-gu, Suwon, Kyounggi-do, Korea
`jhkoo@songgang.skku.ac.kr`
[2] Department of Computer Education, Sungkyunkwan Univ.
Myeongnyun-dong 3-ga 53, Jongno-gu, Seoul, Korea
`sjahn@comedu.skku.ac.kr`

Abstract. The proportional differentiated service (PDS) model has been developed to provide a controllable, consistent, and scalable quality of service (QoS) model. The PDS model enables network operators to get a control knob for the convenient management of services and resources even in a large-scale network. In this paper, we provide an analysis of RED with multiple class-based queues for supporting PDS among different classes using $M/M/1/K$ queueing model and is also verified how to set the main control parameter, max_p, of RED for PDS using ns-2 simulator.

1 Introduction

In the last few years, there has been a wave of interest in providing network services with performance guarantees and in developing algorithms supporting different levels of services. The various solutions that have been proposed can be summarized under the general heading of Quality-of-Service (QoS). The need for such quality assurance, service guarantee, and reliability drove the Internet Engineering Task Force (IETF) to define two QoS solutions - Differentiated Services (DiffServ) and Integrated Services (IntServ). Of these two methods, the DiffServ gained popularity due to its inherent scalability, ease of implementation, and reduced operation complexity at the core of the Internet.

However, the DiffServ can only guarantee that traffic of a high priority class will receive no worse service than that of a low priority class. Recently, it has been further refined into a quantitative scheme: the proportional differentiated service. For this proportional differentiated service is controllable, consistent, and scalable, it provides network operators convenient management of services and resources even in a large-scale network.

The proportional differentiated service (PDS), initially proposed by Dovrolis et al. [1], has been an effort to quantify the differentiation between classes of

---

* Dr. S. Ahn. is the Corresponding Author.

R. Khosla et al. (Eds.): KES 2005, LNAI 3681, pp. 618–623, 2005.

traffic and to enforce that the ratios of delays or loss rates of successive priority classes be roughly constant.

Most mechanisms for PDS have been used independent algorithms for delay and loss differentiation. For example, proportional differentiation of delays has been implemented with appropriate scheduling algorithms [2] [4] [5] and proportional differentiation of loss has been implemented by queue management algorithms [3] [6]. Recently, a joint queue management and scheduling algorithm on proportional differentiation was reported in [7].

In this paper, we limit our discussion to proportional loss differentiation in RED scheme as a representative queue management scheme recommended by the IETF and deployed in many commercial routers and not delay differentiation or the coupling of the two. The rest of the paper is organized as follows. In section 2, we introduce the PDS and its general framework. In section 3, we analyze a main control parameter of RED with multiple class-based queues for PLD. In section 4, we perform network simulation using widely used *ns-2* with Nortel's DiffServ module.

2   General Framework for Supporting PDS

In the proportional differentiation model, the service differentiation can be quantitatively adjusted to be proportional to the differentiation factors that a network service provider sets beforehand. If q_i is the QoS metric of interest and s_i is the differentiation factors for class i, $i = 1, \cdots, n$, in the proportional differentiation model, we should have:

$$\frac{q_i}{q_j} = \frac{s_i}{s_j} \tag{1}$$

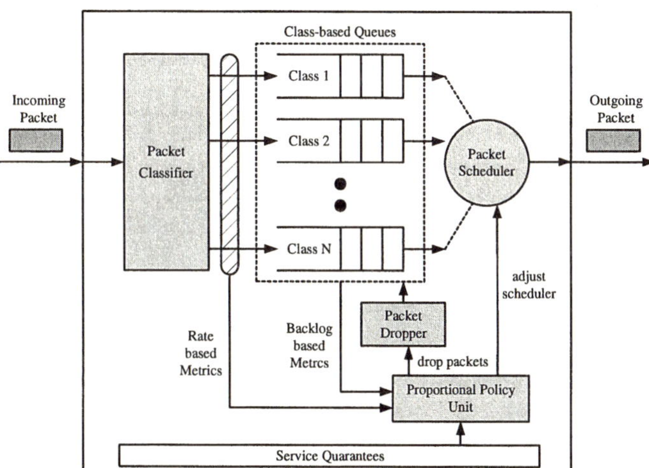

Fig. 1. General framework for supporting proportional differentiated services.

If the desired differentiation among n classes are quantified as the ratios, δ_i, the proportional differentiation of delay(d_i) and loss(l_i) can be rewritten as

$$\frac{d_i}{d_j} = \frac{\delta_i}{\delta_j}, \frac{l_m}{l_n} = \frac{\delta_m}{\delta_n}. \tag{2}$$

The proportional differentiation model has gained attention recently as an effective solution for quantitative service differentiation in IP networks.

A general framework for proportional differentiated service in an IP network consists of four major modules at routers: 1) the packet classifier that can distinguish packets and group them according to their different requirements, 2) the packet dropper that determines both of the following: how much queue space should be given to certain kinds of network traffic and which packets should be discarded during congestion, 3) the packet scheduler that decides the packet service order so as to meet the bandwidth and delay requirements of different types of traffic, and 4) the proportional policy unit that takes the real-time measurement of the QoS metrics and manages the packet dropper and scheduler.

3   Analysis of RED with Multiple Class-Based Queues

For a given resource portion out of common output link through a predestined scheduling, the service rate of each class queue i, μ_i, is determined. With a certain arrival rate, λ_i, the related system utilization factor, $\rho_i = \lambda_i/\mu_i$, can be assumed. Then we can provide an analytical study of RED scheme by deriving the drop probability equations roughly using the traditional $M/M/1/K$ queueing model [8]. With a each class queue i's buffer size of K having tail-drop (TD) scheme, the steady-state probability of finding k packets in the queue is given by

$$\pi_i(k) = \pi_i(0) \prod_{l=0}^{k-1} \rho_i \tag{3}$$

where $\pi_i(0) = \left[\sum_{k=0}^{K} \prod_{l=0}^{k-1} \rho_i \right]^{-1}$. Newly arriving packets will be refused and dropped immediately without service in case when K packets are occupied in the queue system.

For RED queue, incoming packets are dropped with a probability that is an increasing function $d(k)$ of the average queue size k. The average queue size is estimated using an exponential weighted moving average formula, $k = (1 - w_q) * k + w_q * \widehat{k}$, where $\widehat{k}$ is the current queue size and w_q is a weight factor, $0 \leq w_q \leq 1$. RED offers three control parameters: maximum drop probability (max_p), minimum threshold (min_{th}), and maximum threshold (max_{th}) upon averaged queue length (k) with weighted factor (w_q) to tune RED's dynamics.

The typical dropping function of class i, $d_i(k)$, in RED is defined by three parameters min_{th}, max_{th} and $max_{p,i}$ as follows:

$$d_i(k) = \begin{cases} 0, & k < min_{th} \\ \dfrac{max_{p,i} \cdot (k - min_{th})}{max_{th} - min_{th}} & \\ \equiv max_{p,i} \cdot f(k) & min_{th} \le k < max_{th} \\ 1, & k \ge max_{th} \ . \end{cases} \tag{4}$$

Therefore, the drop probability of a packet depending on the dropping function related to each state k can be approximated as follows:

$$P_i^{RED} = \sum_{k=min_{th}}^{K} \pi_i^{RED}(k) d_i(k), \quad min_{th} \le k < K \ . \tag{5}$$

We assumed all max_{th} as same queue size K and same traffic density $\rho_i (= \lambda_i/\mu_i)$. The number of packets in the RED queue is actually a birth-death process with the birth rate in state k equal to $\lambda_i(1 - d_i(k))$ and the death rate equal to μ_i in $M/M/1/K$ queueing model. Accordingly, the stationary distribution of state probability at k, $\pi_i^{RED}(k)$, is derived by modifying Eq. (3) as follows:

$$\pi_i^{RED}(k) = \pi_i(0) \rho_i^k \prod_{l=0}^{k-1} (1 - d_i(l)) \tag{6}$$

where $\pi_i(0) = \left[\sum_{k=0}^{K} \rho_i^k \prod_{l=0}^{k-1} (1 - d_i(l)) \right]^{-1}$.

For proportional loss differentiation, we follow PDS model and quantitative and proportional loss differentiation factor between adjacent classes, δ_L, which is given as network operator's request, described as

$$\frac{l_{i+1}}{l_i} = \delta_L = \frac{P_{i+1}^{RED}}{P_i^{RED}} = \frac{\displaystyle\sum_{k=min_{th}}^{K} \pi_{i+1}^{RED}(k) d_{i+1}(k)}{\displaystyle\sum_{k=min_{th}}^{K} \pi_i^{RED}(k) d_i(k)} \tag{7}$$

where δ_L is the differentiation factor between adjacent classes as a control knob and loss-priorities are decreasing order with class i. The l_i means the desired loss probability of class i.

4   Simulation Results

We perform network simulation using widely used *ns-2* with Nortel's DiffServ module in order to verify the selection method of max_p for PLD at realistic network environments.

The simulation environment consists of a router, a destination, and a set of sources, each source which generates packets with a fixed size of 500 bytes and has the same traffic pattern with CBR exponential distribution over the

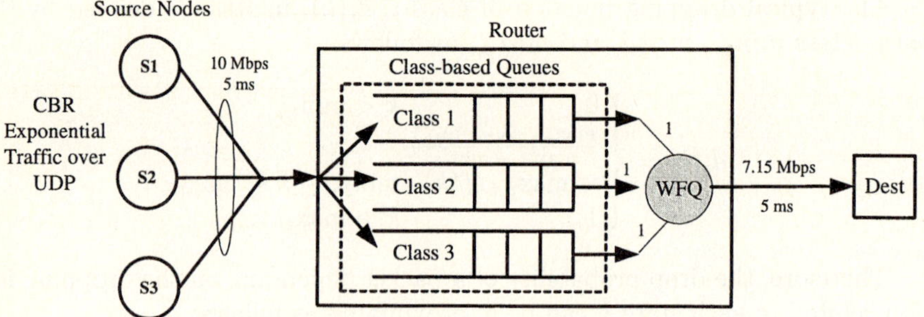

Fig. 2. Network simulation topology.

UDP. And each source chooses only one class type among three service classes and is connected to the router node with a link, which bandwidth and delay are set to 10Mbps and 5ms appropriately. We set the router output link with the capacity of 7.15 Mbps resulting in the offered load as the value of 1.4 and the router has three RED queues in accordance with three classes, each RED queue has a queue size of 120. Each overlapped RED parameters which means the parameters of min_{th} and max_{th} are same across all class's queues were identically set to $min_{th} = 40$, $max_{th} = 120$, and $w_q = 0.002$ except the control parameter of $(max_{p,1}, max_{p,2}, max_{p,3}) = (0.011, 0.021, 0.042)$ which is guided by Eq. (7) in case of δ_L's value of 2. We also use the WFQ scheduler having equal service weight across classes and run 300 seconds of simulated time.

The simulation result in Figure 3 shows the appropriate PLD among different classes after some transient period at initial phase, which is also well-matched with the mathematical analysis.

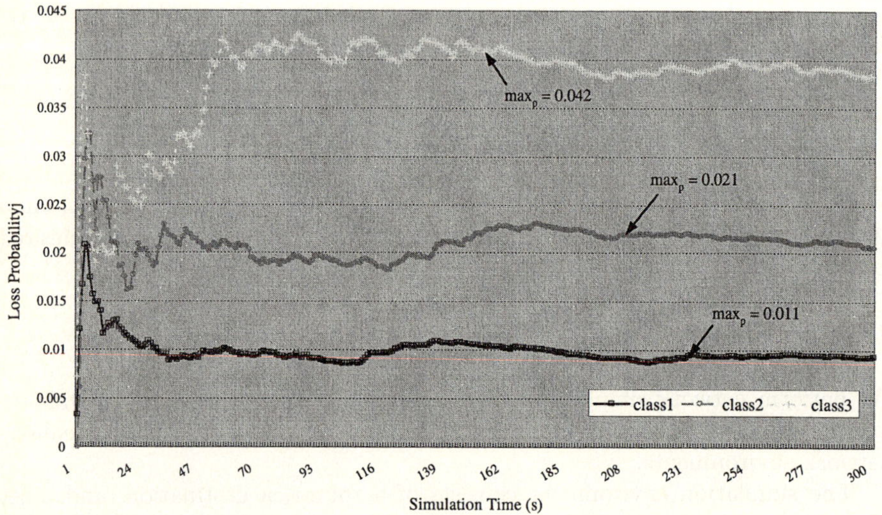

Fig. 3. Simulation result on PLD ($l_1 : l_2 : l_3 = 1 : 2 : 4$).

5   Conclusion

Many mechanisms for differentiated services have been proposed to solve the problems of a high loss rate and inefficient utilization of network resources in the Internet. This work provides an analysis of RED with multiple class-based queues for supporting PDS among different classes using $M/M/1/K$ queueing model and is also demonstrated its loss probability differentiation using ns-2 simulator. This goal is to implement a unified, controllable, and scalable DiffServ router.

References

1. C. Dovrolis and P. Ramanthan, "A Case for Relative Differentiated Services and the Proportional Differentiation Model," IEEE Network, 13(5):26-34, 1999.
2. C. Dovrolis, D. Stiliadis, and P. Ramanathan, "Proportional Differentiated Services: Delay Differentiation and Packet Scheduling," Proc. ACM SIGCOMM, pp. 109-120, 1999.
3. C. Dovrolis and P. Ramanathan, "Proportional Differentiated Services, Part II: Loss Rate Differentiation and Packet Dropping," Proc. IWQoS, pp. 52-61, 2000.
4. Chin-Chang Li et al., "Proportional Delay Differentiation Service Based on Weighted Fair Queueing," Proc. IEEE ICCCN, pp. 418-423, 2000
5. Y. Chen, M. Hamdi, D. Tsang, and C. Qiao, "Proportional QoS provision: a uniform and practical solution," Proc. ICC 2002, pp. 2363-2366, 2002
6. J. Aweya, M. Ouellette, and D. Y. Montuno, "Weighted Proportional Loss Rate Differentiation of TCP Traffic," Int. J. Network Mgmt, pp. 257-272, 2004
7. J. Liebeherr and N. Christin, "JoBS: Joint buffer management and scheduling for differentiated services," Proc. IWQoS 2001, pp. 404-418, Karlsruhe, Germany, 2001.
8. T. Boland, M. May, and J. C. Bolot, "Analytic evaluation of RED performance", in Proc. IEEE Infocom '00, pp.1415-1424, Tel Aviv, Israel, March 2000.

Learning Behaviors of the Hierarchical Structure Stochastic Automata

Operating in the Nonstationary Multiteacher Environment

Norio Baba[1] and Yoshio Mogami[2]

[1] Department of Information Science, Osaka Kyoiku University,
Kashiwara City, Osaka Prefecture, 582-8582, Japan
[2] Department of Information Science and Intelligent Systems, Faculty of Engineering,
University of Tokushima, Tokushima City, Tokushima Prefecture, 770-8506, Japan

Abstract. A new learning algorithm for the hierarchical structure learning automata operating in the nonstationary multiteacher environment is proposed. The efficacy of the proposed algorithm is compared with the modified DGPA algorithm by the several computer simulation studies below.

1 Introduction

The concept of learning automata (LAs) operating in an unknown environment was initially introduced by Tsetlin [1]. He considered the learning behaviors of finite deterministic automata in a stationary random environment. Following his pioneering work, Varshavskii and Vorontsova[2] found that stochastic automata also have learning performances. Since then, the study of learning behaviors of stochastic automata has been done quite extensively by many researchers. The learning automata theory has now reached a relatively high level of maturity, and various successful applications utilizing learning automata have so far been reported[3]-[20].

Although the study of LAs has matured, there are still several problems to be settled. Two of the most important are the low speed of learning, and the insufficient tracking ability to the nonstationary environment.

In this paper, we shall introduce our recent studies regarding learning behaviors of the hierarchical structure learning automata (HSLA) to overcome the second problem. First, we shall propose a new learning algorithm which exploits the idea of relative reward strength algorithm[19]. Then, we shall investigate the efficacy of the proposed algorithm by comparing with the modified DGPA algorithm[13][1].

---

[1] DGPA algorithm is well known as one of the fastest learning automata algorithms today. In this paper, we shall modify the DGPA algorithm to be used as a learning algorithm of the hierarchical structure automata model.

R. Khosla et al. (Eds.): KES 2005, LNAI 3681, pp. 624–634, 2005.

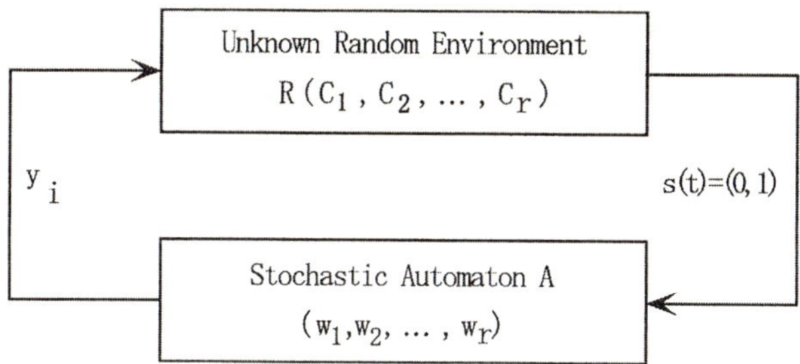

Fig. 1. Basic model of a learning automaton operating in an unknown random environment.

2   Basic Model of a Learning Automaton Operating in an Unknown Environment

The learning behaviors of a variable-structure learning automaton (LA) operating in an unknown random environment have been discussed extensively under the basic model shown in Fig.1. Let us briefly explain the learning mechanism of the learning automaton A under the unknown random environment (teacher environment) $R(C_1, C_2, \ldots, C_r)$.

The learning automaton A is defined by the sextuple $\{S, W, Y, g, P(t), T\}$. S denotes the set of two inputs (0,1), where 1 indicates the reward response from $R(C_1, C_2, \ldots, C_r)$, and 0 indicates the penalty response. (If the set S consists of only the two elements 0 and 1, the environment is said to be a P-model. When the input into A assumes a finite number of values in the closed interval [0,1], it is said to be a Q-model. An S-model is one in which the input into A takes an arbitrary number in the closed line segment [0,1].) W denotes the set of r internal states $(w_1, w_2, \ldots, w_r)$. Y denotes the set of r outputs $(y_1, y_2, \ldots, y_r)$. g denotes the output function $y(t) = g[w(t)]$, that is, one to one deterministic mapping. $P(t)$ denotes the probability vector $[p_1(t), p_2(t), \ldots, p_r(t)]'$ at time t, and its ith component $p_i(t)$ indicates the probability with which the ith state w_i is chosen at time t $(i = 1, 2, \ldots, r)$:

$$p_1(0) = p_2(0) = \cdots = p_r(0) = \frac{1}{r}, \quad \sum_{i=1}^{r} p_i(t) = 1, \tag{1}$$

and T denotes the reinforcement scheme which generates $P(t+1)$ from $P(t)$.

Suppose that the state w_i is chosen at time t. Then, the learning automaton A performs action y_i on the random environment $R(C_1, C_2, \ldots, C_r)$. In response to the action y_i, the environment emits output $s(t) = 1$ (reward) with probability $1 - C_i$; and output $s(t) = 0$ (penalty) with probability C_i $(i = 1, 2, \ldots, r)$. If all of the C_i $(i = 1, 2, \ldots, r)$ are constant, the random environment $R(C_1, C_2, \ldots, C_r)$ is said to be a stationary random environment. (The term "single teacher environment" is also used.) On the other hand, if C_i $(i = 1, 2, \ldots, r)$ are not constant, it is said to be a nonstationary random environment. Depending upon the action of the learning automaton A and the environmental response to it, the reinforcement scheme T changes the probability vector $P(t)$ to $P(t+1)$.

The values of C_i $(i = 1, 2, \ldots, r)$ are not known in advance. Therefore, it is necessary to reduce the average penalty

$$M(t) = \sum_{i=1}^{r} p_i(t) C_i \tag{2}$$

by selecting an appropriate reinforcement scheme. To judge the effectiveness of a learning automaton operating in a stationary random environment $R(C_1, C_2, \ldots, C_r)$, various performance measures such as optimality, ε-optimality, absolute expediency, etc. have been set up. (Details are omitted due to space. The reader is refered to [3]-[6].)

3 Hierarchical Structure Learning Automata(HSLA) Operating in the Nonstationary Multiteacher Environment

In the previous section, we briefly touched upon the basic model of the learning automaton. However, one of the most serious bottlenecks concerning the basic model is that a single automaton can hardly cope with the problems of high dimensionality.

To overcome this problem, Thathachar and Ramakrishnan[7] proposed the concept of HSLA. Since then, the learning behaviors of the HSLA have been extensively studied by many active researchers. In this section, we shall briefly touch upon the learning behaviors of HSLA operating in the nonstationary multiteacher environment.

Fig.2 shows the learning mechanism of HSLA operating in the general nonstationary multiteacher environment. The hierarchy is composed of the single

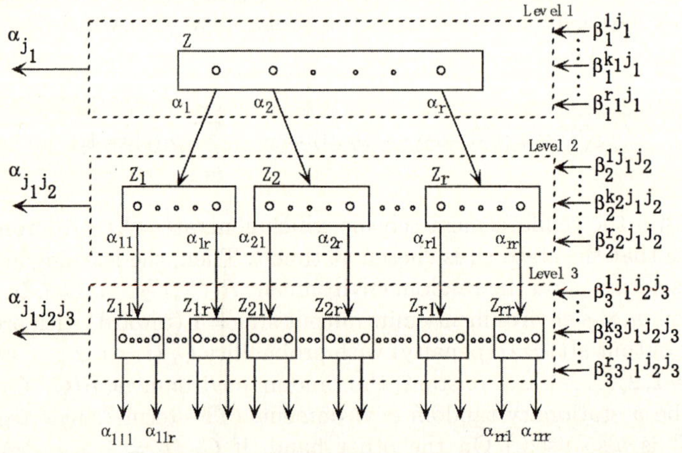

Fig. 2. Learning mechanism of the hierarchical structure learning automata.

automaton Z in the first level, r automata $Z_1, Z_2, \ldots, Z_r$ in the second level, and r^{s-1} automata $Z_{j_1 j_2 \cdots j_{s-1}}$ $(j_i = 1, 2, \ldots, r; i = 1, 2, \ldots, s-1)$ in the sth level $(s = 1, 2, \ldots, N)$. Each automaton in the hierarchichy has r actions.

The operation of the learning system of the HSLA can be described as follows. Initially, all the action probabilities are set equal. Z chooses an action at time t from the action probability distribution $(p_1(t), p_2(t), \ldots, p_r(t))'$. Suppose that α_{j_1} $(j_1 = 1, 2, \ldots, r)$ is the output from Z and $\beta_1^{k_1 j_1}(t)$ $(k_1 = 1, 2, \ldots, r_1)$, where t denotes time, is the reward strength from the k_1th teacher. $(\beta_1^{k_1 j_1}(t)$ can be an arbitrary number in the closed line segment [0,1].) Depending upon the output α_{j_1}, the responses $\beta_1^{k_1 j_1}(t)$ $(k_1 = 1, 2, \ldots, r_1)$, and the responses from the lower levels, the first level automaton Z changes its action probability vector $P(t) = (p_1(t), p_2(t), \ldots, p_r(t))'$. Corresponding to the output α_{j_1} at the first level, the automaton Z_{j_1} is actuated in the second level. This automaton chooses an action from the current action probability distribution $(p_{j_1 1}(t), p_{j_1 2}(t), \ldots, p_{j_1 r}(t))'$. This cycle of operation repeats from the top to the bottom.

The sequence of actions $\{\alpha_{j_1}, \alpha_{j_1 j_2}, \ldots, \alpha_{j_1 j_2 \cdots j_N}\}$ having been chosen by N automata is called the path. Let $\phi_{j_1 j_2 \cdots j_N}$ denote the path. Corresponding to the path, the HSLA recieves reward strengths $\{(\beta_1^{1 j_1}, \beta_1^{2 j_1}, \ldots, \beta_1^{r_1 j_1}), (\beta_2^{1 j_1 j_2}, \beta_2^{2 j_1 j_2}, \ldots, \beta_2^{r_2 j_1 j_2}), \ldots, (\beta_N^{1 j_1 j_2 \cdots j_N}, \beta_N^{2 j_1 j_2 \cdots j_N}, \ldots, \beta_N^{r_N j_1 j_2 \cdots j_N})\}$ from the general multiteacher environment. The HSLA model utilizes these reward strengths in order to update the current recent reward vector. The action probability of each automaton relating to the path is updated by using the information concerning the recent reward vector. After all of the above procedures have been completed, time t is set to be $t + 1$. Let $\pi_{j_1 j_2 \cdots j_N}(t)$ denote the probability that the path $\phi_{j_1 j_2 \cdots j_N}$ is chosen at time t. Then,

$$\pi_{j_1 j_2 \cdots j_N}(t) = p_{j_1}(t) p_{j_1 j_2}(t) \cdots p_{j_1 j_2 \cdots j_N}(t) \tag{3}$$

In this paper, we shall consider the learning behaviors of the hierarchical structure learning automata (HSLA) under the nonstationary multiteacher environment. Before going into details concerning the nonstationary multiteacher environment, we shall define "recent average reward strength":

Definition 1. *Let us assume that the path $\phi_{i_1 i_2 \cdots i_N}$ has been chosen at time t, and, corresponding to $\phi_{i_1 i_2 \cdots i_N}$, reward strengths $\{\beta_s^{1 i_1 i_2 \cdots i_s}(t), \beta_s^{2 i_1 i_2 \cdots i_s}(t), \ldots, \beta_s^{r_s i_1 i_2 \cdots i_s}(t)\}$ have been given from the sth level of the multiteacher environment $(s = 1, 2, \ldots, N)$. Then, "recent average reward strength at the sth level" is defined as follows:*

$$\bar{u}_{i_1 i_2 \cdots i_s}(t) = \frac{1}{r_s}\{\beta_s^{1 i_1 i_2 \cdots i_s}(t) + \beta_s^{2 i_1 i_2 \cdots i_s}(t) + \cdots + \beta_s^{r_s i_1 i_2 \cdots i_s}(t)\} \tag{4}$$

On the other hand, the other recent average reward strength at the sth level $(\exists k \leq s; i_k \neq j_k; s = 1, 2, \ldots, N)$ is defined as follows:

$$\bar{u}_{j_1 j_2 \cdots j_s}(t) = \frac{1}{r_s}\{\beta_s^{1 j_1 j_2 \cdots j_s}(\tau_{j_1 j_2 \cdots j_s}) + \beta_s^{2 j_1 j_2 \cdots j_s}(\tau_{j_1 j_2 \cdots j_s})$$
$$+ \cdots + \beta_s^{r_s j_1 j_2 \cdots j_s}(\tau_{j_1 j_2 \cdots j_s})\}, \tag{5}$$

where $\tau_{j_1 j_2 \cdots j_s}$ is the most recent time when the automaton $Z_{j_1 j_2 \cdots j_{s-1}}$ is actuated and the action $\alpha_{j_1 j_2 \cdots j_s}$ is chosen by the automaton $Z_{j_1 j_2 \cdots j_{s-1}}$, and $\beta_s^{1 j_1 j_2 \cdots j_s}(\tau_{j_1 j_2 \cdots j_s}), \beta_s^{2 j_1 j_2 \cdots j_s}(\tau_{j_1 j_2 \cdots j_s}), \ldots, \beta_s^{r_s j_1 j_2 \cdots j_s}(\tau_{j_1 j_2 \cdots j_s})$ are the reward strengths at the sth level of the multiteacher environment at $\tau_{j_1 j_2 \cdots j_s}$.

4　A New Learning Algorithm of HSLA

In this section, we shall propose a learning algorithm to be used in the HSLA model as shown in Fig.2, where HSLA receive reward strengths from each level of the hierarchy of the nonstationary multiteacher environment.

First, let us define "reward parameter vector":

Definition 2. *Let $\bar{\mathbf{v}}_{i_1 i_2 \cdots i_{s-1}}(t) = (\bar{v}_{i_1 i_2 \cdots i_{s-1} 1}(t),$
$\bar{v}_{i_1 i_2 \cdots i_{s-1} 2}(t), \ldots, \bar{v}_{i_1 i_2 \cdots i_{s-1} r}(t))'$ be the reward parameter vector relating to the sth level LA $Z_{i_1 i_2 \cdots i_{s-1}}$ $(s = 1, 2, \ldots, N)$. Then, each of the components of $\bar{\mathbf{v}}_{i_1 i_2 \cdots i_{s-1}}(t)$ is constructed as follows:*

i) *At the Nth (bottom) level*

$$\bar{v}_{i_1 i_2 \cdots i_N}(t) = a_N \bar{u}_{i_1 i_2 \cdots i_N}(t). \tag{6}$$

ii) *At the sth $(s \neq N)$ level*

$$\bar{v}_{i_1 i_2 \cdots i_s}(t) = a_s \bar{u}_{i_1 i_2 \cdots i_s}(t) + \max_{i_{s+1}}\{\bar{v}_{i_1 i_2 \cdots i_s i_{s+1}}(t)\}. \tag{7}$$

We assume that the following condition holds for all $i_1, i_2, \ldots, i_s$ $(i_q = 1, 2, \ldots, r; q = 1, 2, \ldots, s$ $(s = 1, 2, \ldots, N))$:

$$q_{min} \leq p_{i_1 i_2 \cdots i_s}(t) \leq q_{max}, \tag{8}$$

where q_{min} and q_{max} satisfy the inequalities $0 < q_{min} < q_{max} < 1$ and $q_{max} = 1 - (r-1)q_{min}$, respectively.

Let us now propose a new learning algorithm of HSLA operating in the multiteacher environment.

A. Learning Algorithm

Assume that the path $\phi(t) = \phi_{j_1 j_2 \cdots j_N}$ has been chosen at time t and actions $\alpha_{j_1}, \alpha_{j_1 j_2}, \ldots, \alpha_{j_1 j_2 \cdots j_N}$ have been actuated to the multiteacher environment (MTEV). Further, assume that (corresponding to the actions by HSLA) environmental responses $\{(\beta_1^{1 j_1}, \beta_1^{2 j_1}, \ldots, \beta_1^{r_1 j_1}), (\beta_2^{1 j_1 j_2}, \beta_2^{2 j_1 j_2}, \ldots, \beta_2^{r_2 j_1 j_2}), \ldots, (\beta_N^{1 j_1 j_2 \cdots j_N}, \beta_N^{2 j_1 j_2 \cdots j_N}, \ldots, \beta_N^{r_N j_1 j_2 \cdots j_N})\}$ have been given to the HSLA. Then, the action probabilities $p_{j_1 j_2 \cdots j_{s-1} i_s}(t)$ $(i_s = 1, 2, \ldots, r)$ of each automaton $Z_{j_1 j_2 \cdots j_{s-1}}$ $(s = 1, 2, \ldots, N)$ connected to the path being chosen are updated by the following equation:

$$p_{j_1 j_2 \cdots j_{s-1} i_s}(t + 1) = p_{j_1 j_2 \cdots j_{s-1} i_s}(t) + \lambda_{j_1 j_2 \cdots j_{s-1}}(t)\Delta p_{j_1 j_2 \cdots j_{s-1} i_s}(t) \tag{9}$$

where $\Delta p_{j_1 j_2 \cdots j_{s-1} i_s}(t)$ is calculated by

$$
\Delta p_{j_1 j_2 \cdots j_{s-1} i_s}(t) = \begin{cases} \bar{v}_{j_1 j_2 \cdots j_{s-1} i_s}(t) - \dfrac{1}{|B_s(t)|} \displaystyle\sum_{l_s \in B_s(t)} \bar{v}_{j_1 j_2 \cdots j_{s-1} l_s}(t), & \forall i_s \in B_s(t) \\[2ex] 0, & i_s \notin B_s(t) \end{cases}
\tag{10}
$$

Here, the set $B_s(t)$ is constructed as follows:

1) Place $\bar{v}_{j_1 j_2 \cdots j_{s-1} i_s}(t)$ in descending order.
2) Set $D_{j_1 j_2 \cdots j_{s-1}} = \{ k_s | \bar{v}_{j_1 j_2 \cdots j_{s-1} k_s}(t) = \max_{i_s} \{ \bar{v}_{j_1 j_2 \cdots j_{s-1} i_s}(t) \} \}$.
3) Repeat the following procedure for i_s ($i_s \notin D_{j_1 j_2 \cdots j_{s-1}}$) in descending order of $\bar{v}_{j_1 j_2 \cdots j_{s-1} i_s}(t)$:
   If the inequality $p_{j_1 j_2 \cdots j_{s-1} i_s}(t+1) > q_{min}$ can be satisfied as a result of calculation by (9) and (10), then set

$$
D_{j_1 j_2 \cdots j_{s-1}} = D_{j_1 j_2 \cdots j_{s-1}} \bigcup \{ i_s \}
\tag{11}
$$

4) Set $B_s(t) = D_{j_1 j_2 \cdots j_{s-1}}$.

The learning algorithm proposed above ensures convergence to the optimal set of actions with probability 1 under the rather general nonstationary multi-teacher environment. Due to limitation of space, we don't go into details. Interested readers are kindly asked to attend our presentation.

5 Computer Simulation Results

In this section, we shall compare the learning performance of the proposed algorithm in the nonstatinary multiteacher environment with the algorithm DGPA[13] which is considered to be one of the fastest algorithms today.

Before going into details regarding the computer simulations, we shall try to modify the DGPA[13] algorithm to be used as a learning algorithm in HSLA:

Modified DGPA Algorithm

Let $\phi_{i_1 i_2 \cdots i_N}$ be the path chosen at time t by the HSLA. Further, assume that the reward strength estimate $\hat{\beta}_s^{k_s i_1 i_2 \cdots i_s}(t)$ corresponding to the k_sth teacher at the sth level is calculated by the same procedure as introduced in [13] ($s = 1, 2, \ldots, N$). Then, the average reward strength estimate at the sth level can be obtained by (12):

$$
\hat{u}_{i_1 i_2 \cdots i_s}(t) = \frac{1}{r_s} \{ \hat{\beta}_s^{1 i_1 i_2 \cdots i_s}(t) + \hat{\beta}_s^{2 i_1 i_2 \cdots i_s}(t) + \cdots + \hat{\beta}_s^{r_s i_1 i_2 \cdots i_s}(t) \}
\tag{12}
$$

By using the average reward strength estimate, the reward parameter vector $\bar{\mathbf{v}}_{i_1 i_2 \cdots i_{s-1}}(t)$ relating to the sth level LA $z_{i_1 i_2 \cdots i_{s-1}}$ ($s = 1, 2, \ldots, N$) is constructed as follows:

i) At the Nth (bottom) level

$$
\bar{v}_{i_1 i_2 \cdots i_N}(t) = a_N \hat{u}_{i_1 i_2 \cdots i_N}(t)
\tag{13}
$$

ii) At the sth $(s \neq N)$ level

$$\bar{v}_{i_1 i_2 \cdots i_s}(t) = a_s \hat{u}_{i_1 i_2 \cdots i_s}(t) + \max_{i_{s+1}}\{\bar{v}_{i_1 i_2 \cdots i_s i_{s+1}}(t)\} \qquad (14)$$

By taking advantage of the reward parameter vector, action probabilities of each automaton in HSLA are updated as follows:

Let N, $K(t)$, and Δ be resolution parameter, number of actions with higher reward strength estimates than the reward strength estimate corresponding to the current chosen path, and smallest step size $1/rN$.

Assume that the path $\phi(t) = \phi_{j_1 j_2 \cdots j_N}$ has been chosen at time t and actions $\alpha_{j_1}, \alpha_{j_1 j_2}, \ldots, \alpha_{j_1 j_2 \cdots j_N}$ have been actuated to the multiteacher environment (MTEV). Further, assume that (corresponding to the actions by HSLA) environmental responses $\{(\beta_1^{1j_1}, \beta_1^{2j_1}, \ldots, \beta_1^{r_1 j_1}), (\beta_2^{1j_1 j_2}, \beta_2^{2j_1 j_2}, \ldots, \beta_2^{r_2 j_1 j_2}), \ldots, (\beta_N^{1j_1 j_2 \cdots j_N}, \beta_N^{2j_1 j_2 \cdots j_N}, \ldots, \beta_N^{r_N j_1 j_2 \cdots j_N})\}$ have been given to the HSLA. Then, the action probabilities $p_{j_1 j_2 \cdots j_{s-1} i_s}(t)$ $(i_s = 1, 2, \ldots, r)$ of each automaton $Z_{j_1 j_2 \cdots j_{s-1}}$ $(s = 1, 2, \ldots, N)$ connected to the path being chosen are updated by the following equation:

$$p_{j_1 j_2 \cdots j_{s-1} i_s}(t+1) = \max\{p_{j_1 j_2 \cdots j_{s-1} i_s}(t) - \frac{\Delta}{r - K(t)}, \ 0\} \qquad (15)$$

$$(\forall i_s, i_s \neq j_s) \text{ such that } \bar{v}_{j_1 j_2 \cdots j_s}(t) > \bar{v}_{j_1 j_2 \cdots j_{s-1} i_s}(t)$$

$$p_{j_1 j_2 \cdots j_{s-1} i_s}(t+1) = \min\{p_{j_1 j_2 \cdots j_{s-1} i_s}(t) + \frac{\Delta}{K(t)}, \ 1\} \qquad (16)$$

$$(\forall i_s, i_s \neq j_s) \text{ such that } \bar{v}_{j_1 j_2 \cdots j_s}(t) < \bar{v}_{j_1 j_2 \cdots j_{s-1} i_s}(t)$$

$$p_{j_1 j_2 \cdots j_{s-1} j_s}(t+1) = 1 - \sum_{i_s \neq j_s} p_{j_1 j_2 \cdots j_{s-1} i_s}(t+1) \qquad (17)$$

In order to investigate whether the proposed algorithm can be successfully utilized in various nonstationary environments, we carried out many computer simulations. In the followings, we shall touch upon two of the computer simulation results.

Computer Simulation 1

We have compared the learning performance of the proposed algorithm with that of the modified DGPA algorithm under the following HSLA model and the (S-model) nonstationary multiteacher environment:

A) Hierarchical Structure Learning Automata Model

The hierarchical structure learning automata model is characterized by the following:

1) the number of the levels of HSLA: 5

2) the number of the actions of each automaton in the hierarchy: 2

3) the total number of paths: 32

4) the number of the teachers at each level: 3

5) the optimal path: ϕ_{11111}

B) Nonstationary Multiteacher Environment (NME)

1) Corresponding to the output $\phi(t) = \phi_{i_1 i_2 \cdots i_N}$ from the HSLA, the environmental reward strength $\beta_s^{k_s i_1 i_2 \cdots i_s}(t)$ $(k_s = 1, 2, \ldots, r_s; s = 1, 2, \ldots, N)$ is given at sth level of the hierarchy. $\beta_s^{k_s i_1 i_2 \cdots i_s}(t)$ is characterized by the following equation:

$$\beta_s^{k_s i_1 i_2 \cdots i_s}(t) = a_s^{k_s i_1 i_2 \cdots i_s} + b_s^{k_s i_1 i_2 \cdots i_s} \sin(c_s^{k_s i_1 i_2 \cdots i_s} \pi t + d_s^{k_s i_1 i_2 \cdots i_s}) \quad (18)$$
$$+ e_s^{k_s i_1 i_2 \cdots i_s}(2\xi(t) - 1) \quad (k_s = 1, 2, \ldots, r_s; \; s = 1, 2, \ldots, N)$$

where $\xi(t)$ is the random variable with the uniform probability density function in the closed interval $[0,1]$. Here $a_s^{k_s i_1 i_2 \cdots i_s}, b_s^{k_s i_1 i_2 \cdots i_s}, c_s^{k_s i_1 i_2 \cdots i_s}$, $d_s^{k_s i_1 i_2 \cdots i_s}$ and $e_s^{k_s i_1 i_2 \cdots i_s}$ are the positive scalors which have been chosen in such a way that optimality conditions of the path ϕ_{11111} are not violated.

Table 1 shows each of the values of the parameters $a_s^{k_s i_1 i_2 \cdots i_s}$ and $b_s^{k_s i_1 i_2 \cdots i_s}$ corresponding to the optimal path ϕ_{11111}. (Due to limitation of space, we don't go into details concerning the values of the parameters not corresponding to the optimal path. Interested readers are kindly asked to attend our presentation.)

2) Parameters $a_s^{k_s i_1 i_2 \cdots i_s}$ and $b_s^{k_s i_1 i_2 \cdots i_s}$ not corresponding to the optimal path ϕ_{11111} have been chosen by using the random variables with uniform probability density function in such a way that optimality conditions of the path ϕ_{11111} are not violated.

Table 1. Values of Coefficients a and b for Optimal Path ϕ_{11111}.

reward	1st Teacher		2nd Teacher		3rd Teacher	
	a	b	a	b	a	b
$\beta_1^{k_1 1}$	0.83	0.03	0.85	0.02	0.83	0.03
$\beta_2^{k_2 11}$	0.83	0.01	0.86	0.01	0.82	0.02
$\beta_3^{k_3 111}$	0.89	0.03	0.84	0.02	0.84	0.02
$\beta_4^{k_4 1111}$	0.84	0.03	0.89	0.02	0.86	0.01
$\beta_5^{k_5 11111}$	0.87	0.03	0.83	0.02	0.87	0.01

Table 2. Comparison of the Average Number of Iterations Required for Convergence of the Proposed Algorithm (PA) and the modified DGPA algorithm.

PA	DGPA
375.47	651.97
Parameter λ	Parameter N
$= 0.012$	$= 10$

3) The parameters $c_s^{k_s i_1 i_2 \cdots i_s}, d_s^{k_s i_1 i_2 \cdots i_s}$, and $e_s^{k_s i_1 i_2 \cdots i_s}$ have been chosen by using the random variables with the uniform probability density functions in the closed intervals $[0,1]$, $[0,1]$, and $[0,0.03]$, respectively.

4) Parameters q_{min}, q_{max} and $\lambda_{j_1 \cdots j_{s-1}}(t)$

i) We have used the following values as the parameters q_{min} and q_{max}:
$$q_{min} = 0.01, \; q_{max} = 0.99$$

ii) We have used the same value λ of the parameter $\lambda_{j_1 \cdots j_{s-1}}$ $(s = 1, 2, \ldots, N)$ from the top level to the bottom level.

In each computer experiment, we have carried out 750 simulations where we have set up the following condition: "An algorithm is considered to have converged if each probability of choosing an action in each level corresponding to a path has become greater or equal to a threshold 0.98. If the HSLA have converged to the optimal path, they are considered to have converged correctly."

Before comparing the efficacy of the learning algorithm of HSLA, large number of tests were executed to determine the best value of the parameters for each algorithm. The values of the parameters have been considered as the best if they registered the fastest convergence and the HSLA always converged to the optimal path with 100% accuracy. These best parameters have been chosen as the final parameter values for each algorithm to compare their rates of convergence.

Computer simulations have been carried out. In the above environment, the modified DGPA algorithm required on average 651.97 iterations whereas the proposed algorithm required on average only 375.47 iterations for convergence, being 42% faster than the modified DGPA algorithm. Table 2 summarizes the performances of the two algorithms in the nonstationary multiteacher environment. We have also carried out computer simulations of the learning performances of the two algorithms in the nonstationary multiteacher environment whose characteristic changes suddenly at some time. The simulation results also suggest the efficacy of the proposed algorithm. (Due to limitation of space, we shall omit them. Details will be touched upon in our presentation.)

Computer Simulation 2

Fig. 3 illustrates an intelligent behavior of the robot manipulators going through the maze. S, G, R_i, and C.C. denote the "starting point", the "goal", the ith robot manipulator, and the control center which receives the information concerning whether the robot manipulator has successed in passing the maze or not and decides the action to be taken by the robot manipulator appeared at the starting point S, respectively. The task assigned to the robot manipulators is to find the way which is the easiest to go through the maze. (Each gate closes with some probability which is unknown to the robot manipulators.) To find the best way, the control center exploits the learning performance of the proposed

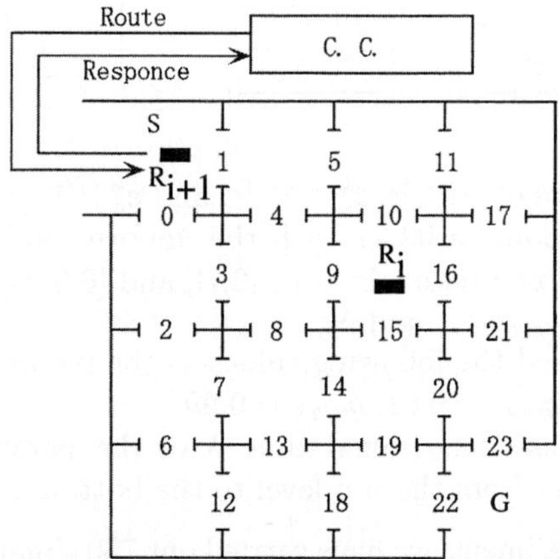

Fig. 3. Robot Manipulators Going Through the Maze.

algorithm (the modified DGPA algorithm). We have carried out many computer simulations concerning this problem. All of these simulation results suggest the efficacy of the proposed algorithm. However, due to limitation of space, we shall omit them. (We shall touch upon them in our presentation.)

6   Concluding Remarks

In this paper, we have proposed a new learning algorithm of the HSLA operating in the nonstationary multiteacher environment. We have presented the several computer simulation results which confirm the efficacy of the proposed algorithm in the nonstationary multiteacher environment. Further research is needed to investigate the learning behaviors of the HSLA under various types of the nonstationary environments.

Acknowledgments

The authors would like to thank the Reviewers for their suggestions and comments. They would also like to express their heartfelt thanks to the Grant-in-Aid for Scientific Research (c) by Ministry of Education, Science, Sports and Culture, Japan and the Foundation for Fusion of Science & Technology (FOST) who have given them partial financial support.

References

1. M.L. Tsetlin, "On the behavior of finite automata in random media", Avtomatika i Telemekhanika, vol.22-10, pp.1345-1354 1961.
2. V.I. Varshavskii and I.P. Vorontsova, "On the behavior of stochastic automata with variable structure", Automation and Remote Control, vol.24, pp.327-333, 1963.
3. K.S. Narendra and M.A.L. Thathachar, "Learning automata - A survey", IEEE Trans. Syst., Man, Cybern., vol.SMC-4, pp.323-334, July 1974.
4. S. Lakshmivarahan, Learning Algorithms Theory and Applications, Springer-Verlag, 1981.
5. N. Baba, New Topics in Learning Automata Theory and Applications, Springer-Verlag, 1985.
6. M.A.L.Thathachar and P.S.Sastry, Networks of Learning Automata Techniques for Online Stochastic Optimization, Kluwer Academic Publisher, 2004.
7. M.A.L. Thathachar and K.R. Ramakrishnan, "A hierarchical system of learning automata", IEEE Trans. Syst., Man, Cybern., vol.SMC-11, pp.236-241, Mar. 1981.
8. K.R. Ramakrishnan, Hierarchical Systems and Cooperative Games of Learning Automata, Ph.D. Thesis, Indian Institute of Science, Bangalore, 1982.
9. N.Baba, "Learning behaviors of hierarchical structure stochastic automata operating in a general multiteacher environment", IEEE Trans. Syst., Man, Cybern., vol.SMC-15, pp.585-587, July/Aug. 1985.
10. M.A.L. Thathachar and P.S. Sastry, "A new approach to the design of reinforcement schemes of learning automata", IEEE Trans.SMC, Vol.SMC-15, pp.168-175, Jan./Feb. 1985.

11. B.J. Oommen and J.K. Lanctôt, "Discretized Pursuit Learning Automata", IEEE Trans. Syst., Man, Cybern., Vol.20, pp.931-938, July 1990.

12. B.J. Oommen and M. Agache, "Continuous and discretized pursuit learning scheme: Various algorithms and their comparison", IEEE Trans. Syst., Man, Cybern. B, Vol.31, pp.277-287, June 2001.

13. M. Agache and B. J. Oommen, "Generalized pursuit learning scheme: New families of continuous and discretized learning automata", IEEE Trans. Syst., Man, Cybern. B, Vol.32, pp.738-749, 2002.

14. G.I. Papadimitriou, M. Sklira and A.S. Pomportsis, "A new class of ε-optimal learning automata", IEEE Trans. Syst., Man, Cybern. B, Vol.34, pp.246-254, Feb. 2004.

15. K.S. Narendra and R. Viswanathan, "A two level system of stochastic automata for periodic random environments", IEEE Trans. Syst., Man, Cybern., Vol.SMC-2, pp.285-289, 1972.

16. N. Baba and Y. Sawaragi, "On the learning behavior of stochastic automata under a nonstationary random environment", IEEE Trans. Syst., Man, Cybern., Vol.SMC-5, pp.273-275, Mar. 1975.

17. K.S. Narendra and M.A.L. Thathachar, "On the behavior of a learning automaton in a changing environment with application to telephone traffic routing", IEEE Trans. Syst., Man, Cybern., Vol.10, pp.262-269, May 1980.

18. P.R. Srikantakumar and K.S. Narendra, "A learning model for routing in telephone networks", SIAM Journal on Control and Optimization, Vol.20, pp.34-57, 1982.

19. R. Simha and J.F. Kurose, "Relative reward strength algorithms for learning automata", IEEE Trans. Syst., Man, Cybern., Vol.19, pp.388-398, Mar./Apr. 1989.

20. N. Baba and Y. Mogami, "A new learning algorithm for the hierarchical structure learning automata operating in the nonstationary S-model random environment", IEEE Trans. Syst., Man, Cybern. B, Vol.32, pp.750-758, Dec. 2002.

Nonlinear State Estimation by Evolution Strategies Based Gaussian Sum Particle Filter*

Katsuji Uosaki and Toshiharu Hatanaka

Department of Information and Physical Sciences,
Graduate School of Information Science and Technology, Osaka University
Suita, Osaka 565-0871 Japan
{uosaki,hatanaka}@ist.osaka-u.ac.jp

Abstract. There has been significant recent interest of particle filters for nonlinear state estimation. Particle filters evaluate a posterior probability distribution of the state variable based on observations in Monte Carlo simulation using so-called importance sampling. However, degeneracy phenomena in the importance weights deteriorate the filter performance. We propose in this paper a novel particle filter, which combines the ideas of Gaussian sum filter based on the Gaussian mixture approximation of the posteriori distribution and Evolution strategies based particle filter using selection process in evolution strategies. Numerical simulation study indicates the potential to create high performance filters for nonlinear state estimation.

1   Introduction

The problem of state estimation of nonlinear stochastic systems based on a sequence of their noisy observations has been ubiquitous in control system science. Central to the nonlinear state estimation problem is estimation of the posterior probability distribution (pdf), which can be obtained by a Bayesian approach combining a prior pdf for the unknown state with a likelihood function relating them to the observations. When observations come sequentially in time, recursive state estimation, which evaluates the evolving posterior pdf recursively in time, is often interested. Unfortunately, the posterior pdf only admits an analytical expression for very restricted cases, including linear Gaussian state space models where well-known Kalman filter [1] can be applied. In many realistic problems, state space models include nonlinear and non-Gaussian elements that preclude a closed form of expression for the optimal state estimate and then several approximations have been proposed. A class of filters called Gaussian filters provide Gaussian approximations to the filtering and predictive pdf's, examples of which include the extended Kalman filter (EKF) and its variation [2] and other filters such as Gaussian sum filters [3], where the posterior pdf's are approximated by the Gaussian mixtures, have been attempted. Recently, "particle filtering," a simulation-based method for Bayesian sequential analysis, attracts much attentions from the massive progress of computing ability [4]. In this approach, a probability density function is represented by a weighted sum based on the discrete grid sequentially chosen by the

---

* This work is partially supported by the Grant-in-Aid for Scientific Research from the Japan Society for the Promotion of Science (C)(2)14550447.

R. Khosla et al. (Eds.): KES 2005, LNAI 3681, pp. 635–642, 2005.

importance sampling and the estimates are obtained based on corresponding importance weights. We propose here a novel particle filter, which combines the ideas of Gaussian sum filter based on the Gaussian mixture approximation of the posteriori distribution and Evolution strategies based particle filter, which modifies the conventional particle filter by using selection rule in evolution strategies. Numerical simulation studies have been conducted to exemplify the applicability of this approach to nonlinear filtering.

2   Gaussian Sum Filter

Consider the following nonlinear state space model.

$$x_{k+1} = f(x_k) + v_k, \tag{1}$$

$$y_k = g(x_k) + w_k \tag{2}$$

where x_k and y_k are the state variable and observation, respectively, f and g are known possibly nonlinear functions, v_k and w_k are independently identically distributed (i.i.d.) system noise and observation noise sequences, respectively. We assume v_k and w_k are mutually independent. Problem to be considered here is to find the best estimate of the state variable x_k in some sense based on the all available data of observations $y_{1:k} = \{y_1, y_2, \ldots, y_k\}$. We can solve the problem by calculating the posterior pdf of the state variable x_k of time instant k based on all the available data of observation sequence $y_{1:k}$.

The posterior pdf $p(x_k|y_{1:k})$ of x_k based on the observation sequence $y_{1:k}$ satisfies the following recursions:

$$\text{Time update:} \quad p(x_k|y_{1:k-1}) = \int p(x_k|x_{k-1})p(x_{k-1}|y_{1:k-1})dx_{k-1}, \tag{3}$$

$$\text{Observation update:} \quad p(x_k|y_{1:k}) = \frac{p(y_k|x_k)p(x_k|y_{1:k-1})}{p(y_k|y_{1:k-1})} \tag{4}$$

with a prior pdf $p(x_0|y_0) \equiv p(x_0)$ of the initial state variable x_0. Here normalizing constant

$$p(y_k|y_{1:k-1}) = \int p(y_k|x_k)p(x_k|y_{1:k-1})dx_k$$

depends on the likelihood $p(y_k|x_k)$, which is determined by the observation equation (2).

Since a closed analytical form solution is not admitted except in very restrictive cases such as linear Gaussian state space models, where the well-known Kalman filter [1] can be applied, some approximations should be introduced. The most popular approximation approach is the extended Kalman filter (EKF) [2], where a linearization technique based on a first order Taylor expansions of the nonlinear system and observation equations about the current estimate is used. Another popular one is the Gaussian sum filter, where the following Gaussian sum representation of the posterior and the predictive pdf's is used [3],

$$p(x_k|y_{1:k}) = \sum_{i=1}^{p} w_{k|k}^{(i)} \mathcal{N}(x_k; \mu_{k|k}^{(i)}, \Sigma_{k|k}^{(i)}), \tag{5}$$

$$p(x_{k+1}|y_{1:k}) = \sum_{i=1}^{p} w_{k+1|k}^{(i)} \mathcal{N}(x_{k+1}; \mu_{k+1|k}^{(i)}, \Sigma_{k+1|k}^{(i)}). \tag{6}$$

Here, $\mathcal{N}(x; \mu, \Sigma)$ is Gaussian pdf with expectation μ and covariance Σ,

$$\mathcal{N}(x; \mu, \Sigma) = \frac{1}{(2\pi)^{\frac{n}{2}} |\Sigma|^{\frac{1}{2}}} \exp\left(-\frac{1}{2}(x-\mu)^T \Sigma^{-1}(x-\mu)\right) \tag{7}$$

The weight $\{w_{k|k-1}^{(i)}, w_{k|k}^{(i)}, (i=1,\ldots,p)\}$ are non-negative and satisfy $\sum_{i=1}^{p} w_{k|k-1}^{(i)} = 1$, $\sum_{i=1}^{p} w_{k|k}^{(i)} = 1$. Then, applying (4), the means and covariances are updated by

$$\begin{aligned}
\mu_{k|k}^{(i)} &= \mu_{k|k-1}^{(i)} + K_t^{(i)}(y_k - g(\mu_{k|k-1}^{(i)})), \\
\Sigma_{k|k}^{(i)} &= (I - K_t^{(i)} \tilde{C}_t^{(i)}) \Sigma_{k|k-1}^{(i)}, \\
K_t^{(i)} &= \Sigma_{k|k-1}^{(i)} \tilde{C}_t^{(i)T} (\tilde{C}_t^{(i)} \Sigma_{k|k-1}^{(i)} \tilde{C}_t^{(i)T} + R_t)^{-1}, \\
\mu_{k+1|k}^{(i)} &= f(\mu_{k|k}^{(i)}), \\
\Sigma_{k+1|k}^{(i)} &= \tilde{A}_{k+1}^{(i)} \Sigma_{k|k}^{(i)} \tilde{A}_{k+1}^{(i)T} + Q_t, \\
\tilde{C}_t^{(i)} &= \left.\frac{dg(x)}{dx}\right|_{x=\mu_{k|k-1}^{(i)}}, \qquad \tilde{A}_t^{(i)} = \left.\frac{df(x)}{dx}\right|_{x=\mu_{k|k}^{(i)}}, \\
w_{k|k}^{(i)} &= \frac{w_{k|k-1}^{(i)} \beta_t^{(i)}}{\sum_{i=1}^{p} w_{k|k-1}^{(i)} \beta_t^{(i)}}, \qquad w_{k+1|k}^{(i)} = w_{k|k}^{(i)}, \\
\beta_t^{(i)} &= \mathcal{N}(y_k; g(\mu_{k|k-1}^{(i)}), \tilde{C}_t^{(i)} \Sigma_{k|k-1}^{(i)} \tilde{C}_t^{(i)T})
\end{aligned} \tag{8}$$

Gaussian sum filter needs the mean and covariance evaluation as in (8), which may be tedious and time-consuming. Furthermore, when severe nonlinearities exist in the models, divergence may occur in Gaussian sum filter due to the linearizations in (8). To resolve these difficulties, Gaussian sum particle filter has been proposed [5], which will be briefly explained in the following.

3   Gaussian Sum Particle Filter

Gaussian sum particle filter assumes the posterior and predictive pdf's by Gaussian sum (5) and (6) as in Gaussian sum filter instead of grid approximation,

$$p(x_k|y_{1:k}) \approx \sum_{i=1}^{p} w_k^{(i)} \delta(x_k - x_k^{(i)}) \tag{9}$$

with Dirac's delta function $\delta(\cdot)$ such that $\delta(x) = 1$ for $x = 0$ and $\delta(x) = 0$ otherwise, in the conventional particle filter. Here, the particles $\{x_k^{(i)}, (i=1,\ldots,p)\}$ are generated and associated weights $\{w_k^{(i)}, (i=1,\ldots,p)\}$ are chosen using the principle of "importance sampling" [6].

Under the assumption

$$w_{k+1|k}^{(i)} = w_{k|k}^{(i)}, \quad i = 1, \ldots, p \tag{10}$$

the mean $\mu_{k+1|k}^{(i)}$ and covariance $\Sigma_{k+1|k}^{(i)}$ of Gaussian pdf $\mathcal{N}(x_k; \mu_{k+1|k}^{(i)}, \Sigma_{k+1|k}^{(i)})$ corresponding to

$$\int \mathcal{N}(x_k; \mu_{k|k}^{(i)}, \Sigma_{k|k}^{(i)}) p(x_{k+1}|x_k) dx_k \tag{11}$$

are estimated through the following Monte Carlo sampling in stead of mean and covariance update (8) in Gaussian sum filter.

For each $i = 1, \ldots, p$, we draw m samples $\{x_k^{(i,j)}, (j = 1, \ldots, m)\}$ from $\mathcal{N}(x_k; \mu_{k|k}^{(i)}, \Sigma_{k|k}^{(i)})$, and then draw $x_{k+1}^{(i,j)}$ from $p(x_{k+1}|x_k^{(i,j)})$. Mean and covariance are then computed by

$$\mu_{k+1|k}^{(i)} = \frac{1}{m} \sum_{j=1}^{m} x_{k+1|k}^{(i,j)},$$

$$\Sigma_{k+1|k}^{(i)} = \frac{1}{m} \sum_{j=1}^{m} (x_{k+1|k}^{(i,j)} - \mu_{k+1|k}^{(i)})(x_{k+1|k}^{(i,j)} - \mu_{k+1|k}^{(i)})^T. \tag{12}$$

On the other hand, observation update (4) with (6) is carried out by importance sampling. We first draw samples $\{x_k^{(i,j)}, (j = 1, \ldots, m)\}$ from the importance function $q(x_k|y_{1:k})$, then compute the corresponding weights by

$$\tilde{w}_{k|k-1}^{(i,j)} = \frac{p(y_k|x_k^{(i,j)}) \mathcal{N}(x_k^{(i,j)}; \mu_{k|k-1}^{(i)}, \Sigma_{k|k-1}^{(i)})}{q(x_k^{(i,j)}|y_{1:k})}. \tag{13}$$

We compute the estimates of the mean and covariance as

$$\mu_{k|k}^{(i)} = \sum_{j=1}^{m} w_{k|k-1}^{(i,j)} x_k^{(i,j)},$$

$$\Sigma_{k|k}^{(i)} = \sum_{j=1}^{m} w_{k|k-1}^{(i,j)} (x_k^{(i,j)} - \mu_t^{(i)})(x_k^{(i,j)} - \mu_t^{(i)})^T, \tag{14}$$

based on the normalized weights $w_{k|k-1}^{(i,j)} = \tilde{w}_{k|k-1}^{(i,j)} / \sum_{i=1}^{m} \tilde{w}_{k|k-1}^{(i,j)}$, and then update the weights by

$$w_{k|k}^{(i)} = \frac{\sum_{j=1}^{m} \tilde{w}_{k|k-1}^{(i,j)} w_{k-1|k-1}^{(i,j)}}{\sum_{i=1}^{p} \sum_{j=1}^{m} \tilde{w}_{k|k-1}^{(i,j)} w_{k-1|k-1}^{(i,j)}}. \tag{15}$$

Compared to the conventional particle filter, this filter can approximate the posterior pdf with smaller number of summands p since the conventional one approximates the pdf with a set of grid points, while this filter approximates it by a weighted sum of continuous functions (Gaussian pdf's), and it leads smaller computation burden. However, there may exist the difficulty [5]. After several updates, the posterior pdf's are approximated by a single Gaussian, and then it leads poor approximation. Moreover, a large computation effort is wasted in updating Gaussians whose contributions to approximate the posterior pdf $p(x_k|y_{1:k})$ is negligible. This is like the degeneracy phenomenon as in the conventional particle filter [4].

4 Evolution Strategies Based Gaussian Sum Particle Filter

In order to reduce the effects of degeneracy in Gaussian sum particle filter, resampling process can be introduced as in the conventional particle filter [7]. The basic idea of resampling is to eliminate the factors with small weights and to concentrate on the factors with larger weights. In particle filter, a new set of particles $\{x_k^{(i)}, i = 1, \ldots, p\}$ are generated by resampling from an approximate representation of $p(x_k|y_{1:k})$ given by (9) so that $\Pr(x_k^{(i)} = x_k^{(j)}) = w_k^{(j)}$. And the weights are reset as $w_k^{(i)} = 1/p$. This filter is called "Sequential importance sampling resampling filter" (SIR) [7].

Resampling process in Sequential importance sampling resampling filter (SIR) can be applied to Gaussian sum particle filter as follows: A new set of Gaussian pdf's $\{\mathcal{N}(x_k; \mu_{k|k}^{(i)}, \Sigma_{k|k}^{(i)}, i = 1, \ldots, p\}$ are generated by resampling from the set of Gaussian pdf's generated by (13)–(15) in the approximate representation of $p(x_k|y_{1:k})$ (5) so that

$$\Pr(\mathcal{N}(x_k; \mu_{k|k}^{(i)}, \Sigma_{k|k}^{(i)}) = \mathcal{N}(x_k; \mu_{k|k}^{(j)}, \Sigma_{k|k}^{(i)})) = w_{k|k}^{(j)} \tag{16}$$

And the weights are reset as $w_k^{(i)} = 1/p$. We can call this filter as "Gaussian sum resampling particle filter" (GSR). The choice of particles in GSR is probabilistic, and hence fluctuation of the computation time may become large.

We propose here a novel particle filter combining the idea of Gaussian sum filter and Evolution strategies (ES) by Rechenberg and Schwefel [8] as in Evolutionary strategies based particle filter (ESP) [9].

ES is one of the Evolutionary Computation (EC) approaches, computational models simulating natural evolutionary processes to design and implement computer-based problem solving systems [10]. It has been applied to continuous function optimization in real-valued n-dimensional space via processes of selection and perturbation such as recombination and mutation depending on the perceived performance (fitness) of the individual structures. Mutation process, which is realized by additive process of random fluctuations, introduces innovation into the population. And then, selection process is carried out as follows: the individuals of higher fitness are chosen deterministically out of the union of parents and offspring or offspring only to form the parents of the next generation in order to evolve towards better search region. Two main selection processes have been proposed, i.e. $(\mu + \lambda)$-selection and (μ, λ)-selection. In these selection processes, λ offspring are created from μ parents and the μ best individuals are selected out of the union of parents and offspring, and offspring only, respectively. These selections are deterministic. Hence, the selection process in ES is employed, we can make the fluctuations of computation time smaller.

This idea leads to a novel filter, Evolution strategies based Gaussian sum particle filter (ESGSP). The filter implements the observation update (4) with (6) as follows:

For each $i = 1, \ldots, p$, $\ell = 1, \ldots, r$, samples $\{x_k^{(i,j,\ell)}, (j = 1, \ldots, m)\}$ are drawn from the importance function $q(x_k|y_{1:k})$, then the corresponding weights are computed by

$$\tilde{w}_{k|k-1}^{(i,j,\ell)} = \frac{p(y_k|x_k^{(i,j,\ell)})\mathcal{N}(x_k^{(i,j,\ell)}; \mu_{k|k-1}^{(i)}, \Sigma_{k|k-1}^{(i)})}{q(x_k^{(i,j)}|y_{1:k})} \tag{17}$$

The estimates of the mean and covariance are computed by

$$\mu_{k|k}^{(i,\ell)} = \sum_{j=1}^{m} w_{k|k-1}^{(i,j,\ell)} x_k^{(i,j,\ell)},$$

$$\Sigma_{k|k}^{(i,\ell)} = \sum_{j=1}^{m} w_{k|k-1}^{(i,j,\ell)} (x_k^{(i,j,\ell)} - \mu_t^{(i,\ell)})(x_k^{(i,j,\ell)} - \mu_t^{(i,\ell)})^T, \tag{18}$$

with the normalized weights $w_{k|k-1}^{(i,j,\ell)} = \tilde{w}_{k|k-1}^{(i,j,\ell)} / \sum_{i=1}^{m} \tilde{w}_{k|k-1}^{(i,j,\ell)}$, $(j = 1, \ldots, m)$. The weights are then updated by

$$w_{k|k}^{(i,\ell)} = \frac{\sum_{j=1}^{m} \tilde{w}_{k|k-1}^{(i,j,\ell)} w_{k-1|k-1}^{(i)}}{\sum_{i=1}^{p} \sum_{j=1}^{m} \tilde{w}_{k|k-1}^{(i,j,\ell)} w_{k-1|k-1}^{(i)}}. \tag{19}$$

We sort the set of $p \times r$ pairs of Gaussian pdf's and corresponding weights, $\{(\mathcal{N}(x_{k|k}; \mu_{k|k}^{(i,\ell)}, \Sigma_{k|k}^{(i,\ell)}), w_{k|k}^{(i,\ell)}\}$, $(i = 1, \ldots, p, \ell = 1, \ldots, r)$ according to the weights $w_{k|k}^{(i,\ell)}$ in descending order, and retain the first p pairs with larger weights, and denote them $\{\mathcal{N}(x_{k|k}; \mu_{k|k}^{(i)}, \Sigma_{k|k}^{(i)}), \tilde{w}_{k|k}^{(i)}\}$, $(i = 1, \ldots, p)$. The weights are normalized as $w_{k|k}^{(i)} = \tilde{w}_{k|k}^{(i)} / \sum_{i=1}^{m} \tilde{w}_{k|k}^{(i)}$. This process corresponds to (p, pr)-selection in ES.

5   Numerical Example

To exemplify the applicability of the proposed ESGSP filter, we carried out a numerical simulation. We consider the following nonlinear state space model with known parameters.

$$x_k = \frac{x_{k-1}}{2} + \frac{25x_{k-1}}{1 + x_{k-1}^2} + 8\cos(1.2k) + v_k = f(x_{k-1}) + v_k,$$

$$y_k = \frac{x_k^2}{20} + w_k \tag{20}$$

and v_k and w_k are i.i.d. zero-mean normal random variates with variance 10 and 1, respectively. The normal distribution with mean $f(x_{k-1}^{(i)})$ and variance 10 is chosen as the importance density $q(x_k|x_{k-1}^{(i)}, y_{1:k})$. Sample paths of the observation process adn the state estimates by the particle filters (SIS ($n = 50$), GSP ($p = 5, m = 10$), and the proposed ESGSP ($p = 5, m = 5, p = 2$)) are given in Fig.1, and that of EKF as well for comparison. Particle filters, especially GSP and proposed ESGSP filters, show well behaviors in nonlinear state estimation, while the estimate by EKF cannot follow the true state.

The other choice of design parameters and evolution processes may improve the performance, and their better choice will be pursued.

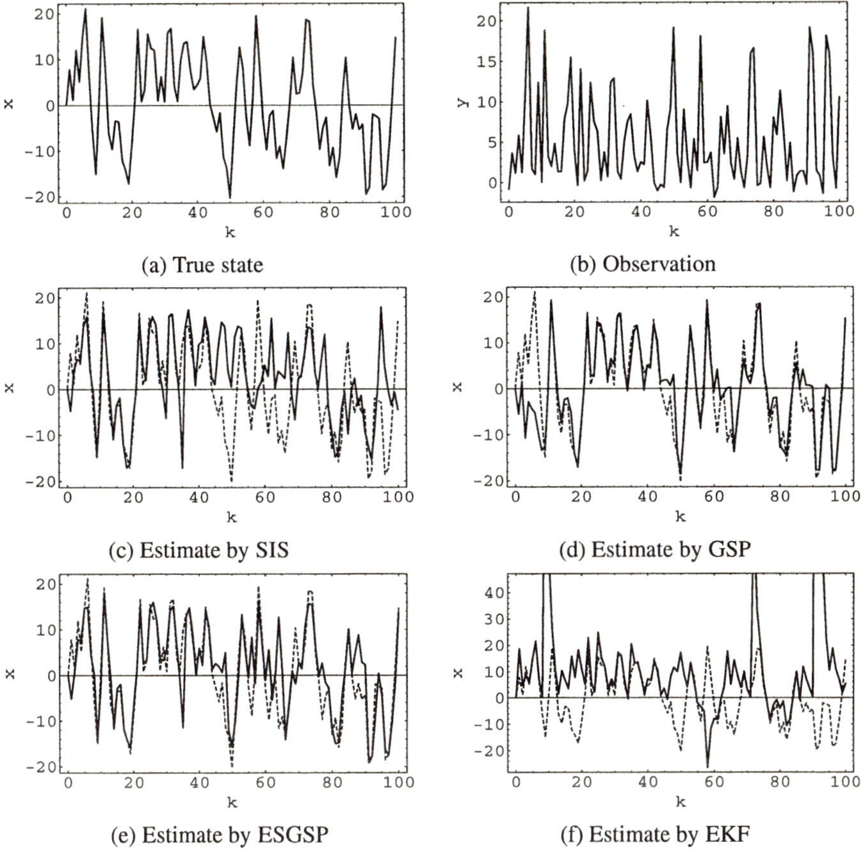

(a) True state

(b) Observation

(c) Estimate by SIS

(d) Estimate by GSP

(e) Estimate by ESGSP

(f) Estimate by EKF

Fig. 1. Sample behavior of state and observation processes, and the state estimates (solid line: estimate, dashed line: true state).

6 Conclusions

Combining the ideas of Gaussian sum filter and Evolution strategies based particle filter (ESP), we have proposed a novel filter, Evolution strategies based Gaussian sum particle filter (ESGSP). Since this filter employs the posterior probability distribution (pdf) approximated Gaussian with the mean and covariance evaluated by Monte Carlo sampling, it can be applied any type of nonlinear systems and non-Gaussian noise cases. Introducing of other evolution operations and better choice of design parameters will lead the particle filters with better performance.

References

1. H. W. Sorenson: *Kalman Filtering: Theory and Application*, IEEE Press (1985)
2. A. H. Jazwinski: *Stochastic Process and Filtering Theory*, Academic Press (1970)
3. D. L. Alspach and H. W. Sorenson: "Nonlinear Bayesian estimation using Gaussian sum approximation," *IEEE Trans. on Automatic Control*, Vol. AC-17, 439–448 (1972)

4. S. Arulampalam, S. Maskell, N. Gordon and T. Clapp: "A tutorial on particle filters for on-line non-linear/non-Gaussian Bayesian tracking," *IEEE Trans. on Signal Processing*, Vol. SP-50, No.2, 174–188 (2002)

5. J. H. Kotecha and P. M. Djurić: "Gaussian sum particle filtering," *IEEE Trans. on Signal Processing*, Vol. SP–51, 2602–2612 (2003).

6. A. Doucet, N. de Freitas and N. Gordon (Eds.): *Sequential Monte Carlo Methods in Practice*, Springer (2001)

7. J. Liu and R. Chen: "Sequential Monte Carlo Methods for Dynamic Systems," *J. of American Statistical Association*, Vol. 93, 1032–1044 (1998).

8. H.-P. Schwefel: *Evolution and Optimum Seeking*, J. Wiley (1995)

9. K. Uosaki, Y. Kimura and T. Hatanaka: "Evolution strategies based particle filters for non-linear state estimation," *Knowledge-Based Intelligent Information and Engineering*, Part. I, 1189–1196, LNAI 3213, Springer (2004)

10. T. Bäck: *Evolutionary Computation*, Oxford Press (1996)

Feature Generation by Simple FLD

Minoru Fukumi[1] and Yasue Mitsukura[2]

[1] University of Tokushima, Dept. of Information Science and Intelligent Systems,
2-1, Minami-Josanjima, 770-8506, Japan
fukumi@is.tokushima-u.ac.jp
[2] Okayama University, Faculty of Education,
3-1-1, Tsushima-Naka, Okayama, 700-8530, Japan
mitsue@cc.okayama-u.ac.jp

Abstract. This paper presents a new algorithm for feature generation, which is approximately derived based on geometrical interpretation of the Fisher linear discriminant analysis. In a field of pattern recognition or signal processing, the principal component analysis (PCA) is often used for data compression and feature extraction. Furthermore, iterative learning algorithms for obtaining eigenvectors have been presented in pattern recognition and image analysis. Their effectiveness has been demonstrated on computational time and pattern recognition accuracy in many applications. However, recently the Fisher linear discriminant (FLD) analysis has been used in such a field, especially face image analysis. The drawback of FLD is a long computational time in compression of large-sized between-class and within-class covariance matrices. Usually FLD has to carry out minimization of a within-class variance. However in this case the inverse matrix of the within-class covariance matrix cannot be obtained, since data dimension is higher than the number of data and then it includes many zero eigenvalues. In order to overcome this difficulty, a new iterative feature generation method, a simple FLD is introduced and its effectiveness is demonstrated.

1 Introduction

In statistical pattern recognition [1], principal component analysis (PCA), Fisher linear discriminant (FLD) analysis, and factor analysis have been widely utilized and their effectiveness has been demonstrated in many applications, especially in face recognition [2]. Recently, many new algorithms have been presented in a field of the statistical pattern recognition and neural networks [3]-[5]. In particular, the simple PCA is a simple and fast learning algorithm and its effectiveness has been demonstrated in face information processing [6][7]. Furthermore, their extension to higher order nonlinear space has been carried out [8][9]. In [2], Eigenface and Fisherface were compared and the effectiveness of Fisherface is shown by means of computer simulations. However, face image size is large compared to the number of image data and therefore many zero eigenvalues are included in a within-class covariance matrix [1]. Then the method could not yield Fisherface directly. In this case, usually PCA is used first to compress data dimension. After data compression, FLD is used to yield eigenvectors. This process requires a huge quantity of matrix computation. FLD is usually better than PCA as a feature generator in pattern recognition [1][2][9]. However zero eigenvalue in the within-class covariance matrix make FLD analysis difficult.

R. Khosla et al. (Eds.): KES 2005, LNAI 3681, pp. 643–649, 2005.
© Springer-Verlag Berlin Heidelberg 2005

On the one hand, simple iterative algorithms for achieving PCA have been proposed [3]-[5]. These algorithms are based on the data method instead of the matrix method and are easy to be implemented. Notice that PCA is based on distribution of all data and is not necessarily effective for pattern classification although PCA has been used in many pattern recognition problems. In this paper, a simple algorithm to achieve FLD analysis is presented. This is carried out by a simple iterative algorithm and does not use any matrix computation. The present algorithm is derived based on geometrical interpretation of maximization of between-class variance and minimization of within-class variance. The matrix method of FLD analysis is due to an eigenvalue problem. In this case, the matrix equation must be solved. Therefore it is difficult to solve its eigenvalue problem owing to a large-sized matrix. Our approach doesn't require matrix computations. We call this the simple-FLD or simple-FLDA.

From results of computer simulations, it is demonstrated that this algorithm is better than PCA in feature generation property.

2 Simple-FLD

The simple-FLD (simple-FLDA) is derived to satisfy FLD properties, which are based on maximization of between-class variance and minimization of within-class variance. These properties are described by the ratio of those in FLD analysis. This is an eigenvalue problem and can be solved by matrix computation including matrix inversion.

To avoid matrix calculation of the eigenvalue problem, the simple-PCA was presented [5]. The simple-PCA is a simple iterative algorithm to obtain eigenvectors without matrix calculation. This iterative learning is carried out using all input data. The objective of the algorithm is to maximize a data variance. After the convergence, the first eigenvector can be obtained. Next, the second eigenvector is also obtained by using the same procedure after subtraction of the first eigenvector component. Then, the third, the fourth and so on can be obtained.

2.1 Maximization of Between-Class Variance

In this paper, FLD is derived in a simple form. First, maximization of between-class variance is introduced by geometrical interpretation. Mean vector h_j in each class is obtained. In Fig.1, h_j is shown as a black dot. Then the mean value for all data is zero. The class-mean vectors are regarded as input vectors in the simple PCA. Therefore maximization process of data variance is the same as that of the simple PCA. The maximization is carried out in the following.

$$y_n = (\alpha_n^k)^T h_j \tag{1}$$

$$\Phi(y_n, h_j) = \begin{cases} h_j & : \text{if } y_n \geq 0 \\ -h_j & : \text{otherwise} \end{cases} \tag{2}$$

In Fig.1, a dotted line shows a line (hyperplane) which adjusts eq.(1) to 0. By using this process, it is hoped that a_n^k converges to n-th eigenvector. a_n^k is the n-th eigenvector after the k-th iteration. This is the same as the simple-PCA in principle. Eeigenvectors are therefore effective up to (c-1), if the number of classes is c. However this is approximated iterative algorithm and therefore more eigenvectors are derived considering numerical error.

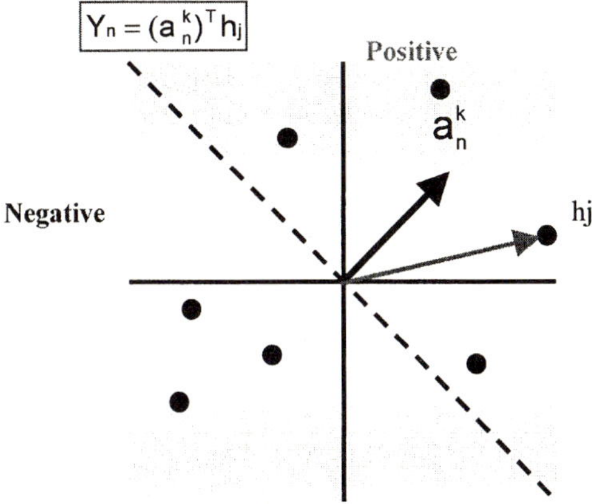

Fig. 1. Maximization of between-class variance

2.2 Minimization of Within-Class Variance

Next, a method for achieving minimization of within-class variance is introduced. In Fig.2, data vector X_j in a class is shown as a black dot. The vector X_j is zero mean in the class. The relationship between the vector X_j and arbitrary vector a_n^k which converges to an eigenvector is considered. The direction to minimize a vector length of the projected X_j is an orthogonal direction to X_j. Therefore such an orthogonal direction is computed and summed to minimize the within-class variance. Let b_j be a direction where X_j direction is subtracted from the vector a_n^k. This calculation is easy to be done and is given by

$$b_j = a_n^k - (\hat{x}_j \bullet a_n^k) \hat{x}_j \tag{3}$$

where

$$\hat{x}_j = \frac{x_j}{\| x_j \|} \tag{4}$$

Actual learning quantity is given as

$$\Phi_i(b_j, x_j) = \frac{\| x_j \|}{\| b_j \|} b_j \tag{5}$$

where b_j is normalized in size. In Fig.2, b_j is the same direction as the dotted line which is perpendicular to X_j. In learning, this quantity is summed and averaged over all sample vectors in the same class. In this case, weighted summation and averaging by the vector length of X_j are carried out. This weighted summation achieves calculation considering an influence of component that vector norm is larger. Then it is hoped that it converges to a direction to minimize the within-class variance.

$$\Phi_n^k = \sum_{i=1}^{c} N_i\, \Phi(y_n, h_i) + \sum_{i=1}^{c} \sum_{j=1}^{N_i} \Phi_i(b_j, x_j) \tag{6}$$

$$a_n^{k+1} = \frac{\Phi_n^k}{\left\| \Phi_n^k \right\|} \tag{7}$$

where N_i is the number of data in the same class i. As shown in eq.(6), the maximization of between-class variance and the minimization of within-class variance are simultaneously achieved and after convergence its eigenvector is obtained.

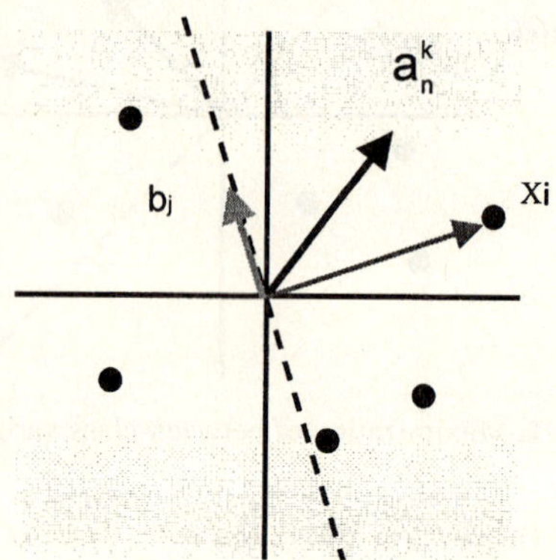

Fig. 2. Minimization of within-class variance

3 Computer Simulation

In order to show the effectiveness of the simple-FLD, computer simulations for a simple 2-class separation problem and a rotated coin recognition problem are carried out. The data distribution example in the 2-class separation is shown in Fig.3. In Fig.3, data of class 1 is generated by

$$y = 0.5\,x - 0.5 + U(-0.01,\, 0.01), \tag{8}$$

where x is an uniform random number. The generated data is illustrated as a black dot in Fig.3. Data of class 2 is generated by

$$y = 2\,x + 1.0 + U(-0.01,\, 0.01). \tag{9}$$

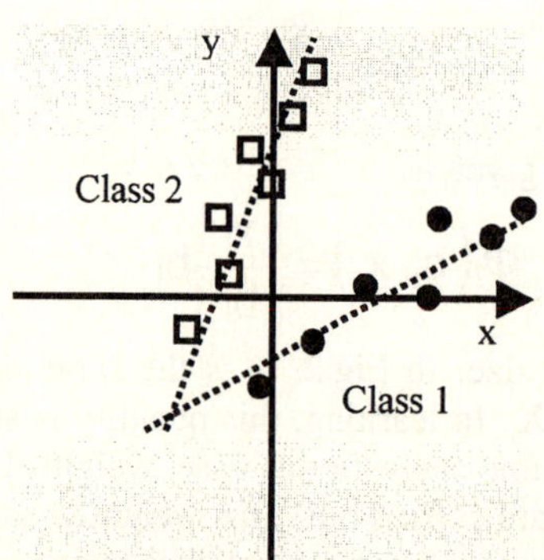

Fig. 3. 2-class separation problem in a 2 dimensional space

In Fig.3, data of class 2 is illustrated as a square. U(-0.01, 0.01) means an uniform random number between –0.01 and 0.01. The number of data is 200 for each class. These data are classified by using the simple-PCA and the simple-FLD followed by the minimum distance classifier. They generate a new feature by performing the inner product (projection to eigenvector) between the eigenvector and data. In this case, the length of the eigenvector is 1. Results obtained by them are illustrated in Figs. 4 and 5. Each figure shows obtained eigenvector and data projection to it. The number of learning iteration of the simple-PCA and the simple-FLD is about 10 to 20 times. The inner product of both the eigenvectors is 0.0044. They are almost orthogonal. As shown in Figs. 4 and 5, the features by the simple-FLD can achieve almost 100 % classification accuracy by using only a simple threshold processing. On the one hand, it is difficult classify the features obtained by the simple-PCA by a threshold processing. The distribution of 2-class data by the simple-PCA is overlapping.

We tested convergence property by changing initial values of eigenvector and then obtained a similar eigenvector. Furthermore a matrix type linear discriminant analysis was computed for the same data. Its eigenvector was almost the same direction as that of the simple-FLD.

Next, we evaluated the feature generation property for the rotated coin recognition problem [10]. The coins are a Japanese 500 yen coin and a South Korean 500 won coin. They have the same size, weight, and color. Their head and tail are also recognized. Therefore four kinds of coin images are recognized by using the simple minimum distance classifier. The learning of eigenvectors is carried out using 10 samples for each class. The rest 40 data of each class is used for evaluation. The total number of learning data and test data is therefore 40 and 160, respectively. The coin image data is transformed by 32x32 points FFT, because they are randomly rotated and therefore rotational invariant features are needed. The data dimension is 544 by symmetry of spectra. The accumulated relevance is greater than 80%, if the number of eigenvectors is 20. In Table 1, EV is the number of eigenvectors obtained. The rows labeled S-FLD and S-PCA are the classification accuracy obtained by feature generations of the simple-PCA and the simple–FLD, respectively.

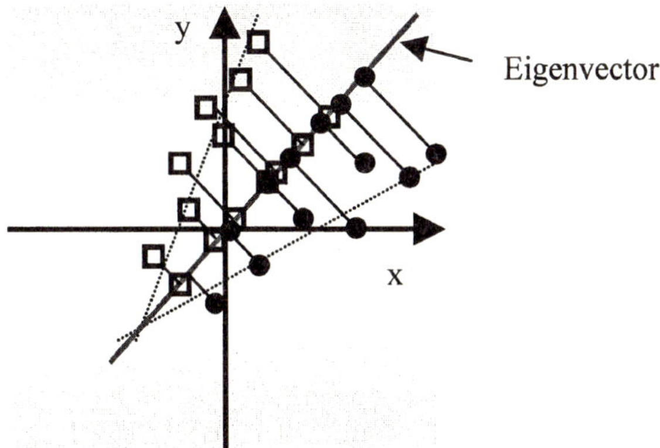

Fig. 4. A resulting eigenvector obtained by the simple-PCA. The vector direction is (0.7115678, 0.702618)

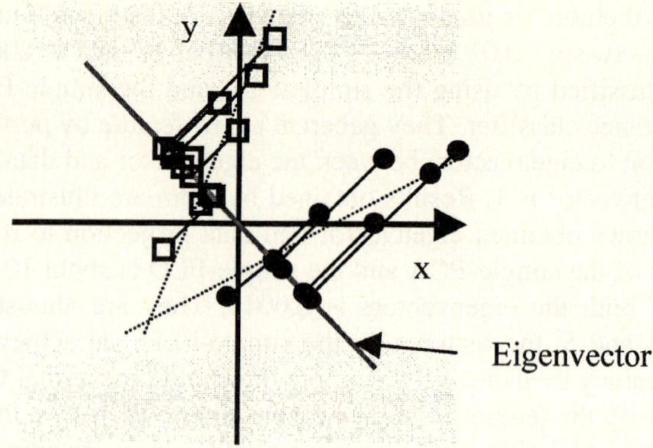

Fig. 5. A resulting eigenvector obtained by the simple-FLD. The vector direction is (0.705755, -0.708456)

Table 1. Recognition accuracy for coin recognition

EV	1	2	3	4	5	10	20
S-FLD	73.4	98.1	98.1	100.0	100.0	100.0	100.0
S-PCA	60.0	84.4	88.4	90.0	96.6	95.6	95.3

As shown in Table 1, recognition accuracy by the simple-FLD is better than that of the simple-PCA. The simple-FLD can achieve 100 % recognition accuracy even by using the minimum distance classifier. More sophisticated classifiers can improve accuracy.

4   Conclusion

We presented the simple-FLD to generate features suitable for pattern recognition. This is derived by geometrical interpretation of the FLD analysis. This method was applied to 2-class separation and rotated coin recognition problems. From the results of computer simulation, it is shown that feature generation property of the simple-FLD is better than that of the simple-PCA.

References

1. R.O.Duda and P.E.Hart, *Pattern Classification and Scene Analysis*, John Wiley & Sons (1973)
2. P.N.Belhumeur, J.P.Hespanha, and D.J.Kriegman, "Eigenfaces vs. Fisherfaces: Recognition Using Class Specific Linear Projection," IEEE Trans. on Pattern Analysis and Machine Intelligence, Vol.19, No.7, pp.711-720 (1997)
3. T.D. Sanger, "Optimal Unsupervised Learning in a Single Layer Linear Feedforward Neural Network," Neural Networks, Vol. 2, No. 6, pp. 459-473 (1989)
4. S.Y. Kung, *Digital Neural Networks*, Prentice-Hall, (1993)
5. M.Partridge and R. Calvo, "Fast dimentionality reduction and simple PCA", IDA, 2, pp.292-298 (1997)

6. M.Nakano, F.Yasukata, and M.Fukumi: "Recognition of Smiling Faces Using Neural Networks and SPCA", International Journal of Computational Intelligence and Applications, Vol.4, No.2, pp.153-164 (2004)

7. H. Takimoto, Y. Mitsukura, M. Fukumi and N. Akamatsu, "A Feature Extraction Method for Personal Identification System by Using Real-Coded Genetic Algorithm", Proc. of 7th SCI'2003, Orlando, USA, Vol. IV, pp.66-70 (2003).

8. B.Scholkopf, et al., "Nonlinear Component Analysis as a Kernel Eigenvalue Problem," Technical Report No.44, Max-Planck-Institute, Germany (1996)

9. M.H.Yang, "Kernel Eigenfaces vs. Kernel Fisherfaces: Face Recognition Using Kernel methods," Proc. of Fifth IEEE International Conference on Automatic Face and Gesture Recognition, pp.215-220, Washington, D.C. (2002)

10. M.Fukumi, S.Omatu, and Y.Nishikawa, "Rotation Invariant Neural Pattern Recognition System Estimating a Rotation Angle", IEEE Trans. on Neural Networks, Vol.8, No.3, pp.568-581 (1997)

Computational Intelligence
for Cyclic Gestures Recognition of a Partner Robot

Naoyuki Kubota[1,2] and Minoru Abe[3]

[1] Dept. of System Design, Tokyo Metropolitan University
1-1 Minami-Osawa, Hachioji, Tokyo 192-0397, Japan
kubota@comp.metro-u.ac.jp
[2] "Interaction and Intelligence", PRESTO, Japan Science and Technology Corporation (JST)
[3] Dept. of Human Artificial Intelligence Systems, University of Fukui
3-9-1 Bunkyo, Fukui 910-8507, Japan
abe@iicx.ia.his.fukui-u.ac.jp

Abstract. This paper proposes a method for cyclic gestures recognition of a partner robot based on computational intelligence. The mobile robot used as a partner robot must make decisions suitable to the human intention in the facing environment. Therefore, it is necessary to recognize the gesture used as a tool for human communication. The proposed method is composed of a fuzzy spiking neural network, a self-organizing map, and a steady-state genetic algorithm. Experimental results show the effectiveness of the proposed method.

1   Introduction

Various robots have been incorporated into human daily life by the progress of robot development in recent years. In general, a human-friendly robot performs social interaction with a human in an environment. Furthermore, such a robot should learn behaviors through the interaction with human. The robot can understand a human behavior, because the robot learns the behavior in the interaction with the human. Especially, gesture and speech recognition play an important role in interaction and communication with a human. It is very natural and useful for a human to use a gesture in order to give the robot a specific task. If the boundary which produces ambiguity by hand motion is regarded as gesture, the boundary will become clear. Gesture is regarded as a symbolic action for communicating intention to the other. Moreover, gesture may include the directionality of the behavior to be reproduced and learning. Therefore, the robot needs to recognize gesture [1,2]. We proposed the gesture recognition of a robot composed of a spiking neural network for extracting a spatio-temporal pattern of human hand motion, self-organizing map for clustering of human hand motion patterns, steady-state genetic algorithm for human detection, but the time length of human hand motion is predefined. Because the time length cannot be exactly fixed beforehand, we propose how to extract a cyclic hand motion.

Section 2 proposes the human detection method based on steady-state genetic algorithm and the gesture recognition method based on a spiking neural network and a self-organizing map. Section 3 shows experiment results of a partner robot based on the proposed method.

R. Khosla et al. (Eds.): KES 2005, LNAI 3681, pp. 650–656, 2005.

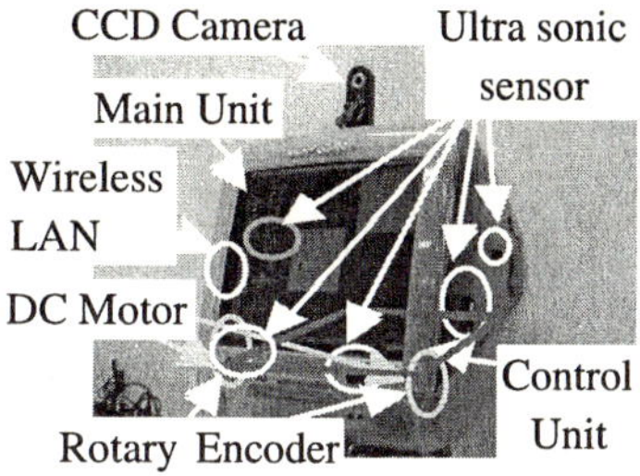

Fig. 1. A partner robot; MOBiMac

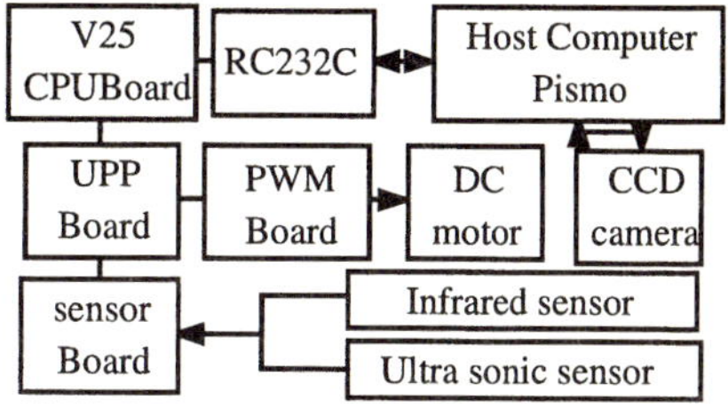

Fig. 2. Construction of MOBiMac

2   A Partner Robot Based on Computational Intelligence

2.1   Human Detection of a Partner Robot

We developed a partner robot; MOBiMac as shown in Figure 1. Two CPUs are used for PC and robotic behaviors. The robot has two servo motors, eight ultrasonic sensors, and CCD camera. Therefore, the robot can take various behaviors such as collision avoiding, human approaching, and line tracing. The robot takes an image from the CCD camera, and extracts a human. If the robot detects the human, the robot extracts the motion of the human hand. Furthermore, the robot expresses the internal or perceptual state by utterance. The image of RGB color space is taken by CCD camera, colors corresponding to human hair and face are extracted by using thresholds. Next, these are detected by using a steady-state genetic algorithm (SSGA) based on template matching. Figure 3 shows a candidate solution of the template used for detecting a human as the target object. A template is composed of numerical parameters of $g_{i,1}$, $g_{i,2}$, $g_{i,3}$ and $g_{i,4}$. (i=1,2, ..., G). In SSGA, only few existing solutions are replaced with the candidate generated by genetic operators in each generation [3]. In this paper, the worst candidate solution is eliminated ("Delete least fitness" selection), and it is replaced with the candidate solution generated by the crossover and the mutation. We use the elitist crossover and adaptive mutation. The elitist crossover randomly selects one individual and generates an individual by combining genetic information from the randomly selected individual and the best individual. Next, the following adaptive mutation is performed to the generated individual,

$$g_{i,j} \leftarrow g_{i,j} + \alpha_j \cdot \frac{f_{\max} - f_i}{f_{\max} - f_{\min}} + \beta_j \cdot N(0,1) . \tag{1}$$

where f_i is the fitness value of the ith individual, $f_{\max}$ and $f_{\min}$ are the maximum and minimum of fitness values in the population; $N(0,1)$ indicates a normal random value; α_j and β_j are the coefficient and offset, respectively. In the adaptive mutation, the variance of the normal random number is relatively changed according to the fitness values of the population. Fitness value is calculated by the following equation,

$$f_i = C^S + C^H + \eta_1 \cdot C^S \cdot C^H - \eta_2 \cdot C^{Other} . \tag{2}$$

where C^S, C^H, and C^{Other} indicate the numbers of pixels of the colors corresponding to human hair, human face, and other colors, respectively; η_1 and η_2 are coefficients. Therefore, this problem result in the maximization problem. By using SSGA, the robot roughly detects a human face candidate. Next, the attention area is extracted according to the central position of the human face (Fig.3 (b)).

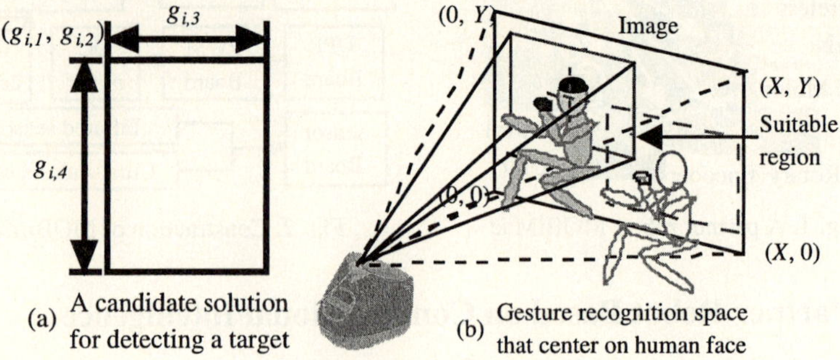

(a) A candidate solution for detecting a target

(b) Gesture recognition space that center on human face

Fig. 3. Human detection by SSGA

2.2 Gestures Recognition

The robot extracts human hand motion from the series of images by using SSGA where the maximal number of images is T. The sequence of the hand positions is represented by $\mathbf{G}(t)=(G_x(t), G_y(t))$ where $t=1, 2, \dots , T$. The sequence of human hand positions is used for recognizing spatio-temporal pattern as a human gesture by spiking neural network (SNN). SNN is often called a pulsed neural network, and is one of the artificial NNs imitating the dynamics introduced the ignition phenomenon of a cell, and the propagation mechanism of the pulse between cells [4]. In this paper, we use a spike response model to reduce the computational cost. First of all, the internal state $h_i(t)$ is calculated as follows;

$$h_i(t) = \tanh\left(h_i^{syn}(t) + h_i^{ref}(t)\right). \tag{3}$$

Here $h_i^{syn}(t)$ including the output pulses from the neurons is calculated by the following equation,

$$h_i^{syn}(t) = \gamma^{syn} \cdot h(t-1) + \sum_{j=1, j\neq i}^{N} w_{ji} \cdot p_j(t-1) + q_i(t). \tag{4}$$

where $h_i^{ref}(t)$ indicates the refractoriness of the neuron; $w_{j,i}$ is a weight coefficient from the jth to ith neuron; $p_j(t)$ is the output of jth neuron at the discrete time t; $q_i(t)$ is the input to the ith neuron from the environment; N is the number of neurons; γ^{syn} is a discount rate satisfying $0<\gamma^{syn}<1.0$. When the neuron is fired, R is subtracted from $h_i^{ref}(t)$ in the following,

$$h_i^{ref}(t) = \begin{cases} \gamma^{ref} \cdot h_i^{ref}(t-1) - R & \text{if } p_i(t-1)=1 \\ \gamma^{ref} \cdot h_i^{ref}(t-1) & \text{otherwise} \end{cases}. \tag{5}$$

where γ^{ref} is a discount rate satisfying $0<\gamma^{ref}<1.0$. When the internal potential of ith neuron is larger than the predefined threshold, a pulse is outputted as follows;

$$p_i(t) = \begin{cases} 1 & \text{if } h_i(t) \geq \theta_i \\ 0 & \text{otherwise} \end{cases}. \tag{6}$$

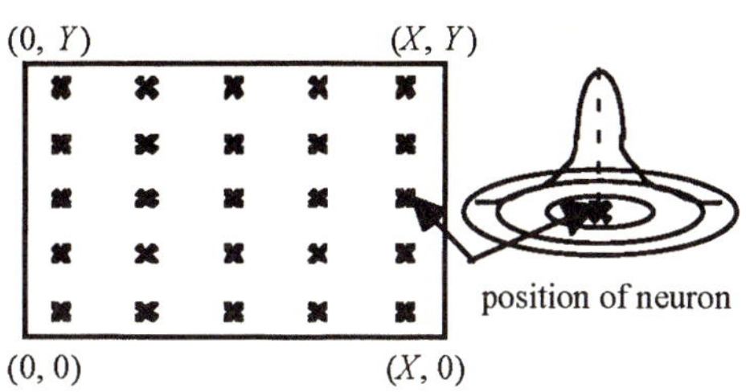

Fig. 4. Spiking neurons arranged on the image

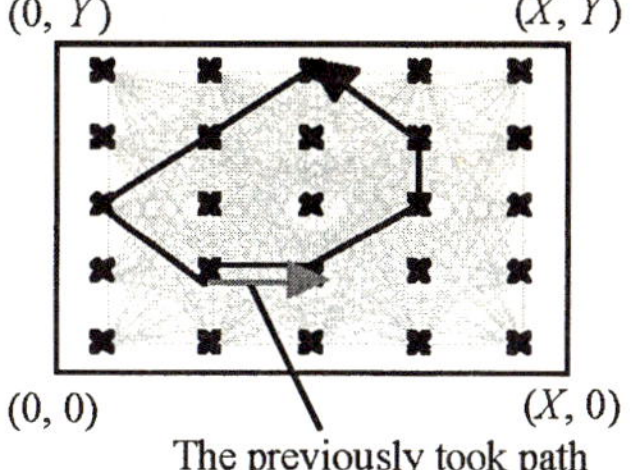

Fig. 5. A cyclic human hand motion

where θ_i is a threshold for firing. Here spiking neurons are arranged on a planar grid (Fig.4) and $N=25$ based on a fuzzy partition. By using the value of a human hand position, the input to the ith neuron is calculated by the radial basis function as follows;

$$q_i(t) = \exp\left(-\frac{\|c_i - G(t)\|^2}{\sigma^2}\right).$$
(7)

where $c_i = (c_{x,i}, c_{y,i})$ is the position of the ith neuron; σ is a standard deviation. The sequence of pulse outputs $p_i(t)$ is obtained by using the human hand positions $G(t) = (G_x(t), G_y(t))$, $t=1, 2, \ldots , T$. The weight parameters are trained based on internal state as follows;

$$w_{j,i} \leftarrow \tanh\left(\gamma^{wgt} \cdot w_{j,i} + \xi^{wgt} \cdot h_i(t) \cdot h_j(t-1)\right).$$
(8)

where γ^{wht} is the discount rate satisfying $0<\gamma^{wht}<1.0$ and ξ^{wgt} is a learning rate satisfying $0<\xi^{wgt}<1.0$. Here we use path value between ith neuron and jth neuron based on weight parameters to memorize a sequential pattern. Because the weight parameters are trained according to the sequential fire of the neurons near the human hand positions, the path value can memorize spatio-temporal patterns of various gestures. If the $w_{j,i}$ is larger than the predefined threshold, the path value between ith neuron and jth neuron is the previously passed one (Fig.5). A pulse is outputted as follows;

$$u_{j,i} = \begin{cases} 1 & if \ w_{j,i} \geq \theta \\ 0 & otherwise \end{cases}.$$
(9)

Here such a human hand motion is assumed as a cyclic gesture pattern, because it is very difficult to display the exactly same hand motion twice. Therefore, when the above condition is satisfied, the human hand motion detection is stopped.

Next, the path values concerning the weight of spiking neurons are as an input for clustering by a self-organizing map (SOM) in order to detect a spatial pattern of a human gesture. SOM is often applied for extracting a relationship among inputs data, since SOM can learn the hidden topological structure from the learning data [5]. The inputs to SOM are the sum of pulse outputs from path values,

$$v = (v_{1,1}, v_{1,2}, \cdots, v_{N,N}) \quad v_{j,i} = \sum_{t=1}^{T} u_{j,i}(t).$$
(10)

When the ith reference vector is represented by $\mathbf{r}_i$, the Euclidian distance between input vector and reference vector is defined as

$$d_i = \|\mathbf{v} - \mathbf{r}_i\|.$$ (11)

where $\mathbf{r}_i=(r_{i,1,1}, r_{i,1,2}, \ldots, r_{i,N,N})$ and the number of reference vectors (output units) is M. Next, the kth output unit minimizing the distance d_i is selected by

$$k = \arg \min_i \left\{ \|\mathbf{v} - \mathbf{r}_i\| \right\}.$$ (12)

Furthermore, the reference vector of the ith output unit is trained by

$$\mathbf{r}_i \leftarrow \mathbf{r}_i + \xi^{SOM} \cdot \zeta_{k,i}^{SOM} \cdot (\mathbf{v} - \mathbf{r}_i).$$ (13)

where ξ^{SOM} is a learning rate satisfying $0<\xi^{SOM}<1.0$; $\zeta_{k,i}^{SOM}$ is a neighborhood function satisfying $0<\zeta_{k,i}^{SOM}<1.0$. In the learning algorithm of SOM, the parameters of ξ^{SOM} and $\zeta_{k,i}^{SOM}$ are gradually reduced toward 0 according to the state of learning. The selected output unit is the nearest pattern among the previously learned gestures.

3 Experiments

This section shows several experimental results of behavior learning using the partner robot MOBiMac. One cycle is defined as following steps; (1) human detection is performed by SSGA (SSGA-1) where the size (X,Y) of an image is (240, 180), (2) human hand detection by SSGA (SSGA-2), (3) spatial and temporal pattern generation by SNN, and (4) gesture clustering by SOM. The size (X,Y) of an image in the steps (2) - (4) is (80, 60). The number of inputs and iterations in SOM are 100 and 1000, respectively. SOM is preliminarily trained by showing some gestures to the robot. Figure 6 shows the experimental result of the human detection by SSGA. The number of individuals and iterations in SSGA are 20 and 30, respectively. In this figure, a green box indicate the position of a candidate solution and red box indicate gesture recognition space, respectively. The human detection is finished when the distance to the human become adequate distance on predifined threshold. The distance to the human is calculated by the area of detected face area as proximity. According to the utterance by the robot, the human stops at the suitable position for gesture recognition. Figure 7 shows the history of the fitness value and proximity. After detected the human, the robot start human hand detection by the seeing space that center on human face (Fig.7(c)). Figure 7 (d) and (e) show the human detection results of the 2nd and the 3rd cycle. Although a human position is changed, the robot traces the central position of the human face. As a result, the robot effieciently recognizes the human gestures in different positions.

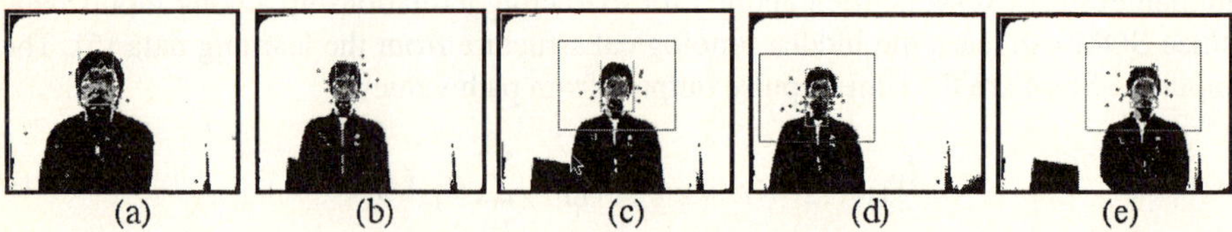

(a)   (b)   (c)   (d)   (e)

Fig. 6. Human detection by SSGA

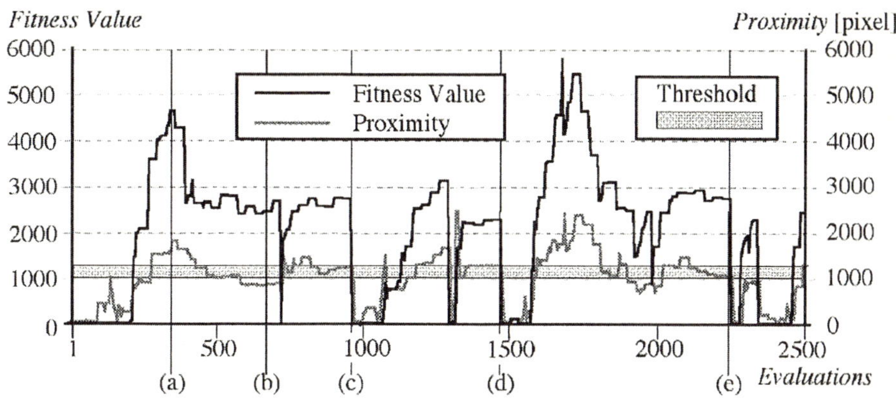

Fig. 7. Change of fitness values in human detection

Figure 8 shows the image processing for extracting human hand motions. This shows the original image in the upper left, the extracted pixels corresponding to the glove color in the upper right, and the neuron paths in the bottom. In this way, SNN can extract the spatio-temporal pattern of human hand motion. Figure 9 shows the history of the nodes selection in SOM. Various nodes are selected in the initial stage, because the human first tried to display various hand motions. After 30th trials, the human performed the same hand motion repeatedly at the right and left of the human face (Fig.9), and SOM was classified using the 3rd node and 6th node. In this way, the proposed method can perform clustering human hand motions.

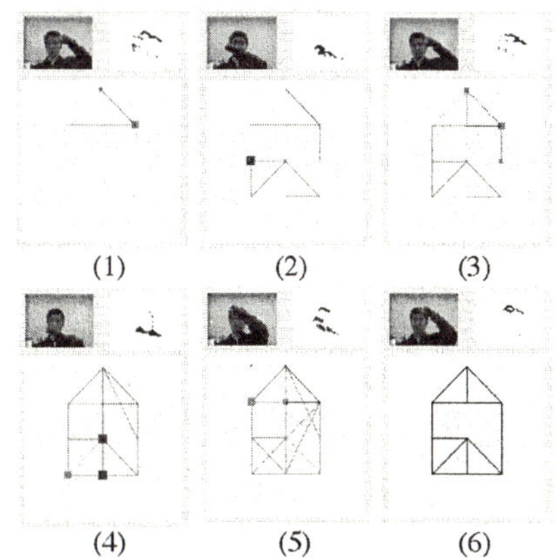

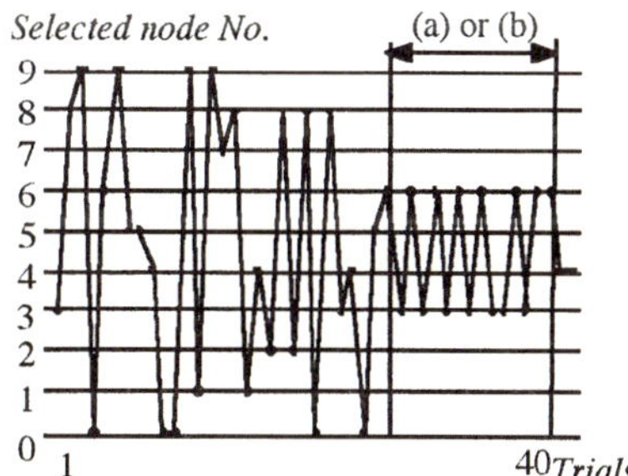

Fig.9. History of the selected node in SOM

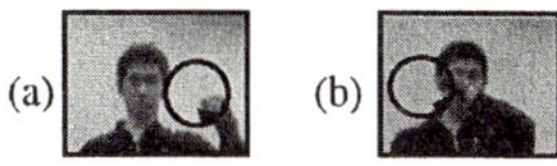

Fig. 8. Image processing for extracting hand motion

Fig. 9. Human hand motion

4 Summary

This paper proposes a method for computational intelligence for cyclic gestures recognition of a partner robot. We applied steady-state genetic algorithm for image processing, a spiking neural network for extracting spatial and temporal patterns of gestures, and self-organizing map for clustering gestures. Experimental results show that the robot can extract and classify the human hand motion by the space that centered on human face. As a future work, we intend to incorporate the obtained action patterns into reactive motions of the robot interacting with human.

References

1. Billard, A.: Imitation. In M. A. Arbib (ed.): Handbook of Brain Theory and Neural Networks, MIT Press. (2002) 566-569
2. Kubota N., Abe M.: Interactive Learning for A Partner Robot Based on Cyclic Gestures, Proc. of SEAL (2004) 336.
3. Syswerda G.:A study of reproduction in generational and steady-state genetic algorithms. In foundations of genetic algorithms. Morgan Kaufmann Publishers, (1991)
4. Takita K., Hagiwara M.: A Pulse Newral Network Learning Algorithm for POMDP environment. Proc. of IEEE (2003)
5. Kohonen T.: Self-organization and associative memory. Springer (1984).
6. Fukuda T., Kubota N.: An intelligent robotic system based on a fuzzy approach, Proceedings of IEEE 87(9). (1999) 1448-1470

Influence of Music Listening
on the Cerebral Activity by Analyzing EEG

Takahiro Ogawa[1], Satomi Ota[2], Shin-ichi Ito[3],
Yasue Mitsukura[2], Minoru Fukumi[1], and Norio Akamatsu[1]

[1] The University of Tokushima, 2-1, Minami-Josanjima, Tokushima, 770-8506 Japan
`{keyroom,fukumi,akamatsu}@is.tokushima-u.ac.jp`
[2] Okayama University,
3-1-1, Tsushimanaka, Okayama, 700-8530 Japan
`{ged17065,mitsue}@cc.okayama-u.ac.jp`
[3] Japan Gain the Summit Co., Ltd.
BLD. 3F, 1-11-10, Sutdio, Bunkyo-ku, Tokyo, 112-0005, Japan
`ito@jgs-g.co.jp`

Abstract. In order to solve a stress problem, researchers have studied
music therapy. It takes the therapist and patient a long time to select
the music. Because the music used in music therapy is of various type.
If the music for it is easily selectable, the music therapy can be carried
out more effectively. In this paper, the purpose is extraction of features
that may be influenced by the music. We pay attention to EEG (electroencephalogram) as an objective and absolute scale. In this paper, we
propose a method that extracts features of the EEG by PCA (principal
component analysis) and CDA (canonical discriminant analysis). Then
we analyze each feature data by NN (neural network). In order to examine whether the proposal system is effective, we try computer simulations
for the EEG classification. According to recognition rate by the NN, it
was considered that the CDA extracted and classified the features of the
EEG better than the PCA.

1 Introduction

Stress is a serious problem in today's society, and has a bad influence on the mind
and body. People lose generous environment and feel daily uneasiness from complicated human relations, economy, and a change of social solution. Moreover
chronic stress causes various illnesses, fatigue and deaths. In order to solve such
a problem, researchers have studied healing, especially music therapy. Music
therapy is a medical treatment method which recovers mind and body by listening to the music. It has been reported that music alleviates pain and anxiety
[1]. The music used in the musical therapy is various, and it takes therapist and
patient a long time to select the music. If the music effective for the musical
therapy is easily selectable, the time of choosing the music can be shortened and
the musical therapy can be carried out effectively. Conventionally, validity of the
musical therapy is estimated by the questionnaire. [2] However, it is difficult to

R. Khosla et al. (Eds.): KES 2005, LNAI 3681, pp. 657–663, 2005.

measure subjective taste because human feeling on stress or relaxed feeling is remarkably complicated. Therefore objective evaluation of the musical therapy is mentioned as an important subject, and the objective evaluation based on physiological change is tried [3]. We think that EEG (electroencephalogram) is very sensitive to mental change because the EEG changes with cerebral activity directly. Therefore we examine the influence of music for humans by using EEG. We propose a method which extracts features of the EEG by a multivariate analysis in this paper. We use PCA (principal component analysis) and CDA (canonical discriminant analysis) as a method of feature extraction. Moreover we verify the effectiveness of the features by NN (neural network). The NN is used in categorization or classification of the EEG because the NN can automatically create a logic without relying on the mechanism of recognition [4].

2 Summary of EEG

The EEG is the biological signal measured by attaching a sensor to a scalp and amplifying the voltage. It is said that the EEG is reflected in a cerebral electric activity. The EEG patterns vary by a measurement point, therefore the measurement point is internationally unified by the 10-20-electrode system. The signal changes by external stimulus as sound and inner stimulus as feelings, therefore the EEG pattern can vary by listening to different music [5]. Moreover the EEG is classified into frequency bands, δ waves (0.5-3 Hz), θ waves (4-7 Hz), α waves (8-13 Hz) and β waves (14-24)Hz. Some researchers have focused on α waves and β waves[6] [7].

3 Proposed System

Figure 1 shows a flow chart of the proposed system. In this paper, first we measure the EEG while listening to the music. Next, we create a data matrix based on appearance rate of frequency ingredients. We try two cases about interval of frequency, which are every 1 Hz and frequency bands. The ingredient of every 1Hz expresses a change finer than that of frequency band. Third, we extract features from the data matrix by PCA and CDA. Finally, we verify the EEG patterns based on the feature vectors by the NN. We compare accuracy obtained PCA and CDA by recognition rate of the NN.

3.1 Measurement of the EEG

In this paper, the EEG is measured using the measurement system by brain-power-development-corporation. The form of the instrument is compact as 120mm (W), 135mm (D), and 35mm (H). Mental and physiological burden to subjects is smaller than conventional instruments. An electrode is fixed to a sensor band and attached in Fp1, left frontal pole (shown in Fig.2). The frontal lobe is an association field to manage feeling such as cognition and judgment[8]

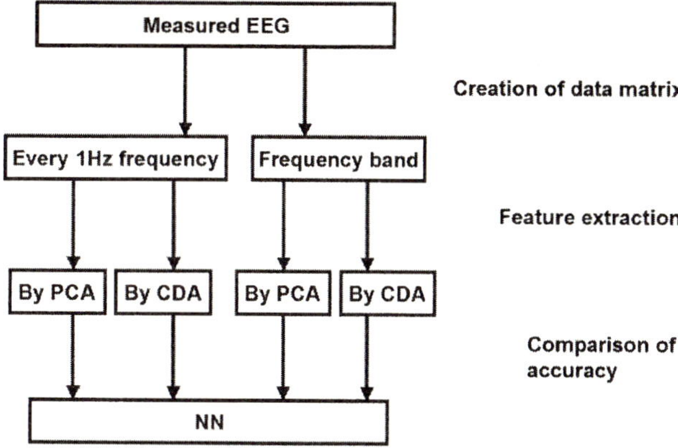

Fig. 1. A flow chart of the proposed system.

[9]. Therefore Fp1 is very sensitive to mental change by hearing stimulus. Moreover, the EEG is measured by the referential derivation method with the left earlobe in Fp1. The method measures the potential difference between the measured part and the reference potential. It is suitable for observing a cerebral activity. The attached EEG analysis software measures the EEG, and results of FFT (fast fourier transform) for each 1-second by 128Hz sampling frequency are yielded. Since this equipment uses the 4-22Hz band path filter, the appearance rate is considered from 4 to 22 Hz. On the one hand, in order to appear strongly influence by the music, the EEG is measured under the following condition. A subject sits on a chair in a PC room that has a little noise. While he measures the EEG, he maintains a quiet closed eye state and relaxed posture. He equips both ears with an earphone for music appreciation, a sensor at Fp1 and a left earlobe for measuring the EEG.

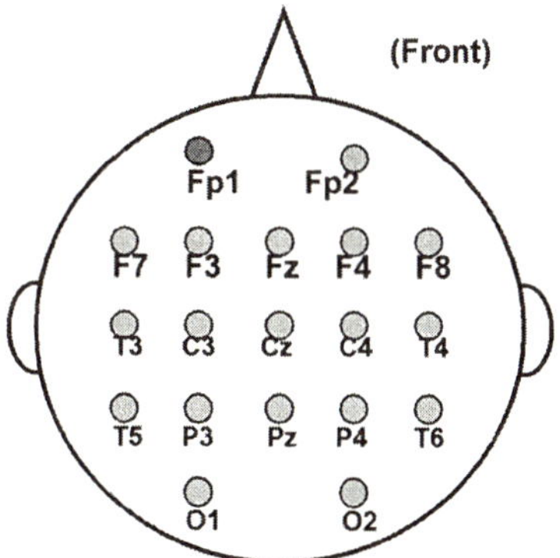

Fig. 2. The measurement points of the EEG.

3.2   Creation of Data Matrix

Data Matrix of Every 1 Hz Frequency. When the EEG was measured m times, the data matrix of every 1 Hz frequency X is defined as

$$
X = \begin{bmatrix} x_{41} & x_{51} & \cdots & & x_{221} \\ x_{42} & & \ddots & & x_{222} \\ \vdots & & x_{ij} & & \vdots \\ & & & \ddots & \\ x_{4m} & & \cdots & & x_{22m} \end{bmatrix}.
\tag{1}
$$

In this matrix, each column $i\ (=4,\ 5,\ \cdots,\ 22)$ expresses frequency bands for every Hz. Each row $j(=1,\ 2,\ \cdots,\ m)$ expresses the status in listening to each music genre. Then x_{ij} means appearance rate, which is the percentage of the spectrum of EEG. When the average of k Hz ingredient in time series data is defines as f_{kj}, the appearance rate x_{kj} is computed by the following formulas.

$$
x_{kj} = \frac{f_{kj}}{f_{4j} + f_{5j} + \cdots + f_{22j}}.
\tag{2}
$$

Data Matrix of Frequency Band. Ingredients of frequency bands are created by compressing that of every 1 Hz. We use 5 frequency band, θ (from 4 to 6 Hz), α_1 (from 7 to 8 Hz), α_2 (from 9 to 11 Hz), α_3 (from12 to 14 Hz), and β (from 15 to 22 Hz). Each ingredient is computed by the average of appearance rate x_{ij}. The data matrix of frequency band Y is defined as

$$
Y = \begin{bmatrix} y_{\theta 1} & y_{\alpha_1 1} & y_{\alpha_2 1} & y_{\alpha_3 1} & y_{\beta 1} \\ & \vdots & & \vdots & \\ y_{\theta m} & y_{\alpha_1 m} & y_{\alpha_2 m} & y_{\alpha_3 m} & y_{\beta m} \end{bmatrix}.
\tag{3}
$$

3.3   Extraction of Features by Multivariate Analysis

The PCA and CDA are one of multivariate analysis methods, which newly make variates in order to clear a difference and a relation between some datas. The features of measured EEG signals are extracted by the methods. The PCA and CDA compute eigenvectors and eigenvalues from the data matrix, and then obtain the feature of the EEG.

3.4   Analysis of EEG Patterns by NN

We detect the EEG by using a three-layer class type neural network shown in Fig.3. Input layer units are given a feature of the EEG extracted by the multivariate analysis. Output layer units show the status in listening to music while measuring the EEG. If the features extracted by the multivariate analysis

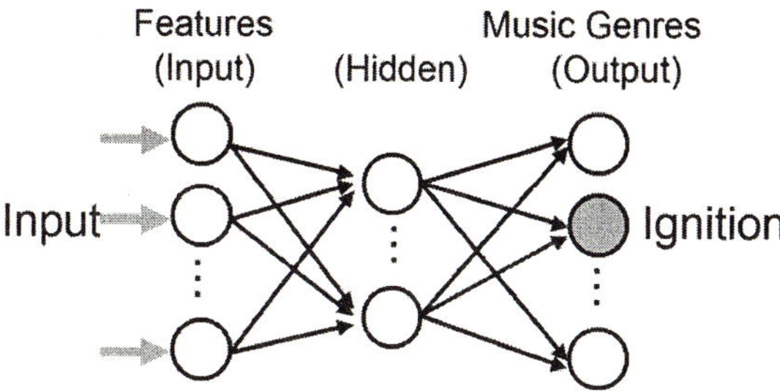

Fig. 3. A neural network for the EEG recognition.

are fed to the NN, a unit of the output layer ignites. The unit corresponds to the music genre. Therefore the NN can detect the music genre. The NN is trained by the back propagation method and verified by the Leave-one-out Cross-validation method.

4 Computer Simulations

We try computer simulations for EEG classification. In this paper the EEGs of three subjects are measured. The subjects are 20-22 years old with hearing normalcy, and have healthy mind and body. 5 kinds of music genre are used, namely jazz, classical, rock, Euro beat and Japanese ballad. It is shown that each genre receives a different impression by the prior questionnaire. The EEG of each subject is measured 5 times for each music genre.

4.1 Result

Tables 1 and 2 show the rate of the EEG patterns recognized by the NN. The number of the input layer units is 5, five components that consist of the 1st to 5th ingredients are fed to the NN. Table 1 shows the recognition rate in the case of every 1 Hz(features extracted from X), and Table 2 shows the recognition rate in the case of frequency bands(features extracted from Y). According to the tables, it is found that the features extracted from Y are recognized lower than that from X. Then according to Table 1, it is found that the average of the

Table 1. The result of recognition rate obtained by the NN, in the case of X (every 1 Hz).

	subject A		subject B		subject C	
Music Genre	PCA	CDA	PCA	CDA	PCA	CDA
jazz	2/5	4/5	4/5	5/5	3/5	2/5
classical	2/5	4/5	3/5	5/5	3/5	5/5
rock	3/5	4/5	1/5	5/5	4/5	5/5
Euro beat	2/5	5/5	3/5	5/5	4/5	5/5
Japanese ballade	4/5	4/5	5/5	5/5	4/5	2/5
Average	52%	88%	64%	100%	72%	76%

Table 2. The result of recognition rate obtained by the NN, in the case of Y (frequency bands).

Music Genre	subject A		subject B		subject C	
	PCA	CDA	PCA	CDA	PCA	CDA
jazz	1/5	1/5	5/5	3/5	3/5	2/5
classical	2/5	2/5	5/5	2/5	2/5	2/5
rock	3/5	3/5	3/5	1/5	4/5	2/5
Euro beat	2/5	3/5	4/5	3/5	3/5	3/5
Japanese ballade	3/5	4/5	4/5	5/5	3/5	2/5
Average	44%	52%	76%	56%	60%	44%

rate by the CDA are recognized better than that by the PCA in all subjects. It is considered that the CDA from the ingredients for every 1 Hz extracted features of the EEG better.

4.2   Consideration

Since a generation source and a transfer system of the EEG have not been clear yet [10], it is complicated and difficult to analyze the EEG. In order to discover correlation and features from complex datas, multivariate analysis has been often used [11] [12]. By the PCA and CDA, an EEG pattern can be expressed by a few ingredients, which emphasize similarity and difference of influence in listening music. It is considered that the features obtained by the PCA and CDA make automatic judgment of the EEG easy. Though the multivariate method has been already reported and effective results were obtained [13], it maybe shows only a rough result for dealing with frequency bands of the EEG. Since recognition rate from frequency bands is low, it is considered that every frequency ingredient is important to improve recognition accuracy in the EEG analysis.

5   Conclusion

In this paper, we extracted features from the measured EEG by the multivariate analysis, then we examined properties of the features by the NN. From simulation results, it was suggested that feature extraction of the EEG by the CDA was effective. Since the proposed method only classified features of the EEG, relation between the EEG and stress or relaxation has not been found yet. We discuss a judgment method of the effect from the classified features in future work. Though we classified the feature of EEG between the music genres in this paper, a different EEG pattern may appear in the same genre. For example, the EEG changes with a type of classical music because many classical music have various impression such as happiness or sadness[14]. Moreover, it is said that the EEG changes with taste[15]. Therefore we should examine not only change of the EEG but also the character of a listening music.

References

1. P.Chiu, and A.Kumar, "Music Therapy: Loud Noise or Soothing Notes?," *International Pediatrics*, Vol. 18, no. 4, pp.204-208, 2003
2. M.Makino, S.Tsutsui, K.Takashima, and M.Yasushi, "An Application of Making an Original CD for Music Therapy," *Japan Biomusic Association*, Vol. 13, no. 2, pp.105-pp.115, 1996 (in Japanese)
3. M.Tanizaki, Y.Kotoh, T.Onishi, and Y.Nakahara, "EEG and Hemo-Dynamics Responses for the Background Noise Using Emotion Spectrum Analyzer and Near-Infrared Spectroscopy(NIRS)," *ISSN*, Vol. 2, no. 1, pp19-26, 2002 (in Japanese)
4. T.Felzer, and Bernd Freisleben, "Analyzing EEG Signals Using the Probability Estimating Guarded Neural Classifier," *IEEE*, Vol. 11, no. 4, pp.361-371, 2003
5. S.Tasaki, T.Igasaki, N.Murayama, and H.Koga, "Relationship between Biological Signals and Subjective Estimation While Humans Listen to Sounds," *T.IEE Japan*, Vol. 122-C, no. 9, pp.1632-1638, 2002 (in Japanese)
6. R.J.Davidson, J.R.Marshall, A.J.omarken, and J.B.Henriques, "While a Phobic Waits: Regional Brain Electrical and Autonomic Activity in Social Phobics during Anticipation of Public Speaking," *Biological Psychiatry*, Vol. 47, no. 2, pp.85-95, 2000
7. J.H.Cruzelier, T.Egner, E.Valentine, and A.Williamon, "Comparing Learned EEG Self-regulation and the Alexander Technique as a Means of Enhancing Musical Performance," *ICMPC7*, pp.89-92, 2002
8. J.J.B.Allen, and J.P.Kline, "Frontal EEG asymmetry, emotion, and psychopathology: the first, and the next 25 years," *Biological Psychology*, Vol. 67, pp.1-5, 2004
9. J.Jung, T.Kobayashi, Y.LI, and S.Kuriki, "Event-Related Potentials during a Target Discrimination Task Based on Texture Cue," *IEICE(D-II)*, Vol. J85-D-II, no. 6, pp.1084-1092, 2002 (in Japanese)
10. O.Yamashita, A.Galka, T.Ozaki, R.Biscay, and P.Valdes-Sosa, "Recursive penalized least squares solution for dynamical inverse problems of EEG generation," *Human Brain Mapping*, Vol. 21, 221-235, 2004
11. H.Arakawa, M.Nakayama, and Y.Shimizu, "A Relation and Pupli Size in the Memorizing Numeral Tasks," *IEICE(D-I)*, Vol. J82-D-I, no. 9, pp.1217-1221, 1999 (in Japanese)
12. O.Sakata, T.Shina, and Y.Saito, "Multidimensional Directed Information and Its Application to Causality Analysis of EEG," *IEICE(A)*, Vol. J83-A, no. 5, pp.466-475, 2000 (in Japanese)
13. S.Yokoyama, T.Shimada, A.Takemura, T.Shina, and Y.Saitou, "Detection of Characteristic Wave of Sleep EEG by Principal Component Analysis and Neural-Network," *IEICE(A)*, Vol. J76-A, no. 8, pp.1050-1058, 1993 (in Japanese)
14. L.A.Schmidt, and L.J.Trainor, "Frontal Brain Electrical Activity (EEG) Distinguishes Valence and Intensity of Musical Emotions," *Cognition and Emotion*, Vol. 15, no. 4, pp.487-500, 2001
15. N.Saji. and R.Saji. "A Study of Relationship between EEG under Music Influence and Music Taste, *Japan Biomusic Association*, Vol. 17, no. 2, pp.226-pp.232, 1999 (in Japanese)

A Goal Oriented e-Learning Agent System

Dongtao Li[1], Zhiqi Shen[2], Yuan Miao[3], Chunyan Miao[1], and Robert Gay[2]

[1] School of Computer Engineering, Nanyang Technological University, Singapore
`lidongtao@pmail.ntu.edu.sg, ascymiao@ntu.edu.sg`
[2] Information Communication Institute of Singapore,
Nanyang Technological University, Singapore
`{zqshen,eklgay}@ntu.edu.sg`
[3] School of Computer Science and Mathematics
Victoria University, Melbourne, Australia
`Yuan.Miao@vu.edu.au`

Abstract. This paper illustrates a goal oriented approach to model agent mediated e-learning system. Most of the traditional e-learning systems are not learner-centric, and they often ignore the diversity of learner population; thus very often their service is not able to directly or effectively match the learner's goal. From the learner's point of view, we propose a goal oriented approach to develop an agent-mediated e-learning system which provides leaner-centric and personalized e-learning services. In this paper, we illustrate how to use Goal Net to model the learner's goal and how to develop such a multi-agent system to provide efficient learning services.

1 Introduction

With the fast growth of computer and communication technology, the traditional learning system is replaced by the convenient e-learning system. However, most of the existing e-learning systems are just using the new technology to wrap the traditional learning style, and did not take the full advantages offered by current technologies [1].

One of the major limitations is that many current e-learning systems are system-centric instead of learner-centric. The system adopts the same learning materials and strategies to different learners without considering their different profiles. The system is not able to take care of the diversity of learner populations. As a consequence, it is also not able to provide effective teaching services to part of the learner group. On the learner side, learners may expect the teaching system to be flexible and the course to be customized to closely fit their background and needs. Thus there is a mismatch between the rigid services provided by the system and flexible goals required by learners.

To fix this problem, we propose a personalized e-learning system based on agent technology. The agent system is designed in a goal-oriented manner, which is modeled using *Goal Net* [2], to provide personalized e-learning service that can effectively match learner's goals. The agent first analyzes a learner's current profile. After that it will generate a suitable learning path for the learner by choosing some specific courses, to adapt the learner's ability and preference. Finally the system will provide the assessment to test the leaner and evaluate the learning process.

R. Khosla et al. (Eds.): KES 2005, LNAI 3681, pp. 664–670, 2005.

Following this introduction, we will first describe the motivation scenario and challenges to develop the above mentioned agent system in the next section. In section 3, we briefly explain *Goal Net* model. Then we will discuss how to build the system model using *Goal Net* in section 4. After that, we illustrate how to convert the model into agent system implementation in section 5. A short conclusion is reached in section 6.

2   Motivation Scenario and Challenges

To make the illustration of the system development clearer, we introduce a typical learning scenario that motivates this research.

Now days IT industry is placed at a state of fast changing due to the knowledge explosion. In order to stay competitive and deal with the dynamic market demands, companies need to frequently train their employees to acquire new knowledge.

Suppose a software company forms a project team for the newly tendered software project, and the project team members need to be trained through an e-learning grid service, to acquire the necessary knowledge and skills for the project. These team members have different background and will take different roles in the new project; thus the new skills they need to acquire and the technical levels of the skills that are required are possibly different. For each learner, a unique learning path needs to be defined based on their learning goals. Further more, different learner may prefer to use different terminals such as laptop or PDA to access the learning materials.

To help the employee having a joyful and efficient learning experience, we design an agent system to provide a learner friendly e-learning system. The system should be able to:

- Dynamically generate a on-going personalized learning path for employee based his/her current skill level and the skills required by the project;
- refine the learning path according to the feedback from the learner in order to optimize the acquisition of the needed competences;
- deliver the selected learning course to the learner site on his/her demand based on the learning path;
- provide self-learning cycle until the assessment of the learning results meets the requirements.

Besides the above explicit requirements, there are also some implicit requirements such as selecting the shortest learning path to reduce the learning cost. All the requirements will be carefully considered in the agent system design.

3   Goal Net Model

Goal Net is a composite goal model which is composed of *states* and *transitions*. The *states*, represented by circles, are used to represent the states that agent need to go through in order to achieve its final goal. The *transitions*, represented by arc and vertical bar, connect one state to anther specifying the relationship between states it joins. Each transition must have at least one *input state* and one *output state*. Each transition is associated with a task list which defines the possible tasks that agent may need to perform in order to transit from the input state to the output state. Figure 1 shows a simple Goal Net.

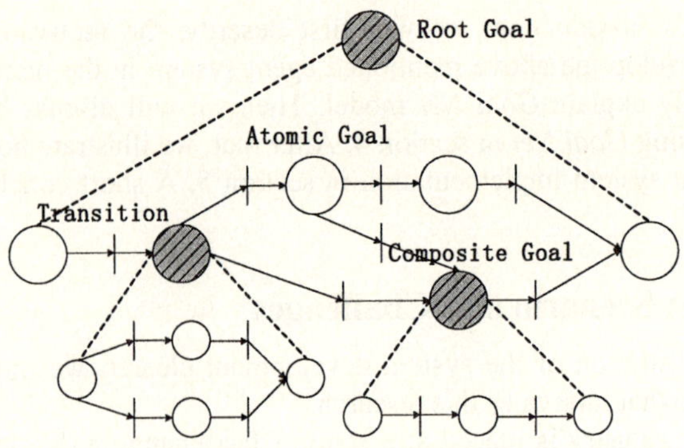

Fig. 1. A Simple Goal Net

There are two types of states in *Goal Net* model, *atomic state* and *composite state*. An *atomic state*, represented by a blank circle, accommodates a single state which could not be split anymore; a *composite state,* represented by a shadowed circle, may be split into states (either composite or atomic) connected via transitions.

In *Goal Net*, there are four types of temporal relations of goals represented by transitions connecting the input state and the output state: *sequence, concurrency, choice* and *synchronization. Sequence* represents a direct sequential relationship between one input state and one output state; *concurrency* has one input state but more than one output states, and all its output states can achieve simultaneously; *choice* specifies a selective connection from one state to other states; *synchronization* specifies a synchronization point from different input state to a single output state. With different combination of the basic temporal relations, *Goal Net* supports a wide range of complicated temporal relations among goals. This is one of the major differences between *Goal Net* and other goal modeling method.

With such a composite state hierarchy and various temporal relations within the hierarchy, a complex system can be recursively decomposed into sub-goals and sub-goal-nets. In such a manner, the system can be easily modeled and simplified.

4 Learning Goal Net

To build an agent system using *Goal Net*, the first task is to construct a learning goal net by identifying goals need to be achieved (*what*), the possible tasks to achieve the goals (*how*), and the environment which may affect the goal achievement (*situation*).

To build the learning goal net, we first identify the overall goal of this goal net, which is *E-Learning Service*. Then we can use goal as a requirement analysis tool to model the problem and acquire the details requirements [6] [7]. Following a system top-down approach, we may split this final goal into a group of sub-goals, of which each sub-goal represents a small problem:

- *Profile Analysis*: collect and analyze the profile of a learner, to assist generating an effective learning path for the learner;
- *Path Generation*: based on the current profile and his/her learning requirements, dynamically select the best learning courses for the learner. Courses can be represented as different goals here;

- *Course Delivery*: negotiate with service provider in the learning grid and deliver the selected course to the learner; customization is also provided according to learner's preference;
- *Assessment*: conduct assessment to a learner, and analyze the assessment results to provide necessary self-learning cycle for learner.

One of the main advantages of *Goal Net* is that the complex goal can be decomposed into easy-to-achieve sub-goals. In the above list, the goals are still too abstract to achieve, we can further decompose these goals until that each goal represents a small problem and can be easily solved or achieved. The high level complex goal will automatically be achieved once all its sub-goals are achieved. Fig.2 shows some of the decomposition process in the learning construction.

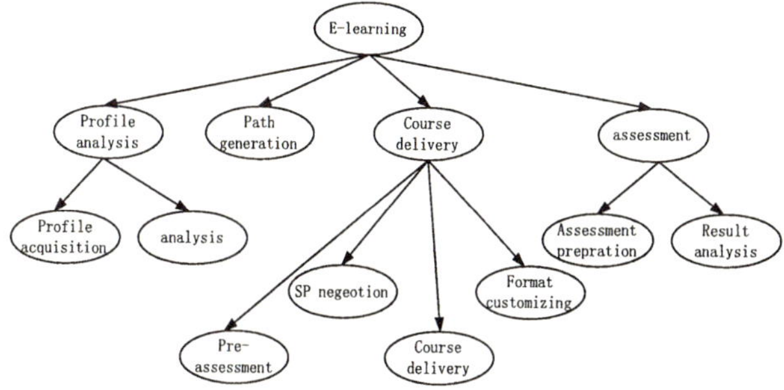

Fig. 2. Learning Goal Net. This goal net shows the high level goals which need to be achieved in order to provide the personalized e-learning service

The *Learning Path Generation* is also a composite goal which is constructed dynamically for each learning application. All the necessary courses or learning objects [3] are identified as goals in the goal net, and all the possible connections are identified as transitions. Property of a course, such as technical grade and cost, are used as the property of the goal. The learner's profile is used as environment information for the learning goal net. Thus we can use the *Goal Selection Algorithm* [4] in *Goal Net* to infer the best path to achieve the final goal. The learning path is dynamically generated during the learning process, which means selecting the next course only when the learner has finished the current course and he/she is ready for the next course, so that the learner's feedback of the previous course can be considered when selecting the next course. Such an on-going learning path generation can easily adapt the change of the learner's requirements.

After all the goals are identified, the next step is to identify the transitions between goals, or work out the tasks to achieve each goal. Each transition may consist of several tasks so that a suitable task can be selected and executed under different situations. For example, to collect a learner's profile, there are three possible ways: getting the learner's profile from the database of human resource; letting the learner fill up a technical survey; or conducting a test to get the learner's profile. Any of these three tasks can achieve the goal *Profile Acquisition*. The system may try to collect the learner's profile from human resource if it is available, otherwise it may try to let the learner to fill in the profile or conduct the test at learner's preference. Such kind of behavior-autonomy gives flexible choices to both the agent and the learner.

To complete the learning goal net model, the environment should be modeled. In goal net, variable-value pairs are used to describe the environment. Some the environment information has already been used in the previous task design. For example, the availability of the learner's profile in human resource can be considered as one environment variable. In this step we formally model all the environment information which may affect the goal pursuit or task execution.

Once the learning goal net is constructed, the skeleton of the system has been clear. The goal net works as a driver to direct the agent system to achieve the learner's goal. The next step is to work out the agent system to execute the learning goal net.

5   Agent Mediated e-Learning System

The learning goal net we designed consists of tens of goals and a number of small tasks. Certainly it is possible that we use one single agent to pursue all the goals from profile collection to assessment analysis. However, it is not efficient to do so. The reason is that the pursuit of different goals requires different types of tasks to be executed; mixture of different level of goals and different types of tasks makes the system difficult to be implemented and maintained. For example, the *Path Generation* goal requires heavy computation to generate a suitable learning path, while *Course Delivery* are more involved in communication and file transfer. The agent will be very heavy-weight and error-prone if it executes the entire learning goal net. It is reasonable that we use several agents and each of them executes part of the goal net.

To derive an agent hierarchy from a goal net, or identify agents from a goal net, we can split the goal net from some of its sub-goals [4]. According to [4], a goal net can be split based on *Modularity*, *Agent Location*, *Load Balancing* and *Agent Organizational Role*. Multiple policies can be applied when identifying agents. Based on the *Modularity* and *Load Balancing* polices, we can derive 5 agents from the learning goal net by splitting the goals in the first level. Fig 3 shows the goal net splitting process, and Fig 4 shows the interaction between different agents.

In this agent hierarchy, the *E-learning Agent* acts as a coordinator of other agents. It controls the general process of the e-learning system. The *Learner Profile Analyzer* is pursuing the *Profile Analysis* goal by processing all the sub-goals and transitions of *Profile Analysis* goal. Once it achieves its goal, that is, it gets the profile analysis results, it will inform the coordinator agent and requests the *Learning Path Generator* to generate the next course or learning object for the learner. The *Learning Path Generator* is trying to find the next learning object the learner should take, that is to find the best next goal in its own goal net. After the learning object is figured out, the *Course Delivery Agent* will start to negotiate with the service provider, customize the course format and deliver it to the learner. The *Assessment Agent* will be informed after the course delivery. Once the learner has finished learning the course, the *Assessment Agent* will automatically starts to assess the learner. If the learner failed the test, the *Assessment Agent* would analyze the assessment results and provide the self-learning cycle until the learner passes the assessment test. When the learner has successfully learned one course (passed the assessment test), his/her profile will be updated and next course will be prepared and started. Once the *Learning Path Genera-*

tor has achieved its final goal, or there is no more next course, the whole learning process is considered accomplished.

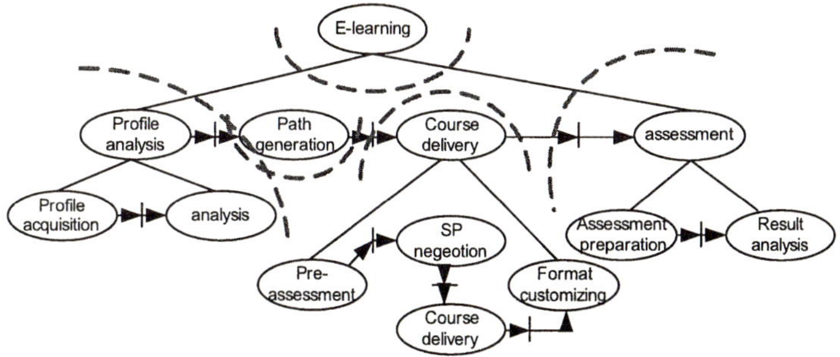

Fig. 3. Learning Goal Net Splitting. The red dash line shows agent identification in the learning goal net

To simplify the work using *Goal Net* in agent system development, we have developed a *Multi-Agent Development Environment* (*MADE*) [4] to assist agent developers. *MADE* provides a graphical user interface to help in *Goal Net* design and agent development. It has been integrated with JADE [5], a popular agent development tool, so that *MADE* provides both mental state development and industry standard compliant agent development environment. *MADE* releases agent developers from the tedious goal net design and agent development work, so that they can focus on *Goal Net* logic and agent task development.

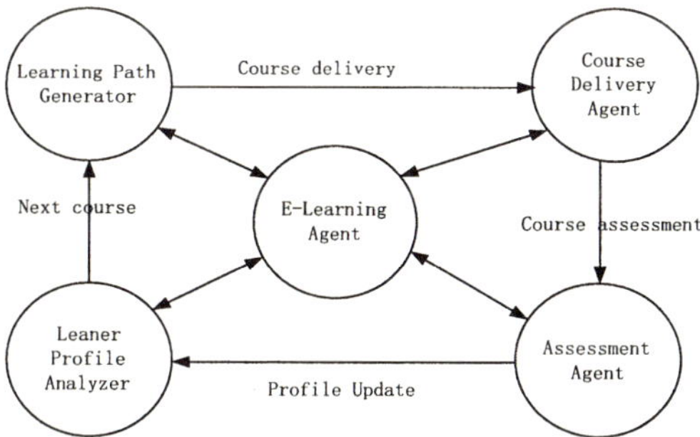

Fig. 4. Learning Agent Hierarchy. Each agent, represented by a circle, takes different responsibility and pursuing a sub-goal. Once it achieves its own goal, it informs the coordinator agent *E-Learning Agent*. The arrow represents the general interaction of the connected agents

6 Conclusion

In this paper, we present a goal oriented approach to provide highly learner-centric and personalized e-learning services through multi-agent system. The system is designed based on *Goal Net*, which models the general learning process and learner's requirements into different goals, and several agents are developed to pursue these goals. The goal-selection and task-selection of *Goal Net* direct agent to dynamically pursue different goals and execute different tasks under different situations; Thus, the system can generate a suitable learning path for the learner based on his/her profile, and provide personalized learning services.

References

1. Albert Angehrn, Thierry Nabeth, Claudia Roda: "Towards personalised, socially aware and active e-learning systems", The Center for Advanced Learning Technologies, November 2001
2. Shen Z. Q. and Gay R., "Goal-Based Intelligent Agents", *International Journal of Information Technology*, Vol 9, No. 1, pp. 19-30, 2003
3. Wiley, D.A. (2003): "Learning Objects: Difficulties and Opportunities." Retrieved April 10, 2003, from the World Wide Web
4. Shen Z.Q.: "Goal-Oriented Modeling for intelligent agents and their applications ", Ph.D. Thesis, Information Communication Institute, NTU, 2003
5. Java Agent Development Framework (JADE). http://jade.tilab.com.
6. A. Dardenne, A. van Lamsweerde, and S. Fickas, "Goal-Directed Requirements Acquisition", *Science of Computer Programming*, Vol. 20, North Holland, pp. 3-50, 1993.
7. Anton, A., "Goal-Based Requirements Analysis", *Second International Conference on Requirements Engineering*, Los Alamitos, California: IEEE Computer Society Press, pp. 136-144, 1996.

iJADE Tourist Guide – A Mobile Location-Awareness Agent-Based System for Tourist Guiding

Tony W.H. Ao Ieong, Toby H.W. Lam, Alex C.M. Lee, and Raymond S.T. Lee

Department of Computing, The Hong Kong Polytechnic University,
Hung Hom, Kowloon, Hong Kong
{cswhai,cshwlam,cscmlee,csstlee}@comp.polyu.edu.hk

Abstract. A mobile location-awareness agent-based tourist guide system called iJADE Tourist Guide is presented. We focus on how to integrate GPS device and agent technology to provide location-awareness tourist information in mobile device. We use GPS device for locating visitor's current position, and adopt iJADE agent to retrieve nearby tourist information. The limitation of computation power and network bandwidth in mobile device can be overcome by utilizing the agent technology. The main objective of the system is to allow visitors to browse the relevant tourist information based on their interest and location in handy device at anytime and anywhere.

1 Introduction

In recent years, the development in Information Technology and communication systems has a profound impact on the tourism industry. These technologies have changed the customer behavior. Tourists would like to have their own tour plan and personalized tourist guide by using these technologies. There are some example developments for tourists such as personalized pricing (priceline.com) and recommendation system. Today, the wireless devices, like PDA, mobile phone, become small in size, light in weight and long run time. Tourists would use the mobile device as their personal tourist guide. In this paper, we propose a new mobile application for tourist which is called iJADE Tourist Guide. iJADE tourist guide is developed based on the intelligent agent-based platform iJADE [1]. The tourist guide is a location awareness application which used the Global Positioning System (GPS) to gather the geographic information. Besides location awareness, iJADE tourist guide is integrated with the intelligent agent to enhance the functionality. The detail about iJADE tourist guide shows in section 3.

The rest of this paper is organized as follows. In section 2, we show the related work about location-awareness applications. In section 3, we show the detail about the iJADE tourist guide. Conclusion and future work appear in section 4.

2 Related Work

GPS equipment is widely used today, as the cost of the receiver is low. There are some consumer products developed for outdoor use, for example, hiking, flying and sailing. In addition, the receivers can be used for driver route guiding. By using GPS and agent technologies, there are some research works on tourist guide system.

A research project CRUMPET, which is funded by European Union, developed a personalized nomadic platform for tourist information systems [2]. The system is

R. Khosla et al. (Eds.): KES 2005, LNAI 3681, pp. 671–677, 2005.

integrated with GPS and agent technology for robust, scalable and seamlessly accessible services. CRUMPET could gather two types of information for tourists: static and dynamic information. The static information is the information that retrieved according to the user's profile and request. The dynamic information is the information that retrieved according to the user's moving route. In addition, the system would learn the user's interest and preferences automatically. This could further filter the irrelevant information to the user.

Cheverst et al.developed a context aware tourist guide system called GUIDE [3]. The system would provide tourist information for the user's mobile device. The system could construct a tour plan for the user by using user's preference and interest. Similar to CRUMPET system, GUIDE also uses wireless communications and GPS to ensure the mobile connectivity and context-awareness. The evaluation shows a high level of acceptability for the system in different types of user.

Georgia Institute of Technology developed another tourist guide system, Cyberguide [4]. The tourist guide provides information to the user based on the information of position and orientation. There are two types of cyberguide: indoor and outdoor. The indoor cyberguide uses infrared (IR) as the positioning component and the outdoor cyberguide uses GPS to capture the user location.

A number of context awareness tourist guide systems has been developed. The main objective of these systems is to select or filter the relevant information to the users based on user's interest and location. However, the existing application not fully utilized the AI technologies. To further improve the functionality, we propose a mobile location-awareness agent-based system for tourist guiding. The detail of iJADE tourist guide is shown in next section.

3   iJADE Tourist Guide

In this paper, we propose a mobile location-awareness agent-based tourist guide system – iJADE Tourist Guide (iJTG). It exploits the intelligent agents in iJADE agent-based framework to provide a context-aware tourist information retrieval system. The information presented to visitors by the system should be tailored to their context. We have identified two broad classes of context to be used, the personal and environmental. Examples of personal context include the visitor's interests, e.g. taste favor, the visitor's current location. Examples of environment context include the time of day, special events for visitors, and the opening of the attractions. iJTG provides the flexibility to enable visitors to explore and learn about the place in their own way, and it also supports interactive services to allow visitors to get the relevant tourist information.

As a mobile application, the narrow bandwidth in wireless connection is a big problem to develop a mobile information retrieval system. iJTG utilizes mobile agent technology to conduct multiple interactions with different information database systems in a single request, the output of the interaction is then sent back to visitor so network loading is reduced as a result. In iJTG, the system consists of two parts, the client side 'iJADE Tourist Guide Mobile Client (iJTGMC)' and the server side 'iJADE Tourist Information Center (iJTIC)'. We will describe the iJADE model and these two parts in the following sections.

3.1 iJADE Model

The aim of iJADE is to provide the aim of iJADE is to provide comprehensive 'intelligent' agent-based APIs and applications for future AI based applications. Figure 1 depicts the two levels of abstraction that are used in the iJADE system: (a) the iJADE system level – ACTS (**A**pplication, **C**onscious, **T**echnology and **S**upporting layers) models, and (b) the iJADE data level – DNA (**D**ata, **N**eural and **A**pplication layers) model. The ACTS model consists of (1) Application Layer, (2) the Conscious (Intelligent) Layer, (3) the Technology Layer, and (4) the Supporting Layer. The DNA model is composed of (1) the Data Layer, (2) the Neural-Network Layer, and (3) the Application Layer. iJADE DNA model provides a comprehensive data manipulation framework that is based on neural-network technology. The "Data Layer" corresponds to the raw data and input "stimuli" (such as the facial images captured from a Web camera and the product information in a cyber store) from the environment. The "Neural-Network Layer" provides the "clustering" of different types of neural networks for the purpose of organization, interpretation, analysis and forecasting operations that are based on the inputs from the Data Layer". The neural networks are used by the iJADE applications in the "Application Layer".

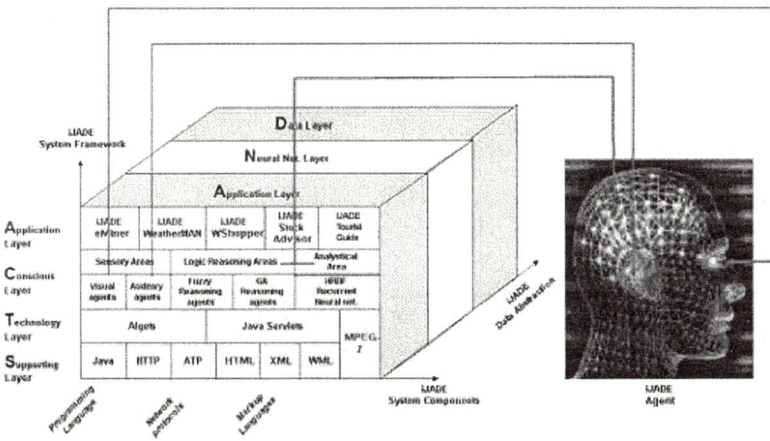

Fig. 1. System architecture of the iJADE (v 2.0) model [1]

3.2 iJADE Tourist Guide Mobile Client (iJTGMC)

As a mobile tourist guide, iJTGMC should promise to free the user from the constraints of stationary desktop computing, so we should focus on what maximally benefit from mobility. Personal digital assistant (PDA) provides the mobile computing environment to implement this kind of mobile application. iJTGMC is a tiny application that runs on the mobile device with limited computation power, it runs without the agent server in order to reduce the computation cost and resources. However, by using the agent client application interface, this tiny application can invoke agent to do the information retrieval and processing, so it benefits from the agent programming. iJTGMC is a context-awareness application, in particular, a location-awareness application which based on the visitor's current location to retrieve the nearby tourist information. Visitors can browse the city's attractions, history, restaurants and transportation information around them from iJTGMC. Visitor will be notified about the special events via iJTGMC as well so that visitors will not miss their favorite events. In order to provide the location-awareness function, we have adopted Global Positioning System (GPS) for positioning in iJTGMC.

3.2.1 Global Positioning System (GPS)

The Global Positioning System (GPS) is a satellite-based navigation system made up of a network of 24 satellites placed into orbit by the U.S. Department of Defense. GPS was originally intended for military applications, but in the 1980s, the government made the system available for civilian use. GPS works in any weather conditions, anywhere in the world, 24 hours a day. There are no subscription fees or setup charges to use GPS [5].

3.2.2 The Client Application

The iJTGMC application consists of 3 major modules: Tourist Guide Manager, GPS Controller and iJADE Runtime Agent. Figure 2 shows the overview of the iJTGMC.

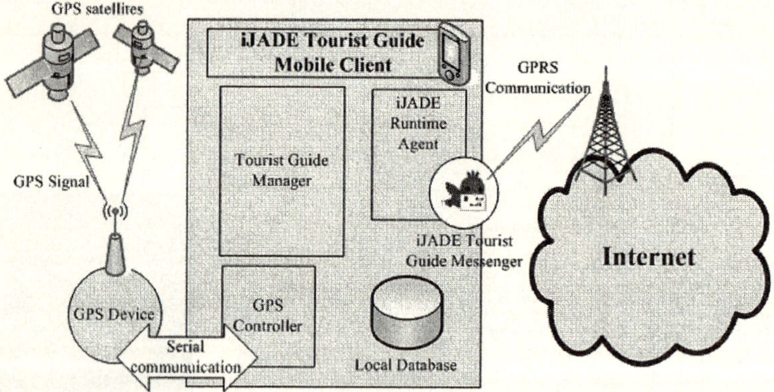

Fig. 2. Overview of the iJTGMC

The Tourist Guide Manager coordinates each module in the application. The manager first gets the current location information from the GPS controller, and then queries the tourist information from the local database. If the relevant information that is nearby the visitor exists in the local database, the manager will retrieve the information from the database and display to the user. If the information does not exist, the manager will ask the iJADE Runtime Agent to retrieve the relevant tourist information by using the iJADE Tourist Guide Messenger (iJTGM). After the iJTGM retrieved the tourist information from the iJADE Tourist Information Center (iJTIC) via wireless GPRS communication, the manager will cache the information in the local database for further use.

The main duty of GPS Controller is to poll the GPS device periodically to get the latest location information. The GPS device can locate its position after receiving adequate GPS signal from various GPS satellites, the position information can be extracted out from the GPS message that collected from the device by using serial communication.

iJADE Runtime Agent is an light-weight agent interface to allow external application to invoke agent in information processing, it invokes the mobile agent iJTGM to retrieve the tourist information from iJTIC via wireless GPRS communication by using agent technology.

3.3 iJADE Tourist Information Center (iJTIC)

In iJADE Tourist Guide system, iJTIC is acting as the server side and providing tourist information to the iJTGMC by using iJTGM. iJTIC consists of iJADE Tourist Information Manager (iJTIM) and Tourist Information Database. iJTIC is a iJADE server which provides the runtime environment for iJADE agents. iJADE Tourist Information Manager is a stationary agent which handles the iJTGM requests, and it returns the relevant tourist information by the position information in iJTGM from its tourist information database. System overview of iJADE Tourist Guide is illustrated in Figure 3.

Visitor can browse the nearby tourist information by using iJTGMC, and iJTGMC retrieves this relevant information by the position information from GPS device. iJTGMC invokes the iJTGM and dispatches it to different iJTICs. The iJTGM will request the iJTIM to get the nearby tourist information by the position information, and then it will dispatch itself to another iJTIC to get other nearby tourist information after finished the information retrieval. As comparing with traditional client-server model, the information retrieval method by using agent can reduce network loading and processing time since only one request can get all the information from different agent servers. For example, a visitor wants to get the nearby tourist information, iJTGMC will send out the iJTGM to do the task. The iJTGM will first locate all the nearby iJTICs, and then visit the iJTICs one by one. The iJTGM gets the shopping mall and discount information from the iJTIC of shopping mall, the catalog from the iJTIC of museum, and the package menu from the iJTIC of hotel. Then the iJTGM carry all the tourist information back to the iJTGMC and display to the visitor. In the meantime, the iJTGMC will notify the visitor that there is a special event in the nearby museum. But for client-server model, it takes at least three requests to the servers in order to finish this task, and it will consume more time and network traffic to accomplish the information retrieval that is not suitable for mobile device.

4 Conclusion and Future Work

In this paper, we have demonstrated a mobile location-awareness agent-based tourist guide system – iJADE Tourist Guide. It integrates GPS device and agent technology to provide a location-awareness tourist information retrieval system. Visitor can access the nearby tourist information anywhere and anytime by a small handheld device with limited computation power and limited network bandwidth in iJTG, these characteristics are benefited by utilizing the iJADE intelligent agent technology. The demonstration of iJTG can be accessed in iJADE website [6]. Further work can focus on how to enhance the functionalities in iJTG, since it is a prototype for demonstration. We can work on fuzzy searching of the tourist information, expert system to answer the questions on tourist information, or tourist route optimization from a set of itineraries with time and money constraints by Genetic Algorithm.

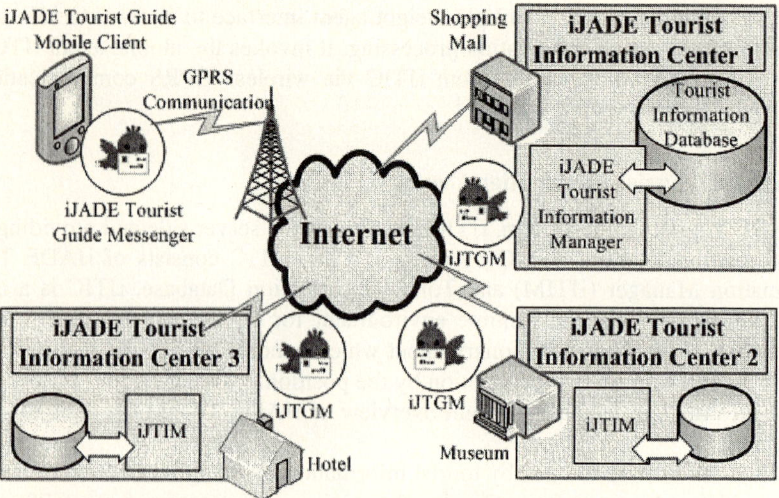

Fig. 3. System overview of iJADE Tourist Guide

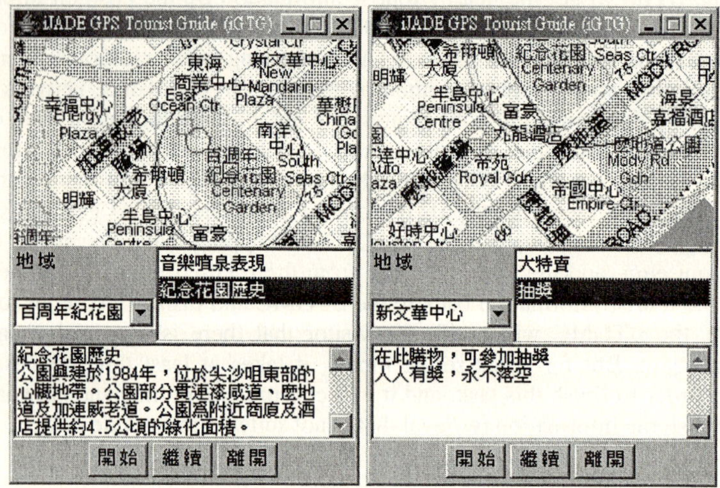

Fig. 4. Screenshots of iJADE Tourist Guide client application

Acknowledgment

This work was partially supported by the iJADE projects B-Q569, A-PF74 and Cogito iJADE project PG50 of the Hong Kong Polytechnic University.

References

1. Raymond S. T. Lee, James N. K. Liu, "iJADE Web-Miner: An Intelligent Agent Framework for Internet Shopping," IEEE Trans. Knowl. Data Eng., vol. 16, no. 4, pp. 461-473, 2004

2. Poslad, S., Laamanen, H., Malaka, R., Nick, A., Buckle, P., Zipl, A., "CRUMPET: creation of user-friendly mobile services personalised for tourism," The Second International Conference on 3G Mobile Communication Technologies, pp. 28-32, 2001
3. K. Cheverst, N. Davies, K. Mitchell, and A. Friday, "Experiences of developing and deploying a context-aware tourist guide: The GUIDE project," The sixth International Conference on Mobile Computing and Networking, pp. 20-31, 2000
4. Abowd, G., Atkeson, C., Hong, J., Long, S., Kooper, R., and Pinkerton, M, "Cyberguide: A mobile context-aware tour guide," ACM Wireless Networks, vol. 3, pp. 421-433, 1997.
5. GPS information: http://www.garmin.com/aboutGPS/
6. iJADE Website: http://www.ijadk.org

The Design and Implementation of an Intelligent Agent-Based Adaptive Bargaining Model (ABM)

Raymond Y.W. Mak and Raymond S.T. Lee

Department of Computing, The Hong Kong Polytechnic University
Hung Hum, Kowloon, Hong Kong
csstlee@comp.polyu.edu.hk

Abstract. In this paper we propose a highly adaptive bargaining model for agent shopping which stimulates two major human bargaining strategies: 1) Payoff-Oriented Strategy and 2) Real-time Adaptive Attitude Switching Strategy. Payoff-Oriented Strategy adjusts the rate of concession by determining the current payoff gained and the eagerness of the adopted attitude at each bargaining round. Also, the buying agent in this work is guided by the Real-time Adaptive Attitude Switching Strategy which comprises a set of attitude switching rules. These rules guide the buying agent to gain higher payoff and prohibit seller from gaining too much payoff. Owing to the substantial experimental results in this work, the two human-like bargaining strategies achieved the adaptive changing eagerness of a particular attitude and adaptive switching attitude in reacting to the opponent's feedback.

1 Introduction

In a typical e-shopping scenario, the main purpose of bargaining is to make profit, which induces the continuous and adaptive changing in the eagerness of a particular attitude with respect to the current payoff gained. However, contemporary product bargaining strategy such as the Market-Driven Negotiation Strategy [1][2][3] and the Adaptive Negotiation Agent [6] uses fixed eagerness value.

In this paper we propose an adaptive bargaining model (ABM) with attitude switching strategy which developed under the iJADE [4][5] platform. To develop a human-like agent with highly adaptability, three bargaining attitudes with different degree of eagerness – including aggressive, neutral and conservative are defined. In summary, the proposed ABM comprises two main modules: 1) The Payoff-Oriented Strategy (POS) – a kind of human-like feature that assumes the price to be proposed is influenced by the current payoff gained and the eagerness of the adopted attitude. 2) The Real-time Adaptive Attitude Switching Strategy (RAASS). RAASS bases on the fact that if only one strategy is adopted during bargaining, the price changing rate will be bounded by that attitude and the amount of payoff that the buying agent can gain is also limited. By adopting RAASS, the amount of concession to be allowed and the payoff gained by the buying agent can be flexibly reduced and maximized respectively.

R. Khosla et al. (Eds.): KES 2005, LNAI 3681, pp. 678–685, 2005.
© Springer-Verlag Berlin Heidelberg 2005

2 Adaptive Bargaining Model (ABM)

A. Modeling the Bargaining Process

i) For modeling the behaviour of the both the selling agent (to decrease its price) and the buying agent (to increase its price) adaptively, we derived the following equations.

The functions for the selling agent to decrease its price and the buying agent to increase its price are given by:

$$p_{seller}(t) = (p_{seller}^{max} - p_{seller}^{min})e^{-st} + p_{seller}^{min} \tag{1}$$

$$p_{buyer}(t) = p_{buyer}^{max} - (p_{buyer}^{max} - p_{buyer}^{min})e^{-bt} \tag{2}$$

ii) For modeling the calculation of payoff gained [1].

The functions for calculating payoff ρ gained by buying agent and selling agent at time t are given by:

$$\rho_{buyer} = \frac{p_{buyer}^{max} - p_{seller}(t)}{p_{buyer}^{max} - p_{buyer}^{min}} \tag{3}$$

$$\rho_{seller} = \frac{p_{buyer}(t) - p_{seller}^{min}}{p_{seller}^{max} - p_{seller}^{min}} \tag{4}$$

iii) For modeling the price changing rate.

The function for the price changing rate:

$$r_{rate} = \alpha * \rho \tag{5}$$

The price increasing (decreasing) rate of the buying agent (selling agent) is determined by the eagerness value of the adopted attitude α and the payoff ρ gained at time t. Fig. 1 shows the internal view of the process:

(I) The buying agent determines the price increasing rate (eqt. 5) with respect to the eagerness value of the current adopted attitude (eqt. 6 - 8) and the current payoff gained (eqt. 3).

(II) Then the buying agent proposes a price (eqt. 2) to the selling agent.

(III) The selling agent calculates the payoff gained (eqt. 4) from this proposed price and determines the eagerness value of the adopted attitude (eqt. 6 - 8). Then the price decreasing rate is determined (eqt. 5).

(IV) After that the selling agent proposes a price (eqt. 1) to the buying agent.

The bargaining process run under this influencing process (by repeating I to IV) until the price is compromised.

B. Payoff-Oriented Strategy (POS)

For modeling the human behavior during bargaining, the properties of the three attitudes are defined. 1) Aggressive attitude: agent with this attitude allows large concession and tends to complete the deal as soon as possible. 2) Neutral attitude: agent with this attitude allows medium concession and tends to complete the deal moderately. 3) Conservative attitude: agent with this attitude allows small concession and tends to complete the deal slowly. To determine the eagerness value of each attitude (the Attitude Factor), the current payoff gained is calculated. The idea of the POS strategy is that the eagerness of the adopted attitude is always changing according to the current payoff gained. Fig. 2 shows the relationship between the attitude factor and the current payoff.

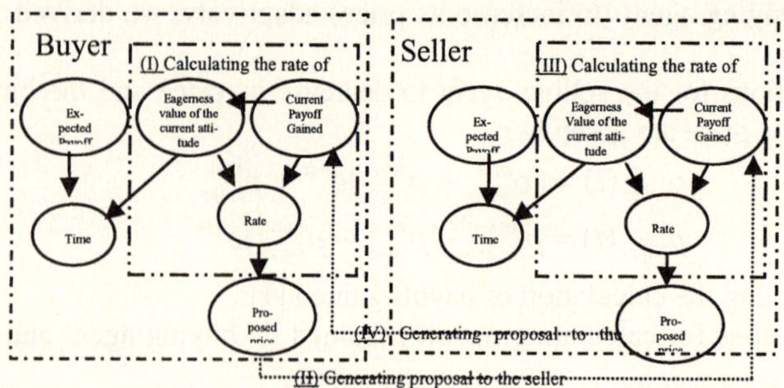

Fig. 1. The Chain of Influences of calculating the rate of concession and generating proposal

The function for calculating the attitude factor α of each attitude:

$$\alpha_{aggressive} = 0.3921 * e^{-2.3041\,\rho} \tag{6}$$

$$\alpha_{neutral} = 0.2548 * e^{-2.3041\,\rho} \tag{7}$$

$$\alpha_{conservative} = 0.1176 * e^{-2.3041\,\rho} \tag{8}$$

For deriving the attitude factor formulae, some experiments were carried out. The experimental setting is described in section 3 – part A.

C. Real-Time Adaptive Attitude Switching Strategy (RAASS)

Since the price changing rate depends both on the attitude factor and the current payoff gained (eqt. 5), in order to encourage the selling agent to decrease the price faster and to prohibit the selling agent from gaining too much payoff, the price increasing rate of the buying agent should be slowed down as it can be seen from eqt. 2 that the higher the value of the price increasing rate, the higher the price is proposed by the buying agent. Once the price increasing rate of the buying agent is slowed down, the price proposed by the buying agent and the amount of concession will then be reduced so the payoff gained by the selling agent will also be reduced. If the payoff gained by the selling agent is reduced, the attitude factor of the selling agent will also be reduced. Then, the price decreasing rate of the selling agent increases and the price proposed by the selling agent becomes lower so that the buying agent can gain higher payoff. In order to slow down the price increasing rate of the buying agent, attitude is switched by following a set of rules stated in Table 1. By comparing the amount of concession ($\delta_{price}^{buyer} > \delta_{price}^{seller}$) allowed by the buying agent at time t and $t+1$ with that of the selling agent, the price decreasing rate of the selling agent can be estimated.

The function to calculate the amount of concession between time t and $t+1$

$$\delta_{price}^{seller} = p_{seller}(t) - p_{seller}(t+1) \tag{9}$$

$$\delta_{price}^{buyer} = p_{buyer}(t+1) - p_{buyer}(t) \tag{10}$$

where δ_{price}^{buyer} (δ_{price}^{seller}) is the difference between the price proposed by the buying agent (selling agent) at time t and $t+1$.

Table 1. The attitude switching rules

Rules	Initial Strategy	Attitude Switching Sequence
Rule 1:	Aggressive	Aggressive → Neutral → Conservative
Rule 2:	Neutral	Neutral → Conservative
Rule 3:	Conservative	Conservative → remain unchanged
Rule 4:	Any	If the current attitude is conservative, then return to its initial attitude.

From Fig 3, it can be seen that the price changing rate of the aggressive attitude is higher than that of the neutral attitude and that of the neutral attitude is higher than that of the conservative attitude. Therefore, the switching sequence should be set in this order - aggressive → neutral → conservative. The Rule 4 controls the agent to return to its initial attitude so that the agent will not follow the properties of that attitude before the price is compromised. From Fig. 4, it can be seen that this property has been proven by experiment.

3   Experimental Results and Evaluation

To illustrate the two human-like features of ABM, a series of experiments were carried out. The experiment is divided into two major sections - i) Observation and Analysis of the Payoff-Oriented Strategy (POS) and ii) Observation and Analysis of the Real-time Adaptive Attitude Switching Strategy.

A. Experimental Setting for Deriving the Relationship Between Attitude Factor and Current Payoff

The experiment described in this section is divided into ten sets of sub-experiments and eqt. 1 is used to model the price changing pattern. In each sub-experiment, ten payoff values (ie. 0.1, 0.2…etc) are used to test against ten price changing rate (ie. 0.01, 0.015, 0.02…etc). Also, the p_{seller}^{min} =5000 and the $p_{seller}^{max} = p_{seller}^{min}(1+x)$ are tested, where {x=10%, 20%...until 100%}. Therefore, totally 100 testing experiments are carried out. In each set of sub-experiment, the first three fastest decay curves are classified into Aggressive attitude, the last three slowest decay curves are classified into Conservative attitude and the four remaining curves are classified into Neutral attitude. Then the attitude factor for a specific attitude is then taken as the average value from the decay values in that patch of curves. From eqt. 5, the value of the attitude factor can be calculated. After carrying out the 10 set of sub-experiments, the ten payoff values are plotted against these attitude factor values. Then the general mathematical model for representing the three human-like attitudes is derived (Fig. 2).

B. Experimental Setting for Evaluating the POS and RAASS Strategy

The individual settings of the buying agent and the selling agent are fixed for every bargaining experiment. One bargaining experiment is defined that totally nine pair of attitudes were tested (Table 2). For the ease of illustrating the bargaining process, long duration and high expected payoff are used.

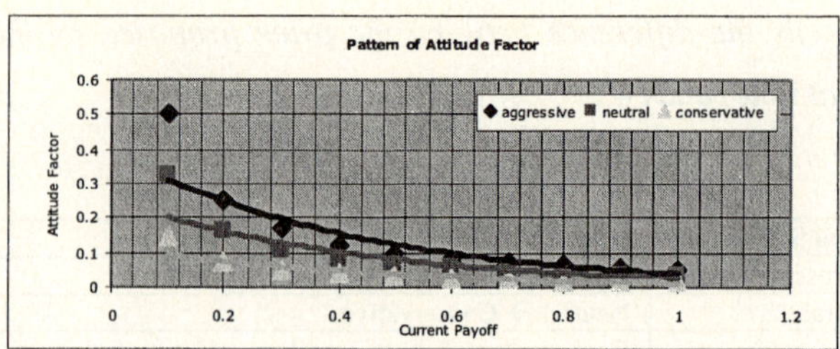

Fig. 2. The relationship between the Attitude Factor and the Current Payoff

- Buying agent $\{p_{buyer}^{max} = 10000, p_{buyer}^{min} = 5000, d = 1000, \varepsilon = 0.97\}$

- Selling agent $\{p_{seller}^{max} = 9000, p_{seller}^{min} = 3000, d = 1000, \varepsilon = 0.97\}$

  (Where d is the expected duration.)

For the experiment of the Payoff-Oriented Strategy, the buying agent and the selling agent adopt one of the three attitudes (aggressive, neutral and conservative) to start the bargaining process. For the experiment of the Real-time Adaptive Attitude Switching Strategy, the buying agent adopts RAASS while the selling agent adopts one of the three attitudes.

Table 2. The record of the bargaining time and the deal price

Buying agent: ε =0.97 Selling agent: ε =0.97			Buying agent					
			aggressive	Time (sec)	neutral	Time (sec)	conserva-tive	Time (sec)
Selling agent	aggressive	Selling agent proposed price	6984.977 [6416.811]	-	6484.098 [6211.125]	10	5835.964	13
		Buying agent proposed price	6879.417 {5701.26}	8	6365.822 {5746.26}	-	5792.19	-
	neutral	Selling agent proposed price	7340.725 [7000.151]	-	6954.995 [6696.119]	-	6044.861	17
		Buying agent proposed price	7289.567 {5825.795}	10	6836.367 {5939.064}	12	6151.303	-
	conserva-tive	Selling agent proposed price	7940.851 [7636.674]	-	7594.458 [7408.268]	-	6827.839	27
		Buying agent proposed price	7916.252 {6154.479}	14	7586.273 {6352.878}	18	6810.881	-

i) Observation and Analysis of the POS Strategy

In this experiment, the properties of the three attitudes are evaluated and illustrated in three different perspectives a) From the perspective of the price decreasing pattern, b) From the perspective of the rate-changing pattern.

a) From the Perspective of the Price Decreasing Pattern

Observation 1: From Fig 4, it can be seen that the pattern of the concession δ_{price}^{seller} reflects the characteristics of the selling agent and the buying agent during the bargaining process. Although the selling agent decreases the price and reduces the amount of concession gradually, there are some differences among the three attitudes. It can also be seen that the attitude adopted by the selling agent can be estimated by checking the initial concession range.

1. $350 < \delta_{price}^{seller} \leq 400$, the attitude is estimated as aggressive.

2. $200 < \delta_{price}^{seller} \leq 250$, the attitude is estimated as neutral.

3. $100 < \delta_{price}^{seller} \leq 120$, the attitude is estimated as conservative.

From Fig. 4 and Table 2, it can be seen that the time spent by the three attitudes is ranked as *Conservative > Neutral > Aggressive*. Also, the order of the ability to gain high payoff is ranked as *Conservative > Neutral > Aggressive*. This also shows the second property of the three defined attitudes.

b) From the Perspective of the Rate-Changing Pattern

Observation 2: From Fig 3, it can be seen that the rate pattern is influenced by both of the attitude adopted by the selling agent and the buying agent. By focusing on the impact of the buying agent's attitude, it can be seen that the rate of the aggressive attitude is the fastest while the conservative attitude is the slowest. The faster the rate, the less time is spent to complete the deal. From Table 2, it can be seen that the ability of the conservative attitude to gain lowest price (high payoff) is the highest while that of the aggressive attitude is the lowest. These also concretize the definition of the three attitudes.

ii) Observation and Analysis of the RAASS

In this experiment, the RAASS is evaluated and illustrated in two different perspectives - a) From the perspective of the price decreasing pattern and b) From the perspective of the rate changes pattern.

a) From the Perspective of the Price Decreasing Pattern

Observation 1: From Fig 4, it can be seen that the RAASS Enabled Aggressive Buying Agent (named as "buyer (smart-agg)") gain more payoff than that of the simple aggressive attitude. The buying agent switches the aggressive attitude to conservative attitude and back to initial attitude to complete the deal. This reflects the second property of the RAASS that it encourages the selling agent to allow more concession so that the buying agent can gain higher payoff. By comparing those prices recorded in table 2, the percentage gained by the RAASS enabled buying agent is shown in table 3.

b) From the Perspective of the Rate-Changing Pattern

Observation 2: From Fig. 3, it can be seen that the RAASS enables buying agent (named as "buyer (smart-agg)") to slow down the rate by switching the aggressive attitude to conservative attitude. This reflects that the change of attitude results in the change of the price increasing rate. This concretizes the concept of the influencing process in Fig 1.

4   Conclusions

In this paper we propose a highly adaptive bargaining model (ABM) for agent shopping. The contribution of this research is that ABM stimulates two major human bargaining strategies: the Payoff-Oriented Strategy (POS) and the Real-time Adaptive Attitude Switching Strategy (RAASS). In summary, the POS is the foundation of this bargaining model which models the changing eagerness of the buyer (seller) in real situation while the RAASS is the enhancement strategy, it adjusts the rate of concession by switching the attitude.

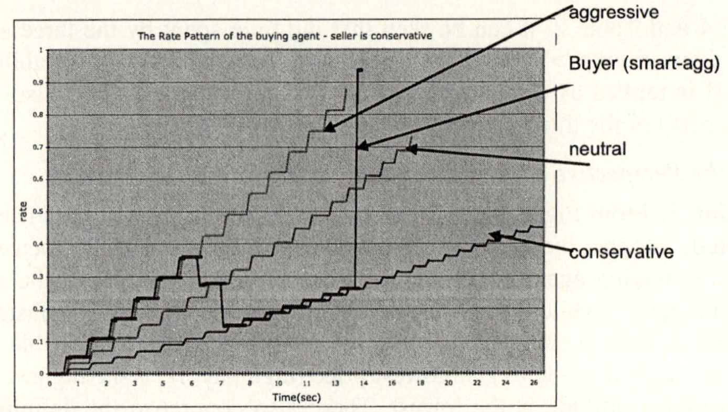

Fig. 3. The rate pattern of the buying agent

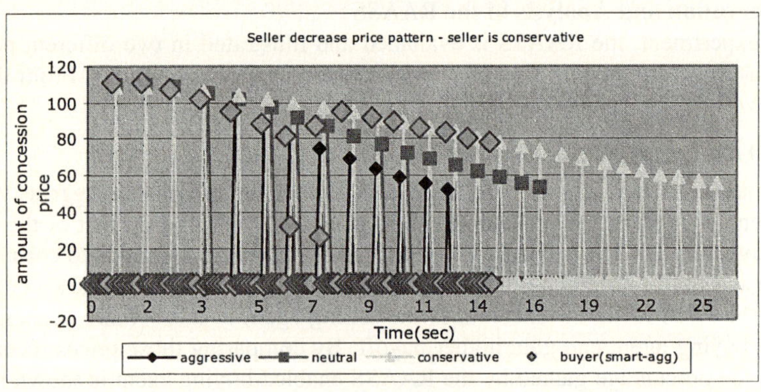

Fig. 4. The concession pattern δ_{price}^{seller} of the conservative selling agent

Table 3. The percentage gained by the RAASS strategy enabled buying agent

		Buying agent
	Selling agent	buyer(smart-agg)
Percent-age Gained	Aggressive	6.72%
	Neutral	3.97%
	Conservative	3.53%

Acknowledgment

This work was partially supported by the iJADE projects B-Q569, A-PF74 and Cogito iJADE project PG50 of the Hong Kong Polytechnic University.

References

1. Kwang Mong Sim and Shi Yu Wang, "Flexible Negotiation Agent With RelaxedDecision Rules," IEEE Trans. Syst., Man, Cybern., B, vol. 34, No. 3, June 2004, p1602-1608.
2. Kwang Mong Sim and Chung Yu Choi, "Agents That React to Changing Market Situations," IEEE Trans. Syst., Man, Cybern., B, vol. 33, No. 2, April 2003, p188-201.
3. Kwang Mong Sim and Eric Wong, "Toward Market-driven Agents for Electronic Auction," IEEE Trans. Syst., Man, Cybern., A, vol. 31, No. 6, November 2001, p474-484.
4. Raymond S.T.Lee and James N.K.Liu, " iJADE Web-Miner: An Intelligent Agent Framework for Internet Shopping", IEEE Trans. On Knowledge and Data Eng., pp.461-473, April 2004.
5. Raymond S.T.Lee and James N.K.Liu, "iJADE IWShopper – A New Age of Intelligent Mobile Web Shopping System Based on Fuzzy-Neuro Agent Technology" Web Intelligence Research and Development (LNAI 2198), pp. 430-412, 2001.
6. Raymond Y.K.Lau, Maolin Tang, and On Wong, "Towards Genetically Optimised Responsive Negotiation Agents", Proceedings of the IEEE/WIC/ACM Inter. Conf. on IAT'04, p295-301

Behavior-Based Blind Goal-Oriented Robot Navigation by Fuzzy Logic

Meng Wang and James N.K. Liu

Department of Computing, The Hong Kong Polytechnic University, Hung Hom, Hong Kong
`{csmwang,csnkliu}@comp.polyu.edu.hk`

Abstract. This paper proposes a new so-called "minimum risk" method, to address the local minimum problem that has to be faced for the goal-oriented robot navigation in an unknown environment. This method is theoretically proved to guarantee the global convergence even in the long-wall, large concave, recursive U-shape, unstructured, cluttered, and maze-like environments. The minimum risk method adopts a strategy of multiple behaviors coordination, in which a novel path-searching behavior is developed by fuzzy logic to recommend the regional direction with minimum risk. This behavior is one of the applications of the proposed memory grid technique. The proposed method is verified by the simulation and real world tests.

1 Introduction

We call it *"blind goal-oriented navigation"* that the robot is required to autonomously reach a desired goal but it does not have a priori known environmental knowledge. Some approaches, e.g. machine learning, neural-fuzzy approach [9], do not guarantee global convergence to the goal because they may get trapped in local minima (or dead ends) of the environment. The local minimum problem has been discussed in the literatures [1-8], some of which call the local minimum problem as deadlock, dead end or limit-cycle problem. We categorize these methods into two types: boundary following type, and virtual subgoal type. These methods have difficulties to guarantee global convergence in the complex environment. For example, the boundary-following strategy is easier to make the robot to be trapped in a wrong wall-following direction when the goal is always at the side of the wall. The unstructured and cluttered environments make the method that recognizes the typical landmarks fail. The recursive U-shape or maze-like environments may cause the robot to regress into the old local minimum.

This paper proposes a new method, minimum risk method, to address the local minimum problem during the blind goal-oriented robot navigation. This method adopts a strategy of multi-behavior coordination, in which a novel Path-Searching (PS) behavior is developed to recommend the regional direction with minimum risk. This paper provides a fuzzy logic framework to implement the behavior design and coordination. Thus errors due to sensor noise and self-localization are effectively handled by our navigation system.

2 Behavior-Based Navigation Strategy

The robot navigation strategy proposed in this paper is comprised of three simple behaviors. These behaviors operate at three different ranges, with the goal-seeking

R. Khosla et al. (Eds.): KES 2005, LNAI 3681, pp. 686–692, 2005.

(GS) behavior at *global*, the path-searching (PS) behavior at *regional* and the obstacle-avoidance (OA) behavior at *local* ranges (see Fig.2). Our behavior coordination strategy is similar but simpler than that of the CDB (context-dependent blending) approach [12]. The CDB approach uses a fuzzy preference combination to carry out command fusion, but we first use a behavior arbitration module to compute all of the defuzzificated weight factors of all behaviors, and then carry out command fusion directly using these weight factors by the equation (1).

$$ v = \frac{\sum v_i \cdot w_i}{\sum w_i} \ , \quad \theta = \frac{\sum \theta_i \cdot w_i}{\sum w_i} \tag{1} $$

where, $i = \{1, 2, 3, \ldots\}$, v and θ are respectively the desired final speed and the delta turn angle values while v_i and θ_i are respectively the speed and turn angle values recommended by each individual behavior. w_i is the defuzzificated weight factors that are output by the behavior arbitration module using fuzzy logic.

As long as the robot is located in a certain situation, (e.g. the robot is close to obstacles), OA, GS, and PS behaviors will at the same time be partially activated. Each behavior is assigned a weighting factor, and these factors are updated continuously and dynamically according to the fuzzy weight rules. One typical example of weight rules is shown as the following: *IF the distance to the front obstacle is VERY NEAR, THEN the weight of OA behavior is HIGH AND the weight of GS behavior is LOW.*

3 Regional Path Searching Behavior

The Path-Searching (PS) behavior is one of the applications of our memory grid techniques. The memory grid is a new type of map to model the robot environment. The main difference from the other maps (e.g. occupancy grid, certainty grid, etc.[10]) is our obstacle and trajectory memory dot model. The memory grid uses the obstacle memory dots to quantify the fuzzy possibility that represents the uncertainty of obstacles detected by sonar sensors, while it uses the trajectory memory dots to save the trajectory traversed by robot. The longer the time for robot to traverse a grid region, the more is the number of the trajectory memory dots saved in this grid. Therefore, the regional direction with minimum obstacle and trajectory memory dots can represent the regional direction with minimum risk. This direction can guide robot trying the best to avoid the previous trajectory, thus choose a new exploration region to escape from local minima and reach the goal. The PS behavior does the job to recommend the direction with minimum risk. Four features are extracted from memory grid data, in which the *iteration risk* α and *collision risk* β are used to infer the Risk Index for the fuzzy navigational rules of PS behavior, and *trajectory dot intensity* κ and *obstacle dot intensity* τ are used to calculate the weight of PS behavior with fuzzy logic.

A) Regional Risk Index

The Fuzzy Rule-Based Risk Index combines two regional risk parameters into a single indicator of safety of traversal of region by mobile robot. The Risk Index r is represented by three linguistic fuzzy sets {DANGEROUS, UNCERTAIN, SAFE}, with the membership functions shown in Fig. 1(a). The Risk Index is defined in terms of both iteration risk and collision risk by a set of intuitive fuzzy logic relations as follows:

1) IF α is HIGH **OR** β is HIGH, THEN r is DANGEROUS.

2) IF α is MEDIUM AND β is MEDIUM, THEN r is DANGEROUS.

3) IF α is MEDIUM AND β is LOW, THEN r is UNCERTAIN.

4) IF α is LOW AND β is MEDIUM, THEN r is UNCERTAIN.

5) IF α is LOW AND β is LOW, THEN r is SAFE.

B) Turn Rules and a Complement Algorithm

The motion control variables of the mobile robot are the translational speed and the rotational turn angle. The robot speed is represented by three linguistic fuzzy sets {STOP, SLOW, FAST}. Similarly, the robot turn angle is represented by five linguistic fuzzy sets {NB, NS, ZE, PS, PB}, where NB is negative-big, NS negative-small, ZE zero, PS positive-small, and PB positive-big. Positive and negative mean that the robot is turned to left and right directions, respectively.

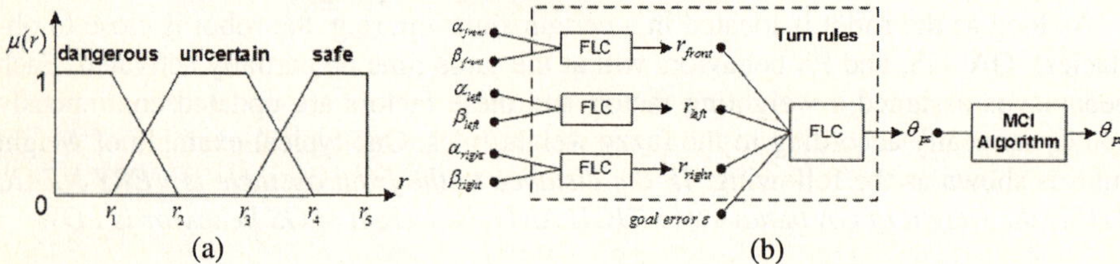

(a)   (b)

Fig. 1. (a) Membership functions for Risk Index; (b) Turn angle determination of PS behavior by both Turn Rules and MCI Algorithm

The Risk Index is used to develop simple fuzzy rules for determination of the robot turn angle as shown in Fig.1(b). FLC implies the fuzzy logic controller. We use a Mamdani model as a fuzzy inference engine, and the Centroid or Center of Gravity (COG) method is used for defuzzification [11]. For complexity reduction, we assume that the speed of the robot is only influenced by the OA behavior, not by the PS or GS behavior. As shown in Fig.2(a), the region available for robot traversal is divided up into three circular side sectors. The radius of the circular sector is a user-defined *regional perception range* of the robot. The Risk Indices for the three regions, r_{left}, r_{front} and r_{right}, are inferred by the Risk Index rules.

The turn rules for the PS behavior are summarized in Fig.2(b). Notice that in Fig.2(b), when the robot needs to turn, but the left and right sectors have the same Risk Indices, then the recommended turn angle is GOAL, where GOAL implies that the recommended turn angle should be toward the direction close to the goal location. Another note, a turn maneuver is not initiated when the three sectors have the same *dangerous* risk indices as shown in the (1,1) element of the top layer in Fig.2(b). The turn rule does not force the robot to choose between left and right directions arbitrarily at this stage, but keeping the turn angle at zero. The final selection will be made by a complement algorithm introduced next.

The kernel idea of the complement algorithm, Minimum Collision & Iteration (MCI), is: if the turn direction recommended by the turn rules does not have LOW iteration risk (by threshold comparison), the collision and iteration risk of all three sectors are compared again (by threshold comparison) to recommend a regional direction that has real minimum collision and iteration risk values.

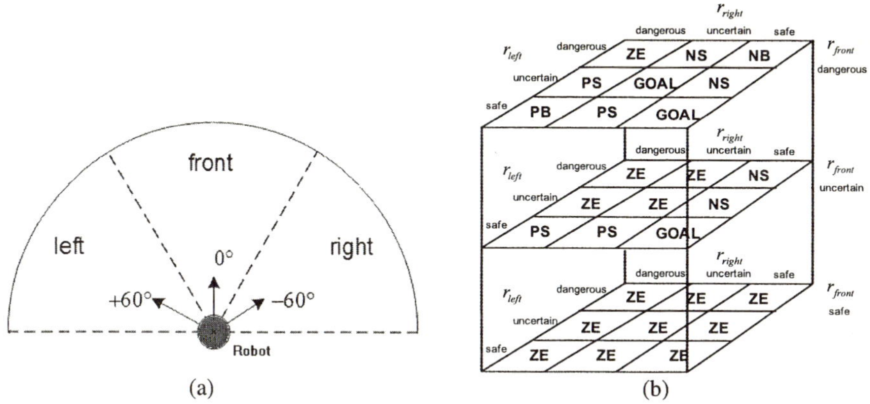

Fig. 2. (a) Three sectors of regions (b) Turn rules for path-searching behavior

C) Weight Rules

The weighting factor w_{ps} represents the strength by which the PS behavior recommendation is taken into account to compute the final motion command. The weight of PS behavior is represented by three linguistic fuzzy sets {SMALL, MEDIUM, LARGE}, and is derived directly from both the trajectory dot intensity κ and obstacle dot intensity τ of a square region, using the following rule set:

1) IF κ is BIG **OR** τ is BIG, THEN w_{ps} is LARGE.
2) IF κ is MEDIUM AND τ is MEDIUM, THEN w_{ps} is LARGE.
3) IF κ is MEDIUM AND τ is SMALL, THEN w_{ps} is MEDIUM.
4) IF κ is SMALL AND τ is MEDIUM, THEN w_{ps} is MEDIUM.
5) IF κ is SMALL AND τ is SMALL, THEN w_{ps} is SMALL.

4 Performance Analysis in Theory

Theorem 1: If there exists a solution path for a blind goal-oriented navigation task, the minimum risk method can guarantee the global convergence.

Prove: Limited by the page numbers, we briefly gave the proof analysis. If there exists a solution path for a blind goal-oriented navigation task, this solution path must have minimum obstacle and trajectory memory dots, i.e. the minimum risk. And the detected regions of the current robot position must have a regional direction that has a minimum risk. The minimum risk method can guarantee to finally recommend the region direction with minimum risk. Thus the solution path must be found and the robot finally reaches the goal. That's global convergence.   ∎

The minimum risk method produces the "trial-and-return" behavior phenomenon as shown in Fig.3(a) and (b). Obviously, this is the Theorem 2.

Theorem 2: If there is not obstacle to block the nearest exit, the minimum risk method must guarantee to find a nearest exit to escape from the local minimum and reach the goal finally.   ∎

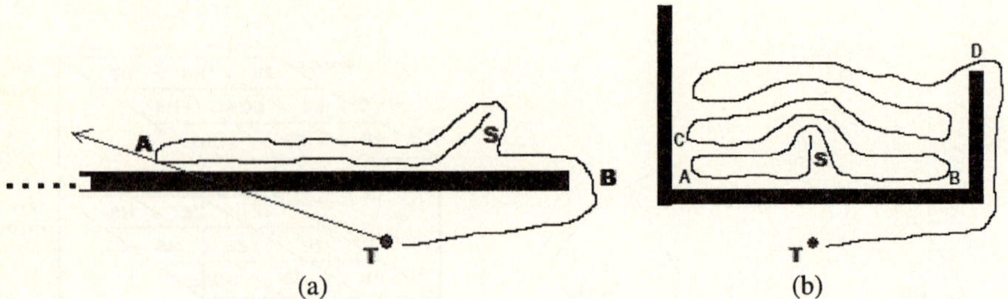

Fig. 3. "Trial-and-return" behavior phenomenon. S is the start of the robot. T is the goal

5   Experimental Results and Comparisons

Our method is tested on a mobile robot (vehicle). The robot has a ring of eight forward ultrasonic sonars that can detect obstacles within three meters. We verify the performance of our minimum risk method that is applied in an unknown long-wall environment with local minimum. The result is shown in Fig.4(a). The "trial-and-return" property enables the robot to be never trapped in a wrong boundary-following direction. That's why the robot leaves the wall at location D and turn toward location E. Fig.4(b) shows the underlying memory grid that is drawn as the spaced horizontal and vertical lines. Fig.4(c) and (d) show the results in the unstructured and cluttered, maze-like environment respectively. More results can be checked on website [13].

We compare our minimum risk method with other related methods. All of them are applied in a concave and recursive U-shape environment as shown in Fig.5. The virtual target method [8] fails to reach the goal as shown in Fig.5(a), because the robot encounters another local minimum at the location "b" and "c" when it is working under the influence of previous virtual subgoal. Krishna and Kalra's method [4] has a good result as shown in Fig.5(b). But it highly depends on the landmark recognition and exact coordination localization. In addition, it is difficult to choose a correct direction to follow the wall boundary as seen in Fig.5(e). Maaref and Barret's method [5] fails to reach the goal in this large concave environment, because it detects the local minimum using a restricted criterion that all sensors must give the small obstacle distances at the same time. Fig.5(c) shows the result of our minimum risk method. The robot exhibits the typical "trial-and-return" behavior phenomenon. This property can help the robot to find the nearest exit to escape from the local minimum and guarantee global convergence. It's further verified in Fig.5(f). Fig.5(d) shows the result of Huang and Lee's method [1]. This method has a conservative leaving criterion that makes the robot traverse a longer path compared with other methods that adopt the boundary-following strategy. More importantly, it is still difficult to choose the correct boundary-following direction. The similar problems occur on the Distbug method [2, 3] and Virtual-target-side method [6]. The main difference among them is that they have different detecting and leaving criteria. Most of related methods [1-8] adopt the analytical models for detecting and leaving criteria, which are not suitable to deal with the uncertainties from sensors and real world, especially from the odometry drift problem.

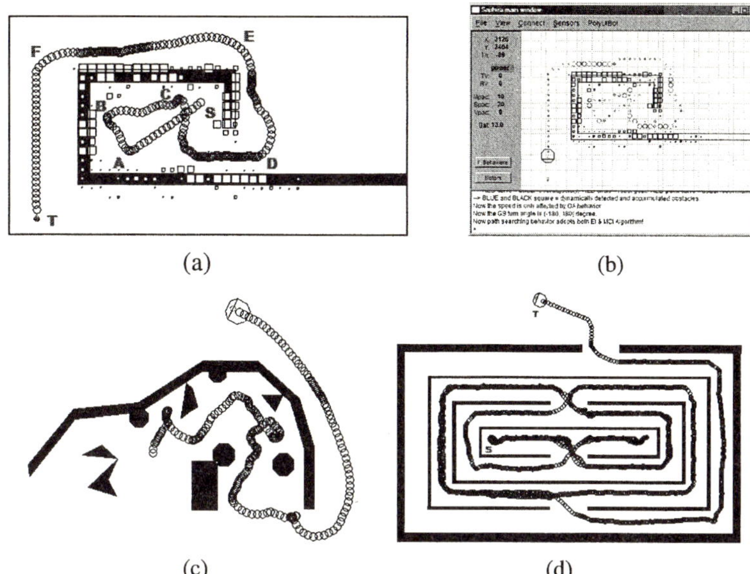

(a)

(b)

(c)

(d)

Fig. 4. Minimum risk approach to address the local minimum problem. (a) in long-wall environment. (b) Underlying memory grid in the control interface. (c) in unstructured and cluttered environment. (d) in maze-like environment

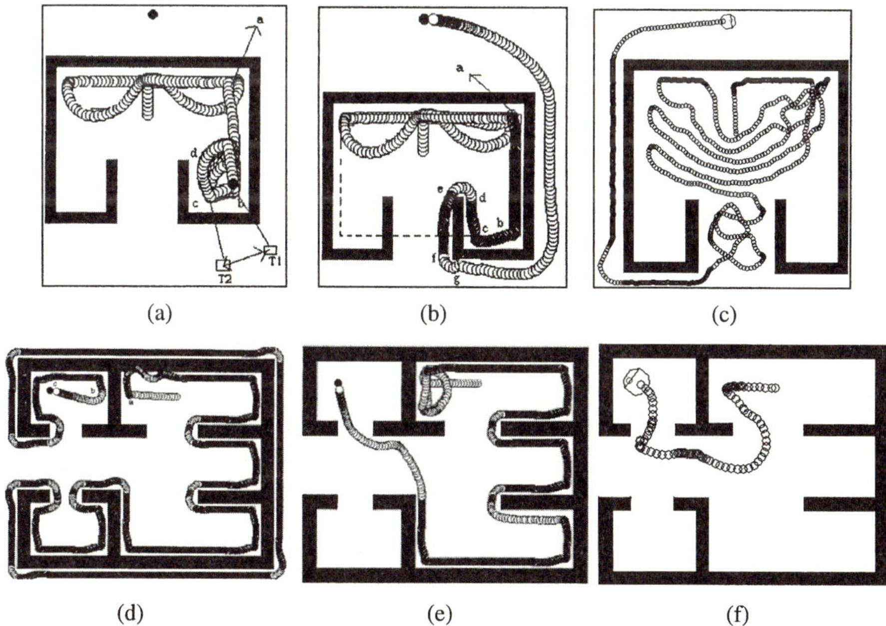

(a)

(b)

(c)

(d)

(e)

(f)

Fig. 5. (a) virtual target method. (b) Krishna and Kalra's method. (c) our minimum risk method. (d) Huang and Lee's method. (e) Krishna and Kalra's method. (f) our minimum risk method

6 Conclusion

This paper proposes a so-called minimum risk method to address the local minimum problem during the blind goal-oriented robot navigation. This method is theoretically proved to guarantee global convergence even in the long-wall, large concave, recursive U-shape, unstructured, cluttered, maze-like environments. A new path-searching behavior is developed to recommend the regional direction with minimum risk. The fuzzy logic framework outperforms the analytical method of the detecting and leaving criterion adopted by other methods.

Acknowledgement

The authors would like to acknowledge the support of the Hong Kong Polytechnic University via RGC grant B-Q515 and departmental grant H-Z87.

References

1. H.P. Huang, P.C. Lee, "A real-time algorithm for obstacle avoidance of autonomous mobile robots", Robotica, Vol.10, (1992) 217-227
2. I.Kamon, E.Rivlin, "Sensory-based motion planning with global proofs", IEEE Transactions on Robotics and Automation, Vol.13, Iss.6, (1997) 814-822
3. J.H.Lim, D.W.Cho, "Sonar based systematic exploration method for an autonomous mobile robot operating in an unknown environment", Robotica, Vol.16, (1998) 659-667
4. K.M. Krishna, P.K. Kalra, "Perception and remembrance of the environment during real-time navigation of a mobile robot", Robotics and Autonomous Systems, vol.37, (2001) 25-51
5. H.Maaref, C.Barret, "Sensor-based navigation of a mobile robot in an indoor environment", Robotics and Autonomous Systems, Vol.38, (2002) 1-18
6. R.Chatterjee, F.Matsuno, "Use of single side reflex for autonomous navigation of mobile robots in unknown environments", Robotics and Autonomous Systems, Vol.35, (2001) 77-96
7. F.G. Pin, S.R. Bender, "Adding memory processing behavior to the fuzzy behaviorist approach: Resolving limit cycle problems in mobile robot navigation", Intelligent Automation and Soft Computing, Vol.5, Iss.1, (1999) 31-41
8. W.L.Xu, "A virtual target approach for resolving the limit cycle problem in navigation of a fuzzy behavior-based mobile robot", Robotics and Autonomous Systems, Vol.30, Iss.4, (2000) 315-324
9. J. Godjevac, N.Steele, "Neuro fuzzy control for basic mobile robot behaviours", Fuzzy Logic Techniques for Autonomous Vehicle Navigation, Physica-Verlag, New York, (2000) 98-118
10. G.Oriolo, G.Ulivi, M.Vendittelli, "Real-time map building and navigation for autonomous robots in unknown environments", IEEE Trans. on SMC, Part B, Vol.28, No.3, (1998) 316-333
11. Meng Wang, James N.K. Liu, "Autonomous Robot Navigation using Fuzzy Logic Controller", IEEE Conference on Machine Learning and Cybernetics, Shanghai, China, (2004) 691-696
12. H.Seraji, A.Howard, "Behavior-based robot navigation on challenging terrain: A fuzzy logic approach", *IEEE Transactions on Robotics and Automation*, Vol.18, Iss.3, (2002) 308-321
13. http://www4.comp.polyu.edu.hk/~csnkliu/polyuibot/

*i*JADE Reporter – An Intelligent Multi-agent Based Context – Aware News Reporting System

Eddie C.L. Chan and Raymond S.T. Lee

The Department of Computing, The Hong Kong Polytechnic University,
Hung Hong, Kowloon, Hong Kong
{c1241986,csstlee}@comp.polyu.edu.hk

Abstract. In this paper, an Intelligent Context – Aware News Reporting System called iJADE Reporter is presented. This system focuses on how context mining techniques are applied on news reporting under a multi-agent architecture, and categorize news content by information retrieval algorithm. This paper also investigates how to improve the similarity measurement between documents by ontology with WordNet graph structure of words. In a web querying case, a common information retrieval algorithm, Term Frequency with Inverse Document Frequency (TFIDF) is used to cluster news contents. The proposed system provides a simple, fast and efficient query in WWW. The proposed system makes use of multi-agent technology to increase the scalability and efficiency of the system. By using TFIDF algorithm and multi-agent based techniques, an online updating news reporter from popular News Website, such as BBC & CNN News Website is developed.

1 Introduction

Retrieving, categorizing and reporting useful news from the web is one of the most challenging problems in machine learning. The common interest among researchers working in diverse fields is motivated by our remarkable innate ability to study and to report news in daily life. The current search engines provided by Google or Yahoo! on news retrieval do not have a logical categorization which is difficult for reading. In this paper, an intelligent multi-agent based context–aware news reporting agents system called iJADE Reporter is presented. For system implementation, iJADE [8] (intelligent Java Agent Development Environment) is adopted to provide an intelligent agent-based platform for the implementation of various AI functionalities.

2 Web Context Mining (WCM) – An Overview

In web content mining, popular search engines (such as Lycos, WebCrawler, Infoseek, and Alta Vista) provide some basic web searching functionalities. However, they fail to provide concrete and structural information [10]. In recent years, interest has been focused on how to provide a higher level (semantic level) organization for semi-structured or even unstructured information on the Web using AI-based Web mining techniques.

Agent-based systems such as Harvest [2], FAQ-Finder [4], Information Manifold [11], OCCAM [9], and Parasite [12] rely either on pre-specified domain specific in-

R. Khosla et al. (Eds.): KES 2005, LNAI 3681, pp. 693–699, 2005.

formation, or on hard-coded information sources to retrieve and interpret documents. For instance, Harvest system [4] relies on semi-structured documents to improve its ability to extract information. Although it "knows" how to find author and title information in Latex documents and how to strip position information from postscript files, it fails to discover new documents or to learn new document structures. Similarly, FAQ-Finder [6] extracts answers to frequently asked questions (FAQs) from FAQ files available on the web with priori knowledge.

Web page ontology [3] can be defined in different ways depending on the objective of the ontology. Most of the web sources have its semantic meaning. Techniques for Ontology Generation, Ontology Mediation, Ontology Population and Reasoning from the Semantic Web have all been major areas of focus. Most web documents are organized in a content hierarchy, with more general nodes placed closer to the root of hierarchy. Each node is labeled by a set of keywords describing the content of documents that are placed in the node. Each document is described by a one-sentence summary including a hyperlink that points to the actual Web document located somewhere on the Web.

3   *i*JADE Reporter – A System Overview

The iJADE Reporter proposed in this paper consists of 4 types of iJADE agents (figure 1):

1) iJADE Search News Agent
2) iJADE Categorize News Agent
3) iJADE Update News Agent
4) iJADE Report News Agent.

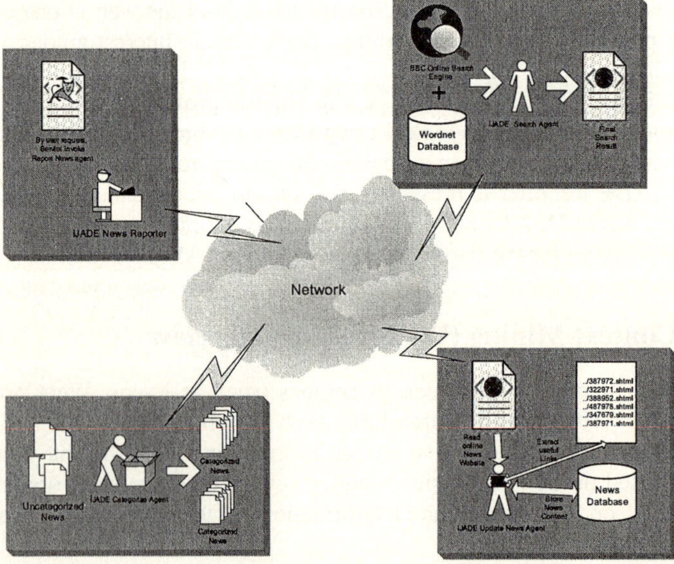

Fig. 1. System Overview of iJADE Reporter

3.1 iJADE Search News Agent

A mobile iJADE agent aims at searching news from popular news websites such as BBC [1] and CNN [3] news websites. It connects to several different popular news search engines; combines the result into search lists and integrates all news search engines using WordNet [14] Dictionary to provide reconstruction and understanding of news query.

3.2 iJADE Categorize News Agent

A stationery iJADE agent aims at categorizing and clustering news into different regions. News categorization is based on calculating the similarity between the web documents by using TFIDF (Term Frequency with Inverse Document Frequency method). TFIDF is a simple but powerful algorithm for machine learning to under-stand semantic document. It exhibits strong characteristics of word frequencies pre-sented in a document. Vector Space Model (VSM) is used for document representa-tion.

3.2.1 Term Frequency with Inverse Document Frequency (TFIDF) Algorithm

TFIDF [13] is an information retrieval algorithm which aims at calculating a specific value of the semantic meaning among words and documents. TFIDF is simple but powerful to express the abstract idea of semantic meaning.

Vector Space Model (VSM) is adopted to represent Web documents. The docu-ments constitute the whole vector space. TFIDF is being used as a weight of term in document.

If a term t occurs in document d,

$$w_{di} = tf_{di} \times \log(N / idf_{di})$$

(1)

where t_i is a word (or a term) in document collection, w_{di} is the weight of t_i, tf_{di} is term frequency (term count of each word in a document) of t_i , N is the number of total documents in the collection and idf_{di} is the number of document in which t_i appears.

3.2.2 Similarity Between Two Documents

Each document d is represented by a vector: $\vec{V}_d = (t_1, w_{d1}; ...; t_i, w_{di}; ...; t_n, w_{dn})$ *where t_i is a word (or a term) in document collection and w_{di} is TFIDF value of t_i in d.*

By calculating the Euclidean distance of two vectors of two documents, the simi-larity can be computed.

$$Sim(d_1, d_2) = \frac{d_1 \bullet d_2}{|d_1||d_2|}$$

(2)

3.2.3 Modified TFIDF by Ontology

In this paper, we proposed ontology-based term frequency (otf) to construct the web ontology on news categorization and retrieval. Assume each word is related to other word, a relationship graph can be constructed as shown in Fig. 3.

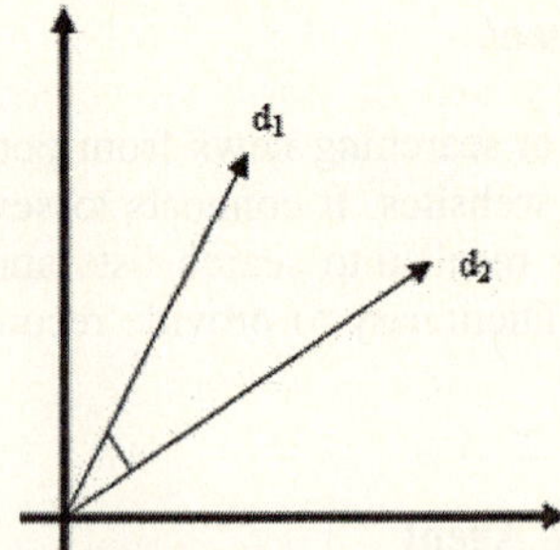

Fig. 2. Euclidean distance of two vectors

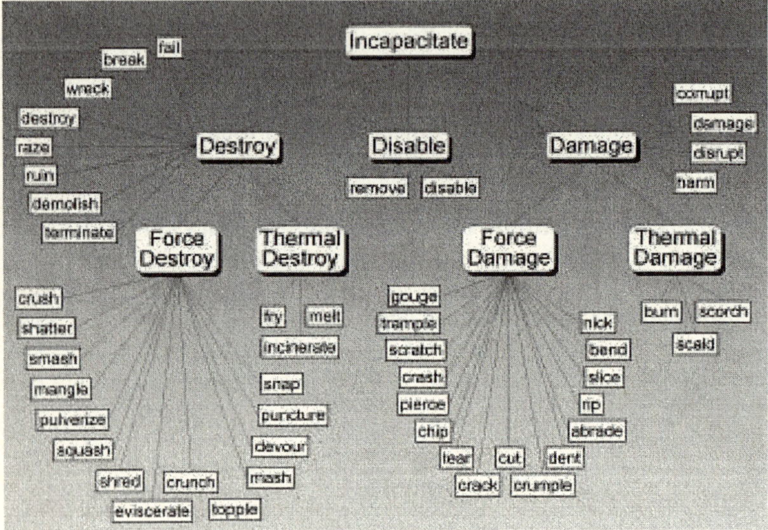

Fig. 3. The word graph example of "Destroy" and "Damage"

In Fig.3., "destroy" and "damage" are similar from the ontology point of view as they are at the same level of the hierarchal structure and have relation between each other. The similarity of two words can be measured by the distance of the tree structure. By comparing the meanings of two terms, the ontology-based term frequency (otf) can be obtained,

$$otf_1 = tf_1 \times (1+(1/D(t_1,t_2)))^{\,tf_2} \tag{3}$$

where t_1, t_2, are different terms; otf_1 is ontology-based term frequency of t_1; tf_1, tf_2 are term frequency respectively to t_1 and t_2; $D(t_1,t_2)$ is the depth between t_1 and t_2. $D(t_1,t_2)$ can be calculated by using WordNet.

For example, assume the term frequencies of "destroy" and "damage" are 3 and 2 respectively and the depth between "destroy" and "damage" is 3. The ontology-based term frequency of "destroy" will be $otf_1 = 3 \times (1+(1/3))^2 = 5.33$. The ontology-based term frequency of "damage" will be $otf_2 = 2 \times (1+(1/3))^3 = 4.67$. After adjusting each term frequency, their term frequency value could be increased, so that two terms will become more significant after computing TFIDF.

3.2.4 News Clustering Technique

In this system, we adopt hierarchical clustering [7] technique to cluster the uncategorized news into the shortest distance of particular news inside same category or region. In hierarchical clustering, there is a set of document $W=\{w1, w2...wi\}$, every document *wi* in W are considered to be a cluster c_i, such that $C=\{c1,c2...ci\}$. Two clusters *ci* and *cj* are randomly chosen and their similarity *sim(ci,cj)* is calculated. They are merged together if the similarity value is greater than the threshold value. Otherwise, this step repeats until reaching the termination condition.

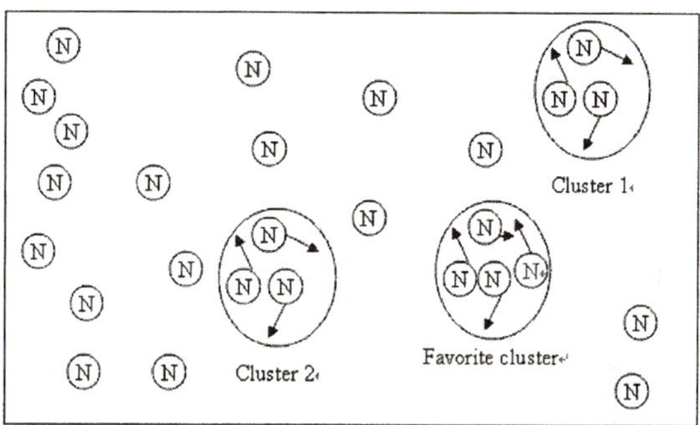

Fig. 4. Cluster News Process

3.3 iJADE Update News Agent

A mobile iJADE agent aims at updating and collecting the news from popular news websites. When user clicks the update button at different categories, news from different news websites can be obtained. Also, the news will be re-categorized and stored in the user local storage. In additional, this agent calculates the preliminary analysis result based on the semantic relationship between words in a document.

The formulation is to simply extract html tag by using web structural mining techniques. The metadata (keywords and title of news) of the html documents is captured, which is used to obtain related news links. After a list of links is given, the news will be further explored to capture related picture links, contents and the term frequency of news is then calculated. In each update process, a region or a category is chosen for information update. Even when the client-side goes offline, the agent will continue to perform its job until the job is finished.

3.4 iJADE Report News Agent

A stationary iJADE agent aims at news reporting. This agent provides a vector list of news with headers, short introduction and content with highlight keywords. Graphics and sound are added to increase the attractiveness in news reporting.

4 Experiments

In this section, the precision rate for news categorization is tested based on the news database consists of over 5000 records. The news articles are subdivided into six categories. They are: Business, Health, Education, Science, Technology and Entertainment. A test set contains over 100 news items without categorization is used to evaluate the performance of the proposed model. To determine whether the categorization or not, human judgment is used.

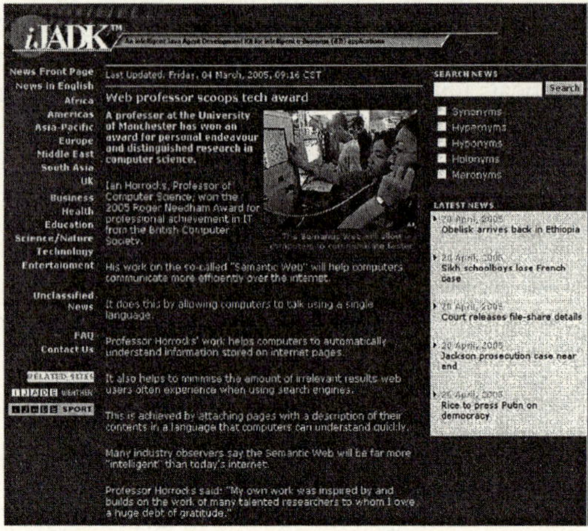

Fig. 5. News Reporting Result

Table 1 and 2 revealed that by using iJADE News Reporter, the precision rate is over 95% with clustering time around 1185 seconds. Compared with other methods, such as FFBP neural network as well as TFIDF with hierarchical clustering technique, iJADE News Reporter gives a better precision rate for categorizing news with reasonable time taken for clustering.

Table 1. Comparison of the Precision Rate

class/category	FFBP NN (3-Layer)	TFIDF+hierarchical clustering	iJADE News Reporter
Business	68.40%	90.30%	95.13%
Health	58.45%	93.37%	96.72%
Education	73.43%	93.88%	95.14%
Science/Nature	67.00%	94.33%	97.44%
Technology	70.22%	93.78%	94.38%
Entertainment	50.23%	91.67%	95.24%
Average	64.62%	93.72%	95.68%

Table 2. Comparison of time taken for clustering 3 different sets of total 100 news

class/category	FFBP NN (3-Layer)	TFIDF+hierarchical Clustering	iJADE News Reporter
Set 1	5277 seconds	1114 seconds	1175 seconds
Set 2	5432 seconds	1123 seconds	1186 seconds
Set 3	5175 seconds	1176 seconds	1195 seconds
Average	5295 seconds	1138 seconds	1185 seconds

5   Conclusion

In this paper, an intelligent agent-based context–aware news reporting system – iJADE Reporter is proposed. Experiments show that iJADE Reporter provides both an effective and efficient solution for news categorization. By integrating with different popular news websites (e.g. BBC, CNN news) to collect, categorize and analysis news, it provides a convenient and semantic-based news retrieval and reporting solution.

Acknowledgment

This work was partially supported by the iJADE projects B-Q569, A-PF74 and Cogito iJADE project PG50 of the Hong Kong Polytechnic University.

References

1. BBC News Website, http://www.bbc.co.uk
2. C. M. Brown and B. B. Danzig, "The harvest information discovery and access system," In Proc. 2nd International World Wide Web Conference, pp. 182-187, 1994.
3. CNN News Website, http://www.cnn.com
4. R. B. Doorenbos, O. Etzioni and D. S. weld, "A scalable comparison shopping agent for the world wide web," Technical Report TR 97-027, University of Minnesiota, 1997.
5. Etzioni, D. S. Weld, and R. B. Doorenbos, "A Scalable Comparison - Shopping Agent for the World Wide Web," Univ. Washington, Dept. Comput. Sci., Seattle, Tech. Rep. TR, P17, 96-01-03, 1996.
6. J. O. Everett, D. G. Bobrow, R. Stolle, R. S. Crouch, V. Paiva, C. Condoravdi, M. Berg, L. Polanyi, "Making ontologies work for resolving redundancies across documents,"Communications of the ACM 45 (2): 55-60, 2002.
7. J. Ham and M. Kamber, "Data Mining Concepts and Techniques," Morgan Kaufmann Publishers, 2002.
8. iJADE official site: http://www.ijadk.org
9. C. Kwok and D. Weld, "Planning to gather information," In Proc. 14th Nat. Conf. AI, pp. 15-21, 1996.
10. H. V. Leighton and J. Srivastava, "Precision among WWW Search Services (Search Engines): Alta Vista, Excite, Hotbot, Infoseek, Lycos," http://www.winona.musus.edu/is-f/library-f/webind2/webind2.htm, 1997.
11. A.Y. Levy, T. Kirk and Y. Sagiv, "The information manifold," AAAI Spring Symposium on Information Gathering From Heterogeneous Distributed Environments, pp. 14-20, 1995.
12. E. Spertus, "Parasite: Mining structural information on the web," In Proc. 6th WWW Conf., pp. 16-22, 1997.
13. C. W. Wen, H. Liu, W. X. Wen and J. Zheng, "A Distributed Hierarchical Clustering System for Web Mining," WAIM2002, LNCS2118, pp. 103-113, Springer-Verlag Berlin Heidelberg, 2001.
14. WordNet, http://www.wordnet.com

Architecture of a Web Operating System
Based on Multiagent Systems

José Aguilar, Niriaska Perozo, Edgar Ferrer, and Juan Vizcarrondo

CEMISID, Dpto. de Computación, Facultad de Ingeniería, Universidad de los Andes
Av. Tulio Febres, Mérida, 5010, Venezuela
`aguilar@ing.ula.ve`

Abstract. The amount of systems, services and applications developed for the Web has grown considerably. The support of the existing operating systems to them, is not the awaited one. Like a solution to this necessity, we propose a model of operating system denominated SOW. It supports and handles a set of services in a heterogenous and dynamic environment like Internet. The SOW is conformed by four subsystems (resource, repositories, web object and communities manager subsystems), where each one carries out a series of coordinated functions, that allow an efficient use of the resources on Internet. In this work we present the design of the SOW based on Agents.

1 Introduction

Given the wide variety of unimaginable services everyday in the web, it's difficult to design an operating system to support each of those services individually. A new domain to solve these needs are the Web Operating Systems (WOSs), which have like main objective provide a platform that allows users to benefit from the computational potential offered in the web, through resource sharing and solving problems of heterogeneity and dynamic adaptability present in it. Thus, in order to reach the best performance in a dynamic environment of distributed resources such as the Internet, the WOS should be configurable and able to adapt to changes related to the availability of software and hardware resources. Taking into account that, the WOS model presented in this paper suggests a series of aspects in order to provide services that can adapt to the web's special features. Hence, our WOS (called SOW) offers a set of services to enable the utilization of the available resources through the Internet.

There are different proposals for management and integration of the computational resources available in a global computer system. Perhaps the most general project is the *WOS* [2] since it allows for management and integration of resources addressing the issue of heterogeneity and volatility in the Web. This project, as well as our proposal, is based on the idea of using versions as a solution to these problems. Other efforts to exploit the resources distributed in global computing include *Jini, Netsolve, Globe, Legion*, and *WebOS* [1, 3, 4, 5]. In contrast with our proposal, most of these systems require entry privileges (*login*) in participating devices and software. Particularly, our model differs from all the mentioned projects, in the sense that no centralized global resource catalog is needed. Our SOW model supports and manages a series of services in a heterogeneous, dynamic and adaptive context, under the application of a reconfiguration approach.

R. Khosla et al. (Eds.): KES 2005, LNAI 3681, pp. 700–706, 2005.

2 General Considerations of the Proposed Architecture

2.1 SOW Features

The main features of our SOW are: **i) Distributed and with different versions:** Different versions of services offer by the SOW are running simultaneously along the network; **ii) Dynamic:** The web is a constantly evolving entity, therefore, the SOW should adapt to unforeseen changes, through the dynamic configuration of its services; the dynamic modification of the information on the resources available in each node; the dynamic grouping of existing nodes according to certain features, and the migration and replication of web objects, **iii) Open:** Our SOW is an open system from two points of view: it accepts different technologies through the network (heterogeneity), and it also allows every node in the Internet to be incorporated into the system; **iv) Intelligent:** Each one of the SOW subsystems will have a certain level of intelligence for the development of some of its functions.

2.2 Integration of the SOW in a Distributed Environment

Our SOW should be at the top of a standard distributed platform (Middleware) in order to use the services provided by this platform. In this way, it coexists with traditional distributed applications, and can be used by a wide community. Particularly, among the distributed services that require our SOW are: a *Name Service* (white pages) for identification and location of resources in the distributed environment by designation; a *Directory Service*; (yellow pages) for identification and location of resources in the distributed environment by attribute; a *Security Service* that provides confidentiality, integrity, and availability; *Synchronization and Coordination Services*; a *Process Management Service* which will be in charge of the processors allocation, internal and global processes scheduling, and processes migration; a *Distributed File System* that will allow the management of different file systems in different nodes and the sharing of information; and a *Communication Services* that allow interaction between existing models (client/server, intermediaries (Proxy, Caches, among others), group communication (multicast), etc.). The interactions can be carried out by messages passing and shared memory.

3 Description of the SOW

The SOW is composed by the resource manager subsystem, the repositories manager subsystem, the web object manager subsystem and the community manager subsystem.

3.1 Resource Manager Subsystem (SMR)

In our proposal, the SMR is made up of mechanisms that allow managing and integrating computational resources present in the Web. The resources may have a physical representation, such as a file or a printer, or abstract representations, such as the CPU time. Due to the dynamic nature of the Web, it's impossible to develop a catalog with every available resource and service. Therefore, instead of having an operating system that provides a set of fixed techniques of process scheduling or caches management, we have an operating system that satisfy the requirement with the particular scheduling technique needed in a given time. In such cases, the concept of versions will be used, and in this way, an action will have the possibility of being coordinated by combining the appropriate versions (which could be physically distributed along the network) in order to achieve an efficient execution of the required services. The SOW has a service configuration mechanism to adequate the utilization of these versions and to have the catalogue of the versions of the different resources, services or strategies. An inference engine handles every requirement to the system. This works as a demand managing reactive system through which heterogeneous computational environment resources provided by the Web are managed. It will particularly carry out the complete search process, determine the service to be provided and assign resources. The SOW will have various inference engines distributed along the network. Whenever a user tries to access a web resource, the SOW gets the request and this request is handled by the local inference engine, which will consult with its repository to determine whether it's possible to satisfy the request locally. Otherwise the request is sent to another inference engine and the previous process is repeated until the request is finally fulfilled. In general, each inference engine has access to its local and remote resource repositories, which store information about the resource versions. The local resource repository is always consulted; the remote resource repository is consulted only if the resource request can't be satisfied locally.

3.2 Local Repositories Manager Subsystem (SMRL)

One of the great challenges of our SOW is implementing an efficient algorithm for the search and discovery of services or resources available in the network. The repositories associated with the SOW nodes provide the necessary information in order to satisfy the service requests. Thus, every node uses its repositories in order to store and continuously update the information about the node, the services and the resources available in it and in its near nodes (a near node is defined by the community manager system). This allows the interaction with different repositories, each one offering different versions of the available services, resource, techniques, applications, etc. The local repositories will have the ability to store local information. When that information is perhaps replicated in other sites, data coherence mechanisms must be at hand in order to keep the information consistent. The SMRL will establish policies to make efficient use of the available capacity in the devices and will use efficient local search procedures. For example, some of the techniques are for the storage optimization, disk access scheduling, among others. On the other hand, since most Internet applications require real time interactions and it's necessary to share available information, the SMRL might establish efficient transaction mechanisms such as the use of parallel I/O scheduling, etc.

3.3 Remote Repositories Manager Subsystem (SMRR)

A problem that our SOW must solve is that it has not an adequate bandwidth to transport web objects. One of the ways to tackle this deficiency is trying to bring users closer to the information they'll require. In our proposal we'll consider the remote repositories as a cache in the web, with the capacity to store objects close to the users, so it's not necessary to move an object from its original location every time it's needed. The most critical aspect in designing these caches in the web is that the storage size is finite. That's why a policy must be used to guarantee optimization of the reduced space, every time old objects have to be replaced with new ones. Other basic elements are the mechanisms to keep the information coherent in these repositories. Among the advantages of having a cache in the web are the reduction of response times (when downloading web objects), the elimination of traffic in the network (by having the information the user would need close to him), among others. The cache in the web works as follows: if a user requests an object, the cache makes a local copy; when the same object is required again, the cache shows its available copy, this way it's not necessary to request it again from the original server.

3.4 Web Object Manager Subsystem (SMOW)

This subsystem manages every kind of web object, particularly the mobile objects and those that have to be replicated. Mobile objects are those that move through Internet sites following certain criteria. Among the advantages offered by the migration of objects in the Internet, we find the reduction of communication costs, and a more fault tolerant system [5, 6]. We use the idea of traces in the definition of mobile object administration policies. When the object moves it leaves a trace of its movements, which is used to locate it when is requested by a user. In addition, the trace can evaporate each certain time. The intention behind the disappearing trail is not to leave confusing traces that might complicate the search of an object at any given time. On the other hand, web objects can be replicated, in this way the system must establish administration policies to guarantee accessibility and best possible location of the replicas in the web.

3.5 Communities Manager Subsystem (SMC)

The nodes will interact in the SOW through mechanisms of request/reply and negotiations. Those nodes exhibiting functional and behavioral affinity are able to associate themselves dynamically to form communities. A node in the SOW is defined according to its abilities and behavior. Those are the elements considered as node features, which as a whole (with its resources and functions) are affiliated to a given community. Our SOW will treat communities from an emerging point of view, that is, communities that emerge and adapt to their environment, that self-organize according to the requirements; without a central control and showing a global order (the group of nodes works very efficiently and with a good structure) [6]. The SMC will have protocols that manage the creation, change and elimination of communities and, in general, the incorporation, elimination or migration of nodes between communities.

4 Design of Our SOW Like a Multiagent System

Each subsystem of the SOW is a MAS. To describe each subsytem like a MAS we will use the MASINA methodology [7].

4.1 Design of the SMR

In the SMR the operations are coordinated by four agents:

- **Interface with the User:** It receives requests of the users, to send them to the reasoning system, and to send the results of the requests to the users.
- **Reasoning System:** It is the main agent of the system. It receives requirements of the interface for processing them, with the purpose of giving an suitable answer to them. It has an inference engine to coordinate the mechanisms of location, configuration and allocation of services required by the users.
- **Services Search:** This agent has like objective the search of services that are required by the inference engine.
- **Versions Manager:** This agent has a *self-configuration module*, which handles the versions of the services found, in order to produce the configuration of the response according to the requirement of the users.

4.2 Design of the SMRL

In the SMRL the operations are coordinated by five agents:

- **Search Coordinator:** It receives the requests of search of services of other subsystems. In addition, It controls the decomposition, classification and distribution of the requests of search of local service.
- **Local Resources Manager:** It manages the resources of hardware and software existing in the node. Also, it updates the catalogue of the local resources.
- **Local Information Manager:** It provides the local information required by the search coordinator. On the other hand, it updates the local information, and also maintains the consistent information after any process of update.
- **I/O Optimizer:** It establises policies to make an efficient use of the capacity available in the devices. In addition, it establishes techniques of optimization and efficient access to the devices.
- **Traces Manager:** it handles the traces at local level. This agent allows the SMRL to participate in the process of creation, reinforcing, evaporation and elimination of traces, altogether with the SMRR and the SMOW.

4.3 Design of the SMRR

In the SMRR the operations are coordinated by two agents:

- **Search:** this agent searches the replicas and the objects that are in the SMRR and that are required by the SMR.
- **Administrator:** this agent maintains the repositories of the SMRR. For this, it has three functions: a) when the SMRR needs to store a new replica or a referenciable object and there are not space in the RR, it eliminates less used objects. b) It enters, eliminates and updates replicas and referenciables objects in the RR. c) It verify the consistency of the information that is in the RR.

4.4 Design of the SMOW

In the SMOW the operations are coordinated by two agents:

- **Replicated Objects Administrator:** this agent transfers the replicas of the objects in the SOW.
- **Migrated Objects Administrator:** this agent migrates the objects Web in the SOW and creates the traces.

4.5 Design of the SMC

In the SMC the operations are coordinated by two agents:

- **Communities Manager:** it manages the existing communities in the SOW. A new community that emerges is characterized before being added to the SOW. In addition, it allows locating communities with certain characteristics or profile, and describing a node before it can be incorporated to a given community.
- **Search Manager:** It receives requests of search of services of a SMR or another SMC, and processes them (It makes the search of a given service).

5 Coordination, Communication and Intelligence Models

The design of each one of the subsystems of the SOW according to the MASINA methodology [7], allows to show the existing communications (speech acts) between the agents. These can be shown through a diagram of interaction of UML. For example, the conversation *to update community* [6], of the agent communities manager of the SMC, is shown in figure 1. In addition, MASINA allows the description of the existing intelligence in some of the agents through the intelligence model. For the same previous agent of the SMC we describe its intelligent model in table 1.

Table 1. Mechanism of Reasoning of agent CA

	REASONING MECHANISM
INFORMATION SOURCE	Obtained results of the group of nodes.
ACTIVATION SOURCE	Request to incorporate or to transfer node.
TYPE OF INFERENCE	Deductive, Inductive.
REASONING STRATEGY	To coordinate activities related to the group of nodes, update and creation of communities, through tools of recognition of patterns based on neuronal networks and fuzzy logic (Engine of fuzzy rules).

6 Conclusions

In spite of the great efforts put on so far, the existing operating system and traditional Middleware architectures aren't prepared to offer an efficient management of web resources. We propose a WOS model that tries to address these issues. The WOS proposed is made up of four subsystems that will carry out a series of coordinated functions that allows an efficient use of Internet resources, in spite of its dynamic features, high mobility and heterogeneity.

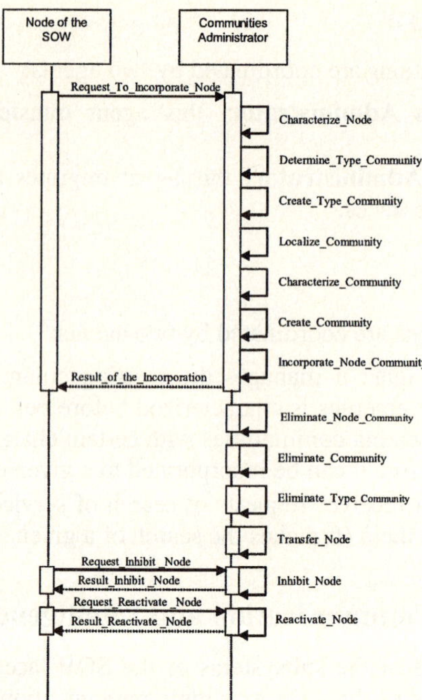

Fig. 1. Conversation: *To Update Community* in the SMC

References

1. Vahdat A., Belani E., Eastham P., Yoshikawa C.: WebOS: Operating System Services for Wide Area Applications". Proc. of Seventh IEEE Symposium on High Performance Distributed Systems (1998) 52-63.
2. Kropf P.: Overview of the WOS Project. Proc. of Advanced Simulation Technologies Conferences, High Performance Computing (1999) 350-356.
3. Casanova, H., Dongarra, J.: NetSolve: A Network-Enabled Server for Solving Computational Science Problems. International Journal of Supercomputer Applications and High Performance Computing, **3** (1997) 212-223.
4. Morgan, S.: Jini to the Rescue. *IEEE Spectrum*, **37** (2000) 44-49.
5. O'Reilly, T.: Building the Internet Operating System. Proc. of O'Reilly Emerging Technology Conference, (2002) 1-2.
6. Aguilar, J., Ferrer, E., Perozo, N., Vizcarrondo, J.: Arquitectura de un Sistema Operativo Web", *Gerencia, Tecnología, Informática (electronic journal: www.cidlisuis.org/aedo),* Instituto Tecnológico Iberoamericano de Colombia, **2** (2003).
7. Aguilar, J.: Especificación Detallada de los Agentes del SCDIA – MASINA CommonKADS. Reporte Técnico Proyecto Agenda Petróleo 9700381, Universidad de los Andes (2003).

A Study of Train Group Operation Multi-agent Model Oriented to RITS*

Yangdong Ye[1], Zundong Zhang[1], Honghua Dai[2], and Limin Jia[3]

[1] Department of Computer Science, Zhengzhou University, Zhengzhou, China, 450052
yeyd@zzu.edu.cn, zhangzundong@gs.zzu.edu.cn
[2] School of Information Technology Deakin University, Melbourne Campus 22
Burwood Highway, Vic 3125 Australia
hdai@deakin.edu.au
[3] School of Traffic and Transportation, Beijing Jiaotong University,
Beijing, China, 100044
jialimin@jtys.bjtu.edu.cn

Abstract. In this paper, aiming at describing interrelationships and communication mechanisms among agents based on a multi-agent framework of Railway Intelligent Transportation System (RITS), we construct the model about stations and trains in the system, which is called Agent-Oriented G-Net Train Group Operation Model (AGNTOM). The framework degrades the complexity of computation and makes the distribution of simulation system easy in design. The simulated experiments prove that the model provides an effective approach for dealing with communication problems in the system.

1 Introduction

Railway Intelligent Transportation System (RITS) is railway transportation (RT) system of new generation, which is characterized by widespread usage of intelligent technologies, such as modern communication, modern information processing, and intelligent automation technologies. Its essential characteristics are integrated with intelligent attributes, openness of architecture/structure, distribution of system structure, etc [1]. The formalized description of train group operation behaviors based on its essential characteristics is one of key issues in RITS. The researches on train group operation model such as TOPNO [2][3], TGOSOPS [4] have established good foundation for the simulation system in RITS study. With the increasing requirements of the simulation system about intelligence and distribution, the models in [2][3][4] have some inadequateness in following aspects: (1) description of the interrelationships among different agents during runtime, such as a train perceiving environments in real time and keeping coordination with environments; (2) effective communications that ensure normal operations or negotiations, such as trains communicating with environments in certain mechanism. Based on those problems, we use agent-oriented G-Nets [5-9] to construct the Agent-Oriented G-Net Train Group Operation Model (AGNTOM) for the multi-agent simulation system in RITS.

The model embodies both object-oriented conceptions and Petri net analysis methods. This model provides an effective approach for dealing with communication

---

* This research was partially supported by the National Science Foundation of China under grant number 600332020 and the Henan Science Foundation under grant number 0411012300

R. Khosla et al. (Eds.): KES 2005, LNAI 3681, pp. 707–713, 2005.
© Springer-Verlag Berlin Heidelberg 2005

problems in the system. The distributed structure makes full use of resources in simulation system and satisfies requirements of distribution on logic and space. Compared with the existing models [2][3][4], AGNTOM has some prominent features including support for asynchronous message-passing, description for the frame of the system, representation of interrelations and communication mechanisms of agents. The simulated experiments show that the model can ensure the interrelations and the effective communication among agents.

Part 2 and Part 3 introduce the framework of the multi-agent simulation system and its communication mechanism. Fourth part denotes the research on the agent-oriented G-Net model of trains and stations in RITS. In part 5 this paper describes the simulated experiments based on AGNTOM. Finally some conclusions and future work are presented.

2   Multi-agent Framework of RITS

In order to satisfy the requirements of essential attributes in train group operation simulation system, the system is designed as a multi-agent system whose framework is shown in Fig. 1. The distributed structure of the system and the interrelationships of the agents oriented to communication are introduced in Fig. 1. The key issue in this system is the description of behavioral characteristics of trains and stations, that is to say, how to construct train group operation model is most important issue in the study on RITS. We use agent-oriented G-Nets [5][9] formal modeling approach to construct train agents and station agents. The detailed explanations of the agents can be found in [5]. We give the descriptions of train agents and station agents below.

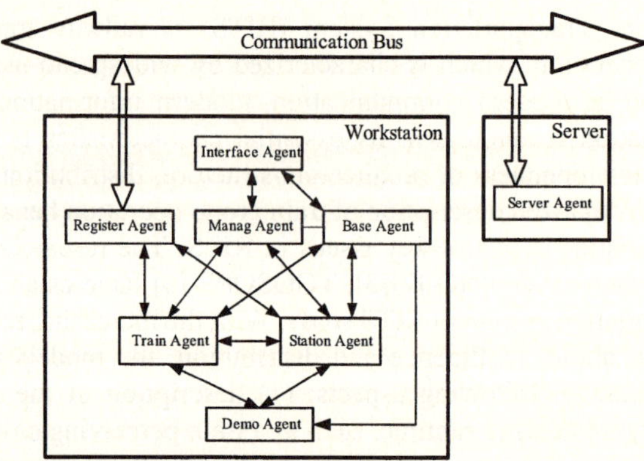

Fig. 1. Framework of train group operation simulation system

A train agent perceives environments, makes decisions by its goals, communicates with station agents that it's relative to and server agent, decides to pull-in, start/stop or not with corresponding information from stations and adjusts dynamically its operations according to itself situation.

A station agent inspects all trains in charge, schedules railway track and resources; analyzes and deals with different type of pull-in trains including pass-by trains, pass-by trains (no stop) and terminal trains.

3   Agent Communication Mechanism

In order to satisfy the specification of agent design, we establish the communication mechanism based on the essential characteristics of RITS in the following figure. Agent communication happens only when it's needed. Agents would not occupy communication resources all the time which means an asynchronous fashion is needed.

We construct agents in agent-oriented G-Net approach that specifies asynchronous and synchronous communication [5]. In the following figure, we describe agent asynchronous and synchronous communication in transport level. Because our model's structure is distributed, the agents communicate with each other by message-passing. Asynchronous communication is implemented by the process in which agents listen to message reception queues and send messages as needed. If a train needs "talk" to a station or its server for an urgency, the communication should not be interrupted or disordered and a clock synchronous module takes effect to deal with this kind of communication.

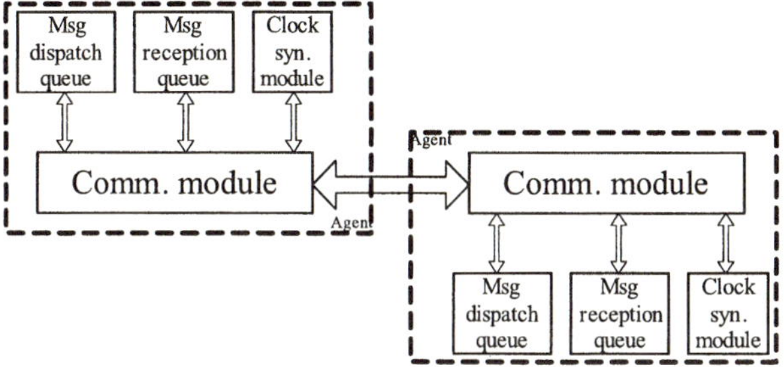

Fig. 2. Communication mechanism between two agents

4   AGNTOM

We construct the agent-oriented G-Net train group operation model (AGNTOM) using agent-oriented G-Net approach, so as to describe the interrelationships and communications among agents in the system. Agent-oriented G-Nets are used to describe agents in the system, which is introduced detailedly in [5][9]. Agent-oriented G-Net integrated with multi methodologies is a formalized approach in agent design. So, the model embodies both OO conceptions and Petri net analysis methods [5].

Train and Station agents are main components of the system. We denote them as the cores of the system. In order to describe their negotiation and cooperation more clearly, validate the correctness of the model, we depict the agent-oriented G-Net model in traditional Petri net, as shown in Fig. 3 (description of transitions and places in [5]). The model can be analyzed by Petri net methods [10] and can narrow the gaps between design and implementation [8][9].

5   Simulated Experiments

Several simulated experiments have been done to show the interrelationships among agents and the communication mechanism based on the railway network graph in Fig. 4. Experiments on Area 1 and 2 are to prove the correct description of the interrelations in the model. Another is to validate the communication among agents.

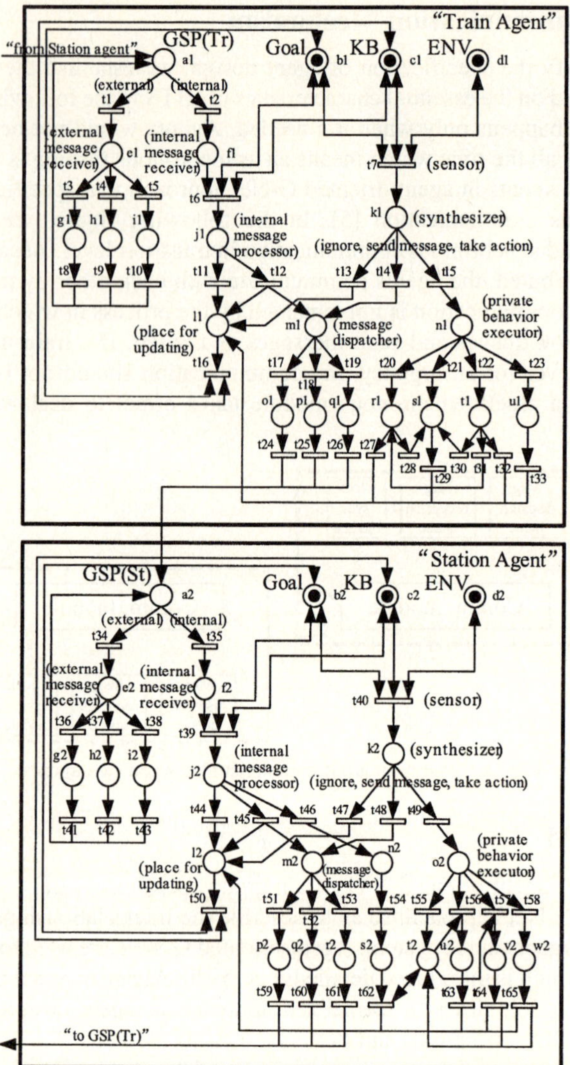

Fig. 3. Agent-Oriented G-Net Train Group Operation Model

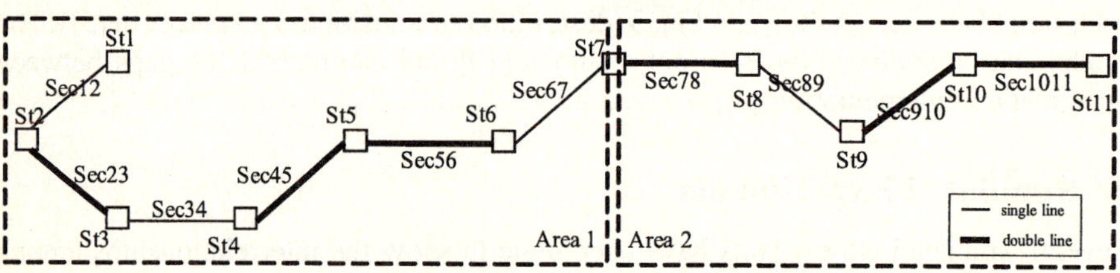

Fig. 4. Railway network graph

5.1   Experiments Based on Area

The first experiment is based on the railway network Area 1 in Fig. 4, and operation goals and plans of all trains are in Fig. 5.

In Fig. 5, T5110 and T3110 will reach and leave Station 4 at same time, which would cause a deadlock. Train agents in the simulation system can detect runtime deadlocks from environments and make adjustment of their behaviors. Just like T5110, its operation plan is adjusted to avoid the deadlock which shows the reactiveness and it adjusts following plans so as to reach its terminal in time which shows the pro-activeness. The whole process can be found in Fig. 6.

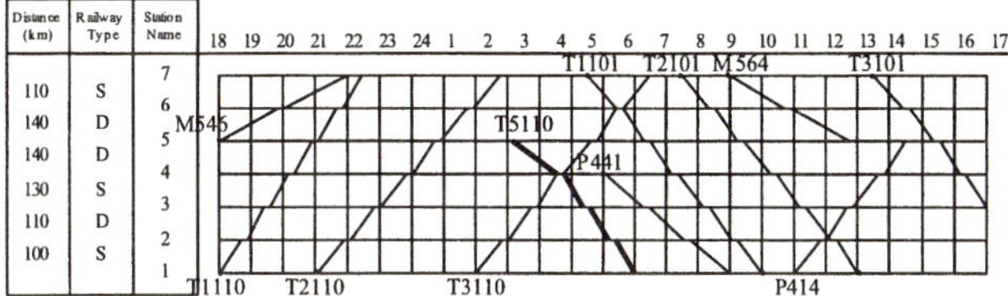

Fig. 5. Plan train operation graph

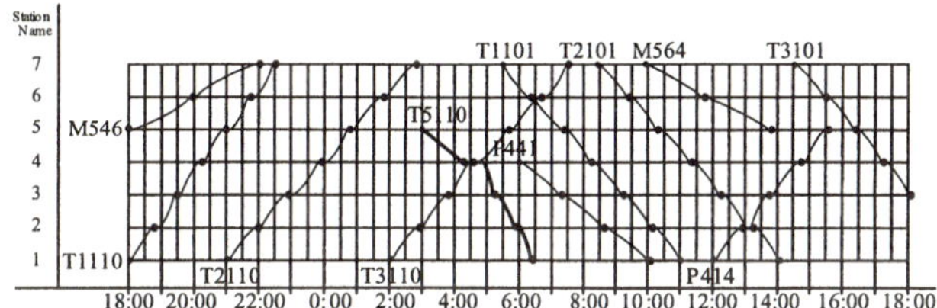

Fig. 6. Simulated results based on Fig. 5

According to the given train operation graph, the simulated results demonstrate the feasibility and the anti-interference ability of operation graphs, which proves the model can describe the interrelation of agents clearly and correctly.

The experiment in Area 2 is similar to the one in Area 1. So, we do not describe it in this paper.

5.2   Experiment of Multi-agent Communication Mechanism

We extend Area 1 with Area 2, and 5 trains are added to Area 2. In order to show the effectiveness of the agent communication mechanism, we run the system again and record the simulated result. In this experiment, we pay more attention to the communication among train agents, station agents, the server agent and the register agent.

Agents communicate with others when negotiation or cooperation. The register agent would not send messages to other agents actively, it replies received messages after completing acts of registering. Train and station agents communicate with the server agent and the register agent for checking in during system initialization. When a train starts, pulls in or stops, it must negotiate with corresponding station, which is supported by effective communications. The detailed statistics of agent communications are list below.

Table 1. Statistics of agent communication

Area 1			Area 2		
Msg dispatch	Msg receive	Count	Msg dispatch	Msg receive	Count
Train Agent	Station Agent	68	Train Agent	Station Agt.	24
Station Agent	Train Agent	68	Station Agt.	Train Agent	24
Train Agent	Register Agt.	69	Train Agent	Register Agt.	25
Register Agt.	Train Agent	69	Register Agt.	Train Agent	25
Station Agt.	Register Agt.	7	Station Agt.	Register Agt.	5
RegisterAgt.	Station Agt.	7	Register Agt.	Station Agt.	5
Server Agt.	Register Agt.	2	Server Agt.	Register Agt.	2
Register Agt.	Server Agt.	2	Register Agt.	Server Agt.	2

This experiment shows that communication happens when one agent need to be reactive to others or take behaviors according to its goals. A communication mechanism is a foundation of agents' reactiveness and pro-activeness. The communication mechanism in the system can support agent communication effectively.

6   Conclusions and Future Work

With a view to the essential attributes of RITS including agents' interrelationships and communications based on the multi-agent framework, this paper studies the Agent-Oriented G-Net Train Group Operation Model (AGNTOM). Because agent-oriented G-Net method is a multi-methodology for dealing with issues in one kind of complex systems like RITS, the model incorporates the object-oriented conceptions, existing Petri net analysis methods and multi-agent techniques. The model describes the interrelation among agents in graphic and formalized fashion and provides the foundation of distributed structure for the system. The simulated experiments have shown the model is with prominence in the description of agents' interrelationships and the effective communication mechanism.

The train group operation simulation system discussed in this article is just a prototype and we will do further research on following aspects.

- More works and researches on formalized description of communication between train and station when negotiation.
- Formalized definitions of interrelationships among independent modules in the train group operation system based on multi-agent framework.
- Further researches on knowledge tokens and environment tokens during system implementing.

References

1. L. Jia, Q. Jiang: Study on Essential Characters of RITS. In: Proc. of 6th Inter. Symposium on Autonomous Decentralized Systems (ISADS 2003). IEEE Comp. Society. Pisa, Italy (2003) 216-221.
2. Y. Ye, L. Jia: Model and Simulation of Train Operation Based on Petri Net with Objects. Journal of System Simulation, Vol. 14, No.2 (2002) 132-135.
3. Y.Ye, L. Zhang, Y. Du, L. Jia: Three-Dimension Train Group Operation Simulation System Based on Petri Net with Objects. In: The IEEE 6th International Conference on Intelligent Transportation Systems (ITSC'2003), Shanghai, China (2003) 1568-1573.

4. Y. Ye, Y. Du, L. Jia: The Object-Oriented Train Group Operation Petri Sub-Net Model. China Railway Science, Vol.23, No.4 (2002) 81-88.
5. Y. Ye, L. Zhang, L. Jia: Research on Agent-Based G-Net Train Group Operation Model. In: Proceeding of 5th World Congress on Intelligent Control and Automation (WCICA 2004), Vol 6, Hangzhou, China (2004) 5227-5231.
6. Y. Deng, S. K. Chang, A. Perkusich, J. de Figueredo: Integrating Software Engineering Methods and Petri Nets for the Specification and Analysis of Complex Information Systems. In: Proc. of the 14th Int. Conf. on App. and Theory of Petri Nets, Chicago (1993) 206-223.
7. A. Perkusich, J. de Figueiredo: G-Nets: A Petri Net Based Approach for Logical and Timing Analysis of Complex Software Systems. Journal of Systems and Software, Vol.39, No.1, Elsevier Science (1997) 39-59.
8. H. Xu, Sol M. Shatz: A Framework for Model-Based Design of Agent-Oriented Software. IEEE Transactions on Software Engineering, Vol.29, No.1 (2003) 15-30.
9. H. Xu, Sol M. Shatz: Extending G-nets to Support Inheritance Modeling in Concurrent Object-Oriented Design. In: Proc. IEEE Int. Conf. on Systems, Man, and Cybernetics (SMC'2000), Vol.4, Nashville, TN, (2000) 3128-3133.
10. T. Murata: Petri Nets: Properties, Analysis and Applications. Proceedings of the IEEE, Vol.77, No.4 (1989) 541-580.

Representation of Procedural Knowledge of an Intelligent Agent Using a Novel Cognitive Memory Model

Kumari Wickramasinghe and Damminda Alahakoon

School of Business Systems, Monash University, Australia
`{kumari.wickramasinghe,Damminda.alahakoon}`
`@infotech.monash.edu.au`

Abstract. To accomplish the autonomous behavior of agents, several top-down and bottom-up agent learning architectures that consist of procedural and declarative knowledge have been implemented. They use mechanisms such as production rules, supervised neural networks and unsupervised neural networks to declare procedural knowledge. An efficient representation of procedural knowledge enhances learning and decision making. Inspired by the representation of rules, facts and knowledge in human memory, this paper presents a novel cognitive memory model, Episodic Associative Memory with a Neighborhood Effect (EAMwNE) as a method to define procedural knowledge.

1 Introduction

The task of an intelligent agent situated in an environment is to accomplish user objectives by mapping sensory inputs to valid output actions. Since the real time environment is dynamic, the optimum mapping between sensors and actuators is a gradual, on going and concurrent learning process for the agent. The cognitive modeling of agents focuses on the strategies of continuous learning varying from "lower level cognitive skills" to "high level cognitive skills" as arithmetic tasks [1].

If specific scenarios are provided humans are capable of generalizing them to similar situations and vice versa. The most fundamental aspects of human behavior include cognition, learning and reaction [2]. This generalization provides much advantage when they encounter unexpected changes in the environment. The cognitive skill acquisition of humans is supported by procedural and declarative knowledge [3].

The two types of knowledge gave rise to two level agent learning architectures termed as top-down and bottom-up [4]. In our early work a conceptual framework for agent learning was proposed through an analogy with biological systems and psychology [5] which is a hybrid of top-down and bottom-up agent learning architectures. The agent uses the bottom layer to make decisions. Therefore the representation of procedural knowledge is a decisive factor for an efficient decision making process. This paper proposes the concept of Episodic Associative Memory (EAM) [6] which is believed to be the main contributor for human autonomy and intelligence as a mechanism to represent procedural knowledge. Due to the limitations in existing episodic memory models in practical use, a novel approach known as Episodic Associative Memory with a Neighborhood Effect (EAMwNE) is proposed.

The paper is organized as follows: an introduction to existing top-down, bottom-up architectures is provided in Section 2; the proposed EAMwNE is elaborated in Sec-

R. Khosla et al. (Eds.): KES 2005, LNAI 3681, pp. 714–721, 2005.

tion 3; the implementation using EAMwNE is experimentally demonstrated in Section 4; Section 5 presents concluding remarks and possible future work.

2   Existing Agent Learning Architectures

In top-down [7] approach explicit declarative knowledge is acquired first and then through practice turns that knowledge into procedural. Bottom-up [1] learning assumes that the declarative can be built after acquiring the procedural knowledge.

In our early work, a hybrid of top-down and bottom-up conceptual framework has been proposed for agent learning [5]. It comprises of *inheritance (Ψ_i)*, *training (Ψ_t)*, *experience (Ψ_e)* and *unexpected (Ψ_u)* layers as shown in Fig. 1. Ψ_i and Ψ_t consist of declarative knowledge and Ψ_e consists of procedural knowledge. It is expected that the implicit knowledge in Ψ_e assists the agent to handle unfamiliar situations in Ψ_u. (Ψ_e) has been implemented using a Growing Self Organizing Map (GSOM) [8].

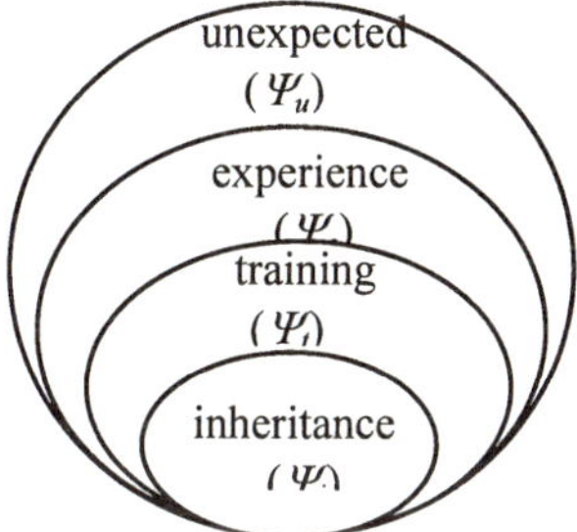

Fig. 1. Proposed 4 tiered agent architecture

Agent architectures use production rules, supervised neural network techniques or unsupervised neural network techniques to implement the procedural knowledge. The neural network techniques are actually mathematical models which represent associative memory. But as procedural knowledge has to be updated continuously implementation using neural network techniques result in the following disadvantages.

- When a new data pattern is added, the whole network has to be retrained.
- Sparsely distributed data could have negatively effect on neural network training. If data patterns consist of varying number of attributes (as the dataset in Section 4) a technique to handle missing values has to be determined.
- Proper training is required to maintain the input-output relationship in data patterns. Results decrease both in over and under training circumstances.
- Parameters such as learning rate, momentum, number of hidden layers and number of hidden layer neurons have to be tuned to obtain accurate results.

The above requirements delay the decision making process of an agent. Agent is always under the time pressure and a decision has to be made in a fraction of a second. In contrast the cognitive models of human memory known as Episodic Associative Memory (EAM) provide following advantages over mathematical models:

- Addition of new data patterns does not affect the existing memory traces.

- Architecture supports sparsely distributed data.
- Data patterns are stored as episodes [9] and do not require training.
- No parameters to be tuned

Therefore, cognitive models provide a light-weight, less time consuming approach. We have already proposed EAM as a technique to model Ψe [10]. But there are some deficiencies in existing EAM models that hinder the practical use. This paper proposes an enhanced EAM, known as Episodic Associative Memory with a Neighborhood Effect (EAMwNE), which eliminates the practical limitations in existing EAMs.

3 Proposed Methodology

Cognitive models of human memory introduce the concept of Episodic Associative Memory (EAM). In EAMs each data pattern gives rise to its own memory trace rather than strengthening a prior representation. Existing cognitive memory models (EAMs) have been briefly described in Section 3.1 highlighting their limitations. An enhanced version, EAMwNE, is proposed in Section 3.2 to provide a practical use. Use of EAMwNE for agent autonomous decision making is described in Section 3.3.

3.1 Existing Episodic Associative Memory (EAM) Models

Convergence-Zone Episodic Memory Model (CZEMM) [9] and Hintzman's Multiple Trace Memory Model (MINERVA) [11] represent a data pattern as a memory trace. Each memory trace consists of a collection of features (known as feature units - Fu), whose value (feature value) defines the pattern. Memory traces are generated as and when knowledge or experience is acquired. When making decisions, all the stored memory traces provide an implicit pool of knowledge.

One of the practical limitations with these models is the low classification accuracy. For example consider the CZEMM representation of data patterns shown in Fig. 2. It consists of three feature units (Fu_1, Fu_2, $Fu3$) and the Fu_3 denote the class to which each data pattern belongs to (array positions 12, 13, 14 in Fig. 2 refer to three classes). The task is to compute the class to which the test pattern t belongs to. The result computed using CZEMM algorithm generates the same number of mappings to classes c1 and c3, indicating t belongs to both c1 and c3.

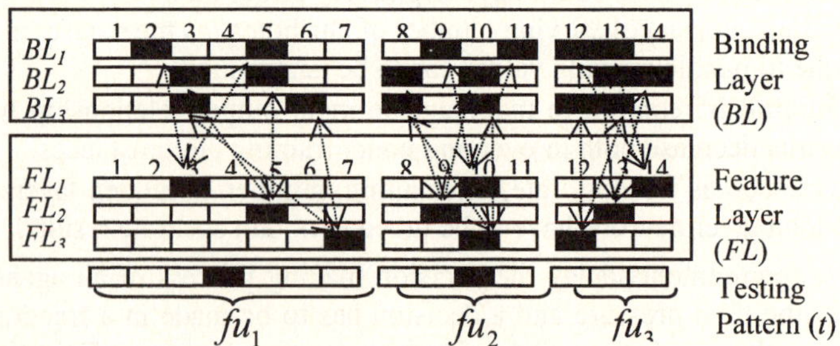

Fig. 2. CZEMM representation of data patterns

The reason for such limitations is mainly due to maintaining only a vertical relationship maintained among feature values in memory traces. They do not consider a lateral relationship among values within a feature, which is more biologically plausible, as there is continuity in the activation level of neurons in the brain. EAMwNE maintains a lateral relationship considering a neighborhood as described below.

3.2 Episodic Associative Memory with a Neighborhood Effect (EAMwNE)

EAMwNE consists of feature units that represent individual features of a data pattern similar to CZEMM. For a data pattern with n distinct features, each feature is represented by a feature unit $fu_i, i = 1....n$. If μ_i denote the number of distinct feature values of $fu_i, i = 1....n$, each fu_i is an array of size μ_i. The neighborhood concept used in EAMwNE is such that the effect from b^{th} location on $(b-1)$ and $(b+1)$ locations is higher than that of $(b-2)$ and $(b+2)$ locations. Therefore, as shown in Fig. 3, a Gaussian distribution is used to calculate the lateral effect on each value position. Lateral effect, $w = \alpha(t)\exp\left(-\left|r_i - r_b\right|/\sigma^2(t)\right)$ where $r_i - r_b$ is the neighborhood size (NH) between i^{th} and b^{th} array positions. Value of w is similar to the amount of learning the neighborhood neurons receive from the winning neuron (b^{th} location according to Fig. 3) in SOFMs. The two functions, $\alpha(t)$ and $\sigma^2(t)$ enhance the neighborhood effect. A neighborhood is considered only within array locations of a given feature unit, but not among feature units.

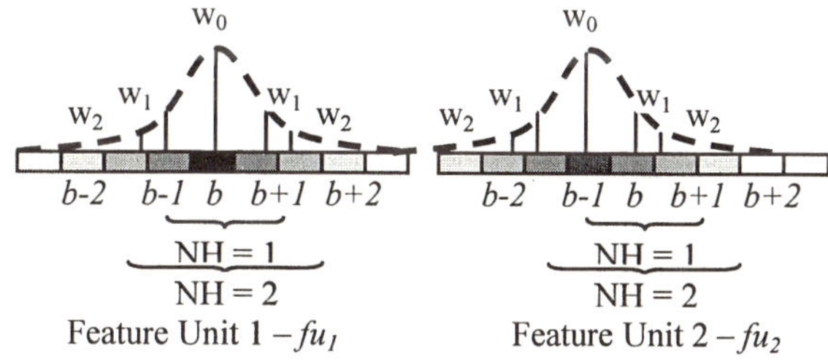

Fig. 3. Neighborhood effect represented by a Gaussian distribution. Gradients of the shadings indicate the magnitude of lateral effect (w) from bth position to the neighboring positions

The algorithm for the creation of memory traces of EAMwNE is as follows:

Step 1: For each training pattern
$P_k, k = 1.......no. of\ training\ patterns$

    Step 2: For each feature unit $fu_i, i = 1.....n$

        Let z = number of distinct feature values of fu_i
        Create an array of size z

  Let $P_k(fu_i) = value\ of\ fu_i\ of\ p_k$

        Let $BL(b)$ denote the array index of $P_k(fu_i)$ in $BL(z)$

  Assign $w = 1$ to $BL(b)$ (value of w when $NH = 0$)

$$\text{Let } c \text{ be the neighborhood size}$$
$$\text{For } NH = 1 \, to \, c$$
$$if\,(0 \le BL(b - NH) < z)$$
$$w_{NH} = \alpha(t)\exp\!\left(-|NH|/\sigma^2(t)\right)$$
$$\text{Create an array of size z}$$
$$\text{Assign } w = w_{NH} \text{ to } BL(b - NH)$$
$$\text{For } NH = 1 \, to \, c$$
$$if\,(0 \le BL(b + NH) < z)$$
$$w_{NH} = \alpha(t)\exp\!\left(-|NH|/\sigma^2(t)\right)$$
$$\text{Create an array of size z}$$
$$\text{Assign } w = w_{NH} \text{ to } BL(b + NH)$$

Repeat Step2

Repeat Step 1

The weight of each array index (w) can be symbolised as w_{ljk} where l = training data pattern number, j = NH and k = array index.

3.3 Use of EAMwNE for Autonomous Decision Making

In agent terminology, feature units – Fu are the set of sensors and actuators available to the agent. Feature values of each Fu correspond to the actual values each sensor or actuator can measure. When the agent perceives a new environment state by its sensors it can determine the action to be taken by its actuators using the pool of existing traces in EAMwNE. Also if the agent encounters an environment situation for which partial knowledge is available it can use EAMwNE to determine the possible values of other sensors and thereupon the output action.

For example, consider an agent with $n1$ number of sensors and $n2$ number of actuators. If $n = n1 + n2$ there will be n number of feature units.

Let μ_i denote the number of distinct feature values of $fu_i, i = 1....n$

Also let $N = \sum_{i=1}^{n} \mu_i$, $N1 = \sum_{i=1}^{n1} \mu_i$ and $N2 = N - N1$

Step 1: Calculate the similarity of existing traces to t. The similarity, e_{lj}, where l, $l=1.....total \ no. \ of \ traces$, $j = \pm NH$

$$e_{lj} = \sum_{k=1}^{N1} w_{ljk} * t(k) \tag{1}$$

where $t(k)$ is the weight of kth position of t

Step 2: Derive the values of actuators using the array indexes $V_k, k = N1.....N2$ as

$$V_k = \sum_{l=1}^{2NH+1} \sum_{j=\pm NH,0} \begin{cases} e_{lj} \ if \ w_{ljk} > 0 \\ 0 \ otherwise \end{cases} \tag{2}$$

Step 3: $\max(V_k)$ is considered as the final value of each actuator.

EAMwNE representation of Fig. 2 is shown in Fig. 4. Applying eq(1) and eq(2) above, $V_{12} = 1.6065$, $V_{13} = 1.9489$, $V_{14} = 2.213$. Applying Step 3, $V_{12} < V_{13} < V_{14}$ and it concludes t belongs to class c3 (no ambiguity). The successful application of this method in a mobile robot navigation task is demonstrated in Section 4.

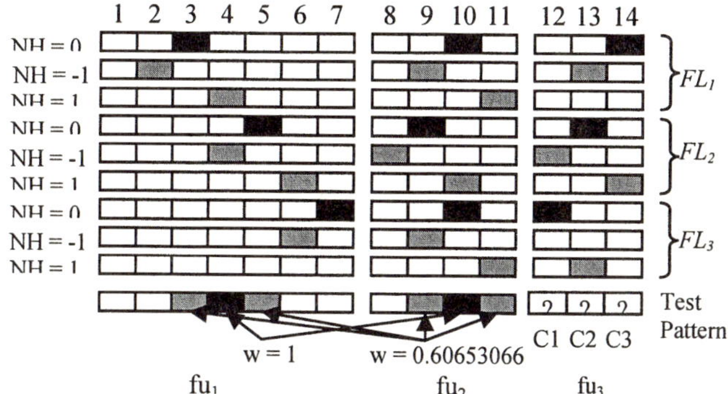

Fig. 4. EAMwNE representation of Fig. 3 with NH=1

4 Demonstration of the Method

EAMwNE representation of procedural knowledge in Ψe was experimented with the "Sensor Data of a Mobile Robot" used by Klingspor [12]. The raw data is classified to: *stable, incr-peak, desc-peak, straight-to, straight-away, no-measurement* and *no-movement* considering the gradient measurement (GT). Basic features are grouped to determine the sensor feature such as *jump, concave, convex* or *line*. Sensor feature is again sub divided into *diagonal, parallel, straight-towards, straight-away*. Sensor features and the sub division provided the procedural knowledge.

Two sets of experiments were carried out as follows:

Case 1: Using the data patterns correspond to a specific sensor feature
Case 2: Using the data patterns for all the sensor features together

In case 1, EAMwNE consists of data patterns which correspond to each sensor feature separately and in case 2 EAMwNE consists of data patterns which correspond to all the sensor features as a combination. A sample of data patterns correspond to case 2 covering all the sensor features are listed in Table 1. Feature units of the data patterns are the gradient measured at each time interval - GT_x (gradient at x^{th} time interval). These data patterns define (Ψ_i) and (Ψ_t) of the conceptual framework.

Table 1. A sample of dataset used in case 2

data pattern	GT_1	GT_2	GT_3	GT_4	GT_5	Outcome
1	-13	-30	-11	70	-11	jump diagonal
2	33	-85	35	-	-	jump diagonal
3	-49	0	-	-	-	jump straight-towards
4	-33	-52	-	-	-	concave parallel
5	27	40	59	-	-	convex straight-away

Using (Ψ_i) and (Ψ_t), EAMwNE traces were formed to represent the procedural knowledge in (Ψ_e). Then (Ψ_e) was used to generate the decision for new data patterns (Ψ_u). For case 1 the decision is whether the new set of sensor data belongs to *diagonal* (d1), *parallel* (d2), *straight-towards* (d3) and *straight-away* (d4) for the considered sensor feature.

Back propagation neural networks (BPNN) were used to compare the results with learning rate = 0.1 and Table 2 summarizes the results obtained for cases 1 and 2.

Table 2. Results obtained for several tests of case 1 and 2

	test	EAMwNE	BPNN
		correct percentage	correct percentage
	jump	93%	80%
Case 1	concave	75%	95%
	convex	80%	85%
	test 1	82%	70%
Case 2	test 2	85%	72%

EAMwNE results are quite promising and also the accuracy of the results for new situations increases if the dataset used to create EAMwNE traces is representative similar to the function of neural networks.

The above formation of Ψ_e belongs to top-down learning. The results for (Ψ_u) are obtained using the bottom layer Ψ_e. In order to support bottom-up learning, if the results are correct the new data item together with the result is added as a new data pattern to the dataset in Ψ_i and Ψ_t to reinforce it in future decision making (that is to use it when creating EAMwNE traces). Similarly if the results are incorrect through human intervention a correct data pattern can be added to Ψ_i and Ψ_t, so that it can enhance the decision making accuracy of (Ψ_u).

5 Conclusion and Future Work

As discussed throughout the paper, cognitive models of human memory provides a light-weight approach to declare procedural knowledge in agent learning architectures. The proposed EAMwNE eliminates the deficiencies of existing EAMs in practical use. Initial indicators show that this is a useful path for further investigations.

As described in Section 2, EAMwNE provides an efficient mechanism for data pattern creation, updation and retrieval of results. It does not require rigorous training or parameter optimization. Most importantly it facilitates the agent to make a decision quickly without compromising the "information processing" task [13]. EAMwNE utilises knowledge as a whole in making a decision. Therefore, EAMwNE has the potential to declare procedural knowledge in agent learning architectures.

The immediate future work will be on further improving the accuracy of EAMwNE. Long term work will mainly be on investigating the potential of EAMwNE to represent cross domain data. This enables the use of a single agent in different domains without heavy changes in the decision making process of it. Humans naturally use cross domain information in making decisions. Similarly, the technique will enhance the accuracy of decisions made by the agent in one domain using relevant knowledge from another domain.

References

1. Sun, R., E. Merrill, and T. Peterson, *From implicit skills to explicit knowledge: a bottom-up model of skill learning.* Cognitive Science, 2001. **25**(2): p. 203-244.
2. Graca, P.R. and G. Gaspar. *Using Cognition and Learning to Improve Agents' Reactions.* in *Adaptive Agents and Multi-Agent Systems.* 2003: Springer-verlag.
3. Sun, R. *An agent architecture for on-line learning of procedural and declarative knowledge.* in *ICONIP'97.* 1997: Springer-Verlag.
4. Sun, R. and L. Bookman, eds. *Computational Architectures Integrating Neural and Symbolic Processes.* 1994, Kluwer Academic Publishers: Norwell, MA.
5. Wickramasinghe, L.K. and L.D. Alahakoon. *Adaptive Agent Architecture Inspired by Human Behavior.* in *IEEE/WIC International Conference on Intelligent Agent Technology IAT-2004.* 2004. Beijing, China: IEEE Computer Society.
6. Miikkulainen, R., *Trace feature map: A model of episodic associative memory.* Biological Cybernetics, 1992. **67**: p. 273-282.
7. Anderson, J.R., *Rules of the Mind.* 1993, Hillsdale, NJ: Lawrence Erlbaum Associates.
8. Alahakoon, L.D., S.K. Halgamuge, and B. Sirinivasan, *Dynamic Self Organizing Maps with Controlled Growth for Knowledge Discovery.* IEEE Transactions on Neural Networks, Special Issue on Knowledge Discovery and Data Mining, 2000. **11**(3): p. 601-614.
9. Moll, M. and R. Miikkulainen, *Convergence-Zone Episodic Memory: Analysis And Simulations.* Neural Networks, 1997. **10**: p. 1017-1036.
10. Wickramasinghe, L.K. and L.D. Alahakoon. *Enhancing Agent Autonomy and Adaptive Behavior by Knowledge Abstraction.* in *The second International Conference on Autonomous Robots and Agents (ICARA 2004).* 2004. Palmerston North, New Zealand: Massey University, New Zealand.
11. Hintzman, D.L., *Schema Abstraction in a Multiple-Trace Memory Model.* Psychological Review, 1986. **93**(4): p. 411-428.
12. Klingspor, V., K.J. Morik, and A.D. Rieger, *Learning Concepts from Sensor Data of a Mobile Robot.* Machine Learning, 1996. **23**: p. 305-332.
13. Sun, R., T. Peterson, and E. Merrill, *A Hybrid Architecture for Situated Learning of Reactive Sequential Decision Making.* Applied Intelligence, 1999. **11**(1): p. 109 - 127.

A Knowledge-Based Interaction Model Between Users and an Adaptive Information System

Angelo Corallo, Gianluca Elia, Gianluca Lorenzo, and Gianluca Solazzo

e-Business Management School - ISUFI
University of Lecce
via per Monteroni sn, 73100 Lecce, Italy
{angelo.corallo,gianluca.elia,gianluca.lorenzo,gianluca.solazzo}
@ebms.unile.it

Abstract. This work approaches the problem of modelling interactions between users and an adaptive system. It is presented the approach adopted in the MAIS[1] project for modelling and designing a recommendation environment for delivering services in a personalized way. The model relies on semantic description of services, on rule-based user profile (part of these studies came from the experience in the KIWI project[2]) and on the use of semantic matching algorithms. We also present the design of the recommendation engine, that is the core subcomponent of the Environment, which performs the task to evaluate each service according to user preferences.

1   Introduction

The current fast evolution in the ICT world suggests that it will be possible to offer access to information systems, in a ubiquitous way, through various kinds of interaction devices (PCs, laptops, palmtops, cellular phones, TV sets, and so on). The MAIS project main research goal is to provide a flexible environment to adapt the interaction and to provide information and services according to ever changing requirements, execution contexts, and user needs.

In the MAIS approach, services can be abstract or concrete. An abstract service is a service with a description completely independent from any kind of implementation. Its function is to represent a service archetype and not a invokable service. A concrete service can be defined as an abstract service with the addition of all the concrete aspects that make it invokable.Web Service orchestration and composition is made by a flexible invocation of available services selected on the basis of their functional properties and quality of service. The MAIS platform uses semantics just for choosing a set of concrete services that referred to a single abstract service which match mainly with functional parameters

---

[1] This work was partially supported by the Italian Ministry of Research - Basic Research Fund(FIRB), within MAIS project, Multi-channel Adaptive Information System. Website at http://www.mais-project.it

[2] Knowledge-based Innovation for the Web Infrastructure -
Web Site http://ra.crema.unimi.it/kiwi

R. Khosla et al. (Eds.): KES 2005, LNAI 3681, pp. 722–729, 2005.

and quality of service. It is not fully considered the opportunity to use semantic structures such as domain ontologies or taxonomies in order to enrich web service description by introducing business characteristics of a service related to a specific domain. In addition to that, it must be considered that user needs and preferences on services are more often implicit, and they should be extracted from the KB of the system. With our research, we aim to define a semantic layer on the service descriptions that can be used to express and explicit user needs in terms of which service is more suitable to his profile. Having these enriched information on services and user profile, we can improve system capabilities to retrieve, select and compose services that are really similar to user needs and preferences contained in his profile. This improvement is intended to be realized by a Recommendation Environment, a component of the MAIS architecture, able to manage a behavioural profiling process and to elaborate recommendation to users. In the following sections we will show how the Recommendation Environment is based on information filtering techniques for adaptive selection and access to e-services, depending on available services and on the service requester (user) profile; we will show also how actually several standard W3C technologies enable to extend web service description using semantics for extra-functional attribute and how is possible to design a recommender environment and more in particular a recommender engine.

2   User-Knowledge Supporting Service Personalization

In a service oriented architecture, interoperability is one of the major issue. It is possible to grant service interoperability at many levels. While the SOA grants a good level of functional interoperability, actually the major efforts aims to grant an acceptable level of semantic interoperability. In this case, semantics can be associated to functional parameters of a service in order to apply match-making rule with users requests[1] . Thus, adding functional or operational (to the pre and post conditions) semantics aims to enhance the service discovery, aggregation and composition phase.

We can also think to add semantics to non functional and to extra-functional parameter such as additional attributes which are specific to a domain and to a service. We want to use this approach in order to select the more appropriate service, given a user profile. From a business point of view, we can achieve differentiation between services provided by different service providers, giving the opportunity to describe a service beyond the capabilities which are tightly technological. From the same point of view, the adding value of the MAIS platform, is to provide to users services in a personalized way, using the user-knowledge held by the system in order to deliver, between all the available services that are also functionally equivalent, that service which has business characteristics closer to the user profile.

The way we want to add semantics is with semantic annotation[2], that provides a richer and more formal service description in order to enable a more efficient discovery mechanism. Semantic annotation aims to create an agreement on a shared vocabulary and a semantic structure for information exchange about a specific domain.

It is necessary to introduce and integrate in the MAIS architecture, an "environment" able to collect and manage the knowledge-base of the user and the service descriptions, and to realize through the use of semantic algorithms the choice of the most suitable service given a user profile. The environment we designed is called "Recommendation Environment", and it has to perform the following tasks:

- to support MAIS platform in choosing the most suitable service through the "Recommendation Engine" component;
- to create and manage user profiles through a "data mining" component;
- to collect all those business events generated by the MAIS platform and that will feed the user profile updating process;
- to have an interface with the MAIS platform in order to obtain all the necessary information to generate a recommendation.

3   Recommendation Environment

The approach used to design and develop the Recommendation Environment is based on component development, that uses business component [3] as fundamental unit for design a software architecture. This approach arise from the strong distributed nature of MAIS architecture, where the units that make up MAIS are distributed in different nodes in the whole infrastructure. We want to grant a loosely coupled interaction and an high degree of interoperability between the architectural components of the MAIS whole system. The exposed interfaces of the components are web services that work with the other component exchanging SOAP Messages. Recommender Environment is a System Level Component (SLC), according to Herzum and Sims [3], that is a group of business components that cooperate to deliver the cohesive set of functionality required by a specific business need with clean interfaces designed so that the system as a whole can be treated as a black box. The Recommender Environment SLC has a declarative interface for communicate with other components.

The Recommender Environment is a composition of Business Component according to the taxonomy in the figure. It's composed by four Process Business Component (PBCs), each of that represent a distinctive business process o business activity. These components, such as the Recommender Engine, the DMPM (Data Mining Profile Module), the Catcher and the Spatial Context Enrichment Component will use to perform a specific task. More in detail, the Catcher is the PBC that is responsible for catch the business events generate by the user interaction with the MAIS platform. These events are sent to DMPM PBC, that is responsible for the elaboration and extraction of a set of association rules, based on event history of the specific user, that describe the user behavior in the interaction with the platform. These rules are the dynamic part of the user profile. The Recommender Engine is the PBC responsible for calculate the affinity degree between user profile and the e-service based on the rules produced

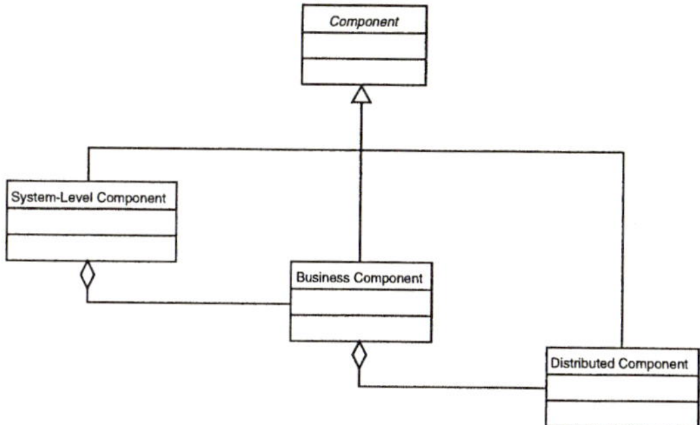

Fig. 1. Component Taxonomy.

by DMPM. In this manner the Recommender Environment perform a recommendation by a ranking of a list of services that expose the same functional interface. The other kind of components are two Entity Business Component that are the main business concepts on which business processes operate. These two EBC, User EBC and e-Service EBC, provide functional interface that expose high level operations that allow to access to information related to the user and e-service. These operations realize an abstraction from the manner in which the information related to the user and e-service are represent and stored. These business components are summarized and represent in the internal view of the Recommendation Environment, that you can see in the figure 2, in which are presented all the relation between the business component

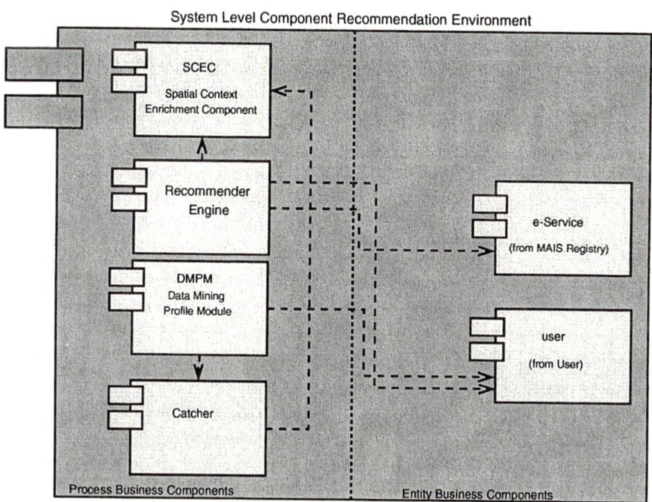

Fig. 2. Recommender Environment Internal View.

4 A Model for the Recommender Engine

Here is shown the model that stands behind the design of the recommender engine[4]

Service description: in this phase, semantic annotation of parameters that describe a service will be performed providing a "link" between these parameters and concepts in a domain ontology. This "semantic markup" will be stored in a service description file that will be accessible over the Internet and retrieved through the use of some Registries. For example, a Provider who wants to publish a service to book rooms in his hotel, whose name is CCIIHotelReservation, will use, in the publishing phase, a Tourism or Accommodation ontology, available in an Ontology Repository, to semantically describe parameters of his service. Annotation is performed browsing the ontology and selecting concepts that have to be linked to additional attributes of the service. We assume that annotation of a parameter, in the CCIIHotelReservation service description, with Restaurant concept means that the Provider can provide all kinds of restaurant defined in the ontology that Restaurant concept subsumes, that is nationalMenuRestaurant, ethnicMenuRestaurant and internationalMenuRestaurant[4].
As an example of service description, we extracted the resulting subset of additional attributes described semantically with OWL-S [5]:

```
<profile:serviceParameter>
    <profile:ServiceParameter rdf:ID="TvInRoom" >
        <profile:serviceParameterName>TvInRoom</profile:serviceParameterName>
        <profile:sParameter rdf:resource="http://localhost.localdomain/ontHotel.rdfs#tv"/>
    </profile:ServiceParameter>
</profile:serviceParameter>
<profile:serviceParameter>
    <profile:ServiceParameter rdf:ID="InternetConnection">
        <profile:serviceParameterName>InternetConnection</profile:serviceParameterName>
        <profile:sParameter rdf:resource="http://localhost.localdomain/ontHotel.rdfs#dsl"/>
    </profile:ServiceParameter>
</profile:serviceParameter>
<profile:serviceParameter>
    <profile:ServiceParameter rdf:ID="TypeOfRestaurant">
        <profile:serviceParameterName>TypeOfRestaurant</profile:serviceParameterName>
<profile:sParameter rdf:resource="http://localhost.localdomain/ontHotel.rdfs#
        internationalMenuRestaurant"/>
    </profile:ServiceParameter>
</profile:serviceParameter>
```

Profile representation: this is the part disposed to contain the information enabling ranking process. This part will contain the information about services extra-functional parameters and can be obtained by the analysis of the events caught by the system during the interaction session with the user. Having a correlation between behavioural data and static attributes of a user, it is possible to extract information on the typologies of user behaviour in the form of rules. Rules are a raw form of knowledge and they will describe in the user profile two or more actions or preferences of the user.
An example of rule is:

IF
*the user is a **student**, requesting from **Rome**, demand for a service belonging to ServiceCategory **HotelReservation***
THEN
*the user prefers **2stars category, room with tv, conditioning air, EthnicMenuRestaurant***

Recommender Engine will verify the premise of the rule (that is the user, who requested for a service belonging to that ServiceCategory, is a student and his request comes from Rome) and will use the preferences contained in the rule consequence to estimate the similarity degree between these preferences and the concrete services. It is important to underline that the Recommender Engine is not able to manage rule with two different ServiceCategory. This is also out of our scope because what we want to obtain is a ranking of concrete services that must belong to the same service category. Each rule can be divided in three part:

1. The Event, that is the service category (element <ServiceCategory>) ;
2. The Conditions, that is the rule premise (element <Rule Premise>), which will consider:
   (a) the User (element <User Promise>),
   (b) the Context (element <Context Promise>),
3. The Actions (element <Rule Consequence>), that is the consequence that contains some values assigned to service extra functional description.

Finally, to each rule can be associated some parameters (element <Rule Parameters>) that will describe how the rule is relevant for an user taking into account either his overall behaviour and overall behaviour of all the MAIS users.

Semantic Matching: The ranking we calculate is a consequence of the application of the algorithm presented in [4] and based on the studies of Paolucci et al.[6].The ranking value of a web service is given by the value of the $SemSimilarity(UP, SP)$ function, defined as follow:

- $SemSimilarity(UP, SP)$ is a function that returns a percentage value representing the similarity degree between its arguments;
- The arguments are a User Profile (UP) and a service description profile (SP);
- The function measures the semantic similarity on all the non-functional parameters contained in the user profile and in the service profile.

More in particular:

$$SemSimilarity(UP, SP) = [\omega_1 QosSS(UP, SP) + \omega_2 AddSS(UP, SP)] \in [0..1]$$

with $\omega_1 + \omega_2 = 1$, in which $QosSS$ is the semantic similarity function calculated on the quality of service[7] parameters, whereas $AddSS$ is the semantic similarity function measured on the additional attributes of a web service. Thus, given a user profile UP and an web service description SP, we define:

$$QosSS(UP, SP) = \frac{\Sigma_i DegreeOfMatch(\alpha_i, \beta_i)}{n}$$

728   Angelo Corallo et al.

and

$$AddSS(UP, SP) = \frac{\Sigma_i DegreeOfMatch(\delta_i, \lambda_i)}{m}$$

The pairs$(\alpha_i, \beta_i) \in \Omega$ and $(\delta_i, \lambda_i) \in \Omega'$ indicate pairs of concepts (to which a pair of non-functional parameters in the user profile and in the service description have been semantically linked) belonging to the ontology Ω and Ω'. The number of QoS parameters is n while the number of additional attributes is m. $DegreeOfMatch$ function calculates the distance between two concepts which belong to an ontology and return a value according to the four cases presented above. That is, the estimation of the semantic similarity is a direct consequence of the $DegreeOfMatch$ applied to pairs of non-functional parameters.

5 Recommender Engine Architecture

The Recommender Engine is composed by two different tier: an enterprise tier (e-tier) and the resource tier (r-tier). The enterprise tier has the task of process the core aspect of the business component. It implements the interaction between business components and it must manage the data integrity. To the e-tier, usually, different users will access at the same time, so it is requested persistence and security. The resource tier manages physical access to shared resource. This tier allows to hide and to separate different aspects related to resources persistence from business logic aspect. The r-tier will be always accessible from its e-tier. The high level architecture of the Recommender Engine will be composed by:

- Enterprise Tier:
    - Facade: realizes the interface with which it is requested a recommendation and it encapsulates the complexity of the interactions between the business objects.
    - Business layer: encapsulates all the business objects that contains applicative logic. In particular it holds the logic for the semantic matching and for the generation of the ranking.

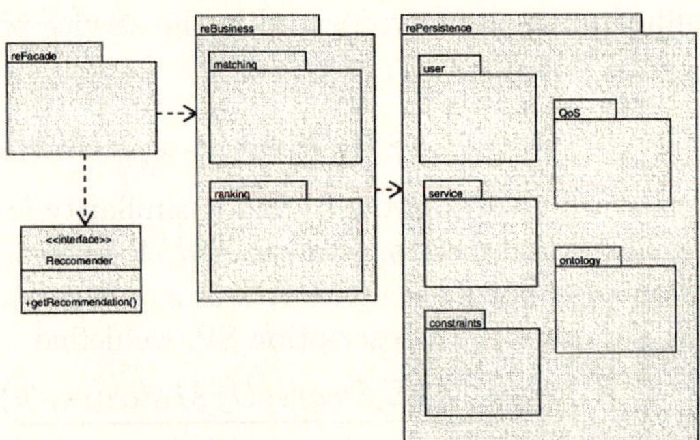

Fig. 3. High Level Architecture.

- Resource Tier:
    - Persistence layer: encapsulates the logic for the data persistence. More in particular, in this layer, we find all the components that manage the PBCs state and define data and business rules for accessing them by exposing their functionality.

6   Conclusion

This work aims to continue to explore the new field for the application of user profiling techniques and information filtering systems to web services and adaptive information systems. Starting from the assumption that service capabilities, that enable delivery in a personalized way, are those information related to the business capabilities of a service, we pointed out the need to add to technological description of a service a semantic layer for the business description. We presented the architecture of the system level component designed for the MAIS architecture, the Recommendation Environment, giving also an insight on the model and on the architecture of the Recommender Engine that is the core component of the Environment. Next steps will be to develop the recommender environment and to test it, by measuring its effectiveness in the generation of recommendations. It is also useful to test the entire environment to a case-study.

References

1. Cardoso, Sheth : Semantic e-Workflow composition. LSDIS Lab., Department of Computer Science, University of Georgia, (2003)
2. Sivashaunmugam K.: The METEOR-S Framework for Semantic Web Service Discovery. LSDIS Lab., Department of Computer Science, University of Georgia, (2002)
3. Herzum P., Sims O.: Business Component Factory. John Wiley & Sons. Chapter 2 (1999)
4. Corallo A., Caforio A., Elia G., Solazzo G.: Service Customization Supporting an Adaptive Information System. Knowledge-Based Intelligent Information and Engineering Systems: 8th International Conference, KES 2004, Wellington, New Zealand, September 20-25, 2004, Proceedings, Part III, p.342-349
5. OWL-S 1.0 Release http://www.daml.org/services/owl-s/1.0/
6. Paolucci et al.: Semantic Matching of Web Services Capabilities. Carnegie Mellon University, Pittsburgh, USA, (2002)
7. Ran S. : A model for Web Service Discovery with QoS, (2003)

Smart Clients and Small Business Model

Phu-Nhan Nguyen, Dharmendra Sharma, and Dat Tran

School of Information Sciences and Engineering,
University of Canberra, ACT 2601, Australia
Phu-Nhan.Nguyen@student.canberra.edu.au
{Dharmendra.Sharma,Dat.Tran}@canberra.edu.au

Abstract. We present in this paper an Asian Grocery server site web application developed for Smart Clients such as Desktop, Notebook, PDA and Pocket PC. The application accesses and retrieves data from an SQL Server 2000 and displays the data in both English and Vietnamese languages. Extensive testing of Smart Clients system and its applications has been performed to gather information and analyze its advantages and inadequacies found using Openwave version 7 Emulator and Microsoft Mobile Explorer version 3.

1 Introduction

Electronic applications in most businesses at present use rich client, e.g., Personal Computer (PC) and thin client, e.g., Personal Digital Assistants (PDA). These are useful but have some limitations. A new technology that has been developed to capture the benefits of both rich and thin client is Smart Client. Smart Client is an internet-connected device that allows the user's local applications to interact with server-based applications through the use of Web services. There are some types of Smart Client such as desktops, workstations, notebooks, tablet PCs, PDAs, and mobile phones. Smart Client is also an evolutionary step over rich client and thin (web) client applications.

In this paper, the applicability of Smart Client Technology is tested for an e-Commerce problem. We have developed an Asian Grocery online shop as a Smart thin client-server application and a Smart Clients Framework (SCF) modeled from the application. The design has been successfully implemented and tested.

2 Smart Clients

Smart Clients are Internet connected devices that allow the user's local applications to interact with server-based applications through the use of Web services. For example, a smart client running a word processing application can interact with a remote database over the Internet in order to collect data from the database to be used in the word processing document. Smart clients are distinguished by the following key characteristics:

- They can be used offline or online, ensuring that user productivity is not compromised even when the user is not connected.
- They consume Web services to provide richer functionality and up-to-date information to the business user.

R. Khosla et al. (Eds.): KES 2005, LNAI 3681, pp. 730–736, 2005.

- They can take advantage of the local processing power of the client device.
- They can be deployed and/or updated from a centralized server.
- Smart client applications support multiple platforms and languages because they are built on Web services [2, 5].

Smart client applications can run on almost any device that has Internet connectivity, including desktops, workstations, notebooks, tablet PCs, PDAs, and mobile phones.

Personal Digital Assistants

PDAs (handheld computers) provide a tremendously flexible and easy way to keep tabs on your appointments. Some PDAs can be used to record voice notes, play videos, display digital photos, listen to music, navigate maps, or read a novel. Using PDAs we could access to the Internet via a wireless fidelity (Wi-Fi) wireless connection, or exchange information from the short distances via Bluetooth device.

The main advantage of a PDA over a notebook is that the PDA has small size and lightweight. Other advantages of PDAs are that they are easy to update – single location update, to deploy – single location update, and to manage. The advantages of PDA are Ease of updating, deploying, and managing.

The PDAs that have the above features also have the following issues: Network dependency, poor user experience – mainly emit HTML, complex to develop, small screen that could not display all information as wish and users have to scroll a lot more to see the whole text, therefore loose the overview, text entry is still too difficult, and limitation of battery life, memory, communication bandwidth, and software applications.

Fat Clients

The Fat clients, Rich clients, Desktop Applications or Client/Server application provide features like rich user experience – by means of better user interface, offline capabilities – need not be connected on a network, high developer productivity, and responsive and flexible.

The Fat clients provide the above features but on the other hand they also have the following issues: tough to update – each location needs modifications, tough to deploy – deployment had to be done at multiple location, and "DLL Hell".

Smart Clients

According to Microsoft website [5], the ten best reasons for developers to start building Smart Client applications are as follows:

1. Improved reliability in heterogeneous network environments
2. Increased performance and scalability
3. Develop applications faster
4. Access to local machine functionality (such as DirectX)
5. Integration with existing desktop applications and systems
6. Ease of deployment and security
7. Mobility support and data synchronization capabilities

8. Native xml and web service support
9. Better user experience and user interface
10. Flexible data access and local caching of data.

Technology to Create Smart Clients

Web services provide the foundation for Smart Client software. Today Microsoft offers a rich set of technologies that provide rich support for Web services and deliver the capabilities for creating Smart Client solutions. These technologies include:

- The .NET Framework, which provides developers with a rich, managed execution environment for creating applications that target a variety of devices.
- Microsoft Visual Studio® .NET, which offers comprehensive tools for rapidly building and integrating Web services and applications, along with an open architecture that enables developers to use any language that targets the .NET Framework.
- Microsoft Windows Server™ 2003, a next-generation enterprise server that provides the platform for building and deploying connected solutions.

Hardware and Operating Systems of PDAs

Most PDAs are built upon CPU technology and only few PDAs are built upon Intel x86 CPU technology [10]. Like a notebook, a PDA has its own operating system, Central Processing Unit (CPU), Random Access Memory (RAM) and storage. The main advantage of a PDA over a notebook is its smaller size and lighter weight. In the current market, there are several manufacturers such as Nokia, Siemens, Sony Ericsson, and Palm creating PDAs, which combine the features of a notebook and a mobile phone together. The main operating system of the PDA is Windows CE or Palm OS, which dominates the market. The other operating systems are Linux 2.4, which uses the Sharp Zaurus Series, and Symbian OS for the Sony Ericsson P800.

3   Unicode

According to [1], Unicode is a universal character encoding required to produce software that can be localized for any language or that can process and communicate data in any language.

The Unicode standard is the product of a joint effort of information technology companies and individual experts. Its encoding has been accepted by ISO as the international standards ISO/IEC 10646. Unicode defines 16-bit codes for the characters of most scripts used in the world's languages. Encoding for some missing scripts will be added over time. The Unicode standard defines a set of rules that help implementers build text-processing and rendering engines. For digital, Unicode represents a strategic direction in internationalisation technology. Many software-producing companies have also announced future support for Unicode.

We have used this advantage by using Vietnamese Unicode to display Vietnamese characters in the Smart Clients. We have experienced that the fastest way to display Vietnamese Unicode properly in Smart Clients using ASP .NET MOBILE 2003 is the use of Vietnamese Profesionals Society (VPS) keyboard and Vietnamese Unicode fonts downloaded from the VPS website. We could install them, manipulate the keyboard symbol, integrate them with ASP .NET MOBILE 2003, and use "advanced saving as" to store the Vietnamese Unicode in the ASPX web pages. Alternatively, we can visit the www.vovisoft.com web site to read information of developing Vietnamese fonts. We can also download the table of Vietnamese symbol, which describes the combination of how to generate the Vietnamese Unicode symbols as shown in Figure 1. However, this method is very time intensive.

Bảng đối chiếu encoding các bộ chữ hiện hành với Unicode

VIQR	VPS	VPS Hex	VISCII	VISCII Hex	VNI	VNI Hex	TCVN	TCVN Hex	Unicode Symbol	Unicode Hex Dec		UTF-8 Hex	
a'	á	E1	á	E1	aù	61 F9	¸	B8	á	00E1	225	C3	A1
a`	à	E0	à	E0	aø	61 F8	µ	B5	à	00E0	224	C3	A0
a?	ả	E4	ả	E4	aû	61 FB	¶	B6	ả	1EA3	7843	E1	BA A3
a~	ã	E3	ã	E3	aõ	61 F5	·	B7	ã	00E3	227	C3	A3
a.	ạ	E5	Ö	D5	aï	61 EF	¹	B9	ạ	1EA1	7841	E1	BA A1
a(	æ	E6	ã	E5	aê	61 EA	¨	A8	ă	0103	259	C4	83
a('	ì	A1	ì	A1	aé	61 E9	¾	BE	ắ	1EAF	7855	E1	BA AF
a(`	¢	A2	¢	A2	aè	61 E8	»	BB	ằ	1EB1	7857	E1	BA B1
a(?	£	A3	Æ	C6	aû	61 FA	¼	BC	ẳ	1EB3	7859	E1	BA B3
a(~	¤	A4	Ç	C7	aû	61 FC	½	BD	ẵ	1EB5	7861	E1	BA B5
a(.	¥	A5	£	A3	aë	61 EB	Æ	C6	ặ	1EB7	7863	E1	BA B7
a^	â	E2	â	E2	aâ	61 E2	©	A9	â	00E2	226	C3	A2
a^'	Ä	C3	¤	A4	aá	61 E1	Ê	CA	ấ	1EA5	7845	E1	BA A5
a^`	À	C0	¥	A5	aà	61 E0	Ç	C7	ầ	1EA7	7847	E1	BA A7
a^?	Å	C4	¦	A6	aå	61 E5	È	C8	ẩ	1EA9	7849	E1	BA A9
a^~	Å	C5	ç	E7	aã	61 E3	É	C9	ẫ	1EAB	7851	E1	BA AB
a^.	Æ	C6	§	A7	aä	61 E4	Ë	CB	ậ	1EAD	7853	E1	BA AD
e'	é	E9	é	E9	eù	65 F9	Đ	D0	é	00E9	233	C3	A9

Fig. 1. Table of Vietnamese Symbols

4 Smart Client Architecture

As a demonstration for our investigation in developing new applications for Smart Clients, we have developed an Asian Grocery Shop for Smart Clients application. This application provides people a very inconvenient way to shop online from anywhere with their smart clients. This is regarded as the main business requirement. The application consisted of two projects, which were Asian Grocery Web and Asian Grocery Mobile.

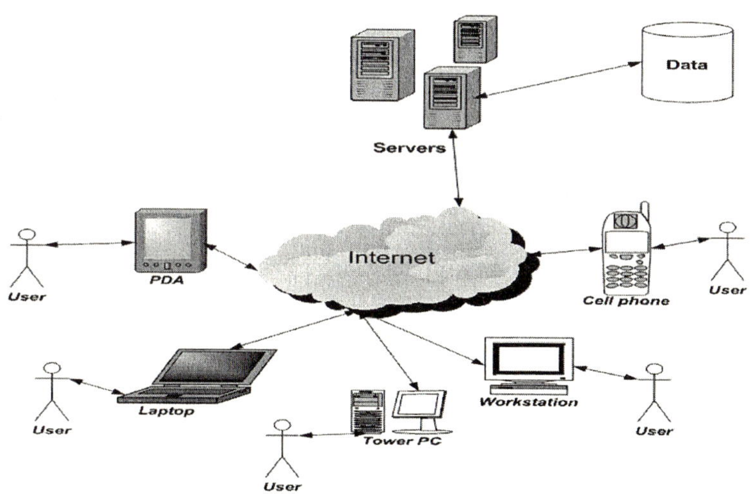

Fig. 2. Summaries of the Smart Clients Framework for small business applications

In order to develop the Asian Grocery Web project, which would be displayed only in PC and Notebook, we used ASP.NET, VB.NET, HTML, XML, JavaScript, Web Services, ADO.NET, Mail Services, IIS Web Server, and SQL Server 2000.

At the beginning, the IIS Web server would send to client a Welcome website, which included Welcome, Menu and Special pages. On the Special page, a DataGrid control was used to display retrieving data, which included information of the product and the link of picture, from the SQL Server database. We combined HTML, JavaScript, VB.NET and ASP.NET tools to get data from the SQL Server database and to display successfully data with picture into the DataGrid control, which allowed the user to enter items he/she would like to buy and to transfer data into the order list. When the user finishes the shopping, he/she is requested to perform payment via the Internet. The system will check his/her credit card number using the LUHM formula. Finally, we used SMTP services to send the order confirmation via email to his/her email address and used ADO.NET to insert data into the SQL Server database.

Asian Grocery Mobile is the second project, which will be displayed only for mobile devices and Pocket PC. We developed the project using ASP.NET Mobile, VB.NET, XML, WML, Web Services, ADO.NET, Mail Services, IIS Web Server and SQL Server 2000. (A tutorial on how an application is developed using the SCF framework is given at http://resources.dmt.canberra.edu.au/sharma/smart-clients.doc.)

5   Testing

We used the Openwave version 7 to test the system. Basic steps for testing are presented in Figure 3. The web mobile application using Openwave 7 does not require a starting or compiling process, just a click on the mobile web page such as Default.aspx from the design window, then select Tools, Openwave Simulator, and View in Openwave 7. For testing with MME, we need to start the project or press F5, then enter the URL http://localhost/conferencetut/ to the URL text box.

6   Conclusions and Further Investigation

PDAs have theirs limitations such as memory, storage, tools, battery life, security issues, very small screen, difficult text entry, software, tools, and programmers. However, PDAs have some very good advantages. They are very small and powerful devices. They combine all the best features of mobile and desktop PC together.

We have presented the Asian Grocery Online application as a demonstration of our Smart Clients Framework. The application is useful for business since it provides good accessibility, saves time and is low effective. For personal use, it is convenient since it allows the user to be able to shop online from anywhere with their Smart Clients. For social uses, the application is a good demonstration for applying new technology to create the best product to serve the society. In particular, the application can serve for Asians who cannot use English.

Some of our further investigations for further developing the SFC framework are as follows

- Security to be supported
- Providing an easier way for data entry, perhaps with voice recognition and dictation

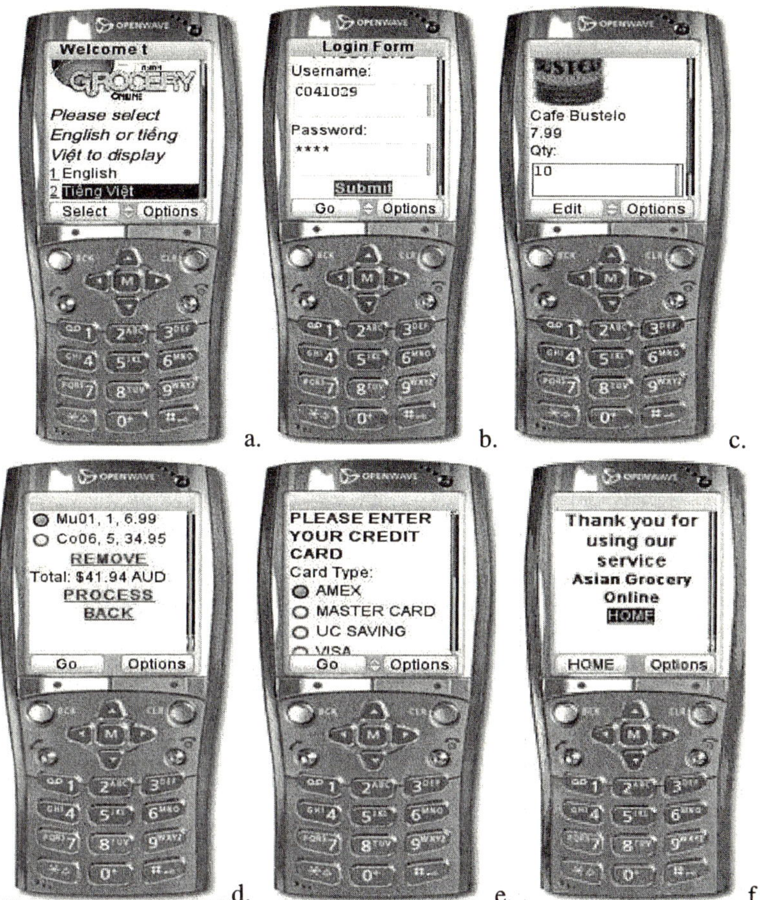

a.   b.   c.

d.   e.   f.

a. Welcome page that user can select English or Vietnamese fonts to display, b. User enters username and password, c. User views product details, d. Display products ordering, e. Select credit card type, f. Thank you

Fig. 3. Basic steps for testing the Asian Grocery Shop system

- Increasing screen size, memory, CPU, storage, and battery life
- Adding more controls and functions to develop web mobile applications such as the ListBox control from which the user can select item and manipulate it; the ObjectList control on which developers can write links that store image and display as the picture not as the text; and the TextBox control using Unicode fonts to which the users can enter their own languages.

References

1. Bettels, J. and Bishop, F. A.: Unicode: A Universal Character Code.
2. Deitel, H. M., Deitel, P.J., Nieto, T. R., and Steinbuhler, K.: Wireless Internet and Mobile Business How to Program. New Jersey (USA), Prentice Hall (2002)

3. Scott, G.: Practical WAP. Cambridge (UK), Cambridge University Press (2001)
4. Jeff, W.: Developing Web Applications with Microsoft Visual Basic .NET and Microsoft Visual C# .NET. Washington (USA), Microsoft Press (2002)
5. http://msdn.microsoft.com/smartclient: Smart Client Developer Center.
6. http://microsoft.com/net/smartclient: Maximize Your Applications with Smart Client Technology
7. http://msdn.microsoft.com/office/understanding/vsto/: Office Developer Center: Visual Studio Tools for the Microsoft Office System
8. http://www.Unicode.org, "What is a Unicode"
9. http://www.openwave.com, "Developer Product"
10. http://corky.net/2600/miscellaneous/pda.shtml

Synthetic Character with Bayesian Network and Behavior Network for Intelligent Smartphone

Sang-Jun Han and Sung-Bae Cho

Dept. of Computer Science, Yonsei University,
134 Shinchon-dong, Sudaemoon-ku, Seoul 120-749, Korea
{sjhan,sbcho}@sclab.yonsei.ac.kr

Abstract. As more people get using mobile phones, smartphone which is a new generation of mobile phone with computing capability earns world-wide reputation as new personal business assistant and entertainment equipment. This paper presents a synthetic character which acts as a user assistant and an entertainer in smartphone. It collects low-level information from various information sources available in smartphone, such as personal information, call log, and user input. Collected information is used for inferring high-level information with Bayesian network and determining agent's goal. This character takes a behavior appropriate to current situation using behavior selection mechanism. The proposed agent is toward a framework for integrating reactive and deliberative behavior.

1   Introduction

Recently mobile phones have become an essential tool for human communication. As more people use mobile phones, various services based on mobile phone networks and high-end devices have been developed. Smartphone which integrates the functions of personal digital assistant (PDA) and mobile phone earns worldwide reputation as new personal business assistant and entertainment equipment because it is all-in-one device: many technologies such as wireless voice/data communication, digital camera, and multi-media player are converged into one device. Current smartphone is just high-end mobile phone, but, they have the potential of being utilized to many novel services by developing innovative applications. However, current mobile devices have constraints of limited processing power, and awkward interaction devices. We are in need of AI techniques specialized in smartphone to cope with these constraints and make intelligent services in real.

There are three major issues in implementing intelligent service in constrained environment.

- To gather information which provides meaningful features for user's state: it is never an intelligent technique to require directly explicit information from user. Desired technique should be able to provide sufficient information while not bothering the user and not invading the user's privacy.

R. Khosla et al. (Eds.): KES 2005, LNAI 3681, pp. 737–743, 2005.

- To infer and predict user's state from collected data: predicting user's state from data can be formulated as conventional classification task. Many AI methods have been successfully applied to this problem.
- Service selection or composition: we can select one service from pre-defined service library or compose novel services appropriate to inferred user's state dynamically.

Here we used personal information, communication log, and the state of smartphone device as the information source, which provides the low-level clues used for inference and action selection. Bayesian network is used for inferring the high-level states from low-level information. In constructing the Bayesian network, we employ commonsense knowledge actively to deal with incomplete, imprecise data. It helps out inferring process by magnifying implicit and hidden clues in that kind of data. In order to generate services appropriate to user's state, we employ the concept of synthetic character based on behavior selection network. A character plays the role of a good medium for entertainment as well as a user assistant with intimate user interface. Its behavior reflects not only user's states such as how he/she feels and how busy he/she is, but device's states such as remaining battery power and things displayed on screen.

2   Intelligent Agent for Smartphone

The proposed agent consists of four components: perception system, emotion system, motivation system, and action selection system. The perception system checks the changes of personal and device information by monitoring address book, call log, user's input, currently running programs, etc. It determines the low-level states such as the category of ongoing scheduled event, the time elapsed after last input, and the number of missed call. The emotion system infers the high-level states such as affect and how busy the user from the low-level state. It uses Bayesian network and the low-level states is used as the evidence for inference. The motivation system determines the goal of the agent using low-level and high-level states. The action selection system chooses agent's action appropriate to the current state. Figure 1 shows an overview of the proposed agent.

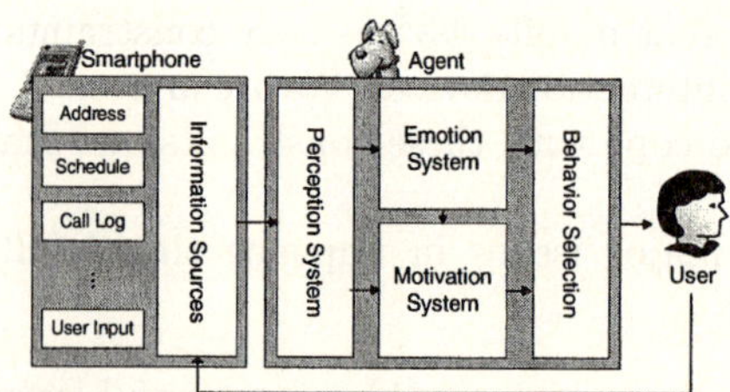

Fig. 1. An overview of the proposed agent.

2.1 Collecting Useful Information

Perception system collects basic information (low-level state) which is used to infer high-level state, determine agent's goals, or select appropriate action. The design methodology of perception system is as follows. First, we discover some events that seem to be related with user's state or agent's goal and define them as low-level state. Next, we specify the possible values that each state has and make concrete conditions for each value of state. For instance, to infer how he/she busy is, we use 'the number of missed calls' as related variable and define 'many', and 'few' as possible states of this variable. The state of 'the number of missed calls' is "many" if the number of calls which were not answered in the last two hours is more than five and 'few' if less than five. It monitors user's input, call/text message log, and changes in device's state or personal information continuously and update the value of the states using defined rules.

2.2 Inferring High-Level States

The low-level state is not enough to recognize user's unspoken needs and respond to it properly. It is needed to mine higher-order information from low-level information. However, there is much uncertainty in attempts to recognize high-level state such as user's object and affect. Bayesian probabilistic inference is one of the famous models for inference and representation of the environment with insufficient information.

The design methodology of Bayesian network is as follows. First, we discover some events that seem to be related user's context and define them as variables in Bayesian network. Next, specify the states that each variable has and make concrete conditions for each state of variable. For instance, to infer how he/she busy is, we use "the number of missed calls" as related variable and define "many", "few" as possible state of this variable. The state of "the number of missed calls" is "many" if the number of calls which did not be answered in the last two hours is over five and "few" if under 5. The accuracy of user modeling hinges upon how we define the variables in Bayesian network and the state of each variable.

After defining variables and its states, we construct the structure of Bayesian network considering the dependency among variables. In this stage, we have to specify the topology and the probability distributions. There are two ways to do this: constructing it automatically from data with learning algorithm and manually using expert's domain knowledge. In this work, we have constructed it manually because it is difficult to gather much amount of personal information data. Figure 2 illustrates the design process of Bayesian network.

The emotion system infers user's high-level state using probabilistic inference technique. We have used user's affect, how busy he/she is, and how close he/she is with someone as the high-level states. Bayesian network is used for inferring high-level states. The low-level states are used as variables in Bayesian network. Perception system traces the changes of personal information and communication log. Then, it sets the values of the variables in Bayesian network.

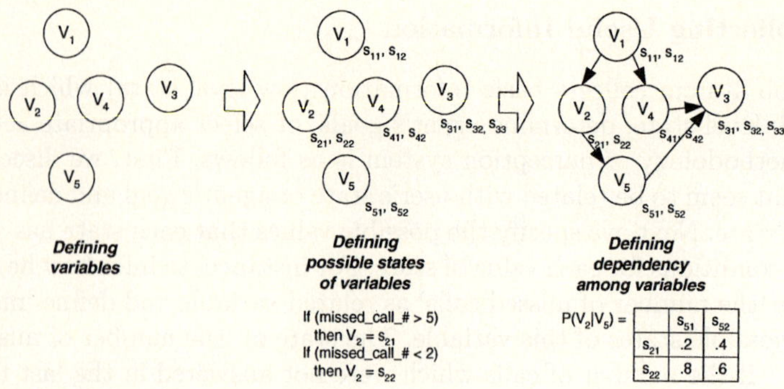

$$P(V_2|V_5) = \begin{array}{c|c|c} & S_{51} & S_{52} \\ \hline S_{21} & .2 & .4 \\ \hline S_{22} & .8 & .6 \end{array}$$

Fig. 2. The design process of Bayesian network.

The structure of Bayesian network is constructed considering the dependency among variables. We have to specify the topology and the probability distributions. There are two ways to do this: constructing it automatically from data with learning algorithm and manually from expert's domain knowledge. In this work, we have constructed it manually because it is difficult to gather enough personal information data.

Bayesian network for this problem has 33 observable variables whose state can be specified by observing user's behavior or personal information and 19 unobservable variables whose state is specified from relationship with other variables. All variables have two states and there are 56 dependencies defined among the variables.

- **Inferring user's affect.** Valence-Arousal (V-A) space has been applied to infer user's affect. V-A space is a simple model which represents affect as the position in the two-dimensional space. It has commonly used in previous studies on affect recognition [2]. It uses two eigenmoods: valence and arousal. Every affect can be described in terms of these eigenmoods. The valence axis ranges from negative to positive. The arousal axis ranges from calm to excited. For instance, "anger" is considered low in valence while high in arousal. In this work, we have used just four simple emotions: joyful, anger, sad, and relaxed depending on the sign of value in each axis.
  General commonsense knowledge has been utilized to define relationships among variables. For example, if there are many business schedules, user's emotion is likely to be negative. Many positive words or emoticons in user's incoming message box imply that user's emotion is likely to be positive.
- **Inferring how the busy user is.** The possible clues for inferring how busy the user is may be the amount of today's schedules, whether it is a holiday, and how frequently he uses the phone. Many business schedules would indicate that user is very busy with his work. No schedule in the evening implies that user has a rest. Busy user is likely to cause many missed calls or unanswered messages. With these commonsense rules, we define causal relationships among the variables.

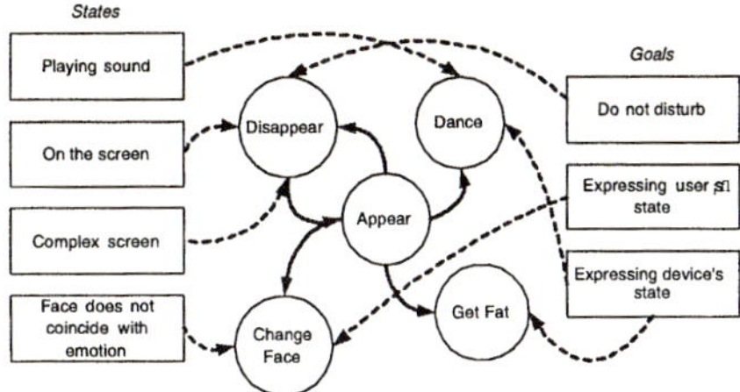

Fig. 3. A part of behavior network (Dotted: External link Solid: Internal link).

- **Inferring how close the user with someone.** In order to infer how close
  the user with people who are registered in address book, we have discovered
  the clues such as the content of incoming text messages, the name of group
  which he/she is added to, the frequency and duration of phone call, etc.
  The commonsense knowledge has been used to construct Bayesian network
  as follows. People who frequently make contact with or are registered at
  "friends" group in address book can be regarded as close. People who never
  got contact with in long time can be regarded as not close.

2.3 Determining Goal of Agent and Selecting Appropriate Action

The ultimate aims of our agents are assisting and entertaining the user in the way
which is appropriate to current situation. Therefore, it is needed to keep it to be
suitable for current situation by changing it from time to time. For this reason,
we determine the goal of the agent considering information about the user and
the device which is collected by perception system. Situation-goal rules are used
for determining current goal. The situation part consists of the low-level states
observed by perception system and high-level states inferred by Bayesian network. Things which are supposed to be need by the used in particular situation
is set as agent's goal. Action that is appropriate to current situation is selected
using behavior network. Whether the agent should make action and what action
should be activated is determined by selection mechanism of behavior network.
Behavior network has the advantage of flexibility with which human can deal in
comparison to simple rule-based selection mechanism. It continuously changes
the activation level of behaviors through selection procedure is adopted for flexibility [3]. The behavior network used in this paper has 26 states, 6 goals, 15
behavior nodes. Table 1 illustrates the basic behaviors of the synthetic character
and figure 3 shows a part of our behavior network.

3 A Working Scenario

To develop the agent prototype, we have used the Microsoft Pocket PC phone
edition SDK 2003. It provides a good implementation of real smartphone device
containing PIMS, voice conversation, text message service, web browser, and

Table 1. Basic behaviors of character.

Behavior	Description
Disappear	Disappears from the screen
Appear	Appears on the screen
Move	Moves to an appropriate position
Sleep	Lies down and sleeps
Turns over	Turns over while sleeping
Wake up	Wake up from sleeping
Get fat	Gets fat (the size of character increases)
Get slim	Gets slim (the size of character decreases)
Change facial expression	Express affect through facial expression
Notice something	Raises a notice board and informs something
Take a rest	Sits down on a chair and takes a rest
Dance	Dances shaking its tail
Vibrate	Vibrates for a while (use vibration motor)
Move to center	Moves to the center of the screen
Happy	Smiles and gets delighted

audio/video player. We have observed the agent's behavior according to user's input in order to demonstrate the feasibility of the agent.

User executes a word processor to read an e-book. Then the screen is filled up with many characters. The value of low-level state, 'complex screen', becomes true and 'do not disturb' is set as the agent's goal. 'disappear' behavior gets additional activation because all of its precondition: 'complex screen', 'awake', and 'on the screen' is true and current goal: 'do not disturb' is in the add list of 'disappear'. As time goes by, the global threshold gets lowered and the activation value of 'disappear' gets higher. After several iteration of activation update procedure, 'disappear' behavior is selected and character is disappeared from the screen as shown in Figure 4(a). User can read the e-book comfortably without manual turning off.

User plays music after reading an e-book. Then, 'complex screen' state becomes false and 'playing sound' becomes true. The motivation system set agent's goal as 'expressing device's state' because 'playing sound' changes from false to true. At first, there are no behaviors of which all preconditions are met because almost behaviors has 'on the screen' as precondition, but they give activation to the 'appear' behavior because 'on the screen' is in the add list of 'appear'. Consequently, 'appear' is selected and the character enters the screen. After that, many behaviors start to compete and the 'dance' behavior has highest activation because all of its preconditions such as 'playing sound', 'awake' and 'on the screen' are met and current goal: 'expressing device's state' is in its add-list. Then, the character begins to dance and entertains the user who is listening to music as shown in Fig. 4(c). Figure 5 shows the agents behavior according to the changes of the low-level and high-level states.

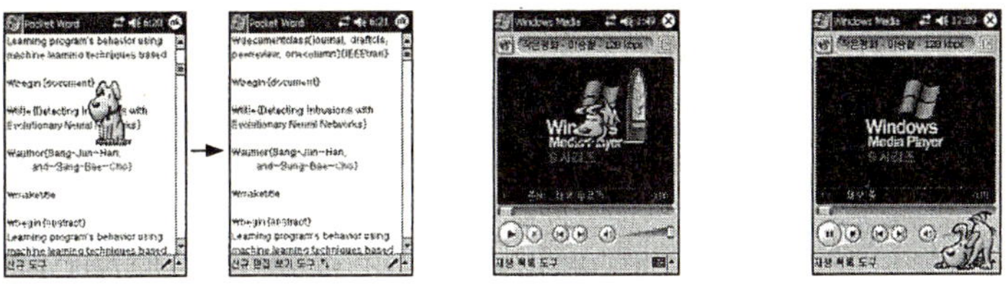

(a) 'disappear' behavior   (b) 'appear' behavior (c) 'dance' behavior

Fig. 4. Agent's behavior.

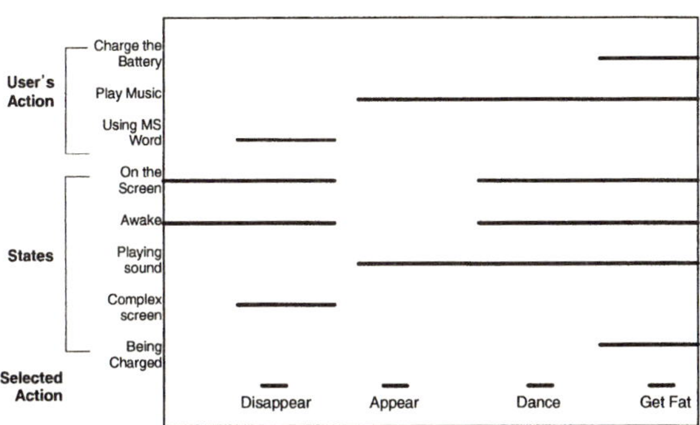

Fig. 5. Changes of states and goals according to user's behavior.

4   Conclusion

In this paper, we proposed a synthetic character as a user assistant and an entertainer. It collects low-level information from various information sources available in smartphone. They are used as clues of high-level information which is inferred by Bayesian network and the basis of determining agent's goal. High/low-level states and agent's goal are inputted into behavior selection mechanism in order to take a behavior appropriate to current situation.

Acknowledgement

This research was performed for the Ubiquitous Computing Research Program funded by the Ministry of Information and Communication of Korea

References

1. E. Charniak, "Bayesian Networks without Tears," *AI Magazine,* vol. 12, no. 4, pp. 50-63, 1991.
2. R. Picard, "Affective Computing," *Media Laboratory Perceptual Computing TR 321,* MIT Media Laboratory, 1995
3. P. Maes, "How to Do the Right Thing," *Connection Science Journal,* vol. 1, no. 3, pp. 291-323, 1989.

The Virtual Technician: An Automatic Software Enhancer for Audio Recording in Lecture Halls

Gerald Friedland, Kristian Jantz, Lars Knipping, and Raúl Rojas

Freie Universität Berlin, Institut für Informatik, Takustr. 9, 14195 Berlin, Germany
{fland,jantz,knipping,rojas}@inf.fu-berlin.de

Abstract. Webcasting and recording of university lectures has become common practice. While much effort has been put into the development and improvement of formats and codecs, few scientist have studied how to improve the quality of the signal before it is digitized. Lecture halls or seminar rooms are not professional recording studios. Good quality recordings require full-time technicians to setup and monitor the signals. This paper describes a tool that eases studioless voice recording by automatizing several tasks usually handled by audio technicians. The expert system measures the quality of the sound hardware used, monitors possible hardware malfunctions, prevents common user mistakes, and provides gain control and filter mechanisms.

1 Introduction

The work presented in this paper has emerged from experiences in developing a lecture recording system, called E-Chalk [1, 2]. E-Chalk produces distance learning as a by-product of traditional chalkboard teaching. The system provides both live transmission and on-demand replay of the lecture from the Web. Remote students follow the lecture looking at the dynamic board content and listening to the recorded voice of the instructor. While the system is now commercially available and in use in several educational institutions in Germany and in other countries, the many possible audio distortion sources in lecture halls and classrooms constitute a usability problem. However, the problems are general and also apply to any other lecture recording system.

The room is filled with multiple sources of noise: Students are murmuring, doors slam, cellular phones ring. In bigger rooms there may also be reverberation that depends on the geometry of the room and of the amount of people in the seats. The speaker's voice has not always the same loudness. Even movements of the lecturer can create noise. Coughs and sneezes, both of the audience and the speaker, result in irritating sounds. The loudness and the volume of the recording depend on the distance between microphone and the speaker's head, which is usually changing all the time. Additional noise is introduced by the sound equipment: Hard disks and fans in the recording computer produce noise, long wires can cause electromagnetic interference that results in noise or humming. Feedback loops can also be a problem if the voice is amplified for the audience. The lecturer's attention is entirely focused on the presentation and technical

R. Khosla et al. (Eds.): KES 2005, LNAI 3681, pp. 744–750, 2005.

problems can be overlooked. For example, the lecturer can just forget to switch the microphone on. In many lectures weak batteries in the microphone cause a drastic reduction of the signal to noise ratio, without the speaker noticing it. Many people have also problems with the operating system's mixer. It differs from sound card to sound card, and from operating system to operating system and usually has many knobs and sliders with potentially wrong settings. Selecting the right port and adjusting mixer settings can take minutes even to experiences users. Another subject is equipment quality. Some sound cards cannot deliver high fidelity audio recordings. In fact, all popular sound cards focus on sound playback but not on sound recording. Game playing and multimedia replays are their most important applications. On-board sound cards, especially those in laptops, have often very restricted recording capabilities. The quality loss introduced by modern software codecs is perceptually negligible compared to the described problems. Improving audio recording for lectures held in lecture halls means first and foremost improving the quality of the raw signal before it is processed by the codec.

For a more detailed discussion of these problems see for example [3]. This article presents a kind of expert system that actively fights audio distortions by simulating the tasks an audio technician would perform using different hardware devices.

2   Related Work

Windows Media Encoder, RealProducer, and Quicktime are the most popular Internet streaming systems. None of them provides speech enhancement or automatized control mechanisms. A possible reason is, that in typical streaming use cases, a high quality audio signal is fed in by professional radio broadcasting stations. An audio technician is assumed to be present. Video conferencing tools, such as Polycom ViaVideo, do have basic filters for echo canceling or feedback suppression. Microsoft's Real-Time Communications API provides an audio tuning wizard [4], but this program only provides a manual input device selection dialog and a dialog that helps a user to adjust the microphone distance. The audio quality needed for a video conference is much lower than what is required for a recording. It is well known that the same noise is less irritating for a listener when it is experienced during a live transmission.

Especially in the HAM amateur radio sector there are several specialized solutions for enhancing speech intelligibility, see for example [5]. Although these solution have been implemented as analog hardware, the underlying ideas are often effective and many of them can be realized in software. Octiv, Inc applies real time filters to increase intelligibility in telephone calls and conferences. They provide hardware to be plugged into the telephone line. Cellular telephones also apply filters and speech enhancement algorithms, but these rely on working with special audio equipment. Generic remastering packets like Steinberg WaveLab need an introductory seminar to be used. Steinberg also provides easy to use software for special purposes like My MP3 Pro, but none is available for live recording.

In academic research many projects seek to solve the Cocktail Party Problem [6]. Most approaches try to solve the problem with blind source separation using extra hardware, such as multiple microphones. Itoh and Mizushima [7] published an algorithm that identifies speech and non-speech parts of the signal before it uses noise reduction to eliminate the non-speech parts. The approach is meant to be implemented on a DSP and although aimed at hearing aids it could also be useful for sound recording in lecture rooms.

3   Enhancing Audio Recordings

Not all of the problems mentioned above can yet be solved only by software, just like a real technician cannot solve every issue without adding hardware. The software described in this paper focuses on the special case of lecture recording and relies on the lecturer using some kind of directional microphone or a headset. This eliminates the influence of room geometry and of cocktail party noise.

3.1   The Expert System Takes Over

A lecture recording system has the advantage that information about speaker and equipment are accessible in advance. The expert system therefore works in two steps: The sound card, the equipment, and the speaker's voice are analyzed before recording. During recording, the expert system uses filters, hardware monitors, and automatic gain control automatically configured using the information collected in the first step.

The expert system is hybrid as it is both model and rule based. The audio signal is analyzed using mathematical models derived from well known standards used in radio and TV broadcasting. A detailed discussion of all the models would go beyond the scope of this article. The authors used mainly the models described in [8]. Based on these standard models, the expert system uses rules to assist a user in assessing the quality of the audio equipment and makes him aware of its influence on the recording. For example, overflows in a ground noise recording will result in the system showing descriptive warnings to the user as hardware problems or handling errors must be the cause.

During the next sections, several models and rules will be discussed along with the user interface.

3.2   Before Recording

Before lectures are recorded, the user creates a so-called audio profile. It represents a fingerprint of the interplay of sound card, equipment and speaker. The profile is recorded using a GUI wizard that guides through several steps, see Figure 1. This setup takes about three minutes and has to be done once per speaker and sound equipment. Each speaker uses his audio profile for all his recordings.

The first screen asks the user to assemble the hardware as it is to be used in the lecture. The wizard detects the sound card and its mixing capabilities. Using

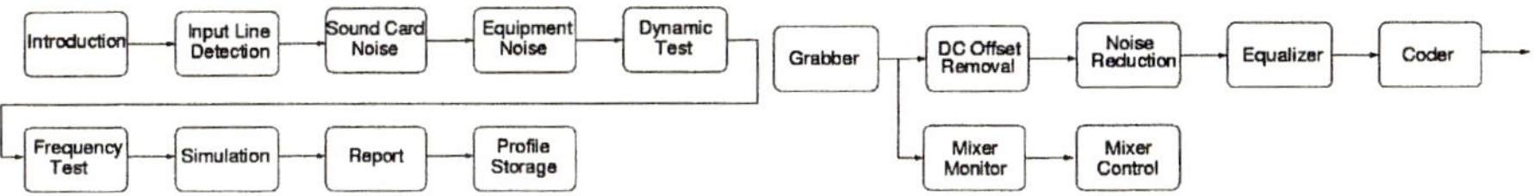

Fig. 1. The steps of the audio profile wizard (left) and the system during recording (right).

the operating system's mixer API the sound card's input ports are scanned to find out the recording devices plugged in. This is done by shortly reading from each port with its gain at a maximum, while all other input lines are muted. The line with the maximum noise level is assumed to be the input source. For the result to be certain, the maximum must differ to other noise levels by a certain threshold, otherwise the user is required to select the line manually. With a single source plugged in, this occurs only with digital input lines because they produce no background noise. At this stage several hardware errors can also be detected, for example if noise is constantly at zero decibel there is a short circuit.

The audio system analyzer takes control over the sound card mixer. There is no need for the user to deal with the operating system's mixer.

The next step is to record the sound card background noise. The user is asked to pull any input device out of the sound card. A few seconds of noise are recorded.

After recording sound card noise level, the user is asked to replug and switch-on the sound equipment. Again, several seconds of "silence" are recorded and analyzed. Comparing this signal to the previous recording exposes several handling and hardware errors. For example, a recording device plugged into the wrong input line is easily detected.

After having recorded background noise, the user is asked to record phrases with special properties. They are language dependent. A phrase containing many explosives is used to determine the range of the gain. In English, repeating the word "Coffeepot" gives good results. This measurement of the signal dynamics is used to adjust the automatic gain control. By adjusting the sound card mixer's input gain at the current port, the gain control levels out the signal. The average signal level should be maximized but overflows must be avoided. If too many overflows are detected, or if the average signal is too low, the user is informed about possible improvements.

During the frequency test, a sentence containing several sibilants is recorded to figure out the upper bound frequency. The system looks at the frequency spectrum to warn the user about equipment anomalies.

The final recording serves as the basis for a simulation and allows fine tuning. The user is asked to record a typical start of a lecture. The recording is filtered (as described in Section 3.3), compressed, and uncompressed again. The user can listen to his or her voice as it will sound recorded. If necessary, an equalizer (according to ISO R.266) allows experienced users to further fine tune the frequency spectrum. The time for filtering and compressing is measured. If this process takes too long, it is very likely that audio packets are being lost during real recording due to a slow computer.

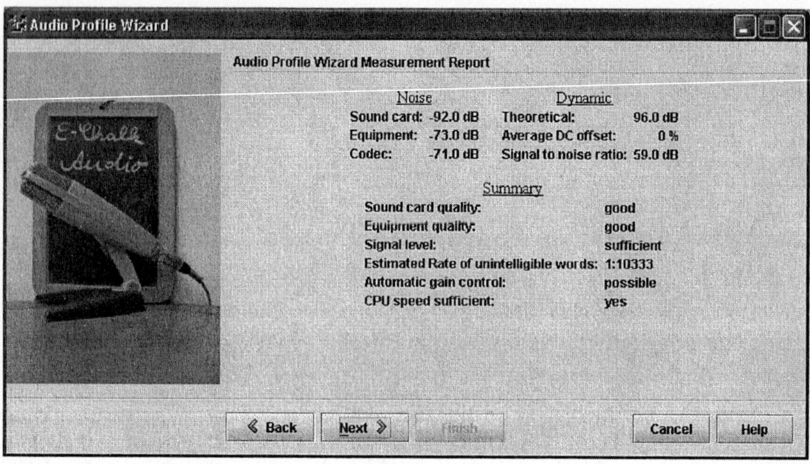

Fig. 2. A report gives a rough estimation of the quality of the equipment. Word intelligibility is calculated according to IEC 268.

At the end of the simulation process a report is displayed, as shown in Figure 2. The report summarizes the most important measurements and grades sound card, equipment, and signal quality into the categories *excellent*, *good*, *sufficient*, *scant*, and *inapplicable*. The sound card is graded using background noise and the card's DC offset calculated from the recordings. The grading of the equipment is based on the background noise recordings and the frequency shape. This is only a rough grading, assisting non expert users to judge the equipment and identify quality bottlenecks. Further testing would require the user to work with loop back cables, frequency generators, and/or measurement instruments.

Among other information, the created profile contains all mixer settings, the equalizer settings, the recordings, and the sound card's identification.

3.3   During Recording

For recording, the system relies on the profile of the equipment. If changes are detected, for example a different sound card, the system complains at start up. This section describes how the recording profile is used during the lecture. Figure 1 illustrates the signal's processing chain.

The mixer settings saved in the profile are used to initialize the sound card mixer. The mixer monitor complains if it detects a change in the hardware configuration such as using a different input jack. It supervises the input gain in combination with the mixer control. A warning is displayed if too many overflows occur or if the gain is too low, for example, when microphone batteries are running out of power. The warning disappears when the problem has been solved or if the lecturer decides to ignore the problem for the rest of the session.

The mixer control uses the values of the dynamic test to level out the input gain using the sound cards mixer. The analog preamplifiers of the mixer channels thus work like expander/compressor/limiter components used in recording

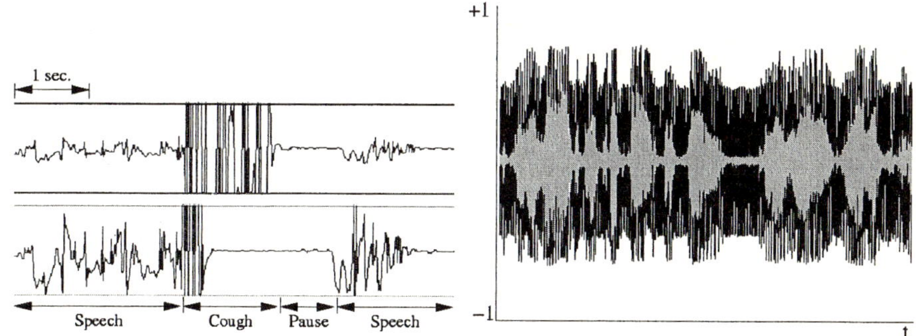

Fig. 3. Left: Without (above) and with (below) mixer control: The speech signal is expanded and the cough is leveled out. Right: Three seconds of a speech signal with a 100 Hz sine-like humming before (black) and after filtering (gray).

studios. This makes it possible to level out voice intensity variations. Coughs and sneezes, for example, are leveled out, compare Figure 3. The success of this method depends on the quality of the sound card's analog mixer channels. Sound cards with high quality analog front panels, however, are becoming cheaper and are getting more popular.

Mixer control reduces the risk of having feedback loops. Whenever a feedback loop starts to grow, the gain is lowered. As in analog compressors used in recording studios, the signal to noise ratio is lowered. For this reason noise filters, as described in the next paragraph, are required.

The signal's DC offset is removed, the sound card background noise level recorded in the profile is used as threshold for a noise gate and the equipment noise as a noise fingerprint. The fingerprint's phase is aligned with the recorded signal and subtracted in frequency space. This removes any humming caused by electrical interference. Because the frequency and shape of the humming might change during a lecture, multiple noise fingerprints can be specified. A typical situation that changes humming is when the light is turned on or off. The best match is subtracted [9]. See Figure 3 for an example. It is not always possible to prerecord the humming, but if so this method is superior to using electrical filters. These have to be fine tuned for a specific frequency range and often remove more than wished.

Equalizer settings are applied before the normalized signal is processed by the codec. The filtering also results in a more efficient compression. Because noise and overflows are reduced, entropy also scales down and the compression can achieve better results. Several codecs have been tested, for example, a modified version of the ADPCM algorithm and codecs provided by the Windows Media Encoder.

4   Summary and Perspective

The system presented in this paper improves the handling of lecture recording systems. An expert system presented via a GUI wizard guides the user through a systematic setup and test of the sound equipment. The quality analysis presented cannot substitute high-quality hardware measurements of the sound equipment but provides a rough guideline. Once initialized, the system monitors and controls important parts of the sound hardware. A range of handling errors and hardware failures are detected and reported. Classical recording studio equipments like graphical equalizers, noise gates, and compressors are simulated and automatically operated. This software system does not replacc a generic recording studio, nor does it make audio technicians jobless. In the special case of on-the-fly lecture recording, however, it eases the handling of lecture recording and improved the overall quality of the recordings without requiring cost-intensive technical staff.

The new automated audio system was tested during an algorithm design course in winter term 2004. Even though having a setup time of three minutes once was at first considered cumbersome, opinion changed when the listeners of the course reported a more pleasant audio experience. The system has now been integrated into the E-Chalk system for deployment and will be in wide use starting this summer term 2005. This will give us the chance for a broader evaluation. In the future the system should be capable of handling multiple microphones and inputs to enable switching between classroom questions and lecturer's voice. One would also like to interface with external studio hardware, such as mixer desks, to enable auto operation.

Freeing users from performing technical setups by automatization, as recently observable in digital photography, is still a challenge for audio recording.

References

1. Friedland, G., Knipping, L., Rojas, R., Tapia, E.: Web Based Education as a Result of AI Supported Classroom Teaching. In: Proceedings of the 7th KES Conference. Volume 2774 of Lecture Notes of Computer Sciences., Oxford, U.K (2003) 290–296
2. Rojas, R., Friedland, G., Knipping, L., Tapia, E.: Teaching With an Intelligent Electronic Chalkboard. In: Proceedings of ACM Multimedia 2004, Workshop on Effective Telepresence, New York, NY, USA (2004) 16–23
3. Katz, B.: Mastering Audio: The Art and the Science. Focal Press (Elsevier), Oxford, UK (2002)
4. Microsoft, Inc: Microsoft Real-Time Communications API. http://www.devx.com/Intel/Article/20008/0/page/4 (2002)
5. Universal Radio, Inc: MFJ-616 Speech Intelligibility Enhancer Instruction Manual. www.hy-gain.com/man/mfjpdf/MFJ-616.pdf (2002)
6. Haykin, S.: Cocktail party phenomenon: What is it, and how do we solve it? In: European Summer School on ICA, Berlin, Germany (2003)
7. Itoh, K., Mizushima, M.: Environmental noise reduction based on speech/non-speech identification for hearing aids. In: Proceedings of the IEEE International Conference on Acoustics, Speech, and Signal Processing, Munich, Germany (1997)
8. Dickreiter, M.: Handbuch der Tonstudiotechnik. 6th edn. Volume 1. K.G. Saur, Munich, Germany (1997)
9. Boll, S.: Supression of acoustic noise in speech by spectral substraction. IEEE Transactions, ASSP **28** (1979) 113–120

Intelligent Environments for Next-Generation e-Markets

John Debenham and Simeon Simoff

Faculty of Information Technology, University of Technology, Sydney
{debenham,simeon}@it.uts.edu.au

Abstract. A complete, immersive, distributed virtual trading environment is described. Virtual worlds technology provides an immersive environment in which traders are represented as avatars that interact with each other, and have access to market data and general information that is delivered by data and text mining machinery. To enrich this essentially social market place, synthetic smart bots have also been constructed. They too are represented by avatars, and provide "informed idle chatter" so enriching the social fabric. The middle-ware in this environment is based on powerful multiagent technology that manages all processes including the information delivery and data mining. The investigation of network evolution leads to the development of new "network mining" techniques.

1   Introduction

Markets play a central role in economies, both real and virtual. One interesting feature of the development of electronic markets has been the depersonalisation of market environments. The majority of on-line trading today is conducted by "filling in pro formas" on computer screens, and by "clicking buttons". This has created a trading atmosphere that is far removed from the vital atmosphere of traditional trading floors. What used to be the trading floor of the Sydney Stock Exchange is now a restaurant, although the trading prices are still displayed on the original board. All of this sits uncomfortably with work in microeconomics that has demonstrated both theoretically and in practice, that a vital trading environment provides a positive influence on liquidity and so too on clearing prices. See, for example, the work of Paul Milgrom, from his seminal "linkage principle" to his recent work [1].

This issue is being addressed by this project in which, virtual worlds technology, based on Adobe Atmosphere is being used to construct an intelligent virtual trading environment that may be "hooked onto" real exchanges. That environment is immersive and complete. It is "immersive" in that: an avatar that may move freely through those virtual trading areas in which it is certified represents each real player. Unstructured data mining tools povide each real player with all the information that she requires, including general information extracted from newsfeeds. The environment also contains synthetic characters that can answer questions like "how is gold doing today?" with both facts and news items. The environment may be seen at: http://research.it.uts.edu.au/emarkets/ — first click on "virtual worlds" under "themes and technologies" and then click on "here". To run the demonstration you will need to install the Adobe Atmosphere plugin. Figure 1 shows a screenshot of a market scenario in a virtual world. The avatar with its "back to the camera" is the avatar representing the agent on the workstation from which the "photo" was taken.

R. Khosla et al. (Eds.): KES 2005, LNAI 3681, pp. 751–757, 2005.
© Springer-Verlag Berlin Heidelberg 2005

Other related projects include the design of trading agents that operate in these information-rich environments [2], and an electronic institution framework that "sits beneath" the virtual environment. The design of this framework is influenced by the Islander framework developed in Barcelona [3]. Unstructured data mining techniques are being developed to tap real-time information flows so as to deliver timely information, at the required granularity [4]. The development of new network mining techniques are presently a major focus.

Fig. 1. A screen shot of one of the eMarkets, including the "chatterbot" avatars

2   The Virtual Trading Environment as a Multi-agent Virtual 3D World

A marketplace is a real or virtual space populated by agents that represent the variety of human and software traders, intermediaries, and information and infrastructure providers. Market activities include: trade orders, negotiations and market information seeking. These activities are *intrinsically social* and *situated*. There is a number of ways to implement an electronic market [5]. The virtual trading environment is *a multi-agent virtual world* [6]. In this environment every activity is managed as a constrained business process by a multiagent process management system. Players can trade securely via the process management system, in addition "chat channels" support informal communication. Representation refers to all aspects of the appearance of objects, avatars, bots and other elements of the virtual world [7].

The virtual trading environment is a multi-agent virtual world (Fig. 1), which integrates a number of technologies: (i) an electronic market kernel (Java-based), which supports the actual market mechanisms; (ii) a virtual world, involving all aspects of the appearance and interactions between market players, bots and other objects in the e-market place (based on Adobe Atmosphere); (iii) a variety of information discovery bots (Java-based), and; (iv) a multi-agent system, which manages all market transactions and information seeking   as industry processes (based on Australian JACK technology).

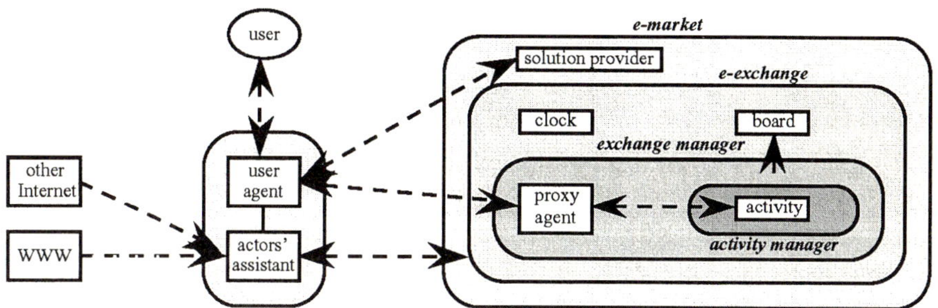

Fig. 2. High-level view of the environment and a user

In the virtual trading environment, all transactions are managed as heavily constrained business processes using a multiagent system. Constraints include process-related (eg: time, value, cost) and player-related (eg: preferences derived from inter-player relationships). The latter are based on a model of 'friendship' development between pairs of actors [8]. Existing process-related constraints will be extended to include measures, (eg. costs) associated with developing and sustaining complex business networks. Research on the evolution of player strategies, with. prior work focusing on highly stylised environments (eg. [9]). The way strategies develop in a virtual marketplace as a result of timely information provision is the subject of current work which will provide the basis for relating strategies to social networks.

Each player in the virtual trading environment, whether a human or a software agent, is represented by an avatar. Generated complex data about interactions between players in different market scenarios, the corresponding transactions and information seeking strategies, reflect all aspects of the market players' behaviour and the interactions between different players involved in electronic market scenarios.

Fig. 2 shows a high-level of the environment and its interaction with a single user. The *e-exchange* is a virtual space in which a variety of market-type *activities* can take place at specified times. Each activity is advertised on a notice *board* which shows the start and stop time for that activity as well as what the activity is and the *regulations* that apply to players who wish to participate in it. A human player works though a PC (or similar) by interacting with a *user agent* which communicates with a *proxy agent* or a solution provider situated in the trading environment. The user agents may be 'dumb', or 'smart' being programmed by the user to make decisions. Each activity has an *activity manager* that ensures that the regulations of that activity are complied with. The set of eight actor classes supported by the trading environment is illustrated in Fig 3.

The following actor classes are supported. *E-speculators* take short term positions in the e-exchange. This introduces the possibility of negative commission charges for others using an e-exchange. Sell-side *asset exchanges* exchange or share assets between sellers. *Content aggregators*, who act as forward aggregators, coordinate and package goods and services from various sellers. *Specialist originators*, who act as reverse aggregators, coordinate and package orders for goods and services from various buyers. The specialist originators, content aggregators and e-speculators represent their presence in the trading environment as avatars that create activities.

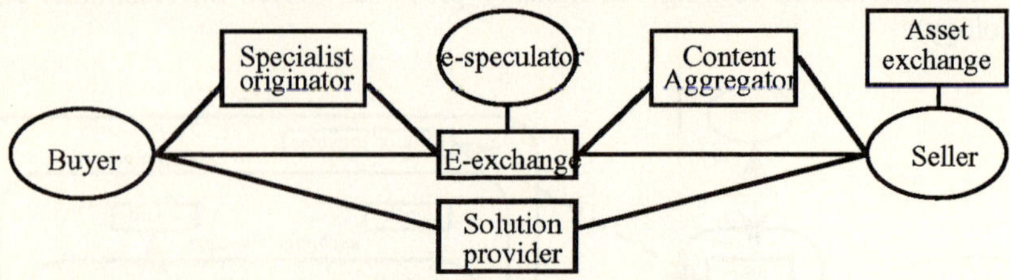

Fig. 3. The eight actor classes

There are valuable studies on the cognitive processes and factors driving alertness to opportunities. There is also a wealth of established work in economic theory, that provides a basis for work in e-markets [10], including: the theory of auctions, theory of bargaining, and theory of contracts. [11] presents mechanisms for negotiation between computational agents. That work describes tactics for the manipulation of the utility of deals, trade-off mechanisms that manipulate the value, rather than the overall utility, of an offer, and manipulation mechanisms that add and remove issues from the negotiation set. Much has to be done before this established body of work may form a practical basis for an investigation into electronic market evolution. Little is known on how entrepreneurs will operate in electronic market places, although the capacity of the vast amount of information that will reside in those market places, and on the Internet generally, to assist the market evolution process is self-evident.

3 Smart Information System

In 'real' problems, a decision to use an e-exchange or a solution provider will be made on the basis of general knowledge that is external to the e-market place. The negotiation strategy used will also depend on general knowledge. Such general knowledge will typically be broadly based and beyond the capacity of modern AI systems whilst remaining within reasonable cost bounds. Once this decision has been made, all non-trivial trading relies on the provision of current information. The *smart information system*, or "actors assistant", aims to deliver "the right information, at the right granularity, to the right person, at the right time".

E-markets reside on the Internet alongside the vast resources of the World Wide Web. The general knowledge available is restricted to that which can be gleaned from the e-markets themselves and that which can be extracted from the Internet in general—including the World Wide Web. The actors' assistant is a workbench that provides a suite of tools to *assist* a trader in the e-market. The actors' assistant does *not* attempt to replace buyers and sellers. For example, there is no attempt to automate 'speculation' in any sense. Web-mining tools assist the players in the market to make informed decisions. One of the issues in operating in an e-market place is coping with the rapidly-changing signals in it. These signals include: product and assortment attributes (if the site offers multiple products), promotions shown, visit attributes (sequences within the site, counts, click-streams) and business agent attributes. Combinations of these signals may be vital information to an actor. A new generation of data analysis and supporting techniques—collectively labelled as *data mining* methods—are now applied to stock market analysis, predictions and other financial and market analysis applications. The application of data mining methods in e-business to date has predominantly been within the B2C framework, where data is mined at an on-line business site, resulting in the derivation of various behavioural metrics of site visitors and customers.

The basic steps in providing assistance to actors are:

- identifying potentially relevant signals;
- evaluating the reliability of those signals;
- estimating the significance of those signals to the matter at hand;
- combining a (possibly large) number of signals into coherent advice, and
- providing digestible explanations for the advice given.

The estimation of the significance of a signal to a matter at hand is complicated by the fact that one person may place more faith in the relevance of a particular signal than others. So this estimation can only be performed on a personal basis. This work does *not*, for example, attempt to use a signal to predict whether the US dollar will rise against the UK pound. What it *does* attempt to do is to predict the value that an actor will place on a signal. So the feedback here is provided by the user in the form of a rating of the material used. A five point scale runs from 'totally useless' to 'very useful'. Having identified the signals that a user has faith in, "classical" data mining methods are then applied to combine these signals into succinct advice again using a five point scale. This feedback is used to 'tweak' the weights in Bayesian networks and as feedback to neural networks [12]. Bayesian networks are preferred when some confidence can be placed in a set of initial values for the weights.

4 Synthetic Characters

If trading environments are to provide the rich social experience that is found in a real, live exchange then they should also contain other classes of actors in addition to essential classes described above. For example, anonymous agents that can provide intelligent casual conversation—such as one might have with a stranger in an elevator. To have a positive role in an eMarket environment, such anonymous agents should be informed of recent news, particularly the news related to the market where the agent is situated (eg. financial news, news from major stock exchanges, news about related companies). If so then they provide a dual social function. First, they are agents to whom one may "say anything", and, second, they are a source of information that is additional, and perhaps orthogonal, to that provided directly to each trading agent as it has specified.

Anonymous synthetic characters have been embedded in the virtual trading environment. Avatars represent these agents visually. They provide casual, but context-related conversation with anybody who chooses to engage with them. This feature markedly distinguishes this work from typical role-based chat agents in virtual worlds. The later usually are used as "helpers" in navigation, or for making simple announcements. They are well informed about recent events, and so "a quick chat" can prove valuable. A demonstration that contains one of these synthetic characters is available at: http://research.it.uts.edu.au/emarkets/. We describe the design of these agent-based characters, and the data mining techniques used to "inform" them.

The following issues have to be addressed in the design of any synthetic character:

- its appearance
- its mannerisms
- its sphere of knowledge
- its interaction modes
- its interaction sequences

and how all of these fit together. Appearance and mannerisms have yet to be addressed. At present the avatars that represent the synthetic characters simply "shuffle around".

An agent's sphere of knowledge is crucial to the value that may be derived by interacting with it. We have developed extensive machinery, based on unstructured data mining techniques, that provides our agent with a dynamic information base of current and breaking news across a wide range of financial issues. For example, background news is extracted from on-line editions of the Australian Financial Review. The agent uses this news to report facts and not to give advice. If it did so then it could be liable to legal action by the Australian Securities and Investment Commission ASIC!

The interaction sequences are triggered by another agent's utterance. The machinery that manages this is designed to work convincingly for interactions of up to around five exchanges only. Our agents are designed to be "strangers" and no more— their role is simply to "toss in potentially valuable gossip".

The aspects of design discussed above cannot be treated in isolation. If the avatars are to be "believable" then their design must have a unifying conceptual basis. This is provided here by the agent's character. The first decision that is made when an agent is created is to select its character using a semi-random process that ensures that multiple instances of the agent in close virtual proximity have identifiably different characters.

The dimensions of character that we have selected are intended specifically for a finance-based environment. They are: politesse, dynamism, optimism and self-confidence. The selection of the character of an agent determines: its appearance (ie: which avatar is chosen for it) and the style of its dialogue. Future plans will address the avatars mannerisms. The selection of agent's character provides an underlying unifying framework for how the agent appears and behaves, and ensures that multiple instances of the agent appear to be different.

The selection on an agent's character does not alone determine its behaviour. Each agent's behaviour is further determined by its moods that vary slowly but constantly. This is intended to ensure that repeated interactions with the same agent have some degree of novelty. The dimensions of moods are: happiness, sympathy and anger. Text mining as opposed to classical data mining techniques is used here to extract knowledge from unstructured text streams.

The agents do not know in advance what will be "said" to them. So they need to be able to access a wide range of information concerning the world's and the market's situation, much of which may not be used. This large amount of information is extracted from different sources on the web, including news pages, financial data and all the exchanges that are part of the virtual environment. Within a particular interaction, the agent's response is driven only by keywords used by his chattermate — just as a human may have done. Despite this simple response mechanism, the agent's dialogue should not appear to be generated by a program. This has partly been achieved by using a semantic network of keywords extracted from the input streams. Words are given a relational weight calculated against others. By analysing which words are related to those used by the chattermate, the agent expresses himself and, to some extent, is able to follow the thread of the discussion. All of this takes place in real time.

In addition to the semantic network, the agent has a personal character and background history, which influences his appearance and the way in which he employs the keywords in his responses. The agent's knowledge is divided into two classes: general and specific. Specific information is contained in a particular data structure. Both classes of information are extracted on demand if they have not previously been pre-fetched.

References

1. Milgrom, P. 2004. *Putting Auction Theory to Work*. Cambridge University Press.
2. Debenham, J.K., 2003. *An eNegotiation Framework*. Proceedings International Conference AI'03. Cambridge, UK.
3. Esteva, Marc; de la Cruz, David; Sierra, Carles, editor. *Proceedings of the First International Joint Conference on Autonomous Agents and Multiagent Systems*, Bologna, Italy.
4. Simoff, S. J. (2002). *Informed traders in electronic markets - mining 'living' data to provide context information to a negotiation process*. In Proceedings of the 3rd International We-B Conference 2002, 28-29 November, Perth, Australia.
5. Barbosa, G.P. and Silva, F.Q.B.: An electronic marketplace architecture based on technology of intelligent agents. *In: Liu, J. and Ye, Y. (eds): E-Commerce Agents*. Springer-Verlag, Heidelberg (2001)
6. Debenham, J. and Simoff, S.: Investigating the evolution of electronic markets. *Proceedings 6th International Conference on Cooperative Information Systems, CoopIS 2001*. Trento, Italy (2001) 344-355
7. Capin, T.K., Pandzic, I.S., Magnenat-Thalman, N. and Thalman, D.: *Avatars in Networked Virtual Environments. John Wiley and Sons, Chichester (1999)*
8. Debenham, J.K.: Managing e-market negotiation in context with a multiagent system. *Proceedings 21-st International Conference on Knowledge Based Systems and Applied Artificial Intelligence, ES'2002: Applications and Innovations in Expert Systems X.* Cambridge, UK (2002) 261-274
9. Tesfatsion, L.: How economists can get alife. In: Arthur, W.B., Durlauf, S. and Lane, D.A. (eds): *The Economy as an Evolving Complex System II.* Addison-Wesley Publishing, Redwood City, CA (1997), 534-564
10. Jeffrey S. Rosenschein and Gilad Zlotkin. *Rules of Encounter*. MIT Press, 1998.
11. Peyman Faratin, Carles Sierra and Nick Jennings. Using Similarity Criteria to Make Issue Trade-Offs in Automated Negotiation. *Journal of Artificial Intelligence.* 142 (2) 205-237, 2003.
12. Kersting, K. and De Raedt, L.: Adaptive Bayesian logic programs. In: Rouveirol, C. and Sebag, M. (eds): *Proceedings of the 11th International Conference on Inductive Logic Programming.* Springer-Verlag, Heidelberg (2001), 104-117.
13. Müller, JP. *The Design of Intelligent Agents: A Layered Approach* (Lecture Notes in Computer Science, 1177). Springer Verlag (1997).
14. Rao, A.S. and Georgeff, M.P. "BDI Agents: From Theory to Practice", in *proceedings First International Conference on Multi-Agent Systems (ICMAS-95)*, San Francisco, USA, pp 312—319.

Autonomous and Continuous Evolution
of Information Systems

Jingde Cheng

Department of Information and Computer Sciences, Saitama University
Saitama, 338-8570, Japan
cheng@aise.ics.saitama-u.ac.jp

"As long as a branch of science offers an abundance of
problems, so long is it alive: a lack of problems foreshadows
extinction or the cessation of independent development."
– David Hilbert, 1900.

"The formulation of a problem is often more essential than
its solution, which may be merely a matter of mathematical
or experimental skill. To raise new questions, new possibilities, to regard old problems from a new angle, requires creative imagination and marks real advance in science."
– Albert Einstein, 1938.

Abstract. This paper presents key characteristics and/or fundamental features
of autonomous evolutionary information systems, the most fundamental issues
in design and implementation of autonomous evolutionary information systems,
our basic ideas and approaches to address the issues, and some challenges for
future research. The notion of an autonomous evolutionary information system,
its actual implementation, and its practical applications concern various areas
including logic, automated reasoning, software engineering, and knowledge engineering, and will raise new research problems in these areas.

1 Introduction

Traditional information systems (without regard to either database systems or knowledge-base systems) are passive in the sense that data or knowledge is created, retrieved, modified, updated, and deleted only in response to operations issued by users
or application programs and the systems only can execute queries or transactions
explicitly submitted by users or application programs but have no ability to do something autonomously by themselves [12, 14, 16, 17]. Although active database systems
allow users to define reactive behavior by means of active rules resulting in a more
flexible and powerful formalism, the systems are still passive in the sense that they
are not autonomous and evolutionary [13, 19]. Future generation of information systems should be more autonomous and evolutionary than the traditional systems in
order to satisfy the requirements from advanced applications of information systems.

An *Autonomous Evolutionary Information System* differs primarily from the traditional information systems in that it can provide its users with not only data or
knowledge stored in its database or knowledge-base by its developers and/or users,
but also new knowledge, which are autonomously discovered by the system itself
based on its database or knowledge-base and the interaction with its users. The new
knowledge may be new facts, new propositions, new conditionals, new inference

R. Khosla et al. (Eds.): KES 2005, LNAI 3681, pp. 758–767, 2005.
© Springer-Verlag Berlin Heidelberg 2005

rules, and so on. Therefore, unlike a traditional information system serving just as a storehouse of data or knowledge and working passively according to queries or transactions explicitly issued by its users and/or application programs, an autonomous evolutionary information system serves as an autonomous and evolutionary partner of its users such that it discovers new knowledge autonomously, communicates and cooperates with its users in solving problems actively by providing the users with advices and helps, has a certain mechanism to improve its own ability of working, and ultimately grows up following its users.

This paper presents key characteristics and/or fundamental features of autonomous evolutionary information systems, the most fundamental issues in design and implementation of autonomous evolutionary information systems, our basic ideas and approaches to address the issues, and some challenges for future research.

2 Autonomous Evolutionary Information Systems

Autonomous evolutionary information systems have the following key characteristics and/or fundamental features: (1) An autonomous evolutionary information system is an information system that stores and manages structured data and/or formally represented knowledge. In this aspect, there is no intrinsic difference between an autonomous evolutionary information system and a traditional database or knowledge-base system. However, this characteristic distinguishes autonomous evolutionary information systems from those "intelligent systems" without a database or knowledge-base. (2) An autonomous evolutionary information system has the ability of reasoning to reason out new knowledge based on its database or knowledge-base autonomously. The ability of knowledge discovery is the most intrinsic characteristic of autonomous evolutionary information systems. (3) An autonomous evolutionary information system has the ability (mechanism) to improve its own ability of working in the sense that it can add new facts and knowledge into its database or knowledge-base autonomously and add new inference rules into its reasoning engine autonomously. It is in this sense that we say that the system is to be autonomously evolutionary. (4) An autonomous evolutionary information system communicates and cooperates with its users in solving problems actively by providing the users with advices and helps that are based on new knowledge reasoned out by itself, and therefore, it acts as an assistant of its users. (5) An autonomous evolutionary information system will ultimately grow up following its users such that two autonomous evolutionary information systems with the same primitive contents and ability may act very differently if they are used by different users during a certain long period of time. Therefore, the evolution of an autonomous evolutionary information system is not only autonomous but also continuous such that its autonomous evolution is a persistently continuous process without stop of interaction with its user.

An autonomous evolutionary information system is different from the so-called "intelligent software agent" [3] in the following aspects: (1) The former is an information system for general-purpose without an explicitly defined specific task in a particular environment as its goal to achieve, while the latter is usually defined as a specific program with an explicitly defined specific task in a particular environment as its goal to achieve. (2) The former in general can serve for the public users, or a specific group of users, or a specific individual user but does not act on behalf of a specific user or a specific group of users in a particular environment, while the latter in general acts on behalf of its principal in a particular environment. (3) The former communicates and cooperates only with its users but does not necessarily communicate and cooperate with other autonomous evolutionary information systems to achieve a task or goal, while the latter often communicate and cooperate with other agents, in the form so-called 'multi-agent system', to achieve a task or goal. (4) The former is autonomously and continuously evolutionary in the sense that it has a certain mechanism to improve its own ability of working, and ultimately grows up following its users, while the latter is only a steady program.

3   Fundamental Issues in Design and Development of Autonomous Evolutionary Information Systems

Reasoning is the process of drawing new conclusions from given premises, which are already known facts or previously assumed hypotheses to provide some evidence for the conclusions. Therefore, reasoning is intrinsically ampliative, i.e., it has the function of enlarging or extending some things, or adding to what is already known or assumed.

Discovery is the process to find out or bring to light of that which was previously unknown. For any discovery, the discovered thing and its truth must be unknown before the completion of discovery process. Since reasoning is the only way to draw new, previously unknown conclusions from given premises, there is no discovery process that does not invoke reasoning. Because the ability to reason out new knowledge is an intrinsic characteristic of autonomous evolutionary information systems, any autonomous evolutionary information systems must have a reasoning engine as an indispensable component which working based on a right fundamental logic system.

The term *'evolution'* means a gradual process in which something changes into a different and usually better, maturer, or more complete form. Note that in order to identify, observe, and then ultimately control any gradual process, it is indispensable to measure and monitor the behavior of that gradual process. The *autonomous evolution* of a system, which may be either natural or artificial, should be a gradual process in which everything changes by conforming to the system's own laws only, and not subject to some higher ones. Making a computing system autonomously evolutionary is very difficult, if not impossible, because the behavior of any computing system itself is intrinsically deterministic (For example, it is well known that any computing system itself cannot generate true random numbers). What we can do is to design and implement some mechanism in a computing system such that it can react to stimuli from its outside environment and then to behave better.

A *reactive system* is a computing system that maintains an ongoing interaction with its environment, as opposed to computing some final value on termination [15]. An autonomous evolutionary information system should be a reactive system with the capabilities of concurrently maintaining its database or knowledge-base, discovering and providing new knowledge, interacting with and learning from its users and environment, and improving its own ability of working.

Measuring the behavior of a computing system means capturing run-time information about the system through detecting attributes of some specified objects in the system in some way and then assigning numerical or symbolic values to the attributes in such a way as to describe the attributes according to clearly defined rules. ***Monitoring*** the behavior of a computing system means collecting and reporting run-time information about the system, which is captured by measuring the system. Measuring and monitoring mechanisms can be implemented in hardware technique, or software technique, or both. For any computing system, we can identify and observe its evolution, i.e. a gradual change process, only if we can certainly measure and monitor the system's behavior. Also, an autonomous evolutionary computing system must have some way to measure and monitor its own behavior by itself.

Learning is the action or process of gaining knowledge of a subject or skill in an art through study, experience, teaching, or training. It usually leads to the modification of behavior of the learner or the acquisition of new abilities of the learner, and results in the growth or maturation of the learner. ***Self-learning*** is the action or process of gaining knowledge or skill through acquiring knowledge by the learner himself (herself) rather than receiving instructions from other instructors. It is obvious that a self-learning process can result in the growth or maturation of the learner only if the learner can correctly valuate the effectiveness of every phase in the self-learning process. ***Valuation*** is the action or process of assessing or estimating the value or price of a thing based on some certain criterion about its worth, excellence, merit, or character. ***Self-valuation*** is the action or process of assessing or estimating the value or price of a thing concerning the valuator himself (herself) that is made by the valuator himself (herself).

For any computing system, its autonomous evolution is impossible if it has no self-learning and self-valuation abilities. However, because the behavior of any computing system itself is intrinsically deterministic and is a closed world, true or pure self-learning and self-valuation is meaningless, if not impossible, to its autonomous evolution. Therefore, the autonomous evolution of a computing system should be accomplished in a gradual process by interaction with its outside environment. On the other hand, because any evolution is a gradual process, and in particular, the autonomous evolution of an autonomous evolutionary information system should be a persistently continuous process without stop of interactions with its user, the self-learning and self-valuation abilities of an autonomous evolutionary information system should grow up in the gradual and continuous process.

Thus, based on the above definitions and discussions, we can understand the following fundamental questions and/or issues about design and development of an autonomous evolutionary information system: What formal logic system can satisfactorily underlie reasoning out new knowledge? How can we formally define that knowledge discovered by the system is new and interesting to its users? How can the system find new and interesting knowledge autonomously? What architecture and/or structure should the system have? What objects in the system should be taken as primary ones such that they have to be measured and monitored in order to identify, observe, and then ultimately control its gradual evolution process? How can the system measure and monitor its own gradual evolution process? How can we design and develop the system such that it can function persistently without stop of interaction with its user? What behavioral changes in the system are intrinsically important to its evolution?

It is obvious that at present we cannot give all answers in detail to all the above problems in a single paper because many studies in this new research direction are still in their early stages and some problems are completely open. Below, we will present our considerations on the fundamental issues and our approaches to the issues, and show some challenges.

4 The Fundamental Logic

The most crucial issue in design and development of an autonomous evolutionary information system is to choose the right fundamental logic system (or systems) to underlie reasoning out new knowledge. The fundamental logic that can satisfactorily underlie autonomous evolutionary information systems has to satisfy at least the following four essential requirements.

First, as a general logical criterion for validity of reasoning as well as proving, the logic must be able to underlie relevant reasoning as well as truth-preserving reasoning in the sense of conditional, i.e., for any reasoning based on the logic to be valid, if its premises are true in the sense of conditional, then its conclusion must be relevant to the premises and true in the sense of conditional. This requirement is crucial to the reasoning engine of an autonomous evolutionary information system because without such a logical criterion for validity, a forward deduction performed by the reasoning engine may produce a lot of useless (even burdensome!) conclusions which are completely irrelevant to given premises or incorrect in the sense of conditional.

Second, the logic must be able to underlie ampliative reasoning in the sense that the truth of conclusion of the reasoning should be recognized after the completion of the reasoning process but not be invoked in deciding the truth of premises of the reasoning. This requirement is obvious to an autonomous evolutionary information system because it should reason out new knowledge whose truth should be unknown at the point of that reasoning is being performed.

Third, the logic must be able to underlie paracomplete and paraconsistent reasoning. In general, our knowledge about the real world (and therefore data and/or knowledge in any autonomous evolutionary information system) may be incomplete or even inconsistent in many ways, i.e., it gives us no evidence for deciding the truth of either a proposition or its negation, or even it directly or indirectly includes some contradictions. Therefore, the abilities of paracomplete reasoning (or reasoning with incomplete information) and paraconsistent reasoning (or reasoning in the presence of inconsistency) are indispensable to an autonomous evolutionary information system.

Finally, the logic must be able to underlie temporal reasoning because autonomous and continuous evolution in an autonomous evolutionary information system is also a time-dependent process of data and/or knowledge update. Not only truth of propositions but also relevant relationships among them may change over time. Without a logic to underlie temporal reasoning, an autonomous evolutionary information system cannot reason about change and update of data and/or knowledge.

Classical mathematical logic and its various classical conservative extensions cannot satisfy any of the above essential requirements [1, 2, 7]. Temporal (classical) logic [4, 18] can satisfy the last requirement but cannot satisfy the first three of the requirements [9]. Traditional relevant logics [1, 2] can satisfy the first requirement partly and the second and third requirements but cannot satisfy the fourth requirement [9]. Strong relevant logics [7] can satisfy the first three of the requirements well but cannot satisfy the fourth requirement [9]. At present, the only hopeful candidate which can satisfy the above all four essential requirements is temporal relevant logics [9], which are obtained by introducing temporal operators and related axiom schemata and inference rules into strong relevant logics.

5 Knowledge Discovery by Relevant and Ampliative Reasoning Based on Strong Relevant Logic

If we represent data and/or knowledge in an autonomous evolutionary information system as a formal theory, then knowledge discovery based on the data /and/or knowledge can be modelled or formalized as the problem of automated theorem finding in the formal theory. However, the problem of automated theorem finding ("What properties can be identified to permit an automated reasoning program to find new and interesting theorems, as opposed to proving conjectured theorems?), originally proposed by Wos in 1988 [20], is still a completely open problem until now.

We have discussed why any of the classical mathematical logic, its various classical conservatives extensions, and its non-classical alternatives is not a suitable candidate for the fundamental logic to underlie automated theorem finding, shown that strong relevant logic is a more hopeful candidate for the purpose, and presented a conceptional foundation for automated theorem finding by entailment calculus based on strong relevant logic [5, 8]. EnCal, an automated forward deduction system for general-purpose entailment calculus [6] can be used for assisting automated theorem finding as well as knowledge discovery.

6 System Architecture Based on Soft System Buses

An autonomous evolutionary information system should be a reactive system with the capability of concurrently maintaining its database or knowledge-base, discovering and providing new knowledge, interacting with and learning from its users, and improving its own ability of working. Moreover, in order to keep a gradual and continuous evolution process without stop of interactions with its user, an autonomous evolutionary information system should be a persistently reactive system such that it functions continuously anytime without stopping its reactions even when it had some trouble, it is being attacked, or it is being maintained, upgraded, or reconfigured.

Conceptually, a *soft system bus*, SSB for short, is simply a communication channel with the facilities of data/instruction transmission and preservation to connect components in a component-based system. It may consist of some *data-instruction stations*, which have the facility of data/instruction preservation, connected sequentially by *transmission channels*, both of which are implemented in software techniques, such that over the channels data/instructions can flow among data-instruction stations, and a component tapping to a data-instruction station can send data/instructions to and receive data/instructions from the data-instruction station [10]. An *SSB-based system* is a component-based system consisting a group of control components including self-measuring, self-monitoring, and self-controlling components with general-purpose which are independent of systems, and a group of functional components to carry out special takes of the system such that all components are connected by one or more SSBs and there is no direct interaction which does not invoke the SSBs between any two components [10].

We considered that an autonomous evolutionary information system can be constructed as an SSB-based system. As a case study to design and develop an autonomous evolutionary information systems, we are developing an autonomous evolutionary information system, named "HILBERT," for teaching and learning logic and various logic systems underlying diverse reasoning forms in scientific discovery as well as everyday logical thinking [11]. Fig. 1 shows the system architecture of HILBERT.

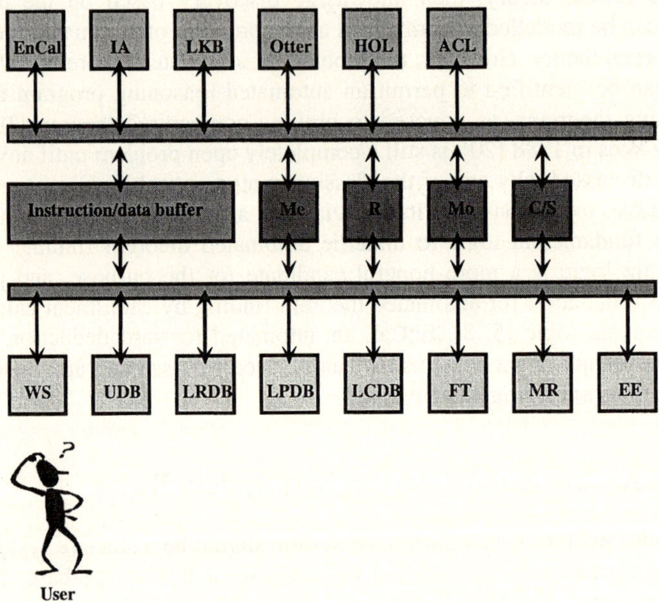

Fig. 1. The system architecture of HILBERT

The central components of HILBERT include a central measurer (**Me**), a central recorder (**R**), a central monitor (**Mo**), a central controller/scheduler (**C/S**), and an instruction/data buffer, all of which are permanent components of HILBERT. The functional components of HILBERT are separated into two groups which are measured, recorded, monitored, and controlled by the central components through two (internal and external) SSBs. One group (named "internal group") of the functional components of HILBERT includes **EnCal** [6], **IA** (intelligent adviser), **LKB** (logic knowledge-base system), and some automated tools for theorem proving and proof checking such as **ACL**, **OTTER**, **HOL**, and so on. These components of internal group can be directly invoked only by the central components since they are general-purpose components serving for the whole of HILBERT rather than individual other functional components or individual users. The functional components of another group can communicate with these components through the SSBs and the instruction/data buffer. While any user of HILBERT cannot invoke these components of internal group directly. Another group (named "external group") of the functional components of HILBERT includes **WS** (Web user interface), **UDB** (user database system), **LRDB** (logic reference database), **LPDB** (logic puzzle database system), **LCDB** (logic course database system), **FT** (formula translator), **MR** (monitoring and recording system), and **EE** (examing and evaluating system). These components of external group can directly communicate with users through an SSB and the instruction/data buffer under the monitor and control of the central components.

7 Self-learning and Self-valuation in Autonomous Evolution

To be an autonomous evolutionary information system, an information system must be able to evolve from a lower form into a better, maturer, or more complete form autonomously. This requires that it must be able to valuate those data, information, or knowledge obtained by self-measuring, self-monitoring, and self-learning and then determine a correct direction or way to improve its own abilities from a lower form into a better, maturer, or more complete form. Since determining a correct direction or way is dependent on the correct valuation of the current state of the system and the current situation of its outside environment, the fundamental question here is: Can the system make the valuation correctly by itself?

On the other hand, since valuation is the action or process of assessing or estimating the value or price of a thing based on some certain criterion about its worth, excellence, merit, or character, any valuator must have a certain criterion before the valuation is made. In fact, the most difficult issue in any valuation is how to establish such a criterion. In general, a human being has different criterion for valuating different things, and his/her various criteria are established in a long time of his/her life by experiencing and learning various things. Moreover, one of the most intrinsic characteristics of human society is that for the same thing, different individual or people may make different valuation based on different criteria. It is this characteristic that leads to the variety and dynamics in human society.

Thus, the fundamental questions are: What is the mechanism to make different criteria for valuation? How can we implement the mechanism? To our knowledge, until now, there is no idea or methodology proposed for addressing these difficult issues.

8 Concluding Remarks

We have presented key characteristics and/or fundamental features of autonomous evolutionary information systems, the most fundamental issues in design and implementation of autonomous evolutionary information systems, our basic ideas and approaches to address the issues, and some challenges for future research. Future work on this new research direction should be focused on designing and developing autonomous evolutionary information systems that work for all sorts of people in various areas of the real world.

The notion of an autonomous evolutionary information system, its actual implementation, and its practical applications concern various areas including logic, theory of computation, automated reasoning and proving, software engineering, and knowledge engineering, and will raise many new research problems in these areas.

References

1. Anderson, A.R., Belnap Jr., N.D.: Entailment: The Logic of Relevance and Necessity, Vol. I. Princeton University Press, Princeton (1975)

2. Anderson, A.R., Belnap Jr., N.D., Dunn, J. M.: Entailment: The Logic of Relevance and Necessity, Vol. II. Princeton University Press, Princeton (1992)

3. Brenner, W., Zarnekow, R., Wittig, H.: Intelligent Software Agents: Foundations and Applications. Springer-Verlag, (1998)

4. Burgess, J. P.: Basic Tense Logic. In: Gabbay, D., Guenthner F. (eds.): Handbook of Philosophical Logic, 2nd edition, Vol. 7. Kluwer Academic, Dordrecht (2002) 1-42

5. Cheng, J.: Entailment Calculus as the Logical Basis of Automated Theorem Finding in Scientific Discovery. In: Systematic Methods of Scientific Discovery − Papers form the 1995 Spring Symposium, Technical Report SS-95-03. AAAI Press, Melon Park (1995).105-110

6. Cheng, J.: EnCal: An Automated Forward Deduction System for General-Purpose Entailment Calculus. In: Terashima, N., Altman, E. (eds.): Advanced IT Tools, Proc. IFIP World Conference on IT Tools, IFIP 96 − 14th World Computer Congress. Chapman & Hall, London Weinheim New York Tokyo Melbourne Madras (1996) 507-514

7. Cheng, J.: A Strong Relevant Logic Model of Epistemic Processes in Scientific Discovery. In: Kawaguchi, E., Kangassalo, H., Jaakkola, H., Hamid, I.A. (eds.): Information Modelling and Knowledge Bases XI. IOS Press, Amsterdam Berlin Oxford Tokyo Washington DC (2000) 136-159

8. Cheng, J.: Automated Knowledge Acquisition by Relevant Reasoning Based on Strong Relevant Logic. In: Palade, V., Howlett, R.J., Jain, L.C. (eds.): Knowledge-Based Intelligent Information & Engineering Systems, 7th International Conference, KES 2003, Oxford, UK, September 3-5, 2003, Proceedings, Part I. Lecture Notes in Artificial Intelligence, Vol. 2773, Springer-Verlag (2003) 68-80

9. Cheng, J.: Temporal Relevant Logic as the Logical Basis of Anticipatory Reasoning-Reacting Systems. In: Daniel M. Dubois (ed.): Computing Anticipatory Systems: CASYS 2003 - Sixth International Conference. AIP Conference Proceedings, Vol. 718. American Institute of Physics, Melville (2004) 362-375

10. 10 Cheng, J.: Connecting Components with Soft System Buses: A New Methodology for Design, Development, and Maintenance of Reconfigurable, Ubiquitous, and Persistent Reactive Systems. In: Proc. 19th IEEE-CS International Conference on Advanced Information Networking and Applications. (2005) Vol. 1, 667-672

11. Cheng, J, Akimoto, N., Goto, Y., Koide, M., Nanashima, K., Nara, S.: HILBERT: An Autonomous Evolutionary Information System for Teaching and Learning Logic. In: Proc. 6th International Conference on Computer Based Learning in Science, Vol. 1. (2003) 245-254

12. Gonzalez, A.J., Dankel, D.D.: The Engineering of Knowledge-Based Systems: Theory and Practice. Prentice Hall, Upper Saddle River (1993)

13. Lausen, G., Ludascher, B., May, W.: On Logical Foundations of Active Databases. In: Chomicki, J., Saake, G. (Eds.): Logics for Databases and Information Systems. Kluwer Academic (1998) 389-422

14. Stefik, M.: Introduction to Knowledge Systems. Morgan Kaufmann Publishers, San Francisco (1995)
15. Pnueli, A.: Specification and Development of Reactive Systems. In: Kugler, H.-J. (Ed.): Information Processing 86. IFIP, North-Holland (1986) 845-858
16. Ullman, J.D.: Principles of Database and Knowledge-Base Systems, Vol. 1. Computer Science Press (1988)
17. Ullman, J.D.: Principles of Database and Knowledge-Base Systems: The New Technologies, Vol. 2. Computer Science Press (1989)
18. Venema, Y.: Temporal Logic. In: Goble, L. (ed.): The Blackwell Guide to Philosophical Logic, Blackwell, Oxford (2001) 203-223
19. Widom, J., Ceri, S. (eds.): Active Database Systems. Morgan Kaufmann (1996) 1-41
20. Wos, L.: Automated Reasoning: 33 Basic Research Problems. Prentice-Hall (1988)

Dynamic Location Management
for On-Demand Car Sharing System

Naoto Mukai[1] and Toyohide Watanabe[1]

Department of Systems and Social Informatics,
Graduate School of Information Science, Nagoya University
Furo-cho, Chikusa-ku, Nagoya, 464-8603, Japan
naoto@watanabe.ss.is.nagoya-u.ac.jp, watanabe@is.nagoya-u.ac.jp

Abstract. Recently, *Car Sharing System* is focused as a new public transportation system in Japan. However, traditional car sharing systems which we call *Pre-Reserve Car Sharing System* require advance reservations, and the number of stations where users get on/off for the systems is small. Thus, the number of users is limited availability and most of users only apply the system to round driving. On the other hand, the advances of global positioning systems and communication techniques improve the system in a new and different way. We call the new system *On-Demand Car Sharing System*. In the new system, users are available for the system on demand, i.e., users can get on/off any time and any stations. Thus, users apply the system to one-way driving in addition to round driving. In this paper, we focus on location balances of cars which is a key factor to improve utilization rate of the system. We propose a relocation algorithm for waiting cars based on virtual spring forces. In the last of this paper, we report our simulation results and show the effectiveness of our algorithm.

1   Introduction

In these years, there's been an increase of research in intelligent transportation system (ITS) [1]. The advances of global positioning systems and communication techniques enable to develop new customer-friendly transportation systems [2] such as Demand Bus System [3]. *Car Sharing System* [4, 5] is one of the systems focused as a new public transportation. The concept of the system is to share a small number of cars among a large number of users. There are three contributions by the system: cost sharing, earth sharing and road sharing. In this way, the system will provide not only non-traditional traffic convenience but also the three sharings for our future traffic environment.

The car sharing system has already been employed in Europe for about fifteen years [6, 7]. Now, the system is in time to the next step with the advancement of the recent technologies. Every traditional car sharing systems, which we call *Pre-Reserve Car Sharing System*, require advance reservations. And, the number of stations, where users get on/off cars, is small in a service area. Thus, the available users for the system are limited in number. Moreover, most of users only apply the system to round driving, i.e., a journey to a place and back again. On the

R. Khosla et al. (Eds.): KES 2005, LNAI 3681, pp. 768–774, 2005.

other hand, we propose a new type of car sharing system based on the advanced technology. We call the new system as *On-Demand Car Sharing System*. In the new system, users are available for the system on demand, i.e., users can get on/off any time and any stations. Thus, users apply the system to one-way driving, i.e., a moving in only one direction, in addition to round driving. The new system can satisfy more meticulous demands (e.g., short-distance and short-time drivings) than existing car sharing systems. However, there are some problems to be solved for realizing the new system. In this paper, we focus on the problem of location balancing for waiting cars. And, we propose a relocation algorithm based on virtual spring forces. The algorithm divides a service area equally for keeping homogeneity of location balances among waiting cars, dynamically.

The remainder of this paper is organized as follows: Section 2 formalizes our intended car sharing problem. In Section 3, we propose a balancing algorithm based on virtual spring forces. In Section 4, we report simulation results and consider effectiveness and possibility of the new system. Section 5 concludes and offers future works.

2   Formalization

2.1   Road Graph

A service area of the system is given by a closed polygon P on x and y coordinates as Equation (1). The position of a vertex p is denoted as $pos(p) = (p_x, p_y)$. The number of vertexes of the polygon is k. We denote the borderline of the polygon as $b(p, p')$.

$$P = \{p_1, p_2, \cdots, p_k\}, B = \{b_1, b_2, \cdots, b_k\}, b_j = (p_i, p_{i+1}) \tag{1}$$

A road network in a service area is given by the concept of graph G which consists of nodes N and edges E as Equation (2). The nodes represent intersections and the edges represent road segments between intersections. The position of node n is denoted as $pos(n) = (n_x, n_y)$. For simplicity, we do not consider one-way traffic. We denote the distance between n and n' as $d(n, n')$. Moreover, we denote the length of the line perpendicular to borderline b from node n as $d(n, b)$.

$$G = (N, E), N = \{n_1, n_2, \cdots\}, E = \{e(n, n') : n, n' \in N\} \tag{2}$$

There are h stations where users get on/off cars in a service area. The stations are corresponded to a subset of nodes in Equation (3).

$$S = \{s_1, s_2, \cdots, s_h\} \in N \tag{3}$$

2.2   Cars

Cars are given by C in Equation (4). The traveling speed of cars is fixed as $|c|$. We assume that the position of car $pos(c)$ corresponds to the position of node

$pos(n) : n \in N$. The state transition of a car is shown in Figure 1(a). In *WAIT*, a car waits for receiving either driving or relocating request at any one of stations. Both requests are given by a pair of ride-on and drop-off stations (s_r, s_d) for a driver. If the car receives a driving request from a user, the car is reserved for the user and waits for his arrival (*RESERV*). The user drives the car from the station to his goal station (*DRIVE*). After the driving, the state of the car transits again to *WAIT*. In different from the way, if the car receives a relocating request from a location controller as described later, the state transits to *RELOC*. However, if an interrupt occurs by a user, the state transits to *RESERV*. The detail of the relocation algorithm is discussed in Section 4.

$$C = \{c_1, c_2, \cdots, c_m\} \tag{4}$$

2.3 Users

Users are given by U in Equation (5). The walking speed of users is fixed as $|u|$. We assume that the position of user $pos(u)$ corresponds to the position of node $pos(n) : n \in N$. A demand of a user is given by $dem(u)$ which contains start node n_r, ride-on station s_r, drop-off station s_d and goal node n_g. The state transition of a user is shown in Figure 1(b). In *FIND*, a customer decides ride-on station s_r and drop-off station s_d. Here, let S_{wait} be the set of stations where there are one or more waiting cars in *WAIT*. The station, which is closest to the start node n_s, is selected from S_{wait} as ride-on station s_r. And, the station, which is closest to the goal node n_g, is selected from S as drop-off station s_d. Next, the user makes a decision on whether he uses the system or not by the condition $T_{use} < T_{nouse}$. The travel time by using the system is given by T_{use} defined in Equation (6). And, the travel time by walking is given by T_{nouse} defined in Equation (7). If the condition is satisfied, firstly the user sends the driving request (s_r, s_d) and walks to the ride-on station s_r (*WALKSTATION*), secondly drives the car to the drop-off station s_d (*DRIVE*), thirdly walks to the goal node n_g (*WALK GOAL*), and finally arrives at his destination (*ARRIVE*). Otherwise, the user walks to the goal node n_g directly by walking (*WALK GOAL*).

$$U = \{u_1, u_2, \cdots\}, dem(u) = (n_s, s_r, s_d, n_g) \tag{5}$$

$$T_{use} = \frac{d(n_s, s_r) + d(s_d, n_g)}{|u|} + \frac{d(s_r, s_d)}{|c|} \tag{6}$$

$$T_{nouse} = \frac{d(n_s, n_g)}{|u|} \tag{7}$$

2.4 Location Controllers

Location controllers, who have a role of relocating cars to more effective positions, are given by L in Equation (8). The moving speed of location controllers is fixed as $|l|$. We assume that the position of location controller $pos(l)$ corresponds to the position of node $pos(n) : n \in N$. The state transition of a location controller is shown in Figure 1(c). In *WAIT*, a location controller waits for receiving a relocating request from a control center. When the location controller receives a relocating request for a car, he checks whether the relocation of the car is possible or not, i.e., *WAIT* or other states. If it is possible, he forwards the relocating request to the car and goes to the ride-station s_r (*RELOC*), then he drives the car to the drop-off station s_d (*DRIVE*). After relocating, he waits for next relocating requests, again (*WAIT*).

$$L = \{l_1, l_2, \cdots, l_o\} \tag{8}$$

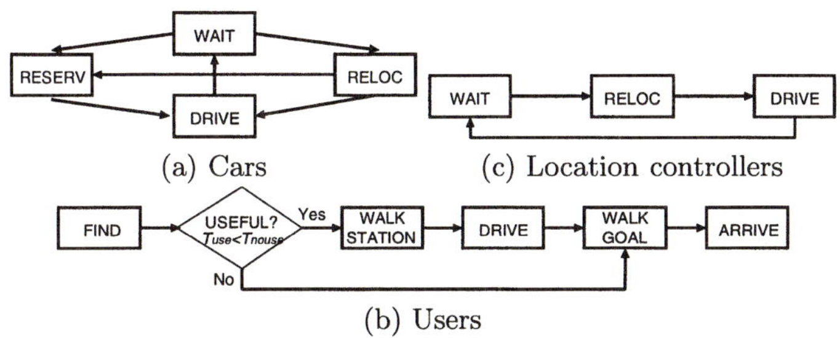

(a) Cars   (c) Location controllers

(b) Users

Fig. 1. State transitions.

3   Relocation Algorithm

A control center issues relocating requests to location controllers by watching positions of cars with a global positioning system. A candidate car for relocating is selected on the basis of virtual spring forces. We attach virtual springs to two places of each vehicle: between the car and other cars and between the car and borderlines of a service area. At the first setup, the length $|vs|$ of a virtual spring vs is calculated by Equation (9). Let $A(P)$ be area of a service area P and $|C_{wait}|$ be the number of waiting cars. The length of springs $|vs|$ represents an appropriate distance between two waiting cars.

$$|vs| = \sqrt{\frac{A(P)}{|C_{wait}|}} \tag{9}$$

The virtual spring force between car c and other waiting cars C_{wait} is given by Equation (10).

$$F_c^{cc} = \sum_{c' \in C_{wait}} (|vs| - d(c, c')) \tag{10}$$

The virtual spring force between car c and borderlines B is given by Equation (11). The spring length is set to half length of original spring length.

$$F_c^{cb} = \sum_{b \in B} (\frac{|vs|}{2} - d(c, b)) \tag{11}$$

The movement energy of car c is given in Equation (12). The waiting factors of virtual spring forces are given by w^{cc} and w^{cb}, respectively. High movement energy shows the off-balance of car distribution.

$$E_c = (w^{cc} \times F_c^{cc}) + (w^{cb} \times F_c^{cb}) \tag{12}$$

A pseudo-code of relocation algorithm is summarized in Figure 2. In the former part of the code, movement energies of all waiting cars are calculated, and a car c_{max} with maximum energy is selected as a ride-on position for location controller. In the latter part of the code, suppose that the car c_{max} moves other stations, a station s_{min} with minimum energy is selected as a drop-off position for a location controller in the same way. Finally, the relocating request (c_{max}, s_{min}) is sent to a location controller closest to c_{max}.

```
//find car with maximum energy
cmax = null; Cwait = cars in WAIT; |vs| = spring length;
WHILE c ∈ Cwait do
    Ec = energy(c, |vs|, Cwait − c); //calculate movement energy
    IF Ec is maximum energy THEN
        cmax = c;
    END
END
//find station with minimum energy
smin = null; Cwait = Cwait - cmax;
WHILE s ∈ S do
    Es = energy(s, |vs|, S − s); //calculate movement energy
    IF Es is minimum energy THEN
        smin = s;
    END
END
l = location controller closest to cmax;
send(l, cmax, smin); //send relocating request to l
```

Fig. 2. Pseudo-code of relocation algorithm.

4   Experiments

There are three experimental patterns shown in Table 1(a). A service area of the system is set to a square (800×800 pixels). A road network in the area is set to a grid network (21×21), i.e., the length of all road segments is 40 pixels. A pair of ride-on and drop-off nodes is selected from the nodes randomly. The service process is repeated until the max time $100000t$. Other parameter setting is shown in Table 1(b). First, we report experimental results related to occurrence rate of driving requests. The occurrence rate of driving requests is set from 1% to 5%. The number of stations is set to a fixed value 20. Figure 3(a) indicates that relocating by driving controller can improve from 20% to 40% compared with PT1:without controller. However, when there are many driving

Table 1. Experimental Setting.

Patterns	$|L|$	$|l|$
PT1: without controller	0	
PT2: controller (driving)	1	$1.0/t$
PT3: controller (walking)	1	$0.2/t$

(a) Patterns

Parameter	Value
$|C|$	5
$|c|$	$1.0/t$
$|u|$	$0.2/t$
$|w^{cc}|$	0.8
$|w^{cb}|$	0.2

(b) Parameters

cars occupied by users, the effect is small. Figure 3(b) indicates that the travel time of users is clearly related to the utilization rate of the system. In other words, high utilization rate shows convenient access in the service area. Figure 3(c) indicates that the number of relocation strongly depends on the speed of location controllers. The relocation by walking reaches the limited number about 1000 even in low occurrence rate. On the other hand, relocation by driving reaches the limited number about 3000. Second, we report experimental results related to the number of stations. The number of stations is set from 10 to 25. The occurrence rate of driving requests is set to a fixed value 1%. Figure 3(d) indicates that a large number of stations in the service area can satisfy more short-distance driving requests. Figure 3(e) shows the same tendency in the previous experimental result. Figure 3(f) indicates that a large number of stations increases the burden on relocation for location controllers. From the experimental results, our relocation algorithm is effective for on-demand car sharing system. Moreover, improving utilization rate of the system can provide more convenient access for users in the service area.

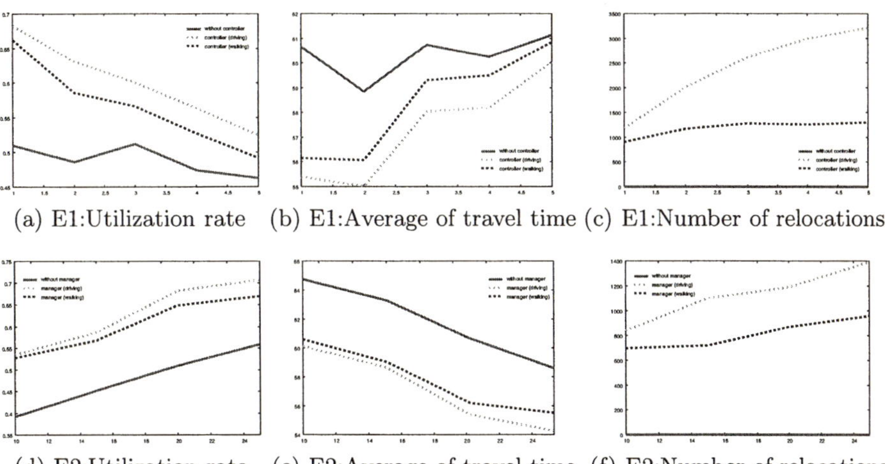

(a) E1:Utilization rate   (b) E1:Average of travel time   (c) E1:Number of relocations

(d) E2:Utilization rate   (e) E2:Average of travel time   (f) E2:Number of relocations

Fig. 3. Experimental results related to occurrence rate and the number of stations.

5   Conclusions

In this paper, we proposed a new transportation system called on-demand car sharing system. The system provides more convenient access, i.e., users can ride-on and drop-off any time and any stations in a service area. The location balance of cars is a key problem to be solved for improving utilization of the system. Therefore, we proposed a relocation algorithm based on virtual spring forces. The effectiveness of the algorithm is estimated by our computer simulation. The results of the simulation show that our algorithm is effective for on-demand car sharing system. In our future work, we must consider a cooperation among location controllers and simulate in real city environment.

References

1. Nakashima, H., Kurumatani, K., Itoh, H.: Supporting a society with ubiquitous computing. IPSJ-MGN450904t **45** (2004)
2. Schaefer, L.A.: Transport applications: architecture using jini technology for simulation of an agent-based transportation system. In: Proceedings of the 33nd conference on Winter simulation. (2001) 1079–183
3. Ohta, M., Shinoda, K., Noda, I., Kurumatani, K., Nakashima, H.: Usability of demand-bus in town area. IPSJ-MBL02023033 **2002** (2002)
4. E., B.: Executive summary in carsharing 2000: Sustainable transport's missing link. The Journal of World Transport Policy & Practice (2000)
5. Abraham, J.E.: Carsharing: A survey of preferences in carsharing. The Journal of World Transport Policy & Practice (2000)
6. Bonsall, P.W.: Car sharing in the united kingdom. a policy appraisal. Journal of Transport Economics and Policy (1981) 35–44
7. Bonsall, P.W.: What makes a car-sharer? Transportation **12** (1984) 117–145

Writer Recognition by Using New Searching Algorithm in New Local Arc Method

Masahiro Ozaki[1], Yoshinori Adachi[1], and Naohiro Ishii[2]

[1] Chubu University, 1200 Matsumoto-Cho, Kasugai, Aichi, Japan 487-8501
ozaki@isc.chubu.ac.jp
[2] Aichi Institute of Technology, Yakusa-Cho, Toyota, Aichi, Japan 470-0392

Abstract. In the previous studies, authors proposed a new local arc method with new similarity evaluation function to the off-line writer recognition, and are obtaining high recognition ratios. However, to calculate the similarity value of the author from the product of the similarity values of three or four kinds of characters, the difference of author similarity value becomes sometimes quite small. Then, it may not be stable recognition method. In this study, the new local arc method was modified to give more information of writers. And by examining the characteristic of the feature of new local arc method, it was found that the feature of each writer was possible to be extracted. Especially, even when the difference between author similarity values is small, stable writer recognition becomes possibl.

1 Introduction

We had proposed the off-line writer recognition systems for Japanese "hiragana" characters [1-5] and Chinese characters [6]. In the papers [2-4], two-dimensional (2D) Fuzzy membership functions which were obtained from simple summation of characters were used for simplicities, and over 98% recognition ratio was obtained. However in those papers, many unstable characters were omitted as inadequate characters for recognition. Therefore, the recognition process happen to be abandoned.

To overcome these problems, a characteristic of each stroke was tested for writer recognition in the previous papers [4-6]. Hand written characters consist of several curved strokes. And curvature distributions in the strokes were able to use as characteristics of each writer. The new local arc method with the new similarity evaluation function was proposed in the previous paper and obtained over 98% recognition ratio without inadequate characters in Japanese characters and 84.6% in Chinese characters.

However there are still two problems to be overcome. One is that the previous algorithm cannot detect stroke in specific directions. Second is that the similarity evaluation function is not suitable for unstable characters.

In this paper, we propose new algorithm to search strokes to avoid loosing feature of writer's specific peculiarity and evaluate similarity values by calculating cosine value.

2 New Algorithm to Search Point on Stroke

To search a local arc in strokes, the algorithm used in the previous studies [4-6] is shown in Table 1.

R. Khosla et al. (Eds.): KES 2005, LNAI 3681, pp. 775–780, 2005.

Table 1. Algorithm to search strokes

Step 1. Search of end point (x0, y0) of chord on a stroke.
Step 2. Search of the other end point (x1, y1) of chord on a stroke for 12 directions. If not exist, go to Step 1.
Step 3. Search a stroke point (x2, y2) on the perpendicular bisector of the chord. If not exist, go to Step 2.
Step 4. Calculate a center (xc, yc) of circumscribed circle of three points (x0, y0), (x1, y1), and (x2, y2).
Step 5. Calculate bisector points (x3, y3) and (x4, y4) on the circumscribed circle of angles (x2, y2)(xc, yc)(x0, y0) and (x2, y2)(xc, yc)(x1, y1).
Step 6. If bisector points are on a stroke, curvature is calculated. If not exist, go to Step 2.

In Step 2, only 12 directions (15 degrees interval) are searched. Therefore in many times, an existence of stroke cannot be detected. In this study, we modified the algorithm of Step 2 as shown in Fig. 1.

In the previous algorithm, searching end point (x1, y1) could not be obtained because of discreteness. Therefore local arc was frequently missed and characteristics of a character were not properly extracted.

By applying the proposed algorithm that is searching end point one dot to one dot on the vector covered in between 15 degree gap as shown in the figure, end point (x1, y1) could be searched appropriately.

However, characteristics of a character is integrated into 12 × 11 dimensional feature matrix (12 directions of chord from 0° to 165° with 15° interval, 11 curvatures from -5 to 5) as same as previous method for simplicity of analysis.

To check the applicability of the proposed algorithm, several simplified characters whose stroke width was 1 dot were tested. Some simplified characters are shown in Fig. 2. The feature matrices obtained from both algorithms are compared in Tables 2 to 5.

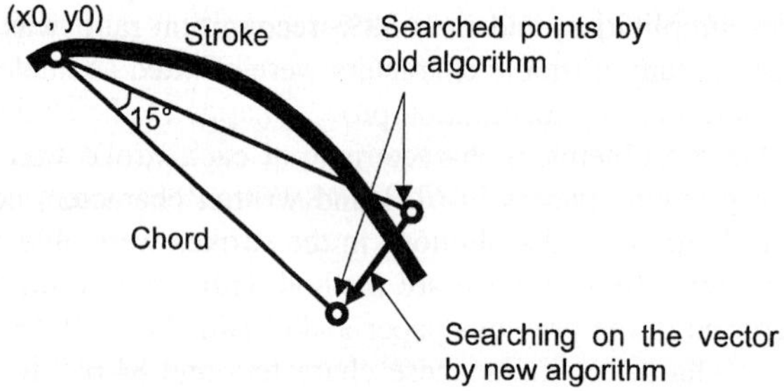

Fig. 1. Searching end point of a chord on a stroke

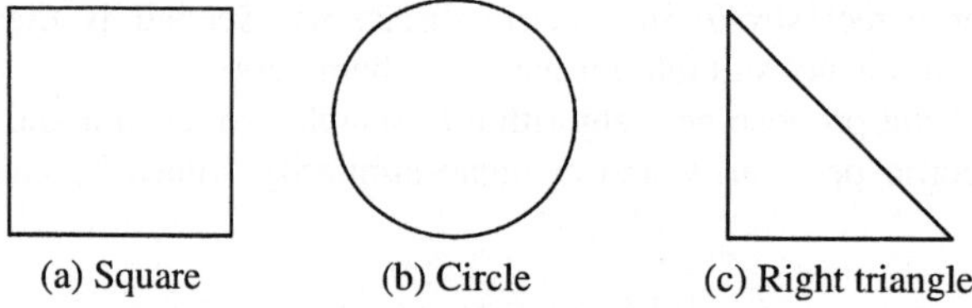

Fig. 2. Tested simple characters

In Table 2, the results of the square whose four sides were equally 99 dots were shown. Old algorithm and new algorithm gave almost same results indicating that

Table 2. Number of local arcs found in the square

Algo-rithm	Curva-ture	12 directions of chord											
		0°	15°	30°	45°	60°	75°	90°	105°	120°	135°	150°	165°
Old	-5	·	·	·	·	·	·	·	·	·	·	·	·
	-4	·	·	·	·	·	·	·	·	·	·	·	·
	-3	·	·	·	·	·	·	·	·	·	·	·	·
	-2	·	·	·	·	·	·	·	·	·	·	·	·
	-1	·	·	·	·	·	·	·	·	·	·	·	·
	0	156	·	·	·	·	·	156	·	·	·	·	·
	1	·	·	·	·	·	·	·	·	·	·	·	·
	2	·	·	·	·	·	·	·	·	·	·	·	·
	3	·	·	·	·	·	·	·	·	·	·	·	·
	4	·	·	·	·	·	·	·	·	·	·	·	·
	5	·	·	·	·	·	·	·	·	·	·	·	·
New	-5	·	·	·	·	·	·	·	·	·	·	·	·
	-4	·	·	·	·	·	·	·	·	·	·	·	·
	-3	·	·	·	·	·	·	·	·	·	·	·	·
	-2	·	·	·	·	·	·	·	·	·	·	·	·
	-1	·	·	·	·	·	·	·	·	·	·	·	·
	0	157	·	·	·	·	·	157	·	·	·	·	·
	1	·	·	·	·	·	·	·	·	·	·	·	·
	2	·	·	·	·	·	·	·	·	·	·	·	·
	3	·	·	·	·	·	·	·	·	·	·	·	·
	4	·	·	·	·	·	·	·	·	·	·	·	·
	5	·	·	·	·	·	·	·	·	·	·	·	·

both algorithms could calculate curvature of vertical and horizontal line in the same manner.

In Table 3, the results of the circle whose diameter was 49 dots were shown. New algorithm gave larger number local arcs indicating that the old algorithm could not detect end point of chord on the circle because of direction interval.

In Table 4, the results of the isosceles right triangle whose two sides were equally 99 dots were shown. The vertical and horizontal lines could be detected as same as in Table 2. But hypotenuse could not be detected by old algorithm because of discrete direction of chord. Furthermore, because of one dot width of the stroke, even new algorithm some times could not detect hypotenuse. This can be understood from small number, 27, compared with other two numbers, 73 and 80.

To know the effect of stroke width, the width of circle was changed to 2 dots, and results were listed in Table 5. Comparing with Table 3, 10 times more local arcs were detected indicating that more than 2 dots width is necessary to obtain appropriate information of curvatures.

From these results, it can be understood that the new algorithm can perform more appropriately.

3   Result and Discussion

a. Compared Chinese Character

48 kinds of well used characters up to 15 strokes shown in Table 6 were chosen and 18 Chinese wrote those for 6 times each (at every more than two days). In total 5184 characters (= 18 subjects × 48 types of characters × 6 times) were collected as an experimental objects. They were taken in a computer through an image scanner, and were cut one by one as 110×110 dots square.

b. Curvature Calculation by the New Local Arc Method

By choosing 5 characters out of 6 collected characters for each type of character, 6 kinds of dictionaries are obtained as same as the previous work for each type of character for each writer.

Table 3. Number of local arcs found in the circle

Algo-rithm	Curva-ture	12 directions of chord											
		0°	15°	30°	45°	60°	75°	90°	105°	120°	135°	150°	165°
Old	-5	·	·	·	·	·	·	·	·	·	·	·	·
	-4	·	·	·	·	·	·	·	·	·	·	·	·
	-3	·	·	·	·	·	·	·	·	·	·	·	·
	-2	·	1	1	·	1	·	·	·	·	·	·	·
	-1	·	·	·	·	·	·	·	·	·	·	·	·
	0	·	·	·	·	·	·	·	·	·	·	·	·
	1	·	·	·	·	·	·	·	·	·	·	·	·
	2	·	·	1	·	·	1	·	·	1	·	·	·
	3	·	·	·	·	·	·	·	·	·	·	·	·
	4	·	·	·	·	·	·	·	·	·	·	·	·
	5	·	·	·	·	·	·	·	·	·	·	·	·
New	-5	·	·	·	·	·	·	·	·	·	·	·	·
	-4	·	·	·	·	·	·	·	·	·	·	·	·
	-3	2	·	·	·	·	1	·	1	·	1	·	1
	-2	·	2	1	·	1	·	·	·	·	·	·	·
	-1	·	·	·	·	·	·	·	·	·	·	·	·
	0	·	·	·	·	·	·	·	·	·	·	·	·
	1	·	·	·	·	·	·	·	·	·	·	·	·
	2	1	·	1	1	1	2	·	·	2	·	1	1
	3	·	·	·	·	1	·	·	·	·	·	·	·
	4	·	·	·	·	·	·	·	·	·	·	·	·
	5	·	·	·	·	·	·	·	·	·	·	·	·

Table 4. Number of local arcs found in the right triangle

Algo-rithm	Curva-ture	12 directions of chord											
		0°	15°	30°	45°	60°	75°	90°	105°	120°	135°	150°	165°
Old	-5	·	·	·	·	·	·	·	·	·	·	·	·
	-4	·	·	·	·	·	·	·	·	·	·	·	·
	-3	·	·	·	·	·	·	·	·	·	·	·	·
	-2	·	·	·	·	·	·	·	·	·	·	·	·
	-1	·	·	·	·	·	·	·	·	·	·	·	·
	0	73	·	·	·	·	·	79	·	·	·	·	·
	1	·	·	·	·	·	·	·	·	·	·	·	·
	2	·	·	·	·	·	·	·	·	·	·	·	·
	3	·	·	·	·	·	·	·	·	·	·	·	·
	4	·	·	·	·	·	·	·	·	·	·	·	·
	5	·	·	·	·	·	·	·	·	·	·	·	·
New	-5	·	·	·	·	·	·	·	·	·	·	·	·
	-4	·	·	·	·	·	·	·	·	·	·	·	·
	-3	·	·	·	·	·	·	·	·	·	·	·	·
	-2	·	·	·	·	·	·	·	·	·	·	·	·
	-1	·	·	·	·	·	·	·	·	·	·	·	·
	0	73	·	·	27	·	·	80	·	·	·	·	·
	1	·	·	·	·	·	·	·	·	·	·	·	·
	2	·	·	·	·	·	·	·	·	·	·	·	·
	3	·	·	·	·	·	·	·	·	·	·	·	·
	4	·	·	·	·	·	·	·	·	·	·	·	·
	5	·	·	·	·	·	·	·	·	·	·	·	·

Table 5. Number of local arcs found in the circle of 2 dots width

Algo-rithm	Curva-ture	12 directions of chord											
		0°	15°	30°	45°	60°	75°	90°	105°	120°	135°	150°	165°
Old	-5	·	·	·	·	·	·	·	·	·	·	·	·
	-4	5	·	·	·	·	·	·	·	·	·	·	·
	-3	·	2	4	4	2	3	4	3	3	3	1	3
	-2	·	7	4	2	5	2	·	3	4	2	4	3
	-1	·	·	·	·	·	·	·	·	·	·	·	·
	0	·	·	·	·	·	·	·	·	·	·	·	·
	1	·	·	·	·	·	·	·	·	·	·	·	·
	2	·	2	5	1	7	7	4	3	7	10	6	1
	3	·	3	2	4	2	2	1	3	1	1	·	3
	4	2	·	·	·	·	·	·	·	·	·	·	·
	5	·	·	·	·	·	·	·	·	·	·	·	·
New	-5	·	·	·	·	·	·	·	·	·	·	·	·
	-4	5	·	·	·	·	·	·	·	·	·	·	·
	-3	20	13	20	10	22	21	30	19	13	14	21	24
	-2	18	22	20	17	17	13	12	8	14	7	15	13
	-1	·	·	·	·	·	·	·	·	·	·	·	·
	0	·	·	·	·	·	·	·	·	·	·	·	·
	1	·	·	·	·	·	2	·	·	·	·	·	·
	2	18	10	23	17	30	28	23	17	28	23	20	22
	3	19	16	13	16	20	22	11	9	10	4	11	7
	4	2	·	·	·	·	·	·	·	·	·	·	·
	5	·	·	·	·	·	·	·	·	·	·	·	·

Table 6. 48 Chinese characters within 15 strokes

了	又	人	几	个	上	大	万
不	心	似	中	用	可	去	他
好	有	在	同	没	但	我	来
的	和	使	受	要	看	是	前
都	能	家	被	起	常	做	得
就	最	然	象	数	想	需	靠

The simple similarity values (cosine values) were calculated between dictionaries and characters. From those values, writers were specified. Some of the recognition ratios obtained by old and new algorithms were compared in Figs. 3 and 4. The new algorithm gives better recognition ratio than the old algorithm as expected. These are indicating that number of local arcs is important to extract a feature of the writer.

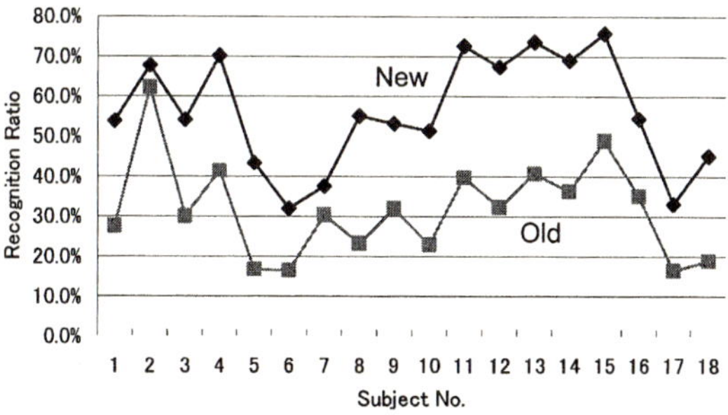

Fig. 3. Comparison of recognition ratio of each writer

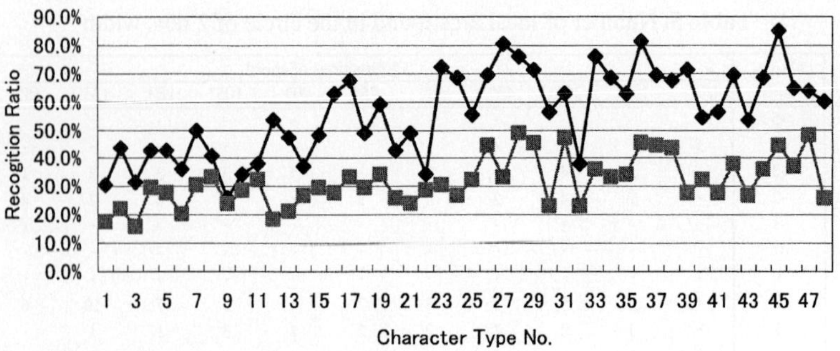

Fig. 4. Comparison of recognition ratio of each type of character

4 Conclusion

New algorithm to search strokes is proposed and examined the applicability of the algorithm. In the case of vertical and horizontal lines, both old and new algorithms gave same results. But in case of circle and leaned line, old algorithm could not search strokes properly. Furthermore, the number of local arcs fairly affected to writer recognition. Therefore new algorithm extract feature of writing habit more accurately.

However, the average recognition ratio is 56.1% (31.7% in old algorithm) and is not high enough to be satisfied. Therefore we have to use the new similarity evaluation function developed in the previous work for practical use.

In the future, we might improve recognition ratio for practical use.

References

1. M. Ozaki, Y. Adachi, N. Ishii, and T. Koyazu: Fuzzy CAI System to Improve Hand Writing Skills by Using Sensuous (1996) Trans. of IEICE Vol.J79-D-Ⅱ NO.9 pp.1554-1561
2. M. Ozaki, Y. Adachi, and N. Ishii: Writer Recognition by means of Fuzzy Similarity Evaluation Function (2000) Proc. KES 2000, pp.287-291
3. M. Ozaki, Y. Adachi, and N. Ishii: Study of Accuracy Dependence of Writer Recognition on Number of Character (2000) Proc. KES 2000, pp.292-296
4. M. Ozaki, Y. Adachi, N. Ishii and M. Yoshimura: Writer Recognition by means of Fuzzy Membership Function and Local Arcs (2001) Proc. KES 2001, pp.414-418
5. M. Ozaki, Y. Adachi and N. Ishii: Development of Hybrid Type Writer Recognition System (2002) Proc. KES 2002, pp.765-769
6. Y. Adachi, M. Liu and M. Ozaki: A New Similarity Evaluation Function for Writer Recognition of Chinese Character (2004) Proc. KES2004, pp.71-76

Development of Judging Method
of Understanding Level in Web Learning

Yoshinori Adachi, Koichi Takahashi, Masahiro Ozaki, and Yuji Iwahori

Chubu University, 1200 Matsumoto-Cho, Kasugai, Aichi, Japan 487-8501
adachiy@isc.chubu.ac.jp

Abstract. In the previous works, we have developed Web education system using the dynamically changing learning material and learning efficiency test have been carried out. Especially, in the new system, it gave priority to the maintenance of the greediness for learning, and three layered learning contents which dynamically changed the learning material by learner himself were proposed. In this study, to maintain the eagerness for self learning by changing learning material based on the understanding level, we developed an understanding level judging system based on the Fuzzy inferences and adequacy of the judging algorithm was tested, and adequate results were obtained.

1 Introduction

Authors have developed education systems using Internet [1-6], and also the dynamically changing learning material according to the intelligibility of the learner has recently been developed, and some research results have been obtained [6, 7]. However, the learning using Web education system is established as a premise of the independent learning. Therefore, Web based learning has not been suitable for the learner who was lacking for the eagerness for learning.

By using dynamically changing learning material which the learner could change the contents according to his understanding level manually, learners' volitions for self learning could be maintained. From the experiments, such results were obtained [6].

Then, in order to promote the learning using the Web education system, it is important to sense fulfillment in the learning. So, the eagerness for learning could maintain. To be obtained such kind of feeling, it is necessary to change the degree of difficulty of contents in proportion to the learner's understanding level, and the appropriate result in proportion to the effort must be obtained. To realize these functions, the hierarchical contents should be prepared, and the algorithm that judges the understanding level and changes the contents in proportion to the intelligibility automatically is necessary.

In this study, the system that judges the learner's understanding level from the learning history will be made by using fuzzy inferences.

2 Judging Algorithm

The motivation of learning is classified into 6 intention groups [8], and is shown in two-dimensional space as shown in Fig. 1.

R. Khosla et al. (Eds.): KES 2005, LNAI 3681, pp. 781–786, 2005.

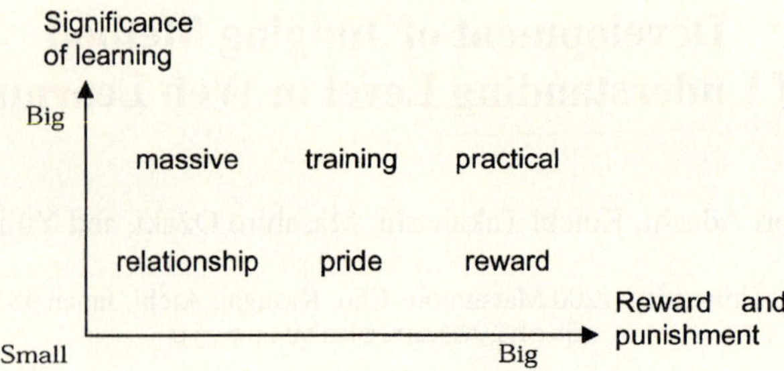

Fig. 1. Classification of learning motivation

The learner whose eagerness for learning is stably strong is mostly belonging to the "massive intention group". The learner belonging to this group has spontaneous motivation based on the intellectual curiosity, apprehensive desire, and ambition. However, the learner whose eagerness for learning is based on the reward initially has strong motivation but soon is used to the reward and loses eagerness for learning. Therefore, "massive intention" is necessary to keep the learning motivation. To realize having "massive intention", it is necessary to present suitable problem to the learner's understanding level and to give fulfillment which can be obtained from correct answer.

To judge the understanding level, we use fuzzy inferences with learner's historical record. Up to now, many systems have been proposed to judge the understanding level by using fuzzy inference. But they only used one kind of record such as number of correct answers or required solving time.

In this work, we used three kinds of record such as number of correct answers, number of referring explanations, and learning response time.

a. Ratio of Right Answer

The ratio of right answer (RRA) is related to the fuzzy inference by the deviation value. The deviation value z_i is obtained from the following equation:

$$z_i = 14\left(\frac{x_i - \mu}{\sigma}\right) + 50 \tag{1}$$

where x_i is the score of the learner i, σ is the standard deviation, and μ is the average score. The fuzzy membership function is defined as follows:

The membership function of RRA is high:

$$\mu_H(z) = \begin{cases} 1 & (z > 70) \\ (z-50)/20 & (50 \leq z \leq 70) \\ 0 & (z < 50) \end{cases} \tag{2}$$

The membership function of RRA is medium:

$$\mu_M(z) = \begin{cases} 1 & (z < 30) \\ (z-30)/20 & (30 \leq z < 50) \\ (70-z)/20 & (50 \leq z \leq 70) \\ 0 & (z > 70) \end{cases} \tag{3}$$

The membership function of RRA is low:

$$\mu_L(z) = \begin{cases} 1 & (z < 30) \\ (50-z)/20 & (30 \le z \le 50) \\ 0 & (z > 50) \end{cases} \tag{4}$$

b. Frequency in Use of Explanation

The frequency in use of explanation (FUE), f_i, is evaluated by the weighted average expressed as follows:

$$f_i = \frac{0.8 \times n_i^H + n_i^M + 0.2 \times n_i^L}{N} \tag{5}$$

where n_i^j is the number of referring explanations of i-th leaner in the j-th level. The deviation value is obtained from Eq. (1). Then, membership functions are defined as follows:

The membership function of FUE high:

$$\mu_H(z) = \begin{cases} 1 & (z > 65) \\ (z-50)/15 & (50 \le z \le 65) \\ 0 & (z < 50) \end{cases} \tag{6}$$

The membership function of FUE is medium:

$$\mu_M(z) = \begin{cases} 0 & (z < 35) \\ (z-35)/15 & (35 \le z < 50) \\ (50-z)/15 & (50 \le z \le 65) \\ 0 & (z > 65) \end{cases} \tag{7}$$

The membership function of FUE is low:

$$\mu_L(z) = \begin{cases} 1 & (z < 35) \\ (50-z)/15 & (35 \le z \le 50) \\ 0 & (z > 50) \end{cases} \tag{8}$$

c. Caution Index of Learning Response Time

Caution index (CI) was proposed by Sato [9] to check the behavior of the learner. Especially, to obtain the caution index of learning response time (CILRT), fuzzy response was introduced by Nagaoka et al. [10]. They defined a membership function for "fast response" by the following equation.

$$\mu_j(t_{ij}) = \begin{cases} 1 & (t_{ij} \le a) \\ \dfrac{b-t_{ij}}{b-a} & (a < t_{ij} \le b) \\ 0 & (b < t_{ij}) \end{cases} \tag{9}$$

where $a = E_j/4$, $b = 2E_j$, $E_j = \sum_{i=1}^{N} t_{ij}/N$, and i is learner number, j is problem number, N is total number of learners, and t_{ij} is response time. Then, Caution index of learner i for learning response time $C.S_i$ was defined as follows:

$$C.S_i(T) = (1 + \alpha_{ti})/2 \tag{10}$$

where α_{ti} is a correlation coefficient between μ_{ij} $(j = 1, \cdots, n)$ of learner i and average response time E_j $(j = 1, \cdots, n)$ of problem j. The membership function of CILRT is defined as follows:

The membership function of CILRT is large:

$$\mu_L(C.S_i) = \begin{cases} 1 & (C.S_i > 0.65) \\ (C.S_i - 0.55)/0.1 & (0.55 \leq C.S_i \leq 0.65) \\ 0 & (C.S_i < 0.55) \end{cases} \tag{11}$$

The membership function of CILRT is small:

$$\mu_S(C.S_i) = \begin{cases} 1 & (C.S_i < 0.55) \\ (0.65 - C.S_i)/0.1 & (0.55 \leq C.S_i \leq 0.65) \\ 0 & (C.S_i > 0.65) \end{cases} \tag{12}$$

These three fuzzy values, ratio of right answer, frequency in use of explanation and caution index of learning response time, are used to make the fuzzy inferences as shown in Table 1.

The min-max-gravity method is used as a defuzzifier method as usual, and understanding level (UL) is judged..

Table 1. Fuzzy inferences to judge understanding level

Rule No.	RRA	FUE	CILRT	UL
1	High	High	Large	Slightly High
2	High	High	Small	High
3	High	Medium	Large	High
4	High	Medium	Small	Very High
5	High	Low	Large	High
6	High	Low	Small	Very High
7	Medium	High	Large	Slightly Low
8	Medium	High	Small	Medium
9	Medium	Medium	Large	Medium
10	Medium	Medium	Small	Slightly High
11	Medium	Low	Large	Medium
12	Medium	Low	Small	Slightly High
13	Low	High	Large	Very Low
14	Low	High	Small	Low
15	Low	Medium	Large	Very Low
16	Low	Medium	Small	Low
17	Low	Low	Large	Low
18	Low	Low	Small	Slightly Low

3   Experimental Results

We applied the above fuzzy inferences to 50 subjects for 39 problems. Some results are depicted in Table 2.

Table 2. Inferred results of understanding level

Subject No.	RRA	FUE	CILRT	UL
5	0.85	0.72	0.43	67.8
6	0.87	0.69	0.54	69.7
7	0.90	0.77	0.38	71.5
10	0.85	0.36	0.48	79.1
13	0.92	0.08	0.41	88.8
16	0.69	0.79	0.43	47.7
25	0.46	0.82	0.48	25.0
30	0.49	0.05	0.83	28.9
34	0.87	0.64	0.40	69.7

Subjects No.7, 16 and 25 have roughly same FUE and CILRT values. Their UL are reasonably proportional to RRA. Then RRA is well reflected to UL.

Subjects No.7, 10 and 13 have roughly same RRA and CILRT values and only have different FUE values. And with increasing of FUE value, UL value is appropriately decreasing.

Subjects No.25 and 30 have roughly same RRA values and have very large difference in FUE and CILRT values. But UL values are roughly same. It is indicating that the effects of FUE and CILRT are canceled out.

Subjects No.5, 6 and 34 have almost same RRA and FUE values, and some difference in CILRT. But UL doesn't have such difference. It is indicating that CILRT doesn't have much influence on UL. CILRT shows that learner whose value is large behaves differently to other learner and need to be marked in the classroom.

In this study, not many learners have CILRT values larger than 0.6 which is the criterion of differentia to others in Eqs. (11) and (12).

4   Conclusion

In this work, we proposed the algorithm to judge understanding level by using fuzzy inferences. In the learning history, we used three kinds of record such as number of correct answers, number of referring explanations, and learning response time. The number of fuzzy inferences is 18 and the min-max-gravity method was used for defuzzification. The followings are obtained from this work:

(1) From the questionnaire to the learner, judged understanding level is adequate.
(2) Caution index of learning response time doesn't have much influence on judging understanding level.
(3) Caution index shows differentia to others and not directly related to the understanding level.
(4) The proposed method requires many learners' records and many problems to calculate correlate coefficient and ratio. Then, it is difficult to apply to the dynamic change of text within learning.

In the future, we have to develop fuzzy inferences which can be used during Web learning and requires no other learner's record. For this purpose, we now are trying to develop fuzzy inferences by using ratio of correct answers, frequency of referring explanations, and average learning response time. These rules can be applied even a small number of problems and only one learner exists on the Web education system. Furthermore, this time caution index of problem was not used. It could be useful to judge understanding level more accurately.

References

1. Ozaki M., Adachi Y., Ishii N. and Koyazu T.: CAI system to improve hand writing skills by means of fuzzy theory (1995) Proc. Intl. Conf. of 4th IEEE Intl. Conf. on Fuzzy Sys. and Intl. Fuzzy Eng. Sym. FUZZ-IEEE/IFES'95, pp. 491-496
2. Ozaki M., Koyazu T., Adachi Y. and Ishii N.: Development of CAI system based on recognition understanding learning model (1995) Proc. IASTED Intl. Conf. Modeling and Simulation, pp.153-155
3. Adachi Y., Kawasumi K., Ozaki M., and Ishii N.: Development accounting education CAI system (2000) Proc. Intl. Conf. on Knowledge-Based Intelligent Eng. Sys. & Allied Tech., pp.389-392
4. Kawada H., Ozaki M., Ejima T., Adachi Y.: Development of the Client/Server System for CAI Education (2002) J. of Nagoya Women's University, No.48, pp.113-120 (in Japanese)
5. Takeoka S., Ozaki M., Kawada H., Iwashita K., Ejima T., Adachi Y.: An Experimental CAI System Based on Learner's Understanding (2002) J. of Nagoya Women's University, NO.48, pp.177-186 (in Japanese)
6. Ozaki M., Koyama K., Adachi Y. and Ishii N.: Web Type CAI System with Dynamic Text Change from Database by Understanding (2003) Proc. Intl. Conf. on Knowledge-Based Intelligent Eng. Sys., pp.567-572
7. Koyama H., Takeoka S., Ozaki M., Adachi Y,: The Development of Authoring System for Teaching Materials – The Dynamically Personalized Hyper Text based on XML – (2003) IEICE Tech. Rep. IEICE Educ. Tech. ET2003-55, pp.23-27 (in Japanese)
8. Ichikawa S,: "Gakushu to Kyoiku no Shinrigaku" (1994) Iwanami Shoten (in Japanese)
9. Sato T.: "Introduction to Educational Information Technology" (1989) Corona Publishing Co. (in Japanese)
10. Nagaoka K. and Wu Y.: An analysis of Learning Response Time Matrix Based on Fuzzy Transformation (1991) Trans. of IEICE D-I, Vol.J74-D-I, No.2, pp.95-100 (in Japanese)

Organising Documents
Based on Standard-Example Split Test

Kenta Fukuoka, Tomofumi Nakano, and Nobuhiro Inuzuka

Graduate School of Engineering,
Nagoya Institute of Technology
Gokiso-cho Showa, Nagoya 466-8555, Japan
`inuzuka@nitech.ac.jp`

Abstract. A purpose of text-mining is to summarise a large collection of documents. This paper proposes a new method to view a summary of large document set. It consists of two techniques, one of which constructs classification trees using a split test called the standard-example (standard-document) split test, and the other is a method to display features in each class of documents classified in the trees. The standard-example split test is a test which divides examples by their distance (or similarity) from a standard-example which is selected by a criterion. This is the first method which applies this test to text mining. The display method exhibits representative words of document classes which emphasise their feature.

1   Introduction

In order to utilise enormous number of documents text-mining is essential. Many methods are studied recently[6–9]. An aim of text mining is to understand structure of collection of documents. For example articles of a magazine of a year consist of some topics and have clusters. Observing the clusters helps to understand the trend of topics in the magazine. An automatic method to show such cluster gives a good overview for such articles.

This paper proposes a method for classifying documents according to provided class labels of documents and representing cluster structure of the document collection. We do not focus accurate classification but to show the cluster structure. The method uses decision tree learning by a novel split approach called standard-example (standard-document) split test. Standard-example split test was first introduced in classification of time series[1]. It splits examples (documents for our aim) by their distance (or similarity) to a standard-example (a standard-documents). This test is suitable to see a representative example of each subclass of examples corresponding to a node of decision tree. A standard-document is placed at the centre of the subclass of documents. For comprehensive presentation of tree, the method displays featured words of standard-documents corresponding to nodes of decision tree. We discuss criteria to choose featured words from documents.

The next section reviews the decision tree induction using standard-example split test, and we modify it for document classification. Section 3 gives a method

R. Khosla et al. (Eds.): KES 2005, LNAI 3681, pp. 787–793, 2005.

Table 1. Decision tree induction by value test.

---

procedure DTVT(E, A, D)

input E: a set of examples, A: a set of attributes, d: a default class;
output a decision tree;

```
1  if E is empty then
2        return a tree with a single node labelled d;
3  elseif all examples are in a class c then
4        return a tree with a single node labelled the class c;
5  elseif A is empty then
6        return a tree with a single node labelled the majority class of E;
7  else
8        a := a best appropriate attribute;
9        T := a tree with only a root node given a test for the attribute a;
10       foreach values v in the range of a do
11             Ev := the set of examples that takes v for a;
12             Tv := DTVT(Ev, A − {a}, the majority class of E);
13             link the root of T to Tv as its child;
14       return T;
```

---

to display a tree induced by affixing featured words of documents to the tree. Section 4 reports a first experiment of the method for classifying net news articles.

2   Decision Tree by Standard-Example Split Tests

For the rest of this paper, we introduce some notations. Let us denote a set of documents by E and an example by e. We denote subsets of it by some modified symbols of E. When examples consist of attribute-value vectors, we use A for a set of attributes and a for each attribute. The value of an attribute a of an example e is referred by $e(a)$.

Decision Tree Induction by Value Test and Standard-Example Split Test. Decision tree learning is a widely used and well-studied classification algorithm. ID3 and its successor C4.5 are representative[4, 5]. Table 1 shows an outline of the algorithm. It constructs a tree recursively. The learning algorithm selects an attribute which divides examples by its value in an evaluation criterion. The attribute makes a node having a test for the attribute. Each of divided examples is processed to construct a subtree and it is connected to the parent node.

The test placed in each node of a decision tree is called value test, which tests the value of an attribute and divides examples according to the value. For an attribute a and its range of values $\{v_1, \cdots, v_n\}$ it divides examples E to

$$E_1 = \{e \in E \mid e(a) = v_1\}, \cdots, E_n = \{e \in E \mid e(a) = v_n\}$$

Table 2. Decision tree induction by standard-example split test.

---

procedure DTSE(E, A, D)

input E: a set of examples, A: a set of attributes, d: a default class;
output a decision tree;

```
1  if E is empty then
2        return a tree with a single node labelled d;
3  elseif all examples are in a class c then
4        return a tree with a single node labelled the class c;
5  elseif A is empty then
6        return a tree with a single node labelled the majority class of E;
7  else
8        e, a, θ := a best appropriate example, attribute and threshold;
9        T :=  a tree with only a root node given a standard-example split test
                    for e, a, θ;
10       E₁ := {e' ∈ E | G(e(a), e'(a)) < θ};
11       E₂ := E − E₁;
12       foreach Eᵢ of E₁ and E₂ do
13             Tᵥ := DTSE(Eᵢ, A − {a}, the majority class of E);
14             link the root of T to Tᵥ as its child;
15       return T;
```

Table 2 gives another decision tree induction algorithm, which does not use value test but uses standard-example split test. A standard-example split test takes an attribute a, example $e \in E$ and a threshold θ and divides E into E_c (close examples) and E_f (far examples),

$$E_c = \{e' \in E \mid \mathrm{dis}(e(a), e'(a)) \le \theta\} \text{ and } E_f = \{e' \in E \mid \mathrm{dis}(e(a), e'(a)) > \theta\}$$

where $\mathrm{dis}(v, v')$ is a distance metric that is defined for each attribute. In the research that applied it to time series classification, the definition of distance metric was a key topic and is discussed carefully.

Division or split is evaluated by the pureness of classes of examples in $E_1, \cdots,$ E_n or E_c and E_f. A common criterion of pureness is the entropy. Appropriateness of selecting attribute is evaluated by the decrease of entropy, which is called information gain. As ID3 and C4.5[4, 5] select the attribute that maximises the information gain, the method by Yamada et al.[1] also selects an example, an attribute, and a threshold maximising the information gain.

Standard-Example Split Test for Document Classification. Table 2 shows the decision tree induction algorithm using standard-example split test.

A reason of introducing standard-example split test in classifying time series was the difficulty of value test for time series attributes. This is because each shape of time-series graph is a different value. Treating a representative time series as a standard-example was an idea. Documents are also similar characteristic. Documents in a collection have a large dimension and each of documents

Table 3. Document decision tree induction by standard-document split test.

---

procedure DTSD(E, D)

input E: a set of documents, A: a set of attributes, d: a default class;
output a decision tree;

1 **if** E is empty **then**
2   **return** a tree with a single node labelled d;
3 **elseif** all documents are in a class c **then**
4   **return** a tree with a single node labelled the class c;
5 **else**
6        $e, \theta :=$ a best appropriate document and threshold;
7        $T :=$ a tree with only a root node given a standard-example split test
                    for e, θ;
8        $E_1 := \{e' \in E \mid G(e, e') < \theta\}$;
9        $E_2 := E - E_1$;
10   **foreach** E_i of E_1 and E_2 **do**
11            $T_v :=$ DTSD$(E_i,$ the majority class of $E)$;
12   link the root of T to T_v as its child;
13   **return** T;

---

can be regarded as a different value. A representative document may represent
the content of documents of a cluster where it belongs. This is also similar to
time-series, we can see a representative shape in representative time-series.

A difference of our document classification algorithm from the decision tree
induction of time series is that we do not think of attributes of documents. It is
assumed that a time series example consists of a couple of time series, such as
movements in X, Y, and Z-coordinates, and each series is treated as an attribute.
Although the original standard-example split test splits examples by the distance
at an attribute, we split documents by the similarity among documents without
thinking any attribute, where we will discuss about the similarity in the following
paragraph. That is, a standard-document split test consists of a document e and
a threshold θ and it splits document collection E as follows.

$$E_c = \{e' \in E \mid \mathrm{sim}(e, e') > \theta\} \text{ and } E_f = \{e' \in E \mid \mathrm{sim}(e, e') \leq \theta\}.$$

2.1   Distance Between Documents

We represent a documents as a vector whose elements correspond to terms
(words) possibly appeared in documents and the values represent the importance of the terms. By representing documents in vectors, we can use the cosine
similarity metric for the similarity.

A common index for importance of words in a document is *tfidf*[2, 3], which
consists of *tf*, term frequency, and *idf*, inverse document frequency.

For a term (a word) w and a document d, $tf(w, d)$ is normally defined as the
number of occurrence of w in d. However, here we define it in a normalised form
as,

$$tf(w,d) = \frac{\log_2(\text{the number of occurrence of } w \text{ in } d + 1)}{\log_2(\text{the number of all words in } d)}$$

Although the word w whose $tf(w,d)$ is large may be important in the document, it is possible the word is very common, such as *is* and *the*. In order to count this possibility, we use the other index $idf(w)$, which is the un-commonality of a word w, which is defined as follows.

$$idf(w) = \log_2 \frac{(\text{the total number of documents})}{(\text{the number of documents that contain the word } w)} + 1$$

By using these indexes a document d is represented by a vector v_d which is defined by,

$$v_d = (tfidf(w_1, d), \cdots, tfidf(w_n, d),$$

where $\{w_1, \cdots, w_n\}$ is all of the words possibly appeared in documents and $tfidf(w,d) = tf(w,d) \cdot idf(w)$. The similarity $\text{sim}(d_1, d_2)$ of two documents d_1 and d_2 is defined as the cosine of their vectors.

3   Representation of Trees

In order to help the user to understand the induced tree, we propose to display featured words of standard-documents of the tree. An expensive alternative is to develop a good interface to show whole documents in nodes of tree. We develop a concise method instead, which gives also concise representation.

As we described in the previous section the *tfidf* is an index of featured words. Featured words of a standard-document, however, does not always contribute split at a node. A split is done by a standard-document d_{stan} and a threshold θ. We assume here this makes a split E_c and E_f. A high *tfidf* value of a word w in d means that the word is not very common in many text and it appeared frequently in d_{stan}. It does not mean that this *tfidf*-value contributes the split. Words that have large *tfidf*-values in E_c may also earn values in E_f. Featured words of the split should be words which contributes the split.

As a candidate of such index, we can consider the following value.

$$F_1(w) = \frac{\sum_{d \in E_c} tfidf(w, d)}{|E_c|} - \frac{\sum_{d \in E_f} tfidf(w, d)}{|E_f|}$$

This index gives a heigh value to a word that is frequent in E_c and rare in E_f. It is still possible that the index F_1 is gives some too general words high values, because such words are highly frequent and even the difference between the counts for E_c and E_f recovers small *tfidf*.

An answer for this problem, we also give the following F_2.

$$F_2(w) = \begin{cases} \dfrac{\sum_{d \in E_c} tfidf(w,d)}{|E_c|} & \text{when } \forall d' \in E_f, w \notin d \\ 0 & \text{when } \exists d' \in E_f, w \in d \end{cases}$$

This suppresses words that appeared in E_f even once.

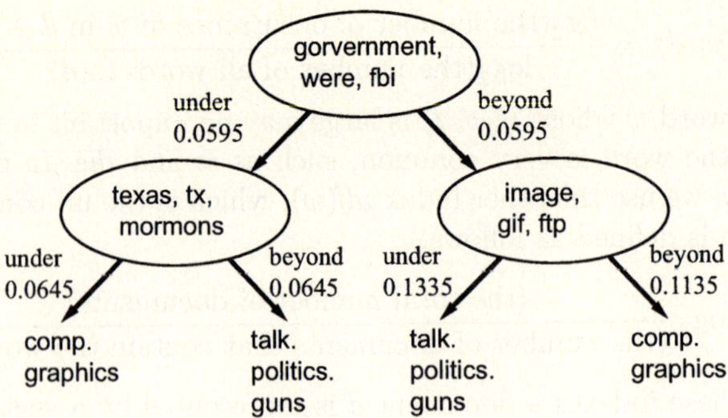

Fig. 1. An example of tree.

4   Experiments

For an explanatory experiment, we used a document set collecting net news, which is taken from UCI machine learning repository. The documents in the collection have pre-classified 20 classes. We used by selecting several classes from the 20 classes. We treated all words in the documents and capital letters are transformed into small.

As the case of two classes the news group `comp.graphics` and `talk.politics. guns` are used. The case of three classes includes `rec.sport.basketball` in addition to them. For these cases 100 documents for each class are used to examines classification accuracy and the numbers of standard-documents used in induced decision trees. They are examined by 10-fold cross validation.

In the case of two class classification the classification accuracy was 76.5% and the number of standard-documents was 4.9. For the case of three classes, the accuracy was 71.0% and the number of standard-documents was 13.0. We should pay attention the conciseness of the tree and the number of the nodes are enough small to keep comprehensibility.

An example of induced tree is shown in Fig.1. For this tree, the candidate of featured words of standard-documents corresponding to the three nodes are listed in Tables 4(a)-(c). Words are listed in the order of *tfidf* and the proposing F_1 and F_2. We think that the words listed by F_1 or F_2 are relatively representative words to express the cluster of documents, although F_1 gives too common words as we discussed already and this problem is dissolved in F_2.

5   Conclusion

This paper proposed a method to give an overview of collection of documents which are already classified. The method includes a new decision tree induction method, standard-document split test. The test was proposed for time-series and we apply it to documents. Our proposal also includes a display method that gives featured words to standard-documents. An explanatory experiment showed that the method gives concise representation for document collection.

Table 4. Featured words to be displayed in nodes of the decision tree in Fig. 1.

(a) For the root node of Fig. 1.

	tfidf	F_1	F_2
1st	weaver	they	government
2nd	spence	gun	were
3rd	prosecution	as	fbi
4th	defense	people	fire
5th	idaho	the	weaver

(b) For the left node of Fig. 1.

	tfidf	F_1	F_2
1st	mormons	texas	texas
2nd	holiness	tx	tx
3rd	texans	mormons	mormons
4th	governor	world	expulsion
5th	tuesday	quoting	quoting

(c) For the right node of Fig. 1.

	tfidf	F_1	F_2
1st	routines	image	image
2nd	routine	gif	gif
3rd	mode	ftp	ftp
4th	wrote	virtual	virtual
5th	include	put	routines

References

1. Y. Yamada, E. Suzuki, H. Yokoi and K. Takabayashi: "Decision tree induction from time serieses data based on a standard-example split test", proc. Twentieth International Conference on Machine Learning (ICML2003), pp.840–847, 2003.
2. D. Harman: "Ranking Algorithms", in Information Retrieval, W. Frakes and R. Baeza-Yates (eds.), Prentice-Hall, 1992.
3. K. Sparck-Jones: "Index Term Weighting", Information Storage and Retrieval, **9**, pp.619–633, 1973.
4. J. R. Quinlan: "C4.5: Programs for Machine Learning", Morgan Kaufmann, 1993.
5. J. R. Quinlan: "Induction of decision trees", Machine Learning, **1**, 1, pp.81–106, 1986.
6. K. Lang: "NewsWeeder: Learning to Filter Netnews", Proc. Twelfth International Conference on Machine Learning, pp.331–339, 1995.
7. T. Joachims: "Text categorization with support vector machines: Learning with many relevant features", Proc. Tenth European Conference on Machine Learning, pp. 262–271, 1998.
8. Y. Yang and X. Liu: "A re-examination of text categorization methods", Proc. SIGIR'99, pp.42–49, 1999.
9. F. Sabastiani: "Machine learning in automated text categorization", ACM Computing Surveys, **34**, pp.1–47, 2004.

e-Learning Materials Development
Based on Abstract Analysis Using Web Tools

Tomofumi Nakano and Yukie Koyama

Nagoya Institute of Technology
Gokiso-cho, Showa-ku, Nagoya 466-8555 Japan
{nakano,koyama}@center.nitech.ac.jp

Abstract. This study includes an original corpus of engineering journals
and is part of the series of E-Learning & English for Specific Purposes
(ESP) researches . Purposes (ESP) researches that includes an original
corpus of engineering journals. In this paper the results of a corpus study
will be presented, and a sample of the ESP e-learning materials being
developed for graduate students in engineering will be shown. Abstracts
were chosen for the corpus this time because students are likely to read
many for their research, and eventually to have to produce their own.
We prepare the 40,000-word corpus that consists of 263 abstracts from
mechanical and electrical engineering journals. The corpus is analyzed
using Wmatrix, which gives part-of-speech tags and semantic tags, and
compares the results with those of the BNC written corpus sampler.
Some special features found in the analysis are frequencies in semantic tags, part-of-speech tags, difference in the use of verbal forms and
multi-words. As an application of the important features, we are developing web-based materials which include the original abstracts with target
items hyper-linked to various pages containing exercises, concordances,
grammar explanations, a bilingual dictionary, etc.

1   Introduction

In the field of English teaching, since Swales claimed in his epoch-making book,
Genre Analysis [11], it has been widely accepted that ESP is one of the most efficient approaches in terms of content appropriateness and students' motivation.
Since we teach at a university of technology, we first started the need analysis
and found that reading, especially reading academic papers, is the most important skill for engineering students. After that, we started to compile an original
corpus of engineering journal papers. This corpus is still growing both in its
discipline coverage and its quantity.

In this study the results of a corpus study and a sample of the ESP e-learning
material will be shown. This material is developed for graduate students in
engineering this time because students are likely to read many articles for their
research, and eventually to have to produce articles of their own. Needless to say,
abstracts play an enormously important role in the academic world, because by
reading abstracts, in many cases, the readers decide whether or not they continue

R. Khosla et al. (Eds.): KES 2005, LNAI 3681, pp. 794–800, 2005.

to read the full papers [4]. Another reason is that, in the English as a Foreign Language (EFL) situation, researchers often write their abstracts in English but the rest of the paper is written in their first language. This also raises the need for abstract analysis in EFL situations such as in Japan.

While reading is the most important language skill for engineering students in Japan [6], they are hindered in reading academic papers by a lack of vocabulary (usually sub-technical or academic) [1] and by difficulty with the grammar of long, often complex sentences [5]. Therefore, this study focuses not only on the word lists but also on part-of-speech and semantic areas. An application introduced in this study also makes it possible to adjust the level of frequency and the degree of specification of the word compared to that in general corpus. As Morton points out, the problem for a student is not technical vocabulary but the difficult words of more general English [7].

Thanks to developments in ICT, E-Learning has become an ideal medium for language learning because of its flexibility and the autonomous learning opportunities it provides inside and outside the classroom. As a new application of the results of relevant analysis, an e-learning material for engineering graduate students will be introduced in the rest of this paper.

2 Corpus Analysis

2.1 The Method of Analysis

The 40,000-word corpus used in this study consists of 263 abstracts from mechanical and electrical engineering journals. This corpus is taken from an originally compiled 1,120,000-word corpus of full papers of these journals. We use Wmatrix [9] for the abstract analysis, not only because this software is very easy to handle but also it has a special function which can determine the characteristics of the corpus. Using the Wmatrix the corpus was automatically tagged, both by part-of-speech tags with CLAWS7 [3] and by semantic tags of USAS (UCREL Semantic Analysis System [8]). Moreover, Wmatrix provides frequency tables and log-likelihood tables of words and these two kinds of tags. Log-likelihood is a measurement which shows the difference in frequencies of two different corpora [2]. Therefore, the information given by log-likelihood is very important for ESP material development in order to grasp the characteristics of the ESP Corpus. The two corpora used in this study are the abstract corpus and BNC written corpus sampler which is the built-in corpus in Wmatrix. In Table 1, the left word list is ordered by the frequency and the right word list is ordered by the log-likelihood. While almost all words are general in the left list, the words in right list are specific to the abstracts or engineering papers.

2.2 Results of the Analysis

The lists of part-of-speech tags and semantic tags are shown in Table 2 and 3 respectively. Both lists show tag names, the frequency in the corpus, its frequency rate, the frequency in BNC written corpus sampler, its frequency rate and the log-likelihood, and these are sorted by the log-likelihood.

Table 1. Left: a word list ordered by the frequency. Right: a word list ordered by the log-likelihood.

rank	word	freq.
1	the	2459
2	of	1205
3	and	945
4	a	683
5	in	525
6	is	522
7	to	521
8	for	396
9	are	262
10	with	260
11	this	213
12	by	190
13	that	183
14	an	172
15	on	156
16	be	149
17	at	128
18	from	127
19	as	120
20	flow	119

rank	word	Abst. freq.	rate	BNC freq.	rate	log-likelihood
1	the	2459	8.38	37283	3.79	1158.11
2	of	1205	4.10	12817	1.30	1068.71
3	flow	119	0.41	10	0.00	772.82
4	model	103	0.35	20	0.00	621.26
5	results	88	0.30	31	0.00	488.40
6	energy	84	0.29	33	0.00	457.50
7	presented	63	0.21	9	0.00	392.35
8	method	59	0.20	16	0.00	340.94
9	fuel	59	0.20	22	0.00	324.30
10	paper	93	0.32	174	0.02	323.55
11	power	71	0.24	67	0.01	315.47
12	using	91	0.31	189	0.02	302.33
13	analysis	50	0.17	10	0.00	300.55
14	combustion	42	0.14	1	0.00	287.94
15	by	190	0.65	1293	0.13	286.05
16	performance	58	0.20	39	0.00	282.24
17	based_on	55	0.19	31	0.00	278.82
18	experimental	43	0.15	4	0.00	277.34
19	conditions	61	0.21	61	0.01	266.37
20	gas	70	0.24	106	0.01	265.30

Table 2. A part-of-speech tag list.

POS tag	Abst. freq.	rate	BNC freq.	rate	log-likelihood
NN1	7297	24.86	147395	15.22	1447.40
JJ	3481	11.86	74927	7.74	533.93
FO	222	0.76	2050	0.21	233.75
VVN	1205	4.10	24675	2.55	226.88
AT	2483	8.46	67521	6.97	84.13
VBZ	522	1.78	11171	1.15	82.10
IO	1204	4.10	30286	3.13	78.32
NN2	2064	7.03	55665	5.75	75.84
IF	398	1.36	8765	0.91	55.09
VVZ	350	1.19	7602	0.79	51.59
VBR	262	0.89	5435	0.56	46.88
⋮					

Examining the results shown in the tables, the findings are as follows:

1. Semantic areas such as objects, mental objects (method and means), substances & materials (gas, solid and general), measurement (length & height, distance, size and volume), comparison, and evaluation occur much more frequently.

Table 3. A semantic tag list.

semantic tag.	Abst. freq.	rate	BNC freq.	rate	log-likelihood	meaning
X4.2	444	1.51	3108	0.32	640.06	Mental object :- Means, method
O1.3	167	0.57	300	0.03	586.57	Substances and materials generally: Gas
O2	610	2.08	6100	0.63	577.74	Objects generally
O3	204	0.69	651	0.07	537.88	Electricity and electrical equipment
A1.5.1	308	1.05	1965	0.20	485.85	Using
N3.1	130	0.44	413	0.04	343.66	Measurement: General
O1	151	0.51	689	0.07	314.64	Substances and materials generally
M4	161	0.55	843	0.09	301.64	Shipping, swimming etc.
O4.6	78	0.27	110	0.01	301.46	Temperature
X2.4	252	0.86	2176	0.22	288.38	Investigate, examine, test, search
N2	143	0.49	760	0.08	264.68	Mathematics
⋮						

2. Parts of speech appearing more often are common nouns, the past participle, general adjectives, the definite article, 'of', 'for', 'is', and 'are'.
3. In the use of verbal forms, the frequency of past participles is significant as found in the journal corpus [5], while the occurrence of past of lexical verbs and infinitive forms is much less compared to BNC written sampler.
4. Multi-words appearing more frequently are 'based on', 'due to', 'used to', 'such as', 'carried out', 'as well', 'in order to', 'in terms of', 'in addition' and 'according to'.

3 Material Development

Through such analysis of corpora data, features of special importance to our students can be selected. Using automatic item generation allows learners to work with different authentic texts each time. Materials underdevelopment include the original abstracts with target items hyper-linked to various pages containing concordances, grammar explanations, a bilingual dictionary, etc.

The outline of material is as follows:

– An abstract is used as the base of this material, whose objective is to enhance the ability of abstract reading comprehension.
– In the course of reading, characteristic words and multi-words are emphasized.
– The materials are designed as a course, which consists of a series of abstracts.

3.1 Design for the Material

Highlighting: Each word in an abstract has a few attributes such as POS tag, semantic tag, and so on though the attributes do not appear directly. When a

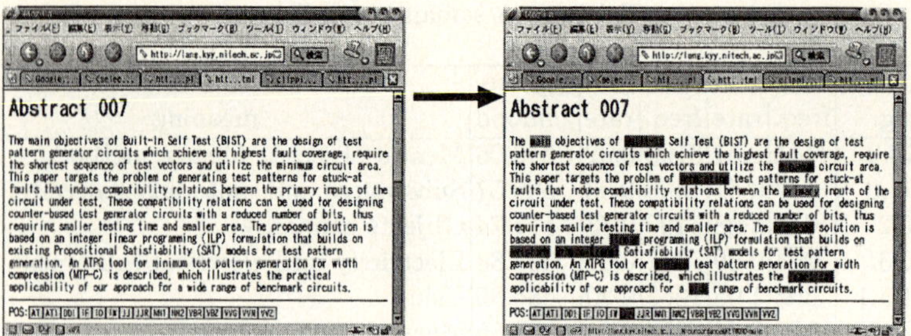

Fig. 1. Highlighting the same tag words.

mouse pointer comes to the word, words which belong to the same tag, they are highlighted. One example is shown in Figure 1. It is important for a reader to understand part of speech in order to understand the structure of the sentence and eventually to comprehend its overall meaning. The highlighting function in this material supports learners to understand the meaning of each POS tag without any previous knowledge. Semantic tags are highlighted in the same way as POS tag. The total number of all POS tags is 137 and the number of semantic tags is 232. The tags are too large in number for a learner to understand all the meanings. Therefore we decided to remove less important tags according to the results of the analysis and reduce the number of tags. On the other hand, there are two parameters in Wmatrix, which are frequency and log-likelihood. With same reason as in the case of tags, we also set lower bound for each parameter, and the two bounds are given before generating materials. It is mentioned in Section 3.2.

A Word List: The material shows the word list which learners should learn. For the selection of the words in the list the lower bounds are applied for frequency rate and log-likelihood as well as tag's ones. Moreover, in order to remove too general words such as 'the' or 'it' , the upper bound for frequency rate in BNC was used for the selection of appropriate words for learners.

The Concordancer: Each word of the list is linked to a concordancer developed by us, which shows part of the sentence that includes the keyword in the several corpora. With this function learners are able to look over usages of the word. An example of the concordancer is shown in Figure 2.

Pop-Up Bilingual Dictionary: When learners keep the mouse pointer on a word for a while,the content of an English-Japanese dictionary appears.We use e_lemma.txt (Ver. 1) [10] to draw the dictionary in this material. This helps learners to concentrate on understanding the structure of each sentence.

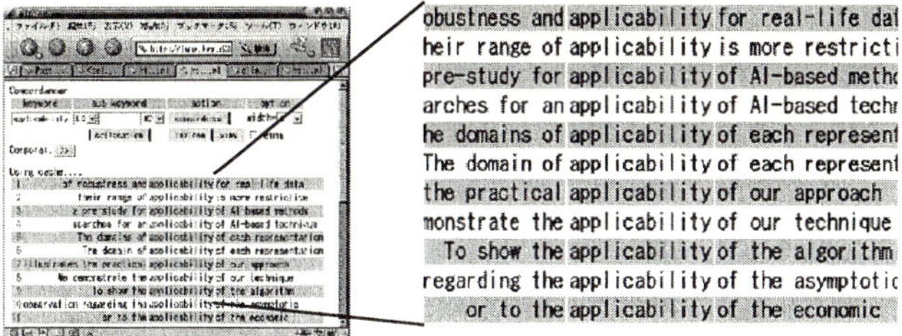

Fig. 2. An example of search by the concordancer.

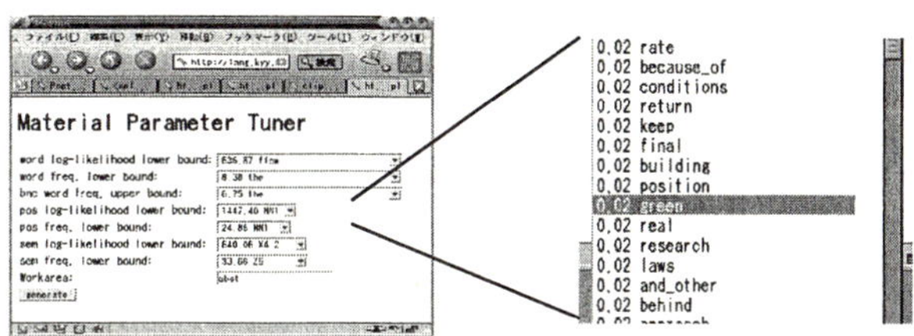

Fig. 3. Material tuner.

3.2   Material Tuner

According to the analysis mentioned above, it is required to set the upper/lower bounds of frequency or log-likelihood, and it is also necessary to avoid redundancy of the words among word lists in a course of the materials. These bounds, or parameters, should be tuned in consideration to the learner's ability and of amount of materials through a trial and error process. Therefore, we developed a material parameter tuner which can adjust these conditions. It is shown in Figure 3. On this page, we can set the following parameters: word log-likelihood lower bound, word frequency lower bound, BNC word frequency upper bound, POS log-likelihood lower bound, POS frequency lower bound, semantic-tag log-likelihood lower bound, and semantic-tag frequency lower bound.

4   Conclusion and Future Works

In this paper, we presented a method of development of e-learning materials based on analysis of an abstract corpus. First, we analyzed the abstract corpus by using Wmatrix. From the results of the analysis it is confirmed that abstracts has specific features compared to general corpus. Therefore, the development of the materials in this study can be seen as meaningful from the ESP perspective.

Next, we designed several functions for the material. Since this is the reading material, these functions are provided for the purpose of better reading comprehension. The functions provided for each material are highlighting of POS and semantic tags, a list of words to learn, links to a concordancer to learn the key word in context in other examples, and a pop-up bilingual dictionary.

Finally, we developed a material tuner to reflect the analysis in materials By using this tuner we can adjust the frequency level or log-likelihood level of the word and tags, which allows a student to learn more comfortably depending on his/her ability.

Future steps in the development of this project include expanding the size of the corpus and the number of disciplines represented; comparison of data for abstracts with those of full papers; and development of materials to guide students in writing abstracts.

References

1. Olsen L. A. and T. N. Huckin. *Technical Writing and Professional Communication.* New York: McGraw-Hill, 1991.
2. Ted Dunning. Accurate methods for the statistics of surprise and coincidence. *Computational Linguistics*, 19(1):61–74, 1994.
3. R. Garside and N. Smith. A hybrid grammatical tagger: Claws4. In R. Garside, G. Leech, and A. McEnery, editors, *Corpus Annotation: Linguistic Information from Computer Text Corpora*, pages 102–121. Longman, 1997.
4. K. Hyland. *Disciplinary Discourses: Social Interactions in Academic Writing.* Pearson Education Ltd., 2000.
5. Yukie Koyama. English for science and technology using corpus based approach. Technical report, Nagoya Institute of Technology, 2003.
6. Yukie Koyama and Robin Nagano. Text analysis based on est corpus and its application to english teaching. Technical report, Nagaoka University of Technology, 2001.
7. R. Morton. Abstracts as authentic material for eap classes. *ELT Journal*, 53(3):177–182, 1999.
8. Scott S. L. Piao, Paul Rayson, Dawn Archer, Andrew Wilson, and Tony McEnery. Extracting multiword expressions with a semantic tagger. In *Proceedings of the Workshop on Multiword Expressions: Analysis, Acquisition and Treatment, (ACL 2003)*, pages 49–56, 2003.
9. P. Rayson. Wmatrix: a web-based corpus processing environment. Technical report, Lancaster University, 2001.
10. Yasumasa Someya. e_lemma.txt (ver.1), 1998.
11. J.M. Swales. *Genre Analysis.* Cambridge University Press, 1990.

Effect of Insulating Coating on Lightning Flashover Characteristics of Electric Power Line with Insulated Covered Conductor

Kenji Yamamoto[1], Yosihiko Kunieda[1], Masashi Kawaguchi[1],
Zen-ichiro Kawasaki[2], and Naohiro Ishii[3]

[1] Suzuka National College of Technology, Shiroko, Suzuka, Mie, 510-0294 Japan
yamamoto@elec.suzuka-ct.ac.jp
[2] Osaka University, 2-1, Yamadaoka, Suita, Osaka, 565-0871 Japan
[3] Aichi Institute of Technology, 1247 Yachigusa, Yagusa-cho, Toyota, 470-0356 Japan
ishii@aitech.ac.jp

Abstract. Both bare conductors and insulated covered conductors are used in overhead distribution line systems. The difference in those conductors will affect striking distances of the conductors to lightning, lightning occurrence probabilities to the conductors and shielding effects of grounding objects on direct lightning. In this research the lightning flashover characteristics and the lightning occurrence probabilities on overhead distribution lines have been studied experimentally at a reduced scale-model. The voltage applied to electrodes in the experiment has a rise time of 1.2 microseconds and a peak to half value decay time of 50 microseconds. This paper presents the experimental data, the lightning flashover characteristics and lightning occurrence probabilities of those conductors.

1 Introduction

Lightning overvoltage which does damage to distribution lines is due to direct lightning strokes as well as indirect ones. A number of studies have been done mainly on indirect lightning strokes to distribution lines [1,2,3]. Recent studies point out that direct lightning hits to the overhead ground wires with the capacity of several tens kA can be prevented by the proper installment of metaloxide arresters [4,5]. The protection of the power distribution system from lightning is emphasized to take into account for direct lightning strokes.

Both bare conductors and insulated covered conductors are used in overhead distribution line systems. The bare conductor is generally used for an overhead ground wire and the bare conductor or the insulated covered one is used for a phase conductor. It is considered that the difference in those conductors will affect striking distances of the conductors to lightning, lightning occurrence probabilities to the conductors and shielding effects of grounding objects on direct lightning. The breakdown tests were carried out by a reduced scale model technique in order to learn about lightning flashover characteristics and occurrence probabilities of lightning impulse voltages for those conductors in this study.

2 Experimental Method

This study investigates lightning flashover characteristics and occurrence probabilities of lightning impulse voltages for the bare conductor and the insulated covered con-

R. Khosla et al. (Eds.): KES 2005, LNAI 3681, pp. 801–807, 2005.
© Springer-Verlag Berlin Heidelberg 2005

ductor on direct lightning to the distribution lines in the final stage of the lightning stepped leader process. The configuration of the experimental equipment is shown in Figure. 1.

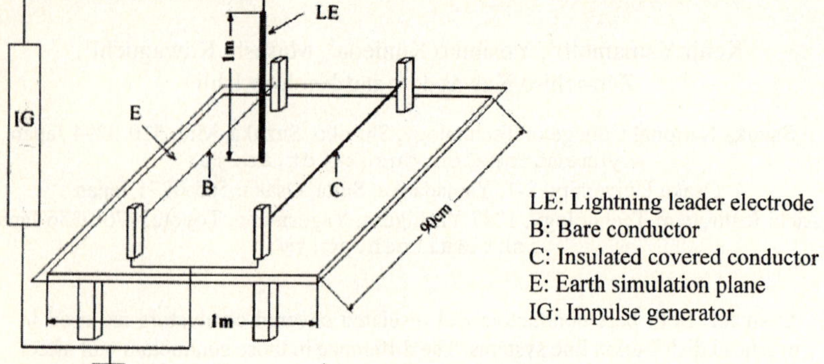

LE: Lightning leader electrode
B: Bare conductor
C: Insulated covered conductor
E: Earth simulation plane
IG: Impulse generator

Fig. 1. Experimental setup

Since this is the first attempt, a simplified, basic model is created. (1)The line conductor: The line conductor on the 6.6kV overhead distribution line in Japan is considered. A line conductor (apprix.0.3mm in diameters, 90 cm in length) of bare annealed copper wire or insulated covered wire described below is put horizontally above a simulated ground, and its end is directly grounded. A Teflon insulated covered wire is used as the insulated covered conductor (A 0.15mm-thick insulated covered wire is used, and the withstand voltage is 1500V by supplying AC voltage for one minute.). (2)Lightning leader: In order to simulate the condition that the lightning leader comes closer to the earth's surface in the final stage of the lightning stepped leader process, the electrode made of 5mm in diameter, 1m-long brass rod with the semi-spherical tip (hereinafter referred to as lightning leader electrode, LE) is placed perpendicular to the ground as shown in Figure 1. (3)Ground: Assuming that the earth is a perfect conductor, 1 mm-thick, 90 cm-long and 1 m-wide aluminum plane was installed parallel to a 75 cm-tall wooden rack. (4) Applied voltage: A standard lightning impulse voltage is applied to the top of the LE electrode by a impulse generator. The voltage   has a rise time of 1.2 microseconds and a peak to half value decay time of 50 microseconds as shown Figure 2.

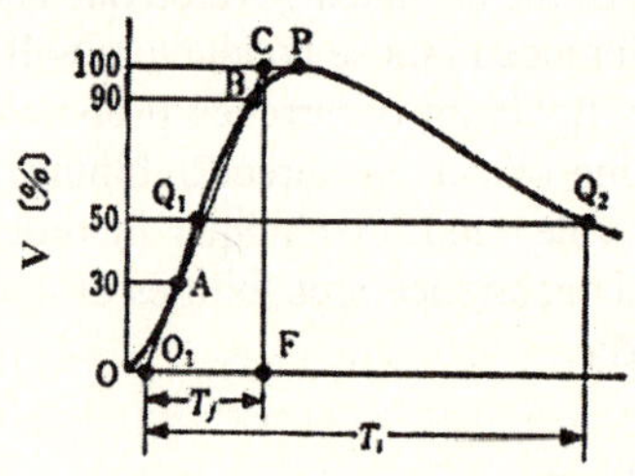

$$T_f = 1.2\ \mu S,\quad T_t = 50\ \mu S$$

Fig. 2. Waveshape of lightning impulse voltage

In addition, to take positive lightning and negative lightning into account, both positive and negative voltage are used respectively. The voltage is applied to the LE up to the fixed numbers of times, and when flashover does occur, the numbers of times of flashover to the bare conductor and the insulated covered conductor are counted. The probability of the lightning stroke occurred to the bare conductor and the insulated covered conductor, k, is defined as numbers of the flashover to those conductors divided by total numbers of the flashover.

Up and down method was used to apply voltage and 50% flashover voltage and occurrence probabilities of lightning strokes of the bare and insulated covered conductors were obtained from 10 times and more discharges.

Figure 1 indicates the configuration of the experimental equipment when a bare conductor was installed parallel to a insulated covered conductor on the earth-simulated aluminum plane.

3   Characteristics of 50% Flashover Voltages

3.1   Flashover Characteristic by Lightning Striking Distance

The study focuses the lightning occurrence probability on each conductor when both bare and insulated covered conductors was installed for distribution line. First of all, characteristics of 50% flashover voltages are investigated when either bare or insulated covered conductor is used alone. That is because it is considered that 50% flashover voltage between the bare conductor and the lightning leader electrode is equal to that between the insulated covered conductor and the lightning leader electrode, when the lightning striking distance is long enough.

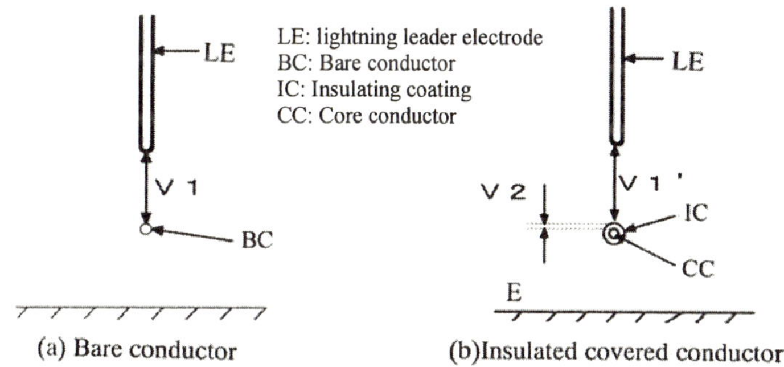

Fig. 3. Electrode configurations

In Figure 3, V1 is the withstand voltage in the air part between the lightning leader electrode and the bare conductor, V1' is the withstand voltage in the air part between the lightning leader electrode and the insulated covered conductor, V2 is the withstand voltage in the insulating coating part of insulted covered conductor and Vi is the withstand voltage between the core conductor of insulated covered conductor and the lightning leader electrode.

When the lightning striking distance is short, that is, the gap length between the lightning leader electrode and the line conductor is short, the following formula is formed;

$$Vi = V1' + V2 > V1$$

That is, 50% flashover voltage is higher between the lightning leader electrode and the insulated covered conductor than between the lightning leader electrode and the bare conductor. On the other hand, when the lightning striking distance is long, that is, the gap length between the lightning leader electrode and the line conductor is long, the following formula is approximately formed;

$$Vi = V1' + V2 \fallingdotseq V1'$$
$$Vi \fallingdotseq V1$$

When the gap length is long, the withstand voltage in the insulating coating part of the insulated covered conductor is much less than that in the air part between the lightning leader electrode and the insulated covered conductor. Consequently, the value of the withstand voltage   between the conductor part of the insulated conductor and the lightning leader electrode is equal to that between the bare conductor and the lightning leader electrode.

3.2   Experimental Results and Discussions

The following experiment was carried out to prove the aforementioned surmise.

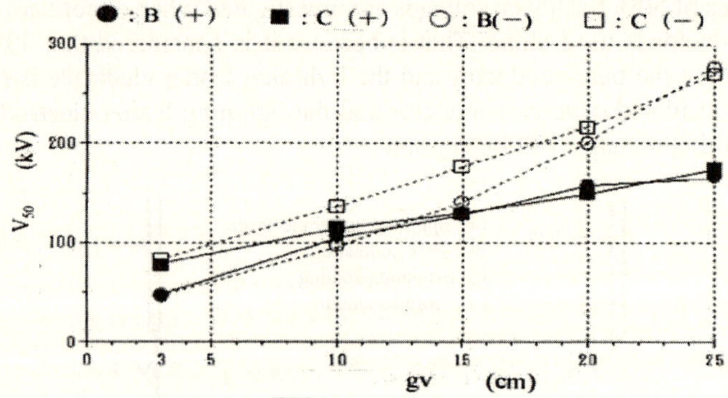

V$_{50}$: 50% Flashover voltage
gv   : Vertical distance between wire and lightning leader electrode

Fig. 4. 50% Flashover voltage for bare and insulated covered conductor

Figure 4 shows 50% flashover voltage of either bare or insulated covered conductor when each conductor is used alone and the distance between the lightning leader electrode and each conductor, gv is changed in the electrode configuration shown in Figures 3 (a) and (b). In Figure 4, the solid line indicates when the positive lightning impulse voltage is applied and the dotted line, the negative lightning impulse voltage. The ● and the ○ are for bare conductors, and the ■ and the □, for insulated covered conductors.

When either positive or negative voltage is applied, 50% flashover voltage tends to be higher when the distance between the lightning leader electrode and the conductor becomes longer. In Figure 4, 50% flashover voltage with the use of the bare conductor for positive applied voltage is equal to that with the use of the insulated covered conductor when the gap length gv is more than 15cm, and that for the negative applied voltage, the gap length gv, more than 25 cm. It was proved that when the lightning striking distance becomes longer, i.e., the gap length becomes longer, the value of the withstand voltage between the conductor part of the insulated conductor and the lightning leader electrode is equal to that between the bare conductor and the lightning leader electrode. It means that the effect of the insulating coating of the covered conductor on the 50% flashover voltage can be neglected when either bare or covered conductor is used alone, and the lightning striking distance is long enough.

4 Lightning Occurrence Probabilities on Conductors

4.1 Experimental Results

Figure 5 shows lightning occurrence probabilities on the bare and insulated covered conductors when the vertical distance between the lightning leader electrode and the conductor is long enough, i.e., gv is 40 cm, and the position of the lightning leader electrode is horizontally moved (See Figures 1). The lightning occurrence probabilities on the conductors are a ratio of discharges on each conductor to the total discharges. Positive lightning impulse voltage is applied, and the height of the conductor from the ground is 10 cm. The distance between the bare and insulated covered conductors is 10 cm.

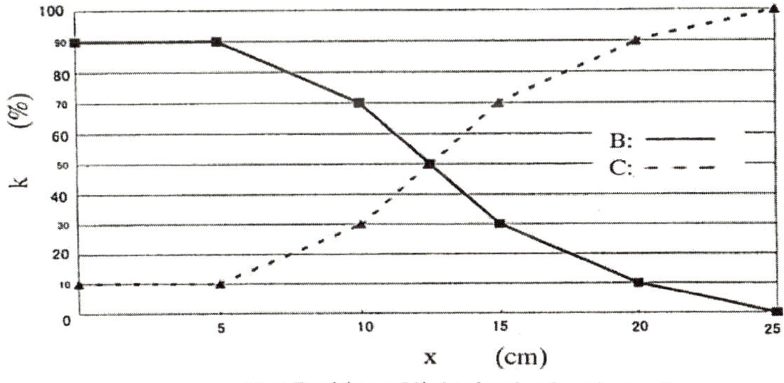

Fig. 5. Lightning occurrence probabilities of bare and insulated covered wires (gv=40cm)

In Figure 5, the horizontal distance from the lightning leader electrode to a middle point between the bare conductor and the insulated covered conductor is indicated as X. The middle point between the bare and insulated covered conductors on the horizontal axis is indicated as x = 0. In this case, an insulated covered conductor is installed at a point where X = +5 cm and parallel to the conductor is installed a bare conductor on the aluminum plane at a height of 10 cm at a point where X = -5 cm. In Figure 5, a solid line indicates the occurrence probability of lightning strokes on the bare conductor, and a dotted line, on the insulated covered conductor.

Figure 5 indicates that the lightning leader electrode moves in the positive direction of X, i.e., on the side of the insulated covered conductor, the lightning occurrence probability on the bare conductor decreases and that on the insulated covered conductor increases. In this case, the gv is 40 cm, and 50% flashover voltage between the bare conductor and the lightning leader electrode is the same as that between the insulated covered conductor and the lightning leader electrode, when the bare or insulated covered conductor is used alone. When $X = 0$, i.e., the distance to the bare conductor is the same as that to the insulated covered one, it is predicted that $k = 50\%$ or its approximate value. However, in Figure 5, when $X = +5$ cm, i.e., the lightning leader electrode is installed right above the insulated covered conductor, the lightning occurrence probability on the bare conductor k is 90% and that on the insulated covered conductor k is 10%, i.e., almost all the lightning strokes occur on the bare conductor.

The results from Figure 5 shows that the position of the lightning leader electrode X50 ($k = 50\%$) is +12.5 cm where the lightning occurrence probability on the bare conductor is equal to that on the insulated covered one. From Figure 5, the lightning occurrence probability on the bare conductor when it is installed parallel to the insulated covered conductor tends to become higher than that on the insulated covered conductor even though the gap length between the lightning leader electrode and the conductor is long enough. It means that the effect of the insulating coating of the insulated covered conductor is not neglected on the lightning occurrence probability to the conductors when the gap length is long enough.

4.2 Discussion

The pictures of discharge channels were taken by a still camera when in the electrode configuration shown in Figure 1 and the flashover was occurred between the lightning leader electrode and the bare or insulated covered conductor. From the aspects of the discharge channels on those pictures, we can make the following assumptions.

An insulated covered conductor has an insulating coating around the conductor. It is supposed that the insulating coating affects discharges from the insulated covered conductor, compared to the bare one. When the applied voltage becomes higher and just before the breakdown, upward discharges occur from both bare conductor and insulated covered conductor. However, the upward discharge from the bare conductor develops so much that it develops very close to the downward discharge from the lightning leader electrode. As a result, even though the insulated covered conductor is close to the lightning leader electrode, the flashover often occurs on the bare conductor, i.e., the gap length is long (gv = 40 cm), it assumes that the lightning occurrence probability on the bare conductor is higher than that on the insulated one. Therefore, when the lightning occurrence probability is considered, it is necessary to grasp downward discharges from the tip of the lightning leader electrode just before its breakdown and upward discharges from the bare and insulated covered conductors as well as flashover voltages.

5 Conclusions

Main results obtained are as follows: (1) When either bare or insulated covered conductor is used alone, the 50% flashover voltage between the simulated lightning leader electrode and the conductor is the same value when the gap length between the lightning leader electrode and the conductor is longer. Thus, the effect of the insulating coating of the insulated covered conductor is neglected. (2) However, when the bare conductor and the insulated covered conductor are installed parallel to each other, the lightning occurrence probability on the bare conductor tends to become higher than that on the insulated covered conductor even though the gap length is long enough, that is, the insulating coating affects the lightning occurrence probabilities of bare conductor and insulated covered conductor. (3) One of the reasons assumed from the aspects of luminosities of the channels of electric discharges is that downward leader from the lightning leader electrode before the electric breakdown, upward leaders from the bare and the covered conductor and an electrode configuration affect the electric breakdown on the conductors.

References

1. S.Rusck, "Induced Lightning Over-voltages on Power-Transmission Lines with Special Reference to the Over-voltage Protection of Low-voltage Networks", Trans. Royal Institute of Technology, Stockholm, Sweden,1958
2. A. J. Ericksson and D.V. Meal, "Lightning Performance and Overvoltage Surge Studies on a Rural Distribution Line", Proc. IEE, 129, Pt.C, No.2, 59, 1982
3. Lightning Overvoltage on Distribution Lines Study Committee, "Lightning Induced Voltages on Overhead Conductors", IEEJ Report No.522 (in Japanese) ,1982
4. S. Yokoyama, "Lightning Protection on Power Distribution Lines against Direct Lightning Hits", IEEJ Trans., vol 114-B(6), pp.568-568(in Japanese) ,1994
5. S.Yokoyama and A. Asakawa, "Experimental Study of Response of Power Distribution Lines to Direct Lightning Hits", IEEE   Trans., vol. PWRD-4(4), pp.2242-2248, 1989

Study on the Velocity of Saccadic Eye Movements

Hiroshi Sasaki[1] and Naohiro Ishii[2]

[1] Fukui University of Technology, Department of Electrical and Electronics Engineering,
Gakuen 3-6-1, Fukui 910-8505, Japan
h-sasaki@fukui-ut.ac.jp
[2] Aichi Institute of Technology, Department of Information Science,
Yakusacho, Toyota 470-0392, Japan
ishii@aitech.ac.jp

Abstract. The velocity of eye movements can be measured by using the corneo-retinal potential of the eyeball. This potential is changed almost in proportion to the eccentricity of the eyeball, and we analyze the detailed motion of the eyeball. Particularly, we measured the velocity of the horizontal saccadic eye movements. Then, we obtained the useful results by investigating experimentally the velocity difference between left eyeball and right eyeball in such following cases. When both eyeballs are opened, the velocity of the adduction is faster than the velocity of the abduction. When one eyeball is covered, the velocity of the covered eyeball is faster than the velocity of the opened eyeball.

1 Introduction

The man's eyeball is a globular form with a diameter of about 2 cm, and is moving by the pair of left and right. The eyeball is moving by extraocular muscles and we can investigate the motion by the electromyogram. However, it is difficult to analyze the detailed motion of the eyeball by this method. On the other hand, the potential difference exists between the cornea and the retina of the eyeball. The cornea side has positive charge and the retina side has negative charge. This potential is called corneo-retinal potential. Since this potential changes by the eye movement, the potential variation can be amplified by the electronystagmography and it can be kept on the record as an electronystagmograph. The advantage of the electronystagmography is that the eye movement of the covered eyeball can also be measured.

In the eye movement, both eyeballs are always moving simultaneously. For example, when we look at the right, the adduction of left eyeball has happened simultaneously with the abdution of right eyeball. Such a movement is called conjugate eye movement. Moreover, the conjugate eye movement can be classified into two types generally. One is the fast eye movement (saccadic eye movement) and the other is the slow eye movement (smooth pursuit eye movement). Saccadic eye movement happens when the point of gazing is changed to look at the object intentionally. Smooth pursuit eye movement happens when the slow moving object is followed by the eye.

Although some results of the study about the physiology of the eye movement have been reported until now, it is thought that there are few results of analysis by measurement data regarding the velocity of the eye movement [1-5]. In this study, we measured by using the corneo-retinal potential, in order to investigate the structure with such interesting character of the eye movement. Particularly, we investigated about the velocity difference between left eyeball and right eyeball at the time of the

R. Khosla et al. (Eds.): KES 2005, LNAI 3681, pp. 808–812, 2005.

horizontal saccadic eye movement. Moreover, we also measured when both eyeballs are opened, and when one eyeball is covered.

2 Method of Measurement

(1) A subject is 24 years old male with his visual acuity as normal.

(2) The electrodes for horizontal component analysis are attached to both of right side and left side 5 mm apart from lateral angle of each eye, and also the center between both eyes respectively. Earth electrode is set on the jaw. (Fig. 1) Silver and silver-chloride electrode is used.

(3) The subject is sat on the chair. In order to prevent swing of the head of the subject during measurement, his jaw is put on a fixed stand.

(4) The central target of the target panel is set to be located in front of the subject correctly. On the panel, targets are attached also in the position of leftward 30 degrees and rightward 30 degrees from the central target respectively. (Fig. 2)

(5) The subject follows two targets which flash mutually for about 10 seconds, in each of the following six patterns. The target changes at interval of 1 second.

(6) The system of measurement is shown in Fig. 2, and the state of the eyeballs at the time of measurement is shown in Table 1.

Measurement (1)　　Mutual gazing between central target and 30 degrees-left target
　　　　　　(1)-A Both eyeballs are opened
　　　　　　(1)-B Right eyeball is covered
　　　　　　(1)-C Left eyeball is covered

Measurement (2)　　Mutual gazing between central target and 30 degrees-right target
　　　　　　(2)-A Both eyeballs are opened
　　　　　　(2)-B Right eyeball is covered
　　　　　　(2)-C Left eyeball is covered

Table 1. State of the eyeballs at the time of measurement

Both eyeballs are opened	Right eyeball is covered	Left eyeball is covered

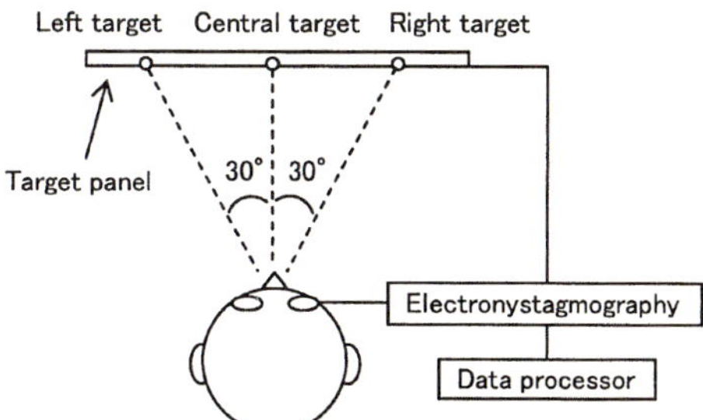

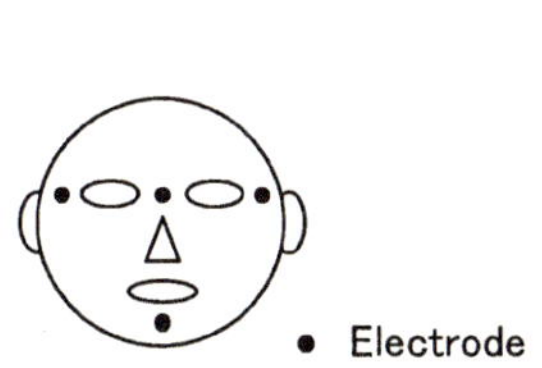

Fig. 1. Position of electrodes　　　　**Fig. 2.** System of measurement

3 Results and Consideration

As an example of data, the result of measurement (1)-A is shown in Table 2. Moreover, in the Figures from number 3 to 8, each data are indicated graphically. The following results are seen from these graphs.

Table 2. Result of measurement (1)-A

Time of turn on [msec] ◇ Central target ◆ Left target		Left eyeball		Right eyeball	
		Latency [msec]	Velocity [deg/sec]	Latency [msec]	Velocity [deg/sec]
◇	340	240	399.4	235	373.0
◆	1355	195	−316.4	200	−391.4
◇	2370	180	409.1	170	373.5
◆	3385	185	−294.0	180	−367.6
◇	4400	185	420.0	185	388.0
◆	5420	230	−290.9	235	−359.0
◇	6435	205	416.7	200	358.0
◆	7450	210	−297.2	210	−360.5
◇	8465	195	418.3	190	376.0

3.1 Case of the Opened Both Eyeballs

From Fig. 3, when the target changes from the center to the left, the velocity of right eyeball which is in the state of the adduction is faster than the velocity of left eyeball which is in the state of the abduction. The difference is about 70 deg/sec. Moreover, since both eyeballs only returned to primary position when the target changes from the left to the center, there is almost no difference in the velocity of both eyeballs. (Primary position … State that stood the head straightly and the visual axis turned to the straight front) On the other hand, from Fig. 6, when the target changes from the center to the right, the velocity of left eyeball which is in the state of the adduction is faster than the velocity of right eyeball which is in the state of the abduction.

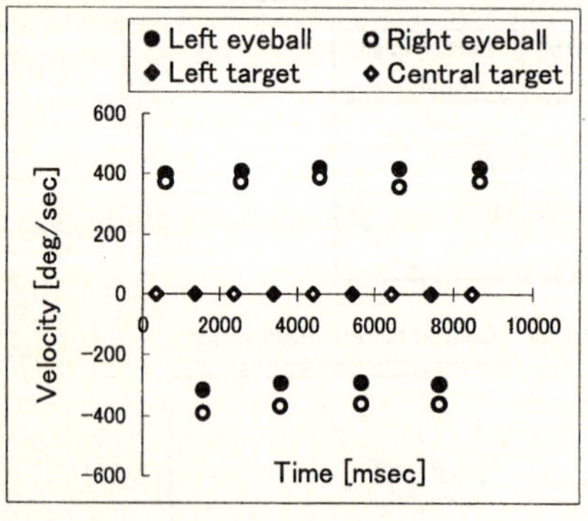

Fig. 3. Result of measurement (1)-A

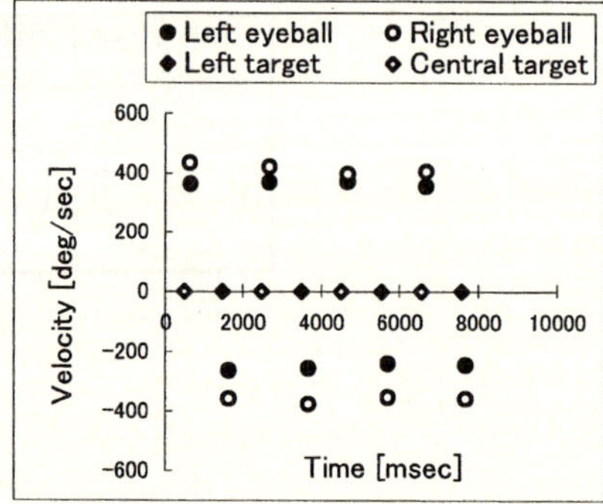

Fig. 4. Result of measurement (1)-B

3.2 Case of the Covered One Eyeball

Fig. 4 and Fig. 5 are the cases of the mutual gazing between central target and left target. In addition, Fig. 4 shows the result of measurement in the case of the covered right eyeball. When the target changes from the center to the left, for the conjugate eye movement, the covered right eyeball is interlocked with the motion of left eyeball and is moving leftward. At this time, the velocity of right eyeball is faster. This is because right eyeball is in the state of the adduction and cannot control its movement by itself. Next, when the target changes from the left to the center, the velocity of right eyeball is slightly faster than the velocity of left eyeball. This is because the covered right eyeball is only moving for the conjugate eye movement, and cannot control its movement by itself.

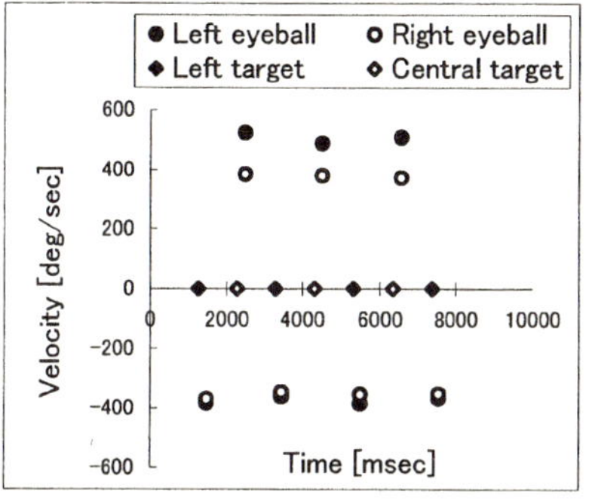

Fig. 5. Result of measurement (1)-C

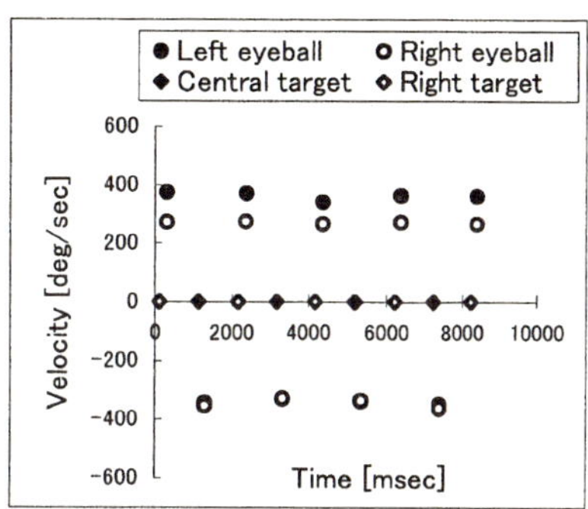

Fig. 6. Result of measurement (2)-A

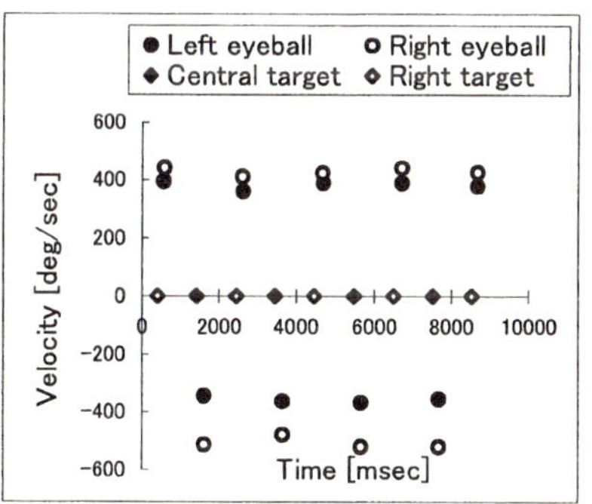

Fig. 7. Result of measurement (2)-B

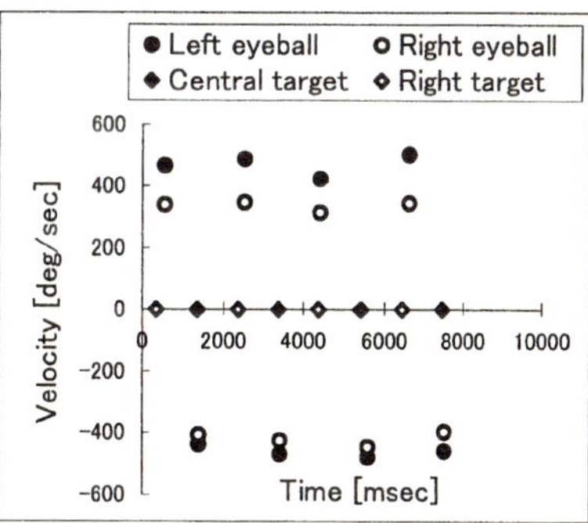

Fig. 8. Result of measurement (2)-C

Fig. 5 shows the result of measurement in the case of the covered left eyeball. Fig. 7 and Fig. 8 are the cases of the mutual gazing between central target and right target. In any cases, it is evident that the velocity of the covered eyeball is faster than the velocity of the opened eyeball. Moreover, also as for the case of the eyeball which cannot actually follow the target, it shows that corneo-retinal potential changes just same as the opened eyeball.

4 Conclusion

In the velocity of saccadic eye movements at the mutual gazing, it turns out that the following difference is between left eyeball and right eyeball.

(1) When both eyeballs are opened, the velocity of the adduction is faster than the velocity of the abduction.
(2) When one eyeball is covered, the velocity of the covered eyeball is faster than the velocity of the opened eyeball.

The knowledge about the velocity of saccadic eye movements was able to be deepened from this study. We would like to continue this kind of measurement further more to obtain significant results from now on.

References

1. Clark, J.J.: "Spatial Attention and Latencies of Saccadic Eye Movements", Vision Res., vol.39, pp.583, 1999
2. Kalesnykas, R.P.: "Retinal Eccentricity and the Latency of Eye Saccades", Vision Res., vol.34, pp.517, 1994
3. Ditchburn, R.W.: "The Function of Small Saccade", Vision Res., vol.20, pp.217, 1980
4. H.Daveson.: Physiology of the Eye (Fifth Edition). Macmillan Press, London, 1990
5. E.R.Berman.: Biochemistry of the Eye. Plenum Press, New York, 1991

Relative Magnitude of Gaussian Curvature from Shading Images Using Neural Network

Yuji Iwahori[1], Shinji Fukui[2], Chie Fujitani[1],
Yoshinori Adachi[1], and Robert J. Woodham[3]

[1] Chubu University, Matsumoto-cho 1200, Kasugai 487-8501, Japan
`iwahori@cs.chubu.ac.jp`
`http://www.cvl.cs.chubu.ac.jp`
[2] Aichi University of Education, Hirosawa, Igaya-cho, Kariya 448-8542, Japan
`sfukui@auecc.aichi-edu.ac.jp`
`http://www.aichi-edu.ac.jp`
[3] University of British Columbia, Vancouver, B.C. Canada V6T 1Z4
`woodham@cs.ubc.ca`
`http://www.cs.ubc.ca`

Abstract. A new approach is proposed to recover the relative magnitude of Gaussian curvature from three shading images using neural network. Under the assumption that the test object has the same reflectance property as the calibration sphere of known shape, RBF neural network learns the mapping of three observed image intensities to the corresponding coordinates of (x, y). Three image intensities at the neighbouring points around any point are input to the neural network and the corresponding coordinates (x, y) are mapped onto a sphere. The previous approaches recovered the sign of Gaussian curvature from mapped points onto a sphere, further, this approach proposes a method to recover the relative magnitude of Gaussian curvature at any point by calulating the surrounding area consisting of four mapped points onto a sphere. Results are demonstrated by the experiments for the real object.

1   Introduction

Surface gradient and curvature are the essential information for the shape representation. Especially, the surface curvature is the invariant and effective feature for the viewing direction, and curvature feature can be used to many applications such as the shape recovery, shape modeling, segmentation, the object recognition and pose determination in the field of computer vision.

Based on the physics based vision approach, Woodham [1] developed a method to get surface curvature using the values of the surface gradients. Using the LUT (Look Up Table), the method obtains the local surface gradients by the empirical photometric stereo using a calibration sphere.

Iwahori has pursued neural network implementations of photometric stereo. In [2] [3], neural network implementation with the PCA (principal component analysis) was proposed.

R. Khosla et al. (Eds.): KES 2005, LNAI 3681, pp. 813–819, 2005.

Angelopoulou and Wolff [4], Okatani and Deguchi [5] proposed a method to recover the local sign of the Gaussian curvature from three images taken under different light source directions without using the values of the surface gradient or the correct light source directions. These methods are applicable to the diffuse reflectance.

While Iwahori *et al.* [6] proposed the method to classify local surface from multiple shading images including the sign of Gaussian curvature using neural network. As with both previous non-parametric, empirical implementations [1, 2] [3], no explicit assumptions need to be made either about light source directions or about the functional model of surface reflectance.

In this paper, we propose a new method to recover not only the classification of local surface but also the relative magnitude of local Gaussian curvarure from shading images using neural network based on the empirical physics based vision approach. The advantage of the method is the direct extraction of surface curvature from shading images. The proposed approach can be applied for non-Lambertian case, rather, can treat the real phenomena of the surface reflectance. Experiments on real data are demonstrated.

2   Principle

2.1   Empirical Constraint

The original purpose of photometric stereo is to determine the surface normal vector (n_1, n_2, n_3) from the observed image irradiances (E_1, E_2, E_3) locally, while, this approach tries to get the local curvature information directly from (E_1, E_2, E_3) at the local five points.

Let the image irradiances at (x_{obj}, y_{obj}) on the test object be $(E_{1obj}, E_{2obj}, E_{3obj})$ and that at (x_{sph}, y_{sph}) on the sphere be $(E_{1sph}, E_{2sph}, E_{3sph})$. Under the condition that the surface material is the same for both of a test object and a sphere, if the following constraint

$$\begin{aligned}
E_{1obj}(x_{obj}, y_{obj}) &= E_{1sph}(x_{sph}, y_{sph}) \\
E_{2obj}(x_{obj}, y_{obj}) &= E_{2sph}(x_{sph}, y_{sph}) \\
E_{3obj}(x_{obj}, y_{obj}) &= E_{3sph}(x_{sph}, y_{sph})
\end{aligned} \tag{1}$$

is satisfied, the corresponding surface normal vector should be the same between a test object and a sphere. This constraint is used to determine the curvature sign and the relative magnitude of a test object using neural network.

2.2   Surface Curvature and Sign

Six kinds of surface curvature exist as shown in Fig.1. Any surface curvature is one of the convex surface, the concave surface, the hyperbolic surface, the convex parabolic surface, the concave parabolic surface, and the plane.

Determining six kinds of surface curvature includes determining the sign of the Gaussian curvature G, further for the case of $G > 0$, it leads the sign of the

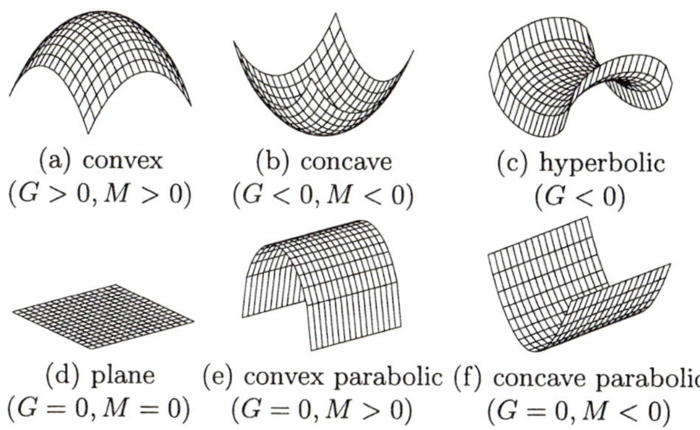

(a) convex
$(G > 0, M > 0)$

(b) concave
$(G < 0, M < 0)$

(c) hyperbolic
$(G < 0)$

(d) plane
$(G = 0, M = 0)$

(e) convex parabolic
$(G = 0, M > 0)$

(f) concave parabolic
$(G = 0, M < 0)$

Fig. 1. Six Kinds of Surface Curvature.

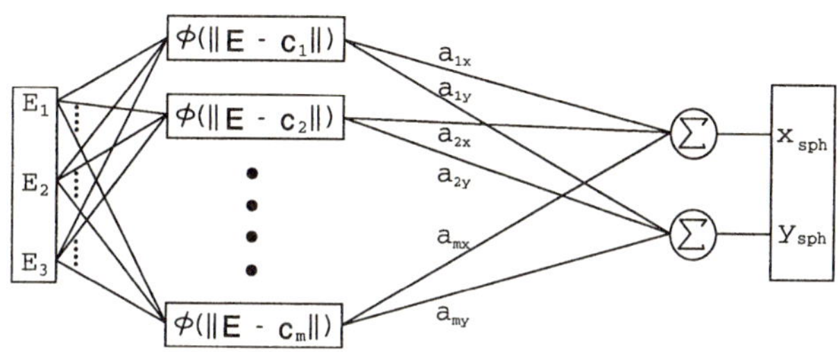

Fig. 2. RBF Neural Network.

mean curvature M, that is, convex or concave curved surface is discriminated uniquely. The sign of G and M for these six surface curvatures is also described for each local surface in Fig 1.

2.3   RBF Networks and OLS Learning

Neural networks are attractive for non-parametric functional approximation. As an algorithm to learn the mapping of the nonlinear functions using neural network, back propagation algorithm is popular and it can often used for this kind of problem. As another candidate and choice except the back propagation, a radial basis function (RBF) neural network [7] is one choice suitable for many applications. In particular, it has been widely used for strict interpolation in multidimensional spaces. In the case that we have many sampled points, RBF neural network has an advantage that the learning is much faster and efficiently done than the standard feed forward neural network with the back propagation algorithm. This is based on the fact that he weights of network can be partially determined for each of outputs and nonlinear mapping problem can be effectively solved using the orthogonal least squared algorithm [7] proposed by Chen *et al.*

3   Curvature Sign and Relative Magnitude

3.1   Mapping onto Sphere by Neural Network

Here, a neural network to do the mapping of $(E_1(x_{sph}, y_{sph}), E_2(x_{sph}, y_{sph}),$ $E_3(x_{sph}, y_{sph}))$ to (x_{sph}, y_{sph}) is constructed for the sphere. With this learning procedure, RBF NN is trained using input/output data from a sphere. Many training vectors are available since data from the calibration sphere are dense and include all possible visible surface normal.

The learning procedure builds an RBF neural network one neuron at a time. Neurons are added to the network until the sum-squared error falls beneath an error goal or a maximum number of neurons has been used. The resulting network generalizes in that it predicts a position (x_{sph}, y_{sph}) on the sphere, given any triple of input values, $[E_1, E_2, E_3]$ of the test object.

3.2   Determination of Curvature Sign

Six kinds of the surface curvature shown in Figure 1 can be discriminated as follows. Local five points on the test object are labeled as ⓪ for the center point, ① for its upper point, ②, ③, ④ for clockwise, respectively.

The method discriminates the kind of surface curvature from this relation of the location of the mapped points onto the sphere. Fig.3 shows that how the mapping of the neighbor five local points is done onto the sphere for each of six kinds of surface curvatures.

Even if the patterns of local points mapped on a sphere are deformed with the mapping of neural network, as far as the regularity of the above properties are kept, the algorithm can determine the kind of local surface curvature among six kinds. For more detailed description of classification, see [6].

3.3   Relative Magnitude of Gaussian Curvature

Following to the classification of the six local surfaces, getting the the relative magnitude of Gaussian curvature is considered. Even if the local surface is in the same classification class, the difference of magnitude has the key to detect the further feature. The magnitude of Gaussian curvature includes the absolute and the relative ones. Absolute value of Gaussian curvature is very small value, while the relative magnitude is sometimes sufficient to recognize the feature points which have the larger values of Gaussian curvature.

The coordinates of four mapped points onto a sphere using NN can be used not only for the classification of local surface but also the relative value of Gaussian curvature. The key idea of this proposed approach is that the area value which is surrounded by the four mapped points onto a sphere corresponds to the relative magnitude of Gaussian curvature.

Let $z = f(x, y)$ be the height distribution of the object, then Gaussian curvature is defined as

$$G = \frac{1}{1 + f_x^2 + f_y^2} \begin{vmatrix} f_{xx} & f_{xy} \\ f_{yx} & f_{yy} \end{vmatrix} \tag{2}$$

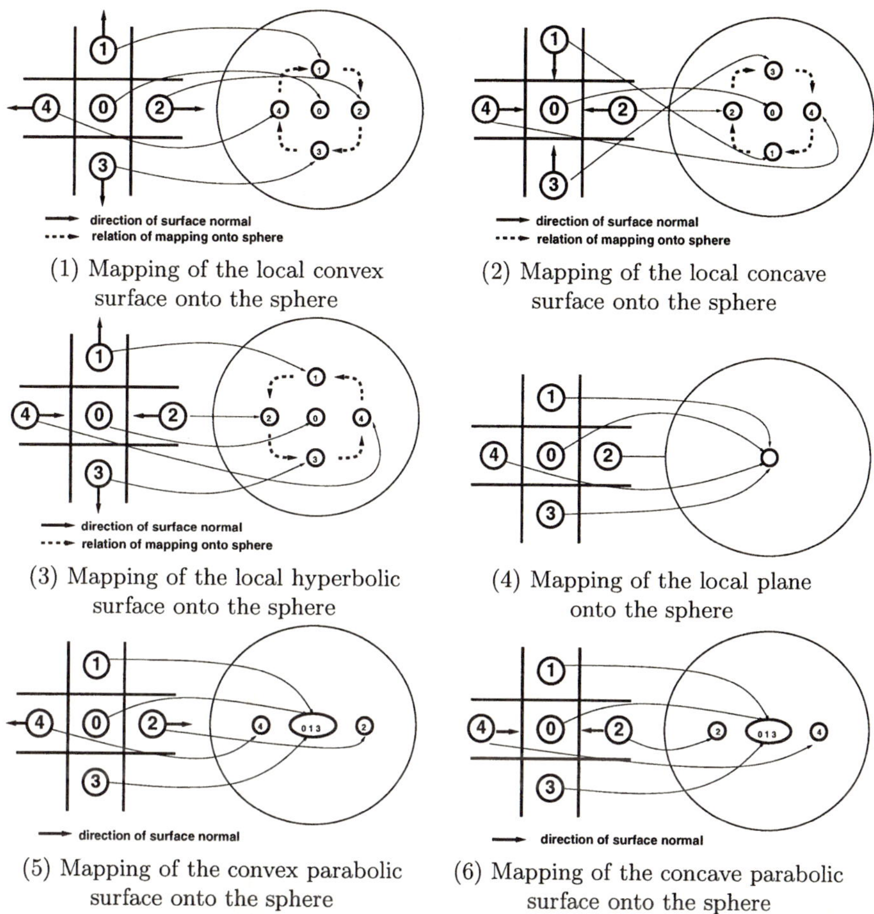

(1) Mapping of the local convex surface onto the sphere

(2) Mapping of the local concave surface onto the sphere

(3) Mapping of the local hyperbolic surface onto the sphere

(4) Mapping of the local plane onto the sphere

(5) Mapping of the convex parabolic surface onto the sphere

(6) Mapping of the concave parabolic surface onto the sphere

Fig. 3. Mapping onto a Sphere.

Here, $f_x = \partial z/\partial x$ and $f_y = \partial z/\partial y$. For the smooth surface, the relation of $f_{xy} = f_{yx}$ holds. Suppose the four mapped points are located as shown in Fig.4-(a), the surrounded area corresponds to the relative magnitude of Gaussian curvature. Here, consider the area value S calculated from the square with four mapped points as shown in Fig.4-(b), then the square area S is given as

$$S = \frac{1}{2}|\vec{a}||\vec{b}|\sin\theta = \frac{1}{2}|\vec{a} \times \vec{b}| = \frac{1}{2}(4\Delta x \Delta y)(f_{xy}^2 - f_{xx}f_{yy}) \qquad (3)$$

This area value corresponds to the relative magnitude of Gaussian curvature, that is, the approximation of the relative value is obtained according to Eq.(3).

4   Experiments

Fig.5-(a)(d) show one image among three input images for two different test objects. Condition of three light sources is the same for a sphere and test objects.

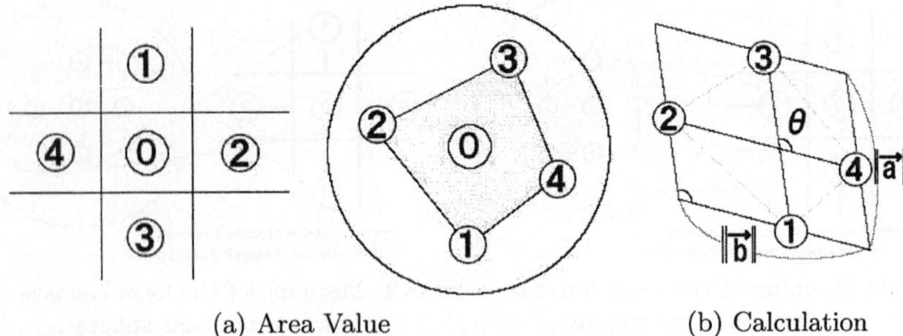

(a) Area Value   (b) Calculation

Fig. 4. Area Value and Calculation.

In the learning of the first RBF neural network of $(E_1, E_2, E_3) \rightarrow (x, y)$, the spread constant was set to be 200 and the number of the learning epochs was set to be 100. For the original images with the size of 512×512, the sampling points are taken 6 dots apart from the test object as the 5 neighbouring points mapped onto a sphere. In addition to the classification of surface curvature, the relative

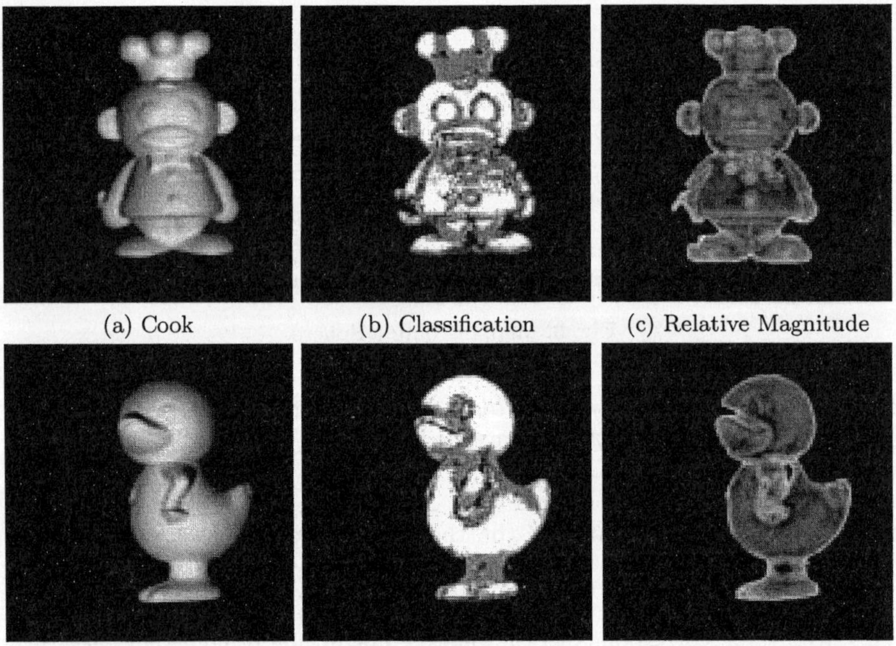

(a) Cook   (b) Classification   (c) Relative Magnitude

(d) Duck   (e) Classification   (f) Relative Magnitude

:Convex   :Plane

:Convex Parabolic   :Concave Parabolic

:Concave   :Hyperbolic

Fig. 5. Example and Result.

magnitude is encoded using the gray scale in Fig.5-(c)(f). Brighter points shows the points with relatively larger value of Gaussian curvature. It is shown that almost robust and reasonable results are obtained.

5   Conclusion

This paper proposed a new method to recover the relative magnitude of Gaussian curvature of a target object. With the neural network implementation, the relative magnitude can be obtained directly from the triple of shading images in addition to the classification by the previous approaches. The entire approach is empirical under the condition that no explicit assumptions are not used for the surface reflectance function nor the illuminating directions. It is shown that the robust results are directly obtained through the experiments. Further subjects are remained for the extension including cast shadow or interreflection problem.

Acknowledgment

Iwahori's research is supported in part by Chubu University Grant, the Kayamori Foundation of Information Science Advancement and the JSPS Grant No. 16500108. Wooodham's research is supported by by the Institute for Robotics and Intelligent Systems (IRIS) and by the Natural Sciences and Engineering Research Council (NSERC).

References

1. R. J. Woodham, "Gradient and curvature from the photometric stereo method, including local confidence estimation," *Journal of the Optical Society of America, A*, vol. 11, pp. 3050–3068, 1994.
2. Y. Iwahori, R. J. Woodham, and A. Bagheri, "Principal components analysis and neural network implementation of photometric stereo," in *Proc. IEEE Workshop on Physics-Based Modeling in Computer Vision*, pp. 117–125, June 1995.
3. Y. Iwahori, R. J. Woodham, M. Ozaki, H. Tanaka, and N. Ishii, "Neural Network based Photometric Stereo with a Nearby Rotational Moving light Source" *IEICE Transactions on Information and Systems*, vol. E80-D, no. 9, pp. 948–957, 1997.
4. E. Angelopoulou and L.B. Wolff: "Sign of Gaussian Curvature From Curve Orientation in Photometric Space" in IEEE Trans. on PAMI, vol.20, NO.10, pp. 1056-1066, Oct, 1998.
5. T. Okatani and K. Deguchi: "Determination of Sign of Gaussian Curvature of Surface from Photometric Data" in Trans. of IPSJ, vol.39, NO.5, pp. 1965-1972, Jun, 1998.
6. Y. Iwahori, S. Fukui, R. J. Woodham, A. Iwata, "Classificaiton of Surface Curvature from Shading Images Using Neural Network," *IEICE Trans. on Information and Systems*, Vol.E81-D, No.8, pp.889-900, 1998.
7. S. Chen, C. F. N. Cowan, and P. M. Grant, "Orthogonal least squares learning algorithm for radial basis function networks," *IEEE Transactions on Neural Networks*, vol. 2, no. 2, pp. 302–309, 1991.

Parallelism Improvements of Software Pipelining by Combining Spilling with Rematerialization

Naohiro Ishii[1], Hiroaki Ogi[1], Tsubasa Mochizuki[1], and Kazunori Iwata[2]

[1] Aichi Institute of Technology, Toyota, 470-0392, Japan
ishii@aitech.ac.jp
[2] Aichi University, Aichi, 470-0296, Japan
kazunori@vega.aichi-u.ac.jp

Abstract. On the instruction level parallelism architecture developed as EPIC, VLIW structure machine et al., the performance is affected by the compiler techniques. The integrated and convergent optimization techniques have been studied for their developments of the parallelism. In this paper, we develop a software pipelining technique for the improvement of the parallel processing in these machine structures. The software pipelining is a loop scheduling technique by overlapping the execution of several consecutive instructions of the program. Then, much registers are needed for the realization of the software pipelining. Here, spilling code and the rematerialization are implemented in the pipelining scheduling. Experimental results of the proposed method are compared with the conventional methos. The results show the improvements of the speedup of the parallel prosecssing in the bench marks.

1   Introduction

High speed processing is expected to realize human computer interactions in the computer systems, such as graphics with high precision and high speed processing ability, high-dimensional image processing, speedy gaming, and integrated media processing [1–3]. Parallel processing by processors, has been a problem to be solved for these purposes. Very long instruction word(VLIW) processors or EPIC structure processors are one of the kind processors developed for the parallel processing[5–7]. In these processors, instruction parallelism processing (ILP) is a key technique to design the high speed processor. Software pipelining is an instruction scheduling technique that exploits the ILP of loops in the program by overlapping the execution of successive iterations of a loop[4, 8]. In this paper, an integrated pipelining algorithm is developed to increase the parallelism for the processing in the VLIW by combining the spilling and rematerialization, which is recomputation of the instruction variable. Here, the spilling and the rematerialization are implemented in the pipelining scheduling. Experimental results of the proposed method are compared with the conventional methods. The results show the improvements of the speedup of the parallel processing in the Livermore kernel, which is a benchmark program.

R. Khosla et al. (Eds.): KES 2005, LNAI 3681, pp. 820–826, 2005.

2   Software Pipelining

Software pipelining is an instruction scheduling technique by overlapping the execution of successive iterations of a loop, which is different from the loop unrolling. There are different approaches to generate a software pipelined schedule for a loop. The most well known scheduling is modulo scheduling. The following is a loop program as an example.

```
for(i = 1; i <= N; i++){
        A;
        B;
        C;
        D;
        E;
},
```

where A, B, C, D, and D are instructions in the loop program. The dependence relation of the instructions is assumed to be A → B → C → D → E. The pipelining of the instructions is expected to perform the execution of the set of instructions as follows,

$$A^{i+4}B^{i+3}C^{i+2}D^{i+1}E^{i},$$

where X^j shows the instruction in the j-th loop. In the realization of the execution of the overlapping instructions, the following prologue and epilogue sections are needed as shown in Fig.1.

(Prologue)
$$A^1$$
$$B^1A^2$$
$$C^1B^2A^3$$
$$D^1C^2B^3A^4$$

(KernelCode)
```
for(i = 1; i <= N - 4; i++ )
```
$$A^{i+4}B^{i+3}C^{i+2}D^{i+1}E^i$$

(Epilogue)
$$E^{N-3}D^{N-2}C^{N-1}B^N$$
$$E^{N-2}D^{N-1}C^N$$
$$E^{N-1}D^N$$
$$E^N$$

Fig. 1. Software pipelining code.

2.1 Initiation Interval (II) in Scheduling

In the modulo scheduling, the initiation interval (II) is a characteristic parameter of the parallelism in the scheduling, which implies the iteration interval for the loop execution. The initiation interval II between two successive iterations is bounded either by loop-carried dependences in the graph (Rec II) or by resource constraints of the architecture (Res II).

Definition 1. *Initiation interval under resource constraints: Res II*
Let the number of the functional units with type r be n_r and the number of the instruction using them be C_r. Then, the initiation interval: Res II, becomes

$$\text{Res II} = \max_{r \in R} \left\lceil \frac{C_r}{n_r} \right\rceil ,$$

where R indicates all functional units and $\lceil\ \rceil$ is Gaussian notation.

Definition 2. *Initiation interval of dependency relation: Rec II*
At the loop of program, cyclic path is generated by adding a dependent feedback arc in the data dependence graph(DDG). Then, the dependency distance δ_e is defined for the dependent feedback arc. Let the summation time of the latency in the DDG, be $\sum l_e$. Then, the initiation interval of dependency relation, is defined as

$$\text{Rec II} \geq \frac{\sum_{e \in E_e} l_e}{\sum_{e \in E_e} \delta_e},$$

where E_e is a set of cyclic pathes.

The lower bound of the modulo scheduling: MII is defined as follows,

$$\text{MII} = \max(\text{Res II}, \text{Rec II})$$

After the determination of MII, the modulo scheduling is carried out in the modulo reservation table. Register assignments are performed by the modulo reservation table.

3 Scheduling Methods to Reduce Registers

Exploiting more parallelism in instruction level, results in a significant increase in the register pressure[8]. The conventional methods and the proposed methods in this paper are shown in the following.

3.1 II++ Method(Increaasing Initiation Interval)

Generally, the registers are fixed in the processor. When registers are needed in their assignment stage, it is natural to increase the initiation interval(II) than it's present value. By increasing the initiation interval(II), necessary registers are reduced. This method is called II++ method. The II++ method decreases the parallelism in the processing, thus results in performance reduction.

3.2 Spilling Method and It's Modification

To solve the register pressure problem, another option is to spill some variables to memory, so that they do not waste registers for a certain number of cycles. Spilling a variable with a long lifetime, was proposed by [8]. The long lifetime variable is selected from the following equation,

$$SP_{order} = \max_{u \in U} \left(\frac{LT_u}{Mem_{traffic}} \right),$$

where SP_{order} shows the order of spilling variable, U is the set of variables in loops and $Mem_{traffic}$ shows the number of instruction access to the variable.

3.3 Multi-variables Method

In this paper, we develop multi-variable method to select the spill code, which is chosen from not only long lifetime but also another variables as the spill codes[6]. The algorithm is as follows,

(1) The set of variables longer than the initiation interval: II, is made as the candidate of spill codes.
(2) Remove variables on the cyclic sub-graph, which satisfy the following equation.

$$\text{Rec II} > \text{II}, \text{where Rec II} = \frac{\sum_{v \in V_c} l_v + l_{load} + l_{store}}{\sum_{e \in E_e} \delta_e},$$

where l_v is a delay time of the instruction v, l_{load} is a delay time of the load instruction and l_{store} is a delay time of the sore instruction. When Rec II > II, the variable of the instruction is not adopted as a spill code.
(3) Variables of the spilling candidate, are computed by the order SP_{order} in subsection 3.2.

3.4 Combined Remateriarization and Spilling Method

The performance of the scheduling often is decreased in the spilling and spilling with lifetime variables. To improve the parallelism of the scheduling, rematerialization which is a recomputation of the variable, is proposed here by the combination of adding spill codes. From the candidate list of variables, the selected variable (from LT/Trf. to be large) is adopted as a rematerialization or spill code. The selected variables are carried out in the computation one by one. The flow chart of the rematerialization is shown in Fig. 2.
In Fig. 2, algorithm of the rematerialization is shown in flow chart. The initiation interval Rec II in Fig. 2 is computed in the following.

$$\text{Red II} = \frac{\sum_{v \in V_e} l_v + l_{recalc}}{\sum_{e \in E_e} \delta_e},$$

where l_{recalc} is a delay time of the variable of the rematerialization. When the rematerialization fails in the computation, the spilling computation is carried out from the flow symbol * in Fig. 2.

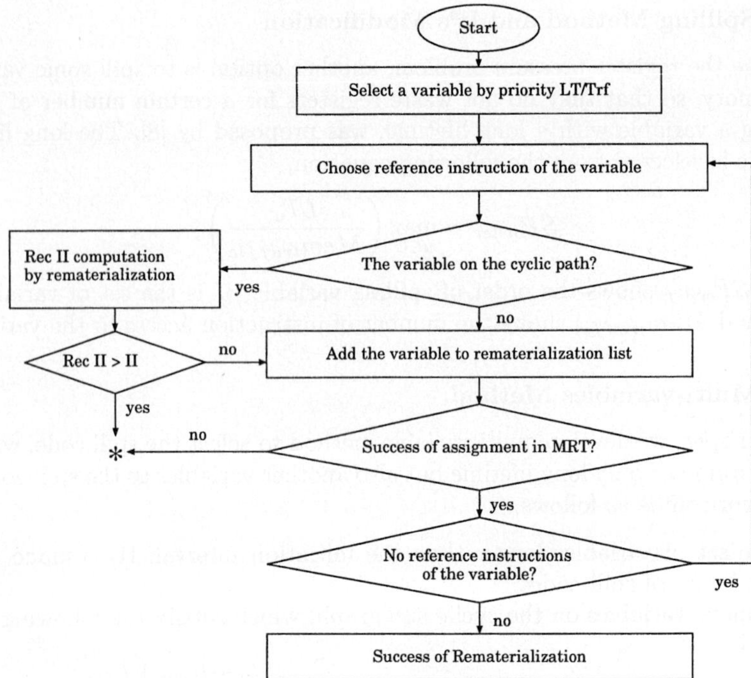

Fig. 2. Flow chart of rematerialization.

4   Experimental Results for Software Pipelining

Experimental simulation was carried out to evaluate the proposed methods. Here, processor models are given in the following Table. 1. There are 4 processor models. P_1 consists of 1 IU unit, 1 FPU unit, 1L/S unit and 1BU unit. The parallelism increases according to P_1, P_2, P_3 and P_4 .

Table 1. Processor models (number of functional unit).

model	IU	FPU	L/S	BU
P_1	1	1	1	1
P_2	2	2	2	2
P_3	3	3	3	3
P_4	4	4	4	4

Experiments by the Livermore kernel as the benchmak, were compared with other conventional methods, which are II ++ in the modulo scheduling without spilling code, the long lifetime method with spilling one variable(Spill-LT in Fig. 3 and Fig. 4), and the long lifetime method with spilling more than two variables(Spill-2LTs in Fig. 3 and Fig. 4). The experimental results by

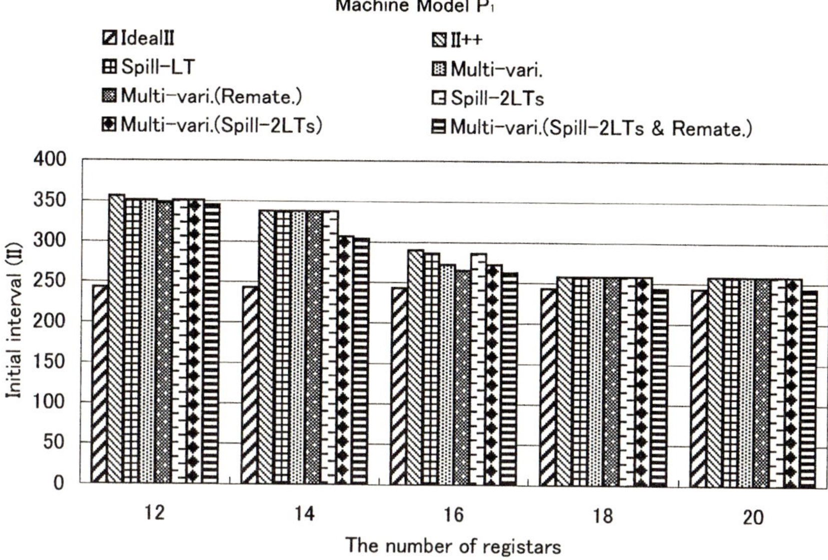

Fig. 3. Initiation intervals for Model P_1.

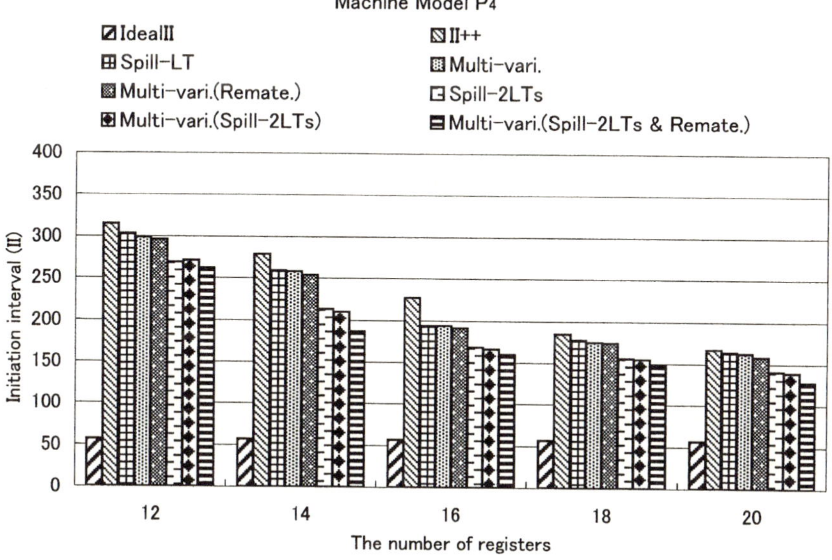

Fig. 4. Initiation intervals for Model P_4.

proposed meth ods here, are shown in Fig. 3 in the P_1 and in Fig. 4 in the P_4. In those figures, the proposed methods are multi-variables method(Multi-vari. in Fig. 3 and Fig. 4), multi-variables method with spilling more than two long lifetime variables (Multi-vari.(Spill-2LTs)), multi-variables method with rematerialization(Multi-vari.(Remate.)), and multi-variables method with spilling more than two long lifetime variables and rematerialization (Multi-vari.(Spill-2LTs & Remate.)). In Fig. 3 and Fig. 4, the proposed multi-variables methods: Multi-vari(Spill-2LTs) and Multi-vari(Spill-2LTs & Remate.) show the short initiation intervals to the number of registers, thus result in faster speed up in the loop calculations. The best method, here is Multi-vari(Spill-2LTs & Remate.). These show the improvement of the parallelism strongly in Multi-vari(Spill-2LTs & Remate.) method. From these experiments, the cooperation of the spilling and the rematerialization show better parallelism results in the performance.

5   Conclusion

In this paper, we discussed the software pipeline techniques in the compiler computations. When the registers are not sufficient in the processors, we have difficult problems in the scheduling of compilers. In this paper, we proposed optimized techniques that collaborate spilling and rematerialization in software pipelining computation. In the experimental results, it was shown that the proposed method generates faster codes than the conventional methods.

References

1. Briggs, P., Cooper, K.D., Torczon, L.: Improvements to graph coloring register allocation. ACM Transactions on Programming Languages and Systems **16** (1994) 428–455
2. Ellis, J.: Bulldog: A Compiler for VLIW Archtectures. The MIT Press (1986)
3. Hennessy, J., Patterson: Computer Architecture: A Quantitative Approach. Morgan Kaufmann Publisheres, Inc, San Mateo CA (1990)
4. J.Zalamea, J.Llosa, E., M.Valero: Register constrained modulo scheduling. ACM Trans. on Parallel & Distributed Systems **15** (2004) 417–430
5. Norris, C., Pollock, L.L.: A scheduler-sensitive global register allocator. Supercomputing '93 Proceedings (1993) 804–813
6. Li, D., Iwahori, Y., Hayashi, T., Ishii, N.: A spill code placement framework for code scheduling. The 11th International Workshop on Languages and Compilers for Parallel Computing (1998) 263–274
7. Shimizu, K., Li, D., Ishii, N.: On the code scheduling of register constraints. AoM/IAoM 17th Annual International Conference (1999) 272–277
8. Llosa, M., Ayguade, E.: Heuristics for register-constrained software pipelining. Proc. of the 1996 Int. Cobf. On Microarchitecture (1996) 250–261
9. N.Ishii, K.Shimizu, D.Li: Cooperation of code scheduling and spilling code placement for the compiler. Proc. Int. Conf. on Soft Eng, Applied to Networking and Parallel/Distributed Computing (2000) 474–481
10. N.Doi, K.Sumiyoshi, N.Ishii: Estimation of earliest starting time for scheduling with communication costs. Proc. Int. Conf. on Soft Eng, Applied to Networking and Parallel/Distributed Computing (2001) 399–406

Content Based Retrieval of Hyperspectral Images Using AMM Induced Endmembers

Orlando Maldonado, David Vicente, Manuel Graña, and Alicia d'Anjou

Dept. CCIA, UPV/EHU, Apdo 649, 20080 San Sebastian Spain
`{manuel.grana,alicia.danjou}@ehu.es`

Abstract. Indexing hyperspectral images is a special case of content based image retrieval (CBIR) systems, with the added complexity of the high dimensionality of the pixels. We propose the use of endmembers as the hyperspectral image characterization. We thus define a similarity measure between hyperspectral images based on these image endmembers. The endmembers must be induced from the image data in order to automate the process. For this induction we use Associative Morphological Memories (AMM) and the notion of Morphological Independence.

1 Introduction

Is well known that the growth in multimedia information, especially images, is the powerful force driving the development of the field of content based image retrieval (CBIR) [1]. In CBIR systems, the images stored in the database are labeled by feature vectors, which are extracted from the images by means of computer vision and digital image processing techniques. In CBIR systems, the query to a database is specified by an image. The query's feature vector is computed and the closest items in the database, according to a similarity metric or distance defined in feature space, are returned as the answers to the query. Of course, there is no universal definition of the feature vectors because of the diversity of image kinds. Hyperspectral images are a special kind of images, in which each pixel contains a fine sampling of the visible and near infrared spectrum, represented by a high dimensional vector. There is a growing need for the maintenance of large collections of hyperspectral images, and for the automated search within these collections. The attempts to define CBIR strategies for them are scarce and partial. The only clear example of an attempt to build up such a system found in the literature searched is [3]. The authors use the spectral mean and variance, as well as a texture feature vector, to characterize hyperspectral image tiles. The approach searches for specific phenomena in the images (hurricanes, fires, etc), using an interactive relevance strategy that allows the user to refine the search.

We propose the characterization of the hyperspectral images by their so-called endmembers. Endmembers are a set of spectra that are assumed as vertices of a convex hull covering the image pixel points in the high dimensional spectral space. Endmembers may be defined by the domain experts (geologists, biologists, etc.) selecting them from available spectral libraries, or induced from the hyperspectral image data using machine learning techniques. In [4, 5] we have proposed an automated procedure that induces the set of endmembers from the image, using AMM to detect the morphological independence property, which is a necessary condition for a collection of endmembers. The goal in [4, 5] was to obtain a method for unsupervised hyperspectral image segmentation.

R. Khosla et al. (Eds.): KES 2005, LNAI 3681, pp. 827–832, 2005.

2 Endmember Induction

Linear mixing models assume the knowledge of a set of endmembers $\mathbf{S} = [s_1, \cdots, s_n]$, where each $s_i \in R^d$ is a d-dimensional vector. Then, one pixel of a hyperspectral image can be expressed as $f(x,y) = \mathbf{S} \cdot \mathbf{a}(x,y) + \eta(x,y)$, where $\eta(x,y)$ is the independent additive noise component and $\mathbf{a}(x,y)$ is the n-dimensional vector of endmember fractional abundance in the pixel. In other words, $\mathbf{a}(x,y)$ are the convex coordinates of the pixel relative to the convex hull defined by the vertices in $\mathbf{S}$. In [4, 5] we were interested in $\mathbf{a}(x,y)$ as unsupervised segmentations of the hyperspectral image. Here we consider that the set of endmembers $\mathbf{S}$ may be *per se* a good characterization of the hyperspectral image, if it has been obtained from it.

The method proposed in [4, 5] to obtain the set of endmembers is based on the notion of morphological independence. Given a vector set $X = \{x_1, ..., x_m\}$, a new vector y is morphologically independent in the erosive sense from X if $\neg \exists x \in X | y \leq x$, and it is morphologically independent in the dilative sense from X if $\neg \exists x \in X | y \geq x$. The partial order defined over the vectors is the one induced from the order of their components: $y \leq x \Leftrightarrow \forall i, y_i \leq x_i$. The vector set X is said to be morphologically independent when all the vectors in the set are independent of the remaining ones in either sense. A set of morphological independent vectors defines a high dimensional box. In [4, 5] we propose the use of AMM's for the detection within a hyperspectral image of a set of morphological independent spectra whose corresponding high dimensional box may be taken as a good approximation to the convex hull of the image data. These spectra are the induced endmembers that will be used to characterize the image. In brief the proposed method consists in the following: (1) A seed pixel is taken as the initial set of endmembers. (2) Erosive and dilative AMM's are built from the current set of endmembers. (3) Each pixel is examined as a candidate endmember testing the response of the erosive and dilative AMM's to it. (4) Morphologically independent pixels are added to the set of endmembers. The process takes into account the variance of the spectra at each band to enhance the robust detection of endmembers, adding and subtracting the per band standard deviation multiplied by a gain parameter to the pixel spectrum before performing the tests with the dilative and erosive AMM´s, respectively. This gain parameter is set by default to 2. The process goes one tim over the image data. If the region is very homogeneous, the process may stop without adding any new endmember, besides the seed pixel taken. Then, the gain parameter is reduced and the process is repeated until the number of endmembers is 2 or more.

3 Similarity Between Images

Let it be $\mathbf{S}_k = \left[s_1^k, \cdots, s_{n_k}^k\right]$ the set of endmembers, obtained as described before from the k-th image $f_k(x,y)$ in the database, where n_k is the number of endmembers detected in this image. Given two images $f_k(x,y)$ and $f_l(x,y)$, we compute the following matrix whose elements are the Euclidean distances between the endmembers of each image:

$$D_{k,l} = \left[d_{i,j}; i = 1, \ldots, n_k; j = 1, \ldots, n_l \right], \tag{1}$$

where $d_{i,j} = \left| s_i^k - s_j^l \right|$. We compute the vectors of the minimal values by rows and columns,

$$\mathbf{m}_k = m_i^k = \min_j \{ d_{i,j} \}, \tag{2}$$

and

$$\mathbf{m}_l = \left[m_j^l = \min_i \{ d_{i,j} \} \right], \tag{3}$$

respectively. Then the similarity between the images f_k and f_l is given by the following expression:

$$d(f_k, f_l) = (|\mathbf{m}_k| + |\mathbf{m}_l|)(|n_k - n_l| + 1). \tag{4}$$

4 Discussion

The endmember induction procedure may give different number of endmembers and endmember features for two hyperspectral images. The similarity measure of eq. 4 is a composition of two asymmetrical views: each vector of minimal distances measures how close are the endmembers of one image to some endmember of the other image. Suppose that all the endmembers $\mathbf{S}_k$ of an image are close to a subset of the endmembers $\mathbf{S}_l$ of the other image. Then the vector of minimal distances $\mathbf{m}_k$ will be very small, not taking into account the unlike endmembers in the second image. However, the vector of minimal distances $\mathbf{m}_l$ will be larger than $\mathbf{m}_k$ because it will take into account the distances of endmembers in $\mathbf{S}_l$ which are unlike to those in $\mathbf{S}_k$. Thus the similarity measure of eq. 1 can cope with the asymmetry of the situation. It avoids the combinatorial problem of trying to decide which endmembers can be matched and what to do in case that the number of endmembers is different from one image to the other. The difference in the number of endmembers is introduced as an amplifying factor. The measure is independent of image size and, as the endmember induction algorithm is very fast, it can be computed in acceptable time. Also the endmember set poses no storage problem.

Our approach does not use spatial features, such as the textures in [3], but the endmembers give a rich characterization of the spectral content of the image. A further work on our approach may be the study of spatial features computed on the abundance images produced by the spectral unmixing, solving the question of band selection or dimension reduction prior to spatial feature computation.

5 Validation Experiment

The validation of CBIR approaches is a subtle issue, because it is not possible in general to ensure a unique response to a query. Our approach to demonstrate the usefulness of the proposed similarity measure is to built up a small database of hyperspectral images which are samples from public well known test hyperspectral images.

The first two images are the ones denominated Indian Pines'92 and Salinas'98. ([6] is an example of research validated with these images). Both are AVIRIS images of 224 spectral bands per pixel, of sizes 145x145 and 512x217 pixels, respectively. The other images are the first chunk of the Cuprite and Jasper images that can be downloaded from the AVIRIS site: http://aviris.jpl.nasa.gov/html/aviris.freedata.html. The precision is 16 bit integers per pixel. All the images correspond to the radiance data cubes, with values ranging up to 10^4. We introduce a normalization step, so that each pixel is a normalized vector of unit magnitude. That accounts for the ordinate ranges in figures 1 and 2.

The samples are obtained cropping rectangular regions of different sizes and locations on these images, some of them overlapping. The validation of our approach will be given by the ability of the proposed measure to obtain as similar images the samples coming from the same image. In other words, the task is to discriminate between Indian Pines and Salinas on the basis of local information given by the samples. This task is not trivial because both Salinas and Indian Pines images cover mainly vegetation areas, as well as some regions of the Jasper and Cuprite image pieces used in the experiment.

We performed this experiment on a set of 20 samples from each image. The success ratio was computed as the percentage of samples coming from the query's image among the 4 closest samples in the database, according to eq. 1. In a 1-leave-out experiment, when each image in the data base is used as a query, the success ratio obtained was 100%. Figure 1 and figure 2 show the response of the system to a Cuprite and an Indian Pines sample query, respectively. The response of the system is the collection of the 4 sample images in the database closest to the query. The hyperspectral images are represented in the figures by the plot of their induced endmembers.

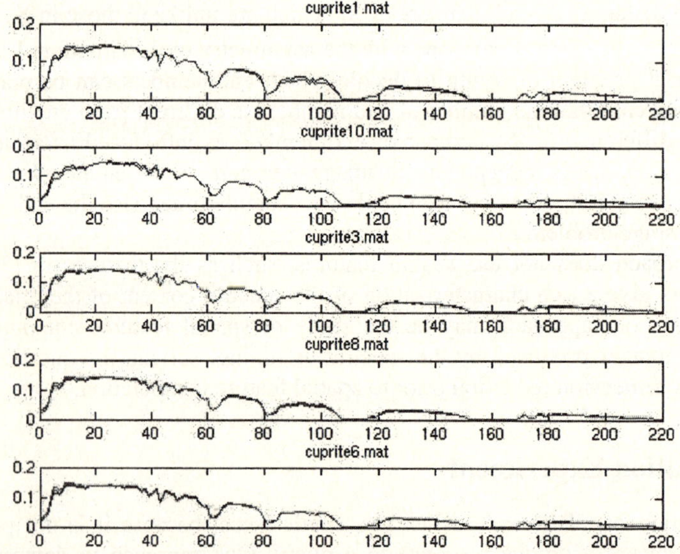

Fig. 1. Answers to the query given by the sample 'cuprite 1' obtained from the database of samples. All the answers come from the same image as the query

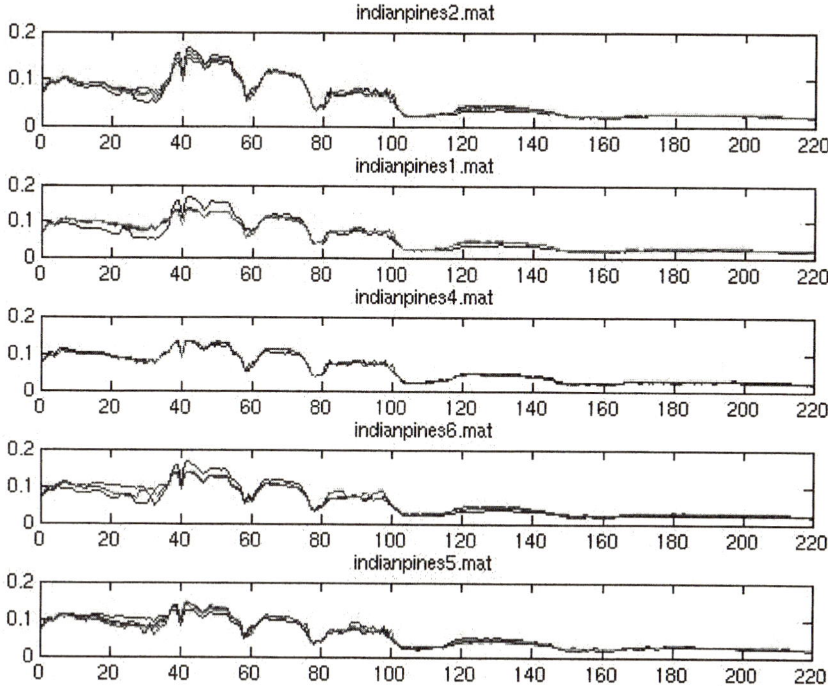

Fig. 2. Answers to the query given by the sample 'Indian Pines 2' obtained from the database of samples. All the answers come from the same image as the query

(Normalized radiance by each spectral band). It can be appreciated that our similarity measure has been able to identify the source images despite the small differences among the endmembers characterizing each query, and the difference the number of endmembers induced. The diversity in the number of endmembers induced from each sample is due to the local per band variance in each sample. In some of them, the extracted region is very homogenous and the process extracts only 2 endmembers, which sometimes appear rather overlapped in the plot. Cuprite samples seem to have less variance than Indian Pines images.

6 Conclussions and Further Work

We propose a similarity measure, based on the endmembers induced from an hyperspectral image, for the construction of CBIR systems of such hyperspectral images. The image characterization is independent of image size and spatial undesired variations, like distortions due to variations in the flying path. Further work may be addressed to the introduction of spatial information in the similarity measure with the aim of detecting specific regions of the images which are being searched for similarity retrieval. This spatial information may come from the abundance images obtained by spectral unmixing.

Acknowledgements

The Spanish Ministerio de Educación y Ciencia supports this work through grant VIMS-2003-20088-c04-04

References

1. A.W.M.Smeulders, M., Worring, S. Santini, A. Gupta, R. Jain, "Content-based image retrieval at the end of the early years", IEEE Trans. Pat. Anal. Mach. Intel., 2000, 22 (12), pp.1349 – 1380
2. D. A. Landgrebe *Signal Theory Methods in Multispectral Remote Sensing*, John Wiley & Sons, Hoboken, NJ, 2003
3. I.E. Alber, Ziyou Xiong; N. Yeager; M. Farber, W.M. Pottenger, "Fast retrieval of multi- and hyperspectral images using relevance feedback", Proc. Geosci. Rem. Sens. Symp., 2001. IGARSS '01, vol.3, pp.1149 – 1151
4. M. Graña, J.Gallego, C. Hernandez, "Further results on AMM for endmember induction", Proc. IEEE Workshop on Adv. Tech. Anal. Remotely Sensed Data, Washingto D.C., Oct. 2003, pp.237 – 243
5. M. Graña, J. Gallego, "Associative morphological memories for endmember induction" Proc. Geosci. Rem. Sens. Symp., IGARSS '03. Tolouse, Jul. 2003, vol.6, pp.:3757 – 3759
6. Gualtieri, J.A.; Chettri, S. 'Support vector machines for classification of hyperspectral data', Proc. Geosci. Rem. Sens. Symp., 2000, IGARSS 2000. pp.:813 - 815 vol.2

Hyperspectral Image Watermarking
with an Evolutionary Algorithm

D. Sal and Manuel Graña*

Dept. CCIA, UPV/EHU, Apdo. 649, 20080 San Sebastian, Spain
`ccpgrrom@si.ehu.es`

Abstract. We propose an evolutionary algorithm for the digital semi-fragile watermaking of hyperspectral images based on the manipulation of the discrete cosine transform (DCT) computed for each band in the image. The algorithm searches for the optimal localization in the support of an image DCT to place the mark image. The problem is stated as a multi-objective optimization problem (MOP), that involves the simultaneous minimization of distortion and robustness criteria. Given an appropriate initialization, the algorithm can perform the search for the optimal mark placement in the order of minutes, approaching real time application restrictions

1   Introduction

The sun transmits to The Earth energy in form of light waves that are reflected, absorbed or transmitted by the materials of the earth's surface. This energy can be picked up by a sensor that registers the energy at different wavelengths reflected from an area discretizing it into image pixels. The hyperespectral sensor performs a fine sampling of the wavelength light energy distribution. Therefore each image pixel corresponds to a high dimensional vector. The whole hyperspectral image is a 3D numerical matrix. This matrix can be read in several ways. One is to asume that each wavelength corresponds to a 2D image, so the whole image is interpreted as a stack of 2D images. Other is to process each pixel as separate plot of energy versus wavelength, thus a reflectance spectrum of the area corresponding to the pixel is obtained. This representation gives the necessary information for the classification and identification of materials, been widely used in a large number of applications, such as detection and identification of the surface and atmospheric constituents present, monitoring agriculture and forest status, and military surveillance[5], [7]. Another way of extracting 2D images from the 3D data volume is to consider the image formed by an image line or column across all the wavelengths. That would correspond to the treatment of the image simultaneously with its acquisition.

Watermarking is a technique for image authorship and content protection [9], [1], [6], [8], [4, 10]. Semi-fragile watermarking tries to ensure the image integrity,

---

* The Spanish Ministerio de Educacion y Ciencia supports this work through grants DPI2003-06972 and VIMS-2003-20088-c04-04.

R. Khosla et al. (Eds.): KES 2005, LNAI 3681, pp. 833–839, 2005.
© Springer-Verlag Berlin Heidelberg 2005

by means of an embedded watermark which can be recovered without modification if the image has not been manipulated. However, it is desirable that the mark is robust to operations like filtering, smoothing and lossy compression which are very common while distributing images through communication networks. A watermarked image must be as indistinguishable from the original one as possible, that is, the watermarking process must introduce the minimum possible visual distortion in the image.

These two requirements (robustness against filtering and minimal distortion) are the contradicting objectives of our work. The trivial watermarking approach consists in the addition or substitution of the watermark image over the high frequency image transform coefficients. That way, the distortion is perceptually minimal, because the watermark is embedded in the noisy components of the image. However, this approach is not robust against smoothing and lossy compression. The robustness can be enhanced placing the watermark in other regions of the image transform, at the cost of increased distortion. Combined optimization of the distortion and the robustness can be stated as a multi-objective optimization.

Multi-objective optimization problems are characterized by a vector objective function. As there is no total order defined in vector spaces, the desired solution does not correspond to a single point or collection of points in the solution space with global optimal objective function value. We must consider the so-called Pareto-Front which is the set of non-dominated solutions. A non-dominated solution is one that is not improved in all and every one of the vector objective function components by any other solution [2]. In the problem of searching for an optimal placement of the watermark image, the trade-off between robustness and image fidelity is represented by the Pareto-Front discovered by the algorithm. We define an evolutive strategy that tries to provide a sample of the Pareto-Front preserving as much as possible the diversity of the solutions. The stated problem is not trivial and shows the combinatorial explosion of the search space: the number of possible solutions is the number of combinations of the image pixel positions over the size of the image mark to be placed. Next section will review multi-objective optimization basics. Section 3 introduces the problem notation. Section 4 describes the proposed algorithm. Section 5 presents some empirical results and section 6 gives our conclusions and further work discussion.

2   Multi-objective Optimization Problem

The general MOP tries to find the vector $x^* = [x_1^*, x_2^*, ..., x_n^*]^T$ which will satisfy the m inequality constraints $g_i(x) \geq 0, i = 1, 2, ..., m$, the p equality constraints $h_i(x) = 0, i = 1, 2, ..., p$ and will optimize the vector function $f(x) = [f_1(x), f_2(x), ..., f_k(x)]^T$.

Vilfred Pareto (1896) generalized the notion of multi-objective optimization, trying to find a good trade-off between solutions.

A vector of decision variables $x^* \in \mathcal{F}$ is Pareto optimal if it does not exist another $x \in \mathcal{F}$ such that $f_i(x) \leq f_i(x^*)$ for all $i = 1, .., k$ and $f_j(x) < f_j(x^*)$ for at least one j. Each solution that carries this property, is called non-dominated solution, and the set of non-dominated solutions is called Pareto optimal set. The plot of the objective functions whose non-dominated vectors are in the Pareto optimal set is called the Pareto-Front.

A vector $\boldsymbol{u} = (u_1, ..., u_n)$ is said to dominate $\boldsymbol{v} = (v_1, ..., v_n)$ (denoted as $\boldsymbol{u} \preceq \boldsymbol{v}$) if and only if $\forall i \in \{1..k\}, u_i \leq v_i \wedge \exists i \in \{1, ..., k\} : u_i < v_i$.

For a given MOP $\boldsymbol{f}(x)$, the Pareto optimal set $\mathcal{P}^*$ is defined as: $\mathcal{P}^* := \{x \in \mathcal{F} \mid \neg \exists x' \in \mathcal{F} : \boldsymbol{f}(x') \preceq \boldsymbol{f}(x)\}$, and the Pareto-Front ($\mathcal{PF}^*$) is defined as: $\mathcal{PF}^* := \{\boldsymbol{u} = \boldsymbol{f} = (f_1(x), ..., f_k(x)) \mid x \in P^*\}$.

3   Watermarking Problem Notation

We have an hyperspectral image X of size m_x x n_x x $nbands$ that we want to protect. To do that, we use a mark image W of size m_w x n_w. The DCT of the image and the mark image are denoted X_t and W_t respectively. X_t is obtained by applying the bi-dimensional DCT to each band. Given two coordinates k, l of the W domain, $1 \leq k \leq m_w$, $1 \leq l \leq n_w$, we denote $x(k, l)$, $y(k, l)$, $z(k, l)$ the coordinates of the X_t domain where the coefficient $W(k, l)$ is added in order to embed the mark.

The algorithm works with a population Pop of P_s individuals which are solutions to the problem. We denote O the offspring population. Let be P_s, P_m and P_c the probabilities of selection, mutation and crossover respectively.

To aboid a possible confusion between the solution vector $(\boldsymbol{x})$ and the original image (X), we will denote the first one as $\boldsymbol{s}^*$. So, the algorithm will try to find the vector $\boldsymbol{s}^* = [s_1^*, s_2^*, ..., s_{P_s}^*]^T$ which optimize $\boldsymbol{f}(\boldsymbol{s}) = [f_1(\boldsymbol{s}), f_2(\boldsymbol{s})]$ where f_1 is the robustness fitness function and f_2 is the distortion fitness function. The algorithm returns a sampling of the Pareto optimal set $\mathcal{P}^*$ of size between 1 and P_s. The user will be able to select the solution which is better adapted to his necessities from the plotted Pareto-Front $\mathcal{PF}^*$.

A solution $\boldsymbol{s}^*$ is represented as a m_w x n_w matrix in which every position $W_t(k, l)$ takes three positive values: $x(k, l)$, $y(k, l)$ and $z(k, l)$. Actually, our mark is a small image or logo. The embedded information is the logo's DCT. So, the corruption of the recovered mark is detected by visual inspection, and can be measured by correlation with the original mark.

4   Algorithm

In this section we will start introducing the fitness functions that model the robustness and distortion of the solutions. Next we define the operators employed. The section ends with the global definition of the algorithm.

4.1   Multi-objective Fitness

Two fitness functions f_1 and f_2, measure the robustness and distortion of the watermark placement represented by an individual solution, respectively. Together, they compose the vector objective function $(\boldsymbol{f})$ to be optimized.

- **Robustness fitness function** f_1: Robustness refers to the property of recovering the mark even when the watermarked image has been manipulated. We are interested in having robustness against lossy compression and smoothing. Both transformations affect the high and preserve the low frequency image transform coefficients. Therefore the closer to the transform space origin the mark is located, the higher the robustness of the mark. As we are embedding the logo DCT, we note that most of the mark information will be in the low frequency coefficients so, they must have priority to be embedded in the positions that are nearer to the low frequencies of X_t. All this requirements are expressed in equations (1) and (2).

Our robustness fitness function is the sum for all the mark pixels of the α-root of the position norm.

$$f_1 = \sum_{k=i}^{m_w} \sum_{l=1}^{n_w} \sqrt[\alpha]{x(k,l)^2 + y(k,l)^2 + k + l} \tag{1}$$

where α is:

$$\alpha = F \frac{\sqrt{x(k,l)^2 + k} + \sqrt{y(k,l)^2 + l}}{x(k,l) + y(k,l) + k + l} + d \tag{2}$$

The robustness does not depend on the band number, because each band DCT has been computed independently. In summary, this function possesses the following properties:

1. As the position in X_t where $s(k,l)$ is embedded is closer to the low frequencies, the function value decreases smoothly.
2. As the pixel of W_t is more important (nearest to the W_t low frequencies), the value of α increases smoothly.

Therefore to maximize the robustness, f_1 must be minimized.

- **Distortion fitness function** f_2: The true distortion is computed as the squared difference between the original image and the inverse of the marked DCT. However, to avoid the computational cost of the DCT inversion, we use as the fitness function of the evolutionary algorithm an approximation that follows from the observation that the distortion introduced adding something to a DCT coefficient is proportional to the absolute value of that coefficient. Thus, the distortion fitness function to be minimized is the following one:

$$f_2 = \sum_{k=i}^{m_w} \sum_{l=1}^{n_w} \mid X_t(x(k,l), y(k,l), z(k,l)) \mid \tag{3}$$

4.2   Evolutionary Operators

Selection Operator: This operator generates O from P. The populations has previously been ordered according to its range and distance between solutions as proposed in [3]. The selection is realized choosing stochastically the individuals, giving more probability to the ones at the beginning of the sorted list.

Crossover operator: This operator is applied with probability P_c and is used to recombine each couple of individuals and obtain a new one. Two points from the solution matrix are stochastically selected as cut points, and the individuals are recombined as in conventional crossing operators.

Mutation operator: Every element of an individual solution s^* undergoes a mutation with probability P_m. The mutation of an element consists of displacing it to a position belonging to the 24-Neighborhood in a 3D grid: given a pixel $W_t(k, l)$ located in the position $x(k, l), y(k, l), z(k, l)$ of X_t, the new placement of $s^*(k, l) \in \{X_t(x(k, l)\pm 1, y(k, l)\pm 1, z(k, l)\pm 1)\}$. The direction of the displacement is chosen stochastically. If the selected position is out of the image, or collides with another assignment, a new direction is chosen.

Reduction operator: At this moment there are two populations: P and O. Now it is time to select the individuals who are going to form the new one. Both populations are joined in a new one of size $2P_s$. This population is sorted according to the range and distance between solutions[3]. This ensures an elitist selection and the diversity of the solutions through the Pareto-Front. The new population P is composed of the best P_s individuals according to this sorting.

4.3   Algorithm

The first step of the GA is the generation of an initial population P and the evaluation of each individual's fitness. The rank and distance of each individual is calculated [3] and is computed to sort P. Once done this, the habitual curl begins: An offspring population O is calculated by means of the selection, crossover, and mutation operators. The new individuals are evaluated before joining them to the population P. Finally, after computing the reduction operator over the new rank and distance of each individual, we obtain the population P for the next iteration.

Since the GA works with many non-dominated solutions, a possible stopping criterion compares the actual population with the best generation, individual to individual, by means of the *crowded_comparison*() [3]. If no individual, or a number of individuals less than a threshold, improve the best solution in n consecutive iterations, the process is finished.

The Pareto-Front is formed by the set of solutions with rank = 1. Once finished the process and chosen a solution, the mark is embedded adding its coefficients to the coefficients of X_t according to the corresponding value of **s**. Before the coefficients are added, they are multiplied by a small value.

5   Results

The results presented in this section concern the application of the algorithm over the well known Indian Pines hyperspectral image of size 145 x 145 x 220. The image transform X_t has been divided in 1452 overlapping quadrants of size 45 x 45 x 110. The initial population is formed by 1452 individuals each one placed randomly in a different quadrant. For the mark has been used an image of size 50 x 50. The GA was executed with $P_s = 20, P_m = 0.05$ and $P_c = 0.9$. We fit the response of the robustness fitness f_1 with $F = 4$ and $d = 3$.

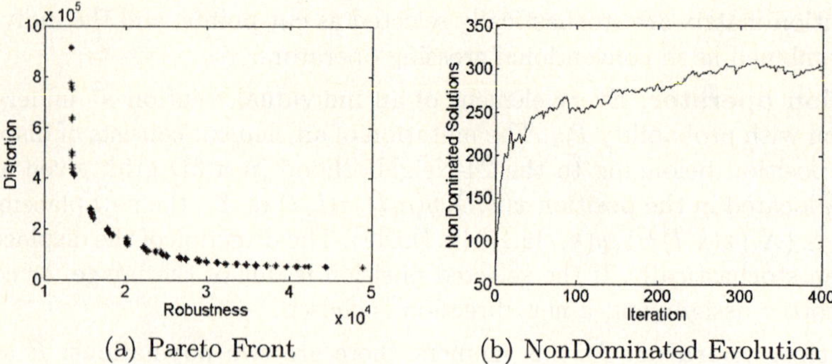

(a) Pareto Front   (b) NonDominated Evolution

Fig. 1. a) Pareto-Front found by GA. b)Evolution of non-dominated solution found by the GA through the iterations.

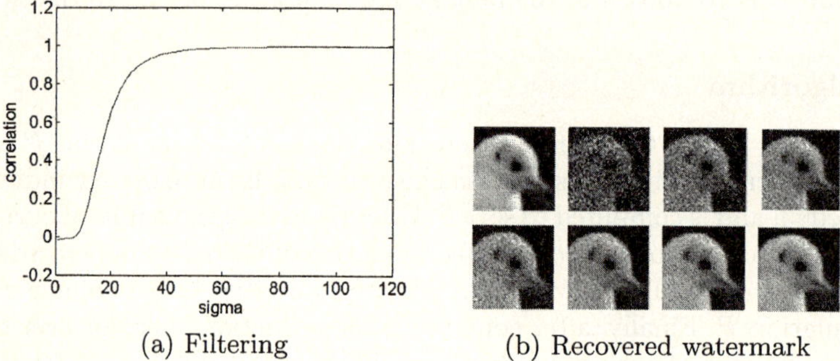

(a) Filtering   (b) Recovered watermark

Fig. 2. a) Robustness level by means of the correlation coefficient of the recovered image mark versus the radius of the smoothing convolution kernel. b) Original watermark and watermark recovered after low pass filtering with sigma $= 10, 20, 30, 40, 50, 60$ and 70 respectively.

Figure 1(a) shows the Pareto-Front consisting of 303 non-dominated solutions, found by the algorithm, following the evolution shown in Figure 1(b). Figure 2(a) plots the correlation coefficient between the original watermark and the corrupted watermark recovered after the marked image has been smoothed bandwise by a low-pass gaussian filter with increasing filter radius applied in the Fourier transform domain. Figure 2(b) shows the visual results of the recuperation of the mark image after smoothing the watermarked image.

Studying each pixel spectrum, experts can know which material form the area represented by this pixel. This is the main objective of hyperspectral imaging, so, it is critical that the watermarking process doesn't disturb these plots. For the noisiest of the solutions shown in Figure 1(a) we computed the correlation of each pixel spectrum with the corresponding one in the original image. The worst value obtained was 0.999. Therefore, this watermarking process is not expected to influence further classification processes.

6 Conclusions

We present an evolutionary algorithm to find a watermark's image placement in an hyperspectral image to protect it against undesirable manipulations. It is desirable that the watermark remains recognizable when the image is compressed or low-pass filtered. We state the problem as a multiobjective optimization problem, having two fitness functions to be minimized. The algorithm tries to obtain the Pareto-Front to find the best trade-off between distortion of the original image in the embedding process and robustness of the mark. The solutions found by the GA provide strong robustness against smoothing manipulations of the image. Because the algorithm works with the entire image DCT, it can be used to hide bigger images or data chunks than other similar approaches. Also it will be more robust than approaches based on small block embedding, experimental verification is on the way to prove this intuition. Furher work must be addressed to the extension of this approach to 3D DCT data. We are studding the robustness against lossy compression.

References

1. D. Augot, J. Boucqueau, J. Delaigle, C. Fontaine, and E. Goray. Secure delivery of images over open networks. *Proceedings of the IEEE*, 87(7):1251–1266, July 1999.
2. C. Coello Coello, G. Toscano Pulido, and E. Mezura Montes. Current and future research trends in evolutionary multiobjective optimization. In M. Grana, R. Duro, A. d'Anjou, and P. Wang, editors, *Information Processing with Evolutionary Algorithms*, pages 213–232. Springer Verlag, New York, 2004.
3. K. Deb, A. Pratap, S. Agarwal, and T. Meyarivan. A fast and elitist muliobjective genetic algorithm: Nsga-ii. *IEEE transactions on evolutionary computation*, 6(2):182–197, April 2002.
4. A. Nikolaidis and I. Pitas. Region-based image watermarking. *IEEE Transactions on image processing*, 10(11):1726–1740, November 2001.
5. C. V. Sanz. Razonamiento evidencial dinamico. un método de clasificación aplicado al análisis de imágenes hyperespectrales. 2002.
6. P. B. Schneck. Persistent access control to prevent piracy of digital information. *Proceedings of the IEEE*, 87(7):1239–1250, July 1999.
7. X. Tang, W. A. Pearlman, and J. W. Modestino. Hyperspectral image compression using three-dimensional wavelet coding.
8. C. Vleeschouwer, J. Delaigle, and M. B. Invisibility and application functionalities on perceptual watermarking- an overview. *Proceedings of the IEEE*, 90(1):64–77, January 2002.
9. G. Voyatzis and I. Pitas. The use of watermarks in the protection of digital multimedia products. *Proceedings of the IEEE*, 87(7):1197–1207, July 1999.
10. R. B. Wolfgang, C. I. Podilchuk, and E. J. Delp. Perceptual watermarking for digital images and video. *Proceedings of the IEEE*, 87(7):1108–1126, July 1999.

A Profiling Based Intelligent Resource Allocation System

J. Monroy, Jose A. Becerra, Francisco Bellas,
Richard J. Duro, and Fernando López-Peña

Grupo Integrado de Ingeniería, Universidade da Coruña
{jmonroy,ronin,fran,richard,flop}@udc.es

Abstract. The work presented here is mainly concerned with the development of an intelligent resource allocation method specially focused in providing maximum satisfaction to user agents tied to resource strapped applications. One of the applications of this type of strategies is that of remote sensing in terms of energy and sensor usage. Many remote sensors or sensor arrays reside on satellites and their use must be economized, while at the same time the agency managing satellite time would like to satisfy the users as much as possible. Here we have developed a cognitive based strategy that obtains models of users and resource use in real time and uses these models to obtain strategies that are compatible with management policies. The paper concentrates in obtaining the user models.

1 Introduction

The resource allocation problem was initially a matter of concern in a variety of areas in operations research and management science [1]. A wide variety of modern applications concerning electric, electronic, communication, computational or sensor network shared systems are compelled to accomplish two conflicting requirements; on one hand increasing in performance and complexity and, on the other, minimizing resource spending. Thus, all these application require a careful resource allocation and planning methodology. To reach a practical implementation in complex cases, intelligent resource allocation appears as a right tool to optimally allocate and reuse resources (examples may be found in [2], [3], and [4]). An intelligent resource allocation method for the distribution of limited resources based on user profiling and aimed at maximizing user satisfaction is presented in this paper. The method addresses the real time profiling of users integrated in resource assignment processes. It is based on evolution and the capabilities of Artificial Neural Networks (ANNs) to model non linear functions. The main idea proposed here is that the problem of assigning resources to tasks and thus, to users, in environments under constraints where the satisfaction of the users is important, may be solved through the implementation of a cognitive process that allows the decision system to directly model in real time the behavior of the users and their satisfaction and use this information as a fitness measure when deciding a resource assignment strategy.

A cognitive strategy usually involves two different tasks: to use the information from the interaction with the environment in order to obtain appropriate models of the environment an of the system itself, and to use the model set as an evaluator so that strategies can be tested on it so that the most adequate one may be chosen before applying it. In the particular case of energy strapped systems that are shared by many users, it is necessary to be able to model the behavior of the users and of the system as

R. Khosla et al. (Eds.): KES 2005, LNAI 3681, pp. 840–846, 2005.

a response to their requests as well as their satisfaction under different circumstances so that these models can be employed to decide the resource allocation sequence that provides the most satisfaction and at the same time preserves the integrity of the system in terms of energy while respecting management policies.

2 Cognitive Approach

One classical way of formalizing the operation of a general cognitive model for a general agent from a utilitarian point of view starts from the premise that to carry out any task, a motivation (defined as the need or desire that makes an agent act) must exist that guides the behavior as a function of its degree of satisfaction. The tools the agent can use to modify the level of satisfaction of its motivation are perceptions through sensors and actions through actuators. In a Cognitive Mechanism, the exploration of actions must be carried out internally and thus a set of mathematical functions, usually denoted as W and S must be somehow obtained. These functions correspond to what are traditionally called:

- *World model (W):* function that relates the external perception before and after applying an action, which in this cases is related to the behavior of the user.
- *Satisfaction model (S):* function that provides de predicted satisfaction from the predicted external perceptions provided by the world model.

There are two processes that must take place in a real non preconditioned operating mechanism: models W and S must be obtained as the agent interacts with the world, and for every interaction of the world, the best possible action must be selected through some sort of optimization using the models available at that time.

When trying to implement this cognitive model computationally, in addition to these basic elements (perceptions, actions, motivations and models) we need to include a new one: the *action-perception pair*. It is just a collection of values from the interaction with the environment after applying an action, that is, data from the real world corresponding to the sensorial data before applying an action, the action applied, the sensorial data after the application of the action and the satisfaction achieved. This action perception pair is a representation of data from the real world and can thus be used to evaluate models in real time.

3 Constructing the MDB

As we have mentioned in the previous section, the actions that must be applied in the environment are selected internally by the agent and this functionality comprises three main elements: a memory that stores the action-perception pairs, a stage to improve the models according to the real information available and finally a stage to select the action to be applied.

Using this scheme we have constructed a Cognitive Mechanism called MDB (Multilevel Darwinist Brain) [5]. The main difference of the MDB with respect to other model based cognitive mechanisms is the way the models are obtained and the actions planned from them. Its functional structure is shown in Fig. 1. The final objective of the mechanism is to provide the action the agent must apply in the environment to fulfill its motivation. Its operation can be summarized by considering that the selected action (represented by the Current Action block) is applied to the Environment through the Actuators obtaining new Sensing values. These acting and sensing values provide a new action-perception pair that is stored in the Short-Term Memory. Then, the Model Learning processes start (for world and satisfaction models) trying to find functions that generalize the real samples (action-perception pairs) stored in the Short-Term Memory. The best models in a given instant of time are taken as Current World Model and Current Satisfaction Model and are used in the process of Action Optimization. The best action obtained from this process is applied again to the Environment through the Actuators obtaining new Sensing values.

These five steps constitute the basic operation cycle of the MDB, and we will call it an iteration of the mechanism. As more iterations take place, the MDB acquires more information from the real environment (new action-perception pairs) so the models obtained become more accurate and, consequently, the action chosen using these models is more appropriate.

There are two main processes that must be solved in MDB: the search for the best world and satisfaction models predicting the contents of the action-perception pair memory and the optimization of the action trying to maximize the satisfaction using the previously obtained models. In the way these processes are carried out lies the main difference of the MDB with respect to other cognitive mechanisms.

In this context, a model is just a non linear function defined in an n-dimensional space that approximates and tries to predict real characteristics. The techniques for obtaining these functions must be found considering that we have samples (action-perception pairs) of the function to model, these samples are known in real time and we want to obtain the most general model possible, not a model for a given set of samples present in a particular instant of time.

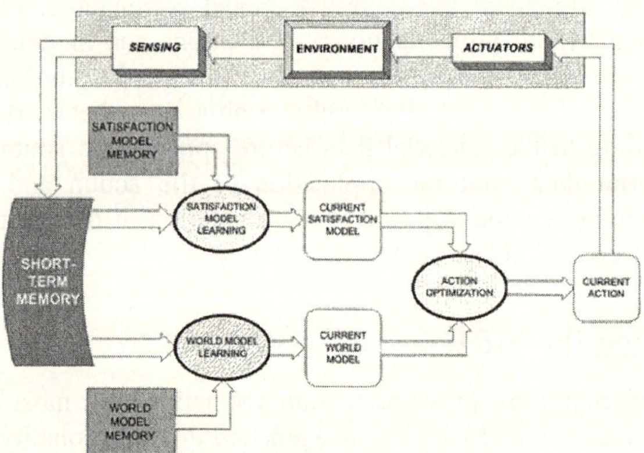

Fig. 1. Block diagram of the MDB

Taking these three points into account, the model search process in the MDB is not an optimization process but a learning process. As commented in [6], learning is different from optimization because we seek the best generalization, which is different from minimizing an error function. Consequently, the search techniques must allow for gradual application as the information is known progressively and in real time. In addition, they must support a learning process through input/output pairs (action/consequence samples) using an error function.

To satisfy these requirements we have selected Artificial Neural Networks as the mathematical representation for the models and Evolutionary Algorithms as the most appropriate search technique. This combination permits an automatic acquisition of knowledge (the models), providing a Darwinist base to the MDB as explained in [5].

After applying an action in the environment and obtaining new sensing values, the search for the models are now evolutionary processes, one for the world models and another for the satisfaction models. The use of evolutionary techniques permits a gradual learning process by controlling the number of generations of evolution for a given content of the action-perception pair memory. This way, if evolutions last just for a few generations (usually from 2 to 4) per iteration, we are achieving a gradual learning of all the individuals. Furthermore, the evolutionary algorithms permit a learning process through input/output pairs using as fitness function an error function between the predicted values provided by the models and the expected values for each action-perception pair

4   Application of the MDB to This Particular Problem

In this particular problem there are two entities for which we must obtain models, the users of the system and the resources of the system. As a first approximation, in this paper we are going to concentrate on the use of the MDB through the production of user behavior models and consider the rest of the elements as fixed.

A user model is implemented as an Artificial Neural Network based agent that processes information corresponding to the amount of resources the user requests, some data on the day of the week and the time of day the resources are requested. As outputs these models provide an estimation of the final use of resources the user really made. This data can then be used by the strategy optimization stage in order to provide the best sequence of tasks in order to maximize the satisfaction of the users while respecting the energy requirements and constraint of the system.

As an experiment to show the way the system works, we have constructed a process queue managing system that contains one model per user and implements a FIFO like short term memory. The user models are ANNs with four input nodes (requested length of process, walltime, day requested, time of day the request was made), two hidden layers with 3 nodes each and an output layer with two nodes (estimated real length of process and estimated walltime).

The user models are updated every time a user requests a process. Initially there is a set of randomly generated models for each user. When the user requests resources, the input data and the results of the running of the processes are used to run a few evolution generations. This evolution uses the corresponding model set and has the aim of minimizing the error between the values predicted by the model and those that were obtained. These evolution processes consist of a standard genetic algorithm with tournament selection (2% window), one point crossover, random mutation and elitist replacement strategy. The best model for each user in a given instant of time is defined through a fitness criterion based on its estimation of the behavior of the user in the previous instants of time. We want the model to provide a good estimation, but we favor overestimation (we do not want to cut a process short) through a non linear fitness function.

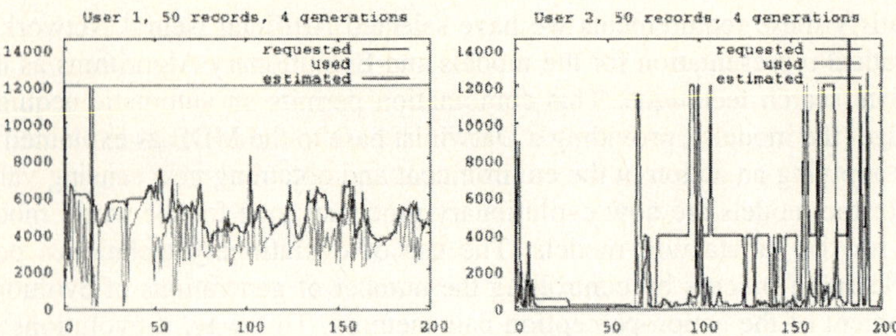

Fig. 2. Some results for two users selected at random. The graphs display the amount of requested resource (in this case computing time), the estimation made by the ANN based user model and the real usage of the resource

The best model for each user in a given instant of time (the model that best predicted the previous task request parameters) is used as the estimator for the strategy optimization block to obtain the most appropriate strategy. This block takes into account all of the requests that are active and uses the user models corresponding to these requests to predict the real resource requirements for each user. With these predictions and considering a fitness criterion that involves user satisfaction through user satisfaction models or management policy rules, and the resource energy constraints the strategy optimization block provides a queue of processes to be run on the system. Once these processes are run, the results obtained can be used to improve the models.

Initially the system will provide poor results as the models are basically random. As time passes and more information is considered, the models will improve and thus, their predictions will be more adequate, leading to a better planning by the system.

5   Experiments

In what follows we present some results of the performance of user models obtained from real data in a resource allocation problem. The users in this problem request computing time and resources up to a maximum allowed value and the system must obtain the best possible allocation strategy in order to maximize resource use and user satisfaction.

The experiments were carried out modeling a number of different users and using different parameters for the MDB, the main two parameters considered were Short Term Memory Length (10, 20 and 50 registers. In all cases the STM behaved as a FIFO) and number of evolution generations for each real action perception pair obtained (4, 10 and 50).

Figure 2 displays some results for two users selected at random. The graphs display the amount of requested resource (in this case computing time), the estimation made by the ANN based user model and the real usage of the resource. This graphs show how the system is able to obtain quite appropriate models of general resource use by a given user in real time, even in cases where the behavior of the user seems to change quite randomly. In fact, when a pattern of behavior changes brusquely, the system lags a little behind in its predictions, but soon it captures user behavior.

In table 1 we display a set of results for different number of registers in the STM and different number of evolution generations in between interaction with the world.

Several aspects can be mentioned. The first one is the improvement achieved with the increase in STM size. Only in one case (second user) the average is worse for a STM size of 50 than for a size of 20, although the variance is still better. However, an increase in short term memory size implies an increase in the computational cost. Consequently equilibrium between these two conflicting aims is desired. In the example presented here, this equilibrium point could be established at 20 as for this value, the results are almost as good as for an STM length of 50 in every case. This fact can be appreciated in Fig. 3 where these two cases are compared for user 3 and the differences are quite slight.

Table 1. Some results for three of the users and different STM sizes and number of generations of evolution per interaction with the World (STM size and generations in the heading of each column). The data in each cell is the sum of the error, its average, variance and standard deviation. Shaded cells are best results for each user

User	10, 4	10, 10	10, 50	20, 4	20, 10	20, 50	50, 4	50, 10	50, 50
1	53.989 0.270 0.141 0.375	56.379 0.282 0.142 0.377	75.095 0.375 0.195 0.442	43.570 0.218 **0.103** 0.321	51.600 0.258 0.130 0.459	75.704 0.379 0.209 0.457	43.432 **0.217** 0.107 0.326	59.553 0.298 0.160 0.400	67.275 0.336 0.193 0.440
2	132.258 0.661 0.554 0.744	120.107 0.600 0.527 0.726	141.994 0.710 0.601 0.775	114.258 **0.571** 0.550 0.742	126.155 0.631 0.551 0.742	143.554 0.718 0.592 0.769	121.040 0.605 0.526 0.725	124.489 0.622 **0.518** 0.720	137.541 0.688 0.587 0.766
3	84.056 0.420 0.344 0.587	87.307 0.437 0.379 0.613	91.172 0.456 0.406 0.637	79.582 0.400 0.346 0.590	77.035 **0.386** 0.346 0.589	87.347 0.437 0.383 0.619	77.330 0.387 **0.308** 0.555	84.216 0.421 0.360 0.600	89.685 0.448 0.399 0.632

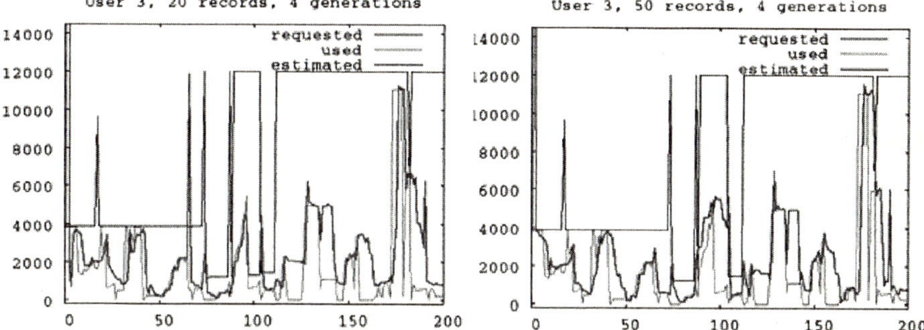

Fig. 3. Some results for obtained for the same user using a STM of 20 records and a STM of 50 records. The graphs display the amount of requested resource (in this case computing time), the estimation made by the ANN based user model and the real usage of the resource

On the other hand, a smaller the number of generations in between interactions with the world leads to better the results. A large number of generations of evolution in between interactions leads to a model that fits the particular contents of the STM in that instant of time but does not generalize well, resulting in larger errors for new estimations. Consequently, it seems that for computational reasons, the best bet is to choose very few generations of evolution in between interactions with the world, between 2 and 4 and a not very large STM of around 20 registers. Notwithstanding these comments, if one compares requested time and estimated time, it is clear that by using this methodology, the profiles obtained would clearly help to improve the planning of the processes.

6 Conclusions

An evolutionary agent based strategy through the incorporation of user profiling is proposed for the optimal assignment of energy strapped limited resources that must accommodate multiple users and require their satisfaction. This strategy provides a mechanism for taking into account the real behavior of the users in terms of resource usage as opposed to resource request and their associated satisfaction when planning the use of the systems.

The approach is based on an on line evolutionary mechanism that adapts sets of artificial neural networks that model the users through their interaction with the world. The adaptation of the models is carried out through a few steps of evolution every time the system interacts with the world, thus leading with time to good models of the real behavior and satisfaction of the users which provide a base for an optimization of resource usage in terms of these models and the policies established by the controlling agency. The results show that the models obtained really improve on the requests made by the users in a problem taking into account elements such as the day the request was made, the time it was made and the amount of resource required as well as its walltime.

Acknowledgements

This work was funded by the MCYT of Spain thorugh project VEM2003-20088-C04-01 and Xunta de Galicia through project PGIDIT03TIC16601PR.

References

1. Ibaraki, T. "Resource Allocation Problems". Algorithmic Approaches. MIT Press, 1988
2. Quijano, N., Gil, A. E. and Passino, K. M. "Experiments for Dynamic Resource Allocation, Scheduling, and Control". IEEE Control Systems Magazine; 63-79, 2005
3. Al agha, K. "Resource Management in Wireless Networks using Intelligent Agents". Int. J. Network Mgmt 2000; 10:29–39, 2000
4. Bhaskaran, S. Forster, B. Paramesh, N. Neal, T. "Remote sensing, GIS and experts systems for fire hazard and vulnerability modelling and dynamic resource allocation". IEEE Geoscience and Remote Sensing Symposium, IGARSS '01. 795-797 vol.2, 2001
5. Bellas, F., Duro, R. J. "Multilevel Darwinist Brain in Robots: Initial Implementation", ICINCO2004 Proceedings Book (vol. 2), pp 25-32, 2004.
6. Yao, X., Liu, Y., Darwen, P., "How to make best use of evolutionary learning". Complex Systems: From Local Interactions to Global Phenomena, pp 229-242, 1996.

Blind Signal Separation Through Cooperating ANNs

Francisco Bellas, Richard J. Duro, and Fernando López-Peña

Grupo Integrado de Ingeniería
Universidade da Coruña
{fran,richard,flop}@udc.es

Abstract. This paper is devoted to proposing and testing a strategy for decomposing compound signals obtained in remote sensing applications through the automatic generation of cooperating ANNs that model it. Each ANN will specialize in one of the primitives that make up the whole. The evolutionary based algorithm that is proposed for this purpose implies that the combination of networks takes place at a phenotypic operational level, this is, the architecture of the networks is not the issue, but rather function they implement. This way, a population of networks that are automatically classified into different species depending on the performance of their phenotype, and individuals of each species cooperate forming a group to obtain a complex output, in this case the signal that is required. The magnitude that reflects the difference between ANNs is their affinity vector, which must be automatically created and modified depending on the actuation of the phenotype of each individual. The main objective of this approach is to model complex functions such as multidimensional signals, which are typical of remote sensing application, providing a decomposition of them into primitive functions.

1 Introduction

Blind Signal Separation (BSS), that is, separating independent sources from mixed observed data is a fundamental problem that is difficult and challenging. It is an intrinsic problem present in most measurement systems and it becomes even more pronounced in remote sensing as there is a lot less control on the mixtures present in the signals acquired. In many practical situations, observations are usually modelled as linear mixtures of a number of source signals, ie. a linear multi-input multi-output system. However, reality is not usually that simple and more ambitious approaches are necessary.

Applications of BSS include fields such as telecommunication systems, sonar and radar systems, acoustics, and any other where multiple signal sources may be present. In the case of its application to remote sensing the main objective is usually to discriminate mixtures in the subpixel range [1][2]. Most authors have resorted to linear models and techniques such as Independent Component Analysis (ICA) whether in the time or in the frequency domain [3][4][5]. Others have considered neural based techniques with some success, especially in linear cases. These methods include the Hebbian learning algorithm [6] or robust adaptive algorithms [7]. In fact, the approaches even seek non linear discriminations, as is the case of nonlinear principal component analysis [8], or Recurrent Neural Networks as employed by Amari et al [9].

In this paper we propose an approach whose emphasis lies on the simultaneous modelling of the underlying signals that make up the sensed whole. A very powerful

R. Khosla et al. (Eds.): KES 2005, LNAI 3681, pp. 847–853, 2005.

technique for the instantiation of the models when the learning process consists in modelling from input-output pairs is the use of artificial neural networks (ANNs). In the case of evolutionary learning, the models adapt to the task through evolution. Thus, the modularity of solutions can be obtained creating complex ANNs as a composition of simpler ones. In this line, a very common approach in the literature [10], [11], [12] is to apply an evolutionary algorithm over two different populations, one made up of simple neuronal units and the other of patterns that indicate how those neural units must be combined. These approaches have been successful, for example, in classification problems.

A different approach is proposed here, where the basic structures participating in the process are complete networks evolving in the same population and they are combined at the phenotypic level into groups that collaborate in the solution of the problem. The work of Xin Yao and Paul Darwen [13] must be mentioned here. They propose an evolutionary learning approach to design modular systems automatically. To do this, they use the concept of speciation introduced by Goldberg [14] through a fitness sharing technique. The main novelty of this work lies in the fact that within each species they use coevolutionary techniques where the concept of just one individual solution does not exist. The authors apply this technique to the prisioner's dilemma and the algorithm works successfully.

This kind of modular approaches imply the existence of collaboration between individuals of the population to achieve a successful result. This is called in the literature symbiotic evolution and is a background problem in multiobjective optimization [15]. In this field, the development of fitness sharing techniques, where the fitness of a given individual is scaled through some similitude measure with respect to the other individuals, is common. These approaches start from the knowledge of the Pareto front to be obtained, this is, they start from the knowledge of the desired objectives and search for individuals specialized in these objectives and that collaborate towards their achievement.

The problem we try to solve in this work is quite different; because we look for an automatic decomposition of the model into simple and basic networks (primitives) that combined provide the desired solution. No knowledge about how to do it (we don't know the best decomposition in primitives) is assumed. Each one of these simple networks is functionally different and, consequently, they belong to different species that evolve together.

Another algorithm that is based on coevolution and species formation is COVNET [16], where authors use a subnetwork population (nodules) that is divided into species that evolve independently and another network population that combine these nodules. The main difference with our approach lies at the species formation level. The number of species in COVNET is prefixed, whereas in our algorithm it arises automatically from the evolution itself.

In the next section we will introduce the main concepts used in this algorithm such as groups and affinity. The remaining sections provide an overview of the operation.

2   Main Concepts of the Algorithm

The main objective of this algorithm is very simple: to develop model a complex function using ANNs. Because of this complexity, we assume it will be simpler to find the solution using more than one ANN. So our approach is based on providing a search algorithm with the capability of obtaining the solution by aggregation of simple ANNs.

Thus, the first important feature of our algorithm is that the solution is provided by a combination of ANNs (*a group*) that combine their outputs to provide a complete solution to the problem.

To conform the groups, we have developed a selection criteria based on a magnitude called *affinity* that classifies the ANNs depending on their phenotype. The main idea is that initially all the networks have a random affinity and through a self organization process these affinities become different automatically. Then, the groups that provide the solution to the problem are made up by combining ANNs with high affinity to each other.

To improve the networks that made up the groups, we propose the use of an evolutionary algorithm as search technique because it is based on a population (a set) of solutions and it is an established technique for adjusting the parameters of ANNs. Furthermore, evolutionary algorithms are very adequate for this approach because of their capabilities to create specialized individuals (*species*) in the population. This way, once some species appear in the population, we can make up the groups that conform the solutions using individuals from these species.

3   Details of the Algorithm

As mentioned in the previous section, individuals will be evaluated (fitness calculation) as a part of a group and not individually. Evolution, on the other hand, takes place over the individuals as usual and the groups are formed in the fitness calculation stage of the algorithm.

The operation is simple and can be summarized in the following 6 steps:

1. Initial creation of a random population of ANNs, as usual in general evolutionary algorithms.
2. Initial creation of a random affinity value (represented by a vector in general) associated to each individual.
3. Fitness calculation of all the ANNs in population. To do this, we apply 5 basic steps:
   a. Each individual in the population creates its own group by choosing other individuals from a window of the population using its and their affinity values.
   b. Each group calculates a fitness value, resulting from the common application of its individuals to the modelling problem.
   c. Each individual creates a new group and the fitness in these new groups is calculated. The best groups remain.
   d. The fitness value obtained in step c) permits adjusting the affinity of the individuals depending on the increase or decrease of fitness.
   e. Steps c) and d) are repeated a given number of iterations. At the end, we assign each individual the fitness of its own group.
4. Selection over the population of ANNs (like in an ordinary evolutionary algorithm) including the affinity value of each individual as a part of the selection criteria favoring reproduction of individuals with similar functional characteristics (same species).

5. Crossover and mutation as usual but including the affinity vector that is transmitted from parents to offspring.
6. Steps 3, 4 and 5 are repeated a given number of generations.

As we can see, this is the operation schema of a simple evolutionary algorithm where the fitness calculation is more complex and where the reproduction is directed towards the formation of species by mixing operationally similar individuals.

Once a group is formed, its fitness is calculated in general as the fitness provided by a set of outputs that are the result of some combination of the individual outputs of its members. In the simplest case the combination is carried out through the addition of these outputs that must be clearly marked to discriminate which ANN affects which output. The group must provide values for all the outputs of the problem, otherwise it is penalized.

After calculating the fitness of all the groups, we carry out a self organization stage (steps from a) to e) in the algorithm), which allows each individual to create more affine groups. On each step of this stage, the individuals form new groups and their fitness is calculated. If the fitness of the new group is better than that of the old one, the group is maintained; otherwise the older group is preserved. After a set number of recombination steps, each individual has formed its most affine group using the technique explained in the next section.

Affinity is used in our algorithm as a label to classify the ANNs depending on the function they perform. It is represented by a numerical vector with fixed upper and lower limits, in which the dimension determines the complexity of the groups that may be formed.

Each element in the affinity vector is represented by a real number. When an individual forms a group it searches for other individuals in the population with *complementary* affinity vectors. Initially, the affinity vectors are random as no knowledge of the characteristics of the individuals and/or their shortcomings is available. Every time a group is evaluated (fitness calculation), if the group improves the fitness of the previous group in which the individual participated, the distance of the affinity vector of the individual (in Euclidean terms) with respect to the affinity vectors of the rest of the individuals in the group is increased. In addition, from this instant, this will be the new group for the individual. On the other hand, if the fitness of the group is worse than that of the previous group, it is the distance with respect to the affinity of the members of the previous group that is increased. This method favours the formation of species, that is, of clusters of individuals with similar affinity vectors.

After each self organization stage, all the individuals have automatically adjusted their affinity vector depending on the function they carry out in the group.

4   Operation Example

Just to display the results the mechanism can produce and the way it works, we consider here a simple problem consisting of an objective function that was constructed by adding 4 different functions (sinusoidal and Gaussian 3D) limited to a range from [-1.5, 1.5] in x and y. In this experiment, the affinity vector considered is two dimensional and the values of each dimension can vary between [-1, 1]. Thus, it seems that an optimal solution for this problem would be the generation of four species, each one specialized in modelling each one of the known functions or any other combination of functions that lead to the perfect modelling of the global function.

The networks that will be evolved in this case are standard multilayer perceptrons which have two inputs (the x and y coordinates) and one output (the z coordinate). The objective is obviously to obtain the network or combination of networks that best approximate the objective function.

The results of modelling obtained with the best group are presented in Fig. 1, where we have represented the prediction (points) and a sampling of the objective function (surface). The results are very satisfactory taking into account the complexity of the expected signal. We are going to take a closer look at what have happened with the affinity values for the individuals and what types of species arose.

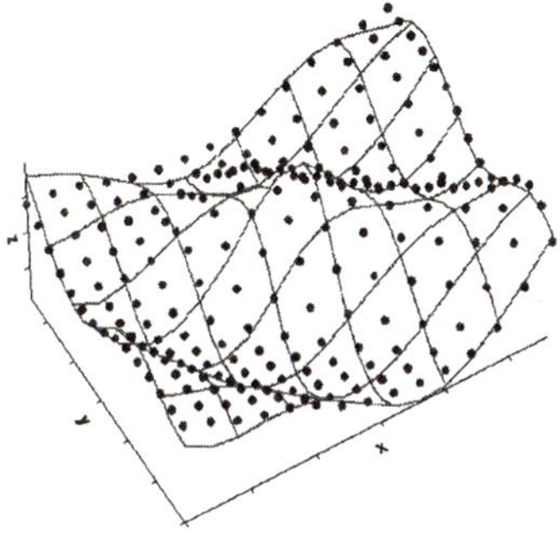

Fig. 1. Modelling result (points) of a 3D function (surface) made up by the addition of 4 basic functions

Fig. 2 displays the evolution of the affinity vectors for the whole population (1000 individuals in this case). In the left graph we have represented the initial random affinities distributed from -1 to 1 in both axes. In the middle graph we show the affinity vectors in generation 2000. As we can see, the values are still not clustered, and some individuals do not belong clearly to a species, although four affinity areas are starting to be delimited. Finally, in the right graph we show the affinity vectors in generation 3600 (final) where 4 clear species have been formed with affinity vectors (-1, 1), (1, -1), (0.5, 0.5) and (-0.5, -0.5).

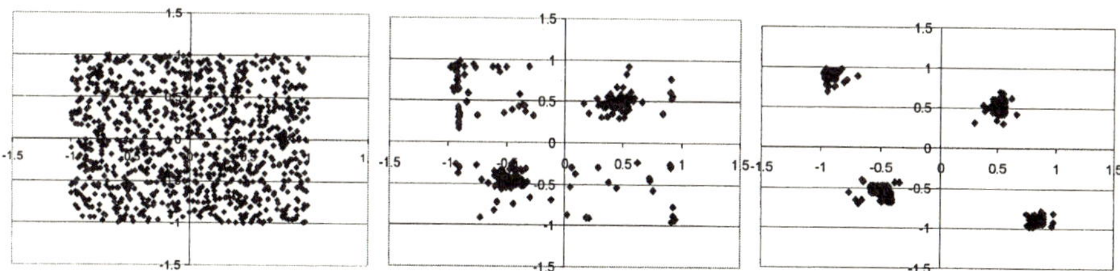

Fig. 2. Evolution of the affinity vectors for the whole population. Left graph represents the initial distribution, middle graph the distribution after 2000 generations and right graph the final distribution

The final group taken as solution of the modelling problem (that provides prediction shown in Fig. 1) is made up of 4 individuals, one from each species. In Fig. 3 we present the functions provided by each one of these individuals in the final group. What is important here is that these primitives have been obtained automatically and combined provide a very successful prediction of the original complex function.

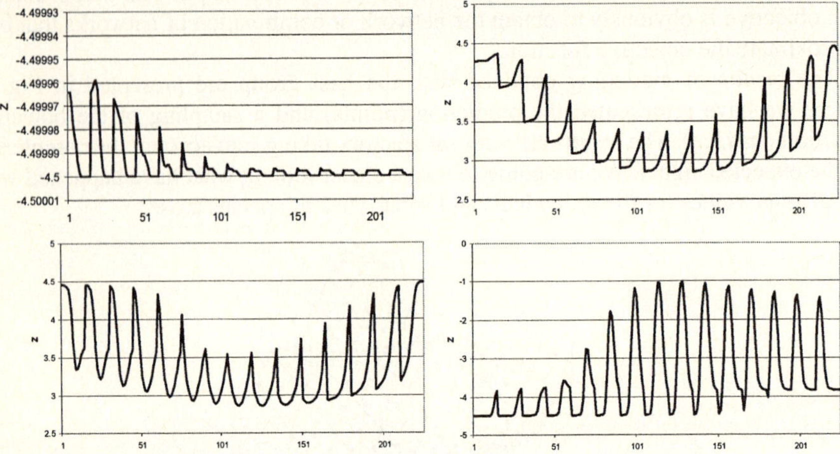

Fig. 3. Functions provided by each individual in the best group (phenotype corresponding to each specie) that combined provide the output shown in Fig 1

5 Conclusions

This work deals with an affinity based strategy for obtaining groups of cooperating artificial neural networks as compound models of complex signals obtained in measuring processes, in particular in remote sensing applications. This strategy may be applied in general to any type decomposition process where the objective is to obtain a group of solution points which, through their combination and cooperation in a phenotypic level produce a modular or decomposable solution to the problem so that parts of these solutions may be later reused in other problems. The main idea behind the phenotypic affinity based strategy is that during the evaluation phase of the evolutionary algorithm the individuals can try out how fit they are through collaboration with other individuals by forming groups. The results of these trials result in modifications of each individual's affinity vector. This process leads to the formation of species of individuals specialized in different aspects or primitives of the global signal, thus allowing for the modularity we desire. The affinity of each individual also participates in the evolution stages as it is assumed that there is a certain correspondence between families or species and affinities for other species in terms of collaboration towards the common task. The results applying this strategy have been very fruitful. Here we have presented a quite complex case with the arising of four species into the population.

Acknowledgements

This work was funded by the MCYT of Spain thorugh project VEM2003-20088-C04-01 and Xunta de Galicia through project PGIDIT03TIC16601PR.

References

1. C. I. Chang, "Target Signature-Constrained Mixed Pixel Classification for Hyperspectral Imagery" IEEE Trans On Geoscience and remote sensing, Vol 40-5, pp 1065-1081, 2002
2. L. Parra, K. Mueller, C. Spence, A. Ziehe, and P. Sajda, "Unmixing Hyperspectral Data," Advances in Neural Information Processing Systems 12, MIT Press, pp 942-948, 2000.
3. T. Nishikawa, H. Saruwatari, and K. Shikano, "Comparison of time-domain ICA, frequency-domain ICA and multistage ICA for blind source separation," in Proc. Eur. Signal Processing Conf. (EUSIPCO), vol. 2, Sep. 2002, pp. 15–18.
4. Marc Castella, Jean-Christophe Pesquet, Athina P. Petropulu "A Family of Frequency- and Time-Domain Contrasts for Blind Separation of Convolutive Mixtures of Temporally Dependent Signals" IEEE Trans. on Signal Processing, vol. 53, n. 1, pp. 107-120, 2005
5. Chia-Yun Kuan, Glenn Healey "Using independent component analysis for material estimation in hyperspectral images" JOSA A, Volume 21, Issue 6, 1026-1034 June 2004
6. C. Jutten and J. Herault, "Blind separation of sources, Part I: An adaptive algorithm based on neurominic architecture", Signal Processing, Vol. 24, pp.1-10, 1991.
7. A. Rummert Cichocki, R. Unbehauen, "Robust Neural Networks with on-line learning for blind identification and blind separation of sources", IEEE Trans Circuits and Systems I: Fundamentals theory and Applications, Vol. 43, No.11, pp. 894-906, 1996.
8. E. Oja and J. Karhunen, "Signal separation by non-linear Hebbian learning, In M. Palaniswami, et al (editors), Computational Intelligence – A Dynamic System Perspective, IEEE Press, New York, NY, pp.83-97, 1995.
9. S. Amari, A. Cichochi and H. H Yang, Recurrent Neural Networks For Blind Separation of Sources", 1995 International Symposium on Nonlinear Theory and its Applications, NOLTA'95, Las Vegas, December, pp.37-42, 1995.
10. Smalz, R., Conrad, M., "Combining evolution with credit apportionment: A new learning algorithm for neural nets", Neural Networks, vol 7, num 2, 1994.
11. Opitz, D.W., Shavlik, J.W., "Actively searching for an effective neural network ensemble", Connection Sci, vol 8, num 3, 1996.
12. Moriarty, D.E., Miikkulainen, R., "Forming neural networks through efficient and adaptive coevolution", Evolutionary Computation, vol 4, num 5, 1998.
13. Darwen, P., Yao, X., "Speciation as automatic categorical modularization", IEEE Transactions on Evolutionary Computation, 1(2), 1997.
14. Goldberg, D., Richardson, J., "Genetic algorithms with sharing for multimodal function optimization", Proceedings of 2nd International Conf. on Genetic Algorithms, 1987.
15. Coello, C., Van Veldhuizen, D., Lamont, G., "Evolutionary algorithms for Solving Multi-Objective Problems", Kluwer Academic Publishers, 2002.
16. García-Pedrajas, N., Hervás-Martínez, C., Muñoz-Pérez, J., "COVNET: A Cooperative Coevolutionary Model for Evolving Artificial Neural Networks", IEEE Transactions on Neural Networks, vol 14, num 3, 2003.
17. F. Bellas, R.J. Duro, "Multimodule ANNs Through evolution using an Affinity Based Operator", Proceedings book JCIS 2003, pp 311-314

Tropical Cyclone Eye Fix Using
Genetic Algorithm with Temporal Information

Ka Yan Wong and Chi Lap Yip

Dept. of Computer Science, The University of Hong Kong, Pokfulam, Hong Kong
{kywong,clyip}@cs.hku.hk

Abstract. Tropical cyclones (TCs) are weather systems with vast destructive power. To give early TC warnings, accurate location of their circulation centers, or "eyes", is required. The pattern matching solution to this TC eye fix problem works by analyzing individual remote sensing images as if they are independent. Temporal information is ignored in such approach and results from individual image are only smoothed afterwards. In this paper, a TC eye fix method using genetic algorithm (GA) with temporal information is proposed. This method gives an improvement of 35% in speed and 20% in accuracy compared to the one without the use of temporal information. Our proposed method can process 6 minutes of radar data in about 10 seconds on a notebook computer, meeting practical real time constraints. It gives an average accuracy within 0.107 to 0.156 degrees in latitude/longitude on the Mercator projected map, well within the relative error of about 0.3 degrees given by different TC warning centers.

1   Introduction

Tropical cyclones (TCs) often cause significant damage and loss of lives in affected areas. Direct economic losses often amount to billions of dollars. The distribution of TCs over the globe from 1986 to 2003 is shown in Fig. 1(a)[1]. To reduce losses, weather warning centers should issue warnings early based on its TC forecast track. This requires the accurate location of the circulation center, or the "eye", of the TC. This is normally done by the analysis of remote sensing data from weather radars or satellites.

Weather radars work by sending out microwave signals to the atmosphere. The reflected signals are then preprocessed to extract the relevant slices suitable for analysis. The radar reflectivity data at 3 km Constant Altitude Plan Position Indicator (CAPPI) (Fig. 1(b)) and the corresponding Doppler velocity data (Fig. 1(c)) are often used for TC eye fix. The former shows the strength of echoes reflected from rain, snow, ice or hail aloft, and the latter shows their radial velocities with respect to the radar.

Since a TC is a weather system with spiraling rainbands whose circulation center is the eye, the line with zero radial velocity with respect to the radar,

---

[1]   As some TC stride more than one basin, the total percentage is greater than 100%.

R. Khosla et al. (Eds.): KES 2005, LNAI 3681, pp. 854–860, 2005.
© Springer-Verlag Berlin Heidelberg 2005

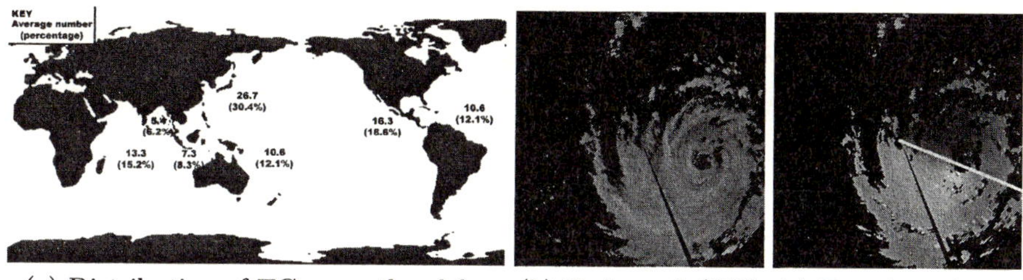

(a) Distribution of TCs over the globe   (b) Radar ref. (RR)   (c) Doppler velocity

Fig. 1. Distribution of TCs over the globe, radar images of Typhoon Yutu.

called zero isodop, is where the TC center should lie. The radars used by the Hong Kong Observatory (HKO) [1] take six minutes to update both types of data. They cover a radius of up to 512 km, with spatial resolution of a few hundred meters. Since TCs in the proximity of hundreds of kilometers from a city fall into the range of radars and pose the greatest threat, we focus on the eye fix process using radar images. To produce timely, accurate and objective TC eye fix results, a number of automated TC eye fix techniques have been proposed in our previous works [2][3]. In this paper, we aim at improving the efficiency and effectiveness of our previous work [3] with the use of temporal information.

2   Related Work

TC eye fix is often done manually in practice. Forecasters estimates the TC center by tracing the movement of spiral rainbands or by overlaying spiral templates on remote sensing images for the best match [4]. However, these techniques are not completely objective. Automated TC eye fix methods, such as pattern matching and wind field analysis, in contrast, employ objective measures.

In wind field analysis, the TC center is found by analyzing the motion field [5] built using the TREC algorithm [6]. Some researchers use mathematical models of vector fields to extract and characterize critical points such as swirl or vortices for TC identification [7]. For pattern matching, the TC eye is fixed by finding the best match of a predefined TC model. Examples include locating the extreme Doppler velocity value using an axisymmetric hurricane vortex flow model [8] and spiral template matching using satellite images [9].

Recently, we have designed an automated eye fix method that makes use of radar data [2]. The spiral rainband of a TC is modeled by the equation $r = ae^{\theta \cot \alpha}$, where a and α are found by image processing and transformation operations. Templates generated are used for matching, and the results are smoothed using Kalman filter to reduce sensitivies to errors and noise.

These eye fix methods require computationally expensive operations such as wind field construction, parameter estimation using search algorithms, and extensive block or object matching. With the large volume and rate of data, high performance computing platforms and clusters are often needed to generate timely results.

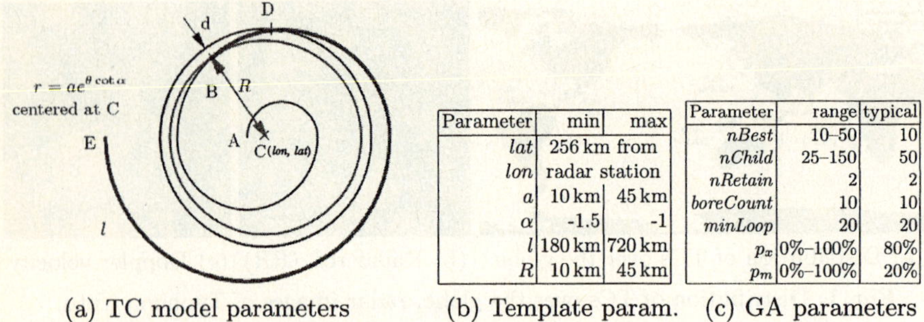

Parameter	min	max
lat	256 km from	
lon	radar station	
a	10 km	45 km
α	-1.5	-1
l	180 km	720 km
R	10 km	45 km

Parameter	range	typical
nBest	10–50	10
nChild	25–150	50
nRetain	2	2
boreCount	10	10
minLoop	20	20
p_x	0%–100%	80%
p_m	0%–100%	20%

(a) TC model parameters   (b) Template param.   (c) GA parameters

Fig. 2. TC model, template and genetic algorithm (GA) parameters.

To speed up the search, genetic algorithm (GA) is applied in our previous work [3][10]. GA makes use of Darwin's idea of "survival for the fittest". The best genes (set of model parameters) that maximizes a fitness function (a quality measure of match) are iteratively generated and selected. The TC model is further modified to a six-parameter one to make the algorithm more adaptive to TCs at various stages. An error of 0.139 to 0.257 degrees was obtained with a ten-fold improvement in speed compared to the original algorithm in [2].

Though effective and efficient, temporal information, which provides information of the development of a TC, has not been exploited in the algorithm. In this paper, we aim at improving the efficiency and effectiveness of the GA-based TC eye fix algorithm using temporal information. A simple template model of TC is defined for matching. Comparison with GA-based TC eye fix method without the use of temporal information is done to evaluate its performance.

In Section 3, a TC template model is introduced. A TC eye fix algorithm that makes use of the model is explained in Section 4. The algorithm is then evaluated in Section 5, followed by a summary in Section 6.

3   A Model of TC

A time-honored technique of manual TC eye fix is to overlay spiral templates on a printout of remote sensing image for the best match [4]. We automate the process by choosing a simple model of TC and doing the match using genetic algorithm. The model is identical to the one in [3] and is briefly explained here.

A TC has a center (point C in Fig. 2(a)) at longitude *lon* and latitude *lat* where a spiral rainband (curve EDBA) with the polar equation $r = ae^{\theta \cot \alpha}$ whirls into. For TCs in the northern hemisphere, $\cot \alpha$ is negative, giving the spiral shape as shown in the figure. TCs in the southern hemisphere have positive $\cot \alpha$ and rainbands swirl in clockwisely. A TC has an eyewall (inner circle of Fig. 2(a)) with radius R (distance BC), which is the boundary between rainy and no-rain areas. Places with a distance slightly larger than R from the center (the outer circle with radius $R + d$) would be rainy. The spiral rainband outside the eyewall (curve BDE) has a length of l, related to the distance of influence of the TC. With this model, six parameters lat, lon, a, α, l and R define the template.

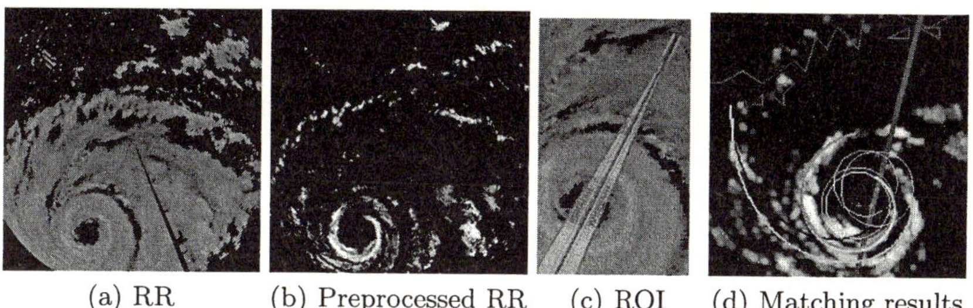

(a) RR   (b) Preprocessed RR   (c) ROI   (d) Matching results

Fig. 3. Input radar image, preprocessed image, ROI and matching results with candidates of GA.

4   The TC Eye Fix Algorithm

A radar reflectivity image is first preprocessed (Fig. 3(a) and 3(b)) before matching is done. It is thresholded with histogram cumulative frequency of 82–86% (corresponding to a reflectivity threshold of about 34 dBZ) to make spiral rainbands stand out. Equalization is then applied to enhance the contrast, while Gaussian smoothing and max filtering are applied to reduce noise in image.

The quality of match is calculated using a fitness function, which is a modified correlation function designed so that high reflectivity areas match the spiral segment BDE and the outer circle of the template, and low reflectivity areas match the inner circle. A more detailed explanation can be found in [10]. Genetic algorithm is used to find the set of parameters for the best match. A number of parameters for GA is defined (Table in Fig. 2(c)).

Initially, $nChild$ template candidates are generated randomly in a Region Of Interest (ROI). To mandate the use of temporal information, at least $nBest$ of them are the fittest candidates of the last iteration of previous frame, if available. In our experiments, the ROI is determined by finding the zero isodop from Doppler velocity data using log-polar transformation [11]. It is an area within $\phi = \pi/45$ radian or $w = 3\,\mathrm{km}$ from the straight zero isodop line (the shaded area in Fig. 3(c)), or a user-defined rectangular area. Domain-specific information is used to limit the values of the six template parameters. lat and lon are limited by the area of coverage of radar, and the limits of the other four parameters are determined by values of typical TCs. Table in Fig. 2(b) summarizes these limits.

After the initial set of candidate templates is generated, the algorithm enters an iterative phase. Here, each template is matched against the preprocessed image for a fitness value, and the fittest $nBest$ candidates are retained as parents. $nChild$ children templates are then generated using the parents with crossover and mutation probabilities of p_x and p_m respectively, with at least $nRetain$ of them verbatim copies of parents. In our experiments, crossover alters at most five template parameters after a randomly selected crossover point. Mutation only alters one of a, α, l, R, or both lat and lon together for better escape from local maxima. The iterative phase ends when the best score does not improve for $boreCount$ iterations after the algorithm runs for at least $minLoop$ iterations, or when the score is over a user-defined threshold $minScore$. This threshold allows the algorithm to give an answer once the fitness score satisfies the user.

The (lon, lat) location found (Fig. 3(d)) is then output and fed to a Kalman filter to smooth out noises caused by bad matches and image defects. Historical TC data, such as average TC speed, are used to determine Kalman filter parameters such as system noise variance. Latitude and longitude values are separately smoothed, with the assumption that they are statistically independent. The Kalman filtered TC center location gives the final output.

5   Evaluation

To evaluate the performance, comparisons with the GA-based TC eye fix algorithm without the use of temporal information [3] is done. A Java-based system prototype is built. A sequence of radar reflectivity images with a range of 256 km captured every 6 minutes (240 images from Typhoon Yutu, 2001-07-24 20:00 to 2001-07-25 19:54), along with their Doppler velocity counterparts, was used for testing. The data covers different stages of a TC life cycle from strengthening to dissipation, allows testing the effectiveness of temporal information.

The efficiency is evaluated by comparing the average number of images the system can process in a minute on a notebook computer with 1.4 GHz Pentium M processor and 256 MB RAM running Windows XP. The effectiveness is evaluated by finding the average Euclidean distance between the algorithm proposed center and the corresponding interpolated HKO best track location. Best tracks are the hourly TC locations determined after the event by a TC warning center using all available data. To allow early return of results, a relatively high *minScore* of 300 is chosen. The table in Fig. 2(c) summarizes the GA parameters used.

5.1   Efficiency

Using a typical parameter set (Table in Fig. 2(c)), the system processes around 6 images a minute on average, a 20% speed improvement with respect to our previous work [3] that did not make use of temporal information. Fig. 4(a) shows results for different p_m and p_x value combinations. Note that the entries for $p_m = 0\%$ is not applicable because fewer than *nChild* distinct children would be generated under this condition. The average speed of the runs are 9.39 images per minute, a 35% speed improvement from 6.95 images per minute in [3].

Fig. 5(a) shows a comparison of the number of iterations for each frame. In general, with the use of temporal information, fewer number of iterations are needed to complete an eye fix exercise. The effect of temporal information is more significant at the late stage of Yutu. Starting from frame 147, the eye can be fixed with only one iteration (around 5 seconds per frame) in most of the frames. An inspection of the radar images shows that Yutu has a very well defined center and spiral structure at that stage. Good matches that give a fitness score higher than our *minScore* threshold happens often, which allows the algorithm to go through fewer iterations per image. These good matches were used as the initial candidate templates of next frame. Hence, there is a higher probability of having a higher fitness score at the first iteration.

Table (a): Number of images per minute

	Crossover probability p_x					
Mutation probability p_m	0%	20%	40%	60%	80%	100%
0%	not applicable, see Sect. 5.1					
20%	5.853	6.908	6.385	6.584	5.853	7.525
40%	7.393	9.364	9.161	8.966	8.104	10.278
60%	8.966	10.033	9.800	9.577	9.577	11.389
80%	9.364	10.805	11.089	10.033	10.535	12.040
100%	10.278	11.389	11.089	10.033	10.805	12.394

(a) Number of images per minute

Table (b): Average error

	Crossover probability p_x					
Mutation probability p_m	0%	20%	40%	60%	80%	100%
0%	not applicable, see Sect. 5.1					
20%	0.112	0.129	0.151	0.135	0.127	0.143
40%	0.107	0.136	0.121	0.107	0.152	0.156
60%	0.152	0.132	0.152	0.152	0.128	0.142
80%	0.125	0.152	0.137	0.139	0.110	0.126
100%	0.123	0.102	0.128	0.150	0.135	0.128

(b) Average error

Fig. 4. Sensitivity study of p_x and p_m on eye fix of Typhoon Yutu.

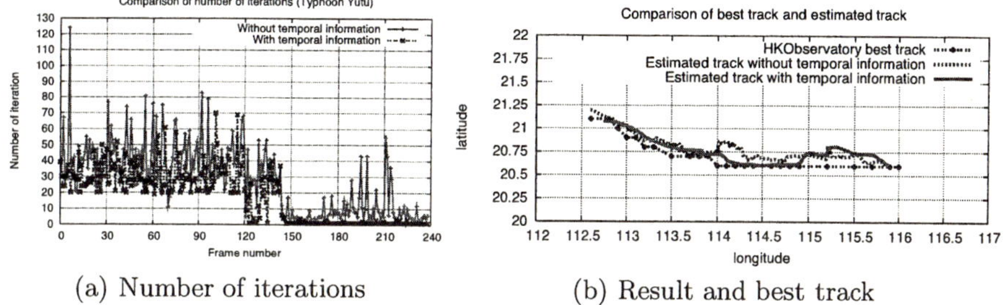

(a) Number of iterations

(b) Result and best track

Fig. 5. Comparison of number of iterations and eye fix results.

5.2 Effectiveness

Using typical parameter values (Table in Fig. 2(c)), our algorithm gives an error of about 0.127 degrees in latitude/longitude on the Mercator projected map. Different p_m and p_x value combinations are tested and shown in Fig. 4(b). The average error values in the table ranges from 0.107 to 0.156 degrees, with an average of 0.133 degrees. The results are much better than that in our previous work [3], which showed an average error of 0.153 degrees using typical parameter values and an accuracy within 0.139 to 0.157 degrees with the use of different p_m and p_x value combinations. The results are also well within the relative error of about 0.3 degrees given by different TC warning centers [12].

Fig. 5(b) shows the comparison between estimated tracks and best track. It can be seen that the estimated track found by the algorithm with temporal information considered is in general closer to the best track. The structure of a TC would not change too much between two frames. The use of the fittest candidates of previous frame in generating the initial set of candidate templates allows information, such as the change of the size of TC eye and the length of spiral rainbands, to be captured. These information provide a good initial set of template parameters for the GA iterations. Better children can be generated with a better initial set of parents, thus increasing the accuracy of the algorithm.

6 Summary

We have introduced a tropical cyclone eye fix method that makes use of genetic algorithm with temporal information. A simple TC template with six parameters is defined, and GA is used to find the template that best fits the image. The fittest templates found in the previous frame are used as part of the initial candidates set for the current frame for eye fix. There is a 35% speed improvement with respect to our previous work [3] that does not use temporal information. The number of iterations needed for fixing the TC eye is fewer in general, and only one iteration is needed when the TC has a well defined structure. The accuracy, in terms of average error on the Mercator projected map, has also been improved from 0.153 to about 0.127 degrees, with an improvement of about 20%.

Acknowledgments

The authors are thankful to the Hong Kong Observatory for the provision of data and expert advices from Dr. LI Ping Wah.

References

1. Hong Kong Observatory. Web page (2005) http://www.hko.gov.hk/.
2. Wong, K.Y., Yip, C.L., Li, P.W., Tsang, W.W.: Automatic template matching method for tropical cyclone eye fix. In: Proceedings of the 17th International Conference on Pattern Recognition. Volume 3., Cambridge, UK (2004) 650–653
3. Yip, C.L., Wong, K.Y.: Efficient and effective tropical cyclone eye fix using genetic algorithms. In: Proc. 8th KES, New Zealand, Springer-Verlag (2004) 654–660
4. Sivaramakrishnan, M.V., Selvam, M.: On the use of the spiral overlay technique for estimating the center positions of tropical cyclones from satellite photographs taken over the Indian region. In: Proc. 12th Radar Meteor. Conf. (1966) 440–446
5. Lai, E.S.T.: TREC application in tropical cyclone observation. In: ESCAP/WMO Typhoon Committee Annual Review. (1998) 135–139
6. Rinehart, R.E.: Internal storm motions from a single non-Doppler weather radar. Technical Report TN-146+STR, NCAR (1979) 262 pp.
7. Corpetti, T., Mémin, E., Pérez, P.: Extraction of singular points from dense motion fields: An analytic approach. J. of Math. Imaging and Vision **19** (2003) 175–198
8. Wood, V.T.: A technique for detecting a tropical cyclone center using a Doppler radar. Journal of Atmospheric and Oceanic Technology **11** (1994) 1207–1216
9. Wimmers, A., Velden, C.: Satellite-based center-fixing of tropical cyclones: New automated approaches. In: Proceedings of the 26th Conference Hurricanes and Tropical Meteorology. (2004)
10. Yip, C.L., Wong, K.Y.: Tropical cyclone eye fix using genetic algorithm. Technical Report TR-2004-06, Dept. of Computer Science, The Univ. of Hong Kong (2004)
11. Young, D.: Straight lines and circles in the log-polar image. In: Proceedings of the 11th British Machine Vision Conference (BMVC-2000). (2000) 426–435
12. Lam, C.Y.: Operational tropical cyclone forecasting from the perspective of a small weather service. In: Proceedings of ICSU/WMO International Symposium on Tropical Cyclone Disasters, Beijing (1992) 530–541

Meteorological Phenomena Measurement System Using the Wireless Network

Kyung-Bae Chang, Il-Joo Shim, Seung-Woo Shin, and Gwi-Tae Park

ISRL, Korea University, 1, 5ga Anam-dong Sungbuk-Gu,
136-713 Seoul, South Korea
{lslove,ijshim,ezshini,gtpark}@korea.ac.kr
http://control.korea.ac.kr

Abstract. In this paper, we proposed a system of the weather disaster prevention in specific areas where the construction of the wire communication infrastructure is hard. This would be able to minimize the damage of men because the system has easiness of materialization with the existing wireless communication network, gets economic profits and can be set up in the back region such as a remote place, island so a person with a mobile terminal can receive a warning matter of the area he is in.

1 Introduction

As the nations influences, and the standards of living increases, industrial classification varies, along with the interest of meteorological demands. As the demand for meteorological information in special purposes increases, recognition and interest toward meteorological phenomenon is elevating. Influenced by these tendencies, the meteorological agency has promoted technology development, following the World Meteorological Organization and with close international cooperation, providing meteorological services for the public, and commercializing meteorological information for special and individual purposes.[1] However, as the nations 70% is covered with mountains, gathering of the meteorological information of particular regions (mountains, canyons, islands etc) by wired transmission systems is difficult and it costs a lot of money. This is one of the obstacles of the construction of an infra for the prevention of disasters caused by meteorological changes.[2] Nevertheless, the existing researches are mostly about Automatic Weather Systems(AWS), which observe meteorological factors. Meanwhile, research on the establishment of networks, which considers geological features, is insufficient.[3] Therefore, in order to solve the problems above, solutions for the following issues must be presented.

1. The existing meteorological data transmitting methods use MODEM or LAN, therefore set up space is limited. Also, selection of the meteorological observation sites becomes limited too.
2. Wide scope forecasting with the use of meteorological satellites are not efficient for the prevention of actual disasters.

This paper proposes a wireless, meteorological phenomena measurement system, which can establish infra with minimum equipment, under bad conditions like this country's geographical features. Recently, wireless communication network industries have been rapidly developing internationally. Especially, our country's wireless communication network establishment has been completed. As seen in the national IT

R. Khosla et al. (Eds.): KES 2005, LNAI 3681, pp. 861–865, 2005.
© Springer-Verlag Berlin Heidelberg 2005

infra establishment status, high speed internet users consists of 70% of the total households, which is 11 million 100 thousand houses in June 2003, which is the biggest ratio per population in the world. Wireless communication network users are up to 32 million 320 thousand people, making a supply rate of 68.3%.[4] Also, high speed wireless network users, that use more than CDMA 2000 1x, are steadily increasing, which tells us that the national wireless communication network is becoming more and more faster. By the use of these, infra establishment completed, existing wireless network equipments, a cheap and easy setting meteorological disaster prevention system can be developed. And by this, we propose the meteorological disaster prevention system based on the web, using wireless communication, which can minimize the enormous amounts of human casualties and property damage by sudden meteorological changes.

2 Proposed Approach

2.1 General System Structure

The existing researches are mostly about meteorological observation algorithms by the use of AWS[5], or about the increase of information transmission speed and efficiency by unification of the networks between Nodes.[6] However, this paper proposes a network integration method using wireless networks.

Fig. 1 shows the general structure of the system, which monitors and collects meteorological information observed from the site's AWS, by the use of the wireless communication network's short message service (SMS). The AWS is designed to be easily set up on specific regions where wired communication is difficult, and by this AWS, meteorological information is gathered. The gathered information is monitored from the disaster prevention agencies by wireless communication networks. Because it isn't necessary for the AWS to always be connected to the wireless network when gathering information, and the gathered information is transmitted on a regular basis, the method of transmitting the data by SMS, which is in real time and has economical advantages, is shown on the dotted line of A.. Meteorological factor observation information is transmitted through a Mobile Terminated Message Delivery service inside the AWS, and in the case of GSM(Group Special Mobile), a maximum of 160 letters(English) can be sent and 52 bytes of information can be transmitted. The meteorological observation data gathered by SMS is united at each disaster prevention agency and is used to prevent meteorological disasters. It is possible for the disaster expected regions to perform immediate arrangements to minimize the damage. Also, by automatically sending disaster warnings to the people carrying mobile terminals (cell phone, PDA) inside the disaster expected regions, human lives can be saved and property damage cam be minimized.

2.2 AWS

Because AWS is set up in mountains or special regions, it must have a self-providing source of energy, and for the efficient use of energy, its system must have low electric power consumption. Accordingly, we manufactured an AWS suitable for special regions, and tested the entire system

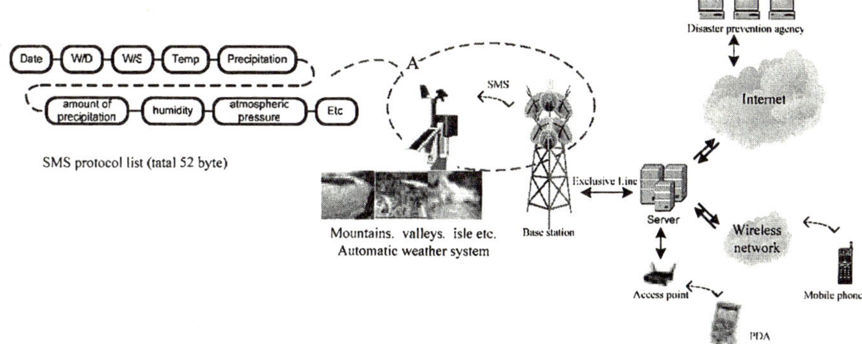

Fig. 1. Automatic Weather Observation system

Fig.2 shows the inner structure of AWS, and Fig.3 is an Embedded System, in which a 400MHz PAX255 ARM Core Chipset is used as the main system of AWS. As shown in Fig.2, this equipment handles, in high speed, each meteorological information data into quantized digital information, through the ADC.

Power is provided by itself from the Solar cells, of which the battery is charged in the daytime, so that independent management without separate equipment support is possible. This enables observation possible from special regions, like mountains, without outer power supply and additional communication line establishments.

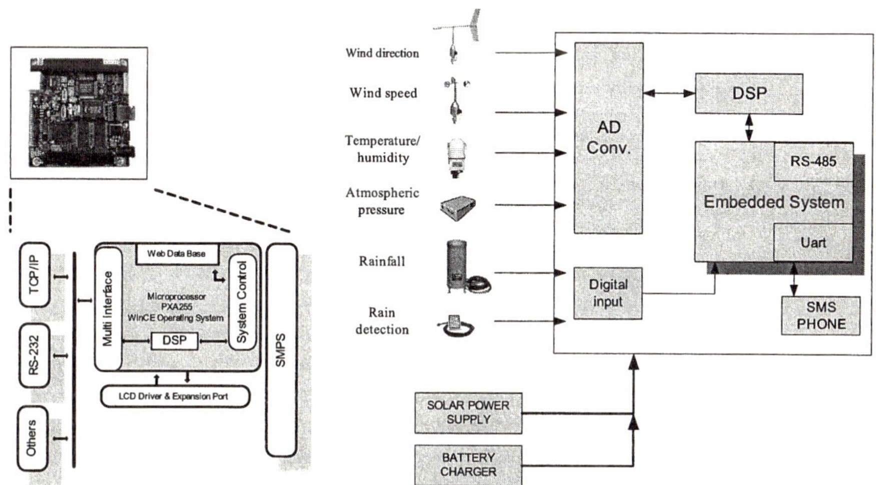

Fig. 2. AWS System Architecture of the AWS Data Logger

Fig. 3. Embedded System

2.3 Experimental Results

In order to evaluate whether if the reliability of the system proposed in Fig.1 is high enough for the use of transmitting meteorological observation information, meteorological information is measured by an individual developed AWS, and tests where performed to monitor the meteorological phenomena with the use of SKT wireless communication networks. If a message fails transmission, in order to secure reliability, the SMSC attempts to re-transmit the short message on time by an algorithm. The re-transmitting algorithm is performed side by side with the notification transmitted from MSC (Mobile Service Center) and HLR (Home Location Register), when a mobile terminal is used. Fig.4 shows the reliability test results of the proposed method, indicating a diagram of the receive delay time of the meteorological information transmission through wireless communication networks. Also, in order to evaluate the reliability of the proposed method, data transmission delay time is analyzed, and tests for the measurement of accuracy is in progress.

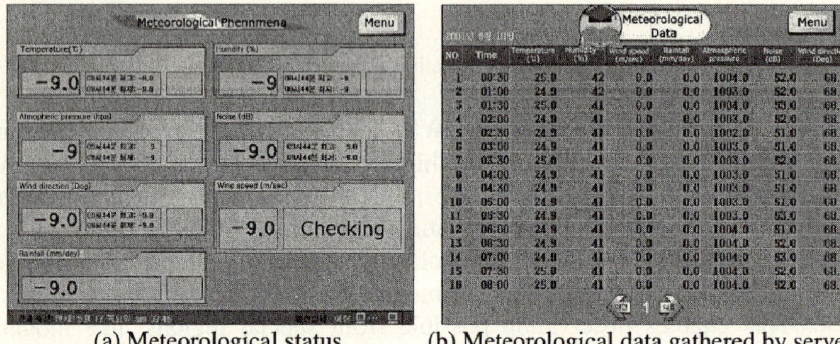

(a) Meteorological status   (b) Meteorological data gathered by server

Fig. 4. A monitoring system screen of the server

Fig.4 shows the actual server's monitoring screen, set up in disaster prevention agencies. (a) shows the meteorological status observed by AWS, and (b) shows the list of observed meteorological data at time intervals.

3 Conclusions

This paper proposed a meteorological disaster prevention system, which can be used at special regions where the establishment of wired communication infra is difficult. By using the existing wireless network, the system's establishment is easier and has economical advantages. Also, the system can be set up on inconvenient regions, like islands or mountains, and an individual carrying a mobile terminal device can transmit the disaster warning status of the region he/she is currently at, therefore minimizing human casualties and property damage, like preventing industrial road and railway loss from rain/snow storms. Thus, this system can be applied at many fields, and can become an important part in the research of minimizing damage from meteorological disasters. For Future work, research about synchronous IMT 2000 technology, like the prevention of disasters by applying photo and sound data of the site's meteorological situation by the use of Multimedia Messaging Service (MMS) through CDMA 2000 1x EVDO (Evolution Data Only), is being planned.

References

1. Korea Meteorological Administration, [Online]. Available: http://www.kma.go.kr

2. Nam-Ho Kyung, Young-Sung Kim, "A Study on the Atmospheric Dispersion of Pollutants in a Complex Terrain," Korea Institute of Geoscience and Mineral Resources(KIGAM), 1990
3. Theagenis J. Abatzoglou, Bruce C. Steakley, "Comparison of Maximum Likelihood Estimators of Mean Wind Velocity From Radar/Lidar Returns," IEEE Conference Record of the Thirtieth Asilomar Conference, vol. 1, pp. 156-160, Nov. 1996
4. Ministry of Information and Communication Republic of Korea, A policy data, "IT conditions of Korea and confronted assignment' [Online]. Available: http://www.mic.go.kr/information/index.jsp
5. Jonathan A.R.Rall, James Campbell, James B.Abshire, James D.Spinhirne, "Automatic Weather Station(AWS) Lidar," IEEE Geoscience and Remote Sensing Symposium vol. 7, pp. 3065-3067, July 2001
6. Arnaud Legrand and Martin Quinson Laboratoire de l'Informatique du Parallelisme Ecole Normale Superieure de Lyon, "Automatic deployment of the Network Weather Service using the Effective Network View," IEEE Parallel and Distributed Processing Symposium, pp. 272-279, Apr. 2004

An Optimal Nonparametric Weighted System for Hyperspectral Data Classification

Li-Wei Ko[1,2], Bor-Chen Kuo[1], and Ching-Teng Lin[2]

[1] Graduate School of Educational Measurement and Statistics,
National Taichung Teachers College, Taichung, Taiwan
`lwko.ece93g@nctu.edu.tw, kbc@mail.ntctc.edu.tw`
[2] Department of Electrical and Control Engineering,
National Chiao-Tung University, Hsinchu, Taiwan
`ctlin@mail.nctu.edu.tw`

Abstract. In real situation, gathering enough training samples is difficult and expensive. Assumption of enough training samples is usually not satisfied for high dimensional data. Small training sets usually cause Hughes phenomenon and singularity problems. Feature extraction and feature selection are usual ways to overcome these problems. In this study, an optimal classification system for classifying hyperspectral image data is proposed. It is made up of orthonormal coordinate axes of the feature space. Classification performance of the classification system is much better than the other well-known ones according to the experiment results below. It possesses the advantage of using fewer features and getting better performance.

Keywords: pattern recognition, feature extraction, feature selection, Hughes phenomenon

1 Introduction

The purpose of this paper is twofold. First, it aims to select the between-class scatter matrix and the within-class scatter matrix in NWFE to take the place of the ones in orthonormal discriminant vector (ODV) in our opinion. Therefore, the orthonormal nonparametric weighted vector (ONV) is made up. Second, combining the ONV with two kinds of well-known feature selections to improve the classification results, one is forward feature selection (FFS) and the other is "plus-k-minus-r" sequential forward selection (SFS).

Hence, the study proposed three methodologies in all for improving the performance of the best individual one. The study is structured as follows. In section 2 and 3, we introduce some well-known feature extraction and feature selection methods which are used to compare with our methodologies by the holdout accuracy. Section 4 describes our methodologies in detail. In section 5, it described experiments design and accuracy results of the study in real data sets. In section 6, we comment about our methodologies and propose some suggestion in the future.

2 Feature Extractions

Main purpose of feature extraction is trying to mitigate the Hughes phenomena [6] when the training sample size is small. Feature extraction is often used for dimension

R. Khosla et al. (Eds.): KES 2005, LNAI 3681, pp. 866–872, 2005.

reduction in classification problems. We described some feature extraction methods which are used in this paper as follows in detail.

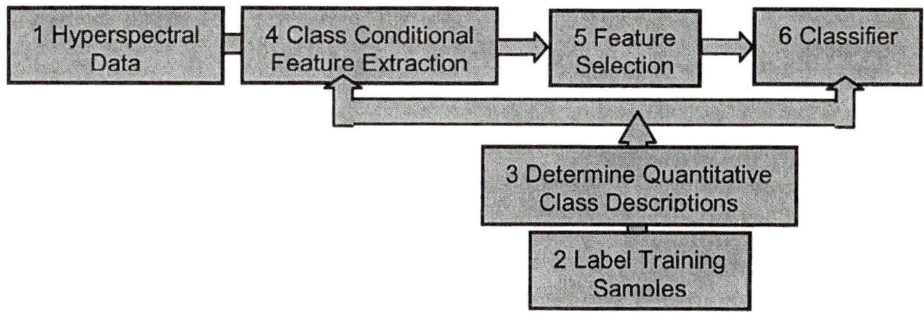

Fig. 1. A schematic diagram for classifying high dimensional data. [9]

2.1 Parametric Feature Extractions

The traditional well-known parametric feature extractions are PCA and LDA [1]. The ODV algorithm [10] is based on LDA in sequential feature extraction. See textbooks on LDA and PCA algorithms [4] and the reference [10] on ODV algorithm for details.

2.2 Nonparametric Feature Extractions

One of limitations of LDA is that it works well when data is normally distributed. A different between-class scatter matrix and a within-class scatter matrix were proposed in nonparametric weighted feature extraction [8] for improving this limitation.

The optimal criterion of NWFE is also by optimizing the Fisher criteria [3], the between-class scatter matrix and the within-class scatter matrix are expressed respectively by

$$S_b^{NW} = \sum_{i=1}^{L} P_i \sum_{\substack{j=1 \\ j \neq i}}^{L} \sum_{k=1}^{n_i} \frac{\lambda_k^{(i,j)}}{n_i} (x_k^{(i)} - M_j(x_k^{(i)}))(x_k^{(i)} - M_j(x_k^{(i)}))^T \tag{1}$$

$$S_w^{NW} = \sum_{i=1}^{L} P_i \sum_{k=1}^{n_i} \frac{\lambda_k^{(i,i)}}{n_i} (x_k^{(i)} - M_i(x_k^{(i)}))(x_k^{(i)} - M_i(x_k^{(i)}))^T \tag{2}$$

In the formula, $x_k^{(i)}$ refers to the k-th sample from class i. The scatter matrix weight $\lambda_k^{(i,j)}$ is a function of $x_k^{(i)}$ and local mean $M_j(x_k^{(i)})$, and defined as:

$$\lambda_k^{(i,j)} = \frac{dist(x_k^{(i)}, M_j(x_k^{(i)}))^{-1}}{\sum_{l=1}^{n_i} dist(x_l^{(i)}, M_j(x_l^{(i)}))^{-1}} \tag{3}$$

$M_j(x_k^{(i)})$ is the local mean of $x_k^{(i)}$ in the class j and defined as:

$$M_j(x_k^{(i)}) = \sum_{l=1}^{n_j} w_{kl}^{(i,j)} x_l^{(j)} \text{ , } L \text{ is the number of classes}$$

$$\text{where } w_{kl}^{(i,j)} = \frac{dist(x_k^{(i)}, x_l^{(j)})^{-1}}{\sum_{l=1}^{n_j} dist(x_k^{(i)}, x_l^{(j)})^{-1}} \tag{4}$$

See reference [8] on NWFE algorithm for details. In NWFE criterion, we regularize S_w^{NW} to reduce the effect of the cross products of between-class distances and prevent singularity by $0.5S_w^{NW} + 0.5diag(S_w^{NW})$. Hence, the criterion of NWFE is to find the first m eigenvectors corresponding to the largest m eigenvalues of $(S_w^{NW})^{-1}S_b^{NW}$.

3   Feature Selections

The best feature subset selection of m features out of n-dimension may be found by evaluating a criterion of class separability for all possible combinations of m features. In this study, we combine the ONV with two kinds of well-known feature selections to improve the classification results, one is forward feature selection and the other is "plus-k-minus-r" sequential forward selection. It will describe in detail as follows.

3.1   Forward Feature Selection

Feature selection techniques generally involve both a search algorithm and a criterion function [4]. Briefly speaking, the forward selection is a stepwise search technique which avoids exhaustive enumeration. It can't guarantee the selection of the best subset. We use forward selection by Duin's PRtools [2] in this paper.

3.2   Plus-k-Minus-r Sequential Forward Selection

The simplest suboptimal search strategy is the sequential forward selection (SFS) technique. The SFS algorithm carries out a "bottom-up" search strategy that, starting from an empty feature subset and adding one feature at a time, achieves a feature subset with the desired cardinality. Unfortunately, the SFS algorithm exhibits a serious drawback, once the features have been selected, they cannot be discarded.

4   Optimal Orthonormal Nonparametric Weighted System

4.1   Orthonormal Nonparametric Weighted Vector

In this paper, orthonormal nonparametric vectors (ONV) takes the S_b^{NW} and S_w^{NW} in NWFE place of ones in ODV algorithm. It also sequentially extracts features which maximize the criterion subject to the orthonormality of features. For using LDA, if the training sample size is very small or the within-class scatter matrix is singular or nearly singular, the performances of LDA will be poor. The method proposed in [7] is time-consuming and only suitable for parametric within-class scatter matrix. For preventing singularity, we also modified the within-class scatter matrix of ONV as $0.5S_w^{ONV} + 0.5diag(S_w^{ONV})$ and the criterion of ONV algorithm is expressed by

$$J_{ONV}(d) = \frac{d^T S_b^{ONV} d}{d^T (0.5S_w^{ONV} + 0.5diag(S_w^{ONV}))d} \tag{5}$$

Kuo and Landgrebe [8] investigate and show that using diagonal parts of the covariance matrix for regularizing is a good strategy.

4.2 ONV System with Plus-k-Minus-r SFS

According to the above section, we know the ONV system algorithm. Hamamoto, et al. [5] propose a new feature extraction method based on the modified "plus-k-minus-r" SFS algorithm. The study proposed the ONV System with plus-k-minus-r SFS is the same as one. We select k = 2 and r = 1 to apply the classification system.

4.3 ONV System with FFS

According to the features extracted by ONV, the study proposed to select f features with FFS to form a new feature space which has the maximum class separability. We name it ONV system with FFS.

5 Experiments and Results

5.1 Data and Algorithms

For evaluating the performance of the proposed system, training and testing data sets are selected from a small segment of a HYDICE (191 bands) hyperspectral image shown in Figure 2. It was collected over the DC Mall maps which have seven classes (Roof, Street, Path, Grass, Trees, Water and Shadow) are selected to form training and testing data. There are 100 training and 100 testing samples in each class. At each experiment, 10 training and testing data sets are randomly selected for estimating system parameters and computing classification accuracies of testing data of different algorithms respectively.

All algorithms studied are listed in Table 1. "Number of classes minus one" features are extracted in each feature extraction algorithms. All classification accuracies are computed by Gaussian classifier (GC). For evaluating the effects of each methodology, testing data accuracy of all algorithms are computed.

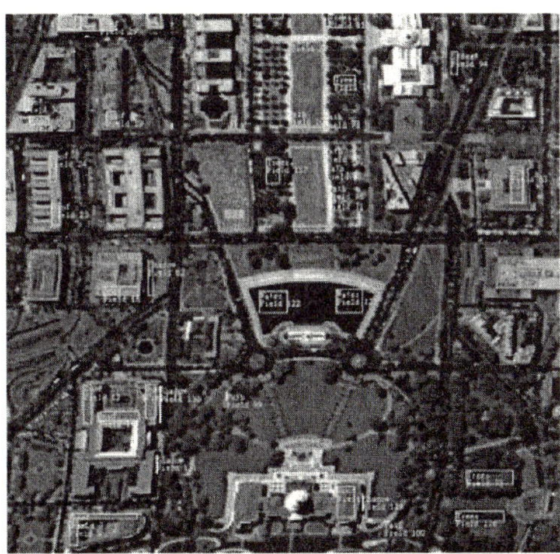

Fig. 2. Seven classes are selected in hyperspectral image of Washington DC Mall

Table 1. Algorithms for Comparison

	Algorithm	Description
Traditional Dimension Reduction Methods (Experiment 1)	PCA	Principal component analysis
	LDA	Linear discriminant analysis
	NWFE	Nonparametric weighted feature extraction
	FFS	Forward feature selection
Similar Methods (Experiment 2)	ODV	Orthonormal discriminant vector
	ODV_SFS	ODV with plus-k-minus-r sequential forward selection
	ODV_FFS	ODV with forward feature selection
Methods Proposed in This Study (Experiment 3)	ONV	Orthonormal nonparametric weighted vector
	ONV_SFS	ONV with plus-k-minus-r sequential forward selection
	ONV_FFS	ONV with forward feature selection

5.3   Results and Findings

The result of experiment 1 is summarized in Fig. 3 and Table 2 and it shows that NWFE uses fewer features and can provide much better accuracy than others. LDA has the best result in experiment 2 shown in Fig. 4 and Table 3 and the ODV, ODV_SFS and ODV_FFS algorithms are not suitable for Washington DC Mall image. The result of experiment 3 is summarized in Fig. 5 and Table 4 and it shows that ONV_FFS with two features outperforms NWFE with two features significantly and other ONV family outperform than NWFE with six features. Figure 5 shows that ONV_FFS has the advantage, using fewer features and getting better performance (93.57%), others algorithms almost have the best accuracy in 6 features.

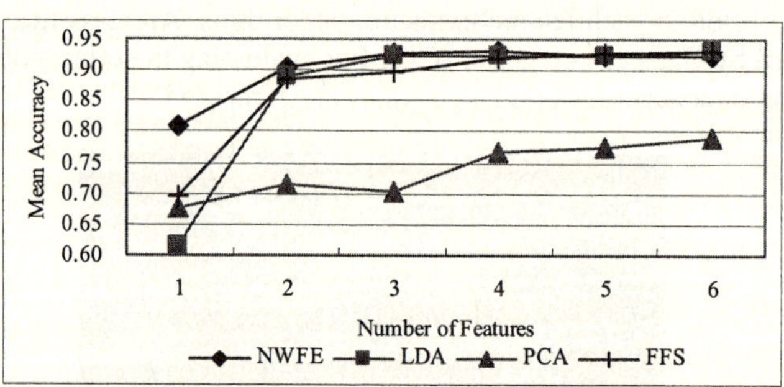

Fig. 3. Mean accuracies of algorithms in experiment 1

Table 2. Mean accuracies of algorithms in experiment 1

Features	NWFE	LDA	PCA	FFS
1	0.8100	0.6157	0.6757	0.6944
2	0.9043	0.8886	0.7157	0.8846
3	0.9271	0.9229	0.7043	0.8973
4	0.9300	0.9243	0.7686	0.9200
5	0.9243	0.9229	0.7743	0.9277
6	0.9229	0.9300	0.7900	0.9311

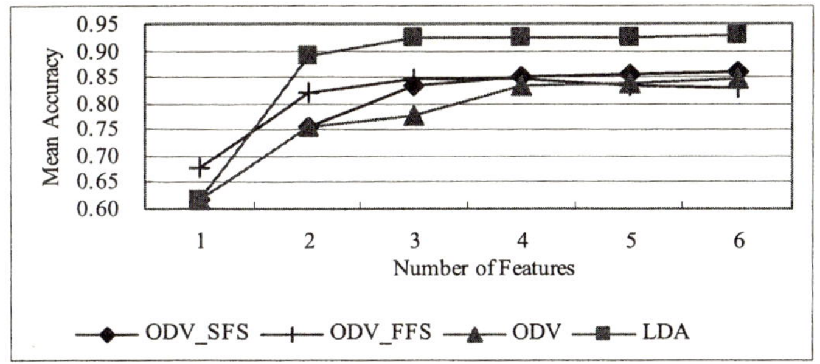

Fig. 4. Mean accuracies of algorithms in experiment 2.

Table 3. Mean accuracies of algorithms in experiment 2

Features	ODV	ODV_SFS	ODV_FFS
1	0.6157	0.6157	0.6771
2	0.7557	0.7557	0.8200
3	0.7786	0.8329	0.8457
4	0.8343	0.8514	0.8443
5	0.8386	0.8557	0.8329
6	0.8457	0.8600	0.8300

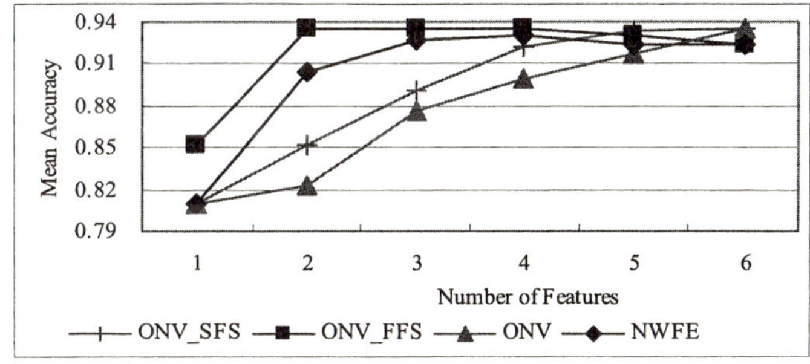

Fig. 5. Mean accuracies of algorithms in experiment 3

Table 4. Mean accuracies for experiment 3

Features	ONV	ONV_SFS	ONV_FFS
1	0.8100	0.8100	0.8514
2	0.8229	0.8514	0.9357
3	0.8757	0.8914	0.9343
4	0.9000	0.9214	0.9329
5	0.9171	0.9329	0.9300
6	0.9343	0.9343	0.9243

6   Conclusions

Traditionally, the feature selection after feature extraction is based on the eigenvalues of the extracted features. In this study, more power selection methods are proposed.

From the experimental results, we can find that the proposed methods (ONV_FFS, ONV_SFS and ONV) outperform others and ONV_FFS can reach the better classification result using fewer features.

Table 5. The best accuracy and used features of each algorithm

Algorithm	PCA	LDA	NWFE	FFS
Best Accuracy	0.7900	0.9300	0.9300	0.9311
# of Used Features	6	6	4	6
Algorithm	**ODV**	**ODV_SFS**		**ODV_FFS**
Best Accuracy	0.8457	0.8600		0.8457
# of Used Features	6	6		3
Algorithm	**ONV**	**ONV_SFS**		**ONV_FFS**
Best Accuracy	0.9343	0.9343		0.9357
# of Used Features	6	6		2

Acknowledgements

Authors would like to thank National Science Council, Taiwan for partially supporting this work under grant NSC 93-2212-E-142-001.

References

1. Belhumeur, P. N., Hespanha, J. P. and Kriegman, D. J., 1997, Eigenfaces vs. Fisherfaces: Recognition Using Class Specific Linear Projection, *IEEE Transaction on Pattern Analysis and Machine Intelligence*, 19(7), 711-720.
2. Duin, R. P. W., 2002, *PRTools, a Matlab Toolbox for Pattern Recognition*, (Available for download from http://www.ph.tn.tudelft.nl/prtools/).
3. Fisher, R.A., 1936, The use of multiple measurements in taxonomic problems, *Annual Eugenics*, Part II, 7, 179-188.
4. Fukunaga, K., 1990, *Introduction to Statistical Pattern Recognition*, San Diego: Academic Press Inc.
5. Hamamoto, Y., Matsuura, Y., Kanaoka T. and Tomita, S., 1991, A Note on the orthonormal discriminant vector method for feature extraction, *Pattern Recognition*, 24, 681-684.
6. Hughes, G. F., 1968, On the mean accuracy of statistical pattern recognition, *IEEE Trans. Information Theory*, IT-14(1), 55-63.
7. Kuo, B-C., Ko, L-W., Landgrebe, D. A. and Pai, C-H., 2003, Regularized Feature Extractions for Hyperspectral Data Classification, *International Geoscience and Remote Sensing Symposium in France*, 21-25 July, 2003, Toulouse.
8. Kuo, B-C. and Landgrebe, D. A., 2004, Nonparametric Weighted Feature Extraction for Classification, *IEEE Trans. on Geoscience and Remote Sensing*, vol. 42, no. 5, pp. 1096-1105.
9. Landgrebe, D. A., 2003, Signal Theory Methods in Multispectral Remote Sensing, John Wiley & Sons Inc., Chapter 7, pp.324.
10. Okada, T. and Tomita, S., 1985, An optimal orthonormal system for discriminant analysis, Pattern Recognition, 18, 139-144.

Regularized Feature Extractions and Support Vector Machines for Hyperspectral Image Data Classification

Bor-Chen Kuo and Kuang-Yu Chang

Graduate School of Educational Measurement and Statistics,
National Taichung Teachers College, Taichung, Taiwan
kbc@mail.ntctc.edu.tw

Abstract. In this study, the performances of using parametric/ nonparametric regularized feature extractions and support vector machine for hyperspectral image classification is explored when the training sample size is small. The classification accuracies of RBF-based SVM using two feature extractions with three regularization techniques are evaluated. The results of two hyperspectral image classification experiments show that the performance of the combination of nonparametric weighted feature extraction and RBF-based SVM outperforms those of others.

1 Introduction

As new sensor technology has emerged over the past few years, high dimensional multispectral data with hundreds of bands has became available. Increasing the number of spectral bands (dimensionality) potentially provides more information about class separability. This positive effect is diluted by poor parameter estimation when statistical classifiers are applied. One of the major issues of hyperspectral image classification is how to mitigate the Hough phenomena [1], [12] or curse of the dimensionality when training sample size is small [1]. Traditionally, feature extraction (FE) or feature selection (FS) techniques are applied to overcome this problem and Fig.1 shows a hyperspectral image classification process [1].

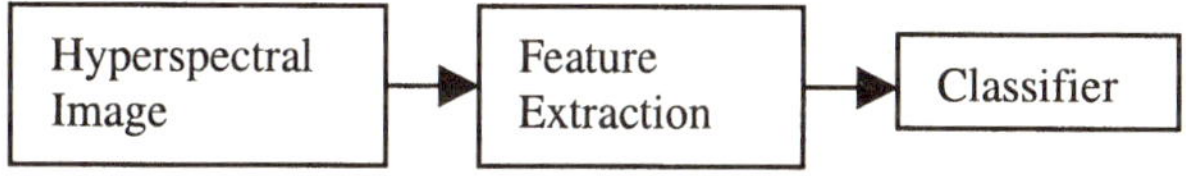

Fig. 1. A hyperspectral image classification process

In small sample size problem, regularization is a kind of technique to overcome it in parametric classifier. Three kinds of regularization techniques are applied in this study, and combine with two kinds of feature extractions.

Support vector machine (SVM) has recently been the subject of intense research activity within the neural networks community [2]-[4], [10]. SVM can be treated as a generalized linear discriminant in a high-dimensional transformed feature space. It is opposite to feature extraction/selection that SVM algorithms try to map data points into a high dimensional feature space [3], then classify them by linear classifiers. The real data experiment results in [2] show that using parametric feature extractions, Fisher linear discriminant analysis (LDA, [6]) and Decision Boundary Feature Extrac-

R. Khosla et al. (Eds.): KES 2005, LNAI 3681, pp. 873–879, 2005.

tion (DBFE, [11]), as a preprocess can not improve the performance of SVM classifier with RBF kernels when training sample size is large enough.

In this study, the relation between parametric/ nonparametric feature extractions and support vector machine is explored when the training sample size is small.

2 Support Vector Machine and Feature Extraction

Given a training set of instance-label pairs $(x_{i,} y_{j}), i = 1,...,l$, where $x_i \in R^n$ and $y \in \{1,-1\}^l$, the support vector machines require the solution of the following optimization problem:

$$\min \frac{1}{2} w^T \cdot w + C \sum_{i=1}^{l} \xi_i$$

$$subject \quad to \quad \begin{array}{l} y_i(w^T \cdot \phi(x_i) + b) \geq 1 - \xi_i, \\ \xi_i \geq 0. \end{array} \tag{1}$$

Here training vectors x_i are mapped into a high dimensional feature space by the function ϕ. Then SVM finds a linear separating hyperplane with the maximal margin in the high dimensional feature space. C > 0 is the penalty parameter of the error term. Furthermore, $K(x_i, x_j) \equiv \phi(x_i)^T \varphi(x_j)$ is called the kernel function and the following RBF kernel is implemented in this study.

$$K(x_i, x_j) = \exp(-\gamma \|x_i - x_j\|^2), \ \gamma > 0. \tag{2}$$

Two feature extraction methods, linear discriminant analysis (LDA, [6]) and nonparametric weighted feature extraction (NWFE, [7], [8]), are applied in this study. One of main purposes of this paper is to show that NWFE is better than LDA for data classification. Since LDA is well known and can be found in [6], only NWFE algorithm is briefly introduced in follow.

NWFE tries to find the feature space in which between-class scatter matrix is maximized and within-class scatter matrix is minimized simultaneously. The between-class scatter matrix is defined as

$$S_b^{NW} = \sum_{i=1}^{L} P_i \sum_{\substack{j=1 \\ j \neq i}}^{L} \sum_{k=1}^{n_i} \frac{\lambda_k^{(i,j)}}{n_i} (x_k^{(i)} - M_j(x_k^{(i)}))(x_k^{(i)} - M_j(x_k^{(i)}))^T \tag{3}$$

The nonparametric within-class scatter matrix is defined as

$$S_w^{NW} = \sum_{i=1}^{L} P_i \sum_{k=1}^{n_i} \frac{\lambda_k^{(i,i)}}{n_i} (x_k^{(i)} - M_i(x_k^{(i)}))(x_k^{(i)} - M_i(x_k^{(i)}))^T \tag{4}$$

where $x_k^{(i)}$ refers to the k -th sample from class i . The scatter matrix weight $\lambda_k^{(i,j)}$ is a function of $x_k^{(i)}$ and local mean $M_j(x_k^{(i)})$, and defined as:

$$\lambda_k^{(i,j)} = \frac{dist(x_k^{(i)}, M_j(x_k^{(i)}))^{-1}}{\sum_{l=1}^{n_i} dist(x_l^{(i)}, M_j(x_l^{(i)}))^{-1}}, \text{ where } dist(a,b) \text{ means the distance from a to b}. \tag{5}$$

If the distance between $x_k^{(i)}$ and $M_j(x_k^{(i)})$ is small then its weight $\lambda_k^{(i,j)}$ will be close to 1; otherwise, $\lambda_k^{(i,j)}$ will be close to 0 and sum of total $\lambda_k^{(i,j)}$ for class i is 1. $M_j(x_k^{(i)})$ is the local mean of $x_k^{(i)}$ in the class j and defined as:

$$M_j(x_k^{(i)}) = \sum_{l=1}^{n_j} w_{kl}^{(i,j)} x_l^{(j)} \text{ , } L \text{ is the number of classes, where } w_{kl}^{(i,j)} = \frac{dist(x_k^{(i)}, x_l^{(j)})^{-1}}{\sum_{l=1}^{n_j} dist(x_k^{(i)}, x_l^{(j)})^{-1}} \tag{6}$$

The weight $w_{kl}^{(i,j)}$ for computing local means is a function of $x_k^{(i)}$ and $x_l^{(j)}$. If the distance between $x_k^{(i)}$ and $x_l^{(j)}$ is small then its weight $w_{kl}^{(i,j)}$ will be close to 1; otherwise, $w_{kl}^{(i,j)}$ will be close to 0 and sum of total $w_{kl}^{(i,j)}$ for $M_j(x_k^{(i)})$ is 1.

To reduce the effect of the cross products of within-class distances and prevent the singularity, some regularized techniques [8], [14], [15], can be applied to within-class scatter matrix. In this study, within-class scatter matrix is regularized by $0.5 S_w^{NW} + 0.5 diag(S_w^{NW})$, where diag(A) means the diagonal parts of matrix A.

The advantages of NWFE are [7], [8]:

1. Using nonparametric form of scatter matrix for non-normal distributed data.
2. Preventing singularity by regularizing within-class scatter matrix.

3 Regularization for Scatter Matrices

Many researches show that feature extraction with regularization can help to improve the classification performances of Gaussian classifier and k-nearest neighbors classifier [1], [7], [8] when training sample size is small. In this study, the performances of SVMs with regularized and non-regularized feature extractions are explored. Three regularization techniques, RFE [8], Maximum Entropy Covariance Selection⎕MECS, [14]⎕, and New LDA (NLDA [15]) are applied in this study.

In RFE, the within-class scatter matrix is rewritten as following

$$S_w^R = \alpha diag(S_w) + \beta \frac{trace(S_w)}{p} + \gamma S_w \text{ , where p is the dimensionality of data} \tag{7}$$

α, β, γ are mixing parameters. $0 \le \alpha, \beta, \gamma \le 1$ and $\alpha + \beta + \gamma = 1$

The MECS [14] approach is a direct procedure that with the singularity and instability of sample group covariance matrices S_i when similar covariance matrices are linearly combined. S_{ps} is the pooled sample covariance matrix.

1. Find the eigenvectors Φ_i^{me} of the covariance given by $S_i + S_{ps}$.

2. Calculate the variance contribution of both S_i and S_w on the Φ_i^{me} basis, i.e.,

$$diag(Z^i) = diag[(\Phi_i^{me})^T S_i (\Phi_i^{me})] = [\xi_1^i, \xi_2^i, \ldots, \xi_n^i] \tag{8}$$

$$diag(Z^{ps}) = diag[(\Phi_i^{me})^T S_{ps} (\Phi_i^{me})] = [\xi_1^{ps}, \xi_2^{ps}, \ldots, \xi_n^{ps}] \tag{9}$$

3. From a new variance matrix based on the large values, that is,

$$Z_i^{me} = diag[\max(\xi_1^i, \xi_1^{ps}), \ldots, \max(\xi_n^i, \xi_n^{ps})] \tag{10}$$

4. Form the MECS estimator $S_i^{mecs} = \Phi_i^{me} Z_i^{me} (\Phi_i^{me})^T$.

The NLDA[16] is in order to expand only the smaller and consequently less reliable eigenvalues of S_p, and keep most of its larger eigenvalues unchanged. S_w is the standard pooled covariance matrix.

1. Find the Φ eigenvectors and Λ eigenvalues of S_p, where $S_p = S_w / [N-g]$;

2. Calculate the S_p average eigenvalues $\overline{\lambda}$ using $\overline{\lambda} = \frac{1}{n}\sum_{j=1}^{n}\lambda_j = \frac{tr(S_p)}{n}$;

3. Form a new matrix of eigenvalues based on the following largest dispersion values

$$\Lambda^* = diag[\max(\lambda_1, \overline{\lambda}), \max(\lambda_2, \overline{\lambda}), \ldots, \max(\lambda_n, \overline{\lambda})]; \tag{11}$$

4. Form the modified within-class scatter matrix

$$S_w^* = S_p^*(N-g) = (\Phi\Lambda^*\Phi^T)(N-g) \tag{12}$$

4 Experiment Design

Two hyperspectral image data sets, Washington DC Mall and Indian Pine Site [1] are used for exploring the effects of combinations of two feature extractions and three regularization techniques in SVM. The DC Mall data set has 191 bands. Indian Pine Site has 220 bands. There are 7 and 9 classes used in Washington DC Mall and Indian Pine Site respectively.

For exploring the impact of training sample size, two cases, each class with 20 and 40 training samples and 100 testing samples are studied. At each experiment, 10 training and testing data sets are randomly selected for estimating system parameters and computing the mean accuracies of testing data of different algorithms respectively.

The Matlab SVM Toolbox, LIBSVM [5], is used in this study. There are two parameters while using RBF kernels: C and γ. It is not known beforehand which C and γ are the best for our problem; consequently some kind of model selection (parameter search) must be done. The goal is to identify good (C, γ) so that the classifier can accurately predict unknown data (i.e., testing data). In v-fold cross-validation, we recommend a "grid-search" on C and γ using cross-validation. Basically pairs of (C, γ) are tried and the one with the best cross- practical method to identify good parameters (for example, $C = 2^{-5}, 2^{-3}, \ldots, 2^{15}, \gamma = 2^{-15}, 2^{-13}, \ldots, 2^3$) [4]. We validation accuracy is picked. It was shown that trying exponentially growing sequences of C and γ is a picked the best parameter C and γ for RBF kernel SVM. In this study, a 5-fold cross-validation is used to find the best parameters. The grid method range for C is from 2^{-5} to 2^{15} step 2^2. The process to pick γ is the same.

5 Experiment Results

The average classification accuracies using regularized feature extraction of two datasets are displayed in Table 1 to Table 4. Table 5 contains the average accuracy of using SVM and the best average accuracy of each case by applying 1 to 10 features.

The followings are some findings:

1. Using non-regularized LDA can not improve the performance of only using SVM.
2. Regularized LDA performs better than non-regularized LDA in SVM.

Table 1. The average accuracy of DC Mall dataset (sample size of each class is 20)

Scatter matrix type	Regularization	Number of features									
		1	2	3	4	5	6	7	8	9	10
LDA	none	0.289	0.199	0.324	0.437	0.479	0.515	0.490	0.540	0.543	0.545
LDA	RFE	0.751	0.243	0.619	0.741	0.844	0.864	0.817	0.724	0.749	0.771
LDA	MECS	0.655	0.208	0.554	0.659	0.729	0.741	0.750	0.751	0.753	0.758
LDA	NLDA	0.698	0.256	0.561	0.677	0.704	0.776	0.763	0.799	0.790	0.783
NWFE	original	0.736	0.257	0.649	0.845	0.860	0.859	0.853	0.873	0.877	0.874
NWFE	RFE	0.743	0.252	0.638	0.842	0.856	0.872	0.874	0.875	0.836	0.858
NWFE	MECS	0.648	0.231	0.520	0.629	0.659	0.718	0.731	0.744	0.734	0.738
NWFE	NLDA	0.698	0.221	0.588	0.701	0.760	0.765	0.784	0.783	0.782	0.782

Table 2. The average accuracy of DC Mall dataset (sample size of each class is 40)

Scatter matrix type	Regularization	Number of features									
		1	2	3	4	5	6	7	8	9	10
LDA	none	0.203	0.240	0.535	0.687	0.711	0.812	0.796	0.756	0.783	0.822
LDA	RFE	0.778	0.228	0.673	0.834	0.884	0.907	0.870	0.806	0.805	0.801
LDA	MECS	0.669	0.220	0.588	0.692	0.782	0.796	0.782	0.801	0.803	0.808
LDA	NLDA	0.741	0.254	0.632	0.762	0.815	0.815	0.831	0.826	0.828	0.819
NWFE	original	0.776	0.266	0.683	0.852	0.908	0.908	0.903	0.910	0.901	0.908
NWFE	RFE	0.784	0.251	0.671	0.842	0.884	0.892	0.904	0.893	0.906	0.904
NWFE	MECS	0.665	0.238	0.554	0.657	0.707	0.742	0.755	0.768	0.783	0.776
NWFE	NLDA	0.735	0.253	0.570	0.721	0.777	0.806	0.800	0.817	0.808	0.808

Table 3. The average accuracy of Indian Pine dataset (sample size of each class is 20)

Scatter matrix type	Regularization	Number of features									
		1	2	3	4	5	6	7	8	9	10
LDA	none	0.259	0.111	0.310	0.413	0.478	0.511	0.536	0.536	0.547	0.542
LDA	RFE	0.460	0.153	0.397	0.590	0.661	0.719	0.721	0.762	0.760	0.770
LDA	MECS	0.455	0.111	0.466	0.568	0.658	0.713	0.731	0.759	0.770	0.774
LDA	NLDA	0.405	0.127	0.418	0.563	0.653	0.699	0.737	0.756	0.768	0.752
NWFE	original	0.441	0.111	0.454	0.601	0.683	0.738	0.743	0.783	0.783	0.798
NWFE	RFE	0.465	0.147	0.429	0.595	0.668	0.731	0.752	0.751	0.762	0.775
NWFE	MECS	0.447	0.111	0.459	0.542	0.622	0.675	0.723	0.735	0.730	0.751
NWFE	NLDA	0.408	0.122	0.402	0.566	0.654	0.693	0.707	0.723	0.737	0.739

Table 4. The average accuracy of Indian Pine dataset (sample size of each class is 40)

Scatter matrix type	Regularization	Number of features									
		1	2	3	4	5	6	7	8	9	10
LDA	none	0.284	0.129	0.329	0.494	0.566	0.607	0.616	0.641	0.647	0.641
LDA	RFE	0.477	0.133	0.383	0.627	0.707	0.739	0.775	0.796	0.801	0.815
LDA	MECS	0.472	0.111	0.495	0.623	0.698	0.755	0.791	0.803	0.807	0.805
LDA	NLDA	0.442	0.164	0.439	0.611	0.702	0.744	0.783	0.799	0.777	0.819
NWFE	original	0.468	0.142	0.501	0.640	0.716	0.781	0.813	0.831	0.834	0.842
NWFE	RFE	0.461	0.144	0.476	0.613	0.696	0.768	0.806	0.822	0.814	0.828
NWFE	MECS	0.446	0.111	0.488	0.636	0.689	0.761	0.749	0.785	0.767	0.799
NWFE	NLDA	0.432	0.140	0.408	0.556	0.689	0.748	0.758	0.760	0.766	0.786

3. NWFE has the best performances, especially when the training sample is small.
4. RFE outperforms other two regularization techniques. The performance of SVM with LDA using RFE in Washington DC Mall is better than that of only using SVM.

Table 5. The best average accuracy

Scatter matrix type	Regularization	Washington DC Mall sample size of each class		Indian Pine Site sample size of each class	
		20	40	20	40
none	none	0.830	0.854	0.785	0.838
LDA	none	0.545	0.822	0.547	0.647
LDA	RFE	0.864	0.907	0.770	0.815
LDA	MECS	0.758	0.808	0.774	0.805
LDA	NLDA	0.799	0.831	0.768	0.819
NWFE	original	0.877	0.910	0.798	0.842
NWFE	RFE	0.875	0.906	0.775	0.828
NWFE	MECS	0.744	0.783	0.751	0.799
NWFE	NLDA	0.784	0.817	0.739	0.786

5. Theoretically extracting only L-1(number of features - 1) is the best choice in LDA [6]. Our experiment result shows that it may not be the best choice for hyperspectral image classification.

6 Conclusions

From the experiment results, we can find that a suitable feature extraction can help classifiers improve their performances especially when the training sample size is small. A proper regularization technique can make LDA more effective in SVM. In [7] and [13], NWFE outperforms other feature extractions when k-nearest-neighbor and Gaussian classifiers are applied. In this study, we also show that NWFE can improve the performances of SVM classifiers.

Acknowledgements

Authors would like to thank National Science Council, Taiwan for partially supporting this work under grant NSC 93-2212-E-142-001.

References

1. D.A. Landgrebe, Signal Theory Methods in Multispectral Remote Sensing, John Wiley and Sons, Hoboken, NJ: Chichester, 2003.
2. C.A. Shah, P. Watanachaturaporn, M.K. Arora and P.K. Varshney, Some Recent Results on Hyperspectral Image Classification, In IEEE Workshop on Advances in Techniques for Analysis of Remotely Sensed Data, NASA Goddard Spaceflight center, Greenbelt, October 27-28, 2003.
3. N. Cristianini and J. Shave-Taylor, Support Vector Machines and other kernel-based learning methods, Cambridge University Press, 2000.
4. C.-W. Hsu, C.-C. Chang and C.-J. Lin, A Practical Guide to Support Vector Classification, 2004. Available at http://www.csie.ntu.edu.tw/~cjlin/libsvm
5. C.-C. Chang and C.-J. Lin, LIBSVM: a library for support vector machines, 2004. Software available at http://www.csie.ntu.edu.tw/~cjlin/libsvm.
6. K. Fukunaga, Introduction to Statistical Pattern Recognition, San Diego, CA:Academic Press, 1990.

7. B-C. Kuo and D.A. Landgrebe, Nonparametric Weighted Feature Extraction for Classification, IEEE Trans. on Geoscience and Remote Sensing, vol. 42, no. 5, pp. 1096-1105, May 2004.
8. B-C. Kuo, D.A. Landgrebe, L-W. Ko and C-H. Pai, Regularized Feature Extractions for Hyperspectral Data Classification, Proceedings of International Geoscience and Remote Sensing Symposium, Toulouse. France, July 2003.
9. B-C. Kuo, K-Y. Chang, C-H. Chang and Y-C. Hsieh, Hyperspectral Image Data Classification Using Feature Extractions and Support Vector Machines. CVGIP 2004.
10. V. Vapnik, The Nature of Statistical Learning Theory. New York, NY: Springer-Verlag. 1995.
11. C. Lee and D.A. Landgrebe, Feature Extraction Based On Decision Boundaries, IEEE Transactions on Pattern Analysis and Machine Intelligence, Vol. 15, No. 4, April 1993, pp 388-400.
12. G.F. Hughes, On the mean accuracy of statistical pattern recognition, IEEE Trans. Inform. Theory, vol. 14, pp. 55 - 63, Jan. 1968.
13. G-S. Chen, L-W. Ko, B-C. Kuo and S-C. Shih, A Two-stage Feature Extraction for Hyperspectral Image Data Classification. Proceedings of International Geoscience and Remote Sensing Symposium, Sep. 20-24, 2004.
14. C.E. Thomaz, D.F. Gillies and R.Q. Feitosa, A New Covariance Estimate for Bayesian Classifiers in Biometric Recognition, IEEE Transactions on Circuits and Systems for Video Technology, Special Issue on Image- and Video-Based Biometrics, vol. 14, no. 2, pp. 214-223, February 2004.
15. C.E. Thomaz and D.F. Gillies, A Maximum Uncertainty LDA-based approach for Limited Sample Size problems - with application to Face Recognition. Technical Report TR-2004-01, Department of Computing, Imperial College, London, UK, January 2004.

An Error Measure for Japanese Morphological Analysis Using Similarity Measures

Yoshiaki Kurosawa, Yuji Sakamoto, Takumi Ichimura, and Teruaki Aizawa

Faculty of Information Sciences, Hiroshima City University, 3-4-1,
Ozuka-higashi, Asaminami-ku, Hiroshima 731-3194, Japan
{kurosawa,ichimura,aizawa}@its.hiroshima-cu.ac.jp
sakamoto@nlp.its.hiroshima-cu.ac.jp

Abstract. The aim of this paper is to propose a Japanese morphological error measure in order to automatically detect morphological errors when analyzing. In particular, we focused on three similarities as the measure and experimented using them. From our experimental results, it was found that the precision of one of measures was 74% and it functioned well.

1 Introduction

Japanese is a language which has no space between written words. When analyzing Japanese sentence, we must appropriately separate it into minimum meaningful units, morpheme, at first. We called this separating processing morphological analysis.

Recently, natural language processing techniques such as morphological analysis have been developed. In particular, the precision of this analysis has exceeded over 90% and the analysis has been used in many tasks.

However, we must check the results whether they are correct or not, because they may remain not a few errors. It is not easy to perform such check precisely by hand. We need effective automatic procedures for finding morphological errors easily.

There is a study for automatically detecting the errors using a statistical measure (Utiyama, 1999). This measure can deal with over-segmentation errors that one morpheme is regarded as two or more morphemes, for example, *Hanako* (personal name) is interpreted as *hana* ("flower" noun) and *ko* ("child" noun).

However, it cannot deal with part-of-speech tagging errors that two morphemes *kara* ("because" conjunctive particle) and *sa* (sentence-final particle) are interpreted as the other *kara* ("spicy" adjective) and *sa* (nominal-suffix) in sentence 1b.

1a: *Doushite ika naka tta no ?*   (Why don't you go there?)
1b: *Kaze wo hii ta <u>kara sa</u> ?*   (<u>Because</u> I caught a cold.)
       vs.   ($^?$<u>Spicy</u> that caught a cold.) … error

In order to detect these errors, we are required to focus on one of Japanese characteristics that various kinds of part-of-speeches emerge repeatedly to add much information such as tense, voice and modality at the end of a sentence in spoken language. For example in sentence 1b, by changing the phrase "*kara sa*" to "*kara kamo* (adverbial particle)," we can make this sentence indicate conjecture. By changing it to "*kara desu* (auxiliary verb)," we can indicate politeness in the same way. These sentences differ from each other from the viewpoint of emerged part-of-speeches. However, they are similar because they have common verbs (catch). If we recognize an error in

R. Khosla et al. (Eds.): KES 2005, LNAI 3681, pp. 880–886, 2005.
© Springer-Verlag Berlin Heidelberg 2005

1b, we consider that we will also recognize the same kinds of errors when seeing these variations as described above.

For this reason, we focus on such similarity of the composed part-of-speeches; see Section 3 for the precise definition of similarity. If a sentence is analyzed incorrectly, the values of the similarity between sentences related to these errors must become higher and the values unrelated to them become lower. On the other hand, a sentence including no errors must make higher value in case of comparing with correct sentences and make lower value in case of doing with incorrect ones. In other words, our aim of this paper is to propose automatic measures for detecting some types of errors using the similarity of the part-of-speeches between sentences.

2 Automatic Error Detect of Morphological Analysis

There are a few studies for using error-correction rules maintaining by hand, and correcting morphological errors (Hisamitsu and Niwa, 1998; Kurosawa et al. 2003, 2005). However, little attention has been given to automatically detecting the errors.

Utiyama calculated a statistical measure for detecting morphological errors, in particular, over-segmentation errors (Utiyama, 1999). He noted the ratio of the appearance probability of the particular string and the partial probability when separated it into two parts. We show his formula below. In the formula, S, A, and B mean a given string and two parts of separated strings of them ($a_1,...,a_k$ and $b_1,...,b_l$), respectively. Moreover, "$\langle w \rangle$" and "$\langle / w \rangle$" are symbols for separate morphemes.

$$L(S, A, B) = \log \frac{\Pr(\langle w \rangle, a_1,...,a_k, b_1,...b_l, \langle / w \rangle)}{\Pr(\langle w \rangle, a_1,...,a_k, \langle / w \rangle)\,\Pr(\langle w \rangle, b_1,...b_l, \langle / w \rangle)} \tag{1}$$

This measure L indicates the difficulty of the separation of a given letter string into two parts (A and B). Therefore, finding larger L points out errors. Certainly the measure may be able to detect over-segmentation, but it cannot detect other types of morphological errors such as under-segmentation and part-of-speech tagging errors (Hisamitsu and Niwa, 1998). In addition, we need to focus on the similarity between two sentences and detect these errors by using it as we explained in Section 1.

3 Similarity Measures

There are, at least, two studies for calculating similarity measures (Yamamoto et al., 2003, 2005; Ichihara et al., 1999). In addition, we make use of these measures for detecting morphological errors, although the aims of these studies are not to detect the errors but to retrieve relevant information. The reason why we use the values is that they are useful on finding the errors, taking the characteristics of Japanese language into account as explained in Section 1. We introduce these measures here briefly.

3.1 Similarity Measure

Yamamoto et al. proposed a similarity measure (SIM) using string weight, which extended the idea of edit distance (Korfhage, 1997). It was defined as follows:

Definition: SIM₃

Let α and β be strings, x and y be a character, and "" be empty string

- If both strings are empty then
$$SIM_3("","") = Score("") = 0.0$$

- Otherwise
$$SIM_3(\alpha, \beta) = MAX(SIM_{3s}(\alpha, \beta), SIM_{3g}(\alpha, \beta)) \tag{2}$$

  - $SIM_{3s}(\xi\alpha, \xi\beta) = MAX(Score(\gamma) + SIM_3(\xi\alpha, \xi\beta))$

    where $\xi (=\gamma\delta)$ is the maximum length string matching from the first character.

  - $SIM_{3g}(x\alpha, y\beta) =$
    $$MAX(SIM_3(\alpha, y\beta) + SIM_3(x\alpha, \beta) + SIM_3(\alpha, \beta))$$

Definition: Score Function

Let ξ be string, $df(\xi)$ the frequency of documents including ξ in the document set for retrieval, and N be the number of documents in the set.
$$Score(\xi) = IDF(\xi) = -\log(df(\xi)/N) \tag{3}$$

As far as the same sequential string is contained between two sentences, this measure makes larger values. Therefore, we can regard these phrases "… *kara sa*," to "… *kara kamo*," and "… *kara desu*" as a sentence of the same sort in case that the elements in "…" are common and their length is long. However, even if a nonidentical letter appears before an error occurs, for example, "… *teru* (subsidiary verb) *kara sa*," the measure makes become lower because of the decrease of common expression. For this reason, we also adopt another similarity.

In addition, we deal with not letter strings but part-of-speech strings because our task is to detect morphological errors including tagging errors.

3.2 Identical Match Rate and Partial Match Rate

There are two heuristics to determine the similarity between sentences (Ichihara et al., 1999). One is whether the sentences contain one or more sets of the same letters, and the other is whether the word orders of them are common. Based on these heuristics, they proposed two similarities: a similarity based on identical match rate (IMR) and a similarity based on partial match rate (PMR).

- **Identical Match Rate**

  This measure uses the number of the same letters of which order is identical. Their definition was shown below. The symbol A and B mean strings. In addition, the value l and onm indicate the number of letters and the maximum number of matches between two sentences.

  $$IMR = sim_1(A, B) = \frac{onm}{l_A} \times \frac{onm}{l_B} \tag{4}$$

- **Partial Match Rate**

  Their definition was shown below. The values nm and ord indicate the number of matches and the partial match rate between two sentences. In addition, $o_{i,j}$ have two values, 0 or 1, according to whether the letter orders are common.

$$ord = \frac{\sum_{j=1}^{n}\sum_{i=1}^{n} o_{i,j}}{\frac{n(n-1)}{2}} \tag{5}$$

$$PMR = sim_2(A, B) = \frac{nm}{l_A} \times \frac{nm}{l_B} \times ord \tag{6}$$

4 Experiment

As we previously mentioned, we attempt to automatically detect morphological errors using three similarities, SIM, IMR and PMR. For this calculation, we deal with part-of-speech information without any surface information of morphologies.

First, we explain our corpora.

– Corpora

We prepared two corpora: a modified corpus (MC) including 1698 sentences, in which morphological errors were correctly modified by hand, and an unmodified corpus (UC) including 1717 sentences. However, not a few sentences consist of large number of part-of-speeches in these corpora,. We restricted the number of them from 4 to 10 to reduce costs for similarity calculation. MC and UC contain 801 and 641 sentences respectively as a result.

4.1 Consideration of Necessary Factor

Our task is to check whether all the sentences in UC are analyzed correctly, that is, to calculate the values of similarity between a sentence in UC and all sentences in MC, to determine an appropriate threshold or formula, and to detect morphological errors including in UC. Because the calculation is based on one-to-many correspondence, in other words, we must calculate 801 similarities per sentence, we need to consider how to deal with these similarities. In this paper, we use two principal representative values: number of rules and variance.

In fact, we checked various representative values whether they are useful. However, it cannot be discussed here lack of space.

We indicate two graphs which described the relation among SIM, IMR and the number of sentences to consider necessary factors for the determination of threshold. Fig.1 shows a part of the result when selecting correct sentences in the UC. Fig.2 shows a part of the result when selecting fault sentences including errors in it.

Fig. 1 and Fig. 2 show different tendency between IMR1 (in case that IMR is 1) and IMR2 (in case that IMR is 2); the number of sentences in IMR2 is higher than IMR1 when the morphological analysis succeeded, on the other hand, the number of sentences in IMR1 is higher than IMR2 when analysis failed. Thus, we make use of two representative values related to IMR values to determine suitable formula: the variance of SIM in case of IMR2 and the number of sentences in case of IMR1.

4.2 Consideration of Threshold or Formula to Detect Morphological Errors

We show the graph of scatter plot between them (Fig.3). In it, we marked only 20 dots (sentences) selected at random from UC. A half of them (square) indicates sentences including errors. The other half (triangle) indicates sentences including no errors.

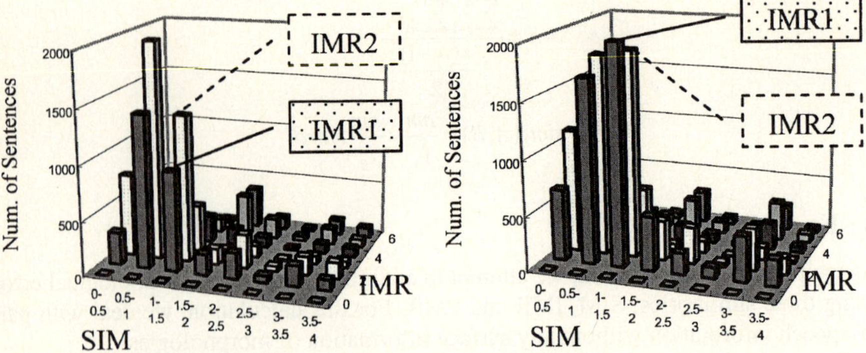

Fig. 1. Relation between SIM and IMR in case of CORRECT Result

Fig. 2. Relation between SIM and IMR in case of ERROR Result

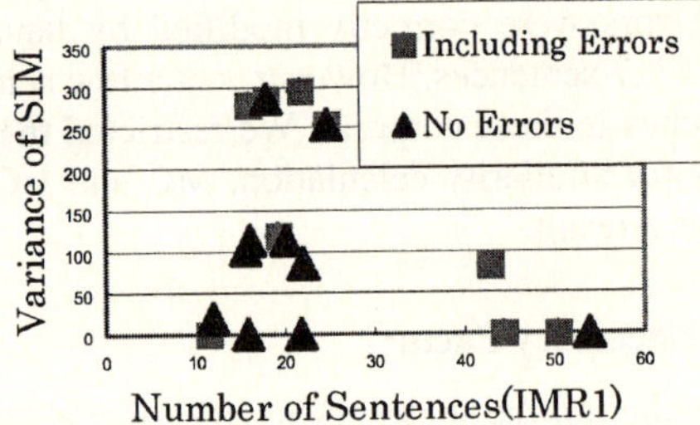

Fig. 3. Scatter Plot between Variance of Similarities and Number of Sentences (IMR1)

As a result, we tentatively got this formula to detect morphological errors (7). The variable x in it means the number of sentences in case of IMR1 explained above.

$$SV1 = 35/4x + 350 \tag{7}$$

If values which we assign the calculated x and get are larger than the variances of SIM, we judge that the analysis is incorrect.

Next, we got another formula (8) focusing on PMR, although we do not show some graphs indicating the relation between SIM and IMR due to lack of space.

$$SV2 = -1/2x + 180 \tag{8}$$

In the formula, the variable x means the number of sentences that the PMR value is in the range from 0 to 0.5 and the SIM value is in the range from 1 to 1.5. If calculate values are larger than the number of sentences which the SIM is in the range from 1 to 1.5, we judge that some errors occur.

4.3 Result

We experimented after implementing our system based on the factor previously mentioned. We confirmed that this system could detect morphological errors such as under-segmentation and part-of-speech tagging errors that *karasa* is interpreted as *kara* ("spicy" adjective) and *sa* (nominal-suffix) as explained in Section 1.

In the following subsections, we discuss the efficiency of our formulas when adopting only one formula and two formulas at a time.

- **Efficiency of single formula**

  The result of these experiments is described below (see SV1 and SV2 in Table 1).

Table 1. Precision and Recall

	SV1	SV2	SV1+SV2
Sentences (correct/false)	801 (744 / 57)		
Detected Sent. (correct/false)	23 (17 / 6)	25 (10 / 15)	5 (5 / 0)
Precision	74,0%	40,0%	100,0%
Recall	29,8%	17,5%	8,8%

The formula of SV2 does not function well because the values of SV2 are worse (Precision 40% and Recall 17.5%). Thus, this measure is not effective to detect morphological errors.

On the other hand, the precision of SV1 is relatively better (74%) although the recall (29.8%) is much lower. Our formulas, however, are not most appropriate because the precision means that we remain to need a considerable check of the result of morphological analysis. We consider this measure still has room for improvement.

- **Efficiency of multiple formula**

  Next, we experimented using new system which adopt two formulas at a time (SV1+SV2 in Table 1). In a sense, this is a kind of hybrid measure.

This precision (100%) is good and this measure function well. Certainly, the number of this detection (5) is very small and the detection is required to improve its recall, but it has an advantage that we do not need to perform a post-check procedure after morphological analysis because its precision reaches 100%.

5 Conclusion and Future Works

We proposed three measures related to similarities to detect various types of morphological errors without surface information of morphologies. From our experimental results, we found a useful measure because of 74% Precision. In addition, the other hybrid measure functions well (100% Precision) although the detection adapts to only 5 cases.

In future works, we need to adopt machine-learning techniques such as ADG, Automatically Detect Groups (Hara et al., 2004). ADG is one of genetic programming method for rule extraction and it has an advantage that rules for exceptional data can be extract in case of being missed by only a single rule. From this viewpoint of the characteristics of ADG, we consider ADG is useful because the reasons of morphological errors are not simple and the factors of them are more complex. Therefore, we will experiment with this approach on a large scale.

References

1. Asahara, M. and Matsumoto, Y. (2000), "Extended models and tools for high-performance part-of-speech tagger," In *Proceedings of The 18th International Conference on Computational Linguistics*, pp.21-27.

2. Hara, A. et al. (2004), "Discovery of cluster structure and the clustering rules from medical database using ADG; automatically defined groups," In "Knowledge-Based Intelligent System for Healthcare," Advanced Knowledge International, pp.51-86.

3. Hisamitsu, T. and Niwa, Y. (1998), "Post-processing of Japanese morphological analysis using transformation rules and contextual information," *IPSJ SIGNotes Natural Language Abstract*, 126, pp.55-62 (in Japanese).

4. Ichihara, H. et al. (1999), "*Youso no junnjokannkei kara mita ruijibunn saiteki syougou kennsaku* [Similar sentence retrieval using sequence relations of its elements]," In *Proceedings of the 5th Annual Conference of Natural Language Processing Society*, pp.521-524 (in Japanese).

5. Korfhage, R. (1997), "Information Storage and Retrieval," Wiley Computer Publishing, John Wiley & Sons, Inc., New York.

6. Kurosawa, Y., Ichimura, T., and Aizawa, T. (2003), "A description method of syntactic rules on Japanese filmscript," In *Proceedings of the 7th International Conference on Knowledge-Based Intelligent Information and Engineering Systems*, pp.446-453.

7. Kurosawa, Y., Ichimura, T., and Aizawa, T. (2005), "A description method of syntactic rules on filmscripts," *Journal of Natural Language Processing*, 12(2), pp.25-62 (in Japanese).

8. Matsumoto, Y. et al. (2000), "Morphological analysis system ChaSen version 2.2.1 manual," http://chasen.aist-nara.ac.jp/.

9. Utiyama, M. (1999), "Statistical measure for detecting errors in results of Japanese morphological analysis," *IPSJ SIGNotes NL Abstract*, 129, pp.71-77 (in Japanese).

10. Yamamoto, E. et al. (2003), "An IR similarity measure which is tolerant for morphological variation," *Journal of Natural Language Processing*, 10(1), pp.63-80 (in Japanese).

11. Yamamoto, E. et al. (2003), "Dynamic programming matching for large scale information retrieval," In *Proceedings of the 6th International Workshop on Information Retrieval with Asian Language*, pp.100-108.

Distributed Visual Interfaces
for Collaborative Exploration of Data Spaces

Sung Baik[1], Jerzy Bala[2], and Yung Jo[1]

[1] Sejong University, Seoul 143-747, Korea
`{sbaik,joyungki}@sejong.ac.kr`
[2] Datamat Systems Research, Inc., 8260 Greensboro Drive, Suite 120
McLean, VA 22102, USA
`jbala@dsri.edu`

Abstract. In our knowledge-based society, collaboration environments define a
key Knowledge Management trend. The critical step in the implementation of
this step is to capture the collaboration surrounding visual reports (e.g., charts)
used in decision-making. Indeed, the ability to provide an accurate and integrated visualization of the data spaces in order to support organizational decision-making is a crucial component of modern information-rich enterprises.
This ability can be significantly improved through the introduction of collaborative decision support tools tailored for visualization. This paper presents the development of a class of such tools, called DIVI (Distributed Intelligent Visual
Interfaces). The DIVI approach is based on the application of distributed visualization nodes and a mechanism for visual information abstrac-
tion/generalization. DIVI technology is envisioned to have a major impact on
those applications requiring information to be collected from different data
sources for collaborative group decision-making. Typical applications include
planning systems, logistics, computer-aided medical group diagnosis, and educational services.

1 Introduction

Today's decision makers rely heavily on the availability of information anywhere, at
anytime, and for any task and mission. However, as the amount of information available for them explodes and its availability comes from physically and logically distinct sites, the need arises for distributed and compatible access [1-3]. To accomplish
this, new computing tools are required that are capable of integrating data from different sources, distributing them throughout global communication systems, and presenting them in an optimal way. Since today's information exchange support infrastructures were designed and built to accommodate the requirements for voice and
low speed data, their collaborative capabilities are limited to electronic emailing,
electronic meeting rooms, and forms of video-conferencing. Furthermore, they are
characterized by:

- Too many interface scenarios with unclear functions,
- A flood of individual data streams to visual interfaces,
- Disjointed decision processes due to inappropriate information exchange, and
- Fragmented visualizations of the battle-space rendered by static interfaces.

R. Khosla et al. (Eds.): KES 2005, LNAI 3681, pp. 887–892, 2005.

Specifically, disjointed decision processes cause negative situational awareness, increase the time needed to comprehend the significance of information, cause incomplete and inaccurate understanding of the data-space, and delay decisions while waiting for more data. They do not have the capability to tailor the presentation of information to support decision-relevant details. As a result, initiatives to develop new classes of collaborative visual interfaces are needed, and they are specifically motivated by the following observations:

- As information network and Human Computer Interface (HCI) technologies converge, the use of distributed interactive computer-interfaces will experience a dramatic increase in terms of the use, scope ('functionality') and diversity of the users,
- Dynamic data-spaces change in unpredictable ways. Therefore, it is impossible, in principle, to pre-set all the HCI specifications into the global information exchange systems for the information visualization, and
- Manually setting HCI specifications into a distributed information exchange system is difficult, time-consuming, and often impossible; adaptability provides a fundamental vehicle for simplifying this process.

However, the development of such tools presents design and implementation challenges and requires further research efforts on the dynamics of distributed collaborative information exchange systems. This paper presents the development of a class of such tools called DIVI (Distributed Intelligent Visual Interfaces).

2   Visualization for Collaboration

Visualization as a support tool for decision-making has emerged in the last few years as a recurrent topic and a catchword. The purpose of distributed visualization is to support collaborative visualization in which multiple users can work together using multiple perspectives on the information [4]. Since most significant decision-making is often, by nature, collaborative with several decision makers, collaborative visualization allows geographically separate users to both share data and co-operate with the use of visualization and analysis [5]. If the data directly from the simulation can be shared then users can apply different visualization techniques at each location rather than looking at a static display provided by one of the participants [6]. In order to utilize visualization as knowledge, diverse forms of information are needed. It is thus clear that a considerable amount of knowledge for visualization should be stored and made available to the visualization engine to support the intelligent rendering of data designed to facilitate the decision-making process. Such knowledge should also be updated in an incremental fashion using both self-adaptation systems and Human Computer Interface (HCI) as new need and awareness. Adaptive visualization is a technology that can enhance the power of program visualization. The idea of adaptive visualization is to adapt the level of details in a visualization to the level of user knowledge about information [7]. In the domain of collaborative workspaces, it makes good sense to explore the possibilities of distributed visualization by letting each interface deal with the specifics of its own visualization tasks. At the same time, truly collaborative visualization networks should also allow for the seamless exchange of views, raw data, and all the relevant control parameters. Collaborative workspaces should involve not only data sharing but also sharing of the decision-making and visualization control processes. It is through the weighting of evidence, its corresponding certainty (or uncertainty) and responsibility levels that the overall collaborative visualization structure can make decisions on rendering and requests for further data sources.

3 Distributed Visualization Framework

The traditional metaphors associated with HCI, those of manipulation, locomotion and communication, have to be augmented by new metaphors such as collaborative information exchange for communities. The new metaphors require that visualization provide the users with the ability to conduct meaningful team interactions in order to enhance capabilities for situation awareness. The same metaphors also require that the Computer element be driven both by explicit users' requests and implicit ('hidden') users' motivations and pragmatics. Toward that end, the Interface element provides an intelligent medium where users and computers can conduct a meaningful dialogue. Users' strength of motivations (interested vs. bored) and beliefs (random vs. sure answer) are assessed based on new interface modalities such as facial expression, gesture recognition, speech intonation, and haptic stress.

The intelligent aspect of this distributed visualization paradigm is exhibited by its ability to bridge over semantic misalignments that occur between the Human (e.g. commander, analyst) and Computer components while gaining relevant feedback about users' beliefs and pragmatics. Alignment processes (e.g. semantic) should be available through an intelligent medium where users and computers can conduct a meaningful dialogue.

The goal for the Interface element is to facilitate the dialogue between the Human and Computer elements by bridging their apparent conceptual gaps using communication alignment at the semantic, pragmatic and motivational level. Semantic alignment is usually the only alignment current visualization techniques presently use, and even this is done only to a limited extent. The DIVI approach goes beyond semantic alignment and would assess users' pragmatics using decision-making tools and uncertainty analysis. Users' pragmatics, motivations and beliefs can then further facilitate the semantic alignment process as they provide meaningful feedback to the overall interpretation process. The novel DIVI metaphor for collaborative information exchange draws from the framework of distributed and behavior-based Artificial Intelligence, can adapt and reconfigure itself according to users' needs, and is developed and implemented using evolution and machine learning techniques.

Figure 1 presents an overall DIVI Framework for collaborative decision making. The framework consists of four components to support information sharing and collaborative decision making in distributed environments. For the first component, the agent analyzer is responsible for analyzing distributed information at each site. A global model at each site is generated by this analyzer with the distributed information. As a connective step for customizing the information, the bridging module is responsible for converting the information to a form of collaborative environment. In alignment module the elements that delay the decision making with awareness of incorrect situation and malfunction of meaning are removed. Lastly, as a collaborative workspace consisting of an interface, there are two components: the customized view and adaptive visualization. The customized view is one of the visualization steps for cognitive representation. Adaptive visualization is also one of the visualization steps, consisting of the rendering process and the conferencing module. The rendering process is offered to models for collaborative interaction using the Model-View-Controller Paradigm [8] (MVCP) at each Visualization Interface. The conferencing module effectively enables remote video conferencing by streaming video, voice and analyzed information through the network. Therefore, users can be offered semantic networks effectively.

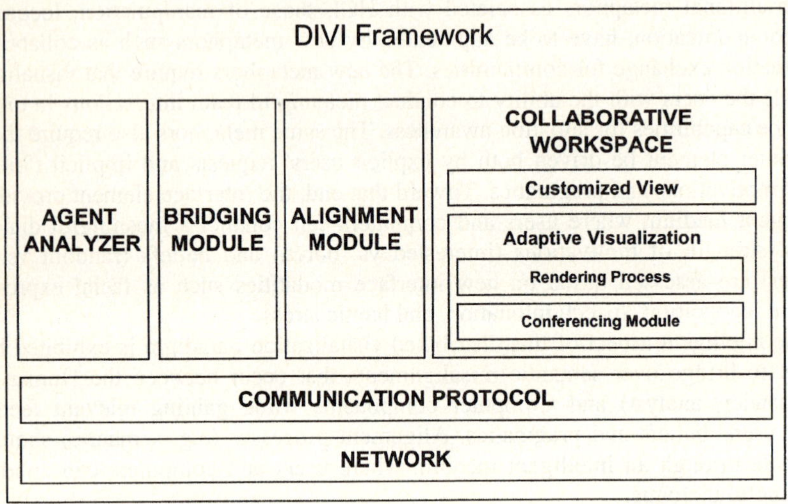

Fig. 1. DIVI Framework for Collaborative Decision Making

4 DIVI Implementation

The DIVI architecture applies advanced distributed 3D visualization and knowledge-based techniques, and it is specifically characterized by functional components capable of

- Tessellating Collaborative Workspaces through a network of Adaptive Visualization Interface (AVI) to enhance visualization tasks geared towards global decision making,
- Endowing each AVI with a Local Knowledge Database that provides relevant information to efficiently perform its duties in a distributed visualization environment,
- Endowing the Collaborative Workspace with software tools to augment the belief reasoning tools and thus to anticipate events and create virtual scenarios for increasing the efficiency and robustness for both the data collection and the decision-making processes (in order to achieve alignment),
- Generating customized information landscapes based on data models and control parameters obtained from local as well as from remote AVI data streams (Model/Control Layer),

- Distributing partial views and/or landscapes in a convenient format for ease of transmission through a Visual Layer,
- Exploiting the Visualization Layer for assemblage of complex, piece-wise defined, multi-resolution landscapes without excessive computational burden, and
- Using a highly modular and re-configurable implementation of AVI that will promptly react to changes in network connectivity, communication disruption, and global data decisions related to new data flow and/or visualization needs. Reconfiguration allows for adaptive reallocation of tasks and network connectivity at the global level and for AVI changes at the local level, respectively.

The DIVI software modules consist of a set of distributed objects (components) that can be reused in different network-based scenarios. To support the DIVI Framework; Distributed Agent Technology [9], JAVA 3D and JMF (Java Media Framework) are used to implement the components:

- As an analyzer for collaborative decision making, distributed agent technology is introduced for helping users offer consistent model information as well as clear model patterns. The technology introduced to our DIVI Framework is applied on the component of agent analyzer.
- JAVA 3D, which was introduced to our rendering process on visual interface, offers augmented and variable views for manipulating modeling information and transforming perspective. As one of the techniques providing variable rendering, the JAVA 3D technique is applied to the collaborative workspace. Therefore, we can produce 3D Space and offer effective perspectives. The JAVA 3D technique is very compatible with agent analyzer which is implemented by java programming. Because of its multimedia support and adaptation within different platforms, the technique was applied to the DIVI framework.
- JMF was used in the conferencing module for collaborative decision making and applied to effectively interact in a distributed collaboration environment. In this environment, using RTP, video, audio, and analyzed information of an individual data stream can be transferred in real-time. So it supports information sharing for decision making.

The implementation effort follows a component-based software paradigm. The term component-based software defines a software model, consisting of an architecture and a set of application programming interfaces (API), within which one can define modular software components that can be dynamically combined to form an application. The implemented components allow users to tailor visualization landscapes to achieve highly realistic situation-awareness of data-space, generate optimal plans for future operations, fully support the execution of current operations, and formulate efficient and effective real-time controls of the data pipeline. The DIVI enterprise visualization system maintains a proper balance between data collection, HCI, visualization and rendering decisions, and the decision-making process, through the development of benchmark tools, measures the overload information level on the human users and their ability to capture information and react to it in a proper and timely fashion. The DIVI graphical rendering software implements the Model-View-Controller Paradigm (MVCP) at each Visualization Interface. In MVCP, the rendering data is stored in models and transformed from models into graphical views through controllers. The controller can achieve this transformation with a broad variety of actions, including filtering and multi-resolution, routing data to different views within the local or remote nodes, and selecting which part of the view geometry and/or graphical appearance is affected by incoming data.

5 Conclusions

The initial experiments with DIVI have validated the integration of distributed visual interfaces and intelligent controls for supporting human decisions in collaborative environments. This integration extends the traditional metaphors associated with HCI, those of manipulation, locomotion and communication, towards new metaphors such as collaborative information exchange for communities. The current version of the DIVI architecture utilizes distributed rule-based representations to facilitate control mechanisms for interface visualization rendering. The database of rules evolves as users interact with the system. This process of communication alignment bridges initial conceptual gaps at the semantic, pragmatic, and motivational levels.

Summarizing DIVI research, we envision a class of decision-making tools that integrates distributed visualization and knowledge engineering techniques and creates an assembly of dynamic re-configurable interfaces. The interfaces render a coherent picture of the dynamic data spaces and support creative processes for the collaborative development of new knowledge.

References

1. Molina, A. I., Redondo, M. A., Bravo, C., Ortega, M., Using Simulation, Collaboration, and 3D Visualization for Design Learning: A Case Study in Domotics, LNCS(3190), pp. 164-171, 2004
2. Brezillon, P., Adam, F., Pomerol, J. C., Supporting Complex Decision Making Processes with Collaborative Applications - A Case Study, LNCS(2806), pp. 261-276, 2003
3. Cera, C. D., Regli, W. C., Braude, I., Shapirsten, Y., Foster, C. V., A collaborative 3D environment for authoring design semantics, IEEE Transactions on Computer Graphics and Applications(22), pp. 43-55, 2002
4. Schonhage, B., Ellens, A., Information exchange in a distributed visualization architecture: the shared concept space, Proceedings of Distributed Objects and Applications, pp. 335-334, 2000
5. Manuel, C., Jonice, O., Julia, S., Jano M. S., Decisio-Epistheme: An Integrated Environment to Geographic Decision-Making, LNCS(2569), pp. 126-136, 2002
6. Mark, W., Jason, W., Ken, B., A Distributed Co-Operative Problem Solving Environment, LNCS(2329), 2002
7. Peter, B., Hoah-Der, S., Adaptive Visualization Component of a Distributed Web-Based Adaptive Educational System, LNCS(2363), 2003
8. Baik, S. W., Bala, J., Hadjarian, A., Pachowicz, P., A Naïve Geography Analyst System with Cognitive Support of Imagery Exploitation, LNAI(2972), 2004.
9. Baik, S. W., Bala, J., Cho, J. S., Agent based Distributed Data Mining, LNCS(3320), pp. 42-45, 2004

Attribute Intensity Calculating Method
from Evaluative Sentences by Fuzzy Inference

Kazuya Mera[1], Hiromi Yano[2], and Takumi Ichimura[1]

[1] Faculty of Information Sciences, Hiroshima City University,
3-4-1, Ozuka-Higashi, Asa-minami-ku, Hiroshima 731-3194, Japan
{mera,ichimura}@its.hiroshima-cu.ac.jp
http://www.ariake.jp/mera/index.html
[2] Graduate School of Information Sciences, Hiroshima City University,
3-4-1, Ozuka-Higashi, Asa-minami-ku, Hiroshima 731-3194, Japan
yano@nlp.its.hiroshima-cu.ac.jp

Abstract. We propose a method to calculate the attribute's intensity of an object from multiple opinions. The intensity is calculated by min-max inference and the membership value of each degree group is obtained from the rate of collected opinions. The rate is calculated considering to the reliability of each opinion which are extracted from evaluative sentences on the Internet. Extracting the opinions from sentences, classifying the opinions into four degree groups, and applying the reliabilities of the opinions are done based on the grammatical features. Furthermore, in order to deal with such degree expression and reliability expression, we define a quintuplet dataset which consists of "evaluative subject," "focused attribute," "orientation expression," "degree expression," and "reliability expression."

1   Introduction

There is innumerable valuable information for various objects such as merchandises, services, organizations, individuals, and the information will strongly help to evaluate the objects. When we evaluate something new, we consider the attributes of the object such as price, design, taste, and so on.

There are two types of attributes; quantitative attributes like prize and size, and qualitative attributes like taste and impression. It is not so difficult to evaluate quantitative attribute because it has one specific value. On the other hand, it is hard to deal with qualitative attributes. When we deal with such qualitative information, fuzzy inference is used. Fuzzy inference can deal with ambiguous linguistic expression such as "a little" and "very."

Mamdani proposed a min-max inference to deal with such data. The method firstly obtains the membership value of each degree group about an attribute from the point of view of one person. Next, the fuzzy sets of the groups are calculated based on the fuzzy membership functions which are customized for the person. Then, the fuzzy sets are united and its gravity point is calculated as an integrated value of the attribute.

Now, we can obtain variable opinions about the target attributes from the Internet. However, we have to consider some more situations for the opinions from the point of view of the general public. Especially for qualitative attributes, people often de-

R. Khosla et al. (Eds.): KES 2005, LNAI 3681, pp. 893–900, 2005.
© Springer-Verlag Berlin Heidelberg 2005

clare different opinions because people often have different impressions for the same statement. Furthermore, people often choose different expressions even if they have same impressions.

In spite of such situation, people can integrate such conflicting opinions. When they deal with the opinions, they rely on the rate of the each opinion's group. In the former method, the rate is calculated only from the number of the opinions. However, when people express their opinion, they append reliability information of their opinions grammatically.

Therefore, we propose a method to calculate the integrated value of attribute from multiple opinions based on min-max inference method. Firstly, we extract the evaluative opinions from natural sentences on the Internet. The opinions are expressed by using quintuplet dataset which we developed. Next, the opinions are classified into four degree groups (a little, relatively, medium, and very) based on the degree expression. Then, the rates of the degree groups are calculated considering to the reliability expression in each opinion. Successively, the membership values of the degree groups are calculated from the opinions' rates. Then, the membership values are applied into the membership functions which we prepared to calculate the fuzzy set, and the gravity value is calculated as an integrated value of the collected evaluative opinions.

2   Process of Integrated Value Calculation

2.1   Mamdani's Min-Max Inference Method

In this paper, we calculate the integrated value of a target attribute from collected evaluative sentences based on Mamdani's min-max inference. The Mamdani's method calculates the integrated value of an attribute from the individual feeling. The integrated value is calculated from the membership values of the degree groups from the point of view of the person. Next, the fuzzy sets of the groups are calculated based on the fuzzy membership functions which are customized for the person. Then, the fuzzy sets are united and the gravity point is calculated as an integrated value of the attribute. Figure 1 shows an example of the method. Firstly, we research the correspondence of the membership function for each degree expression with the individual feeling about the degree expression. In this example, three fuzzy sets (small, medium, large) are prepared. Next, the subject indicates the membership value of each degree against the target attribute. In this example, the subjects indicates the membership values as "small: 0.2, medium: 0.1, and large: 0.7," respectively. Based on the membership values, the fuzzy sets are calculated by cutting of the head. Then, all the fuzzy sets are united and the gravity point is calculated as a integrated value of the subject's feeling.

2.2   Development for Multiple Opinions

When we deal with multiple opinions obtained from multiple subjects, we often meet some conflicts among the collected opinions. There are two significant reasons; the impressions for the same intensity are often different among the people and the expressions for the same impressions are also often different.

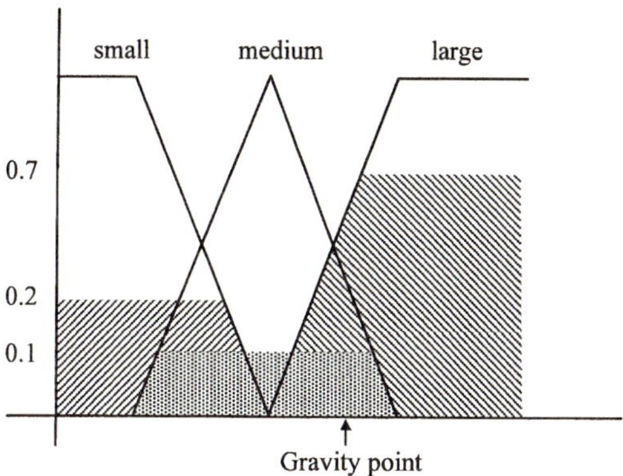

Fig. 1. Min-max inference method

When people evaluate qualitative attribute like spiciness and appearance, the criteria are quite different among the people because of their preferences, their experiences, and so on. However, if we can collect multiple opinions, the effect from the minor opinions will be weakened.

On the other hand, people may choose different expressions even if they have same impressions. Some people often use big words and the other seldom uses big words. The correspondence of the impression with the expression is different among the people. In order to avoid such problem, it is the best way to construct the fuzzy sets of the expressions based on the feeling of all the subjects. However, it is almost impossible if we deal with the evaluate sentences explored from the Internet. Therefore, we propose a method to calculate the integrated value from multiple opinions written in natural language expressions. We show our process as follows;

0. Prepare the membership functions for the general public

Generally, when we deal with the opinions of multiple subjects, we have to use the membership functions customized for the subject. However, it is almost impossible if we deal with the evaluate sentences explored from the Internet. Hersh researched the membership functions of some expressions [1] for 19 subjects as shown in Figure 2. Therefore, we prepare simpler membership functions based on Hersh's result as shown in Figure 3.

1. Extract the quintuplet datasets (as described in section 3.1) from evaluative sentences and classify them into groups which indicate the same object and attribute

The former method cannot deal with the information about degree and reliability. Then, we define the quintuplet dataset which consists of "evaluative subject," "focused attribute," "orientation expression," "degree expression," and "reliability expression for the extracted information." Each element is expressed in natural language words. These words are extracted by using grammatical templates as described in section 3.2.

2. Classify the datasets into groups which express same degree of attribute

"Degree expression" in a dataset is expressed by an adverb. However, there are many adverbs in natural language and it is impossible to define a group for every words.

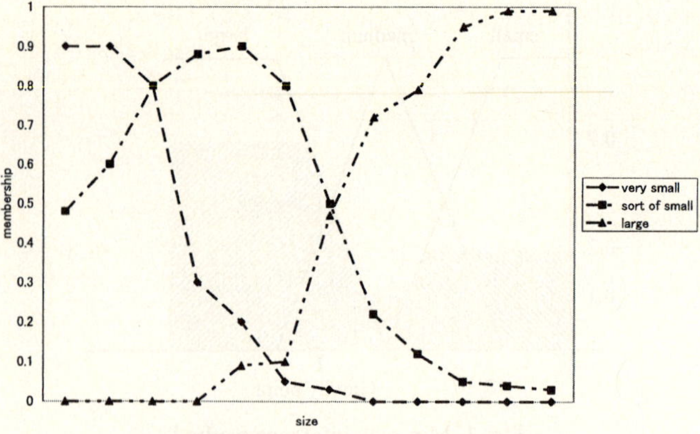

Fig. 2. Membership functions by Hersh

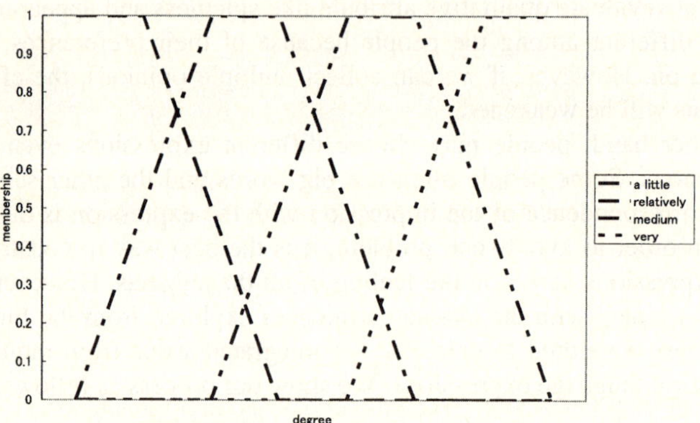

Fig. 3. Membership functions for our method

Therefore, we classify the adverbs into four groups based on the intensity of the attribute (a little, relative, moderate, very). The classification method is described in section 3.3.

3. Calculate the rate of the degree groups' opinions

In order to apply min-max inference, the membership values of the degree groups are required. In this paper, we calculate the membership values from the rate of the collected opinions. In general, the rate of the opinions are calculated the number of the opinions in each degree group. However, when we meet some different opinions such as "food-A is **surely** bad" and "food-A **might be** good," we usually believe the opinion whose reliability feels higher. Then, we calculate the groups' rate considering to the opinion's reliability which is guessed from the reliability expression in the sentence. The membership for group A is calculated as follows;

$$R_A = \frac{\sum_i \omega \times \rho_i}{\sum_j \omega \times \rho_j} \qquad (i \in A, j \in X)$$

R_A means the rate of the group A. i and j are the opinion which is classified into group A and all opinions, respectively. ρ_i indicates the reliability value of the opinion i. ω means the weight of an opinion. We define ω as 1.0. The relationship between the reliability expression and the value is described in section 3.3.

4. Calculate the membership values for degree groups

The degree groups' rate indicates the tendency of the collected opinions. We recognize the values as the membership degree which the subjects feel, because the rate is calculated based on their opinions and the result reflects the degree that they feel each degree.

$$\mu_A = R_A$$

5. Calculate the integrated value of the attribute

By applying Mamdani's min-max inference, we calculate the integrated value from the gravity point of the united fuzzy sets.

3   Quintuplet Dataset for Attribute Information

3.1   Definition of Quintuplet Dataset

In order to express the evaluative information obtained from natural language sentences, Kobayashi proposed the triplet dataset which consists of "evaluated subject (ES)," "focused attribute (FA)," and "orientation expression (OE)." She defined these elements as the fillers for the following template;

<Focused Attribute> of <Evaluated Subject> is <Orientation Expression>

For example, when the template is applied into an example sentence "The price of food-A is very expensive," the ES is "food-A," the FA is "price," and the OE becomes "expensive." When we apply another example "food-B is spicy," the ES is "food-B," the OE is "spicy," and the FA is not extracted [2].

However, this type of dataset cannot distinguish the intensity of the attribution between "food-A is **a little** hot" and "food-B is **extremely** hot." Furthermore, when we meet some different information such as "food-A is **surely** bad" and "food-A **might be** good," most of us pay attention to the reliability of the information. Therefore, we add two more elements of data for "the degree of the attribute's intensity" and "the reliability of the dataset" into the former triplet dataset. We will explain these new elements in section 3.3 and 3.4.

3.2   Quintuplet Dataset Extracting Method

We extract the quintuplet dataset from collected natural language sentences about the evaluated subject from the Internet. At first, the raw sentences are analyzed morphologically using ChaSen which is a Japanese morphological analyzer [3]. Next, the sentences are divided into clauses. Successively, the "quintuplet extracting rules" are applied to the clauses which contain "orientation expression."

The quintuplet extracting rules are described in BNF format as shown in table 1. At first, the system checks whether or not there is "sentence-final expression" at the beginning and/or end of the sentence. Then, the system extracts the phrase which will contain the information about the evaluated subject. The extracting rules are provided as templates for the phrases. They show the word order of the extracted elements and some particles in Japanese. When "evaluated subject" or "focused attribute" contains multiple elements in a clause, multiple datasets are obtained from the clause.

3.3 Classifying Dataset for the Degree Group

There are a lot of expressions to indicate the degree in Japanese. Mera et al. investigated the effects of the modifications for 31 adverbs (positive: 18, negative:13) in order to analyze the degree of the affirmative/negative intention in the user's reply [4]. Table 2 shows the adverbs' groups and their values to change the positive/negative degrees. It shows that the adverbs like *"hijouni"* and *"totemo"* enhance the expressed attribute, and the adverbs like *"sukoshi"* and *"shoushou"* deflate the attribute. We classify the quintuplet dataset into four positive groups and three negative groups based on these groups and we prepare the membership functions for each degree.

Table 1. Quintuplet extracting rules for Japanese sentences

	Rules
clause::=	(<SFE> | φ) <phrase> (<SFE> | φ)
phrase::=	<Subj> *no* <Attr> (*ga* | *wa* | *mo* | *ni* | *wo*) <OrieP> |
	<Subj> *no* <OrieP> *na* <Attr> |
	<Subj> *no* <OrieP> |
	<OrieP> (*no* | *na* | φ) (<Subj> | <Attr>) |
	<OrieP> *no aru* <Subj> |
	<Attr> (*ga* | *wa* | *mo* | *ni*) <OrieP> |
	<OrieP> <Suffix> *no* <Attr> |
	<OrieP> <Subj> |
	<OrieP> <Attr> *no* <Subj>
Subj::=	<Evaluative Subject> (<Para> <Subj> |φ)
Attr::=	<Focused Attribute> (<Para> <Attr> |φ)
OrieP::=	(<Degree>|φ) <Orientation Expression>
Suffix::-	*teki* | *fuu* | *sei*
Para::-	*ya* | *to* | *mata* | *matawa*
SFE::-	(shown in Table 2)
Degree::-	(shown in Table 1)

3.4 Reliability Expressions of Evaluated Sentences

Aoyama researched the effect of the sentence's reliability using Japanese sentence-final expressions. In his questionnaire, he presented the sentences with sentence-final expressions and the subjects replied the reliability of the contents in the range [-100, 100]. 0 indicates "the speaker does not know about the content at all" and 100 indicates "the speaker is certain of the content." He experimented with 26 positive expressions and 42 negative expressions [5]. Table 3 shows a part of the expressions. In this paper, we apply 22 positive expressions and 16 negative expressions whose standard deviation of the reliability value is less than 20.00.

Table 2. Japanese adverbs to express the degree of predicates

Group	Adverbs
positive	*Hijouni, Zuibun* (greatly, extremely), *Kekkou, Kanari, Daibu* (quite, very well), *Totemo* (very, really), *Jyuubun, Yoku* (enough, sufficiently), *Soutou* (considerable)
	Warito, Wariai, Warini (comparatively, relatively), *Maamaa* (moderate)
	Sukoshi (slightly), *Chotto, Shoushou, Tashou* (a little), *Ikuraka* (somewhat)
	(no modified expression by adverb)
negative	*Mattaku, Zenzen* (quite, entirely), *Sukoshimo, Sappari, Chittomo* (not at all), *Toutei* (hardly), *Mettani* (rarely), *Totemototemo* (not at all)
	Sonnani, Amari, Anmari (not very), *Sahodo* (not so much), *Taishite* (not very much)
	(no modified expression by adverb)

Table 3. Some of Japanese sentence-final expressions and their reliability values

sentence-final expression	Reliability
Taro wa kekkon suru. (Taro gets married.)	93.10
*Taro wa kekkon suru **darou**.* (Taro will get married.)	67.80
*Taro wa kekkon suru **kamosirenai**.* (Taro may get married.)	49.40
***Kanarazu** Taro wa kekkon suru.* (Surely Taro gets married.)	93.40
Taro wa kekkon sinai. (Taro doesn't get married.)	-92.86
***Osoraku** Taro wa kekkon sinai **darou**.* (Probably, Taro won't get married.)	-52.14
*Taro ga kekkon suru **ka douka sirimasen**.* (I don't know whether Taro gets married or not.)	2.14

4 Conclusive Discussion

We proposed a method to calculate the attribute's intensity of an object from multiple opinions. We developed min-max method by calculating the membership values from the rate of classified opinions. We prepared membership functions to deal with the opinions from general public. Furthermore, our method considered the reliability expressions in the sentences. We assigned reliability values for 38 Japanese sentence final expressions, and we classified 31 modifiers into four positive/negative groups. In order deal with degree and reliability expressions, we developed the format of opinions and we proposed the dataset extracting rules by using 17 templates based on the Japanese grammatical features.

Our method will be applied not only evaluating subjects but also learning the attributes of subjects automatically. The result can be employed as "favorite degree" of the subjects in "Emotion Generating Calculations Method [6]."

References

1. Hersh H. M. and A. Caramazza: A fuzzy set approach to modifiers and vagueness in natural language, J. Exp. Psych.: General, 105, 3.
2. Kobayashi, N., Inui, K., et al.: Collecting Evaluative Expressions by A Text Mining Technique, Technical Report of IPSJ, Vol. 2003-NL-154, (2003) 77–84 (in Japanese)

3. Matsumoto, Y., Kitauchi, A., et al.: Japanese Morphological Analysis System ChaSen version 2.2.1, http://chasen.naist.jp/hiki/ChaSen/ (in Japanese)
4. Yoshie, M. Mera, K., et al.: Analysis of affirmative/negative intentions of the answers to yes-no questions and its application to a web-based interface, Journal of Japan Society for Fuzzy Theory and Systems, Vol.14, No. 4, (2002) 393-403 (in Japanese)
5. Aoyama, H.: A Mathematical Analysis of Epistemic Modalities and the logic of Conversation, Journal of Logical Philosophy, No. 2 (2001) 1–19 (in Japanese)
6. Mera, K.: Emotion Orientated Intelligent Interface, Doctoral Dissertation, Tokyo Metropolitan Institute of Technology, Graduate School of Engineering.

Proposal of Impression Mining
from News Articles

Tadahiko Kumamoto[1] and Katsumi Tanaka[1,2]

[1] National Institute of Information and Communications Technology,
Kansai Science City, Kyoto 619-0289, Japan
`kuma@nict.go.jp`
[2] Kyoto University,
Kyoto, Kyoto 606-8501, Japan

Abstract. Each word in a language not only has its own explicit meaning, but also can convey various impressions. In this paper, we focus on the impressions people get from news articles, and propose a method for determining impressions of these news articles. Our proposed method consists of two main parts. One part involves building an "impression dictionary" that describes the relationships among words and impressions. The other part of the method involves determining impressions of input news articles using the impression dictionary. The impressions of a news article are represented as scale values in user-specified impression scales, like "sad – glad" and "angry – pleased". Each scale value is a real number between 0 and 1, and is calculated from the words (common nouns, action nouns, verbs, adjectives, and katakana characters) extracted from an input news article using the impression dictionary.

1   Introduction

In the technological fields of human properties, studies for inferring information providers' emotions have been actively done in recent years [1, 2, 8, 9]. The approach of mining impressions from text, however, has not been exhaustively explored. This paper, therefore, focuses on the impressions people get from news articles, and proposes a method for automatically determining impressions of news articles. Our proposed method consists of two main parts. One part involves the construction of an "impression dictionary" that describes the relationships among words and impressions, and the other part determines the impressions which a journalist put in a target news article using the impression dictionary.

We suggest that our proposed method will satisfy the specifications listed below from the perspective of cost and practicality.

(1) Automatic construction of impression dictionaries.
In this paper, an electronic dictionary that describes relationships among words and impressions is called an impression dictionary. Generally, methods that require a worker's judgment in constructing so-called dictionaries are not only expensive, but also cause many of the problems discussed below. i) There is

R. Khosla et al. (Eds.): KES 2005, LNAI 3681, pp. 901–910, 2005.

individual variation among the impressions people get from a news article. ii) Criteria of the judgment may change according to worker's mental states such as personality, physical condition, and emotion. iii) Reconstructing and correcting impression dictionaries is not easy in terms of the time and labor required. Hence, the fully automated construction of impression dictionaries is indispensable.

(2) Extraction of arbitrary kinds of impressions.
The limitation of impressions to be extracted narrows the application range of the method. Since impressions to be extracted vary according to application domains and situations, impression mining methods should be designed so as to be able to extract arbitrary kinds of impressions. This causes a new problem, however. That is, numerically expressing the strength of each impression is required in order to determine characteristic impressions from two or more impressions extracted.

(3) Data describing accurate relationships among words and impressions is not needed.
It is not easy to create such data in advance because the problems described above (1) occur. In addition, it seems impossible to create data that describes an accurate relationship among words and impressions against all arbitrary kinds of impressions. Hence, an unsupervised learning strategy should be adopted.

In this paper, we verify the effectiveness of our proposed method by comparing the impressions which our proposed method determined about one hundred news articles [1] and the impressions which fifty people gave of the same articles. In this comparison, we adopt four impression words that denote basic impressions and have distinctive meanings, and compose two impression scales; "sad – glad" and "angry – pleased". The impressions of a news article are represented as scale values in these two impression scales. Each scale value is a real number between 0 and 1, and is calculated from the words (common nouns, action nouns, verbs, adjectives, and katakana characters) extracted from its input news article using the impression dictionary which our proposed method created from the Nikkei Newspaper Full Text Database [6] including two million news articles or more over a 12-year period, from 1990 to 2001.

The remainder of this paper is organized as follows. In Sect. 2, we summarize our related work. In Sect. 3, we present a method for building an impression dictionary, and for determining impressions of news articles using the impression dictionary. In Sect. 4, we verify the effectiveness of our proposed method. Finally, we conclude this paper, and discuss challenges to be overcome in the future.

2   Related Work

Studies for inferring information providers' emotions are the most prevalent in the technological fields of human properties. The goal of these studies has been enabling smoother communication with robots and avatars and smoother communication between people using e-mail. In order to infer speakers' emotions, the

---

[1] These articles were all collected from the Yahoo News site
(http://dailynews.yahoo .co.jp/fc/).

studies have proposed a system that extracts vocal features from their utterances [9], a system that recognizes their emotionally charged or offensive utterances, like "wow" and "oh my god"[1], and a system that measures their facial expressions with a three-dimensional measuring instrument, and matches the facial expressions with five typical facial expressions [2], where their target emotions have been limited to several specific reactions, like "anger", "joy", "sadness", "normal", "laugh", and "excitement". Since these studies do not use common nouns, action nouns, verbs, adjectives, and katakana characters that our proposed method uses, the studies and our study complement one another. On the other hand, a system that infers speakers' emotions using positive feelings [4] of the words appearing in their utterances has been proposed [8], while positive feelings associated with a word was determined by workers based on their experiences. This means that specification (1) is not satisfied. Since several specific kinds of emotions are inferred by "IF-THEN" logic, re-writing all the rules is necessary in order to enhance to other kinds of emotions. Hence, it is difficult to satisfy specification (2).

Methods of extracting writers' reputations and evaluation from movie reviews, book reviews, and production evaluation questionnaires also have been studied. For example, Turney proposed a method of classifying various genres of reviews into "recommended" or "not recommended" [7]. The primary characteristics of his method are the following: One is that the class into which an input text is classified is determined based on a relationship between each of the words extracted from the text and the predefined reference words, i.e. "excellent" and "poor", and the other is that the relationship is determined based on the number of hits obtained using the AltaVista Advanced Search engine [2]. This method satisfies specification (1) described above, but it is difficult to satisfy specification (2) because the two reference words were designed only for a specific impression scale; "recommended – not recommended".

We hit on the idea of classifying input documents into two or more impression classes using a text classification method. Many researchers have previously tackled the problem of creating more accurate classifiers using less accurate answer data [5, 10]. Specification (3), however, is not satisfied because their methods have still required a large amount of correct answer data. On the other hand, we submit that the documents with impression tags which can be produced by our proposed method would have a partially bad relationship, but they can be reused as correct answer data, which may contribute to the creation of classifiers that are accurate enough. This problem will be the subject of future research.

3 Impression Mining Method

In this section, we present a procedure for calculating the scale value of a word in an impression scale, and creating an impression dictionary that describes a relationship between words and scale values in an impression scale, and then we present a procedure for determining the scale value of a news article in an impression scale.

---

[2] http://www.altavista.com/sites/search/adv

Table 1. Numbers of articles including each impression word.

Editions	sad (kanashii)	glad (ureshii)	angry (okoru/ikaru)	pleased (yorokobu)
1990	213	844	238	985
1991	217	835	263	906
1992	111	664	187	694
1993	126	638	199	643
1994	114	648	185	593
1995	180	885	268	790
1996	179	958	298	766
1997	168	924	252	779
1998	162	930	289	833
1999	140	898	248	755
2000	156	951	226	739
2001	146	945	199	746
Total	1,912	10,120	2,852	9,229

3.1   Design Concept

As described in specification (1), automatic construction of an impression dictionary is the most important task to be solved. In addition, as described in specification (3), an unsupervised learning method that does not use any correct answer data should be proposed. We, therefore, design some heuristic rules as Turney did. Also, the proposed method must be independent of the kind of impression scales so that the method can deal with any arbitrary kinds of impressions, rather than with specific kinds of impressions, as described in specification (2). First, we set up the assumption that articles including an impression word e transmit an impression denoted by the impression word. Next, we use a very large newspaper text database to refine this assumption, and formulate which of two comparable impression words a word co-occurs more often with in the database. Consequently, an impression dictionary can be automatically constructed for arbitrary kinds of impression scales.

3.2   Parameters of Newspaper Text Database

The text database we use is the Nikkei Newspaper Full Text Database [6] from 1990 edition to 2001 edition, which has two million or more articles in total. Each edition consists of about 170,000 articles (about 200MB). The numbers of articles including the impression words "sad (kanashii)", "glad (ureshii)", "angry (okoru/ikaru)", and "pleased (yorokobu)" in each edition are listed in Table 1, where the Japanese translation of each impression word is written by the Roman alphabet in parentheses.

3.3 Automatically Constructing Impression Dictionaries

Each entry in an impression dictionary corresponds to a word whose part of speech is a common noun, action noun, verb, adjective, or katakana character [3], and has a scale value and a weight in each of target impression scales. The scale value in an impression scale of a word shows which of two impression words composing the impression scale the word co-occurs more often with, and is calculated using the following procedure.

In the Y edition, let the number of articles including an impression word e be $DF(Y, e)$, and let the number of articles including both e and target word w be $DF(Y, e\&w)$. Joint probability $P(Y, e\&w)$ of e and w is calculated as follows.

$$P(Y, e\&w) = \frac{DF(Y, e\&w)}{DF(Y, e)} \tag{1}$$

Interior division ratio $R(Y, e_1, e_2, w)$ of $P(Y, e_1\&w)$ and $P(Y, e_2\&w)$ shows which of two impression words e_1 and e_2 composing the impression scale $e_1 - e_2$ a target word w co-occurs more often with, and is calculated as follows.

$$R(Y, e_1, e_2, w) = \frac{P(Y, e_1\&w)}{P(Y, e_1\&w) + P(Y, e_2\&w)} \tag{2}$$

where $R(Y, e_1, e_2, w) = 0$ when the denominator is 0. The values of $R(Y, e_1, e_2, w)$ in all the editions are calculated, and the scale value $S(e_1, e_2, w)$ in $e_1 - e_2$ of w is calculated as follows.

$$S(e_1, e_2, w) = \sum_{Y=1990}^{2001} R(Y, e_1, e_2, w) \bigg/ \sum_{Y=1990}^{2001} T(Y, e_1, e_2, w) \tag{3}$$

if $DF(Y, e_1\&w) + DF(Y, e_2\&w) > 0$ then $T(Y, e_1, e_2, w) = 1$
otherwise then $T(Y, e_1, e_2, w) = 0$

The $T(Y, e_1, e_2, w)$ term is used as a token for excluding $R(Y, e_1, e_2, w)$ of the edition in which w did not co-occur with e_1 and e_2. Since we know the words (for example, the words related to the Olympics) that do not appear every year but are strongly related to specific impression words, the $T(Y, e_1, e_2, w)$ term was introduced with such words. In the meantime, the values of $\sum_{Y=1990}^{2001} T(Y, e_1, e_2, w)$ and $\sum_{Y=1990}^{2001}(DF(Y, e_1\&w) + DF(Y, e_2\&w))$ may be large or small. Therefore, we define the weight $M(e_1, e_2, w)$ in $e_1 - e_2$ of w as follows.

$$M(e_1, e_2, w)$$

$$= log_N \sum_{Y=1990}^{2001} T(Y, e_1, e_2, w) \times log_{N^2} \sum_{Y=1990}^{2001} (DF(Y, e_1\&w) + DF(Y, e_2\&w)) \tag{4}$$

---

[3] Negative forms of verbs and adjectives such as "do not erase (kesanai in Japanese)" and "be not happy (tanoshikunai)" are dealt with as one word in this paper.

Table 2. Words that co-occur most frequently with "sad" and "glad".

Part of speech	Target word	Scale value	Weight	Target word	Scale value	Weight
common noun	death	0.839	1.170	smile	0.162	1.083
	sacrifice	0.804	1.048	Olympic	0.159	1.212
action noun	divorce	0.805	0.877	manufacture	0.197	1.086
	mourning	0.875	0.861	scream	0.049	1.228
verb	lose	0.801	0.997	can win	0.097	1.060
	die	0.822	0.880	sell	0.189	1.104
adjective	sad	0.986	1.535	good	0.189	1.144
	unhappy	0.801	0.949	glad	0.068	1.858
katakana	Revi	0.857	0.346	variety	0.175	0.806
	Spielberg	0.811	0.302	par	0.129	0.843

Table 3. Words that co-occur most frequently with "angry" and "pleased".

Part of speech	Target word	Scale value	Weight	Target word	Scale value	Weight
common noun	anger	0.926	1.135	repute	0.194	1.104
	opposition party	0.817	1.015	parting with	0.016	1.128
action noun	remonstrance	0.850	0.958	welcome	0.190	1.060
	prosecution	0.803	0.799	attracting	0.173	1.072
verb	angry	0.985	1.611	open	0.175	1.088
	scold	0.812	1.012	pleased	0.048	1.840
adjective	ridiculous	0.817	0.901	approvingly	0.139	0.827
	irresponsive	0.801	0.706	cheap	0.152	0.879
katakana	booing	0.861	0.625	birdie	0.150	0.645
	Wahid	0.938	0.416	bio	0.154	0.697

where N denotes the total number of editions used in constructing its impression dictionary, and is equal to 12 in this example.

A sample of the impression dictionary constructed using the method mentioned here is shown in Tables 2 and 3, where, in each part of speech, the heaviest two words in weight are selected and presented out of the words whose scale values are 0.8 or more or are 0.2 or less. The number of target words registered in the impression dictionary is shown in Table 4, together with their total co-occurrence frequencies with each impression word.

3.4   Determining Scale Values of News Articles

First, an input article $TEXT$ is segmented into words by using the Japanese morphological analyzer called "Juman" [3], and the words whose parts of speech are common nouns, action nouns, adjectives, verbs, and katakana characters are extracted from the input article. Negative forms of verbs and adjectives, however, are not segmented and are regarded as one word. For example, the negative form "be not sad (kanashikunai in Japanese)" of the adjective "sad (kanashii)"

Table 4. Parameters of impression dictionary.

	Different number	Co-occurrence freq. sad	glad	Different number	Co-occurrence freq. angry	pleased
common noun	23,114	122,899	715,782	23,675	127,919	799,663
action noun	6,742	35,909	195,510	7,013	39,125	238,390
verb	13,503	66,969	483,990	14,228	71,995	544,829
adjective	3,244	18,824	153,212	3,263	19,599	157,231
katakana	17,683	29,322	37,624	17,750	29,723	38,413
Total	64,286	273,923	1,586,118	65,929	288,361	1,778,526

Table 5. The number of articles in each genre used for performance evaluation.

Society	Oversea	Sports	Economy	Politics	Industrial	Region	Total
27	11	19	7	16	7	13	100

is not segmented into "sad (kanashii)" and "not (nai)", but is dealt with as one adjective "be not sad (kanashikunai)". Next, the scale value $S(e_1, e_2, w)$ and weight $M(e_1, e_2, w)$ of each word are obtained by consulting the impression dictionary, and the scale value $O(e_1, e_2, TEXT)$ of the input article is calculated using (5).

$$O = \sum^{TEXT} (S \times |2S - 1| \times M) \bigg/ \sum^{TEXT} (|2S - 1| \times M) \qquad (5)$$

where the $|2S - 1|$ term denotes an inclined distribution depending on the scale value S. When the scale value S is 0.5, the $|2S - 1|$ term is 0. When the scale value S is 0 or 1, the $|2S - 1|$ term is 1. Many of the words that appear in the articles seem to be independent of the impression of the articles. The inclined distribution described here has been introduced to remove the adverse effect that such general words cause in calculating (5).

4 Verification of Effectiveness

In order to verify the effectiveness of our proposed method, we use one hundred articles collected from the Yahoo News site on the World-Wide Web, and compare the evaluation results which our proposed method determined and the evaluation results which fifty subjects (30 women and 20 men ranging in age from their twenties to their sixties) gave for each article. Here, the number of articles in each genre is shown in Table 5.

First, we asked the subjects the following question. "Suppose that you are an announcer. What do you feel when you read these articles out loud?" At

Table 6. Article with the highest standard deviation in score distribution and its score distribution (This article was translated into English by the authors).

(a) Article ID No. 62 (Genre: Society)

On national road No.175 of Shidaka, Maizuru, Kyoto Prefecture, a sightseeing bus came to a standstill because the Yura river was overflowing, and 37 employees of the Toyooka Branch of the Pensioners' Union for Hyogo Local Government took shelter on the roof of the bus. Their rescue started at 6:00AM on the 21st, with the first helicopter sighting of the bus passengers atop the bus occurring at 6:12AM. The Maizuru antidisaster headquarters said that the Maritime Self-Defense Force confirmed the safety of all 37 people from the bus at 6:30AM, and rescued everyone at 8:24AM. A Navy self-defense member who had been approximately five kilometers away went to the site in a rubber dinghy and rescued one woman at 6:20AM. The rescue team of the Japan Coast Guard found the bus passengers hanging on to utility poles. These people were rescued by the helicopter at 6:30AM. In addition, area homeowners got on the roofs of their homes because the flooding was so serious. The rescue team rescued eleven people in total by 8:15AM. At that time, the water was waist-high. The people rescued by the helicopter were all carried to the West playing field in Maizuru, and the two men who first arrived at 6:45AM were taken to the Maizuru Red Cross hospital.

By Yomiuri Shimbun, Inc.   Issued at 10:07AM, October 21, 2004

(b) Score distribution in the impression scale "sad – glad"

Score	5 points	4 points	3 points	2 points	1 point
Number of subjects	4	13	16	15	2

the same time, we advised, "You may feel various kinds of feelings. But please focus on sadness, gladness, anger, and pleasure. If you experience ambivalent feelings after reading an article, you should put a priority on the most strongly felt emotion." They read each article in turn, and evaluated it by five steps using each of the two impression scales, "sad (5 pts) – somewhat sad (4 pts) – neutral (3 pts) – somewhat glad (2 pts) – glad (1 pt)" and "angry (5 pts) – somewhat angry (4 pts) – neutral (3 pts) – somewhat pleased (2 pts) – pleased (1 pt)". Here, as a sample of the evaluation results, is the article that had the highest standard deviation (1.03) in score distribution. The article is shown in Table 6, together with its score distribution.

Next, an impression dictionary was constructed from the Nikkei Newspaper Full Text Database for a 12-year period using the method described in 3.3, which describes the relationship among the target words extracted from all the articles in the database and each of the two impression scales, i.e. "sad – glad" and "angry – pleased". The scale value of each article was calculated using the method described in 3.4. We compared the scale value which our proposed method determined for each article and the evaluation results of fifty subjects' reactions to this article. Note that, if the scale value of an article is 0.570 or more, the tag "sad/somewhat sad" was attached to the article against the impression scale "sad – glad" and the tag "angry/somewhat angry" was attached to the article

Table 7. Results of performance-evaluation.

	sad – glad	angry – pleased
Number of agreement cases	2,614	3,046
Agreement ratio	52.3%	60.9%
Chance ratio	48.3%	48.9%
Maximum ratio	74.6%	77.1%
Minimum ratio	1.6%	0.7%

against the impression scale "angry – pleased". Similarly, if the scale value of an article is 0.343 or less, the tags "glad/somewhat glad" and "pleased/somewhat pleased" were attached to the article. Otherwise, the tag "neutral" was attached to the article. When the tag that our proposed method attached to an article was the same with the tag that one subject attached to the article, the number of agreement cases increased by 1. The number of agreement cases and its ratio are shown in Table 7, together with chance ratio (ratio obtained when an impression mining method always attaches the "neutral" tag which were most attached by the subjects) and maximum/minimum ratio (ratio obtained when an impression-mining method always attaches the tag which was most/least attached by the subjects against each article). The threshold values were experimentally set.

We can see from Table 7 that the agreement ratio is 12 points higher than the chance ratio against the impression scale "angry – pleased", and our proposed method is comparatively simple but has high accuracy. On the other hand, the agreement ratio is just slightly higher than the chance ratio against the impression scale "sad – glad", and we do not say that the accuracy is excellent. Anyway, it is certain that our proposed method has not reached a practical accuracy, i.e. 90% or more. However, the theoretical maximum ratios are 74.6% and 77.1%. This indicates that it is obviously necessary not only to complicate an impression mining method but also to deal with users' knowledge, sensibility (personality, preference, interest), mental state (emotion, physical condition), and subscription environment (place, time, subscription history) in order to achieve practical accuracy. We would like to address this problem in the near future.

5   Conclusion

In this paper, we have proposed a method for determining the scale values (real numbers between 0 and 1) of news articles. Our proposed method consists of two main parts. The first part automatically builds an impression dictionary from a very large newspaper text database. The impression dictionary describes a relationship between words (common nouns, action nouns, adjectives, verbs, and katakana characters) and impression scales. The other part of the method automatically determines the scale value of an input news article in a user-specified impression scale. In order to achieve automatic and unsupervised construction of impression dictionaries and use arbitrary kinds of impressions, we formulated which of two impression words that compose an impression scale each of the words extracted from an input news article co-occurs more often with.

Future research will focus on a more detailed analysis of the results of the performance evaluation obtained in this experiment, and develop an impression mining method that achieves a higher accuracy. We will also explore the factors that influence the relationship between articles and impressions toward personal adaptation. We will place particular focus on the order in which articles are read and the hotness of topics, and develop a method for managing users' subscription histories. Now, we are applying our proposed impression mining method to Web retrieval, user modeling, and Web page production systems, and are developing the systems which effectively use impressions of Web pages and news articles. By evaluating these systems, we will be able to investigate what kind of impressions should be extracted and how such impressions can be extracted more accurately.

References

1. Fukui, M., Shibazaki, Y., Sasaki, K., Takebayashi, Y.: Multimodal personal information provider using natural language and emotion understanding from speech and keyboard input. Information Processing Society of Japan SIG Notes, HI64-8 (1996) 43–48
2. Kuraishi, H., Shibata, Y.: Feeling communication system by facial expression analysis/synthesis using individual models. IPSJ SIG Notes, DPS74-14 (1996) 79–84
3. Kurohashi, S., Nagao, M.: Manual of Japanese morphological analysis system Juman version 3.61, http://pine.kuee.kyoto-u.ac.jp/nl-resource/juman.html (1999)
4. Mera, K., Ichimura, T., Aizawa, T., Yamashita, T.: Invoking emotions in a dialog system based on word-impressions. Trans. Japanese Society for Artificial Intelligence, **17**, 3 (2002) 186–195
5. Nagata, M., Taira, H.: Text classification — Trade fair of learning theories. IPSJ Magazine, **42**, 1 (2991) 32–37
6. Nikkei Newspaper Full Text Database DVD-ROM, 1990 to 1995 editions, 1996 to 2000 editions, 2001 edition, Nihon Keizai Shimbun, Inc.
7. Peter D. Turney,: Thumbs up or thumbs down? Semantic orientation applied to unsupervised classification of reviews. Proc. Conference on Association for Computational Linguistics (2002)
8. Ren, F., Matsumoto, K., Mitsuyoshi, S., Kuroiwa, S., Lin, Y.: Researches on the emotion measurement system. Proc. IEEE International Conference on System, Man and Cybernetics (2003) 1666–1672
9. SGI Japan Ltd., http://www.sgi.co.jp/newsroom/press_releases/ 2004/sep/st.html
10. Tsukamoto, K., Sassano, M.: Text categorization using active learning with AdaBoost. IPSJ SIG Notes, NL146-13 (2001) 81–89

An Experimental Study on Computer Programming with Linguistic Expressions

Nozomu Kaneko and Takehisa Onisawa

University of Tsukuba, 1-1-1, Tennodai, Tsukuba, Ibaraki 305-8573, Japan
`{nozom,onisawa}@fhuman.esys.tsukuba.ac.jp`

Abstract. This paper describes the end-users programming system that generates computer programs through interaction using linguistic expressions with users who have little knowledge of computer programming. The present system employs the idea of *computer programming by paraphrasing*, which is used as the useful method of human's everyday communication. The system also uses the case-based reasoning method for paraphrasing. Therefore users can add the meanings of linguistic expressions that the system does not have. Two subject experiments are carried out in order to confirm the usefulness of the present system. The one task is simple text editing and the other task is more complex one. Ten subjects in all perform the experiments and the experimental results are discussed in this paper.

1   Introduction

Conventional computer programming is performed by a programming language. Therefore, computer programming is difficult for users without programming language knowledge. For such a problem, many studies on the end-user programming are progressed. One of them is famous as the programming by demonstration (PBD) [1], which generates a computer program by user's illustration. However, it is difficult for users to tell their intentions to the system precisely since the PBD system infers the user's intention only from the illustrations. There are also some end user programming systems with linguistic expressions. For example, NaturalJava [2] can generate a syntax tree of Java language from linguistic expressions, and CLEPE [3] performs the problem-solving on the conceptual level using the ontology. However, these systems still have open problems. The former system deals with knowledge only about the limited world, and the latter one requires a large amount of knowledge about the real world.

This paper applies paraphrasing to computer programming using linguistic expressions [4] so that even end users with little knowledge of computer programming can make programs since paraphrasing is efffective in human's everyday communication. The present computer programming system uses the case-based reasoning method for paraphrasing. Therefore users can add the meanings of linguistic expressions that the system does not know. Two subject experiments in which some text editing tasks are performed using the present system are carried out and the usefulness of the present system is experimentally discussed.

R. Khosla et al. (Eds.): KES 2005, LNAI 3681, pp. 911–917, 2005.

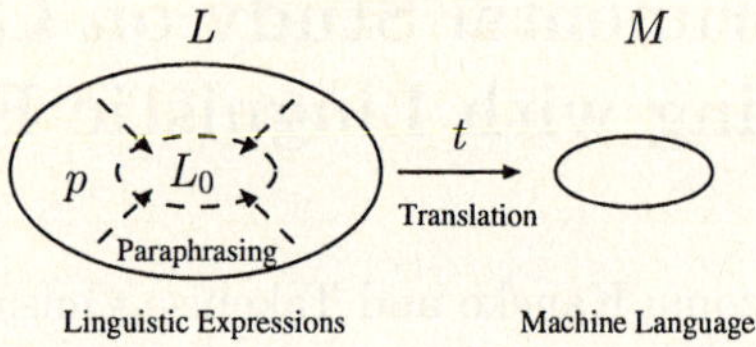

Fig. 1. Translation and paraphrasing.

This study is an example of *everyday language computing* [5], in which the importance of natural language in computing is emphasized.

2 Programming by Paraphrasing

Computer programming using linguistic expressions is regarded as the conversion process from linguistic expressions into a machine language. The conversion process is divided into two parts, the translation part and the paraphrasing part as shown in Fig. 1. Set L of linguistic expressions has quite a lot of elements. On the other hand, set M of a machine language has a fewer elements than set L. Let L_0 be the set of elements each of which has an approximately one-to-one correspondence to an element of M. The translation t is a map from specific linguistic expressions to some commands in a machine language. The paraphrasing p is a map between two linguistic expressions in L. The sequence of paraphrasing maps linguistic expressions in L into some expressions in L_0. End users can understand the paraphrasing without any knowledge of computer programming since the paraphrasing is the map defined within L.

3 System Structure

The structure of the present system is shown in Fig. 2. The system interacts with the user by linguistic expressions, and generates a program through the paraphrase and the translation. The case-based reasoning method is used for the paraphrase and the translation. Users can add paraphrase cases into the case database. On the other hand, the translation cases are required to be prepared beforehand because users are hard to understand them. The system retrieves a similar case for the inputted linguistic expressions, and outputs a translation or a paraphrase based on the selected case. The system is implemented with Lisp on the GNU Emacs.

3.1 Translation Part

The translation part generates computer programs from inputted linguistic expressions using the case-based reasoning method. This part has the translation case database. The translation case is the description of the relation between

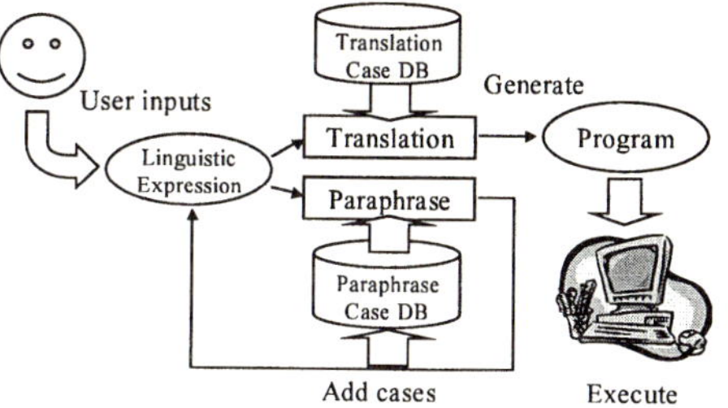

Fig. 2. System structure.

linguistic expressions and instructions in a computer programming language. Inputted lingustic expressions are compared with linguistic expressions in the case database based on the phrase structure. The phrase structure of linguistic expressions is obtained as the parse tree by the morphological and the syntactic analyses. Especially, ChaSen [6], the morphological analysis system is used to obtain a morphological sequence from Japanese sentences.

3.2   Paraphrase Part

The paraphrase is performed by the case-based reasoning as well as the translation, except that outputs are not computer programs but only linguistic expressions. The paraphrasing is performed in the following way. The cases partially corresponding to the inputted linguistic expressions are selected as candidates. The similarity between two parse trees is defined by Eq. (1) based on the Convolution Kernel [7, 8].

$$K(T_1, T_2) = \max_{n_1 \in N_1} \max_{n_2 \in N_2} C(n_1, n_2) \ , \tag{1}$$

where $C(n_1, n_2)$ is the similarity between nodes n_1 in the tree T_1 and n_2 in the T_2, and is defined recursively in the following way.

Case 1. If the part-of-speech of n_1 is different from that of n_2, $C(n_1, n_2) = 0$.
Case 2. If the part-of-speech of n_1 is the same as that of n_2, and n_1 and n_2 are pre-terminals, then $C(n_1, n_2) = 1$.
Case 3. Else if the part-of-speech of n_1 is the same as that of n_2, and n_1 and n_2 are not pre-terminals,

$$C(n_1, n_2) = \prod_{k \in ch(n_1)} \prod_{l \in ch(n_2)} (1 + C(k, l)) \ , \tag{2}$$

where $ch(n_1)$ and $ch(n_2)$ are the sets of children of n_1 and n_2, respectively.

Next, the system generates new paraphrases in order of the similarity according to the procedure shown in Fig. 3. (1) The system compares the parse tree of inputted linguistic expressions with that of linguistic expressions before paraphrasing in the selected paraphrase case, and then detects the difference between the two parse trees. (2) The difference found out in (1) is applied into

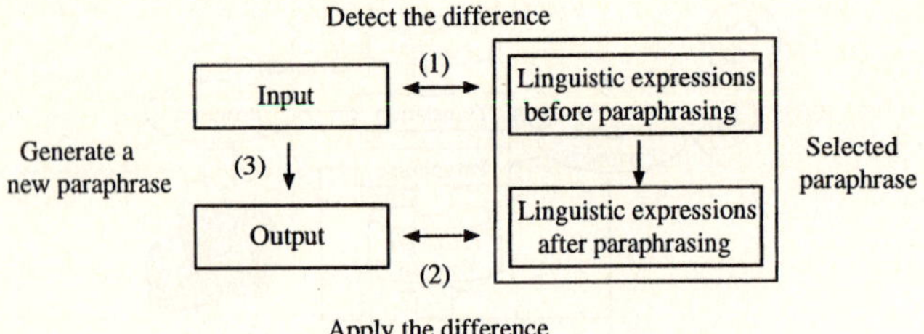

Fig. 3. Procedure of paraphrase generation.

タイトル：言語理解の構造
著者：テリー・ウィノグラード
出版社：産業図書
出版年：1976
ISBN：4-7828-5236-3

テリー・ウィノグラード：『言語理解の構造』，産業図書（1976）．

Fig. 4. Text before editing (Exp. 1).   **Fig. 5.** Text after editing (Exp. 1).

the linguistic expressions after paraphrasing in the case. As a result, (3) the new paraphrase is generated. Finally, the system asks users if the paraphrase is appropriate to inputted linguistic expressions or not. The suitable paraphrase is stored in the database.

3.3   Addition of a Paraphrase Case

If no cases are found in the database corresponding to inputted linguistic expressions, the system asks users to paraphrase them into other expressions. Added linguistic expressions are available in the next interaction.

4   Subject Experiments

4.1   Experiment 1

Five subjects make computer programs to edit the text data shown in Fig. 4 into that shown in Fig. 5 in this experiment. As a result of this experiment, all five subjects make computer programs completely for editing the text data. Table 1 classifies obtained paraphrase cases into some categories. "Concreting" is to paraphrase abstract linguistic expressions into other more concrete ones. Even if there are conceptual gaps between users' intensions and commands which the system can execute, "concreting" paraphrase bridges the gaps. For example, "edit the title, move the author and the publisher, and write the publication year" is the concreting paraphrase of "change the format of the bibliography data." "Re-wording" is a kind of paraphrase that adds users' favorite words into system's vocabulary, for example, the paraphrase between "select it" and "begin the selection." There are also some added cases due to users' mistakes or system's bugs.

Table 1. Classification of obtained paraphrases (Exp. 1).

Subjects	A	B	C	D	E	Total
Concreting	10	6	16	18	16	66
Re-wording	1	5	10	1	2	19
Mistake	1	0	1	2	5	9
Bug	0	0	1	2	0	3
Total	12	11	28	23	23	97

Table 2. Results of questionnaire (Exp. 1).

Subjects	A	B	C	D	E
Ease of use[1]	3	4	4	2	4
Intelligibility[2]	4	4	4	3	5
Intuitiveness[3]	4	5	5	4	4

[1] 1:very difficult $\sim$ 5:very easy
[2] 1:hardly intelligible $\sim$ 5:very intelligible
[3] 1:hardly intuitive $\sim$ 5:very intuitive

Table 3. Tendency to paraphrase (Exp. 1).

Subjects	A	B	C	D	E	Total
Total number of linguistic expressions after paraphrasing($= X$) (Average)	52 (4.33)	49 (4.45)	88 (3.14)	113 (4.91)	96 (4.17)	398 (4.10)
Number of reparaphrased linguistic expressions ($= Y$)	10	15	34	8	11	78
Y/X	19%	31%	39%	7%	12%	20%

After the experiment, the subjects answer three questions by five grade evaluation: "Ease of use of the system", "Intelligibility of obtained linguistic expressions" and "Whether programming by paraphrasing is intuitive or not". The results of the questionnaire are shown in Table 2. It is found that these subjects can easily understand the programming with linguistic expressions.

Table 3 shows subjects' tendencies to paraphrase. As for subjects B and C, it is noticed that approximately 30% of all linguistic expressions after paraphrasing are paraphrased into other linguistic expressions again. In contrast, subject D seldom reparaphrase linguistic expressions and has many linguistic expressions after paraphrasing since his linguistic expressions are concrete rather than abstract. Various tendencies to paraphrase are observed among subjects.

4.2   Experiment 2

Five other subjects carry out experiment 2 in which four kinds of tasks are given and the text data in Fig. 6 are editted into the HTML form shown in Fig. 7.

Task (1): Make the first line a title.
Task (2): Make the third line with "*" a headline.
Task (3): Itemize the part where the line starts by "-".
Task (4): Edit the text data shown in Fig. 6 to the one in a web browser as shown in Fig. 7, completely. This task includes tasks (1) $\sim$ (3) and the additional task to complete the HTML form.

Although three of five subjects achieve all tasks, two subjects G and I do not achieve the task (4) since they often misunderstand the meanings of linguistic

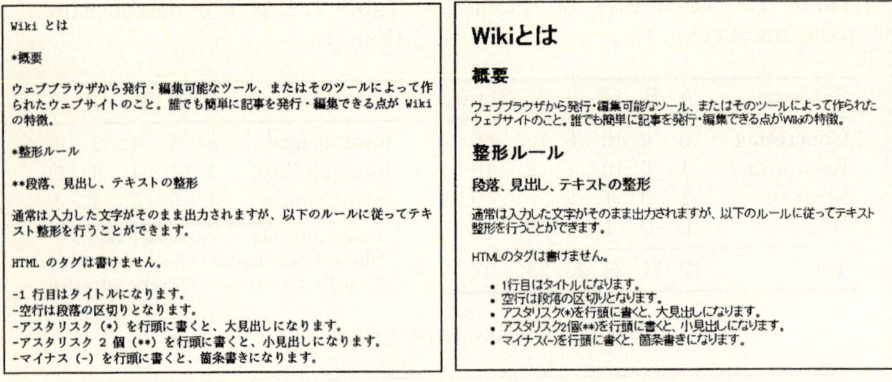

Fig. 6. Text before editing (Exp. 2).   **Fig. 7.** Text after editing (Exp. 2).

<table>
<tr><td>if the beginning of a line fits "-", then insert the item tag ⇒</td><td>

```
(if (save-excursion (beginning-of-line) (looking-at "-"))
  (progn
    (beginning-of-line)
    (delete-char 1)
    (insert "<li>")))
```

</td></tr>
</table>

(a) Linguistic expressions   (b) Computer program

Fig. 8. Example of generated computer program.

expressions in the interaction with the system and then they perform the task without noticing such misunderstanding.

Fig. 8 shows an example of generated computer programs in task (3). Linguistic expressions in the left side of the arrow are translated into the program.

The same questions as the ones in experiment 1 are given to the subjects after the experiment. The results of the questionnaire are shown in Table 4. Although experiment 2 is more complex than experiment 1, many subjects answer that obtained linguistic expressions are intelligible and that programming by paraphrasing is intuitive. However, it is found that the usability of the system is insufficient to the complexity of the task. The subjects are also asked to answer the question about the difficulty of each task by five grades from 1: very easy to 5: very difficult. Table 5 shows the results. Especially, subjects G and I, who do not achieve the task (4), answer "very difficult" for task (3).

Table 4. Results of questionnaire (Exp. 2).

Subjects	F	G	H	I	J
Ease of use	2	2	2	2	2
Intelligibility	4	2	5	4	4
Intuitiveness	4	4	2	5	4

Table 5. Difficulty of each task in experiment 2.

Subjects	F	G	H	I	J
Task (1)	1	3	1	3	1
Task (2)	1	3	2	3	1
Task (3)	3	5	4	5	3
Task (4)	4	–	4	–	5

The important comments are given by many subjects. The present system does not consider users' trial and error. However, subjects prefer to perform the task by trial and error when they do not follow the procedure for a given task completely. It is necessary for the system to support users' trial and error for a complex task such as task (4) in this experiment.

5 Conclusion

This paper describes the end-users programming system using linguistic expressions and usefulness of the system is experimentally confirmed. The results of the experiments show that the proposed *programming by paraphrasing* method is intuitive and that obtained linguistic expressions are intelligible for users. However, the experimental results also show that the usability of the present system is insufficient for a complex task. Furthermore, the present system cannot deal with ungrammatical input. Therefore, it is necessary to correct grammatical errors.

References

1. Henry Lieberman: "Your Wish is My Command: Programming by Example", Morgan Kaufmann Publishers (2001).
2. David Price et al.: "NaturalJava: A Natural Language Interface for Programming in Java", Proc. of 5th International Conference on Intelligent User Interfaces, New Orleans, Louisiana, pp. 207-211 (2000).
3. Kazuhisa Seta, Mitsuru Ikeda, Osamu Kakusho, and Riichiro Mizoguchi: Capturing a Conceptual Model for End-User Programming: Task Ontology As a Static User Model, Proc. of 6th International Conference on User Modeling, Chia Laguna, Sardinia, pp. 203-214 (1997).
4. Nozomu Kaneko, Takehisa Onisawa: "End-User Programming by Linguistic Expression employing Interaction and Paraphrasing", Proc. of SCIS & ISIS 2004, Keio Univ, Japan (2004).
5. S. Iwashita, N. Ito, T. Sugimoto, I. Kobayashi and M. Sugeno, "Everyday Language Help System Based on Software Manual", Proc. of IEEE International Conference on Systems Man and Cybernetics (SMC2004), pp. 3635-3640, Hague, Netherlands, Oct. (2004).
6. Yuji Matsumoto, et al.: Morphological Analysis System ChaSen Version 2.2.1 Manual, Nara Institute of Science and Technology (2000).
7. M. Collins, N. Duffy: "Convolution Kernels for Natural Language", Neural Information Processing Systems, Vol. 14, pp. 625-632 (2002).
8. Tetsuro Takahashi, Kozo Nawata, Kentaro Inui, Yuji Matsumoto: "Effects of Structural Matching and Paraphrasing in Question Answering", IEICE Transactions on Information and Systems (2003).

Enhancing Computer Chat:
Toward a Smooth User-Computer Interaction

Calkin A.S. Montero and Kenji Araki

Graduate School of Information Science and Technology, Hokkaidō University,
Kita 14-jo Nishi 9-chome, Kita-ku, Sapporo City, 060-0814 Japan
{calkin,araki}@media.eng.hokudai.ac.jp
http://sig.media.eng.hokudai.ac.jp

Abstract. Human-computer interaction (HCI) has fundamentally changed computing. The ubiquity of HCI can be seen in several kinds of application areas, such as text editing, hypertext, gesture recognition, and the like. Since HCI is concerned with the joint performance of tasks by humans and machine, human-computer conversational interaction plays a central role when trying to bring the computer behavior closer to the human conversational behavior. In this paper we introduce our idea of modeling computer *chat* based on the observed "critical" behavior of the structure of human chat.

1 Introduction

Theoretical and practical research on the structure of the human dialogue has been a topic of investigation for decades in Artificial Intelligence (AI). Since Alan Touring [1] presented his philosophical dissertation regarding the question "Can machines think?", conversational understanding has been considered the mean to determine whether the machine is intelligent. Turing proposed an "imitation game" in which an interrogator poses questions to an unseen entity (machine or person) then based on the responses the interrogator judges which is which. The main issue for Turing was that using language as humans do is sufficient, by itself, to determine intelligence.

Since, human-computer conversation (HCC) as part of human-computer interaction (HCI) has been growing until becoming nowadays one of the most important areas in AI, reaching a similar grade of development like that of a better-known area of language processing, i.e., machine translation (MT). One of the most famous examples of HCC systems is ELIZA [2] in which many of the issues raised by Turing became relevant. ELIZA is a pattern-matching natural language program that performs the role of a Rogerian therapist during its conversation with a user. The program worked by making use of a cascade of regular expressions substitutions that each matched certain part of the user's input and changed it into the system's reply.

A best performance was PARRY program [3]. PARRY attempted to simulate a paranoid patient, implementing a model behavior based on conceptualizations and beliefs - i.e., accept, reject, neutral: judgments of conceptualization. PARRY

R. Khosla et al. (Eds.): KES 2005, LNAI 3681, pp. 918–924, 2005.

introduced some important ideas for the development of conversational agents: topic and context understanding and conversational strategy or dialogue management.

In recent years a considerable amount of research has been carried out on conversational spoken systems. In this regard Jupiter [4], system that provides a telephone-based conversational interface for international weather information, stands as a good example. Other examples of conversational systems include airline travel information [5], speech-based restaurant guides [6], and telephone interfaces to emails or calendars [7]. The dialogue manager is the core of such systems, regulating the flow of the dialogue, deciding how the system's side of the conversation should proceed, what to ask or what assertions to make and when to ask or make them, being the dialogue constrained to certain specific domain.

Several approaches to modeling constrained dialogue have been proposed. Among those, dialogue grammar approach [8], based on the observation that there are regularities of sequence during the dialogue, as a question is follow by an answer, proposal by acceptances, and so forth. However, this approach assumes that only one state results from a transition, when actually utterances are multi-functional, i.e., an utterance can be both, a rejection and an assertion, which makes the approach unsatisfying. Plan-based approach [9], based on the observation that the user's utterances are not simple string of words but communicative actions or speech acts. This approach assumes that the user's speech acts are part of a plan and the goal is to uncover and respond appropriately to it [10]. However, one of the limitations of the plan-based approach is that the illocutionary act recognition, needed in order to infer the user's plan, has shown to be redundant [11]. Finally, joint action or collaborative approach, based on the assumption that both parties to a dialogue are responsible for sustaining it, existing a collaboration between the interlocutors in order to achieve mutual understanding [12]. However, all of those dialogue modeling approaches focused on creating dialogue agents in order to achieve specific goal, being their dialogues domain restricted, failing when it comes to hold a human-like dialogue.

Recently, the development of non-specific-goal[1] dialogue agents has exponentially increased. Those agents are called chat robots or chatbots. A chatbot is defined as a computer program that simulates human conversation through AI. In fact, ELIZA has been considered the first created chatbot. The dialogue modeling of a chatbot is based on certain words combinations that it finds in a phrase given by the user, and in the case of sophisticated chatbots, like ALICE [2] [13], AIML[3] is used to transcribe pattern-response relations. Chatbots are wide spread on the Web, being applied for electronic commerce customer service. An ELIZA-clone chatbot application to question answering has been proposed by C.S. Montero and K. Araki [14, 15]. However, it is highly improbable to create all the possible patterns-response that could appear during a dialogue, a shortcoming that makes the chatbot to easily lose the conversation flow.

---

[1] Open domain dialogue.

[2] Artificial LInguistic Computer Entity.

[3] Artificial Intelligence Markup Language, base on eXtended Markup Language, XML.

In this research a deep analysis of the human chat has been performed in order to apply its characteristics of criticality, described in Sect. 2, to modeling computer chat. Experiments show an improvement of the chatbot performance after providing its database with "critical categories," as it is described in Sect. 3. A conclusion with reflections on future directions of the research is given in Sect. 4.

2   Characteristics of Criticality in Human Chat

A human chat section has been considered *critically self-organized* [16]. Self organized criticality states that large interactive systems naturally evolve toward a critical state in which a minor event can lead to catastrophe. The critical state is established entirely due to the dynamical interactions among individual elements of the system, implying self-organization. Considering the system to be a human chat section, we argue that the chat is a dynamic system whose environment is formed by both interlocutors utterances. The chat evolves from an initial flat state to a state where a single utterance can completely change the direction of the conversation - catastrophe -, introducing a new topic, repeating the cycle again. This behavior has been called *Critically Self-Organized Chat* [16]. It could be argued that being aware of the "critical behavior" of the human chat allows to create more human-like dialogue flows when modeling computer chat, therefore its importance.

The critical behavior of the human chat can be observed in the following fragment of a transcription of a real chat between two friends[4]:

1A: Have you had your lunch over lah

1B: No ...

2A: Uhm I haven't had my lunch also

2B: Uh busy uh

3A: Uhm busy playing with my nieces

3B: Both of them are in their own place eh at your home now

4A: Hah The younger one is always at my home mah

4B: Both of them – Orh

5A: My mum look after her

5B: Ah

6A: The baby is only two three months

6B: Um so weekends the parents come

7A: You hear my little niece crying

7B: Heh uh Uhm so are you spring cleaning already

8A: Ah intend to do it today lor

8B: Orh ....

9A: Ah – Then weekdays definitely can not – Weekdays I'm so busy in the office you know

9B: Ah hah

10A: This whole this week itself ah

10B: Ah hah

In this chat, the utterances 1A to 2B form the flat initial state of the system: topic-introductory utterances. At this point there are not big changes in the chat, hence it could be said the system is in equilibrium. The next utter-

---

[4] University of South California, Dialogue Diversity Corpus, Dialogue 91, http://www-rcf.usc.edu/~billmann/diversity/DDivers-site.htm.

ance - 3A - initiates the chat toward a determined topic: A's nieces. This topic is remaining the main one, although small "avalanches" occur, it is to say, the focus of the topic is slightly changed, until the utterance 7B (the number of the utterances does not reflect the number of turns taken by the speakers) that crashes the main topic (catastrophe) giving birth to a complete new direction of the chat: laundry time. The utterance that leads to a catastrophe - changing the direction of the chat - has been called "critical utterance" [16]. It is worth noticing how the new topic rises from a flat equilibrium state to a state where a unique utterance - 9A - changes again the direction of the chat, starting a new cycle.

Taking in consideration the critical self organization of the human chat when modeling computer chat, we argue, a more human-like, smooth performance could be achieved. In Sect. 3 it is presented the basis of the application of criticality to modeling computer chat along with a design experiment.

3   Method: Modeling Computer Chat

In order to model computer chat, taking in consideration the properties of criticality of the human chat, it is needed to observe the behavior of the human chat in terms of the interrelationship between the speakers utterances. This interrelationship can be considered as the utterance co-occurrence during the chat. In order to achieve this goal, we applied a data mining tool called KeyGraph. The KeyGraph [17] identifies relationship between terms in a document, particularly focusing on co-occurrence relationships of both high-probability and low-probability events.

We analyzed the chat section[5] shown in Sect. 2. Each speaker's utterance was segmented into words, being those words filtered - eliminating common ones, i.e., I, you, is, and the like - and a sentence co-occurrence relationship was determined as:

$D=$

$\quad w_1 :: S_1, S_2, S_4 \dots$

$\quad w_2 :: S_9, S_{25} \dots$

$\quad w_3 :: S_1, S_3, S_{10} \dots$

$\quad \dots$

$\quad w_n :: S_{24}, S_{25}, \dots S_m$

where:

w_k $(k = 1, 2, 3, \dots, n)$, represents a word in a sentence.

S_l $(l = 1, 2, 3, \dots, m)$, represents a sentence (utterance transcription).

Feeding the KeyGraph with this preprocessed document (preprocessed chat section), the visual result showed clusters of interrelated sentences, where one critical utterance was leading to the other, and the links of the clusters were showing the shifting between topics during the chat.

---

[5] During the analysis the complete chat section was taken in consideration.

3.1   Design Experiment

In the experiment a native English speaker performed a chat with ALICE chatbot. The performed chat was analyzed, following the above method, in order to find co-occurrence between the user's utterance and the chatbot replies. The graphical view of this chat section could be seen in Fig. 1.

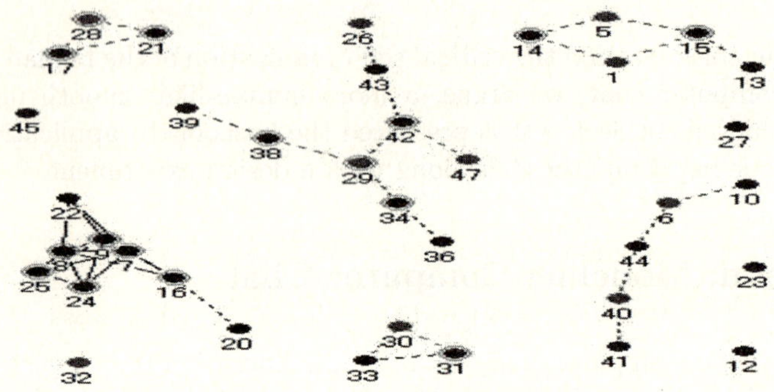

Fig. 1. Graphical View: Non Critical Chat.

In this figure, the clusters represent the relationship between the interlocutors utterances (user and computer) and the links between clusters represent the transition between one topic and the other. It can be observed that the main clusters are not interconnected, meaning the chatbot in many cases could not keep a natural flow of the chat, giving as a result vagueness during the dialogue. This chat section had 48 turns of the interlocutors (user - chatbot).

After analyzing this graph, in order to enhance the computer chat, we try to add criticality to the dialogue by making the chatbot to ask "intelligent questions" at certain points of the conversation, so as to make the shifts from one topic to the other, i.e., interconnecting the clusters shown in the graph, more naturally and in a human-like way. The chatbot database is made of "categories", each one including a pattern to match, i.e., user input, and a template, i.e., chatbot reply. Adding "critical categories" to the database the performance of the chatbot has shown improvement. For instance, if there is an utterance the chatbot does not know how to reply to, a general pattern for smoothly shifting the topic, by asking a question at this point, as to create a link in the graph, gives to the dialogue the desired criticality.

The same user was requested to performed a chat with the enhanced chatbot. The result of the analysis is shown in Fig. 2. This chat section was described by the user as "more interesting" than the previous one, having 82 turns of the interlocutors. In Fig. 2, the shown clusters are more interrelated than in Fig. 1 indicating a better flow of the chat section. For example, a fragment of the performed chat between the user and the chatbot is as follow:

Table 1. Improvement of the Chatbot

	No. of Turns	% Vague Replies	User's Opinion
Baseline Chatbot	48	14.6%	"loosy"
Enhanced Chatbot	82	3.7%	interesting to talk with

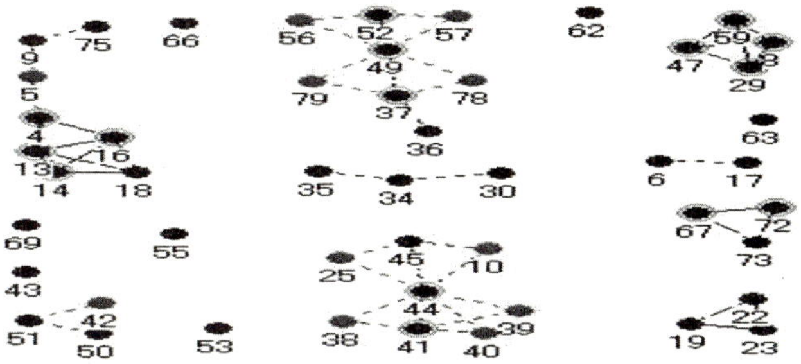

Fig. 2. Graphical View: Critical Chat.

Chatbot: What is like living in Sapporo?
User: sweet as
Chatbot: In the context of living in Sapporo I don't understand "what is like living in Sapporo?"

In this case there was not a pattern for matching the user's utterance "sweet as", being the rules for non-matching patterns of the chatbot a failure. After adding criticality to the chatbot, a reply for the expression is:

User: sweet as
Chatbot: "sweet as"! what is the weather like over there?
User: is a nice day so far

Table 1 shows a resume of the chatbot performance. The success of the method it is difficult to major mathematically due to its novelty, however, the satisfaction of the user while chatting with the enhanced chatbot indicates its betterment. Although the performance of the chatbot is not completely perfect, as it can be seen in Fig. 2 not all the clusters are linked to each other, a remarkable improvement in its performance can be observed, having less vague replies, showing the validity of the approach.

4   Conclusion

In this paper a novel method for enhancing computer chat has been proposed. Taking in consideration the observed critically self-organized behavior of the human chat when modeling computer chat has shown to be an useful approach. By making the chatbot to ask "intelligent question" at certain points of the chat, the criticality increases obtaining as a result a more natural flow of the dialogue, being the chatbot able to keep the chat section going longer time. Future works are oriented toward the application of the enhanced chatbot to Web-based question answering (QA) system in open domain.

References

1. Turing A.M.: Computing Machinery and Intelligence. Mind, Vol. 59, No. 236, pp. 433-460. (1950)
2. Weizenbaum J.: ELIZA-A Computer Program for the Study of Natural Language Communication Between Man and Machine. Communications of the ACM 9, No.1, pp. 36-45. (1966)
3. Colby K., Hilf F., Weber S.: Artificial Paranoia. Artificial Intelligence, Vol. 2, pp. 1-25. (1971)
4. Zue V., Seneff S., Glass J., Polifroni J., Pao C., Hazen T.J., and Hetherington L.: JUPITER: A Telephone-Based Conversational Interface for Weather Information. IEEE Transactions on Speech and Audio Processing, Vol. 8, No. 1, pp. 85-96. (2000)
5. Bratt H., Dowding J., Hunicke Smith J.: The SRI Telephone-based ATIS System, in Proceedings of the ARPA Spoken Language System Technology Workshop, pp. 22-25. (1995)
6. Lau R., Flammia G., Pao C., Zue V.: WebGALAXY: Beyond Point and Click - A Conversational Interface to a Browser, in Proceedings of the 6th International WWW Conference, pp. 119-128. (1997)
7. The University of Texas at Austin (2004) SmartVoice, http://www.utexas.edu/its/smartvoice/
8. Sacks H., Schegloff E., Jefferson G.: A Simplest Systematics for the Organization of Turn-taking in Conversation. In: Schenkein J, (editor) Studies in the Organization of Conversational Interaction. Academic Press, New York. (1978)
9. Searle R.J.: Speech Acts: An essay in the Philosophy of Language. Cambridge University Press, Cambridge. (1969)
10. Allen F.J., Perrault R.C.: Analyzing Intention in Dialogues. Artificial Intelligence, 15(3):143–178. (1980)
11. Cohen P.R., Levesque H.J.: Speech Acts and the Recognition of Share Plans, in Proceedings of the Third Biennial Conference, pp. 263-271. Canadian Society for Computational Studies of Intelligence. (1980)
12. Clark H.H., Wilkes-Gibbs D.: Referring as a Collaborative Process. Cognition, 22:1–39. (1986)
13. Wallace R.S.: A.L.I.C.E. Artificial Intelligence Foundation. http://www.alicebot.org
14. Montero C.A.S., Araki K.: Information Acquisition Using Chat Environment for Question Answering. KES 2004, Lecture Notes in Artificial Intelligence (LNAI) 3214, pp. 131-138, 2004. Springer-Verlag Berlin Heidelberg 2004.
15. Montero C.A.S., Araki K.: Improvement of a Web-Based Question Answering System Using WordNed-Upgraded Chat. Joint Convention Record, The Hokkaido Chapters of The Institutes of Electrical and Information Engineers, pp. 282-283, IEEE Sapporo Section, Japan. (2004)
16. Montero C.A.S., Araki K.: Discovering Critically Self-Organized Chat. The Fourth IEEE International Workshop on Soft Computing as Transdisciplinary Science and Technology. To appear. (2005)
17. Ohsawa Y.: *KeyGraph*: Visualized Structure Among Event Clusters. In: Ohsawa Y., and Mcburney P. (eds) Chance Discovery. Springer, Berlin Heidelberg New York.

Effect of Direct Communication in Ant System

Akira Hara, Takumi Ichimura,
Tetsuyuki Takahama, Yoshinori Isomichi, and Motoki Shigemi

Faculty of Information Sciences, Hiroshima City University,
3-4-1, Ozuka-higashi, Asaminami-ku, Hiroshima, 731-3194, Japan
{ahara,ichimura,takahama,isomichi}@its.hiroshima-cu.ac.jp
http://www.chi.its.hiroshima-cu.ac.jp/

Abstract. In multi-agent systems, communication among agents plays an important role for cooperative behavior. In this paper, we investigate the effect of direct communication in a multi-agent system, Ant System (AS). Ant agents in AS usually perform indirect communication through pheromones. We introduce the mechanism that ants communicate directly with neighboring ants so that each ant's information can be exchanged. As experimental results, the AS with direct communication acquired better solutions than conventional AS. If the ants have excessive range of direct communications, however, the search performance becomes worse. We showed that local communications are effective in order to search solutions while striking a balance between centralization and diversification of the ants' action. For the effective search, it is important to set the range of the direct communication appropriately.

1   Introduction

Due to the advance of computer networks and web applications, the quantity of data on networks has increased in recent years. Search engine is a human computer interaction system, which computer users utilize for getting the necessary information from the vast data in the internet. Agents who collect the information automatically are indispensable for the system on the networks. Recently, retrieval support systems also have been studied. For example, multiple agents partition the search space and collect appropriate information for users cooperatively. In another system, each user's agent utilize the past retrieval results of other users in a community for effective search. We have to investigate the methodology by which the artificial agents instructed by humans cooperate with one another effectually. In this paper, we pick up swarm intelligence that is one kind of multi-agent systems, in order to examine component technique of the cooperative systems. We investigate the effect of direct communication among agents.

Swarm intelligence means that multiple simple programs interact with each other and intellectual behavior is emerged as a whole. The characteristics of swarm intelligence are autonomy of agents, adaptability for dynamic environments, and robustness for faults. Ants or bees in the real world are examples

R. Khosla et al. (Eds.): KES 2005, LNAI 3681, pp. 925–931, 2005.

of the swarm intelligence. Ants generate the volatile chemical substance that is called pheromone in their body, and ants secrete the pheromone in the route that hey passed in foraging. Moreover, ants trace the route with pheromones that other ants secreted. Thus, ants perform communication asynchronously through the pheromones on the field. This phenomenon is called the pheromone communication. Ants just trace pheromone trail. However, a shorter route to food is formed, and effective foraging behavior is realized.

Traveling salesman problem (TSP) is one of the typical combination optimization problems. Ants' ability that they can select a shorter route is very useful for this problem. Dorigo *et al.* proposed a search algorithm based on this idea. This search algorithm based on the pheromone communication by ants is called Ant Colony Optimization (ACO)[1–3].

Artificial ants' communication in ACO is indirect communication through the pheromone on the field as well as the real world. In this research, we introduce direct communication between ants. By the direct communication, an ant can exchange information with other ants, which are in the communication range, every time the ant moves from one city to another. The ants make a comparison between the both best tours, and then the ant with a longer tour refers to the other's best tour. This mechanism helps the concentration of ants to the vicinity of good tours. Moreover, the direct communication occurs locally. Each ant encounters different ants. Therefore, each ant has original route information. This mechanism also helps the diversification of ants' action. In this research, we aim to investigate the effectiveness of the direct communication between ants.

2   Ant System (AS)

Ant System (AS)[2] is the most basic ACO algorithm. The AS can solve the TSP efficiently. The goal of TSP is to find a closed tour of minimal length connecting n given cities. This problem can be defined on a graph (N, E) where the cities are the nodes N and the connections between the cities are the edges E. The amount of the pheromone on each edge is expressed by a nonnegative real value. We use $\tau_{ij}(t)$ as an expression of the amount of the pheromone on the edge between city i and city j. The initial amount of pheromone on edges is assumed to be a small positive constant τ_0.

In AS, multiple artificial ants build solutions to the TSP by moving on the problem graph form one city to another until they complete a tour. During an iteration of the AS algorithm, each ant $k(k = 1, \ldots, m)$ builds a tour executing $n = |N|$ steps. The ants move while referring to the inter-city distances and the amount of pheromone on the edges connecting respective cities. The transition rule is the stochastic selection of the next visiting city. When ant k is in city i and N^k is the set of cities that the ant still has to visit, the probability for the ant to go from city i to city j is given by:

$$p_{ij}^k(t) = \frac{[\tau_{ij}(t)]^\alpha [\eta_{ij}]^\beta}{\Sigma_{l \in N^k} [\tau_{il}(t)]^\alpha [\eta_{il}]^\beta} \tag{1}$$

where η_{ij} is the characteristic information of problem domain. In TSP, $\eta_{ij} = 1/d_{ij}$ (d_{ij} is the distance between city i and city j). This information η_{ij} represents the heuristic desirability of choosing city j when in city i. By using this parameter, ants can avoid searching useless routes, and they can approach the optimal solution early. Two nonnegative real values α and β are adjustable parameters that control the relative weight of pheromone intensity and inter-city distance.

After every ant completes its tour, each ant lays pheromone on the respective edges included in its tour. Each ant lays pheromone in proportion to its tour length. If the tour length is short, the ant lays much pheromone. On the other hand, if the tour length is long, the ant lays a little pheromone. The pheromone in each edge evaporates gradually. Concretely, the amount of pheromone on each edge is updated by the following rules.

$$\tau_{ij}(t+1) = (1-\rho)\tau_{ij}(t) + \sum_{k=1}^{m} \Delta\tau_{ij}^{k}(t) \tag{2}$$

$$\Delta\tau_{ij}^{k}(t) = \begin{cases} 1/L^k & \text{if } (i,j) \in T^k \\ 0 & \text{otherwise} \end{cases} \tag{3}$$

where T^k is the round tour done by ant k, and L^k is its length. As shown in equation (3), the quantity of the pheromone which laid by each ant is the inverse of the length of its tour. Therefore, a large quantity of pheromone is laid on the routes which consist of a short tour, and vice versa. Pheromones based on the search results by all ants are accumulated, and the information of pheromone intensities is shared by all ants. $\rho(0 < \rho < 1)$ in equation (2) represents an evaporation rate of pheromone. The quantity of old pheromone decays gradually by this parameter. Therefore, recent pheromone have a greater influence on ants' action. This pheromone evaporation enables ants to avoid local minimum solutions, and to search for various solutions.

Then, each ant starts the search for a new round tour based on the updated pheromone information. This procedure is repeated until the iteration t reaches the maximum number t_{max}.

A large amount of pheromone accumulates in the edges, which compose a shorter round tour or where a lot of ants passed by. Each ant selects the route stochastically according to the distribution of these accumulated pheromones. This nature of ants causes a positive feedback loop, and more ants concentrate on the discovered better route. This mechanism improves the possibility of acquiring the optimal solution.

3   Direct Communication Between Agents

Pheromone communication of ants is indirect communication through the field. In this paper, in addition to the indirect communication, we introduce direct one-to-one communication into AS. By the direct communication, an ant can exchange information with other ants, which are in the communication range, every time the ant moves from one city to another. The ants make a comparison between each other's best tours, and then the ant with a longer tour refers to the other's better tour. The outline of direct communication that each ant takes every move is as follows:

1. Ant searches for other ants in the communication range.
2. The ant communicate with the discovered ant, and receive information of the best tour which the discovered ant has generated before.
3. If the length of the received tour is shorter than its own best tour, the ant lay virtual individual pheromone for its exclusive use.
4. The ant searches for the other ant in the communication range. Return 2, if the other ant exists in the communication range.

We can change the range of direct communication. When the range is the minimum, ants communicate only with the ants which are in the same city. As the range becomes large, ants can communicate with the ants which are in other neighboring cities. By changing the communication range, it is possible to prompt to concentrate on the vicinity of good routes, or to avoid excessive concentration to the same route.

In addition to the shared pheromone information, each ant has individual pheromone information. When an ant performs direct communication with another ant, the ant lay the virtual individual pheromone that only itself can recognize. The probability for selecting next visiting city is calculated by using both the shared pheromone information and individual pheromone information. The probability is as follows:

$$p_{ij}^k(t) = \frac{[\tau_{ij}(t) + \tau_{ij}^k(t)]^\alpha [\eta_{ij}]^\beta}{\sum_{l \in N^k} [\tau_{il}(t) + \tau_{il}^k(t)]^\alpha [\eta_{il}]^\beta} \tag{4}$$

where $\tau_{ij}^k(t)$ is the individual pheromone information that the only ant k can use. In the conventional AS, all ants share the same pheromone information. Therefore, a lot of ants concentrate on the same route with much pheromone. In AS with direct communication, however, each ant has both individual pheromone and shared pheromone information. It is considered that the individual pheromone information is useful to keep the diversification of ants' action.

When ant k performs communications with ant h, individual pheromone information of ant k is updated based on equations (5), (6), and (7). σ is evaporation rate of the virtual individual pheromone information. T^k is the round tour generated by ant k, and L^k is its length.

$$\tau_{ij}^k(t+1) = (1 - \sigma)\tau_{ij}^k(t) + \Delta\tau_{ij}^k(t) \tag{5}$$

$$T^k = \begin{cases} T^k & (L^k \leqq L^h) \\ T^h & \text{otherwise} \end{cases} \tag{6}$$

$$\Delta\tau_{ij}^k(t) = \begin{cases} 1/L^k & \text{if } (i,j) \in T^k \\ 0 & \text{otherwise} \end{cases} \tag{7}$$

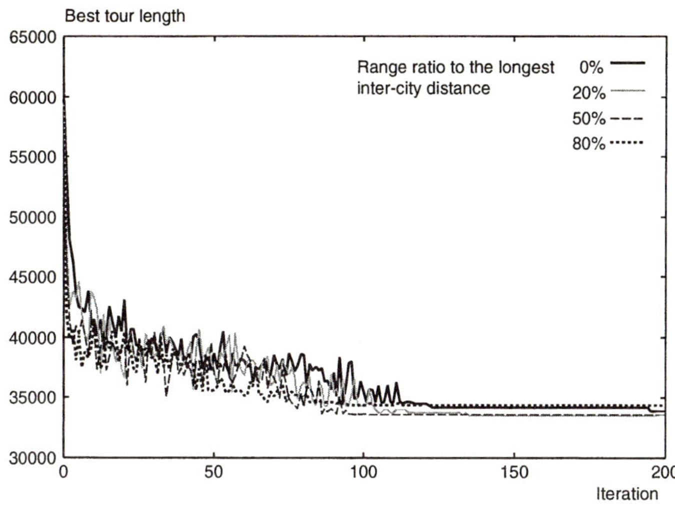

Fig. 1. The length of the best tour in 48 cities map.

4   Experimental Results

4.1   Parameter Settings

We aim to examine the effect of the direct communication on search performance. We investigate how the performance changes when the range of direct communication is changed. The respective parameters in AS are $\alpha = 1.0$, $\beta = 2.0$, $\rho = 0.5$, $\sigma = 0.5$, and $\tau_0 = 1.0$. The following results show average values of 5 trials. We use city maps of TSP which are opened for the benchmark by the TSPLIB site[4]. Here, we show the results by a map of 48 cities, att48. We use 48 ants as many as the number of cities.

4.2   Relation Between Communication Range and Convergence Speed

We investigated how convergence speed is influenced by changing the communication range. The curves in Fig.1 shows the length of the best tour at every iteration. The range parameters in the respective curves are 0%, 20%, 50%, 80% of the longest inter-city distance in the map.

Fig.1 shows that the curve of range=0% converged at about 200 iterations. Similarly, the curves of range 20%, 50%, 80% converged at about 150 steps, 110 steps, 95 steps respectively. The wider the range of communication becomes, the earlier the convergence time is. If the communication range is wider, an ant communicates with more other ants. Therefore, individual pheromone informations of most ants become the same state. As a result, a lot of ants concentrate on the same route, and the search converged early.

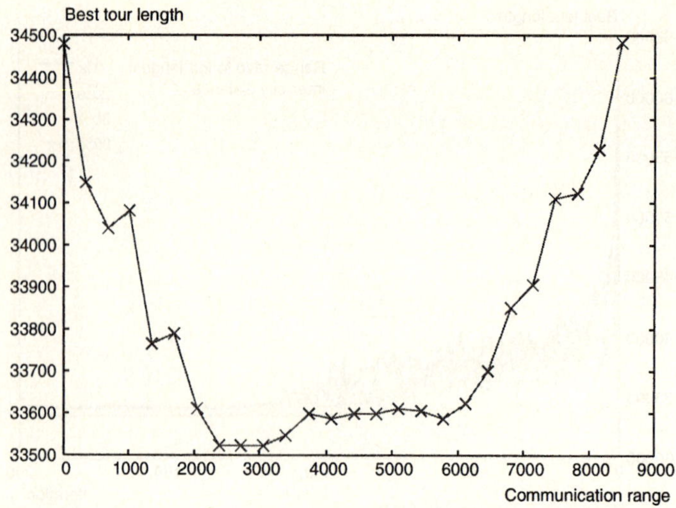

Fig. 2. The best tour's length in various range of direct communication.

4.3   Relation Between Communication Range and the Best Tour's Length

Next, we investigated how quality of solutions is influenced by changing the communication range. Fig.2 shows the length of the best tour in the various ranges of direct communication. This figure shows that we can get the best performance when the range is 2400 $\sim$ 3100. This range corresponds to the 30% $\sim$ 35% of the longest inter-city distance in the map. When the communication range was expanded more than the value, quality of acquired solutions became worse. This is because a lot of ants concentrate on the same route early and the diversity of solutions is lost. This results reveals that there is an appropriate range for direct communication, and excessive wide range has a bad influence upon the performance. It is considered that a balance between centralization and diversification of search is kept by setting the range from 30% to 35% of the longest inter-city distance.

Moreover, Table 1 shows the lengths by the conventional AS and the proposed method. This table also shows the correct answer announced by TSPLIB[4]. In the proposed method, we set the range at 30% of the longest inter-city distance. This table shows the performance of the proposed method is better than that of the conventional AS, and the solution by the proposed method is very close to the correct answer.

Table 1. Comparison of the search performance.

conventional AS	AS with direct communication	correct answer
35169	33546	33522

Thus, ants can search for solutions with keeping a balance between centralization and diversification of their action by using the local direct communication.

5   Conclusions and Future Work

In this research, we introduced direct communication into AS in addition to indirect pheromone communication, and examined the influence on the search. By some experiments, we showed that if the range of direct communication is set properly, a good balance between centralization and diversification of the search can be held.

From now on, we will investigate the technique for dynamic control of the balance between centralization and diversification of search. Moreover, we have to apply this component technique to intelligent human computer interaction systems.

References

1. M.Dorigo and G.D.Caro: "Ant Algorithms for Discrete Optimization", *Artificial Life*, Vol.5, No.2, pp.137-172 (1999).
2. M.Dorigo, V.Maniezzo and A.Colorni: "The ant system: Optimization by a colony of cooperating agents", *IEEE Transactions on Systems, Man, and Cybernetics - Part B*, Vol.26, No.1, pp.29-41 (1996).
3. M.Dorigo and L.M.Gambardella: "Ant colony system: A cooperative learning approach to the traveling salesman problem", *IEEE Transactions on Evolutionary Computation*, Vol.1, No.1, pp.53-66 (1997).
4. TSPLIB
   URL http://www.iwr.uni-heidelberg.de/groups/comopt/software/TSPLIB95/

Multi-agent Cluster System for Optimal Performance in Heterogeneous Computer Environments

Toshihiro Ikeda[1], Akira Hara[1], Takumi Ichimura[1], Tetsuyuki Takahama[1],
Yuko Taniguchi[1], Hiroshige Yamada[2], Ryota Hakozaki[2], and Haruo Sakuda[2]

[1] Hiroshima City University, Japan
`{ikeda,ahara,ichimura,takahama}@chi.its.hiroshima-cu.ac.jp`
`taniguchi@nlp.its.hiroshima-cu.ac.jp`
[2] Hiroshima Prefectural Hiroshima Kokutaiji High School, Japan
`t-yamadak953257@hiroshima-c.ed.jp`

Abstract. In an environment where various personal computers exist together, computers may have different operating systems (OS) and various specifications. In this case, an OS which is not suitable for constructing a PC cluster may be included in that environment and the performance of PC cluster is influenced by personal computers with low specification. In this research, automatic OS decision function, grouping function and automatic start scheduling function are suggested to reduce the influence given by these problems. These functions are realized by multiple agents. In addition, by showing the difference of performances between diskless nodes and non-diskless nodes, we verified the effectiveness of automatic OS decision function.

1 Introduction

We need huge calculation resource to realize intelligent behavior on personal computers. In recent years, PC cluster systems are used to get huge calculation resource. The systems are useful for graphics calculations, data mining and so on [1] [2]. However, there might be a case that it is not possible to prepare enough number of personal computers to get such high performance, for reasons such as lack of budget. In this case, a PC cluster is constructed by personal computers that we are usually using. When we use such personal computers, the computers may be unsuitable to construct a PC cluster. For example, in the environment, there exist various OS and personal computer with different specifications. The former environment makes it difficult to construct a PC cluster, because the environment includes OS that is not suitable to construct a PC cluster. The latter environment also makes it difficult to construct a PC cluster, because a PC cluster is influenced by the personal computers with low performance.

The purpose of this research is to reduce the influence given by these problems, and draw out the maximum performance of the PC cluster in heterogeneous environments. This paper proposes the following functions realized by multiple agents to achieve this purpose.

- Automatic OS decision function in order to construct a PC cluster by using diskless boot in the environment where different kinds of OS exist.
- Grouping function in order to group into the same performance computers in environment where personal computers with different specs exist together.

R. Khosla et al. (Eds.): KES 2005, LNAI 3681, pp. 932–937, 2005.

- Automatic start scheduling function in order to realize effective use of starting time by using the group as a calculation node as soon as it is started.

In addition, by showing the difference of performances between diskless nodes and non-diskless nodes, we verified the effectiveness of automatic OS decision function.

2 PC Cluster

PC cluster is a kind of decentralized memory parallel computer. It can be used as one calculation resource by connecting two or more personal computers with a high-speed network [3] [4].

A PC cluster is consisted of management nodes, calculation nodes and user nodes. Management nodes are personal computers that manage the entire PC cluster. Calculation nodes are personal computers that execute a parallel program. User nodes are personal computers set to be able to use the PC cluster. One personal computer might play two or more roles.

A PC cluster that uses personal computers with different specs as calculation nodes is called a hetero cluster. Hetero clusters have an advantage that a personal computer of any spec can be used as a calculation node. However, the performance of hetero cluster is influenced by the personal computer with low specification. Therefore they might not be able to demonstrate the performance proportional to the number of calculation nodes.

As a similar technology to the PC cluster, there is grid computing. The purpose of grid computing is to make it available for everyone, anytime, anywhere by connecting the calculation resource to the network. It is basically different from PC clusters in the sense that grid computing consists of different types of computers, where PC clusters usually consist of only one type. Moreover, grid computing is used across the firewall and the number of nodes is dynamic, where PC clusters are inside the firewall and the number of nodes is static.

3 Diskless Node

A diskless node is a calculation node that does not use its own hard disk. There are following advantages of diskless nodes:

- It is possible to use PC with any OS as a calculation node.
- Computer failure does not occur easily.

The PC cluster consisted of only diskless nodes are called a diskless cluster. A diskless node can be achieved by using CD-ROM boot or network boot. Network boot and CD-ROM boot is called diskless boot.

3.1 CD-ROM Boot

In CD-ROM boot, the OS image recorded on the CD-ROM is used to start a personal computer. The diskless node achieved by CD-ROM boot uses RAM instead of hard disk.

The advantage of diskless nodes achieved by CD-ROM boot is that a PC cluster can be constructed without changing the setting of the network or constructing any troublesome servers. On the other hand, RAM space that can be used for calculation decreases.

3.2 Network Boot

Network boot is a method of starting a PC by using the OS image received from a server computer which has the OS image.

The advantage of diskless nodes achieved by network boot is that it is easy to consolidate the management since the server holds the kernel and the files. However, the calculation nodes have to refer files via the network, therefore the performance decreases due to network loads

4 Multi-agent Cluster System in Heterogeneous Computer Environments

The management agent exists in the management computer and client agents exist in the other personal computers. The management agent gets information of personal computers connected to network by communicating with client agents. Information that the management agent got from client agents are about the kind of OS, the performance of central processing unit, capacity of RAM, speed of network interface card, and transmission speed between server and client. Automatic OS decision function, grouping function, and automatic start scheduling function are realized by using the information.

Fig.1 shows the composition chart of a PC cluster constructed by the proposal methods

4.1 Automatic OS Decision Function

Automatic OS decision function is a function which inspects the OS of personal computers and decides on which starting method to use.

Before starting the PC cluster, the management agent inspects the OS of PCs which are connected to the network and decide whether or not the OS can be used as a calculation node. Based on this decision, a starting method is chosen. When a personal computer has OS which can be used as a calculation node, it starts with local boot. If not, it starts with network boot.

Using this function, it is possible to construct a PC cluster in the environments where different kinds of OS exist.

4.2 Grouping Function

Grouping function is a function which groups personal computers connected to the network depending on the performance of central processing unit, capacity of RAM, speed of network interface card and the transmission speed between server and client. In addition, if personal computers providing services such as DHCP server and NFS server are detected by analyzing packets or scanning ports, grouping function adds them to a group which disapproves start-up of these nodes.

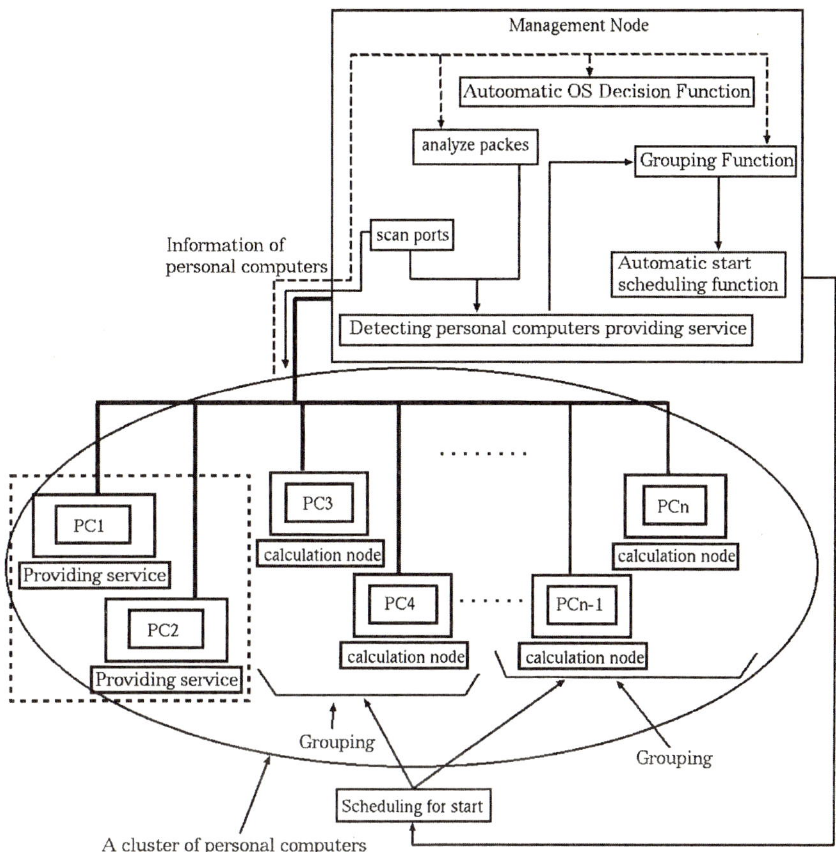

Fig. 1. Composition chart of PC cluster which be constructed by proposal method

Personal computers included in the group disapproving start-up are not used as calculation nodes. Therefore, it avoids damage induced by using personal computers providing services as calculation nodes.

By grouping personal computers, priority of starting the computers is assigned. Also, when executing the program, influence from the difference of specifications can be held down by using one group as a virtual PC cluster.

4.3 Automatic Start Scheduling Function

Start scheduling function is a function which decides the starting order of groups and uses the group as calculation nodes as soon as it is started. As for the starting order, groups with lower evaluation value calculated by equation (1) are started before the groups with higher evaluation value.

$$V_k = t_k + \alpha \frac{w}{\sum_{i=1}^{Nk} P_{k,i}} \tag{1}$$

Where V_k is the evaluation value of group k. t_k is the boot time of group k. w is the size of program. N_k is the number of personal computers included in group k. $P_{k,i}$ is the performance of personal computer i included in group k. This value is obtained by the performance of CPU, the capacity of RAM, the speed of network interface card and transmission speed between server and client. α is a weight parameter. We can consider difference of boot time between groups by t_k.

Using this function will make effective use of starting time, without having to wait for all the groups to complete their start-up.

5   Experiments

In this research, we timed the calculation time of π calculated by equation (2) in the environment which is shown in Table.1. The calculation of π is accelerated by splitting the integral areas and calculating each area with different PCs. In this experiment, the management node also assumes the role of the user node.

$$\pi = \int_0^1 \frac{4}{1+x^2}\, dx \tag{2}$$

Table 1. Experimental environment

	Management node	Calculation nodes
CPU	Intel Pentium 4, 3.2GHz	Intel Pentium4, 3.2GHz
RAM	1G	2G
Network Speed	100MBase-T	100MBase-T

The purpose of this experiment is to compare the efficiency of the calculation node which is started with diskless boot and the calculation node which is started with local boot.

A program which is converted into parallel program by using MPI is used in this experiment. In the experiment, first we measured the calculation time of PC cluster, changing the number of calculation nodes of the PC cluster. We also measured the calculation time, changing the ratio of the diskless nodes in the four calculation nodes. These experiments assume that all calculation nodes have already been started.

All calculation times shown in the Tables 2 and 3 are average of twenty calculations.

Table 2. The calculation time of PC cluster when changing the number of calculation nodes

	One node	Two nodes	Three nodes	Four nodes
calculation time by diskless nodes (sec)	48.249	22.681	17.007	13.832
calculation time by non-diskless nodes (sec)	39.865	20.215	13.646	10.305

Table 3. The calculation time of PC cluster when changing the ratio of the diskless nodes in the four calculation nodes

Diskless nodes	0	1	2	3	4
Non-diskless nodes	4	3	2	1	0
calculation time (sec)	10.305	11.236	12.805	13.317	13.832

6 Discussion and Conclusion

Table 2 shows that the performance of PC cluster with all diskless calculation nodes is lower than that of PC cluster with all non-diskless calculation nodes. In addition, Table 3 shows that the performance of PC cluster decreased by raising the ratio of the diskless nodes in calculation nodes.

Therefore, if personal computers having the OS that can be used as calculation nodes are started by local boot, the PCs should be used as non-diskless nodes for keeping high performance. By our proposed function, the biggest possible number of PCs can be used as non-diskless nodes automatically.

From these facts, we show that deciding the appropriate starting method by automatic OS decision function is effective to construct a PC cluster in environments where different OS exist together.

In this research, the following three functions are proposed to achieve the best performance in environments where different models exist together:

- Automatic OS decision function
- Grouping function
- Automatic start scheduling function

Results of experiments proved that Automatic OS decision function is an effective function in such environment. In experiments, we assumed that all calculation nodes have already been started, but in the future, we will conduct the experiment considering the time to boot calculation node and verify the effectiveness of the other two functions

We will use this PC cluster as an important calculation resource to do data mining or to learn cooperative behavior of soccer agents in RoboCup.

References

1. Sigeru Muraki,Eric. B. Lum, Kwan-Liu Ma, Masato Ogata, Xuezhen Liu: A PC Cluster System for Simultaneous Interactive Volumetric Modeling and Visualization. IEEE Symposium on Parallel and Large-DataVisualization and Graphics, pp. 95-102, October 2003.
2. Wylie, B., Pavlakos, C., Lewis, V., Moreland, K: ScalableRendering on PC Clusters. IEEE CG&A, Vol. 21, No. 4, pp.62-70,2001.
3. Sterling, T., Savarese, D., Beeker, D.J., Dorband, J.E., Renawake, U.A., Packer, C.V.: BEOWULF: A parallel workstation for scientific computation. In Proceedings of the 24th International Conference on Parallel Processing, 11-14, 1995
4. Tomoyuki Hiroyasu, Mitsumori Miki, Kenzo Kodama, Junichi Uekawa, and Jack Dongarra: A Simple Installation and Administration Tool for the Large-Scaled PC Cluster System: DCAST. ClusterWorld Conference and Expo, June 23-26, 2003

Exploring the Interplay Between Domain-Independent and Domain-Specific Concepts in Computer-Supported Collaboration

Christina E. Evangelou and Nikos Karacapilidis

Industrial Management and Information Systems Lab
MEAD, University of Patras, 26500 Rio-Patras, Greece
{chriseva,nikos}@mech.upatras.gr

Abstract. Communities of practice need the appropriate means to collaborate in order to reach decisions by exploiting all possible knowledge resources. To address this issue, we have developed a web-based platform that enables members of such communities collaborate through carrying out well-structured argumentative discourses. Our approach comprises a variety of concepts, methods, models and techniques, deriving among others from the decision making, knowledge management and argumentation fields, and properly interweaves them with the aid of an ontology model. This paper explores the interplay of the domain-independent and domain-specific concepts coming from the above fields, and comments on their embodiment in the above platform. Our multidisciplinary approach provides the foundations for developing a platform for brainstorming and capturing of the organizational knowledge in order to augment teamwork in terms of knowledge elicitation, sharing and construction, thus enhancing decision quality.

1 Introduction

Decision making at a strategic level comprises the identification of the basic goals of an enterprise, the consideration of alternative courses of action, and the allocation of resources necessary for carrying out these goals [1]. These tasks are usually undertaken by a group of managers working together to solve the related issues. According to Simon [2], decision making comprises three principal phases: identifying problematic situations or opportunities that call for decisions (intelligence phase), inventing or developing possible courses of action and testing of their feasibility (design phase), and selecting a certain course of action to be followed (choice phase). In the strategic decision making case, the quality of a formulated strategy (course of action) depends on the quality of the knowledge used during all these phases [3]. Moreover, it is strongly associated with the level of communication and coordination established between the decision makers during their collaboration.

In order to provide contemporary business organizations with the necessary means to develop successful strategic plans, an integration of tools and approaches coming from diverse disciplines is required. Of particular importance is the provision of a platform supporting collaboration, which aids decision makers attain a synthesis (and reflect on the convergence) of their highly specialized state-of-the-art knowledge. Our aim is to furnish communities of practice engaged in strategic decision making with the appropriate means to collaborate via a properly structured argumentation platform, so as to reach a decision by exploiting all possible knowledge resources. This plat-

R. Khosla et al. (Eds.): KES 2005, LNAI 3681, pp. 938–945, 2005.

form can be used for bringing together people holding complementary knowledge of a specific domain, which can be unified, revised and improved while it is being used for decision support. Our overall approach comprises a variety of concepts, methods, models and techniques deriving from the fields of Decision Support Systems, Knowledge Management Systems, Multicriteria Decision Aid, Argumentation and Semantic Web, as well as from the particular knowledge domain considered each time (the Manufacturing Management domain is considered in this paper).

This paper explores a series of issues related to the seamless integration of the concepts coming from the above fields. The proposed integration is achieved through the development of an ontology model that exploits their interrelations and rules holding among them [4]. Always referring to computer-supported collaboration, we distinguish between domain-independent and domain-specific processes to be provided, and we comment on the interplay between the associated concepts. The conceptual framework of our approach is presented in Section 2. Its embodiment in an already developed web-based collaboration platform is discussed in Section 3, through an example concerning argumentative discourse on the issue of building additional capacity for the needs of a specific company. Final remarks and future work directions are sketched in Section 4.

2   The Proposed Conceptual Framework

To identify the prevailing concepts and parameters to be taken into account in our overall approach towards computer-supported collaboration, we conducted a comprehensive survey of research and practical issues related to the abovementioned disciplines. Our main conclusion is that the concepts of *knowledge*, *argument*, and *decision*, together with the associated processes of *knowledge gathering* and *sharing*, *argumentation* and *decision making* are ubiquitous in a collaboration setting, independently of the knowledge and problem domain under consideration. Thus, they constitute the core of our conceptual framework[1]. More specifically, the concept of knowledge refers to the experience, values, contextual information, and expert insights that enable stakeholders (e.g. decision makers, domain experts) evaluating and incorporating new experiences and information. Arguments are considered as knowledge items ("pieces of knowledge") that are asserted during a discourse about a specific problem. Decision is synonymous to choice, and refers to the desired outcome of a discourse. On the other hand, knowledge gathering refers to the actions related to the extraction of knowledge from the discourse participants in order to share it with their peers. Knowledge sharing involves all the actions related to the dissemination of knowledge to the discussion participants in order to aid them reach a decision in an efficient and effective way. Argumentation refers to the process of introducing, supporting or defeating alternative courses of action, based on well-defined rules and procedures. Finally, decision making refers to the cognitive process of shaping a position, opinion or judgment in order to resolve a problem, attain a goal or seize an op-

---

[1] The term *conceptual model* is used to characterize a model of a knowledge area (often called domain), that represents the primary entities, relationships between these entities, attributes and attribute values of entities and relationships, as well as the rules that associate entities, relationships and attributes [5]

portunity. Issues to be addressed in a decision making setting concern the decision making process *per se*, the decision makers involved and the outcome of the overall process.

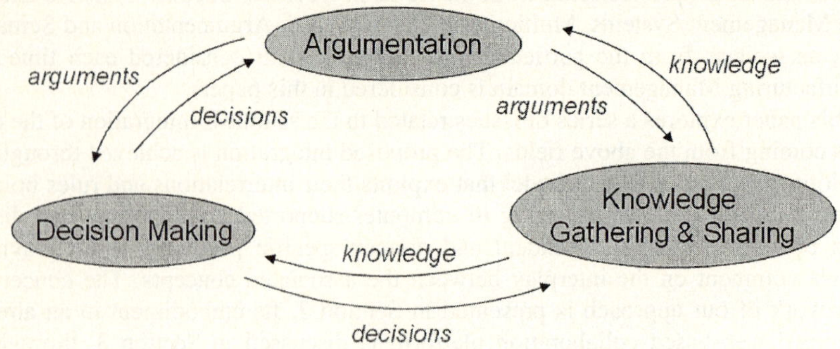

Fig. 1. The interplay of the fundamental concepts of our approach

The core of our approach's conceptual model is founded on the interplay of decision making, argumentation and knowledge management during discourses carried out by decision makers aiming at reaching a decision. As sketched in Figure 1, appropriate knowledge is required as an input to the decision making activities we consider. This knowledge can be expressed explicitly by using structured arguments to feed the discourses carried out for collaborative decision making purposes. The final outcome of these discourses is a set of decisions, resulted out of appropriate evaluation and reasoning mechanisms, which may then constitute new knowledge. With respect to the knowledge-based view of Decision Support Systems (see, for instance, [6]), a decision is a piece of knowledge indicating the nature of an action commitment. In other words, decision making results in production of new knowledge. If this new knowledge is properly made explicit (e.g. in the form of a structured argument), then it can be reused for argumentation during a new (related) decision making activity. As derives from the above rationale, the concepts of knowledge management, argumentation and decision making are the processes through which knowledge, arguments and decisions are interrelated.

As widely argued in the relevant literature, the efficient exchange of knowledge amongst the decision makers and thus the facilitation of their communication should rely on the establishment of a common language ("terms of reference"), as far as the representation of the issue, the assessment of the current situation and the objectives to be attained are concerned. The use of ontologies is valuable for such purposes. From an information science point of view, ontologies are the hierarchical structures of knowledge about things, by subcategorising them according to their essential or relevant cognitive qualities. They are a means to accomplish a shared understanding of different knowledge domains and allow for sharing and reuse of bodies of knowledge across groups and applications. Moreover, they figure prominently in the emerging Semantic Web as a way of representing the semantics of documents and enabling these semantics to be used by web applications and intelligent agents [7, 8].

An ontology model for managing the knowledge that is embedded in collaborative decision making processes and articulated through argumentative discourse has been already introduced in [9]. This model is generic, in that it can serve any application facilitating collaboration among individuals and groups, independently of the knowledge or problem domain considered. However, when taking into account a particular domain, two additional issues are of major importance. These refer to the way the abovementioned argumentative discourse is modelled and evaluated. The former corresponds to the method according to which the problem will be structured, and the form according to which the argumentation-based graph will be visualized, while the latter to the procedures according to which alternative courses of action will be weighed. These two issues delineate the domain-specific part of the ontology model to be used in a particular collaborative setting (see Figure 2).

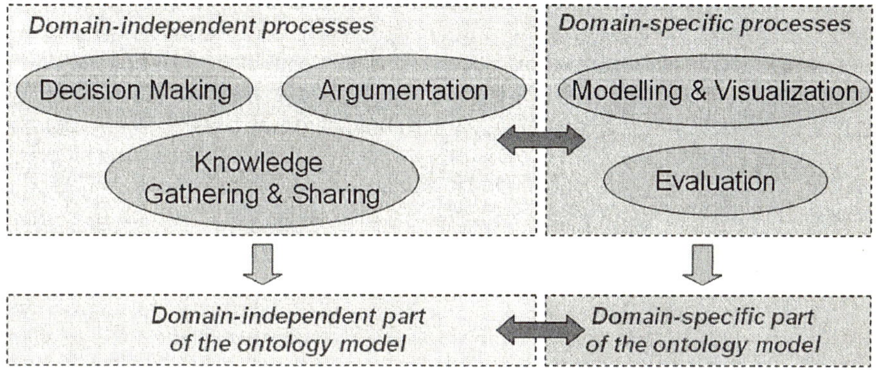

Fig. 2. Domain-independent and domain-specific processes

For instance, in the case considered in this paper, the knowledge domain is that of manufacturing management, while the problem domain concerns strategy development. The domain-specific concepts of our approach stem from methods, models, and techniques coming from the Strategy Development and the Multicriteria Decision Aid disciplines. More specifically, we have employed a series of decision making approaches that apply to the specific problem domain (this corresponds to the design phase of decision making). These are the *SWOT* (Strengths, Weaknesses, Opportunities, and Threats) framework [10], the *Resource Based View* (RBV) of the firm [11], and the *PIMS* (Profit Impact of Marketing Strategy) approach [12]. The above decision making approaches provide insights about the modelling and visualization of the problem under consideration. Furthermore, in order to evaluate the outcome of an argumentative discourse, we adopted as alternative scoring mechanisms the *Analytic Hierarchy Process* [13], the *Outranking Relations* [14], and the *SMART* [15] techniques (with the aid of them, our framework provides support during the choice phase of decision making). Each of these techniques provides different means for disaggregating a complex problem, by measuring the extent to which the alternative solutions meet the objectives set by the stakeholders, and by sorting the proposed courses of action.

Finally, common terms of reference should be established each time about the particular domain. In our case, the domain-specific part of the ontology model used was built on top of semantics concerning product design and development, process selection, plant location and design, capacity management, manufacturing planning and control, quality control, workforce organization, equipment maintenance, product distribution, and inter-plant coordination.

3   The Proposed Platform for Collaboration

Managers, being they decision makers or knowledge workers, need to communicate in order to collaborate, especially when they work in a distributed environment both in time and space terms. The concept of collaboration is based on the existence of common goals, objectives, or criteria which are either defined prior to or negotiated through the collaborative process [16]. The conceptual framework presented in the previous section is embedded in a web-based collaboration platform that enables stakeholders carry out argumentative discourses, the aim being to solve issues through knowledge-based decision making. According to our approach, stakeholders can argue/discuss in a structured way, by exchanging their assessments and beliefs through an *argumentation graph* that serves visualization purposes. Stakeholders, provided that they have the rights to do so, may initiate a new discourse, contribute to an ongoing one, and retrieve pieces of knowledge from past (closed) discourses (their actions are generally defined by a set of rules reflecting their level of authority and expertise). They can contribute to the overall collaborative process by asserting discourse items (goals, alternative solutions, criteria, preferences, constraints, etc.), which reflect their opinion about the problem under consideration. Discourse items are appropriately placed in the argumentation graph (according to the model selected for the particular domain). Each discourse item is treated as a piece of knowledge with a specific semantic value, according to the place of the graph where it has been inserted, and the set of meta-data that it carries. In such a way, the argumentation graph is also considered as a problem-specific knowledge graph.

After a discourse initiation request and the definition of the problem under consideration, a set of rules are triggered in order to define the structuring of the argumentation graph. These rules associate the problem domain with the past declarations of preference and usage of the available decision making approaches and scoring mechanisms (these are stored in the platform's model base). For instance, in a strategy development problem, two alternative decision making approaches are these of RBV and SWOT. These approaches formalize the context of the discourse, but they do not limit participants in the expression of their alternative views. On the other hand, the provided scoring mechanisms are methods and techniques coming from the Multicriteria Decision Aid discipline that serve as a guide for selecting the most acceptable alternative solution. In this way, our platform is able to propose the appropriate structure of the graph, while the final selection of the decision making approach and the scoring mechanism to be adopted is up to the discourse initiator (in this case the new preference is recorded and triggers the related models).

Figure 3 illustrates an instance of a discourse that concerns resolving an issue of building additional capacity for the needs of a specific company. In this case, the argumentation graph is structured according to criteria apt to the RBV decision making framework and a scoring mechanism that is based on the Analytical Hierarchy Process. When a discourse item is inserted into the argumentation graph by one of the discussion participants, it is automatically assigned with a specific semantic value and treated accordingly. This is achieved through a set of meta-data that every discourse item carries (maintaining information about its type and creator, and the place of the graph where it is inserted), and a set of properly defined rules. By referring to a discourse item type, we point to a part of the ontology that describes all possible kinds of statements a decision maker can make. These are explicitly defined by the `domain`, `goal`, `alternative` (solution), `criteria`, `support` (in favour or against) and `constraint` classes, fully described in the generic ontology model we have defined [9]. As mentioned in the previous section, this model comprises diverse elements, which have been properly interwoven for efficient and effective decision support and knowledge management during argumentative discourses.

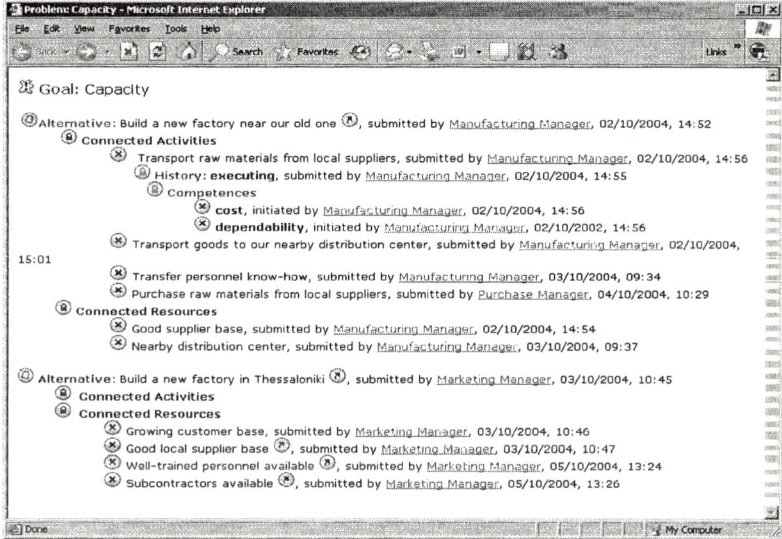

Fig. 3. An instance of the argumentation graph

Due to the emergence of the Semantic Web, the physical model of our ontology model has been described with the Web Ontology Language (OWL), by employing the Protégé ontology editor [17]. All the above concepts, represented as elements in the ontology model, formulate OWL classes (in the OWL ontology representation language). The construction of each class comprises the definition of the class name, properties, and instances, as well as a set of asserted conditions. Besides interrelating such classes, the declaration of properties in the form of rules serves for making explicit the concepts interconnections. In this way, whenever a discourse item insertion takes place, the related ontology class is dynamically updated, including the inserted item as its instance. Furthermore, all arguments expressed during the discourse are classified with respect to the structure of the ontology model, in order to be reused in a future discourse of the same problem domain.

4 Conclusions

In this paper we have explored a series of issues related to the interplay of domain-independent and domain-specific concepts of an ontology model developed to be embodied in a web-based collaboration platform. Our multidisciplinary approach provides the foundations for developing a platform for brainstorming and capturing of the organizational knowledge in order to augment teamwork in terms of knowledge elicitation, sharing and construction, thus enhancing decision quality. This is due to its structured problem-specific language for conversation and its mechanism for evaluation of alternatives. During the implementation of our approach, much attention was paid to openness and extensibility issues. Thus, we exploited XML technologies to guarantee an interoperable, generic, extensible and neutral information modelling. Our future work directions concern the development of more domain-specific ontology concepts, as well as the integration of existing ontology models in order to further support decision making activities.

Acknowledgements

The authors thank European Social Fund (ESF), Operational Program for Educational and Vocational Training II (EPEAEK II), and particularly the Program HERAKLEITOS, for funding the above work.

References

1. Chandler, A.D. Jr: Strategy and structure: Concepts in the History of the Industrial Enterprise. MIT Press, Casender, MA (1962).
2. Simon, H.: The new Science of Management Decision. Prentice Hall, Englewood Cliffs, NJ (1977).
3. Feurer, R. and Chaharbaghi, K.: Strategy Development: Past, Present and Future. Management Decision Vol. 33(6), 11-21 (1995).
4. Davies, J., D. Fensel and F. van Harmelen (eds.): Towards the Semantic Web. Ontology-driven Knowledge Management. Wiley, Chichester, UK (2003).
5. Daconta, M.C., Obrst, L.J. and Smith, K.T: The Semantic Web: A Guide to the Future of XML, Web Services, and Knowledge Management. John Wiley & Sons (2003).
6. Holsapple, C. and Whinston A.: Decision Support Systems-A knowledge Based Approach. West Publishing Company (1996).
7. Chandrasekaran, B., Josepheson, J. and Benjamins, V.R.: Ontologies: What Are They? Why Do We Need Them? IEEE Intelligent Systems Vol. 14(1), 20-26 (1999).
8. Duineveld, A.J., Stoter, R., Weiden, M.R., Kenepa, B. and Benjamins, V.R.: Wonder-Tools? A Comparative Study of Ontological Engineering Tools. International Journal of Human-Computer Studies Vol. 52, 1111-1133 (2000).
9. Evangelou, C., Karacapilidis, N. and Abou Khaled, O.: Interweaving Knowledge Management, Argumentation and Decision Making in a Collaborative Setting: The KAD Ontology Model. International Journal of Knowledge and Learning Vol. 1(1/2), 130-145 (2005).
10. Porter, M.E.: Competitive Strategy. The Free Press, New York (1980).
11. Wernerfelt, B.: A Resource-Based View of the Firm. Strategic Management Journal Vol. 5, 171-180 (1984).
12. Buzzell, R.D. and Gale, B.T.: The PIMS Principles-Linking Strategy to Performance. The Free Press, New York (1987).
13. Saaty, T.L.: The Analytic Hierarchy Process, McGraw-Hill, New York (1980).
14. Roy, B.: The outranking approach and the foundations of ELECTRE methods. Theory and Decision Vol. 31, 49-73 (1991).

15. Edwards, W.: SMARTS and SMARTER: Improved simple methods for multiattribute utility measurement. Organizational Behaviour and Human Decision Processes Vol. 60, 306-325, (1994).
16. Nunamaker, J.F. Jr., Applegate, L. and Konsynski, B.: Facilitating Group Creativity: Experience with a Group Decision Support System. Journal of Management Information Systems Vol. 3(4), 5-19 (1987).
17. Protégé: The Protégé Project, Stanford Medical Informatics [Available online at: http://protege.stanford.edu].

Using XML for Implementing Set of Experience Knowledge Structure

Cesar Sanin and Edward Szczerbicki

Faculty of Engineering and Built Environment, University of Newcastle
University Drive, Callaghan, NSW 2308, Australia
{Cesar.Maldonadosanin,Edward.Szczerbicki}@Newcastle.edu.au

Abstract. Among all knowledge forms, storing *formal decision events* in a knowledge-explicit way becomes an important development. Set of experience knowledge structure can help in achieving this purpose. Set of experience has been shown as a shape able to acquire explicit knowledge of formal decision events. However, to make set of experience knowledge structure practical, it must be worldwide transportable and understandable. The purpose of this paper is to show an effective form of transformation of the set of experience into a shareable and understandable shape able to travel among different systems and technologies.

1 Introduction

Knowledge is a valuable asset of incalculable worth and has been considered as the only true source of competitive advantage of a company [2]. Thus, the focus of managers has turned to knowledge administration and companies want technologies that facilitate control of all forms of knowledge. Knowledge management can be considered as the key for the success or failure of a company.

Lin et al. [7] describe the concept of knowledge as an organized mixture of data, integrated with rules, operations, and procedures, and it can be only learnt through experience and practice. One of the most complicated issues about knowledge is its representation, because it determines how knowledge is acquired and how knowledge is transformed from tacit knowledge to explicit knowledge.

Decision-makers, when making decisions, extract the most significant characteristics from the current circumstances, and relate them to comparable situations and actions that have worked well in the past. In consequence and understanding that a *formal decision event* is a decision occurrence that was made following procedures that make it structured and formal [12], tools for representing and store *formal decision events* in a knowledge-explicit way are evidently necessary.

Set of experience knowledge structure has been developed to facilitate representation of formal decision events inside computers. It is a structure developed to support a platform named Knowledge Supply Chain System (KSCS) [11]. In brief, the KSCS takes information from different technologies that make formal decision events, integrates them and transforms them into knowledge making use of sets of experience. Because different technologies can be both source and target of sets of experience, it must be possible to construct them in a wide-reaching language.

R. Khosla et al. (Eds.): KES 2005, LNAI 3681, pp. 946–952, 2005.

Making set of experience a tool able to be universally useful requires translating it into a language to which every technology can be accommodated easily. Thus, many languages for exchanging information have emerged. Such languages give specific format and meaning to the data the document contains. Multiple technologies using a shareable language to exchange documents would provide stability across generations of technologies [13], and, they would give long-term life to set of experience.

The purpose of this paper is to show an effective transformation of set of experience into a shareable and understandable shape able to travel among different systems and technologies. The developed transportable set of experience can be applied in many technologies, and more specifically in the KSCS. Set of experience knowledge structure has the potential to improve the way knowledge is managed as an asset in current decision making environments.

2 Background

2.1 Set of Experience

In this paper a succinct idea of set of experience knowledge structure and its components is given, for further information Sanin and Szczerbicki [12] should be review.

Set of experience has been developed to store formal decision events in an explicit way [12]. It is a model based upon existing and available knowledge, which must adjust to the decision event is built from. Four basic components surround decision-making events, and are stored in a combined dynamic structure that comprises set of experience. These four components are *variables, functions, constraints,* and *rules*.

Variables usually involve representing knowledge using an attribute-value language [8]. This is a traditional approach from the origin of knowledge representation, and is the starting point for set of experience. Variables that intervene in the process of decision-making are the first component of the set of experience, and they are the origin of the other components.

Based on the idea of Malhotra [9] who states that "to grasp the meaning of a thing, an event, or a situation is to see it in its relations to other things", variables are related among them in the shape of functions. Functions, the second component, describe associations between variables; moreover, functions can be applied for reasoning optimal states. Therefore, set of experience uses functions, and establishes links among the variables constructing multiobjective goals.

According to Theory of Constraints (TOC), Goldratt [3] maintains that any system has at least one constraint; otherwise, its performance would be infinite. Thus, constraints are another way of relationships among the variables. A constraint, as the third component of set of experience, is a restriction of the feasible solutions in a decision problem, and limits the performance of a system with respect to its goals.

Finally, rules are suitable for representing inferences, or for associating actions with conditions under which the actions should be performed. Rules, the fourth component of set of experience, are another form of expressing relationships among variables. They are conditional relationships of the universe of variables.

In conclusion, sets of experience represent *formal decision events* that can be used in platforms to support new decisions decision-making.

2.2 XML for Exchanging Information and Knowledge

Web technologies have developed tools necessary for integration of distributed applications. These technologies include the development of standards and protocols that have led to the creation of some hundreds of different 'languages'. Up to date we have found almost 390 different languages; most of them are based upon XML. These languages present defined vocabulary, structure and constraints for expressing information and knowledge. Among them, we have distinguished some important languages that can be useful for our purposes: XML, MATHML, XRML, RULEML, RFML, OptML, SNOML, LPFML, AIML and PMML.

EXtensible Markup Language (XML) was developed under the basis of simplicity, and communicating format and meaning. XML, under the sponsorship of the W3C, has grown into a family of standards integrating technologies. It describes structured data in one document; subsequently it can be used by another application or document. It is easy to understand, read, and author [11].

XML is a mechanism for creating markup languages, allowing users to create their own set of markup-tags to represent data relationships. These tags can be chosen to reflect the domain specific semantics of the information or knowledge. XML allows to structure information and knowledge as labeled trees. The XML attribute-value mechanism allows encoding graphs as XML trees.

In addition, XML allows defining restrictions on the set of tags that can be used in documents. This is done by a XML Schema Definition (XSD). XSD communicates the structure of XML, specifies the validity of each tag, and expresses in a grammar-like formalism what kinds of permitted sequences of tags are allowed in a document [5]. XSD provides critical information that allows XML processors to parse the coding and make certain that it contains all the information the application needs.

MATHML appeared as a language to display specialized mathematical information. It is derived from XML, and like XML, it does have content about the information it transports. However, even the goals of MATHML were to have a system for encoding flexible and extensible mathematical material for the web, without involving some complexity, its structure does not present the simplicity of XML at all. Consequently, MATHML is not primarily intended for direct use by authors, MATHML is mainly to be used by machines [14]. MATHML, despite of its no simplicity, seems to be a good option to transmit pure mathematical information; however it shows some complicatedness on exchanging rules.

Other mathematical languages are more specialized in optimization models. Languages such as OptML, SNOML, and LPFML can transmit models for different optimization methods such as linear and non-linear optimization, and stochastic programming. They can carry variables, functions, and constraints, but they cannot configure rules, creating a big restriction for our purposes.

PMML, another XML-based language, allows defining predictive models. One of its major goals is to facilitate moving predictive models across applications and systems [1]. Currently, PMML 3.0 is supporting predictive models such as: regressions, DT, clustering, association rules, NaiveBayes, neural nets, rule sets, SVM, text models, and sequences. Even though all this predictive models help in decision-making, and could make part of set of experience, the impossibility of including optimization models, limits its use for our implementation. Set of experience is based upon constraints as one of its components, and PMML does not allow its use yet.

Knowledge representation for software agent communication has also adopted the XML platform [4]. Additionally, the W3C [15] defends XML and agrees in terms of its use on knowledge representation saying: "XML provides a universal storage and interchange format for distributed knowledge representation. XML offers new general possibilities from which AI knowledge representation (KR) can profit". For instance, specialized languages for the use of knowledge have been developed based upon XML. AIML, RFML, RULEML and XRML are examples of this kind of technologies that exchange ontological concepts. They are most focused on the representation of objects and rules in an XML format [6]. Even though AIML, RFML, RULEML and XRML are originated on XML, they are excessively specialized and are not worldwide used. Hence, XML remains as the preferred worldwide option for exchanging information and knowledge in a simple and not specialized form.

Finally and most important, XML has emerged as the leading method for application integration [13]. It has gradually materialized as the preferred data exchange across different computer platforms and systems, as well as a standard for the development and delivery of Internet content.

3   Structure of Set of Experience in XML

XML permits the representation of structured, hierarchical data, capable of representing not just the values of individual information and knowledge items, but also the relationships between them. Transforming set of experience into a XML configuration makes necessary to represent it in a tree shape.

Each set of experience is identifiable by the whole set of elements that comprises it. There are some kinds of fields, which are considered the head of set of experience. Because set of experience keeps a formal decision event that has occurred at a defined moment, date (<date>) and hour (<hour>) of occurrence are stored. Category (<category>) encloses fields about the areas (<area>, <subarea>) and objective functions (<subject>) involved in the decision event. An example of a head of set of experience in XML-configuration is shown:

```xml
<?xml version="1.0" encoding="UTF-8" standalone="no"?>
<set_of_experience xmlns:xsi="http://www.w3.org/2001/XMLSchema-instance"
   xsi:noNamespaceSchemaLocation="set of experience model.xsd">
      <date>2004-11-11</date>
      <hour>14:10:00</hour>
      <category>
          <area>Human Resources</area>
          <subarea>Salary Office</subarea>
          <subject>Payment Level</subject>
          <subject>Worker's Morale</subject>
      </category>
```

3.1   Variables

Describing the variables includes the fields of name of the variable (<var_name>), cause value of the variable (<var_cvalue>), effect value of the variable (<var_evalue>), the unit of the values (<unit>), and the boolean value of internal variables (<internal>). An example of a variable in XML can be seen:

```
<variable>
      <var_name>X1</var_name>
      <var_cvalue>2</var_cvalue>
      <var_evalue>5</var_evalue>
      <unit>quantity product 1</unit>
      <internal>true</internal>
</variable>
```

3.2 Functions

Four of the elements are presented just once in the set of experience: objective (<obj>), name of function (<fn_name>), symbol (<sym>), and unit (<unit>). The other elements can be present more than once; they are factors (<factor>). Factors comprise: associated factors (<assofactor>) and simple factors (<simfactor>), and those contains different elements such as parenthesis (<lpar>, <rpar>), but most important the term (<term>), which is a joint of operator (<oper>), coefficient (<coef>), variable (<variable>), and potency (<poten>). An example of a function is shown:

```
<function>
      <obj>Min</obj>
      <fn_name>Payment Level</fn_name>
      <sym>=</sym>
      <factor>
            <assofactor>
                  <lpar>(</lpar>
                        <term>
                              <coef>3</coef>
                              <variable>X1</variable>
                        </term>
                        <term>
                              <oper>+</oper>
                              <variable>X2</variable>
                              <poten>2</poten>
                        </term>
                  <rpar>)</rpar>
                  <poten>2</poten>
            </assofactor>
      </factor>
      <unit>money</unit>
</function>
```

3.3 Constraints

Constraints are functions that bound related variables, and they are expressed similarly except for the use of comparative relationships. Two of the components of the constraints are presented just once: value (<value>) (that is, the independent value) and symbol (<sym>) (that is, $\leq, \geq, <, >, =$). The other element, factor, can be more than once; they are expressed in the same way as in functions.

3.4 Rules

Rules contain two basic elements: the one that compound the conditions named joint (<joint>) and the one that correspond to the consequence named as itself (<conse-

quence>). Joint could be presented more than once and it comprises: conjunction form of conditions (<jnt>) (that is, AND / OR) and condition (<condition>). Condition and consequence, each one on its branch, can be presented more than once. Condition and consequence are comprised by: factors, already described, symbol (<sym>), variable (<variable>), and value (<value>). An example of a rule in XML is shown:

```
<rule>
    <joint>
        <condition>
            <factor>
                <simfactor>
                    <term>
                        <coef>1/1.2</coef>
                        <variable>Firing</variable>
                    </term>
                </simfactor>
            </factor>
            <sym>&lt;=</sym>
            <variable>Competitor's Firing</variable>
        </condition>
    </joint>
    <consequence>
        <variable>Status of Firing</variable>
        <sym>=</sym>
        <value>VERY GOOD</value>
    </consequence>
</rule>
```

4 Conclusion

Marvin Minsky [10], referring to knowledge representation says "each particular kind of data structure has its own virtues and deficiencies, and none by itself would seem adequate for all the different functions involved". Set of experience knowledge structure is a suitable representation for formal decision events. Additionally, if it is shaped in a wide-reaching understandable and transportable language as XML, it will advance the notion of administering knowledge in the current decision making environment.

XML enables us to create set of experience documents and in that form, companies which are expanding the knowledge management concept externally, can explore new ways to put explicit knowledge in the hands of employees, customers, suppliers, and partners.

Further research will lead to the extension of set of experience XML-configuration that allows transmitting any kind of decision models such as heuristic algorithms.

References

1. DMG, Data Mining Group: PMML Version 3.0. Viewed April 2005, <http://www.dmg.org/v3-0/> (2004).
2. Drucker, P.: The Post-Capitalist Executive: Managing in a Time of Great Change. Penguin, New York (1995).

3. Goldratt, E. M. and Cox, J.: The Goal. Grover, Aldershot, Hants (1986).
4. Grosof, N.: Standardizing XML Rules: Rules for E-Business on the Semantic Web. Invited talk in: Workshop on E-Business and the Intelligent Web at the International Joint Conference on Artificial Intelligence, August, Seattle, (2001).
5. Harmelen, F. Fensel, D.: Practical Knowledge Representation for the Web. In proceedings: IJCAI'99 Workshop on Intelligent Information Integration, July (1999).
6. Lee, J.K. and Sohn, M.M.: The Extensible Rule Markup Language. Communications of the ACM, Vol. 46 (5). (2003) 59-64.
7. Lin, B., Lin, C. et al.: A Knowledge Management Architecture in Collaborative Supply Chain. Journal of Computer Information Systems, Vol. 42 (5). (2002) 83-94.
8. Lloyd, J. W.: Logic for Learning: Learning Comprehensible Theories from Structure Data. Springer, Berlin (2003).
9. Malhotra, Y.: From Information Management to Knowledge Management: Beyond the 'Hi-Tech Hidebound' Systems. In Srikantaiah, K. and Koening, M. E. D (Eds.): Knowledge Management for the Information Professional, Information Today, Inc., (2000) 37-61.
10. Minsky, M.: AI Topics. Viewed December 2004, <http://www.aaai.org/AITopics/html/#minsky> (2004).
11. Sanin, C. and Szczerbicki, E.: Knowledge Supply Chain System: A Conceptual Model. Knowledge Management: Selected Issues. Szuwarzynski, A. (Ed), Gdansk University Press, Gdansk, (2004) 79-97.
12. Sanin, C. and Szczerbicki, E.: Set of Experience: A Knowledge Structure for Formal Decision Events. Foundations of Control and Management Sciences, In print (2005).
13. Singh, S.: Business-to-manufacturing integration using XML. Hydrocarbon Processing, Vol. 82 (3). (2003) 62-65.
14. W3C: MathML 2.0 Recommendation 21 October 2003. Viewed August 2004, <http://www.w3.org/TR/2003/REC-MathML2-20031021/> (2003).
15. W3C: XML. Viewed April 2004, <http://www.w3.org/XML/> (2004).

A Framework of Checking Subsumption Relations Between Composite Concepts in Different Ontologies[*]

Dazhou Kang[1], Jianjiang Lu[1,2], Baowen Xu[1,2], Peng Wang[1], and Yanhui Li[1]

[1] Department of Computer Science and Engineering, Southeast University,
Nanjing 210096, China
[2] Jiangsu Institute of Software Quality, Nanjing 210096, China
bwxu@seu.edu.cn

Abstract. Ontology-based applications always need checking subsumption relations between concepts. Current subsumption checking approaches are not appropriate to check subsumption relations between composite concepts in different ontologies. This paper proposes a framework, which is practical and effective to check subsumption relations between composite concepts in different ontologies. The framework is based on learning the mutual instances of different ontologies. It analyses the problems when generating and checking of mutual instances, and provides effective processes to solve them. The framework is fault tolerance and can deal with large-scale mutual instances; it can greatly reduce the manual annotation work, which is the main limitation in mutual instances approaches; intensions are also considered to avoid unnecessary instance checking and hit-and-miss results.

1 Introduction

Ontology is the basic of knowledge sharing and reusing on the Semantic Web [1]. One of the most important issues in ontology-based applications is to find the subsumption relations between concepts in different ontologies to gain interoperability. However, checking subsumption relations between composite concepts in different ontologies is still an open problem.

Current logic checking approaches [2] can only checking subsumption relations between concepts in the same ontology. Ontology mapping approaches mainly find relations between atomic concepts [3]. Logic reasoning is an exact approach of subsumption checking [4]. It can deal with composite concepts and the result can be believable. However, different ontologies often have few inter-relations; then the reasoning results will always be uncertain. Learning methods use machine-learning methods based on the similarity measures between concepts to find mappings between concepts [5]. The concept similarity can include both the lexical, structural features and the instances of concepts. However, the methods can mostly find the one-to-one concept mappings. When checking subsumption relation, this approach can only deal with atomic concepts and the results are not precise [6].

---

[*] This work was supported in part by the NSFC (60373066, 60425206, 90412003), National Grand Fundamental Research 973 Program of China (2002CB312000)

R. Khosla et al. (Eds.): KES 2005, LNAI 3681, pp. 953–959, 2005.

The ontologies may be populated by instances. If a set of instances is classified in both ontology O_1 and O_2, then we call it a set of mutual instances of O_1 and O_2. Using the mutual instances, the subsumption relation can be checked easily by learning their corresponding sets of instances. If there are no or few mutual instances, then the experts annotate more instances and classify them into concepts [7]. Generally, mutual instances can lead to a simple and efficient checking of subsumption relations between different ontologies, but faces several problems.

This paper proposes a framework of checking subsumption relations between concepts in different ontologies based on mutual instances. Section 2 defines the problem. Section 3 analyses the problems when generating and checking of mutual instances, and provides effective processes to solve them. Section 4 discusses the intensions. Section 5 proposes the framework. Section 6 gives the conclusions.

2   Concepts Subsumption Between Different Ontologies

Definition 1. A source is a set of instances. If every instance in a source S has been classified into one or more concepts in an ontology O, i.e. the instances are individuals in O, then S is called a source with respect to O. Let the concepts in ontology O form the set $C = \{c_1, c_2, \ldots, c_n\}$. For each c_i, we introduce $S(c_i)$ means the set of instances in S such that belong to concept c_i; and for any instance a in S, let

$$C(a)=\{c_i \in C \mid a \in S(c_i)\}. \tag{1}$$

Definition 2. A composite concept is a Boolean expression on the set of concept names in an ontology, and is called a query for short in this paper. The queries in concept set C and their expected answers in source S are defined as:

1. Every concept name in C is a query, the answer to the query c_i is $S(c_i)$;
2. If Q is a query, then $\neg Q$ is also a query, the answer $S(\neg Q) = S - S(Q)$;
3. If Q_1 and Q_2 are queries, then $Q_1 \wedge Q_2$, $Q_1 \vee Q_2$ are queries, the answers are $S(Q_1 \wedge Q_2) = S(Q_1) \cap S(Q_2)$, and $S(Q_1 \vee Q_2) = S(Q_1) \cup S(Q_2)$.

Definition 3. Let P, Q be queries, if for any source S, it is true that $S(P) \subseteq S(Q)$, then Q subsumes P, notated as $P \sqsubseteq Q$. If it is true that $P \sqsubseteq Q$ and $Q \sqsubseteq P$, then $P \equiv Q$. If $S(P) \cap S(Q) = \varnothing$, then P is disjoint with Q, else P and Q are overlapped.

Let O_1 and O_2 be ontologies, C and D be the set of concepts in O_1 and O_2, S and T be the sources with respect to O_1 and O_2 respectively, $M=S \cap T$ be the mutual instances. The problem is checking the subsumption relation between P and Q, where P is a query in ontology O_1 and Q is a query in ontology O_2; the result of the checking can be $P \equiv Q$, $P \sqsubseteq Q$, $Q \sqsubseteq P$, P is disjoint with Q, P and Q are overlapped, or unknown.

Our framework is based on the assumption that checking subsumption relations between concept queries is equivalent to checking subsumption relations between the corresponding sets of mutual instances, i.e.

$$M(P) \subseteq M(Q) \rightarrow P \sqsubseteq Q. \tag{2}$$

This assumption is not proper true, but it is acceptable when adequate mutual instances are considered.

3   Checking Based on Mutual Instances

3.1   Weighted Mutual Instances

In practice, there may be some mistake in the mutual instances. We should estimate checking results considering fault tolerance. For example, let X, Y be two sets of instances, if $X \cap Y$ have thousands of instances, $X-Y$ have only one instance, then it is likely that $X \subseteq Y$, and the latter instance is annotated mistakenly. However, it is not always the case, some instances may be important than others. Weighted mutual instances are proposed to denote such importance.

Definition 4. When annotating instances, the experts can give a weight value $w(a)$ for any instance a. The higher $w(a)$ denotes the more importance of a. If there is no expert's opinion on a, then we set $w(a)=1$ by default. If the experts give different weight values for the same instance, then we consider $w(a)$ be the average value.

Then when checking $X \subseteq Y$, it is equivalent to checking

$$\frac{\sum_{a_i \in X-Y} w(a_i)}{\sum_{b_j \in X \cap Y} w(b_j)} \leq T_k, \tag{3}$$

where T_k is a given threshold that is very small. From now on, we assume all instance are weighted, and we always use Equation 3 to check whether $X \subseteq Y$.

3.2   Large-Scale Mutual Instances

It is intuitionistic that the more the mutual instances, the better the checking results. However, the large-scale mutual instances will make the checking process time-consuming: when $M(P)$ and $M(Q)$ are both large sets, it is hard to check whether $M(P) \subseteq M(Q)$. The basic idea to solve the problem is to partition mutual instances into sets, and deal with the sets instead of individual instances.

Definition 5. The instances can be partitioned into instance classes. Let C be the set of concepts in the ontology O_1, and S be a source with respect to O_1. For any instance a, b, if $C(a)=C(b)$, i.e. a and b are classified into the same concepts, then a and b are in the same instance class identified by $C(a)$. Let

$$I(E, S)=\{a \in S \mid C(a)=E\}, \tag{4}$$

where E is a subset of C. We call $I(E, S)$ an instance class in S with respect to E.

Although there may be numerous possible instance classes, we only consider non-empty instance classes, and the number of them is far smaller than the instances.

We change the checking of $M(P) \subseteq M(Q)$ into two parts: a preparative offline process and an online process. In the offline process, for each nonempty instance classes $I(E, M)$ such that E is a subset of C, the subsumption relation between $I(E, M)$ and every $M(d_i)$ such that $\{d_i \mid d_i \in D$ or $\neg d_i \in D\}$ is checked and recorded. In the online process, the basic problem of checking whether $P \sqsubseteq Q$ is further simplified.

Firstly, both queries can be transformed into negation normal form in which negations only apply to concept names by using the following two equations:

$$\neg(e_1 \wedge e_2) = \neg(e_1 \vee e_2);$$
$$\neg(e_1 \vee e_2) = \neg(e_1 \wedge e_2). \tag{5}$$

Then the problem can be split into sub problems using the following equations:

$$P \sqsubseteq Q_1 \wedge Q_2 \;\Leftrightarrow\; P \sqsubseteq Q_1, P \sqsubseteq Q_2;$$
$$P_1 \vee P_2 \sqsubseteq Q \;\Leftrightarrow\; P_1 \sqsubseteq Q, P_2 \sqsubseteq Q. \tag{6}$$

In the end, each sub problem is in the form as

$$\bigwedge_{c_i \in G} c_i \sqsubseteq \bigvee_{d_j \in H} d_j; \tag{7}$$

where G is a subset of $\{c_i \mid c_i \in C$ or $\neg c_i \in C\}$, and H is a subset of $\{d_j \mid d_j \in D$ or $\neg d_j \in D\}$.

From Definition 3, we can deduce that

$$M(\bigwedge_{c_i \in G} c_i) = \bigcup_E I(E, M); \tag{8}$$

where $E \sqsubseteq C$ satifies $\forall c_i \in C, c_i \in G \rightarrow c_i \in E, \neg c_i \in G \rightarrow c_i \notin E$.

The process of checking whether Equation 3 holds:

1. If for any $I(E, M)$, $I(E, M)$ is subset of any $M(d_i)$ in $\{d_i \mid d_i \in D$ or $\neg d_i \in D\}$, then Equation 3 is true. The results recorded in the offline process are used.
2. For all $I(E, M)$ not subset of any $M(d_i)$, put their instances into a set X. Then Equation 3 is equivalent to $X \subseteq M(\bigvee_{d_j \in H} d_j)$.

The result of whether $P \sqsubseteq Q$ holds can be integrated by results of sub problems.

In this process, it is not necessary to get whole $M(P)$ and $M(Q)$. The number of nonempty instance classes is far smaller than the number of instances in $M(P)$. The size of X is much smaller than the original instances in most cases. Therefore, the process can deal with the large-scale mutual instances.

3.3   Annotation of Mutual Instances

Not any two ontologies have many mutual instances. When there are no or few mutual instances, we should annotate new mutual instances. However, the annotation cannot be completely automatic, but needs manual work. On the one hand, the more the mutual instances, the better the checking results; on the other hand, more mutual instances lead to more manual annotation work. Selecting adequate and compact set of instances for manual annotation is an important task.

The following process assumes that S and T are both abundant, but M is not adequate. We want to annotate some instances in S and T to get more mutual instances. It is obvious that for any instances a, b not in $S \sqcap T$, if $C(a)=C(b)$ and $D(a)=D(b)$, then only one of them needs to be annotated manually. However, before the annotation, either $C(a)$ or $D(a)$ is unknown. Our idea is using learning method to do a tentative annotation for all instances, and then remove redundant instances based on the learning results. The human experts will annotate the residual instances.

Firstly, train learners for tentative annotation. For each concept c_i in C, partition S into two sets of instances that do and do not belong to c_i: $Pc_i=\{a \in S(c_i)\}$, $Nc_i=S-Pc_i$.

Then use Pc_i and Nc_i as the sets of positive and negative training examples respectively to train a learner Lc_i for instances of c_i. It outputs a value over $[0, 1]$ as the confidence of the input instance belonging to c_i. For each concept d_i in D, a learner Ld_i for instances of d_i can also be trained. Then we get two sets of learners:

$$LC=\{Lc_i \mid c_i \in C\}, LD=\{Ld_i \mid d_i \in D\}. \tag{9}$$

Secondly, apply the learners to do the tentative annotation, and remove redundant instances. Let S', T' be empty sets. For each nonempty instance class $I(E, S)$ in S,

1. Orderly apply the learners in LD to every instances in $I(E, S)$. For each instance a in $I(E, S)$, we can get a vector over $[0, 1]^D$ showing the confidence of a belonging to each d_i in D.
2. Using these vectors, apply a clustering algorithm [8] to partition $I(E, S)$ into several subsets. The clustering algorithm should ensure that if the distance between two vectors is greater than a predefined threshold, then the corresponding instances are clustered into different subsets.
3. Because if instances a and b are clustered into the same subset, then it is likely that $D(a)=D(b)$; and we know $C(a)=C(b)$. One of them is probably redundant. So for each subset, only preserving one representative element (the one with highest weight value or closet to the clustering mean), and add it into S'. If the representative element is instance a, then define $w(a)= \sum_i w(a_i)$ be the weight of a, where a_i is all the instances in the subset.
4. Afterward, we get a new class $I(E, S')$ much smaller than $I(E, S)$. An example of this process is shown in Fig.1.

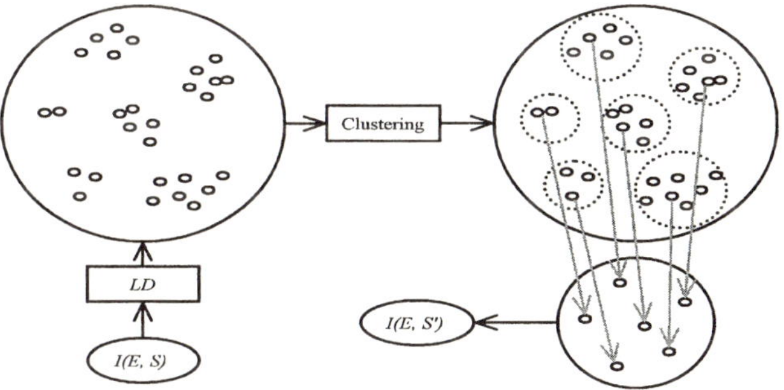

Fig. 1. The process

In the end, we can get a new source S', which is much smaller than S. We can get a new source T' from T using the learners in LC in the same way.

In $S' \cup T'$, probably redundant instances have been removed using the learning results. So the members of $S' \cup T'$ are potential to be mutual instances which are useful in checking inter-relations. The experts annotate the instances in S' using O_2, and annotate the instances in T' using O_1. Then we can expand the two sources to $S \cup T'$ and $T \cup S'$. The increased mutual instances are now $M'=M \cup S' \cup T'$.

Not all ontologies are populated by many instances. If one of O_1 and O_2 has no or few instances, then the above process should be modified. Without the loss of generality, let O_2 have no instances, i.e. T is empty. We can randomly select some instances form S, annotate them with O_2, and use them as T. Then S' can be further selected by learning T with respect to O_2.

4   Intension Annotation

The mutual instances only show the extensions of concepts. It takes no account of intensions of the concepts. Two concepts can share extension without sharing intension. For example, the concepts *creatures with a kidney* and *creatures with a heart* have the same extension but different intensions. It may lead to even absurd results when mutual instances are not complete. For example, if M is selected in a women's school, then it may be true that $M(student) \subseteq M(schoolgirl)$; considering the intensions of concepts *student* and *schoolgirl* can clearly exclude this result.

Definition 6. For each concept c, the experts annotate c by a set of intensions $A(c)$. Let $A(Q)$ be the intensions of query Q and for any queries Q_1 and Q_2, it is true that

$$A(Q_1 \wedge Q_2) = A(Q_1) \cup A(Q_2);$$
$$A(Q_1 \vee Q_2) = A(Q_1) \cap A(Q_2). \tag{10}$$

When the intension annotation is complete, we consider for any queries P and Q:

$$A(Q) \subseteq A(P) \ \rightarrow \ P \sqsubseteq Q. \tag{11}$$

However, intension annotation is only an assistant tool. It can estimate obvious relations between queries: if $A(Q)-A(P)$ is a large set, then it is impossible that $P \sqsubseteq Q$. This can avoid much checking work on the mutual instances. In the above example, since $A(schoolgirl)-A(student)$ at least contains many intensions of girls, we can clearly deduce that $M(student) \subseteq M(schoolgirl)$ is only a hit-and-miss result.

5   The Framework

In the framework, we assume all instance are weighted using Definition 4, and when checking subsumption relations between sets of instances, we use Equation 3 for fault tolerance.

Beforehand, for the ontologies O_1 and O_2, the intensions of the concepts are annotated, and the instances are annotated using the method in Section 3.3. The mutual instances are then organized into instance classes and the offline process in Section 3.2 is processed.

When checking subsumption relation between queries P and Q in O_1 and O_2:

Firstly, we can divide the checking problem into several basic problems: $P \sqsubseteq Q$, $Q \sqsubseteq P$, $P \sqsubseteq \neg Q$. The basic problems are all in the form $P \sqsubseteq Q$.

Then for each basic problem $P \sqsubseteq Q$:

1. check $A(Q)-A(P)$: if $A(Q)-A(P)$ is a large set, then return *false*;
2. check $M(P) \subseteq M(Q)$ using the offline process in Section 3.2: if it is true that $M(P) \subseteq M(Q)$, then return *true*, else return *false*.

Finally, get the result of the original problem using:

$$P \equiv Q \ \Leftrightarrow \ P \sqsubseteq Q, Q \sqsubseteq P;$$
$$P \text{ is disjoint with } Q \ \Leftrightarrow \ P \sqsubseteq \neg Q;$$
$$P \text{ and } Q \text{ are overlapped} \ \Leftrightarrow \ \neg(P \sqsubseteq \neg Q). \tag{12}$$

6   Conclusions

This paper proposes a framework of checking subsumption relations between composite concepts in different ontologies based on mutual instances. We discuss three key problems about mutual instances: how to deal with the large-scale mutual instances, how to estimate checking results considering fault tolerance, and how to reduce the manual annotation work. Then we give a process to deal with the large-scale mutual instances, provide the checking based on the weighted mutual instances considering fault tolerance, and propose a way greatly reducing the manual annotation work to get mutual instances. Intensions are also considered to avoid unnecessary instance checking and hit-and-miss results. Using these methods, we propose a framework, which is practical and effective to check subsumption relations between composite concepts in different ontologies. The framework is practical and effective.

References

1. Guarino, N.: Formal Ontology and Information Systems. In: Proceedings of the 1st International Conference on Formal Ontologies in Information Systems, IOS Press, Trento, Italy (1998) 3–15
2. Deen S.M., Al-Qasem, M.: A Query Subsumption Technique. In: Proceedings of the 10th International Conference Database and Expert Systems Applications, Florence, Italy (1999) 362–371
3. Kalfoglou, Y., Schorlemmer, M.: Ontology mapping: the state of the art. The Knowledge Engineering Review, vol. 18, no. 1 (2003) 1–31
4. Baader, F., Calvanese, D., McGuinness, D.L., Nardi, D., Patel-Schneider, P.F. (Eds.): The Description Logic Handbook: Theory, Implementation, and Applications. Cambridge University Press (2003)
5. Doan, A., Madhavan, J., Domingos, P., Domingos, P., Halevy, A.: Learning to map between ontologies on the semantic web. In Proceedings of the eleventh international conference on World Wide Web, Honolulu, Hawaii, USA (2002) 662–673
6. Rodríguez, M.A., Egenhofer, M.J.: Determining Semantic Similarity Among Entity Classes from Different Ontologies. IEEE Transactions on Knowledge and Data Engineering, Vol. 15 (2003) 442–456
7. Erdmann, M., Maedche, A., Schnurr, H.-P., Staab, S.: From manual to semi-automatic semantic annotation: About ontology-based text annotation tools. In: Proceedings of the COLING 2000 Workshop on Semantic Annotation and Intelligent Content, Luxembourg (2000) 34–48
8. Everitt, B.S., Landau, S., Leese, M.: Cluster Analysis (4th edition). Oxford University Press (2001)

A Knowledge Acquisition System
for the French Textile and Apparel Institute

Oswaldo Castillo Navetty and Nada Matta

University of Technology of Troyes, 12 rue Marie Curie
10010 Troyes, France
{Oswaldo.castillo,nada.matta}@utt.fr

Abstract. The management of knowledge and know-how becomes more and more important in organizations. Building corporate memories for conserving and sharing knowledge has become a rather common practice. However, we often forget that the efficiency of these activities is strictly connected to the appropriation capacities and learning of the organizational actors. It is through this learning that new skills can be acquired. In this paper, we propose general guidelines facilitating the process of creation and appropriation of professions memories built by means of methods from the knowledge engineering and from the educational engineering techniques.

1 Introduction

The French Textile and Apparel Institute (IFTH) is leading a project on knowledge capitalization for "pullovers 3D knitting" by utilizing a framework for defining a training platform for workers in the knitting industry. The objective is to train the textile companies of a French region to this type of knitting in order to improve the manufacturing process. Our role in this project is to define and conceive a knowledge acquisition system on the pullovers 3D knitting. The learning system we defined is based on knowledge engineering and educational techniques.

2 Knowledge Management

In order to take highest advantage of their intellectual capital, nowadays, the IFTH and most of the others companies develop and apply knowledge strategies. Thereby it becomes increasingly easier to realize an appropriate management, innovation and creation of their knowledge.

Dieng-Kuntz [1] presents some definition of knowledge management. "Knowledge capitalization in an organization has as objectives to promote the growth, the transmission and the preservation of knowledge in this organization" [2]. "It can carry both theoretical knowledge and know-how of the company. It requires the management of company knowledge resources to facilitate their access and their re-use" [3]. "It consists of capturing and representing knowledge of the company, facilitating its access, sharing and re-use. This very complex problem can be approached by several points of view: socio-organizational, economic, financial, technical, human, and legal" [4].

R. Khosla et al. (Eds.): KES 2005, LNAI 3681, pp. 960–966, 2005.

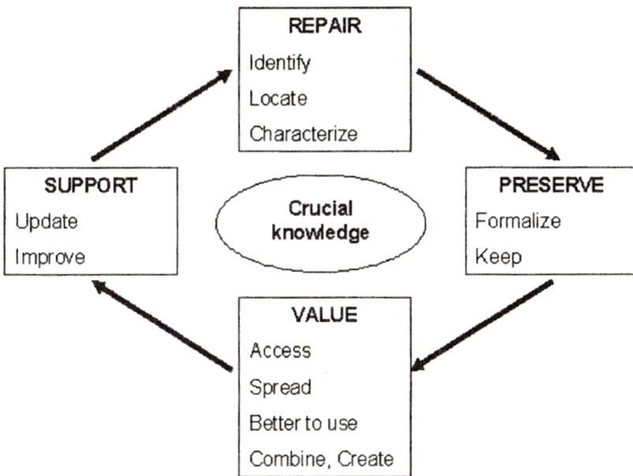

Fig. 1. Knowledge management life-cycle

We adopted the knowledge management life-cycle proposed by Grundstein (figure 1), where, according to him, "in any operation of knowledge capitalization, it is important to identify the strategic knowledge to be capitalized" [5].

3   Corporate Memory and Profession Memory

Implementation in organizations of knowledge management cycle-life assures knowledge formalization. This formalized knowledge is more and more represented in the form of corporate memory. We can define this memory as the "explicit, persistent and disembodied representation of knowledge and information in an organization" [6]. A corporate memory should supply "the good knowledge or information to the good person at the right time and at the good level" [1]. One of the corporate memory types [7] is the profession memory, which clarifies the repository, the documents, the tools and the methods employed in a given profession. Our study is based essentially on the appropriation of knowledge from a profession memory.

For knowledge capitalization in the IFTH about pullover 3D knitting, we are going to focus on the design and the use of knowledge based profession memory. The latter is based on collecting and explicitly modeling knowledge. This knowledge can come from experts and specialists of the company as well as from documents [1].

4   Appropriation of a Profession Memory

One of the main motivations for building a memory in a company is the improvement of employee's learning. This learning can be at an individual, group or organizational level [1]. In the case of our work in the IFTH, learning will be approached in an individual way for their employees. By looking at our preliminary experiences of constructing a profession memory, and more specifically at the learning from such memories, we noticed that the learning from a profession memory is not easy. These memories are generally presented under several points of view (classifications, constraints, processes, problem solving strategies, etc.). The links between these views are put in background because the knowledge formalization shows the nature of the knowledge [8]. Learning and following the learning progress in such a memory can be easy for a "cognitician" (cognition specialist) or a knowledge engineer but it is complex for an employee who is specialist on his profession and who wants to learn a know-how formalized by an expert in his domain.

To facilitate the learning from a profession memory, we adapted techniques from educational engineering by modifying the way of building the profession memory, and especially, by showing this memory, to the employee in a different way.

4.1 Educational Engineering

According to Paquette [9], educational engineering or training engineering contains all principles, procedures and tasks that allow to:

- define the contents of a training by means of a structural identification of the knowledge and the recounted,
- realize an educational scenario of the training activities and to define the context of use and the structure of the learning material,
- define infrastructures, resources and services necessary for distributing lessons and preserving their quality.

We use the general architecture proposed for the educational engineering, constituted by four interdependent modules:

- Expert module: represents the knowledge of the taught subject;
- Student model: represents the students needs, his/her level and his/her difficulties, which represents, his/her knowledge state, as well as the history of his/her interactions with the system;
- Educational module: contains the educational system, its purpose is to plan and to pass the learning;
- Interface module: concerns the support management and the system's communication modes with the student.

In other terms, the educational engineering allows us to develop a learning system [10] [9]. Our particular interest is the construction of a system based on a profession memory. By taking into account the practical knowledge (problem solving) of the training contents, our system becomes a practical learning system.

5 CPLS: Construction of a Practical Learning System

In this article, we are going to present our proposition of the practical learning system's expert module for the pullover 3D knitting in the IFTH.

5.1 Proposition of the Expert Module

Our expert module is represented as a guide named "activity process". It is also used to explain on detailed each step of the process. Expert module stores teaching materials consisting of problem solving strategy, exercises and formative assessment (quizzes), help/hints, answers and solution's prototypes to each assessment and exercise.

5.1.1 Representation of the Activity Process

For the definition of the expert module, the knowledge engineer has to find and emphasize a guiding thread to assure the progress in learning, with the expert's help. This guide has to show the aims, the deep knowledge and the links inside the knowledge. Based on MASK [11], the proposed guide (figure 2) in the practical learning system on the pullovers 3D knitting at the IFTH, is an arrangement that allows understanding of the succession of stages to be realized within the activity.

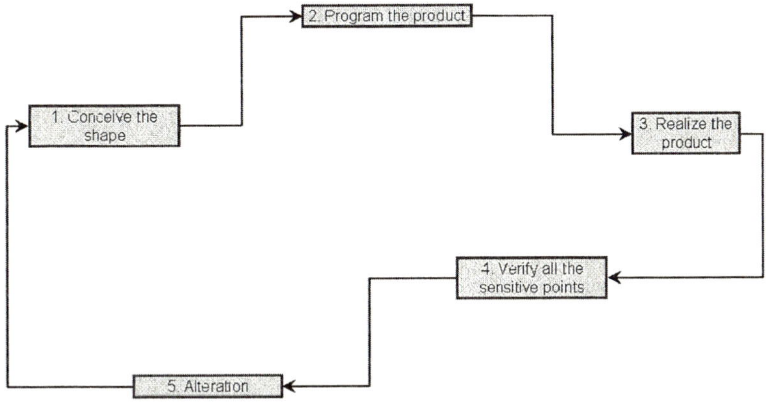

Fig. 2. Representation of the activity process

From this process, the knowledge engineer, always working with the expert, has to represent the interactions in every stage of the activity process and the problem solving strategy.

5.1.2 Representation of the Interactions at Each Activity Process Stage

The knowledge engineer and the expert have to represent the interactions of each step of the activity process with their environments. Still based on MASK [11], each step proposed in the practical learning system is described according to its inputs, outputs, resources and its actors.

5.1.3 Representation of the Problem Solving Strategy

Every step in the activity process must be detailed, by showing how the expert behaved to reach his/her objectives. The problem solving strategy is generally represented by the objectives to reach and the sequence of the steps followed by the expert. Every problem solving strategy (figure 3) in the practical learning system on the pullovers 3D knitting of the IFTH is represented by a list of targets. Each target contains a list of steps and tasks that the student has to develop.

5.1.4 Representation of Exercises

To allow training and making practical experiences, the practical learning system offers the possibility for realizing exercises. On one hand the exercises allow learning in several phases of the practical knowledge, and on the other hand an evaluation of the student's progress. These exercises mainly teach problem solving strategies.

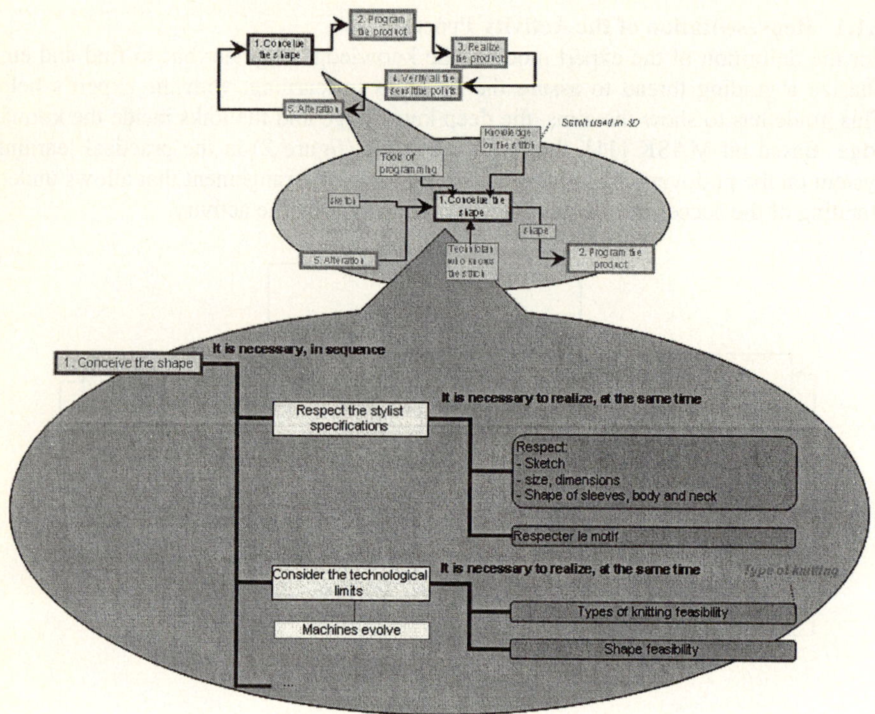

Fig. 3. Representation of the problem solving strategy

The learning system enables students to realize exercises of various complexity levels. Execution of exercises is also guided by the corresponding problem solving strategy. Indeed, this strategy is revealed to the student, step by step with these objectives and the control structure at each step. The student is then asked to execute the instructions one by one.

An exercise in the practical learning system on the pullovers 3D knitting at the IFTH can be represented with: a statement, indications and observations, a problem solving strategy, estimated time, a result's prototype (illustrations) and the result characteristics (list to tick and links with the corresponding skills).

5.1.5 Progress Process for Skills Acquisition

The activity process can be a guide to the practical knowledge learning. It allows to supply a global view as well as to guide a student to focus on every step of the process and to learn the strategies as well as the difficulties of the activity. We show student the global process and we ask him/her to explore the system step by step. For every stage, interactions with the environment are shown at first, followed by the corresponding problem solving strategy (figure 4).

The link between skills and characteristics of the results of the exercises allows progress in skills acquisition. These skills are related to each activity level. More global exercises are finally presented to a student to assure a global understanding of the activity.

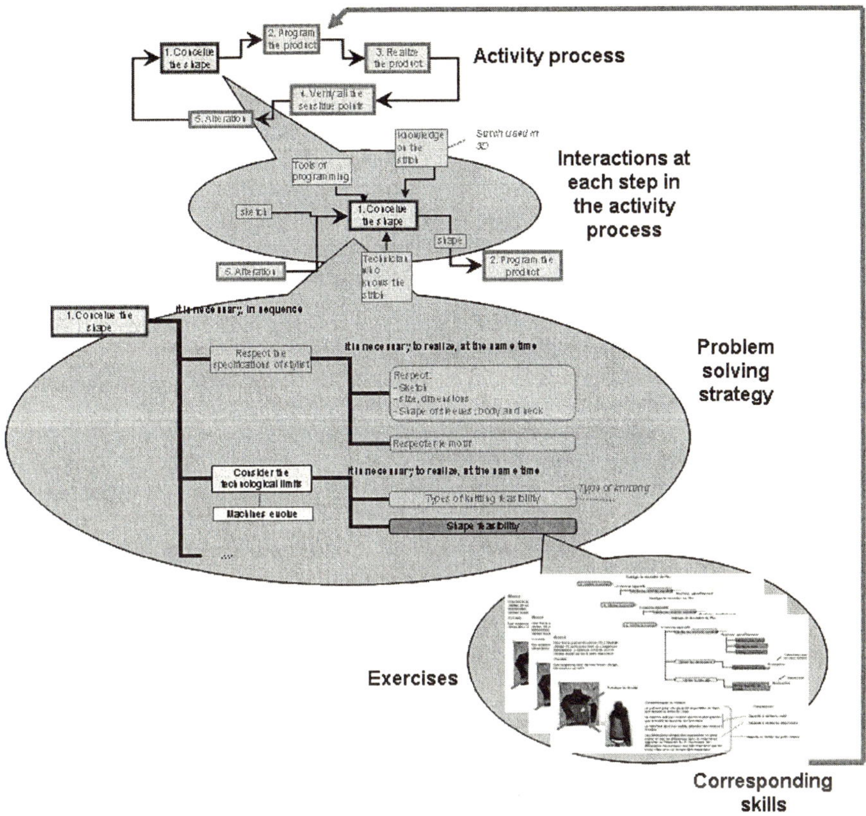

Fig. 4. Process illustration of knowledge learning progress

6 Conclusions

We are defining a tool to support this type of learning system. Together with the IFTH, an application for the textiles is being developed. Experiences on this type of domain will allow to deepen our studies and to enrich our learning system by other techniques.

The learning system on the pullovers 3D knitting will be integrated into a training platform which will be integrated into the textile industry in our region. We intend to study the feedback of this type of training.

We showed the relations which can exist between knowledge appropriation and skills acquisition. We aim at studying the representation formalisms and techniques of skills acquisition in a more detailed way.

References

1. Dieng-Kuntz, R., Corby, O., Gandon, F., Giboin, A., Golebiowska, J., Matta, N., Ribière M.: Méthodes et outils pour la gestion des connaissances: une approche pluridisciplinaire du knowledge management, 2ᵉ édition, Dunod (2001)

2. Steels, L.: Corporate knowledge management, Proc. ISMICK'93, Compiègne, France (1993), 9-30
3. O'Leary, D.E.: Entreprise Knowledge Management, Computer, 31, 3 (1998), 54-61
4. Barthès, J.-P.: ISMICK and Knowledge Management, In J.F. Schreinemakers (éd.), Knowledge Management: Organization, Competence and Methodology, Proc. ISMICK'96, Würzburg, Ergon Verlag, Rotterdam (1996), 9-13
5. Grundstein, M., Barthès, J.-P.: An industrial view of the process of capitalizing knowledge, Proc. ISMICK'96, Rotterdam (1996), 258-264
6. Van Heijst, G., Van der Spek, R., Kruizinga, E.: Organizing Corporate Memories, Proc. KAW'96, 10th Banff knowledge acquisition for knowledge-based system workshop, Banff, Canada (1996), 42-1/42-17
7. Tourtier, P.-A.: Analyse préliminaire des métiers et de leurs interactions, Intermediate report of the project GENIE, INRIA-Dassault-Aviation (1995)
8. Castillo, O., Matta, N., Ermine, J.-L.: De l'appropriation des connaissances vers l'acquisition des competences, Proc. C2EI'04, Nancy, France (2004)
9. Paquette, G.: L'ingénierie pédagogique: pour construire l'apprentissage en réseau, Presses de l'université de Québec (2002)
10. Rolland, M.: Bâtir des formations professionnelles pour adultes, Editions d'Organisations (2000)
11. Ermine, J-L.: La gestion de connaissances, Hermès sciences publications, Paris (2002)

Two-Phase Path Retrieval Method
for Similar XML Document Retrieval

Jae-Min Lee and Byung-Yeon Hwang

Department of Computer Engineering, Catholic University, Korea
{likedawn,byhwang}@catholic.ac.kr

Abstract. The existing three-dimensional bitmap indexing techniques have a performance problem in clustering the similar documents because they cannot detect the similar paths. The existing path clustering method which is based on the path construction similarity is one approach to solve the problem above. This method, however, consumes too much time in measuring the similarities between the similar paths for clustering. This paper defines the expected path construction similarity and proposes two-phase path retrieval method which effectively clustering the paths using it. The proposed method solved the performance degrade problem in path clustering by filtering the paths to be measured using the expected path construction similarity.

1 Introduction

As XML has been the standard for the content storage and exchange, the clustering and retrieval of XML documents become the key basis technique in various areas[1, 2, 3]. Due to this, various methods for XML retrieval have been proposed [4, 5, 6, 7]. Among these, three-dimensional bitmap indexing [8, 9, 10, 11] has been the representative method for XML retrieval which is based on the paths. This method first maps the XML to three-dimensional bitmap index which consists of document, path, and word. Then it extracts the information fastly through the bitwise operation. The method, however, caused the performance problem in clustering and retrieval of XML documents since it cannot detect the similar paths. The method in [12] first clusters the paths using the path construction similarity before actually clustering and retrieving the XML documents. This method also has the problem of too much time consuming because it has to measure the similarities between the similar paths. The accuracy of the clustering and retrieval is surely improved but the speed of the clustering and retrieval is the problem in the method. The paper proposes two-phase path retrieval method which can detect the similar path to overcome the problems mentioned above. The proposed method employs the expected path construction similarity to measure the similarity between the paths. The method can determine the need for measuring the path construction similarity. Due to this, the number of operations of measuring the similarity is decreased and consequently the performance of clustering and retrieval is improved.

2 The Related Research

This section discusses the three-dimensional bitmap indexing which is a representative path-based indexing method. The path construction similarity is also discussed which is proposed to solve the problem of three-dimensional bitmap indexing.

R. Khosla et al. (Eds.): KES 2005, LNAI 3681, pp. 967–971, 2005.

2.1 Three-Dimensional Bitmap Indexing

In order to fastly extract information, the three-dimensional bitmap indexing [8, 9, 10, 11] dissimilates the XML documents to documents, paths, and words and maps those to three-dimensional index. This three-dimensional index consists of bitmaps or linked lists in which the bitwise operation is possible so as to obtain the result for user's queries effectively. The three-dimensional bitmap indexing also determines how evenly the similar paths are distributed over the documents to cluster the similarly-structured documents based on the paths.

Unfortunately the existing three-dimensional bitmap indexing cannot detect the similar paths resulting in the performance degrade in document clustering and retrieval.

2.2 The Path Clustering Methods Based on the Path Construction Similarity

The path construction similarity was proposed to solve the problem of three-dimensional bitmap indexing which cannot detect the similar paths. The similarity is based on how many nodes are matched, but more importantly how many nodes are sequentially matched. The path construction similarity measures the degree of sequential node match between two paths.

The path construction similarity is defined as the average of node values in the path which has more nodes between the two. The path construction similarity ($P.C.Sim(P_1, P_2)$) between the paths P_1 and P_2 is defined as follows:

$$P.C.Sim(P_1, P_2) = \frac{\sum_{i=1}^{NodeNum(P_2)} NodeValue(P_2, N_i)}{NodeNum(P_2)}$$

The path clustering method [8] based on the path construction similarity solves the problem of existing three-dimensional bitmap indexing, but there is a performance degrade problem in speed by using too much time in the path clustering.

3 Two-Phase Path Retrieval

This section discusses the two-phase path retrieval method which solves the problem of existing path clustering method which is based on the path construction similarity.

3.1 The Expected Path Construction Similarity

If it is known how many nodes are in the paths and how many paths match between two paths, the optimal path construction similarity can be calculated without actually measuring the path construction similarity. The optimal path construction similarity means the virtual path construction similarity in which the order of nodes is arranged ideally, regardless of the real order of nodes in the two paths.

For instance, in Figure 1, suppose the path A_B_E is the criterion path and an arbitrary path consisting of A, B, C, D, E, F, G, H, I is the target path to be compared. The values of nodes A, B, and E are 1's since they match the nodes in the criterion path. Since the number of matching nodes is three, other unmatched nodes can be inserted into four places randomly. The sum of node values of the inserted nodes in each place is $n/2^n$ when the number of nodes to be inserted in n. For the unmatched six nodes to be inserted into four places maintaining the optimal path construction similarity, they should be distributed evenly over the places. In other words, when the number of places with n being 1 is two and the number of places with n being 2 is two, the path construction similarity is the highest $(((3+2\cdot1/2^1+ 2\cdot2/2^2)/9=0.55)$.

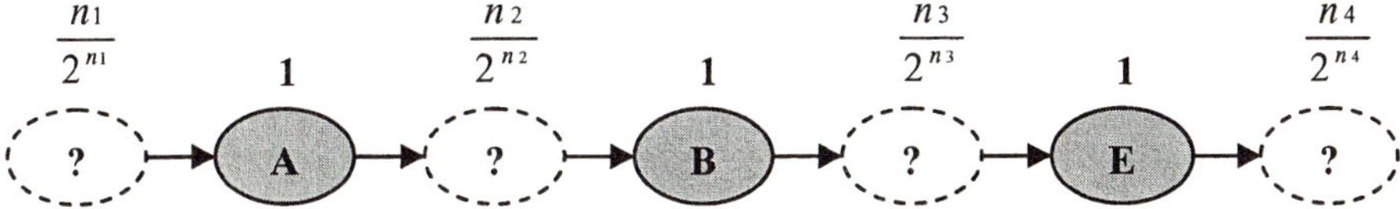

Fig. 1. An example of measuring the expected path construction similarity

This is generalized as follows. Suppose t is the number of nodes in the path to be compared and m is the number of matching nodes with the criterion path. When the number of unmatched nodes $(t-m)$ is divided by the number of spaces $(m+1)$, let t_1 be the quotient and t_2 be the remainder. First, $t_1 X (m+1)$ nodes are inserted into $(m+1)$ places with each place having t_1 nodes. Therefore the sum of node values included in each places becomes $t_1/2^{t_1}$. Then the remaining t_2 nodes are inserted into the places randomly. The t_2 places with sum of node values of $t_1/2^{t_1}$ now changed to the sum of node values to be $(t_1+1)/2^{t_1+1}$. Therefore the expected path construction similarity is defined as the Definition 1.

Definition 1. The expected path construction similarity ($EPCSim(P_1, P_2)$) within a cluster is defined as follows (Here P_1 is the criterion path and P_2 is the target path to be compared when $NodeNum(P_1)$ is always less then or equal to $NodeNum(P_2)$. m is the number of matching node among the nodes in the paths P_1 and P_2, t is $NodeNum(P_2)$. t_1 is $\lfloor (t-m)/(m+1) \rfloor$ while t_2 is $(t-m)\bmod(m+1)$.)

$$EPCSim(P_1, P_2) = \dfrac{m + \dfrac{t_1}{2^{t_1}}\cdot(m+1) - \dfrac{t_1}{2^{t_1}}\cdot t_2 + \dfrac{t_1+1}{2^{t_1+1}}\cdot t_2}{t}$$

$$= \dfrac{2^{t_1+1}\cdot m + 2t_1\cdot(m+1) - t_1\cdot t_2 + t_2}{2^{t_1+1}\cdot t}$$

3.2 Measuring the Expected Path Construction Similarity

Measuring the expected path construction similarity is simple. The nodes in the paths P_1 and P_2 are read sequentially first, and the number of matching nodes m, and then the number of all nodes t in longer path. Then those are simply plugged into the Definition 1. Figure 2 shows the algorithm measuring the expected path construction similarity.

```
double GetExpectedPCSimilarity(Path path, int PCID, double limit)
{
  int i; double b, m, t, t1, t2, sim = 0;
  t = GetTotalNodeNumber(center, PCID);
  t1 = System.Math.Floor((t-m)/(m+1)) ;
  t2 = (t-m) % (m+1);
  b = System.Math.Pow(2, t1 + 1);

  do{
    if(MF.indexes.index[PCID].center[path.ID] == true) M++;
    path = path.nextNode;
  }while(path != null)

  sim = ( b*m + 2*t1*(m+1) – t1*t2 +t2 ) / b*t;
  if(sim > limit) return GetPCSimilarity(path, PCID, limit);
  else return sim;
}
```

Fig. 2. The algorithm measuring the expected path construction similarity

3.3 Two-Phase Path Retrieval Algorithm

In a nutshell, the proposed method in this paper improves the performance of the similar path retrieval by diminishing the target path clusters which are used to measure the path construction similarity. Figure 3 shows the two-phase path retrieval algorithm. The phase one measures the expected path construction similarity using the centroid of the specific path cluster and the centroid of the path to be compared. When the measured similarity is more than the threshold, the phase two measures the actual path construction similarity. The phase two is omitted when the similarity is below the threshold. This step is iterated over all the path clusters.

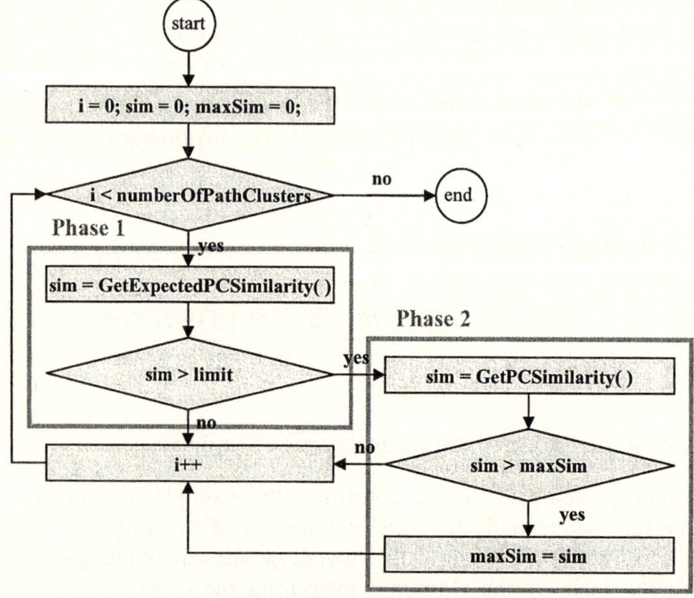

Fig. 3. The two-phase path retrieval algorithm

4 Conclusion

The existing three-dimensional bitmap indexing has the problem of performance degrade in clustering similar documents due to its lack in detecting similar paths. The existing clustering methods based on the path construction similarity are those to overcome this problem. These methods, however, consumes too much time in measuring the similarities between similar paths to cluster the paths. This paper defines the expected path construction similarity and proposes the two-phase path retrieval method which effectively clusters the paths using the new similarity measure. The proposed method determines target paths first to measure the similarity using the expected path construction similarity. The method consequently solves the problem of performance degrade in path clustering.

References

1. Chen, Y., Davidson, S.B., Zheng, Y.: BLAS: An Efficient XPath Processing System. Proc. of the ACM SIGMOD (2004) 47-58
2. Delobel, C., Reynaud, C., Rousset, M.: Semantic Integration in Xyleme: A Uniform Tree-based Approach. Data & Knowledge Engineering, Vol.44 (2003) 267-298
3. Cooper, B., Sample, N., Franklin, M., Shadmon, M.: A Fast Index for Semistructured Data. Proc. of the 27th VLDB Conference, Roma, Italy (2001)
4. McHugh, J., Abiteboul, S., Goldman, R., Quass, D., Widom, J.: Lore: A Database Management System for Semistructured Data. ACM SIGMOD Record, Vol.26 (1997) 54-66
5. Zhang, C., Naughton, J., Dewitt, D., Luo, Q., Lohman, G.: On Supporting Containment Queries in Relational Database Management Systems. Proc. of the ACM SIGMOD (2001) 425-436
6. Chung, C., Min, J., Shim, K.: APEX: An Adaptive Path Index for XML Data. Proc. of the ACM SIGMOD (2002) 121-132
7. Sayed, A., Unland, R.: HID: An Efficient Path Index for Complex XML Collections with Arbitrary Links. Lecture Notes in Computer Science, Vol.3433 (2005) 78-91
8. Yoon, J.P., Raghavan, V., Chakilam, V., Kerschberg, L.: BitCube: A Three-Dimensional Bitmap Indexing for XML Documents. Journal of Intelligent Information System, Vol.17 (2001) 241-254
9. Lee, J.M., Hwang, B.Y.: xPlaneb: 3-Dimensioal Bitmap Index for XML Document Retrieval. Journal of Korea Information Science Society, Vol.31, No.3 (2004) 331-339
10. Lee, J.M., Hwang, B.Y., Lee, B. J.: X-Square: Hybrid 3-Dimensional Bitmap Indexing for XML Documents Retrieval. Lecture Notes in Computer Science, Vol.3307 (2004) 221-232
11. Yoshikawa, M., Amagasa, T.: XRel: A Path-based Approach to Storage and Retrieval of XML Documents using Relational Databases. ACM Transactions on Internet Technology, Vol.1, No.1 (2001) 110-141
12. Lee, J.M., Hwang, B.H.: A Search Method of Similar XML Documents based on Bitmap Indexing. Proc. of the 21th Korea Information Processing Society Spring Conference, Vol.11, No.1 (2004) 16-18

MEBRS: A Multiagent Architecture
for an Experience Based Reasoning System

Zhaohao Sun[1] and Gavin Finnie[2]

[1] School of Economics and Information Systems,
University of Wollongong, Wollongong NSW 2522 Australia
zsun@uow.edu.au

[2] Faculty of Information Technology, Bond University, Gold Coast Qld 4229 Australia
gfinnie@staff.bond.edu.au

Abstract. This paper reviews eight different inference rules for experience-based reasoning (EBR), and proposes a multiagent architecture for an EBR system, which constitutes an important basis for developing any multiagent EBR systems (EBRS). The proposed architecture consists of a global experience base (GEB), and a multi-inference engine (MIE), which is the mechanism for implementing eight reasoning paradigms based on eight inference rules for EBR. The proposed approach will facilitate research and development of experience management, knowledge-based systems, and recognition of fraud and deception in e-commerce.

Keywords: Experience-based reasoning, experience management, knowledge-based system, multiagent system.

1 Introduction

Experience-based reasoning (EBR) is a reasoning paradigm based on logical arguments [2]. EBR as a technology has been used in many applications [13]. For example, EBR has been used in help desk systems to adapt to new business situations by "learning" from experience, tailoring a help desk to effectively maintain critical business systems [17][18]. However, the existing applications and studies have not been based on any firm theoretical foundations, because there are few studies on EBR from a logical viewpoint [9][12], although there are a lot of empirical works on EBR mainly in the business fields. How to automate experience remains a big issue. Taking into account research and development of case-based reasoning (CBR), Sun and Finnie [13][12] proposed eight different inference rules for EBR from a logical viewpoint, which cover all possibilities of EBR. However, how to implement an EBR system, and what is an experience-based inference engine based on the proposed eight inference rules for EBR [11][13] are still open problems. This paper attempts to fill the above gap by providing a multiagent architecture for an EBR system (for short, MEBRS), which is a core part for any multiagent EBR system. More specifically, this paper reviews eight different inference rules for EBR. Then the paper examines the multiagent architecture, MIE, and some other agents in the MEBRS. The rest of this paper is organized as follows: Section 2 reviews inference rules for EBR from a logical viewpoint. Section 3 proposes an architecture for EBR systems. Section 4 examines the multiagent framework for experience based inference engine. Section 5 looks at some intelligent agents in the MEBRS. Section 6 ends this paper with some concluding remarks.

R. Khosla et al. (Eds.): KES 2005, LNAI 3681, pp. 972–978, 2005.

2 Inference Rules for Experience Based Reasoning

From a logic viewpoint, there are eight basic inference rules for performing EBR [13], which are summarized in Table 1, and cover all possible EBRs, and constitute the fundamentals for all EBR paradigms [9][11][13]. The eight inference rules are listed in the first row, and their corresponding general forms are shown in the second row respectively. Because four of them, *modus ponens* (MP), *modus tollens* (MT), *abduction* and *modus ponens with trick* (MPT) [10] are well-known in AI and computer sciences [8][7][11], we do not go into them any more, and focus on reviewing the other four inference rules in some detail. First of all, we illustrate *modus tollens with trick* (MTT) with an example. We have the knowledge in the knowledge base (KB):

1. If Klaus is human, then Klaus is mortal
2. Klaus is immortal.

What we wish is to prove "Klaus is human". In order to do so, let

- $P \rightarrow Q$: If Klaus is human, then Klaus is mortal
- P: Klaus is human
- Q: Klaus is mortal.

Therefore, we have P: Klaus is human, based on MTT, and the knowledge in the KB (note that $\neg Q$: Klaus is not mortal). From this example, we can see that MTT is a kind of EBR.

Abduction with trick (AT) can be considered as a "dual" form of abduction, which is also the summary of a kind of EBR [13]. Abduction can be used to explain that the symptoms of the patients result from specific diseases, while abduction with trick can be used to exclude some possibilities of the diseases of the patient [12]. Therefore, abduction with trick is an important complementary part for performing system diagnosis and medical diagnosis based on abduction.

Inverse modus ponens (IMP) is also a rule of inference in EBR [13]. The "inverse" in the definition is motivated by the fact that the "inverse" is defined in mathematical logic: "if $\neg p$ then $\neg q$", provided that if p then q is given [3]. Based on this definition, the inverse of $P \rightarrow Q$ is $\neg P \rightarrow \neg Q$, and then from $\neg P$, $\neg P \rightarrow \neg Q$ we have $\neg Q$ using modus ponens. Because $P \rightarrow Q$ and $\neg P \rightarrow \neg Q$ are not logically equivalent, the argument based on IMP is not valid in mathematical logic. However, the EBR based on IMP is a kind of common sense reasoning, because there are many cases that follow IMP. For example, if John has enough money, then John will fly to China. Now John does not have sufficient money, then we can conclude that John will not fly to China.

The last inference rule for EBR is *inverse modus ponens with trick* (IMPT) [13]. The difference between IMPT and IMP is again "with trick", this is because the reasoning performer tries to use the trick of "make a feint to the east and attack in the west"; that is, he gets Q rather than $\neg Q$ in the *inverse modus ponens*.

So far, we have reviewed eight different inference rules for EBR (see Table 1) from a logical viewpoint, four of them have been thoroughly used in computer science, mathematics, mathematical logic, philosophy and other sciences. The rest have not been appeared in any publications except [11][13], to our knowledge. However, they are all the abstraction and summary of experience or EBR in real world problems. Therefore, any research and development of each listed inference model is significant for understanding of intelligence, logic, EBR.

Table 1. Experience-based reasoning: Eight inference rules

MP	MT	abduction	MTT	AT	MPT	IMP	IMPT
P $P \rightarrow Q$ $\therefore Q$	$\neg Q$ $P \rightarrow Q$ $\therefore \neg P$	Q $P \rightarrow Q$ $\therefore P$	$\neg Q$ $P \rightarrow Q$ $\therefore P$	Q $P \rightarrow Q$ $\therefore \neg P$	P $P \rightarrow Q$ $\therefore \neg Q$	$\neg P$ $P \rightarrow Q$ $\therefore \neg Q$	$\neg P$ $P \rightarrow Q$ $\therefore Q$

It should be noted that the inference rules "with trick" such as MTT, AT, MT and IMPT are non-traditional inference rules. However, they are really abstractions of some experience-based reasoning, although few have tried to formalize them. The "with trick" is only an explanation for such models. One can give other explanations for them. For example, one can use fraud or deception to explain them [13]. One can also use inference rules "with exception" to explain them.

3 An Architecture of EBR Systems

The architecture of an EBRS consists of experience users, U, which are either human customers or intelligent agents delegated by the human users or other systems, as shown in Fig. 1. The EBRS mainly consists of a user interface, a global experience base (GEB) and a multi-inference engine (MIE). The user interface consists of some kind of natural language processing systems that allows the user to interact with the EBRS [7] (p. 282). The GEB, like the experience factory [1] and case base [11], consists of all the experiences that the system collects periodically, and the adapted new experiences retained when the system is running. The MIE is a multi-inference engine which consists of the mechanism for implementing eight reasoning paradigms based on eight inference rules for EBR with manipulating the GEB to infer experience requested by the experience user. The remarkable difference between the mentioned EBRS and the traditional KBS lies in that the latter's inference engine is based on a unique reasoning paradigm (or inference rule), while the MIE is based on many different reasoning paradigms.

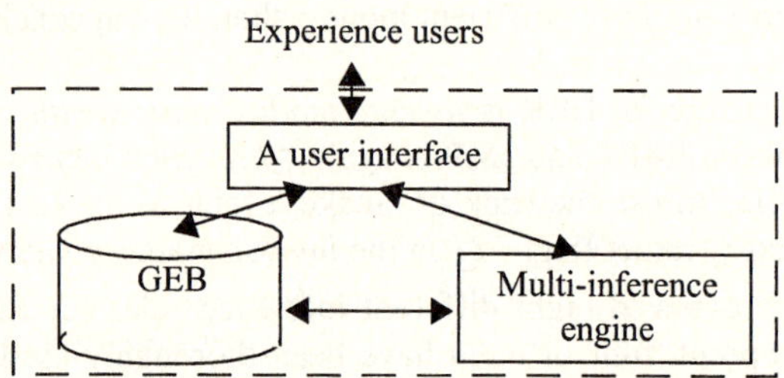

Fig. 1. A general architecture of an EBRS

4 MEBIE: A Multiagent Framework for Experience Based Inference Engine

As mentioned in the previous section, the MIE is a multi-inference engine for EBR. MIE could automatically adapt itself to the changing situation and perform one of the mentioned inference rules for EBR. However, any existing intelligent system has not reached such a high level. The alternative strategy is to use multiagent technology to implement the MIE.

Multiagent systems (MAS) have been studied in the field of distributed AI (DAI) for more than 20 years [11][15]. Today these systems are not simply a research topic, but are also an important subject of academic teaching and industrial and commercial application. Recently, the term MAS is used for all types of systems composed of multiple agents showing the following characteristics:

- Each agent has incomplete capabilities to solve a problem
- There is no global system control over agents
- Data are decentralized
- Computation is asynchronous.

Rationality is a compelling notion in MAS. An ideal rational agent is defined as follows: for each possible percept sequence, it acts to maximize its expected utility, on the basis of its knowledge and the evidence from the percept sequence [5] (pp. 3-4).

Autonomy is also an important property of intelligent agents [5], in particular for mobile agents, which can move from one location to another while preserving their internal state [11]. Autonomy is the ability of agents to handle human user-defined tasks independently of the user and often without the user's guidance or presence [11].

Based on the above discussion, we propose a multiagent framework for an experience-based inference engine (for short MEBIE), which is a core part of a multiagent EBR system (MEBRS), as shown in Fig. 2. In this framework, eight rational agents (from MP agent to AbT agent) are semi-autonomous [11]. These eight agents are mainly responsible for performing EBR corresponding to eight inference rules in the EBRS respectively. In what follows, we discuss each of them in some detail.

1. The *MP agent* in the MEBIE is responsible for manipulating the GEB based on *modus ponens* to infer the experience requested by the experience user. This agent can be considered as an agentization of an inference engine in a traditional KBS in the EBRS. The function of the MP agent can be extended to infer the experience in the GEB based on *fuzzy modus ponens* [16] and *similarity-based modus ponens* [11]. It should be noted that the function of following mentioned agents can be also extended to infer the experience in the GEB based on corresponding fuzzy or similarity-based inference rules for EBR [13]

2. The *MT agent* manipulates the GEB to infer the experience requested by the experience user based on *modus tollens,* which is an inference rule discussed in AI

3. The *Ab agent* is responsible for manipulating the GEB to infer the experience requested by the experience user based on *abductive reasoning* [6]. This agent can generate the explanation for the experience inferred by the MEBIE

4. The *MPT agent* is responsible for manipulating the GEB to infer the experience requested by the experience user based on *modus ponens with trick*. This agent can perform reasoning with trick [9], therefore the understanding of the experience provided by this agent requires special attention under some situations

5. The *IMPT agent* manipulates the GEB to infer the experience requested by the experience user based on *inverse modus ponens with trick*. This agent can also perform reasoning with trick [10], therefore the understanding of the experience provided by this agent requires special attention

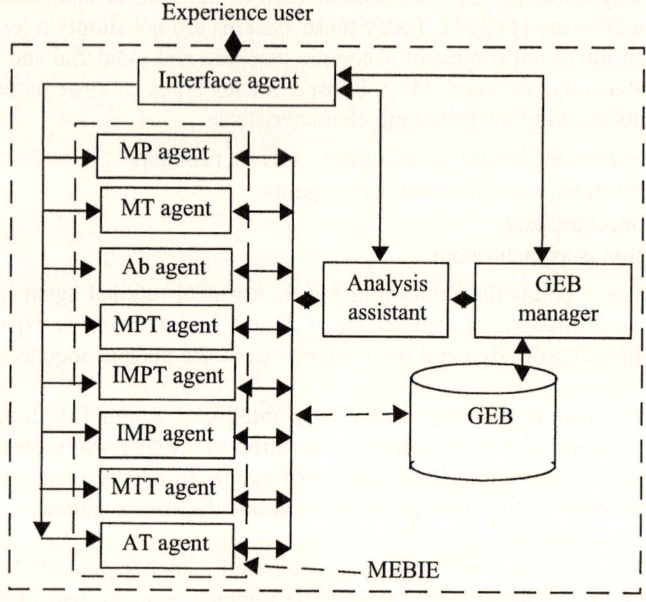

Fig. 2. MEBIE in a MEBRS

6. The *IMP agent* is responsible for manipulating the GEB to infer the experience requested by the experience user based on *inverse modus ponens* (see Section 2)

7. The *MTT* agent manipulates the GEB to infer the experience requested by the experience user based on *modus tollens with trick*. This agent can also perform reasoning with trick [13], therefore the understanding of the experience provided by this agent also requires special attention

8. The *AT* agent manipulates the GEB to infer the experience requested by the experience user based on *abduction with trick*. This agent can also perform reasoning with trick [13], therefore the understanding of the experience provided by this agent requires special attention.

5 Some Other Agents in MEBRS

For the proposed MEBRS there are some other intelligent agents, shown in Fig. 2. These are an interface agent, an analysis assistant and a GEB manager. In what follows, we will look at them in some depth.

The *Interface agent* is an advisor to help the experience user to know which rational agent s/he likes to ask for help. Otherwise, the interface agent will forward the problem of the experience user to all agents in the MEBIE for further processing.

The output of the experience provided by the MEBIE can be considered as a sub-output, the final output as the solutions to the problem of the experience user will be processed with the help of the *Analysis agent*. Since different agents in the MEBIE use different inference rules, and then produce different, conflicting results with knowledge inconsistency. How to resolve such knowledge inconsistency is a critical issue for the MEBRS. This issue will be resolved by the Analysis assistant of the MEBRS. The Analysis assistant will

- Rank the degree of importance of the suboutputs from the MEBIE taking into account the knowledge inconsistency
- Give an explanation for each of the outputs from the MEBIE and how the different results are conflicting
- Possibly combine or vote to establish the best solutions
- Forward them to the interface agent who then forwards them to U.

The *GEB manager* is responsible for administering the GEB. Its main tasks are GEB creation and maintenance, experience evaluation, experience reuse, experience revision, and experience retention. Therefore the functions of the GEB manager are an extended form of the functions of a CBR system [11], because case base creation, case retrieval, reuse, revision and retention are the main tasks of the CBR system [4].

Now let us have a look at how the MEBRS works. The experience user, U, asks the interface agent to solve the problem, p. The interface agent asks U whether a special reasoning agent should be need. U does not know. Thus, the interface agent forwards p (after formalizing it) to all agents in the MEBIE for further processing. The agent i in the MEBIE manipulates the experience in the GEB based on $p,$ and the corresponding reasoning mechanism, and then obtains the solution s_i, which is forwarded to the Analysis assistant. After the Analysis assistant receives all solutions s_i, $i = \{1, ..., 8\}$ to p, it will rank the degree of importance of the solutions, give an explanation for each of the solutions and how the results are conflicting, and then forward them (with p) to the interface agent who would then forward them to U. If U accepts one of the solutions to the problem, then the MEBRS completes this experience transaction. In this case, the GEB manager will look at whether this case is a new experience. If yes, then it will add it to the GEB. Otherwise, it will keep some routine record to update the GEB. If U does not accept the solution provided, the interface agent will ask U to adjust some aspects of the problem p, which is changed into p', then the interface agent will once again forward the revised problem p' to the MEBIE for further processing.

6   Concluding Remarks

This paper first examined eight different inference rules for EBR. Then the paper proposed a multiagent architecture for an EBR system (MEBRS), and looked at the MEBIE and some other agents in the MEBRS. The proposed approach will facilitate research and development of EBR systems, knowledge based systems.

In future work, we will develop a prototype system for a multiagent system development environment for EBR system. To this end, we will further examine the internal mechanisms of the Analysis assistant such as how the agent ranks the degree of importance of the solutions produced by agents. We will also examine the communication, coordination, and cooperation among the agents with the proposed multiagent architecture.

References

1. Bergmann R. Experience Management: Foundations, Development Methodology and Internet-Based Applications. LAIN 2432. Berlin: Springer, 2002
2. Bosch. http://www.cs.tut.fi/~ohar/Slides2001/Bosch/tsld027.htm, 2001
3. Epp SS. Discrete Mathematics with Applications, Brooks/Cole Publishing Company Pacific Grove, 1995, 801 p.
4. Finnie G, Sun Z. R^5 model of case-based reasoning. Knowledge-Based Syst. 16(1) 2003, 59-65
5. Huhns MN, Singh MP (eds) Readings in Agents. San Francisco: Morgen Kaufmann Publishers, Inc. 1998
6. Magnani L. Abduction, Reason, and Science, Processes of Discovery and Explanation. New York: Kluwer Academic/Plenum Publishers, 2001
7. Nilsson NJ. Artificial Intelligence. A New Synthesis. San Francisco, California: Morgan Kaufmann Publishers, Inc. 1998, 513 p.
8. Russell S, Norvig P. Artificial Intelligence: A modern approach. Upper Saddle River, New Jersey: Prentice Hall, 1995
9. Sun Z, Finnie G. Brain-like architecture and experience-based reasoning, In: Proc. 7th Joint Conf on Information Sciences (JCIS), September 26-30, 2003 Cary, North Carolina, USA. 1735-1738
10. Sun Z, Weber K. Turing test and intelligence with trick. In: Proc 8th Ireland Conf on AI (AI-97), Londonderry, Ireland, 1997, 217-224
11. Sun Z, Finnie G. Intelligent Techniques in E-Commerce: A Case based Reasonng Perspective, Heidelberg: Springer, 306 p
12. Sun Z, Finnie G. Experience Based Reasoning: A similarity-based perspective, IEEE Trans on Knowledge and Data Engineering, 2004, under review
13. Sun Z, Finnie G. Experience based reasoning for recognising fraud and deception. In: Proc. Inter Conf on Hybrid Intelligent Systems (HIS 2004), December 6-8, Kitakyushu, Japan, IEEE Press, 2004, pp 80-85
14. Torasso P, Console L, Portinale L, Theseider D. On the role of abduction. ACM Computing Surveys, 27 (3), 1995, 353-355
15. Weiss, G (ed.). Multiagent Systems: A modern approach to distributed artificial intelligence, Cambridge, Massachusetts, London: The MIT Press, 1999
16. Zimmermann HJ. Fuzzy Set Theory and its Application (3rd Edn). Boston/Dordrecht/London: Kluwer Academic Publishers, 1996
17. http://wm2003.aifb.uni-karlsruhe.de/workshop/w06/GWEM2003_CfP_engl-ish.html
18. http://www.iese.fraunhofer.de/experience_management/.

Experience Management in Knowledge Management

Zhaohao Sun[1] and Gavin Finnie[2]

[1] School of Economics and Information Systems, University of Wollongong,
NSW 2522 Australia
zsun@uow.edu.au
[2] Faculty of Information Technology, Bond University Gold Coast Qld 4229 Australia
gfinnie@staff.bond.edu.au

Abstract. This paper examines experience and knowledge, experience management and knowledge management, and their interrelationships. It also proposes process perspectives for both experience management and knowledge management, which integrate experience processing and corresponding management, knowledge processing and corresponding management respectively. The proposed approach will facilitate research and development of knowledge management and experience management as well as knowledge-based systems.

Keywords: Knowledge management, experience management, knowledge-based systems, e-commerce.

1 Introduction

While knowledge management (KM) has become well-established in information systems (IS), business and management, artificial intelligence (AI), and information technology (IT) with books, conferences, commercial tools and journals on the topic [12], experience management (EM) has received a small amount of research attention. There have been two German workshops on EM [18]. In addition, research on a tool called the experience factory that has been used to store and retrieve experience in software development projects has led to some commercial product development [19].

While knowledge has received considerable attention in the above mentioned areas, experience has not drawn similar attention [13]. In particular, how to automate experience based on intelligent techniques and software engineering methodology is still a big issue.

However, without any doubt there is a close relationship between experience and knowledge. For example, experience could be considered a refinement of knowledge or a special instance or form of knowledge [14]. Therefore, it is significant to examine the relationship between experience and knowledge, EM and KM and their interrelationships. To this end, the remainder of this paper is organised into the following sections: Section 2 examines knowledge, experience and their relationships, Section 3 and 4 looks at KM, EM and their interrelationships. The final section concludes the paper with some concluding remarks.

2 Knowledge and Experience

Knowledge and experience are both intelligent assets of human beings. They have been emphasised in a different way although they have a close relationship.

R. Khosla et al. (Eds.): KES 2005, LNAI 3681, pp. 979–986, 2005.

There is no consensus on what knowledge is. Over the millennia, the dominant philosophies of each age have added their own definition of knowledge to the list. Knowledge as a construct or an atom is defined as "understanding the cognitive or intelligent system possesses that is used to take effective action to its system goal" [16]. In computer science, knowledge is usually defined as the objects, concepts and relationships that are assumed to exist in some area of interest [13]. Various knowledge exists in encyclopaedias, handbooks, manuals, other reference materials, speeches, lectures, the head of human beings, and the Internet/WWW, in particular.

Knowledge became an important construct in AI in the 1970's, although there are no definitions of knowledge in some books on AI such as Ref. [9]. At that time, AI researchers believed that more powerful intelligent systems required much more built-in knowledge about the domain of application [9] (p 10). With knowledge-based systems (KBS) becoming an important application of AI in the 1980's, knowledge is a central part of KBS, in which the knowledge base and inference engine are main parts [13] (p 13).

Knowledge has also played a pivotal role in business management and information management (IM) [6]. How to find useful knowledge from a large database or from the WWW to assist business decision making has become one of the most important issues in data mining and business management.

However, any investigation into knowledge without taking into account experience seems to be less meaningful, although it is not easy to define what experience is, just as it is hard to define what knowledge is. Generally speaking, however, experience can be taken as previous knowledge or skill one obtained in everyday life [13] (p.13). For example, Peter avoided a traffic tragedy on Pacific highway yesterday, because he drove carefully and focused on the drive. This is a typical experience for driving. In other words, experience is previous knowledge which consists of problems one has met and the successful solution to the problems. Therefore, experience can be taken as a specialization of knowledge.

In CBR terminology, a piece of experience is denoted as a case [13]. All cases are stored in a case base. Therefore, a case base is essentially an experience base. A previous experience, which has been captured and learned in a way that it can be reused in the solving of future problems, is referred to as a past case, previous case, stored case, or retained case. Correspondingly, a new case or unsolved case is the description of a new problem to be solved and its possible solution.

So far, we have discussed knowledge, experience and their relationships. What the difference is between them is an interesting topic, although few in AI, Information Systems, KM and EM give a deep insight, because it is the basis for differentiating KM from KM. In what follows, we try to use Q-A-R (Question-Answer-Remark) method to differentiate them with some comments [12].

Q1: What are you going to study in your school?
A1: I am going to study to gain knowledge.
R1: Few say that "I am going to study to gain experience"
Q2: Why did you visit that old doctor?
A2: Because he has rich experience in diagnosing and treating the disease that I suffered.
R2: In this case, the knowledge of the doctor in diagnosing and treating the mentioned disease is not sufficient to attract the customer to see the doctor. This is a com-

mon sense. That is, experience is a more important asset than knowledge in some field. Therefore, in diagnosis, a doctor's experience can be considered the kernel of his possessed knowledge.

Q3: What are you doing?

A3: I am collecting experience

R3: One seldom says "I am collecting knowledge"

Q4: Have you drawn some lessons from that experience?

A4: Yes, I have

R4: One seldom asks "Have you drawn some lessons from that knowledge?"

From the Q-A-R consideration, we see that experience and knowledge are two different concepts. Furthermore, possessing knowledge is only one necessary condition for a field expert [13]. Experience may be more important than knowledge for a field expert to deal with some tough problems. Accumulation of knowledge is the necessary condition of accumulating experience for a field expert. However, knowledge and experience are abstractions at two different levels. Experience is at a higher level, because because experience is a meta-knolwdge in some cases [13]. From a historical viewpoint, transforming the experience of a human being into knowledge has always been an important topic in science and technology. On the other side, knowledge accumulation and distillation might lead to new experience.

3   Knowledge Management

As mentioned previously, knowledge management (KM) has become well-established in IS, IM, and AI [3][6]. However, KM is also a new and exciting research topic in these and other fields such as business management and business decision making. KM is being viewed by organisations as providing the capability of improving competitiveness and increasing internal efficiency by facilitating the leveraging of existing knowledge within the organisation. This is achieved by providing a centralised repository for knowledge which is then accessible as needed. More specifically, KM is a discipline that focuses on knowledge processing and corresponding management that permeates each of following process stages [6][7][12]:

- Understand knowledge
- Discover knowledge
- Capture, and acquire knowledge from a variety of sources
- Select, Filter and classify the existing knowledge
- Define storage structures for saving knowledge
- Design ontology of knowledge
- Generate, adapt and/or create new knowledge
- Measure and/or evaluate knowledge according to the objective of an organisation development
- Visualise knowledge
- Distribute and/ transfer knowledge to other organisations or individuals commercially or non-commercially
- Recommend, share, utilise/apply and sell knowledge
- Retain and maintain knowledge as an asset.

It is significant to separate management function from knowledge processing functions, and then integrate them in KM. This is also one of the characteristics of this paper, as shown in Fig. 1.

The management of knowledge processing for each process stage includes analysis, planning, organisation, support, collaboration [7], coordination and possible negotiation [12]. Generally speaking, management issues related to each or some of the knowledge processing stages include:

- Organisation of team
- Knowledge processing or management task assignment to specific person or teams
- Member and team communication and coordination.

Knowledge understanding includes knowledge learning, which is an important part for students in schools. In most cases, knowledge understanding is the goal of knowledge learning, while knowledge understanding can promote further knowledge learning.

Knowledge creation is sometimes also a consequence of knowledge sharing, which is valid, in particular, in some organisations, in which the opportunities for the creation of new idea or knowledge that have the potential to add value to it increases because of a developing knowledge sharing environment [8].

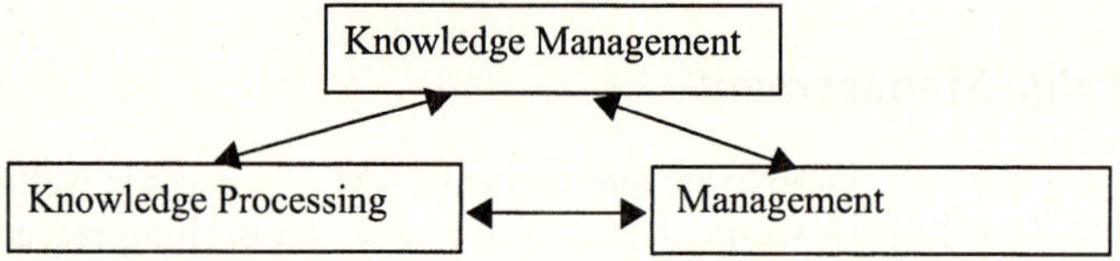

Fig. 1. KM as integration of knwoledge processing and managment

In a most general sense, the history of human civilization (culture) can be considered as the history of knowledge management, at least since the invention of paper-making (in AD 105) and printing technology (around 1041- 48)[1]. Therefore, a real KM would cover a majority of human activities in culture development. This is the reason why it is difficult for a process model to cover all possible activities in KM [12]. However, from a narrower sense, modern KM just began after the inception of modern computers after World War II in 1945 although the term KM was introduced at the end of last century [2] (p 9). Since then, all the mentioned process stages have been examined and developed based on modern information technology. With the dramatic development of the Internet and the WWW at the end of last century, KM has been drawing increasing attention from both researchers and companies, because "the basic economic resource is no longer capital, nor natural resources. It is and will be knowledge" [4].

There are two perspectives for KM: One is from a researcher viewpoint, another is from an organisational viewpoint [12]. From a researcher viewpoint, each of the mentioned process stages can be considered as a research field or research topic. That is, each of them still requires further systematic investigation, and optimisation and automation based on information technology. Most of researchers in KM usually

---

[1] See http://www.printersmark.com/Pages/Hist1.html

focus on a few process stages of the proposed model. Few researchers have studied all the process stages thoroughly or in parallel. From an organisational viewpoint, in particular, from a business viewpoint, an organisation or a company also focuses on one stage or some process stages in order to maximize the profits from the activities in KM. For example, one company only develops the knowledge visualization software and then sell its software in order to help its customer with facilitating knowledge visualization and knowledge distribution and utilization.

In order to formalize the above consideration, we assume that

$$KM = \{<KU, M>, <KD, M>, <KC, M>, <SSD, M>, <KOD, M>,$$
$$<KGC, M>, <KME, M>, <KV, M> <KDi, M>, <KR, M>, <KRM, M>\}$$

Where, KM is considered as a set consisting of eleven elements, each of which corresponds to a process stage in the knowledge processing and its management (M for short). For example, $<KDi, M>$ denotes knowledge distribution and its corresponding management. From a set theoretical viewpoint, 2^{KM} consists of all possible subsets of KM, each of the subsets in 2^{KM} corresponds to the research interests of a researcher, or business activities of a company. For example, $\{<KV, M>, <KDi, M>\} \in 2^{KM}$ consists of knowledge visualization, knowledge distribution and their management, which are the business activities of a publisher.

It should be also noted that from the history of modern computing, any reasonable abstraction from data has facilitated the research and development of IT. For example, the abstraction from data to information led to the fast development of information engineering and IM [13]. Based on this idea, we can see that the abstraction process from data to experience requires corresponding processing technology such as data processing, information processing, knowledge processing and experience processing which further involve data management, IM, KM (including intelligent agents and ES) and EM respectively. That is, human-level experience processing also requires EM. Just as data management, IM, and KM have played an important role in IT, IS, and AI, EM will also play an important role in IS and e-commerce.

4   Experience Management

In this section we examine experience management (KM) and look at the interrelationship between KM and EM. At the same time, we will also discuss inheritance, *specialization and generalization* from knowledge to experience, which will lead to some characteristics of EM different from KM.

From an object-oriented viewpoint [11], a subclass Y inherits all of the attributes and methods associated with its superclass X; that is, all data structures and algorithms originally designed and implemented for X are immediately available for Y [10] (p 551). This is the inheritance or reuse of attributes and operations. As we know, experience can be considered as a special case of knowledge (see Section 2), methodologies, techniques and tools for KM can be directly reused for EM, because EM is a special kind of KM that is restricted to the management of experience. On the other hand, experience has some special features and requires special methods different from that of knowledge, just as a subclass Y of its superclass X usually possesses more special attributes and operations. Therefore, two issues are very important for EM:

- What features of experience management (EM) are different from that of KM?
- Which special process stages does EM require?

In what follows, we will try to resolve these two issues. First of all, we define that EM is a discipline that focuses on experience processing and corresponding management which is in each of following process stages [2] (pp 1-14):

- Discover experience
- Capture, gain and collect experience
- Model experience
- Store experience
- Evaluate experience
- Adapt experience
- Reuse experience
- Transform experience into knowledge
- Maintain experience.

In these process stages, "maintain experience" includes update the available experience regularly, while invalid or outdated experience must be identified, removed or updated. Transform experience into knowledge is an important process stages for EM, which is the unique feature of EM different from those of KM. In the history of human beings, almost all invaluable experiences are gradually transformed through induction, summary and publication into knowledge, which then is spread widely in a form of books, journals and others means.

It should be noted that transformation between experience and knowledge is still an open topic from an intelligent system viewpoint. Further, from an IT viewpoint, discovery of knowledge from a huge database has become an important research field, that is, data mining and knowledge discovery, while discovery of experience from a collection knowledge or social practice is also a big issue for EM, because discovery of experience from knowledge or social practice is at a higher level than discovery of knowledge from data.

Similar to the discussion in the previous section, we assume that

$$EM = \{<ED, M>, <EC, M>, <EM, M>, <ES, M>, <EE, M>, <EA, M>,$$
$$<ER, M>, <ET, M>, <EM, M>\}$$

Where EM is considered as a set consisting of ten elements, each of which corresponds to a process stage in the experience processing and its own management (M for short). For example, $<ED, M>$ is for experience discovery and its own management, and $<EA, M>$ denotes experience adaptation and its own management. Based on set theory, 2^{EM} consists of all possible subsets of EM. Each of the subsets in 2^{EM} corresponds to a kind of business activities or research interests of some researchers. For example, $\{<EA, M>, <ER, M>\}$ can be considered as the activities of novice in his/her work.

5   Concluding Remarks

This paper examined experience and knowledge, EM and KM, and their interrelationships. It also examines EM and KM as integration of experience processing and corresponding management, knowledge processing and corresponding management respectively from a process modelling viewpoint.

EM and experience-based reasoning (EBR) research [15] will provide a new way of looking at data, knowledge, experience and their management for organisations. This will include experience structures, experience retrieval, experience similarity, experience processing and experience adaptation, and EBR. Successful solution of these problems could provide the basis for new advances in KM and EM. There is also significant potential for EBR in opening up a broad new range of applications, not only in business and e-business but in a number of domains such as deception recognition [15].

In future work, we will develop a prototype system for multiagent EBR systems, which can be used for business negotiation and brokerage. We will also develop a spiral model for experience management.

References

1. Abramowicz W, Kowalkiewicz M, Zawadzki P. Ontology frames for IT courseware representation 2003, pp 1-11. In Cokes E. [3]
2. Bergmann R. *Experience Management: Foundations, Development Methodology and Internet-Based Applications*. LNAI 2432. Berlin: Springer 2002
3. Cokes E. *Knowledge Management: Current issues and challenges*, Hershey, PA: IRM Press, 2003
4. Drucker P. Post-Capitalist Society. New York: Harper Business, 1993
5. Finnie G, Sun Z. A logical foundation for the CBR Cycle. *Int J Intell Syst* 18(4) 2003, 367-382
6. Hasan H. Handzic M (eds.). *Australian Studies in Knowledge Management*. Australia: University of Wollongong Press, 2003, 568 p.
7. McManus DJ, Snyder CA. Knowledge management: The missing element in business continuity planning, in Cokes [3], 2004, pp 79-91,
8. Mitchell HJ. Technology and knowledge management: Is technology just an enabler or does it also add value? In: Coakes E: *Knowledge management: Current issues and challenges. Hershey*: IRM Press, 2003, 66-78
9. Nilsson NJ. *Artificial Intelligence. A New Synthesis*. San Francisco, California: Morgan Kaufmann Publishers, Inc. 1998, 513 p.
10. Pressman RS. *Software Engineering: A Practitioner's Approach* (5th Edn), Boston: McGrawHill Higher Education, 2001, 860 p
11. Satzinger JW, Jackson RB, Burd SD. *Systems Analysis and Design in a Changing World* (3rd Edn). Course Technology, 2004
12. Sun, Z. A waterfall model for knowledge management and experience management. In: *Proc. Inter Conf on Hybrid Intelligent Systems* (HIS 2004), December 6-8, Kitakyushu, Japan, IEEE Press, 2004
13. Sun Z and Finnie G. *Intelligent Techniques in E-Commerce: A Case-based Reasoning Perspective*. Heidelberg: Springer-Verlag, 2004
14. Sun Z, A waterfall model for knowledge management and experience management.In: *Proc. Inter Conf on Hybrid Intelligent Systems* (HIS 2004), December 6-8, Kitakyushu, Japan, IEEE Press, 2004, pp 472-475
15. Sun, Z. Finnie G. Experience based reasoning for recognising fraud and deception. In: *Proc. Inter Conf on Hybrid Intelligent Systems* (HIS 2004), December 6-8, Kitakyushu, Japan, IEEE Press, 2004, pp 80-85

16. Voss A. Towards a methodology for case adaptation. In: Wahlster W (ed.): *Proc 12th European Conf on Artificial Intelligence (ECAI'96)*, John Wiley and Sons; 1996. pp 147-51
17. Zimmermann HJ. *Fuzzy Set Theory and its Application*. Boston/Dordrecht/London: Kluwer Academic Publishers; 1991.
18. http://wm2003.aifb.uni-karlsruhe.de/workshop/ w06/GWEM2003_CfP_english.html.
19. http:// www.iese.fraunhofer.de/experience_management/.

Rule Based Congestion Management –
Monitoring Self-similar IP Traffic in Diffserv Networks

S. Suresh and Özdemir Göl

School of Electrical and Information Engineering, University of South Australia
Mawson Lakes Campus, Adelaide, South Australia-5095, Australia
pessuresh@hotmail.com

Abstract. In this paper, we discuss employing a server in a router, the architecture of which is based on policy, for congestion management of IP traffic in Diffserv networks. We exploit the congestion information and employ that information to manipulate the network level parameters to adapt the traffic according to the current congestion state of the network. The congestion monitors interact with the policy server, which, depending on the network state, decides the policy/rules that should be enforced by the Diffserv network. The implementation of the selected policies typically leads to accepting, remarking or dropping the IP traffic packets entering the network. We have also indicated how these modules could functionally be used to generate and store appropriate rules based on policies for congestion management. Since the system will now be rule based, there is a need for policing action to monitor traffic and enforce these rules when congestion sets in, and we propose mobile agents accomplishing this job.

Index Terms: Quality of service, IP Networks Congestion Management, Self-similar input, Policy based rules, Mobile Agents.

1 Introduction

Policy-based networking (PBN) is emerging as a popular approach to automating network management as evident by numerous research and development efforts in academia, industry and the IETF [1]-[7]. PBN relies on the use of policies that configure dynamically a network node in a vendor-independent and interoperable manner. At present efforts in this area are getting focused on the Differentiated Services (Diffserv) architecture [8],[9]. Policies in this respect have two related purposes, viz., i) to define the goals of the organisation, in our case congestion management and ii) to allocate the resources, in our case buffer space in the router and the link bandwidth, to achieve the goals. For this purpose, it is possible to represent congestion management policy as a parameter pertaining to the communication network management, which can be interpreted by automated managers, particularly in routers. In this paper, we discuss a policy-based framework for congestion management of IP traffic having application in quality of service (QoS) requirements. The traffic congestion management policies are dynamically decided and enforced in the Diffserv network through the PBN infrastructure based on the information provided by congestion monitors installed in the network nodes. The motivation for our research is aimed at bringing out the salient features of the policy and consequently a rule based congestion monitoring and managing server architecture. This paper presents our approach in this direction and it is organised as follows. Section 2 presents the background and related

R. Khosla et al. (Eds.): KES 2005, LNAI 3681, pp. 987–995, 2005.
© Springer-Verlag Berlin Heidelberg 2005

works. Section 3 presents our framework for dynamic congestion management policy decisions for the present day IP traffic, and the rules framed. In section 4, an architecture for a policy based congestion management server system, which can dynamically adapt to changes in policy depending on the traffic, is presented along with the details on the traffic congestion monitoring agent algorithm. Finally, section 5 concludes this paper.

2 Related Background

We discuss in this section some related background works on Diffserv architecture and policy-based network management.

2.1 Differentiated Services

In the Internet QoS schemes, the Differentiated Services (Diffserv) approach [8],[9] has gained importance both in the industry and in academic research. Diffserv networks are generally divided into domains and each domain has routers as shown in Fig.1. IP packets are tagged with distinct labels before entering an IP Diffserv domain and will receive particular forwarding treatment at each router along the path. The main reason that a Diffserv domain would fail to achieve the QoS guarantees may be due to the edge routers making admission decisions without knowledge of congestion inside the domain, and such admitted flows will use and also waste network resources towards just getting dropped at downstream-congested routers.

The three main activities that are used in a Diffserv domain are the following. i) Traffic Classification (CL), ii) Traffic Conditioning (TC) and iii) Queue Controlling (QC). Based on these three activities, the following are the two main policy types [10] that are applicable in a Diffserv domain.

- *Edge router policy*: At the Edge router, the packets are classified according to packet flows and assigned labels called CIDs (Classifier Identifiers). After getting classified, they are metered and based on the metering they are either marked and scheduled or dropped based on say, the RED algorithm. Thus the edge policy incorporates the Traffic classifier, Traffic Conditioner and Queue control activities
- *Core router policy*: At the Core router the classified packets are metered and based on the metering they are either marked and scheduled or dropped based on say, the RED algorithm. In core routers there is no classifying activity. The Core policy incorporates only traffic Conditioning and Queue Control activities.

With this as background, we will now analyse congestion-monitoring problem associated with the modern IP networks.

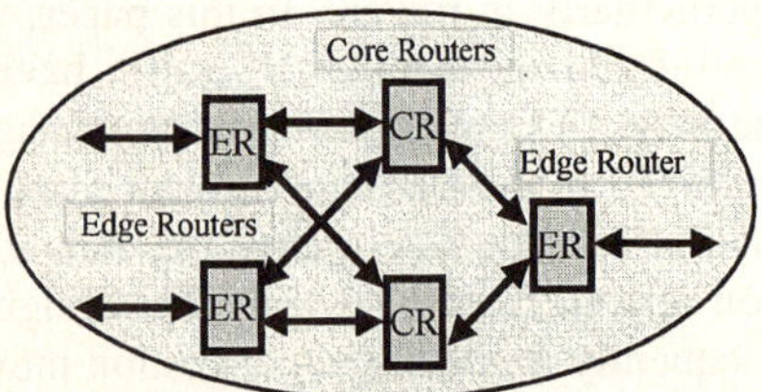

Fig. 1. Routers in the Diffserv domain

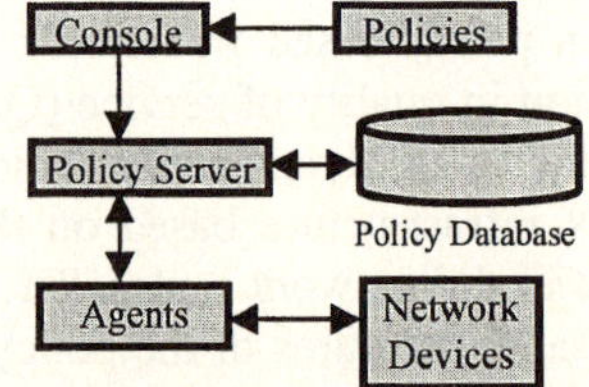

Fig. 2. Schematic of Policy Architecture

2.2 Policy-Based Network Management (PBNM)

PBNM [1]-[7] is a software tool used for managing network resources to provide QoS in IP networks. In the Diffserv framework, it is used for configuring Diffserv routers in an administrative domain. The tool provides the means to allocate resources to a particular user as specified in the Service Level Agreement (SLA) based on certain policy. Thus, policies for classification, policing and queuing/scheduling must cooperate to assure QoS. A schematic of Policy based architecture is illustrated in Figure 2. The details of one such architecture that we proposed [11], will be discussed in a later section for congestion management. With reference to Fig.2, the administrators or operators using the console define policies. They are fed to the policy server and stored into the policy database. They are deployed through agents into the network devices. The server also manages network interfaces of the devices by using the information managed and sent by these agents.

3 Congestion Management Policy

Traffic congestion in a computer network refers to degradation in the performance of the network when the network is heavily loaded [12]. It can cause data loss, large delays in data transmission and also large variations in these delays thus affecting the QoS. It is also widely accepted that Poisson model is not sufficient to characterize the traffic in the current IP network. Fowler and Leland [13] and Leland and Wilson [14], [15] have provided convincing evidence for the significance of self-similar or fractal IP network traffic in IP network. Taking these aspects into consideration, we discuss the importance of policies, particularly for managing congestion in the above scenario and also propose a possible architecture for a policy server based on rule modules.

3.1 Congestion Monitoring

We will now explore, the technique of monitoring and reporting on congestion level in a Diffserv domain. Edge nodes perform appropriate conditioning on the incoming flows so that the arrival rate of aggregated flows of same class at each interior router will not be more than expected. However, network dynamics cause traffic belonging to the same class to be directed via a particular interior router thus causing congestion. The congestion not only will waste lots of resources in the network, but may also cause more serious problem, such as not meeting the accepted QOS, and getting the whole network domain to crash. This means that congestion monitoring is a necessity. Before going into the details on congestion monitoring, the role of a congestion monitoring agent (CMA) and the implementing algorithm, we will discuss about the nature of self similar IP traffic and the corresponding rule base for congestion management.

3.2 Self-similar IP Traffic

It has been reported [13]-[15], that Self-Similar models were found to describe traffic well, over time scales ranging from milliseconds to hours and can, therefore, be expected to provide a very accurate picture of the nature of congestion. In the context, the self-similar IP traffic is characterized by the four parameters m, a, H and C. The parameter m>0 represents the mean input packet arrival rate, a > 0 is a variance coefficient, H $\in$ {1/2, 1} the Hurst self-similarity parameter of the IP input traffic and C > m is the service rate. The following approximate formula [16]-[18] then holds good.

$$P(L)=\exp\{(-1/2am)[(C-m)/H]^{2H}[L/(1-H)]^{2(1-H)}\} \tag{1}$$

where P(L) represents the probability for the queue size in the buffer to exceed the size L. For a given value of H, from eqn. (1), substituting for C = m/ρ, where ρ is the utilization factor,

$$\log P(L) = -K\,L^{2(1-H)} \tag{2}$$

$$\text{where } K = \{-[m^{(2H-1)}/2a][(1-\rho)/\rho H]^{2H}/(1-H)^{2(1-H)}\} \tag{3}$$

$$\text{hence } L = [-\log P(L)/K]^{(r)} \tag{4}$$

$$\text{where } r = 1/2(1-H) \tag{5}$$

We now show, how we employ these equations for generating appropriate rules for use in the CMA.

3.3 Derivation of Rule Base for Self-similar Input

Random Early Detection (RED) algorithm as proposed by Floyd [19], [20] for congestion avoidance, enables controlling the average queue size. In this scheme, if congestion is detected well before the router buffer is full, the edge router notifies the source to reduce its transmission window size i.e. the average rate of transmission, for that link. The router starts dropping packets only when the average queue size in the buffer exceeds some fixed threshold. RED has several shortcomings [21],[22],[23]. Since the IP input traffic from the source is self-Similar, it is characterized by the parameters a, m, ρ and H. Due to the filtering action of the first algorithm used in the RED, they may not be fully represented/conveyed in the average queue length. We have therefore proposed [22],[23] that the first thing to do is to assess the average queue length for sensing congestion, and then from it assess the drop probability (PM$_a$) value for quick management action, but taking into consideration the respective self similar parameters. So we consider that this modified RED algorithm based on probabilities [22],[23] corresponding to the queue lengths, would be appropriate for self-similar inputs. Fixing the values for the queue thresholds L_{min} and L_{max} however, depends on the desired average queue size to be maintained. A useful rule-of-thumb [19], [20] is to set L_{max} to at least twice L_{min}. With this back ground we will now propose the rules to be followed, based on the probability-based algorithm for self-similar inputs.

We take the same queue length based algorithm proposed by Floyd, but instead of the actual queue lengths, we compute the appropriate probability values through equation (4) taking into consideration the H $\in$ {1/2,1} and then, we fix the marking probability P(M$_a$) as under [22],[23].

$$P(M_a) = 0 \text{ for } L_{min} > L_a \tag{6}$$

$$P(M_a) = F\{[\log P_{min}]^r - [\log P(L_a)]^r\}/\{[\log P_{min}]^r - [\log P_{max}]^r\}$$
$$\text{for } L_{max} > L_a \geq L_{min} \tag{7}$$

$$P(M_a) = 1 \text{ for } L_a \geq L_{max} \tag{8}$$

In equation (7), F can be any value between 0 and 1 and this is the same as maxP in the equation of Floyd [19].

3.4 Policies and Rules for Congestion Management

Based on the above discussion, the policies to be followed for implementing the proposed congestion management scheme in IP networks would then be as below.

- The incoming IP traffic is self-similar (Fig.3)
- Packet marking for congestion management will be employed at the edge routers using modified RED algorithm [22]

  Based on these two policies, the following rules are framed for execution.

- For a given self similar IP traffic, the queue size in the router is calculated using exponential weighted moving average [19], [20] technique (Red-EWMA Queue in Fig.3)

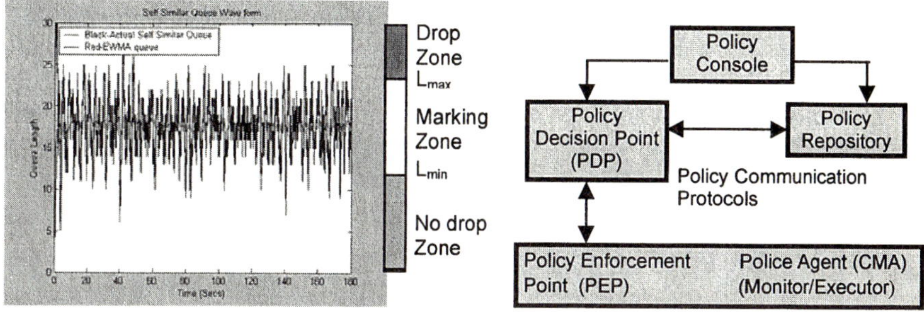

Fig. 3. Rule base using modified RED   **Fig. 4.** Basic Architecture of Policy

- The self similar properties as evidenced by the parameters a, m, ρ and H of the IP input traffic are computed by taking samples of the instantaneous queue lengths periodically for a given duration and these are used to compute the overflow probability [22],[23] corresponding to the average queue length.
- The modified RED algorithm (as above) is used on this average queue value by selecting two thresholds for the queue size, viz., minimum queue size L_{min} and maximum queue size L_{max}.
- The dropping probabilities for this self similar IP input, corresponding to these two thresholds are computed
- Till the average queue size reaches the minimum threshold level L_{min}, do not mark packets
- When the average queue size is between the two bounds, viz., minimum threshold L_{min} and the maximum threshold L_{max}, mark each packet with a probability, which is a function of the current average queue size (eqn. 7). This is an indication for the source to cut down the transmitting window size in a suitable manner, to prevent congestion from starting.
- When the average queue size is greater than the maximum threshold value L_{max}, start dropping packets. This means the transmission rate gets slowed down automatically.

4   Server Architecture for Rule Based Congestion Management

We have proposed an architecture for building the policy based server [11] for congestion management, as shown in Fig.4. The role of each module in this architecture is explained below.

- Policy Console: This is a user interface to store policies, and monitor status of the policy-managed environment.
- Policy Decision Point (PDP): This module enables making decisions based on policy based rules and the state of the services those policies manage. Policy enforcement points enforce these decisions.
- Policy Repository: A directory and/or other storage service where policy and related information are stored. This is basically in the present case, the Congestion Management Information Base (CMIB).
- Policy Enforcement Point (PEP): a software based mobile agent running on or within a resource that enforces a policy decision and/or makes a configuration change. The "Police" in our case will be a mobile agent, which will function, based on the policy-based rules. One more functionality added to the police agent is the role of CMA (congestion monitoring agent).
- Policy Communication Protocols: A protocol to read/write data from the policy repository and a protocol to communicate between PDP and PEP. Structured Query Language (SQL) could be used as Policy communication protocol.

4.1   Congestion Managing Agent (CMA)

The major task of any Congestion Monitoring Agent (CMA) is to monitor the congestion status in the Diffserv domain. CMA is just one more functionality added at the router where another agent is already residing in that router for that purpose. In this context, to support dynamic resource management, the routers in the domain should have some additional capability to monitor the congestion status and the consequent resource utilisation. For getting this information, we can make the agent at the routers report their queuing status for every class of traffic they handle. As an example, we can monitor the network queue system status for every class of traffic and the agent in the router sends out a report to Server for action. The server can then use this information to estimate the buffer utilisation for each class in the router and the proportion of the traffic for each flow in one class and take appropriate action.

With regard to congestion monitoring, in this paper, we follow the method of probability based RED [22],[23]; particularly for the IP traffic. The functionalities needed for this at a router include monitoring the average queue size of same traffic class, determining the congestion status as Nil, Low, Medium or High, and then based on the Probability RED, notify the congestion data to Server within the Diffserv domain. Using this Congestion information it is possible to control the window size appropriately, the details of which is beyond the scope of this paper.

We define three policy rules: Rule1, Rule2 and Rule3 for handling IP traffic (see the algorithm shown below). The PDP selects one of these three rules depending on information periodically received from the congestion monitors (CMA). The measurements processes by these monitors pertain to congestion status, as well as bandwidth usage and link utilization in the network consequent on the congestion. The level of traffic shaping is determined using an *Exponentially Weighted Moving Average* (EWMA) approach [19],[20]. The algorithm described below uses the predefined policy rules to make a decision depending on bandwidth availability in the network based on congestion information.

4.2 Congestion Managing Agent Algorithm

The Algorithm is as follows.

Initialization:
Start Congestion Monitor CMi for each Router i to calculate the congestion status.
Fix alpha = 0.002
// Fixed value of alpha for historical data
//Compute EWMA of queue length Le as:
Le=(1-alpha)*Le+(alpha)*L(j);
//Where L(j) is the current sample value of the queue length of the buffer of Router i.
//Fix the buffer length as Lmin, Lmax and B.
//Lmin is the lower threshold, L max is the upper threshold and B is the buffer capacity.
if Le < Lmin then
Rule1: Accept as in-profile traffic
Mark traffic with DSCP1 (Diffserv Code Point 1)
else if Lmin<=Le<Lmax, then
Rule2: Mark traffic with DSCP1(Diffserv Code Point 1)
Else if Le>Lmax then
Rule3: Drop out-of-profile Traffic
Mark traffic with DSCP2 (Diffserv Code Point 2)
End.

The implementation of the algorithm is being done using Java (RMI) on mobile agent-based system. Mobile agents being active entities, move from place to place to access the places where they have to serve. In our model, mobile agents would be multi-threaded entities, whose states and codes would be transferred to the new place when the mobile agent migration has to take place. Each mobile agent would be identified by a globally unique identifier, which would be generated by the system at the time of its creation. The identifier would be independent of the mobile agent's current place and hence does not change when the agent moves from node to node.

5 Conclusion

In this paper we have discussed employing a server in a router whose architecture is based on policy, for congestion management of IP traffic in Diffserv networks. We exploit the congestion information and employ that information to manipulate the network level parameters to adapt the traffic according to the current congestion state of the network. In this scheme, congestion monitors interact with the policy server, which, depending on the network state, decides the policy/rules that should be enforced by the Diffserv network. Since the system will now be rule based, there is a need for policing action to monitor traffic and enforce these rules when congestion sets in. Thus we propose mobile agents accomplishing this job. We have discussed the policy server architecture for Congestion Management and also the role of mobile agent for congestion management/control in IP networks.

References

1. Blight, D.C. and. Liu, C.G.,"Policy Based Network Architecture for End to End QoS", 1999, www.ee.umanitoba.ca/~blight/papers/ icc99.pdf
2. Fukuda, K. ., et al., "Policy-based Networking Service over Heterogeneous Public IP Service (DynaServ). Proc. of ICCC'99, Tokyo, June, 1999.
   http://www.cse. unsw.edu.au/~cs9333/ mini-conf/preg4t3.ppt
3. Moffett, J.D. and Sloman, M.S., "Policy Hierarchies for Distributed Systems Management", IEEE-JSAC Special Issue on Network Management, Vol.11, No.9, pp.1404-1414, December, 1993
4. Sloman, M.S., "Policy Driven Management for Distributed Systems., Journal Of Network and Systems Management, 2, (4), pp.333-360., 1994.
5. Suresh, S., "Applications of policy and mobile agents in the management of computer networks"., M.S. (By Research) Thesis., Anna University, Chennai., India., pp. 22-26., July, 2001.
6. Yi Pan., et.al., "Multi-domain Policy Based Network Architecture",
   www.ics.uci.edu/~ypan/QoSppt/WebPPT/PS/multidomain/MDPS.ppt
7. CISCO System Inc., "Cisco QoS Policy Manager Solution",
   www.cisco.com/warp/ public/cc/pd/wr2k/qoppmn/prodlit/ qpm2_ov.pdf
8. Blake, S., et.al, "RFC 2475: An Architecture for Differentiated Services", December 1998.
9. Bernet, Y., et.al, "An Informal Management Model for Diffserv Routers", draft ietf - Diffserv-model-06.txt, *Internet Draft*, IETF, February 2001.
10. Bernet, Y., et.al., "Diffserv Policies and Their Combinations in a Policy Server Called PolicyXpert", *IEICE SIG on Information Networks & SIG on Network Systems*, Technical Report, March 2002.
11. Suresh, S. et.al., "Policy Server Architecture for Traffic Congestion Management in IP Networks", Proc. 7th International Symposium on Digital Signal Processing and Communication Systems (IEEE sponsored) Coolangatta, Gold Coast, Australia, Dec, 8-11, 2003.pp.211-216.
12. Jain,R. and Ramakrishnan, K.K., "Congestion Avoidance in Computer Networks with a Connectionless Network Layer: Concepts, Goals and Methodology", Proceedings Computer Networking Symposium, Washington,D.C., pp.134 -143., April, 1988.
13. Fowler, H.J, and Leland,W.E., "Local area network traffic characteristics with implications for broadband network congestion management", IEEE J. Sel. Areas in Communic., 9(7), pp.1139-1149., 1991.
14. Leland,W., et.al., "On the Self-Similar Nature of Ethernet Traffic (Extended Version)"., IEEE/ACM Transactions on Networking, 2(1), February 1994.
15. Leland. W.E. and Wilson. D., "High time-resolution measurement and analysis of LAN traffic: Implications of LAN interconnections"., IEEE INFOCOM-91, 1991.
16. Agapie.M., "Fractal Traffic in high speed data networks", isis.pub.ro/iafa2003/ files/8-1.pdf
17. Garroppo, R.G., "Performance Analysis of a Single Server Queue Loaded by Long Range Dependent Input Traffic", 2nd Working-Conference on Optical Network Design and Modeling, Roma, February 9-11, 1998 isis.pub.ro/iafa2003/files/8-1.pdf
18. Norros, I., "On the use of Fractional Brownian Motion in the theory of connectionless Networks" IEEE-JSAC, Vol.13, No.6, pp.953-962, 1995.
19. Floyd, S. and Van Jacobson., "Random Early Detection Gateways for Congestion Avoidance", IEEE/ ACM Transactions on Networking, August 1993

20. Floyd, S., "Random Early Detection (RED): Algorithm, Modeling and Parameters Configuration"., www.ece.poly.edu/aatcn/pres_reps/ JTao_RED_report.pdf.
21. Wu-chang Feng, et.al., "The BLUE Active Queue Management Algorithms"., IEEE/ACM Transactions on Networking, Vol. 10, No. 4, August 2002.
22. Suresh, S., and Ozdemir Gol, "Congestion Management of Self Similar IP Traffic - Application of RED Scheme", Proc. 2nd IFIP/IEEE International Conference on Wireless and Optical Communication Networks., Dubai, UAE, March 6-8, 2005.
23. Suresh, S, and Ozdemir Gol, "Congestion Management of Self Similar IP Traffic using Normal and Exponential Marking", Accepted for presentation in the 3rd International conference on Information Technology: Research and Education (ITRE –2005 -IEEE Communications Society), 27-30 June, 2005, Hsinchu, Taiwan.

A Parallel Array Architecture
of MIMO Feedback Network
and Real Time Implementation

Yong Kim and Hong Jeong

Department of Electronic and Electrical Engineering
POSTECH, Pohang, Kyungbuk, 790-784, South Korea
{ddda,hjeong}@postech.ac.kr
http://isp.postech.ac.kr

Abstract. Blind source separation(BSS) of independent sources from
their convolutive mixtures is a problem in many real-world multi-sensor
applications. However, the existing BSS solutions are more often than
not based upon software and thus not suitable for direct implementation
on hardware. In this paper, we present a new FPGA architecture for
the blind source separation of a multiple input mutiple output(MIMO)
measurement system. The algorithm is based on feedback network and is
highly suited for parallel processing. The implementation is designed to
operate in real time for speech signal sequences. It is systolic and easily
scalable by simple adding and connecting chips or modules. In order to
verify the proposed architecture, we have also designed and implemented
it in a hardware prototyping with Xilinx FPGAs.

1 Introduction

Blind source separation is a basic and important problem in signal processing.
BSS denotes observing mixtures of independent sources, and, by making use
of these mixture signals only and nothing else, recovering the original signals.
In the simplest form of BSS, mixtures are assumed to be linear instantaneous
mixtures of sources. The problem was formalized by Jutten and Herault [1] in
the 1980's and many models for this problem have recently been proposed.

In this paper, we present K. Torkkola's feedback network [2, 3] algorithm
which is capable of coping with convolutive mixtures, and T. Nomura's extended
Herault-Jutten method [4] algorithm for learning algorithms. Then we provide
the linear systolic architecture design and implementation of an efficient BSS
method using these algorithms. The architecture consists of forward and update
processor. We introduce in this an efficient linear systolic array architecture that
is appropriate for FPGA implementation. The array is highly regular, consisting
of identical and simple processing elements(PEs). The design very scalable and,
since these arrays can be concatenated, it is also easily extensible. We have designed the BSS chip using a very high speed integrated circuit hardware description language(VHDL) and fabricated Field programmable gate array)FPGA.

R. Khosla et al. (Eds.): KES 2005, LNAI 3681, pp. 996–1003, 2005.

The organization of this paper is as follows. Section 2 derives an algorithm for the feedback network, and Section 3 shows the detailed architecture of the feedback network. The simulation results and conclusions are, respectively, presented in Sections 4 and 5.

2 Background of the BSS Algorithm

Real speech signals present one example where the instantaneous mixing assumption does not hold. The acoustic environment imposes a different impulse response between each source and microphone pair. This kind of situation can be modeled as convolved mixtures. Assume n statistically independent speech sources $s(t) = [s_1(t), s_2(t), \ldots, s_n(t)]^T$. There sources are convolved and mixed in a linear medium leading to m signals measured at an array of microphones $x(t) = [x_1(t), x_2(t), \ldots, x_m(t)]^T (m > n)$,

$$x_i(t) = \sum_p \sum_{j=0}^{n} h_{ij,p}(t)s_j(t-p), \quad for \ \ i = 1, 2, \ldots, m, \tag{1}$$

where $h_{ij,p}$ is the room impulse response between the jth source and the ith microphone and $x_i(t)$ is the signal present at the ith microphone at time instant t.

The feedback network algorithm was already considered in [2, 6]. Here we describe this algorithm. The feedback network whose ith output $y_i(t)$ is described by

$$y_i(t) = x_i(t) + \sum_{p=0}^{L} \sum_{j \neq i}^{n} w_{ij,p}(t)y_j(t-p), \quad for \ \ i, j = 1, 2, \ldots, n, \tag{2}$$

where $w_{ij,p}$ is the weight between $y_i(t)$ and $y_j(t-p)$. In compact form, the output vector $y(t)$ is

$$y(t) = x(t) + \sum_{p=0}^{L} W_p(t)y(t-p) = [I - W_0(t)]^{-1}\{x(t) + \sum_{p=1}^{L} W_p(t)y(t-p)\}. \tag{3}$$

The learning algorithm of weight W for instantaneous mixtures was formalized by Jutten-Herault algorithm [1]. In [4], the learning algorithm was the extended Jutten-Herault algorithm and proposed the model for blind separation where observable signals are convolutively mixed. Here we describe this algorithm. The learning algorithm of updating W has the form

$$W_p(t) = W_p(t-1) - \eta_t f(y(t))g(y^T(t-p)), \tag{4}$$

where $\eta_t > 0$ is the learning rate. One can see that when the learning algorithm achieves convergence, the correlation between $f(y_i(t))$ and $g(y_j(t-p))$ vanishes. $f(.)$ and $g(.)$ are odd symmetric functions. In this learning algorithm, the weights are updated based on the gradient descent method. The function $f(.)$ is used as the signum function and the function $g(.)$ is used as the 1st order linear function because the implementation is easy with the hardware.

3 Systolic Architecture for a Feedback Network

We present a parallel algorithm and architecture for the blind source separation of a multiple input multiple output(MIMO) measurement system. The systolic algorithm can be easily transformed into hardware. The overall architecture of the forward process and update is shown first and then follows the each detailed internal structure of the processing element(PE).

3.1 Systolic Architecture for Forward Process

In this section, we introduce the architecture for forward processing of the feedback network. The advantage of this architecture is spatial efficiency, which accommodates more time delays for a given limited space. The output vector of the feedback network, $\boldsymbol{y}(t)$ is

$$\boldsymbol{y}(t) = [\boldsymbol{I} - \boldsymbol{W}_0(t)]^{-1}\{\boldsymbol{x}(t) + \sum_{p=1}^{L} \boldsymbol{W}_p(t)\boldsymbol{y}(t-p)\}. \tag{5}$$

Let us define $\boldsymbol{C}(t) = [\boldsymbol{I} - \boldsymbol{W}_0(t)]^{-1}$ where the element of $\boldsymbol{C}(t) \in \boldsymbol{R}^{n \times n}$ is $c_{ij}(t)$.

$$\boldsymbol{y}(t) = \boldsymbol{C}(t)\{\boldsymbol{x}(t) + \sum_{p=1}^{L} \boldsymbol{W}_p(t)\boldsymbol{y}(t-p)\} = \boldsymbol{C}(t)\boldsymbol{x}(t) + \sum_{p=1}^{L} \hat{\boldsymbol{W}}_p(t)\boldsymbol{y}(t-p). \tag{6}$$

Applying (6) and the above expressions to (2), we have

$$y_i(t) = \sum_{j=1}^{n} c_{ij}(t)x_j(t) + \sum_{p=1}^{L}\sum_{j=1}^{n} \hat{w}_{ij,p}(t)y_j(t-p), \quad for \ \ i = 1, 2, \ldots, n. \tag{7}$$

Let us define the cost of the pth processing element $f_{i,p}(t)$ as

$$f_{i,0}(t) = 0,$$

$$f_{i,p}(t) \equiv f_{i,p-1}(t) + \sum_{j=1}^{n} \hat{w}_{ij,p}(t)y_j(t-p), \quad for \ \ p = 1, 2, \ldots, L. \tag{8}$$

Combining (7) and (8), we can rewrite the output $\boldsymbol{y}(t)$ as

$$y_i(t) = \sum_{j=1}^{n} c_{ij}(t)x_j(t) + f_{i,L}(t), \quad for \ \ i = 1, 2, \ldots, n. \tag{9}$$

We have constructed a linear systolic array as shown in Fig. 1. This architecture consists of $L+1$ PEs. The PEs have the same structure, and the architecture has the form of a linear systolic array using simple PEs that are only connected with neighboring PEs and thus can be easily scalable with more identical chips.

During $p = 1, 2, \ldots, L$, the pth PE receives three inputs $y_j(t-p)$, $\hat{w}_{ij,p}(t)$ and $f_{i,p-1}(t)$. Also, it updates PE cost $f_{i,p}(t)$ by (8). The $L+1$th PE calculates

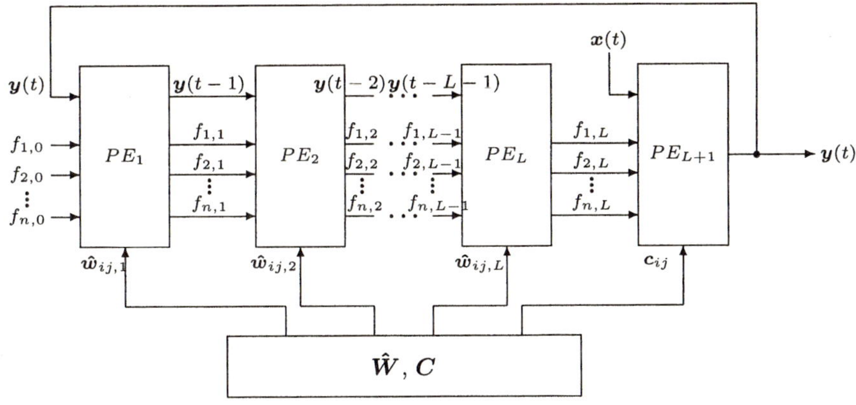

Fig. 1. Linear systolic array for the feedback network.

$y(t)$ according to (9) using inputs $x(t)$ and $c_{ij}(t)$. In other words, we obtain the cost of PE using recursive computation. With this method, computational complexity decreases plentifully. The computational complexity for this architecture is introduced in the last subsection.

The remaining task is to describe the internal structure of the processing element. Fig. 2 shows the processing element. The internal structure of the processing element consists of signal input part, weight input part, calculation part, and cost updating part. The signal input part and weight input part consist of the FIFO queue in Fig. 2, and take signal input $y(t)$ and weight $w(t)$ respectively. The two data move just one step in each clock. As soon as all inputs return to the queue, the next weight is loaded. When all the weights are received, the next input begins. The calculation part receives inputs $y(t)$ and w at two input parts, then updates PE cost $f_{i,p}(t)$ according to (8). This part consists of adder, multiplier, and register.

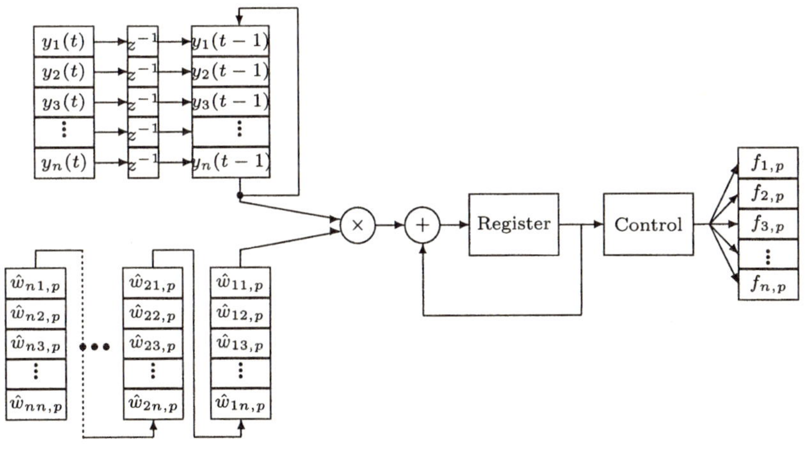

Fig. 2. The internal structure of the processing element.

3.2   Systolic Architecture for the Update

This section presents the architecture that updates the weights. This architecture also consists of the processing elements that operate in the same method. We define the processing element which the used in the update as the Update Element(UE). Efficient implementation in systolic architecture requires a simple form of the update rule. The learning algorithm of updating has the form

$$w_{ij,p}(t) = w_{ij,p}(t-1) - \eta_t f(y_i(t))g(y_j(t-p)),$$
$$i, j = 1, 2, \cdots, n, (i \neq j) \quad p = 1, 2, \cdots, L. \tag{10}$$

In this architecture, the function $f(.)$ is used as the signum function $f(y_i(t)) = sign(y_i(t))$ and the function $g(.)$ is used as the 1st order linear function $g(y_j(t)) = y_j(t)$ because the implementation is easy with the hardware.

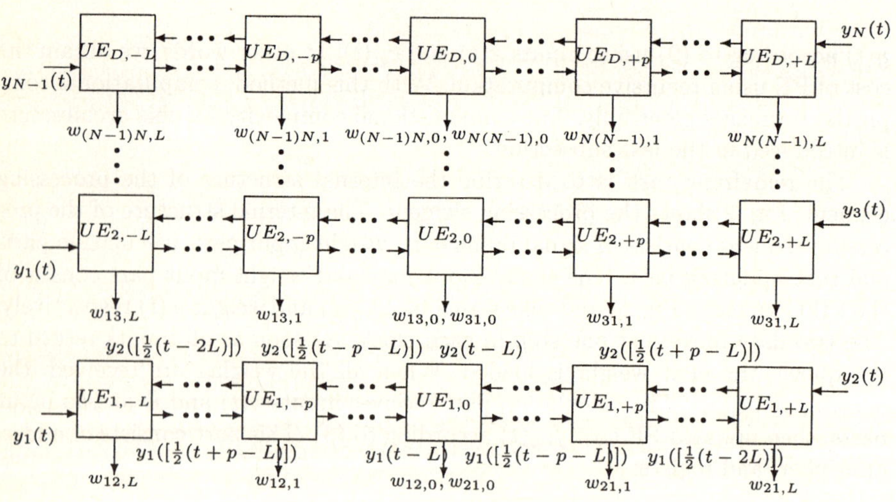

Fig. 3. Linear systolic array for update.

Fig. 3 shows the systolic array architecture of the update process. If the number of signals is N, then the number of rows D is $(N^2+N)/2$. All arrays have the same structure and all weights can be realized by using $y(t)$ simultaneously. In a row in Fig. 3, if the number of PE is L, the number of columns is $2L+1$. In other words, the architecture of the update consists of $D \times (2L + 1)$ UEs. The cost of (d,p)th UE has the form

$$u_{d,p}(t) = u_{d,p}(t-1) - \eta_t f(y_i(1/2(t-p-L)))y_j(1/2(t+p-L)),$$
$$d = 1, 2, \cdots, D. \tag{11}$$

Fig. 4 shows the internal structure of the UE of the feedback network. The processing element performs simple fixed computations resulting in a simple

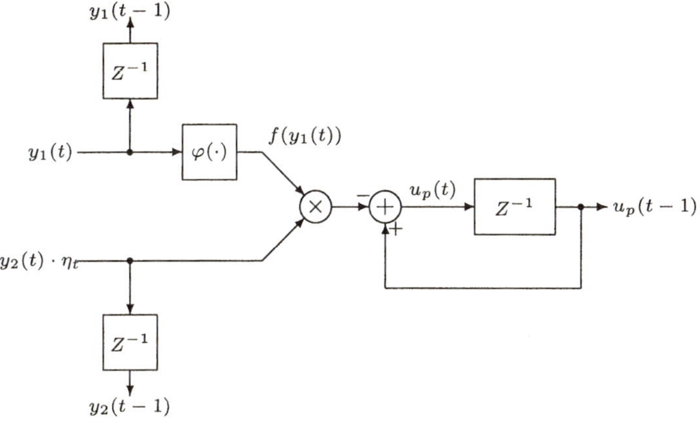

Fig. 4. The internal structure of update element.

design and low area requirements. The pth UE receives two inputs y_i and y_j, then one input becomes $f(y_i)$. The cost of UE is added to the accumulated cost of the same processor in the previous step.

3.3   Computational Complexity

We assumed that auditory signals from n sources were mixed and reached n microphones far from the sources. The forward process of the feedback network algorithm uses two linear arrays of $(L+1)$ PEs. (nL) buffers are required for output $\boldsymbol{y}(t)$,nL buffers for cost of PE, and (DnL) buffers for weight $\boldsymbol{W}(t)$. Since each output $\boldsymbol{y}(t)$ and weight $\boldsymbol{W}(t)$ are B_y bits and B_w bits long, a memory with $O(nLB_y + DnLB_w)$ bits is needed. Each PE must store the partial cost of B_p bit and thus additional $O(nLB_p)$ bits are needed. As a result, total $O(nL(B_y + B_p + DB_y))$ bits are sufficient.

The update of the feedback network algorithm uses a linear array of $D(2L+1)$ UEs. $2D(2L+1)$ buffers are required for output $\boldsymbol{y}(t)$ and $D(2L+1)$ buffers for cost of UE. If UE stores the partial cost of B_u bits, total $O(DL(4B_y + 2B_w))$ bits are sufficient.

4   System Implementation and Experimental Results

The system is desiged for an FPGA(Xilinx Virtex-II XC2V8000). The entire chip is designed with VHDL code and fully tested and error free. The following experimental results are all based upon the VHDL simulation. The chip has been simulated extensively using Cadence simulation tools. It is designcd to interface with the PLX9656 PCI chip.

As a performance measure, Choi and Chichocki [5] have used a signal to noise ratio improvement, defined as

$$SNRI_i = 10 \log_{10} \frac{E\{(x_i(k) - s_i(k))^2\}}{E\{(y_i(k) - s_i(k))^2\}}. \tag{12}$$

Two different digitized speech signals $s(t)$, as shown in Fig. 5, were used in this simulation. The received signals $x(t)$ collected from different microphones and recovered signals $y(t)$ using a feedback network are shown in Fig. 6, 7. In this case, we obtained $SNRI_1 = 2.7061$, and $SNRI_2 = 4.9678$.

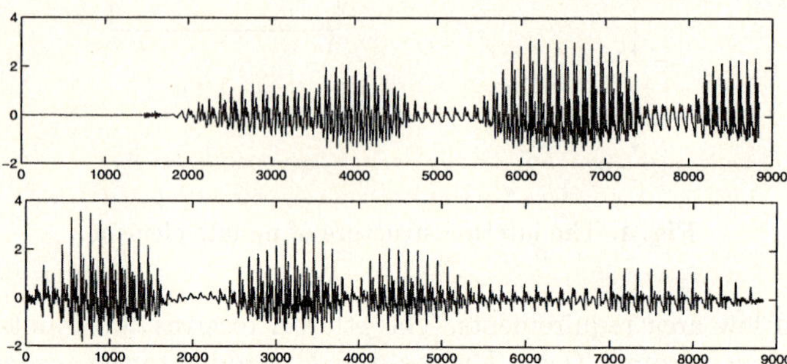

Fig. 5. Two original speech signals.

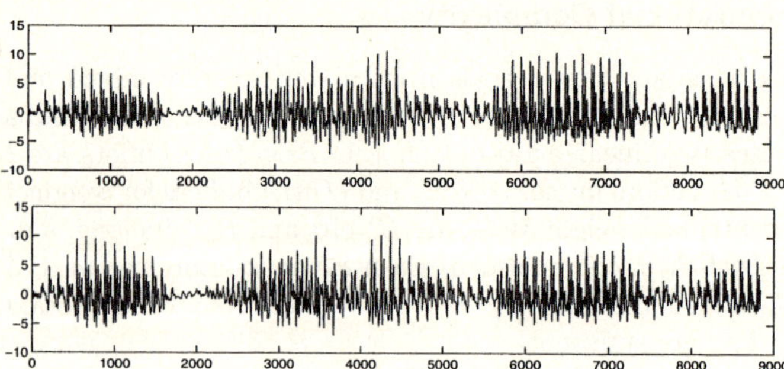

Fig. 6. Two mixtures of speech signals.

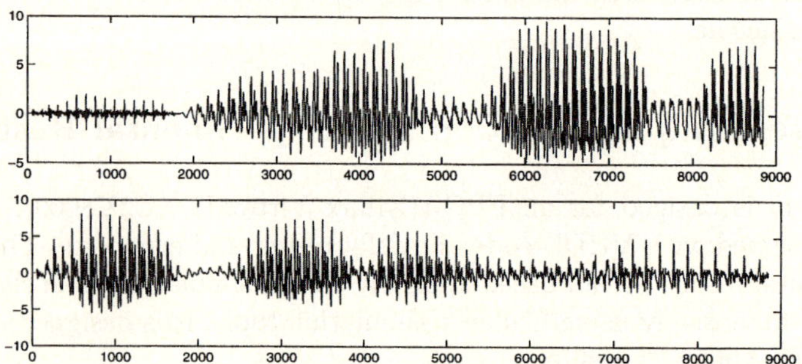

Fig. 7. Two recovered signals using the feedback network.

5   Conclusion

Because the algorithms used for hardware and software implementation differ significantly it will be difficult, if not impossible, to migrate software implementations directly to hardware implementations. The hardware needs different algorithms for the same application in terms of performance and quality. We have presented a fast and efficient FPGA architecture and implementation of BSS. In this paper, the systolic algorithm and architecture of a feedback network have been derived and tested with VHDL code simulation. This scheme is fast and reliable since the architectures are highly regular. In addition, the processing can be done in real time. The full scale system can be easily obtained by the number of PEs, and UEs.

References

1. C. Jutten and J. Herault, Blind separation of source, part I: An adaptive algorithm based on neuromimetic architecture. *Signal Processing* 1991; vol.24, pp.1-10.
2. K. Torkkola, Blind separation of convolved sources based on information maximization. *Proc. IEEE Workshop Neural Networks for Signal Processing* 1996; pp.423-432.
3. Hoang-Lan Nguyen Thi, Christian Jutten, Blind Source Separation for Convolutive Mixtures. *Signal Processing* 1995; vol.45, pp.209-229.
4. T. Nomura, M. Eguchi, H. Niwamoto, H. Kokubo and M. Miyamoto, An Extension of The Herault-Jutten Network to Signals Including Delays for Blind Separation. *IEEE Neurals Networks for Signal Processing* 1996; VI, pp.443-452.
5. S. Choi and A. Cichocki, Adaptive blind separation of speech signals: Cocktail party problem. *in Proc. Int. Conf. Speech Processing (ICSP'97)* August, 1997; pp. 617-622.
6. N. Charkani and Y. Deville, Self-adaptive separation of convolutively mixed signals with a recursive structure - part I: Stability analysis and optimization of asymptotic behaviour. *Signal Processing* 1999; 73(3), pp.255-266.
7. K.C. Yen and Y. Zhao, Adaptive co-channel speech separation and recognition. *IEEE Tran. Speech and Audio Processing* 1999; 7(2),pp. 138-151.

Wavelength Converter Assignment Problem in All Optical WDM Networks

Jungman Hong[1], Seungkil Lim[2], and Wookey Lee[3]

[1] Consulting Division, LG CNS
Seoul, 100-768, Korea
jmhong@lgcns.com
[2] Division of e-businessIT, Sungkyul University
Anyang, 430-742, Korea
seungkil@sungkyul.edu
[3] Division of Computer Science, Sungkyul University
Anyang, 430-742, Korea
wook@sungkyul.edu

Abstract. This paper considers the problem of locating the wavelength converter in all optical Wavelength Division Multiplexing (WDM) network. The problem is formulated as an integer programming problem by using path variables. In order to solve the associated linear programming relaxation which has exponentially many variables, a polynomial time column generation procedure is exploited. Therewith, an LP-based branch-and-price algorithm is derived to obtain the optimal integer solution for the problem. Computational results show that the algorithm can solve practical size problems in a reasonable time.

1   Introduction

Wavelength division multiplexing (WDM) is a promising technique to utilize the enormous bandwidth of the optical fiber. Multiple channels can be operated along a single fiber simultaneously in WDM. Thus, WDM divides the huge bandwidth of a single-mode optical fiber into channels whose bandwidths are compatible with peak electronic processing speed. Such WDM networks can be classified into two broad categories, one representing broadcast-and-select networks, and the other one representing wavelength-routed WDM networks. In the wavelength-routed WDM networks, the use of wavelength to route data is referred to as wavelength routing, and the network which employs this technique is known as a wavelength-routed network. Such a network consists of wavelength-routing switches interconnected by using optical fibers. Some routing nodes are attached to access stations where data from several end-users could be multiplexed on to a single optical channel. The access station also provides optical-to-electronic conversion and vice versa to asterface the optical network with conventional electronic equipment. A wavelength-routed network which carries data from one access station to another without any intermediate optical-to-electronic conversion is referred to as an all-optical wavelength-routed network.

R. Khosla et al. (Eds.): KES 2005, LNAI 3681, pp. 1004–1013, 2005.

In order to transfer data from one access station to another, a connection needs to be set up. This operation is performed by determining a path in the network connecting the source station to the destination station and by allocating a common free wavelength on all of the fiber links in the path. Such an all-optical path is referred to as lightpath. To establish any lightpath, we require that the same wavelength be allocated on all of the links in the path. This requirement is known as the wavelength-continuity constraints and wavelength-routed networks with this constraints are referred to as wavelength-continuous networks [4–6].

The capacity (bandwidth) on a wavelength channel is another valuable resource in a WDM-based optical network. When a station is to be connected directly to another in a wavelength-routed optical network, a lightpath is to be established from the source to the destination on a particular wavelength along all the links in the corresponding route. This wavelength-continuity constraint in all-optical networks is that a signal is forced to remain on the same wavelength from source to destination, and it may lead to blocking some connections even when adequate capacity is available on all of the links [7]. There are some studies that consider routing and wavelength assignment in WDM networks for better utilizing limited bandwidth [8–11]. In order to eliminate or reduce the effects of the wavelength-continuity constraints, a device called a wavelength converter is utilized. A wavelength converter is a device that takes a signal on an input wavelength and converts it to a specific output wavelength [12, 13].

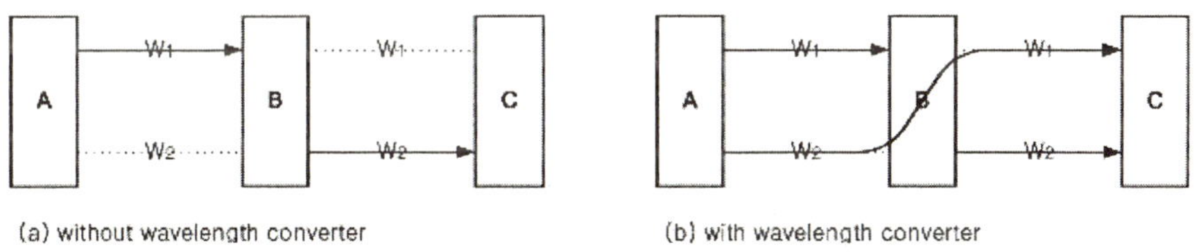

Fig. 1. Wavelength-continuity constraint and wavelength converter.

Consider the network in Figure 1. Two lightpaths have been established in the network, between nodes A and B on wavelength W_1 and between nodes B and C on wavelength W_2. Suppose a lightpath between nodes A and C needs to be set up. If there are only two wavelengths available in the network, establishing such a lightpath is impossible even though there is a free wavelength on each of the links. But, if a wavelength converter at node B is employed to convert data from wavelength W_2 to W_1. Then, new lightpath between nodes A and C can be established [5].

Wavelength converters have been shown as improving the utilization of the network bandwidth in high-speed optical networks [5, 14–17]. A simulation-based study on the performance evaluation of full and limited wavelength conversion was done by Lee et al. [12]. In the study, it was shown that the wavelength conversion can improve the call blocking performance. In addition, it was indicated that a limited number of converters might provide most of the benefits of a network with full conversion capabilities. An analytical model for evaluating the path blocking performance in such networks is presented in [14]. On the other hand, Yates et al. [18] have found that a limited-range converter can achieve most of the benefits of a full-range converter.

In this paper, a wavelength converter location problem is considered. This wavelength converter assignment problem is now specifically described. First, it is given the set of nodes that represent optical switches and the set of undirected links between nodes. Second, it is given the input data of the expected traffic requirements between node pairs. Third, the number of wavelength that can be used is given initially too. Forth, the number of wavelength converters that can be installed is also given. Finally, it is given the range of wavelengths over which each converter can change. The described problem can be considered as maximizing the expected traffic requirements that can be served, given the number of employed wavelength. Accordingly, three kinds of decisions have to be made subject to the associated objective, namely, determining which nodes have to be selected to locate the wavelength converter, finding a route for each traffic requirement, and assigning a wavelength for each route.

2   Notations and Formulation

In this section, we introduce some notation and definitions and present an integer programming formulation of the suggested problem. It is assumed that the physical network $G = (V, E)$, a set of available wavelengths, a set of selected pairs of nodes, and the available wavelength converters are all given.

2.1   Notations

Some notation and definitions are introduced, which are used in the integer programming formulation.

V : set of all the nodes,
E : set of all the undirected links,
W : set of all the available wavelengths,
K : set of all the selected pairs of nodes of traffic requirement,
o_k, d_k : two selected node pairs of $k \in K$,
r_k : integer value that represents the expected traffic requirement that is estimated as the number of wavelength to sever the traffic , $k \in K$,
a_k : the revenue by severing the unit traffic of node pairs of $k \in K$,
C : number of all the available wavelength converters,
$P(k)$: set of the paths between o_k and d_k, for all $k \in K$,
$P(k, e, w)$: set of the paths between o_k and d_k that pass through link e with wavelength w, for all $k \in K$, $e \in E$ and $w \in W$,
$P(k, v)$: set of the paths between o_k and d_k that use the wavelength converter at node v, for all $k \in K$ and $v \in V$,

The decision variables are specified as follows;

x_p = the portion of the traffic r_k, $k \in K$ that is severed by path $p \in P(k)$

$$y_v = \begin{cases} 1 & \text{if node } v \text{ is selected to locate wavelength converter} \\ 0 & \text{otherwise} \end{cases}$$

To discriminate link types, all the undirected and directed links are represented by $\{i, j\}$ and (i, j), respectively and $e \in E$ represents the undirected link.

2.2   Mathematical Formulation

The proposed problem is mathematically expressed as in the integer programming, called Problem AC. Problem (AC)

$$\max \sum_{k \in K} \sum_{p \in P(k)} a_k x_p$$

s.t.

$$\sum_{p \in P(k)} x_p \leq r_k, \quad \forall k \in k, \tag{1}$$

$$\sum_{k \in K} \sum_{p \in P(k,e,w)} x_p \leq 1, \quad \forall e \in E, \forall w \in W, \tag{2}$$

$$\sum_{k \in K} \sum_{p \in P(k,v)} x_p \leq M y_v, \quad \forall v \in V, \tag{3}$$

$$\sum_{v \in V} y_v \leq C, \tag{4}$$

$$0 \leq x_p \leq 1, \quad \forall p \in P(k), \forall k \in K$$

$$y_v \in \{0, 1\}, \quad \forall v \in V$$

Constraints (1) ensure that the severed traffic for the demand k is less than r_k between nodes o_k and d_k for $k \in K$ in the solution. Constraints (2) describe that at most one path can use one wavelength on one link. Constraints (3) imply that the wavelength of a path can be converted if there is a wavelength converter at node $v \in V$. Constraints (4) ensure that the total number of wavelength converter is less than equal to C. In the formulation AC, variables x_p ($p \in P$, $k \in K$) is real value that range from 0 to 1. If x_p has a fractional solution in the optimal solution, this indicates the path p ($p \in P$, $k \in K$) dynamically shares the wavelength with other node pairs at some links of the path in the long time periods. Problem AC has exponentially many variables. However, it will be shown in the next section that its LP relaxation of problem AC can be solved efficiently by the column generation method, which has successfully been used in solving many communication network configuration problems [19–22].

3 Solution Algorithm

3.1 Column Generation

Let LPAC be the LP relaxation of Problem AC and Z_{LPAC} be the optimal objective function value of Problem LPAC , which is called as the master problem. As shown in the previous, LPAC has exponentially many columns corresponding only to x_p variables, while the y_v variables are all included in LPAC, so that a column generation procedure is desired to implement only on x_p variables. Such a newly generated column enters into the basis of the restricted master problem, composed of some columns of the master problem.

Let α_k, β_{ew} and γ_v be the dual variables associated with constraints (1), (2) and (3), respectively. Also, let α_k^*, β_{ew}^*, and γ_v^* be the dual variables corresponding to any current Problem LPAC.
For a path $p \in P(k)$, define the followings;

E_p : set of the links of which path p consists,
$S(w)$: set of the wavelengths convertible from $w \in W$ with wavelength converter, $S(w)=\{(w-s+|W|) \mod |W|, \cdots, w-1, w+1, \cdots, (w+s) \mod |W|\}$

$$f_{uv}^p = \begin{cases} 1 & \text{if } e = (u, v) \in E_p \\ 0 & \text{otherwise.} \end{cases},$$

$$z_{ew}^p = \begin{cases} 1 & \text{if link } e = \{u, v\} \in E_p \text{ with wavelength } w \\ 0 & \text{otherwise.} \end{cases},$$

$$y_v^p = \begin{cases} 1 & \text{if node } v \in V \text{ is selected to locate the wavelength converter} \\ & \text{in path } p \in P(k) \\ 0 & \text{otherwise.} \end{cases},$$

Then, the subproblem that must be optimized to generate the most attractive column (i.e., the column contributing to the maximization of the objective function of the master problem) is now derived as Problem (ACSP (k)).
Problem (ACSP (k))

$$\min \ \alpha_k^* + \sum_{w \in W} \sum_{e \in E} \beta_{ew}^* z_{ew}^p + \sum_{v \in V} \gamma_v^* y_v^p - a_k$$

s.t.

$$\sum_{u \in V} f_{uv}^p - \sum_{u \in V} f_{vu}^p = \begin{cases} 1 & \text{if } v = o_k \\ -1 & \text{if } v = d_k \\ 0 & \text{otherwise} \end{cases}, \forall e = \{u, v\} \in E, \tag{5}$$

$$f_{uv}^p + f_{vu}^p \leq \sum_{w \in W} z_{ew}^p, \qquad \forall e = \{u, v\} \in E, \tag{6}$$

$$f_{iv}^p + f_{vj}^p + z_{\{iv\}w}^p - z_{\{vj\}w}^p \leq y_v^p + 2, \qquad \forall (i, v), (v, j) \in E, \forall w \in W, \tag{7}$$

$$f_{iv}^p + f_{vj}^p + z_{\{iv\}w}^p - \sum_{\lambda \in S(w)} z_{\{vj\}\lambda}^p \leq 2, \qquad \forall (i, v), (v, j) \in E, \forall w \in W, \tag{8}$$

$$\sum_{v \in V} y_v^p \leq C, \qquad\qquad (9)$$

$$f_{uv}^p, z_{ew}^p, y_v^p \in \{0,1\}, \qquad \forall e = (u,v) \in E, \forall w \in W, \forall v \in V.$$

Constraints (7) describe that wavelength of the path can be convertible if there exists a wavelength converter, constraints (8) ensure that the range of wavelength that the wavelength converter can change is restricted, and constraints (9) ensure that the total number of wavelength converter is less than equal to C.

The column generation problem $ACSP(k)$ associated with demand node pairs k represents a shortest path problem between node o_k and node d_k, to be solved with link weights (β_{ew}^*, for every $e \in E$) and node weights (γ_v^*, for every $v \in V$). For the subproblem, a modified version of shortest path algorithm, MSP, to minimize the sum of the link and node weights along the resulting path is to be exploited. It is noticed, referring to Ahuja et al. [24], that since the link and node weights have nonnegative values, the MSP algorithm can solve the column generation problem optimally in polynomial time.

Let $dis_c(u, \dot{w}, v, \tilde{w})$ be the length of the minimum cost path from node u with wavelength $\dot{w}$ to node $v \in V$ with wavelength $\tilde{w}$ when c number of converters are used along the path. It is shown in Figure 2. Note that the minimum cost path with no wavelength converter, $dis_0(u, \dot{w}, v, \tilde{w})$, can be easily constructed with shortest path algorithm for each wavelength. Then, the MSP algorithm of constructing a shortest paths between all pair of nodes is now derived in a step-by-step procedure.

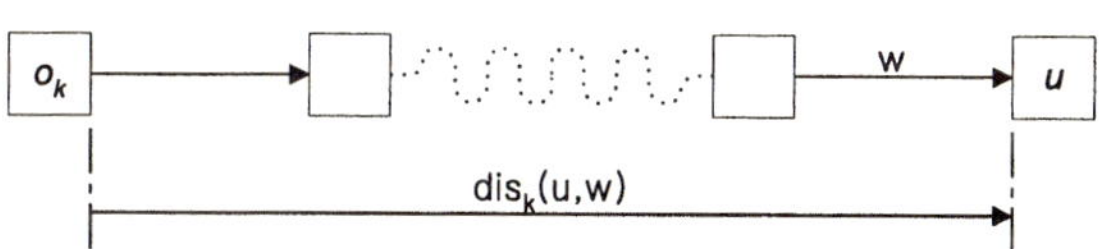

Fig. 2. Example of $dis_k(u, w)$.

Algorithm (MSP)

Step 1. set $c = 1$,

Step 2. set $d_c(u, \dot{w}, v, \ddot{w}) = \infty, \forall u, v \in V, \dot{w}, \ddot{w} \in W$,

set $d_c(u, \dot{w}, u, \ddot{w}) = 0, \forall u \in V, \dot{w}, \ddot{w} \in W$,

Step 3. For $\forall u, v, i \in V, \ \forall \dot{w}, \ddot{w}, \tilde{w}, w \in W$

if for $\tilde{w} \neq w, \ w \in S(\tilde{w})$,

$d_c(u, \dot{w}, v, \ddot{w}) > d_0(u, \dot{w}, i, \tilde{w}) + d_{c-1}(i, w, v, \ddot{w}) + \gamma_v^*$,

then $d_c(u, \dot{w}, v, \ddot{w}) = d_0(u, \dot{w}, i, \tilde{w}) + d_{c-1}(i, w, v, \ddot{w}) + \gamma_v^*$,

Step 4. if $c < C$, then $c = c + 1$ and goto Step 2.

Step 5. set $d^*(u, v) = \min_{c \leq C} \left\{ \min_{\dot{w} \in W} \left\{ \min_{\ddot{w} \in W} \{d_c(u, \dot{w}, v, \ddot{w})\} \right\} \right\}, \forall u, v \in V.$

Proposition 1. *At termination of the algorithm MSP(k), the resulting path from o_k to d_k is the optimum solution of the problem ACSP(k), if given β_{ew}^* and γ_v^* are all non-negative values.*

Proof. See [23] for a complete proof of this proposition.

If the resulting length of a shortest path, dis_k, plus α_k^* is less than a_k, then the path can be added to the current restricted master problem. Otherwise, no column is generated with respect to $k \in K$. The coefficients of the newly generated column are determined by the links and nodes included on the path. If the new column is generated for the node pairs k and its corresponding path is p, then the coefficient of the kth row of constraints (1) is 1, and the coefficients of constraints (2) and (3) are the values of their corresponding variables z_{ew}^p and y_v^p, respectively.

3.2   Branching Rules

When no more columns can be generated, the associated solution finding for the proposed problem is completed at the root node of the enumeration tree, giving the final solution. Then it is necessary to check if the obtained solution is integral. If it is integral, then the solution process is done with the final solution as the optimal solution of Problem AC. Otherwise, it is required to initiate the branch-and-bound procedure to find an integer solution, for which the node variable $y_v \in V$ is used as a branching variable. Consider a node $v \in V$ for branching,

$$\text{where } v = \arg\left\{ \max_{v \in V} \left\{ \sum_{\{p|p \in P(k,v), k \in K\}} x_p | y_v \text{ is fractional} \right\} \right\}.$$

At the root node, the associated subproblem is a shortest path problem. After branching, the subproblem remains as a shortest path problem. A few extra steps are then added to deal with the branching restrictions. Suppose that we need to branch on node variable y_v, and that two new branches are generated; one of them forcing y_v to be 1 and the other one forcing y_v to be 0. In the branch with $y_v = 1$, the branching process is done by adding the linear constraint $y_v = 1$. In the branch with $y_v = 0$, the linear constraint $y_v = 0$ is added, and the subproblem being under branching is a shortest path problem in the network with the cost of branching variable (node cost) provided with a sufficiently large value.

4   Computational Results

In this section, all the above mentioned solution procedures are put together into the proposed branch-and-bound algorithm. This branch-and-bound algorithm is then applied to solve numerical examples. In all the experimental results presented here, the initial restricted Problem LPAC has the columns corresponding to all the shortest paths between the node o_k and the nodes d_k with randomly selected wavelength. In addition to these initial columns, columns are generated iteratively by solving the associated shortest path problems ACSP(k) until

LPAC is optimized. When no more column can be generated to be added to the problem, it is necessary to check if the obtained solution is integral. If it is integral, then the solution process is done with the final solution as the optimal solution of Problem AC. Otherwise, it is required to initiate the branch-and-bound procedure to find an integer solution.

In implementing the proposed algorithm, the objective function value of an LPAC is used as the bound for each branch node and the best bound rule is used for selecting the next node for further branching in the enumeration tree. In the experiment, the CPLEX callable mixed integer library is used as an LP solver and some other routines in the library are also used for adding constraints and variables. All the numerical problems are solved on a PC by the proposed algorithm coded in C language.

The performance of the proposed algorithm is evaluated for its efficiency and effectiveness with randomly generated graphs. The nodes in the graphs are randomly placed in a 10000-by-10000 square grid. Links connecting pairs of nodes are placed at probability $P(u,v) = Aexp\left(\frac{-d(u,v)}{BL}\right)$, where $d(u,v)$ represents the distance between nodes u and v, and L is the maximum possible distance between two nodes [25]. The parameters A and B are real values defined in the range $(0,1]$. The nodes have the minimum degree of 2 and the maximum degree 5. The node pairs (o_k, d_k) are selected randomly and r_k is randomly generated individually from the uniform distribution defined over the interval $[1,3]$, and the value of a_k is set to the value of the minimum hop distance in the given network. In this method, we can guarantee the fairness in the resulting solution. The set of the wavelengths convertible from $w \in W$ with wavelength converter, $S(w)$, is $\{(w - s + |W|) \mod |W|, \cdots, w - 1, w + 1, \cdots, (w + s) \mod |W|\}$, where s was set 10. For each of the problem cases, 20 problems are generated and solved. Table 1 summarizes the experimental results.

Table 1. Computational Results with $|V| = 20$.

$|V|$	$|W|$	C	Z_{LPAC}	CPU (sec.)	# of generated variables	network degree	objective value	percentage gap
		2	112.1	5.2	1052.5	2.9	110.7	1.3
		4	119.9	5.2	1093.7	3.0	117.8	1.8
	5	6	122.7	6.1	1250.6	3.0	118.8	3.2
		8	117.4	5.5	1105.5	3.0	114.2	2.8
		10	116.8	5.0	1005.6	3.0	113.7	2.7
		2	207.2	115.4	3083.3	2.9	203.2	1.9
		4	214.6	205.6	2847.4	2.9	209.2	2.5
20	10	6	224.5	201.6	2788.9	3.0	217.6	3.1
		8	215.4	320.0	2162.2	2.9	209.5	2.8
		10	211.7	428.1	2437.9	2.9	207.1	2.2
		2	286.8	485.8	4432.2	2.9	279.5	2.6
		4	279.2	309.0	2674.6	2.9	275.0	1.5
	15	6	283.7	300.7	3077.0	3.0	277.0	2.4
		8	299.9	598.7	3920.6	2.8	291.1	2.9
		10	318.4	632.7	4221.3	2.9	308.2	3.2

5   Conclusion

This paper considers an optimal problem of locating the wavelength converter in all optical Wavelength Division Multiplexing (WDM) network. The problem is formulated as an integer programming problem by using path variables. In order to solve the associated linear programming relaxation which has exponentially many variables, a polynomial time column generation procedure is exploited. Therewith, an LP-based branch-and-bound algorithm is derived to obtain the integer solution for the problem. Computational results show that the algorithm can solve practical size problems in a reasonable time.

References

1. Green, P.E.: Fiber-optic networks. Prentice-Hall, Englewood Cliffs, NJ. (1993)
2. Mukherjee, B.: Optical communication networks. McGraw-Hill. (1998)
3. Ramaswami, R., Sivarajan, K.N.: Optical networks: a practical perspective. Morgan Kaufmann, San Francisco, California. (1998)
4. Ramaswami, R.: Multi-wavelength lightwave networks for computer communication. IEEE communications magazine. **31** (1993) 78-88
5. Ramamurthy, B., Mukherjee, B.: Wavelength conversion in wdm networking. IEEE journal on selected areas in communications. **16** (1998) 1061-1073
6. Ramaswami, R., Sivarajan, K.N.: Routing and wavelength assignment in all-optical networks. IEEE/ACM transaction on networking. (1995) 489-500
7. Kaminow, I.P. et al.: A wideband all-optical wdm network. IEEE journal on selected areas in communications. **14** (1996) 780-799
8. Zang, H., Ou, C., Mukherjee, B.: Path-protection routing and wavelength assignment (RWA) in WDM mesh networks under duct-layer constraints. IEEE/ACM Transactions on networking. **11** (2003) 248-258
9. Koyama, A., Barolli, L., Okada, Y., Kamibayashi, N., Shiratori, N.: A wavelength assignment method for WDM networks using holding strategy. AINA. (2003) 337-342
10. Sivakumar, M., Subramaniam, S.: Wavelength conversion placement and wavelength assignment in WDM optical networks. HiPC. (2001) 351-360
11. Talay, C., Oktug, S.F.: A GA/Heuristic hybrid technique for routing and wavelength assignment in WDM networks. EvoWorkshops. (2004) 150-159
12. Lee, K.C., Li, V.: A wavelength-convertible optical network. IEEE/OSA journal of lightwave technology. **11** (1993) 962–970
13. Subramaniam, S., Azizoglu, M., Somani, A.: All-optical networks with sparse wavelength conversion. IEEE/ACM transactions on networking. **4** (1996) 544-557
14. Kovacevic, M., Acampora, A.: Benefits of wavelength translation in all-optical clear-channel networks. IEEE journal on selected areas in communications. **14** (1996) 868–880
15. Barry, R., Marquis, D.: Models of blocking probability in all-optical networks with and without wavelength changers. IEEE journal on selected areas in communications. **14** (1996) 858-867
16. Birman, A.: Computing approximate blocking probabilities for a class of all-optical networks. IEEE journal on selected areas in communications. **14** (1996) 852-857

17. Jia, X.-H., Du, D.-Z., Hu, X.-D., Huang, H.-J., Li, D.-Y.: On the optimal placement of wavelength converters in wdm networks. Computer Communications **26** (2003) 986-995
18. Yates, J., Lacey, J., Everitt, D., Summerfield, M.: Limited-range wavelength translation in all-optical networks. IEEE INFOCOM'96, San Francisco, CA. (1996)
19. Clarke, L.W., Gong, P.: Capacited network design with column generation. The Logistics Institute, Georgia Tech. (1995)
20. Lee, K., Park, K., Park, S.: Design of capacitated networks with tree configurations. Telecommunication systems. **6** (1996) 1-19
21. Park, K., Kang, S., Park, S.: An integer programming approach to the bandwidth packing problem. Management science. **42** (1996) 1277-1291
22. Parker, M., Ryan, J.: A column generation algorithm for bandwidth packing. IEEE journal on selected areas in communications. **2** (1994) 185-195
23. Hong, J.: Some Optimization Issues in Designing Multimedia Communication Networks. Ph.D. Thesis, Department of Industrial Engineering, KAIST (2000)
24. Ahuja, R.K., Magnanti, T.L., Orlin, J.B.: Network flows: theory, algorithms, and applications. Prentice-Hall. (1993)
25. Waxman, B.: Routing of multipoint connections. IEEE journal on selected areas in communications. **6** (1998) 1617-1622

Real-Time System-on-a-Chip Architecture for Rule-Based Context-Aware Computing

Seung Wook Lee[1], Jong Tae Kim[1], Bong Ki Sohn[2], Keon Myung Lee[2], Jee Hyung Lee[1], Jae Wook Jeon[1], and Sukhan Lee[1]

[1] School of Information and Communication Engineering, Sungkyunkwan University, Korea
jtkim@skku.ac.kr
[2] School of Electrical and Computer Engineering, Chungbuk National University, Korea

Abstract. A rule-based system can be a solution for context reasoning in context-aware computing systems. In this paper we propose new flexible SoC (System-on-a-Chip) architecture for real-time rule-based system. The proposed architecture can match up values and variables of the left-hand sides of *'if-then rules'* (rule's LHS) in parallel. Compared to previous hardware rule-based system, we reduce the number of constraints on rule representations and combinations of condition terms in rule's LHS by using a modified contents addressable memory and a crossbar switch network (CSN). The modified contents addressable memory (CAM), in which the match operation of the system is processed in parallel, stores the rule-base of the system. The crossbar switch network is located between the input buffer that stores external raw input data and the working memory, and can freely configure condition operation of rule's LHS and working memory with stored data within the input buffer. The proposed SoC systems architecture has been designed and verified in a SoC development platform called SystemC.

1 Introduction

The context-aware computing system can be seen as a service system that becomes aware of current context using acquired information from various sensors, and provides appropriate service to the user. Since context-aware applications depend on sensors to get context information, efficient sensor management is required[1]. The sensor management method depends on the applications or on the sensors used. In a context-aware computing system, the proper reasoning method is needed to become aware of current context. If the method is not affected by the application and the used sensors, an identical context reasoning method can be used in different applications. One of the methods is rule-based systems that compare facts in working memory, which is continually updated, to a stored rule-base, which is represented by *if-then* rules, and infer system response. However, the rule-based systems based on conventional software algorithms (RETE[2], etc.), which are computation intensive and run slowly, are inadequate for real-time applications such as context-aware computing systems. For speeding-up rule-based systems, a parallel implementation architecture based on a multiprocessor platform had been proposed by A. Gupta[3]. But, considering its application area such as portable device, this approach is not suitable to context-aware computing systems. Namely context-aware computing systems must consider power consumption and size for mobility.

R. Khosla et al. (Eds.): KES 2005, LNAI 3681, pp. 1014–1020, 2005.
© Springer-Verlag Berlin Heidelberg 2005

In this paper, we propose a real-time rule-based system architecture that can be applied to a mobile intelligent robot. The intelligent robot must provide appropriate service to elderly people by reasoning current context with acquired information from external sensors. The reasoning speed is important factor to measure robot capacity. In K. M. Lee[6], we have proposed SoC based whole system architecture for context-aware system which includes communication unit, processing unit and rule-based system unit. So, this article will give details of the real-time rule-based system architecture used in K. M. Lee[6]. This proposed architecture does not require reconfiguration of hardware architecture as in other research[4-5]. Also, it can match up values and variables of rule's LHS in parallel. By using modified CAM and CSN, the number of constraints on rule representations and combinations of condition terms in rule's LHS is reduced. The real-time rule-based system architecture is described and verified using SystemC ver. 2.0 and simulated with 56 rules to assist elderly people at home. This paper is organized as follows. In section 2, the proposed hardware architecture is presented in detail and the expression of rule's LHS and working memory (WM) is discussed. Section 3 explains the fast variable binding method based on a hardware structural approach. In section 4, the experimental results are shown. Finally, the conclusions are drawn in section 5.

2   Real-Time SoC Architecture for Rule-Based Systems

In this section, we discuss distinctive features of hardware real-time rule-based system architecture in detail

2.1   Overview of Hardware Architecture

The real-time rule-based system architecture can be schematized as in Fig. 1.

Host-processor. The host-processor is in charge of the software part of the whole system. The inferred rule (i.e., the rule selected by the conflict resolution module) of the real-time rule-based system is executed on the host-processor. For detailed discussion of the software part, see K. M. Lee[6].

Interface module. The operations of real-time rule-based system start with input information from external sensors. The input information from external sensors is processed to take it into the real-time rule-based system on the host-processor. The interface module communicates with the host-processor.

Input buffer. The input buffer, which is a register bank with unique attributes for each register, stores data from the interface module. The data stored in this module is used for configuring WM using CSN. For improving system performance, the input buffer is updated by the conflict resolution module directly, rather than by the host-processor.

This is, in case some executed rules must change WM; there would not be overhead to transfer data between the host-processor and interface module.

Crossbar switch network. CSN, which includes pre-processing modules (PPMs), is located between the input buffer and WM. The PPM is an ALU (arithmetic logic unit) and it can compute such operator as +, -, $\geq$, $\leq$, $\neq$ and ==.

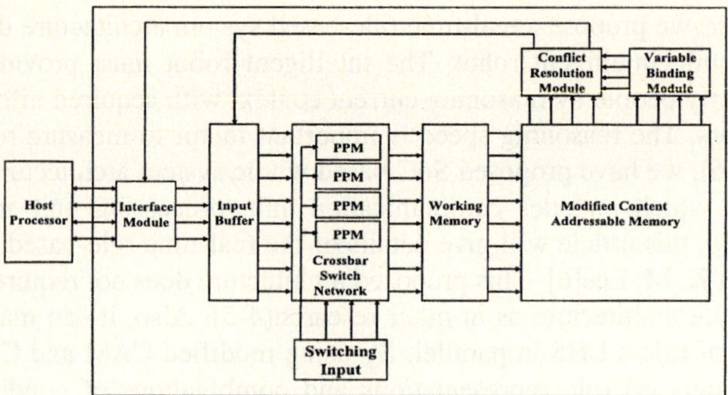

Fig. 1. Overview of Real-Time Rule-Based System Architecture

Modified content addressable memory. This memory stores *if*-parts of all rules and executes matching operations in parallel with CAM. Also, we propose modified architecture of general CAM for efficient matching operation.

Variable binding module. We propose a new variable binding method which can process variables equivalence test in rule's LHS using a hardware structural approach in parallel.

Conflict resolution module. In this module, the conflict set is resolved. The applied resolution strategies are priority-based resolution and LRU (least recently used)–based resolution.

2.2 Expression of Rules

In this section, we discuss the expression method of rules for using the real-time rule-based system.

The applicable rule's LHS is classified according to the following: (1) with an attribute, (2) AND relation between attributes, (3) with a condition term and (4) AND relation between attributes and condition terms.

Given that a rule is expressed as follows,

$$If <E> \text{ AND } <F> \text{ } Then \text{ } <Action>$$

We will use the term "expression" to refer to $<E>$, $<F>$ can represent attributes or results of a condition term in the rule. In other words, each expression indicates unique attribute in the whole rule-base. For example, in the case of Rule 4, $<E>$ is $<wait\_response>$ which is taken as an attribute value from external sensors of the robot, $<F>$ is $<clock\_time - event\_time\_1 \geq specific\_time\_1>$ which is computed using PPMs with stored data in the input buffer for configuring WM. Assuming that all rules can be constructed by 6 expressions in the whole rule-base and each rule can include a maximum of 2 expressions, 4 rules in Fig. 2 are stored in rule-base memory as shown in Fig. 3. The same applies to WM. In this case, the matching operation between rule-base and WM can be easily implemented using modified CAM which will be discussed in section 2.4 in detail.

Rule 1
> *If*   <OK_response>
> *Then*   reset abnormal_pulse_rate_flag
>   reset abnormal_temperature_flag
>   reset OK_response_flag

Rule 2
> *If*   <abnormal_pulse_rate> AND <abnormal_temperature>
> *Then*   set abnormal_state_flag
>   ask the master "How do you feel?"
>   set event_time_1

Rule 3
> *If*   <blood_pressure ≥ upper_normal_pressure>
> *Then*   set abnormal_blood_pressure_flag
>   prepare hypotensive drug

Rule 4
> *If*   <wait_response> AND <clock_time – event_time_1 ≥ specified_time_1>
> *Then*   set abnormal_state_flag
>   reset wait_response_flag

Fig. 2. Example of Rules

Expression <A> : <OK_response>
Expression <B> : <abnormal_pulse_rate>
Expression <C> : <abnormla_temperature>
Expression <D> : <blood_pressure ≥ upper_normal_pressure>
Expression <E> : <wait_response>
Expression <F> : <clock_time – event_time_1 ≥ specified_time_1 >

	<A>	<B>	<C>	<D>	<E>	<F>
Rule 1	(value)	empty	empty	empty	empty	empty
Rule 2	empty	(value)	(value)	empty	empty	empty
Rule 3	empty	empty	empty	(value)	empty	empty
Rule 4	empty	empty	empty	empty	(value)	(value)

Fig. 3. Stored Rule's LHS in Rule-Base

2.3 Modified CSN

The modified CSN is located between the input buffer and WM, which is used for configuring WM. In the general case, the cross point switch, which can be only one intersection of row bus line and column bus line, can configure dynamic connection of a network, which is a dedicated bi-directional connection between an input and an output. In our system, the dedicated connection between the input buffer and WM is uni-directional. As Fig. 1 indicates, stored data within the input buffer is only transferred to WM by CSN, but the opposite transfer does not occur. The modified CSN can support multi-point switching on an input bus line, so that the same input data can be transferred to multiple outputs simultaneously. In the case that multiple outputs need of an input data, several switching operations are required in general CSN. But, in our CSN, it can be done with one switching operation. This multi-point switching can reduce constraints of data fetch scheduling which incur performance degradation.

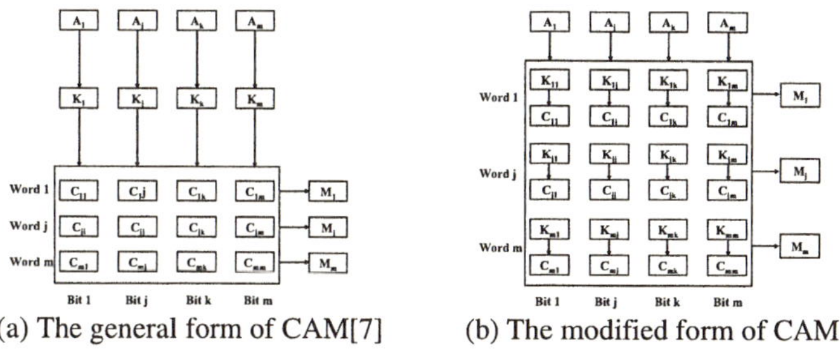

(a) The general form of CAM[7]   (b) The modified form of CAM

Fig. 4. The Content Addressable Memory

The results of condition terms (i.e. <blood_pressure> ≥ <upper_normal_pressure> in Fig. 2 are required by WM. In Fig. 1, the PPMs are used to compute these results of condition terms. These PPMs are arithmetic logic units that can compute such opera tor as +, -, ≥, ≤, ≠ and ==. For selection of operator, the operator selection signal is input to the CSN switching simultaneously.

2.4 Modified CAM

When the rule-base is composed in the rule-based systems, the use of CAM as the storage device can speed up the matching operation of rules.

We also used CAM for storing rules, but, the required operation of CAM differs from general CAM. As shown in Fig. 4(a), the key register can mask a part of the input bits in general CAM. Thus, these masked bits are only compared with stored contents. In our real-time rule-based system, WM has all expressions that can configure all rule's LHS, but only some of them are used for making a rule. Thus CAM in our system requires the opposite masking operation of general CAM. For example, if WM has 6 expressions and rule's LHS is stored as shown in Fig. 3, empty expressions in rule's LHS should be excluded from the matching operation. Namely, only expressions with values should be masked in memory cells not on the input side. Thus, we modified the architecture of CAM as shown in Fig. 4(b). The key register location differs from that of general CAM.

3 Variable Binding Method

Due to the slow processing speed, using a rule based systems in real-time application has been inadequate. The main reason was that the matching operation consumes around 80% of the total processing time (and about 80% of the operation time is spent on the variable binding process). There are several algorithms to reduce processing time of the matching operation like the RETE algorithm. The RETE algorithm solves the problem using the RETE network, in which repeated computing is removed. However, software algorithms have inherent drawback by their sequential characteristic. Thus, we propose a new hardware structural approach by which the processing time for variable binding can be considerably reduced. Beginning from an example, WM and rule's LHS are shown in Fig. 5, where X0, X1, Y0 and Y1 are variables. A vari able in rule's LHS can address unique memory location of its own variable CAM as shown in Fig. 5. Then the variable binding module copies corresponding expression values in WM to the variable CAM as shown in Fig. 5. And the stored values in its variable CAM for each variable are checked whether they are all the same or not. In the case of the example in Fig. 5, due to the fact that the values of X0 and X1 are not the same, variable binding of the rule fails. The required size of hardware logic for determining whether all stored values in a variable CAM are the same or not can be neglected.

4 Experimental Results

The real-time rule-based system is designed for application to a mobile intelligent robot to help elderly people. This mobile intelligent robot requires real-time processing for awareness of current context. We used SystemC ver. 2.0 for description, simulation and verifying the real-time rule-based system.

Expression <A>	Expression <B>	Expression <C>	Expression <D>	Expression <E>	Expression <F>	Expression <G>	
red	yellow	green	gray	white	black	black	Working Memory

Expression <A>	Expression <B>	Expression <C>	Expression <D>	Expression <E>	Expression <F>	Expression <G>	
red	X0	green	X1	white	Y0	Y1	Rule's LHS

Variable CAM (X)

X3	empty
X2	empty
X1	gray
X0	yellow

Variable CAM (Y)

Y3	empty
Y2	empty
Y1	black
Y0	black

Not Match !

Fig. 5. Example of Unsuccessful Variable Binding

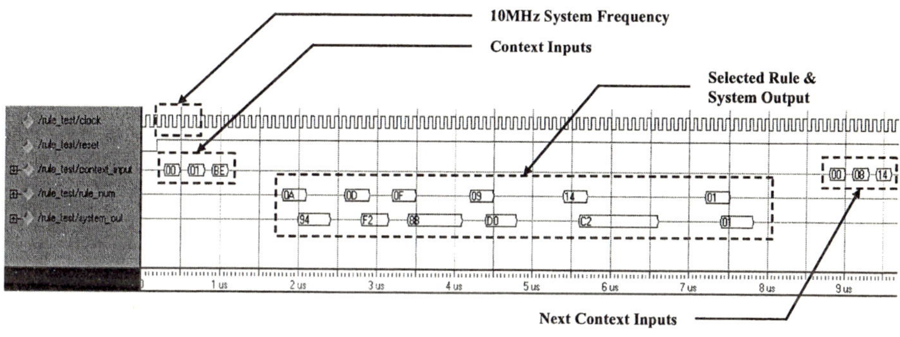

Fig. 6. Simulation Results of SystemC ver. 2.0

The programmed code written with SystemC can be transformed to synthesizable HDL code which is independent of the technology library used by the Synopsys SystemC compiler and the Design compiler. Also, the code can be extended by only changing parameters in the header files. Fig. 6 and Table 1 shows the simulation results. For the experiment, the designed real-time rule-base system has 100 words in the input buffer, 50 words in WM and a CAM size of 2.8 kbytes for the rule-base. In the case of system operation with a 10MHz system clock, it can reason an average of 102,000 times per second.

Table 1. Performance estimation results with 56 rules

Operation environment		CPU usage		Elapsed time
Pentium-4 2.4GHz	JESS (Java Expert System Shell)	(OS + JESS)	7 %	1.0 [sec]
		(assumed)	100 %	70.0 [msec]
	Optimized C-code		100 %	3.2 [msec]
Our system 10MHz	SystemC ver. 2.0 Platform	-		9.8 [μsec]

5 Conclusion

In this paper, we proposed an SoC architecture for rule-based system that can be applied to real-time applications like a intelligent mobile robot. This system was described using SystemC ver. 2.0 and tested with 56 rules to reason current context. According to experimental results, it has sufficient processing power to manage current context. And the use of modified CAM and CSN PPMs can improve system performance compared to previous architectures.

Acknowledgement

This research was performed for the Intelligent Robotics Development Program, one of the 21st Century Frontier R&D Programs funded by the Ministry of Commerce, Industry and Energy of Korea.

References

1. Kanter, T.G.: Attaching context-aware services to moving locations Internet Computing. IEEE, Volume: 7, Issue: 2, March-April 2003 Pages:43 – 51
2. C.L. Forgy: RETE: A Fast Algorithm for the Many Pattern/Many Object Pattern Match Problem. Artificial Intelligence (1982) vol. 19 17-37.
3. A. Gupta, C. Forgy, A. Newell: High Speed Implementations of rule-based systems. ACM Transactions on Computer Systems, Volume 7 Page 119-146, 1989
4. Pratibha and P.Dasiewicz: A CAM Based Architecture for Production System Matching. reprinted in VLSI for Artificial Intelligence and Neural Networks, edited by J.G. Delgado-Frias and W.R. Moore, Plenum Press (1991) 57-66.
5. Dou, C.: A highly-parallel match architecture for AI production systems using application-specific associative matching processors. Application-Specific Array Processors, Proceeding, International Conference (1993) 180 – 183
6. K. M. LEE, et al.: An SoC-based Context-aware System Architecture. Lecture Notes in Computer Science, Volume 3215 / 2004, KES 2004, Wellington, New Zealand, September 20-25, 2004, Proceedings, Part III
7. M. Morris Mano: Computer System Architecture 3rd ed. Prentice Hall (1997) 456-768

Mobile Agent System for Jini Networks Employing Remote Method Invocation Technology

Sang Tae Kim, Byoung-Ju Yun, and Hyun Deok Kim

Dept. of Electronic Engineering, Kyungpook National University,
Daegu 702-701, South Korea
hyundkim@ee.knu.ac.kr

Abstract. A mobile agent system employing a Remote Method Invocation(RMI) technology has been demonstrated for a resource-limited mobile device. The agent system facilitates the mobile device to support Jini services without any additional client program installation in it. It also provides a dynamic service list, which facilitates the users to search and to utilize the various services supported on a Jini network in a real time through a web browser of the mobile device.

1 Introduction

As the mobile devices such as Portable Digital Assistants(PDAs) and cellular phones become more powerful enough to run much more sophisticated applications, the internetworking services using the devices become more important. An intelligent mobile device should be able to interact with other mobile devices or even with non-mobile devices anytime and anywhere in the world. Especially, it becomes more important to interact a mobile device with electronic devices or appliances in a house for a seamless home network services [1]. The middlewares such as Jini and UPnP play a key role for the internetworking between the devices with different operation systems and different protocols [2],[3].

Based on the Jini network technology, the users can explore the whole advantages of the home network where the various appliances such as a refrigerator, a TV, and a washing machine and the peripherals of a personal computer such as a printer and a scanner are interconnected on a single network. However, it is not easy to apply the middlewares developed for the home network directly to the mobile devices since the resources of the mobile device such as a processing power and a memory size are inherently insufficient. Namely, most of the middlewares of the home networks have been developed originally as a part of the desktop applications and it cannot be applicable for the resource-limited mobile devices. Thus, it is important to develop middlewares suitable for the resource-limited mobile devices to realize a networking between a mobile device and other electronic devices and home appliances.

In this paper, we demonstrate a mobile agent system for a Jini network by employing a Remote Method Invocation(RMI) technology. The agent system enables the resource-limited mobile device to participate in a Jini network without

R. Khosla et al. (Eds.): KES 2005, LNAI 3681, pp. 1021–1027, 2005.

any additional client program installation in it. A couple of examples such as a file search service and a printing service have been demonstrated based on the agent system.

In the section 2, we will briefly review the Jini network technology. After that we will describe our mobile agent system and the experimental results in section 3 and section 4, respectively. The conclusion will be given in section 5.

2   Jini Network Technology

The Jini network technology is the next-generation connection technology developed by Sun Microsystems. It is built around the Java programming language and environment and provides an object-oriented distributed network where the devices and the applications are treated as services. The users can easily utilize the services through the Jini network without any changing the configurations of the client. In addition, there is no need to install any additional programs such as device drivers into the client. The Jini network supports a plug and work operation when a device or an application is added to the network.

The Jini network component is composed of a service provider connected to Jini network, a service client utilizing the service and a lookup service as shown in Fig. 1. The service provider can provide a Java object also known as a proxy and the lookup service manages, operates and controls the Jini network [4].

If a new service provider participates in the network, the service provider searches the lookup service by using a discovery protocol. At the end of a successful discovery, the service provider requests a registration to the lookup service and then the lookup service returns a registrar, a Java object, to the ser-

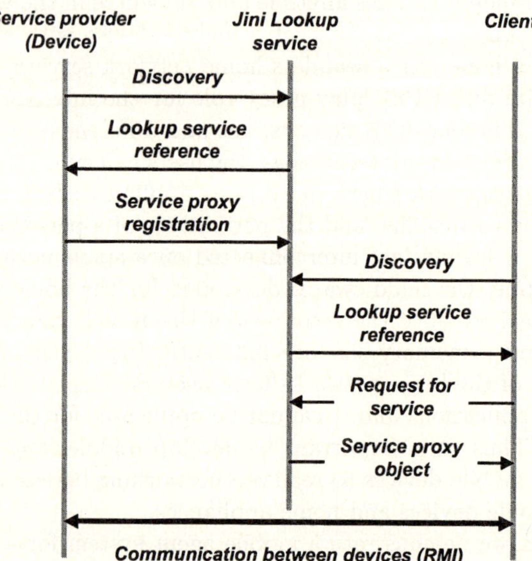

Fig. 1. Jini network components

vice provider. When the client wants to utilize a service provided by a service provider, it directly communicates with the service provider through a Java RMI and then the client utilizes the Java service [5]-[7].

The Jini network technology is originally developed for a desktop environment and implemented by using a Java Virtual Machine(JVM). However,the mobile devices such as cellular phones and PDAs has insufficient resources and thus have several problems to perform the operations related a floating point, a reflection and a error processing. Namely, it is difficult to apply the Jini technology to a mobile device since the mobile device may not support the JVM. Thus, it is important to develop a middleware which enables a mobile device to participate in the Jini network even when the resources of the devices is insufficient.

3 Implementation

To utilize a Jini network services with a resource-limited mobile devices, we have implemented a mobile agent system. The agent system supports a Jini network technology to the mobile devices without any additional client program installation in it. It also provides a dynamic service list, which facilitates the users to utilize the Jini network services in a real time through a web browser of the mobile device.

The configuration of Jini network with the mobile agent system is shown is in Fig. 2. The Jini network consists of four parts, a lookup service, mobile agent, a service provider and a client and the agent system consist of a dynamic list and a servlet.

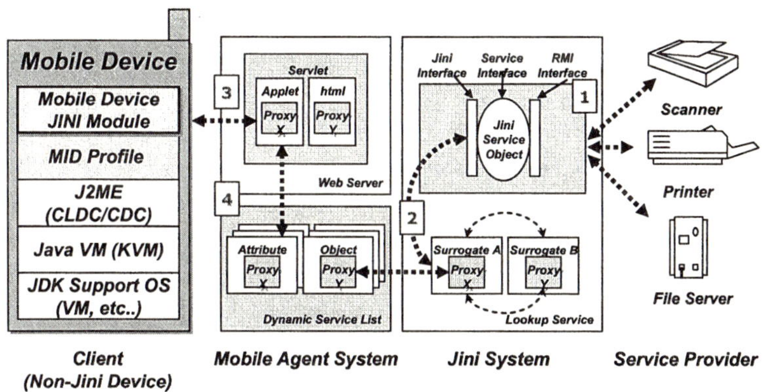

Fig. 2. Configuration of the Jini network with the mobile agent system

The operation principle is as follows. The service registration procedure of the service provider is same as that of the conventional Jini network technology. If any service is registered in a lookup service, however, the service is also registered in the dynamic list of the agent system. When the client wants to use a service provided by the service provider, it searches the dynamic list of the agent system through the servlet of the agent system and then the client is connected to the service provider via the agent.

The agent communicates with the service provider through a Java RMI and the service provider can provide the Java service to the client via the agent system. The agent system sends and receives data based on Java. Thus the mobile devices requires a virtual machine to communicate with the agent system. We used Java Micro Edition(J2ME) on the mobile devices and Kilobyte virtual machine(KVM) or Classic virtual machine(CVM) as a virtual machine [8]. It is notable that the J2ME is designed for the devices with a smaller memory size and a lower processing power and does not support Jini network service. Thus the role of the J2ME on the mobile device is only to process the data based on Java and the enabler of the Jini network service with the mobile is the mobile agent system.

The Jini network technology provides a plug and play operation in the distributed network, which is a very important feature since it facilitates the interworking between devices. Namely, the user can easily add or remove devices without any information on the device in Jini network. The mobile agent system is designed to support a plug and play operation. Thus, the agent system provides a dynamic service list and the users can explore the services available in a real time through a web browser of the mobile device.

4   Result

To evaluate the validity of the proposed mobile agent system we implemented a Jini network with a mobile agent system. The client(mobile device) was a PDA(*Compaq iPAQ PocketPC H*3600) and the service providers were a printer, a scanner and a ftp server.

The user interface of the mobile agent system is shown in Fig. 3. The user interface is a web browser and the web server is implemented by using a servlet, which enables the client to participate in the Jini network through a web browser. It is notable that we have the client to support a Java applet by using Jeode VM though the client was a non-Jini device and the web browser of the client did not support a Java applet [9].

The mobile agent system is developed to provide a dynamic service list in the client screen and thus one can easily searches services available in the Jini network.

The source code to implement a dynamic service list is as follows:

```
import java.util.ArrayList;
public class Client implements Runnable
{

  // The Vector declaration for stores a service list

    private java.util.List vec = new ArrayList();
```

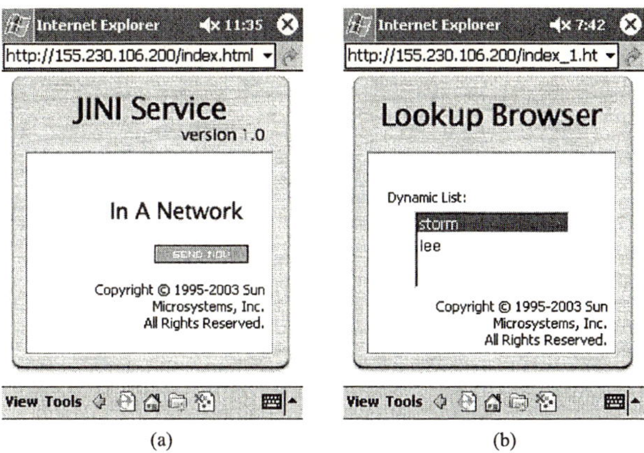

(a)   (b)

Fig. 3. The user interface of the mobile agent system (a) The initial state, and (b) The dynamic list describing available Jini services

```
class Discoverer implements DiscoveryListener
{

  // A list of the registered services in the Lookup service

  vec.add(newregs[i].getLocator().getHost()+" ");
   for(int j = 0; j < vec.size(); j++) {
       dic += (String)vec.get(j);
   }
 }
}
```

We have demonstrated two exemplary services, a file search service and a fiber printing service.

The result of the file search service is shown in Fig. 4. The service provider was a desktop computer and we can search files stored in the service provider from the client(mobile device). After reading a particular file(test.txt), we have the file to be printed by another service provider(a printer). The result of file printing service is shown in Fig. 5. When the file printing service has been completed the printer server sends a message to the Jini Lookup service to inform the client of the completion of the service.

Servlet receives Proxy and Service ID of Print service from Jini-Agent registered Jini Lookup Service and search Print service matched service which is user wanted. It gets an appropriate attribute about printer service and delivers the contents of specific document file with printer server. When the data is delivered successfully with the printer and printing is accomplished from the printer, and then printer server delivers a response message to Jini Lookup server. Finally, it prints contents of selected file by activating Printing Service. Fig. 5 shows screen which receives a response message.

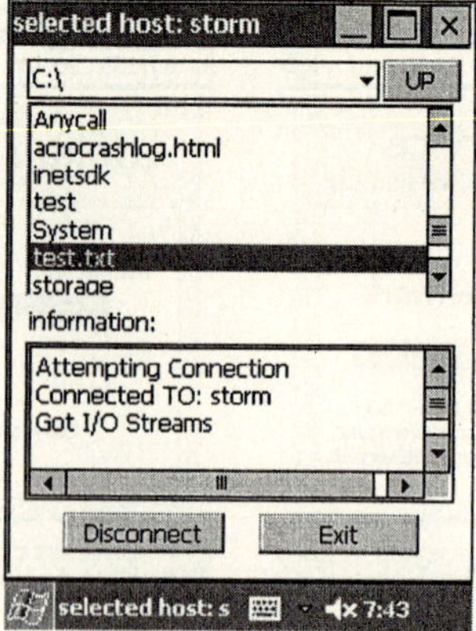

Fig. 4. File search service

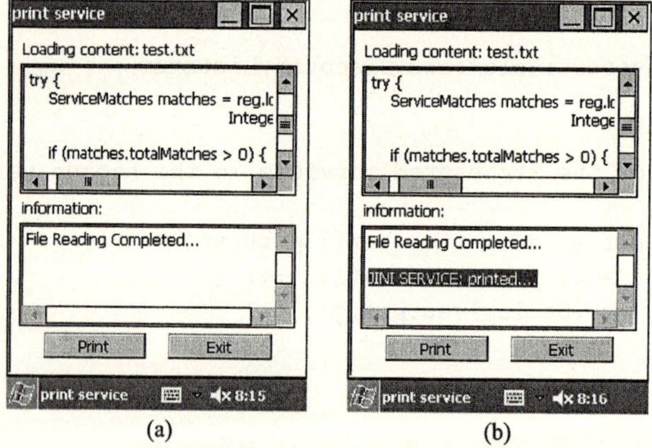

(a)   (b)

Fig. 5. Printing service (a) Loading content of file (b) contents of selected file by activating printing service

5   Conclusion

A mobile agent system employing a Remote Method Invocation(RMI) technology has been demonstrated to provide Jini network services to the resource-limited mobile device. The Jini network technology support a plug and play operation in a distributed network, which facilitates the interworking between devices based on different operating systems and different protocols. However, since the Jini network technology is originally developed for a desktop environment application, it is difficult to apply the technology to a mobile device with an insufficient resource.

The mobile agent system enables the mobile device to participate in a Jini network without any additional client program installation in it. It provides a dynamic service list, which facilitates the users to search and to utilize the various services available on a Jini network through a web browser of the mobile device.

The the agent system communicates with the service provides through a Java RMI, while it sends/receives only service identities(IDs) and commands to/from the mobile devices. Sice the mobile devices communicate with the devices on the Jini network via the agent system, therefore, the agent system does not require any wrapper code. The agent system enables an internetworking between a resources-limited mobile device and other electric devices one a Jini network.

Acknowledgement

This work was partially supported by the Mobile Technology Commercialization Center(MTCC) and the Brain Korea 21(BK21) Project.

References

1. Landis, S.: Reaching out to the cell phone with Jini. System Sciences (2002)
2. Stang, M.: Enterprise computing with Jini technology. IT Professional (2001)
3. Miller, B.A.: Home networking with Universal Plug and Play. IEEE (2001)
4. Sing, Li.: Professional Jini. Wrox Press Ltd (2000)
5. Campadello, S., Koskimies, O., Raatikainen, K.: Wireless Java RMI. IEEE (2001)
6. Courtrai, L.: Java objects communication on a high performance network (2001)
7. Gupta, R., Talwar, S., Agrawal, D.P.: Jini home networking. IEEE (2002)
8. Helal, S.: Pervasive Java, Part II, Pervasive Computing. IEEE (2002)
9. Insignia.: Jeode Platform White Paper (2001)

Incorporating Privacy Policy into an Anonymity-Based Privacy-Preserving ID-Based Service Platform[*]

Keon Myung Lee[1], Jee-Hyong Lee[2], and Myung Geun Chun[1]

[1] School of Electrical and Computer Engineering,
Chungbuk National University, Korea
[2] School of Information and Communication Engineering,
SungKyunKwan University, Korea
kmlee@cbnu.ac.kr

Abstract. Hiding users' real ID is one of promising approaches to protecting their privacy in ubiquitous computing environment. Policy-based privacy protection is also known as another useful approach. In the previous work, we have proposed a privacy preservation platform, called AP^3(the Anonymity-based Privacy-Preserving ID-based service Platform), which allows users to use pseudonyms for hiding real IDs and enables ID-based services like buddy service at the same time in the ubiquitous environment. AP^3 does not yet support privacy policy mechanism which allows users to control the accessibility of others to themselves according to their preference. In this paper, we propose an extended version of AP^3, named PAP^3(the Policy-incorporated Anonymity-based Privacy-Preserving ID-based service Platform), which incorporates a privacy policy mechanism to control the accessibility about which application or user can get access to whom at what time and at which place. This paper presents the PAP^3 architecture and describes how to enforce privacy policy for the access control in PAP^3.

1 Introduction

Emerging ubiquitous computing environment promises *anytime, anywhere* computing services by deploying embedded sensors, sensor networks, and contextual information processing. For ubiquitous computing service, it is essential to use contextual information (such as the identity, location of user, service time, neighboring objects, and so on) about the user in order for her not to specify the details about how to operate the facilities. Lots of contextual information about users are collected through sensors embedded in the environment and stored somewhere in the environment. Some pieces of contextual information are privacy-sensitive, and thus if someone can gain access to such information, he may figure out *when*, *where*, and *who* did *what*. Among the contextual information, the users' identity(ID) and location are most sensitive. If it is possible

---

[*] This work was supported by the Regional Research Centers Program of the Ministry of Education & Human Resources Development in Korea.

R. Khosla et al. (Eds.): KES 2005, LNAI 3681, pp. 1028–1035, 2005.

to completely hide real IDs of users, we can lessen lots of privacy concern. In practice when they use some computing services, users frequently encounter with the situations in which their IDs are asked. In the meanwhile, the locations of users may imply some aspects of privacy-sensitive information.

In the literature of ubiquitous computing, we can find several approaches to protecting users' privacy, especially, location privacy.[1-9] The policy-based approaches and the anonymity-based approaches are the representative ones for privacy protection. In the policy-based approaches, a designated server takes charge of handling access control to privacy-sensitive information based on the privacy policies.[2] In the anonymity-based approaches, applications have their own spatial service area and take care of only users who enter into their service area.[3] To protect their own privacy, users go under pseudonyms when they join in a service area. Thanks to pseudonyms, attackers come to have difficulty in associating users with their real IDs.

In the previous work[1], we introduced an anonymity-based privacy-preserving ID-based service architecture hereafter called AP^3 which enables to hide real IDs by using pseudonyms, and allows to provide ID-based services and location-based services for the authorized users and applications. In the AP^3 architecture, once a user or an application is authorized to access a user's current pseudonym, he can always access the user as long as the user does not revoke the authorization. However, it would be desirable for the users to control the accessibility to themselves according to their preference. For example, *when she is off-duty, Alice wants not to be accessed by her company colleagues even though they are authorized to be able to communicate with her while she is online. Bob wants not to be accessed by others but his family while he is in a hospital to see a doctor.* In order to provide this kind of access control, some privacy policy mechanism has to be implemented. However, AP^3 has not yet considered such a privacy policy mechanism.

In this paper, we propose an extension of AP^3, called PAP^3 (Policy-incorporated Anonymity-based Privacy-Preserving ID-based service Platform). The PAP^3 architecture enables users to define their own privacy policy to control the accessibility to themselves. This paper is organized as follows: Section 2 presents some related works on privacy preservation in ubiquitous computing. Section 3 briefly explains the AP^3 architecture and Section 4 introduces the proposed PAP^3 architecture. Finally, Section 5 draws conclusions.

2   Related Works

Privacy concern in ubiquitous computing environment has been addressed in various research works.[1-9] Myles et al.[2]'s policy-based privacy preservation method depends on a server which makes access control decision on privacy-sensitive information. For the access control, the server refers to the applications' privacy policy and the users' privacy preference. Users register to the server their privacy preference about who can use which data of them. When an application requests data from the server, it also sends its privacy policy for the data along with the request. The server contains a set of validators used to check the conformity of an application's privacy policy against the corresponding user's privacy preference. The privacy policy-based control method allows flexible access control based on various criteria such as time of the request, location, speed, and identities of the located objects. But, there are no guarantees that the server adequately protects the collected data from various attacks. The users are also asked to blindly trust the server which controls the access to their privacy-sensitive data.

Beresford et al.[3] proposed an anonymity-based privacy preservation method which is an alternative of the policy-based method. The method tries to protect the individual's privacy by depersonalizing user data. In this scheme, when a user enters into a service area of an application, she uses a pseudonym instead of her real ID. The use of pseudonyms makes it difficult for malicious applications to identify and track individuals. Due to the anonymity-based nature, it is not easy to incorporate deliberate access control like privacy policy and to provide ID-based services like buddy services, safety alert service, callback service, etc. In Lee et al.[1], we proposed AP^3 which is an anonymity-based architecture to use pseudonyms to hide real IDs but to allow ID-based services by running a white page service agent.

Grutester et al.[4] proposed a privacy preservation method to use a distributed anonymity algorithm that is implemented at the sensor network level. In their architecture, the sensors can keep track of the number of users in an area and monitor changes in real-time, and a location server collects the sensor data and publishes it to applications. Their anonymity algorithm controls the resolution of users' IDs to guarantee users to be k-anonymous, which means that every user at the moment is indistinguishable from at least $k - 1$ other users. To enable this, the method employs the special naming scheme for locations in which names are encoded into a hierarchically organized bit stream. It reports only some upper part of the bit stream when it wants to increase the level of anonymity. This approach uses pseudonyms for locations instead of using pseudonyms for users. As a matter of fact, the pseudonyms for locations are the blurred IDs of the real location IDs.

3   The AP^3 Architecture

This section briefly explains the AP^3 architecture into which we incorporate a privacy policy mechanism. The AP^3 architecture works on the following environment[1]: Two counterparts want to communicate with each other in an anonymity-based platform despite they keep changing their pseudonyms. They do not disclose their real IDs to the sensor networks to protect their privacy. While they communicate with each other, they use their pseudonyms. Their devices are assumed not to have unique IDs which the sensor networks could use to associate them with specific users. That is, there are no ways to directly bind user IDs with device IDs. There are special servers called *zone agents*, each of which takes care of a spatial service area, collects data from the sensors in their own area, communicates with user devices (i.e., user agents) joined in the area,

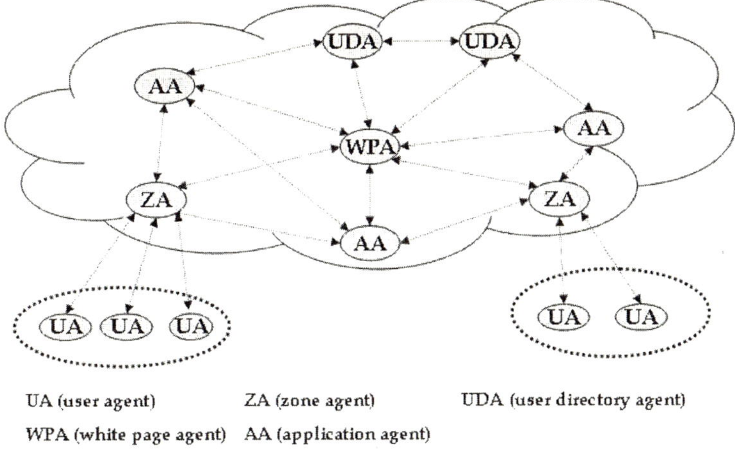

Fig. 1. The AP^3 architecture.

and relays the communication messages between user devices and other users' devices or other applications outside the area. One of the most crucial capabilities for the ID-based services in the anonymity-based platforms is to enable communicating partners to know their corresponding partners' pseudonyms in a secure way.

Figure 1 shows the AP^3 architecture which is comprised of several agents : a *white page agent*, a set of *zone agents* to take charge of physical zones, a set of *user agents* to provide user-interfaces, a set of *user directory agents* to update the pseudonym records residing in the white page agent on behalf of users, and a set of *application agents* which correspond to applications. Users register their encrypted pseudonyms with the white page agent when they change their pseudonym. Both users and applications can get the pseudonyms of correspondents from the white page agent once they have proper key so-called *friend key*. The database of the white page agent contains the records each of which is made of the tuples (real ID of user i, a list of user i's current pseudonyms encrypted with different friend keys). With the help of her user directory agent, each user comes to give unique friend keys to her authorized users and applications. In order to enable a user to update the encrypted pseudonym field of her record in the white page agent without revealing her identity, the architecture asks the i's user directory agent to register new pseudonym on behalf of user i in the following way: The i's user agent sends to the white page agent the pseudonym update request message encrypted in a secret key shared with its directory agent. Then the white page agent broadcasts the message to all directory agent without figuring out its contents. Upon the broadcasted message, multiple directory agents including the i's directory agent come to update their records residing in the white page agent. To hide who actually updates her pseudonym, some directory agents other than the i's directory agent also update their records by inserting meaningless noise in the records.

When a user (or an application) i wants to communicate with a user j, i needs to know the current pseudonym of j. At the moment, user i is assumed to know the real ID of user j. The i's user agent asks the white page agent for the pseudonym of user j. After obtaining the pseudonym, user i communicates with user j using the pseudonym. To prevent eavesdropping, the pseudonym request message is encrypted using the public key of the white page agent. The request message contains a secret key as well as several IDs including the counterpart's ID. The secret key is later used to encrypt the response message which consists of a collection of tuples (a real ID, a list of encrypted pseudonyms). In the list, each encrypted pseudonym element is a pair(friend key f_k, pseudonym) encrypted using the friend key f_k. If user i is a friend of user j, i must have a friend key for j's pseudonym. Therefore, the i's user agent can recover the pseudonym for user j from the response message.

Only friends of a user are allowed to get her pseudonym. When a user or an application i establishes a friendship with user j, the following protocol is exercised: The i's user directory agent sends the i's friendship request message encrypted by the key shared by user directory agents to the white page agent which later on broadcasts it to all user directory agents. Then j's user directory agent gets the information about i from the i's user directory agent, and asks the user j whether she accepts. If she accepts it, the j's user directory agent creates a friend key for i, informs i of the friend key through the j's user directory agent, and then updates the j's record in the white page agent so as to take care of the friend key.

4   The Proposed PAP³ Architecture

In order to provide the accessibility control for users on the AP³ architecture, we incorporate the privacy policy scheme into AP³ and propose its extended architecture named PAP³(the Policy-incorporated Anonymity-based Privacy-Preserving ID-based service Platform). The PAP³ architecture has the same architecture with AP³ except the component for handling privacy policies.

The AP³ architecture protects the privacy by hiding users' real ID but enables the authorized users and applications (i.e., friends) to access users. The PAP³ architecture is proposed to enable users to control the accessibility of their friends to themselves by specifying *who* can access them *at what time* and *at which place*. The basic mechanism employed in PAP³ for privacy policy enforcement is to make spoiled the pseudonym fields for the restricted friends.

4.1   The Privacy Policy Enforcement Architecture

To enforce privacy policies in PAP³, the user directory agents maintain their users' privacy policy and spoil the pseudonym fields for the restricted friends according to the privacy policies, if needed. It is not enough to invalidate the pseudonym fields kept in the white page agent, because the friends corresponding to the fields still have the valid pseudonym unless the user changes her pseudonym. Therefore, each time a privacy policy is exercised, the user is supposed to change her pseudonym. PAP[3] causes a user agent to initiate the policy enforcement and then its user directory agent takes action to enforce the privacy policy.

The users are supposed to express their privacy preference in the form of privacy policy which consists of a set of access control rules. The privacy policy is registered with their user directory agent and at the same time compiled into their user agent in the form of policy event profile. The user agents generate *pseudonym change events* whenever they change their pseudonym to follow the imposed policy. Due to the insufficient computing resource for a user agent running on user devices, the user agents maintain only the compiled policy event profile which tells when they should generate pseudonym change event. Regardless of privacy policy, once a user's service zone has been changed and thus her pseudonym has been changed, then a pseudonym change event is forwarded along with the new assigned pseudonym to its user directory agent. On each pseudonym change event message, the user directory agent checks whether there is some privacy policy to apply. In the meanwhile, the policy event profile consists of information about times at which some privacy policy should be checked. A policy event profile for user agents is expressed in the list of policy check times e.g., (9:00, 12:00, 17:00).

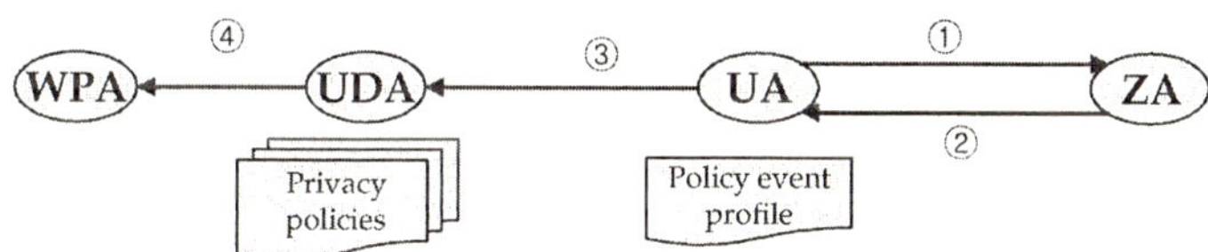

Fig. 2. The Privacy Policy Enforcement Flow.

Figure 2 shows how the privacy policy is enforced in PAP[3]. User agents periodically check their policy event profile to see whether there is a policy to follow at the moment. If any, it finishes all its communication connection to others and asks its current zone agent for a new pseudonym.(Step 1) Then, the zone agent assigns a new pseudonym to the user agent.(Step2) After that, the user agent sends a pseudonym change event along with the new pseudonym to its user directory agent.(Step 3) When a user directory agent receives a pseudonym change event from its user agents(Step 4), it performs the following tasks:

1. Check the user's privacy policy and determine the friends to be blocked from the pseudonym access.
2. Place the *access denied* message in the corresponding field of the blocked friends instead of the current pseudonym.
3. Encrypt each field with the corresponding friend key and build a record consisting of those fields.
4. Send the updated record to the white page agent according to the AP[3]'s pseudonym update protocol[1].

Table 1. An Example of Privacy Policy.

	Rule 1	Rule 2	Rule 3	Rule 4
Who	Bob	family	boss	relative
When	weekday 9:00-17:00	always	weekday on-duty	weekday daytime
Where	public place	anywhere	public place	public place, home

When a friend is prohibited from accessing the user, he cannot get the user's valid pseudonym from the white page agent even though he has a valid friend key.

4.2   Privacy Policy Representation

A user's privacy preference is expressed in privacy policy consisting of access control rules which specify who can get the user's valid pseudonym at which place and at what time. Table 1 shows an example of privacy policy for Alice, where Rule 1 tells that *Bob can access Alice when she is in some public place from 9:00 to 17:00 on weekday.* Rule 2 tells that *her family members can access Alice any time, any place.* At the communication setup phase, each user is supposed to send her real ID to the communicating counterpart so as to be used for the privacy policy check, even though she masquerades under a pseudonym. In the meanwhile, it is not straightforward to use class terms like *family, relative* and so on to designate a group of people in policy description. As in Table 1, the usage of these class terms make it easy for users to describe their privacy preference. Such a convenience is also expected in the description of locations and times for access control rules. Sometimes there are special relationships among class terms like *subclass, instance of*, and the like and it would be very useful to use such information in the policy description. Figure 3 shows a part of relationships among partner classes. This kind of information can be regarded as an ontology for the domain. Therefore, it is expected that the ontology for *partners, locations*, and *times* would give great convenience in policy description.

Each user has her own personal relationship with others. Hence, she needs to write down such information for herself when she writes her own privacy policy. Suppose that *Alice is the spouse of Daniel.* From this fact along with

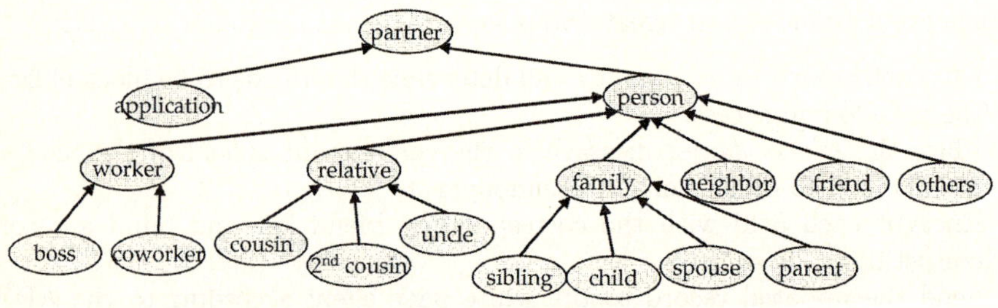

Fig. 3. A part of class relationship for partners.

the relationship information shown in Fig. 3, it can be inferred that *Daniel is a family member of Alice*. Therefore, when Daniel wants to access Alice, he can get the valid pseudonym of Alice by the Alice's access control rule shown in Rule 2 of Table 1. To support this kind of inference, the user directory agents should have an inference mechanism.

5   Conclusions

The proposed PAP^3 architecture is an extension of the AP^3 architecture, which allows users to use the privacy policy for deliberate access control based on the communicating partners, user's location, and access times. To provide such service, PAP^3 takes the strategy to assign a new pseudonym to user and to spoil pseudonym fields of the white page agent for restricted friends each time there needs to enforce some privacy policy. It is expected that the use of ontology for partners, locations, and times is very convenient in privacy policy description. Due to lack of any available physical anonymity-based ubiquitous computing environment, the proposed architecture is not yet implemented. As the further works, therefore, there remain to develop a simulation environment for the assumed ubiquitous computing environment and to implement the PAP^3 architecture on it, and then to define the protocols for PAP^3 in a detail specification. It is also an interesting issue to define the comprehensive set of an ontology for the partners, locations, and times for privacy policy description.

References

1. K.M. Lee, S.H. Lee. A Multiagent Architecture for Privacy-Preserving ID-based Service in Ubiquitous Computing Environment. *Lecture Notes in Artificial Intelligence*. 3339. (2004). 14-25.
2. G. Myles, A. Friday and N. Davies. Preserving Privacy in Environments with Location-Based Applications. *IEEE Pervasive Computing* 2(1). (2003). 56-64.
3. A. R. Beresford and F. Stajano. Location Privacy in Pervasive Computing. *IEEE Pervasive Computing* 2(1). (2002). 46-55.
4. M. Gruteser, G. Schelle, A. Jain, R. Han and D. Grunwald. Privacy-Aware Location Sensor Networks. http://systems.cs.colorado.edu/Papers/Generated /2003PrivacyAwareSensors.html (2003).
5. X. Jiang and J. Landay. Modeling Privacy Control in Context-aware Systems. *IEEE Pervasive* 1(3). (2002).
6. M. Langheinrich. A Privacy Awareness System for Ubiquitous Computing Environments. In *Ubicomp 2002*. (2002).
7. S. Lederer, A. K. Dey, J. Mankoff. Everyday Privacy in Uniquitous Computing Environment. In *Ubicomp 2002 Workshop on Socially-informed Design of Privacy-enhancing Solutions in Ubiquitous Computing*. (2002).
8. A. R. Prasad, P. Schoo, H. Wang. An Evolutionary Approach towards Ubiquitous Communications: A Security Perspective. In *Proc. of SAINT 2004: The 2004 Symposium on Applications & Internet*. (2004).
9. P. Osbakk, N. Ryan. Expressing Privacy Preferences in terms of Invasiveness. In *Position Paper for the 2nd UK-UbiNet Workshop(University of Cambridge, UK)*. (2004).

A Quantitative Trust Model
Based on Multiple Evaluation Criteria

Hak Joon Kim[1] and Keon Myung Lee[2,*]

[1] Division of Multimedia Information, Howon University, Korea
[2] School of Electric and Computer Engineering,
Chungbuk National University, Korea
`kmlee@cbnu.ac.kr`

Abstract. This paper is concerned with a quantitative computational trust model which takes into account multiple evaluation criteria and uses the recommendation from others in order to get the trust value for entities. In the proposed trust model, the trust for an entity is defined as the expectation for the entity to yield satisfactory outcomes in the given situation. Once an interaction has been made with an entity, it is assumed that outcomes are observed with respect to evaluation criteria. When the trust information is needed, the satisfaction degree, which is the probability to generate satisfactory outcomes for each evaluation criterion, is computed based on the outcome probability distributions and the entity's preference degrees on the outcomes. Then, the satisfaction degrees for evaluation criteria are aggregated into a trust value. At that time, the reputation information is also incorporated into the trust value. This paper presents in detail how the trust model works.

1 Introduction

Along with the widespread Internet applications such e-commerce, P2P services and so on, the users have to take some risks in doing transactions with unknown users or systems over the Internet. In everyday life, we estimate the trust degree in the others by collecting the past interaction experience with them and sometimes referring to the reputation, i.e., word of mouth. By the same token, an on-line entity could reduce the risks to run with the help of the trust information for the interacting entity. Even though there have been proposed various trust models[1-9], there are no models yet generally accepted. Some models are qualitative models[1] and others are quantitative models[2, 6–8]. Some models depend only on users' ratings to compute the trust value, and others get the trust values by observing the behaviors of the entity over some period.

The trust has been defined in a various way because there are no consensus on what constitutes the trust[1-9]. The following is the Gambetta's[4] which is a well-known definition of trust: *Trust(or, symmetrically, distrust) is a particular*

---

* This work was supported by the Regional Research Centers Program of the Ministry of Education & Human Resources Development in Korea. Corresponding author

R. Khosla et al. (Eds.): KES 2005, LNAI 3681, pp. 1036–1043, 2005.

level of the subjective probability with which an agent will perform a particular action, both before [we] can monitor such action (or independently of his capacity of ever to be able to monitor it) and in a context in which it affects own action. In the trust models, the following three types of trust are usually considered: situational trust, dispositional trust, and general trust. Situational trust (a.k.a., interpersonal trust) is the trust that an entity has for another entity in a specific situation. Dispositional trust (a.k.a, basic trust) is the dispositional tendency of an entity to trust other entities. General trust is the trust of an entity in another entity regardless of situations. On the meanwhile, the reputation is valuable information for estimating the trust for an entity.

The trust of an entity can be differently measured according to which aspects we evaluate. For example, suppose that we are interested in the trust for a restaurant. Some people may be concerned about the taste of the food. Some people may pay more attention to the waiting time to get a table. Others may also have interest in the availability of their favorite food. They may also be concerned for all these things and put probably different weights on them. Therefore we proposed a trust model to consider multiple evaluation criteria and to enable entities to reflect their preference on the outcomes. In the model, we define the trust for an entity as the probability of the entity to produce satisfactory outcomes in the given situation. Once an interaction has been made with an entity, it is assumed that outcomes are observed for evaluation criteria. When the trust information is needed, the satisfaction degree which is the probability to generate satisfactory outcomes for each evaluation criterion, is computed. Then, the satisfaction degrees for evaluation criteria are aggregated into a trust value. At that time, the reputation information from others is also combined into the trust value.

The remainder of this paper is organized as follows: Section 2 briefly presents several related works on trust models. Section 3 introduces the proposed trust model and Section 4 draws the conclusions.

2   Related Works

Trust and reputation have gained great attention in various fields such as economics, distributed artificial intelligence, agent technology, and so on. Various models for trust and reputation have been suggested as a result[1-9]. Some models just give theoretical guidelines and others provide computational models.

Abul-Rahman et al.[1] proposed a qualitative trust model where trust degrees are expressed in four levels such as *very trustworthy, trustworthy, untrustworthy,* and *very untrustworthy.* The model has somewhat ad-hoc nature in defining the trust degrees and the weights.

Azzedin et al.[2] proposed a trust model for a peer-to-peer network computing system, which maintains a recommender network that can be used to obtain references about a target domain. The model is specialized for the well-structured network computing system and thus there are some restrictions on applying the model to general cases.

Derbas et al.[6] proposed a model named TRUMMAR which is a reputation-based trust model that mobile agent systems can use in order to protect agents from malicious systems. The model pays special attention to use reputation for trust modeling, but does not consider the multiple evaluation criteria.

Shi et al.[7] proposed a trust model which uses the statistical information about the possible outcome distribution for actions. In the model, trust is described as an outcome probability distribution instead of a scalar value. When choosing a candidate, it is used to compute the expected utility value for the candidate entities' actions.

Wang et al.[8] proposed a trust model based on Bayesian networks for peer-to-peer networks. In the model, a Bayesian network is used to represent the trust between an agent and another agent. Such a Bayesian network represents the probabilities to trust an entity in various aspects. The recommendation values from other entities also are incorporated into the model.

3   The Proposed Trust Model

In the literature, there is no consensus on the definition of trust and on what constitutes trust management. For the purpose of getting a computational trust model, however, we define the situational trust as follows: *Situational trust is the expectation for an entity to provide satisfactory outcomes with respect to the evaluation criteria in a given situation.* This section proposes how to evaluate the situational trust based on the above definition of situational trust , how to handle the dispositional trust and the general trust, and how to use recommandation from others and to adjust the recommanders' trust.

3.1   Situational Trust

The situational trust is the trust assigned to an entity for a specific situation. So, the trust value depends on the situations. Most existing approaches have interest in how much the considered entity's behaviors are satisfactory. It is assumed in their methods to rate the satisfaction degree in a single perspective and it is somewhat ad-hoc on how to rate the satisfaction degrees for an entity to other entities in a specific situation. In order to provide a comprehensive computational model, we propose a method to evaluate situational trust which takes into account several evaluation criteria and the trusting entity's preference. In the proposed method, the situational trust is computed as follows: First, an entity α accumulatively constructs the empirical probability distribution of possible outcomes for the interacting entities β with respect to each evaluation criterion in the given situation. Each time the entity α needs to measure the trust in another entity β, she computes the satisfaction degrees with β over each evaluation criterion in the situation. After that, the situational trust of α in β is determined by aggregating the satisfaction degrees in terms of evaluation criteria.

Let $TS_\alpha(\beta, \delta; EC)$ be the situational trust of entity α in entity β in the situation δ with respect to evaluation criteria $EC = \{ec_1, ec_2, ..., ec_n\}$, where

ec_i is an evaluation criterion. It expresses the degree of expectation for trusted entity β to yield satisfactory actions with respect to the evaluation criteria in the given situation.

In order to get the situational trust, whenever entity α has an interaction with β, α keeps the records about the evaluation outcomes with respect to the evaluation criteria. The evaluation outcomes are given in either continuous values or categorical attributes. In the case of continuous outcomes, the outcome domain is quantized into several prespecified intervals and outcome values are expressed in the corresponding interval labels.

Empirical Outcome Probability Computation. Let $PG^t(\alpha, \beta, \delta, o_i; ec_k)$ be the α's observational probability for entity β to make outcome o_i in the situation δ up to time t with respect to an evaluation criterion ec_k. The probability is basically computed as follows: Here, $N^t(o_i)$ is the number of outcome o_i for ec_k so far time t and n is the number of total interactions.

$$PG^t(\alpha, \beta, \delta, o_i; ec_k) = \frac{N^t(o_i)}{n} \tag{1}$$

The situational trust measure of Eq. (1) accumulates all past experience and thus it is possible to retain low level of trust in an entity due to her old bad behaviors, and vice versa. Therefore, it would be desirable to forget the very old experience with entities. In order to put the forgetness effect for the old experience, it maintains a sequence of observations happened within the recent time window $[t - dt, t]$ at time t. The observation sequence is constructed in an ordered list $OL(ec_k)$ of pairs of an observation value o_i and its timestamp t_i, $OL(ec_k) =< (o_i, t_i)|\ t_i \in [t - dt, t] >$.

Let $PL^t(\alpha, \beta, \delta, o_i; ec_k)$ be the probability of outcome o_i for ec_k within time window $[t - dt, t]$ which is computed as follows: Here $n_{[t-dt,t]}$ denotes the total number of α's interactions with β in the time window $[t - dt, t]$, and $N^{[t-dt,t]}(o_i)$ is the number of outcome o_i in the window $[t - dt, t]$.

$$PL^t(\alpha, \beta, \delta, o_i; ec_k) = \frac{N^{[t-dt,t]}(o_i)}{n_{[t-dt,t]}} \tag{2}$$

At each interaction, the new empirical probability value $p^t(\alpha, \beta, \delta, o_i; ec_k)$ is determined by taking the weighted sum of the global empirical probability value PG^t and the local empirical probability value $PL^t(\alpha, \beta, \delta, o_i; ec_k)$ as follows: Here ρ indicates the weighting factor to reflect the ignorance effect on the past experience.

$$p^t(\alpha, \beta, \delta, o_i; ec_k) = \rho * PG^t(\alpha, \beta, \delta, o_i; ec_k) + (1 - \rho) * PL^t(\alpha, \beta, \delta, o_i; ec_k) \tag{3}$$

Then these quantities are normalized to preserve the property of probability measure as follows:

$$P^t(\alpha, \beta, \delta, o_i; ec_k) = \frac{p^t(\alpha, \beta, \delta, o_i; ec_k)}{\sum_{o_j} p^t(\alpha, \beta, \delta, o_j; ec_k)} \tag{4}$$

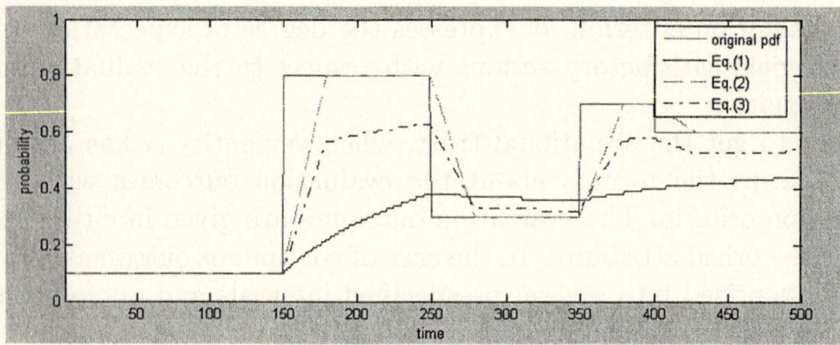

Fig. 1. Empirical Outcome Probability.

Figure 1 shows the relationship between the imaginary original probability density function(original pdf) for an outcome, its global empirical probability computed by Eq.(1), its local empirical probability computed by Eq.(2) and its combined probability with the ignorance effect computed by Eq.(3). In the simulation, it was assumed that the probability is measured at each unit time interval and the local empirical probability is computed by averaging the most recently measured 30 probability values. Figure 1 indicates that the combined probability could give more meaningful information about the outcome probability than the global empirical probability or the local empirical probability alone.

Satisfaction Degree Computation. The trusting entity α makes her mind on which outcomes are satisfactory for her own preference with respect to each evaluation criterion. For an evaluation criterion ec_i, suppose that its possible outcome is $PO(ec_i) = \{o_{1i}, o_{2i}, ..., o_{ni}\}$. Then, the entity α specifies earlier on her satisfactory outcome set $SO(\alpha, ec_i)$ along with the relative preference for each outcome as follows: $SO(\alpha, ec_i) = \{(o_1, ow_1), ..., (o_j, ow_j)\}$ where $o_i \in PO(ec_i)$ and $wo_k \in [0, 1]$ is the α's relative preference for the outcome o_k. The satisfaction degree $SD_\alpha(\beta, \delta; ec_i)$ of α with β with respect to ec_i is determined as follows:

$$SD_\alpha(\beta, \delta; ec_i) = \sum_{(o_k, wo_k) \in SO(\alpha, ec_i)} wo_k \cdot P^t(\alpha, \beta, \delta, o_k; ec_i) \qquad (5)$$

Situational Trust Computation. In the proposed method, the situational trust is measured by the satisfaction degrees of an entity α with other entity β with respect to multiple evaluation criteria EC. For example, when a user determines the trust in a restaurant, she considers her satisfaction degrees with it in the point of her own criteria such as taste of food, waiting time to take a table, availability of her favorite food, and so on. The situational trust of α in β in situation δ with respect to evaluation criteria EC is computed as follows: Here, wc_i is the relative importance that α weighs for the evaluation criterion ec_i.

$$TS_\alpha(\beta, \delta; EC) = \sum_{ec_i \in EC} wc_i \cdot SD_\alpha(\beta, \delta; ec_i) \qquad (6)$$

3.2   Dispositional Trust

The dispositional trust TD_α represents the dispositional tendency of trusting entity α to trust other entities. Each entity is supposed to assign her own dispositional trust value. It could be used as the initial general trust when an entity starts an interaction with a new entity.

3.3   General Trust

The general trust $TG_\alpha(\beta)$ of entity α in entity β is the trust that α has on β regardless of situations. It plays the role of initial situational trust for β in a new situation. It can be used as the reputation weight for β at the beginning. It can be also used as the situational trust value while enough interactions have not yet made. The general trust of α in β is obtained by averaging the situational trusts for the experienced situations Φ as follows:

$$TG_\alpha(\beta) = \frac{\sum_{\delta \in \Phi} TS_\alpha(\beta, \delta; EC)}{|\Phi|} \tag{7}$$

3.4   Reputation

When an entity decides whether it starts an interaction with another entity, it is valuable to refer to available reputation of the entity. A reputation is an expectation about an entity's behavior which is formed by the community having interacted with the entity based on the information about or the observations of its past behaviors.

When the recommenders γ are available for an entity β in a situation δ, the entity α might take into account the recommendation for β. Each entity could have different preference on the outcomes with respect to the evaluation criteria and thus it is not so meaningful to directly use the trust values for β from the recommenders γ. Therefore we take the approach to take the recommender γ_j's empirical outcome probabilities $P^t(\gamma_j, \beta, \delta, o_k; ec_i)$ for β instead of γ_j's trust value $TS_{\gamma_j}(\beta, \delta; EC)$ on β. With the received outcome probabilities, the satisfaction degrees $SD^r_{\gamma_j}(\beta, \delta; ec_i)$ from γ_j are computed as follows: Here wo^α_k is the preference degree of the entity α for the outcome o_k.

$$SD^r_{\gamma_j}(\beta, \delta; ec_i) = \sum_{o_k \in SO(\alpha, ec_i)} wo^\alpha_k \cdot P^t(\gamma_j, \beta, \delta, o_k; ec_i) \tag{8}$$

Then the situational trust value $TS^r_{\gamma_j}(\beta, \delta; EC)$ from the recommender γ_j is computed as follows: Here wc_i indicates the relative importance degree for the evaluation criterion ec_i that the entity α has.

$$TS^r_{\gamma_j}(\beta, \delta; EC) = \sum_{ec_i \in EC} wc_i \cdot SD^r_{\gamma_j}(\beta, \delta; ec_i) \tag{9}$$

The reputation value $TR_\alpha(\beta, \delta; EC)$ of β for α is computed by taking the weighted sum of the situational trust $TS^r_{\gamma_i}(\beta, \delta; EC)$ from recommenders γ_j

as follows: Here the weighting factor wr_j is the recommendation trust value for the recommender γ_j. That is, wr_j is the degree to which α believes the recommendation from γ_j. These weights are updated through the interaction with the entities.

$$TR_\alpha(\beta, \delta; EC) = \frac{\sum_j wr_j \cdot TS^r_{\gamma_j}(\beta, \delta; EC)}{\sum_j wr_j} \tag{10}$$

3.5   Combination of Situational Trust and Reputation

When an entity starts to work in a community, she assigns her own dispositional trust value. The dispositional trust is used as the initial general trust when she interacts with an entity for the first time. Until sufficient number of interactions has made for a given situation, the general trust is used as the situational trust. Once the situational trust $TS_\alpha(\beta, \delta; EC)$ and the reputation $TR_\alpha(\beta, \delta; EC)$ are obtained, the final trust value $TS_\alpha(\beta, \delta; EC)$ is computed by their weighted aggregation as follow: Here, w is the relative weighting factor for the situational trust, $w \in [0, 1]$.

$$TS_\alpha(\beta, \delta; EC) = w \cdot TS_\alpha(\beta, \delta; EC) + (1 - w) \cdot TR_\alpha(\beta, \delta; EC) \tag{11}$$

3.6   Update of the Recommender Trust

The recommender's trust wr_i is updated according to how much their recommendation score is similar to the final computed trust value $TS_\alpha(\beta, \delta; EC)$. If the recommendation score is similar to the final trust value, the recommender's recommendation trust is increased by a small amount. Otherwise, the recommender's recommendation trust is decreased by an exponential factor term. The following shows the update rule for the recommender trust.

Let $\Delta = |TS_\alpha(\beta, \delta; EC) - TS^r_{\gamma_i}(\beta, \delta; EC)|$. If $\Delta < \epsilon$, $wr_i(t+1) = \min\{wr_i(t) \cdot (1 + \eta), 1\}$ where ϵ and η are small values such that $0 \leq \epsilon, \eta \leq 1$. Otherwise, $wr_i(t + 1) = wr_i(t)(1 - e^{-\lambda\Delta})$ where λ is a small value such that $0 \leq \lambda \leq 1$.

4   Conclusions

The trust information for the online entities are valuable in reducing the risks to take on doing some transactions. We proposed a comprehensive computational trust model which has the following characteristics: The model allows to look at entity's trust in the point of multiple evaluation criteria. It maintains the empirical outcome distributions for evaluation criteria and enables the trusting entities to express their preference on the outcomes when estimating trust in other entities. In addition, the model makes it possible for the entities to put different weights on the evaluation criteria. When it uses the recommendations from others, it takes the outcome distributions instead of their recommending trust values. Thereby, it allows to reflect the trusting entity's preference and her own weighting on the evaluation criteria in the evaluation of the recommendation. As the further study, there remains to develop some techniques to automatically learn the entities' preference on the outcomes and weighting on the evaluation criteria.

References

[1] A. Abdul-Rahman, S. Hailes. Supporting Trust in Virtual Communities. *Proc. of the Hawaii Int. Conf. on System Sciences.* (Jan.4-7, 2000, Maui Hawaii).

[2] F. Azzedin, M. Maheswaran. Trust Modeling for Peer-to-Peer based Computing Systems. *Proc. of the Int. Parallel and Distributed Processing Symposium.* (2003).

[3] U. Hengartner, P. Steenkiste. Implementing Access Control to People Location Information. *Proc. of SACMAT'04.* (Jun.2-4, 2004. New York).

[4] D. Gambetta. Can We Trust Trust?. In *Trust: Making and Breaking Cooperative Relations.*(Gambetta. D (ed.). Basil Blackwell. Oxford. (1990).

[5] D. H. McKnight, N.L. Chervany. The Meanings of Trust. *Technical Report 94-04.* Carlson School of Manangement, University of Minnesota. (1996).

[6] G. Derbas, A. Kayssi, H. artial, A. Cherhab. TRUMMAR - A Trust Model for Mobile Agent Systems Based on Reputation. In *Proc. of ICPS2004.* IEEE. (2004).

[7] J. Shi, G. v. Bochmann, C. Adams. A Trust Model with Statistical Foundation. In *FAST'04.* Academic Press. (2004).

[8] Y. Wang, J. Vassileva. Bayesian Network Trust Model in Peer-to-Peer Network. In *WI'03.* IEEE. (2003).

Groupware for a New Idea Generation with the Semantic Chat Conversation Data

Takaya Yuizono[1], Akifumi Kayano[1], Tomohiro Shigenobu[2],
Takashi Yoshino[2], and Jun Munemori[2]

[1] Department of Mathematics and Computer Science,
Interdisciplinary Faculty of Science and Engineering, Shimane Univ.,
1060, Nishikawatsu, Matsue, Shimane 690–8504, Japan
`yuizono@cis.shimane-u.ac.jp`
[2] Department of Design and Information Science,
Faculty of Systems Engineering, Wakayama Univ.,
930, Sakaedani, Wakayama-shi, Wakayama 640–8510, Japan
`{s020055,yoshino,munemori}@sys.wakayama-u.ac.jp`

Abstract. The GUNGEN groupware for a new idea generation has been developed and applied to the KJ method with chat conversation data. First, Remote Wadaman V, which is a seminar system equipped with the semantic chat function, supports collecting the chat data as knowledge including explicit and implicit knowledge via cooperative interaction. The semantic chat function allows user to assert semantics corresponding to chat statements with one-selection interface. Next, KUSANAGI supports cooperative idea creation from the collected chat data leaded by the KJ method.

The Remote Wadaman V with the semantic chat function has been applied to seminar over half year. The availability of chat data as an ideal label for an idea generation method was investigated.

1 Introduction

Knowledge is thought more valuable in these days and an information technology for knowledge management attracts many organizations [2, 3]. It is considered that knowledge consists of tacit knowledge and explicit knowledge. Many researchers study knowledge management system [2, 3] based on Semantic Web [1]. The Web system is progressive system of ordinary information systems. Those systems are considered to be efficient for the explicit knowledge transformation.

Otherwise, it is considered that the tacit knowledge has an important role in the knowledge creative process called as SECI model [6]. And the tacit knowledge is transmitted from human to human by daily communication. The technology to capture various type of knowledge from conversation data is necessary to support the knowledge creative process. Therefore, we have developed GUNGEN groupware [11] to support knowledge creative process based on SECI model.

The GUNGEN that consists of two components; first is Remote Wadaman V for collecting conversation chat data including tacit knowledge and explicit

R. Khosla et al. (Eds.): KES 2005, LNAI 3681, pp. 1044–1050, 2005.

knowledge, and second is KUSANAGI (Knowledge Units Summary Assistants with Networking Associated Grouping Interface) for supporting the KJ method.

An idea generation, as it were, is a kind of a knowledge creation. The idea generation supported groupware, such as Cognoter [8] and GDSS etc., seems mainly to support brainstorming for capturing an idea only in face-to-face meeting. The idea generation method named KJ method [4] is well known in Japan. Some KJ method tools, such as KJ Editor [7] and D-Abductor [5], have been developed in Japan but support carry out the KJ method including brainstorming only. Our study focuses on creation of some knowledge from chat conversation data collected in cooperative work.

The purpose of this paper is to describe the groupware for a new idea generation from chat conversation data. In second section, the seminar system equipped with semantic chat is described. Next, groupware to support the KJ method for an idea generation is described. Then, application of semantic chat to remote seminar is explained and those results are discussed.

2   Remote Wadaman V

2.1   Overview of the Seminar System

The seminar system is a conferencing system based card system after intelligent productive technique [9]. User writes his/her report on a card window. The seminar system named Remote Wadaman has supported sharing the card window, which displays a seminar report and both a speaker and a teacher can use shared cursors over networked computers. The system has been applied to a usual seminar since 1996. In recent, the seminar system has been improved and named Remote Wadaman V. The system has semantic chat function described in 2.2 and five senses interface using many type sensors [10].

The example screen of seminar system is shown in Fig. 1. The seminar information window indicates the state of a seminar, which includes a participant's list, a report list, and a current speaker, etc. Few participants use voice communication at the same time, but all participants can use a text-based chat always. The text-based chat supports image characters [10] and the semantic chat. The chat data is saved as log text and converted to XML format file.

2.2   The Semantic Chat Function for a Seminar Setting

The semantic chat is that a user adds a tag as semantics to conventional chat. The expected effect is increasing interaction, increasing quality of data for idea generation, and progressing user's participation by driving their interaction.

Semantics corresponding to tag should be considered for the usage context and interaction. The nine kinds of tags are decided for semantic tags. The nine tags and those semantics are shown in Table 1. In particular, the "Opinion" tag is expected to encourage the effect of positive contribution to a seminar contens, and the "Question" tag is expected to be the trigger of the reply with another participants using the "Answer" tag.

Report shared window   Seminar information window   Log window

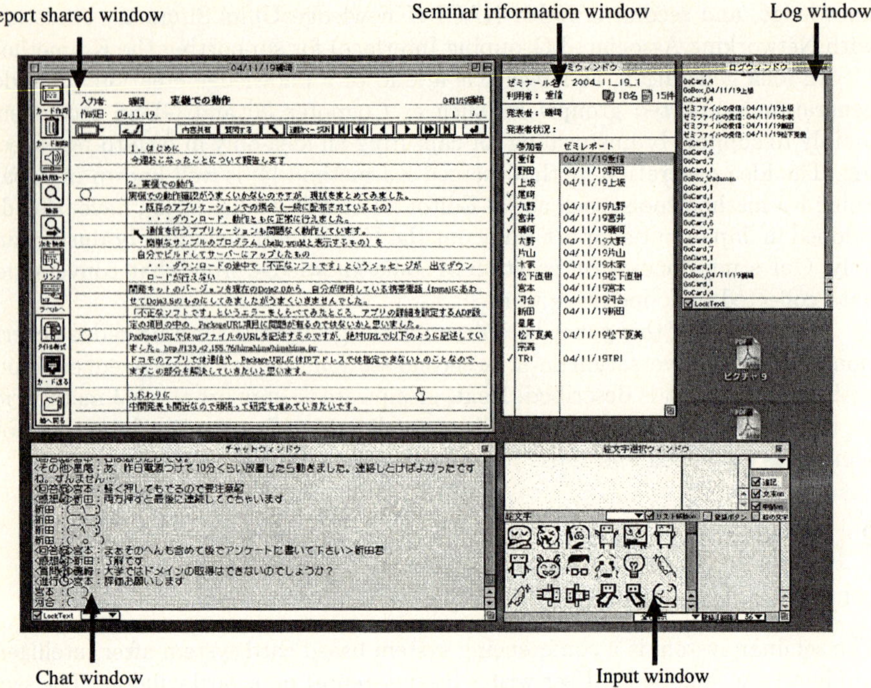

Chat window   Input window

Fig. 1. An example screen of Remote Wadaman V.

The user interface for the semantic chat should support as possible as simple operation to avoid disturbing user's thinking. The selection interface of the input window is shown in Fig. 2 (a). The interface allows us mouse selection with a popup menu or a key selection only.

First, a user inputs chat data into the input window with keyboard. Next, he mouse-selects the popup button on the top right side of the input window to select the semantic tag for the chat data. The operation of selection causes sending chat data for all clients. Then the chat data indicated the selected tag is appeared in the chat window shown in Fig. 2 (b).

3   KUSANAGI

KUSANAGI has been developed to support the KJ method execution with multi PC's screen and to support the KJ method via networked personal computers (Fig. 3). To support the creative knowledge process [11], KUSANAGI supports the KJ method with the chat data collected in the seminar with the semantic chat function. Therefore, KUSANAGI has a graphical user interface for choosing the chat data as an idea label.

The groupware supports three steps of the KJ method; those steps are (1) brainstorming for showing ideas, (2) grouping ideas for a concept formation, and

Table 1. Items of semantic tag for seminar system.

Items	Semantics
Opinion	Idea, Suggestion
Question	Question
Answer	Answer to a question
Impression	Impression, Comment
Explanation	Explanation
Memo	Memorandum
Progress	Direction statement for session
Greeting	Greeting, for example as hello, etc.
Other	Others above.

(a) Semantic chat interface.

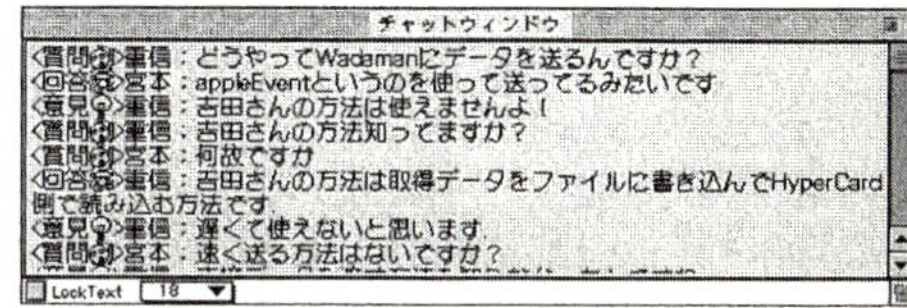

(b) Chat window.

Fig. 2. Semantic chat interface of Remote Wadaman V.

(3) writing a conclusion sentences from those results of preceding two steps. The system screen is shown in Fig. 3; idea labels, the groups of ideas as islands and the conclusion sentences are shared with each computers.

The system has been implemented with the JAVA programming language. The virtual window supports to enlarge a shared screen with a scrolling function and allows us to use a multi-screen sized window space. When two computers, each computers has one screen, are set for the KJ method, the window allows multi-screen sized view to empower the user's look-ability.

4   Usage

The semantic chat function has been used in the seminar with Remote Wadaman since Novenber 2004. The seminar has been weekly held at a room in Wakayama University. The seminar member is one professor and seven-teen students.

In next section, the semantic chat data obtained from seven seminars is compared with the conventional chat data from seven similar seminars in 2003.

5   Discussion

5.1   Usability of Semantic Chat Function

The mean value of seminar data with the semantic chat function was compared with those of a conventional chat in Table 2. The number of chats with the

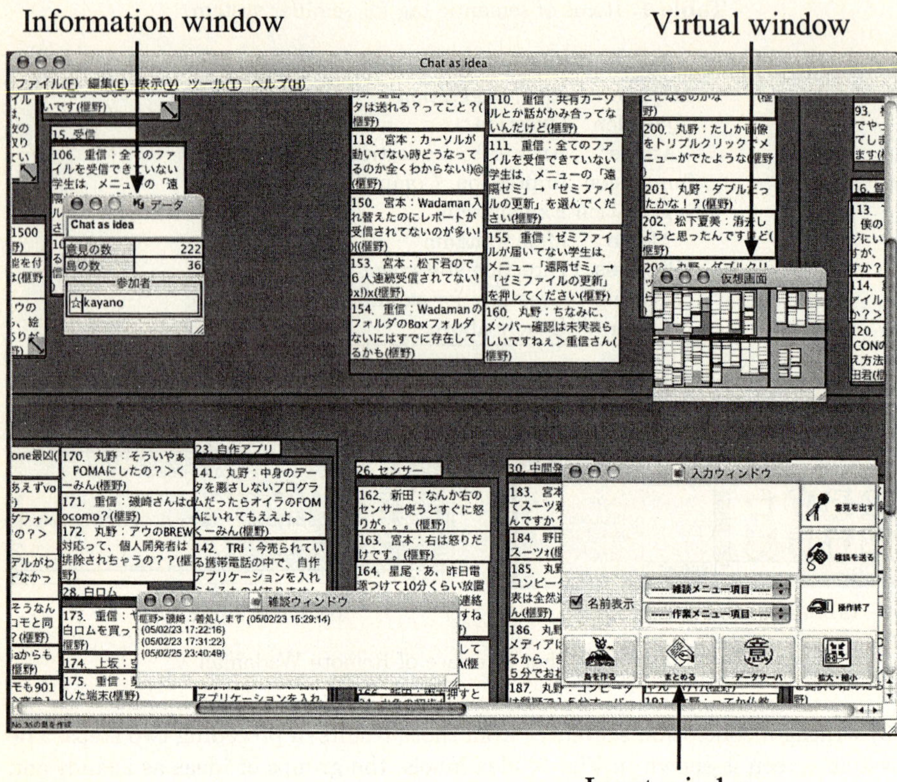

Fig. 3. An example screen of KUSANAGI.

Table 2. Performance of the seminar system with the semantic chat function.

As idea label	Semantic chat in 2004	Conventional chat in 2003
Number of chats	94.3	80.4
Time (min.)	66.1	59.1
Number of report cards	24.1	29.7
Number of speakers	12.9	12.7

semantic function, which has the tag choice interface for one-selection operation, is not differing from those of the conventional chats. Therefore, the semantic chat function is not burden for user chatting in the seminar.

5.2   Availability of Chat Conversation Data

The quality of chat data with the semantic chat function is considered in this section. It is estimated how many numbers of chat data may be utilized as an

Table 3. Availability of chat data as an idea label.

As idea label	Semantic chat in 2004(N=660)			Conventional chat in 2003(N=563)		
Person	X	Y	Z	X	Y	Z
Usage able+ Maybe usage able	231 (35%)	143(21%)	224 (34%)	200(35%)	66 (12%)	145(26%)
Nonsense	429(65%)	525(79%)	436 (66%)	363(65%)	497 (88%)	418(74%)

idea label for the brainstorming session by three person X, Y, Z. The person Y and Z did not take part in the seminar. The result is shown in Table 3. The approximately 30 percent of chat data with semantic chat may be available as an idea label. In the view of availability of an idea label, the chat data using the semantic chat is more adaptive for the person, who has no participation in the seminar, than the chat data using conventional chat.

6   Conclusion

We have developed a groupware GUNGEN that consists of two components; Remote Wadaman V for collecting conversation data with the semantic chat function and KUSANAGI for supporting the cooperative KJ method with the virtual window.

The semantic chat function of Remote Wadaman V has been expected to utilize for production of explicit knowledge from the interaction data. The function was applied to weekly seminar, and compared with the seminar employing conventional chat. The result shows that the chat data with the semantic chat may be more adaptive to an idea label for idea generation method than those with conventional chat function.

In future, we will employ the semantic chat data for some idea generation, such as brainstorming, and concept formation, and explore the effective usage of the semantic chat data. In other words, the collected data has been applied to the idea generation method with KUSANAGI to support a concept creation including explicit knowledge generation from implicit data.

References

1. Bosak J. and Bray Tim: XML and the Second-Generation Web , Scientific American (1999).
2. Daconta, M. C., Obrst, Leo J. and Smith, K.T: The Semantic Web: A Guide to the Future of XML, Web Services, and Knowledge Management, Wiley, (2003).
3. Davies, J., Fensel, D. and Harmelen, F.V. edit.: TOWARDS THE SEMANTIC WEB: Ontology-driven Knowledge Management, Wiley (2003).
4. Kawakita, J.: KJ Method, Chuokoron-sha, Tokyo (1986) (In Japanese).
5. Misue, K. and Sugiyama, K.: Evaluation of a thinking support system D-ABDUCTOR from an operational point of view , Trans. IPS Japan, Vol. 37, No. 1 (1996) 133–145 (In Japanese).

6. Nonaka, I. and Takeuchi, H.: The Knowledge-Creating Company, Oxford University Press (1995).
7. Ohiwa, H. et al.: Idea processor and the KJ method , Journal of Information Processing, Vol.13, No.1 (1990) 44–48 (In Japanese).
8. Stefik, M., Foster, G, Bobrow, D.G., Kahn, K., Lanning, S., Suchman, L.,: Beyond the Chalkboard: Computer Support for Collaboration and Problem Solving in Meetings , Comm. ACM, Vol. 30, No. 1, ACM PRESS (1987) 32–47.
9. Umesao, T.: Technique for Intellectual Productive Works, Iwanami-shoten, Tokyo (1969)(In Japanese).
10. Yoshida, H., Yoshino, T. and Munemori, J.: Data Processing Method of Small-Scale Five Senses Communication System , KES 2004, LNAI 3214 (2004) 748–755.
11. Yuizono, T. et.all: A Proposal of Knowledge Creative Groupware for Seamless Knowledge , KES 2004, LNAI 3214 (2004) 876–882.

Dual Communication System Using Wired and Wireless with the Routing Consideration

Kunihiro Yamada[1], Takashi Furumura[2], Yoshio Inoue[2], Kenichi Kitazawa[2], Hiroki Takase[2], Yukihisa Naoe[2], Toru Shimizu[3], Yoshihiko Hirata[2], Hiroshi Mineno[4], and Tadanori Mizuno[4]

[1] Tokai University
[2] Renesas Solutions Corporation
[3] Renesas Technology Corporation
[4] Shizuoka University

Abstract. As the actual testing result without any countermeasure in the communication environments in the house, it has gained that the communication success rate is 82% in the current wireless type 802.15.4 and is 70% in the wired PLC, and is 90% in the dual communication system with the mutual complement of them. Moreover, in order to improve the communication quality, the routing system through the nodes is investigated. As the result, it is found that the routing system in which the features of the wireless and wired communications are considered will considerably reduce the number of the routes and the number of the communication times per route. Thus, we think that one of the methods to put the measures into practice will be gained.

1   Introduction

In restricted areas such as the home, the ability to communicate quality is improved using node relays combining wired and wireless communication to achieve a network without having to change the communication environment excessively.

We found that an effective communication network can be configured if it is known which floors have nodes in order to effectively use the wireless correspondence (IEEE802.15.4) and it is known whether the power distribution in Japan is in the same or different phase in order to effectively use the Power Line Carrier Communication (PLC).

We simulated 6 simple nodes and found that correspondences per route averaging 4.7 times were reduced to 1.8 times if it is known which floors have wireless nodes and could be reduced to 1.4 times if it is known whether the PLC uses the same or a different phase. We verified that our proposal is feasible for practical applications.

The average home generally has limited, independent information networks. To unify them, communication is required that does not require that the communication environment be excessively changed.[1]

We planned to use PLC using electric power lines as the wired medium and IEEE802.15.4 as the wireless medium. However, PLC and IEEE802.15.4 are already in practical use or are about to be, as independent communication media [2], [3], [4].

R. Khosla et al. (Eds.): KES 2005, LNAI 3681, pp. 1051–1056, 2005.
© Springer-Verlag Berlin Heidelberg 2005

1052   Kunihiro Yamada et al.

2 Communication Quality in Wired or Wireless Single Communication

2.1 Quality of Wired Communication

The result of measurement of PLC communication quality in a common family house is shown in Fig. 1. X-axis indicates Packet Error Rate (PER), and the value of 0% means the best condition. Y-axis shows the channel frequency distribution (%).A frequency distribution of 70% achieves the best PER 0%. PER 100% where the correspondence is completely impossible is 20% present. The cause of impossible communication was improper phase combination in the power cable placement [3], [5]

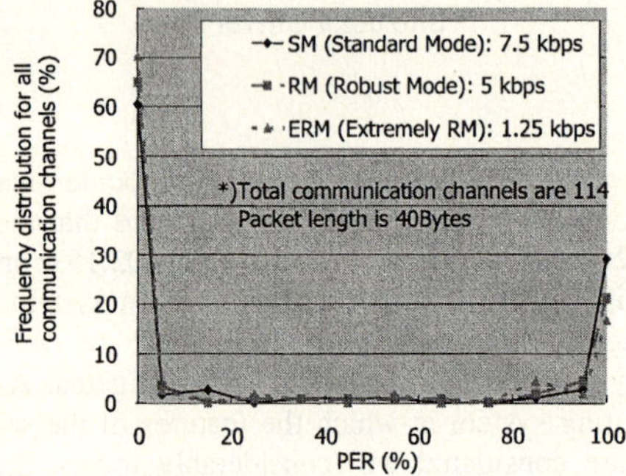

Fig. 1. Testing result of PLC (100kHz–400kHz) in typical Japanese home

2.2 Quality of Wired Communication

Fig. 2 shows the result of a simplified experiment to measure how the distance changes the level of received signals and PER in a certain packets, using IEEE 802.15.4.

When the relative distance is 30m, PER was found to be 1.5% or so. However, IEEE 802.15.4 is easily affected by obstacles such as metal pieces or reinforced concrete (rebar) as it is so in other wireless communication methods, and sometimes it is not possible to provide stable communication. In structures such as homes made of steel-reinforced concrete, communication between floors may be very difficult.

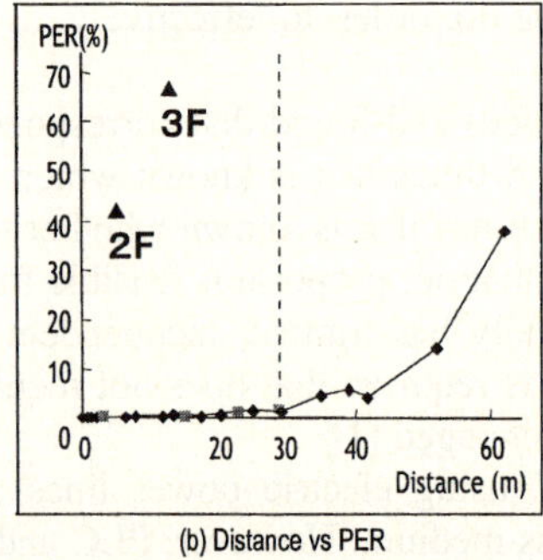

Fig. 2. IEEE802.15.4 test result data

3 Wired and Wireless Mutually Supplemental Network System

Our intention is to develop networks for relatively small spaces such as a detached three-story house with a 300m^2 floor area, which do not require any changes in the communication conditions or assistance from engineers to operate the networks.

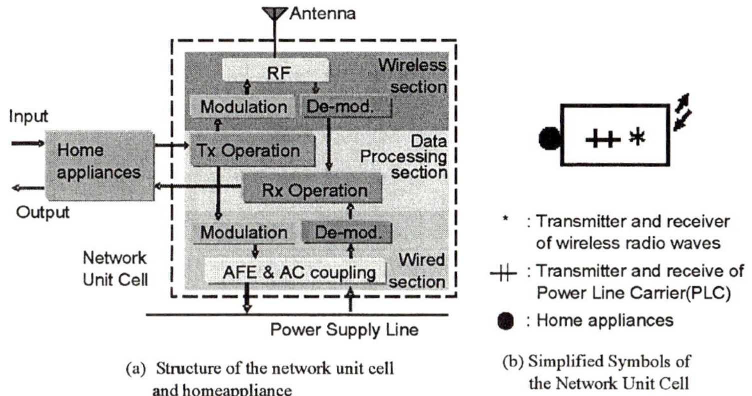

(a) Structure of the network unit cell
and homeappliance

(b) Simplified Symbols of
the Network Unit Cell

Fig. 3. Network Unit Cell

We called the communication system having the wired and wireless media a "network unit cell". Fig.3(a) shows how the home appliances are interfaced with the network unit cell and Fig.3(b) explains the symbols used. The network unit cell has three sections, the wireless section, the wired section, and the data processing section.

We defined the evaluation criteria for the communication quality in modeling the networks as follows. When the packet data of about 40 bytes is sent 10 times successively, we can classify the received data into the correct data (Truth: The same data is received in the correct condition two times successively.), possibly the correct data (Truth?: The correct data is received two times, but not successively.), and the false data (False: conditions other than above). The communication quality in the data transmitting and receiving is classified into three ranks in the following examination.

Table 2 shows the quality of communication when wired and wireless systems are combined. Individual communication quality is improved from 70% for wired and 82% for wireless to 94.6% overall. When "Truth" and "Truth?" is set at "Truth", the value of 96% is achieved.

Table 1. The communication quality in % when the wired and wireless communication media are used simultaneously

		Wireless		
		Truth 82	Truth? 14	False 4
Wired	Truth 70	Truth 57.4	Truth 9.8	Truth 2.8
	Truth? 10	Truth 8.2	Truth?□Truth? 1.4	Truth□□False 0.4
	False 20	Truth 16.4	Truth?□False 2.8	False□False 0.8

4 Improvement of Communication Quality with Node Relay

Although combining wired and wireless communication is estimated to achieve a correspondence of 96%, inserting several relay nodes (routing nodes) to improve quality more if communication is not possible between two points [6]. We studied the use of six nodes, providing one node per two power phases per floor for floors 1, 2, and 3 (Fig. 4).

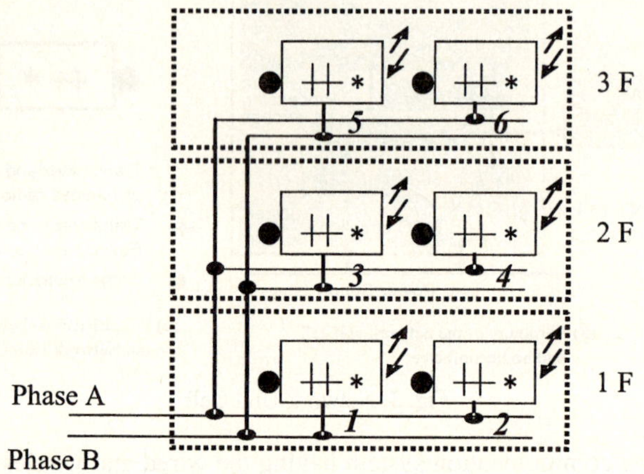

Fig. 4. Routing node example at three-story house

Fig. 5 shows whether communication is possible between nodes. PLC is possible for the same phase but not for a different phase (Fig. 1). Under these conditions, the number of communications per route is 1 to 7 (Table 1) if the floor number of a node or the power supply phase (same or different) is not known. If the floor number is known, the number of communications becomes a maximum of 3 per route using an algorithm to route the node of the same floor if communication is not possible once. If the power supply phase is known, the number of communications per route becomes 2 or fewer between all nodes.

		Receiving node No.					
		1	2	3	4	5	6
Sending node No.	1	—	Z	P	x	P	x
	2	Z	—	x	P	x	P
	3	P	x	—	Z	P	x
	4	x	P	Z	—	x	P
	5	P	x	P	x	—	Z
	6	x	P	x	P	Z	—

P: Success by PLC communication
Z: Success by 802.15.4 communication
x: Communication impossible

Fig. 5. Communication between two nodes

Table 2. Comparison of communications per route when the floor number and power supply phase are known

Floor	unknown	known	known
Same/different phase	(unknown)	(unknown)	(known)
Number of routes	Number of communications per route		
1 time	18/18 times	18/18 times	18/18 times
2 times	–	–	12/24 times
3 times	24/ 72 times	12/36 times	–
4 times	24/ 96 times	–	–
5 times	24/120 times	–	–
6 times	48/288 times	–	–
7 times	24/168 times	–	–

5 Conclusions

A routing algorithm that uses the features of the wired and wireless communications considerably reduces the number of routes and average number of communications. In wired communication PLC, the power supply phase, i.e., whether it is the same or different, greatly affects communication, as does whether the floor is the same or not greatly affects wireless communication [7]. Using this algorithm reduced 162 routes and an average of 4.7 communications to 30 routes and an average of 1.8 communications per route. Knowing whether the power supply phase is the same or different improves the number of communications to an average of 1.4 times. The routing that uses these features is thus effective in a mutually complementary system.

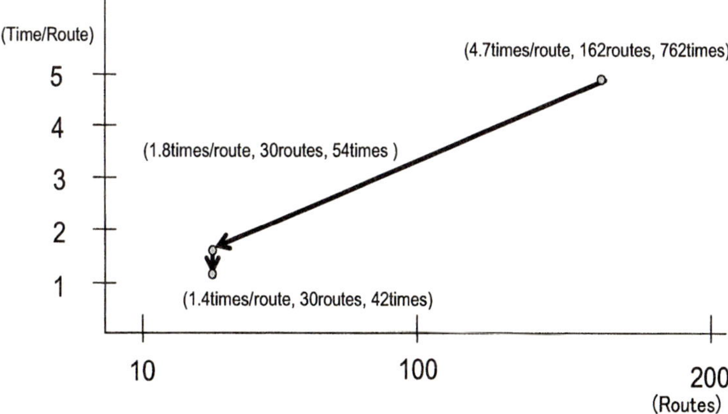

Fig. 6. Verification result of algorithm

References

1. H. Fujimura, Y.Ogawa, H.Kawana, A.Kuwayama, "Home Safety Station "Beruhi"" Matsushita Technical Journal Vol. 47 No 47 Feb. 2001
2. G. Evans. "The CEBus Standard User's Guide" section 5. The Training Dept. Publications. 1996
3. URL http:/www.itrancom.com
4. E. Takagi, Y. Takabetake, "Current Status of and Future Trends in Wireless Technologies" Toshiba Review Vol. 58 No 4 Apr. 2003
5. 5. T. Furumura, Y.Hirata, S.Kuroda, K.Yamada, M.Kojima, T.Mizuno, "Power Line communication System for Home Network" pp93-96 Multimedia, Distributed, Cooperative and Mobile Symposium, IPSJ Symposium Series Vol. 2003 No 9 ISSN 1344-0640 June 2003
6. K.Yamada□etc "Dual Communication System Using Wired and Wireless Correspondence in a Small Space" KES2004, Proc., pp898-904.
7. URL http://www.zigbee.org/

Development and Evaluation
of an Emotional Chat System Using Sense of Touch

Hajime Yoshida[1], Tomohiro Shigenobu[1], Takaya Yuizono[2],
Takashi Yoshino[1], and Jun Munemori[1]

[1] Wakayama University, Japan
{s029015,s020055,yoshino,munemori}@sys.wakayama-u.ac.jp
[2] Shimane University, Japan
yuizono@cis.shimane-u.ac.jp

Abstract. We have developed a prototype of a chat system using sense of touch. Depending on the power of grasping a mouse, two or more kinds of face marks ((^ ^), (> <), etc.) are inputted into a chat window. We apply this system to a kind of an electronic conferencing system, named a remote seminar support system. From the results of the experiments, we found that the input method using sense of touch was useful for busy user who had no time to type keyboards. And we found that the users of this system were interested in this system.

1   Introduction

PC and the Internet have become widespread, and therefore we can do real-time communication easily. Real-time communication functions are a text-based chat, a videoconference, and so on. We have developed the remote seminar support system with a real-time communication functions since 1996 [1] [2].

One of the problems of communication via network is the difficulty of maintaining the tension of participants. This problem arises from the lack of information about each other's feeling. We try to solve this problem by exchanging information about "five senses information" in network communication application [3] [4].

Ambe and Ohmura have proposed to use the sense of touch for telecommunication [5]. However, they focus on a communication between 2 persons, and they used a specific hardware, named "Hearty Egg". Therefore we cannot apply this approach to the electronic conferencing system (i.e. the remote seminar support system).

We assume that the power of fingers grasping is a parameter of feeling of tension, so therefore we have developed the prototype of a chat system using sense of touch.

In the other research, a conference state estimation method by biosignal processing has been proposed [6]. In their paper, the heart rate is used for a parameter of feeling. Their approach is also helpful for us to conduct our research.

This paper describes the prototype of an emotional chat system using sense of touch and its application to seminars.

2   Remote Seminar Support System

We have been using the remote seminar support system, named "RemoteWadaman", in usual seminars (Fig.1). RemoteWadaman has mainly two windows, the shared

R. Khosla et al. (Eds.): KES 2005, LNAI 3681, pp. 1057–1063, 2005.

document window with shared cursors, and the text-based chat window with pictograph. If necessary, we can also use video communication software with RemoteWadaman. RemoteWadaman has been developed on HyperCard.

When we hold a remote seminar, students write their report using RemoteWadaman in advance. A professor comments on these reports and coaches them.

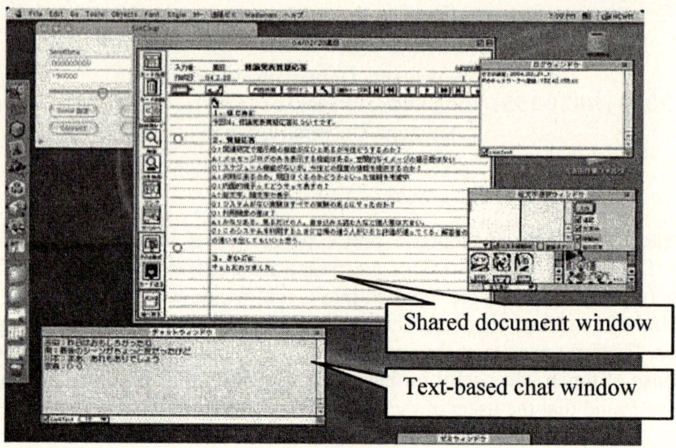

Fig. 1. A screenshot of RemoteWadaman

3 System Overview

We have developed a new remote seminar support system with a chat system using sense of touch. Depending on the power of grasping a mouse, a face mark is inputted to the chat window. The components of this system are shown as in Table 1. A pressure sensor is attached to a mouse, and a digital multimeter reads the value of the sensor. The digital multimeter connected to a PC will read the value of the sensor and this value to the PC (Fig.2).

Table 1. Sytem components

Category	Detail
Hardware	Macintosh PowerBook G4 (Apple Computer) PowerPC G4 500Mhz / 512M Memory
	Sensor: FlexiForce (Tekscan)
	Digital Multimeter: PC500 (Sanwa Electric Instrument)
Software	Remote seminar support system: RemoteWadaman (Original Software)
	Receive and process data from sensor software

This system consists of two software, as shown in Table 1. We have made software that receives and processes data from the digital multimeter. If the value of the sensor is kept in a range for 0.2 seconds, the counter increases one by one. And if the value of the counter became 5, this program sends relevant information about face marks to RemoteWadaman (Fig. 3).

The system can handle two sensors. These are attached to the left side and the right side of the mouse. The relation between sensor value and face marks is shown in Fig. 4.

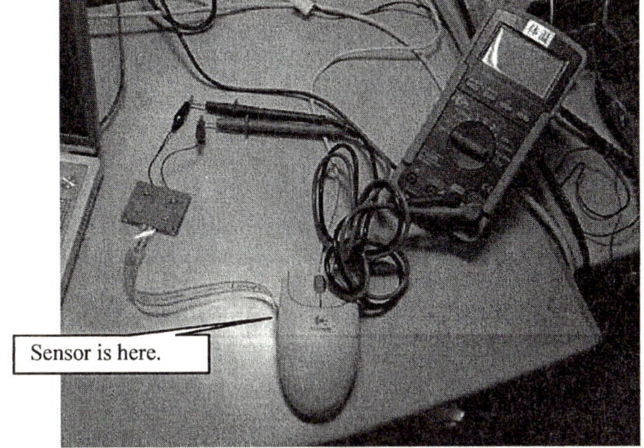

Fig. 2. The sensor and the Digital multimeter

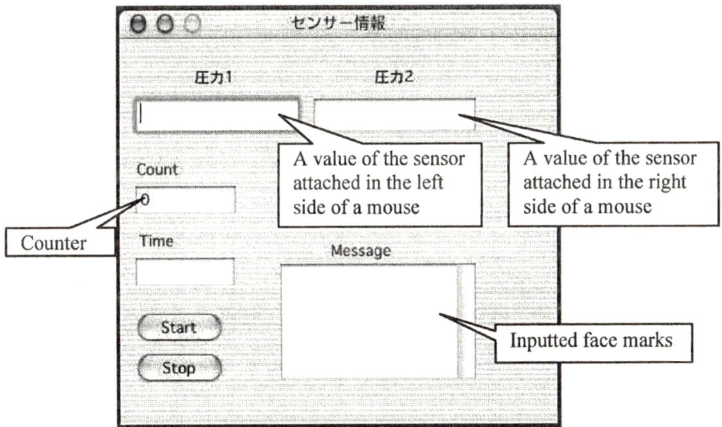

Fig. 3. A screenshot of data processing software

Combination of sensor value		Face mark	Meaning
Left side	Right side		
Softly	-	(^_^)	Joy
Strongly	-	\(^_^)/	Happy
-	Softly	(`o´)	Angry
Softly	Softly	(; ;)	Sad
Strongly	Strongly	(￣ ^ ￣ ;)	Confuse

Fig. 4. Relation between a pressure and five face marks

4 Experiments

4.1 Experiments in Actual Seminars

We performed three experiments in actual seminars. Overview of each experiment is shown in Table 2. These experiments were performed by about 15 subjects.

Experiment 1

In this experiment, only the professor used this input function. Students used text-based chat, but they did not use this input function. The results of the questionnaire are shown in Table 3.

From the result of Question 2 (Table 3), the participants felt that the professor had participated in the chat. And the professor felt that he had participated in the chat. Therefore, this input method is useful for participation. And From the result of Question 4 (Table 3), these 5 face marks are enough to communicate each other.

Experiment 2

In this experiment, four students used the sensor-attached mouse. They inputted the rating of each report using this input method.

The results of the questionnaire are shown in Table 4. From the result of the questionnaire, the approach to evaluate reports using this prototype did not work adequately. The reason is that it is difficult for them to input a face mark on their own. They worried about incorrect input. The result of the Question 5 (Table 4) shows that the users of this prototype rated it higher than other participants. We found that more participants should be able to use this input method.

Table 2. Experiments overview

No.	NUMBER OF THE SENSOR	NUMBER OF THE VARIATION OF FACE MARKS	USER
1	2	5	A Professor
2	2	5	4 students

Table 3. Result of experiment 1

No.	Question	Score (average)
Q1	Was this experiment interesting?	3.2
Q2	Did you think that the professor is in text-based chat?	3.0
Q3	Did you think that this function is useful for remote seminar?	2.9
Q4	Are these 5 face marks enough for communication?	3.3

[Score: 1(worst)-5(best)]

4.2 Experiment with Input Method of Face Marks

Some of the experimental subjects pointed out difficulty of the input using this method. We measured the difficulty of the input. We have made a simple test application (Fig.5).

Table 4. Result of experiment 2

No.	Question	Score (Average)	Score (Only the users)
Q1	Was this experiment interesting?	3.5	4.0
Q2	Did you think that this function is useful for remote seminar?	2.5	3.0
Q3	Could you input face mark on your own?	2.8	2.8
Q4	Did you think that the report evaluation using face mark came good?	2.7	3.0
Q5	Was it good that two or more users participate in a seminar using this method?	3.5	4.3

[Score: 1(worst)-5(best)]

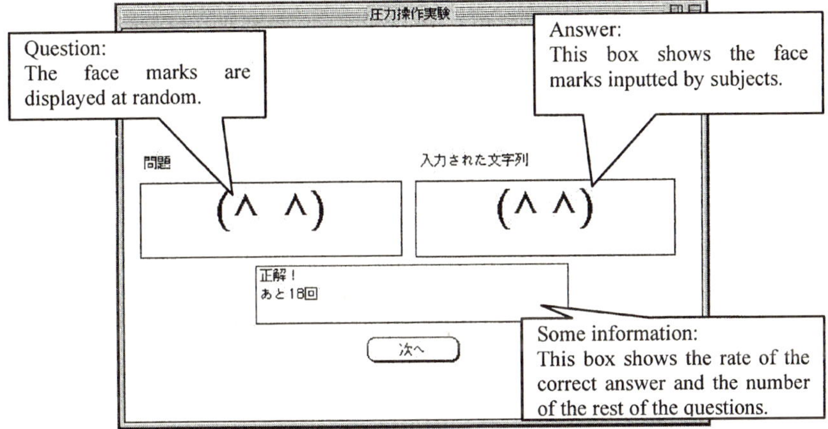

Fig. 5. A screenshot of an input test application

The test application displays face marks in the left side box at random. When an experimental subject inputs a face mark using the sensors, the relevant face mark is displayed.

We experimented using the test application (Fig.5). Each subject inputted 20 face marks. The application logged the rate of a correct answer. Three subjects had experienced this input method in other experiment of this paper, and five subjects used this method for the first time.

The results of the questionnaire are shown in Table 5. From the result of questionnaire, it is difficult for a person who uses this input method for the first time to input face marks on his own. In this experiment, the average rate of the correct answers is 54.4 percent. However, from the result of Question 2 and 3 in Table 5, most subjects feel that this input method is very interesting. We think most users will use this method delightfully when the input method can be improved in RemoteWadaman.

We found that the input method needs to be improved in two ways. One is that a grasping power depends on the person. We think that it is necessary to add an adjustment function to this system.

Table 5. Result of experiment with input method

	Scores of each subjects								AVERAGE
	A	B	C	D	E	F	G	H	
Q1	2	3	2	2	2	2	4	2	2.4
Q2	4	4	4	2	3	4	5	4	3.7
Q3	3	5	4	2	3	3	5	4	3.6
D1	35%	55%	40%	55%	55%	60%	50%	50%	50%
D2	12	9.2	6.1	5.8	7.4	7.5	11.1	6.7	8.2
D3	0	0	1	0	0	0	2	2	-

Items of table 5 are below;   [Score: 1(worst)-5(best)]
Q1: Could you input face mark on your own?
Q2: Was this input method interesting?
Q3: Was this experiment interesting?
D1: Rate of correct answer
D2: Average of the time required to input face marks
D3: Number of times of experience

The other is that it is difficult to press only one side. For example, if the user wants to input an anger mark, he needs to press only the sensor attached to the right side of the mouse. If we use two or more sensor, we think that it is necessary to consider where a sensor should be attached. In this experiment, there is no difference between a person with experience and a newcomer.

5 Discussion

From the result of Experiment, this input method is useful for busy user who had no time to type keyboards. In remote seminar, this system is useful for the professor.

And we found that the users of this method appreciated it more highly than other participants. Therefore, this input method is effective for not only busy participants but also other participants. We think that this chat system will be the most effective when all the members will use this input method.

6 Conclusion

We propose a new concept to keep tension for serious participation in a remote seminar support system. We have developed a prototype system and applied it to actual seminars. We found that the input method using sense of touch is useful for participation and all the users feel that this method is quite interesting from the experiments. Therefore, it is effective for keeping tension of participants. Next, we measured the difficulty of input. We found that it is difficult for a person who uses this input method for the first time. However, they are interested in this input method. We think that this method is effective for active participation.

References

1. J. Munemori, H. Yoshida et al., "Remote Seminar Support System and Its Application and Estimation to a Seminar via Internet," Transaction of Information Processing Society of Japan, Vol.39, No.2, pp.447-457 (1998)

2. T. Yoshino, J. Munemori, "Application and Evaluation of Distributed Remote Seminar Support System RemoteWadaman II over two years," Transaction of Information Processing Society of Japan, Vol.43, No.2, pp.555-565 (2002)
3. J. Munemori, "Modified Brain Model Hyper Communication Mechanisms," Proceedings of 2001 IEEE International Conference on Systems, Man, and Cybernetics (SMC2001), pp. 622-627, 2001.
4. J. Munemori, T. Yoshino, "The Prototype of Brain Model Hyper Communication Mechanisms," IEEE International Conference on Consumer Electronics (ICCE 2002), pp.232-233, 2002.A. N. Netravali and B. G. Haskell, Digital Pictures, 2nd ed., Plenum Press: New York, 1995, pp. 613-651.
5. M. Ambe, K. Ohmura, "The role of haptic communication channel for the listeners in informal tele-communication," Human Interface 2000 symposium.
6. M. Hosoda, A. Nakayama et al, "Conference State Estimation by Biosignal Processing," Transaction of VRSJ, Vol.9, No.2, pp151-160 (2004)

Theory and Application of Artificial Neural Networks for the Real Time Prediction of Ship Motion

Ameer Khan[1], Cees Bil[1], and Kaye E. Marion[2]

[1] RMIT University, School of Aerospace, Manufacturing and Mechanical Engineering
GPO Box 2476V, Melbourne, Victoria, 3001 Australia
[2] RMIT University, School of Mathematical and Geospatial Sciences
GPO Box 2476V, Melbourne, Victoria, 3001 Australia

Abstract. Due to the random nature of the ship's motion in an open water environment, the deployment and the landing of vehicles from a ship can often be difficult and even dangerous. The ability to predict reliably the motion will allow improvements in safety on board ships and facilitate more accurate deployment of vehicles off ships. This paper presents an investigation into the application of artificial neural network methods for the prediction of ship motion. Two training techniques for the determination of the artificial neural network weights are presented. It is shown that the artificial neural network based on the singular value decomposition produces excellent predictions and is able to predict the ship motion in real time for up to 10 seconds.

1 Introduction

An algorithm capable of predicting the motion of a ship is required for the successful deployment of a ship system currently used on ships that operate in open sea environments. The predicted motion and attitude of the ship will be transmitted to the ship system to ensure successful activation. With the predicted ship motion, the correct flight conditions can be calculated thereby allowing successful and safe deployment of the ship system.

A key requirement of the algorithm is to predict in advance by x, x+2 and x+4 seconds (where x=3-4 seconds) whether the angles are likely to exceed the "launch lockout value" which is the condition where the system cannot be activated. When it is predicted that the angles exceed the launch lockout value the algorithm will automatically select another ship system that is not in lockout. Otherwise, the algorithm allows activation of the ship system. Therefore it is necessary to predict the maximum and minimum roll angles or the turning points in the motion. It is important that the predicted angles are of a high accuracy as the batteries for the system are "one shot" batteries, which means that the process of deployment once activated cannot be reversed.

The motion of a ship in an open water environment is the result of complex hydrodynamic forces between the ship, the water and unknown random processes. This leads to the necessity to use statistical prediction methods for the prediction of this motion rather then a deterministic analysis, which would lead to a ship specific model that involves highly complex calculations [1].

R. Khosla et al. (Eds.): KES 2005, LNAI 3681, pp. 1064–1069, 2005.

2   Rationale for Performing the Research

The ability to predict the ship motion reliably in any sea state will enable better control of systems operated off ship platforms. For example, the landing and take off of helicopters and aircraft whether manned or unmanned from ship decks in rough sea conditions can be difficult and at times dangerous.

If the motion of a ship can be predicted with reasonable error bounds and communicated to the aircraft or helicopter, touchdown dispersion can be improved on landing and a smoother aircraft trajectory can be achieved on take off. Prediction of ship motion is also important for the deployment of missiles and remote piloted vehicle from ship platforms for the correct trajectory calculation [2]. In some cases there is a launch "lock out" condition where the missile or remote piloted vehicle cannot be launched safely.

Most statistical techniques used for time series prediction have difficulty dealing with noisy data, do not have much parallelism and fail to adapt to circumstances. Artificial Neural Networks in contrast promise to produce predictions with high accuracy as well as high efficiency due to their ability to learn and adapt according to the conditions present. There is also the possibility that the methodology developed can be applied to other scenarios that require the prediction of a stochastic harmonic process in real time. In particular the paradigm could be used in areas such as economics, acoustics and other areas of interest which have stochastic harmonic motion present.

3   Artificial Neural Networks

Artificial neural networks (ANN) form a class of systems that are inspired by biological neural networks [3]. A three layered feed forward ANN consisting of an input layer, a hidden layer and an output layer was utilized in this investigation.

A single neuron has n inputs including a bias term, which has been set to 1 for this investigation. The inputs are each multiplied by their corresponding weight value, which are summed together and subsequently entered into an activation function. The output of the activation function will correspond to the output of the neuron. Mathematically, the output of a neuron is given as:

$$out = f\left(net\right) = f\left(\sum_{i=0}^{n-1} I_i w_i + w_n\right) \tag{1}$$

where the inputs are $\{I_i,\ i = 0,\ldots,\ n\text{-}1\}$. In this investigation the activation function shown below was used:

$$f\left(net\right) = \tanh\left(net\right) \tag{2}$$

The entire ANN process is depicted in Fig. 1.

3.1   Training the Network

The training of the network can be viewed as a minimization process where the weights in the ANN are systematically adjusted in a manner that reduces the error between the output of the ANN and the desired output. The algorithm used to determine the minimum must ensure that global minimum is achieved and has not discovered a local minimum.

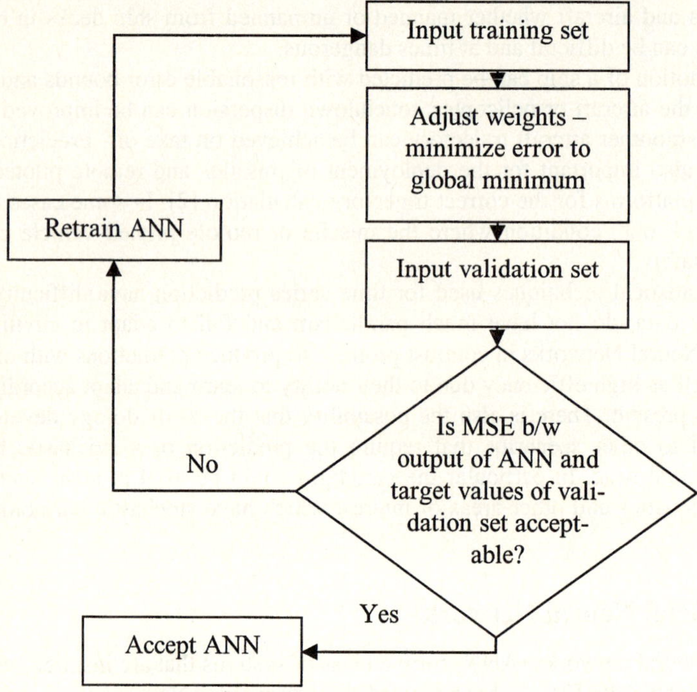

Fig. 1. The artificial neural network process flow chart

Conjugate-Gradient Algorithm

The conjugate gradient (CG) algorithm created for the ANN in this investigation was based on the Polak-Ribiere algorithm. The mathematical justifications for the algorithm are beyond the scope of this paper but a detailed description can be found in Polak [4]. In a general sense, the algorithm generates a sequence of vectors and search directions. It can be shown that the exact minimum will be obtained if the multi-dimensional function can be expressed as a quadratic. The ANN error function is quadratic close to the minimum so it is expected that once close to a minimum, convergence to the local minimum will be very rapid [5].

Genetic Algorithm

The genetic algorithm (GA) is part of a rapidly growing field of artificial intelligence called evolutionary computing and can be used to find the global minimum through the use of genetics and natural selection [6]. It uses the principle of survival of the fittest where the 'fittest' (best) 'survivor' (solution) evolves to create the next population [7]. To ensure maximum efficiency, the GA is used to establish an approximate location of the global minimum and then the conjugate gradient method is used to rapidly ascertain the exact values of the weights.

Singular Value Decomposition

The aim of this investigation is to develop a methodology to predict ship motion in real time. Singular value decomposition (SVD) is an important tool of matrix algebra that has been applied to a number of areas [8]. The application of SVD method showed a significant increase in the speed at which the weights for the ANN are calculated. A detailed description of the SVD technique is beyond the scope of this paper but essentially the matrix X which satisfies the function

$$A.X = B \tag{3}$$

when A and B are known can be calculated efficiently using SVD. When applying it to the ANN process the weights between the input layer and the hidden layer are initially randomly generated. The training samples are then inserted into the ANN and the hidden layer activation functions are calculated creating a matrix equivalent to A. Also, the values for the inverse transfer function of the output are also calculated creating a matrix equivalent to B. Applying SVD and solving Eq (3), the approximate optimal weights X are found.

3.2 Validation of ANN Model

To ensure that the weights in the ANN have been correctly set and that the output of the ANN is sufficiently reliable, a validation process is applied after training has been completed. Essentially the data used to test and evaluate the ANN model is divided into a training set and a validation set. The training set is exclusively used for training the ANN while the validation set is used to test the accuracy of the ANN model.

4 Application of ANN to Ship Motion Prediction

The algorithms developed were applied to measured ship roll angle data taken from Frigate class vessels operating in sea states 5-6. The ANN algorithms were applied to nine separate data sets which were each between 300-600 seconds in length. The training data was set to two thirds of the data sets and the validation set was designated as the final third of the data sets. All results shown are the predictions made using the validation set only.

The results are shown in Table 1. The emphasis of the results is on the accuracy and the speed of the ANN. The accuracy was assessed using two criteria. Criterion 1 is the percentage of predictions accurate within the 95% confidence interval. This criterion essentially assesses the ability of the predictive algorithm to represent every aspect of the ship's motion. As stated in the introduction, the success criterion for this investigation is to develop an algorithm that can predict when a predefined angle is exceeded. The second criterion is percentage forecasts that correctly predicted that the roll angle would exceed a predefined lock-out angle of 7°.

Two ANN training regimes were used. The first used a combination of the GA and CG algorithms to calculate the weights for the ANN. The GA and CG methods were used alternatively until no further improvements were produced. The second used the SVD method to calculate the weights for the ANN. A range of different ANN structures were used in this investigation. The number of input nodes varied from 4 to 6 and the number of hidden layer nodes varied from 1 to 3. The results shown in Table 1 are the average results found by applying the ANN to the nine data sets. Ten trials were used for every ANN trained using the SVD node configuration and the best result chosen. The processing times for the SVD method shown in Table 1 represent the total time to complete the ten trials. A portion of a ten second in advance prediction of the ANN trained using SVD is shown in Fig 2 and shows that predicted motion essentially mirrors the actual motion.

Table 1. Results of the application of artificial neural networks trained using GA/CG and SVD methods for determination of weights

Prediction Interval (sec)	Combined GA and CG method			SVD method		
	Criterion 1 (%)	Criterion 2 (%)	Processing time (sec)	Criterion 1 (%)	Criterion 2 (%)	Processing time (sec)
1	73.34	97.83	106.2	99.73	100	0.116
2	64.41	99.56	167.79	99.72	100	0.105
3	61.76	99.4	297.10	99.57	100	0.118
4	66.42	99.54	38.73	99.7	100	0.123
5	63.19	99.27	47.46	99.51	100	0.115
6	62.76	99.56	34.68	99.53	100	0.118
7	66.56	99.57	108.14	99.68	100	0.113
8	59.21	98.34	231.06	99.5	100	0.126
9	65.74	97.58	44.25	99.35	100	0.137
10	63.37	99.77	61.77	99.61	100	0.114

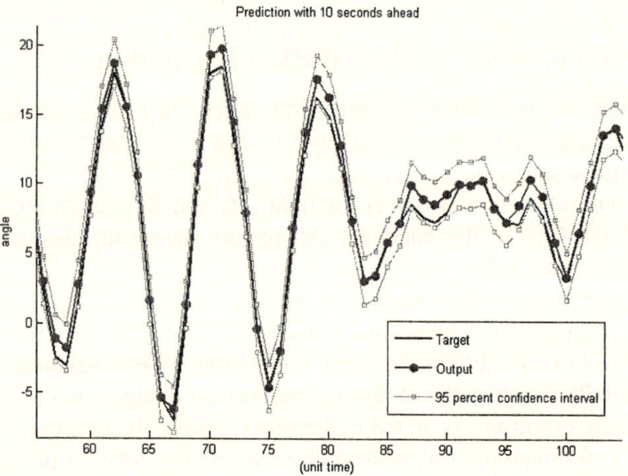

Fig. 2. Portion of 10 second prediction of SVD trained ANN

5 Discussion of Investigation Results

Table 1 shows the effectiveness of ANN for the prediction of ship motion. Using the SVD training scheme far superior standards of accuracy end efficiency were achieved compared to the ANN trained using the GA/CG training scheme. The SVD trained ANN was able to achieve accuracy levels of above 99% based on criterion 1 for predictions up to 10 seconds in advance. In contrast the GA/CG trained ANN only achieved mediocre accuracies of only 59% for some prediction intervals based on criterion 1.

The SVD trained ANN also performed better on criterion 2. The results show that the SVD trained ANN produced accuracy levels of 100% for criterion 2. The GA/CG method also produced good results of greater than 97% accuracy for prediction up to 10 seconds. This demonstrates that both training regimes are capable producing adequately accurate predictions of the lock-out condition however one of the primary stated objectives is to predict the lock-out condition in real time.

From Table 1 it is clear that the computational time of the GA/CG trained ANN is too excessive. In some cases it can take up to four minutes to find the best solution. In comparison, the SVD trained ANN was able to produce the results in approximately one tenth of a second. This implies that the SVD trained ANN is suitable for real time prediction applications.

6   Conclusion

In this paper artificial neural network based methods for the prediction of the ship motion was presented. The weights for the artificial neural network were determined using a combination of the genetic algorithm and conjugate gradient techniques and separately using the singular value decomposition technique. It was found that the singular value decomposition based algorithm outperformed the genetic algorithm and conjugate gradient technique and was able to produce a set of weights that allowed very accurate forecasts in fractions of a second which implies that this method has the real time capability required in operational circumstances. The ANN architecture is therefore sufficiently capable of representing the motion of ship in an open sea environment subjected to complex hydrodynamic forces.

Acknowledgements

The authors would like to thank BAE Systems Australia for supporting the project.

References

1. Price, W. G., Bishop, R. E. D.: Probabilistic theory of ship dynamics. Chapman and Hall Ltd, Salisbury (1974)
2. Sidar, M. M., Doolin, B. F.: On the feasibility of real-time prediction of aircraft carrier motion at sea. IEEE Transactions on Automatic Control. 28 (1983) 350-356.
3. Suykens, J. A. K., Vandewalle, J. P. L., Moor, B. D.: Artificial neural networks for modeling and control of non-linear systems. Kluwer Academic Publishers, Dordrecht (1996)
4. Polak, E.: Computational methods in optimisation. Academic Press, New York (1971)
5. Press, W. H., Flannery, B. P., Vetterling, W. T., Teukolsky, S. A.: Numerical recipes in Fortran: The art of scientific computing. Cambridge University Press, New York (1992)
6. Haupt, R. L., Haupt, S. E.: Practical Genetic Algorithms. John Wiley and Sons, (1998)
7. Obitko, M.: Genetic algorithms. http://cs.felk.cvut.cz/~xobitko/ga/ (Accessed: 16 March 2004),
8. Gass, S. I., Rapcsák, T.: Singular value decomposition in AHP. European Journal of Operational Research. 154 (2004) 573-584.

Soft Computing Based Real-Time Traffic Sign Recognition: A Design Approach

Preeti Bajaj[*], A. Dalavi, Sushant Dubey, Mrinal Mouza,
Shalabh Batra, and Sarika Bhojwani[**]

G.H Raisoni College of Engineering, Nagpur
preetib123@yahoo.com, dubey_sushant@rediffmail.com
mrinalmouza@indiatimes.com, shalabhbatra100@yahoo.com

Abstract. The traffic sign detection and recognition system is an essential module of the driver warning and assistance system. During the last few years much research effort has been devoted to autonomous vehicle navigation using different algorithm. Proposed work includes a neural network based drivers assistance system for traffic sign detection. In this paper authors have implemented a high-speed color camera to enhance its performance in real time scanning. The proposed algorithm increases the efficiency of the system by 7 to 10% as compared to conventional algorithms. The system includes two main modules: detection module and recognition module. In the detection module, the thresholding is used to segment the image. The features of traffic signs are investigated and used to detect potential objects. In recognition module, we use complimenting and then ANDing techniques. The joint use of classification and validation networks can reduce the false positive rate. There liability demonstrated by the proposed method suggests that this system could be a part of an integrated driver warning and assistance system based on computer vision technology.

1 Introduction

Traffic signs provide the driver information about safe and efficient navigation. Automatic recognition of traffic signs is, therefore, important for automated driving or driver assistance systems [1-2]. During the last few years much research effort has been devoted to autonomous vehicle navigation using digital image processing. Most has been aimed to road boundary detection and obstacle avoidance, and several very robust and reliable systems have been implemented. Several different techniques have been applied, as shown in [3-4], but lately the use of neural networks has produced promising results because of their robustness and computational simplicity [5-6]. There exists a system based on Global Position System (GPS) and digital road maps with speed limits included [7]. Only a few systems have been designed for the purpose of autonomous vehicle navigation by traffic sign detection and recognition, [8].

Since these signs are usually painted (specially in most of Asian countries) by distinctive colors, they may be easily detected using color information [10]. However, color information is sensitive to the change of weather or lighting condition and, therefore, it is sometime difficult to extract traffic signs reliably only by color.

---

[*] Professor & Head, ETRX Department
[**] Research Scholars

R. Khosla et al. (Eds.): KES 2005, LNAI 3681, pp. 1070–1074, 2005.

2 Methodology

In addition, in urban cluttered scenes, other signboards or buildings with the similar color to traffic signs make it more difficult to extract an appropriate region for the target signs in the image. To cope with such problems, it was proposed to use shape information [9]. To recognize traffic signs, it is often necessary to recognize their characters and symbols. For reliable recognition of characters and symbols, it is desirable to capture a sign in a large size in the image. To do this, it is proposed to use a camera with a telephoto lens and direct it to the target. The camera direction is automatically determined based on the prediction of the target motion in the image. This paper describes an active-vision system that can recognize various traffic signs in real-time based on the above-mentioned approach. The recognition algorithm is designed intensively using built-in functions of an off the-shelf image processing board in order to realize both easy implementation and fast recognition.

Fig1 illustrates the system configuration. The system has two cameras; one is equipped with a wide-angle lens (called *wide-camera*) and the other with a telephoto lens (called *tele-camera*), and a PC with an image processing board. The wide-camera is directed to the moving direction of the vehicle. The tele-camera changes the viewing direction to focus the attention to the target sign.

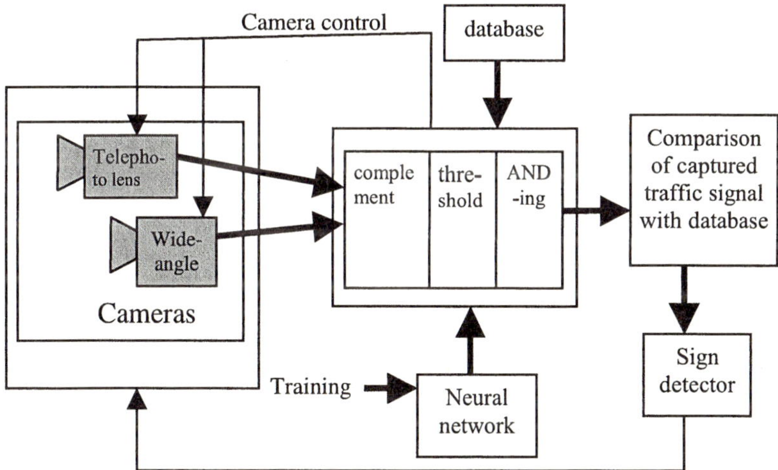

Fig. 1. The Block Schematic for Traffic Sign Detection

The Design Includes the Following Steps
1. Neural Network training for all traffic signs.
2. Detect candidates of traffic signs with threshold and complementing techniques.
3. Screen candidates using edge information.
4. Predict the motion of a screened candidate in the image and direct the tele camera to it.
5. Extract characters and symbols (if necessary).
6. Identify the sign by Neural Network tool.
7. Alerting the driver.

3 Design Approach

3.1 Complementing the Image

Converts object of interest as the watershed transform detects intensity "valleys" in an image, the program uses the imcomplement function on the enhanced image to highlight the intensity valleys. Conversion of the input image to grayscale format is accomplished by imcoplement function (if it is not already an intensity image), and then converts this grayscale image to binary by thresholding. The output binary image BW has values of 0 (black) for all pixels in the input image with luminance less than level and 1 (white) for all other pixels.

3.2 Thresholding the Image

To decide the threshold for an image authors has created morphological structuring element using a technique called structuring element decomposition. The principle is that dilation by some large structuring elements can be computed faster by dilation with a sequence of smaller structuring elements. For example, dilation by an 11-by-11 square structuring element can be accomplished by dilating first with a 1-by-11 structuring element and then with an 11-by-1 structuring element. This results in a performance improvement of a factor of 5.5. After thresholding the smaller elements gets filtered out.

3.3 ANDing of Images

The binary converted image and the image from which the smaller elements have been filtered out are ANDED to get final image, which is then compared with the image, which is in the data base.

Sample Coding for Algorithm

```
   Ic = imcomplement(image);
%(IM) computes the complement of the image IM. IM can be %a bi-
nary,intensity, or RGB image. IM2 has the same class %and size
as IM.
   BW = im2bw(Ic, graythresh(Ic));
%Convert an image to a binary image,based on threshold
   se = strel('disk', 4);
%a structuring element, SE, of the type specified by %shape Flat
Structuring elements
   BWc = imclose(BW, se);
%IM2 = imclose(IM,SE) performs morphological closing on the
grayscale or binary image IM, returning the closed image,IM2
   BWco = imopen(BWc, se);
%IM2 = imopen(IM,SE) performs morphological opening on the gray-
scale or binary image IM with the structuring element SE
   mask = BW & BWco;
%search or cell command ..go to celldisp(i1) %i=input('Enter the
File Name in the current %directory\n','s');
```

```
    dt = ismember(mask, image2)
% [c,i] = setdiff(image2(:),mask(:));print(c);
%tf = ismember(A,S) returns a vector the same length as A %con-
taining logical true (1) where the elements of A are %in the set
S, and logical false (0) elsewhere. In set %theoretic terms, k
is 1 where A belongs to S. A and S %can be cell arrays of
strings.
```

4 Implementation and Results

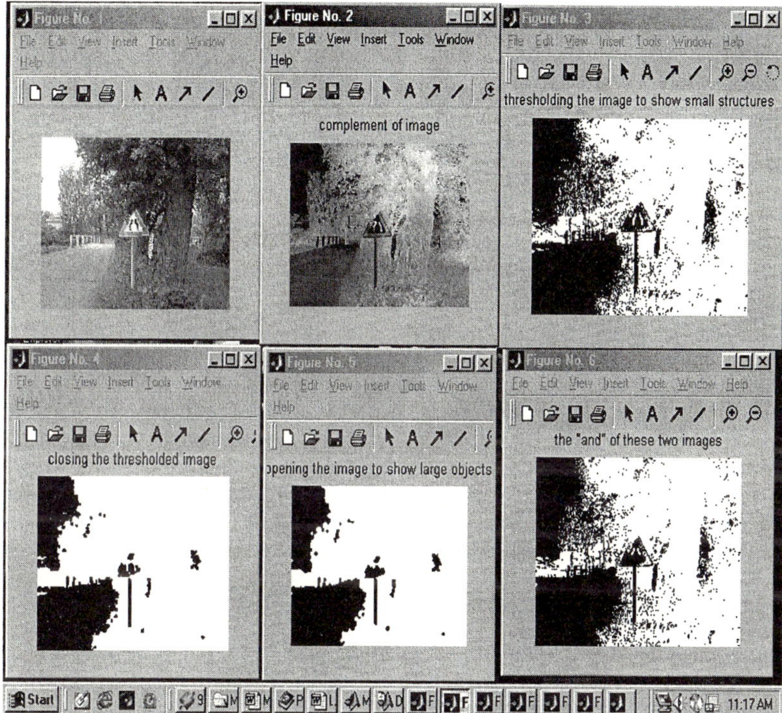

Fig. 2. (a-f)- (a) Traffic sign (b) Complemented image (c) Thresholded image (d) Closed thresholded image (e) Image with large objects (f) ANDed image

Fig. 3. System alerts with a message "narrow road ahead"

5 Conclusion

Proposed work includes a neural network based drivers assistance system for traffic sign detection. In this paper authors have implemented a high-speed color camera to enhance its performance in real time scanning. The proposed algorithm increases the efficiency of the system by 7 to 10% as compared to conventional algorithms. If the driver is not attentive then the proposed strategy will help the car to be attentive so that accidents can be avoided and it will also see that no road sign should be missed.

References

1. S. Azami, S. Katahara, and M. Aoki. "Route guidance sign identification using 2-d structural description". *Proceedings of the 1996 IEEE Symp. on Intelligent Vehicles*, pp. 153–158, 1996.
2. U. Franke et al. "Autonomous driving goes downtown". *IEEE Intelligent Systems & Their Applications*, Vol. 13, No. 6, pp. 40–48, 1998.
3. Charles Thorpe and Tadeo Kanade. *"1987 Year End Report for Road Following at Carneige Mellon"*, CMU-RI-TR-88-4. The Robotics Institute. Carnegie Mellon Uiversity. April 1988.
4. Graefe, V.,B16chl, B. *"Visual Recognition of Traffic Situations for an Inteligent Automatic Co-pilo't"*. PROMETHEUS Workshop, Proceedings of the 5th workshop, Munich, 1991,pp.98-108.
5. Pomerleau, D.A. *ALVINN: "An Autonomous Land Vehicle in a Neural Network"*, Technical Report CMU-CS-89-107. School of Computer Sience. Carnegie Mellon University. 1989.
6. P. Viola and M. J. Jones. "Robust Real-Time face detection". International Journal of Computer Vision 57(2), pages 137–154, 2004
7. R. Thomas. Less is more "intelligent speed adaptation for road vehicles". *IEE Review*, 49(5):40–43, May 2003.
8. Austermeier H., Biiker U., Mertsching B., Zimmermann S. *"Analysis of Traffic Scenes by Using the Hierarchical Structure Code"*. Advances in Structural and Syntactic Pattern Recognition, proc. of the International Workshop on Structural and Syntactic Pattern Recognition, Bern, Switzerland, August 1992. Bunke and Wang, Series in machine perception and artificial inteligence, Vol. 5, pp.561-570.
9. G. Piccioli et al. "Robust method for road sign detection and recognition. *Image and Vision Computing"*, Vol. 14, pp. 209–223, 1996.
10. L. Priese et al. "Traffic sign recognition based on color image evaluation". In *Proceedings of the 1993 IEEE Symp. on Intelligent Vehicles*, pp. 95–100, 1993.

A Soft Real-Time Guaranteed Java M:N Thread Mapping Method

Seung-Hyun Min[1], Kwang-Ho Chun[2], Young-Rok Yang[1], and Myoung-Jun Kim[1]

[1] Dept.of Computer Sicence, Chungbuk National Univ., Korea
imturtle@iita.re.kr
[2] Dept. of Electronic & Information Eng. Chonbuk National Univ., Korea

Abstract. A 1:1 mapping model between a user thread and a system thread has a merit that supports parallelism, and a M:N mapping has a merit that supports parallelism and fast context exchange. It is necessary to improve performances in order to guarantee soft real-time by minimizing the context exchange between threads. This paper proposes a Java thread model, which supports a M:N mapping in a Linux Java virtual machine, by introducing the concept of LWP Light Weight Process (LWP) in a Java application level. The proposed model is able to maintain a fast processing speed, and independence from a Java platform. In addition, this model guarantees the soft real-time of a Java application program by applying a MTL-LS algorithm to LWP scheduling, and guarantees an excellent performance and soft real-time of the proposed model, which is introduced using a comparison between previous 1:1 model and previous M:1 mapping model.

1 Introduction

Java is a kind of programming language, which has the merits of platform independence, high security levels, multi-threads, and garbage collecting. This language is used in the field of Internet applications, and application programs of an internal system. However, it also has several problems supporting real-time functions in a Java application program level. For instance, a number of properties, which include a delay time occurring during the management of garbage collector memory, inability to access a physical memory, and absence of a standardized real-time API, suggest that Java is unsuitable for use in real-time application programs.

In spite of these linguistic characteristics, studies supporting soft real-time in a Java application program level have been conducted. The RTSJ (Real-Time Specification for Java) is a type of technical document to support real-time, and provides several APIs for real-time application programs [1]. There are many studies on RMS (Rate Monotonic Scheduling) analysis in a Java platform proposed by Nilsen, Static Cycle Scheduling, and a real-time garbage collector, but these studies are not easy to apply in practical applications, due to their hardware dependency [2].

QJVM is a type of virtual machine, which uses a green thread library, where thread generation, resources management, scheduling, synchronization, and other functions are performed in a user level [3]. This QJVM only considers M:1 mapping using a Move-to-Rear List Scheduling (MTR-LS) algorithm, which satisfies soft real-time. Because a M:1 mapping model performs context exchange in a use level regardless of the type of kernel, the cost of overhead is very low. However, its response time may

R. Khosla et al. (Eds.): KES 2005, LNAI 3681, pp. 1075–1080, 2005.

be lowered by a pending state, in which all threads wait until a thread enters a kernel during input or output processes. In addition, the QJVM cannot support parallelism in a kernel because multiple user threads are mapped with a single kernel, and only support concurrency at a user level. In the present version of the Java virtual Linux machine and Linux used a native thread mapping model in order to support parallelism. This method can support parallelism in a kernel level due to the fact that user threads and single kernel are mapped by a 1:1 model. However, this method presents many overhead problems due to the entrance into a kernel when context exchange occurs at a user level. In order to satisfy soft real-time, an algorithm, which fairly assigns resources when multiple user thread scheduling supported by Java is performed in a certain level of application program, is required [4-6]. Thus, a mapping model, which supports a fast context exchange and parallelism, is necessary to guarantee soft real-time.

This paper proposes a Java M:N thread model in order to guarantee soft real-time by supporting a fast context exchange and parallelism. In the proposed model, a user can map multiple green threads using a single LWP.

This paper consists of five Sections. Section 2 investigates a MTR-LS algorithm performed in a Java virtual machine. Section 3 describes the Java M:N mapping model proposed in this paper. Section 4 examines the performance of the proposed 1:1 and M:1 model in a Linux virtual machine through a test. Finally, Section 5 concludes this paper.

2 MTR-LS Algorithm

A MTR-LS algorithm can reserve CPU resources as much as it is required at an application program level, and makes a schedule based on the CPU quantum, which is a type of service time input from users. The priority of all threads can be assigned by the value of quantum, and the thread can preempt CPU according to the value of quantum. Thus, the MTR-LS algorithm can provide a guaranteed service in soft real-time. Fig. 1 presents a thread flow in the MTR-LS algorithm implemented in a Java virtual machine. In addition, the characteristics presented in each thread are as follows.

A. Thread creation
1. A quantum item is added to the thread material structure of a thread.
2. The value of the quantum is clearly input by the user.
3. The priority is defined at a queue according to the value of the quantum.

B. Wait queue
1. It waits until that the condition variable becomes true.
2. An arrange is not analyzed.

C. Runnable queue
1. As an arrange of the priority, a hip material structure, which has a lower time complexity than that of the linked list, is used.
2. A proper priority can be assigned by the value of quantum.
3. CPU resources can be serviced as much as the amount of quantum that a thread has using a MTR-LS method.

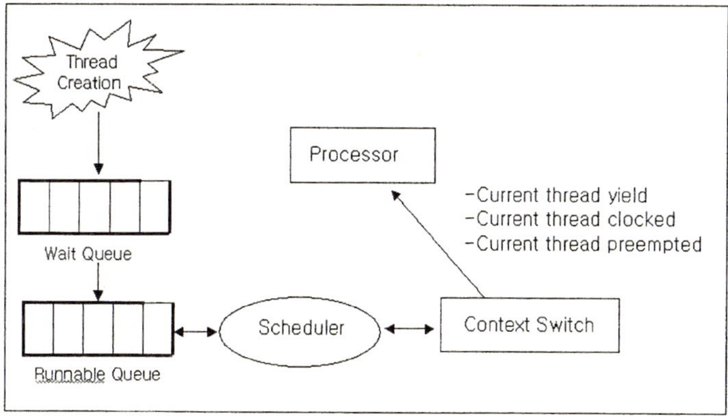

Fig. 1. Thread flowchart

D. Context switch
1. Periodical calculation of the value of quantum
    - The first in-line function
    - The second in-line function

In order to apply the soft real-time guaranteed algorithm proposed in this study, a number of changes to the MTR-LS are required. First, it is necessary to simultaneously generate a green thread and LWP, and the value of the quantum must be assigned to the LWP. Second, a green thread must be scheduled using a cooperative type of scheduling, and the LWP must be scheduled using a MTR-LS preemptive algorithm.

3   M:N Thread Mapping Model

A user thread used in a Java application program level should be mapped using a system thread of an operating system. Then, it can perform calculations by preempting a CPU. Because the supporting methods for a thread used in Unix, Windows NT, Windows 95, and Linux are different to one another, a system thread mapping model and user thread mapping model may different in each operating system. A thread used in an operating system is classified as a user thread and a kernel thread. A user thread is produced using user libraries, and a kernel thread is a unit object of the actual scheduling in an operating system. Although 1:1, M:1, and M:N models are used as thread mapping models, the M:N mapping model presents the best performance for this case, which required a fast context exchange and parallelism, among them. Because a Linux system can only support a 1:1 mapping model, it is possible to improve the performance by supporting a M:N mapping model in an application program level. Fig. 2 illustrates the structure of the proposed model, and the model is applied using a LWP at a Java application program level.

A single or multiple green threads can be mapped using a LWP, and the mapping is clearly achieved by a user. The priority for a LWP is properly assigned according to the value of time quantum, which is input by a user. The value of time quantum assigned to a LWP is shared by green threads, and changed by the priority together with a fast context exchange. Because multiple LWPs are processed by a native thread in a Linux user level, and are mapped using a 1:1 mapping model, a parallel process in a kernel level is possible. A LWP follows a MTR-LS type of preemptive scheduling method, and the priority is properly assigned using the value of quantum, which is input by the user as shown in Fig. 2. This model has the advantage that it can manage CPU resources in the context of an application by inputting the value of quantum from the viewpoint of the user, and can estimate the possibility of scheduling.

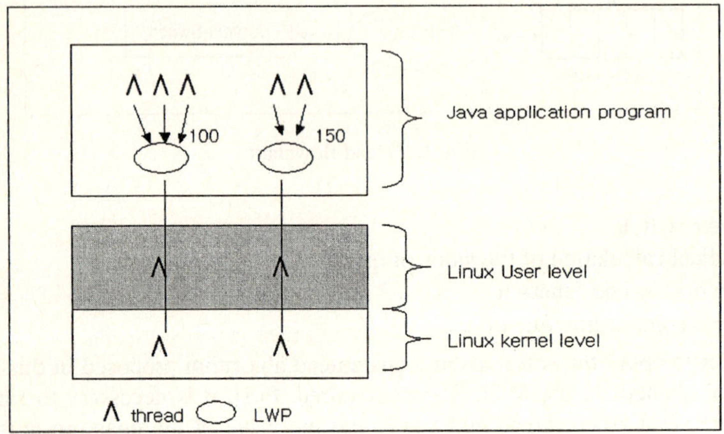

Fig. 2. M:N Thread Mapping Model

4   Implementation and Test

We investigates a certain consideration to implement the model proposed in this study, and compares the performances of a 1:1 and M:1 mapping model, respectively, through a test.

A Runnable array object, which is used as a green thread, generates a green thread, and is scheduled as a storage object using a type of cooperative method. The Runnable object can be implemented as a circular structure, and a data sharing issue in the multiple thread can be solved using a synchronizing method [7]. A code for generating and storing a green thread for a Runnable object can be expressed as follows.

```
Private Runnable[] task =
{
        new Runnable() { public void run() { ⊢ },
        new Runnable() { public void run() { ⊢ },
        new Runnable() { public void run() { ⊢ }
}
```

A LWP can be produced by transferring multiple Runnable object arrays to a thread generator. The green thread is executed by sharing all data areas of a LWP. The LWP process generation codes are as follows.

```
new Thread(new Runnable[] {new by_runnable_Object(),
                    new My_other_runnable_Object[]});;
```

The LWP process technology consists of various items. A quantum item is to be applied when a MTR-LS algorithm is used.

```
public class Thread
{
                Object value ;
                String Id ;
                long quantum ;
                int priority ;
                Node next ;
                public Node(Object  o ) {
                                value = 0 ;
                                next = null
                }
}
```

A test for the comparison between the M:N thread mapping model and the 1:1 and M:1 mapping models applied in a Linux Java virtual machine was performed sing a Pentium-III 600MHz computer, Linux paran 7.3, and JDK 2.2 glibc (Native Version). The test generated 9 equal threads, and measured the finishing time of each thread during the given 1,000 looping times. The 1:1 model didn't use a scheduling policy. The LWP generates 3 LWPs (three green threads per LWP), and didn't apply a MTR-LS algorithm. The LWP+MTR-LS is a test, which is conducted by generating a LWP, using a MTR-LS algorithm. The test revealed that the method applied by a M:N mapping model and MTR-LS algorithm performed better than that of the other two models in a Java application program level.

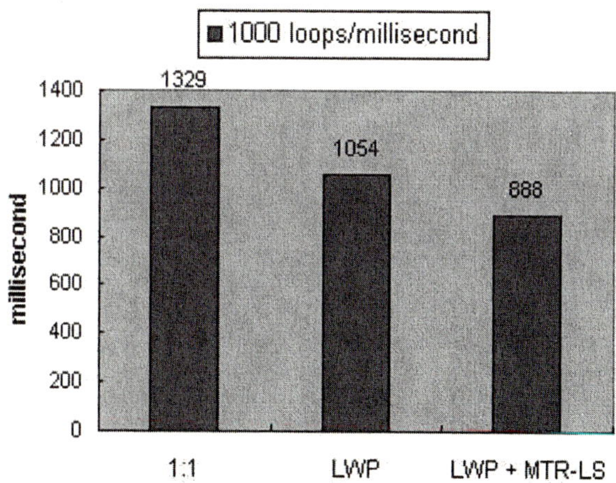

Fig. 3. The test results

5 Conclusions

This paper proposed a M:N thread mapping model, which guarantees soft real-time, in a Java application program level. The proposed model generates a green thread and LWP in a Java application program level, and the mapping is achieved using a M:N mapping model. In addition, a Java application program can guarantee soft real-time by applying a MTR-LS algorithm to a LWP, which is generated by a M:N mapping model. Consequently, the proposed model, which is implemented through a test performed better than that of the 1:1 and M:1 models.

References

1. G. Bollella, et al., The Real-Time specification for Java, Addison-Wesley, June 2000.
2. K. Nilsen, Java for Real-time, In the journal of Real-Time System, pp.197-205, February 1996.
3. James C. Pang, Gholamali C. Shoja, Eric G. Manning, Providing Soft Real-time QoS Guarantees for Java Threads, Concurrency and computation: Practice and Experience, pp.521-538, March-April 2003.
4. P. Goyal, H.M. Vin, H. Cheng, Start-time Fair Queueing: A Scheduling Algorithem for Integrated Service Packet Switching Networks, In Proceedings of ACM SIGCOMM'96, pp. 157-168, August 1996.
5. A. Demers, S. Keshav, S. Shenker, Analysis and simulation of a Fair Queueing Algorithm In Proceedings of ACM SIGCOMM, pp. 1-12, september 1989.
6. J. Bruno, E. Gabber, B. Ozden, A. Silberschatz, Move-To-Rear List Scheduoing: A New scheduling Algorithem for Providing Qos Guarantees, Proceedings of the Conference On Multimedea, pp. 63-73, November 1997.
7. Allen I. Holub, Taming Java Threads, Apress, Jun 2000.

A Genetic Information Based Load Redistribution Approach Including High-Response Time in Distributed Computing System

Seong Hoon Lee[1], Dongwoo Lee[2], Wankwon Lee[3], and Hyunjoon Cho[3]

[1] Department of Computer Science, Cheonan University
115, Anseo-dong, Cheonan, Choongnam, Korea, 330-180
`shlee@cheonan.ac.kr`
[2] Department of Computer Science, Woosong University
17-2 Jayang-dong dong-ku Daejon, Korea 300-718
`dwlee@woosong.ac.kr`
[3] School of Information Technology & Engineering, Jeonju University 1200,
3rd Street Hyoja-dong Wansan-Koo, Jeon-Ju, Chonbuk, Republic of Korea
`{wklee,chohj}@jj.ac.kr`

Abstract. Load redistribution algorithm is a critical element in distributed system. We propose a new load redistribution algorithm based on genetic information to obtain high response time in distributed systems. Under sender-initiated load redistribution algorithms, the sender continues to send unnecessary request messages for load transfer until a receiver is found while the system load is heavy. Because of these unnecessary request messages, it results in inefficient communications, low cpu utilization, and low system throughput. This algorithm decreases response time and increases acceptance rate.

1 Introduction

An objective of load redistribution in distributed systems is to allocate tasks among the processors to maximize the utilization of processors and to minimize the mean response time. Load redistribution algorithms can be largely classified into three classes: static, dynamic, adaptive. Our approach is based on the dynamic load redistribution algorithm. In dynamic scheme, an overloaded processor(sender) sends excess tasks to an underloaded processor(receiver) during execution. Dynamic load redistribution algorithms are specialized into three methods: sender-initiated, receiver-initiated, symmetrically-initiated. Basically our approach is a sender-initiated algorithm.

Under sender-initiated algorithms, load redistribution activity is initiated by a sender trying to send a task to a receiver[1, 2]. In sender-initiated algorithm, decision of task transfer is made in each processor independently. A request message for the task transfer is initially issued from a sender to an another processor randomly selected. If the selected processor is receiver, it returns an accept message. And the receiver is ready for receiving an additional task from sender. Otherwise, it returns a reject message, and the sender tries for others until receiving an accept message. If all the request messages are rejected, no task transfer takes place. When a distributed system becomes to heavy system loads, it is difficult to find a suitable receiver because most processors have additional tasks to send. So, many request and reject mes-

R. Khosla et al. (Eds.): KES 2005, LNAI 3681, pp. 1081–1087, 2005.

sages are repeatedly sent back and forth, and a lot of time is consumed before execution. Therefore, it causes low system throughput, low cpu utilization.

To solve these problems in sender-initiated algorithm, we use a new genetic algorithm. A new genetic algorithm evolves strategy for determining a destination processor to receive a task in sender-initiated algorithm. The rest of the paper is organized as follows. Section 2 presents the Genetic Algorithm-based sender-initiated approach. Section 3 presents several experiments to compare with conventional method. Finally the conclusions are presented in Section 4.

2   Genetic Algorithm Based Load Redistribution Approach

2.1   Load Measure and Coding Method

We employ the CPU queue length as a suitable load index because this measure is known the most suitable index[5]. This measure means a number of tasks in CPU queue residing in a processor. We use a 3-level scheme to represent a load state on its own CPU queue length of a processor. T_{up} and T_{low} are algorithm design parameters and are called *upper* and *lower thresholds* respectively.

The transfer policy uses the threshold policy that makes decisions based on the CPU queue length. The transfer policy is triggered when a task arrives. A node identifies as a **sender** if a new task originating at the node makes the CPU queue length exceed T_{up}. A node identifies itself as a suitable **receiver** for a task acquisition if the node's CPU queue length will not cause to exceed T_{low}.

Each processor in distributed systems has its own population which genetic operators are applied to. We use binary encoding method in this paper. So, a string in population can be defined as a binary-coded vector $<v_0, v_1, ..., v_{n-1}>$ which indicates a set of processors to which the request messages are sent off. If the request message is transferred to the processor P_i(where $0 \leq i \leq n-1$, n is the total number of processors), then $v_i=1$, otherwise $v_i=0$. Each string has its own fitness value.

2.2   Load Redistribution Approach

2.2.1 Overview

In sender-based load redistribution approach using genetic algorithm, Processors received the request message from the sender send accept message or reject message depending on its own CPU queue length. In the case of more than two accept messages returned, one is selected at random. Suppose that there are 10 processors in distributed systems, and the processor P_0 is a sender. Then, genetic algorithm is performed to decide a suitable receiver. It is selected a string by a probability proportional to its fitness value. Suppose a selected string is $<-, 1, 0, 1, 0, 0, 1, 1, 0, 0>$, then the sender P_0 sends request messages to the processors (P_1, P_3, P_6, P_7). After each processor(P_1, P_3, P_6, P_7) receives a request message from the processor P_0, each processor checks its load state. If the processor P_3 is a light load state, the processor P_3 sends back an accept message to the processor P_0. Then the processor P_0 transfers a task to the processor P_3.

2.2.2 Fitness Function

Each string included in a population is evaluated by the fitness function using following formula in sender-initiated approach.

$$F_i = \left(\frac{1}{\alpha \times TMP + \beta \times TMT + \gamma \times TTP} \right)$$

α, β, γ used above formula mean the weights for parameters such as TMP, TMT, TTP. The purpose of the weights is to be operated equally for each parameter to fitness function F_i. Firstly, TMP(Total Message Processing time) is the summation of the processing times for request messages to be transferred. This parameter is defined by the following formula. The $ReMN$ is the number of messages to be transferred. It means the number of bits set '1' in selected string. (where, x={i□v_i=1 for $0 \leq i \leq n$-1})

$$TMP = \sum_{k \in x} \left(\mathrm{Re}MN_k \times TimeUnit \right)$$

Secondly, TMT(Total Message Transfer time) means the summation of each message transfer times($EMTT$) from the sender to processors corresponding to bits set '1' in selected string. The objective of this parameter is to select a string with the shortest distance eventually. So, we define the TMT as the following formula.

$$TMT = \sum_{k \in x} EMTT_k$$

Lastly, TTP(Total Task Processing time) is the summation of the times needed to perform a task at each processor corresponding to bits set '1' in selected string. This parameter is defined by the following formula. The objective of this parameter is to select a string with the fewest loads. Load in parameter TTP is the volume of CPU queue length in the processor.

$$TTP = \sum_{k \in x} \left(Load_k \times TimeUnit \right)$$

So, in order to have a largest fitness value, each parameter such as TMP, TMT, TTP must have small values as possible as. That is, TMP must have the fewer number of request messages, and TMT must have the shortest distance, and TTP should have the fewer number of tasks.

2.2.3 Algorithm

This algorithm consists of five modules such as Initialization, Check_load, Individual_evaluation, Genetic_operation and Message_evaluation. Genetic_operation module consists of three sub-modules such as Local_improvement_operation, Reproduction, Crossover. These modules are executed at each processor in distributed systems. The algorithm of the proposed method for sender-initiated load redistribution is presented as fig 1.

An Initialization module is executed in each processor. A population of strings is randomly generated without duplication. A Check_load module is used to observe its own processor's load by checking the CPU queue length, whenever a task is arrived in a processor. If the observed load is heavy, the load redistribution algorithm performs the following modules. A Individual_evaluation module calculates the fitness value of strings in a population. A Genetic_operation module is executed on the population in such a way as follows. Distributed systems consist of several groups with autonomous computers. When each group consists of many processors, we can suppose that there are p parts in a string corresponding to the groups. The following genetic operations are applied to each string, and new population of strings is generated:

<table>
<tr><td align="center">Algorithm: GA-based sender-initiated load redistribution algorithm</td></tr>
</table>

```
Procedure Genetic_algorithm Approach
 { Initialization();
   while ( Check_load() )
     if ( Load_i > T_up ) { Individual_evaluation(); Genetic_operation();
                       Message_evaluation();  }
    Process a task in local processor;
 }
Procedure Genetic_operation()
 { Local_improvement_operation(); Reproduction(); Crossover();
 }
```

Fig. 1. Proposed algorithm

(1) Local_improvement_operation
String 1 is chosen. A copy version of the string 1 is generated, and part 1 of the newly generated string is mutated. This new string is evaluated by proposed fitness function. If the evaluated value of the new string is higher than that of the original string, replace the original string with the new string. After this, the local improvement of part 2 of string 1 is done repeatedly. This local improvement is applied to each part one by one. When the local improvement of all the parts is finished, new string 1 is generated. String 2 is then chosen, and the above-mentioned local improvement is done. This local_improvement_operation is applied to all the strings in population.

(2) Reproduction
The reproduction operation is applied to the newly generated strings. We use the "*wheel of fortune*" technique[4].

(3) Crossover
The crossover operation is applied to the newly generated strings. These newly generated strings are evaluated. One-point crossover used in this paper differs from the pure one-point crossover operator. In pure one-point crossover, crossover activity generates based on randomly selected crossover point in the string. But the boundaries between parts(p) are used as an alternative of crossover points in this paper. So we select a boundary among many boundaries at random. And a selected boundary is used as a crossover point. This purpose is to preserve an effect of the Local_improvement_operation of the previous phase. Therefore, the crossover activity in this paper is represented as fig 2.

The Genetic_operation selects a string from the population at the probability proportional to its fitness, and then sends off the request messages according to the contents of the selected string. A Message_evaluation module is used whenever a processor receives a message from other processors. When a processor P_i receives a request message, it sends back an accept or reject message depending on its CPU queue length.

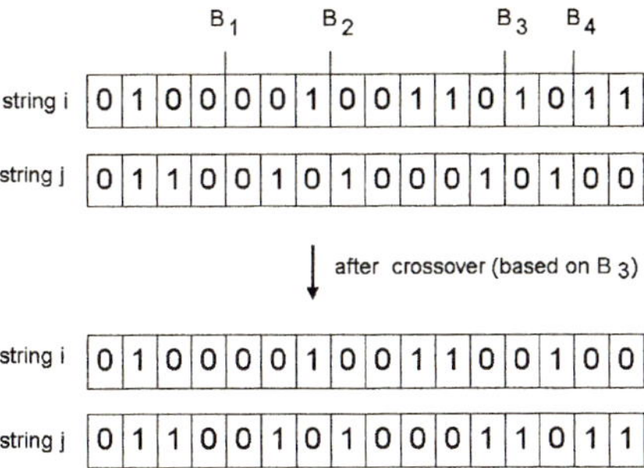

Fig. 2. Crossover Activity

3   Experiments

We executed several experiments on the proposed genetic algorithm approach to compare with a conventional sender-initiated algorithm. Experiments in this paper have the following assumptions. Firstly, each task size and task type is the same. Secondly, the number of parts(p) in a string is four. The values of these parameters P_c, P_m were known as the most suitable values in various applications[3]. Table 1 shows the detailed contents of parameters used in our experiments. Weights for *TMP, TMT, TTP* are 0.025, 0.01, 0.02. The load rating over systems supposed about 60 percent.

Table 1. Contents of parameter

number of processor	24
P_c	0.7
P_m	0.1
Number of strings	50
number of tasks to be performed	5000

[**Experiment 1**] We compared the performance of proposed method with a conventional method by using the parameters on the table 2 and table 3. The experiment is to observe change of response time when the number of tasks to be performed is 5000.

Fig 3 shows result of the experiment 1. In conventional methods, when the sender determines a suitable receiver, it select a processor in distributed systems randomly, and receive the load state information from the selected processor. The algorithm determines the selected processor as receiver if the load of randomly selected processor is T_{low}(light-load). These processes are repeated until a suitable receiver is searched. So, the result of response time shows the severe fluctuation. In the proposed algorithm, the algorithm shows the low response time because the load redistribution activity performs the proposed genetic_operation considering load states when it determines a receiver.

[**Experiment 2**] This experiment is to observe the performance when the probability of crossover is changed. And is to observe the performance when the probability of mutation is changed.

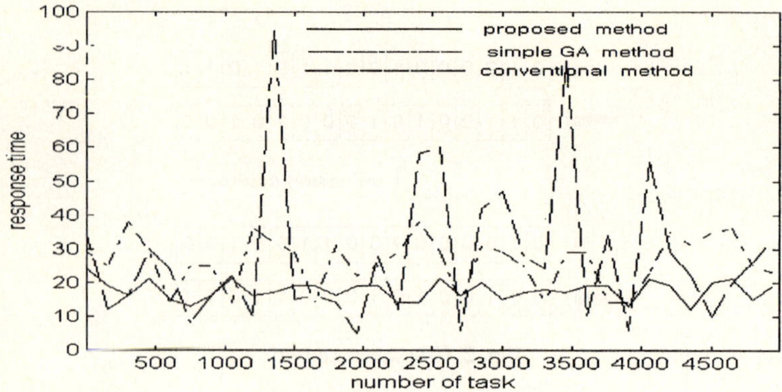

Fig. 3. Result of response time

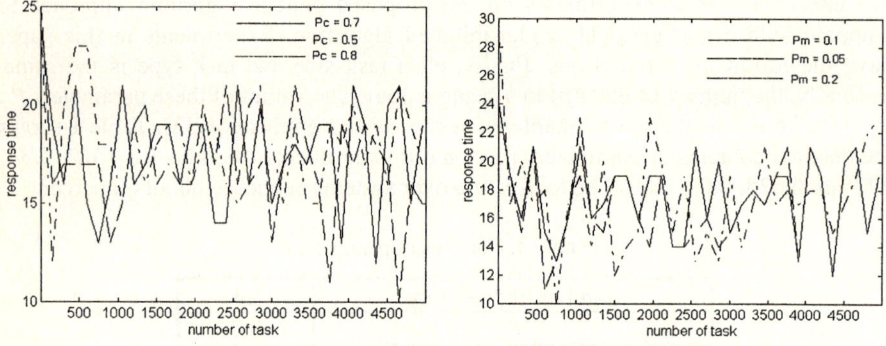

Fig. 4. Result depending on the changes of P_c, P_m

Fig 4 shows the result of response time depending on the changes of P_c when P_m is 0.1. In accordance with value of P_c, it shows a different performance. But the proposed algorithm shows better performance than that of conventional algorithm and simple genetic algorithm approach. And shows the result of the response time depending on the changes of P_m when P_c is 0.7. In accordance with value of P_m, it shows a different performance. But the proposed algorithm shows better performance than that of conventional algorithm and simple genetic algorithm approach.

4   Conclusions

We propose a new dynamic load redistribution scheme in a distributed system, which is based on the new genetic algorithm with a local improvement operation. Through the various experiments, the performance of the proposed scheme is better than that of the conventional scheme and the simple genetic algorithm approach on the response time and the mean response time. But the proposed algorithm is sensitive to the weight values of *TMP*, *TMT* and *TTP*. In future, we will study on method for releasing sensitivity of the weight values.

References

1. A. Y. Zomaya and Y.H. the, "Observations on Using Genetic algorithms for Dynamic Load Balancing", IEEE Tr. On Parallel and Distributed Systems, vol.12 no. 9, pp.899-911 Sep. 2001.
2. A. Hac and X. Jin, "Dynamic Load-Balancing in a Distributed System Using a Sender-Initiated Algorithm", Proc. 13th Conf. Local Computer Networks, pp. 172-180, 1988.
3. J.Grefenstette, "Optimization of Control Parameters for Genetic Algorithms," *IEEE Trans on SMC*, vol.SMC-16, no.1, pp.122-128, January 1986.
4. J.R. Filho and P. C. Treleaven, "Genetic-Algorithm Programming Environments," *IEEE COMPUTER*, pp.28-43, June 1994.
5. T. Kunz, "The Influence of Different Workload Descriptions on a Heuristic Load Balancing Scheme," *IEEE Trans on Software Engineering,* vol.17, No.7, pp.725-730, July 1991.
6. T.Furuhashi, K.Nakaoka, Y.Uchikawa, "A New Approach to Genetic Based Machine Learning and an Efficient Finding of Fuzzy Rules," *Proc. WWW'94*, pp.114-122, 1994.
7. J A. Miller, W D. Potter, R V. Gondham, C N. Lapena, "An Evaluation of Local Improvement Operators for Genetic Algorithms," *IEEE Trans on SMC*, vol.23, No 5, pp.1340-1351, Sept 1993.
8. Terence C. Fogarty, Frank Vavak, and Phillip Cheng, "Use of the Genetic Algorithm for Load Balancing of Sugar Beet Presses," *Proc. Sixth International Conference on Genetic Algorithms*, pp.617-624, 1995.
9. Garrism W. Greenwood, Christian Lang and steve Hurley, "Scheduling Tasks in Real-Time Systems using Evolutionary Strategies," *Proc. Third Workshop on Parallel and Distributed Real-Time Systems*, pp.195-196, 1995.

An Architecture of a Wavelet Based Approach for the Approximate Querying of Huge Sets of Data in the Telecommunication Environment

Ernesto Damiani[1], Stefania Marrara[1],
Salvatore Reale[2], and Massimiliano Torregiani[2]

[1] Università degli Studi di Milano, Dipartimento di Tecnologie dell'Informazione
via Bramante 65, 26013 Crema (CR), Italy
{damiani,marrara}@dti.unimi.it
[2] Siemens Mobile Communications S.p.A.
Via Monfalcone 1 - Cinisello Balsamo, 20092 (MI), Italy

Abstract. Third generation mobile networks were designed to satisfy raising requests of performance, reliability and availability on behalf of customers all over the world. These networks have a new architecture where new Network Entities carry out complex functionalities and can serve many calls at the time. In order to grant high efficiency and the best performance of the network, such a complex scenario must be monitored and controlled by a certain number of Operation & Maintenance Centres (OMCs). Typically, these centres pose queries to the underlying DBMS that require complex operations over Gigabytes or Terabytes of disk-resident data, and thus, take a very long time to execute to completion and produce exact answers. Due to the *exploratory nature* of these applications, an exact answer may not be required, and a user may in fact prefer a fast, approximate answer. In this paper we propose an architecture for approximate query processing to be integrated in the OTS architecture of Siemens Mobile S.p.A. The main purpose of this paper is to propose a practical application of wavelet-based synopses techniques in order to improve the performances of the management system of real telecommunication network.

1 Introduction and Related Work

In the last years, *approximate query processing* has emerged as a viable solution for dealing with the huge amount of data, the high query complexities, and the increasingly stringent response-time requirements that characterize today's Decision-Support Systems (DSS) applications. Typically, DSS users pose very complex queries to the underlying Database management System (DBMS) that require complex operations over Gigabytes or Terabytes of disk-resident data, and thus, take a very long time to execute to completion and produce exact answers. Due to the *exploratory nature* of many DSS applications, there are a number of scenarios in which an exact answer may not be required, and a user may in fact prefer a fast, approximate answer.

R. Khosla et al. (Eds.): KES 2005, LNAI 3681, pp. 1088–1093, 2005.

One of the answers to the problem of optimizing the execution of queries to massive data sets is offered by the use of small data structures called *synopses*. A good starting point, for an overview on synopsis data structures for relational data warehouse, is [BDF$^+$97], which describes the state of the art in *data reduction* techniques, for reducing massive data sets down to a "big picture" and for providing quick approximate answers to queries. The data reduction techniques surveyed by the paper are singular value decomposition, wavelets, regression, log-linear models, histograms, clustering techniques, index trees, and sampling. Each technique is described briefly and then evaluated *qualitatively* on to its effectiveness and suitability for various data types and distributions, on how well it can be maintained under insertions and deletions to the data sets, and on whether it supports answers that progressively improve the approximation with time.

Wavelets are a mathematical tool for hierarchical decomposition of functions/signals. A survey of their use in databases is [VWI98]: in this paper we can find an overview of the use of Haar wavelets for approximate query processing. The key idea is to apply wavelet decomposition to the input data collection (attribute column(s) or OLAP cube) to obtain a compact data synopsis that comprises a select small collection of *wavelets coefficients*. The results in this paper show that wavelets can be effective in handling aggregates over high-dimensional OLAP cubes, while avoiding the high construction costs and storage overheads of histogramming techniques.

1.1   Wavelet-Based Representations

Wavelets are a useful mathematical tool for hierarchically decomposing functions in ways that are both efficient and theoretically sound. Broadly speaking, the wavelet decomposition of a function consists of a coarse overall approximation together with detail coefficients that influence the function at various scales. Our approach is based on the *multi-dimensional Haar wavelet* decomposition. There are two common methods in which Haar wavelets can be extended to transform the data values in a *multidimensional array* as presented in the survey paper [CGRS01]:

- *Standard* Haar decomposition [SDS96, VW99]
- *Non-Standard* Haar decomposition [MVW00, CGRS00]

In the first method, they first fix an ordering for the data dimensions (e.g., $1, 2, \ldots, n$), and then proceed to apply the complete one-dimensional wavelet transform for each one-dimensional *row* of array cells along dimension k, for all $k = 1, \ldots, n$.

Instead, the non-standard decomposition alternates between dimensions during successive steps of pairwise averaging and differencing: given an ordering for the data dimensions (e.g.,$1, 2, \ldots, n$), they perform *one step of pairwise averaging and differencing* for each one-dimensional row of array cells along dimension k, for each $k = 1, \ldots, n$. (The results of earlier averaging and differencing steps are treated as data values for larger values of k.) This process is then repeated recursively only on the quadrant containing averages across all dimensions.

An basic work for our architecture is [CGRS01]: in this work, they propose an approach to general-purpose approximate query processing that consists of two basic steps. First, multi-dimensional Haar wavelets are employed to efficiently obtain effective, compact synopses of general relational tables. Second, using novel query processing algorithms, standard (aggregate and non aggregate) SQL operators are applied *directly* over the wavelet-coefficient synopses of data to obtain fast and accurate approximate query answers.

Aim of this work is to prove that these techniques can be effectively applied inside a company environment, providing benefits to a real case of data management.

The structure of the paper is as follows. Section 2 introduces a general overview of environment for which we provide the synopsis architecture, and presents an example of query that could receive good benefits from an approximate querying approach; Section 3 presents the architecture of the approximate querying system we propose to integrate to Siemens' OTS, and, finally, in Section 4 we outline our conclusions.

2   The Environment and the OTS System

Third generation mobile networks were designed to satisfy raising requests of performance, reliability and availability on behalf of customers all over the world. To meet these requirements, a brand new architecture has been built where, with particular reference to the UMTS Radio Access Network (UTRAN), new Network Entities carry out more complex functionalities and can serve many more calls at the time than the previous networks did. On the other hand, in order to grant high efficiency and the best performance of the network, such a complex scenario must be monitored and controlled by a certain number of Operation & Maintenance Centres (OMCs).

In this situation, a few thousands of NodeBs together with few tens of Radio Network Controllers (RNCs) give a snapshot of the amount of Network Entities directly managed by a unique OMC whilst, in GSM system, the same analysis was done by less than an hundred of Base Station Systems (BSSs). Finally, an important role to optimize the usage of UMTS networks is played by the evaluation of performance measurements periodically given from NodeBs and RNCs. These measurements can be mainly grouped into:

- Physical Link Measurements (e.g. Bit Error Rate, Uplink/Downlink SNR);
- Message Flow Measurements (e.g. Packet Drop Rate);
- Call Measurements (e.g. call drop, call establishments, total calls, handover requests, handover failures).

These huge amount of raw measurement data are difficult to evaluate and, in general, do not give a clear idea of the actual course of the network health. Therefore, starting from these huge amount of raw measurement data got directly from controlled Network Elements, many performance indicators can be computed applying, for instance:

- primitive mathematic operators like addition, subtraction, multiplication and division;
- trends like average, maximum and minimum applied to the entire set of data or to well-defined period of time;
- time and object aggregations;
- telecommunication operators.

Performance Indicators can be split in three classes, which are: number of attempts, number of successes and number of failures. Conversely, any of the three indicator classes can be calculated from the two others:

\# of Success + \# of Failures = \# of Attempts

To analyze network performance, such classes can be also reported as rates. With respect to the classes defined above, it is possible to define success and failure rates. In case of hard handover or Inter-System handover, a drop rate is added, which indicates a failed handover procedure with the consequence that the connection is lost. As in the formula above, one of the rates can be calculated if the other value(s) are known.

Among all the indicators, those which are mainly exploited to analyse system performance are:

- *Call setup success/failure rate*
- *Connection release rate*, which describes the ratio of CN initiated connection releases related to the total number of CN initiated connection releases;
- *Connection drop rate* expressing the ratio between the number of abnormal call releases and the number of released calls;
- *Outgoing/Ingoing Handover success/failure rate*. This indicator can be furthermore specialized in Inter-NodeB, Intra-RNC, Inter-RNC via Iur and Inter-system (GSM and UMTS).

This work, which has the aim to provide fast approximate results to the analysis described above, was born from a cooperation between the University of Milan, Department of Computer Technologies, and Siemens Mobile Communications S.p.A., a multinational company in the field of telecommunication networks. The software developed by Siemens for mobile networks performance analysis, and basis of this work, is *OTS* (O&M ToolSet). OTS is composed by several tools that collect informations from the elements of the net (e.g., BTS and BSC) and produce several analysis aimed to optimize the net itself. The main software is *RadioCommander*, a software used to check the network in real time. If we look inside OTS, we can see that the main tool composing the system is *PM+*, which computes streams of data coming from the net, and produces reports of the behavior of the system. Data coming from the physical components of the net are collected inside an Oracle database and then used by PM+ to create its reports, which are stored on disk. Then another tool, the *Proposer*, uses these reports to create summarized performance profiles by means of aggregate queries.

The *Proposer* is one of the performance bottlenecks of OTS; computing aggregate queries over the data stored in the Oracle database requires long computation times and huge memory resources basically for two reasons:

1. Proposer is not designed to query directly the database but the communication is carried by file exchanges;
2. each analysis requires the aggregation of huge sets of data.

In such a application, the company is often more interested to obtain an approximate answer computed in a short time rather than an exact one obtained in some minutes or, at the worst, hours. In this work we propose an architecture to apply approximate query processing techniques to obtain approximate answers in very short computation times. The basic idea for approximate answers is to store pre-computed summaries of the Oracle database, also called *synopses* (concise data collections), and to query them instead of the original database, thus saving time and computational costs. In our scenario, the Proposer should pose a query directly to the DBMS, which should use the synopsis instead of the original data to answer the query. Consequently, the answer should come within a much shorter time to the price of a loss in precision. The research work's purpose involves techniques for such transparent query transformation, as well as evaluation methods for estimating the error produced by the evaluation of the query on the synopsis instead of the original data. A query (simplified) example of the analysis required on telecommunication data is:

```
select sum(shoAddInterRncFail_6_7_2)
from MEAS_RNC_CELL_APPL_35
where (suspect = 0)
   and (TO_NUMBER(TO_CHAR(date_time,'HH24MI')) between 800 and
   1899) and (date_time between TO_DATE('2004-04-03 15:00',
   'YYYY-MM-DD HH24:MI') and TO_DATE('2004-04-03 4:MI')) and
   RNC_ID =: id<int>
```

Typically, queries in this environment do not have complex structures and joins that can compromise the performances of using synopses. Hence, telecommunication data are an optimal benchmark to test the benefits of using synopses for querying huge data sets and compute new complex analysis that are too expensive, in terms of computational costs, on the original data.

3   Architecture

ApproSQL is designed as a module that sits on top of the DBMS. ApproSQL precomputes statistical summaries on the relations in database. In our approach, the statistics take the form of wavelets, and are stored as regular relations inside the database; they are also incrementally maintained up-to-date as the base data is updated. ApproSQL answers user queries using the precomputed summaries. Approximate answers are provided by rewriting the Proposer's query over the summary relations and executing the new query. The rewriting involves suitably scaling the results of certain operators within the query. Finally, the query and

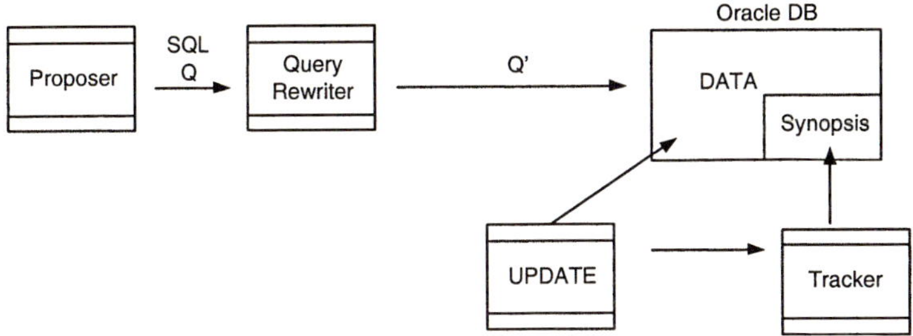

Fig. 1. ApproSQL architecture.

the approximate answer are analyzed to provide guarantees on the quality of the answer, and report error bounds. The high-level architecture of ApproSQL is depicted in Figure 1, along with the steps taken during query processing. As new data arrives, the *ApproSQL Tracker* maintains the synopses up to date, with few or no accesses to the original data.

4   Conclusions

In this paper we propose an architecture for approximate query processing to be integrated in the OTS architecture of Siemens Mobile S.p.A.

The main purpose of this paper is to propose a practical application of wavelet-based synopses techniques in order to improve the performances of the OTS system and test the benefits of this approach on real telecommunication data.

References

[BDF⁺97] D. Barbarà, W. DuMouchel, C. Faloutsos, P. J. Haas, J. M. Hellerstein, Y. Ioannidis, H. V. Jagadish, T. Johnson, R. Ng, V. Poosala, K. A. Ross, and K. C. Sevcik. The new jersey data reduction report. In *Bulletin of Technical Committee on Data Engineering*, pages 20(4): 3–45, 1997.

[CGRS00] K. Chakrabarti, M. Garofalakis, R. Rastogi, and K. Shim. Approximate query processing using wavelets. In *Proceeding of Very Large Data Bases Conference, VLDB*, 2000.

[CGRS01] Kaushik Chakrabarti, Minos Garofalakis, Rajeev Rastogi, and Kyuseok Shim. Approximate query processing using wavelets. *VLDB Journal: Very Large Data Bases*, 10(2–3):199–223, 2001.

[MVW00] Y. Matias, J. S. Vitter, and M. Wang. Dynamic maintenance of wavelet-based histograms. In *Proceeding of Very Large Data Bases Conference, VLDB*, 2000.

[SDS96] E. J. Stollnitz, T. D. DeRose, and D. H. Salesin. Wavelets for computer graphics. In *Morgan-Kauffman Publisher Inc.*, 1996.

[VW99] J. S. Vitter and M. Wang. Approximate computation of multidimensional aggregates of sparse data using wavelets. In *ACM Sigmod*, 1999.

[VWI98] J. S. Vitter, M. Wang, and B. Iyer. Data cube approximation and histograms via wavelets. In *Proc. the 7th Int. Conf. on Information and Knowledge Management.*, 1998.

Performance Comparison
of M-ary PPM Ultra-wideband Multiple Access System Using an Intelligent Pulse Shaping Techniques

SungEon Cho[1] and JaeMin Kwak[2]

[1] Professor of Dept of Computer & Communication
Sunchon National University, Sunchon-si, Chonnam, 540-742, Korea
chose@sunchon.ac.kr
[2] Korea Electronics Technology Institute
kjm@keti.re.kr

Abstract. Ultra-wideband(UWB) system is recently becoming very attractive candidate for low cost short range communication, due to recent development in high speed intelligent switching technology. In this paper, we theoretically analyze the probability of error for M-ary pulse position modulation (PPM) ultra-wideband (UWB) multiple access system using intelligent pulse shaping techniques. We first consider the optimum detection of UWB signals using M-ary orthogonal PPM in additive white Gaussian noise (AWGN). Then we derive receiver signal to noise power ratio (SNR) and bit error rate (BER) in multiple access interference (MAI) environment. Numerical results considering some practical parameters are presented and the BER performance according to different pulse shapes are compared.

1 Introduction

Ultra-wideband (UWB) system is recently becoming very attractive candidate for low cost short range communication, due to recent development in high speed intelligent switching technology [1], [2]. This technology seems to be able to offer low emissions, high bit rates, and spectrum reuse. UWB technology is defined as any radio technology having a spectrum that occupies a bandwidth greater than 20 percent of the center frequency of at least 500MHz [3]. Its characteristics of extremely short pulse duration, typically less than few nanoseconds, results in spreading the energy of the radio signal very thinly from near d.c. to a few gigahertz. It typically uses time hopping scheme for communication, in which data modulation is accomplished by additional pulse position modulation (PPM) at the rate of many pulses per data bit [4], [5]. There are some previous studies on the performance and capacity of the UWB multiple access system. In [5] multiple access performance assuming free space propagation conditions and additive white Gaussian noise (AWGN) are studied. In [6], [7] the information theoretic capacity of M-ary PPM UWB multiple access system using rectangular pulse waveform is derived. However, to the best of the authors' knowledge, no results have been published on the probability of error for M-ary PPM UWB multiple access system considering various pulse shapes. Therefore we derive the probability of error for M-ary PPM UWB multiple access systems using some different pulse shapes, which are rectangular pulse, Gaussian monocycle, and Rayleigh monocycle, by an approach similar to [7].

R. Khosla et al. (Eds.): KES 2005, LNAI 3681, pp. 1094–1103, 2005.

The rest of this paper is organized as follows. In Section 2, the system model of UWB multiple access system is described. The SNR and BER considering pulse shapes are analyzed in section 3. Numerical results and some conclusions are drawn in section 4 and section 5, respectively.

2   System Model

A typical time hopping format of the k-th UWB transmitter output signal $s_{tr}^{(k)}(t)$ is given by

$$s_{tr}^{(k)}(t) = \sum_{j=-\infty}^{\infty} A^{(k)} p_{tr}(t - jT_f - c_j^{(k)}T_c - d_{\lfloor j/N_p\rfloor}^{(k)}) \tag{1}$$

where $p_{tr}(t)$ is transmitted waveform, and k is user index. T_f is frame interval, $A^{(k)} = \sqrt{E_p}$, $c_j^{(k)}$, $d_j^{(k)}$, and N_p are respectively the amplitude, time hopping sequence, time shift modulation for user k, and the number of pulses that forms an UWB data symbol.

The notation $\lfloor q \rfloor$ denotes the integer part of q. For PPM time shift $d_{\lfloor j/N_p\rfloor}^{(k)} \in \{\delta_1, \delta_2, \ldots, \delta_M\}$, we assume $\delta_1 = 0 < \delta_2 = T_c/M < \ldots < \delta_M = T_c(M-1)/M$. Therefore the signal transmitted by k-th transmitter consists of large number of pulses shifted to different times.

The received signal from k-th transmitter is given by

$$s_{rec}^{(k)}(t) = \sum_{j=-\infty}^{\infty} A^{(k)} p_{rec}(t - jT_f - c_j^{(k)}T_c - d_{\lfloor j/N_p\rfloor}^{(k)}) \tag{2}$$

where $p_{rec}(t)$ is the received pulse waveform with duration T_p from the k-th transmitter.

If we assume N_u active users are in the UWB multiple access system, the total received signal is expressed by

$$r(t) = \sum_{k=1}^{N_u} s_{rec}^{(k)}(t - \tau^{(k)}) + n(t) \tag{3}$$

where $\tau^{(k)}$ is time delay from user k and $n(t)$ is AWGN with double sided power spectral density $N_0/2$.

In this paper, we assume $c_j^{(k)}T_c$ are iid discrete uniform random variables over $[0, T_c, \ldots, (N_h-1)T_c]$ and N_h is an integer value equal to T_f/T_c. Also $d_{\lfloor j/N_p\rfloor}^{(k)}$ are iid discrete uniform random variables over $[0, T_c, \ldots, T_c(M-1)/M]$. And $\tau^{(k)}$ are iid continuous uniform random variables over frame time $[0, T_f]$.

In general, M, N_p, N_h are system design parameters. The specific values of these parameters depend on performance requirement and implementation considerations, among other criteria. In this paper, the number of pulse N_p for an UWB symbol is assumed to be one for simplicity, so that symbol duration $T_s = T_f$.

For the performance analysis in next section, we define correlation function $h(\tau)$ of a pulse $p_{rec}(t)$ as

$$h(\tau) = \int_0^{T_f} p_{rec}(t) p_{rec}(t - \tau)dt. \tag{4}$$

where the notation τ is the time difference between user 1 and k given by

$$\tau = (c_j^{(k)} - c_j^{(1)})T_c + (d_j^{(k)} - d_j^{(1)}) + (\tau^{(k)} - \tau^{(1)}) \tag{5}$$

Pdf of τ is equal to the convolution of pdfs from six independent random variables $(c_j^{(k)}, c_j^{(1)}, d_j^{(k)}, d_j^{(1)}, \tau_j^{(k)}, \tau_j^{(1)})$ and its range is $[-(2T_f - T_c/M), 2T_f - T_c/M]$.

In our analysis we consider UWB PPM signals as orthogonal signal time separated, hence M-ary PPM orthogonal optimum receiver for AWGN channel is used for receiver structure and it is assumed that receiver is perfectly synchronized to user1.

As shown in Fig. 1, in the receiver, the received signal r(t) is cross-correlated with each of the M signal waveforms and the correlator outputs are sampled at $t = T_f$. The optimum detector observes the M correlator outputs r_i, $i = 1, 2, \ldots, M$ and select the signal corresponding to the largest correlator output. Without loss of generality, we assume $d_j^{(k)} = \delta_1$ (this assumption is also used in next section for performance analysis), hence the first correlator output r_1 among M correlators at receiver has statistically highest probability of showing maximum output [8].

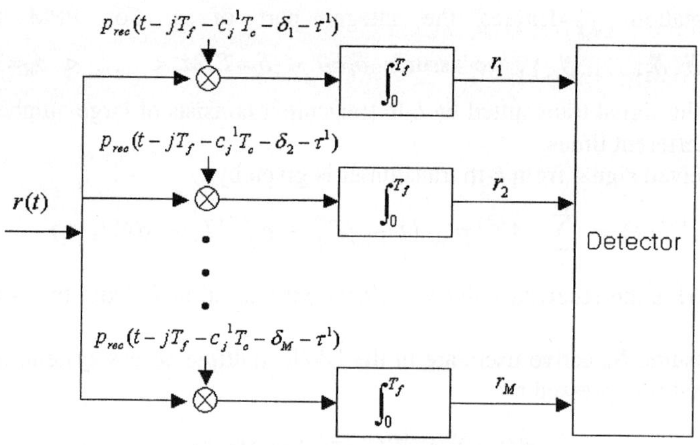

Fig. 1. Optimum Receiver for AWGN

3 Performance Analysis

In our analysis, it is assumed that waveform $p_{rec}(t)$ at the receiver antenna is known and have the following formats [9]

$$p_{rec\_RE}(t) = \sqrt{\frac{1}{T_p}} \quad 0 \le t \le T_p \tag{6.1}$$

$$p_{rec\_GA}(t) = A_G[1 - (\frac{t}{\sigma} - 3.5)^2] \cdot \exp[-0.5(\frac{t}{\sigma} - 3.5)^2] \tag{6.2}$$

$$p_{rec\_RA}(t) = A_R[\frac{t - 3.5\sigma}{\sigma^2}] \cdot \exp[-0.5(\frac{t}{\sigma} - 3.5)^2] \tag{6.3}$$

where RE, GA, and RA mean rectangular pulse, Gaussian monocycle, and Rayleigh monocycle, respectively. A_G and A_R are normalization factor for $p_{rec\_GA\_}(t)$ and $p_{rec\_RA\_}(t)$, respectively, to have unit energy. The pulse width T_p is assumed to be 7σ to have 99.99% energy in the Gaussian monocycle[10].

The energy of any pulse $p_{rec}(t)$ is defined as

$$E_p = \int_0^{T_p} p_{rec}(t)^2 \, dt. \tag{7}$$

Therefore pulse energy normalization factor A_G and A_R are obtained as following

$$A_G \approx \frac{1}{\sqrt{1.3291\sigma}} \tag{8.1}$$

$$A_R \approx 1.0623\sqrt{\sigma}. \tag{8.2}$$

The correlation function $h(\tau)$ in (4) considering pulse shapes (6) can be solved by following expressions according to the range of τ

$$h_{left}(\tau) = \int_0^{T_p+\tau} p_{rec}(t)p_{rec}(t-\tau)dt, \quad -T_p \le \tau \le 0 \tag{9.1}$$

$$h_{right}(\tau) = \int_\tau^{T_p} p_{rec}(t)p_{rec}(t-\tau)dt, \quad 0 \le \tau \le T_p \tag{9.2}$$

where $h_{left}(\tau)$ is left part of $h(\tau)$ with minus value of τ and $h_{right}(\tau)$ is right part of $h(\tau)$ with plus value of τ.

By using (9), following correlation functions are obtained.

Case 1: Rectangular Pulse

$$h_{left\_rec\_RE}(\tau) = \frac{T_p + \tau}{T_p}, \quad -T_p \le \tau \le 0 \tag{10.1}$$

$$h_{right\_rec\_RE}(\tau) = \frac{T_p - \tau}{T_p}, \quad 0 \le \tau \le T_p. \tag{10.2}$$

Case 2: Gaussian Monocycle

$$h_{left\_rec\_GA}(\tau) = 0.047/\sigma^4 \begin{bmatrix} (-658\sigma^4 - 290\tau\sigma^3 - 14\tau^2\sigma^2 + 2\tau^3\sigma) \\ \cdot \exp(-\dfrac{49\sigma^2 + 2\tau^2 + 14\sigma\tau}{4\sigma^2}) \\ + (12\sigma^4\sqrt{\pi} - 12\sqrt{\pi}\tau^2\sigma^2 + \tau^4\sqrt{\pi}) \\ \cdot erf(\dfrac{7\sigma+\tau}{2\sigma}) \cdot \exp(-\dfrac{\tau^2}{4\sigma^2}) \end{bmatrix} \tag{11.1}$$

$$h_{right\_rec\_GA}(\tau) = 0.047/\sigma^4 \begin{bmatrix} (-658\sigma^4 + 290\tau\sigma^3 - 14\tau^2\sigma^2 - 2\tau^3\sigma) \\ \cdot \exp(-\dfrac{49\sigma^2 + 2\tau^2 - 14\sigma\tau}{4\sigma^2}) \\ + (12\sigma^4\sqrt{\pi} - 12\sqrt{\pi}\tau^2\sigma^2 + \tau^4\sqrt{\pi}) \\ \cdot erf(\dfrac{7\sigma-\tau}{2\sigma}) \cdot \exp(-\dfrac{\tau^2}{4\sigma^2}) \end{bmatrix}. \tag{11.2}$$

Case 3: Rayleigh Monocycle

$$h_{left\_rec\_RA}(\tau) = -0.2821/\sigma^2 \begin{bmatrix} (14\sigma^2 + 2\sigma\tau) \cdot \exp(-\dfrac{49\sigma^2 + 2\tau^2 + 14\sigma\tau}{4\sigma^2}) \\ + \sqrt{\pi}(-2\sigma^2 + \tau^2) \cdot erf(\dfrac{7\sigma+\tau}{2\sigma}) \cdot \exp(-\dfrac{\tau^2}{4\sigma^2}) \end{bmatrix} \tag{12.1}$$

$$h_{right\_rec\_RA}(\tau) = -0.2821/\sigma^2 \begin{bmatrix} (14\sigma^2 - 2\sigma\tau) \cdot \exp(-\dfrac{49\sigma^2 + 2\tau^2 - 14\sigma\tau}{4\sigma^2}) \\ + \sqrt{\pi}(-2\sigma^2 + \tau^2) \cdot erf(\dfrac{7\sigma-\tau}{2\sigma}) \cdot \exp(-\dfrac{\tau^2}{4\sigma^2}) \end{bmatrix}. \tag{12.2}$$

From (6), (10), (11), and (12), plots of pulse shape $p_{rec}(t)$ and correlation function $h(\tau)$ are obtained and shown in Fig. 2.

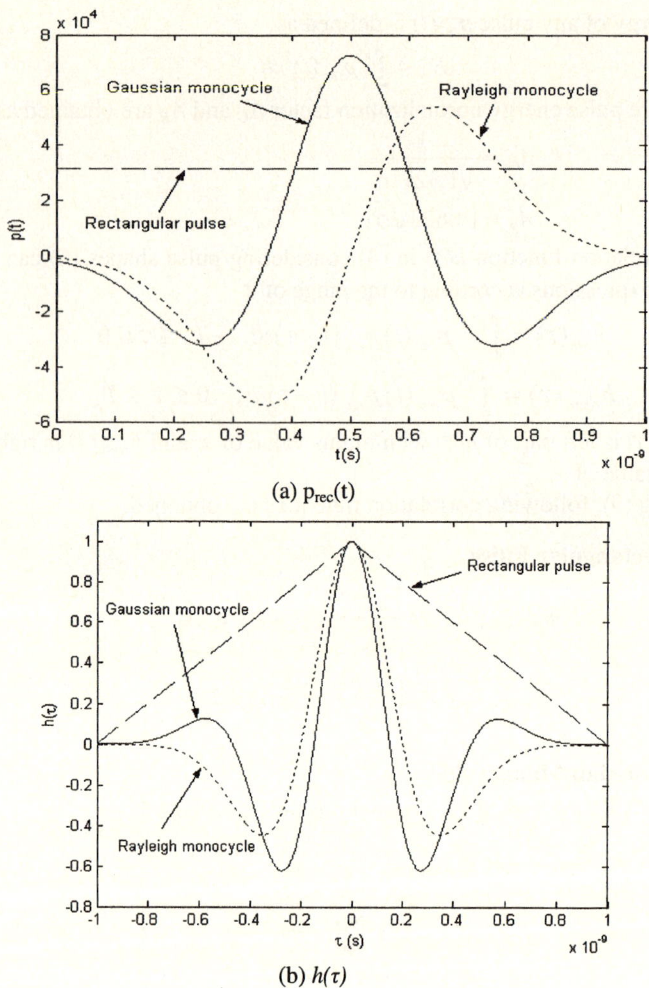

Fig. 2. Optimum Receiver for AWGN

From above equation we can know that $h_{left}(\tau)$ and $h_{right}(\tau)$ are symmetric to each other. Expectation and variance of $h(\tau)$ are generally expressed as

$$E[h(\tau)] = \int_{-T_p}^{T_p} h(\tau) f_\tau(\tau) d\tau \tag{13}$$

$$\sigma_h^2 = \int_{-T_p}^{T_p} (h(\tau) - E[h(\tau)])^2 f_\tau(\tau) d\tau$$
$$= E[h^2(\tau)] - (E[h(\tau)])^2. \tag{14}$$

Where $f_\tau(\tau)$ is probability density function (pdf) of random variable τ. Since (13) and (14) can not be solved to simplified form, so in our work, expectation and variance of $h(\tau)$ are obtained by averaging equations (13) and (14) after generation of random number τ 500,000 times where τ are obtained by summation of six random numbers $(c_j^{(k)} T_c, c_j^{(1)} T_c, d_j^{(k)}, d_j^{(1)}, \tau^{(k)}, \tau^{(1)})$ generated.

The receiver correlator output r_i at the sample time is given by

$$r_i = \begin{cases} A^{(1)} + N_M + N, & i = 1 \\ N_M + N, & i \neq 1 \end{cases}$$

(15)

where N_M and N are respectively, the multiple access interference (MAI) and AWGN component.

The MAI component N_M and it's variance σ_M^2 are written by

$$N_M = \sum_{k=2}^{N_u} A^{(k)} h(\tau)$$

(16)

$$\sigma_M^2 = \sum_{k=2}^{N_u} (A^{(k)})^2 \sigma_h^2$$

(17)

Where σ_h^2 is the variance of $h(\tau)$. Strictly speaking, N_M is not Gaussian random variable. However it is assumed that there are sufficiently many active users and therefore N_M is assumed to have Gaussian pdf by central limit theorem.

Therefore the pdf of receiver correlator output r_i is given by

$$f_{r_i}(r_i) = \begin{cases} Normal(A^{(1)} + E[N_M],\ \sigma_M^2 + N_0/2) & signal \\ Normal(E[N_M],\ \sigma_M^2 + N_0/2) & no\ signal \end{cases}$$

(18)

where $Normal(x, y^2)$ is Gaussian pdf with mean x and variance y^2

Considering AWGN and MAI, the signal to noise power ratio (SNR) at the output of first correlator is given by

$$SNR = \frac{(A^{(1)})^2}{\sigma_M^2 + N_0/2} = \frac{(A^{(1)})^2}{\sum_{k=2}^{N_u} (A^{(k)})^2 \sigma_h^2 + N_0/2}.$$

(19)

With perfect power control, $A^{(1)} = A^{(k)} = \sqrt{E_p}$, SNR is rewritten by

$$SNR = \frac{E_p}{(N_u - 1)E_p \sigma_h^2 + N_0/2} = \frac{1}{(N_u - 1)\sigma_h^2 + 1/(SNR_0)}$$

(20)

where $SNR_0 = 2E_p/N_0 = 2E_s/N_0 = 2kE_b/N_0$.

For M-ary orthogonal signaling using PPM, the probability of a symbol error, P_e, is simply obtained by

$$P_e = 1 - P(r_1 > r_2, r_1 > r_3, \ldots, r_1 > r_M).$$

(21)

It can be shown the P_e is expressed in integral form as [8]

$$P_e = \frac{1}{\sqrt{2\pi}} \int_{-\infty}^{\infty} \{1 - [1 - Q(y)]^{M-1}\} \exp(-\frac{(y - \sqrt{SNR})^2}{2}) dy$$

(22)

where $Q(x) = \int_x^{\infty} (1/\sqrt{2\pi}) \exp(-z^2/2) dz$.

Since no closed form expression exist for P_e except when $M = 2$, it is useful to find simple tight bounds for P_e for system feasibility studies. By using $P_e = P[\bigcup_{i \neq 1} r_i > r_1]$, it can be upper bounded using the union bound

$$P_e \leq (M-1)P[Y_i > Y_0] = (M-1)Q(\sqrt{\frac{SNR}{2}}). \qquad (23)$$

For M equally likely orthogonal PPM signals, all symbol errors are equi-probable and the average number of bit errors per k-bit symbol is

$$\sum_{n=1}^{k} n \binom{k}{n} \frac{P_e}{2^k - 1} = k \frac{2^{k-1}}{2^k - 1} P_e. \qquad (24)$$

So, the bit error rate is given by

$$P_b = \frac{2^{k-1}}{2^k - 1} P_e. \qquad (25)$$

Using (10), (11), (12), (14), (22), and (25), bit error rate (BER) of M-ary PPM UWB multiple access system according to pulse shapes can be evaluated.

4 Numerical Results

The performance of M-ary PPM UWB multiple access system considering various pulse shapes will be evaluated under AWGN channel. The basic system parameters used to provide numerical results are shown in Table 1.

Table 1. System Parameters

Parameters	VALUE
Pulse type	Rectangular pulse Gaussian monocycle Rayleigh monocycle
Pulse width, T_p	1 ns
M	2, 4, 8
E_b/N_0	5, 10 dB
Processing gain, T_f/T_p	32 ~ 960
The number of users	1 ~ 31

The SNR performance using Gaussian monocycle for M=2, 4, 8 versus processing gain in AWGN channel are plotted in Fig. 3 The N_h is adjusted to have the same processing gain according to different M. It was found that at the same processing gain, although M and N_h values are different, SNR performance shows almost same performance, and as processing gain increase, the SNR is converged to 10 dB which is equal to SNR0.

In Fig. 4, SNR performances are compared according to different pulse shapes. It is clear from the figure that SNR performance of UWB multiple access system using Gaussian sund Rayleigh monocycle show almost the same curves, and outperform the SNR of UWB system using rectangular pulse. This follows from the fact that although all pulses used in our analysis have the same energy, the variance of the pulse autocorrelation is higher for the rectangular pulse than for the Gaussian and Rayleigh monocycles, so MAI component for rectangular pulse has large value. In addition, in case of using Gaussian or Rayleigh monocycles, the SNR variation according to the processing gain is

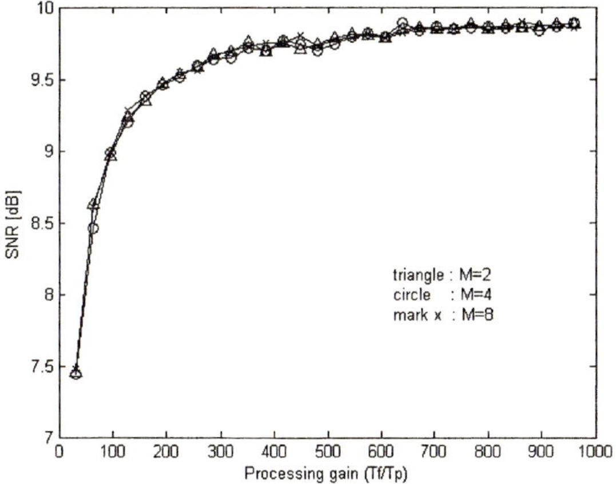

Fig. 3. SNR versus processing gain, T_f/T_p (Gaussian monocycle, SNR0 = 10 dB, # of users = 30)

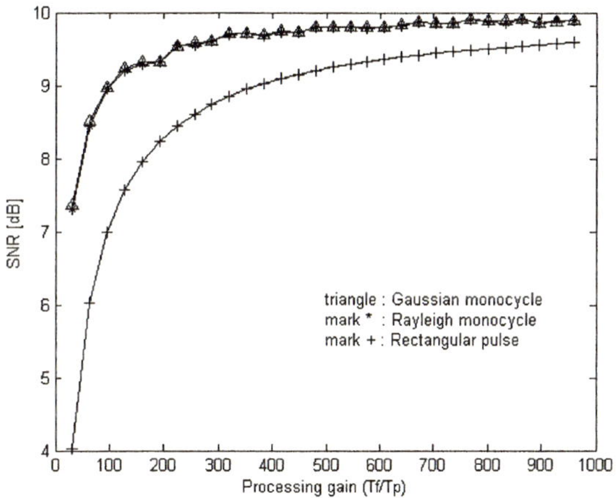

Fig. 4. SNR versus processing gain, T_f/T_p (M=8, SNR0 = 10 dB, # of users = 30)

noticeable in the range of [0, 200]. Therefore, to achieve good performance the pulse types and processing gain should be carefully selected.

Fig. 5 and Fig. 6 show the BER performance of M-ary PPM UWB multiple access system according to pulse types. The pulse width is set to $T_p = 1$ ns and processing gain is 128, so that the transmission symbol rate, $R_s = 1/T_f = 7.8125$ Msps.

In the Fig. 5, the BER of M-ary PPM UWB multiple access system assuming $E_b/N_0 = 5$ dB and 7.8125Msps symbol rate versus the number of users in AWGN channel is plotted. As in the case of M-ary orthogonal signaling, BER of M-ary PPM UWB multiple access system becomes better as the modulation level M becomes larger. And as expected, UWB system using Gaussian and Rayleigh monocycles show almost the same BER curves, and BER performance of UWB multiple access system using rectangular pulse shows worst performance.

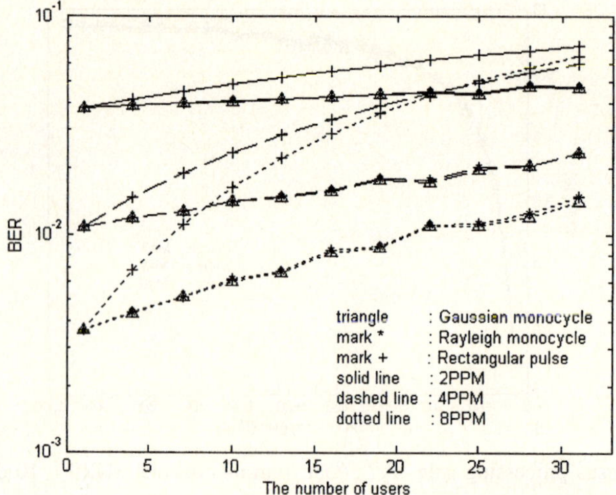

Fig. 5. BER versus the number users (processing gain = 128, E_b/N_0 = 5 dB)

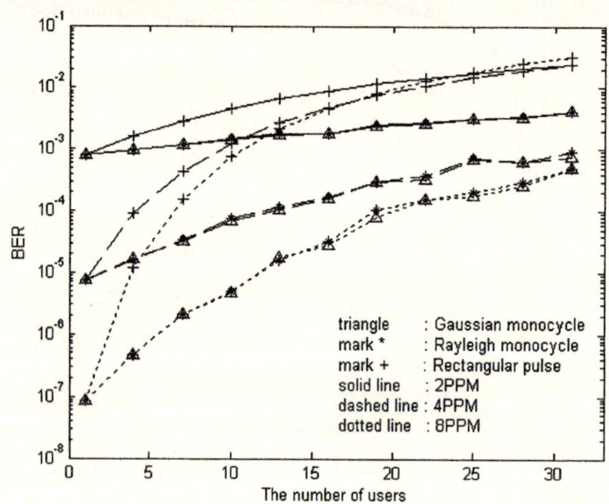

Fig. 6. BER versus the number users (processing gain = 100, E_b/N_0 = 10 dB)

Fig. 6 represents the BER performance of UWB multiple access system with E_b/N_0 = 10 dB. From Fig. 6, for the BER reference 10^{-3}, 4 and 8-ary PPM using Gaussian and Rayleigh monocycles can achieve more than 31 users whereas 4-ary and 8-ary PPM using rectangular pulse support at most 8 users. In 2-ary PPM case, no UWB multiple access system using rectangular pulse, Gaussian, or Rayleigh monocycles can achieve BER reference 10^{-3}.

5   Conclusion

In this paper, we derived and compared the SNR and error rate performance of M-ary PPM UWB multiple access system using an intelligent pulse shaping techniques. It is notable that UWB multiple access system using rectangular pulses shows the worst BER performance. This is because variance of receiver correlation value between the reference user signal and the multiple access interference user signal for the rectangular pulse has higher value than for Gaussian or Rayleigh monocycles so that the MAI component for the case of rectangular pulse is increased. Therefore, in designing a high rate UWB multiple access system, the pulse type should be carefully selected for maximizing the system performance. In future work, the effect of adaptive pulse shapes on the UWB system performance in practical multi-path fading environment should be studied.

References

1. Foster, E. Green, S. Somayazulu, and D. Leeper "Ultra-wideband technology for short- or medium-range wireless communications," Intel Technology Journal, May 2001.
2. Mattew L. Welborn, "System considerations for Ultra Wideband Wireless Networks," 2001 IEEE Radio and Wireless Conference (RAWCON), August 2001, Boston, MA, USA.
3. FCC Notice of Proposed Rule Making, "Revision of Part 15 of the Commission's Rules Regarding Ultra-wideband Transmission Systems," ET-Docket 98-153.
4. R. A. Scholtz, "Multiple access with time-hopping impulse modulation," in Proc. IEEE MILCOM Conf., vol. 2, Oct. 1993, pp. 447-450.
5. M. Z. Win and R. A. Scholtz, "Ultra-wide bandwidth time-hopping spread-spectrum impulse radio for wireless multiple-access communications," IEEE Transactions on Communications, vol. 48, Apr. 2000, pp. 679-691.
6. L. Zhao and A. M. Haimovich, "Capacity of M-ary PPM ultra-wideband communications over AWGN channels," in Proc. IEEE VTC Conf., vol. 2, Oct. 2001, 2, 1191-1195
7. L. Zhao and A. M. Haimovich, "The Capacity of an UWB multiple-access communication system," in proc. IEEE ICC., vol. 3, 2002, pp. 1964-1968.
8. J.G. Proakis, Digital Communications. 3rd edition, New York, NJ: McGraw-Hill, 1995.
9. J. T. Conroy, J. L. Lociero, and D. R. Ucci, "Communication techniques using monopulse waveforms," in Proc. IEEE MILCOM, vol. 2, 1999, pp. 1191-1185.
10. L. Zhao and A. M. Haimovich, "Performance of ultra-wideband communications in the presence of interference," IEEE Journal on Selected Areas in Communications, vol. 20, dec. 2002, 1684-1691.

Implementation of a Network-Based Distributed System Using the CAN Protocol

Joonhong Jung[1], Kiheon Park[1], and Jae-Sang Cha[2]

[1] School of Information and Communication Engineering, SungKyunKwan University,
300 Cheoncheon-Dong, Suwon, Kyung-Gi-Do, 440-746 Korea
nelcast@skku.edu, khpark@ece.skku.ac.kr
[2] Department of Information and Communication Engineering, SeoKyeong University,
16-1 Jungneung-Dong, Sungbuk-Ku, Seoul, 136-704 Korea
chajs@skuniv.ac.kr

Abstract. Network-based systems are widely used in many application areas such as intelligent systems and distributed systems. To design distributed systems, distributed information processing and intelligent terminals are required. This paper deals with the design considerations and implementations of a distributed system using network architecture. Network nodes, called smart nodes, are designed for supporting the CAN(controller area network) protocol and a new message scheduling method is proposed in order to transfer data between these nodes. An experimental setup is built to perform velocity control of four DC-Motor subsystems. The experimental results verify that the proposed distributed system provides an acceptable control performance.

1 Introduction

In a centralized system, mainly implemented in traditional small scale systems, all control functionalities reside in a single main controller. The point-to-point communication channels are set to connect the central controller to system components(i.e. supervisory devices, controllers, I/O devices, etc.). A problem with centralized architecture is that the system performance is generally decreased when the system is scaled to increasingly large dimensions. In addition, another problem is that reliability and safety of the system is too sensitive to the fault of the main controller. Distributed control architecture is completely opposite to the centralized control architecture. Distributed system contains a large number of interconnected devices and subsystems to perform desired control operations. This can provide robust reliability and can improve control performance by distributing the main controller load over a number of subsystems.

In distributed systems, point-to-point connection is no longer suitable as a communication method. Recently, there has been a great deal of interest in common network architecture for control applications[1],[2],[3]. In a network-based control system, intelligent system components are connected to a common bus. Integrating a control system into a network has great advantages over the traditional control system, which uses point to point connection: it allows remarkable reduction in wiring, provides an easy to install and maintain system, and improves compatibility[3],[4]. However, a network-based control system has the critical defect that network uncertainties, such as time delay and data loss, can degrade the control system's performance because

R. Khosla et al. (Eds.): KES 2005, LNAI 3681, pp. 1104–1110, 2005.

feedback loops are established through a common bus [5],[6],[7]. Therefore, the important issue in development a network-based control system is analyzing the effect of network uncertainties.

The main objective of this paper is to propose a design methodology and to analyze control performance with respect to network uncertainties for network-based control system. Firstly, we select the CAN as a network protocol and design smart network nodes for supporting this CAN protocol. These nodes have four types in all, which include sensor nodes for sensing system outputs, controller nodes for computing control inputs, actuator nodes for actuating control inputs, and monitor nodes for supervising entire systems. Secondly, we propose a new scheduling method to transfer data between nodes. Thirdly, experimental setup is built to perform velocity control of four DC-Motors using these nodes. The experimental results verify that the proposed distributed control system shows acceptable control ability. Also, these results show that network uncertainties, such as time delay and data loss, are the primary factors of decreasing performance of a network-based control system.

2   Design of Smart Network Nodes Using the CAN

The CAN was originally developed for automotive applications in the early 1980's and is also being used in a broad range of industrial applications as well as automation control systems[8],[9],[10]. The CAN protocol comprises a physical, data link, and application layer of the seven layer ISO/OSI reference model and uses a CSMA/BA (carrier sense multiple access/bitwise arbitration) medium access method. The physical layer defines how signals are actually transmitted. In this paper, the Philips PCA82C250 CAN transceiver[11] is used in order to interface the CAN controller with the physical bus. It controls the logic level signals from the CAN controller into the physical levels on the bus and vice versa. The data link layer is divided by a transfer layer and an object layer. The transfer layer is responsible for bit timing and synchronization, message framing, arbitration, error detection and signaling, and fault confinement etc. The object layer is concerned with message filtering and message handling. In this paper, the Philips SJA1000 CAN controller[12] is used and implements the complete CAN protocol defined in the data link layer.

In order to build a distributed control system using the CAN, network nodes, called smart nodes[4], which transmit and receive various information through the CAN bus are needed. The block diagram of the CAN node designed in this paper is shown in Fig. 1.

The smart network node is classified as a sensor node, a controller node, an actuator node, and a monitor node according to the functions of the application layer. Sensor node converts a plant output into the CAN message and transmits this message to the controller node. Controller node calculates a control input by using the output data transmitted from the sensor node and transmits it to the actuator node. Actuator node generates PWM(pulse width modulation) signal according to the control input transmitted from the control node. Monitor node supervises state of entire system and node. In order to communicate between the nodes, the CAN controller is initialized according to the data transmission speed, the bit timing, and the identifier etc. The environment setting and data handling of the CAN controller used in this paper refers to the literature cited [12].

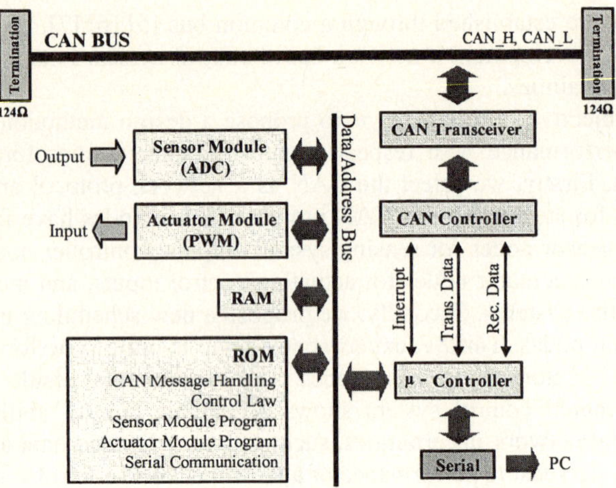

Fig. 1. Block diagram of the CAN node

3 Development of Network-Based Distributed Control System

In this paper, the CAN based distributed control system, which can control four DC-Motors is constructed. Fig. 2 shows a block diagram of the entire system(The block named T indicates the bus termination).

The subsystem for controlling the N'th DC-Motor comprises a sensor node N, a controller node N, and an actuator node N(N=1,2,3, and 4). The Tamagawa DC-Motor TS679N10E63 is used and the parameters provided by vendors are referred to, so that the second order system is modeled as follows[13].

$$\dot{x}(t) = \begin{bmatrix} -629.63 & -23.70 \\ 2264.09 & 0 \end{bmatrix} x(t) + \begin{bmatrix} 370.37 \\ 0 \end{bmatrix} u(t), \quad y(t) = \begin{bmatrix} 0 & 1 \end{bmatrix} x(t) \tag{1}$$

where $x(t) = [i_a(t) \quad \omega_m(t)]^T$ the state vector, $i_a(t)$ the armature current, $\omega_m(t)$ the angular velocity, and $u(t)$ the input voltage. Each subsystem executes a servo control tracking the reference speed, which is transmitted from the monitor node.

We choose the sampling period by h=20[ms] for discretization and design a controller where an integral action is added in the forward path[14]. The controller gains are selected such that the performance index $J = \sum \left\{ \bar{x}(k)^T Q \bar{x}(k) + u(k)^T R u(k) \right\}$ is minimized(where $\bar{x}(k)$ is the modified state vector $[x^T(k) \quad r(k-1)-y(k-1)]^T$, $r(k)$ is the reference speed, Q is positive semi-definite symmetric matrix, and R is positive constant).

Also, a state observer is designed easily using pole placement method.

The flow of the messages in one sampling cycle, is described by Fig. 3, showing a total time spent sending a control data from a sensor node to an actuator node.

Network uncertainties in the network-based control system are roughly regarded as time delay and data loss. Time delay occurs at the sensor, actuator, and controller node due to not only computation and capsulation of the message but also exchange of control messages through the network[3],[6]. This time delay depends on the bus speed and performance of the hardware and application program at each node. In this paper, we calculate time delay experimentally using counter values obtained from each node's micro-controller(Intel 80C196KC [15]). The measured counter values are T_0, T_1, T_2, T_3 in Fig. 3, and the computation delay $\tau_{com,a}$ is transmitted from the actuator node, the total time delay in the N'th DC-Motor is calculated by the eq(2).

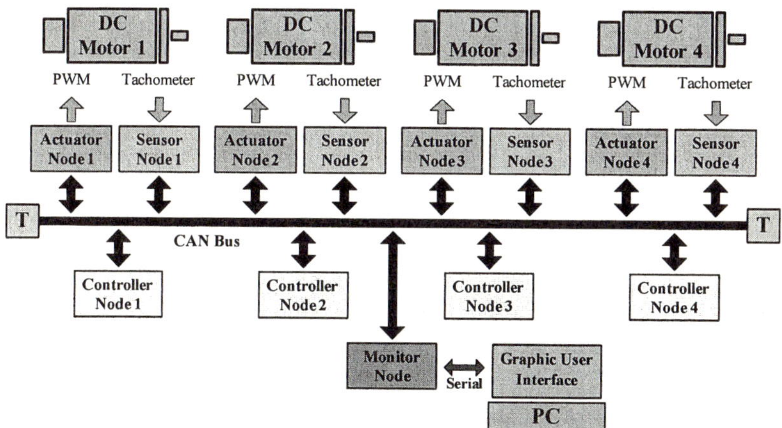

Fig. 2. Distributed Control System for speed control of DC-Motor using the CAN

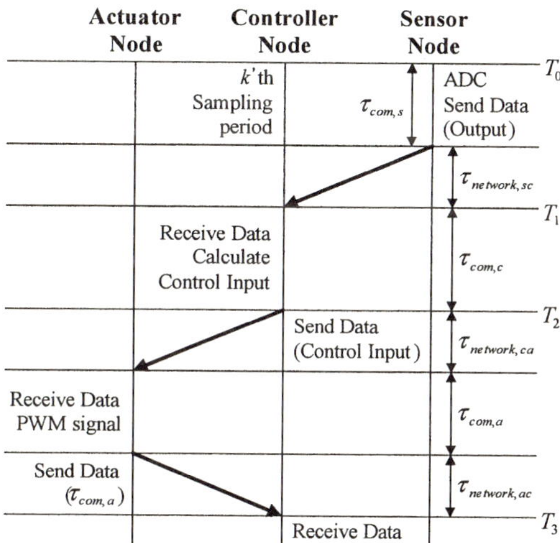

Fig. 3. Timing diagram of the N'th DC-Motor subsystem.

$$\tau = \tau_{com,s} + \tau_{network,sc} + \tau_{com,c} + \tau_{network,ca} + \tau_{com,a} = \frac{T_3 + T_2 + \tau_{com,a}}{2} - T_0 \tag{2}$$

Now, let us consider the situation that the data loss is occurred over the network. In this case, it is impossible to control the DC-Motor speed correctly. Without loss of generality, we assume that messages in the k'th sampling period are similar to the (k-1)'th sampling periods. In this reason, if data fails to transmit from one node to another node, the data is discarded promptly and past data is used for the control processing. This scheduling concept is more efficient than a retransmission method which requires additional waiting time and diminishing control performance. We set the maximum waiting time as deadline and the data, which exceeds this bound, is classified as data loss. When a message is received without errors, the CAN controller transmits receive interrupt signal to the micro-controller. If a data fails to transmit, it is discarded promptly and past data is used in the control processing.

4 Experimental Results

The experimental setup is shown in Fig. 4.

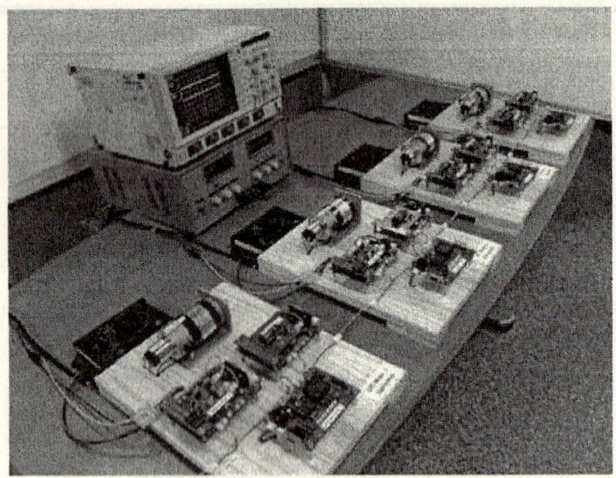

Fig. 4. Experimental setup

All nodes are connected through the CAN bus and the transmission rate of the CAN bus is set at 1[Mbps].

Fig. 5 shows the comparison of measurement output and simulation output of the first DC-Motor subsystem.

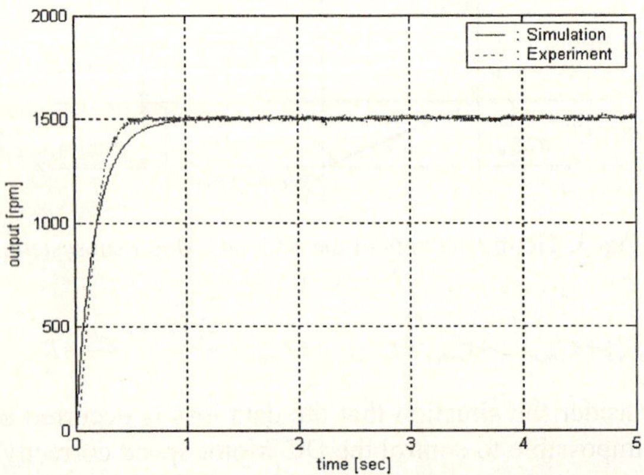

5. Comparison of measurement output and simulation output. The difference between iment and simulation results from DC-Motor modeling errors

A DC-Motor is generally a stable system, the motor's stability is not altered the time delay or data loss[3],[4],[7]. However, the control performance can degrade when occurrence of these uncertain factors increases. To confirm this issue, we observed output changes of the DC-Motor as varying time delay and data loss in the first subsystem. The counter values of the first subsystem are measured and the time delay is calculated as 2.388[ms] by using the eq(2). After setting the maximum waiting time as 3[ms], we cannot find the error or loss of transmission data with a number of times of the experiment. Thus, we conclude that the data transmission rate of experimental setup is nearly 1. When the time delay and the transmission rate are simultaneously changed, Table 1 shows the results of the control performance calculated by comparing the output of Fig. 5 as a reference with the output of the DC-Motor.

Table 1. Performance comparison of DC-Motor outputs due to time delay and data loss(%)

	$\tau=2.388[ms]$	$\tau=10.388[ms]$	$\tau=18.388[ms]$
r=1	100	97.07	94.55
r=0.8	95.21	94.82	93.54
r=0.6	93.52	91.45	90.92
r=0.4	89.77	88.99	83.26
r=0.2	83.76	76.16	72.76

5 Conclusion

In this paper, design considerations for a network-based distributed control system are proposed and practical analysis of control performance with respect to network uncertainties is done experimentally. Four types of the CAN smart nodes are designed and a new scheduling method is proposed in order to transfer data between these nodes. The experimental setup is built to perform velocity control of four DC-Motors using these nodes. Through the experimental results, it is verified that the proposed distributed control system provides an acceptable control performance. Also, this results show that network uncertainties are the primary factors of decreasing performance of a network-based system.

It is expected that the results of this paper will find applications in the core techniques of intelligent systems, sensor network, smart home, building automation, factory automation, and other application areas.

References

1. Feng-Li Lian, James Moyne, and Dawn Tilbury: Network Design Consideration for Distributed Control Systems. IEEE Trans. Control Sys. Tech., Vol. 10. No. 2. (2002) 297–307
2. Yodyium Tipsuwan and Mo-Yuen Chow: Control Methodologies in Networked Control Systems. Control Engineering Practice. Vol. 11. Issue 5. (2003) 483–492
3. Special Section on Networks and Control. IEEE Control Sys. Mag., vol. 21. no. 1, (2001) 22–99
4. Gregory C. Walsh, Hong Ye, and Linda G. Bushnell: Stability Analysis of Networked Control Systems. IEEE Trans. Control Sys. Tech. Vol. 10. No. 3. (2002) 438–446
5. Michael S. Branicky, Stephen M. Phillips, and Wei Zhang: Stability of Networked Control Systems-Explicit Analysis of Delay. Proc. Amer. Control Conf. Chicago. IL. (2000) 2352–2357

6. Feng-Li Lian, James Moyne, and Dawn Tilbury: Modeling and Optimal Controller Design of Networked Control Systems with Multiple Delays. Int. J. Control, Vol. 79. No. 6. (2003) 591–606
7. Arash Hassibi, Stephen P. Boyd, and Jonathan P. How: Control of Asynchronous Dynamical Systems with Rate Constraints on Events. Proc. 38th Conf. on Decision & Control. Phoenix. AZ. (1999) 1345–1351
8. Marco Di Natale: Scheduling the CAN Bus with Earliest Deadline Techniques. Proc. of IEEE Real-Time Sys. Symp. (2000) 259–268
9. BOSCH GmbH.: CAN Specification Part A & Part B. (1991)
10. Wolfhard Lawrenz: CAN System Engineering-From Theory to Practical Application. 1st edn. Springer-Verlag, Berlin Heidelberg New York (1997)
11. Philips: PCA82C250 CAN Controller Interface-Data Shcet. (2000)
12. Peter Hank and Egon Johnk: SJA1000 Stand-Alone CAN Controller-Application Note (97076). Systems Laboratory Hamburg Germany. (1997)
13. Benjamin C. Kuo and Farid Golnaraghi: Automatic Control Systems. 8th edn. John Wiley & Sons, Third Avenue New York (2003)
14. Kazuo Aida and Toshiyuki Kitamori: Design of a PI-type State Feedback Optimal Servo System. Int. J. Control. Vol. 52. No. 3. (1990) 613–625
15. Intel: 8XC196KC/8XC196KD User's Manual (1992)

Implementation of Adaptive Reed-Solomon Decoder for Context-Aware Mobile Computing Device

Seung Wook Lee[1], Jong Tae Kim[1], and Jae-Sang Cha[2]

[1] School of Information & Communication Eng. Sungkyunkwan Univ., Korea
`jtkim@yurim.skku.ac.kr`
[2] Department of Information and Communication Eng. Seokyung Univ., Korea

Abstract. In the context-aware mobile computing environment, the context-aware mobile computing devices adapt their behavior to available services which require different communication protocols by the type of transmitted data (e.g. video, voice, text and etc.). In this paper, we present a novel architecture for an adaptive Reed-Solomon decoder which can apply to context-aware mobile computing device to support various code specifications for heterogeneous communication protocols. This decoder can vary the length of the codeword, generate polynomials and correct erasures. For this decoder, we use a modified Euclidean algorithm based on a $GF(2^8)$ finite field for the error locator polynomial. Also, this decoder can support any specification in the range of $(1 \leq n \leq 255)$ for the codeword and $(1 \leq t \leq 16)$ for the errors.

1 Introduction

Context-aware mobile computing network is a wireless network which consists of service provider nodes and user's context-aware mobile computing device. In Context-aware mobile computing network, a context-aware mobile computing device has to process various communication protocols between heterogeneous service provider nodes. That is, the context-aware mobile computing device has to cognize a type of protocol used service delivery, and adapts their behavior to available service. In this process, these communication protocols require characterized error-correction mechanism by wireless channel state such as bandwidth, error probability, signal-to-noise ratio(SNR) and so on. So, adaptive error-correction scheme is required in context-aware mobile computing device.

In this paper, we propose the architecture of adaptive Reed-Solomon(RS) decoder for context-aware mobile computing device. RS code has been widely used in various protocols for forward error correction(FEC). This decoder can adjust code specification dynamically for supporting context-aware mobile computing device and operate in high-speed using pipeline processing. In this decoder architecture, we used modified Euclidean algorithm based on $GF(2^8)$ finite field for error locator polynomial. So, the decoder can support any specification $(1 \leq n \leq 255)$ for codeword and $(1 \leq t \leq 16)$ for errors which can apply to most standard for wireless network systems.

This paper is organized as follows: In section 2, we briefly review some related works on the subject of the adaptive RS decoder. Section 3 presents the proposed adaptive RS decoder and Section 4 explains the performance and a synthesis of the results for the decoder. Finally, in Section 5, we draw our conclusions.

R. Khosla et al. (Eds.): KES 2005, LNAI 3681, pp. 1111–1117, 2005.

2 Related Works

There are several architectures that have been implemented with the purpose of supporting any RS code specification and these have involved the use of several different techniques.

Christof Paar[2] proposed an efficient architecture involving the use of the Galois field multiplier based on reconfigurable hardware. In this architecture, by using a $GF((2^n)^m)$ composite field instead of the regular $GF(2^m)$ finite field, the number of I/O bits needed for the multiplier is reduced. This approach can speed up the primitive arithmetic operation of the decoder, but it can also cause more hardware overhead because of the requirement for fully reconfigurable hardware.

Sourav Roy[3] used a hardware-software co-design method for the implementation of an RS decoder accelerator based on a relatively small amount of reconfigurable hardware and a DSP processor. This hardware accelerator speeds up the primitive arithmetic operations, while the DSP can change the code specification by means of software parameters. The RS decoding algorithm is a computation intensive algorithm, however this decoder architecture used fine-grain granularity for the co-design, which can cause serious data transmission overhead to be incurred between the DSP and the reconfigurable hardware. Hence, the operation speed of this decoder depends on the data transmission speed between the DSP and the reconfigurable hardware logic.

Huai-Yi Hus[4] used a module-based architecture. The modules are configured by means of reconfigurable hardware by selecting the appropriate code specification. This architecture provides a high-speed data rate and a small change of the generate polynomial. However, the configuration process based on the reconfigurable hardware has to be performed for every change of code specification.

M. K. Song[5] implemented an adaptive RS decoder which can decode any RS code of any block length and any message length, without the use of reconfigurable logics. This decoder allows for 4-step pipeline processing and reduces the number of required MEA cells by using a clock rate for the recursive MEA cells which is several times faster than that used for the other part of the decoder. Indeed, in order to reduce the required number of MEA cells, the clock rate of the other parts need to be decreased.

In summary, the above-mentioned architectures tend to be one-sided, in that they privilege either the performance aspect or the implementation aspect. Therefore, we propose a more efficient architecture, which takes into consideration both the performance and implementation aspects, by using a modified Euclidean algorithm. In our decoder, no reconfigurable hardware is used. Therefore, changing the code specification does not require the hardware architecture or size to be changed. Instead, control registers, which store the decoder set-up values and control operations of the decoder for each code specification, are utilized.

Adaptive Reed-Solomon Decoder Architecture

shows the architecture of the adaptive RS decoder. This decoder can receive deword length up to n_{max} and correct error symbols up to t_{max}, where n_{max}

and t_{max} are the maximum codeword length that is decodable and the maximum error correction capability, respectively. The host-processor can adjust the specification by inputting the set-up values, n_c and t_c, and the roots of the general polynomial for the RS code specifications, where n_c and t_c are the desirable codeword length and the error correction capability respectively. This operation, for the purpose of changing the specification, can be performed immediately, by writing to the control registers via the general purpose I/O of the host processor.

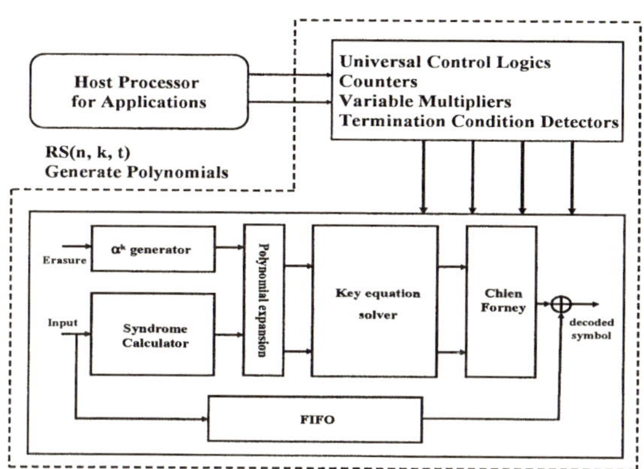

Fig. 1. The adaptive Reed-Solomon decoder

3.1 Syndrome Calculator

The syndrome calculator computes the syndrome equation using the roots of the generate polynomial. The syndrome polynomial, $g(x)$, and the generate polynomial, S_k, are as follows[7]:

$$g(x)=\prod_{j=0}^{2t-1}(x-\alpha^{h+j})=\sum_{j=0}^{2t-1}g_j x^j \,,\ S_k =\sum_{i=0}^{N-1}r_i X^i\Big|_{x=\alpha^k} =\sum_{i=0}^{N-1}r_i\alpha^{ki}\ \text{for } 1\le k\le d-1 \tag{1}$$

where h is an integer constant, which is called the power of the first of the consecutive roots of $g(x)$ in eq(1), r_i represents the coefficient of the received polynomial, $r(x)$, and α^{ki} represents the roots of the generate polynomial, $g(x)$, used for data encoding in eq(1). By carefully choosing the integer, h, the circuit complexity of the encoder and decoder can often be reduced[7]-[8]. The integer constant, h, is usually, but not always, chosen to be 1 or 0. Thus, the roots of the generate polynomial are fixed in the typical architecture used for supporting a fixed RS code specification. In this case, considering the logic delay and area, α-constant multipliers based on $GF(2^8)$ are more suitable than variable $GF(2^8)$ multipliers. However, in the case where the code specification is susceptible to be changed, the generate polynomial can also be changed.

Thus, variable $GF(2^8)$ multipliers are more appropriate for the adaptive RS decoder, as shown in Fig. 2. Also, by using the systolic array structure to compute the syndrome values, the syndrome calculator for the RS(n_{max}, k, t_{max}) code can compute the syndrome for RS(n_c, k, t_c) by means of the control logics, which count the number of input code symbols and the required number of cells.

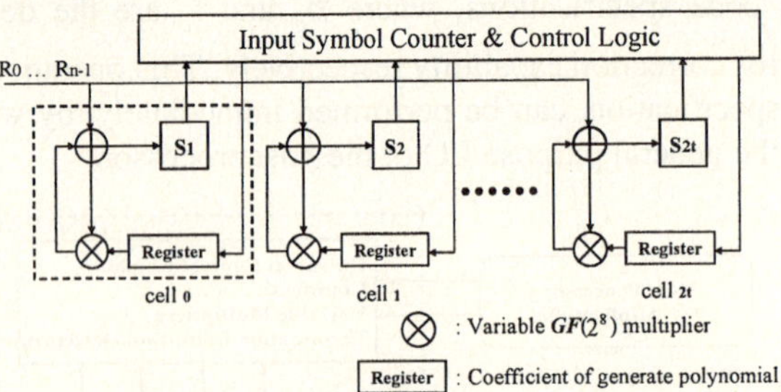

Fig. 2. The modified syndrome calculator

3.2 Error Locator Polynomial

In order to compute the error locator polynomial and error evaluator polynomial in this decoder, the MEA cells are used in a recursive manner. The MEA cell used in our adaptive RS decoder is based on the architecture that we proposed in [6]. The MEA cell architecture performs well for a fixed code specification and the feature which distinguishes it from the conventional architecture used to find the error locator polynomial is described in [6]. In this paper, we modified the previously proposed architecture with the goal of supporting any RS code specification. Figs. 3 shows the architectures of the recursive MEA cells using the (a) register shifting [7] and (b) register addressing techniques[6]. In the register shifting technique, as shown in Fig. 3(a), it takes t_{max} recursions to compute the algorithm, where each recursion requires t_{max} codeword symbol times. Therefore, using a single cell recursively requires a total symbol time of $(t_{max})^2$ to compute the MEA. As shown in Fig. 3(a), the number of registers which store the coefficients of the $R(x)$, $Q(x)$, $\lambda(x)$ and $\mu(x)$, depends on the error correcting capability, t_{max}. The values stored in each register are shifted synchronously with the system clock. Thus, regardless of the number of coefficients in the $R(x)$, $Q(x)$, $\lambda(x)$ and $\mu(x)$ registers, the values are shifted out after an identical number of clock ticks. That is, in the case of small t_c in the RS(n_c, k, t_c) specification, degradation of the decoding speed can occur. In the case of Fig. 3(a), 16 clock ticks are always required in every iterations. On the other hand, in the register addressing technique, the values stored in each register can be referred to immediately the control logic. Therefore, the required computation clock time can be reduced to the error correcting capability, t_c, as follows[6]:

$$The\ required\ clocks = (n_c - k)^2 - (t_c^2 + 2t_c) + 1 \qquad (2)$$

$(t_c^2 + 2t_c)$ is the reduction in the symbol times afforded by the register addressing technique[6].

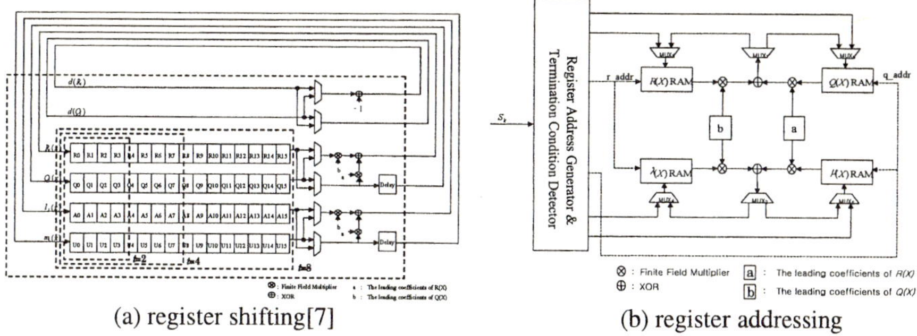

(a) register shifting[7]   (b) register addressing

Fig. 3. The structure of an MEA cells

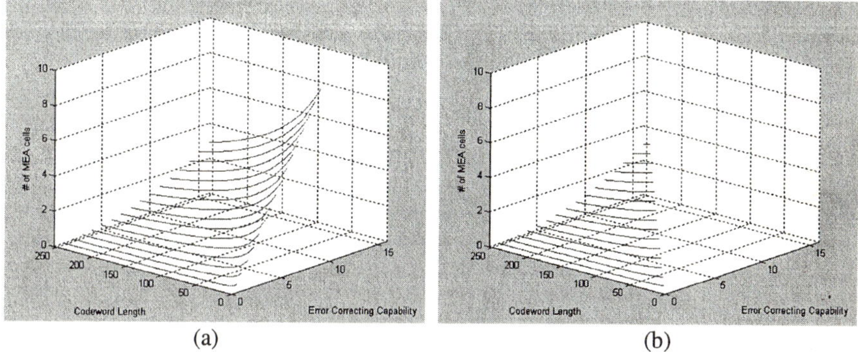

(a)   (b)

Fig. 4. Analysis of the relation for the # of MEA cells, codeword length and error correction capability for a code rate of 0.5(a) and 0.85(b)

In general, an RS decoder for a fixed specification is designed in the pipeline manner. Therefore, the appropriate number of MEA cells required is as follows:

$$\#\ of\ required\ MEA\,cells = \left| \frac{(n_c - k)^2 - (t_c^2 + 2t_c) + 1}{n_c} \right| \tag{3}$$

According to eq(3), in the case where $n = 64$ and $t = 16$, the number of required MEA cells can be 10 for pipeline processing. The MEA cells occupy the largest area of the whole RS decoder. So, in order for the decoder to be adaptive, too many MEA cells cannot be integrated when considering the decoder area efficiency. Thus, we analyze the number of MEA cells required for the code rate (k/n) from a practical viewpoint, as shown in Figs. 4. As the code rate decreases, the error correction capability is improved, but the number of MEA cells required increases. However, the use of too low a code rate can decrease the bandwidth efficiency[8]. As shown in Table 2, the existing industrial standard is a code rate of 0.87 ~ 0.92.

4   Implementation

The proposed adaptive RS decoder was synthesized using ANAM a 0.25-um, four-metal, 2.5v CMOS technology cell library. Table 2 shows the synthesized results.

Table 1. The synthesized results

The Adaptive Reed-Solomon decoder	
Syndrome calculator	16,128 gates
MEA cells (i.e. 4 MEA cells)	83,028 gates
Chien and Forney module	7,651 gates
Total gate counts	145,315 gates
Clock rate	100 MHz
Adaptability	$GF(2^8)$, $(1 \leq n \leq 255)$, $(1 \leq t \leq 16)$
Erasure correcting	Yes

Table 2. Comparison of our decoder with individual decoders

Frequency 100 MHz	Individual decoder k/n	Gate size	Data rate	Our decoder Gate size	Data rate
DVB	0.92	32,114			
CCSDS	0.87	107,826	800 Mbps	145,315	800 Mbps
ATSC	0.90	95,242			
IMT2000	0.87	19,832			

Table 3. Comparison of our decoder with the versatile RS decoder developed in [5]

	M. K. Song[5]	Our decoder
Adaptability	$GF(2^8)$ $(13 \leq n \leq 255)$ $(1 \leq t \leq 10)$	$GF(2^8)$ $(1 \leq n \leq 255)$ $(1 \leq t \leq 16)$
Latency	$3n - 1 + (4t + t + 20)/pk$ $(k = 2)$ k: # of MEA cells	$4n$ 4-MEA cells
Frequency	30 MHz	100 MHz
Date rate	240 Mbps	800 Mbps
Total # of gates	211,296 gates	145,315 gates

Table 2 shows a comparison of our decoder with various individual decoders for the industrial standard specifications. As shown in Table 3, our decoder satisfies the listed industrial standard specifications with fixed hardware architecture. Table 3 shows that our decoder is more efficient than the versatile decoder developed by M. K. Song[5]. Although our decoder has 2 more MEA cells, its total area is smaller, and it can correct more errors for the same codeword length.

5 Conclusion

In this paper, we proposed a novel architecture for a adaptive Reed-Solomon decoder, for context-aware mobile computing device, which can accept any RS code specifications within the range of $(1 \leq n \leq 255)$ and $(1 \leq t \leq 16)$. Also, the proposed decoder can correct erasures and adopt various generate polynomials. To accomplish this, we analyzed the number of MEA cells required as a function of the code rate (k/n). Thus, the proposed decoder includes only 4 MEA cells for a code rate of more than 0.85. This decoder can change the specifications by writing the set-up values to the control registers. The set-up values consist of n_c and t_c and the roots of the generate polynomial of the specification. Also, the performance of the decoder satisfies the industrial standard specifications.

Acknowledgement

This work was supported by IT SoC Promotion Group, Korea IT Industry Promotion Agency(Human Resource Development Project for IT SoC key Architect).

References

1. Gruia-Catalin Roman et al. "Network abstractions for context-aware mobile computing" International Conference on Software Engineering Proceedings of the 24th International Conference on Software Engineering
2. Paar C. and Rosner M., "Comparison of arithmetic architectures for Reed-Solomon decoders in reconfigurable hardware", FPGAs for Custom Computing Machines, Proceedings, The 5th Annual IEEE Symposium on, 16-18 April 1997, pp 219 -225
3. Roy S., Bucker M., Wilhelm W. and Panwar B.S., "Reconfigurable hardware accelerator for a universal Reed Solomon codec", Circuits and Systems for Communications, 2002. Proceedings. ICCSC '02. 1st IEEE International Conference on, 26-28 June 2002, pp. 158 -161
4. Huai-Yi Hsu and An-Yeu Wu, "VLSI design of a reconfigurable multi-mode Reed-Solomon codec for high-speed communication systems", ASIC, 2002. Proceedings. 2002 IEEE Asia-Pacific Conference on, 6-8 Aug. 2002, pp. 359 -362
5. M. K. Song, E. B. Kim, H. S. Won and M. H. Kong, "Architecture for decoding adaptive Reed-Solomon codes with variable block length", Consumer Electronics, IEEE Transactions on, Vol. 48, Issue: 3, Aug. 2002, pp. 631-637
6. Dong Hoon Lee, Jong Tae Kim, "Efficient Recursive Cell Architecture for the Reed-Solomon Decoder", Journal of the Korean Physical Society, Vol. 40, Number 1, 2002, pp. 82-86
7. Howard M. Shao, T. K. Truong, Leslie J. Deutsch, Joseph H. Yuen, Irving S. Reed, "A VLSI Design of a Pipeline Reed-Solomon Decoder", IEEE Trans. Computers 34(5), 1985, pp. 393-403
8. Irving S. Reed and Xuemin Chen, "Error-Control coding for Data Network", Kluwer Academic Publishers, 1999

Authenticated IPv6 Packet Traceback Against Reflector Based Packet Flooding Attack

Hyung-Woo Lee[1] and Sung-Hyun Yun[2]

[1] Dept. of Software, Hanshin University, Osan, Gyunggi, 447-791, Korea
`hwlee@hs.ac.kr`
[2] Div. of Information and Communication Engineering, Cheonan University,
Anseo-dong, Cheonan, Chungnam, 330-704, Korea
`shyoon@cheonan.ac.kr`

Abstract. IPv6 can provide end-to-end security services such as access control, confidentiality, and data integrity with less impact on network performance. However, we can also prospect that there will be much more dangerous and tremendous type of attack on IPv6 than that on IPv4. So, we propose new IP traceback mechanism for tracing the spoofed real source on IPv6 packet using authentication mechanism. Proposed authentication mechanism supports the key disclosure mechanism on IPv6 packet marking with ICMPv6 on the IP traceback process.

1 Introduction

The global need for IP addresses has even added political force to the drive for IPv6 implementation. Aside from the increased address space, IPv6 offers a number of other key design improvements over IPv4. (1) IPv6 provide improved efficiency in routing and packet handling. And (2) IPv6 format allows plug-and-play deployment of Internet-enabled devices such as cell phones, wireless devices, and home appliances. (3) By providing globally unique addresses and embedded security, IPv6 can provide end-to-end security services such as access control, confidentiality, and data integrity with less impact on network performance. Finally, (4) IPv6 multicast saves network bandwidth and improves network efficiency because IPv6 multicast completely replaces IPv4 broadcast functionality[1]. Although IPv6 is powerful as above compared with existing IPv4 mechanism, we can also prospect that there will be much more dangerous and tremendous type of attack on IPv6 than that on IPv4[2].

In this paper, we overviewed the possible attack mechanism on IPv6 by considering of the vulnerability of proposed IPv6 structure[3,4]. IPv6 can provide end-to-end security services such as access control, confidentiality, and data integrity using IPSec(AH/ESP) functions[1]. However, *spoofing attack on IPv6 is also possible by forging the real source*[2,3]. The reason is that we can commonly use IPv6 over IPv4 tunneling mechanism for encapsulating IPv6 traffic within IPv4 packets, to be sent over an IPv4 backbone. Therefore, this enables similar diverse DDoS attack[6,9] on IPv6 to be applicable through an existing IPv4 infrastructure.

R. Khosla et al. (Eds.): KES 2005, LNAI 3681, pp. 1118–1124, 2005.
© Springer-Verlag Berlin Heidelberg 2005

As a solution, we suggest *authenticated IP traceback mechanism on IPv6* against DDoS attack. IPv6 improves productivity by enabling network connectivity via a wider range of media and delivery mechanisms. But for general acceptance, the new IPv6 networks must demonstrate responsiveness at least equal to that of IPv4. Therefore, we considerate on the authenticated IP Traceback mechanism with hashed *MAC* function on IPv6 header.

2   IPv6 and Its Possible Attacks

2.1   IPv6 Protocol: Overview on Header Format

The features which IPv6 protocol can be summarized as follows: (1) New header format, (2) Large address space, (3) Efficient and hierarchical addressing and routing infrastructure, (4) Stateless and stateful address configuration, (5) Security, (6) Better Quality of Service(QoS) support, (7) New protocol for neighboring node interaction, (8) Extensibility[1,5].

IPv6 Header: A new format is designed to minimize header overhead. The IPv6 header is only twice the size of the IPv4 header. This is achieved by moving both nonessential and optional fields to extension headers that are placed after the IPv6 header.

The IP version number for IPv6 packets will be the binary representation of number 6 instead of 4. This information is contained in the "version" field, which is the first part of the IP packet header. And the "time-to-live" field has been renamed "hop limit". The 8-bit IPv4 type of service (TOS) field has been renamed for IPv6 purpose and is now called the "traffic class" field. It is expected that traffic classes will closely mirror the IPv4 usage of the TOS bits for differentiated services.

A new 20-bit "flow label" field has been added to the packet header to facilitate the identification of a specific traffic stream. This flow can be associated with a particular traffic class. The *20-bit Flow Label field* in the IPv6 header may be used by a source to label sequences of packets for which it requests special handling by the IPv6 routers, such as non-default quality of service or advanced "real-time" IPv6 packet control services such as flooding and DDoS control[5].

2.2   Possible Attacks on IPv6

Within RFC 2462 it is defined that if during the neighbor discovery process IPv6 device which received response that some other particular device is already using its proposed address - it must not use it. Therefore, it is fairly easy for a malicious user to craft the address which already exists on the local segment and hence achieve Denial of Service(DoS) attack on particular IPv6 device trying to initially obtain the stateless IPv6 address and start the IPv6 communication[2,3,4].

Packet Flooding Attack in IPv6: IPv6 is enabled by default, but the problem can only be exploited remotely against hosts which are reachable through IPv6. *Additionally reflector based packet flooding attack cause DDoS situation also possible on IPv6 as IPv4 network.*

In IPv6 network, the operator of a reflector cannot easily locate the slave that is pumping the reflector same as in IPv4, because the traffic sent to the reflector does not have the slave's source address, but rather the source address of the victim[11]. The attacker first locates a very large number of reflectors on intermediate IPv6 hosts. They then orchestrate their slaves to send to the reflectors spoofed traffic purportedly coming from the victim, V. The reflectors will in turn generate traffic from themselves to V.

Reflector based Flooding Attack on IPv6: *The flood at V arrives not from a few hundred or thousand sources, but from a million sources, an exceedingly diffuse flooding likely clogging every single path to V from the rest of the IPv6 Internet.*

2.3   Against Reflector Packet Flooding Attack on IPv6

Traffic generated by reflectors may have sufficient regularity and semantics that it can be filtered out near the victim without the filtering itself constituting a denial-of-service to the victim ("collateral damage"). Therefore, the filtered traffic could then be rate-limited, delayed, or dropped[14].

Traceback Mechanism with Packet Classification: *In principle it could be possible to deploy traceback mechanisms that incorporate the reflector end-host software itself in the traceback scheme, allowing traceback through the reflector back to the slave[8,10,11].*

In addition, traceback may not help with traceback in practice if the traceback scheme cannot cope with a million separate Internet paths to trace back to a smaller number of sources. *So we need an advanced new mechanism against reflector-based packet flooding attack(DDoS) by using combined technique both packet classification and advanced traceback mechanism.*

2.4   DDoS Traffic Identification/Control Mechanism on IPv6

In IPv6, *ECN(Explicit Congestion Notification) field* was proposed as an implementation mechanism of congestion manager. The ECN field in the IPv6 consists of two bits after the Differencial Service field[5]. The four possible different value (00, 01, 10, 11) indicate whether the end-nodes (sender and destination) are using en ECN-Capable Transport as well as whether there is congestion at the sender. If ECN is 11, it mean that router begins to experience congestion.

So, we can use these field to control DDoS traffic at routers as *ACC (aggregate-based congestion control)*. Additionally, as hacking attacks are extremely diverse, it evaluates traffic based on *congestion signature*, which is corresponding to the congestion characteristic traffic.

ACC with ECN field: *If traffic shows congestion exceeding a specific bandwidth based on the characteristic of DDoS attack network traffic, the ACC module judges based on congestion signature that a hacking attack has happened and, working with a filtering module on ECN field on IPv6 packet, provides a function to block the transmission of traffic corresponding to the congested DDoS attack.*

3 Proposed IPv6 Traceback Against Reflector Packet Flooding Attack

3.1 IP Traceback on IPv6 Against Reflector Packet Flooding Attack

Step 1: Marking with Flow Label and Hop Limit Fields. IPv6 headers consists of 8 fields spread over 40 bytes[1,2,5]. CLASS field consists of 6-bit *Differentiated Services(DS) field* and 2-bit *ECN* value. Aggregates or aggregated flows may also be referred to as classes of packets. This study use 2 bits out of 6 DS field as *TF(traceback flag)* and *PF(pushback flag)*. *Flow Lable(FL) field* is a 20-bit value used to identify packets that belong to the same flow. We use router's information on FL field for IPv6 packet traceback. And we also use *Hop Limit(HL) field* value for calculating traceback distance. The router extracts 20-bit Flow Label field in the IPv6 header for tracing the packet's real transmission path, which is used by a source to label sequences of packets for requesting special traceback handling module on the IPv6 router.

Notation and Definition: Let's say A_x is the 128 bit IPv6 address of R_x, P_x is IPv6 packet arrived at router R_x, and M_x is 28 bits on the header of P_x in which marking information can be stored. In packet P_x, M_x is composed of 8-bit *CLASS* field, and 20-bit Flow Label field. CLASS field has been defined is not used currently. Thus the use of CLASS field does not affect the entire network.

3.2 Authenticated IPv6 Packet Traceback

We need a mechanism to authenticate the packet marking. We propose a much more efficient technique to authenticate the IPv6 traceback packet(the Authenticated Marking Scheme). This technique only uses one cryptographic MAC (Message Authentication Code) computation per marking, which can be adapted so it only requires the 20-bit overloaded IP flow label field for storage and ICMPv6 Traceback Mechanism.

Step2: Authentication Mechanism on Traceback. Two parties can share a secret key K. When party A sends a message M to party B, A appends the message with the MAC of M using key K. When B receives the message, it can check the validity of the MAC. A well-designed MAC guarantees that nobody can forge a MAC of a message without knowing the key.

Basic Assumption: Let f denote a MAC function and f_K the MAC function using key K. If we assume that each router R_i shares a unique secret key K_i with the victim V, then instead of using hash functions to generate the encoding of a router's IP address, R_i can apply a MAC function to its IP address and some packet-specific information with K_i.

The packet-specific information is necessary to prevent a replay attack, because otherwise, a compromised router can forge other routers markings simply by copying their marking into other packets. We could use the entire packet content in the MAC computation, i.e. encode R_i with its IP address A_i on packet P_i as $f_{K_i}(<P_i; A_i>)$.

We extend the scheme by using the *time-released keys authentication scheme* based on [10]. The basic idea is that each router R_i first generates a sequence of secret keys, $\{K_{j;i}\}$ where each key $K_{j;i}$ is an element of a hash chain. By successively applying a one-way function g to a randomly selected seed, $K_{N;i}$, we can obtain a chain of keys, $K_{j;i} = g(K_{j+1;i})$. Because g is a one-way function, anybody can compute forward (backward in time), e.g. compute $K_{0;i}$; ... ; $K_{j;i}$ given $K_{j+1;i}$, but nobody can compute backward (forward in time) due to the one-way generator function.

Step 3: Authenticated Traceback Path Marking at Routers. When informed of the occurrence of abnormal traffic, router R_x performs marking for packet P_x corresponding to congestion signature classified by decision module.

Router R_x commits to the secret key sequence through a standard commitment protocol, e.g. by signing the first key of the chain $K_{0;x}$ with its private key, and publish the commitment out of band, e.g. by *posting it on a web site and also sending it on iTrace packet*. We assume that each router has a certified public key. The time is divided into intervals. Router R_x then associates its key sequence with the sequence of the time interval, with one key per time interval. It is also sent by ICMPv6[5,13] Traceback message to the victim V. Therefore, in time interval t, the router R_x uses the key of the current interval, $K_{t;x}$, to mark packets in that interval. The router uses $K_{t;x}$ as the key to compute the hashed MAC. R_x will then reveal/disclosure the key $K_{t;x}$ after the end of the time interval t. And the interval value t can also sent by ICMPv6 traceback packet.

4   Advanced iTrace for Authenticated Reconstruction

4.1   Authenticated iTrace Structure

ICMPv6 traceback element contains the time, in NTP timestamp format (eight octets) for interval value. The timestamp can be consistent with the starting and ending time of validity of the applicable hash key as reported in *Key Disclosures* in subsequent ICMPv6 traceback packets.

An attacker may try to generate fake Traceback messages, primarily to conceal the source of the real attack traffic, but also to act as another form of attack. We thus need authentication techniques that are robust but quite cheap to verify. In iTrace message, hash code(the HMAC Authentication Data element) is used, supported by signed disclosure of the keys most recently used (the Key Disclosure and Public Key Information elements). The primary content of the Key Disclosure element consists of a key used to authenticate previous traceback messages and the starting and ending times between which that key was used.

4.2   Generate iTrace Message Against Reflector Attack

We can generate the suspicious packet into ICMPv6 traceback packet and send it by iTrace module to the victim host. In an IP header excluding the option and the padding, the length of the unchanging part to the bit just prior to the option is 128 bits excluding HLEN, Hop Limit and checksum. The 128 bits in a packet can be used to represent unique characteristics, so a router can use them in generating a ICMPv6 traceback message for the packet. This iTrace packet will be sent to the victim hosts. The detailed steps will be as follows.

We can extract unique information on packet P_x and set it Q_x. Q_x is composed of 128-bit information representing the characteristics of network packets and providing the uniqueness of the packet. The 128 bit information can be divided into two 64-bit sub-blocks as follows. $Q_x = B_{x1}|B_{x2}$. And notation $|$ means common concatenation function. And 64-bit B_x can be obtained from these two 64-bit sub-blocks. $B_x = B_{x1} \oplus B_{x2}$.

Now the router R_x is aware of its own 128-bit IP address A_x and $t; x$ as the timestamp information. Then the router calculates $A_x' = (A_x|t; x) \oplus K_x$ by generating 64-bit information with 64-bit key timestamp(key identifier) information. Then the the router R_x can generate $B_x' = B_x \oplus A_x'$. And then B_x' is included in an ICMPv6 traceback packet and sent to the victim host. And then the victim system authenticate this information.

4.3 Analysis for the Proposed Mechanism

In order to evaluate the performance of the proposed method(AIPv6PM: Authenticated IPv6 Packet Marking against Reflector Packet Flooding Attack), we analyzed the performance using ns-2 Simulator[16]. In this study, a classification technique is adopted in classifying and packet control on flooding traffic and as a result the number of marked packets has decreased. As a result, we can reduce the malicious DDoS traffic in victim host.

Table 1. Comparison of performance with existing IPv4 traceback methods.

	Net.load	Sys.load	Mem.	Traceback	Sec.	Auth.	pkt #	Protocol
Filter	$\times$	$\times$	$\times$	$\times$	$\times$	$\times$	n	IPv4
Logging	$\times$	$\times$	$\uparrow$	$\bigtriangledown$	$\times$	$\times$	n	IPv4
APPM	$\downarrow$	$\uparrow$	$\uparrow$	$\Diamond$	$\Diamond$	$\triangle$	$n \cdot p$	IPv4
iTrace	$\downarrow$	$\uparrow$	$\uparrow$	$\Diamond$	$\Diamond$	$\times$	$n \cdot p$	IPv4
AIPv6PM	$\downarrow$	$\downarrow$	$\uparrow$	$\triangle$	$\triangle$	$\triangle$	n	IPv6

$\times$:N/A $\downarrow$:low $\uparrow$:high $\triangle$:good $\Diamond$:moderate $\bigtriangledown$:bad n:traffic p:sampling ratio

Table 1 shows the comparison of the performance of the proposed method with that of existing 'IPv4 traceback-related technologies. Existing packet marking methods and iTrace methods based on node and edge sampling cause low load on the management system and the network but create heavy load when a victim system restructures the traceback path. These methods are considered suitable in terms of traceback function and scalability. However, they are vulnerable to DDoS attacks. Proposed method improves the bandwidth of the entire network and can restructure the path to the source of DDoS attacks with a small number of marking packets. In the method, the path to the source of attack can be restructured only with n traceback messages if the packet has been transmitted via n routers on the network. As a disadvantage, the method requires additional memory at routers for the DDoS-related identification function.

5 Conclusions

This study proposed a method for a victim system to trace back the actual IPv6 address of the attacker for spoofed traffic when a reflector DDoS attack happens. We proposed a new authenticated marking technique that provides the functions of identifying/controlling DDoS hacking attacks on the network and at the same time enables victim systems to trace back the spoofed source IPv6 packets. The proposed method is superior to existing ones in load, performance, stability and traceback function.

Acknowledgements

The first author was supported by University IT Research Center(ITRC) Project. The second author was supported by Cheonan University Research Foundation.

References

1. S. Deering, R. Hinden, "Internet Protocol, Version 6 (IPv6) Specification", RFC2460, IETF drafe, December 1998.
2. Franjo Majstor, "Does IPv6 protocol solve all security problems of IPv4?", Information Security Solutions Europe, 7-9 October 2003.
3. Timo Koskiahde, "Security in Mobile IPv6", 8306500 Security protocols, Tampere University of Technology, 2002.
4. Henry C.J. Lee, Miao Ma, Vrizlynn L.L. Thing, Yi Xu, "On the Issues of IP Traceback for IPv6 and Mobile IPv6", URL: http://citeseer.ist.psu.edu/689796.html
5. Pete Loshin, "IPv6: Theory, Protocol, and Practice", Second edition, Morgan Kaufmann, 2003.
6. John Elliott, "Distributed Denial of Service Attack and the Zombie and Effect", IP professional, March/April 2000.
7. Tatsuya Baba, Shigeyuki Matsuda, "Tracing Network Attacks to Their Sources", IEEE Internet Computing, pp. 20-26, March, 2002.
8. Hassan, Aljifri, "IP Traceback: A New Denial-of-Service Deterrent?", IEEE SECURITY & PRIVACY, pp.24-31, May/June, 2003.
9. L. Garber, "Denial-of-Service attacks trip the Internet", Computer, pages 12, Apr. 2000.
10. Andrey Belenky, Nirwan Ansari, "On IP Traceback", IEEE Communication Magazine, pp.142-153, July, 2003.
11. Vern Paxson, "An Analysis of Using Reflectors for Distributed Denial-of-Service Attacks", ACM SIGCOMM, Computer Communication Review, pp.38-47, 2001.
12. Chang, R.K.C., "Defending Against Flooding-based Distributed Denial-of-Service Attacks: A Tutorial", IEEE Communications Magazine, Volume: 40 Issue: 10 , Oct 2002, pp. 42 -51.
13. Steve Bellovin, Tom Taylor, "ICMP Traceback Messages", RFC 2026, Internet Engineering Task Force, February 2003.
14. P. Ferguson and D. Senie, "Network ingress Filtering: Defeating denial of service attacks which employ IP source address spoofing", May 2000. RFC 2827.
15. D. X. Song, A. Perrig, "Advanced and Authenticated Marking Scheme for IP Traceback", Proc, Infocom, vol. 2, pp. 878-886, 2001.
16. K. Fall, "NS Notes and Documentation", The VINT Project, 2000.

A Relationship Between Products Evaluation
and IT Systems Assurance

Tai-hoon Kim[1] and Seung-youn Lee[2]

[1] San-7, Geoyeo-Dong, Songpa-Gu, Seoul, Korea
taihoon@empal.com
[2] SKK Univ., Dept. of Information & Communication Eng. Kyonggi, 440-746, Korea
syoun@ece.skku.ac.kr

Abstract. IT systems consist on very many components and very complex, so the implementation of the security countermeasures needs more critical considerations. Indeed, IT systems contain many subsystems, and most of subsystems consist of one or more IT products. In this hierarchy structure, the security characteristics of each IT product may affect the total assurance of the IT systems. Therefore, the security should be considered at the base level of the IT systems, in other words, IT product the base of the IT systems. For this work, this paper presents our research results about the security and assurance relationship between IT products and IT systems.

1   Introduction

In general case, the aims of IT systems are the protection of information and information systems. For more precisely, IT security is the protection of information and information systems from many types of threats such as unauthorized disclosure and modification arising from human or systems-generated activities.

Most IT systems are built using other systems. When we built IT systems, hardware platforms, operating systems and other products are often purchased and used as the foundation on which applications are implemented.

It is very important that the developers of designers of IT systems must determine the security properties of these components, whether COTS or government-furnished. For the more assurance, the testing, evaluation and certification approaches should be recursively applicable to all the integrated components. In the best of all possible scenarios, all products included in the IT systems will have been evaluated under successful evaluation and certification scheme.

If unevaluated components are used, the developers must perform some amount of testing on the unevaluated components to determine the security properties of the components and the way these security properties affect the integrated systems. The assurance derived from the documentation accompanying the unevaluated components will affect the testing program required. When available, the actual code should be reviewed since documentation frequently cannot be relied upon. Unavailability of code limits the assurance that can be placed in the component. Testing the security properties of the integrated components and the components developed by the developer is all part of the developer's testing program [1].

R. Khosla et al. (Eds.): KES 2005, LNAI 3681, pp. 1125–1130, 2005.

This paper presents our research results about the security and assurance relationship between IT products and IT systems. The result of this paper is the start point of IT systems assurance framework.

2   Evaluation of IT Products by CC

The Common Criteria (CC) philosophy is to provide assurance based upon an evaluation of the IT product or system that is to be trusted [2]. Evaluation has been the traditional means of providing assurance. In fact, there are many evaluation criteria. The Trusted Computer System Evaluation Criteria (TCSEC), the European Information Technology Security Evaluation Criteria (ITSEC), and the Canadian Trusted Computer Product Evaluation Criteria (CTCPEC) existed, and they have evolved into a single evaluation entity, the CC. The CC is a standard for specifying and evaluating the security features of IT products and systems, and is intended to replace previous security criteria such as the TCSEC.

The CC is presented as a set of distinct but related 3 parts. Part 3, Security assurance components, establishes a set of assurance components as a standard way of expressing the assurance requirements for Target of Evaluation (TOEs), catalogues the set of assurance components, families, and classes, and presents evaluation assurance levels that define the predefined CC scale for rating assurance for TOEs, which is called the Evaluation Assurance Levels (EALs). But about the evaluation with the CC, for example, the evaluation for EAL 3, whenever each product or system is evaluated, same evaluation process is required for all component included in EAL 3.

2.1   Common Criteria

The multipart standard ISO/IEC 15408 defines criteria, which for historical and continuity purposes are referred to herein as the Common Criteria (CC), to be used as the basis for evaluation of security properties of IT products and systems. By establishing such a common criteria base, the results of an IT security evaluation will be meaningful to a wider audience.

The CC will permit comparability between the results of independent security evaluations. It does so by providing a common set of requirements for the security functions of IT products and systems and for assurance measures applied to them during a security evaluation. The evaluation process establishes a level of confidence that the security functions of such products and systems and the assurance measures applied to them meet these requirements. The evaluation results may help consumers to determine whether the IT product or system is secure enough for their intended application and whether the security risks implicit in its use are tolerable.

The CC is presented as a set of distinct but related parts as identified below.

Part 1, Introduction and general model, is the introduction to the CC. It defines general concepts and principles of IT security evaluation and presents a general model of evaluation. Part 1 also presents constructs for expressing IT security objectives, for selecting and defining IT security requirements, and for writing high-level specifications for products and systems. In addition, the usefulness of each part of the CC is described in terms of each of the target audiences.

Part 2, Security functional requirements, establishes a set of functional components as a standard way of expressing the functional requirements for TOEs (Target of Evaluations). Part 2 catalogues the set of functional components, families, and classes.

Part 3, Security assurance requirements, establishes a set of assurance components as a standard way of expressing the assurance requirements for TOEs. Part 3 catalogues the set of assurance components, families and classes. Part 3 also defines evaluation criteria for PPs (Protection Profiles) and STs (Security Targets) and presents evaluation assurance levels that define the predefined CC scale for rating assurance for TOEs, which is called the Evaluation Assurance Levels (EALs).

In support of the three parts of the CC listed above, it is anticipated that other types of documents will be published, including technical rationale material and guidance documents.

2.2 Protection Profile

A PP defines an implementation-independent set of IT security requirements for a category of TOEs. Such TOEs are intended to meet common consumer needs for IT security. Consumers can therefore construct or cite a PP to express their IT security needs without reference to any specific TOE.

The purpose of a PP is to state a security problem rigorously for a given collection of systems or products (known as the TOE) and to specify security requirements to address that problem without dictating how these requirements will be implemented. For this reason, a PP is said to provide an implementation-independent security description. A PP thus includes several related kinds of security information (See the Fig. 1).

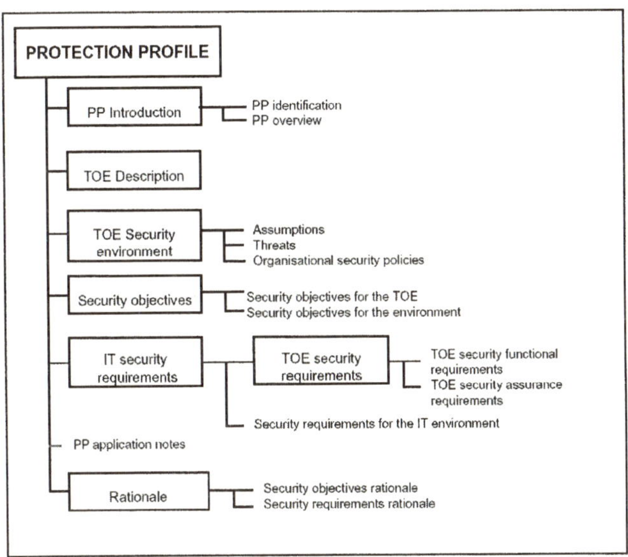

Fig. 1. Protection Profile content

A description of the TOE security environment which refines the statement of need with respect to the intended environment of use, producing the threats to be countered and the organizational security policies to be met in light of specific assumptions.

3 Building and Evaluation of IT Systems

Implementation of any security countermeasure may require economic support. If your security countermeasures are not sufficient to prevent the threats, the existence of the countermeasures is not a real countermeasure and just considered as like waste. If your security countermeasures are built over the real risks you have, maybe you are wasting your economic resources.

Building of security countermeasures is started from measuring the level of security assurance or robustness of IT systems.

For considering and deciding the security level needed to an IT systems, we are researching some models.

First step is dividing IT systems we will protect and making block (See the Fig.2):

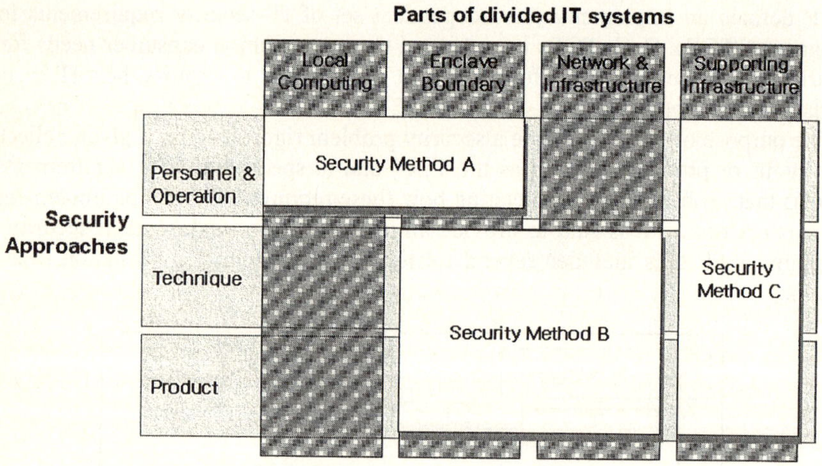

Fig. 2. Security Approaches and Methods for divided IT systems

Second step is building security countermeasures by using Block Region (See the Fig.3):

4 Relationship Between IT Products and IT Systems Assurance

In this chapter, we will prove the research results about the relationship between IT products and IT systems assurance.

Because the IT systems contain very many components for achieve their objectives, attackers may exploit weaknesses existed somewhere in the IT systems components. Weakness may be existed in the database which stores some privacy files, or in the person who have the responsibilities for managing accounts of employees. But in this paper, we only focused on the IT products included in the IT systems.

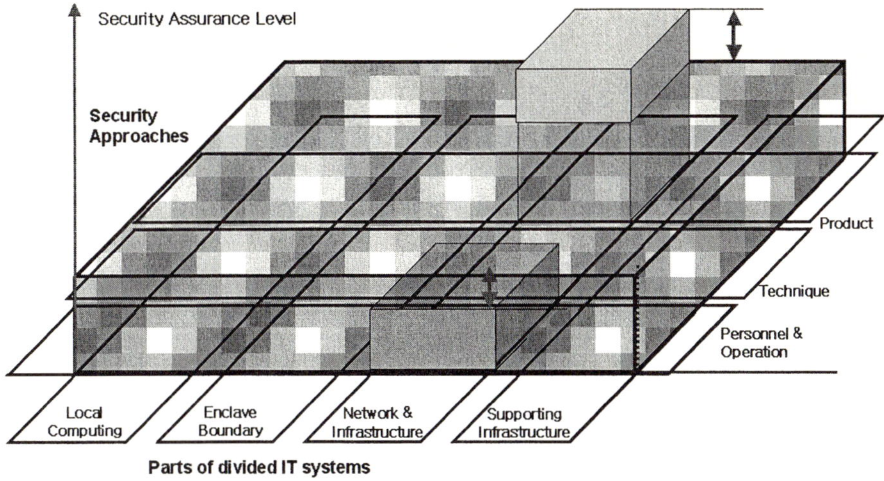

Fig. 3. Building security countermeasures by using Block Region

IT systems consist of various products such as database, personal computer, network switch, firewall and IDS. If any threat agent wants to compromise the IT systems, he or she attacks a certain product for making his or her base post. If the threat agent may compromise a PC in the IT systems, some times later, he or her may compromise whole IT systems.

In the past, IT systems designers focused on installing some Security Products such as FW, IDS and AVS (Anti-virus systems) to protect their systems. But as you know, though there are cop systems, some accidents are occurred.

Most important thing in the design of the IT systems, all components of the systems should have capabilities for protecting themselves. Database system should be not only the system for storing some data but also the system for securing the data stored in them. Network switch should be not only the product for transforming the data but also the product for hiding and monitoring the data transformed in them.

This concept is depicted in next figure:

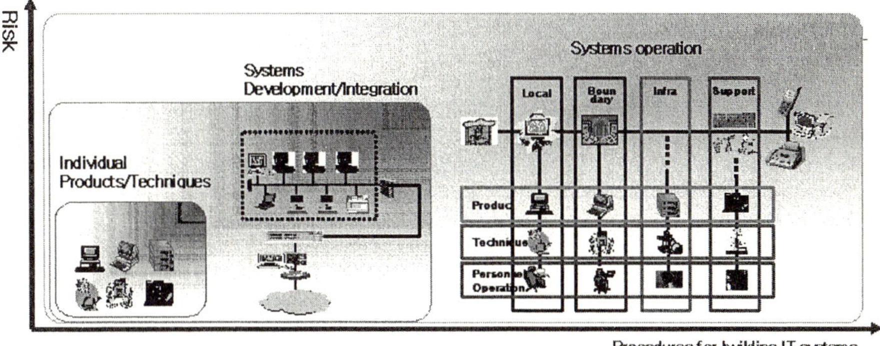

Fig. 4. IT products and IT systems

Because the IT products are the base component for IT systems security, IT system designer should select the products with careful. And the robustness strategy should be considered.

An integral part of the process is determining the recommended strength and degree of assurance for proposed security services and mechanisms that become part of the solution set. The strength and assurance features provide the basis for the selection of the proposed mechanisms and a means of evaluating the products that implement those mechanisms.

Robustness strategy should be applied to all components of a solution, both products and systems, to determine the robustness of configured systems and their component parts. It applies to commercial off-the-shelf (COTS), government off-the-shelf (GOTS), and hybrid solutions. The process is to be used by security requirements developers, decision makers, information systems security engineers, customers, and others involved in the solution life cycle. Clearly, if a solution component is modified, or threat levels or the value of information changes, risk must be reassessed with respect to the new configuration.

Various risk factors, such as the degree of damage that would be suffered if the security policy were violated, threat environment, and so on, will be used to guide determination of an appropriate strength and an associated level of assurance for each mechanism. Specifically, the value of the information to be protected and the perceived threat environment are used to obtain guidance on the recommended strength of mechanism level (SML) and evaluation assurance level (EAL).

5 Conclusions and Future Work

This paper will contain our research results about the security and assurance relationship between IT products and IT systems. The result of this paper is the start point of IT systems assurance framework.

To make IT systems more securable, IT systems designers should be recommended to adopt the step and block model suggested by Dr. Tai-hoon Kim [6-7].

References

1. Tai-Hoon Kim, Byung-Gyu No, Dong-chun Lee: Threat Description for the PP by Using the Concept of the Assets Protected by TOE, ICCS 2003, LNCS 2660, Part 4, pp. 605-613
2. Tai-hoon Kim and Haeng-kon Kim: The Reduction Method of Threat Phrases by Classifying Assets, ICCSA 2004, LNCS 3043, Part 1, 2004.
3. Tai-hoon Kim and Haeng-kon Kim: A Relationship between Security Engineering and Security Evaluation, ICCSA 2004, LNCS 3046, Part 4, 2004.
4. Tai-hoon Kim, Tae-seung Lee, Kyu-min Cho, Koung-goo Lee: The Comparison Between The Level of Process Model and The Evaluation Assurance Level. The Journal of The Information Assurance, Vol.2, No.2, KIAS (2002).
5. Eun-ser Lee, Kyung-whan Lee, Tai-hoon Kim and Il-hong Jung: Introduction and Evaluation of Development System Security Process of ISO/IEC TR 15504, ICCSA 2004, LNCS 3043, Part 1, 2004
6. Tai-hoon Kim and Seung-youn Lee: Design Procedure of IT Systems Security Countermeasures, ICCSA 2005, LNCS 3481, 2005.
7. Tai-hoon Kim and Seung-youn Lee: Security Evaluation Targets for Enhancement of IT Systems Assurance, ICCSA 2005, LNCS 3481, 2005.

Knowledge Acqusition for Mobile Embedded Software Development Based on Product Line*

Haeng-Kon Kim

[1]Department of Computer Information & Communication Engineering,
Catholic University of Daegu, Kyungbuk, 712-702, South Korea
`hangkon@cu.ac.kr`

Abstract. To achieve the goal of assuring short development cycles for new products, they are developing a process-based knowledge-driven product development environment by employing information technology. The success of a KBS critically depends on the amount of knowledge embedded in the system. Expert knowledge, which results from an individual's extensive problem-solving experience, has been described as *unconscious knowledge*. Besides shortening development time, properly handled reuse will also improve the reliability since code is executed for longer time and in different contexts. In this paper, We propose a design process suitable for developing product lines for mobile embedded systems with KBS. The process starts in a requirement-capturing phase where the requirements from all products in the line are collected. The contributions of this work with respect to mobile embedded systems are an outline of a development process, focusing on the special considerations that must be taken into account when designing a PLA for the systems.

Keywords: Knowledge Based Systems, Intelligent Agents, Mobile Embedded Systems, CBD, Domain Engineering

1 Introduction

A core idea in product line software engineering is to develop a reuse infrastructure that supports the software development of a family of products. It is of key importance to the success of the product line development to spend the resources in a way that promises a high return on investment and reduced time-to-market.

Analysis phase defines the estimated benefits from applying the product line approach and the portfolio of products included into the product line, the domains that have potential for reuse, and software artifacts that are considered as reusable assets of the product line [4].

Business, architecture, process and organization are the important keys to effective product lines. The business strategy defines what organizations, customers and participants the company deals with regarding the product line and the basic lines, how the software development work has to be organized. Software production can be based on in-house components, fixed-price component deliveries, commercial software components, and software development controlled by subcontracts or participation contracts. All these co-operation manners should be allowed to use in the devel-

---

* This work was supported by Korea Research Foundation Grant, (KRF-2004-041-D00626)

R. Khosla et al. (Eds.): KES 2005, LNAI 3681, pp. 1131–1138, 2005.

opment of a product line. Product line development aims at the reuse of software requirements, architecture, components and processes. Analysis of the commonality and variability of functional and structural properties of systems is the main difference between the development of a product line and the development of a single software system [1]. This paper presents evaluation criteria for the development of a product line and the state-of-the practices in application of the product line approach. Research results give an overview of the situation in the mobile embedded software development. A question form was sent to automation, electronics, telecommunications and software business companies that were known to use software components in their software production.

2 Mobile Embedded SW Development Based on Product Line

2.1 Embedded Software Development Process on PL

In this section we discuss the development process in which a PLA for mobile embedded products is constructed. The process is iterative and includes architectural analysis of properties that are of vital importance for a PLA, e.g. flexibility. Moreover, the derivation of products from a PLA is dealt with. The development process proposed in this paper is shown in Figure 1 where the process is divided into *domain engineering*, *PLA development*, and *application engineering*. Domain engineering consists of requirements analysis in embedded software domain, architecture analysis in embedded software and developments of assets for PL. Application engineering consists of requirements analysis in embedded software, system design and implementation.

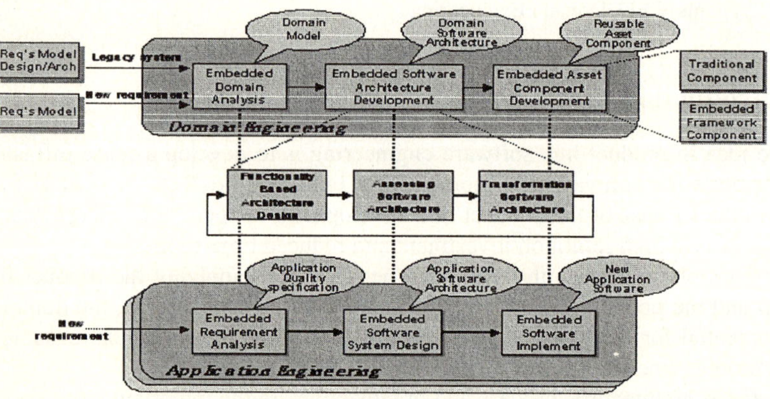

Fig. 1. Architecture for Embedded Software Development Process on PL

The analysis of business drivers (Table 1) includes an analysis from three points of view organization culture and processes, product line architecture, and assets management. The first view analyses the state of software development and conformance between the described and applied software development and conformance between the described and applied software development practices. The second view estimates the existing and predicted properties and quality of the products inserted into the product line. The third view estimates the potential to maintain assets by the developed and applied supporting mechanisms such as software and product configuration management (SCM, PDM). Although we concentrate on the development of product line architectures and assets, we must always remember the reality: a good architecture needs well-defined processes and supporting mechanisms for assets management in order to be successful.

Table 1. Defining the Domain drivers of a product line

Elements	Organization culture and processes	Product line architecture	Assets Management
Product assortment Expected sales The number of products The value of reuse for products Similarity and diversity of products **Business manners** In-house components Commercial off-the-shelf components Original component manufactures Contracted software Appropriateness of the product-line approach	Cultural defferences in multi-site software development Current state of development processes Component development, acquisition and utilization processes	Commonality and variability of the functional features of the products Quality requirements of the products Artfacts and their representations Quality of artefacts Documentation of artefacts Business rationale: cost, time-to-market, Partnerships	Mechanisms to manage variability inside the product line Mechanisms to share the properties of the product line with business stakeholders

Domain definition gives the answers to the second question of where, when and how the product line approach could be applied. The phase includes the analysis of problem and solution domains, technologies and personnel (Table 2). Scope, commonality and variability, recent representations and their quality are the primary focus of the domain analysis that is considered from the three points of view of a product line development. If only common features are included and implemented as a platform, the minimalist approach is used. If maximalist approach is used, all product features are included in the product line and tools support automated instantiation of a product. Product line development also requires a new organizational culture, and therefore, commitment of all software engineers involved in the product line is needed before the initiation of a product line. There are two approaches to develop a product line: evolutionary and revolutionary.

2.2 Features Modeling for Mobile Embedded Software on PL

Features are rarely independent from each other. A feature can for instance depend on other features in order to deliver the desired functionality. Another example of a relation between features is the mutually exclusive relation, implying that only one of the related features can appear in the final product. If mutually exclusive features must co-exist in the product, effort has to be made to resolve the conflict. Features in mobile embedded software systems will also exhibit dependencies related to the

Table 2. Defining domains: preconditions and targets of a product line

Elements	Organization culture and processes	Product line architecture	Assets Management
Domain analysis Domains and their scopes Commonalities and variability within domains Stability and maturity of domains Representations of reusable domain knowledge Quality of these representations Evolutionary or revolutionary approach	Support for domain analysis Collecting and ordering domain knowledge Representations for knowledge management Individual and organizational readiness in product line development	Quality requirements of the application domains Diversity of domains and implementation technology Standards Mapping domains to a product line architecture Domain rationale	Conceptual classification of the requirements and product features Models for representing and storing requirements and architecture into the core assets repository Models for storing documents of core assets
Personnel Management Domain expertise Technical skills	Commitment Decision making Roles, rights and responsibilities Expertise of software processes and metrics	Domain analysts Domain and product architects Dissemination of architecture Expertise of design and analysis of software architectures and components	Asset managers Distributed and virtual assets management infrastructure Expertise of data modeling and management

temporal domain in the multiple view as in figure 2. It For instance, consider the lock-free break feature and the anti-slid feature in automotive vehicles. Both features need information about the wheels' velocity, thereby having a shared temporal dependency related to the freshness of the sensor data. Such relations on features are specified in a *feature graph*. The feature matrix and the feature graph constitute the basis for deciding what features to include in the PLA. Identifying the commonality among the features for real-time systems is more complicated since we also have to take the temporal domain into consideration. When the scope for the PLA has been decided, the features should be mapped to components that, together with their interrelations, constitute the actual architecture.

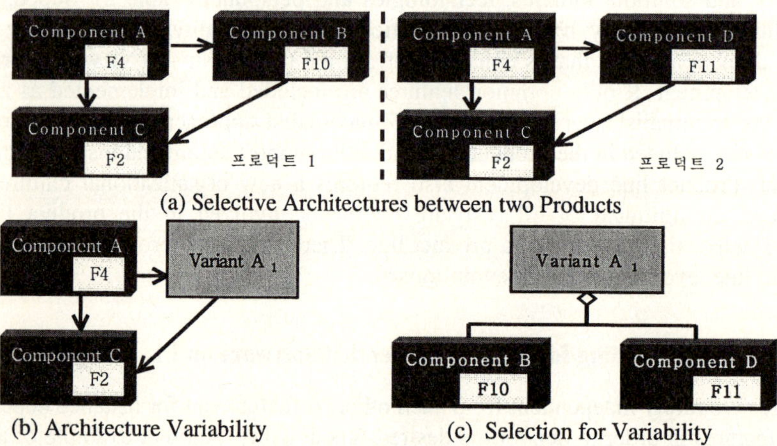

(a) Selective Architectures between two Products

(b) Architecture Variability   (c) Selection for Variability

Fig. 2. Relationship for Multiple View on PL

Both approached can be applied to an existing product line and a new product line. The evolutionary approach is based on re-engineering and evolution of existing software architectures and components. Some of shared artifacts fall in and some may fall out of the scope of a product line. In the revolutionary approach, architectural styles and design patterns are applied to meet the requirements of a super-set of product line members or all expected product line members. It has 3 different views in the PL, as use case, collaborations and futures as in figure 2. The analysis of models, methods and tools in this paper, as well as earlier developed artifacts and applied practices answers to the question of what new practices should be developed in order to have success in the product line development.

The parts of software that have most potential for reuse are selected for core assets development. That is why the analysis of core assets evolution is also considered from the three viewpoints of a product line processes, architecture and assets management.

2.3   The Practice in Mobile Embedded Software

The success of a product line is based on the trade-off of all of the product line related and domain-related quality attributes. Especially in the initiation phase, while estimating how new features and components affect the product line architecture, the amount of the required modifications is tried to be minimized while the reusability of the product line architecture is aimed to be maximized. Figure 3 shows overall architecture for developing mobile embedded software based on product line, which is consists of development process assistor, component repository, assets manager and integrator. Although the architecture of products and their qualities is not the cure-all, it is the best place to start the initiation of a product line. In order to get an understanding of whether a product line was appropriate for different technologies and businesses, application domains were classified into five categories according to the

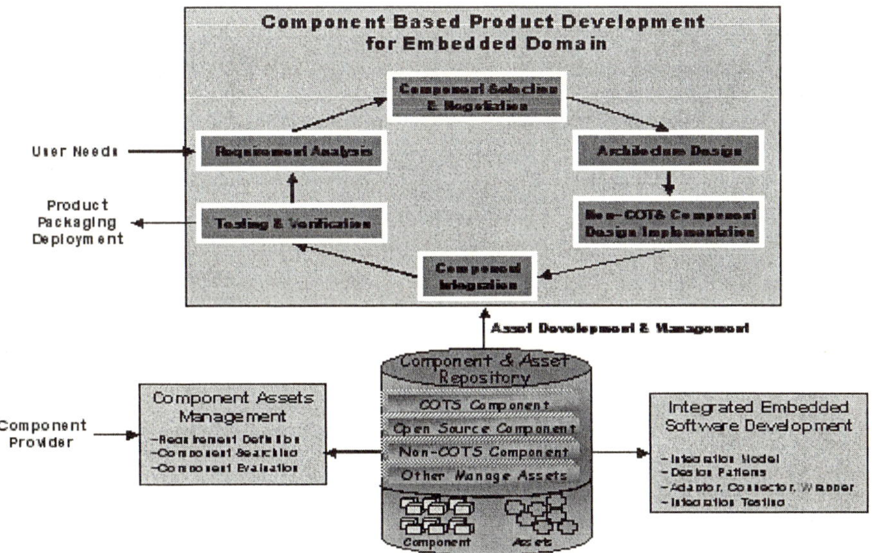

Fig. 3. Architecture for Embedded Software Development Process on PL

architectural layers that all systems normally embody. Although a system can consist of all these five technology domains, a domain itself may also from a product. All domains have their own organization culture, if they are managed by separate organizations. However, the technology domains have a culture of their own as well, although they are developed in the same site of a company. Therefore, a domain should be considered a separate software unit that has its own life cycle in the domain engineering and product engineering.

All domains that use different technologies, i.e. languages, modeling methods, testing techniques etc., should be organized as the sub-domains of a product line, if they belong to the core competitiveness of the company and they are integrated into the same products. Otherwise, they should be considered as separate product lines.

However, the variability was more obvious in the user interface domain and the data processing and management domain than in the other domains. On the other hand, standards had more effect on telecommunications and hardware-related software than the real-time and data processing and management domains. In the case of the standard being stable, stability and maturity of the application domain was also considered very high. However, the diversity in the product assortment seems to decrease the estimated expertise in the domain and the possibility of reusing software. Although the domain knowledge was estimated to be very high and all domains had potential for reuse, the domain knowledge was purely documented or it was not documented at all. Representations of design artifacts, that is the way, how requirements, software architecture and software components were defined and documented, were best done in the real-time software domain as in figure 4. They were weakest in the user interface and hardware-related software domains. However, these differences were not linear. In some companies, requirements and/or components were exactly defined, but software architecture was inadequately defined.

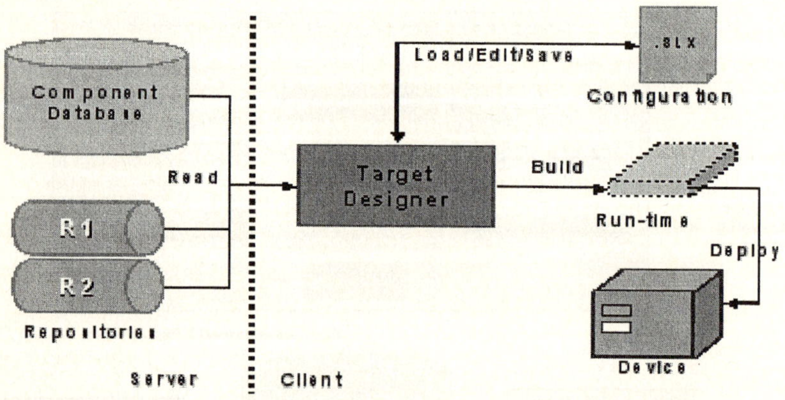

Fig. 4. Architecture for Embedded Software Development on Distributed Environments

A defined change procedure was followed up in every company, except in one this was just under construction. Architecture analysis methods were not used at all. However, component and integration testing could be improved only a little in the mobile embedded software domain contrary to telecommunications, data processing and management, and hardware-related software domains in which the domains component and integration testing was ranked to be at the level 0 or 1 (scale 0…5).

3 Conclusions

For mobile embedded software systems we also have to consider temporal behavior. A typical example of how the temporal requirements influence the functional design is the following. Consider features that are functionally equivalent between a set of products. If these products will run on different infrastructures, i.e. operating system and hardware platform, the functionality may be partitioned among components in different ways to fulfill the timing requirements. In the high-end product we can partition a feature in such a way that it will be easy to maintain while in a low-end product we have to make an architecture that is focused on performance to be able to fulfill the timing requirements due to the limited resources in the low-end product. After the functional design the PLA must be analyzed in order to verify that the architecture is flexible enough to facilitate all products in the product line.

The use of architectural patterns is best learned by working as mentor-apprentice couples but, nevertheless, this work-organization model is rarely used. At the top level in the learning process, a domain engineer learns how to develop a domain component framework. At this stage, a domain architect's expertise has been achieved. The domain architect, however, is not responsible only for the development of a product line but also for the maintenance of the product line architecture. After the domain engineer has the skills of a domain architect, he or she could deploy his or her knowledge to a project team by working as a product architect. The other possibility is to continue as a domain architect in the asset engineering team and act as a mentor in a mentor-apprentice couple. In some organizations, the domain knowledge is deployed in this way, moving between the domain and product engineering teams, but without acting as a mentor-student couple.

The product line approach requires new organization culture. A product line should be shared among all projects engineering staff and their full commitment is needed. Advantaged solutions achieved in small product engineering groups should also be imported to the product line, otherwise they are not exploited and developers' commitment is lost. Furthermore, uniform practices should be defined and followed in all sub-domains. In this way, the integration of sub-domains might be less problematic in practice.

There are two kinds of evolution: the horizontal evolution of domains and the vertical evolution of products. The management of the reusable assets, such as component libraries, product line architecture, and test cases, has to be controlled by a domain engineering team, contrary to the product versions and releases that are controlled by a product engineering team with PDM tools. This kind of work-organization is just being started but it will take a couple of years to get thorough information of how successful the development of the product lines has been. Therefore, the analysis of product line development practices presented here is insufficient, but it gives an estimation of which areas need more investigation and development.

A product line produces several benefits to the business. First, trying to catch the balance between sub-domains forces to consider the set of products instead of a product, and therefore, the important properties of a product line are most invested in.

Secondly, software engineers might need to see the wood for the trees: the product line architecture is an investment that must be looked after. Thirdly, the sub-domains of a product line architecture may use, change and manage their own technologies, and evolve separately but in a controlled way.

References

1. 2002 Embedded Software Tools Worldwide Forecast, Gartner Dataquest Market Statistics 110850, 2002.
2. Worldwide Embedded Software Tools Outlook, 2002, Gartner Dataquest Alert, 2002.
3. Technical report, "STARS ProductLine Concept", http://www.asset.com/stars/darpa/prdd-line.html, 2003.
4. Bosch, Design & Use of Software Architectures, Addison-Wesley, 2000.
5. Technical report, "A Fraemwork fot Software Productline Practce Version 3.0", Carnegie Mellon SEI, 2000 http://www.sei.cmu.edu/;plp/framework.html
6. Desmond F. D'Souza, Alan c. Wills, Objects, Components, and Frameworks with UML, Addison-Wesley, 1998.
7. Peter Herzum, Oliver Sims, Business Component Factory: A Comprehensive Overview of Component-Based Development for the Enterprise, OMG press, December, 1999.
8. "The Embedded Software Strategic Market Intelligence Program 2002/2003", VDC(Venture Development Corporation), Feb., 2003.
9. Rick Lehrbaum, "EDC: Embedded Linux remains #1 choice of developers - despite tools dissatisfaction", http://www.linuxdevices.com/articles/AT4787985721.html.
10. Rosenblum,D.S and Natarajan,R. "Supporting Architectureal Concerns in Component Interoperability Standards," IEEE Proceedings Software Special Issue on Component-Based Software Engineering, Vol. 147, No .6, pp.215-223, Dec., 2000.
11. Clements, P., Kazman, R., and Klein, M., Evaluating Software Architectures, Addison - Wesley, 2002

Gesture Recognition by Attention Control Method for Intelligent Humanoid Robot

Jae Yong Oh[1], Chil Woo Lee[1], and Bum Jae You[2]

[1] Department of Computer Engineering, Chonnam National University,
300 Yongbong-dong, Buk-gu, Gwangju, 500-757, Korea
`ojyong@image.chonnama.ac.kr`, `leecw@chonnam.ac.kr`
`http://image.chonnam.ac.kr`
[2] Intelligent Robotics Research Center, Korea Institute of Science and Technology,
39-1 Hawolgok-dong, Seongbuk-gu, Seoul, 136-791, Korea
`ybj@kist.re.kr`

Abstract. In this paper, we describe an algorithm which can automatically recognize human gesture for Human-Robot interaction by utilizing attention control method. In early works, many systems for recognizing human gestures work under many restricted conditions. To solve the problem, we propose a novel model called *APM(Active Plane Model)*, which can represent 3D and 2D gesture information simultaneously. Also we present the state transition algorithm for selection of attention. In the algorithm, first we obtain the information about 2D and 3D shape by deforming the *APM*, and then the feature vectors are extracted from the deformed *APM*. The next step is constructing a gesture space by analyzing the statistical information of training images with PCA. And then, input images are compared to the model and individually symbolized to one of the pose model in the space. In the last step, the symbolized poses are recognized with HMM as one of model gestures. The experimental results show that the proposed algorithm is very efficient to construct intelligent interface system.

1 Introduction

Recently, various applications of robot system become more popular accordance with rapid development of computer hardware and software, artificial intelligence, and automatic control technology. So far, robot mainly have been used for industrial field, however, nowadays it is said that the robot will do an important role in home service application in the near future. To make the robot more useful, we require further researches on natural communication method between human and the robot system, and autonomous behavior generation. The gesture recognition technique is one of the most convenient methods for natural human-robot interaction, so it is to be solved for implementation of intelligent robot system. Briefly speaking, that we understand the human behavior and gesture is to track the temporal sequence of movement of human body automatically, and then to estimate some meanings of the temporal movement from previously trained and arranged motion data. That means, for understanding human behavior in a sequence of image, we have to construct a model space for human behaviors previously and analyze temporal changes of human body. However, it

R. Khosla et al. (Eds.): KES 2005, LNAI 3681, pp. 1139–1145, 2005.

is very difficult to analyze temporal changes, namely historical meaning of bodily motion automatically because a human body is a three-dimensional object with very complicated structure and flexibility. As knowing from the facts mentioned above, most algorithms are work under many restricted conditions. In this paper, we describe the algorithm which can automatically recognize human gesture without such constraints by utilizing three-dimensional features extracted from stereo images.

The paper structured as follows: Section 2 presents a definition of gestures. Section 3 presents an overview of the recognition system. Section 3.1 and 3.2 describe how the bodily region and features can be extracted from the stereo images. And in section 3.4, we explain how the gesture model space is constructed by analyzing the statistical information of training images with the Principle Component Analysis (PCA) method. And then, section 3.5 presents how the sequential poses are classified into model gestures with Hidden Markov Model (HMM) algorithm. Section 4 shows the efficiency of the proposed method with experimental results. Finally the paper is closed with mentioning the conclusion and further works.

2 Define Recognition System

2.1 Definition of Gestures

Human use many kinds of gestures in daily life. But it is very difficult to implement the recognition algorithm which can understand all the gestures. For controlling the robot system, we can assume several prominent gestures. Therefore, here we define six different gestures for human-robot interaction as shown in Fig.1. These gestures are mutually isolated from each other in meaning, and are classified into static or periodic type.

Fig. 1. This shows definition of basic gestures for humanoid robot. [Normal], [I love you], [Stop], [Point] are static gestures, [Good bye] and [Come here] are periodic gestures

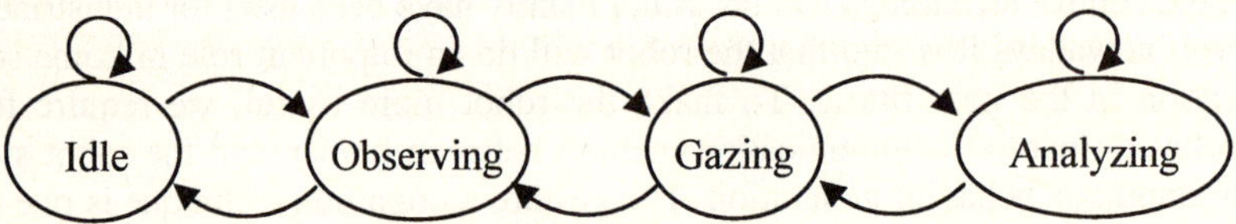

Fig. 2. This shows the State transition for recognition

2.2 State Transition for Recognition

Human gesture could be interpreted into different meanings according to the situation and actor's intention even though they use the completely same motion. That means gesture recognition is very subject and situation dependent task, consequently we need to analyze the context of environment where the gestures is adopted. Here, we define four states of behaviors assuming that the purpose is to understand the gesture which is used in our daily life. And we can recognize the context of gestures and catch the intention of a doer by analyzing the state transition model. So gestures are defined using transition of these states. The [Idle] is the ready state that there is no motion and no change in the field of view. And [Observing] is the state in which we can see something related to motion of object, but we do not know whether the motion is induced by human or not. We can proceed with the state if we can find the facial region; [Gazing] state is activated. [Gazing] is the state in which we try to find the intention of a person; to estimate significance of the motion. If the motion has a specific meaning, the process proceeds to the [Analyzing] state. In the [Analyzing] state, the motion is interpreted by the recognition algorithms into abstract meanings which are previously defined. Fig. 2 shows the transition between each state.

3 Gesture Recognition System

3.1 Preprocessing

In the first step of gesture recognition, we need to extract the foreground image having bodily motion. In the paper, we use a simple foreground extraction method using depth information. If we concern the situation of visual interaction with a robot, we can assume that human body is relatively closer to the robot than other background objects. The procedure is as follows; first, we detect a face region of the frontal person [1], and then estimate the distance from the robot camera to the face by calculating stereo geometry. Finally, we can obtain the closer region of input image as the foreground. For the face detection algorithm, we adopt CMU's face detection algorithm which uses Gaussian Mixture Model (GMM) of color distribution of the face region and the background region [1]. And, for calculating the stereo geometry, we use a commercialized vision system; Bumblebee Camera set of Point Grey Research. This system estimate correspondence between stereo images with Sum of Absolute Differences [2].

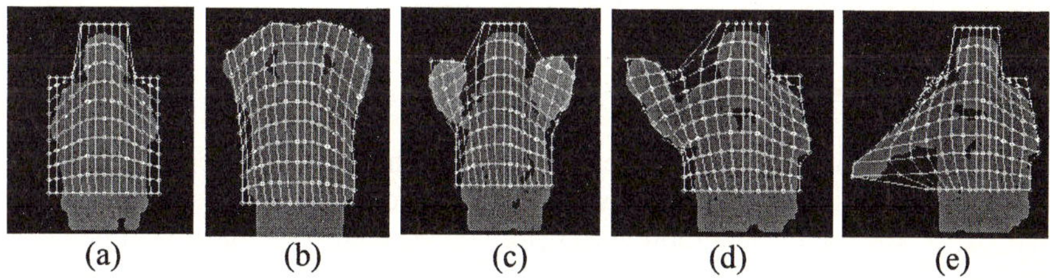

(a)   (b)   (c)   (d)   (e)

Fig. 3. This shows deformed APM examples, (a) Normal, (b) I love you, (c) Stop, (d) Goodbye, (e) Point

3.2 Feature Extraction

Once bodily region is extracted, we need to find the features of the body configuration that can efficiently represent the posture and motion of the body [3][4]. In this paper, we propose an efficient model, *Active Plane Model* (APM), for representing the posture and motion information simultaneously. The APM is characterized by that it can represent three-dimensional depth information and two-dimensional shape information at the same time. And the APM is less sensitive to noise or appearance variation since it uses the standardized model as shown in Fig 3. (a).

APM consist of several nodes connected to each other in grid shape and it is deformed by the shape and depth information. For the first step to deform the APM, we find the contour of foreground region. The contour is searched from the minimal rectangular boundary surrounding foreground region into center of rectangle. If we find the contour, outer nodes of APM are fixed, then, inner nodes are moved to new location in order to keep the same distance between neighbor nodes. That means the inner nodes are rearrange by calculating the average position between the most outer nodes as shown in equation (1).

$$N_i^{t+1} : \text{Average of neighbor nodes of } N_i^t \tag{1}$$

$$N_i^{t+1} - N_i^t \leq T \;, \; {}_{(t \geq 0)} \tag{2}$$

In equation (2), T expresses a suitable threshold distance. By adopting the criterion of equation (2), we can stop the deformation quickly. Fig. 3 shows the deformation shape of APM for several gestures. From this deformed APM, we can calculate a displacement of nodes of grid. Therefore, multi-dimensional features are extracted for every frame by using the equation (3).

$$F_t = \{N_1, N_2, ..., N_n\}$$
$$N_i = \{x_i, y_i, z_i\} \tag{3}$$
$$(0 \leq t \leq T, \;\; 1 \leq i \leq n, \;\; n \geq 4)$$

In the equation (3), F_t is a feature vector set at time t, N_i is i-th nodal position of APM, and n is the total number of APM nodes. Also, direction vector of pointing gesture; for the case of 'Come here', can be calculated using APM. The most outlier point in an APM can be assumed as pointing point, and then direction vector is obtained by connecting the point to the face region.

3.3 Attention Control

Understand of human intention is the most difficult things in gesture recognition field because human gestures have various meaning according to the time or environment. In this paper, we assume that motion ROI(Region of Interest) is the area of behavior's intentions If the ROI is defined by changes of images, this ROI includes many error factors. Therefore, we introduce an attention control method using APM. The motion ROI can be supposed using most outlier point in the APM and this motion ROI have majority of behavior's intentions. *Motion History Image*(MHI) is calculated to extract a features for periodic gestures like 'good bye' or 'come here' in the motion ROI. In the MHI, the most has the highest value, earlier motions have decreasing values, and we can estimate the Orientation of motion easily. And then, we estimated the orientation in the ROI of Motion using APM, and this orientation information is used for feature vector.

3.4 Pose Symbolization

A common method for linear dimension reduction is Principal Components Analysis (PCA). The method is performed by applying eigen-value decomposition on the co-variance matrix of feature data. A small subset of resulting eigenvectors, covering a desired amount of the data variance, forms a new linear subspace for representing the data set [5]. Using the *Principal Component Analysis*, a small number of component vectors are chosen which can represent the whole space, and we call it dimension reduction. Fig. 4 shows an example of the reduced space, *Gesture Space*.

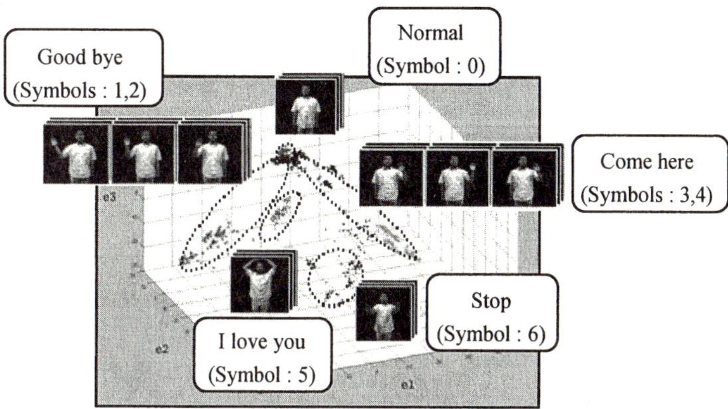

Fig. 4. Projection of model gesture sequences into the gesture space. Each gesture can be represented by sequence of pose symbols

3.5 Symbolic Gesture Recognition

HMM is a stochastic recognition method of probabilistic network with *hidden* and *observable* states. This method is well suited to time-series data with both spatial and temporal dependencies. If image sequences can be classified into several gesture patterns, the images are converted into a sequence of symbols by a pose symbols and this sequence can be used as the input of HMM. HMM λ is represented by the following variables. The state transition probability a_{ij} indicates that state of HMM will change from i to j. The probability $b_{ij}(y)$ indicates that the output symbol y will be observable in the transition of state j from state i. Also, π_i is the probability of initial state. Learning of a HMM is equal to estimating the parameters $\{\pi, A, B\}$ of the HMM.

4 Experimental Results

For the training and evaluation of the algorithm, we construct an image database about different gestures. The resolution of the image is 320x240 in pixels and capturing speed is about 10 frames per a second. We recorded six kinds of gesture image for ten different persons. We made an experiment for recognizing gestures with test images. In the experiment, an APM of 15x10 in resolution was utilized, so that 450-

dimensional feature vector was extracted for every frame. We used only five eigen-vectors, and this low-dimensional vector can represent the whole gesture space well. We used six different HMMs for the six different gestures. The HMM parameters were estimated with the Forward-backward algorithm [6] under assuming that each HMM had 5 states. Table 1 shows the recognition result for each gesture. "Good bye" and "Stop" gestures have recognition rates that are lower than the average rate respectively. We suppose that the gestures contain little motion or 3D information comparing to other gestures. Fig. 5 shows the failed cases of recognition.

Table 1. Recognition rate for the test image set

	Correct (%)	Incorrect (%)
Normal	98	Come here (2)
I love you	95	Normal (5)
Good bye	85	Normal (10), Come here (5)
Stop	80	Come here (20)
Come here	85	Normal (15)
point	100	-
Average	90.5	-

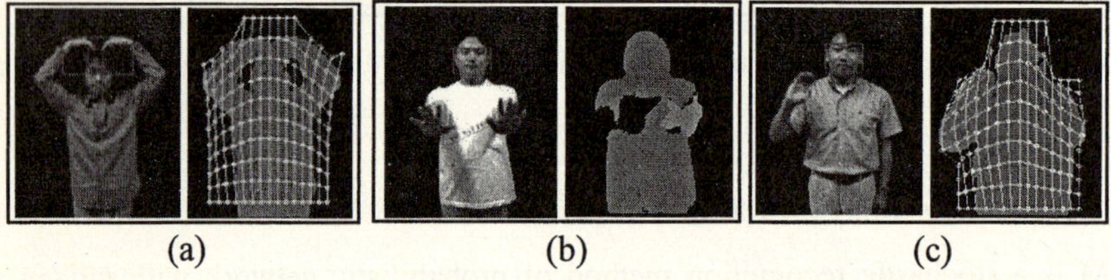

(a)   (b)   (c)

Fig. 5. This shows examples of failed gesture recognition. (a) Failure to detect face, (b) Failure to calculate disparity, (c) Too little motion

5   Conclusions

In this paper, the novel gesture recognition method using 2D shape and 3D depth information simultaneously is proposed. Also we propose the state transition model for recognition and novel attention control method. This algorithm works very well since it is robust to shape and depth variation. There are many gesture recognition algorithms using visual information. However, most algorithms work under many restricted conditions. To eliminate such constraints, we have proposed *Active Plane Model* (APM) to recognize human gestures. This model comes from structured model approach instead of geometric feature model approach using edge or corners. Therefore, we can make a general model for recognition of human gesture in real world. Also we can estimate the direction vector of pointing gesture easily by using the APM. However, there are some problems to be solved. One problem is that it is difficult to classify all gestures with several models obtained from training image. That means the algorithm become unstable under little information of ambiguous 2D shape and 3D depth. So, in order to improve the method for real world application, we must gather all kind of gesture images and analyze them statistically. Also our further work includes analyzing gestures of multiple persons concurrently.

References

1. F. J. Huang and T. Chen: Tracking of Multiple Faces for Human-Computer Interfaces and Virtual Environments, IEEE Intl. Conf. on Multimedia and Expo., July (2000) 1563-1566
2. Point Grey Inc. (http://www.ptgrey.com)
3. James Davis: Recognizing Movement using Motion Histograms, MIT Media Lab. Technical Report No. 487, March (1999)
4. Ross Cutler, Matthew Turk: View-based Interpretation of Real-time Optical Flow for Gesture Recognition, Third IEEE International Conf. on Automatic Face and Gesture Recognition, (1998)
5. Chil-Woo Lee, Hyun-Ju Lee, Sung H. Yoon, and Jung H. Kim: Gesture Recognition in Video Image with Combination of Partial and Global Information, in Proc. of VCIP, Lugano, July, (2003) 458-466
6. Shun-Zheng Yu; Kobayashi, H.: An efficient forward-backward algorithm for an explicit-duration hidden Markov model, Signal Processing Letters, IEEE, Volume: 10, Issue: 1, Jan. (2003), 11-14

Intelligent Multimedia Service System
Based on Context Awareness in Smart Home*

Jong-Hyuk Park[1], Heung-Soo Park[1], Sang-Jin Lee[2], Jun Choi[2], and Deok-Gyu Lee[3]

[1] R&D Institute, HANWHA S&C Co. Ltd.,
Hanwha Blg. #1, Jangyo-Dong, Jung-Gu, Seoul, Korea
{hyuks00,parkhs}@hanwha.co.kr
[2] Center for Information Security Technologies, Korea University,
5-Ka, Anam-Dong, Sungbuk-Gu, Seoul, Korea
{sangjin,jun}@korea.ac.kr
[3] Divison of Information Technology Engineering, SoonChunHyang University,
Eupnae-ri, Shinchang-myun, Asan-si, ChoongNam, Korea
hbrhcdbr@sch.ac.kr

Abstract. According to the rapid change of a computing environment, it is coming an age of ubiquitous computing that we can make use of information service at any time and anywhere as the paradigm of digital convergence such as broadcasting, communication, and appliances. Consequently, Smart Home environment is constructed in the house. Various services such as security/automation, health monitoring, medical examination and treatment, entertainment, education, and e-business are realized at home. Both affective, systematic managements and services that correspond with the above environment are required in multimedia services as well. In this paper, we propose a Intelligent Multimedia Service System(IMSS) based on context awareness in Smart Home which provides multimedia interoperability among incompatible multimedia play devices, intelligently processes context awareness information, transparent, and provides safe services.

1 Introduction

As a digital convergence paradigm caused by the rapid growth of computing environments, various contents such as image and voice data are freely implemented with regard to appliance, service, and form of network. The many needs for pleasant and convenient life like human, machine, robot, intelligent device, facility, building, and medical system grow larger everyday. In accordance with these needs, the request that services various multimedia among incompatible multimedia play devices is going to increase in the side of multimedia service[1][6][10].

In this paper, we propose IMSS which provides multimedia interoperability among incompatible multimedia play devices, intelligently processes context awareness information, provides transparent, and safe services.

---

* This research was supported by the MOCIE (Ministry of Commerce, Industry and Energy)-DCP Project and the MIC (Ministry of Information and Communication), Korea, under the ITRC (Information Technology Research Center) support program supervised by the IITA (Institute of Information Technology Assessment)

R. Khosla et al. (Eds.): KES 2005, LNAI 3681, pp. 1146–1152, 2005.

The rest of this paper is organized as follows. In section 2, we explain Related Works which is core technologies of Smart Home and technologies for active multimedia distribution. In section 3, we discuss IMSS which proposed system in this paper and scenario. In section 4, we consider the proposed system and then come to conclusion.

2   Related Works

In this section, we discuss the core technologies for Smart Home and technologies for active multimedia distribution. Ubiquitous Computing is a computing environment that various kinds of computers soak into people, objects, and environment are connected with each other so that this enables us to implement a computing at anytime and anywhere. As a meaning of "it is commonplace", "it exits at the same time in every time and every where", it is more advanced computing environment than existence home-network and mobile computing. Xerox PARC of America describes the requirements of ubiquitous computing throughout the research about computer and network as follows. The interface of it has to be invisible and embodied virtuality of real computer has to be possible at everywhere and the network has to be connected always. This ubiquitous computing is the main factor to organize Smart Home[2]-[4], and related technologies for Smart Home have WSN[5], Context Awareness[7][8], and MPEG-21 DIA/Video transcoding[11]-[13].

Furthermore, related technologies for active multimedia distribution has MPEG-21 IPMP and REL. For the active multimedia distribution to Home, It must be provided method which Contents Provider or Right-holder trusted. We will explain MPEG-21 IPMP and REL which are one of the these methods. MPEG-21 Intellectual Property Management and Protection(IPMP) will be discussed as MPEG-n with special emphasis on [11][16] and [17] as it was discussed so in the days when it was being re-established and at present(n=2,4,21). In incompatible network and device, all users express an agreement about their copyrights of digital items, protect, and manage the copyrights about valuable multimedia as a intellectual character. In MPEG-n IPMP, they propose a digital contents distribution business model. Right Expression Language(REL) is a language that expresses the right to read machine-read expressions, the vocabularies stated in RDD(Rights Data Dictionary)[14][15].

3   IMSS

3.1   IMSS Architecture

The IMSS which proposed in this paper consists of two parts. One is to create, process, and distribute multimedia to Smart Home. The other is to consume multimedia in Smart Home. In this paper, we discuss briefly the front part based on [18], and then discuss in detail multimedia service in Smart Home(Refer to Fig.1). A front part of IMSS consists of CA, DOI-RA, CP, MMC, CS, PG etc. Certificate Authority(CA) issues a certificate to authenticate that user is authorized to creators distributors, etc. It provides authentication functions of user or device etc. and system validation. The Digital Object Identifier Registration Authority(DOI-RA) provides a function that generates identifier in order to manage the multimedia in efficient and systematic way.

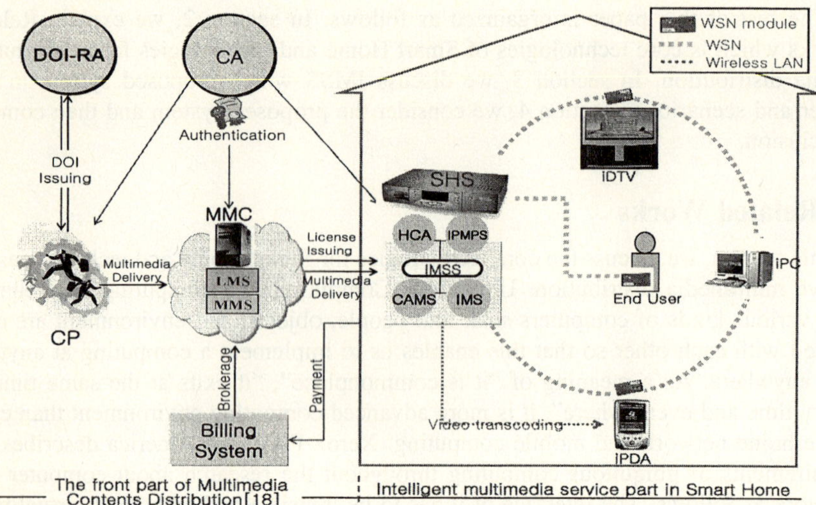

Fig. 1. IMSS Model

Contents Provider(CP) is a copyright holder who provides multimedia, creating safe contents by downloading packager. CP creates digital item by adding metadata which describes raw multimedia file in the process of packaging and usage rule which sets up the right of usage. Multimedia Mall Center(MMC) manages safe multimedia files which are distributed by copyright holder, and provides interface that users can consume. It consists of MMS and LMS. Multimedia Management System (MMS) manages multimedia which is safely packaged by the CP. It provides related information by using multimedia metadata. License Management System(LMS) manages license used to unlock the user's file lock system in order to play packaged multimedia, and gives a license to consumed-subject when the right payment from consumed-subjet is accepted.

3.2 IMSS Design

The IMSS proposed in this paper consists of four objects: SHS, IMPD, WSN module and End User.

Smart Home Station(SHS) is a main element of IMSS and is composed of HCA, IPMPS, CAMS, and IMS. It composed of 5 layer as follows: communication layer, operating system layer, middle-ware layer, core engine layer, and user interface layer. Home Certificate Authority(HCA) takes authentication functionality for Smart Home user and IMPD(Intelligent Multimedia Play Device) like private CA certified by public CA. Intellectual Property Management and Protection System(IPMPS) take security functionality for downloaded multimedia into Intelligent Multimedia Server(IMS) through external networks. It provides minimum security requirements such as right authorization, delegation and management suitable for Smart Home in order to support a transparency for the user. Table 1 is part of license management included usage right and user authorization. It is expressed by XML form.

Table 1. Usage right management & authorization

```
<!—Usage right management -->
    <grant>
      <keyHolder licensePartID="UserA"><info><dsig:KeyValue>
        <dsig:RSAKeyValue><dsig:Modulus>ZaDKoQQyzS==</dsig:Modulus>
          <dsig:Exponent>kDABSS==...</dsig:Exponent></dsig:RSAKeyValue>
          </dsig:KeyValue></info></keyHolder>
      <mx:play/><mx:diReference><mx:identifier>urn:grid:a1-aseffde-574812
        6845-e</mx:identifier></mx:diReference>
    <!—Usable Date -->
        <notBefore>2005-02-23T14:30:00</notBefore>...</validityInterval>
        <validityIntervalDurationPattern>
    <!—Usable date count from Issue day -->
          <duration>P6D</duration></validityIntervalDurationPattern>
    </grant>
    <issuer><timeOfIssue>2005-02-23T14:05:23</timeOfIssue></issuer>.........
  <!—Usage right authorization part  -->
   <entry><subject>......
    <!-- user ID -->
        <may-not-delegate/>
    <!— right authorization to another user -->
        <using access>all</using access><valid><notBefore>2005-02-
          25T01:00:00</notBefore><notAfter>2005-02-25T23:59:59</notAfter>
   </entry>
<!—validate date -->
   <not-before>2005-02-26T23:59:59</not-before>
   <not-after>...</not-after><name>...</name>
</license>
<!-- XML Signature part -->
<Signature xmlns="http://www.w3.org/2000/09/xmldsig#"><SignedInfo>..</KeyInfo>
</Signature> ...
```

Context Awareness Management System(CAMS) stores and manages a context information(user location, user preferences, user multimedia start and end time etc.) collected via WSN into database, determines a multimedia service by receiving a user context information. IMS(Intelligent Multimedia Server) applies a MPEG-21 DIA and real time video transcoding to downloaded multimedia in order to provide adapted resolution for each IMPD. It provides interoperability among incompatible multimedia play devices.

WSN module takes functionality for network communication among WSN modules, and context collection functions for each object attached on end user and IMPD. WSN module consists of low power manager, sensor part, operating system, and the protocol module, middleware part, a application program.

IMPD takes functionality to display multimedia such as iDTV, iPDA, iPC etc., and embeds WSN module, so multimedia device context is delivered to SHS through WSN("i" meaning: intelligent).

End user is authenticated by Biometrics such as voice, fingerprint, face recognition, etc., and ultimately consume the multimedia in Smart Home. He or She always takes mobile phone, ring or watch which WSN Module is embedded. He or She sends context(user location, user mood, current user status of body including temperature, blood press, etc.) information to SHS through WSN module in realtime.

3.3 Scenario

In this section, we will discuss scenario as follows: User and device are authenticated and then user moves from Living Room to Room 2. At this time, the user contexts are sent to SHS through WSN, and this context is used to determine multimedia service for IMPD(Refer to Fig. 2)

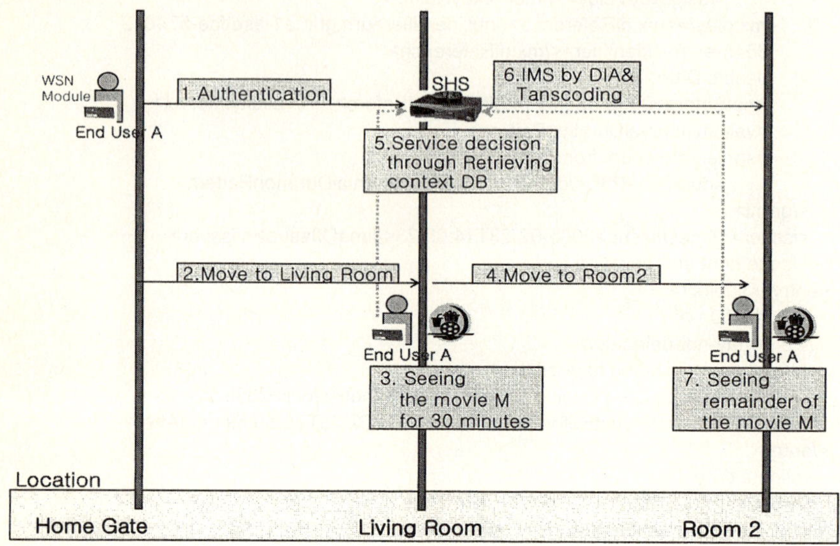

Fig. 2. Scenario Flow

User and IMPD Authentication: When an user comes in home gate, he or she is authenticated by Biometric and communication with SHS. If a new IMPD in home is added, user registers it at SHS, it is automatically authenticated by HCA as well.

Multimedia Consumption and Adapted Streaming Service: User A downloads multimedia from external Multimedia Mall Center by means of the IMS of SHS. A SHS sends this updated information through WSN to all IMPDs in home as soon as the download is completed. While he or she sees a movie "M" for 30 minutes as IMPD (iDTV), he or she moves to Room 2. At this time, the current location context(Room 2) of User A is awared through WSN module. When he or she wants to see remainder of seen the movie, he or she simply gesticulates. At that moment, start and end time of recent seen the movie M is retrieved in context DB of SHS, IMS determine to streaming service. And then provides multimedia streaming service to IMPD(iPDA) in Room 2 through Wireless LAN.

With the consideration against information flowing, The iPDA context(Resolution) in Room 2 is sent to SHS in Living Room through WSN, It is provided streaming service which is suitable adapted resolution of iPDA by DIA and transcoding. Table 2 shows some part of context DB about above scenario.

Table 2. Some part of Context DB about Scenario

User	Location	IMPD	Recent Movie	Play time (s)		Preference Movie	IMPD Resolution
				Start	End		
.	.	.	.	.	.	.	.
.	.	.	.	.	.	.	.
.	.	.	.	.	.	.	.
A	Living Room	iDTV	M	0	1800	Action	1280x720
A	Room 2	iPDA	M	1801	-	Action	320x240
.	.	.	.	.	.	.	.
.	.	.	.	.	.	.	.

4 Conclusion

We consider the analysis of the performance of proposed system. The first, It provides a user-oriented multimedia service based on context-awareness suitable for Smart Home environment without the passive intervention of the user. The second, It offers a multimedia interoperability between incompatible devices with a uni-formatted contents by DIA and real-time video transcoding. The third, in proposed system, the convenience of user and the efficiency of user rights management can be increased by using a license system capable of switching user rights through IPMPS of SHS.

The IMSS provides intelligent multimedia service, which is grafted WSN technology and context awareness computing. It is the core technology of ubiquitous times, as one generation advanced type from general multimedia service in existing home network. It solves problem of interoperable service among incompatible multimedia service devices in home as well. It adaptively grasps the resolution for each device and solves the problem by means of optimized video trans-coding for each corresponded device. Finally, we increased efficiency multimedia use right management in Smart Home.

References

1. Mark Weiser: The Computer for the 21st Century, Scientific American (1991), 94-104
2. Buxton, W.: Ubiquitous Media and the Active Office, http://www.billbuxton.com/ubicom- p.html
3. M. Satya: IEEE Pervasive Computing Magazine, http://www.computer.org/ pervasive.
4. Mark Weiser: Hot topic: Ubiquitous Computing IEEE Computer(1993), 71-72
5. I.F. Akyildiz, W. Su, Y. Sankarasubramaniam, and E. Cayirci: A survey on sensor networks, IEEE Communication Magazine(2002), vol. 40, no. 8, 102-114
6. Enyi Chen, Degan Zhang, Yuanchun Shi, Guangyou Xu: Seamless Mobile Service for Pervasive Multimedia. PCM (2004), 754-761
7. Cliff Randell, Henk Muller: Context Awareness by Analyzing Accelerometer Data, http://www.cs.bris.ac.uk/Tools/Reports (2000)
8. Rye: A Survey of Context-Awareness, MobiCom, ACM Press(1996), 97-107
9. A.K. Dey and G.D. Abowd: Towards an understanding of context and context-awareness, HUC99.
10. MPEG-21 Part1: Vision, Technologies and Strategy, ISO/IEC JTC1/SC29/WG11 N4333 (2001)

11. MPEG-21 Overview v.5, ISO/IEC JTC1/SC29/WG11 N5231 (2002)
12. MPEG-21 Digital Item Adaptation AM (v.5.0), ISO/IEC JTC1/SC29/WG11 N5613 (2003)
13. W. X. Guo, Z. W. Guo, and Ahmad I: MPEG-2 To MPEG-4 Trans-coding, Proceeding of workshop and Exhibition on MPEG-4/2002 (2002), 83-86
14. MPEG-21 Part 5: Right Expression Language FDIS, ISO/IEC JTC1/ SC29/WG11 N5599, (2003)
15. The MPEG-21 Rights Expression Language, Rights com Ltd., White Paper (2003)
16. Draft requirements for MPEG-21 IPMP, ISO/ IEC JTC1/SC29/WG11 N6271, MPEG meeting (2003)
17. MPEG IPMP Extensions Overview, ISO/IEC W6338, MPEG Munchen Meeting (2004)
18. Jong-Hyuk Park, Sung-Soo Kim, Jae-Won Han, Sang-Jin Lee: Implementation of the H-IPMP System Based on MPEG-21 for secure Multimedia Distribution Environment, WSEAS TRANSACTIONS on INFORMATION SCIENCE and APPLICATIONS, Issue 5, Volume 1 (2004), 1301-1308

Analysis of Opinion Leader in On-Line Communities

Gao Junbo, Zhang Min, Jiang Fan, and Wang Xufa

Department of Computer Science and Technology,
University of Science and Technology of China, Hefei, Anhui, P.R. China, 230027

Abstract. In the 1940's, Paul proposed the concept of "opinion leader", and with the development of information technology and network, the importance of on-line community in information spread has become more and more greater. So the information analysis on Internet is becoming quite important. We construct the social network with the replying relations between comments mapped to the comment authors' relations. Then we validate its characteristics of small world networks and find out the opinion leaders in on-line community.

Keywords: Opinion leader, on-line community, small-world network

1 Introduction

In the 1940's, Paul from Colombia University proposed the concept of "opinion leader". He said that the "opinion leader" is sensitive to the trend and having a great influence on peoples' decision making. And with the fast development of information technology, Internet has become the fourth media of information release. The analysis and disposal of information have been focused. The aim of setting such places is to make communities where people share the common context [1] by activating the interaction among people. Based on theses fundaments, we construct the social networks in on-line community from the replying relations between comments and authors. After the analysis of the networks, we have found its small-world characteristics and have found out the opinion leaders from the on-line community with examples.

The left part of this paper is organized as follows: in Section 2 the characteristics of small world networks will be introduced; and in Section 3 we will give the method of the social network construction and analysis this networks. The analysis of opinion leader and examples will given in Section 4. And in Section 5 we will have a conclusion.

2 Small World Characters

Watts and Strogalz [2,3] defined a network as small-world if it has high clustering coefficient and short characteristic path length. Small world networks always show a tendency of clustering and clustering Coefficient (C) can evaluate numerically such a tendency. For example, in social network it shows the possibility of two persons that have common friends also to be friends, which can be calculated using the following formula

$$C = \frac{1}{N} \sum_{i=1}^{N} \frac{e_i}{k_i(k_i - 1)/2} \tag{1}$$

R. Khosla et al. (Eds.): KES 2005, LNAI 3681, pp. 1153–1159, 2005.

Where, N is the total number of nodes, k_i is the number of the neighbors for the ith nodes, e_i is existing links between node i's k neighbors. C is an average over all nodes. Characteristic Path Length (L) represents the average shortest distance of all pairs of nodes, which is the average minimal links between the pairs. The high clustering coefficient and short characteristic path length properties of small world network are evaluated in compare with random graph.

According to Watts' model, small world network is constructed in between the completely regular and completely random graph, which is created by rewiring a certain percentage p of all edges in regular graph. As shown in Fig 1, with p increasing from 0 to 1, the network evolves from Regular network to Small World Network and finally when p equals 1: the network becomes completely random. So, for small world network, the adding in of shortcut dramatically decreased the characteristic path length of the graph and the keeping of some of the regular graph properties guarantee the high clustering coefficient.

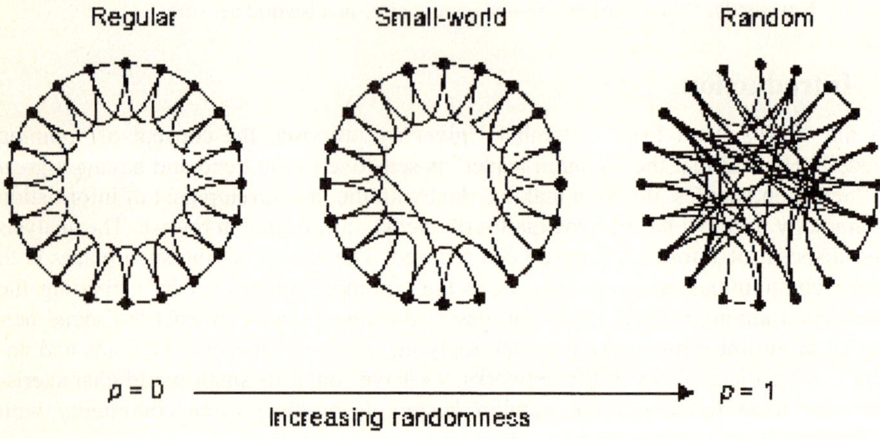

Fig. 1. Evolution of Small World Network

3   Construct and Analysis of Community Network

3.1   Construct of Community Network

By observing BBS member's behavior, we find that it is not perfect to judge member's behavior only by words in his comments [4]. Matsumura and Matsuo had done many works in the context of chance discovery. They used the IDM to mining the opinion leaders. Influence Diffusion Model (IDM) is a method for discovering influential comments, people, and terms from threaded online discussions, such as BBS [5]. But this method doesn't consider the association between topics, interaction between characters and topic's time series, etc. It can't precisely analyze character s' behavior. Since the investigation of individual behavior by single character has some fault, we have to consider whether we can discovery the feature of community's organization and structure in the whole community. Further more, we can investigate the interpersonal relationship in community and achieve the aim of investigating character's behavior. Here, we mainly adopt social network analytical method. This method origins from social metrology method founded by American social psychologist Renault. Social metrology method is the base of social network analysis in metrological analysis. The essential thought of social metrology method used by Renault to analyze human relations is still influencing the development of social network analysis. According to BBS's characteristic, we define the community network constituted by BBS members as follow:

Node in network: BBS member (including who post new comment and who reply to the comment);

Edge in network: replying relation between members (adding a edge between two nodes which have replying relation);

Now, we construct a community network. It is a complex network constituted by characters and their interaction relation. (In brief, we define this network as an undirected and unweighted graph).

3.2 Experiment of Small-World Feature

We select china military forum (http://bbs.china.com/military/bbs.jsp) as data source to validate that our community network fits in with small-world network's feature *(high clustering coefficient and short average path length)*. Aiming at mass data, we select one day's data as a dataset and use it to construct an undirected and unweighted graph. We have selected 41 days' data from Nov 20 to Dec 30,2004. We have experimented on these dataset separately for many groups and show some experiment results as follow:

Since G isn't always a connected graph, we consider its' connected sub-graph instead (the biggest connected sub-graph's nodes amount to 90% of G). Experiment data is shown in table 1:

Table 1.

Date	$|V(G)|$	$|V(g_M)|$	$|V(g_M)|/\,|V(G)|*100\%$
2004-12-01	1878	1779	94.73%
2004-12-02	1849	1740	94.10%
2004-12-03	2146	2041	95.12%
2004-12-04	1496	1367	91.38%
2004-12-05	1511	1441	95.37%

Hence, we will validate whether g_M fits in with small-world network's feature. We adopt *JUNG(Java Universal Network/Graph Framework)* as experiment platform. The results are shown in table 2.

Where C and L is separately clustering coefficient and average path length of g_M, C_{rand} and L_{rand} is separately clustering coefficient and average path length of corresponding random graph.

We can easily conclude from table 2 that g_M fits in with the feature as follow:

$$C \gg C_{rand}, L \approx L_{rand}$$

According to the definition of small world (Watts & Strogatz), we find that g_M fits in with small-world network's feature. It is validated that our community network has the feature of small world.

Table 2.

Date	$|V|$	$|E|$	C	C_{rand}	L	L_{rand}
2004-12-01	1779	3132	0.55816	0.00198	6.27783	5.79712
2004-12-02	1740	3217	0.54608	0.00248	6.27813	5.77337
2004-12-03	2041	3529	0.59823	0.00124	6.20001	5.90808
2004-12-04	1367	2145	0.58937	0.00106	6.62903	5.55563
2004-12-05	1441	2367	0.57191	0.00359	6.48281	5.60829
2004-12-06	1414	2146	0.63825	0.00374	6.65321	5.60516
2004-12-07	1394	2237	0.57598	0.00403	6.59011	5.60423
2004-12-08	1772	3002	0.56289	0.00126	6.47697	5.79443
2004-12-09	1512	2337	0.59713	0.00108	6.64479	5.64972
2004-12-10	1439	2433	0.57811	0.00200	6.29426	5.61001

4 Opinion Leader Analysis and Experiment

4.1 Opinion Leader's Characteristic

Rogers (a famous Communication scholar) has concluded opinion leader's character-istic and pointed out that opinion leader must give demonstration to others. [6] He must have encyclopedia knowledge, rich social resource, many adherents, compre-hensive ability and larruping charm. Also, people who are core in BBS community must be wide field, actively participate discussion and present remarkable individual opinion. We believe that opinion leaders can provide fascinating topics which trigger the activation of the community.

Objectively, these members can always conveniently access Internet and have lots of time to access Internet. Subjectively, they accumulate plenty of experience on net-work and establish personal authority by posting valuable comments. Depending on their excellent character and broad information source, they can launch remarkable topic, attract many members to take part in discussion and intensively influence other members. It makes them have a lot of adherents in forum. They can rapidly gain re-source predominance and become the core of forum. These members are good at communication and they can often communicate with others, send and receive great amount of information, express own opinion by communication. Whether a forum member can become opinion leader depends on his inherent rallying point. This influ-ence seems to have no obvious restriction, but actually can often restrict others while compulsive restriction can't do. In many time, forum administrator's compulsive restriction (such as deleting comment, forbidding ID) works not good as opinion leader's influence effect. Usually, natural influence also includes some factors about compulsive influence, such as seniority factor. In daily life, one's seniority reflects what has happened in his life and his experience. Also, logging on forum and posting new comments will accumulate member's experience point. More experience points mean more time on forum, deeply participating discussion and abundant network experience. On certain condition, seniority factor can also influence leader's validity. Generally speaking, forum member will more trust in senior leader.

4.2 Mining Opinion Leader

Based on above community network, we can obtain a network that has forum author as its node and replying relation as its edge. The procedure of finding opinion leader is the procedure of finding key node in network. We try to find these key nodes by analyzing average path length. Based on above community network, we give definition as follow:

Definition 1: CN is a community network and L denotes CN's average path length;

Definition 2: CN_i is a community network without the i-th node (all edges connecting i-th node are removed). L_i denotes CN_i's average path length.

According to above two definitions, we define ΔL as average path length's change value.

$$\Delta L_i = L - L_i \tag{2}$$

Now, we can judge a node's importance by the change value from L to L_i. That is to say, the node with high ΔL is the key node in network, it plays a very important role in the whole network' connectivity. For our community network, the node with high ΔL can connect nodes, even clusters, which seem irrelevant. Therefore, this kind of node can be deemed as key node that is the opinion leader we are looking for.

4.3   Experiment

In the above section, we select one day's data in forum as a dataset and construct an undirected and unweighted grapy G. We only select partial data because the amount of data in forum is too large and we want to illuminate more intuitively. The community network is shown in Fig 2 (opinion leader is represented by solid node in Fig 2).

At first, we validate this community network fits in with small-world network's feature. The result is shown in table 3.

Table 3.

$|V|$	$|E|$	C	C_{rand}	L	L_{rand}
54	110	0.3950	0.0803	2.9139	2.8539

Table 4.

No.	node	ΔL	Origin comments	Out-degree	In-degree
1	洪浩	0.904584	3	16	33
2	坚决抵制	0.903564	2	10	23
3	zzzzzzzzza	0.902865	2	14	18
4	qinchenxv	0.902166	1	8	16
5	hrb–浦汪	0.902136	1	6	16

Then, we find opinion leader in forum by computing each node's ΔL that is shown in table 4 (one's out-degree denotes the sum of his comments replying to others' comments, one's in-degree denotes the sum of other's comments replying to his comments).

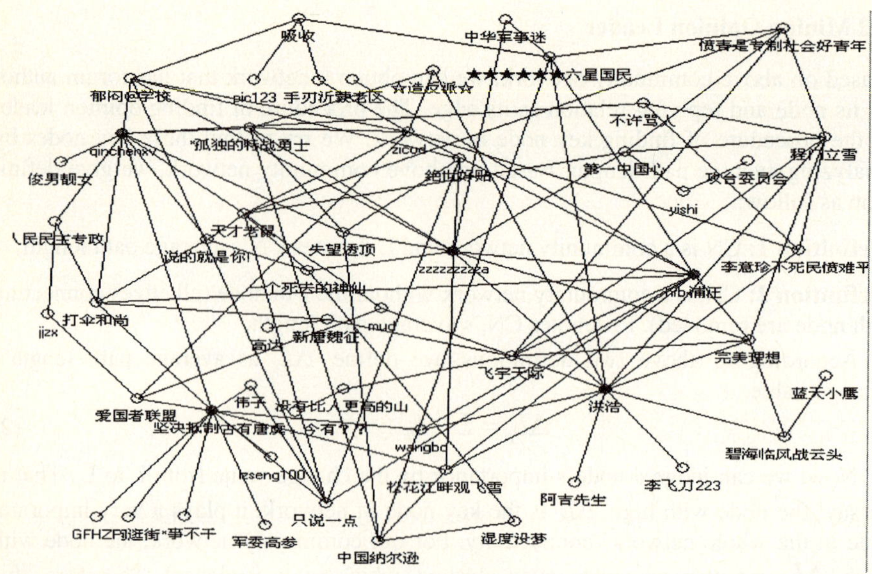

Fig. 2. Community network

From table 4 we find that opinion leaders we have selected are all core members of network community. For member No.1, he is no doubt the most important character in network. The majority of members in community network has paid great attention to his comments, at the same time, he also actively replies to other members' comments and communicates with many members. His status is beyond other opinion leaders. Several other opinion leaders' comments also draw attention from community members. Otherwise, we find out that these opinion leaders are all senior members in forum. They are well known by the others and have posted plenty of comments in the past several years, which attract abroad attention from forum members. They completely accord with the feature of opinion leader. Therefore, we believe that the method using small-world network to find opinion leader is correct and the author we have found is assuredly opinion leader in forum.

5 Conclusion and Perspectives

The development of Internet is rapidly changing our real world, it provides to us a virtual community that is a new space and communication surroundings. Based on the feature of communication in virtual community, interpersonal mutual relationship is a hotspot that attracts attention from domestic scholars and overseas scholars. Small World structure is popular in both natural networks and man-made systems. But the discovery and utilizing of Small World structure and its properties in the field of mining the opinion leader in on-line communities is just beginning. In this article, we have investigated an online forum and constructed a community network in forum. We analyze this network's feature and find it fits in with small-world network's feature. At last, we have found opinion leader in online forum based on small-world network. Our future works will be based on the following ideas. The community network can give not only the information of which peoples are important but also the correlations of peoples. And thus, the idea represented by the article can also be got from the network, which might help some in mining the opinion leaders in on-line community. In addition, by combining the community network of several related dataset, we could find out the latent opinion leaders. We will further analyze network member's dynamics behavior in network, try to find their network feature and do further research to analyze character's behavior in complex network.

References

1. Yukio Ohsawa: "Chance Discovery for Making Decision in Complex Real World", New Generation Computing, Vol. 20, No. 2, 2002.
2. D.J. Watts and S.H. Strogatz, "Collective dynamics of 'small-world' networks", Nature 393, 440-442(1998)
3. D.J. Watts, "Small World", Princeton University Press, Princeton (1990)
4. Matsumura, Ohsawa, "Mining and Characterizing Opinion Leaders from Threaded Online Discussions". KES2002
5. Naohiro Matsumura, Yukio Ohsawa, and Mitsuru Ishizuka: Influence Diffusion Model in Text-based Communication, WWW02, 2002.
6. E.M. Rogers: Diffusion of Innovations, The Free Press, 1962.

Optimum Pricing Strategy for Maximization of Profits and Chance Discovery

Kohei Yamamoto and Katsutoshi Yada

Faculty of Commerce, Kansai University, 3-3-35, Yamate, Suita, 564-8680 Osaka, Japan
{da20699,yada}@ipcku.kansai-u.ac.jp

Abstract. The objective of this paper is to present and discuss methods for formulating optimum pricing strategies for maximizing store profit levels using consumer purchase data. In order to maximize outlet profit levels, it is necessary to seek such pricing strategies after achieving an in-depth understanding of the various features of consumer purchase behavior concerning a wide range of products and their related prices. We authors have used a very large amount of consumer purchase behavior data in order to clarify the effects on store-level profits of a wide range of specific products to develop a support system for pricing specialists. At the end of this paper, we analyze the framework that has been presented for the system from the standpoint of its utility for discovery of new business chances and discuss the need to clarify the problems connected with the use of data mining-based pricing strategies.

1 Introduction

During the current deflationary period, there has been continuous and growing price competition among retailers aimed at capturing and keeping customers. Thus, special sales are being carried out everyday at super markets. However, in the case of food retailing in Japan, it is very rare that a retailer will scientifically analyze such sales in advance, using data from previous sales, for the purpose of establishing pricing strategies. In most cases, previous experience and the sales strategies of other rival retailers are used for the purpose of setting the prices of the products that are featured during such special sales. This paper proposes using very large amounts of customer purchase data to develop optimum pricing strategies for supermarkets.

Consumers have different needs and, therefore, have different standards [6] for judging the attractiveness of the special sales prices for different products. So, it is only natural that the reactions of different consumers to specific sales will differ. Thus, lowering prices may not, in given situations, result in increases in sales. The store marketing personnel that decide on product item prices must utilize suitable pricing strategies [5] based on an in-depth understanding of consumer purchase behavior. For this purpose, it is important to pay attention to the implications of accumulated data collected with the use of customer membership cards for the purpose of understanding consumer purchase behavior. However, regrettably, there are numerous studies on the sales and profit levels pertaining to certain product items, but virtually no studies on the creation of pricing strategies from the standpoint of optimizing profits for each given store as such. The price of a given item can often affect the sales and profits of other items so that there is a need for each store to clarify these complicated relationships between items in order to maximize the profit levels of each store.

R. Khosla et al. (Eds.): KES 2005, LNAI 3681, pp. 1160–1166, 2005.

The objective of this paper is to discuss analytic methods for formulating optimum pricing strategies for maximizing outlet profit levels using consumer purchase data. This study is a typical example of using data mining methods for marketing applications. Therefore, looking that this process from the standpoint of chance discovery [4], there are several problems that must be dealt with. For example, in this paper, the suggested methods of analysis tend to be based on statistical evaluation criteria. Thus, in the case of relatively small quantity phenomena in the pricing data, i.e., hard to notice, exception-related phenomena concerning related products, the important potential effects of these phenomena may not be noticed, or may be ignored. In reviewing the various cases in this study, it is necessary to keep in mind the possible problems related to the development of data mining oriented methods of pricing, and we wish to present and discuss the possibilities connected with using chance discovery techniques when formulating pricing strategies.

2 A Framework for Optimum Pricing Strategy for Maximization of Profits

Our development of pricing strategy for maximizing profit is based on a framework that consists of 4 stages. The first 2 stages are not concerned directly with analysis of pricing as such. Rather, they consist of the general content required for the analysis of customer purchase data. At the first stage, the data that is to be subject to analysis is cleaned and error data and noise are processed. At the second stage, the basic analysis of the specific outlets and customers included in the research is carried out, and the data is inspected to check whether or not the overall data includes areas or specific outlets that have unique features widely differing from the other areas and outlets that appear in the data.

Step three consists of discovering the specific groups of related products or product categories that will be effective for use for special sales aimed at increasing the sales of the given outlet. Within the vast number of different items handled by a retail outlet, there are categories of products that are effective for raising sales and other categories that are not. In order to make selections of product items that are effective for building optimum pricing strategies, we must select groups of products for special sales based on indices of customer response to specific sales. Stage 4 is the stage where pricing strategies for the specific product items (Or, item categories) are chosen for maximizing profit. If the sales of a single given product are to be maximized, it is sufficient to sell that product at the lowest possible price. However, this approach cannot be said to be acceptable for the goal of maximizing profits of the given outlet itself. Lowering the price for a given product will not only have the effect of lowering the contribution of the related products to total store profits, but will also have an effect on the sales of competing product items that are being sold at their usual price levels. For the purpose of maximizing the profit of the individual store, we think that it is necessary to include all groups of products that contain product items that are likely to be affected by the lowering of prices of the special sale items in the analysis for the formulation of the pricing optimization strategies to be used [1]. For this purpose, the past purchase records of individual customer respondents are used and this makes it possible to provide a proper framework for the formulation of the pricing optimization strategies.

Regarding the first stages, since there are many factors contained in the data for individual consumers that are similar to standard consumer research operations, this paper will be focused on the last two stages of the optimization process.

3 Discovering the Optimum Product Items

For achieving optimum pricing, before prices as such are considered, we think that there is a need to determine whether lowering the prices for a given group of products will actually cause an increase in sales or not. Within the vast array of products being sold, it is necessary to clarify the aims of the strategy and on which products the lowered pricing is to be focused. Concerning this point, it is necessary to carefully classify all products to determine which products are more suited for weekend special sales and those that are more suited for weekday special sales.

3.1 Price Indicating the Rise/Fall Effects of Lowering the Price in Increments of One Yen

Here, we examine a specific product "A" as the possible focus of a special sale. In the case of this paper, the most frequent price of the product "A" (The price that appears most frequently on register receipts) is defined as the "normal price" of the product and prices lower than this price are "special sale prices". The specific days upon which these lower prices are used are defined as "sales days". Further, in order to discover whether product "A" is more suitable for a weekday or for the weekend, the 1-yen increment Product Indexes are used. PI value is the amount of the product sold per 1000 visitors to the store. The reason for the use of this index is that the rise and fall of sales volume (PI) in increments of a single yen makes it possible to remove the effects of the size of the particular outlet being examined from the calculations. The calculation that is used is as follows:

$$\frac{p_a(d) - p_a(r)}{r - d}$$

Pa= sales quantity per 1000 visitors
Pa (x) = product "a" x yen/time PI value
r = Normal price in yen
d = special sales price in yen

These calculations are carried for each product in each product category, making it possible to discover the products that are the most suited for special sales.

3.2 Example of Discovering Suitability of a Category for Weekends vs. Weekdays

Using the method indicated above, from customer purchase data for supermarkets located in Japan's Kanto Region, effectiveness of specific product categories for both weekdays and weekends were calculated. The results of the calculations showed with relatively great degrees of significance that there were many examples of lack of effect in changing product prices. In other words, there cases when lowering the price actually caused sales to decrease rather than increase. The definition of a product that is suitable for special sales is a product the sales of which will increase when the price is lowered. Likewise, on a specific category basis, more than 50% of the products in the category must be suitable for special sales for the category to be a "suitable category for special sales". Using these definitions, it became clear that on weekdays, vegetables and confectionary products are a suitable category for special sales and that instant food and processed meats are effective product categories for weekends. With these results in hand, the marketing staff can decide whether specific categories are suitable or not.

4 Optimized Pricing for Maximized Profits

In this section, we examine in detail the subject of product optimize prices. In order to discuss optimized pricing, we aim for the maximization of total profit for the entire outlet. If we try to optimize pricing of each product from the point of view of the marketing staff of the maker of each product, we would probably have to sell each product for the lowest possible price. However, each outlet is handling many different products, and the prices of some products have an influence on the sales than other products. Thus, to comprehend the situation for the entire store, it is necessary to take into consideration many very complicated factors. Therefore, we present a framework for optimizing prices that clarifies the complicated relationships between the different products by using the purchasing histories of individual customers.

4.1 Predictions of Purchasing Probability by Means of the Logit Model Using Customer Data

First, in the case of the logit model, the model for forecasting the purchase probability for purchase ratios is derived for the product "A" using individual customer purchase histories for product "A" for the desired variables. Consumer purchase behavior is affected not only by price, but also such factors as purchase loyalty. In addition, the prices of competitive products and other factors also affect purchase. Since it impossible to include all of these variables in a single model, for the basis of the statistical indices, the factors included were narrowed down to 10–20 significant factors and the variables were chosen according to the stepwise method to build the forecasting model. The result was that we became able to forecast, based on certain conditions, whether an individual customer will buy a given product or not and using these predicted figures, we are able to estimate the sales for a given product. In addition, using the purchase probability prediction model to analyze competing products as well, we are able identify product groups that will affect the pricing for product "A".

4.2 Sales Forecasts and Maximization of Store Profit

Using the predictive model referred to above, base on certain conditions, in other words price levels set for the different prices set for items contained in a specific product item category, it is possible to calculate the level of probable purchase for individual customers and using these estimations, it becomes possible to forecast the sales for a given store. In the category, "soy sauce", in the case where the sales of the leading brand "K" were predicted, the prices of the all other brands of K were fixed and when the price of K was varied, the results were as shown in Figure 1. The x-axis indicates prices for K in 10-yen increments. The y-axis indicates the predicted sales quantities for K and indicates the rate of growth in sales. As can be seen from the

Figure, the more the price of K is lowered, the more sales increase. However, with each 10-yen cut in the price, the rate of sales increase slows and the sales increase effect of price-cutting decreases.

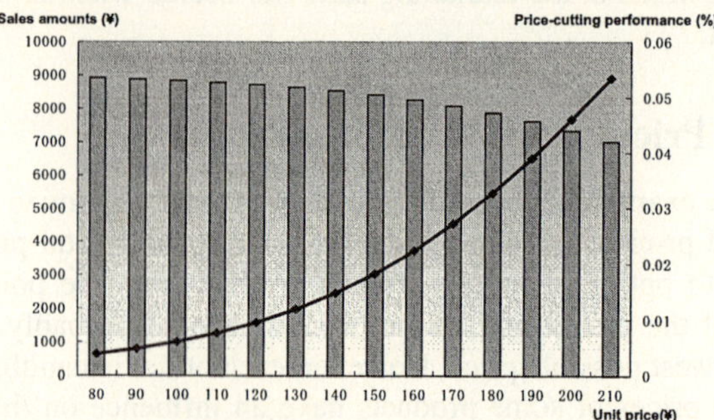

Fig. 1. Changes in price and the effects on the sales of Brand K soy sauce

In this paper, the logit model was used as the purchase probability prediction model, but depending on the specific products for which this model is used, there are many cases where sufficiently useful predictions cannot be made. As an example of this, this model assumes the existence of a "reference price" [2] [3] for category (or product) below which the consumer will make a purchase and that there exist certain prices levels where very significant increases in sales will occur.

4.3 The Case of Chinese Cabbage-Related Product Prices

As an example, we examined the situation for the pricing of the product group consisting of Chinese cabbage and closely related product items. At the store where the data was collected, the category also included related items consisting of half of a Chinese cabbage (Below, "H") and a quarter (Q). To maximize the profit of the store, it is necessary to set different prices for all of these items and, in practice, this is difficult. The related variables consisted of prices for a Chinese cabbage half (H) and (Q) and the prices for the other related products other than Chinese cabbage such as the prices for tofu and ponsu (A citrus-based liquid flavoring). The individual purchasing histories regarding frequency of purchase of seafood and vegetables were also extracted from the data for the calculations. When the model, using these variables, was used for the calculations, the optimum prices for Chinese cabbage H and Q were 198 yen and 100 yen, respectively and the profit levels obtained were also at the maximum level.

It was possible to discover the price levels for the two Chinese cabbage items that maximized profit, but since the Chinese cabbage item prices affected the sales of the other products, in order to maximize store profits, it was necessary to formulate an optimal pricing strategy that included the other products as well. For example, the price of the Chinese cabbage was found to a strong influence on the sales of ponsu. In the case of the pattern encountered for Chinese cabbage prices shown above, the price for ponsu was 158 yen. However, when the level of profit contributed by the ponsu was included in the calculations and the calculations for the maximization of profit for the store were run, as shown in Figure 2, when the price of ponsu was 248 yen, the maximum profit contributed by the two Chinese cabbage items and ponsu was the point of maximum store profit. In order to carry out maximization of store profit pricing strategies, it is necessary to analyze the effects of not only a given product and its 0direct competitors, but also to sufficiently investigate the effects on the prices of other related products on the calculations.

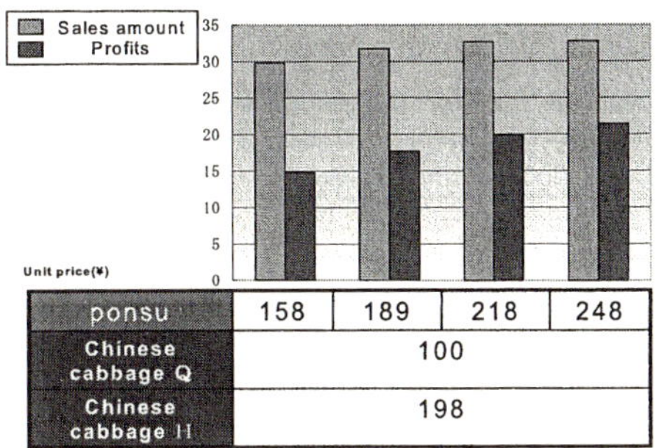

ponsu	158	189	218	248
Chinese cabbage Q	100			
Chinese cabbage H	198			

Fig. 2. The pattern for prices for Chinese cabbage (H,Q) and ponsu aimed at the maximum level of store profit

5   Future Topics and the Study of Business Chance Discovery

In this paper, methods were proposed to create optimized pricing strategies for the maximization of store profit. The paper contained concrete procedures and methods for achieving these aims. By means of simulations using actual retail store data, it was possible to demonstrate the effectiveness of using these methods. Store marketing personnel, using our methods, will be able to create optimal pricing strategies and will probably able to maximize store profits.

We are, as yet, left with many unsolved problems for which solutions must be found. In this paper, in order to predict purchasing probability, the model that was used was the logit model. When actual purchase data is used for applications using this model, there are occasions when the calculations do not produce sufficiently usable predictions. In the case of certain special products and categories, the calculations are strongly influenced by noise and the effects of abnormal values and the level of accuracy obtained was unstable. In this paper, aside from the example involving the logit model, regarding data mining algorithms, sufficient comparisons were not carried out. Therefore, we must find other algorithms that can attain a high and stable level of accuracy not affected by noise or subject to abnormal values.

All of the content of this paper is based entirely on knowledge gained from the use and analysis of data to obtain results related to pricing strategies and, because of that, the paper does not contain sufficient content having to do with chance discovery. The reason for this is that human beings (Experts) were not using their own perceptions and their abilities to notice factors and to integrate these inputs into the strategy creation process. In addition, the largest problems concerning the methods proposed in this paper related to creating optimized pricing are related to the selection of the variables for use with the prediction model and the reliance solely on statistical indices. Considering the matter of the discovery of new business chances, the chances are relatively high that the use of these methods alone will seldom result in such discoveries and that the important "chances" will get away. Regarding the product items pricing, there is a need, in the future, to find methods that lead to discoveries of new opportunities that would not be discovered under ordinary circumstances and to develop an integrated framework for such methods.

References

1. Blattberg, R.C., Wisniewski, K.J.: Price-induced Pattern of Competition, Marketing Science, Vol.8 (1989) 291-309.
2. Mayhew, G.E., Winer, R.S.: An Empirical Analysis of Internal and External Reference Prices Using Scanner Data, Journal of Consumer Research, Vol.19 (1992) 62-70.
3. Mazumdar, T., Papatla, P.: An Investigation of Reference Price Segments, Journal of Marketing Research, Vol.37, Issue 2 (2000) 246-258.
4. Ohsawa, Y.: Chance Discovery for Making Decisions in Complex Real World, New Generation Computing, Vol.20, No.2 (2002) 143-163.
5. Tellis, G.J.: Beyond the Many Faces of Prices: An Integration of Pricing Strategies, Journal of Marketing, Vol.50 (1986) 146-160.
6. Urbany, J.E., Dickson, P.R., Kalapurakal, R.: Price Search in the Retail Grocery Market, Journal of Marketing, Vol.60 (1996) 91-104.

Risk Management by Focusing
on Critical Words in Nurses' Conversations

Akinori Abe[1], Futoshi Naya[1], Hiromi Itoh Ozaku[1], Kaoru Sagara[2],
Noriaki Kuwahara[1], and Kiyoshi Kogure[1]

[1] ATR Intelligent Robotics and Communication Laboratories
2-2-2, Hikaridai, Seika-cho, Soraku-gun, Kyoto 619-0288 Japan
{ave,naya,romi,kuwahara,kogure}@atr.jp
[2] Seinan Jo Gakuin Univ.
1-3-5, Ibori, Kokura, Kitaku, Kitakyushu, Fukuoka 803-0835 Japan
sagara@seinan-jo.ac.jp

Abstract. This paper proposes another dynamic prediction procedure
that observes nurses' conversations to determine critical points in nursing
accidents or incidents. The points would be be suggested by words that
have a certain feature. We regard such words that might cause accidents
or incidents as chances. This paper analyzes these types of words and
describes their features to dynamically determine the point at which an
accident or incident occurs.

1   Introduction

Because of malpractice in medical care (nursing), hospitals have been sued and
sometimes forced to pay huge amounts of money. Therefore, some hospitals in-
sure against medical malpractice to protect themselves from bankruptcy. Of
course, medical malpractice is not only a problem for hospitals but also has
serious effects on hospital patients. Taking out insurance might protect hospitals from bankruptcy, but it cannot save the lives of patients. Thus, it is
quite a passive measure for preventing malpractice in medical care. Recently,
it has been recognized that medical risk management is very important both
for hospitals and their patients. Medical risk management aims to reduce medical accidents and minimize the costs of medical care, including insurance fees
and financial losses. For this reason, there exists an accident or incident report
database [HiyarihattoDB]. Risk-management experts can refer to this type of
database and generate generalized patterns for frequently occurring accidents.
Thus, it is important to compile a textbook on nursing accidents to avoid them.

It is more important, however, to perform dynamic nursing risk management that does not need any compiled data on accidents. In previous papers
[Abe 2004a, Abe 2004b, Abe 2005a], we proposed a model that can dynamically
predict nursing accidents or incidents. In the proposal, nursing accidents or incidents can be logically determined. It is not a static process, but it does require
a set of model of accidents or incidents. To build the models, it is necessary to
survey many example sets of human errors or nursing accidents, and if we cannot

prepare proper models, it is impossible to determine the causes of nursing accidents or incidents because in a logical process models are required for reasoning. Therefore, it is necessary to establish a dynamic prediction process that does not require prepared models.

In this paper, we propose another dynamic risk-prediction procedure that analyzes nurses' conversation type to determine possible critical points of nursing accidents or incidents. The points will be suggested by words or phrases that have certain features. We regard such a word or phrase that might cause accidents or incidents as a chance, because such a word or phrase is usually unconsciously used and can be a trigger of accidents or incidents.

The remainder of this paper is structured as follows. In Section 2 we analyze the dialogue data set from nurses and show the features of words and phrases that might cause accidents or incidents. In Section 3 we propose a prototype risk-management system that determines critical words or phrase to dynamically predict accidents or incidents. Section 4 includes our conclusions.

2   Dialogue Data from Clinical Meetings

Kuwahara et al. proposed an integrated monitoring system for nursing activity that couples ubiquitous apparatus with fixed apparatus [Kuwahara 2004]. The system was proposed to monitor the workflow of nurses' activities and report on their entire workflow in real-time. In the current (experimental) system, nurses carry wearable sensors that record their (self-)conversations or narration, the number of footsteps, body angle etc. In addition, video monitoring systems are placed in each room to visually record part of their activities (no sound), while radio frequency (RF) stations are installed to record their locations. Furthermore, digital pens are used by nurses to electronically record their worksheets. These data are useful, after statistical analyses, to estimate types of nursing activities from the live data.

In this paper, however, we focus on nurses' conversation data. We previously proposed the concept of scenario violation by analyzing nurses' narrations [Abe 2005a], and we manually analyzed the set of narrations to determine and model types of nursing activity or processes such as injection. In this paper, though, we analyze the features of nurses' conversations to determine critical words or phrases that might lead to accidents or incidents.

2.1   Example from Dialogues in Nursing Meetings

As reported in [HiyarihattoDB], there are many factors associated with nursing accidents or incidents. Since, during the experiments conducted by [Kuwahara 2004], we collected narration and conversation data from nurses, in this paper we focus on *oral* miscommunications that usually lead to serious accidents or incidents by nurses. To identify symptoms or possibilities for miscommunication during nursing activities, we survey dialogues between nurses, especially focusing on dialogues during handovers (clinical meetings) where important information on nursing activities is discussed and exchanged. Figure 1

> A. Well....
> A. The patient's name is Mr. HK, he is 59 years old and a male.
> A. Well, on June 23, he was hit in his face, mandible and body by his subordinate. He was transported to the hospital by an ambulance. He got seriously injured in his face. In addition, he got injured in his elbow. I have not fully checked the ENCAL[a]...
> A. He is a patient for the third department.
> C. He is also a patient for the dental surgery department.
> A. His mandible is well fixed and has gradually recovered. He was prescribed antibiotics and an analgesic. He had finished all the medicine before Aug. 4.
> ......
> A. Well, he has a chronic with around 3cm thick and... Though the doctor did not say at first.... but in the MunThera[b], the doctor said that his brainstem was bent and he was in quite a dangerous situation.....
> B. What do you mean by "bend"? Is it "pressured?"
> A. Oh! yes, his brainstem was pressured......
> ....
> A. He also has a headache. During an Anamnese[c], he said that, in 2003, he was diagnosed with a traumatic subarachnoid hemorrhage and was operated on with neurosurgical clipping....
>
> ---
> [a] Abbreviation of clinical record at an admission?
> [b] Mund (mouth) Therapy; explanation of condition
> [c] Inquiry before the admission (in German).

Fig. 1. Example of dialogue between nurses (translated from Japanese).

shows an extraction set from a set of dialogue data during a clinical meeting. In the dialogues nurses seem to frequently use such abbreviations as ENCAL and MunThera or such specialized terms as medicine names and Anamnese. One reason is that if they use such abbreviations or specialized terms, it is slightly difficult for their patients to understand their conversations. In fact, for those us who are neither nurses nor medical doctors, it is quite difficult to take comprehensive dictation of their conversations not only because of the poor quality of recordings but also their specialized terms. In addition, it is stylish and fashionable to use such abbreviations and specialized terms originating from German, Greek or Latin.

2.2   Analysis of the Dialogue Data

As explained above, we assume that one reason for accidents or incidents is miscommunication. In addition, our main purpose is to build a (dynamic) nursing risk-management system. In this paper, therefore, we analyze dialogues between nurses particularly from the aspects of accidents or incidents.

Neither an accident nor an incident was reported on the day when the above dialogue set was recorded, so it is not possible to directly analyze reasons for accidents or incidents from the dialogue data. However, we can find some symptoms for incidents from the above data by applying natural language analysis techniques.

In the above, we briefly analyzed the dialogues. Some aspects are repeated below, but there we summarize features of the dialogues during clinical meetings to identify some typical features of them.

- It is impossible to guess from the dictated data, but they spoke very fast. Accordingly, part of their dialogue was very unclear for us and perhaps even for them. However, they responded quickly. Also they might have only briefly checked the information by reading printed materials. Otherwise, they could supplement missing information or understand it before the end of the report.
- They frequently used abbreviations and jargon. Such specialized terms are useful for shortening the length of meetings. In addition, such terms work as cyphers against patients. Of course, using specialized terms is not a serious problem for nurses, and in fact, it seems better to use such terms to avoid miscommunication.
- Many topics that seem to have little or no relation to each other were discussed during one meeting, because the nurses have patients suffering from various diseases.
- During the meeting, possibilities for other accidents or incidents were sometimes discovered. Some of these phenomena can be regarded as associate remembrance. They can be regarded as Chance Discovery, and it is important to analyze and utilize such phenomena to avoid accidents or incidents, though this paper does not deal with them. We plan to deal with this type of chance in future work.

The above lists include quite typical features, and usually such features do not lead nurses to cause accidents or incidents, since they are accustomed to using abbreviations and jargon, even if they fail to catch the words, they can properly supplement the missing information by considering the situation. In addition, doctors say that medicines must not be confused with those that have a similar name when they know the patient's disease or condition. In the case of medicine, one of the most serious problems is observed in confusing the unit of medicine. For instance, confusion between "Anginal (25 mg)" and "Anginal (100 mg)" or "2% Xylocaine" and "10% Xylocaine."

In addition, we observed other features in their conversations.

- A lot of filler was observed. This is evidence that they needed to think over or change their words for better understanding.
- Difficulty in choosing neutral or commonly used terms was observed. Such terms as jargon and medicine names that they usually use seemed to easy to use. However, it appeared slightly difficult to use common words to express the status or situation of patients. Since they did not seem to have a dictionary for these types of expression, it appeared quite difficult to generate adequate words in a short time.
- Sometimes the oral report was neither concrete nor clear. This might have occurred due to time limitations. For the same reason mentioned above, they could not prepare words before their report, so sometimes their reports tended to become quite abstract and unsatisfactory.

We think the above features present opportunities for error, thus we analyze the feature of such words and phrases in the next section.

2.3   Chance for Error Due to Words and Phrases

Iwahara reported that people sometimes carefully selects fonts or pictographs to avoid miscommunication in e-mail communication [Iwahara and Hatta 2004]. In fact, he analyzed miscommunication in emotional aspects — para-linguistic information. In other communication such as simple conversation, we usually select adequate words or change intonation to avoid miscommunication not only for para-linguistic aspects, but also for verbal aspects. Since we have common understandings for roles of words and intonation, if we use them according to the common understandings, few miscommunications will occur.

Thus if we have enough time to select adequate words, very few problems occur in communication. However, since nurses often do not have enough time for handing over, sometimes miscommunication indeed seems to occur. This is probably due to inadequate or insufficient usage of words or phrases.

Here we highlight the problem by using examples from dialogue data gathered during clinical meetings. We especially analyze the data from the aspect of the possibility of nursing accidents. In Fig. 1, we observe a very interesting interaction that includes "What do you mean by bend? Is it pressured?". To express patients' situations, nurses should usually use everyday words because there is no such dictionary that deals with irregular cases. Then nurses, by using their own vocabulary and background knowledge, generate expressions. In fact, because of time limitations, they sometimes fail to generate adequate words, or even when they do successfully generate words, those words may be sometimes too abstract or vague for listeners. Moreover, often the background knowledge between speaker and listener differs. In the above case, since a listener asked about the word's meaning, the misunderstanding seems to have been solved and no accident occurred. However, if he/she had failed to confirm and proceeded with the task with his/her own understanding, an accident might have occurred. In another case (not included in Fig. 1), we observed such an abstract phrase as, "Please arrange an intravenous drip to be most effective after dinner." This is quite an abstract phrase, so another nurse confirmed as it meaning, "It is best to arrange 30 min. intravenous drip just after dinner." If she/he had not made a confirmation, an accident might have occurred. The shown examples did not cause any nursing accidents, but if there had been no confirmations, accidents might have occurred.

By surveying conversations among nurses, we can categorize words and phrases used in the conversation as follows:

- abstract / concrete
- specialized term, jargon / everyday word, general word, commonly used word

From the above examples, for safe conversation among nurses, concrete terms, specialized terms, and jargon seem to better to use because nurses generally know the terms well. and thus can be regarded as common sense for them. Furthermore, in conversations among nurses, words regarded as general or commonly used, and abstract phrases play important roles in miscommunication though of course not all of them do so.

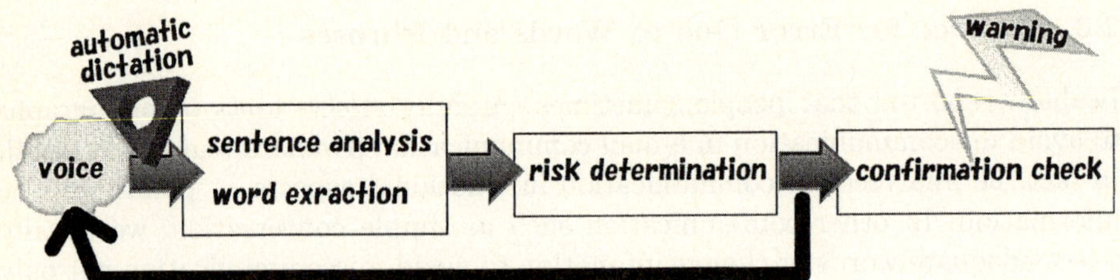

Fig. 2. A procedure for the accident or incident determination system.

From the viewpoint of Chance Discovery [Ohsawa 2002], such words and phrases that can lead us to accidents or incidents can be regarded as chance, because this process is an unconscious one and hidden or latent reasons for accidents or incidents can be included such words or phrases (keyword), becoming possible triggers for accidents or incidents.

3   Dynamic Nursing Risk Management

In the previous section we discussed that in conversations among nurses, words that are regarded as general or commonly used, and abstract phrases, play important roles in miscommunication. We also pointed out that such words and phrases leading to accidents or incidents can be regarded as chances. Since, in general, this type of miscommunication is an unconscious process and is dominated by an individual's background knowledge, it is quite difficult to detect by normal methods. The problem is everybody believes that they are correct. This section proposes a simple prototype for an accident or incident determination system that observes conversations and determine the chance (words or phrases) of avoiding accidents or incidents. The system's procedure is shown in Fig. 2.

The most important point of the system is the $\boxed{\text{risk determination}}$ component that discovers a risk (chance) that might lead nurses to cause accidents or incidents. It can determine the possibility of accidents or incidents. In addition, another check point is necessary. This is the $\boxed{\text{confirmation check}}$ component, which checks whether nurses confirm ambiguous information. If not, the system should warn the nurse of the possibility of accidents or incidents. If they confirm the meaning of the information, the system needs not sound an alarm. If the system issues a warning even in a safe case, it might be harmful because nurses will ignore such meaningless messages in the future.

As we have already pointed out, the most important function is determination of risk. We assume that we can determine the point by analyzing nurses' conversations, but of course, before analysis, real-time automatic dictation of nurses' conversations is necessary. However, a more important matter is to generate a dictionary that can distinguish abstract from concrete or specialized terms and jargon from general terms. In the previous section, we presented a couple of examples to illustrate features of possible risk-leading terms as chances. Actually, it is not so difficult to distinguish abstract and concrete information and published dictionaries can be applied to generate the special dictionary. However, to generate a dictionary that can distinguish specialized terms and jargon from general terms is quite difficult, though, specialized terms and jargon can be extracted from various dictionaries or term sets. However, regarding general words, we need to collect critical words that may cause miscommunication. Although, some communication theory can be applied, we still need to collect more conversation examples for analysis to generate the dictionary.

4   Conclusions

In this paper we analyzed dialogues from clinical meetings where critical information on patients was discussed and exchanged. We categorized the words and phrases used in the conversations as [abstract/concrete] and [specialized terms, jargon/everyday word, general words, commonly used word] and demonstrated that those regarded as general or commonly used, and abstract phrases, play important roles in miscommunication. Considering the results, we proposed a prototype of an accident or incident determination system. In this paper, since we dealt with a small amount of dialogue data, we need to collect more data to perform more detailed analysis in order to determine more detailed reasons for accidents or incidents. We also need to generate a dictionary that can be used to determine critical words or phrases.

Acknowledgements

This research was supported in part by the National Institute of Information and Communications Technology (NICT).

References

[Abe 2004a] Abe A., Kogure K., and Hagita N.: Determination of A Chance in Nursing Risk Management, *Proc. of ECAI2004 Workshop on Chance Discovery, pp. 222–231* (2004)

[Abe 2004b] Abe A., Kogure K., and Hagita N.: Nursing Risk Prediction as Chance Discovery, *Proc. of KES2004, Vol. II, pp. 815–822* (2004)

[Abe 2005a] Abe A., Naya F., Ozaku H., and Kogure K.: Scenario Violation in Nursing Activities, *Proc. of ICML05 the 4th Int'l Workshop on Chnace Discovery* (2005) submitted

[HiyarihattoDB] Database of accident or incident reports, `http://www.hiyari-hatto.jp/`. (in Japanese)

[Iwahara and Hatta 2004] Iwahara A. and Hatta T.: Can We Encode Emotional Semantic Information in Written Message? — A basic study towards a development of a new miscommunication-free e-mail system —, *Proc. of PRICAI2004 Workshop on Language Sense on Computer, pp. 10–17* (2004)

[Kuwahara 2004] Kuwahara N. et al.: Ubiquitous and Wearable Sensing for Monitoring Nurses' Activities, *Proc. SCI2004, Vol. VII, pp. 281–285* (2004)

[Ohsawa 2002] Ohsawa Y.: Chance Discovery for Making Decisions in Complex Real World, New Generation Computing, *Vol. 20, No. 2, pp. 143–163* (2002)

Extracting High Quality Scenario for Consensus on Specifications of New Products

Kenichi Horie and Yukio Ohsawa

University of Tsukuba

Abstract. In this study, analysis has been executed on the scenarios for consensus on specifications of new products, which were extracted with the refinement of Double Helix Loop process for Chance Discovery. The aim of this analysis was to confirm that SCENARIO MAP, which refines KeyGraph by attaching macro photography of defects on the undefined names of the defects on its nodes, helps examinees to extract high quality scenarios through group discussion and to find the roles of SCENARIO MAP and Group discussion with its effect. 11 high quality scenarios for future proposal were extracted successfully after 104 scenarios. We found that SCENARIO MAP played the role to help examinees to identify the undefined defects and interpret the context of objective data and subjective data easily and the group discussion accelerated them to exchange their inherent experiences, know how and knowledge. The human centric process of SCENARIO MAP and group discussion can be regarded to give effect to extract these high quality scenarios.

1 Introduction

There is a growing interest to utilize an abundance of text files described in free format for a variety of purposes in enterprises so as to find new knowledge. It is emphasized to utilize the interaction of people to find new knowledge with data mining process because expected new knowledge has not been created yet only from the data mining tools with the text file described in free format. In this study, the Charged-Couple Device inspection equipment for defects appeared on the surface of related films for Flat Panel Display (FPD) was chosen because new proposals for new specifications have been demanded by our customers to improve the current performance of the equipment. The difficulty that confronted us is that non of new proposals have been created though a group discussion among a manager and technical sales persons has been held frequently based on common text files of customer call report and test report on the server in our company [Evans 87].

The Double Helix Loop process for Chance Discovery [Ohsawa 03] is human centric process with data mining tool. (Ref. Fig1) One helix is the process undertaken by human who forms the interpretation cycles, such as understanding of subject's own concerns and awareness and understanding of chances from the subject data (DM-a) and understanding of the chance in the environment from the objective data (DM-b). The other helix is a process undertaken by computer, receiving and mining the data. So, we adopted the Double Helix Loop process for the case at first but found that it was neither easy for us to interpret the context of KeyGraph [Ohsawa 98] nor to extract scenarios through group discussion. Because, the objective defects on 16 kinds

R. Khosla et al. (Eds.): KES 2005, LNAI 3681, pp. 1174–1180, 2005.

of related films for FPD are so many and each of them are slightly different in shape and size and are undefined in our sales group. Furthermore, the names of defects were determined between technical sales people and his customer and/or similar names for different defects sometimes were used in the customer call reports and test reports. It is also said that Double Helix Loop process takes a long time because both of objective data and subjective data shall be shown to people by turn and needs its efficiency for business application [Ohsawa 03].

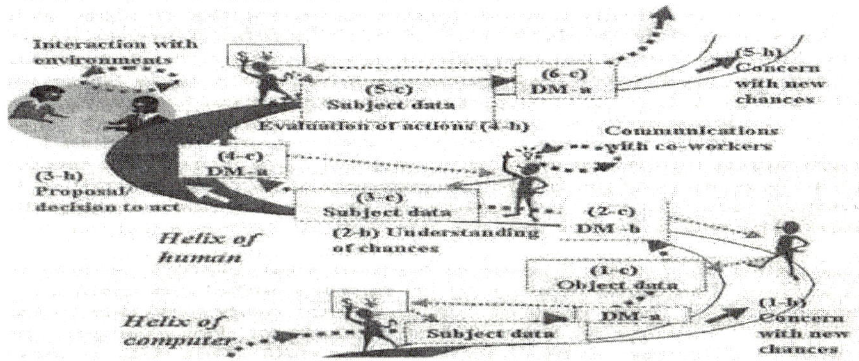

Fig. 1. Double Helix Loop process [Ohsawa 03]

The analysis is executed for the purpose of the followings.

(1) Confirming application of the following refined process to the case can be helpful to extract high quality scenarios
(2) Finding roles of SCENARIO MAP [Ohsawa 03][Usui 03], which is made by attaching macro photography of defects on the corresponded names of defects on nodes of KeyGraph and Group Discussion to extract high quality scenarios
(3) Finding effects of the refinements of the process to extract high quality scenario

2   The Refinements of Double Helix Loop Process

We consider modifying the Double Helix Loop process and adopt the following refinements of the process.

(1) Separation of all customer reports and test reports into 16 sets of the text data, each of which is written by one sales person for one kind of film.
(2) Creation of "SCENARIO MAPs" before group discussion, which identify the undefined defects in common by attaching macro photography of defects to the corresponded names of defect on nodes of KeyGraph.
(3) Skip individual scenario creation process for 16 sets of subjective data but execute after accumulating them in order to shorten the process time [Brooks 86].
(4) Govern the rule that all members in the group must articulate their own opinion and/or proposal per case to prevent inclining, connivance, and silence during group discussion [Festinger 57]. The extraction of scenarios created per case shall

be made by show of hands of all members by the end of set up time per case in order to control total planned process time.

Furthermore, Polaris [Okazaki 03], which is improved from KeyGraph, is used for data mining tool. Because it is capable visualizing word relations in a document, which aids user in creating scenarios. Scenario is defined as a context with time series. In KeyGraph, the word, which serves in an *island*, is extracted first on its appearance frequency and the degree of collocation with words to come in the same island. Further the degree of collocation of each word in the document and foundations and is calculated, and the words with high value are extracted and defined as *bridges* between *islands*.

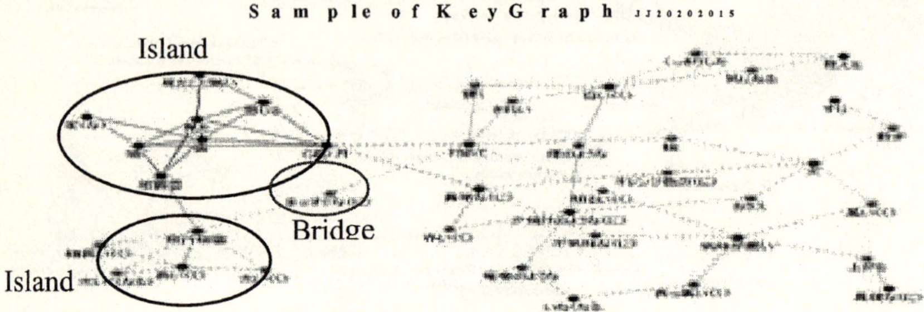

Fig. 2. KeyGraph sample

3   Experimentation of Refined Double Helix Loop Process

Firstly, we chose one (1) sales manager(SM), two (2) experienced technical sales person (ETS) and three (3) inexperienced technical sales people (ITP) in total of six (6) people. Secondly, 322 sheets of macro photography for 20 groups 62 kinds of defects were collected from customer call report and test report for 16 kinds of related films for FPD. Finally, the following process flow was adopted for the experiment.

(1) Objective data Pretreatment: Separate one (1) set of customer call report and test report for one (1) kind of films and prepare 16 sets of them individually.
(2) Data Mining: Output 16 charts of Polaris from 16 sets of them.
(3) SCENARIO MAP: Prepare 16 SCENARIO MAPs by attaching macro photography of defects to nodes with these objective names on the chart of Polaris.
(4) Showing SCENARIO MAP: Show 16 SCENARIO MAPs to all examinees on the screen one by one.
(5) Group Discussion: Utter individual interpretation of SCENARIO MAP or his own proposal and write the scenarios on white board. Delete, add or amend the scenarios while discussions till the end of set up time.
(6) Extraction of Scenarios (Subjective data): Select the scenarios by show of hands at the end of set up time and extract the scenarios approved by all of examinees.
(7) Data Mining: Output one (1) chart of Polaris from these subjective data.
(8) SCENARIO MAP: Prepare one (1) SCENARIO MAPs by attaching macro photography of defects to nodes of these objective names on the chart of Polaris.
(9) Repeat (4) ~ (6)

4   Analysis of Extracted Scenarios in the Experimentation

We analyzed the classification of scenarios first and the roles of SCENARIO MAP to create these scenarios secondly and evaluation them in the view of Usefulness, Recognition, and Consensus last.

4.1   Analysis of Classifications of Extracted Scenarios

We executed the experiment of 16 cases with refined Double Helix Loop Process.

104 scenarios were extracted from 16 SCENARIO MAPs. These 104 scenarios are classified to 85 present situations and 19 future proposals. (Ref. Table 1.) It was observed that 82 of 104 scenarios contain words of red nodes of these SCENARIO MAPs. (78.8%).

Table 1. Classfication of scenarios

Experiment	Present situations	Future proposals	Total
Scenarios from 16 SCENARIO MAPs	85	19	104
Scenarios from 104 scenarios		11	11

11 future proposals (Ref. Table 2) were extracted from SCENARIO MAP (Ref Fig 3), which is made from the time series of these104 scenarios. It was observed that only future proposals were created and extracted after 104 scenarios (subjective data).

Table 2. Example of 11 Scenarios extracted from 104 scenarios

- Use scattered fluorescent lumps with high frequency power supply in regular reflection and in transmission at two stages in order to inspect defects such as Chink, Pustule, Black dot, White dapple, Foreign body in general on PET, PS, Special film and coated film.
- Use metal halide lump/ or halogen lump in regular reflection and in transmission at 2 stages in order to inspect defects such as White dapple, Air bubble, Crenellation
- Use filter software together with lighting conditions in order to inspect defects such as Scratch, Spottiness, Line, in case it is difficult to adjust lighting conditions.
- Develop and propose a sheet feed conveyor with the inspection equipment to our customers in accordance with strong requirement from the market.
- Develop the software for image processing such as filter and macro together with lighting conditions in accordance with the tendency of customer's requirement to inspect micro defects such as Scratch, Spottiness, Line.

4.2   Analysis of Roles of SCENARIO MAP to Extract Scenarios

We analyzed 104 scenarios and measured Correlation matrix and Multivariate transformation figure for number of characters of scenarios and that of words of black nodes and red nodes contained in the scenarios per SCENARIO MAP.

Number of characters of 85 present situations has strong correlation 0.800++ with that of words of black nodes contained in these scenarios. (Ref. Table 3) But the number of characters of 19 future proposals does not have correlation with that of words of black nodes contained in the scenarios. Neither number of characters of 85 present situations nor that of 19 future proposals has correlation with each number of words of red nodes contained in these scenarios.

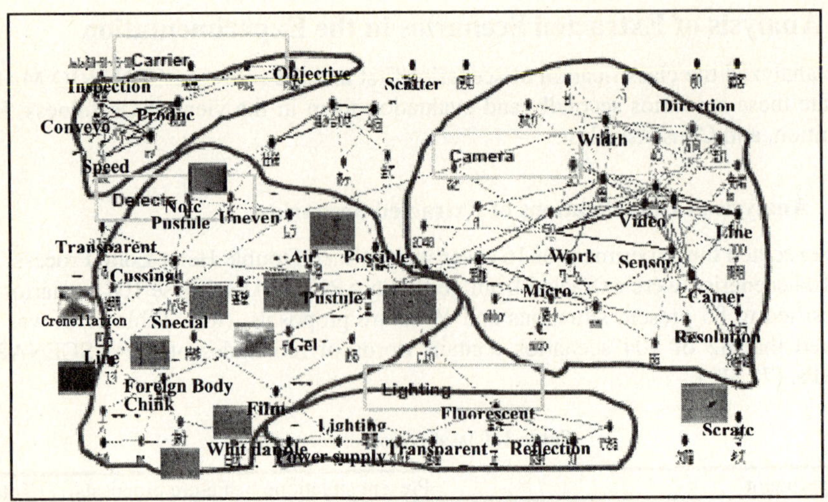

Fig. 3. SCENARIO MAP made from 104 scenarios

Table 3. 85 present situations correlation matrix

Variances	No. of Characters	No. of Black Nodes	No. of Red Nodes
No. of Characters	1.000	0.800++	0.529
No. of Black Nodes	0.800++	1.000	0.293
No. of Red Nodes	0.529	0.293	1.000

The analysis was executed for time series variance among number of characters of scenarios and numbers of words of black nodes and red nodes contained in 104 scenarios, which were extracted from each SCENARIO MAPs.

It was observed that the following common features in all of 10 SCENARIO MAPs, which extracted 19 future proposals. (100%) (Ref. Fig.4)

(1) The number of characters of the present situations decreases just before extracting a future proposal.

(2) The number of words of red nodes, which are contained in the present situations, decreases just before extracting a future proposal.

(3) In addition, we observed a following feature in 8 out of 10 SCENARIO MAPs, which extracted 19 future proposals. (80%)

(4) The number of words of red nodes, which are contained in the present situations, increases just after extracting a future proposal.

We confirmed from above analysis result that the black nodes in *Island* refer examinees to most of topics of SCENARIO MAP and help them to extract 85 present situations. We infer that examinees first paid attention to defects, which are defined in common on SCENARIO MAP, and moved their attention to *island*, which was linked to them. They referred to *bridges* and interpreted the topic of the *island* from the relation with these defects roughly. So, the number of characters of first scenario for present situations was low. Then, they started interpreting the topic deeply referring to the *island* with red nodes, which linked in isolation. So, the number of characters of the scenarios for present situations turned to be increased with that of red nodes once after interpreting the topic of another *island* deeply.

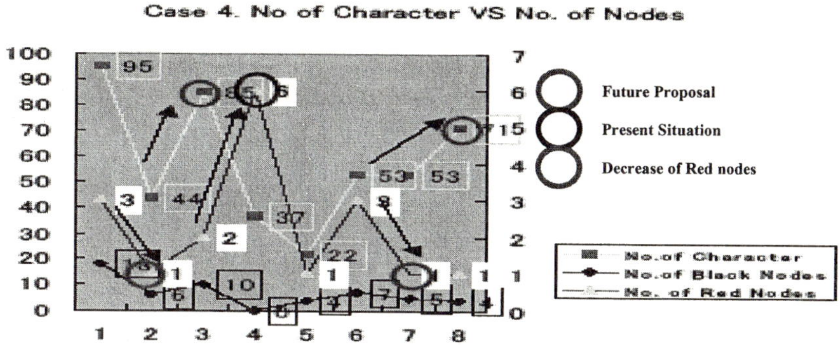

Fig. 4. Number of Character and Black & Red nodes in scenarios

Repeating the above process, examinees fully interpreted the context of SCENARIO MAPs and extracted scenarios on present situation first and then future proposal. After data mining 104 scenarios (subjective data) with Polaris, SCENARIO MAP helped all examinees to project their subjective thoughts clearly and to externalize future proposals with group discussion. .

4.3 Analysis of Evaluation of Extracted Scenarios

The evaluation was executed with full marks of 10 points by all examinees in the view of Usefulness, Recognition, Consensus and Contribution.

We use Mean Value (Mean) and Standard Deviation (SD) of Statistic and observed the following features.

(1) Usefulness: Sales manager marked the lowest score 4.7 points to 85 present scenarios but the highest score 7.2 points to 19 and 11 future proposals. The average score for present situations was marked low 5.4 points but high 7.0 and 6.8 points for 19 and 11 future proposals.

(2) Recognition: Experienced technical sales people marked the highest score between 8.8 and 9.3 points for all proposals. But inexperienced people marked the lowest score between 5.2 and 5.9 points for them.

(3) Consensus: All examinee marked higher score between 8.7 and 8.9 points for all proposals.

(4) Contribution: Sales manager marked the highest score 5.2 points for 85 present situations, 6.9 points for 19 future proposals and 8.1 points for 11 future proposals. The average score of all examinee increases 4.3 points for 85 present situations and 5.3 points for 11 future proposals

In addition, we observed that all SD for 11 future proposals are equal or lower than that of 19 future proposals among all categories.

5 Conclusion

In this study, 11 future proposals are regarded as high quality scenarios, which are evaluated highly (6.8 point) for usefulness and are practical for the real business proposal. These high quality scenarios were extracted from the refined Double Helix

Loop process:(1) Objective data Pretreatment,(2) Objective data on SCENARIO MAP after data mining with Polaris.(3) Group Discussion, (4) Extraction of Scenarios (Subjective data), (5) Subjective data on SCENARIO MAP after data mining with Polaris, (6) Group Discussion, (7) Extraction of Scenarios (Subjective data)

We concluded that SCENARIO MAP of 104 scenarios helped all examinees to project their subjective thoughts clearly. In addition, the group discussion stimulated their interest in knowing experiences; know how, knowledge, each other and awoke the consciousness of the possibility for several scenarios for future proposal. Thus, 11 future proposals were externalized and extracted successfully [Polanyi 83] [Nonaka 95]. It is, however, necessary for us to execute an action for 11 future proposals in business and evaluate the result in order to evaluate the performance of the refined Double Helix Loop process in business environment. Furthermore, it shall be our next theme to identify the mechanism that both number of characters of scenarios and that of words on red nodes decrease before extracting future proposals.

References

[Brooks 86] Brooks, R.A.," A Robust Layered Control Systems for a Mobile Robot", IEEE Transaction on Robotics and Automation 2, No.1, pp.14-23 (1986)

[Evans 87] Evans, J., "Bias in Human Reasoning: Caused and Consequences", Hillsdale (1987)

[Festinger 57] Festinger, L., "A theory of cognitive dissonance, Evanston, IL: Row, Peterson", (1957)

[Nonaka 95] Nonaka, I., and Takeuchi, H., "The Knowledge-Creating Company", Oxford University Press (1995)

[Ohsawa 98] Ohsawa, Y., Benson E. N., and Yachida, M., KeyGraph: Automatic Indexing by Co-occurrence Graph based on Building Construction Metaphor, Proc. Advanced Digital Library Conference (IEEE ADL`98), pp.2-18 (1998)

[Ohsawa 03] Ohsawa, Y., Modeling the Process of Chance Discovery, Ohsawa, Y. and McBurney P., Eds, *Chance Discovery*, Springer Verlag pp.2-15 (2003)

[Okazaki 03] Okazaki, N. and Ohsawa, Y.: "Polaris: An Integrated Data Miner for Chance Discovery" In proceedings of The Third International Workshop on Chance Discovery and Its Management, Crete, Greece (2003)

[Polanyi 83] Polanyi, M., *Tacit Dimension*, Peter Smith Pub Ind. (1983)

[Usui 03] Usui, M. and Ohsawa, Y.: Chance Discovery in Textile Market by Group Meeting with Touchable Key Graph, On-line Proceedings of Social Intelligence Design International Conference, Page
http://web.rhul.ac.uk/Management/News-and-Events/conferences/SID2003/tracks.html
(2003)

How Can You Discover Yourselves During Job-Hunting?

Hiroko Shoji

Faculty of Science and Engineering, Chuo University,
1-13-27 Kasuga, Bunkyo-ku, Tokyo, 112-8551, Japan
`hiroko@indsys.chuo-u.ac.jp`

Abstract. The authors collected the cases of job-hunting by female university students and analyzed their thinking process. The findings obtained is that reviewing the job-hunting log is sometimes effective in prompting self-discovery. Many previous studies have already shown that inspiration via interaction plays an important role in creative thinking process such as learning and problem solving. The cases presented in this study suggest that creative thinking process is observed also in job-hunting as an everyday activity. Therefore, the framework of creativity support studies can be applied to the support for job-hunters' thinking process as well.

1 Introduction

There are frequent government announcements of the upswing in business activities and the employment scene is seemingly getting better to some extent, however, university students often still have difficulty in their job-hunting process. Recently, the employment rate of new graduates of universities and high schools has decreased, resulting in a growing number of young part-time jobbers. The scarcity of employment for the younger generation has become a social problem. Coupled with the decrease in younger population, many universities are actively trying to survive the inter-university competition, by focusing on the support and guidance for job-hunters to feature their high employment rate. As an academic, the author has many opportunities to have close contact with students in job-hunting. Taking a closer look at their process, you find that some students can get an official job offer relatively soon regardless of the difficulty in finding employment, and others cannot for a long time. In spite of the same affiliation and the similar capabilities, their job-hunting processes have ended with very different results. This can hardly be attributed only to the difference in personal capability. In addition, some students can hardly convince themselves of a job offer obtained as a result of compromise, and others can be more positive toward job hunting after they have found another goal different from an initial one. Where does this difference come from? If you can give them a good clue, you could support their job-hunting process. The authors originally have been working on a microscopic observation of thinking process [2]. We thought that the findings to be obtained from an investigation of how individual student's activity and thinking process is different by person may provide a catalyst to know the approach and/or attitude to suitable job-hunting process for an individual person.

With this as a trigger, the authors have decided to have an interview with them about their job-hunting and log their activities, and have been working on it since

R. Khosla et al. (Eds.): KES 2005, LNAI 3681, pp. 1181–1185, 2005.

January of 2002. Targeted at students graduating in March of 2003, the first batch of investigations conducted since January of 2002 collected their job-hunting process log and made an analysis of their thinking process. The findings obtained are as follows: (1) The process characterized as "concept articulation" is sometimes observed in the mental world of students during job-hunting, and (2) Such concept articulation can lead them to find their own identity and therefore help them succeed in job-hunting [3]. This paper reports the analysis result of the data from the second batch of investigations conducted subsequently.

2   Collection of Job-Hunting Process Cases

This series of investigations aims at observing the thinking process of individual students in job-hunting. This first requires them to externalize their own thinking about job-hunting process. This study has had an individual interview with each student in job-hunting to log what they say and analyze the content. The second batch of investigations were targeted at eight third-year students in Kawamura Gakuen Women's University as of January of 2003 and conducted in January, April, July, and October. The five out of them had already got a job lined up during the investigations (by the end of October of 2003). The limitations of the number and selection method of the subjects prevent us from clearly stating the statistical significance of these investigations, however, tracing their behavior and/or thinking can serve as a useful reference to know the average job-hunting process and/or thinking of the university's students (or the female students in general).

The method used for the first batch of investigations was simple in that only listening to and logging what the subjects say were conducted every time. The second batch chose to have the subjects record (or log) their job-hunting process and think referring to their activity log during interviewing. The activity logging started one month before the first interview and had been recording their activities as appropriate. The second (conducted in April) and subsequent interviews in the second batch presented the subjects with what they said in the previous interviews and allowed them to express their view referring to the content as well as their activity log. This change in investigation method is based on the assumption that reviewing activity log and the content of previous interviews is effective in prompting concept articulation during job-hunting.

3   The Effect of Concept Articulation via Introspection

Among eight cases collected, one for a certain subject is described below. Due to limitations of space, only a part of it is included here.

1. The excerpt from the first survey interview (conducted in January of 2003):
   - I want to get a job that involves using computers. For example, software engineer who has contact with customers, or something. But now, no matter what type of job you are engaged in, you will use personal computers, won't you?
   - All I have done so far is requesting company brochures. I haven't had an interview with any company yet. So, initially I was confused as I had nothing to write in the log. Then, I chose to add to my log what I wrote in my "entry sheets" (i.e. a company-specific inquiry sheet about applicants themselves).

2. The excerpt from the second survey interview (conducted in April of 2003):
   - I have already submitted my entry sheets to many companies, participated in placement seminars, and had interviews with several companies. I'm working really hard, if I say so myself.
   - At school, I thought I excel at using computers, but comparing to other job-hunters in the science and engineering majors, I realized that's not true. In the meantime, I will get qualification as a system administrator.
   - On the day when I had an interview with a company, I often wrote my impression of the interviewer in the log. I'm ashamed that most of the content is what has got me and complaints.
   - Reviewing my self-introductory documents made for a variety of companies brought me to finding out that the content varies with the company. I feel that I wrote something I think desirable to the company. This may be necessary for passing a screening by the company, but I feel that I may be wrong.
   - From now on, I want to stop saying a different thing depending on the company targeted and be vocal about what I really think.
3. The excerpt from the third survey interview (conducted in July of 2003):
   - After the second survey interview, I found myself at a loss as to what to write in my self-introductory document and job-hunting log. Rereading my log has made it clear that "computer" is a keyword. The other dominant content is my impression of persons I met. Especially, the words "fostering" and "teaching" frequently appear.
   - I inclined to be engaged in a job that involves not only computers but also education and training. When I talked about this during an interview with a company in May, the interviewer suggested to me that I be an instructor as an example, and I got serious about it.
   - I got an official job offer as an instructor from two companies (between May and June).

At the second survey interview, this subject reviewed her own job-hunting log to find out that "the content written in her entry sheets varies with the company" and think why she cannot write in self-advertisement and/or the job she wants although she should do so (see Figure 1). This provided her with a clue to get conscious of what she want to do and what her aptitude and strength are, however, the unclarity of the concept temporarily put her at a loss as to what to write in the document. After reviewing the log to check for the frequently used words and topics, she got conscious that "computer" and "fostering and teaching" are important topics. In an interview she had after gaining an understanding of this concept, a job as an "instructor" came up in conversation and consequently she decided that she should aim at being an instructor (see Figure 2). Later on, she was hunting a job with a focus on an instructor to finally get an official offer as she wish without delay.

This case shows that viewing job-hunting process log which is an externalization of your thought promotes the dialogue with yourself and leads you to concept articulation of yourself, i.e. self-discovery.

Many previous studies have already shown that inspiration via interaction plays an important role in creative thinking process such as learning and problem solving. For example, Schoen mentioned through an analysis of architect and psychotherapist cases that the dialogue with professionals themselves is important for their thinking process [1]. He also showed that viewing externalizations of your thought, such as sketches and documents, can prompt the dialog with yourself, i.e. introspection. Suwa, on the other hand, scrutinized how architects interact with their own sketches to find out that the interaction with sketches is effective in prompting them to reinterpretation and unexpected discovery [4]. Interestingly, the observation in this study suggests that a similar phenomenon to the ones addressed by previous studies with a focus on professional tasks, including Schoen's and Suwa's, does occur also in a non-professional task of university students' job-hunting.

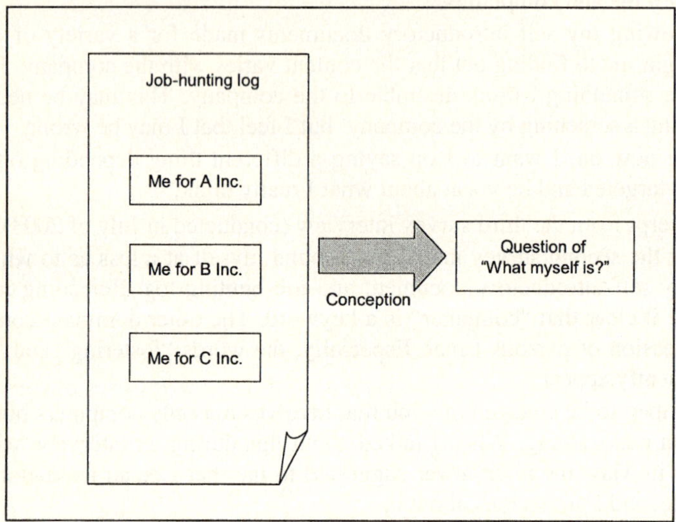

Fig. 1. The effect of introspection via activity log (1)

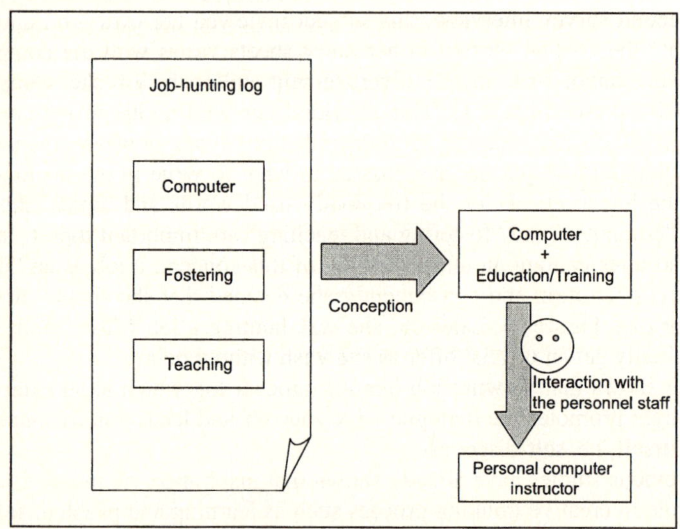

Fig. 2. The effect of introspection via activity log (2)

4 Conclusion

The authors collected the cases of job-hunting by female university students and analyzed their thinking process. The findings obtained is that reviewing the job-hunting log is sometimes effective in prompting self-discovery. The authors want to address an effective method for supporting self-discovery and concept articulation during job-hunting as well as to make a more detailed analysis of the cases in the future.

Not a few job-hunters are forced to continue their activity for a long time and/or make a compromise on employment conditions. This is definitely a bitter experience for them, however, this process with much frustration for compromise involved may, on the contrary, provide them with an opportunity for concept articulation. Hopefully, job-hunters who have only vague concept or requirements will take advantage of an interaction process called "job-hunting" to lead to self-discovery. Those who could successfully do concept articulation (i.e. self-searching or self-discovery) during job-hunting are expected to be able to address their job aspiringly with higher motivation, even if they work for a company that doesn't match their initial wishes. That event is exactly an example of value creation realized through interaction.

Actual employment assistance provided by most of employment bureaus in universities, on the other hand, includes only uniform services intended for all the students (job-hunters), such as providing corporate information, recommendation of taking qualification exams, cultivation of academic capability for passing written exams, and conveyance of know-how. Any one of these support services may be useful for job-hunting as problem-solving with a problem to be solved made clear, however, is not necessarily likely to help job-hunters who need concept articulation during self-searching. Employment assistance in the future should introduce concept articulation, i.e. a framework for supporting job-hunting as a creative thinking process. The findings from this study, which suggest a possibility of applying a framework of creativity support study to the support for job-hunters' thinking, can serve as a reference to explore a new job-hunting support method.

References

1. Schoen, D.A., The Reflective Practitioner: How Professional Think in Action, Basic Books, 1983.
2. Shoji, H., and Hori, K., "Creative Communication for Chance Discovery in Shopping," New Generation Computing, Vol.21, No.1, pp.73-86, 2003.
3. Shoji, H., "How Can You Facilitate Concept Articulation during Job-hunting?" The 2005 International Conference on Active Media Technology, 2005. (to appear)
4. M. Suwa, T. Purcell, and J. Gero, "Macroscopic analysis of design processes based on a scheme for coding designers' cognitive actions," Design Studies Vol.19, No.4, pp.455-483, 1998.

An Extended Two-Phase Architecture for Mining Time Series Data

An-Pin Chen, Yi-Chang Chen, and Nai-Wen Hsu

Institute of Information Management, National Chiao Tung University,
Hsinchu, Tiawan, 1001 Ta Hsueh Road, Hsinchu, Taiwan, R.O.C.
{apc,ycchen,sosorry}@iim.nctu.edu.tw

Abstract. Time series data vary with time. In the past, most of the researches focused on the matching of feature points or measuring of the similarities. They could successfully represent the feature patterns in a visualized way. In the mean while, those researches did not sufficiently describe the results in simple and understandable words. In this research, a two-phase architecture for mining time series data is introduced. By combining some different mining techniques, the difficulties mentioned above may be overcome. This architecture mainly consists of Exploratory Data Analysis (EDA) and techniques related to mining association rules. After the phase I analysis, quantitative association rules are obtained by phase II. Meanwhile, the rules of the architecture are able to be verified by accuracy analysis. Finally, a result of comparison with the traditional data mining techniques and this architecture shows that the two-phase architecture is superior to traditional techniques to the time series data.

1 Introduction

In recent year, mining time series database becomes more and more important in database research field. Many methods have been proposed to query time series data for similarity by their representation, including Fourier transforms, wavelets transforms and symbolic mappings [7]. Most of them are emphasized on the pattern recognition or the similarity measure [1], [11]. Until the data mining development, mining association rules is a new technique to identify the frequent patterns of information in a database. Most researches on mining time series data focused on indexing, clustering, classification and segmentation [9]. Although they could improve the accuracy and efficiency of execution, they did it in a classical analysis way. Typically, the classical analysis approach was aimed to build a model for analyzing data, but exploratory data analysis was instead to analyze data for constructing a model. For instance, Bezdek et al. mentioned that the advantage of exploratory data analysis (EDA) is to infer what model would be appropriate in advanced. The clustering is just one step of EDA procedure [2]. In order to aid human analysts, the result of EDA is usually used for visualization. Consequently, it is difficult to obtain an intuitive interpretation in a high-dimensional space. Therefore, association rules mining forms a simple expression to represent the relation between items.

The mining technique is traditionally used to find the relation among items in the supermarket. Latterly, some algorithms have been proposed to deal with quantitative attributes to adapt more generalized date set [10], [13], [16]. According to them, inter-transaction association rules, a concept reversed to mine the association among

R. Khosla et al. (Eds.): KES 2005, LNAI 3681, pp. 1186–1192, 2005.

items within the same transaction, extracting from different transaction records could reveal more useful information [17]. However, it may be limited only in categorical data. By association rules mining on inter-transaction records and interval data, the method has been generally provided to mine different kinds of data-sets such as time series data.

As regards time series data, especially the stock data, many researches doing analysis just are focused on only one variable, price change. However, their results are not sufficient to conclude something but proven that some information content might be hidden according to the trend of trading volume [4]. Even such as Karpoff and Moosa et al. have merely provided evidences on supporting positive correlation between the absolute value of price change and the trading volume [8], [14]. Therefore, we tried to investigate a two-phase architecture to mine more accuracy rules from stock trading information to help investors making better decisions.

2   The Concept of Two-Phase Architecture

In this research, a two-phase architecture is proposed to explore rules relating the behaviors of patterns within a sequence of time series data. EDA is adapted in the first phase to aggregate sequence with the same properties. In order to minimize the impact of the noise data, kernel smoothing is used for extracting feature points of each sequence. The second phase is to find the quantitative association rules from sets of feature points in the same cluster. And the two-phase architecture will be introduced in the following section. It includes phase I: Exploratory Data Analysis and phase II: Mining Quantitative Inter-Transaction Rules (ITARs), shown as Figure 1.

2.1   Phase I: Exploratory Data Analysis

In this phase, the relevant hiding information would be extracted and quantized from the raw data, and following steps are continuities of EDA which are different with traditional clustering [2]:

a. Assessment of clustering tendency attempts to evaluate raw data, if it has structure. Before evaluating, the initial number of clusters has to be decided. Some statistic indexes could be adopted to determine the initial number of clusters in the hierarchical or crisp clustering.
b. Clustering is a method for grouping unlabeled data which is set of units with the same properties. By clustering, the difference between clusters could be easily distinguished.
c. Clustering validity makes that clustering methods could be quantized. A good clustering is a unit set with the high homogeneity intra cluster and the high heterogeneity inter clusters. With the different kind of analysis, the different clustering validities are used.

Owing to the noise data or considering the performance of model, smoothing method could be utilized after phase I. In general, pattern recognition techniques are used to extract the nonlinear patterns from noise data and to identify the movement of time series data. In order to automatically detect the patterns, a classical statistical estimator, called smoothing estimator, is suited to find the extremes. Such as, kernel

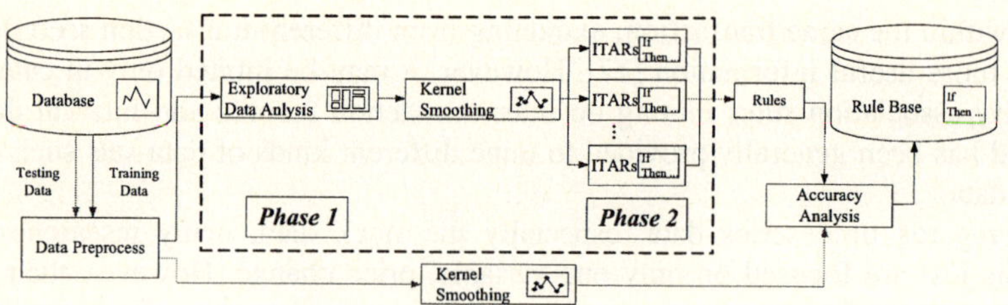

Fig. 1. Two-phase architecture

smoothing is the general way [11]. Therefore, the original data sequence could be represented by feature points, and next the complexity would be reduced.

After the phase I, the units with the same properties will be aggregated in the same cluster, and each cluster will be different. Each cluster still represents a set of time series data. As regards kernel smoothing, the purpose of applying it and extracting feature points is to reduce the computing time and to get the more accurate result.

2.2 Phase II: Mining Quantitative Inter-transaction Rules

In classical mining methods, mining association rules is just focused on interesting relations between categories in transaction database. The relation from x to y is usually represented as x->y to describe the condition and inference result by simple words [5]. It is not sufficient so that this phase combines two mining concepts, mining quantitative association rules and mining inter-transaction rules.

a. Quantitative association rule mining is adapted to the database with quantitative attributes. It could be solved with Apriori method by mapping the quantitative association rule problem into the Boolean association rule problem. It works by partitioning the value of quantitative attributes into interval, but the problem of deciding the interval width and the number of partitions is produced in the same time. Generally, using a common statistics measure is a distance-base approach to the problem. The well clusters could be formed by finding a set of N clusters that minimize a given distance metric. The number of units in the cluster is defined as the frequency. By deciding the density threshold d, and the frequency threshold s, the clusters are ensured sufficiently dense and frequent.

b. Mining inter-transaction rules is the technique to break the barriers of the transaction and the scope. Mining intra-association rules is concluded to the relations among items within the same transaction. Following the existence of temporal relations between transactions, the prediction or any extended analysis could be realized.

3 Experiment

As Figure 1 shows, the original dataset would be separated into two parts, one part is training set which is used to generate rules drawn inside the dotted line rectangle, and the others, testing one will be utilized to test the accuracy of the result rules. Finally, the valid rules would be stored into rule base. In the following experiments, the traditional architecture would be compared with the proposed architecture.

The simulation dataset is collected from Taiwan Stock Exchange Co., Ltd (TSEC). There are totally 480 records of the transaction volume during Jul. 2002 and Jun. 2004. Each of those records includes 270 daily ticks. The first 378 records are utilized as the training set, and the remainder is the testing set.

3.1 The Detail Methodologies of Proposed Architecture

In phase I, Self-Organizing Map (SOM) is employed for cluster analysis because of its excellent exploration in financial field [3][6]. The well-known validation method, Dunn's index [2], is to identify whether the clusters are compact and well-separated. RMSSTD is adopted to determine the number of clusters existing simultaneously in the data set [15].

While iterating SOM in clustering records, the clustering result in each number of clusters having maxima Dunn's index is chosen to be the representation. Figure 2.a shows Dunn's index of the representative cluster in each number of clusters. To determine the best number of cluster, the 48th representative cluster having maxima change rate of RMSSTD is recognized as the best result, shown as Figure 2.b.

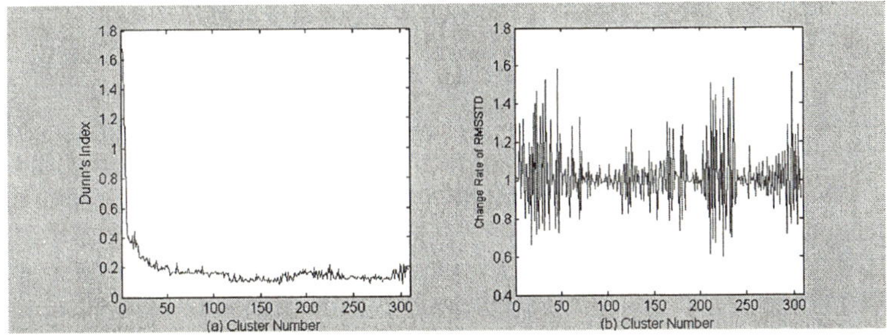

Fig. 2. (a) Dunn's index in each cluster. (b) The change rate of RMSSTD in each cluster

In the clustering result, kernel smoothing is applied to extracting feature points. A meaningful pattern could be represented by five points [12]. This argument is realized by sliding window and breaks the barrier of transaction. Hence, a unit with n feature points could be expressed as (n-5) sliding windows. Mining inter-transaction rules is applied to figure out the relations between each two feature points of sliding window. Therefore, all sliding windows in each unit of the same cluster are selected.

In order to mine the association rules from quantitative data, the data should be partitioned into the corresponding interval in advance. While mapping into the interval, Apriori algorithm is proper one to find quantitative association rules. If intervals are too large, the rules are meaningless. If intervals are too small, the few rules may not have the sufficient support. The distance-based interval could be decided by two criteria to evaluate whether the cluster is sufficiently dense and frequent. Afterward, the result is shown as Figure 3.a.

Some rules are found by mining inter-transaction association rules over interval data, shown as Figure 3.b. However, the robustness of these rules should be tested. A unique rule may occur only a few times, so this kind of rules is eliminated in this experiment. For this reason, the remainder rules are verified the accuracy by the testing set.

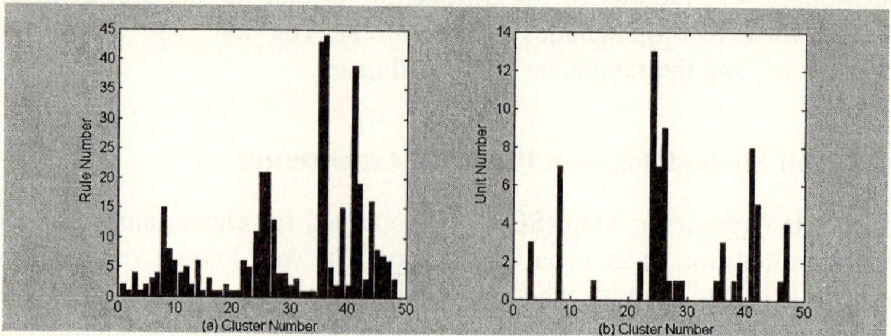

Fig. 3. (a) The number of units in each cluster. (b) The number of rules in each cluster

The accuracy rate is defined as a percentage to measure the occurrence of rules in the testing data. Figure 4 shows the accuracy of rules which are found from two-phase architecture and traditional architecture separately.

3.2 Result Comparison and Discussion

In this section, three criteria are introduced to evaluate the ability of analyzing time series data: (i) the number of rules, (ii) the accuracy of rules, and (iii) the effectiveness of rules. An intuitive measure of the number of rules is to compare with how many rules are found. The accuracy of rules is to evaluate how well the rules would be in the test data quantitatively. How effectiveness of the rules is to explore what information in the rules.

The first criterion is illustrated by Figure 4, which shows that 70 rules are figure out by two-phase architecture in one cluster. But in traditional architecture, only 7 rules are found while the training data are separated into different cluster, shown by the different color in Figure 4.b. According to the criterion (ii), 22 rules that were exceeded 70% accuracy that were discovered by two-phase architecture in the Figure 4.a. However, only 2 rules exceed the same threshold by traditional architecture, shown as Figure 4.b. Finally, criteria (iii) examined the rules which were found from these two kinds of architecture. Two example rules are illustrated as table 1 and 2:

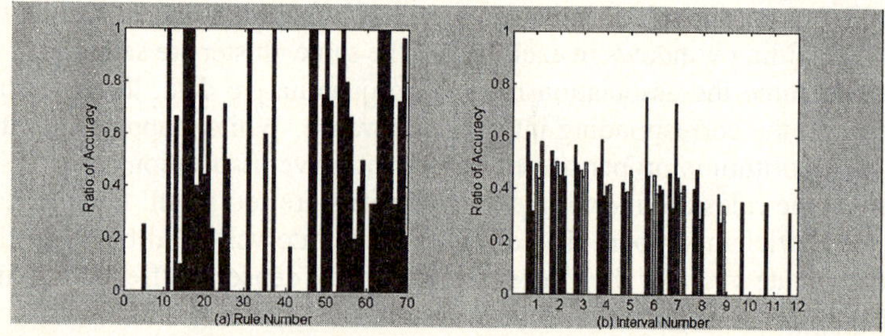

Fig. 4. (a) The ratio of accuracy of each rule in two-phase architecture. (b) The ratio of accuracy of each rule in traditional architecture

Table 1. The presented rule 1 from the training data is separated into 13 intervals

Output rule	Rule explanation
If (FP_1=4 and FP_2=1 and FP_3=5 and FP_4=2) Then FP_5=4	If ($0.0766<FP_1<0.0938$ and $0<FP_2<0.0439$ and $0.0938<FP_3<0.1122$ and $0.0439<FP_4<0.0595$) Then ($0.07665<FP_5<0.0938$)

Table 2. The presented rule 2: the training data is separated into 7 intervals

Output rule	Rule explanation
If (FP_1=1 and FP_2=1 and FP_3=1 and FP_4=2) Then FP_5=1	If ($FP_1<0.0609$ and $FP_2<0.0609$ and $FP_3<0.0609$ and ($0.0609<FP_4<0.1014$) Then ($FP_5<0.0609$)

FP_i is denoted to the ith feature point. $FP_i = j$ means the ith feature point that is mapped into the jth interval. $p < FP_i < q$ is the interval of normalization range.

The interval size of rule 1 is small, compared with rule 2, so the inference would more precise than rule 2. Rule 1 displays a "W" shape which infers the high trading volume accompanying a positive price trend in the next feature point. Therefore, rule 1 provides more information content and more precise than rule 2 does. The rules found by two-phase architecture are the same as rule 1, while the rules found by traditional architecture are like rule 2. According to the result of experiments, the two-phase architecture is more superior to the traditional architecture by three criteria.

4 Conclusion

The proposed two-phase architecture could discover the interesting rules which are quantitative, valid and predictable. Time series data, especially the stock data, could be analyzed by the techniques of EDA, the feature extraction and the association rules mining. In the comparison, the proposed architecture could be more outperformed to the traditional architecture by generally three-criterion evaluation. In short, meaningful analysis or prediction on the trend by the proposed two-phase architecture would be better. As regards the detailed methodologies, more efficient ones could be superseded in the future.

References

1. Agrawal, R., Lin, K., Sawhney, H. S., Shim, K.: Fast Similarity Search in the Presence of Noise, Scaling, and Translation in Time-Series Databases. In Proc. of the 21st VLDB Conf. Zurich, Switzerland (1995) 490-501
2. Bezdek, J.C., Pal, N.R.: Some New Indexes of Cluster Validity. IEEE Transactions on Systems, Man and Cybernetics – part B: Cybernetics, Vol. 28, No. 3. (1998) 301-315
3. Blazejewski, A., Coggins, R.: Application of Self-Organizing Maps to Clustering of High-Frequency Financial Data. In Proc. of AWDM&WI2004. Dunedin, New Zealand. CRPIT, 32. Purvis, M., Ed. ACS (2004) 85-90
4. Blume, L., Easley, O'Hara, D.M.: Market Statistics and Technical Analysis: the Role of Volume. J. of Finance, Vol. 49, No. 1. (1994) 153-181

5. Das, G., Lin, K., Mannila, H., Renganathan, G., Smyth, P.: Rule Discovery from Time Series. In Proc. of the 4th Int'l Conf. on KDDM1998, New York (1998) 16-228
6. Deboeck, G., Kohonen, T.: Visual Explorations in Finance with Self-Organizing Maps., Springer Finance, London (1998)
7. Han, J., Kamber, M.: Data Mining: Concepts and Techniques. The Morgan Kaufmann Series in Data Management Systems, Morgan Kaufmann Publishers. (2000)
8. Karpoff, J.M.: The Relation between Price Changes and Trading Volume: a Survey. J. of Financial and Quantitative Analysis, Vol. 22, (1987) 109-126
9. Keogh, E., Kasetty, S.: On the Need for Time Series Data Mining Benchmarks: a Survey and Empirical Demonstration. In Proc. of SIGKDD2002, Vol. 7, No. 4. (2002) 349-371
10. Lent, B., Swami, A., Widom, J.: Clustering Association Rules. In Proc. of 13th Int. Conf. on Data Engineering. (1997) 220-231
11. Lo, A.W., Mamaysky, H., Wang, J.: Foundations of Technical Analysis: Computational Algorithms, Statistical Inference, and Empirical Implementation. J. of Finance, Vol. 55, No. 4. (2000) 1706–1765
12. Levy, R.A.: The Predictive Significance of Five-Point Chart Patterns. J. of Business, Vol. 44, No. 3, (1971) 316-323
13. Miller, R.J., Yang, Y.: Association Rules over Interval Data. In Proc. of SIGMOD Conf. 97 Tucson, Arizona, USA (1997) 452–461
14. Moosa, I. A., Al-Loughani, N. E.: Testing the Price-Volume Relation in Emerging Asian Stock Markets. J. of Asian Economics, Vol. 6, No. 3. (1995) 407-422
15. Sharma, S.: Applied Multivariate Techniques. John Wiley & Sons, Inc., New York (1995)
16. Srikant, R., Agrawal, R.: Mining Quantitative Association Rules in Large Relational Tables. In Proc. of Int. ACM-SIGMOD Conf. Management of Data, Montreal Canada (1996) 1-12
17. Tung, A.K.H., Lu, H., Han, J., Feng, L.: Breaking the Barrier of tTransactions: Mining Inter-Transaction Association Rules. In Proc. of Int. Conf. on KDD'99, San Diego, USA (1999), 297-301

A Performance Comparison
of High Capacity Digital Watermarking Systems

Yuk Ying Chung[1], Penghao Wang[1], Xiaoming Chen[1], Changseok Bae[2],
Ahmed Fawzi Otoom[1], and Tich Phuoc Tran[1]

[1] School of Information Technologies, University of Sydney, NSW2006, Australia
[2] Post-PC Platform Research Team, Electronics and Telecommunication Research Institute,
Daejeon, Korea

Abstract. Material in music, film, book and software publishing industries can be easily duplicated on a large scale by personal computers, and unauthorised or illegal copies have become a serious problem in recent years. In this situation the digital watermark can be used as a direct method of preventing unauthorised copying. This paper presents a novel digital watermarking system based on Vector Quantisation (VQ) and Discrete Cosine Transform (DCT), which can embed a grey level image. The proposed new system can embed 16 times more information than other traditional DCT based watermarking systems. The recovered images have high Peak Signal to Noise Ratio (PSNR) value and good visual quality. This system is robust in terms of both JPEG compression and other signal processing attacks.

1 Introduction

A digitally watermarked image is obtained by invisibly concealing a signature information within the host image. The signature is recovered using an appropriate decoding process. The challenge is to ensure that the watermarked image is perceptually indistinguishable from the original and that the signature is recoverable even when the watermarked image has been compressed or transformed by standard image processing operations. There are two types of digital image watermarking schemes: one based on spatial domain and the other on frequency domain. Here the frequency domain approach is used. The method by which the watermark is embedded into the DCT coefficient is described.

Many studies have discussed one-dimensional watermarks [1][2] and two-dimensional watermarks [3][4]. A one-dimensional watermark is a set of ID numbers which has less perception and information about copyright ownership when compared to two-dimensional watermark. This paper describes a novel grey-level digital watermark embedding algorithm that provides more perceptual information compared with a binary digital watermark. A grey-level digital watermark is first decomposed into a series of binary digital images for implementing multiple watermarking. The multiple watermarks can be inserted into any region of an original image. Discrete cosine transform (DCT) [5] coding is one of the common algorithms used in watermarking, based on one-dimensional watermarks or two-dimensional binary digital watermarks. The multiple watermarks are then inserted into the DCT transform of the original images.

R. Khosla et al. (Eds.): KES 2005, LNAI 3681, pp. 1193–1198, 2005.

In this paper, the signature image was first quantised using vector quantisation with the Linde-Buzo-Grey (LBG)[6] algorithm in order to hide larger amounts of watermark data. Binary bits of watermark were embedded directly into the Least Significant Bit (LSB) [7][8] of DCT coefficients in an image. Using the duplication of copies of the watermark provides further robustness against the attacks of signal processing techniques. To extract the watermark, the LSB can be taken out and the VQ encoded watermark data is decoded by the VQ decoder using source codebook. This method is self-extractable.

Section 2 gives a brief description of watermarking procedures. Section 3 and 4 give the experimental result and conclusion respectively.

2 The Embedding and Extraction of Grey-Level Digital Watermarking

a. Embedding Digital Watermark Procedure

In this paper, the vector quantisation scheme is used for compressing the signature image. The compressed indices are injected into the DCT coefficients of the host image. The host data is first DCT [5]-transformed. To hide large amounts of data, the signature image is first quantised using vector quantisation with the Linde-Buzo-Grey (LBG) [6] algorithm. Vector quantisation transforms the vectors of data into indices that represent clusters of vectors.

A grey-level digital watermark is decomposed into a series of binary bit planes by bit decomposition. The 8-bit grey-level image has been divided into 8 planes (b7, b6, b5, b4, b3, b2, b1 and b0), each binary bit plane of the watermark can implement multiple watermarking. The embedding procedure first segments the eight original images (I_1, I_2, I_3, I_4, …I_8) into 8 x 8 pixels blocks and each block is DCT transformed independently. The eight bits grey-level digital watermark is first quantised using Vector Quantisation (VQ); the VQ encoded signal will be embedded into the eight host transformed images. The output eight images from multiple watermarks embedding process will have the inverse DCT transformation. The output eight images (I_1', I_2', I_3', I_4', …I_8') are the watermarked images.

The embedding procedure is as follows:

1. Eight images (I_1, I_2, I_3,….I_8) will be divided into 8x8 blocks, the DCT being transformed independently;
2. The eight bit grey-level digital watermark is first quantised using Vector Quantisation (VQ);
3. The eight bit VQ encoded signal is divided into 8 planes (b7, b6, b5,…b0);
4. The LSB of each modified DCT coefficient is replaced by the watermark bit streams in each block;
5. The output images from step 4 will have IDCT transform and
6. The output images from step 5 are the watermarked images (I_1', I_2', I_3', … I_8').

b. Extracting Digital Watermark Procedure

For recovery of the hidden data, the watermarked image is first segmented into 8 x 8 blocks, and the segmented image will then have the DCT transformation. The transformed image is subtracted from the coefficients. The algorithm can extract the watermark in self-extractable way. The extracting method is as follows:

1. Eight images (I_1', I_2', I_3',….I_8') will be divided into 8 x 8 blocks and have the DCT transform independently;
2. Copy the LSB of each modified DCT coefficients in each transformed image from step1;
3. The 8 bits VQ encoded watermark data from step 2 will be combined together to form 8 bit VQ indices;
4. The grey-level watermark can be decoded by a VQ decoder using the VQ codebook and the encoded 8 bit indices and
5. The host images (I_1, I_2, I_3, … I_8) can be obtained after the IDCT transform process.

3   Results and Discussion

The Peak Signal to Noise Ratios (PSNR) is used to measure the distortion of watermarked image.

$$PSNR = 10\log_{10}[\frac{255^2}{\frac{1}{M \times N}\sum_{i=1}^{M}\sum_{j=1}^{N}(x_{i,j} - x_{i,j}')^2}]$$

Where $x_{i,j}$ and x_{ij}' denote the pixels of the original and reproduced images and the images are the size $M \times N$.

Different images and grey level signature images were used in testing this algorithm. The test images used in the work are shown in Figure 1. The host image *Boat* is 512 x 512 grey level, and the watermark image *Elaine* is 256 x 256 grey level. Two set of testing were done. Part (a) shows the testing results from the DCT digital watermarking system without using VQ algorithm; part (b) shows the testing results of DCT and VQ digital watermarking system.

(a) Boat 512 x 512

(b) Elaine 256 x 256

(c) Recovered host image Boat after DCT and VQ watermark with (PSNR) 46.36dB replacing 5th LSB in Table 1.

(b) Extracted watermark image Elaine with 28.12dB replacing 5th LSB in Table 1.

Fig. 1. Host image **Boat** with 512x512 grey-level, watermark image **Elaine** with 256x256 grey-level

(a) Results for the DCT Digital Watermarking System (Without Using VQ)

Table 1 and Table 2 show the result of replacing different bits (2nd lsb, 3rd lsb, 4th lsb, 5th lsb and 6th lst) in host image by watermark information. Table 1 shows the result for the recovered host image and Table 2 shows the result for the extracted watermark image.

Table 1. The recovered host image with different bit replacement

Position	PSNR (in dB)	RMSE Root Mean Square Error	MSE Mean Square Error	NMSE Normalized Mean Square Error
2nd	54.518	0.479	0.230	0.0109
3rd	50.855	0.730	0.534	0.0252
4th	**45.834**	**1.300**	**1.689**	**0.0796**
5th	40.165	2.500	6.260	0.2950
6th	34.245	4.946	24.46	1.1530

Table 2. The extracted watermark image with different bit replacement

Position	PSNR (in dB)	RMSE Root Mean Square Error	MSE Mean Square Error	NMSE Normalized Mean Square Error
2nd	24.345	15.460	239.11	10.517
3rd	34.383	4.869	23.703	1.043
4th	**44.060**	**1.598**	**2.553**	**0.112**
5th	34.610	4.740	22.50	0.989
6th	51.839	0.653	0.426	0.0187

(b) Results for the DCT and VQ Digital Watermarking System

Table 3 shows the test results from replacing different bits (1st Least Significant Bit *LSB* to 8th *LSB*) in the host image with watermark information. From Table 3, it can be seen that the best result can be obtained by replacing the 5th LSB.

The quality of the recovered host image and the extracted watermark can be found in Figures 1(c) and 1(d): Figure 1(c) is the recovered host image *Boat* with 46.36dB and Figure 1(d) is the extracted *Elaine* watermark with 28.12dB.

Table 3. The test results from replacing different bits (1st Least Significant Bit *LSB* to 8th *LSB*) in the host image with watermark information

Least Significant Bit *LSB*	Recovered host image with VQ and DCT		Extracted watermark image	
	PSNR in dB	*MSE* Mean Square Error	*PSNR* in dB	*MSE* Mean Square Error
1	63.09	0.128	18.7	882
2	60.48	0.234	24.5	232
3	56.94	0.53	27.12	127
4	52.14	1.6	28.03	101.8
5	**46.36**	**6.06**	**28.12**	**100**
6	40.49	23.4	27.58	114
7	34.51	92.06	23.52	291
8	28.57	364.9	18.62	900

In order to test and verify the robustness of the new watermarking algorithm [9], the JPEG compression and other signal processing attacks were tested. In this work, it was found that the use of Pseudo Random Number Generator *PRNG* in the watermarking system not only increased the security but also increased the quality of the recovered information after some distortion. Figure 2, Figure 3 and Figure 4 show the result of cropping distortion test. The results of Figure 3 (c) with VQ and PRNG is better than Figure 2(c), Figure 4(c) with VQ and PRNG is better than Figure 10(b).

(a). watermarked image

(b). watermarked image cropped

(c). watermark image retrieved

Fig. 2. Cropping test for watermarked image without VQ and PRNG

(a). watermarked image with PRNG + VQ

(b). watermarked image cropped with PRNG + VQ

3(c). watermark image retrieved with PRNG + VQ

Fig. 3. Cropping test for watermarked image with VQ and PRNG

Original host image

Watermark image
Figure 4 (a)

Modified watermarked image

(b). Retrieved watermark without VQ

(c). Retrieved watermark with VQ

Fig. 4. Cropping test for grey-scale watermarked image with VQ and PRNG

4 Conclusion

In this paper, a novel digital watermarking technique was presented which can embed the grey level image. In the proposed method, the digital watermark is first quantised using VQ with LBG [6] algorithm. VQ transforms the vectors of data into indices that represent clusters of vectors. Then the vector is decomposed into a series of binary digital images for implementing multiple watermarks. The decomposed watermark image is embedded into the DCT[5] domain by modifying DCT coefficient values. The technique presented here is a new technique that has the ability to hide up to 25 per cent of host image size data in the host image—16 times more than other traditional DCT based watermark systems.[7][8] Section 3 shows the result of embedding watermarked image into different bits. The recovered images Boat and Elaine have high PSNR value and good visual quality. The test results prove that this digital watermarking system is robust in countering both JPEG and other signal processing attacks.[9]

References

1. R. B. Wolfgang, C. I. Podilchuk, I. J. Delp, "Perceptual Watermarks for Digital Images and Video", Proceedings of the IEEE, Vol. 87, No.7, pp.1108-1126, July 1999.
2. G. Voyatzis, I. Pitas, "The Use of Watermarks in the Protection of Digital Multimedia Products", Proceedings of IEEE, Vol.87, No.7, pp.1197-1207, July 1999.
3. C.T. Hsu, J. L. Wu, "Hidden Digital Watermarks in images", IEEE Trans. On IP, Vol.8, No.1, pp58-68, Jan. 1999.
4. C.T. Hsu, J. L. Wu, "Multiresolution Watermarking for Digital Images", IEEE Trans. On Circuits and System (II), Vol.45, No.8, pp1097-1101.
5. N.Ahmed, T. Natarajan, and K.R.Rao, "Discrete cosine transform,",IEEE Trans. Comput., Vol, C- 23,pp90-93,Jan 1974.
6. Y.Linde, A.Buzo, and R.M.Gray,"An algorithm for vector quantizer design," IEEE Trans. Communications., Vol, Com-28,pp.84-95,Jan. 1980
7. G.Voyatzias and I.Pitas, "Chaotic mixing of digital images and applications to watermarking," in Proceedings of European Conference on Multimedia Applications, Services and Techniques (ECMAST'96), vol. 2, pp 687-695, May 1996.
8. R.G. van Schyndel, A.Z. Tirkel, and C.F. Osborne, " A Digital Watermark," in Proc. 1994 IEEE Int. Conf. On Image Proc,. Vol II, (Austin, TX), pp86-90, 1994.
9. F. A. P. Petitcolas, R. J. Anderson, "Evaluation of copyright marking   systems", Proceedings of IEEE Meltimedia Systems (ICMCS'99), vol.I,pp574-579, June 1999, Florence, Italy.

Tolerance on Geometrical Operation as an Attack to Watermarked JPEG Image

Cong-Kha Pham[1] and Hiroshi Yamashita[2]

[1] The University of Electro-Communications, Tokyo, Japan
[2] ROHM CO., LTD., Japan

Abstract. A method of embedding binary data into JPEG bitstreams has been reported in [1]. However, attacks as geometrical operations to the watermarked JPEG image data have not been analyze. In this work, we propose a method for improvement the tolerance on the geometrical operations as attacks to the watermarked JPEG image.

1 Introduction

Recently, digital networks and multimedia systems grow up rapidly. In order to protect multimedia contents from illegal uses, electronic watermark technologies have been developed. These technologies have been researched for a purpose to embed information to the digital contents. A method of embedding binary data into image data which reported in [1] is an example among the electronic watermark technologies. In this technology, JPEG is targeted as an image compression method. And, binary data have been embedded into the JPEG bitstreams. The embedded data will completely extracted in the decoding stage due to JPEG is a nonreciprocal encoding. However, since the embedding positions are fixed to specific DCT coefficients, deletions of the embedded data are easily performed for purpose of illegal uses. Also, in this techology, attacks as geometrical operations to the watermarked JPEG image data have not been analyze.

We have proposed a method to enable selection of non-fixed embedding positions [2]. In the proposed method, we have been tried to control the image quality deterioration due to the non-fixed embedding positions have been adopted. In this work, we propose a method for improvement the tolerance on the geometrical operations as attacks to the watermarked JPEG image based on the proposed method using non-fixed embedding positions.

2 Previous Work

In JPEG encoding, processing of DCT, quantization, and Huffman encoding are given at every sub-block of 8x8 of the input image. After quantization, DCT coefficients are zigzag scanned, and an end of block (henceforth EOB) sign that shows the end of the block is added at the AC coefficient end. The previous work proposed to embed data after this EOB sign [1]. Fig. 1 shows the encoding procedure of the previous work. It also shows that from the experiment, for the embedding position at P=63 a best result can be obtained. Here, P is an embedding position after the zigzag scan of the DCT table. And,

R. Khosla et al. (Eds.): KES 2005, LNAI 3681, pp. 1199–1204, 2005.

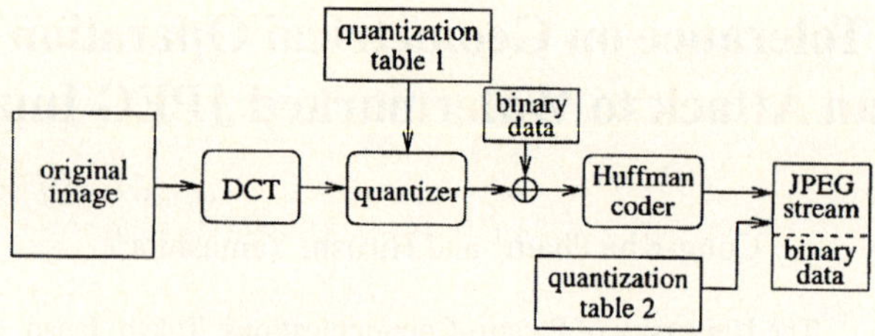

Fig. 1. Conventional method.

(a)   (b)   (c)

Fig. 2. Gray scale images to which binary data is embedded (256x256).

in order make an influence in a reverse-quantization by the embedded data becomes to minimum during decoding stage, the quantization table in which the values are changed and stored in the JPEG file header as a conventional JPEG encoding. Namely, the value of quantization table which is corresponding to the embedding position P=63 is changed to 1.

3 Proposed Method

The purpose of the previous work is to control the image quality deterioration because of the embedded data to minimum [1]. However, since the embedding positions are fixed to specific DCT coefficients, deletions of the embedded data are easily performed for purpose of illegal uses. Also, in this techology, attacks as geometrical operations to the watermarked JPEG image data have not been analyze.

In our work, the method of which the non-fixed embedding position has been proposed to improve this respect [2]. As a result, the embedded data can not easily done when some body try to remove these embedded data. However, there is a possibility that the image quality deterioration will increase when compare with the previous work. Therefor, it is needed to control the image quality deterioration while making the embedding positions become non-fixed. Also, it is needed to verify the influence to the image quality deterioration because of changing of the embedding positions.

3.1 Verification of Image Quality

The embedding procedure for verifications has been done using the same encoding system configuration as shown in Fig. 1. 1-bit of data is embedded to each of 8x8 block of MCU (Minimum Coded Unit). Therefor, in the case of 256x256 size of the gray scale image as shown in Fig. 2, 1024-bit of information can be embedded in to the image. 32x32 size of binary image as shown in Fig. 3 has been used as embedding data

Fig. 3. Binary image used as embedding data (32x32).

in the verification. The PSNRs of the gray scale images which embedded by the binary data at each embedding position in DCT coefficient after quantization are calculated. The above operation has been done at each JPEG compress rate (Q value: 0-100).

3.2 Verification Results

Fig. 4 shows the PNSRs at each embedding position in each Q value of the gray scale image as shown in Fig. 2(a).

PSNR is increasing along to the increasing of Q value at each embedding position. On the other hand, there is an uneven to PSNR at each embedding position as shown in Fig. 5 (Q=70) for the same value of Q. Similar feature has been found for the result of the other two images as shown in Fig. 2(b) and Fig. 2(c). Therefore, the influence of these unevens due to the characteristics of the images is small, and it seems to be that the influence due to the value of the quantization table used in quantization and reverse-quantization is large.

4 Decision of Embedding Positions

When thinking about non-fixed embedding positions, it is possible to decide the embedding positions from a plural candidates using a random or a constant conditions. However, the PSNR value of the embedding position which is selected from the plural candidates is almost equals to the averaged value of the PNSR values at the embedding position selected by one place in the above-mentioned verification. Therefore, it is necessary to decide the embedding position by selecting the position in which a high

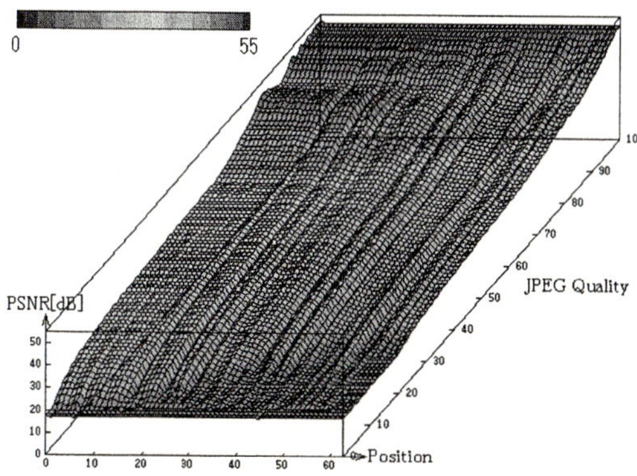

Fig. 4. PSNR of each embedded position for each Q value (without of operation to quantization table).

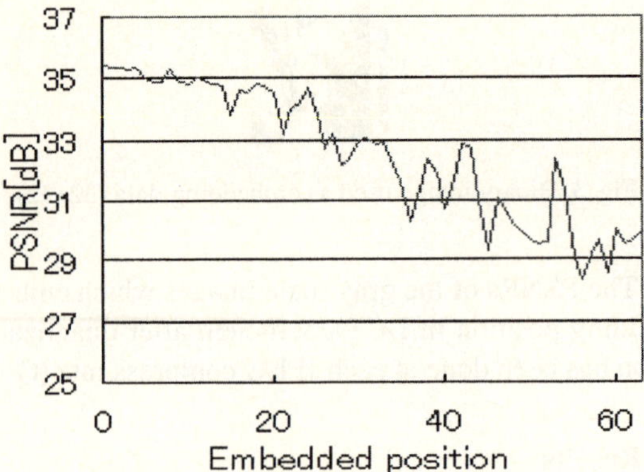

Fig. 5. PSNR of each embedded position for Q=70.

PNSR value was obtained among the candidates. Moreover, for the selected candidate, the influence will appear to the result for the number of candidates because there is the uneven of PSNR values. The average of PSNR value becomes small when a large number of candidates are selected due to the candidate is selected in order in which PSNR value is high, and it causes an increase of the image quality deterioration. On the other hand, when the number of candidates becomes small, non-fixation of the embedding position is ruined. The embedding positions have been decided in consideration of the above criterion.

From verification results, the top position of the uneven is selected as the embedding position candidate. It was weak to the attack of the geometrical operation etc. for the embedding positions in high frequency region. However, it becomes visually remarkable for the embedding positions in low frequency region. Therefor, five points of P=23, 32, 42, 43, and 53 for the embedding positions in middle frequency region have been selected as candidates. The first propose selection method is to select one at random from these five candidates [2]. The second propose selection method is to select the embedding position of two places in one MCU in consideration of the tolerance to the geometrical operation, and 1-bit of information data is embedded in these two places. In addition, the embedding position candidate are assumed to be 12 points for the selected two places. And in order to decrease the influence by the characteristic of the frequency region, we propose to divide the candidates into two groups according to the frequency region and select one point respectively at random in each group.

On the otherhand, it is possible to change the values of the quantization table like the previous work as mentioned above [1]. This method is effective to control the image quality deterioration. However, this technique is effective only when the embedding positions are fixed. Therefore, we do not to change the values of the quantization table because of the embedding positions are non-fixed in our work.

5   Tolerance of Geometrical Operation

In order to evaluate the tolerance to the geometrical operation, we calculate the extraction rate of the embedded data which were embedded to the JPEG bitstreams and were geometrically operated. 32x32 size of binary images as shown in Fig. 6(a)-(e) have been used as embedding data which are embedded to 256x256 size of the gray scale images as shown in Fig. 2(a)-(c).

The geometrical operations such as enlargement (2 times) and reduction (1/2 times) were done to the embedded JPEG images. The embedded data were extracted as the restoration images, and the extraction rate for each restoration image were calculated using the equation (1).

$$\text{Extraction rate} = \frac{\text{Data bits which exactly extracted}}{\text{Data bits which embedded to JPEG image}} [\%] \tag{1}$$

Figs. 7 and 8 show the example of the restoration images from the extracted data in the case of the target image shown in Fig. 2(a), using the propose method. Fig. 9 shows the example of the restoration image from the extracted data in the case of the target image shown in Fig. 2(a), using the previous method [1]. Fig. 10 shows the calculation result of the extraction rate. From the results, we can see that, in both of the previous method and the proposed methods, the data lossed by quantization are eased because of increasing of Q value, and the extraction rate goes up. An increase at the extraction rate was seen from Q value 40 as for the propose method was about 40%-50% when comparing with the previous method, and 92% was confirmed as the maximum in case of Q=100. Moreover, it seems that it is easy to recognize the image of the restoration images with the proposed methods than that of the previous method as shown in Fig. 7(a), Fig. 8(a) and Fig. 9(a)-(b).

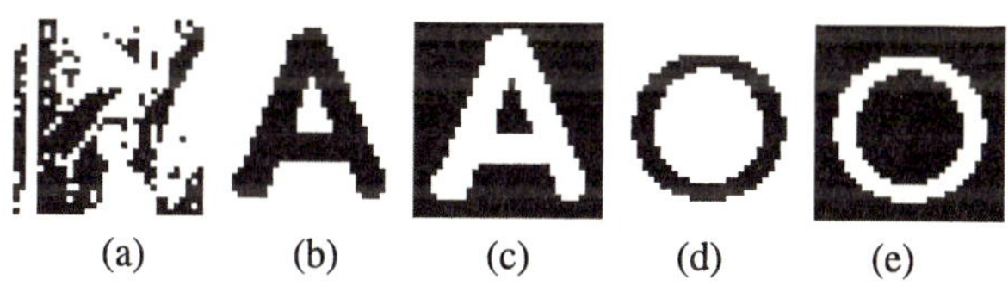

(a)   (b)   (c)   (d)   (e)

Fig. 6. Image data for embedding.

(a)   (b)   (c)   (d)   (e)

Fig. 7. Restoration image form extracted embedding data using the propose method (2 times enlargement as the geometrical operation).

(a)   (b)   (c)   (d)   (e)

Fig. 8. Restoration image form extracted embedding data using the propose method (half reduction as the geometrical operation).

(a)   (b)

Fig. 9. Restoration image form extracted embedding data using the previous method (2 times enlargement (a); half reduction (b)).

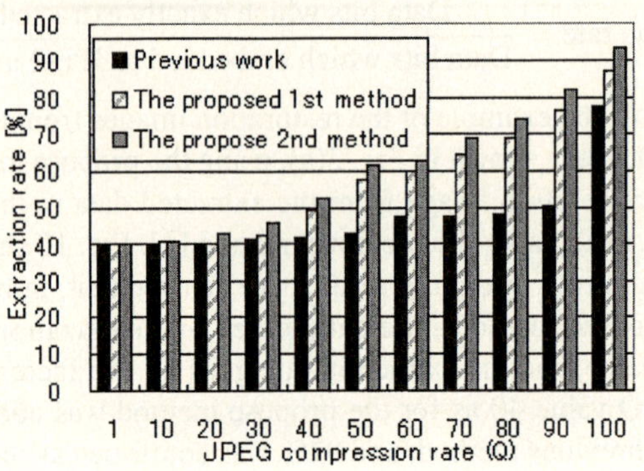

Fig. 10. Extraction rate of each method.

6   Conclusions

We have proposed the method to enable selection of non-fixed embedding positions. In the proposed method, we have been tried to control the image quality deterioration due to the non-fixed embedding positions have been adopted. Also, we have proposed the method for improvement the tolerance on the geometrical operations as attacks to the watermarked JPEG image using non-fixed embedding positions. The tolerance to the geometrical operation has been improved from recognized respect as the extraction rate and the extrated image. However, the information after the attack by the geometrical operations have not received with 100%, a further improvement becomes future tasks.

References

1. H. Kobayashi, Y. Noguchi and H. Kiya, "A Method of Embedding Bianary Data into JPEG Bitstreams," IEICE, Vol.83-D-II, No.6, pp. 1469-1476, 2000.
2. H. Yamashita, C-K. Pham, "Improvement of Robustness on Embedding of Binary Data to JPEG Image", 2004 RISP International Workshop on Nonlinear Circuit and Signal Processing (NCSP'04), pp. 65-68, Hawaii, Mar. 2004.

A High Robust Blind Watermarking Algorithm in DCT Domain

Ching-Tang Hsieh and Min-Yen Hsieh

Department of Electrical Engineering, Tamkang University
151, Ying-Chuan Rd. Tamsui, Taipei County Taiwan 25137, R.O.C.
hsieh@ee.tku.edu.tw, 692350092@s92.tku.edu.tw

Abstract. In this paper, we propose an effective algorithm of gray-level watermarks based on the characteristics of image. The DCT coefficients of the watermark image are embedded into the position of DCT domain of host image by considering the similarity of DCT blocks energy of both images. The algorithm recovers the watermark without any reference to the original image and we propose a post-processing procedure, which can correct the errors by the survived information, to obtain higher quality watermark. The experimental results show that the proposed algorithm adapted to the original image produces the most imperceptible and the most robust watermarked image under print-and-scanner and various attacks.

1   Introduction

Motivated by the overwhelming urge for internet data security, digital watermarking has recently emerged as an important area of research in multimedia data processing. The technology of digital watermarking has gained prominence and emerged as a leading candidate that could solve the fundamental problems of legal ownership and content authentications for digital multimedia data. The purpose of digital watermarking is not to restrict use of multimedia resources, but to resist attack from unauthorized users. It is an invisible mark inserted into the digital multimedia data so that it can be detected in the later stage for evidence of rightful ownership. To achieve a high performance watermarking, the algorithm must possess properties of invisibility, security and robustness.

The techniques proposed so far can be divided into two main groups according to the embedding domain of the container image. One is to modify the intensity value of the luminance in the spatial domain [1]. However, this technique has relatively low information hiding capacity and can be easily erased by lossy image compression. The other is to change the image coefficients in a frequency domain (e.g. DCT, DFT, DWT). Cox *et al.* [2] uses spread spectrum to embed watermark in the discrete cosine transform (DCT) domain. Hsu and Wu embedded an image watermark into selectively modified middle frequency of DCT coefficients of container image [3]. Joseph et al. developed a digital image watermarking using the Fourier-Mellin transform that is invariant to image manipulations or attacks due to rotation, scaling and translation [4]. Lu et al. uses cocktail watermark to improve the robustness and used human visual system (HVS) to maintain high fidelity of the watermarked image[5]. Several other methods used discrete wavelet transform (DWT) to hide the data to the frequency domain [6,7,8].

R. Khosla et al. (Eds.): KES 2005, LNAI 3681, pp. 1205–1211, 2005.

In recent literatures, most authors concentrated on using binary data watermarking [9-13]. The gray-level image watermarking for protecting multi-media data has rarely been discussed [14,15,16]. In order to reserve the visual quality of host image and improve the robust of watermark, the scheme of watermark can be roughly categorized to two kinds. The early technique is that embedding the watermark into the medium or high frequency domain and another is that using JND or HVS to embedding the watermark in the suitable position dispersing. In this paper we proposed an adapted approach considering the similarity of DCT blocks energy between watermark and original images and by that the watermark can be dispersed in DCT domain without using JND. So the watermark will have both the robust and invisible properties.

In the watermark extraction we also propose a post-processing procedure, which can correct the errors by the survived information using a filter in the DCT domain, to obtain higher quality watermark.

2   Embedding Process

In the proposed method we embed the grey watermark W in the grey original image Y and the flowchart is shown in Fig. 1.

We segment the original image Y into 8*8 blocks without overlap and define $f_k(x',y'), 0 \leq x', y' < 8, k = 0,1...,K-1$, here k denote the k th block and K is the amount of the blocks. Then we estimate the complexity of each DCT of $f_k(x',y'), F_k(u,v)$. The complexity of each $F_k(u,v)$ is defined by the function given in (1).

$$E = \sum_i \sum_j |f_v(i,j)| + |f_h(i,j)| + |f_{high}(i,j)| \tag{1}$$

where $f_v(i,j)$, $f_h(i,j)$ and $f_{high}(i,j)$ are the blocks of each component in DCT as shown in Fig. 2. We decide one location of each area from horizontal energy, vertical energy and high frequency in the chosen $F_k(u,v)$. The M horizontal energy mark up the serial h(i); the vertical energy and the high frequency energy make up the serial v(i) and high(i), respectively. We obtain the serial $h_{sort}(i), v_{sort}(i)$ and $high_{sort}(i)$ by sorting the serial h(i), v(i) and heigh(i) and the generated serials, $h_{index}(i)$, $v_{index}(i)$ and $high_{index}(i)$ reserve the original position information of the component of sorted serial. For example, if h={15,28,41,9,3}, $h_{sort}(i)$={3,9,15,28,41} and $h_{index}(i)$={5,4,1,2,3}. Similarity, we obtain $dc_{index}(k)$ and in $dc_{sort}(k)$ from dc(k), which is composed of DC value, in $F_k(u,v)$ and where k=1,2,...,K. We decide one segment of M dimension from $dc_{sort}(k)$ represented by $sub\_dc_{sort}(i)$, here i =1,2,..., M, which is used for embedding the dc component of the watermark DCT.

The watermark image is divide by 2*2 block without overlap, which is defined as $w_m(x',y'), 0 \leq x', y' < 2, m = 0,1...,M-1$, here m denote the mth block and M is the amount of all blocks. We obtain $W_m(u,v)$, which is the DCT of each $w_m(x',y')$, that generate the four serials, wdc(i), wh(i), wv(i) and whigh(i) from the four position. Then we can get the sorted serials, $wdc_{sort}(i)$, $wh_{sort}(i)$, $wv_{sort}(i)$ and $whigh_{sort}(i)$, and the serials recording location information of original, $wdc_{index}(i)$, $wh_{index}(i)$, $wv_{index}(i)$ and $whigh_{index}(i)$. The proposed method of embedding is: (1) the part of ac: $wh_{sort}(i)$, $wv_{sort}(i)$ and $whigh_{sort}(i)$ substitute for $h_{sort}(i)$, $v_{sort}(i)$ and $high_{sort}(i)$ directly and to restore them by the information of index serials, $h_{index}(i)$, $v_{index}(i)$ and $high_{index}(i)$. And then we place the serials in the chosen position from $F_k(u,v)$. (2) the part of dc: we estimate n_dc(i), which is the similarity of chosen $sub\_dc_{sort}(i)$, using $wdc_{sort}(i)$ by the following function.

$$\text{n\_dc(i)} = (\frac{wdc_{sort}(i) - minw}{maxY - minY}) * (maxY - minY) + minY \tag{2}$$

where maxY and minY is the maximum and minimum of $sub\_dc_{sort}(i)$ and minw is the minimum of $wdc_{sort}(i)$. We replace $sub\_dc_{sort}(i)$ with n_dc(i) and restore the sequence by $dc_{index}(k)$. Then we place them back in $F_k(u,v)$ and compute the IDCT of $F_k(u,v)$. We estimate PSNR to compare with the original image and if PSNR is less than a predefined threshold T1 we choose $sub\_dc_{sort}(i)$ from $dc_{sort}(k)$ again until $R \geq T1$. Lastly we get the image including the water marking and we storage the key file including the location information of embedding and the ratio of approximation. Embedding the watermark to dc can improve the robust of image, but that will influence the quality of vision when dc is changed. In the paper the proposed method can improve robust without influencing quality by choosing segment to math repeatedly.

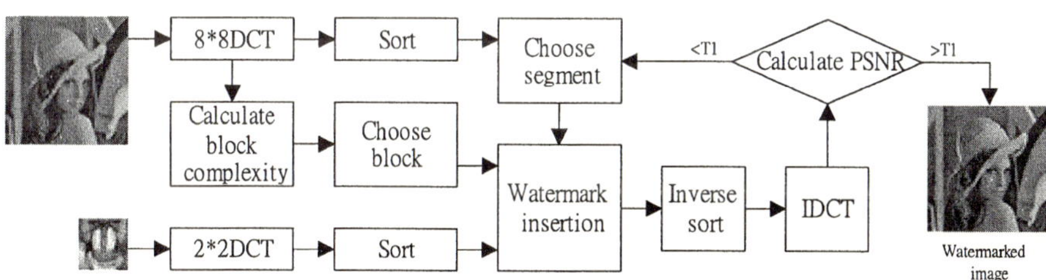

Fig. 1. The flowchart of proposed embedding process

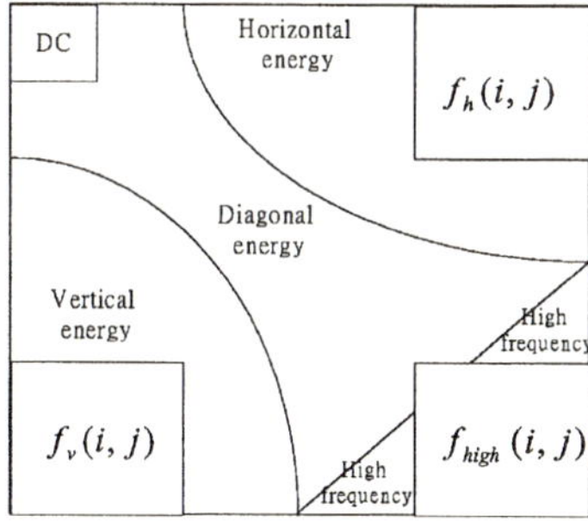

Fig. 2. DCT energy distribution map

3   Extraction Process

In general, the method to extract the watermark is the inverse of embedding. In the proposed method, we extract the watermark without original image and we implement a post-processing to improve the image quality efficiently.

We divide the image Y', which is including watermark, by 8 *8 block without overlap. After DCT of each block, we find out the blocks including watermark by the location information serial m(i) in ac part of watermark and extract the serials of horizontal energy, vertical energy and high frequency area by $h_{index}(i)$, $v_{index}(i)$ and $high_{index}(i)$. Then we can obtain the serials, $wh_{sort}(i)$, $wv_{sort}(i)$ and $whigh_{sort}(i)$, and the dc component of each block make up one serial. By $dc_{index}(k)$ we can find out the segment, n_dc'(i), including the watermark and we get $wdc_{sort}'(i)$ by restoring the ratio as shown in Fig. 3(a). Because of the sorting of $wdc_{sort}(i)$, even there is noise or error produced by attacking in the image the roughly trend will not be changed. The post-processing of proposed method is that we filter $wdc_{sort}'(i)$ by medium filter to find out its trend and adjust the serial to the original starting point by $wdc_{sort}(0)$ as shown in Fig.3(b). By such, we can correct the error information by the surviving watermark. Lastly, we combine with the four serials to $W_m(u,v)$ and we can get the watermark image W' by implementing 2*2 IDCT.

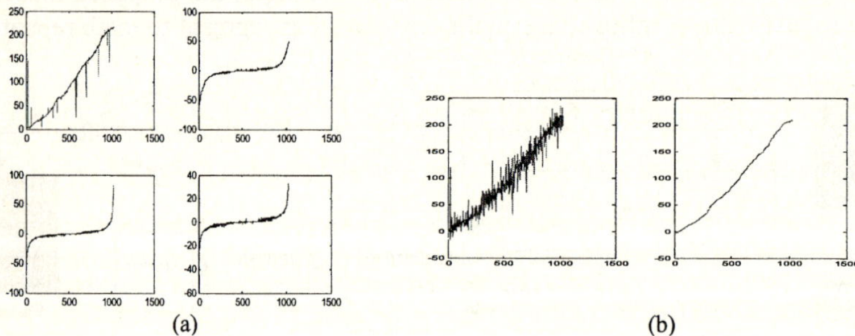

(a)   (b)

Fig. 3. (a) the extracted serials of embedded watermark, (b)finding the trend of attacked DC component using the medium filter

4   Experimental Results

In the experiment, we embed the 64*64 gray watermark image to the 512*512 original image. The evaluation of image quality of embedded image and extracted image is evaluated by PSNR. In the experiment T1 is defined as 40dB and the PSNR of embedded image is 42 dB. Fig.4 (a) and Fig.4 (b) is the original image Lena and watermark image Baboon and Fig. 4 (c) is the image including watermark. Fig. 4 (d) is the extracted watermark from the image that is not attacked and its PSNR is 45dB. We can note that Fig.4 (a) is so similar to Fig.4 (c) that the watermark is invisible.

To test the watermark robust we introduce 14 attacks and 12 compressions of different qualities, as shown in Fig.5, including the destructive print-scan attack. Fig. 7(a) shows the PSNR of watermark extracted from the attacked image.

Against to the attack of rotation and scaling we only have to manually restore the image by image processing software and then we can extract the watermark by proposed method. A gray image can prove the more perceptual information so that the watermark can easily verify by human eye as shown in Fig. 6. JPEG is the popular tool for image compression, so any watermark must have capacity to resist JPEG washing for real applications. We test the image on 12 different JPEG compression qualities as shown in Fig. 7(b). Fig.8 shows the difference results between applying the post-processing and not.

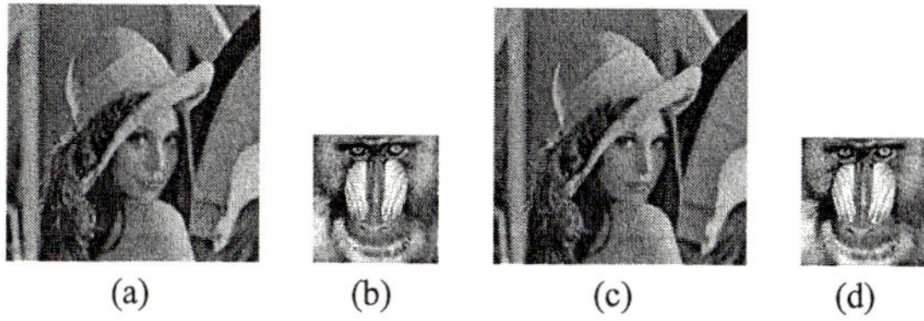

(a)   (b)   (c)   (d)

Fig. 4. (a)Original image Lena, (b)watermark Baboon, (c)watermarked image, (d)watermark Extraction

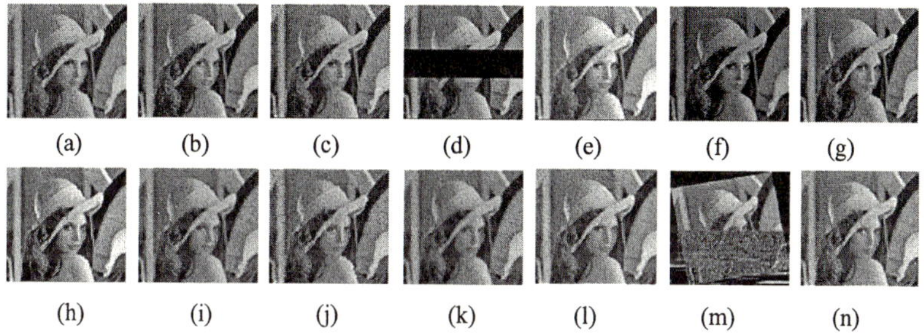

(a) (b) (c) (d) (e) (f) (g)

(h) (i) (j) (k) (l) (m) (n)

Fig. 5. (a)-(t) attacked watermarked images: (a) low-pass filter (mask size 5_5) (b) sharpened (c) noise (d)cropping25% (e) increase brightness (f) decrease brightness (g) scaling (h) contrast enhancement (i) mosaic (j) smart blurred (k) Gaussian blurred (l) median filtered (mask size 7_7 (m) rotation (n) print-and-scan

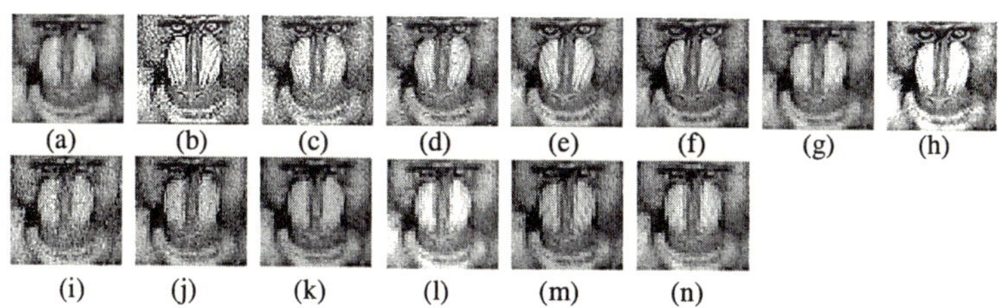

(a) (b) (c) (d) (e) (f) (g) (h)

(i) (j) (k) (l) (m) (n)

Fig. 6. Watermark extracting from fig. 7

5   Conclusion

In this paper, we propose a new block DCT-based digital watermarking scheme, which insert an adapted gray level watermark in the original image by DCT. In extraction processing we propose a post-processing procedure to correct the error on

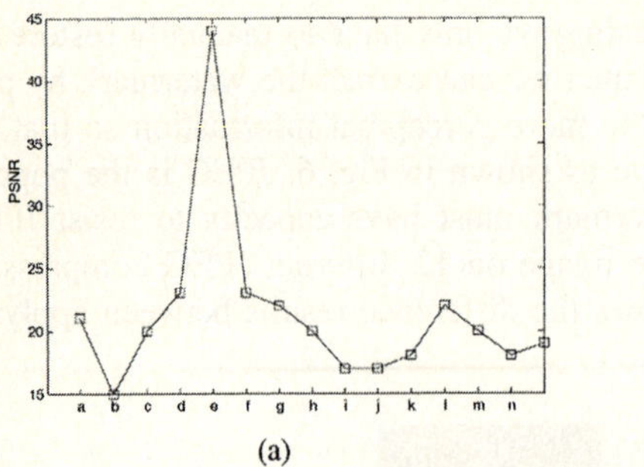
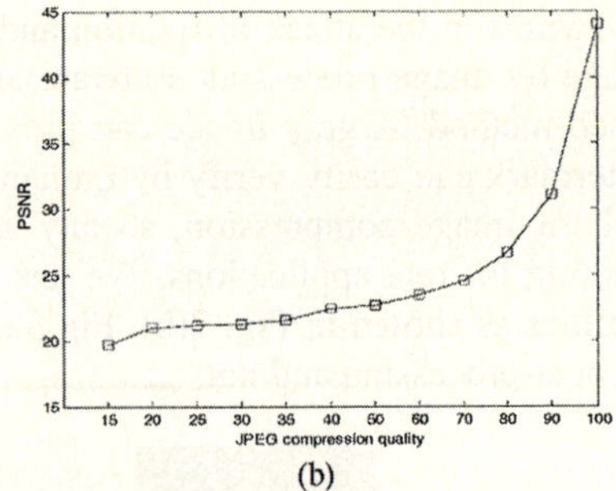

(a)   (b)

Fig. 7. (a) 12 PSNR of Fig.8 and (b) PSNR of JPEG compression attack

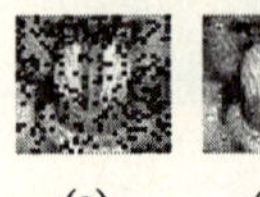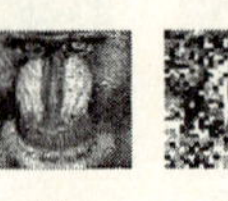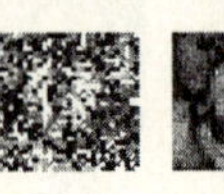

(a)   (b)   (c)   (d)

Fig. 8. Extracted watermark process (a)(c) without post-processing, (b)(d) with post-processing

attacked image. The experimental result shows that the algorithm can against the various attacks effectively and can extract the watermark correctly.

References

1. Van Schyndel, R.J., Tirkel, A.Z., Osborne, A.F.: A digital watermark., Proceedings of the IEEE International Conference on Image Processing, Vol. 2, (1994) 86-90.
2. I. J. Cox, J. Kilian, T. Leighton, and T. Shamoon: Secure spread spectrum watermarking for multimedia, IEEE Transaction on Image Processing, Vol. 6, (1997) 1673–1687.
3. Hsu, C.T., Wu, J.L.:Hidden digital watermarks in images, IEEE Transaction on Image Processing, Vol. 8, (1999) 58-68.
4. Joseph, J.K., O' Ruanaidh, Pun T.: Rotation, Scale and Translation Invariant Digital Image Watermarking, Signal Processing, Vol. 66, No. 3, (1998) 303-317.
5. C. S. Lu, S. K. Huang, C. J. Sze, and H. Y. M. Liao: Cocktail watermarking for digital image protection, IEEE Transaction on Multimedia, Vol. 2, (2000) 209–224.
6. Wei, Z.h., Qin, P., Fu, Y.Q.: Perceptual digital watermark of image using wavelet transform, IEEE Transaction on Consumer Electronics, Vol. 44, No. 4, (1998) 1267-1272.
7. Dugad, R., Ratakonda, K., Ahuja, N.: A new wavelet-base for watermarking image, Proceedings of the IEEE International Conference on Image Processing, Vol. 2, (1998) 419-429
8. Hsu, C.T., Wu, J.L.: Multiresolution watermarking for digital images, IEEE Transaction on Consumer Electron, Vol. 45, (1997) 1097-1101.
9. Fengsen Deng, Bingxi Wagn: A novel technique for robust image watermarking in the DCT domain, IEEE International Conference on Neural Networks & Signal Processing, (2003) 1525-1528.
10. Peter H. W. Wong, Oscar C. Au, Y. M. Yeung: A novel blind multiple watermarking technique for images, IEEE Transaction on Circuits and Systems for Video Technology, Vol. 13, No. 8, (2003) 813-830.
11. Tay R. J., Havlicek J. P.: Image watermarking using wavelets, IEEE International Conference on Circuits and Systems, Vol. 3, (2002) 258-261.

12. Bao P. J., Xiaohu Ma: Image adaptive watermarking using wavelet domain singular value decomposition, IEEE Transaction on Circuits and Systems for Video Technology, Vol. 15, (2005) 96-102.

13. M. Wu, B. Liu: Watermarking for image authentication, IEEE International Conference on Image Processing, Vol. 2, (1998) 437-441.

14. Shih-Chang Hsia, I-Chang Jou: A high robust watermarking technique using sub-band filtering, IEEE International Conference on Multimedia Modelling, (2004) 72-78.

15. Ho A.T.S., Jun Shen, Chow A.K.K., Woon J.: Robust digital image-in-image watermarking algorithm using the fast Hadamard transform, ISCAS '03, Circuits and Systems, Vol. 3, (2003) 826-829.

16. Yong Duk Chung, Chung Hwa Kim: Robust image watermarking against filtering attacks, SICE Annual Conference, Vol. 3, (2003) 3017-3020.

Improved Quantization Watermarking with an Adaptive Quantization Step Size and HVS

Zhao Yuanyuan and Zhao Yao

The Institute of Information Science, Beijing Jiaotong University, Beijing, China
zyykxp@hotmail.com, yzhao@center.njtu.edu.cn

Abstract. This paper proposes a new image-adaptive watermarking technique which utilizes a new combination of an adaptive quantization step size and a HVS(human visual system) model in the wavelet domain. Here we use Quantization Index Modulation(QIM) method with an adaptive quantization step size to realize the embedding scheme. The HVS masking is accomplished pixel by pixel by take into account the luminance and the frequency content of all the image subbands. The watermarking consists of a pseudorandom sequence which is adaptively embedded into the subbands. As usual, the watermark bits are detected by a minimum distance detector. Experimental results prove the effectiveness of the new algorithm.

1 Introduction

As a result of the rapid development of digital technology, image watermarking is finding more and more support as a possible solution for the protection of intellectual property rights. Many techniques have been proposed in the literature over the last few years. One of the most important approaches proposed so far is Quantization Index Modulation (QIM) [1]. QIM methods are multi-bits watermark scheme. It can achieve very efficient trade-offs among the amounts of embedded information (rate), the amounts of embedding-induced distortion to the host signal, and the robustness to intentional and unintentional attacks.

To the aim of effectively image compression without degrading subjective image quality, theoretical models of the human visual system (HVS) have been deeply studied. Similarly, it is today widely accepted that robust image watermarking techniques should largely exploit the characteristics of the HVS, for more effectively hiding a robust watermark.

In this paper, a novel blind watermarking algorithm, which embeds the watermark in the DWT domain by exploiting the adaptive QIM method, is presented. The main novelty of the algorithm resides in the adaptive quantization. Here, the adaptive quantization is accomplished through the values of the pixels in adjacent domain and a mask giving a pixel by pixel measure of the sensibility of the human eye to local image perturbations. As said above, to effectively hide the watermark, each bit is quantized by one of the two non-intersect quantizations with an adaptive step size. Mask construction relies on a work by Lewis and Knowles [2], in which the authors propose a method to evaluate the optimum quantization step for each DWT coefficient according to psychovisual considerations. Some modifications used here are in order to make it suitable to the computation of the maximum visibly tolerable watermark energy that can be used for each DWT coefficient [3][4]. For watermark detection, we use the minimum distance decoder. Extensive experiments aimed at assessing the performance of the new system both from the point of view of watermark invisibility and from the point of view of robustness; the system has demonstrated to be resistant to JPEG compression and Gaussian noise.

This paper has five parts. The first part is the introduction. The second part is the watermarking embedding scheme, the adaptive quantization step size will be introduced particularly. The third part is watermarking extraction scheme. In the fourth part we will show the experimental results. The last part is the conclusion.

2 Perceptual Watermark Embedding Scheme

Since its excellent spatio-frequency localization properties, the DWT is very suitable to exploit the adaptive watermark embedding scheme: if a DWT coefficient is modified, only the region of the image where the particular frequency corresponding to that coefficient is present will be modified.

2.1 Watermark Embedding

The watermark embedding scheme is shown as Fig.1:

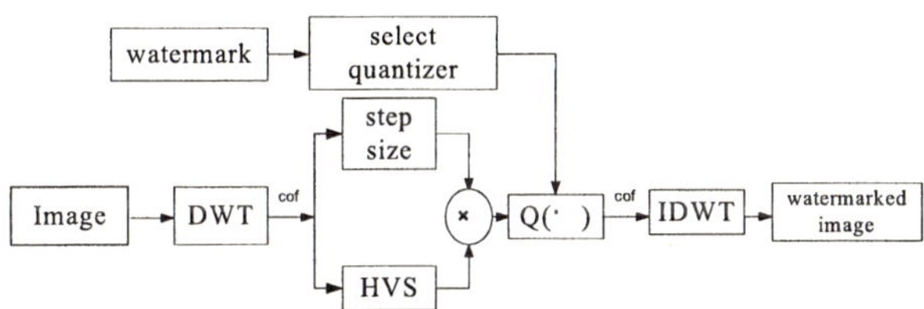

Fig. 1. Watermark embedding scheme

The image to be watermarked is first decomposed through DWT in four levels.

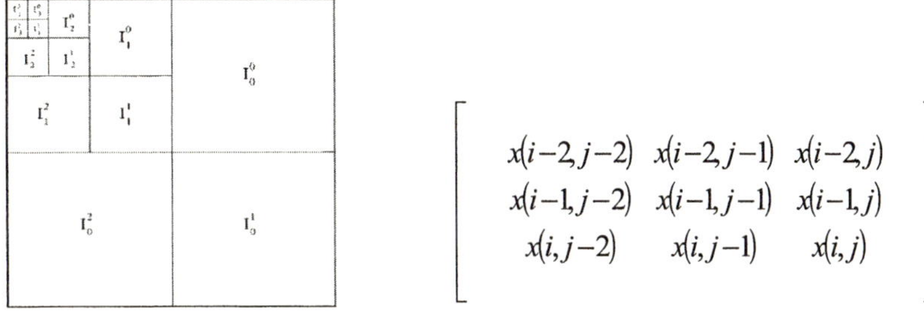

$$\begin{bmatrix} x(i-2,j-2) & x(i-2,j-1) & x(i-2,j) \\ x(i-1,j-2) & x(i-1,j-1) & x(i-1,j) \\ x(i,j-2) & x(i,j-1) & x(i,j) \end{bmatrix}$$

Fig. 2. Sketch of the decomposition of an image

Fig. 3. The adaptive quantization step size's in calculation

We call I_l^θ the subbands at resolution level $l = 0,1,2,3$ and with orientation $\theta = \{0,1,2,3\}$ (see Fig.2). The watermark, consisting of a pseudorandom binary (0,1)

sequence, is inserted by modifying the wavelet coefficients belonging to the detail bands at level 1 and level 2, i.e., I_1^θ, I_2^θ.

The choice of embedding the watermark into this middle detail subbands was motivated by experimental tests, as the one offering the best trade-offs between robustness and invisibility. Inserting the watermark into these middle frequency subbands could give a higher robustness (e.g., compression or noise), and give the low visibility of disturbs at the same time. In more detail, subbands coefficients are quantized with two non-intersect quantizers indexed the pseudorandom binary sequence (0 or 1) as follows:

$$X_I^\theta = \begin{cases} round\left(\dfrac{x_I^\theta - \Delta_I^\theta/4}{\Delta_I^\theta}\right) \cdot \Delta_I^\theta + \Delta_I^\theta/4, w=0 \\ round\left(\dfrac{x_I^\theta + \Delta_I^\theta/4}{\Delta_I^\theta}\right) \cdot \Delta_I^\theta - \Delta_I^\theta/4, w=1 \end{cases} \tag{1}$$

The adaptive quantization step size are modified according to the rule

$$\Delta_I^\theta(i,j) = \Delta_I^{\theta\,\prime}(i,j) \cdot P_I^\theta(i,j) \tag{2}$$

where $\Delta_I^{\theta\,\prime}$ is the adaptive quantization step size lying on the adjacent pixels for watermark embedding, and P_I^θ is a weighing function considering the local sensitivity of the image to noise. It is this weighing function that allows to exploit the masking characteristics of the HVS.

2.2 The Adaptive Quantization Step Size

The mainly advantage when using the adaptive quantization step size is that the embedding strength is more or less proportional to the perceptual sensitivity to distortions. Because for bright image areas a larger quantization step size is chosen, a larger robustness is achieved in those areas; information embedded in bright areas can be retrieved at the detector with larger reliability. The overall robustness will gain from this adaptive quantization, while the watermark is as perceptible as in the case of a fixed quantization step size.

Here we use a model for determining the adaptive quantization step size: there is a linear relation between Δ and the group of pixel values $x(i,j)$. Such a linear relationship can be defended by referring to Weber's law which gives a linear relationship between the sensitivity of the human eye and the luminance value. So the quantization step size is determined by

$$\Delta'(x) = \frac{\alpha}{(2M+1) \cdot (2N+1)} \sum_{i=i-M}^{i+M} \sum_{j=j-N}^{j+N} x(i,j) \tag{3}$$

where $(2M+1) \cdot (2N+1)$ is the number of pixels in the aforementioned group and α is embedding strength parameter. It can easily be seen that this model is brightness scale invariant.

For a fixed quantization step size, both at the embedder and the detector, Δ is known. But for an adaptive quantization step size Δ is a function of $x(i,j)$. At the embedder the quantization step size is calculated by $\Delta'(x)$, but at the detector Δ has to be calculated from the received signal $(y+n)$, so the quantization step size is $\Delta(y+n)$. This Δ is only an estimation of the quantization step size used at the embedder and therefore may not be completely accurate. Because the detection depends on the adaptive quantization step size, the estimation error causes bit errors in estimating the received message. Based on this errors, we have modified this step size as follows:

$$\Delta_l^{\theta'}(i,j) = \frac{\alpha}{(M+1)\cdot(N+1)-1}\left(\sum_{i=i-M}^{i}\sum_{j=j-N}^{j}x_l^{\theta}(i,j) - x_l^{\theta}(i,j)\right) \tag{4}$$

This equation means that we only calculate the pixels on the top left corner of $x_l^{\theta}(i,j)$. In order to reduce the estimation error, $x_l^{\theta}(i,j)$ is not included in the calculation as Fig.3.

2.3 Perceptual Weighing

In order to embed into the images the maximum, but still unperceptible, level of watermark, the weighing function has to consider how the eye perceives disturbs. In particular, the eye is less sensitive to noise in high resolution bands, and in those bands having orientation of 45; the eye is less sensitive to noise in those areas where brightness is high or low; the eye is less sensitive to noise in highly textured areas.

Based on these considerations, we computed the quantization step of each coefficient as the weighted product of two terms where the meaning of each term in this equation is explained below.

Let us start the analysis of by the first of the expression in (2):

$$P_l^{\theta}(i,j) = F(l,\theta)\cdot L_l(i,j) \tag{5}$$

To take into account how sensitivity to noise changes depending on the band (in particular depending on the orientation and on the level of detail), we let

$$F(l,\theta) = \begin{cases} \sqrt{2}, & if \ \theta=1 \\ 1, & otherwise \end{cases} \cdot \begin{cases} 1.00 & if \ l=0 \\ 0.32 & if \ l=1 \\ 0.16 & if \ l=2 \\ 0.10 & if \ l=3 \end{cases} \tag{6}$$

The second term takes into account the local brightness based on the graylevel values of the low pass version of the image. Since the eye is less sensitive in the regions with high brightness, we can compute this factor in the following way:

$$L_l(i,j) = \frac{1}{256}I_3^3\left(1+round\left(\frac{i}{2^{3-l}}\right), 1+round\left(\frac{j}{2^{3-l}}\right)\right) \tag{7}$$

Based on the consideration that the human eye is less sensitive to changes in very dark regions as well, the modified factor can be as follows:

$$L_l'(i,j) = \begin{cases} 1-L_l(i,j), & if \ L_l(i,j)<0.5 \\ L_l(i,j), & otherwise \end{cases} \tag{8}$$

In our scheme, we use the normalization of this factors.

3   Watermark Detection

Watermark detection is accomplished without referring to the original image. In general, we use minimum distance decoder in QIM watermark detection.

With the received sequence, we can easily find the nearest reconstruction sequence of each quantizer (the 0-quantizer and the 1-quantizer) with the adaptive quantization step.

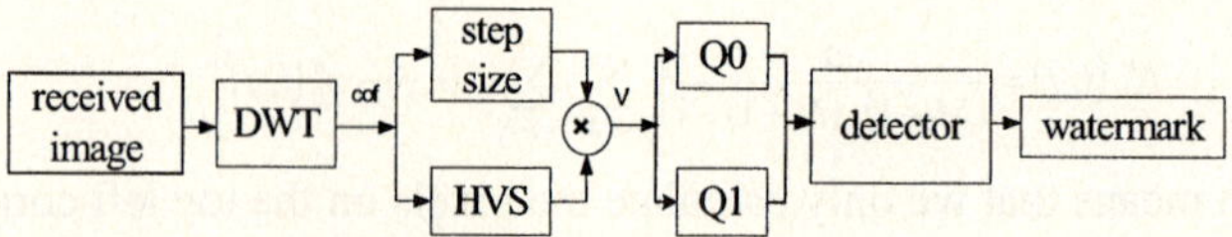

Fig. 4. Watermark detection scheme

At first, we should calculate the received image's adaptive quantization step according to the formula (2); then with using this step, we can decide the corresponding quantizer by the minimum distance decoder; from the quantizer, we can estimate the embedded watermark bits. Watermark detection process is shown in figure 4.

4   Experimental Results

The algorithm has been extensively tested on various standard images and attempting different kinds of attacks, in this section some of the most significant results will be shown. For the experiments presented in the following, the Daubechies-1 filtering kernel has been used for computing the DWT.

In our experiments we embed 32×32 bits binary sequence into 256×256 test image namely Lena and Boat as Fig.5 and Fig.6. PSNR value is 42.7262dB and 41.0642dB. Here $\alpha = 5$.

First, watermark invisibility is evaluated: in Fig. 5(a) and Fig. 6(a), the original "Lena" and "Boat" image is presented, while in Fig. 5(b) and Fig. 6(b), the watermarked copy is shown: the images are evidently undistinguishable, thus proving the effectiveness of DWT watermarking and the masking procedure. In particular, it is evident that the watermark is mainly hidden into high activity regions and around edges from Fig. 7 which $\alpha = 7$ (see, such as Lena, the high level of watermark at the borders of the hat and the shoulders of the girl, and over the feathers). Fig.8 shows the Lena image with the mixed quantization step size which use the same quantizers. The mixed step size is 16 and it's robustness is close to Fig. 5(b). But the PSNR is 39.7538dB, less than our scheme with the adaptive quantization in Fig. 5(b). And we can easily found that Fig.5(b) is better than Fig.8.

A set of distortions is applied to the watermarked image and the watermark is extracted from the distorted image. The corresponding bit error rate (BER) is calculated to measure the robustness of the algorithm to that particular distortion. Here the attacks are JPEG compression and additive noise.

As a first experiment, JPEG coding with decreasing quality was applied to the watermarked image, and 1024 different watermarks were tested for presence. In fact, we have embedded 1024 watermark bits repeatedly into 6 detail subbands. So we have quantized almost 15360 pixels. Fig. 9 illustrates the bit error rate when the applied distortion is JPEG compression with quality factors on the image Lena and Boat. In Fig. 9, the quality factor is in the range of 50-100.

 (a)
 (b)

Fig. 5. (a) Lena original image, (b) Lena watermarked image ($\alpha = 5$)

 (a)
 (b)

Fig. 6. (a) Boat original image, (b) Boat watermarked image ($\alpha = 5$)

Fig. 7. Lena watermarked image ($\alpha = 7$)   **Fig. 8.** Lena watermarked image with fixed step size

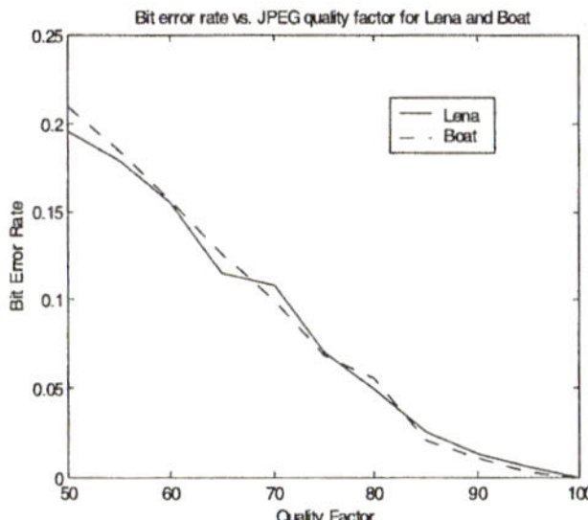

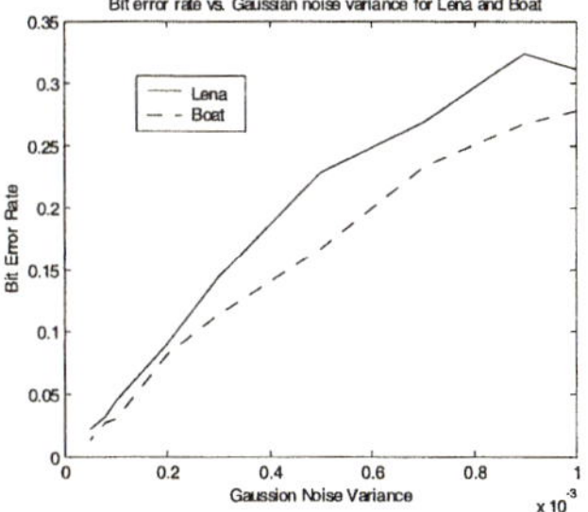

Fig. 9. Effect of JPEG compression   **Fig. 10.** Effect of additive noise

Additive Gaussian noise of zero mean and variance in the range 0.0001-0.0015 is added to the watermarked image. Fig.10 show the plot of BER versus noise variance.

From the above results, we can conclude that our scheme can get good performance with different attacks. With HVS and the adaptive quantization step size, we can get the watermarked image which has invisible distortion and good robustness.

5 Conclusion

In this paper, a novel algorithm for image watermarking has been presented. The algorithm embeds the watermark code by quantized the DWT coefficients of the image. With exploiting an adaptive quantization step size and a model of the HVS, the performances of the novel algorithm are good. Experimental results, in fact, supported the suitability of DWT watermarking schemes for hiding watermarks into images.

References

1. Brian Chen and Georgy W. Wornell, "Quantization Index Modulation Methods for Digital Watermarking and Information Embedding of Multimedia", Journal of VLSI Signal Processing 27, 7–33, 2001.
2. A. S. Lewis, G. Knowles, "Image Compression Using the 2-D Wavelet Transform", IEEE Transactions on Image Processing, Vol. 1, No. 2, April, 1992.
3. Christine I. Podilchuk, and Wenjun Zeng, "Image-Adaptive Watermarking Using Visual Models", IEEE Journal on Selected Areas in Communications, Vol. 16, NO. 4, May 1998.
4. Zhenghua Yu and H. R. Wu,"Human Visual System based Objective Digital Video Quality Metrics", IEEE Proceedings of ICSP2000.

Energy-Efficient Watermark Algorithm
Based on Pairing Mechanism

Yu-Ting Pai[1], Shanq-Jang Ruan[1], and Jürgen Götze[2]

[1] Department of Electronic Engineering, National Taiwan
University of Science and Technology, Taipei, Taiwan
{M9302101,sjruan}@mail.ntust.edu.tw
http://lps.et.ntust.edu.tw
[2] University of Dortmund, Germany
juergen.goetze@uni-dortmund.de
http://www-dt.e-technik.uni-dortmund.de

Abstract. In recent years, digital watermarking is a technique for labeling digital images by hiding secret information which can protect the copyright. The goal of this paper is to develop a DCT-based watermarking algorithm for low power and high performance. Our energy-efficient of technique focus on algorithm level processing. We improve Hsu and Wu's algorithm by using DCT coefficients to pairing blocks directly. The experimental results show our approach not only can reduce a halt of pairing operations required, but also increase the PSNR around 0.2 db.

1 Introduction

Due to the rapid and extensive growth of Internet, images can be distributed much easier and more convenient. On the contrary, it is easy to make unjust replica and interpolation, so the copy right protection becomes a momentous theme. Digital watermarking is the technique that can embed secret information in general binary images, like scanned text, figures, and signatures. In the literature, several researchers have investigated digital watermarking with different contributions. Cox *et al.* argued inserting watermark in DCT domain and making it robust to signal processing operations [1]. Fei *et al.* proposed an algorithm for improving resistance to compression [2]. Kii *et al.* used patchwork to endure cropping of the image [3]. Wu and Hsieh researched the watermarking by using zerotree of DCT that extract the watermark form watermarked image without using original image [4].

Digital image processing in portable device has become a popular technique nowadays, such as laptop computer, personal digital assistant (PDA) and cellular phone. Because these applications are battery powered, reducing power dissipation in a process or a system is practical. Many investigator have discussed the various low power technologies on digital image processing, such as low power CMOS image sensors [5], coding and decoding [6], and compression [7] *et al.*. Although low power image process is a popular topic, few writers have

R. Khosla et al. (Eds.): KES 2005, LNAI 3681, pp. 1219–1225, 2005.

reported on low power watermarking. Mohanty *et al.* implemented watermarking by hardware instead of software [8]. Dand low power consumption result. The hardware implementation has advantages of low power , high performance, and reliability. The disadvantage of his algorithm is that the processing needs to be done pixel by pixel. Hsu and Wu investigate a DCT-based embedding technique which focus on robustness. It's experimental results showed their approach can survive the cropping of an image, image enhancement, and JPEG compression [12]. However, it need a plenty of computation. Hence, it has high power dissipation.

The goal of this paper is to develop a DCT-based low power watermarking algorithm. Due to hide data in frequency domain image has robust characteristic to avoid attacks, watermarking has been discussed generally in frequency domain. Hiding watermarking in frequency domain can be divided into DCT-based [9] or DWT-based [10]. DWT has higher performance on peak signal-to-noise ratio (PSNR) however the block-based transform of DCT is easier to implement on hardware [11]. Hence we focus on low power dissipation of watermarking under DCT domain. We present a energy-efficient watermarking approach which is an improvement of Hsu and Wu's algorithm [12]. We perform block-based permutation based on frequency coefficients instead of using spacial coefficients proposed by Hsu and Wu to reduce the number of arithmetic operations. Our experimental results show that our pairing mechanism not only reduce 94.14% arithmetic operations of original pairing mechanism, but also increase the PSNR by 0.414 db for the best case.

The rest of this paper is organized as follows. The proposed watermarking scheme is presented in Section 2. This is followed by the performance experimental results in Section 3. In the last part, the conclusions are presented in Section 4.

2   Energy-Efficient Mechanisation for Watermarking

Pairing mechanism proposed in [12] used variance to define which block has higher frequency to hide more information. However, the computation of variance operation is complicated. In this paper, we directly exploit frequency coefficients as the criterion for pairing blocks.

We first discuss the characteristic of DCT, and introduce pairing mechanism in Section 2.1. Section 2.2 represents the analysis and comparison between the original work [12] and the proposed model.

2.1   Pairing Mechanism

In order to deduce our model, we introduce DCT first . The DCT used by JPEG, H.264, and MPEG-4 standards are performed on blocks of 8×8. This size gives a good compromise between compression efficiency and computational efficiency. Therefore, our work is based on the blocks of 8×8. Note that it can be applied to any block size without loss of generally. The two-dimensional DCT is defined by

$$C(0,0) = \frac{1}{N} \sum_{x=0}^{N-1} \sum_{y=0}^{N-1} f(x,y) \tag{1}$$

$$C(u,v) = \frac{1}{2N^3} \sum_{x=0}^{N-1} \sum_{y=0}^{N-1} f(x,y)[cos(2x+1)u\pi][cos(2y+1)v\pi] \tag{2}$$

for $u,v = 0,1,...,N\!-\!1$ where N is the block size. The inverse DCT (IDCT) can be written as

$$f(x,y) = \frac{1}{N} C(0,0) + \frac{1}{2N^3} \sum_{u=0}^{N-1} \sum_{v=0}^{N-1} C(u,v)[cos(2x+1)u\pi][cos(2y+1)v\pi] \tag{3}$$

for $x,y = 0,1,...,N\!-\!1$ where $[cos(2x+1)u\pi] \times [cos(2y+1)v\pi]$ of the second term represented the set of basis functions.

As can be seen in Eq. (3), a digital signal can be represented as a superposition of basis functions. Now we consider a set of single term in Eq. (3) for particular u,v

$$\{C(u,v)cos[(2x+1)u\pi]cos[(2y+1)v\pi] \,|$$
$$x = 0,1,...,N\!-\!1 \text{ and } y = 0,1,...,N\!-\!1\} \tag{4}$$

where $u,v \in \{0,1,...,N\!-\!1\}$. $|C(u,v)|$ is the gain of basis functions. Based on the observation, the spacial signal becomes more complexity with increasing $|C(u,v)|$, u, and v. We let S be the signal complexity in spacial domain.

$$S = \sum_{u=0}^{N-1} \sum_{v=0}^{N-1} k(u)k(v)|C(u,v)| \tag{5}$$

where $k(u)$ and $k(v)$ are ascending function. An image block can be hidden more information with a large value of S. Hence, high-frequencies are significant parameters for pairing mechanism. However, the middle-frequencies are also important elements, because watermark is hidden in middle-frequency. Therefore, we only consider middle- or high-frequency coefficients for determining pairing mechanism in the proposed approach. Note that $k(u)$ and $k(v)$ can be written as a two dimension filter $H(u,v)$ that passes high- and middle-frequencies and rejects low ones. In order to simplify Eq. (5), we replace $|C(u,v)|$ with $C(u,v)^2$ for reducing arithmetic operation. We let S' be our proposed model to determine sorting sequence as shown below.

$$S' = \sum_{u=0}^{N-1} \sum_{v=0}^{N-1} C(u,v)^2 \times H(u,v) \tag{6}$$

$$H(u,v) = \begin{cases} 1 & , \omega_l < \sqrt{u^2 + v^2} < \omega_h \\ 0 & ,\text{otherwise} \end{cases} \tag{7}$$

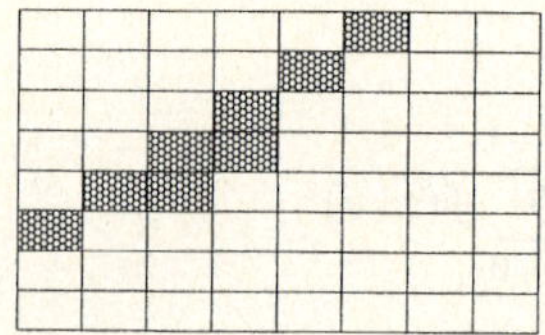

Fig. 1. The $H(u, v)$ filter by selecting middle-coefficients.

where $0 < \omega_l < \omega_h \leq (N-1)^2$. We assume that there are T frequency coefficients will pass the $H(u, v)$ filter. We call this approach as pairing blocks by sum of square mechanism (PBSM). In order to reduce the computation complexity, our objective is to use few coefficients to determine PBSM. A $H(u, v)$ filter example shown in Fig. 1, and the shadow lattices are the coefficients selected. According to the experimental results shown in the next section, we can hold performance with few computation, thereby lowering power.

2.2   Analysis and Statistics

In order to compare PBSM with original pairing mechanism, we discuss the computation complexity of S' and variance. The equation of variance can be represented as follow

$$V_{ar} = \sigma^2 = \frac{1}{N^2} \sum_{x=0}^{N-1} \sum_{y=0}^{N-1} [f(x, y) - \mu]^2 \tag{8}$$

$$\mu = \frac{1}{N^2} \sum_{x=0}^{N-1} \sum_{y=0}^{N-1} f(x, y) \tag{9}$$

where μ is the mean of a block and N^2 is the number of pixel in a block. Due to $H(u, v)$ of PBSM is used to determine which coefficient is selected, we implement $H(u, v)$ filter by placing the index of selected coefficients into memory. Hence, we reduce S' to a simpler form

$$S' = \sum^{T} C(u, v)^2. \tag{10}$$

In Table 1, we analyze the the number of arithmetic operations of Eqs. (8), (9), and (10). As observed in Table 1, the proposed method can reduce computation significantly when $T = 8$. Obviously, low computation complexity not only reduce

Table 1. The number of arithmetic operations required for $512 * 512$ pixel image.

	+	−	÷	Square	Total
Var	516096	262144	8192	262144	1048576
PBSM	28672	0	0	32768	61440

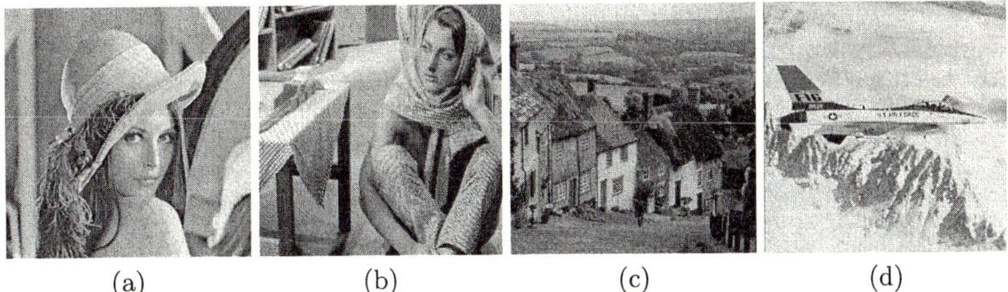

Fig. 2. The test images of size 512×512, (a) Lenna, (b)Barbara, (c)Goldhill, and (d)Jet.

Fig. 3. (a) The test image "Stream" of size 512×512, (b) the watermark of size 256×256, (c) the watermarked image, and (d) the extracted watermark.

power consumption, but also decrease operation time, hence energy-efficient. In next section, we will experiment original and proposed models on several images to tradeoff computation complexity and image quality.

3   Experimented Results

Although low power watermark processing is our target, the noise made by watermark must be unapparent in our approach. In this reason, we use peak signal to noise ratios (PSNR) to evaluate the distortion of watermarked image.

$$\text{PSNR(db)} = 10 \log_{10} \frac{255^2}{\text{MSE}} \qquad (11)$$

$$\text{MSE} = \frac{1}{N^2} \sum_{x=0}^{m-1} \sum_{y=0}^{m-1} (\alpha_{xy} - \beta_{xy})^2, \qquad (12)$$

where MSE is mean square error between a watermarked image and its original one. Since the extracted watermark is a visually recognizable pattern, we define a similarity measurement between the referenced watermark W and extracted watermark W' as normalized correlation (NC)

Table 2. Experimental results of original and proposed architecture.

	PSNR					NC				
	Lenna	Barbara	Goldhill	Jet	Stream	Lenna	Barbara	Goldhill	Jet	Stream
Var	39.0549	31.1323	35.9062	37.5904	32.1727	0.9960	0.9967	0.9981	0.9964	0.9975
PBSM	39.4690	31.1678	36.1699	37.9535	32.3882	0.9965	0.9974	0.9979	0.9965	0.9974
Diff	0.4142	0.0355	0.2637	0.3632	0.2155	0.0005	0.0007	-0.0002	0.0001	-0.0001

$$NC = \frac{\sum_x \sum_y W(x,y)W'(x,y)}{\sum_x \sum_y [w(x,y)]^2} \tag{13}$$

where $0 < NC \leq 1$ and the best qulaity of extracted watermark is $NC = 1$. It means the extracted watermark is the same as the original one.

To contrast the analytical results derived in the previous section with empirical data, we experiment on various images. The $H(u,v)$ filter was implemented by Fig. 1. The experimental results of PSNR and NC value are shown in Table 2. The second row of Table 2 are test images with size $512 \times 512 \times 8$ bit shown in Fig. 2 and Fig. 3(a). In the first column, "Var" represents the original algorithm, PBSM represents the proposed algorithm, and the "Diff" is the difference of PSNR computed as $(PSNR_{original} - PSNR_{proposed})$. In this experiment, we employed a binary watermark image with a size 256×256 shown in Fig. 3(b). We can observe that the performance (PSNR) can be held with only few sample coefficients. Particularly, the performance of PBSM is superior to original approach. Therefore, we can implement pairing mechanism by only using eight middle-frequency coefficients to determine pairing mechanism. Fig. 3 is an example for watermarking by PBSM. Fig. 3(c) is watermarked image from Fig. 3(a) and Fig. 3(b). As can be seen, Fig. 3(a) and Fig. 3(c) are so alike, that we can hardly distinguish one from the other. Fig. 3(c) and Fig. 3(d) are watermarked image and extracted watermark respectively. We can see that retrieval of watermark can be identify easily. Table 3 shows the number of arithmetic operations required in our experiment. Compared to the original variance approach [12], our pairing mechanism reduce number of operations form 1048576 to 61440 for $T = 8$. It is worth noting that we do not have to use complicated division operation.

4   Conclusion

This paper presented a new technique for embedding watermark into a still image based on DCT. A low power mechanism was proposed for pairing blocks. We directly exploit frequency coefficients as the criterion for pairing blocks. According to low computation reduce power consumption and operation time, we devised a watermark technique targeted at reducing arithmetic operation. Our experimental results showed that proposed method not only reduce 94.14% of pairing operations required, but also increase the PSNR 0.414 db for the best case. Therefore, our approach perform an useful mechanism successfully.

The power reduction is achieved in algorithm level, so that it can be applied on different kinds of hardware such as DSP, MIPS, ARM, etc. As the multimedia is widely used nowadays, low power watermark is practical in the future.

References

1. Cox, I. J., Kilian, J., Leighton, F. T., Shamoon, T.: Secure spread spectrum watermarking for multimedia. IEEE Transactions on Image Processing, Vol. 6, pp. 1673 - 1687, Dec. 1997.
2. Fei, C., Kundur, D., Kwong, R. H.: Analysis and design of watermarking algorithms for improved resistance to compression. IEEE Transactions on Image Processing, Vol. 13, pp. 126- 144, Feb. 2004.
3. Kii, H., Onishi, J., Ozawa, S.: The digital watermarking method by using both patchwork and DCT, IEEE International Conference on Multimedia Computing and Systems, Vol. 1, pp. 7 - 11, June 1999.
4. Wu, C.-F., Hsieh, W.-S.: Digital watermarking using zerotree of DCT. IEEE International Conference on Consumer Electronics, Vol. 46, pp. 87 - 94, Feb. 2000.
5. Sohn, S.-M., Kim, S.-H., Lee, S.-H., Lee, K.-J., Kim, S.: A CMOS image sensor (CIS) architecture with low power motion detection for portable security camera applications. IEEE Transactions on Consumer Electronics, Vol. 49 , pp.1227 - 1233, Nov. 2003.
6. Masselos, k., Merakos, P., Stouraitis, T., Goutis, C. E.: Novel vector quantization based algorithms for low-power image coding and decoding. IEEE Transactions on Circuits and Systems II: Analog and Digital Signal Processing, Vol. 46 , pp.193 - 198, Feb. 1999.
7. Jeong, H., Kim, J., Cho, W.-k.: Low-power multiplierless DCT architecture using image correlation. IEEE Transactions on Consumer Electronics, Vol. 50 , pp.262 - 267, Feb. 2004.
8. Mohanty, S. P., Ranganathan, N., Namballa, R. K.: VLSI implementation of invisible digital watermarking algorithms towards the development of a secure JPEG encoder. IEEE Workshop on Signal Processing Systems, 27-29 , pp.183 - 188, Aug. 2003.
9. Hernandez, J. R., Amado, M., Perez-Gonzalez, F.: DCT-domain watermarking techniques for still images: detector performance analysis and a new structure, IEEE Transactions on Image Processing, Vol. 9, pp. 55 - 68, Jan. 2000.
10. Hsieh, M.-S., Tseng, D.-C., Huang, Y.-H.: Hiding digital watermarks using multiresolution wavelet transform. IEEE Transactions on Industrial Electronics, Vol. 48, pp. 875 - 882, Oct. 2001.
11. Iain E. G. Richardson: Video Codec Design, John Wiley & Sons, LtD, pp. 127 - 161, 2002.
12. Hsu, C.-T., Wu, J.-L.: Hidden Digital Watermarks in Image. IEEE Transactions on Image Processing, Vol. 8, pp. 58 - 68, Jan. 1999.

Implementation of Image Steganographic System Using Wavelet Zerotree

Yuk Ying Chung[1], Yong Sun[1], Penghao Wang[1],
Xiaoming Chen[1], and Changseok Bae[2]

[1] School of Information Technologies,
University of Sydney, NSW2006, Australia
[2] Post-PC Platform Research Team,
Electronics and Telecommunication Research Institute, Daejeon, Korea

Abstract. With the development of mobile communication and internet technologies, digital media can be transmitted conveniently over the network. In this process the algorithms used to protect secret messages during transmission become an important issue. This paper presents a new high capacity image steganograpic model based on wavelet transform [1] and modified adaptive Wavelet zerotrees [2]. The proposed new system can embed more information than traditional algorithms without compression. The recovered hiding image has high Peak Signal to Noise Ratio (PSNR) value and good visual quality. The system is robust in terms of both JPEG compression and other signal processing attacks.

1 Introduction

Steganography is an ancient art of conveying messages in a secret way such that only the receiver knows the existence of message. The techniques of steganography are classified into linguistic steganography and technical steganography. The former consists of linguistic or language forms of hidden writing. The disadvantage is that users need to have a working knowledge of linguistics.

The encrypted message using the classic cryptography cannot pass the checkpoint on the network as a simple level of protection. Steganography provides another layer of protection on the secret message, which will be embedded in another media in such a way that the transmitted data will be meaningful to everyone. Steganography conceals the existence of the secret messages when compared with cryptography techniques which attempt to conceal the content of messages.

For the blind wavelet watermarking system [5][6], the nature of the picture can affect the capacity of information which can be embedded [3]. In Table (3), the second column (*Pixels*) indicates the number of bits which can be embedded in each different images which are named in the first column (*Image*). For example, *pepper* can only embed 5828 bits but *baboon* can embed 29462 bits in wavelet watermarking system. Therefore, it is advantageous to have an adaptive compression algorithm which can compress the hiding image in variable compression ratio according to the capacity of different images. The wavelet zero tree [2] is one of the adaptive compression algorithms. The subband coefficients [4] can be chosen according to the capacity size. However the coefficients of wavelet zero tree are inter-dependent to each other and cannot be robust to any attack. The hiding image cannot be recovered

R. Khosla et al. (Eds.): KES 2005, LNAI 3681, pp. 1226–1232, 2005.

if any data has distorted. In this paper, a new modified wavelet zero tree algorithm is proposed to implement this high capacity image steganographic system which can have both adaptive compression and robust to JPEG and signal processing attack.

2 Implementation of High Capacity Image Steganographic System

Our new high capacity image steganographic system is based on wavelet watermarking model using Vector Quantization (VQ) and modified wavelet zerotree to compress the hiding image before being embedded into the cover image. Figure 1 shows the detail of how to implement this steganographic system.

The steps of our new proposed modified wavelet zerotree algorithm are:

1. Transform the cover image with 9-7 tap Atonini biorthogonal filters;
2. Transform the hiding image with integer wavelet, output the transformed coefficients in zerotree structure;
3. Find the hiding position in the cover image with threshold values;
4. Embed LL2, HL2 and LH2 first;
5. Then randomised LL2 and embed individually;
6. Embed HL2 and LH2 together;
7. Repeat to embed LL2 if any more hiding position is available and
8. Inverse wavelet transform to the cover image to obtain the steganographic image.

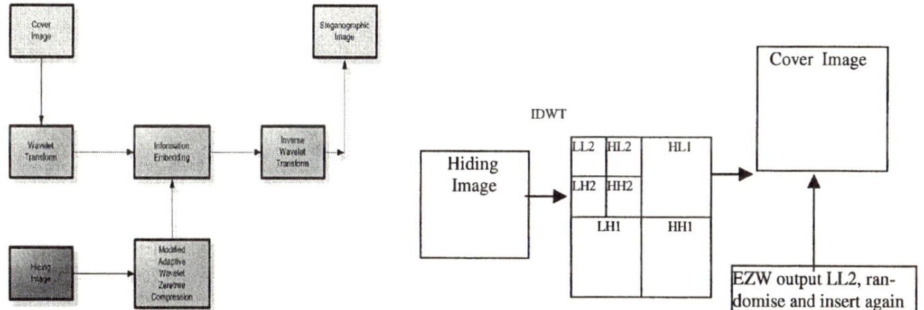

Fig. 1. Image steganography using wavelet transform and modified adaptive wavelet zerotree

Fig. 2. Modified wavelet zerotree LHLL

3 Results

3.1 Results of Fractal, Wavelet Zerotree and VQ

(i) Three common compression algorithms were tested: Fractal, Wavelet zero tree and VQ without any signal processing attack. From Table 1, Fractal has the best result if a signal processing attack is not applied to the watermarked image.

(ii) Three compression algorithms were tested with some signal processing attacks. The attacks are as following:

    i) ¼ HH part of the watermarked image has been cropped;

    ii) Apply JPEG compression factor 8 to the watermarked image;

iii) Add 1% Gaussian noise to the watermarked image and
iv) Adjust the contrast and brightness of the watermarked image using factor 10 and 10.

Table 1. Test result for the existing compression algorithms without any attack

	Baboon	Barbara	Lena
Fractal			
Wavelet Zerotree			
VQ			

Table 2. Barbara test with watermark repeat three times

	No Attack	1/4 Cropping	Low jpeg compression	1% noise	Adjustment 10 10
Fractal					NA
Wavelet Zerotree					
VQ					

From Table 2, the existing image compression algorithms Fractal, Wavelet zero tree and VQ are not suitable for the high capacity image steganographic system because they cannot robust to the signal processing attack.

3.2 Results of the New LHLL Algorithm

From Section 3.1, the existing image compression algorithms are not suitable for the high capacity image steganographic system. In this paper, we have proposed the new modified wavelet zero tree algorithm LHLL. The results show that the LHLL algorithm is robust to the signal processing attacks and suitable for the high capacity image steganographic system.

Different host images and grey level hiding images were used in testing this algorithm. The original hiding image *Elaine* used in this work is shown in Figure 3. The cover image airplane is 512x512 grey level, and the hiding image *Elaine* is 256 x 256 grey level. Figure (4) shows the result of recovered hiding image *Elaine* using VQ and LHLL. The output from VQ (Figure 4a) is clearer than Figure 4b but has blocking effect, while the output from LHLL (Figure 4b) is more smooth.

Fig. 3. Original hiding image Elaine

(a)   (b)

Fig. 4. (a) VQ PSNR: 28.0893db, (b) LHLL PSNR 28.519db

Table (3), (4) and (5) compares the recovered images using VQ compression and modified wavelet zerotree compression. From Table (3), VQ has better results in a no distortion test. From Table (4) and (5), the proposed LHLL system has better result in the cropping test and the JPEG compression test. Therefore LHLL system is more robust against the JPEG compression and signal processing attacks. Table (6) and Table (7) show the good visual quality of recovered hiding image *Elaine* in different cropping tests and JPEG compression tests.

Table 3. No distortion comparison

Image	Pixels	VQ (in dB)	LHLL (in dB)
splash	3425	15.7564	22.6995
512elaine	4858	28.0893	26.4694
pepper	5828	27.7617	26.7765
yongslena	5941	28.0893	26.8098
airplane	8237	28.0893	27.3933
truck	8420	28.0893	27.4392
tankacp2	12848	28.0893	28.1476
man	13566	26.2491	21.394
airport	15722	28.0893	28.2726
bridge	18092	27.0837	22.5966
aerial	18882	27.5116	23.3567
baboon	29462	28.0893	28.5191
MEAN		26.74897	25.82286

Table 4. Cropping test comparison

Image	Pixels	VQ (in dB)	LHLL (in dB)
splash	3425	9.97564	12.417
512elaine	4858	11.6062	14.5929
pepper	5828	12.2836	14.7415
yongslena	5941	8.96975	10.3613
airplane	8237	8.83446	11.5996
truck	8420	11.072	14.284
tankacp2	12848	14.4947	11.9262
man	13566	12.4626	11.6377
airport	15722	16.0601	13.6516
bridge	18092	16.9804	14.1874
aerial	18882	14.9279	12.9251
baboon	29462	17.398	15.7352
MEAN		12.92211	13.17163

Table 5. JPEG compression comparison

Image	Pixels	VQ (in dB)	LHLL (in dB)
splash	3425	8.36437	14.2653
512elaine	4858	9.16166	14.4313
pepper	5828	8.55078	12.919
yongslena	5941	8.46289	13.1005
airplane	8237	8.18773	11.9466
truck	8420	8.43636	11.7939
tankacp2	12848	9.76477	10.8904
man	13566	9.27771	10.3959
airport	15722	9.04062	11.2073
bridge	18092	9.23814	11.4947
aerial	18882	9.4127	11.859
baboon	29462	10.2476	12.9694
MEAN		9.012111	12.27278

4   Conclusion

High capacity image steganographic model based on VQ, modified wavelet zerotree
(LHLL) and wavelet watermarking were investigated. In the proposed method, the
hiding image is compressed by the adaptive Wavelet zerotrees. The recovered Elaine
has high PSNR value and good visual quality. The test results prove that this new
high capacity Steganographic system is robust in countering both JPEG and other
signal processing attacks.

Table 6. Cropping test result for LHLL system

Cover image (hiding capacity)	Crop1	Crop2	Crop3	Crop5 (1/4)
512elaine (4858) in dB	25.6435	25.0461	19.9841	14.5929
pepper (5828) in dB	21.2855	19.1966	16.5246	14.7415
airplane (8237) in dB	23.4297	22.686	21.8207	11.5996

Table 7. JPEG compression test result for LHLL system

Cover image (hiding capacity)	Factor5	Factor7	Factor9	Factor11
512elaine (4858) in dB	19.1581	21.1557	26.547	26.4716
pepper (5828) in dB	16.7882	18.2918	26.181	26.7765
airplane (8237) in dB	13.8485	15.1261	27.1091	28.2755

References

1. M. Atonini, M. Barlaud, P. Mathieu and I. Daubechies, "Image coding using the wavelet transform" IEEE Transactions on Image Processing, 1, 205-220 (1992)
2. C. Valens c.valens@mindless.com "Embedded Zerotree Wavelet Encoding" 1999

3. S. Servetto and C. Podilchuk, "Capacity issues for digital watermarking", in IEEE Int. Conf. Image Processing '98

4. Algazi, V.R. and R.R. Estes, "Analysis based coding of image transform and subband coefficients", Proceedings of the SPIE, vol.2564(1995), p.11-21

5. Pan R, Gao YX, "Image watermarking mehod based on wavelet transform", Journal of Image and Graphics, 2002, 7(7): 667~671.

6. M. Barni, F.Bartolini, V. Cappellini, A. Lippi, and A. Piva, "A DWT-based technique for spatio-frequency masking of digital signatures", in Proceedings of the 11th SPIE Annual Symposium, Electronic Imaging '99, Security and Watermarking of Multimedia Contents, P. W. Wong, ed., vol. 3657, (San Jose, CA, USA), January 1999

7. http://www.data-compression.com/vq.shtml

8. R. M. Gra, "Vector Quantization", IEEE ASSP Mug., p4-29, April 1984

A RST-Invariant Robust DWT-HMM Watermarking Algorithm Incorporating Zernike Moments and Template

Jiangqun Ni*, Chuntao Wang, Jiwu Huang, and Rongyue Zhang

School of Information Science and Technology, SunYat-Sen University
Guangzhou 510275, P.R. China
Phn: 86-20-84036167
issjqni@zsu.edu.cn

Abstract. In this paper, a coarse to fine strategy is presented to achieve the geometrical synchronization incorporating Zernike moments and the template. At the coarse stage, the rotation and scaling parameters are estimated with Zernike moments of the translation-normalized image, while the translation parameters are estimated with the centroid's increments between the original image and the image that is corrected for rotation and scaling. At the fine stage, the accurate Rotation, Scaling and Translation (RST) parameters are obtained by matching the template around the roughly estimated RST values. A watermarking scheme based on the vector Hidden Markov Model in wavelet domain (DWT-HMM) is also adopted. High robustness is observed against geometric manipulations and other StirMark attacks in our simulations.

1 Introduction

Robustness against geometric attacks is the key issue [1~2] in designing the watermarking algorithm. In this paper, we propose a coarse to fine strategy to achieve geometrical synchronization. At the coarse stage, the rotation and scaling parameters are estimated with Zernike moments [3] of the translation-normalized image, while the translation parameters are estimated with the centroid's increments between the original image and the image that is corrected for rotation and scaling. At the fine stage, the accurate Rotation, Scaling and Translation (RST) parameters are obtained by matching the template around the roughly estimated RST values, which thus greatly reduces the searching space. Finally, the geometrically attacked image is recovered with the accurate RST values.

In our previous work [4], an algorithm based on vector HMM in wavelet domain is presented for robust multi-bit image watermarking. In this paper, the coarse to fine geometrical resynchronization scheme is combined with the HMM-based watermarking scheme and gives rise to the RST-invariant robust DWT-HMM watermarking algorithm.

The remainder of this paper is organized as follows. Section 2 and 3 introduce HMM-based watermarking scheme and embedding/extraction of the template, respec-

---

* Corresponding author

R. Khosla et al. (Eds.): KES 2005, LNAI 3681, pp. 1233–1239, 2005.
© Springer-Verlag Berlin Heidelberg 2005

tively. Section 4 explains the strategy for geometrical synchronization. Section 5 shows the experimental results. Finally, we draw the conclusion in section 6.

2   Watermarking Scheme Based on DWT-HMM

The watermarking algorithm in [4] employs a vector DWT-HMM, which takes into account both the energy correlation across the scale and the different subbands at the same scale of the wavelet pyramid [5]. To incorporate the vector DWT-HMM, the image wavelet coefficients are represented with the structure of vector tree. In the interest of resisting against JPEG attack, the watermark is embedded only to the coarsest 2 wavelet scales. Therefore, each vector tree includes only 15 nodes. The process of watermark embedding can be described as follows: (1) the watermark **m** is coded to $\mathbf{m}_c$ with 1/3 convolution code; (2) $\mathbf{m}_c$ is proceeded with DSSS (Direct Sequence Spread Spectrum) and interleaving to generate **w**; (3) **w** is grouped into M-bit segments; (4) each M-bit (M=5 in our work) segment is encoded with a 15-bit long code word and embedded into the 15-node vector tree. The process is shown in Fig.1, which is the integrated diagram for geometrical synchronization and HMM-based watermarking.

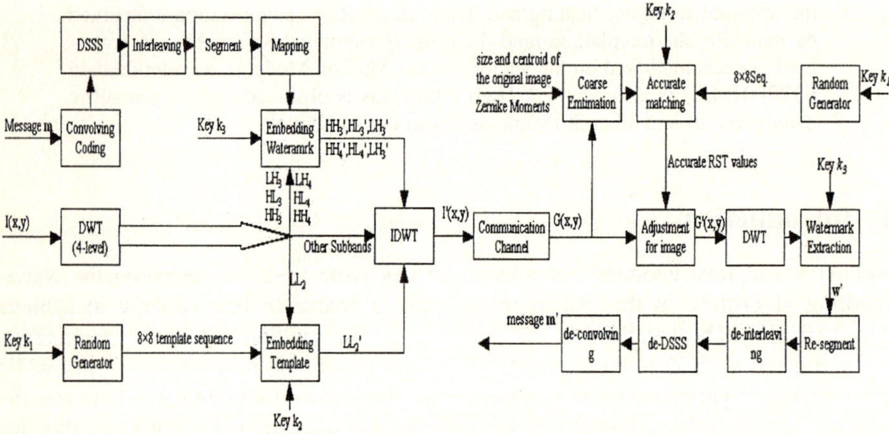

Fig. 1. Diagram for geometrical synchronization and watermarking

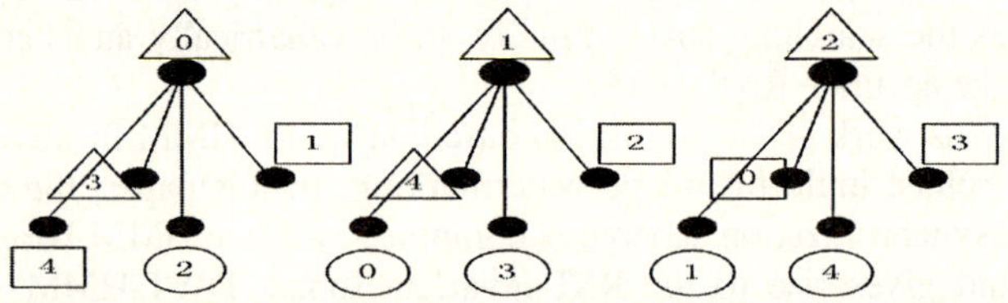

Fig. 2. Optimal mapping pattern from M=5 bits to a 15-node vector tree. The triangle and circle represent the repeated versions of the M=5 bits and the rectangle represents the inverse version

Base on the robustness performance analysis in [4], the 15-bit long code word should be designed so that the parent node is selected to embed data with priority and the distance among the $2^M = 32$ code words is kept large. With these principles, the optimal mapping/coding strategy is developed, as shown in Fig. 2. The watermark extracting is the inverse process of the watermarking embedding (See Fig.1).

3 Embedding and Detection of the Template

A set of random sequence $\mathbf{S}\{s_i; s_i \in \{-1, +1\}, i = 1, 2, \quad ,64.\}$ is first generated with the key k_1 as a seed. Then they are reshaped into an 8*8 rectangle to use as the template, which is embedded into an 8*8 area selected from the LL_2 subband with the key k_2. The template embedding strategy is described in [6].

In template detection, an 8*8 area is similarly chosen from the LL_2 subband with the same key k_2, where a sequence $\mathbf{T}\{t_i; t_i \in \{-1, +1\}, i = 1, 2, \quad ,64\}$ is extracted with the method in [6]. For the extracted sequence $\mathbf{T}$ and the original one $\mathbf{S}$, their correlation is computed with the following formula:

$$\rho_{T,S} = (1/64)\sum_{n=1}^{64} T_n \cdot S_n . \tag{1}$$

If $\rho_{T,S}$ is not less than a predefined threshold, the template detection is successful. Both the embedding and detection processes are shown in the Fig.2.

4 Coarse to Fine Geometrical Resynchronization

4.1 Coarse Estimation of RST Parameters

If an image is rotated for α, then the rotated angle α can be estimated with the Zernike moments of the two images before and after rotation [7].

If an image is scaled for β, it's easy to deduce the expression (2) according to the definition of Zernike moments in [3]:

$$Z_{nm}^s = \beta^{2+n} Z_{nm} \quad (|m| = n), \tag{2}$$

where Z_{nm}^s is the $(n, m)th$ Zernike moment of the scaled image. With the method similar to rotation estimation [7], the scaling factor β can be estimated using (2). In order to make the estimation of rotation and scaling be independent of the translation estimation, a translation-normalized strategy [3] is employed before the estimation.

If an image is translated for Δx and Δy rightward and downward respectively, then Δx and Δy can be estimated with the following formula:

$$\Delta x = \bar{x}_1 - \bar{x}_0 \ and \ \Delta y = \bar{y}_1 - \bar{y}_0 , \tag{3}$$

where $(\bar{x}_0, \bar{y}_0)$ and $(\bar{x}_1, \bar{y}_1)$ are the centroids of the images before and after translation respectively. Note that translation estimation with (3) should be applied after the correction of rotation and scaling, since they would also contribute to the centroid's increments.

4.2 Accurate Recognition of RST Parameters

To recognize the RST parameters accurately, the template is matched around the roughly estimated RST values in the section 4.1. The matching algorithm is as follows: 1) the related searching spaces are constructed with the roughly estimated RST values; 2) the constructed spaces are searched element by element and the attacked image is corrected for every searching; 3) the template sequence is detected and the correlation $\rho_{T,S}$ is computed with the formula (1); 4) the searching is converged if the 3 $\rho_{T,S}$ s from the consecutive 3 matching are not less than a predefined threshold and the 2nd $\rho_{T,S}$ is larger than or equal to the other two. Otherwise, the searching is continued until the convergence is reached or the spaces are completely searched. Finally, the accurate RST parameters corresponding to the maximum $\rho_{T,S}$ are obtained when the matching is successful. For the separately geometrical attack (e.g., rotation, scaling and translation), the searching spaces for the accurate recognition of rotation, scaling, and translation parameters are as follows:

1. $S_r = \{r; r = Step_r * k + \alpha_0 - DA, k \in [0, 2DA / Step_r]\}$ and $S_t = \{(x, y); x, y \in \{-1, 0, 1\}\}$;

2. $S_s = \{s; s = Step_s * k + \beta_0 - DS, k \in [0, 2DS / Step_s]\}$;

3. $S_t = \{(x, y); x = Step_t * k_1 + \Delta x - DX, y = Step_t * k_2 + \Delta y - DY, k_1 \in [0, 2DX / Step_t], k_2 \in [0, 2DY / Step_t]\}$.

The α_0, β_0 and $\Delta x / \Delta y$ are the estimated RST parameters; the DA, DS and DX / DY are all positive constant values determined by experiments; the $Step_r$, $Step_s$ and $Step_t$ are the searching steps. Therefore, the average searching time is $(2DA / Step_r + 1) * 9 / 2$ for rotation, $(2DS / Step_s + 1) / 2$ for scaling, and $(2DX / Step_t + 1) * (2DY / Step_t + 1) / 2$ for translation, respectively.

4.3 Accurate Recognition of Joint RST

Under the joint RST attacks, the spaces of S_r, S_s and S_t should be searched jointly and orderly. Considering the accuracy of RST estimation, the space S_s should be searched first, then S_r and S_t. However, the estimated RST parameters would deviate from the true values to some extent under the joint RST attacks. In order to match the template efficiently and accurately, the coarse to fine matching strategy can be further deployed, i.e., the approximate RST parameters are first estimated coarsely by increasing the RST searching step and enlarging the RST searching spaces, then the searching spaces of S_r, S_s and S_t are reconstruct based on the coarsely estimated results and these 3 new spaces are searched again orderly and jointly.

5 Experimental Results

In our experiments, 5 standard 512*512*8b images with different texture characteristics are tested. The images are first decomposed into 4-level pyramids with the 9/7 biorthogonal wavelet; the 8*8 template and a 60-bit watermark are then embedded in the LL_2 and $HH_i, HL_i, LH_i (i = 3, 4)$ subbands, respectively. The watermarked images with template are shown in Fig. 3.

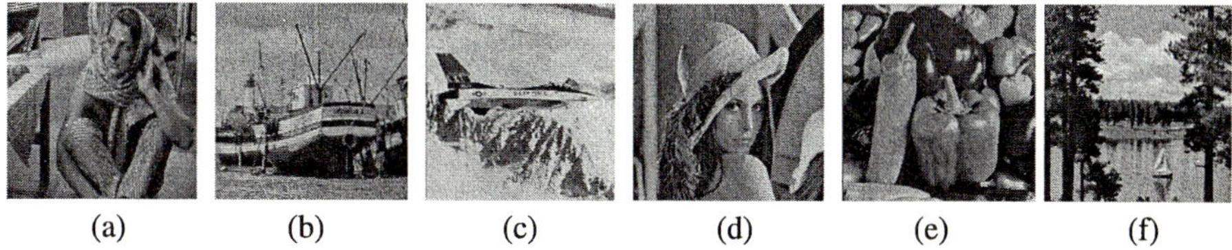

(a)   (b)   (c)   (d)   (e)   (f)

Fig. 3. Watermarked images with template: (a) barb (PSNR=40.26db); (b) boat (PSNR=39.74db); (c) f16 (PSNR=38.83db); (d) lena (PSNR=42.72db); (e) peppers (PSNR=40.33db); (f) sailboat (PSNR=39.84db)

In rotation and scaling estimation, the higher order the Zernike moments are used, the higher estimation accuracy can be. With the usage of template registration, however, the Zernike moments can be reduced to $(0,0)th, (2,2)th, (3,3)th$ and $(4,4)th$ order on the condition of $n = |m|$ and $|m| > 0$. Note that the $(1,1)th$ Zernike moment is always zero [3]. Fig. 4 gives the rotation and scaling estimation results for "lena" image, where $\sigma_r = \pi/4m$ and $\sigma_s = 1/20$ are set for the rotation and scaling estimation respectively, which is also the case for other images.

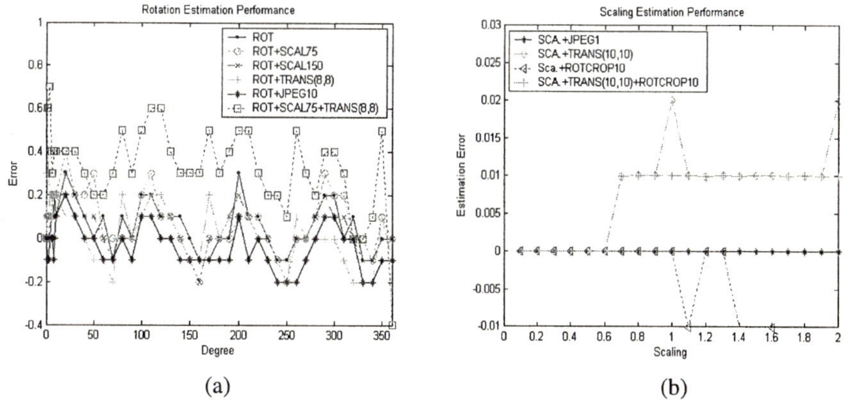

(a)   (b)

Fig. 4. Performance of rotation and scaling estimation with Zernike moments: (a) Performance of rotation estimation; (b) Performance of scaling estimation

The images in Fig.3 are then attacked with StirMark 4.0. For template matching, the related parameters are set as $DA = 0.3$, $DS = 0$, $DX = DY = 1$, $Step_r = 0.1$ and $Step_t = 1$. Thus the average searching time is (7*3*3/2)=31.5 for rotation, 1 for scaling and (3*3/2)=4.5 for translation. The threshold $Thresh$ is determined to be 0.59 given the false alarm probability $1*10^{-6}$. The related experimental results are included in the Table 1.

Table 1. Watermark Extraction under attacks in StriMark 4.0. The type of MedianCut is 2*2, 3*3, 4*4, 5*5 and 7*7, the detail parameter 1_0.01_0_0_1_0 is for X-Shearing, 1_0_0_0.01_1_0 for Y-Shearing, 1_0.01_0_0.01_1_0 for XY-Shearing and 1.013_0.008_0.011_1.008 for G_Shearing that stands for General_Shearing

Image / Attacks	barb	boat	f16	lena	sailboat
Rot._Crop*(°)	-20~20	-20~20	-20~20	-20~20	-20~20
Scaling	≥ 0.3	≥ 0.3	≥ 0.3	≥ 0.3	≥ 0.3
X-Shearing	OK	OK	OK	OK	FAIL
Y-Shearing	OK	OK	OK	OK	OK
XY-Shearing	FAIL	FAIL	OK	OK	FAIL
G_Shearing	OK	OK	OK	OK	FAIL
AddNoise	1~2	1~2	1~7	1~3	1~3
JPEG	19~100	15~100	16~100	18~100	19~100
MedianCut	$\leq 7*7$	$\leq 5*5$	$\leq 5*5$	$\leq 3*3$	$\leq 5*5$
Sharpening	OK	OK	OK	OK	OK

In addition, the images in Fig.3 are implemented with a joint attack: scaling to 1.2, translation for (5,5), rotation 10° and JPEG compression with quality factor 20. One of the examples is shown in Fig. 5(a); Fig. 5(b) is the corrected version. The related results are shown in Table 2.

(a)   (b)

Fig. 5. The jointly attacked "lena" image and its corrected version. (a) The jointly attacked image; (b) The image recovered from the joint attack

Table 2. Template matching and watermark extraction under the joint attacks, where $DS = 0$, $DA = 0.5$, $DX = DY = 1$, $Step_r = 0.1$ and $Step_t = 1$ are set

Image / Parameter		barb	boat	f16	lena	sailboat
Scaling	Estimated	1.2	1.2	1.2	1.2	1.2
	Matched	1.2	1.2	1.2	1.2	1.2
Rotation	Estimated	10.7	10.3	9.7	9.6	9.7
	Matched	10.1	10.1	10.1	10.1	10.1
Translation		(4,3)	(4,3)	(4,3)	(4,3)	(4,3)
Correlation		0.875	0.875	0.875	0.875	0.875
Extraction		OK	OK	OK	OK	OK

6 Conclusion

In this paper, a coarse to fine geometrical synchronization strategy is proposed to counter geometrical attacks, which reduces the searching spaces and improves the searching efficiency greatly. A watermarking scheme based on the vector DWT-HMM is also employed for multi-bit image watermarking. High robustness is observed against StirMark attacks and their joint attacks in our experimental results. Consequently, combining the HMM based watermarking scheme with the coarse to fine geometrical synchronization strategy gives rise to a RST-invariant robust watermarking algorithm.

References

1. J. Cox, M. L. Miller, J. A. Bloom, "Digital Watermarking," *Electronics and Industry Publishing Company, PRC*, 2003.5.
2. V. Licks and R. Jordan, "Geometric Attacks on Image Watermarking Systems: A Survey," *Submitted to Journal Publication*, 2003.
   [online]. http://www.eece.unm.edu/~vlicks/texts/publications.htm
3. Khotanzad and Y. H. Hong, "Invariant Image Recognition by Zernike Moments," *IEEE Trans. on Patter Analysis and Machine Intelligence*, vol. 12, no. 5, May 1990.
4. R. Zhang, J. Ni, J. Huang, "A robust multi-bit image watermarking based on HMM in the wavelet domain," accepted by the *Chinese Trans. on Software*.
5. M. N. Do and M. Vetterli, "Rotation Invariant Texture Characterization and Retrieval using Steerable Wavelet-domain Hidden Markov Models," *IEEE Transactions on Multimedia*, Volume: 4 Issue: 4, Dec. 2002 Page(s): 517 -527.
6. X. Kang, J. Huang, Y. Q. Shi, and Y. Lin, "A DWT-DFT composite watermarking scheme robust to both affine transform and JPEG compression," *IEEE Trans. On Circuit and Systems for Video Technology*, vol. 13, no. 8, Aug. 2003.
7. W.-Y. Kim and Y.-S. Kim, "Robust rotation angle estimator," *IEEE Trans. on Patter Analysis and Machine Intelligence*, vol. 21, no. 8, Aug. 1999.

Security Evaluation of Generalized Patchwork Algorithm from Cryptanalytic Viewpoint

Tanmoy Kanti Das[1], Hyoung Joong Kim[2,*], and Subhamoy Maitra[1]

[1] Applied Statistics Unit, Indian Statistical Institute
203 B T Road, Kolkata 700 108, India
{das_t,subho}@isical.ac.in
[2] Department of Control and Instrumentation Engineering,
Kangwon National University, Chunchon 200-701, Korea
khj@kangwon.ac.kr

Abstract. In this paper we present a cryptanalysis of the generalized patchwork algorithm under the assumption that the attacker possesses only a single copy of the watermarked image. In the scheme, watermark is inserted by modifying randomly chosen DCT values in each block of the original image. Towards the attack we first fit low degree polynomials (which minimize the mean square error) on the data available from each block of the watermarked content. Then we replace the corresponding DCT data of the attacked image by the available data from the polynomials to construct an attacked image. The technique nullifies the modification achieved during watermark embedding. Experimental results show that recovery of the watermark becomes difficult after the attack.

Keywords: Cryptanalysis, Digital Watermarking, Information Security, Multimedia Systems, Discrete Cosine Transform.

1   Introduction

Serious interest has been growing in past few years in the field of Digital Watermarking towards copyright protection of electronic contents. Thousands of watermarking schemes are being proposed and in the state of the art works, the schemes are robust against signal processing attacks and employ complicated signal processing techniques to gain user confidence. However, even the well known schemes are generally not evaluated against serious cryptographic analysis other than routine signal processing attacks like rotation, cropping, random geometric bending (Stirmark) [12, 13], etc. In [6], it has been argued that the correlation based watermarking schemes are vulnerable against collusion attack, but that may require large number of watermarked copies which may not be available in practice. Recently the most successful attacks are the ones that use a single watermarked copy for cryptanalysis [2, 4, 5, 7, 9–11]. In the same line, here we study the security of the Generalized Patchwork Algorithm [14] with respect to single copy attack.

---

* This work was in part supported by the Media Service Research Center (MSRC-ITRC), Ministry of Information and Communication, Korea.

R. Khosla et al. (Eds.): KES 2005, LNAI 3681, pp. 1240–1247, 2005.

Any invisible watermarking scheme inserts a signal $s^{(i)}$ to the host image I to get the watermarked copy $I^{(i)}$ which is visually indistinguishable from I. This watermarked copy $I^{(i)}$ is sold to the ith buyer. The watermark $s^{(i)}$ can be made buyer-specific for forensic tracking if required. Now the watermark retrieval algorithm works in the following way. From the available image (which may be attacked using image processing or cryptanalytic techniques) a signal $s^{\#}$ is recovered. From $s^{\#}$, the buyer i is suspected if $s^{(i)}$ possesses some significant correlation with $s^{\#}$.

It is well accepted in the field of cryptology and information hiding [1] that the algorithm will be known to the attacker but not the secret information (key or watermark). Hence, a watermarking strategy is considered to be secured, if a cryptanalyst, who has access to the algorithmic principle of the strategy but has no access to the secret key, should not be able to erase the watermark [1]. This principle had been introduced by Kerckohoffs [8] in as early as 1883. There are many instances to justify that "Security by Obscurity" (the assumption that opponent will stay ignorant about the system being used) can't work. One of the recent examples is the Secure Digital Music Initiative (SDMI) challenge [2]. In the challenge, the algorithmic principles of watermarking strategies were kept secret. In spite of that, authors of [2] were able to successfully attack one of the schemes. Thus, while analyzing a watermarking scheme, it is assumed that the scheme is known to the attacker but the keys are unknown.

The existing watermarking models need to be analyzed using cryptanalytic techniques as it is done for standard cryptographic schemes. We here look into the watermarking scheme as a cryptographic model and provide an attack which can be considered as a cipher-text only jamming attack since the proposed attack removes/invalidates the secret watermark using a watermarked copy only. In the process of cryptanalysis one needs to construct an attacked image $I^{\#}$ from the available image $I^{(i)}$ in a manner that there is no significant correlation between $s^{\#}$ and $s^{(i)}$. That is, if the watermark is a binary string, then the strings $s^{\#}$ and $s^{(i)}$ will match at approximately half of the places and will not match at approximately in the rest half of the places. Thus, it will be impossible to identify the malicious buyer i anymore. Moreover, $I^{(i)}$ and $I^{\#}$ need to be visually indistinguishable. Note that to the attacker only a single copy of the image is available and he has no knowledge about the watermark signal $s^{(i)}$ or the key.

In Section 2 we present the scheme [14] that we attempt to cryptanalyse. Exact cryptanalysis strategy is presented in Section 3 with necessary background material and experimental results. Section 4 concludes the paper.

2   The Generalized Patchwork Algorithm

Take an image of size $M \times M$. Consider the image in DCT domain, where the Discrete Cosine Transform has been done using block size $N \times N$. There is a secret key k and a secret watermarking signal w. The watermarking signal $w = [w_1, \ldots, w_t]$ is $t = \frac{M \times M}{N \times N}$ bits binary pattern.

2.1 Bit Embedding

Now we consider each DCT block where a bit is inserted. The key k is used as a seed of pseudo random number generator to get random index values from $[Z_1, Z_2]$, $1 \leq Z_1 < Z_2 \leq N \times N$, where the index values are matched with orders of the JPEG-like zigzag pattern in the DCT block. Now two index sets I^0 and I^1, each containing $2n$ elements, are generated using the secret key k as follows:

$$I^0 = I^{0+} \cup I^{0-}, \ I^{0+} = \{I_1^0, \ldots I_n^0\}, \ I^{0-} = \{I_{n+1}^0, \ldots I_{2n}^0\},$$
$$I^1 = I^{1+} \cup I^{1-}, \ I^{1+} = \{I_1^1, \ldots I_n^1\}, \ I^{1-} = \{I_{n+1}^1, \ldots I_{2n}^1\}.$$

I^0 and I^1 are used to embed the watermark bits 0 and 1, respectively. Define $A^j = \{A_1^j, \ldots A_n^j\}$ as the DCT coefficients corresponding to I^{j+} and $B^j = \{B_1^j, \ldots B_n^j\}$ as the DCT coefficients corresponding to I^{j-} for $j = 0, 1$. Suppose that bit 0 is the watermark signal. Then we calculate the means $\overline{A^0} = \frac{1}{n}\sum_{i=1}^{n} A_i^0$, $\overline{B^0} = \frac{1}{n}\sum_{i=1}^{n} B_i^0$ and the sample variances $S_{A^0}^2 = \frac{1}{n-1}\sum_{i=1}^{n}(A_i^0 - \overline{A^0})^2$, $S_{B^0}^2 = \frac{1}{n-1}\sum_{i=1}^{n}(B_i^0 - \overline{B^0})^2$. The embedding function is as follows:

$$A_i^{0*} = (1 + sign(S_{A^0}^2 - S_{B^0}^2)P_1)A_i^0 + sign(\overline{A^0} - \overline{B^0})\sqrt{P_2}\frac{S_{E0}}{2},$$

$$B_i^{0*} = (1 - sign(S_{A^0}^2 - S_{B^0}^2)P_1)B_i^0 - sign(\overline{A^0} - \overline{B^0})\sqrt{P_2}\frac{S_{E0}}{2}, \qquad (1)$$

where

$$S_{E0} = \sqrt{\frac{\sum_{i=1}^{n}(A_i^0 - \overline{A^0})^2 + \sum_{i=1}^{n}(B_i^0 - \overline{B^0})^2}{n(n-1)}}.$$

Thus, in total $4n$ DCT coefficients are chosen for inserting the watermark and out of them $2n$ coefficients are modified.

2.2 Bit Extraction

For decoding and detecting the watermark, the following strategy is used. Note that here we work on the watermarked image and the scheme is oblivious, so the original image is not required. Only the secret key k need to be known to generate the index sets. Take

$$T_0^Q = \beta \max\{\frac{S_{A^0}^2}{S_{B^0}^2}, \frac{S_{B^0}^2}{S_{A^0}^2}\} + (1-\beta)\frac{(\overline{A^0} - \overline{B^0})^2}{S_{E0}^2},$$

$$T_1^Q = \beta \max\{\frac{S_{A^1}^2}{S_{B^1}^2}, \frac{S_{B^1}^2}{S_{A^1}^2}\} + (1-\beta)\frac{(\overline{A^1} - \overline{B^1})^2}{S_{E1}^2}. \qquad (2)$$

If $P_1 = 0$ and $P_2 > 0$, then $\beta = 0$. On the other hand, if $P_1 > 0$ and $P_2 = 0$, then $\beta = 1$. If $T_0^Q > T_1^Q$, then we extract bit 0 else we extract bit 1.

The basic idea behind the bit extraction is as follows.

Case 1. Consider a specific DCT block and the case $P_1 > 0$ and $P_2 = 0$, i.e., $\beta = 1$. If 0 has been inserted, then the values in DCT coefficients A^0 and B^0

will change, but A^1 and B^1 will not change. Thus, the sample variances $S^2_{A^1}$ and $S^2_{B^1}$ of the watermarked image will be the same as the sample variances of the original image and statistically the values will be equal (as they both are sample variances). So T^Q_1 will stay close to 1. On the other hand, though in the original image $S^2_{A^0}$ and $S^2_{B^0}$ are statistically equal (both are sample variances), in the watermarked image one of those two values will increase and the other will decrease. Thus, T^Q_0 will be away from (considerably larger than) 1.

Case 2. Consider a specific DCT block and Now we explain the case $P_1 = 0$ and $P_2 > 0$, i.e., $\beta = 0$. If 0 has been inserted, then the values in DCT coefficients A^0 and B^0 will change, but A^1 and B^1 will not change. Thus, $(\overline{A^0} - \overline{B^0})^2$ will be away from 0. On the other hand, $\overline{A^1}$ and $\overline{B^1}$, both being the sample mean, $(\overline{A^1} - \overline{B^1})^2$ will be close to 0. Thus, T^Q_0 will be away from (considerably larger than) 0 and T^Q_1 will be close to 0.

In [14], both $P_1 > 0$ and $P_2 > 0$ case has not been considered in detail. Due to space constraint we also omit that here, however, the attack strategy will be similar as explained in the following section. That will be presented in the full journal version of this paper.

3   Cryptanalysis of the Generalized Patchwork Algorithm

As we have discussed in the last section, due to the embedding of the bits one of T^Q_0 and T^Q_1 becomes much larger than 1 (case 1) or 0 (case 2) (due to the modification of DCT coefficients) and the other one stays close to 1 (case 1) or 0 (case 2) (as it is sample data from the DCT values of the original image). Towards the attack, we need to modify the DCT data such that both T^Q_0 and T^Q_1 becomes close to 1 (case 1) or 0 (case 2), i.e., the effect of the modification in some specific DCT coefficients can be nullified.

The main tool for implementing this attack is fitting a polynomial over the DCT data of the watermarked image. The strategy of polynomial fitting in DCT domain has been used in [5] for cryptanalysis of the CKLS scheme [3]. However, in that case the DCT data had been sorted first and then the polynomial fitting has been attempted. In this case the polynomial fitting is done without sorting the data (we will explain the reason later in Remark 1) and further it has been done on every $N \times N$ DCT block as the watermark insertion algorithm [14] works over each DCT block to insert a bit.

For each $N \times N$ DCT block, we take the DCT coefficients in the index range $[Z_1, Z_2]$. Corresponding to the data we fit a polynomial of degree d, such that the mean square error is minimized. Then we replace the DCT values in the index range $[Z_1, Z_2]$ by the estimated values available from the polynomial. This provides the basis of our attack. Since we are approximating the DCT coefficients by a polynomial on the watermarked image, the effect of modifying the original DCT data will be approximated and nullified depending on the degree of the approximating polynomial.

It is clear that as we increase the degree d of the polynomial, the mean square error is less and the quality of the image will stay good, but the effect of nullification of the watermark will be less. On the other hand, if the degree is decreased, the mean square error will be more and the extracted image will not have good image quality, but the removal of watermark information will be achieved in a much better way. Thus, we need to make a compromise over this based on the DCT data.

Remark 1. Now we explain why we do not sort the DCT data before fitting the DCT polynomial. If the data get sorted then it becomes monotonically increasing or decreasing (according to the sorting). Hence the data can be very well modeled with low degree polynomial with very small mean square error. Thus, when one gets back to the DCT coefficients from the polynomial, the modified data (and also the data that are not modified) are not disturbed much and consequently the watermark is not removed.

3.1   Experimental Results

We consider the parameters used in [14] for a concrete description of the attack. Image size of 512×512 has been considered. During the embedding procedure $N \times N$ DCT blocks are taken, where $N = 64$. The index range in each block is chosen in $[Z_1 = 200, Z_2 = 700]$. The value of n has been taken to be 30. The watermark is of length $\frac{512 \times 512}{64 \times 64} = 64$ bits.

The attacker buys a single copy and then modifies it in such a manner such that the buyer-specific watermark cannot be extracted from the attacked images. Thus, the main motivation of the attacker is removing the watermark completely. Towards that direction we made extensive experiments and found that one can use polynomials of degree 3, 4 or 5 by which the watermark is removed completely and the attacked image quality also stays good (> 30 dB PSNR with respect to the attacked image).

Let us now briefly mention what we exactly mean by complete removal of the watermark [4, 5]. Consider that the watermark w is a binary string of length t. Let the recovered watermark from the attacked image is $w^{\#}$. Then the correlation between w and $w^{\#}$ is number of places they are matching minus the number of places they are not matching divided by t. Thus, the correlation between w and $w^{\#}$ is 0 when they match at half of the places and do not match in the rest half of the places. By complete removal of watermark we mean that the correlation will be close to zero, i.e., the number of bit errors is half the length of watermark string, i.e., bit error rate is close to 50%. In this case, from $w^{\#}$, no inference on w can be done as the probability that one bit of $w^{\#}$ will be equal (or unequal) to the corresponding bit of w is $\frac{1}{2}$. Note that very high bit error is not good in terms of cryptanalysis, as in that case the bitwise complement of $w^{\#}$ will be very close to w. Thus, significant negative and or positive correlations are both undesirable for cryptanalysis. What we need is a correlation very close to 0 and then we can ascertain that the watermark is removed and the attack is successful.

We present the experimental results in Table 1, for three different 512×512 images. We insert a 64 bit watermark considering two cases: (i) $P_1 = 0.5, P_2 = 0$

and extract the watermark taking $\beta = 1$, (ii) $P_1 = 0, P_2 = 9$ and extract the watermark taking $\beta = 0$, with the proper key after the attack and calculate how many bits are matching with the original watermark. That around half of the bits are matching (around 32 out of 64) demonstrates the success of the attack. The correlation can be calculated as $\frac{x-(64-x)}{64}$, when x bits are matching. That is, if $x = 32$, then the correlation is 0. In our experimentation, the maximum absolute value of correlation after the attack is 0.125, when $x = 36$ or 28. This clearly indicates that the cryptanalysis presented here is successful.

Also we provide the PSNR of the attacked image with respect to the watermarked image to present the image quality. One can check that we get PSNR (of the attacked image with respect to the watermarked image) values greater than 30 dB. In Figure 1, we present the original, watermarked and attacked images which are visually indistinguishable. This visual representations are for 512×512 size images and we use a polynomial of degree 3 for attack.

Fig. 1. Left: Original Image, Middle: Watermarked Image, Right: Attacked Image.

We present the "number of bits matching, the PSNR value in dB" in each cell corresponding to the degree of the polynomials which we take as $3, 4$, and 5.

Table 1. Experimental results.

	$P_1 = 0.5, P_2 = 0$			$P_1 = 0, P_2 = 9$		
	3	4	5	3	4	5
Lena	31, 30.11	30, 30.13	30, 30.14	31, 30.29	30, 30.31	29, 30.34
Goldhill	32, 30.13	30, 30.14	33, 30.16	36, 30.18	31, 30.27	34, 30.42
Fishboat	28, 30.56	33, 30.68	31, 30.86	29, 30.58	31, 30.64	33, 30.81

To improve the PSNR of the attacked image with respect to the watermarked image further we can modify the attack in the following way. Instead of fitting one polynomial in the range of $[Z_1, Z_2]$, we fix number of polynomials in the range. For the experimental purpose, where $Z_1 = 200, Z_2 = 700$, we can fit five polynomials in the intervals $[200, 300], [301, 400], [401, 500], [501, 600], [601, 700]$. We have checked that this increases the PSNR by an average of 1 dB than what presented in Table 1.

4   Conclusion

In this paper we have presented a cryptanalysis of the Generalized Patchwork Algorithm presented in [14]. To embed the watermark signal, some randomly chosen DCT coefficients are modified. We used the technique of polynomial fitting on the DCT coefficients to erase that modification. The audio watermarking scheme presented in [15] uses the similar strategy as mentioned in [14] for images. Thus, it is expected that a similar strategy like the attack presented in this paper may be successful against the audio watermarking scheme [15].

References

1. R. J. Anderson and F. A. P. Petitcolas, On the limits of steganography, *IEEE Journal of Selected Areas in Communications. Special Issue on Copyright and Privacy Protection*, vol. 16, no. 4, pp. 474-481, May 1998.
2. J. Boeuf and J. P. Stern, An analysis of one of the SDMI candidates, *Information Hiding 2001*, vol 2137, pp. 395-410, in *Lecture Notes in Computer Science*, Springer-Verlag, 2001.
3. I. J. Cox, J. Kilian, T. Leighton and T. Shamoon, Secure spread spectrum watermarking for multimedia, *IEEE Transactions on Image Processing*, vol. 6, no. 12, pp. 1673-1687, 1997.
4. T. K. Das, S. Maitra and J. Mitra, Cryptanalysis of optimal differential energy watermarking (DEW) and a modified robust scheme, *IEEE Transactions on Signal Processing*, PART II, vol. 53, no. 2, pp. 768-775, February 2005.
5. T. K. Das and S. Maitra, Cryptanalysis of correlation based watermarking schemes using single watermarked copy, *IEEE Signal Processing Letters*, vol. 11, no. 4, pp. 446-449, April 2004.
6. F. Ergun, J. Kilian and R. Kumar, A note on the limits of collusion-resistant watermarks, In *Eurocrypt 1999*, vol. 1592, pp. 140-149, in *Lecture Notes in Computer Science*, Springer-Verlag, 1999.
7. T. Kalker, J. P. M. G. Linnartz and M. v. Dijk, Watermark estimation through detector analysis, In *International Conference on Image Processing*, 1998.
8. A. Kerckhoffs, La cryptographie militaire, *Journal des Sciences Militaires*, 9th series, IX, pp. 5-38, January 1883, and pp. 161-191, February 1883.
9. D. Kirovski and H. S. Malvar, Embedding and detecting spread-spectrum watermarks under the estimation attack, In *IEEE International Conference on Acoustics, Speech, and Signal Processing*, 2002.
10. D. Kirovski and F. A. P. Petitcolas, Replacement attack on arbitrary watermarking systems, In *ACM Workshop on Digital Rights Management*, 2002.
11. M. K. Mihcak, R. Venkatesan, M. Kesal, Cryptanalysis of discrete-sequence spread spectrum watermarks, In *Workshop on Information Hiding 2002*, vol. 2578, pp. 226-246, in *Lecture Notes in Computer Science*, Springer-Verlag, 2003.
12. F. A. P. Petitcolas, R. J. Anderson, M. G. Kuhn and D. Aucsmith, Attacks on copyright marking systems, In *2nd Workshop on Information Hiding*, vol. 1525, pp. 218-238, in *Lecture Notes in Computer Science*, Springer Verlag, 1998.
13. F. A. P. Petitcolas and R. J. Anderson, Evaluation of copyright marking systems, In *IEEE International Conference on Multimedia Computing and Systems*, Florence, Italy, vol. 1, pp. 574-579, June 1999.

14. I.-K. Yeo and H. J. Kim, Generalized patchwork algorithm for image watermarking, *Multimedia Systems*, vol. 9, pp. 261-265, 2003.
15. I.-K. Yeo and H. J. Kim, Modified patchwork algorithm: A novel audio watermarking scheme, *IEEE Transactions on Speech and Audio Processing*, vol. 11, no. 4, pp. 381-386, July 2003.

Reinforcement Learning by Chaotic Exploration Generator in Target Capturing Task

Koichiro Morihiro[1,2], Teijiro Isokawa[2,3],
Nobuyuki Matsui[2,3], and Haruhiko Nishimura[4]

[1] Hyogo University of Teacher Education, Hyogo 673-1494, Japan
`mori@info.hyogo-u.ac.jp`
[2] Himeji Institute of Technology, Hyogo 671-2201, Japan
[3] Graduate School of Engineering, University of Hyogo, Hyogo 671-2201, Japan
`{isokawa,matsui}@eng.u-hyogo.ac.jp`
[4] Graduate School of Applied Informatics, University of Hyogo,
Hyogo 650-0044, Japan
`haru@ai.u-hyogo.ac.jp`

Abstract. The exploration, that is a process of trial and error, plays a very important role in reinforcement learning. As a generator for exploration, it seems to be familiar to use the uniform pseudorandom number generator. However, it is known that chaotic source also provides a random-like sequence as like as stochastic source. Applying this random-like feature of deterministic chaos for a generator of the exploration, we already found that the deterministic chaotic generator for the exploration based on the logistic map gives better performances than the stochastic random exploration generator in a nonstationary shortcut maze problem. In this research, in order to make certain such a difference of the performance, we examine target capturing as another nonstationary task. The simulation result in this task approves the result in our previous work.

1 Introduction

In reinforcement learning[1, 2], it is necessary to introduce a process of trial and error designed to maximize rewards obtained from environment. This trial and error process is called an exploration. Because there is a trade-off between exploration and exploitation (avoiding bad rewards), balancing of them is very important. This is known as the exploration-exploitation dilemma. The schema of the exploration is called a policy. There are many kinds of policies such as ε-greedy, softmax, weighted roulette and so on. In these existing policies, exploring is decided by using stochastic random numbers as its generator, according to the reference value and the provided criterion[2, 3].

As the generator for exploration, it seems to be general to use the uniform pseudorandom number generator. However, it is known that chaotic source also provides a random-like sequence as like as stochastic source[4, 5]. From this point of view, we propose an application of the feature of deterministic chaos for a generator of the exploration. In practice, we use the logistic map that is

R. Khosla et al. (Eds.): KES 2005, LNAI 3681, pp. 1248–1254, 2005.

known as the simplest chaotic function. In order to see the effects of generator for exploration, we previously examined a little complex problem so-called shortcut maze[6] known as a nonstationary problem and found that the deterministic chaotic generator for the exploration based on the logistic map gives better performances than the stochastic random exploration generator[7]. In this paper, we look into a target capturing problem as another nonstationary task to make certain such a difference of the performance.

2 Learning Algorithm and Exploration Generator

2.1 SARSA: Modified Q-Learning

Machine learning that gives a computer system an ability to learn has been developed and used in various situations. A lot of learning algorithms and methods are proposed for the system to acquire step by step the desired function. Reinforcement learning is originated in experimental studies of learning in psychology. In the reinforcement learning, the system gets only an evaluative scalar feedback from its environment, not an instructive one as in supervised learning. Q-learning is known as the best-understood reinforcement learning algorithm[8]. It forms a Q-mapping from state-action pairs by rewards obtained from the interaction with the environment.

Furthermore, from the point of policy control method, there are two types of approaches. One is called on-policy method and the other is off-policy method[2]. In on-policy method, the behavior policy of agent and the estimation policy are completely the same. On the other hand, in off-policy method, these two policies are different each other. Q-learning is an off-policy method learning algorithm. In this respect, Q-learning is not suitable to examine effects of the exploration generator's nature on learning, because that the exploration generator constitutes only the behavior policy. Then, as the learning algorithm in this research, we adopt SARSA which is one of the on-policy method called modified Q-learning[9].

To enhance the learning efficiency, we also combine the eligibility trace $e_t(s, a)$ with SARSA. The learning rule is written as follows, where α is the learning rate, γ is the discount rate and λ is the eligibility trace parameter. In this paper, $\alpha = 1, \gamma = 0.8, \lambda = 0.5$ are used for the simulations.

$$Q_{t+1}(s, a) = Q_t(s, a) + \alpha \delta_t e_t(s, a) \tag{1}$$

$$\delta_t = r_{t+1} + \gamma Q_t(s_{t+1}, a_{t+1}) - Q_t(s_t, a_t) \tag{2}$$

$$e_t(s, a) = \begin{cases} \gamma \lambda e_{t-1}(s, a) + 1 & (s = s_t, a = a_t) \\ \gamma \lambda e_{t-1}(s, a) & (otherwise) \end{cases} \tag{3}$$

2.2 Exploration Generator

In the reinforcement learning, many kinds of exploration policies have been proposed as a process of trial and error such as ε-greedy, softmax, weighted roulette. Then, exploring is decided by using the stochastic random numbers as its generator in these existing policies. As the policy, we use ε-greedy which decides exploration or exploitation based on the given threshold value $\varepsilon \in [0, 1]$. In this paper, if there are k kinds of possible actions, an action which has maxmium Q-value is selected in $1 - \varepsilon$ and the other actions are selected in $\varepsilon/(k - 1)$, respectively.

It has been general to use the uniform pseudorandom number as the stochstic exploration generator. In our research, we deal with two kinds of exploration generators that are such a conventional generator and a chaotic deterministic generator. As the chaotic deterministic generator, we use a logistic map which generates a value in the closed interval [0,1] as follows.

$$x_{t+1} = ax_t(1 - x_t), \qquad (a = 4) \tag{4}$$

3   Simulation and Result

3.1   Target Capturing Task

The original environment is a 16×12 grid world shown as (a) in Fig. 1. The problem for an agent is capturing a targt "T_1" in the shortest path from start cell "S". The target moves within the shaded ragen of 4 cells in each time step. The agent selects one of the action in up, down, left, right in each time step and gets the reward -1 in every moving step except for capturing. Capturing the target (this is attained by taking an adjacent position to the target), the agent gets the reward +1 and retuns to the start cell "S". If there is no free cell in a selected way, the agent stays where it is and gets the reward -1.

In this paper, we call the journey of the agent from "S" to the target an episode. After the first $n_c(= 70)$ episodes, one more target "T_2" appears in the environment as shown in (b) in Fig. 1. In this environmental change, the number of steps in the shortest way to the target to capture changes from 15 into 5. The agent keeps searching the new shortest way up to 900 episodes in total. We call this 900 episodes an event. Within one event, each Q-value is never initialized and the agent iteratively use the updated Q-value. When a new event begins, all Q-values are initialized to 0. To evaluate the learning performance, we take the average of 1000 events with different inital conditions for the exploration generator.

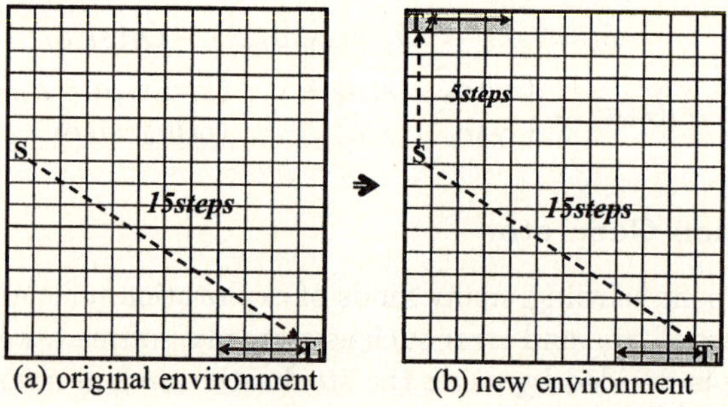

Fig. 1. Target capturing task used for the learning evaluation.

3.2   Results and Considerations

Let t_n be the number of actual steps from "S" to the target at episode n. The minimal step t_{min} is 15 for $0 \leq n < 70$ and 5 for $70 \leq n < 900$. As a measure of learning performance, we take $q(n) = t_{min}/t_n$. In the case the agent captures the target in the shortest path, this measure gives $q(n) = 1$. If the agent takes much more steps, it gets close to 0. Figure 2 shows the transition of $\langle q \rangle$ which is the average of every $q(n)$ in 1000 events. In the original environment ($0 \leq n \leq 69$), the stochastic random generator gains $\langle q \rangle = 0.65$ and the deterministic chaotic generator gains $\langle q \rangle = 0.67$. The $\langle q \rangle$ once descends by the environmental change at n=70, and $\langle q \rangle$ ascends again by adaptation. At n=899, the stochastic random generator and the deterministic chaotic generator regains $\langle q \rangle$ =0.76 and 0.86, respectively.

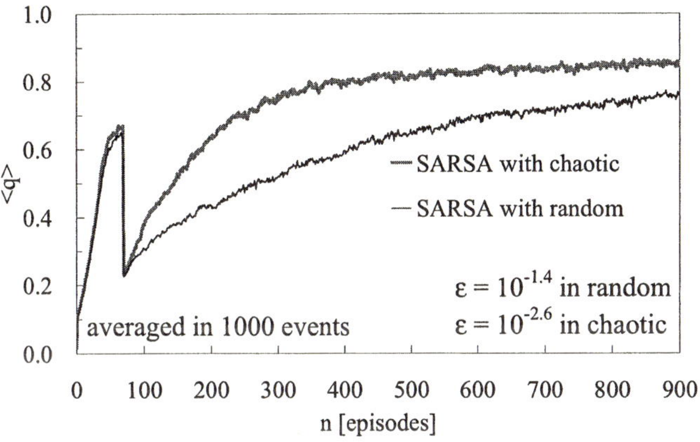

Fig. 2. Evolution of the performance measure $\langle q \rangle$ in the target capture learning.

The evolution of measure $\langle q \rangle$ depends on ε which is the criterion of ε-greedy policy. We show the dependece of final $\langle q(n = 900) \rangle$ on ε in Fig. 3. In the case of random generator, the maximum $\langle q(n = 900) \rangle = 0.76$ is given at $log(\varepsilon) = -1.4$. In the chaotic generator, the maximum $\langle q(n = 900) \rangle = 0.86$ is obtained at $log(\varepsilon) = -2.6$. The chaotic generator shows better performance than the random generator. Moreover, the chaotic generator case maintains near the best performance stably within the wider range of ε.

Indeed, Fig. 2 corresponds to the learning curves in the cases of the best performance with each generator. Through the all episodes, the measure $\langle q \rangle$ shows that the chaotic generator case is better than the random generator case.

We further examine that how many episodes each exploration generator needs to find the shortest path for the first time in each case of the best performance at log(ε) =-1.4 and -2.6. Figure 4 shows the cumulative frequency on 1000 events $R_{q>0.8}$ that the agent discovers almost shortest ($\langle q \rangle$ grows more than 0.8) path until episode n. Before the environment changes, both exploration generators discover such path with about the same performance of 99%. After the environ-

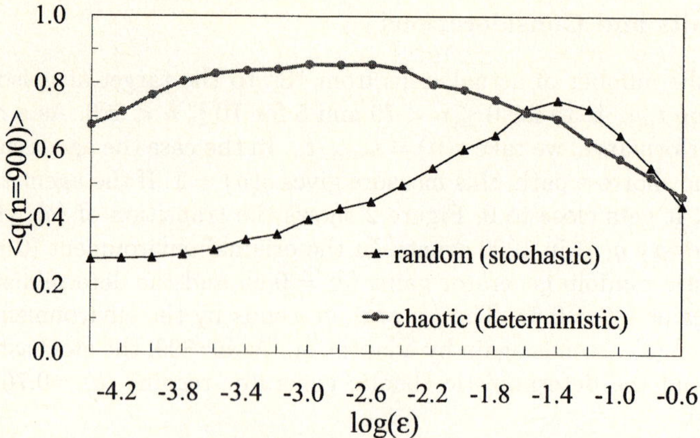

Fig. 3. Dependence of the final $\langle q \rangle$ on the parameter in ε-greedy policy.

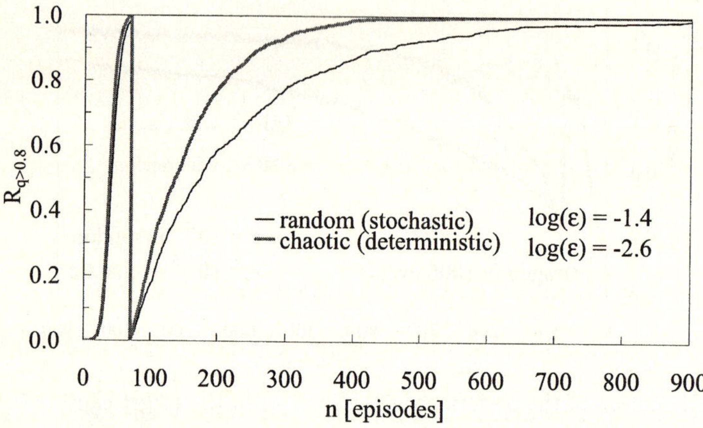

Fig. 4. Evolution of the cumulative frequency $R_{q>0.8}$ in discovery of the almost shortest ($\langle q \rangle$ grows more than 0.8) path.

ment changes, the random generator case needs 248(=318-70) episodes and the chaotic generator case needs 140 episodes to attain 80% performance. Also, the random generator case and the chaotic generator case take 369 and 192 episodes to 90% performance, respectively. This indicates high adaptability of the chaotic generator to the environmental change compared with the random generator.

To understand the above difference of perfromance, we pay attention to the relation between q(n) and e(n) which is the frequency of explorations in episode n. In general, the decrease of q(n) causes the increase of e(n). Figure 5 shows the relation between $q(n)$ and $e(n)$ in the case of one of the best performance events in each generator. In the both generator cases, the circle and triangle points are distributed dropping on the right of them. The circle points spreading wider than triangle points indicate that more explorations are occurred in many episodes in the chaotic generator case.

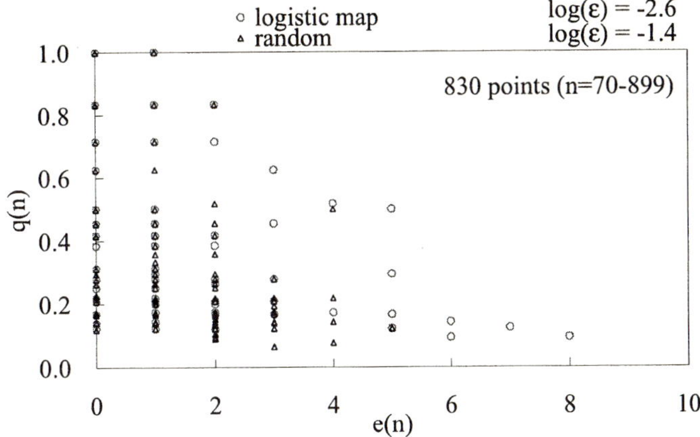

Fig. 5. Relation between q(n) and e(n) in the case of one of the best performance events in each generator.

Explorations are necessary for improvement of learning and it appears in a later exploitative episode. Concretely, if q(n) increases after its decreasing while e(n) decreases after its increasing, then learning is expected to be improved by the exploration. So, we probe the differences of second order for q(n) and e(n) as follows.

$$\delta^2 q(n) = (q(n+1) - q(n)) - (q(n) - q(n-1)) \tag{5}$$

$$\delta^2 e(n) = (e(n+1) - e(n)) - (e(n) - e(n-1)) \tag{6}$$

Figure 6 shows the relation between $\delta^2 q(n)$ and $\delta^2 e(n)$ in this case. Many circle points spreads in the 2nd quadrant region ($\delta^2 q(n) > 0$ and $\delta^2 e(n) < 0$). This means that learning is frequently improved by exploration with the chaotic generator. Comparing with the circle points, triangle points have the

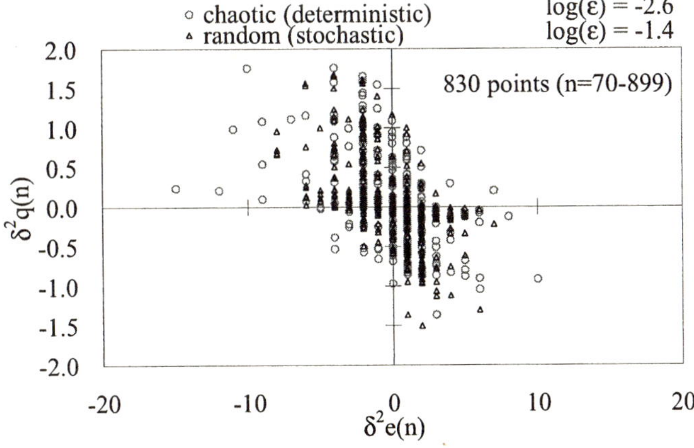

Fig. 6. Relation between the differences of 2nd order for q(n) and e(n) in Eqs. (5), (6).

tendency gathering around the origin. This reflects the gradual learning progress by exploration with the random generator.

4 Conclusion

We find that the deterministic chaotic generator for the exploration based on the logistic map gives better performances than the stochastic random exploration generator in this target capturing task. Moreover, the stochastic random generator is sensitive to the parameter ε, so it performs well only near the best ε. Contrary to this, the deterministic chaotic generator is stable to ε and work well in much wider range of ε. These tendencies are similar to the results obtained in the cases of shortcut maze problem, approving the fact in our previous work[7]. In order to understand such a difference of the performance of stochastic random generator and chaotic generator, we proceed further investigation about other learning tasks and other chaotic generators, such as the tent map which is homeomorphic to the logistic map.

References

1. L. P. Kaelbling, M. L. Littman and A. W. Moore: Reinforcement Learning: A Survey, Journal of Artificial Intelligence Research, Vol.4, pp.237–285 (1996)
2. R. S. Sutton and A. G. Barto: Reinforcement Learning, The MIT press, Cambridge, MA, (1998)
3. S. B. Thrun: Efficient Exploration In Reinforcement Learning, Technical report CMU-CS-92-102, Carnegie Mellon University, Pittsburgh, PA, (1992)
4. T. S. Parker and L. O. Chua: Practical Numerical Algorithms for Chaotic Systems, Springer-Verlag, (1989)
5. E. Ott, T. Sauer and J. A. Yorke: Coping with Chaos: Analysis of Chaotic Data and The Exploitation of Chaotic Systems, John Wiley & Sons, Inc., New York, NY, (1994)
6. A. B. Potapov and M. K. Ali: Learning, Exploration and Chaotic Policies, International Journal of Modern Physics C, Vol. 11, No.7, pp.1455–1464 (2000)
7. K. Morihiro, N. Matsui and H. Nishimura: Effects of Chaotic Exploration on Reinforcement Maze Learning, in Knowledge-Based Intelligent Information and Engineering Systems, Part I, (KES2004), LNAI3213, pp.833-839, Springer (2004)
8. C. Watkins and P. Dayan: Q-learning, Machine Learning, 8, pp.279–292, (1992)
9. G. A. Rummery and M. Niranjan: On-line q-learning using connectionist systems, Technical Report CUED/F-INFENG/TR 166, Cambridge University Engineering Department, (1994)

Prediction of Foul Ball Falling Spot in a Base Ball Game

Hideyuki Takajo[1], Minoru Fukumi[2], and Norio Akamatsu[2]

[1] Department of Information Engineering, Takuma National College of Technology,
551 Kouda, Takuma-cho, Mitoyo-gun, Japan
`takajo@di.takuma-ct.ac.jp`
[2] Department of Information Science and Intelligent Systems, Faculty of Engineering,
Tokushima University, 2-1 Minami-Josanjima, Tokushima, Japan
`{fukumi,akamatsu}@is.tokushima-u.ac.jp`

Abstract. In baseball games, foul balls sometimes fly into the spectators seats. Some person can be seriously wounded for those balls. This research aims at developing a system which computes the orbit of a foul ball on real time using camera, predicts of the spot where foul balls fall and informs spectators about its danger. Detection of balls consists of the following three steps: At first each frame is processed using a smoothing filter for noise reduction. Next, temporal Laplacian is carried out for a series of those filtered images. Finally, whether a part of a ball or not is judged for each pixel by translating the resultant images into polar coordinates. The pitching motion is also considered. We report a method for prediction of foul ball falling spot with fuzzy inference.

1 Introduction

There are many cases where a net for protection from foul balls isn't arranged in front of the audience in a baseball field except for the back of the home base. This makes the audience and the players feel a sense of closeness, and has an effect to raise liveliness. Whereas, it increases foul balls which jump into the seat and hurt an audience. Actually, in Japan, the accident that a foul ball hit a student's eye in a high school baseball game was happened. Furthermore, in the practice of daily club activities of baseball, there is danger that a foul ball hits people outside the ground as well. At present, high school baseball federation in Japan takes the following measures against this problem. Some persons in the audience sound whistle to draw attention when a foul ball goes into the gallery.

This research aims to forecast the landing point of foul balls before they fall into the audience by analyzing the images of baseball games. In this paper, we report the method to find out the ball in the images and forecast the landing point by fuzzy reasoning.

2 The Present Method

Although setting up the camera in ballpark and processing images in real time are necessary for practical use, we takes a picture beforehand with the DV(Digital Video) camera and analyze the image preserved on the tape at present. The DV camera has 680 million pixels. In the standard of the DV camera, 30 frames/second (Interrace is considered, it is essentially 60 frames/s) can be obtained by the size of the 720×480

R. Khosla et al. (Eds.): KES 2005, LNAI 3681, pp. 1255–1260, 2005.

pixels. The record format of the image is a National Television System Committee standard.

We are developing programs on IBM PC/AT compatible machine and Linux (Kernel 2.4) as this system must be built-in type one for practical use. Besides this, DV file operation library (libdv) of free software is used.

The detection of the ball consists of three steps. To begin with, various processes are carried out for the each frame image, and next, the Laplacian is taken for the image which is the result of the previous step. Finally, in the image which is taken by the Laplacian, the parts with the big change are plotted on polar coordinates. The direction in histogram of the plotting number which exceeded a given threshold is judged as the flying direction of the ball. These three steps are described below in detail.

2.1 Pre-processing of Images

First of all, each frame image is processed by the following procedures.

(a) Reading color images of DV format from a movie file
(b) Decomposition of interlaced image
(c) Translation from color image to monochrome one
(d) Application of spatial filter

In this study, we simply resolves an interlaced image into the odd number line and even line, and only one piece of the two images is used.

The translation to the monochrome image in (c) is because the detection of the ball is mainly based on the change of the luminance in our method.

The filter used in (d) is for smoothing and is expressed as the following matrix:

$$Smoothing\ filer\ matrix = \begin{bmatrix} 1 & 1 & 1 \\ 1 & 1 & 1 \\ 1 & 1 & 1 \end{bmatrix} \tag{1}$$

Fig. 1 shows the result obtained by applying from (a) to (d) procedures.

Fig. 1. The ball which the batter hit is flying to the left on the screen

2.2 Time-Directional Processing

In this phase, Laplacian of the successive two images is calculated. Laplacian filter is popular to detect the difference of two images. As the velocity of a ball movement is enough fast in comparison with the frame rate of DV camera, when the ball passed a point, the brightness of the pixel rapidly increases, and it rapidly drops to original brightness. In this study, we use the filter to detect the ball in images. Figure 2 is the one that Laplacian was performed for the image of Figure 1. The parts such as the ball, bats, and batters are extracted well as an area with a large change in brightness.

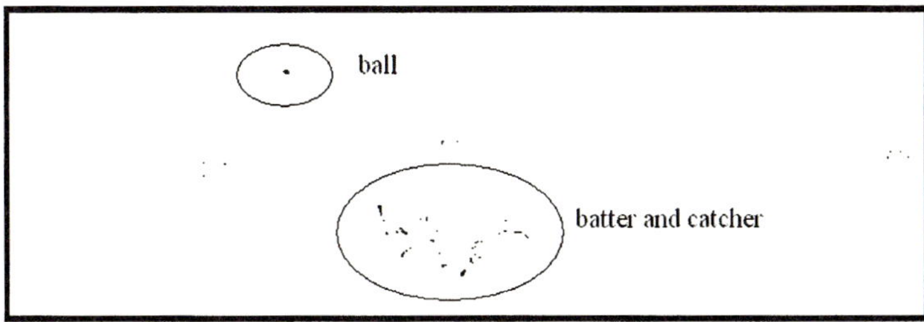

Fig. 2. Application of Laplacian filter

In actual processing, a certain threshold is set, and if the difference of brightness is smaller than this threshold, we consider a given image has no change. Though the region of the ball becomes easy to be detected, if this threshold is decreased, the influence of the noise becomes easy to be received. As shown in Figure 1, since those areas such as the right side of the pitcher and the grass in the outfield are whitish, when the ball passed on these parts, it isn't detected as a ball for little difference in brightness. Therefore, it is necessary to set this threshold to an appropriate value in consideration of these trade-offs.

2.3 Histogram of the Ball Candidate for Each Direction

In this phase, coordinates of the points in the image obtained as a result of processing in section 2.2 are converted into polar coordinates as shown in Figure 3.

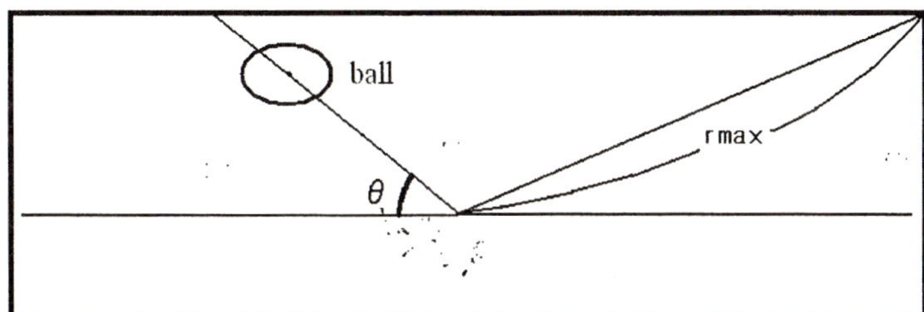

Fig. 3. Conversion into polar coordinates

For each pixel(r,θ) on the image of Figure 2, according to whether it is bright more than before, we define the function δp(r,θ) as follows:

$$\delta_p(r,\theta) = \begin{cases} 1, \textit{if the pixel turned brighter.} \\ 0, \textit{otherwise} \end{cases} \qquad (2)$$

In the following, θi *and* ri stand for areas as follow:

$$\frac{\pi}{N_\theta}i \le \theta_i \le \frac{\pi}{N_\theta}(i+1), i = 0,1, \quad ,N_\theta -1 \tag{3}$$

$$\frac{r_{max}}{N_r}i \le r_i \le \frac{r_{max}}{N_r}(i+1), i = 0,1, \quad ,N_r -1 \tag{4}$$

where r_{max} is the distance between the catcher and the furthest point on the image, N_θ and N_r are the number of regional divisions in the direction of θ and r, respectively. $N_\theta = N_r = 30$ in Figure 4.

We define NL(ri,θj) as the number of the pixels where δp(r,θ) = 1 in the region (ri,θj), and δL(ri,θj) as follows:

$$\delta_L(r_i,\theta_j) = \begin{cases} 1, if\ N_L(r_i,\theta_j) \ge threshold \\ 0, otherwise \end{cases} \tag{5}$$

The purpose of setting the threshold to NL(ri,θj) is to remove the influence of some noise. The areas where δL(ri,θj) = 1 become the candidates of the ball.

In general, δL(ri,θj) is a function of time. Then, it can be written as δL(ri,θj,t) t=0,1,2 … , where t=0 corresponds to the moment when the pitcher throw a ball out.

Next, we define functions L and LT as follows:

$$L(r_i,\theta_j,t_k) = \sum_{t=0}^{t_k} \delta L(r_i,\theta_j,t) \tag{6}$$

$$L_T(r_i,\theta_j,t_k) = \begin{cases} 0,\ if\ L(r_i,\theta_j,t_k) = 0 \\ 1,\ if\ 1 \le L(r_i,\theta_j,t_k) \le 2 \\ 0,\ if\ L(r_i,\theta_j,t_k) > 2 \end{cases} \tag{7}$$

The value of function LT(ri,θj,tk) is zero when L(ri,θj,tk) is larger than two. This is due to the following reason. It seems to be about two times at most that the ball exists in some regions, as the speed of the ball is enough rapid. Therefore, this is disregarded as not ball but result of detecting moving human.

Figure 4 shows the area of LT(ri,θj,tk)=1, where the ball is at the position shown in Figure 2.

Finally, Ls(θj,tk) that totals LT(ri,θj,tk) on the r direction is defined as follows.

$$Ls(\theta_j,t_k) = \sum_{r_i} L_T(r_i,\theta_j,t_k) \tag{8}$$

We decide that the ball is detected if the value of Ls(θj,tk) exceeds some threshold. Figure 4 shows the detected direction to which the ball is flying.

3 Fuzzy Inference

3.1 Fuzzy If-then Rules

Our inference method is based on both direction and velocity of the hit ball. Since we use only one camera, it is impossible to decide the direction to which hit ball flies accurately. Though it is possible to increase reliability by using two cameras, the speed to decide the falling spot slows down. In this study, we must make decision as soon as possible to inform spectators about danger. In this paper we infer the direction with only one camera by fuzzy inference.

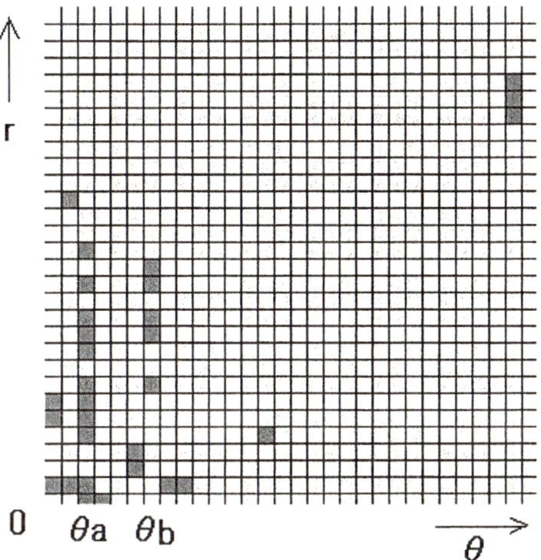

Fig. 4. Two balls are detected: θa and θb correspond to pitched ball and hit one, respectively

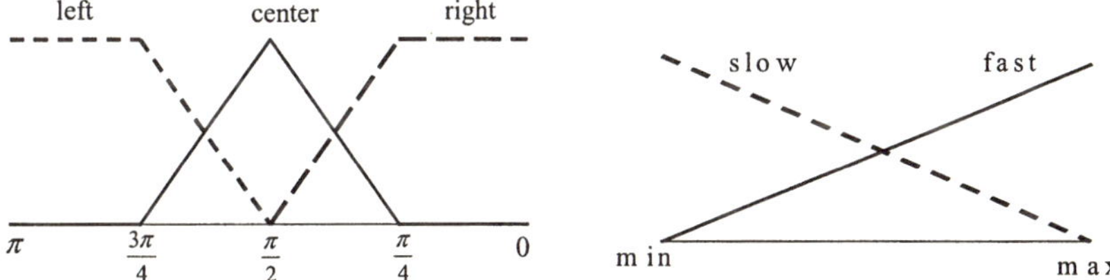

Fig. 5. Membership function about direction   **Fig. 6.** Membership function about velocity

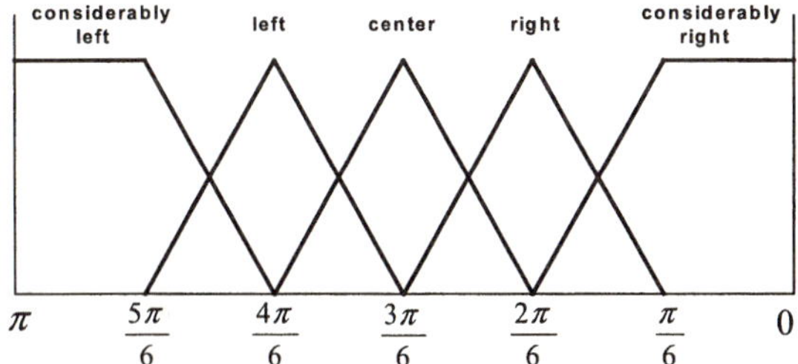

Fig. 7. Member functions of the consequent division

In general, a distant object moves slowly compared with a nearby object on the screen, even if they are moving in the same velocity. Therefore, if the movement of the ball is fast, it can be thought that it is flying not toward the outfield but to the horizontal direction. From this fact, we use the following if-then rules:

Rule1: if direction is **left** and velocity is **fast** then direction is **considerably left**.
Rule2: if direction is **left** and velocity is **slow** then direction is **left**.
Rule3: if direction is **center** and velocity is **fast** then direction is **center**.
Rule4: if direction is **center** and velocity is **slow** then direction is **center**.
Rule5: if direction is **right** and velocity is **slow** then direction is **right**.
Rule6: if direction is **right** and velocity is **fast** then direction is **considerably left**.

Each member-ship function is shown in Figures 5, 6 and 7.

Note that as shown in Table 1, the direction of balls (d), (e) and (f) is equivalent, but the velocity is different. The result of fuzzy inference shows that the angle of the direction is smaller in the order (f), (e) and (d). It implies that those balls fall down nearer in the same order and this result agrees with the actual data. On the other hand, actual falling point of the ball (g) is near in spite of the speed being slow. The velocity of the ball (g) is small as it is a simple fly, so that it falls in near place.

The direction of the ball (i), (j) and (k) is significantly different from the actual direction. This is because the balls which fly to the right side of the pitcher on the image is made to fly to the right direction in present if-then rule.

Table 1. The inferred directions with fuzzy reasoning is shown. Each direction is displayed in degrees. The direction converted as right side is widely spread on the image compared with left one owing to left-side camera position. The velocity is the ratio to average of pitching ball speed

	direction on image	converted direction	velocity on image	inferred direction	actual direction
(a)	15	46	2.01	27	35
(b)	153	86	2.14	90	138
(c)	45	55	1.8	33	0
(d)	57	58	0.73	66	80
(e)	57	58	0.92	63	48
(f)	57	58	1.14	60	0
(g)	63	60	0.71	69	43
(h)	111	74	1.18	72	80
(i)	129	79	1.18	81	174
(j)	141	82	1.42	81	185
(k)	141	82	1.57	81	113

4 Conclusions

In section 2, we presented the method for efficiently detection of balls.

Fuzzy inference is useful for prediction of the foul ball falling spot without plural cameras as shown in section 3. On the other hand, our method has problems that a low speed ball such as a trivial fly can be mistook as one flying toward the outfield. It is considered that these problems are improved by changing the rule of fuzzy inference.

References

1. E.H.Mamdani: Application of Fuzzy Algorithms for Control of a Simple Dynamics Plant. Proc. IEE 121 (12). (1974) 1585-1588

Automatic Extraction System of a Kidney Region Based on the Q-Learning

Yoshiki Kubota[1], Yasue Mitsukura[2], Minoru Fukumi[1],
Norio Akamatsu[1], and Motokatsu Yasutomo[3]

[1] The University of Tokushima, Tokushima, Japan
{ktyosh,fukumi,akamatsu}@is.tokushima-u.ac.jp
[2] Okayama University, Okayama, Japan
mitsue@cc.okayama-u.ac.jp
[3] Higashi Tokushima National Hospital, Tokushima, Japan

Abstract. In this paper, a kidney region is extracted as a preprocessing of kidney disease detection. The kidney region is detected based on its contour information that is extracted from a CT image using a dynamic gray scale value refinement method based on the Q-learning. An initial point to extract the kidney contour is decided by training gray scale values along horizontal direction with Neural Network (NN). Furthermore the kidney contour is corrected by using the snakes more accurately. It is demonstrated that the proposed method can detect stably the kidney contour from CT images of any patients.

1   Introduction

In these years, kidney disease patients are increasing. A diagnosis using the X-ray CT images as an inspection method of the kidney disease is widely used. The inside state of a patient's body could be seen without damaging it. However, the CT image data has a huge quantity. Therefore the doctor needs a long time and a heavy labor for diagnosis. In order to overcome these issues, the purpose of this research is to develop an automatic extraction system of a kidney region as a preprocessing of automatic detection of a kidney disease in the next step.

Until now, various researches as a technique that extracts a kidney contour have been reported. However, it is difficult to perform edge extraction stably only with a single threshold from various images and to extract the contour of the kidney [1]. Moreover, in automatic contour extraction by a dynamic contour model [2,3], it is necessary to specify a rough form of internal organs that serves as an initial candidate for extraction [4,5]. Therefore, in this method, reinforcement learning [6~10] is used to extract the kidney contour. In particular, this paper regards a kidney contour extraction as a benchmark problem, a maze learning, used in the field of reinforcement learning, and a kidney contour is extracted by solving this maze. But it is necessary for solving the maze to decide the initial points (start and goal). Therefore, they are decided by training gray scale values along a horizontal direction with Neural Network (NN). The contour of the kidney is automatically extracted by this method, however this extracted contour has a little accidental error. Therefore, it is corrected by using the snakes. Above the kidney region can be extracted accurately by it.

R. Khosla et al. (Eds.): KES 2005, LNAI 3681, pp. 1261–1267, 2005.

2 Medical Images

2.1 Abdominal X-Ray CT Images

An image treated in this research is an abdominal X-ray CT image. A CT image apparatus irradiates an X-ray beam with narrow band to a certain sectional view, and constructs an image how many X-rays were absorbed. That is, the gray scale values of CT images display the Hounsfield number for every pixel that constitutes the image.

2.2 A Contrast Medium

A X-ray absorptivity is different in each part of a human body and is displayed on an X-ray image. Therefore, when the X-ray absorptivity is equal in internal organs to be detected and their neighboring ones, it is difficult to discriminate the internal organs to be detected with the naked eye. Then, in the clinical spot, the doctor carries out classification between internal organs to be extracted and circumference internal organs by making contrast artificially using a contrast medium.

In this study, since a kidney region is pinpointed easily, the image at 80 seconds after injection of a contrast medium for the patient is used. Fig. 1 shows an image without a contrast medium. The image at 80 seconds after injection of a contrast medium is shown in Fig. 2.

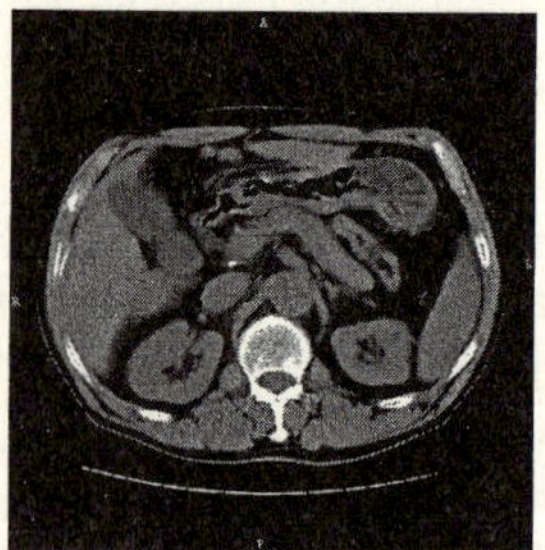

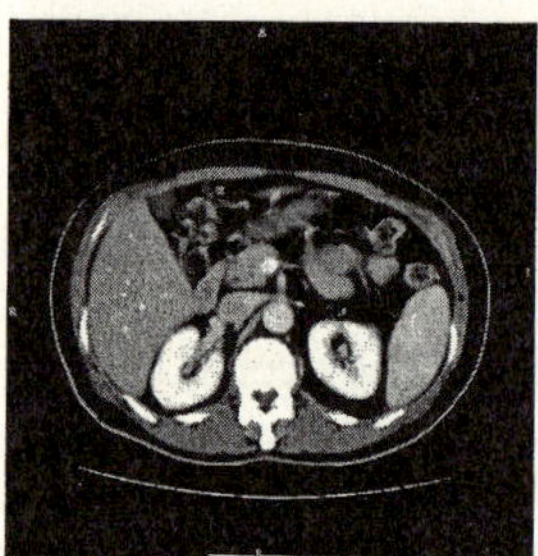

Fig. 1. Before injection of a contrast medium   **Fig. 2.** After injection of a contrast medium

3 The Flow of The Proposed Method

The fundamental flow of our method consists of the following procedures.

Step1: Backbone domain extraction
Step2: Specification of rough kidney position
Step3: Contour extraction of the kidney
Step4: Sector of calibration by using the snakes

First, the position of the backbone is pinpointed in Step1. Next, in Step2, a rough kidney position is estimated using the position of this backbone. Next, in Step3, Q-learning is used within the rough kidney region, and kidney contour edge is detected. Finally, in Step4, by using the snakes, the kidney contour is corrected more accurately.

4 Backbone Domain Extraction

As a preprocessing before extracting a kidney region, since there is a little individual difference even in a human body, the backbone that is computationally easy to be extracted is pinpointed [11~13]. The procedure of backbone extraction is shown below.

1) A binary image.
   Pixels with 230 or greater gray scale values in an image are displayed white as a binarized image. An original CT image is shown in Fig. 3.
2) Domain limitation
   A diagonal line is drawn from a center in the image and only pixels in the part below the diagonal line are displayed.
3) Labeling
   Labeling is performed, and a region with the largest area is extracted. The image after extraction of the backbone is shown in Fig. 4.

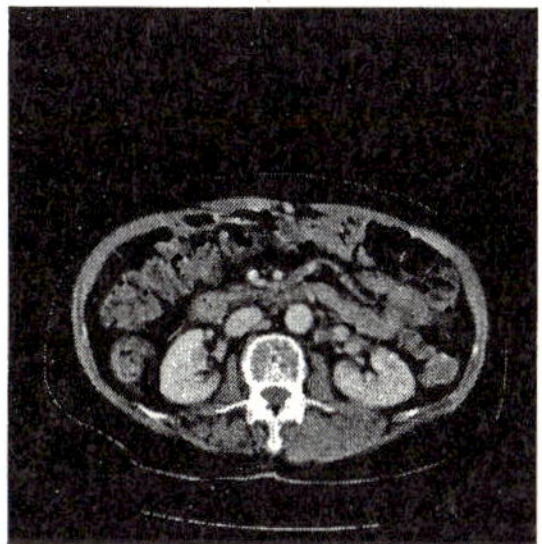

Fig. 3. An original CT image

Fig. 4. After labeling

The position of the backbone is pinpointed by the above processing. On the structure of a human body, since the kidneys are in the both sides of this backbone, a kidney contour can be extracted on the basis of this backbone.

5 Specification of Rough Kidney Position

5.1 The Reference Point of the Backbone

First, to decide the position of the kidney, we decide the reference point of the backbone as shown in Fig. 5.

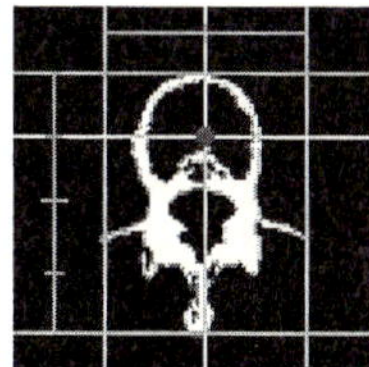

Fig. 5. The reference point

5.2 Initial Point Decision by NN

Next, we decide the beginning point (start) and end the point (goal) to search for the kidney contour. These generally have an individual difference and a slice position difference. Then, we decide rough positions of initial points (start and goal) by using NN. Because NN can deal with a nonlinear problem, it is suitable for this problem. Gray scale values of pixels of the reference points with x=100 to 256 are fed to the input layer, and distances from x=100 to the start and the goal points are put in the output layer as teacher data, where x means a coordinate value along the horizontal direction. The NN is shown in Fig. 6.

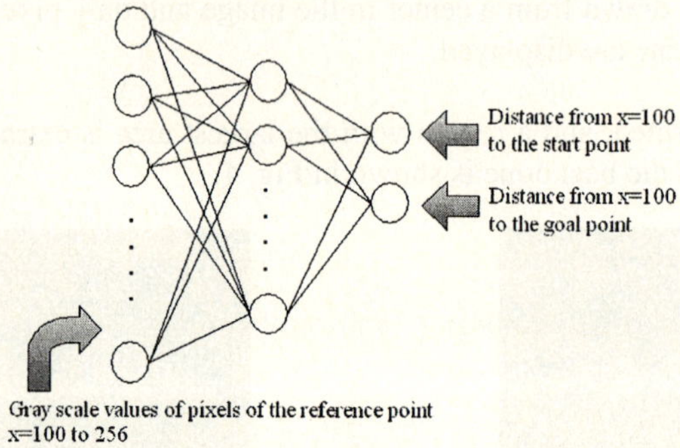

Fig. 6. NN

We evaluated out method by the leave one out method. The examples of result are shown in Figs. 7 and 8.

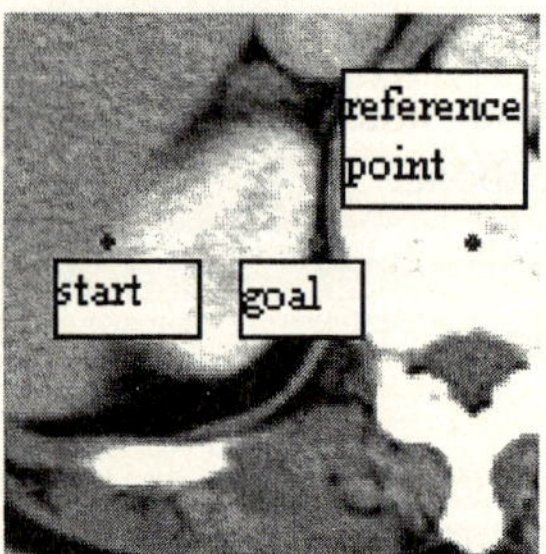

Fig. 7. The 14th slice

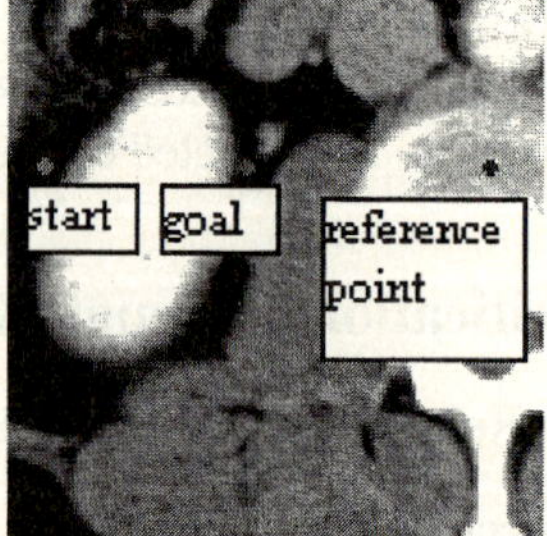

Fig. 8. The 26th slice

6 Edge Extraction Based on Q-Learning

6.1 The Basic Concept of Q-Learning

In Q-learning, a subject which solves a problem is called agent and all things other than an agent are called environment [14,15]. An agent observes environment, takes an action and receives the remuneration according to state changes. Furthermore, a Q value is updated by the following formulas using the received remuneration.

6.2 An Agent's Action Pattern

An agent can move to a pixel in neighboring eight, and has Q values (Q table) according to each moving direction. An agent's action pattern is shown below.

1) If an agent passes for pixels of a gray scale value range ($0 < x^* < T1$ or $T2 < x < 255$), it will receive a penalty (reward: -0.1).
2) If an agent passes for pixels of a gray scale value range ($T1 < x < T2$), it will receive no reward (reward: 0).
3) If an agent reaches the goal, it will receive reward (reward: 10).
4) The reward according to an action is received and Q value is updated.

*x: a gray scale value that an agent passed

6.3 Computer Simulation

We extracted the kidney contour by using the initial point requested in Section 5. The numbers of success images are 8/15 (53.3%). The reason for the failing images are that learning did not end. Therefore a further improvement for it is necessary. The examples of extraction are shown in Figs. 9 and 10.

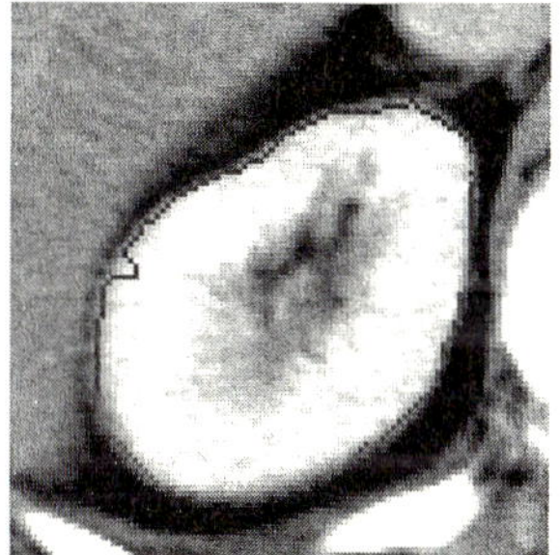

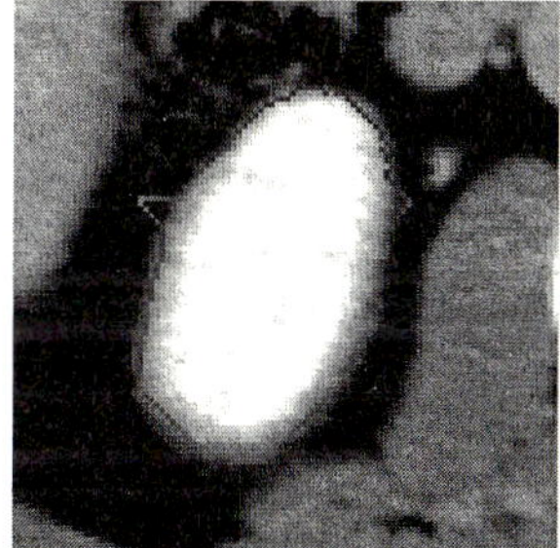

Fig. 9. The 17th slice   **Fig. 10.** The 26th slice

An excellent result was obtained though there were a few error margins.

7 Sector of Calibration by Using the Snakes

7.1 Snakes

The snakes [2,3] is one of the contour extraction methods. It is a method of extracting the contour of an object body by transforming an initial curve based on edge and contour shape information [4,5]. The snakes is a method that the energy shown by the equation (1) is minimized.

$$E_{snakes}(k) = \omega_1 \cdot E_{cont}(k) + \omega_2 \cdot E_{curv}(k) + \omega_3 \cdot E_{image}(k) \tag{1}$$

$E_{cont}(k), E_{curv}(k)$: Internal energy (first, second)

$E_{image}(k)$: External energy

7.2 Computer Simulation

In the result of section 6, there are a few error margins with an actual kidney contour. Therefore, it is corrected by using the snakes. Each parameter in equation (1) was decided experimentally. The results of experiment are shown in Figs. 11 and 12. They are improved compared to those shown in Figs. 9 and 10.

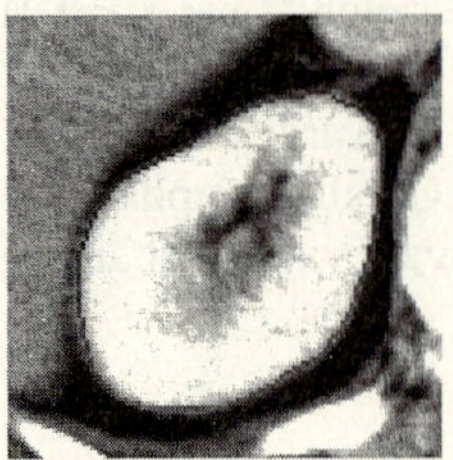

Fig. 11. The 17th slice

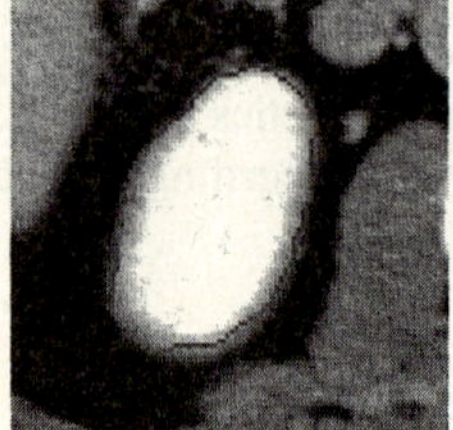

Fig. 12. The 26th slice

The error margins were corrected, and resulting contours are better for kidney region extraction.

8 Conclusion

In this paper, the technique of the automatic extraction of a kidney region that used Q-learning was proposed as a pretreatment of automatic detection of a kidney disease. The success probability was low as 53%, and therefore a further improvement is necessary.

References

1. M. Inoue, N. Yagi, M. Hayashi, H. Nakasu, K. Mitami, M. Okui, "Practice image processing studied by the C language", Ohmsha, 1999, In Japanese
2. T. F. Cootes, C. J. Taylor, A. Lanitis et al: Building and Using Flexible Models Incorporation Grey-Level Information. Proc of ICCV: 242-246, 1993
3. T. F. Cootes, C. J. Taylor, D. H. Cooper et al: Active Shape Model – Their Training and Application. CVIU61(1): 38-59, 1995
4. M. Kass, A. Witkin, D. Terzopoulos, "Snakes: Active contour models," Int. J. Comput. Vision., vol. 1, no. 3, pp.321-331, 1988.
5. D. Terzopoulos, A. AWitkin, M. Kass, "Symmetry-seeking models and 3D object reconstruction," Int. J. Comput. Vision, vol. 1 no. 2, pp.211-221, 1987.
6. M. Errecalde, M. Crespo y C. Montoya. "Aprendizaje por Refuerzo: Un estudio comparativo de de sus principales metodos". Proc. del 2 Encuentro Nacional de Computacion (ENC'99) Sociedad Mexicana de Ciencia de la Computacion. Mexico, 1999.
7. T. Mitchell. "Machine Learning". Capitulo 13. (Version preliminar).
8. S. Russell y P. Norvig. "Artificial Intelligence. A modern Approach". Prentice -Hall- 1995
9. R. Sutton y A. Barto. "Reinforcement Learning: an introduction". The MIT Press, 1998
10. G. Kimura, K. Miyazaki, S. Kobayashi, The design indicator of Reinforcement Learning system, Instrument and Control, Vol38, No10, pp. 618-623, 1999, The Society of Instrument and Control Engineers, In Japanese

11. Y. Morita, The 2nd volume of new edition science-of-nursing complete works – Physiology-, Mejikaru Furenndo company (incorporated company), 1992, In Japanese

12. S. Akira, The 18nd volume of new edition science-of-nursing complete works –Adult science of nursing 3-, Mejikaru Furenndo company (incorporated company), 1992, In Japanese

13. T. Ohsawa, The volume 3 accorging to system science-of-nursing lecture –Clinical radiology-, Igaku-Shoin (incorporated company), 1970, In Japanese

14. C. J. C. H. Watkins. "Learning from Delayed Rewards". PhD thesis, Cambridge University, 1989.

15. C. J. C. H. Watkins y P. Dayan. "Q-leaning". Machine Learning, 8: 279-292, 19921. Baldonado, M., Chang, C.-C.K., Gravano, L., Paepcke, A.: The Stanford Digital Library Metadata Architecture. Int. J. Digit. Libr. 1 (1997) 108–121

Drift Ice Detection
Using a Self-organizing Neural Network

Minoru Fukumi[1], Taketsugu Nagao[1], Yasue Mitsukura[2], and Rajiv Khosla[3]

[1] University of Tokushima, Dept. of Information Science and Intelligent Systems,
2-1, Minami-Josanjima, 770-8506, Japan
`fukumi@is.tokushima-u.ac.jp`
[2] Okayama University, Faculty of Education,
3-1-1, Tsushima-Naka, Okayama, 700-8530, Japan
[3] La Trobe University, Melbourne, Australia

Abstract. This paper proposes a segmentation method of SAR (Synthetic Aperture Radar) images based on a SOM (Self-Organizing Map) neural network. SAR images are obtained by observation using microwave sensor. For teacher data generation, they are segmented into the drift ice (thick and thin), and sea regions manually, and then their features are extracted from partitioned data. However they are not necessarily effective for neural network learning because they might include incorrectly segmented data. Therefore, in particular, a multi-step SOM is used as a learning method to improve reliability of teacher data, and carry out classification. This process enable us to fix all mistook data and segment the SAR image data using just data. The validity of this method was demonstrated by means of computer simulations using the actual SAR images.

1 Introduction

In recent years, abnormal weather in an earth scale is increasing rapidly with growing environmental destruction. At present, various observation equipments are used as detailed data acquisition means on the earth [1]-[3]. Detection of the drift ice treated in this paper is carried out for prevention of the marine peril in winter. As for the present drift ice observation, the vessel and the airplane are used well. However, it is difficult to observe the earth surface by them as a continuous data acquisition means from receiving restrictions of weather conditions heavily. Moreover, since drift ice has the feature that floats on the sea surface and move easily, it always has the necessity for observation. Therefore, in this paper, drift ice is detected using the SAR (Synthetic Aperture Radar) images obtained from the observation satellite. SAR that can observe any region without being dependent on day and night has the merit that is not influenced of weather conditions. This means that it can always observe the drift ice.

This paper tries to classify the drift ice using a self-organizing (SOM) neural network. The objective is to classify SAR images into three regions, sea, thick-ice, and thin-ice. "Land" can be detected easily from its location. In this case, correct drift ice information on the sea is necessary to evaluate classification accuracy. However, there is no correct information that can be used as teacher data. Only information is the rough drift ice observation data (sea ice condition chart) obtained from the Hydrographic and Oceanographic Department of the 1st Regional Coast Guard in Japan. This data show concentration levels of drift ice in numerical scale and is usually ob-

R. Khosla et al. (Eds.): KES 2005, LNAI 3681, pp. 1268–1274, 2005.

tained by visual observation. This scale roughly uses 5 kinds of concentration degrees, 0, 1-3, 4-6, 7-8, and 9-10. For instance, "1-3" shows that drift ice covers over 10 to 30 % of sea. If we use these data as teacher, there may be some errors on class label in clipping teacher image data, because this is rough data. This fact can decrease their classification reliability.

This paper therefore presents a method to improve classification reliability by using a multi-step SOM. A popular single-step SOM cannot correct misclassified teacher data. From results of computer simulations, it is shown that the multi-step SOM is better in classification than a usual single-step SOM.

2 SAR Image and Features

The sensors installed in a satellite for remote sensing are divided into OPS and SAR. The observation range of OPS is a wavelength band from visible light (0.4 - 0.7μm) to the heat infrared rays (8 - 14μm). The observation range of SAR is in the area of the microwave. In SAR, in addition to passive sensors of OPS, the other sensor launches the microwave, and an active type sensor that measures its reflection wave is used. The SAR images are obtained in gray scale levels. From these SAR images, an image with 16×16 pixels is clipped out as teacher and evaluation data. These images are used in training and testing neural networks. However SAR images have a noise which appears by interference of microwave, called Speckle noise. To remove this noise on images, in this paper, the median filter is used. In addition, 1 pixel size of SAR image is 12.5m square. The size of 16x16 pixels is therefore 200m square. Hereafter every processing is carried out using this image size.

The following three kinds of features are used in this paper:

  (i) Fourier power spectrum
 (ii) Higher-order autocorrelation function [4][5]
(iii) Image features based on a run length method [6][7]

Besides these features, we can use another image features. For instance, Fractal and images features based on the cooccurrence matrix of gray scale levels [9] can be used.

2.1 Fourier Power Spectra

This feature can be obtained by calculating 2-dimensional Fourier transform. Fourier power spectra of three kinds, sea, thick ice, and thin ice are shown in Fig. 1. They have a 16x16 points window size and a different features to classify the drift ice.

2.2 High-Order Autocorrelation Function

This feature is invariant to parallel translation. It does not change if the position of the object in an image is changed. Let an object in an image be f(r). A feature can be obtained to calculate the formula defined by

$$x_n(a_1, a_2, ..., a_N) = \int f(r)f(r+a_1) \cdots f(r+a_N)dr$$

In this paper, a direction is only 3x3 pixels area around a reference point "r". Order "N" is fixed at 2 as the maximum value. On this condition, 25 patterns can be obtained without equivalent patterns when a reference point is moved collimation. 25 local patterns are shown in Fig.2.

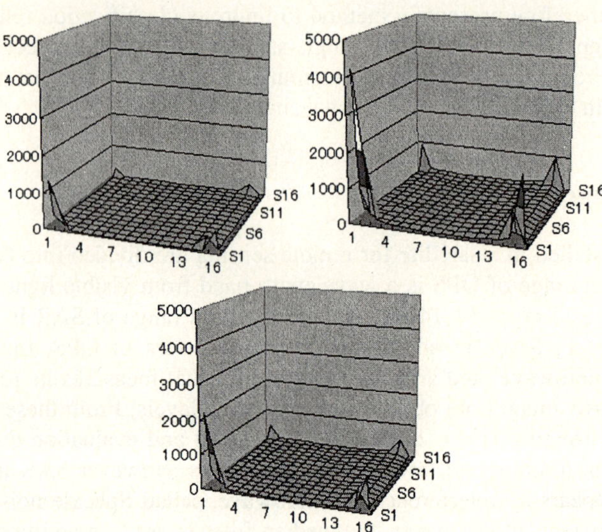

Fig. 1. Fourier Power Spectra. Horizontal axis: index of spectra. Vertical axis: amplitude. The top left is sea, the top right is thick ice, and the bottom is thin ice

2.3 Run Length Method

Run length method is often used as a coding method for compressing information. In this paper, a feature is extracted by finding run length of a texture image. The run length of an image indicates the length where pixels with the same gradation continue on any angle. The run length matrix is defined as

$$P(i, j, \theta) : \quad \text{gray level i, length j, direction } \theta$$

The sum of the run length matrix is computed as

$$T = \sum_{i=1}^{g} \sum_{j=1}^{N} P(i, j; \theta)$$

and then 5 kinds of features are computed. For instance, one of five features is

$$f_1 = \frac{1}{T} \sum_{i=1}^{g} \sum_{j=1}^{N} \frac{P(i, j; \theta)}{j^2}$$

This is a short runs emphasis. The other four functions are also used to generate run length features. Those are computed for four directions. The number of total run length features is 20.

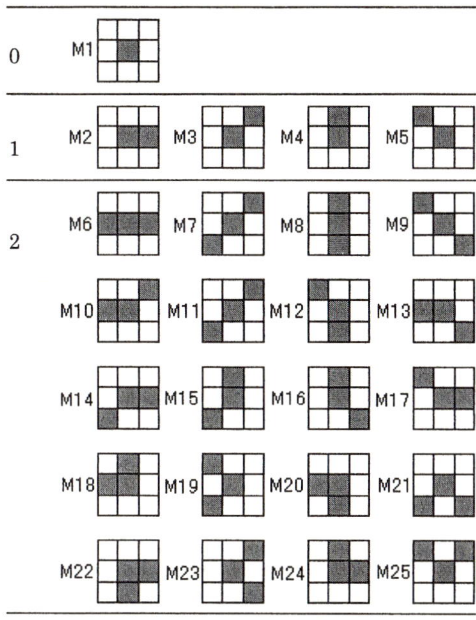

Fig. 2. 25 local patterns based on higher-order autocorrelation function

3   A Multi-step SOM

The SOM (self-organizing map) was proposed by T.Kohonen [8] and is a competitive learning network model. In this model, a winner unit which is nearest to an input feature can carry out selective learning. This paper tries to map input features into a 2-dimensional space and classify them into three classes, sea, thick ice, and thin ice. The input features used in this paper are Fourier spectra with 137 components by symmetry, 25 higher-order autocorrelation functions, and 20 run length features. The number of input features to SOM is therefore 182.

However sea and thin ice features are similar and there is a possibility where they can be misclassified in data clipping. In order to correct this situation, we present a multi-step SOM algorithm. This can correct a misclassified class based on data similarity by an unsupervised learning. Concrete algorithm is summarized in Fig.3. After SOM learning, when a unit has two class labels of inputs, a label with fewer input data and its data are eliminated from the unit. Then SOM is carried out again without eliminated data. After relearning of SOM, the eliminated data are fed to the SOM and its class label is corrected. This process is repeated several times.

The multi-step SOM stops if the number of class correction after SOM relearning is less than a given number. Fig. 4 shows the resulting classification obtained by the multi-step SOM. A unit labeled "reject" shows that it has no data assigned by the multi-step SOM or has the same number of data for two classes. A single-step SOM included 9 units labeled "reject". However only 7 "reject" units left after the multi-step SOM. This decreased the number of rejected data less than the half.

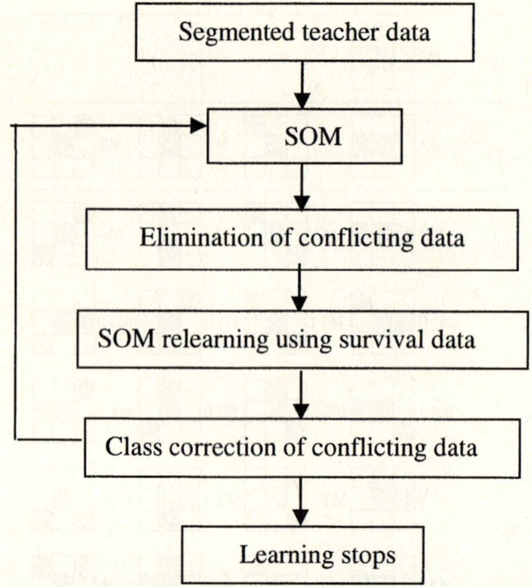

Fig. 3. A multi-step SOM algorithm

thick	thick	thick	thin	reject
reject	thick	thick	reject	sea
sea	reject	thick	reject	sea
reject	thick	thick	thin	sea
thick	thick	thin	thin	reject

Fig. 4. Classification result of teacher data obtained using the multi-step SOM

4 Simulation Result

The number of training data is 80 for each class. After the multi-step SOM, the number of sea data is 87 including 7 thin ice data before learning. After the learning, the number of thick ice and thin ice data are 80 and 73, respectively. The SAR image classification is carried out using this result.

Figure 5 shows comparison of two kinds of SOM learning results. Fig.5 (iv) shows ice concentration degrees in Okhotsk sea. The data size is 1,600 x 1,600 pixels (20,000m x 20,000m). The concentration degrees in this area are 7-8. In the figure, blue pixels are sea, light blue pixels are thin ice, and green pixels are thick ice. By using the multi-step SOM, sea and thin ice regions are corrected.

5 Conclusion

In this paper, a multi-step SOM is presented to correct misclassified ambiguous teacher data. This can improve reliability of teacher data and is better in classification accuracy for drift ice detection using real SAR image data.

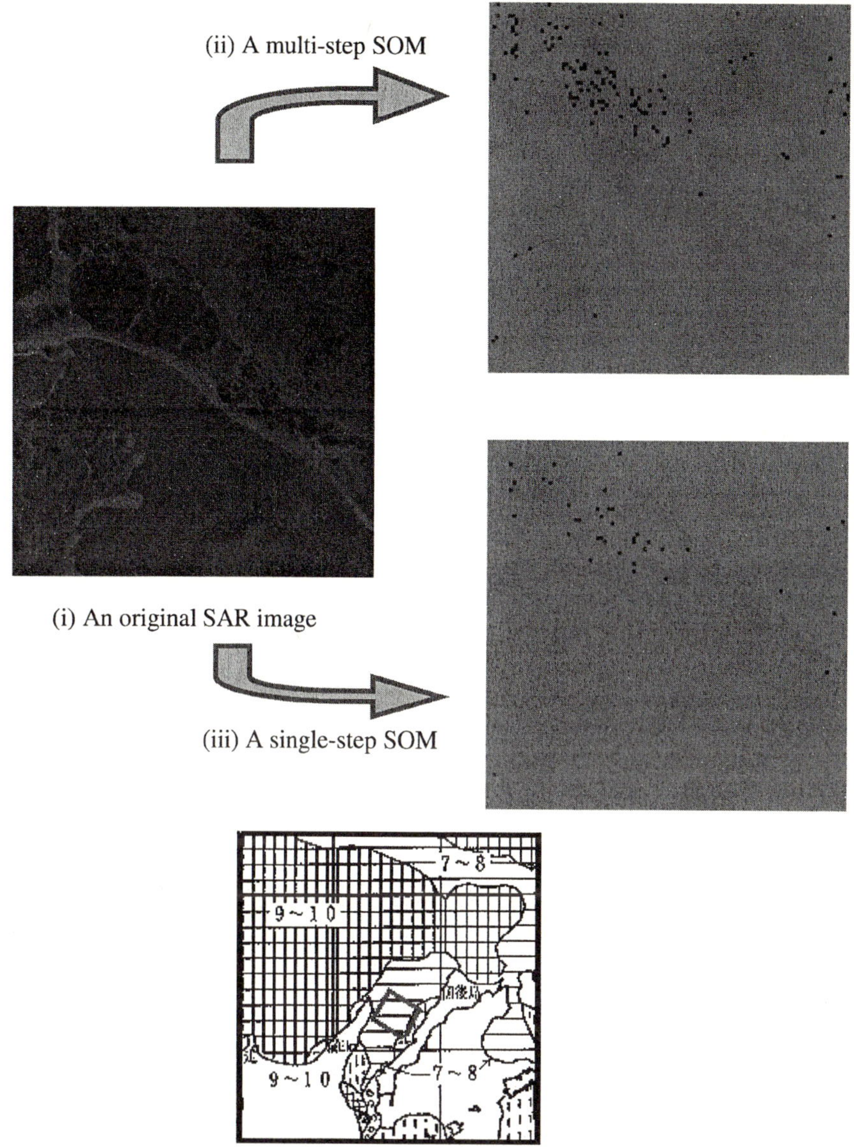

(iv) Drift ice concentration degrees in Okhotsk sea.

Fig. 5. Comparison of two kinds of SOM learning

References

1. B. Bhanu and G. Jones III, "Object Recognition Results Using MSTAR Synthetic Aperture Radar Data", Proc. of the IEEE Workshop CVBVS 2000, Session C: Synthetic Aperture Radar Image Analysis (2000).

2. A. Bors, E. Hancock and R. Wilson, "3-D Terrain from Synthetic Aperture Radar Images", Proc. of the IEEE Workshop on CVBVS 2000, Session C: Synthetic Aperture Radar Image Analysis (2000).
3. Betti, M. Barni, and A. Mecocci, "Using a wavelet-based fractal feature to improve texture discrimination on SAR images", Proc. of the 1997 International Conference on Image Processing (1997).
4. T. Kurita, K. Hotta, and T. Mishima, "Scale Invariant Recognition of Face Image Using Higher Order Local Autocorrelation Features of Log-Polar Image", IEICE D-II Vol.J80-D-II No.8 pp.2209-2217 (1999) in Japanese.
5. M.Kreutz, B. Volpel, and H.Janben: "Scale-invariant image recognition based on higher-order autocorrelation features, Pattern Recognition", Vol.29, No.1 (1996).
6. Galloway,M.M.: "Texture Analysis Using Gray Level Run Lengths," Computer Graphics and Image Processing, 4, pp.172-179 (1975).
7. T.Nagao, Y.Mitsukura, M. Fukumi and N. Akamatsu: "Drift ICE Recognition using Remote Sensing Data by Neural Networks", CD-ROM Proc. of ICONIP'2002, Singapore, #1325, pp.1-5 (2002)
8. Teuvo Kohonen, The Self Organizing Map, Proc. of The IEEE, Vol.78, 9, pp.1464-1480 (1999).
9. S. Tadokoro, T. Nagao, M. Fukumi, and N. Akamatsu, "Research on Drift Ice Detection from SAR Images Using Bio-Information Processing", Proc. of SIP'2003, Hawaii, pp.167-172 (2003).

Word Sense Disambiguation of Thai Language with Unsupervised Learning

Sunee Pongpinigpinyo and Wanchai Rivepiboon

Computer Engineering Department, Faculty of Engineering
Chulalongkorn University, Bangkok, 10330 Thailand
Sunee.Po@Student.chula.ac.th, Wanchai.R@Chula.ac.th

Abstract. Many approach strategies can be employed to resolve word sense ambiguity with a reasonable degree of accuracy. These strategies are: knowledge-based, corpus-based, and hybrid-based. This paper pays attention to the corpus-based strategy that employs an unsupervised learning method for disambiguation. We report our investigation of Latent Semantic Indexing (LSI), an unsupervised learning, to the task of Thai noun and verbal word sense disambiguation. We report experiments on two Thai polysemous words, namely หัว /hua4/ and เก็บ /kep1/ that are used as a representative of Thai nouns and verbs respectively. The results of these experiments demonstrate the effectiveness and indicate the potential of applying vector-based distributional information measures to semantic disambiguation. Our approach performs better than a baseline system, which picks the most frequent sense.

1 Introduction

Word Sense Disambiguation (WSD) is the process of resolving natural language ambiguities, which is one of the most important problems in the computational linguistics. It refers to the process of selecting the most appropriate meaning or sense to a given ambiguous word within a given context. Resolving the word ambiguity is considered as the major bottleneck for large scale language understanding applications and their associate tasks such as machine translation (MT), information retrieval (IR), natural language understanding (NLU) and others. These various range applications of natural language processing need knowledge of word meaning to select the correct word sense in a context. For example, the Thai word หัว /hua4/ has many different senses, one of which can be translated into English as head, and another as chief.

As with other languages, working with Thai single words and Thai compound words also deals with the ambiguity of word meanings. Polysemy refers to a word that has more than one related meaning or sense, which is derived from the same word form and listed in the same lexical entry in a dictionary [17]. Homonymy is a type of word that has a completely different meaning or sense, which accidentally has the same word form and is listed in the same lexical entry in a dictionary [17].

Many approaches have been proposed for eliminating the ambiguous. Most of the word sense researching is to assess in English sentence and to assist in English language translation. Most Thai words, especially Thai compound words, have several

R. Khosla et al. (Eds.): KES 2005, LNAI 3681, pp. 1275–1283, 2005.

meanings or senses and depend on context words which ambiguous between these senses. The Thai compound words make ambiguity of word meanings more complex problem. This type of word is composed of a combination of different words where each part of the combined word has a complete meaning by itself. The Thai new compound word may retain a partial meaning of the original combined words or completely different meaning from each word. For example, the compound word หางเสือ (หาง (tail), เสือ (tiger)) has two senses which mean tiger tail and rudder. The word หาง (tail) maintains its meaning that is an end body part of animal. When it is combined with เสือ (tiger), it still retains its relation meaning in the equivalent level as the end body part of a tiger. But the compound word sometimes has a completely different meaning. For example, หางเสือ (หาง (tail), เสือ (tiger)) which means rudder has the meaning which is completely different from each word.

Three main approaches have been applied in the WSD field.

1. Machine Readable Dictionaries (MRD) rely on information provided by Machine Readable Dictionaries (MRD) [1], [11], [12].
2. Supervised learning approaches use information gathered from training on a corpus that has sense-tagged for semantic disambiguation [13]. A major obstacle of this approach is the difficulty of manual sense-tagged in a training corpus that impedes the applicability of many approaches to domains.
3. Unsupervised leaning approaches determine the class membership of each object to be classified in a sample without using sense-tagged training examples [14], [21]. These approaches are considered to have an advantage over supervised learning approach as they do not require costly hand-tagged training data.

This paper describes an investigation of the use of Latent Semantic Indexing (LSI) to solve the semantic ambiguity of two Thai polysemous words, namely หัว /hua4/ and เก็บ /kep1/, which are used as a representative of nouns and verbs respectively. In this research, we focus on the data that are created by combining an individual word หัว /hua4/ and เก็บ /kep1/ with other words to be compound words, reduplicative and repetitive words with transparent meanings หัว /hua4/ and เก็บ /kep1/ which have parts of speech other than noun and verb respectively are excluded in our study.

Latent Semantic Indexing (LSI) [3] is a corpus-based statistical method for inducing and representing aspects of the meanings of words and passages (of natural language) reflective in their usage. The method generates a real valued vector description for documents of text. Basically, the central concepts of LSI is that the information about the contexts in which a particular word appears or does not appear provides a set of mutual constraints to determine the similarity of meaning of sets of words to each other. The advantage of LSI is that it is a fully automatic corpus based statistical procedure that does not require syntactic analysis.

The remainder of this paper is organized as follows. Section 2 describes the related work using corpus statistics to disambiguate word sense meaning. Section 3 briefly describes the Latent Semantic Indexing. Section 4 explains the experimental methodology to Thai word sense disambiguation. Section 5 reports and discusses the results. In section 6 some conclusions are dawn and some suggestion for future work on the area of WSD offered.

2 Related Work

A wide rang of approaches have been investigated and a large amount of effort devoted to tackle WSD. Currently one of the most successful lines of research is the corpus-based approach using statistical or Machine Learning (ML) algorithms. Both supervised learning and unsupervised learning are applied to learn statistical models or classifiers from corpora in order to perform WSD. For the research reported in this paper, the focus will be on the use of unsupervised methods for WSD.

The method adopted by Schutze and Zernik [18], [22] avoids tagging each occurrence in the training corpus and associates each sense of a polysemous word with a set of its co-occurring words. If a word has several senses, then the word is associated with several different sets of co-occurring words, each of which corresponds to one of the senses of the word.

Yarowsky [21] used an unsupervised learning procedure with noun WSD. The proposed algorithm starts with a set of labeled data (seeds) and builds a classifier which is then applied on the set of unlabeled data. Only those instances that can be classified with a precision exceeding a certain minimum threshold are added to the labeled set. The result of Yarowsky [21]'s method shows that the average percentage attained was 96.1% for 12 nouns when the training data was a 460 million word corpus, although Yarowsky used only nouns and did not discuss more than two senses of a word.

Pedersen and Bruce [14] presented three unsupervised learning algorithms to distinguish the sense of an ambiguous word in untagged text. These were McQuitty's similarity analysis, Ward's Minimum-Variance method and the EM algorithm. These algorithms assign each instance of an ambiguous word to a known sense definition based solely on the values of automatically identifiable features in text. Pedersen and Bruce reported that the disambiguating of nouns is more successful than adjectives or verbs. The best result for verbs was provided through the use of McQuitty's method (71.8%), although they tested only 13 ambiguous words of which only 4 were verbs. Purandare and Pedersen [15], [16] have developed methods of clustering multiple occurrences of a given word into senses based on their contextual similarity. In this paper we adapt those techniques to the problem of name discrimination.

3 Latent Semantic Indexing

Latent Semantic Indexing (LSI) is a psychological model and computational simulation intended to help explain the way that humans learn and represent the meaning of words, text, and other knowledge.

LSI is a vector based model of semantic based on word co-occurrences [4]. Words that occur together in the same or similar contexts are considered to be semantically similar. Likewise, words that occur together co-occurrence and have a syntagmatic and/or thematic connection are considered to be semantically similar. The LSI algorithm is trained on a corpus of documents. Documents here are any semantically cohesive set of words, such as sentences, paragraphs, any articles, etc. To build the LSI model for the experiments in this paper, a large co-occurrence matrix of documents is then created, with rows corresponding to words in the vocabulary, and columns to documents. Each entry in the matrix is a weighted frequency of the corresponding term in the corresponding document.

LSI relies on a Singular Value Decomposition (SVD) [19] of a matrix (word ⊓ context) derived from a corpus of natural context that pertains to knowledge in the particular domain of interest. SVD is a form of factor analysis and acts as a method for reducing the dimensionality of a feature space without serious loss of specificity. Typically, the word by context matrix is very large and quite often sparse. SVD reduces the number of dimensions without great loss of descriptiveness. SVD is the underlying operation in a number of applications including statistical principal component analysis [8], text retrieval [2], [6] pattern recognition and dimensionality reduction [5] and natural language standing [10].

The next step is to reduce the very large sparse matrix into a compressed matrix which based on Singular Value Decomposition (SVD). The result of the SVD is a k-dimensional vector space containing a vector for each term and each document.

$$M = T \times S \times D'$$

The original $t \times d$ matrix M is decomposed into a reduced rank $t \times k$ term matrix T, a diagonal matrix of singular values, S and a $d \times k$ document matrix D.

Decreasing k, the number of dimensions retained, reduces the accuracy with which M can be recreated from its component matrixes, but importantly, it reduces the noise from the original matrix. As empirical results in Information Retrieval (IR), we chose 300, 400 and 500 respectively as a value of k in our experiments. We find that best value of k is 500 in this our Thai word senses disambiguation experiment although 300 is generally chosen to be a good value of k in IR experiments [7]. In this task, we are only interested in the term matrix, T. Each row of T is a vector representation of the semantics of a particular word, in a k dimensional space. We can now compare the semantic distance between any two words by looking at the cosine of the angle of the two corresponding rows (vectors) in the matrix T. In this research, the cosine of the angles between the context vectors is used to calculate the correlation between words. The cosine measure for two vectors $\vec{x}$ and $\vec{y}$ can be calculated as follows:

$$Similarity = cos(\vec{x}, \vec{y}) = \frac{\vec{x} \cdot \vec{y}}{|\vec{x}||\vec{y}|} = \frac{\sum_{i=1}^{n} x_i y_i}{\sqrt{\sum_{i=1}^{n} x_i^2} \sqrt{\sum_{i=1}^{n} y_i^2}}$$

The advantages of using LSI with word meaning are that it is a fully automatic, corpus based statistical procedure that does not require syntactic analysis. LSI uses a fully automatic mathematical/statistical technique to extract and infer semantic relations between the meanings of words from their contextual usage in large collections of natural discourse. The analysis yields a mathematically well-defined representation of a word that can be thought of as a kind of average meaning.

4 Experimental Methodology

We evaluate our method using sources of sense-tagged corpus. In supervised learning sense-tagged corpus is used to induce a classifier that is then applied to classify test data. Our approach, however, is purely unsupervised and the sense-tagged corpus is used to carry out an evaluation of the discovered sense groups.

Since there is no available Thai corpus-based, which contains Thai polysemous words in public, so in this research, we use a Thai corpus-based, which contains Thai polysemous words หัว /hua4/ and เก็บ /kep1/, were created by [9]. According to [9], the polysemous words and their contexts were randomly extracted from the corpus of "Bangkok Business" newspaper from November 1st, 1999 to October 31st, 2000 with the total size of 132 MB. The corpus contains sentences, which have sense of หัว /hua4/ and เก็บ /kep1/. The data contain 2,200 samples of หัว /hua4/ and เก็บ /kep1/ of each word. Each instance of หัว /hua4/ and เก็บ /kep1/ was hand-tagged with its sense defined in the Thai Royal Institute Dictionary BE (Buddhist Era) 2525. There were some senses found from the data that were not provided in the dictionary, so we had to analyze these additional senses. The characteristic of Thai text language is that there is no word boundary in Thai written text. Therefore, the collected data which contained the polysemous words หัว /hua4/ and เก็บ /kep1/ must be word-segmented. The segmentation was processed automatically by SWATH [20] which is a Thai word segmentation program from the NECTEC. The error correction was verified manually based on the context. The distributions of senses of หัว /hua4/ and เก็บ /kep1/ are presented in Table 1 and Table 2 respectively.

Table 1. Definitions of sense หัว /hua4/

Senses	Definitions
Head	Body part, which contains the brain.
Head of coin	Side of coin where a person's profile is represented, opposite of tails.
Intelligence	Ability of a person's brain.
View point	The way of a person views or thinks about as issue.
Talent	Talent or special ability to do something.
Top	Top part or pointed ends of and object.
Front	Front or pointing part of an object.
Early hours	The early hours or part of a time
Bulb	Globular base of stem of some plants sending roots downward and leaves upwards.
Concentrate	Concentrated substance.
Entity	Metonyms use of head to refer to an individual; extended metaphorically to refer to an organization.
Hair	Hair on the head of a human or an animal; hairstyle.
Brain	Brian. It also refers to the seat of consciousness, thought, memory and emotion.
Emotion	Emotional and psychological state.
Machine part	Vital part of machine. For example, part which pulls the rest of an engine, part that cuts or emanates sound or energy.
Chief	Top position of leadership, importance and honor, an individual holding these positions.
Heading	Information shown at the top of a page; title, heading; letterhead.
Headline	Headlines in newspaper.
Topics	Information represented in headlines, titles, and headings.
Titles or Names	Titles or names of newspapers, book and magazine.

Table 2. Definitions of sense เก็บ /kep1/

Senses	Definitions
To pick up	To pick something up from the ground or the floor.
To arrange	To put away; to arrange objects in a cabinet.
To take	To collect; to harvest; to take under one's care.
To keep	To keep or store, to prevent loss or damage.
To gather	To gather; to save.
To charge	To collect or to charge a fee.
To hide	To keep out of sight; to keep hidden from others; to hide.
To kill	To get rid of; to eliminate.
To purchase	To acquire; to buy in stock markets.

5 Results

The input of the testing component is the testing corpus, which is already segmented. The output is the most likely senses of words given by the WSD systems. The performance of the method was computed as precision rates by applying the following formula:

$$Precision\ Rates = \frac{Total\ number\ of\ correct\ answers}{Total\ number\ of\ answered\ senses} * 100$$

5.1 Baseline System

As a baseline system, the most frequent sense (MFS) of a word is chosen as the correct sense. The frequency of word senses is calculated from the occurrences of the word senses in the corpus, with ties broken randomly.

5.2 Experimental Results

Table 3 and Table 4 show experimental results compare with the baseline system for the disambiguation of หัว /hua4/ and เก็บ /kep1/ respectively. The first column of Table 3 and Table 4 are sense definitions. The number of sentences of occurrences of each sense is shown in the second column of Table 3 and Table 4. The final column shows the total number of correct answers that could be estimated correctly. Table 3 shows that the polysemous word หัว/hua4/ has 20 senses the percentage attained at 71.27 %. Table 4 shows that the polysemous word เก็บ /kep1/ has 9 senses the percentage attained at 75.58 %. As a result, it can be pointed out that less polysemous words are the better performances of the method. The reasons why less polysemous words have clearer sense indicators as is that their senses are not closely related. Different senses occur with a totally different context. We compare our experimental results of both polysemous word หัว /hua4/ and เก็บ /kep1/ with the baseline system. It can be seen that our approach outperforms that of the baseline system.

Table 3. Precision rate of disambiguation of หัว/hua4/ and the baseline

Sense Definitions	Frequency of Senses	Precision (%)
Brain	138	77.5
Bulb	159	75.9
Chief	30	76.5
Concentrate	55	76.5
Early hours	41	70.4
Emotion	13	72.8
Entity	460	61.2
Front	133	69.8
Hair	37	78.7
Head	506	62.4
Head of coin	6	60.1
Heading	7	70.2
Headline	41	74.5
Intelligence	88	73.2
Machine part	50	77.2
Talent	5	63.7
Titles or names	56	70.9
Top	77	71.3
Topics	60	77.2
View point	238	65.4
	Average	**71.27**
Baseline = 23.00 %		

Table 4. Precision rate of disambiguation of เก็บ /kep1/ and the baseline

Sense Definitions	Frequency of Senses	Precision (%)
To arrange	41	80.7
To charge	627	72.1
To gather	295	79.8
To hide	61	76.5
To keep	832	70.2
To kill	10	75.8
To pick up	7	79.8
To purchase	20	73.5
To take	322	71.8
	Average	**75.58**
Baseline= 37.81 %		

6   Conclusion

In this paper, we have presented a novel method of Thai word sense disambiguation by using Latent Semantic Indexing (LSI). LSI has one property that is very attractive for processing in solving ambiguity word semantic. No knowledge sources are required for the analysis. Thus it eliminates the need for dictionaries. It needs neither any training or uses word sense tags from corpus. The use of unlabelled data is especially important in corpus-based natural language processing because raw corpora are ubiquitous while sense tags data are expensive to obtain. This is shown that it is a suitable method of further developing a Thai word sense disambiguation program.

The results from the research on word sense disambiguation of Thai polysemous word หัว/hua4/ and เก็บ/kep1/ are promising. The data are free running text and have large number of senses per word (twenty senses for หัว /hua4/ and nine senses for เก็บ /kep1/). Our approach performs better than the baseline system that chooses the most frequent sense.

In future work, we will intend to use larger and more diverse corpora. Furthermore, our long term goal is that we will develop this method combining with other learning methods to enhance the ability of solving word sense ambiguity problem

References

1. Eneko Agirre and German Rigau.: A proposal for word sense disambiguation using conceptual distance. In Proceedings of the International Conference Recent Advances in Natural Language Processing, Tzigov Chark, Bulgaria, 1995.
2. Michael W. Berry, Susan T. Dumais and Gavin W. O'Brien.: Using Linear Algebra for Intelligent Information Retrieval. SIAM: Review, 37(4):573-595, 1995.
3. Michael W. Berry.: Large Scale Singular Value Computations. International journal of Supercomputer Applications, Vol.6, pages 13-49, 1992.
4. Scott Deerwester, Susan T. Dumais, George W. Furnas, Thomas K. Landauer and Richard Harshman.: Indexing by Latent Semantic Analysis. Journal of the American Society for Information Science, Vol. 41, pages 391-407, 1990.
5. Richard O. Duda,. Peter E. Hart and David G. Stork.: Pattern Classification, 2ndEd., Wiley, 2000.
6. Susan T. Dumais.: Latent Semantic Indexing (LSI) and TREC-2. In Proceedings of The 2nd Text Retrieval Conference (TREC-2), March, pages 105-115, 1994.
7. Peter W. Foltz.: Latent Semantic Analysis for text-based research. Behavior Research Methods, Instruments and Computers, 28(2), pages 197-202, 1996.
8. I T Jolliffe.: Principal Component Analysis, Springer Verlag, 1986.
9. Wipharuk Kanokrattanukul.: Word Sense Disambiguation in Thai Using Decision List Collocation. Master Thesis, Department of Linguistics, Faculty of Arts, Chulalongkorn University, 2001.
10. Thomas K. Landauer and Susan T. Dumais.: A Solution to Plato's Problem: The Latent Semantic Analysis Theory of the Acquisition, Induction, and Representation of Knowledge. Psychological Review, 104(2), pages 211-240, 1997.
11. Claudia Leacock, Martin Chodorow and George A. Miller.: Using Corpus Statistics and WordNet Relations for Sense Identification. Computational Linguistics, 24(1):147-165, 1998.
12. George A. Miller, Martin Chodorow, Shari Landes, Claudia Leacock and Robert G. Thomas.: Using a semantic concordance for sense identification. In Proceedings of the ARPA Human Language Technology Workshop, 1994.
13. Hwee Tou Ng and Hian Beng Lee.: Integrating Multiple Knowledge Sources to Disambiguate Word Sense: An Examplar-Based Approach. In Proceedings of the 34th Annual Meeting of the Association for Computational Linguistics, Santa Cruz, 1996.
14. Ted Pedersen and Rebecca Bruce.: Distinguishing word senses in untagged text. In Proceedings of the 2nd Conference on Empirical Methods in Natural Language Processing, pages 197-207, 1997.
15. Amruta Purandare.: Discriminating among word senses using McQuitty's similarity analysis. In Companion Volume to the Proceedings of HLT-NAACL 2003 – Student Research Workshop, pages 19–24, Edmonton, Alberta, Canada, May 27 - June 1 2003.

16. Amruta Purandare and Ted. Pedersen.: Word sense discrimination by clustering contexts in vector and similarity spaces. In Proceedings of the Conference on Computational Natural Language Learning, pages 41–48, Boston, MA, 2004.
17. J I Saeed.: Semantics, The United Kingdom, Blackweel Publishers, 1997.
18. Hinrich Schutze.: Dimensions of Meaning. In Proceedings of Supercomputing, pages 787-796, 1992.
19. Gilbert Strang.: Algebra and its applications, 2nd ed., Academic Press, 1980.
20. Smart Word Analysis for Thai. National Electronics and Computer Technology Center (NECTEC), Information Research and Development Division. Available from: http://www.links.nectec.or.th/, 2002.
21. David Yarowsky.: Unsupervised Word Sense Disambiguation Rivaling Supervised Methods. In Proceedings of the 33rd Annual Meeting of the Association of Computational Linguistics, Cambridge, Massachusetts, 1995.
22. Uri Zernik.: Train1 vs. Train2: Tagging Word Sense in Corpus. Lexical Acquisition: Exploiting on-line Resources to Build a Lexicon. In Proceedings Recherche d'Informations Assistée par Ordinateur, pages 91-112, 1991.

Differential Evolution
with Self-adaptive Populations

Jason Teo

AI Research Group, School of Engineering and Information Technology,
Universiti Malaysia Sabah, Kota Kinabalu, Sabah, Malaysia
jtwteo@ums.edu.my

Abstract. In this paper, we present a first attempt at self-adapting the population size parameter in addition to self-adapting crossover and mutation rates for the Differential Evolution (DE) algorithm. The objective is to demonstrate the feasibility of self-adapting the population size parameter in DE. Using De Jong's F1-F5 benchmark test problems, we showed that DE with self-adaptive populations produced highly competitive results compared to a conventional DE algorithm with static populations. In addition to reducing the number of parameters used in DE, the proposed algorithm performed better in terms of best solution found than the conventional DE algorithm for one of the test problems. It was also found that that an absolute encoding methodology for self-adapting population size in DE produced results with greater optimization reliability compared to a relative encoding methodology.

1   Introduction

In working towards parameterless evolutionary algorithms, significant progress has been made through adaptation and self-adaptation of crossover and mutation rates [3, 5]. In fact, it has been found to actually improve the quality of solutions resulting from the search process [1]. However, significantly less work has been done in terms of adapting population size. It has been highlighted that an appropriate population size is critical to both run-time efficiency and optimization effectiveness [6, 7]. Although a number of different approaches have been used in adapting the sizes of dynamic populations [2, 4, 7, 8], they are based on adaptation and not self-adaptation — that is the changes made to the population size parameter are based on some form of feedback from the search [1]. Hence, population size is not included as an element of the genetic encoding that is subjected to evolutionary processes and pressures as in self-adaptation. Since self-adaptation has been shown to be a very effective method for dynamically adjusting evolutionary parameters [5], we thus propose a similar treatment for population size. The main objective of this study is to implement and investigate a dynamic approach to the population sizing problem through self-adaptation. It must be pointed out from the outset that our aim in this particular study is simply to demonstrate the feasibility of implementing a self-adaptive population size for reducing the number of user-defined parameters and not to propose a new algorithm which can outperform a conventional evolutionary algorithm.

R. Khosla et al. (Eds.): KES 2005, LNAI 3681, pp. 1284–1290, 2005.

2   The Proposed Algorithms

An evolutionary algorithm based on *Differential Evolution* (DE) [9] was used in this study for testing the viability of self-adapting the population size parameter. The base DE algorithm used in this study is a slightly modified version which uses the randomly chosen main parent to also represent the trial vector and a Gaussian mutation operator, similar to [10, 11] where this DE scheme was found to be highly beneficial for vector optimization and data mining. Our proposed algorithm is called **D**ifferential **E**volution with **S**elf-**A**dapting **P**opulations (DESAP). The main contribution of this proposed algorithm is that the population size is automatically determined from initialization right through the evolutionary search process to completion through self-adaptation. A self-adaptive strategy is also used for controlling the crossover and mutation rates for both our proposed DESAP as well as conventional DE algorithms so that the only difference between the conventional and proposed algorithms is contributed by the self-adaptation of the population size parameter. Investigating the effects of self-adapting crossover and mutation against a canonical DE would also be interesting but is beyond the scope of this paper. Two different versions of DESAP are proposed: (1) DESAP-Abs which uses an absolute encoding methodology, and (2) DESAP-Rel which uses a relative encoding methodology for encoding the population size parameter as follows:

1. Automatically create a random initial population of $(10 * n)$ individuals Ω, where n represents the number of design variables. The crossover rate δ and mutation rate η are assigned random values according to a uniform distribution between $[0, 1]$. In DESAP-Abs, the population size parameter π is initialized to *round(initial population size $+ N(0, 1)$)*, where the initial population size is perturbed by adding to it a Gaussian value $N(0, 1)$. In DESAP-Rel, the population size parameter π is initialized to a random value according to a uniform distribution between $[-0.5, 0.5]$ which represents a relative population growth rate, which can be negative as well as positive to allow for variations in population size that can either decrease or increase.

2. Repeat
   (a) Evaluate the individuals in the population.
   (b) Repeat
        i. Select at random an individual as the main parent α_1, and two individuals, α_2, α_3 as supporting parents.
        ii. Select at random a variable j.
        iii. **Crossover:** With some probability
             $Uniform(0, 1) > \delta^{\alpha_1}$ or if $i = j$, do

$$\omega^{child} \leftarrow \omega^{\alpha_1} + F(\omega^{\alpha_2} - \omega^{\alpha_3}) \tag{1}$$

$$\delta^{child} \leftarrow \delta^{\alpha_1} + F(\delta^{\alpha_2} - \delta^{\alpha_3}) \tag{2}$$

$$\eta^{child} \leftarrow \eta^{\alpha_1} + F(\eta^{\alpha_2} - \eta^{\alpha_3}) \tag{3}$$

DESAP-Abs:
$$\pi^{child} \leftarrow \pi^{\alpha_1} + int(F(\pi^{\alpha_2} - \pi^{\alpha_3})) \tag{4}$$

1286   Jason Teo

DESAP-Rel:

$$\pi^{child} \leftarrow \pi^{\alpha_1} + F(\pi^{\alpha_2} - \pi^{\alpha_3})) \tag{5}$$

otherwise

$$\omega^{child} \leftarrow \omega^{\alpha_1} \tag{6}$$

$$\delta^{child} \leftarrow \delta^{\alpha_1} \tag{7}$$

$$\eta^{child} \leftarrow \eta^{\alpha_1} \tag{8}$$

$$\pi^{child} \leftarrow \pi^{\alpha_1} \tag{9}$$

where each variable in the main parent is perturbed by adding to it the difference between the two values of this variable in the two supporting parents. The amplification factor F for the difference between the two values of the supporting parents is set to 1 in this set of experiments.

iv. **Mutation:** With some probability
$Uniform(0,1) > \eta^{\alpha_1}$, do

$$\omega^{child} \leftarrow \omega^{child} + N(0, \eta^{\alpha_1}) \tag{10}$$

$$\delta^{child} \leftarrow N(0,1) \tag{11}$$

$$\eta^{child} \leftarrow N(0,1) \tag{12}$$

DESAP-Abs:

$$\pi^{child} \leftarrow \pi^{child} + int(N(0.5, 1)) \tag{13}$$

DESAP-Rel:

$$\pi^{child} \leftarrow \pi^{child} + N(0, \eta^{\alpha_1}) \tag{14}$$

(c) Until the population size is M

(d) Calculate new population size

DESAP-Abs:

$$M_{new} = round(\sum_1^M \pi/M) \tag{15}$$

DESAP-Rel:

$$M_{new} = round(M + (\pi * M)) \tag{16}$$

(e) if $M_{new} <= M$ then :
carry forward only the first M_{new} current individuals into the next generation;

otherwise :
carry forward all current individuals into the next generation and add the remaining $(M_{new} - M)$ individuals by cloning the best individual from the current population.

3. Until maximum number of generations is reached.

The initial population size for DESAP-Abs and DESAP-Rel are both set at 10 times the number of design variables as recommended by the authors of the original DE algorithm [9]. For DESAP-Abs, the population size of subsequent generations is taken as the average of the population size attribute from all individuals in the current population (Eq. 15) whereas for DESAP-Rel, it is taken to be the current population size plus a percentage increase or decrease according to the population growth rate (Eq. 16). It should be pointed out also that elitism and culling is achieved naturally in both DESAP-Abs and DESAP-Rel. Elitism is only executed when the population size of the next generation exceeds the population size of the current generation ($M_{new} > M$). In this case, all of the current population will survive into the next generation and on top of that, the best solution in the current population is copied into $M_{new} - M$ individuals to make up the required number of individuals in the next generation. In the case where the population size of the next generation is less than the population size of the current generation ($M_{new} < M$), then only the best M_{new} individuals will survive into the next generation and the remainder of the current population will be culled. No elitism or culling will occur if the population size of the next generation equals the population size of the current population ($M_{new} = M$).

3   Experimental Setup

To investigate the effects of a dynamic self-adaptive population, both versions of our proposed DESAP algorithm are compared against a conventional DE algorithm, which is similar to DESAP in all aspects except that the population size is static and non-adaptive. Five different population sizes of 10, 20, 30, 50, and 100 were used for the conventional DE algorithm. As an initial explorative study, these algorithms were compared using De Jong's F1-F5 test functions [12] only. In future work, we intend to extend the range of test suites as well as to obtain some statistical information. We treat these benchmark problems in their original form, that is as function minimization problems. Each evolutionary setup was run for 100 generations and repeated 50 times using different seeds. The discussion of results will focus only on the solutions found at the end of the evolutionary runs. Due to the inherent nature of dynamic population sizing in DESAP, a comparison based on similar number of objective evaluations is not straightforward and would require introducing artificial constraints on the evolutionary search of the DE algorithms with fixed population sizes.

4   Results and Discussion

Table 1 lists the overall best solution found over 50 runs. Both versions of our proposed algorithm DESAP produced highly similar results as the conventional DE algorithm. In fact, the proposed DESAP algorithm obtained the best over-all solution for F4. It converged to the best possible solution for F3 and F5 as did the DE algorithm with different population sizes. DESAP obtained slightly

Table 1. Overall best solution found over 50 runs.

Algorithm	Test Function				
	F1	F2	F3	F4	F5
DESAP-Abs	8.8E-06	9.2E-05	0	2.7598	0.9980
DESAP-Rel	8.4E-06	1.1E-04	0	3.1400	0.9980
DE-10P	1.7E-05	5.4E-05	0	3.3556	0.9980
DE-20P	7.0E-05	3.6E-05	0	4.0818	0.9980
DE-30P	6.9E-05	1.5E-04	0	4.4530	0.9980
DE-50P	1.5E-05	1.7E-06	0	5.2456	0.9980
DE-100P	1.6E-06	1.6E-05	0	4.8230	0.9980

better solutions than DE for F1 except when DE was set to largest population size (DE-100P). For F2, DESAP performed marginally better than DE-30P and marginally worse against the remaining setups for DE. Comparing between DESAP-Abs and DESAP-Rel, the results were highly similar between both versions except that the absolute encoding produced marginally better results for F2 and F4 and vice-versa for F1. Thus, in terms of the overall best solution found, DESAP produced highly competitive results compared to the conventional DE algorithm.

Table 2 lists the average of the best solutions found over 50 runs. The results again indicate that across the 50 different evolutionary runs, DESAP was able to perform better on average compared to DE with smaller population sizes of 10, 20, and 30 for F1, F3 and F4 and better than DE with population sizes of 10 for F2 and F5. Again, DESAP produced the most favorable average best solution for F4. DE with the larger population sizes of 50 and 100 were able to produce slightly better solutions on average compared to DESAP with the exception of F4 where DESAP outperformed all configurations of the conventional DE. Comparing between DESAP-Abs and DESAP-Rel, both versions had the same average of the best solutions for F1 but was better using the absolute encoding for F2, F3 and F5, whereas the opposite held true for F4. In terms of standard deviation, the results indicate that the performance of DESAP is quite stable compared to the conventional DE algorithm. The dynamic nature of the population size does not appear to adversely affect the stability of the evolutionary search process and was in fact more stable than DE when DE was set to use the smaller population sizes. For DE with the larger population sizes, the standard

Table 2. Average and standard deviation of best solutions found over 50 runs.

Algorithm	Test Function				
	F1	F2	F3	F4	F5
DESAP-Abs	0.0005 ± 0.0004	0.0113 ± 0.0133	0.48 ± 0.81	7.7899 ± 2.1918	5.2351 ± 4.6760
DESAP-Rel	0.0005 ± 0.0006	0.0156 ± 0.0153	1.14 ± 1.41	7.0239 ± 2.1012	5.8397 ± 5.0260
DE-10P	0.0017 ± 0.0020	0.0271 ± 0.0325	3.74 ± 2.52	21.6080 ± 8.6251	7.9802 ± 4.9501
DE-20P	0.0009 ± 0.0012	0.0129 ± 0.0198	2.36 ± 2.22	16.7103 ± 7.0218	5.5710 ± 4.7626
DE-30P	0.0005 ± 0.0003	0.0098 ± 0.0114	1.52 ± 1.59	14.3968 ± 4.9793	3.2273 ± 3.9585
DE-50P	0.0003 ± 0.0003	0.0034 ± 0.0037	0.54 ± 0.88	12.9318 ± 5.0388	1.8220 ± 2.3722
DE-100P	0.0001 ± 0.0001	0.0018 ± 0.0018	0.16 ± 0.42	10.4214 ± 3.0276	0.9980 ± 0.0000

deviation of the best solutions of DESAP were not very different compared to DE. Comparing between the two different methodologies of self-adapting the population size parameter, the absolute encoding appeared to generate less variation in terms of the best solutions found in all the test functions except for F4. This suggests that the evolutionary search process is more stable using the absolute encoding compared to the relative encoding.

Table 3. Final population sizes for the absolute encoding over 50 runs. P.S. refers to population size and S.D. refers to standard deviation.

	DESAP-Abs					DESAP-Rel				
	F1	F2	F3	F4	F5	F1	F2	F3	F4	F5
No. of Variables	3	2	5	30	2	3	2	5	30	2
Minimum P.S.	22	11	42	297	13	14	9	37	251	20
Maximum P.S.	54	25	52	300	42	210	24	120	355	26
Average P.S.	28.92	18.72	49.08	299.16	21.52	36	19.62	53.18	303.46	20.2
S.D. of P.S.	4.44	2.53	1.95	0.71	6.33	27.20	2.11	13.034	17.17	1.01

Table 3 provides an analysis of the dynamic population size reached after 1000 generations for the 50 evolutionary runs of DESAP. One clear observation that can be made here is that there is a fair amount of dynamics present in DESAP when attempting to solve these five problems. The population size analysis for DESAP-Abs is discussed first. There is a noticeable difference between runs that achieved a smaller population size at the end of the search process and those that achieved a higher population size. For F2-F4, the final population sizes appear to be limited within a much narrower range as compared to F1 and especially F5. Surprisingly, the problem with the most number of design variables, F4, produced the least amount of variation in terms of final population sizes with only a standard deviation of 0.71 and a range difference of only 3 individuals between the minimum and maximum population size. It should also be pointed out that although DESAP-Abs utilized a higher population size than those used by the conventional DE algorithm in the F4 experiment, in which it obtained superior results, this supports our motivation in proposing a parameterless evolutionary algorithm — an evolutionary algorithm should be able to automatically determine, adjust and self-adapt its population size appropriately to the problem at hand, which is exactly that achieved by our proposed DESAP-Abs algorithm. For DESAP-Rel, there is a much larger variation in terms of population size compared to DESAP-Abs except for F5. This observation corresponds to the larger standard deviation that was also detected for the best solutions found over 50 runs. Therefore, running DESAP using the absolute population size encoding provided more stability than using the relative population size encoding. Coupled with the fact that the best solutions obtained did not differ significantly between the two types of encoding, it would appear that using the absolute population size encoding may be preferable from a stability point of view.

5   Conclusion and Future Work

Two versions of DESAP were implemented using absolute and relative encoding methodologies respectively for dynamically self-adapting the population size parameter. The results show that DESAP performed similarly well when compared against the conventional DE algorithm. Moreover, it outperformed the conventional DE algorithm for one of the test problems. Additionally, since no significant advantages could be observed using the relative encoding, it is concluded that the use of an absolute encoding for the self-adaptation of population size may be more favorable by virtue of providing more stability over the course of the evolutionary search process compared to the relative encoding. For future work, it would be beneficial to test DESAP in computationally expensive evolutionary applications, such as in evolutionary robotics. If a dynamic and self-adjusting population size produced sufficiently good results, then time-consuming preliminary test runs for hand-tuning the evolutionary parameters could be eliminated entirely and the time better utilized for the actual evolutionary runs.

References

1. Eiben, A., Hinterding, R., Michalewicz, Z.: Parameter control in evolutionary algorithms. IEEE Transactions on Evolutionary Computation **3** (1999) 124–141
2. Harik, G., Lobo, F.: A parameter-less genetic algorithm. 1999 Genetic and Evolutionary Computation Conference. Morgan Kaufmann (1999) 258–265
3. Lis, J.: Parallel genetic algorithm with dynamic control parameter. 3rd IEEE Conference on Evolutionary Computation. IEEE Press (1996) 324–329
4. Shi, X., Wan, L., Lee, H., Yang, X., Wang, L., Liang, Y.: An improved genetic algorithm with variable population size and a pso-ga based hybrid evolutionary algorithm. 2nd International Conference on Machine Learning and Cybernetics. IEEE Press (2003) 1735–1740
5. Spears, W.: Adapting crossover in evolutionary algorithms. 4th Annual Conference on Evolutionary Programming. MIT Press (1995) 367–384
6. Sarker, R., Kazi, M.: Population size, search space and quality of solution. 2003 Congress on Evolutionary Computation. IEEE Press (2003) 2011–2018
7. Smith, R.: Population size. Handbook of Evolutionary Computation. Oxford University Press (1997) 1–5
8. Tan, K., Lee, T., Khor, E.: Evolutionary algorithms with dynamic population size and local exploration for multiobjective optimization. IEEE Transactions on Evolutionary Computation **5** (2001) 565–588
9. Price, K.: An introduction to differential evolution. In Corne, D., Dorigo, M., Glover, F., eds.: New Ideas in Optimization. McGraw-Hill (1999) 79–108
10. Abbass, H., Sarker, R.: The Pareto differential evolution algorithm. International Journal on Artificial Intelligence Tools **11** (2002) 531–552
11. Abbass, H.: Speeding up backpropagation using multiobjective evolutionary algorithms. Neural Computation **15** (2003) 2705–2726
12. De Jong, K.: An Analysis of the Behaviour of a Class of Genetic Adaptive Systems. PhD thesis, Univesity of Michigan (1975)

Binary Neural Networks – A CMOS Design Approach

Amol Deshmukh[1], Jayant Morghade[2], Akashdeep Khera[2], and Preeti Bajaj[3]

[1] Lecturer & Research Associate, Electronics Dept,
G.H.Raisoni College of Engineering, Nagpur, India
amolydeshmukh@yahoo.com
[2] Research Associate, G.H.Raisoni College of Engineering, Nagpur, India
[3] Professor & Head, ETRX Dept, G.H.Raisoni College of Engineering, Nagpur, India
preetib123@yahoo.com

Abstract. The proposed work includes CMOS design of a Neuron to generate binary logic Artificial Neural Network (ANN) as well as Multilayer Neural network. Several neural net chips exist on the market today. Some of these chips operate as analog devices by running below threshold on the transistors thereby gaining continuous properties instead of discrete properties afforded by CMOS transistor logic. In the current paper, authors have proposed a weighter circuit. It is designed with the help of NAND & XOR gates & binary connections are stored in flops. Both the gates provide more flexibility than the way the neuron deals with the input. Two-phase clocking with no overlap is used to ensure that all weights are properly shifted in without any data corruption. The same concept is extended to multilayer network.

1 Introduction

An artificial neural network is a mathematical or computational model for information processing based on a connectionist approach to computation. It involves a network of relatively simple processing elements, where the global behavior is determined by the connections between the processing elements and element parameters. In a neural network model, simple nodes or neurons are connected together to form a network of nodes hence the term neural network [1]. Existing neural net chips operates as analog devices by running below threshold on the transistors, thereby gaining continuous properties instead of discrete properties afforded by CMOS transistor logic. Other chips, however, do use purely the digital capacity of CMOS to execute a network; and yet others use hybrids between the two to create a more robust design [2-3]. The distributed neural computing is typically performed within the entire array composed of neurons and weights. This indicates that parallel distributed computing makes efficient use of the stored data (weights) and of the input data (inputs)[4].

2 Methodology

The neurons are usually arranged in architectures that support the function that they are expected to perform. On a basic level each neuron's connection to another neuron or input carries with it some strength, or weight. Whenever an input arrives at a neuron, it is multiplied by this connection weight. Typical neurons multiply each input by a connection weight and add up all of these weighted inputs. This sum is usually sent through an activation function to determine whether or not a neuron will fire. While

R. Khosla et al. (Eds.): KES 2005, LNAI 3681, pp. 1291–1296, 2005.

such networks exist primarily as software and computer models, some have been hardwired into circuits as well as ICs.

A neural network consists in an array of elementary processors called neurons and a coupling network formed by resistive elements called synapses. Through this network, each neuron can be connected to each other. The connections can be positive or negative. The positive is called as excitatory connection, a negative one is called as inhibitory. Each synapse can either source or sink current to the input line of the connected neuron [5]. The direction of the current is determined by a combination between two values: the connected neuron output value and a connection weight programmed into the synapse. It is always recommended to use the XOR function between two values.

3 Design Approach

The key circuit component in VLSI is the transistor, a three-or four-terminal device that can behave as a switch in digital circuits and an amplifier in analog ones. Transistors may be fabricated using a variety of technologies: BJT (bipolar junction transistor), MOS (metal oxide semiconductor), CMOS (complementary MOS), and others. For neural network implementations, MOS and BJT devices have been used the most; when both occur together the process is called BiCMOS and is the most prevalent present day analog VLSI technology. Fig. 1 shows the circuit symbols for those transistors of most interest to ANN VLSI.

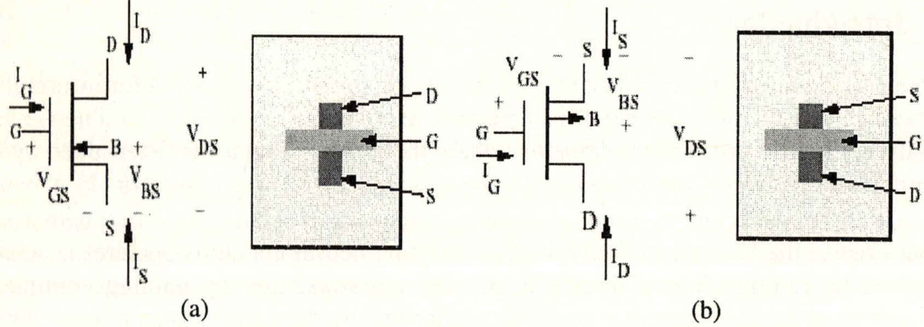

Fig. 1. (a) Basic Modeling of nMOS, (b) Basic Modeling of pMOS

In the MOS transistor, the drain current, I_D, which flows from the outside into the drain, D, and then through the device to the source, S, is controlled by the voltage at the gate, G, with respect to the source, VGS, when the latter is above threshold voltage, Vth. As the threshold voltage can be used as a fine control on the ANN weights, it is dependent upon the voltage of the bulk, B, to source voltage, VBS, where the bulk material is that of the substrate into which the transistor is embedded. The two types of MOS transistors, NMOS and PMOS, are distinguished by their internal conduction mechanisms with the currents and voltages of the latter being ideally the negative of the former in the complementary case desired for CMOS fabrications [6].

To accomplish the weighting function performed by typical neurons, logical operations on each of the inputs to the neuron with known values (weights) have been performed. This is accomplished by storing binary connection weights in flops. These connection weights have been implemented at the individual neuron because it eliminated the need to have one central memory holding all weights for the whole network. Each input to each neuron has two weights associated with it. The first weight is NAND-ed with the input, and the output of this operation is XOR-ed with the second weight [7]. The reason for including both of these operations is that they provide us with more flexibility in how the neuron deals with the input. Fig. 2 shows the hardware used to weight each input to each input. The Table 1 illustrates the advantages of using two weights instead of one.

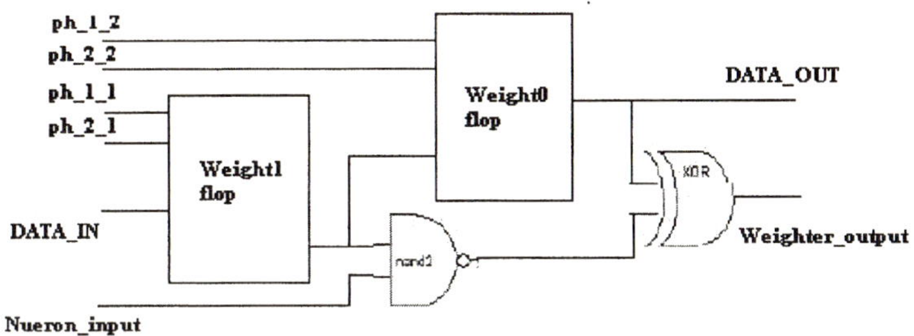

Fig. 2. Weighter circuit

Each independent set of two weights passes a different input to the neuron's activation function. Neuron's activation function is the NOR5. Each neuron weights each of its five inputs separately and then NOR5-s them all together. Neurons can be weighted in order to produce a desired output from a given input. Each flop (storing one weight) in addition to having its output connected to a logic gate also has its output connected to the next flop. In this way, all of the flops all over the chip form a distributed shift register. Since the clock is spread over almost the entire chip, 2-phase clocking with no overlap is used to ensure that all weights are properly shifted in without any data corruption.

Table 1. Truth Table for Weights

Weight1	Weight0	Input	Output
0	0	0	1
0	0	1	1
0	1	0	0
0	1	1	0
1	0	0	1
1	0	1	0
1	1	0	0
1	1	1	1

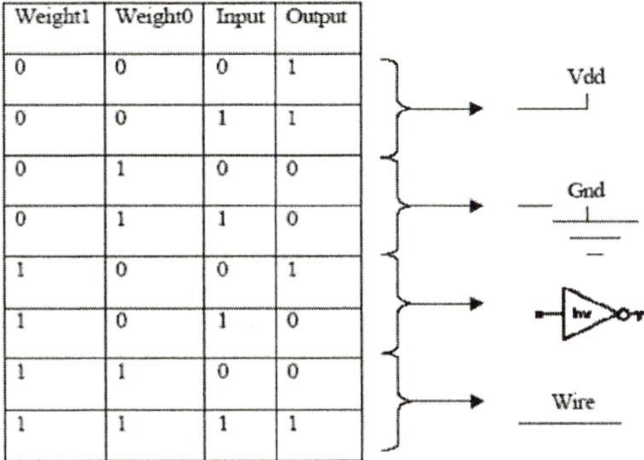

4 Design and Simulation

The design & simulation is performed with Microwind Elite, a tool for CMOS design [8] & Tanner Tools [9]. Fig. 2 shows the basic weighter circuit with neuron input &

weighter output. This facet takes two ph_1 clocks and two ph_2 clocks to match the layout. The layout had two clocks so that each weighter can snap together with the weighter above, making the clocks run as four parallel lines over each layer. Fig. 3 shows the basic circuit of a neuron with five input blocks.

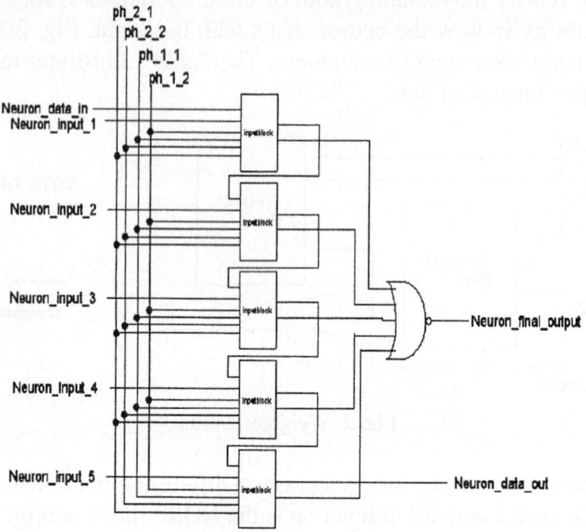

Fig. 3. Neuron circuit

5 Simulation Results

Fig. 4 shows the verification of Truth Table 1 with simulation. The circuit consumes power of 0.4 mW. Fig 5 shows the neuron layer generated using Tanner Tools. Fig.6 & 7 shows the Neural Network generated using Tanner Tools and its waveforms. The waveform shows that the neuron input1 recovered at output as it is with a delay of 4 ns, which is very less, compared to other systems.

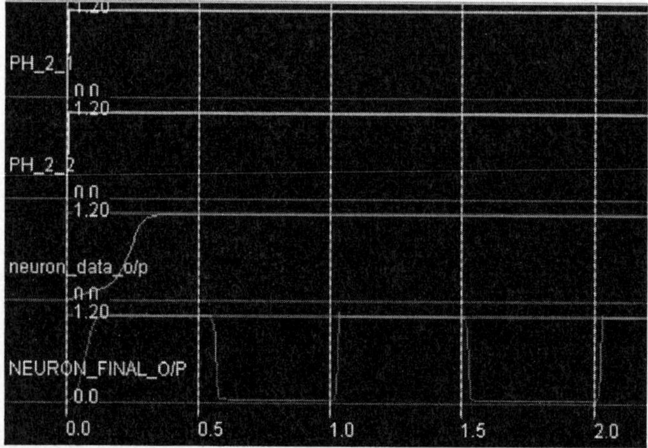

Fig. 4. Neuron output

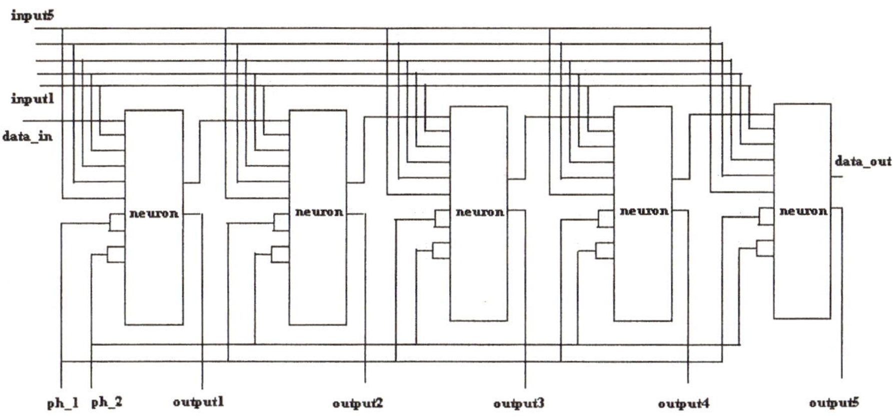

Fig. 5. Neuron layer

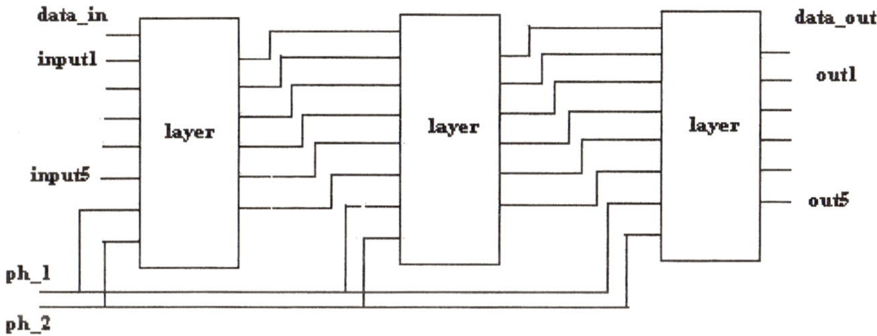

Fig. 6. Neural Network generated through Tanner Tools

Fig. 7. Waveforms of Neural Network

6 Conclusion

A Neuron simulation is successfully implemented and tested with CMOS Technology. The proposed design of Neuron can be used for variety of knowledge based intelligent applications. Authors have extended this work to Multilayer neural network. It can also be extended to closed loop supervised neural networks.

References

1. Hammerstrom, D., 1995, A Digital VLSI Architecture for Real-World Applications, San Diego, CA, Academic Press, pp.335-358.
2. M. Verleysen, et al, A Large VLSI Fully Interconnected Neural Network, 1988 Symposium on VLSI circuits, Tokyo (Japan), 22-24 August 1988, pp.27-28
3. Valeriu Bieu, How to Build VLSI-Efficient Neural Chips, Proceedings of the Intl. ICSC Symposium on Engineering of Intelligent Systems EIS'98.
4. Jacek M. Zurada, Introduction to Artificial Neural Systems, West-publishing company.
5. C. Diorio, et al, Adaptive CMOS: From Biological Inspiration to Systems–on–a-Chip, Proceedings of the IEEE, Vol. 90. No.3, March 2002.
6. C. Diorio, P. Hasler, B. A. Minch, C. A. Mead, A Single- Transistor Silicon Synapse, IEEE Tran. on Electron Devices, vol. 43, no. 11,pp. 1972-1980, Nov. 1996.
7. Daniel Smith & Shamik Maitra, Design and Implementation of a Binary Neural Network E158 Intro to CMOS VLSI Design
8. Microwind User's Manual, INSA Toulouse, National Institute of Applied Sciences, Department of Electrical & Computer Engineering.
9. Tanner Tools Manual, Tanner Research INC, USA.

Video Rate Control
Using an Adaptive Quantization
Based on a Combined Activity Measure

Si-Woong Lee[1], Sung-Hoon Hong[2], Jae Gark Choi[3],
Yun-Ho Ko[4], and Byoung-Ju Yun[5]

[1] Div. of Information Communication and Computer Engineering,
Hanbat National University, Daejeon, Korea,
[2] Dept. of Electronics, Computer and Information Engineering,
Chonnam National University, Gwangju, Korea
[3] Dept. of Computer Engineering, Dongeui University, Busan, Korea
[4] Dept. of Mechatronics Engineering, Chungnam National University, Daejeon, Korea
[5] Dept. of Information and Communication, Kyungpook National University,
Daegu, Korea

Abstract. A new rate control algorithm for videos is presented. The method comes from the MPEG-2 Test Model 5(TM5) rate control, while a buffer constraint and a new measure for the macroblock (MB) activity based on spatio-temporal sensitivity are introduced. Experimental results show that the proposed method outperforms the TM5 rate control in picture quality.

1   Introduction

The video coding standards, such as MPEG, H.263 and H.264 generate variable bit rates intrinsically. In the applications where the compressed video bit streams with variable bit rates are transmitted through a limited channel bandwidth, video buffers at the encoder and the decoder are necessary to smooth out the bit rate variations. In addition, a proper bit-rate control algorithm at the encoder is necessary to prevent the buffers from overflowing and underflowing. Although the rate control itself is not specified in the standards, it strongly affects the performance of video coding.

The conventional rate control algorithms can be classified into two classes according to whether they use the rate-quantizer (R-Q) model or not. The model based approaches, such as H.263 TMN Ver.10 (TMN10) [1] and MPEG-4 Verification Model Ver.8 (VM8) [2] have been considered as rate control algorithms maximizing the video quality at a given bandwidth. However, they are inappropriate for use in a wide variety of the bit rates. Among model-free approaches, the MPEG-2 TM5 rate control [3] is the most representative. The TM5 rate control consists of three steps to adapting the MB quantization parameter for controlling the bit rates. Among them, the adaptive quantization step modulates the quantization parameter adaptively according to the spatial activity of the MB, considering the characteristic of the human perception. The human perception,

R. Khosla et al. (Eds.): KES 2005, LNAI 3681, pp. 1297–1302, 2005.

however, depends not only on the spatial activity but also on the temporal activity of the MB. Based on this observation, we propose a new spatio-temporal activity measure for the adaptive quantization in this paper. In addition, we introduce a buffer constraint in the target bit allocation step. The constraint controls the buffer status to operate within a proper range, and prevents buffer underflowing and overflowing.

2 Proposed Rate-Control Algorithm

The proposed method adopts the three-step structure of the TM5 rate control. However, it redesigns the first and third steps of the TM5 to enhance the performance of the rate control, which are described precisely in the following subsections.

2.1 Step 1: Target Bit Allocation Using Buffer Constraint Condition

The target bit allocation is first performed at picture level as the same in the TM5 as follows,

$$T_i = \max\{\frac{R}{1 + \frac{N_p X_p}{K_p X_i} + \frac{N_b X_b}{K_b X_i}}, \frac{bit\_rate}{8 \times picture\_rate}\}$$

$$T_p = \max\{\frac{R}{N_p + \frac{N_b K_p X_b}{K_b X_p}}, \frac{bit\_rate}{8 \times picture\_rate}\}$$

$$T_b = \max\{\frac{R}{N_b + \frac{N_p K_b X_p}{K_p X_b}}, \frac{bit\_rate}{8 \times picture\_rate}\} \tag{1}$$

Readers can refer to [3] for more detailed descriptions of the parameters in Eq. 1.

The target bit calculated according to Eq. 1 is then adjusted to prevent the buffer underflow and overflow. In the constant-bit-rate (CBR) transmission case, if the output bit rate of the encoder is controlled to ensure no encoder buffer overflow or underflow, then the decoder buffer will never underflow or overflow when using the same buffer size at the encoder and decoder[5]. Thus, for the CBR case, it is sufficient to control the output bit rate of the encoder to avoid encoder buffer underflow and overflow. When a coded bit stream is stored in the encoder buffer and then transmitted over a CBR channel, the condition to prevent encoder buffer underflow and overflow is

$$0 \leq E(i) \leq E^{max}, \text{ for } \forall i \tag{2}$$

$$E(i) = E(i-1) + B(i) - T_{CBR} \tag{3}$$

where E^{max} is the encoder buffer size and $E(i)$ is encoder buffer fullness after encoding the ith frame. $B(i)$ is the bit amount generated from the ith frame and T_{CBR} is the bit amount transmitted during a frame period. From Eq. 2 and Eq. 3, the range of $B(i)$ to prevent encoder buffer underflow and overflow is

$$B(i) \in [min\_B^e(i), max\_B^e(i)]$$

$$= [max(T_{CBR} - E(i-1), 0), E^{max} + T_{CBR} - E(i-1)] \tag{4}$$

Now, we set the range of target bit $B^T(i)$ to be allocated to the ith frame such that

$$B^T(i) \in [min\_B^T(i), max\_B^T(i)]$$
$$= [min\_B^e(i) + B_U^e, max\_B^e(i) - B_O^e] \qquad (5)$$

where B_U^e and B_O^e are margins to prevent the buffer fullness from being too close to "0" and to E^{max}, respectively. In the experiments, they are set as follows

$$B_U^e = B_O^e = 0.05 \times E^{max} \qquad (6)$$

Finally, the target bit for the current frame is allocated as follows

$$B^T(i) = \begin{cases} T_i, T_p, T_b & \text{if } min\_B^T(i) \leq T_i, T_p, T_b \leq max\_B^T(i) \\ min\_B^T(i) & \text{if } min\_B^T(i) \geq T_i, T_p, T_b \\ max\_B^T(i) & \text{if } T_i, T_p, T_b \geq max\_B^T(i) \end{cases} \qquad (7)$$

2.2   Step 2: Bit-Rate Control

In bit-rate control step, a quantizer parameter Q_j for the macroblock to be coded is derived according to the virtual buffer fullness. The algorithm for this step in the proposed method is the same with the TM5, and thus the detailed description of this step is omitted here.

2.3   Step3: Adaptive Quantization Using Spatio-temporal Activity Measure

The adaptive quantization in the TM5 considers only the spatial characteristic of the human visual perception. Thus, the quantization step is modulated according to the spatial activity measure of the current MB. There still, in fact, exists another feature in the human visual system (HVS) called temporal masking, which can be suitably used to enhance the performance of the adaptive quantization. Temporal masking states that it takes a while for the HVS to adapt itself to the scene when the scene changes abruptly. During this transition the HVS is not sensitive to details. Considering both spatial and temporal masking properties of the HVS, an adaptive quantization based on the spatio-temporal MB activity is introduced in the proposed method.

The modulated quantization parameter for the jth MB is obtained by

$$mquant_j = Q_j \cdot N\_act_j \qquad (8)$$

where, Q_j is the quantization parameter obtained in the bit-rate control step, and N_act_j is the MB activity factor, which is calculated as

$$N\_act_j = \alpha \cdot N\_act_j^s + (1 - \alpha) \cdot N\_act_j^t \qquad (9)$$

The above equation indicates that, differently to TM5, the normalized activity factor of the current MB is calculated as the weighted sum of the spatial activity factor, $N\_act_j^s$, and the temporal activity factor, $N\_act_j^t$, which are calculated as

$$N\_act^s_j = \frac{2 \cdot act^s_j + avg\_act^s}{act^s_j + 2 \cdot avg\_act^s}$$

$$N\_act^t_j = \frac{2 \cdot act^t_j + avg\_act^t}{act^t_j + 2 \cdot avg\_act^t} \tag{10}$$

where avg_act^s and avg_act^t are the average values of the spatial activity measure act^s_j, and the temporal activity measure act^t_j of the last picture to be encoded, respectively. act^s_j and act^t_j for the jth MB are calculated as

$$act^s_j = 1 + \min_{sblk=1,8} (var\_sblk)$$

$$act^t_j = 1 + |Mvp(j)| \tag{11}$$

where var_sblk is the variance of each spatial 8×8 block, and $Mvp(j)$ is the predicted motion vector of the jth MB. For the case where the motion vector for the current MB is not available in rate control process, such as the H.264 video encoder, we use the predicted motion vector instead of the motion vector in Eq. 11.

The parameter α in Eq. 9 is the weight of the spatial activity in the MB activity factor. For intra-pictures, $\alpha = 1.0$ is used, and for inter-pictures $\alpha = 0.5$ is used. Note that $\alpha = 1.0$ means the TM5 rate control algorithm.

3   Experimental Results

To evaluate the performance of the proposed algorithm, 4 standard test sequences of "Carphone", "Container", "Coastguard", and "Foreman" are used, and the results are compared with those of the TM5. The rate control algorithms are implemented on the H.264/AVC JM2.0 video codec [4]. The frame rate is 30frames/sec, and the input sequences of 4:2:0 QCIF format are coded at the channel rate of 128kbps.

Fig. 1 and table 1 show the results of the buffer constraint condition. In this experiment, GOP structure of $N_{GOP} = 10$, $M_B = 0$ is applied and the encoder and decoder buffer sizes are set to $E^{max} = D^{max} = T_{CBR} \times 150ms$ assuming the overall delay of $150ms$. From the results, it is shown that, by applying the buffer constraint condition, the proposed method keeps the buffer status within the proper boundary, while the TM5 rate control suffers overflows and underflows.

Fig. 2, Fig. 3 and table 2 show the results for the adaptive quantization using the proposed spatio-temporal activity measure. In this experiment, except for the first frame, which is intra-coded, all the subsequent frames are predictive-coded. Fig. 2 shows the number of generated bits over 100 frames. Note that the two methods show very similar results excepting a little bit larger inter-frame variation in bit-rates of the proposed method. This indicates that the proposed method modulates the number of bits so that more bits are allocated to stationary frames. Fig.2 shows the PSNR values of the reconstructed frames. Note that the proposed method is superior to the TM5 in PSNR over all the frames. Table 2 shows the average performance over 100 frames of both methods.

Table 1. # of overflows and underflows (150ms).

Input Seq.	TM5		Proposed	
	Overflow	Underflow	Overflow	Underflow
Carphone	0	6	0	0
Container	15	5	0	0

Table 2. Performance comparison.

Input Seq.	TM5			Proposed		
	PSNR	Rate	QP	PSNR	Rate	QP
Carphone	37.26	127.9	13.85	37.99	128.0	14.00
Coastguard	30.23	127.9	20.02	31.11	128.1	19.52
Container	39.15	128.0	8.48	40.19	128.1	8.71
Foreman	35.18	128.0	16.39	35.90	127.9	16.16

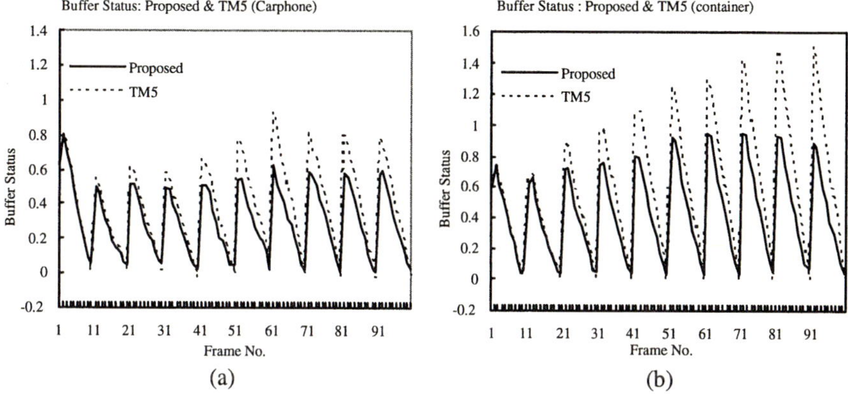

Fig. 1. Buffer fullness (a) "Carphone" (b) "Container".

It is noted that both methods provide exact rate control results. In the respect of average PSNR, the proposed method shows improvements about 0.7 ∼ 1.1 dB over the TM5. The results demonstrate that the spatio-temporal activity measure for the adaptive quantization is superior to the spatial activity measure in TM5 in picture quality.

4   Conclusions

This paper proposed a TM5-like video rate control algorithm. To avoid buffer overflow and underflow, we introduced the buffer constraint in the target bit allocation step. In addition, considering the temporal characteristic of the human perception, a spatio-temporal MB activity measure for the adaptive quantization is proposed. By evaluating the MB temporal activity according to the magnitude of the motion vector, the quantization step can be modulated so that more bits are assigned to the more perceptible still areas.

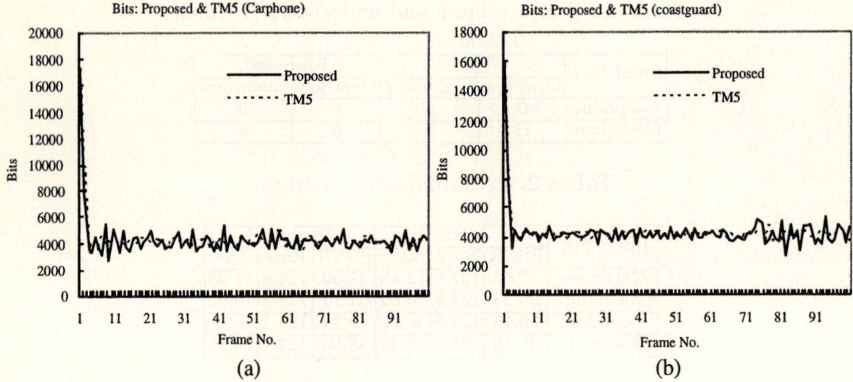

Fig. 2. Generated bits (a) "Carphone" (b) "Coastguard".

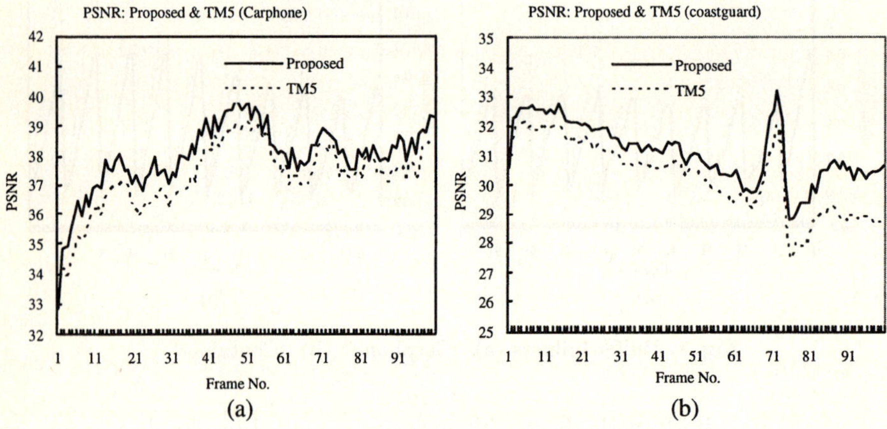

Fig. 3. PSNR (a) "Carphone" (b) "Coastguard".

References

1. ITU-T/SG16, "Video codec test model near-term version 10," April 1998.
2. ISO/IEC JTC1/SC29/WG11, "Text of ISO/IEC 14496-2 MPEG video VM-version 8.0," July 1997.
3. ISO/IEC JTC/SC29/WG11, "Test Model 5," Draft, April 1993.
4. ISO/IEC JTC1/SC29/WG11, "Working draft of reference software for advanced video coding," Mar 2003.
5. A. R. Reibman and B. G. Haskell, "Constraints on variable bit rate video for ATM networks," *IEEE Trans. Circuits Syst. Video Technol.*, vol. 2, no. 4, pp. 3612 – 3372, December, 1992.

Author Index

Lecture Notes in Artificial Intelligence (LNAI)

Vol. 3476: J. Leite, A. Omicini, P. Torroni, P. Yolum (Eds.), Declarative Agent Languages and Technologies II. XII, 289 pages. 2005.

Vol. 3464: S.A. Brueckner, G.D.M. Serugendo, A. Karageorgos, R. Nagpal (Eds.), Engineering Self-Organising Systems. XIII, 299 pages. 2005.

Vol. 3452: F. Baader, A. Voronkov (Eds.), Logic for Programming, Artificial Intelligence, and Reasoning. XI, 562 pages. 2005.

Vol. 3451: M.-P. Gleizes, A. Omicini, F. Zambonelli (Eds.), Engineering Societies in the Agents World V. XIII, 349 pages. 2005.

Vol. 3446: T. Ishida, L. Gasser, H. Nakashima (Eds.), Massively Multi-Agent Systems I. XI, 349 pages. 2005.

Vol. 3445: G. Chollet, A. Esposito, M. Faundez-Zanuy, M. Marinaro (Eds.), Nonlinear Speech Modeling and Applications. XIII, 433 pages. 2005.

Vol. 3438: H. Christiansen, P.R. Skadhauge, J. Villadsen (Eds.), Constraint Solving and Language Processing. VIII, 205 pages. 2005.

Vol. 3430: S. Tsumoto, T. Yamaguchi, M. Numao, H. Motoda (Eds.), Active Mining. XII, 349 pages. 2005.

Vol. 3419: B. Faltings, A. Petcu, F. Fages, F. Rossi (Eds.), Constraint Satisfaction and Constraint Logic Programming. X, 217 pages. 2005.

Vol. 3416: M. Böhlen, J. Gamper, W. Polasek, M.A. Wimmer (Eds.), E-Government: Towards Electronic Democracy. XIII, 311 pages. 2005.

Vol. 3415: P. Davidsson, B. Logan, K. Takadama (Eds.), Multi-Agent and Multi-Agent-Based Simulation. X, 265 pages. 2005.

Vol. 3403: B. Ganter, R. Godin (Eds.), Formal Concept Analysis. XI, 419 pages. 2005.

Vol. 3398: D.-K. Baik (Ed.), Systems Modeling and Simulation: Theory and Applications. XIV, 733 pages. 2005.

Vol. 3397: T.G. Kim (Ed.), Artificial Intelligence and Simulation. XV, 711 pages. 2005.

Vol. 3396: R.M. van Eijk, M.-P. Huget, F. Dignum (Eds.), Agent Communication. X, 261 pages. 2005.

Vol. 3394: D. Kudenko, D. Kazakov, E. Alonso (Eds.), Adaptive Agents and Multi-Agent Systems II. VIII, 313 pages. 2005.

Vol. 3392: D. Seipel, M. Hanus, U. Geske, O. Bartenstein (Eds.), Applications of Declarative Programming and Knowledge Management. X, 309 pages. 2005.

Vol. 3374: D. Weyns, H. V.D. Parunak, F. Michel (Eds.), Environments for Multi-Agent Systems. X, 279 pages. 2005.

Vol. 3371: M.W. Barley, N. Kasabov (Eds.), Intelligent Agents and Multi-Agent Systems. X, 329 pages. 2005.

Vol. 3369: V. R. Benjamins, P. Casanovas, J. Breuker, A. Gangemi (Eds.), Law and the Semantic Web. XII, 249 pages. 2005.

Vol. 3366: I. Rahwan, P. Moraitis, C. Reed (Eds.), Argumentation in Multi-Agent Systems. XII, 263 pages. 2005.

Vol. 3359: G. Grieser, Y. Tanaka (Eds.), Intuitive Human Interfaces for Organizing and Accessing Intellectual Assets. XIV, 257 pages. 2005.

Vol. 3346: R.H. Bordini, M. Dastani, J. Dix, A.E.F. Seghrouchni (Eds.), Programming Multi-Agent Systems. XIV, 249 pages. 2005.

Vol. 3345: Y. Cai (Ed.), Ambient Intelligence for Scientific Discovery. XII, 311 pages. 2005.

Vol. 3343: C. Freksa, M. Knauff, B. Krieg-Brückner, B. Nebel, T. Barkowsky (Eds.), Spatial Cognition IV. XIII, 519 pages. 2005.

Vol. 3339: G.I. Webb, X. Yu (Eds.), AI 2004: Advances in Artificial Intelligence. XXII, 1272 pages. 2004.

Vol. 3336: D. Karagiannis, U. Reimer (Eds.), Practical Aspects of Knowledge Management. X, 523 pages. 2004.

Vol. 3327: Y. Shi, W. Xu, Z. Chen (Eds.), Data Mining and Knowledge Management. XIII, 263 pages. 2005.

Vol. 3315: C. Lemaître, C.A. Reyes, J.A. González (Eds.), Advances in Artificial Intelligence – IBERAMIA 2004. XX, 987 pages. 2004.

Vol. 3303: J.A. López, E. Benfenati, W. Dubitzky (Eds.), Knowledge Exploration in Life Science Informatics. X, 249 pages. 2004.

Vol. 3301: G. Kern-Isberner, W. Rödder, F. Kulmann (Eds.), Conditionals, Information, and Inference. XII, 219 pages. 2005.

Vol. 3276: D. Nardi, M. Riedmiller, C. Sammut, J. Santos-Victor (Eds.), RoboCup 2004: Robot Soccer World Cup VIII. XVIII, 678 pages. 2005.

Vol. 3275: P. Perner (Ed.), Advances in Data Mining. VIII, 173 pages. 2004.

Vol. 3265: R.E. Frederking, K.B. Taylor (Eds.), Machine Translation: From Real Users to Research. XI, 392 pages. 2004.

Vol. 3264: G. Paliouras, Y. Sakakibara (Eds.), Grammatical Inference: Algorithms and Applications. XI, 291 pages. 2004.

Vol. 3259: J. Dix, J. Leite (Eds.), Computational Logic in Multi-Agent Systems. XII, 251 pages. 2004.

Vol. 3257: E. Motta, N.R. Shadbolt, A. Stutt, N. Gibbins (Eds.), Engineering Knowledge in the Age of the Semantic Web. XVII, 517 pages. 2004.

Vol. 3249: B. Buchberger, J.A. Campbell (Eds.), Artificial Intelligence and Symbolic Computation. X, 285 pages. 2004.

Vol. 3248: K.-Y. Su, J. Tsujii, J.-H. Lee, O.Y. Kwong (Eds.), Natural Language Processing – IJCNLP 2004. XVIII, 817 pages. 2005.

Vol. 3245: E. Suzuki, S. Arikawa (Eds.), Discovery Science. XIV, 430 pages. 2004.

Vol. 3244: S. Ben-David, J. Case, A. Maruoka (Eds.), Algorithmic Learning Theory. XIV, 505 pages. 2004.

Vol. 3238: S. Biundo, T. Frühwirth, G. Palm (Eds.), KI 2004: Advances in Artificial Intelligence. XI, 467 pages. 2004.

Vol. 3230: J.L. Vicedo, P. Martínez-Barco, R. Muñoz, M. Saiz Noeda (Eds.), Advances in Natural Language Processing. XII, 488 pages. 2004.

Vol. 3229: J.J. Alferes, J. Leite (Eds.), Logics in Artificial Intelligence. XIV, 744 pages. 2004.